MARINE FISH CULTURE

MARINE FISH CULTURE

by

John W. Tucker, Jr., Ph.D.
Harbor Branch Oceanographic Institution
and
Florida Institute of Technology, Melbourne

KLUWER ACADEMIC PUBLISHERS
Boston / Dordrecht / London

Distributors for North, Central and South America:
Kluwer Academic Publishers
101 Philip Drive
Assinippi Park
Norwell, Massachusetts 02061 USA
Telephone (781) 871-6600
Fax (781) 871-6528
E-Mail <kluwer@wkap.com>

Distributors for all other countries:
Kluwer Academic Publishers Group
Distribution Centre
Post Office Box 322
3300 AH Dordrecht, THE NETHERLANDS
Telephone 31 78 6392 392
Fax 31 78 6546 474
E-Mail <services@wkap.nl>

Electronic Services <http://www.wkap.nl>

Library of Congress Cataloging-in-Publication Data

Tucker, John W., 1948-
Marine fish culture / by John W. Tucker, Jr.
p. cm.
Includes bibliographical references (p.) and index.
ISBN 0-412-07151-7 (alk. paper)
1. Marine fishes. 2. Fish-culture. I. Title.
SH163.T835 1998
639.3'2--dc21 98-42062
CIP

Printed on acid-free paper.

Printed in the United States of America

CONTENTS

Acknowledgments ix

The Author x

1 Introduction 1
- Purposes 1
- Recent status 1
- Organization of the book 12
- Overview 12
- Summary 16

2 Characteristics of Marine Fish 17
- Reproduction 17
- Development 19
- Critical periods 32
- Growth 33
- Wild vs hatchery fish 33
- Summary 34

3 The Rearing Environment 49
- Temperature 50
- Salinity 55
- Solids 59
- Light 59
- Color and texture of hatchery tanks and fittings 66
- Green-water culture 67
- Structured habitat 67
- Disturbances 67
- Microbial environment 68
- Water chemistry 69
- Waste products 72
- Gases 81
- Gas-related disorders 90
- Effects of seawater collection and usage 96
- Toxic metals 97
- Pesticides and other pollutants 122
- Toxic construction materials 142
- Toxicants accidentally introduced by the culturist or the building 144
- Therapeutic compounds 145
- Contaminant effects on beneficial bacteria 145
- Testing and electronic monitoring of water quality 145
- Predators 145
- Noxious microorganisms 146

Environmental problems caused by aquaculture . . . 147
Summary . . . 148

4 **Water Sources** . . . 149
Criteria . . . 149
Major types . . . 150
Summary . . . 152

5 **Water Treatment** . . . 155
Requirements . . . 155
Materials . . . 155
Treatment options . . . 156
System design . . . 169
System monitoring and control . . . 172
Environmental considerations . . . 174
Summary . . . 174

6 **Culture Units** . . . 175
Considerations in choosing culture units . . . 175
Characteristics of culture units . . . 175
Applications of culture units . . . 191
Hatchery design . . . 208
Summary . . . 210

7 **Obtaining Fish for Stocking** . . . 211
Stock from the wild . . . 211
Stock from the hatchery . . . 211
Spermatogenesis (sperm formation) . . . 232
Oogenesis (egg formation) . . . 232
Oocyte maturation . . . 233
Endocrine control of oocyte maturation and ovulation . . . 237
Induced ovulation . . . 238
Timing and egg quality . . . 257
Artificial fertilization . . . 265
Care of eggs . . . 267
Storage of gametes . . . 269
Natural ovulation . . . 270
Care of broodfish . . . 289
Egg collection . . . 290
Induced vs natural ovulation . . . 290
Broodfish adaptability . . . 291
Examples . . . 291
Genetic considerations . . . 295
Hybridization . . . 296
Sex control . . . 296
Summary . . . 298

8 **Nutrition of Larval Fish** . . . 299
Feeding criteria . . . 299
Choice and culture of foods . . . 307
General feeding practices . . . 336
Specific feeding practices . . . 352
General methods used in our hatchery . . . 372
Industrial-scale larval food processing in Italian hatcheries . . . 373
Summary . . . 374

9 **Nutrition of Juvenile and Adult Fish** . . . 375
Requirements and components . . . 375
Broodstock nutrition . . . 407
Nutritional disorders . . . 408
Environmental considerations . . . 411
Feed studies . . . 411
Suggested feed formulas . . . 460
Making and storing feeds . . . 461
Feeding methods . . . 464
Summary . . . 467

10 **Energetics** . . . 469
Energy budget components and influencing factors . . . 469
Specific examples . . . 472
Energy budgets for farms . . . 473
Summary . . . 473

11 **Fish Health** . . . 481
Disease prevention . . . 481
Therapeutic agents . . . 486
Biological disease agents and their control . . . 491
Neoplasias . . . 491
Viruses . . . 491
Bacteria . . . 495
Fungi . . . 500
Protozoans . . . 501
Coelenterates . . . 504
Free-living flatworms . . . 504
Parasitic flatworms . . . 504
Cestodes . . . 506
Nematodes . . . 506
Acanthocephalans . . . 506
Crustaceans . . . 506
Leeches . . . 507
Lampreys . . . 507
Other pathological conditions . . . 508
Genetic disorders . . . 508

Environmentally-induced disorders . . . 508
Overexposure to therapeutic chemicals . . . 510
Nutritional disorders . . . 511
Anaphylactic shock . . . 513
Gas bladder disorders . . . 513
Susceptibility . . . 514
Environmental and human health considerations . . . 518
Summary . . . 520

12 Handling and Transporting Marine Fish . . . 521
Sensitivity of different stages . . . 521
Handling marine fish . . . 521
Grading marine fish . . . 525
Transporting marine fish . . . 526
Summary . . . 532

13 Culture of Established and Potential Species—Food Fish . . . 533
Status . . . 533
Family accounts . . . 533
Environmental considerations . . . 571
Summary . . . 571

14 Culture of Established and Potential Species—Bait Fish . . . 575
Family accounts . . . 575
Environmental considerations . . . 576
Summary . . . 576

15 Culture of Established and Potential Species—Ornamental Fish . 578
Family accounts . . . 578
Environmental considerations . . . 586
Summary . . . 586

16 The Future of Marine Fish Culture . . . 587
Obtaining fish for stocking . . . 587
Environment, water treatment, culture units . . . 587
Nutrition . . . 588
Health . . . 588

Literature Cited . . . 589

Glossary . . . 719

Appendix—Conversion Factors . . . 727

Species Index, by Family . . . 731

ACKNOWLEDGMENTS

It took nearly all of my spare time for the last eight years to write this book. I thank the following for their contributions prior to and during that time.

For encouraging me to raise fish: Stan Blum, Blain Kennedy, Gil and Caroline Kennedy, Jim McVey, John Ryther, and my parents Betty and John Tucker, Sr.

For providing practical and inspirational advice during graduate school: Bill D. Anderson, Bill Hettler, John Merriner, John Miller, Dave Peters, and Allyn Powell.

For collaborating on fish culture research: Bill FitzGerald, Grant Gilmore, Jon Hateley, Blain Kennedy, Bill Lellis, Gene Parsons, Mark Subramanyam, and others.

For being outstanding research assistants: Blake Faulkner, Dan Sennett, and Peter Woodward.

For obtaining hard-to-find references: Kris Metzger, Marilyn Morris, and Sandra Murray.

For providing photographic and darkroom assistance: Tom Smoyer (Harbor Branch Oceanographic Institution).

For allowing use of unpublished material: Phil Bush, LeRoy Creswell, Harry Grier, Blain Kennedy, Gary Loverich (Ocean Spar Technologies), Carole Neidig, Gary Rumsey, and Bob Winfree (U.S. National Park Service).

For reviewing chapter 11: Blain Kennedy. For reviewing pages 486-490: Rosalie Schnick.

For providing places to investigate and raise fish: Grice Marine Biological Laboratory (South Carolina), Pamlico Marine Laboratory (North Carolina), Virginia Institute of Marine Science, National Marine Fisheries Service (North Carolina), Queensland Department of Primary Industries (Australia), University of the West Indies (Jamaica), Cayman Islands Natural Resources Laboratory, Cayman Islands Turtle Farm, Guam Aquaculture Development and Training Center, Micronesian Mariculture Demonstration Center (Palau), Bahamas Department of Fisheries, Bermuda Division of Fisheries, and Harbor Branch Oceanographic Institution.

For providing funding for investigation and culture of fish: North Carolina State University, National Marine Fisheries Service, College of William and Mary, Harbor Branch Oceanographic Institution, Florida Department of Environmental Protection, Allen Johnson, U.S. Department of the Interior, Sea Grant College Program (National and Puerto Rico), Bermuda Division of Fisheries, and others.

Cover: Two female Nassau groupers (foreground) and one male in a 26-m^3 cage, during a spawning session (also see Fig. 7-16). (photo by Phillipe G. Bush, Cayman Islands Department of the Environment)

THE AUTHOR

Beginning with home aquariums in 1971, the author has raised 40 species and hybrids of fish. He has degrees in forestry, zoology (development of marine fish larvae), and marine science (energetics of marine fish eggs and larvae). He has done fish culture research at the National Marine Fisheries Service Laboratory in Beaufort, North Carolina, and government laboratories in Australia, the Bahamas, Bermuda, Cayman Islands, Guam, and Palau. His work has focused on spawning, larval rearing, nutrition, and health of marine fish, with the goal of improving technology for both farming and stock enhancement. He has been a technical advisor for many research and commercial projects. He was Fulbright Professor for a year at the University of the West Indies in Kingston, Jamaica, and has taught nine marine fish culture courses at the Florida Institute of Technology, Bermuda Biological Station for Research, University of the West Indies, National Polytechnic Institute (Mexico), and University of Puerto Rico. He also has made scientific visits to Belize, Canada, England, Hong Kong, Japan, Martinique, New Zealand, Okinawa, Saudi Arabia, Singapore, St. Croix, St. Thomas, Taiwan, Thailand, and Wales. He is a scuba diving instructor for the National Association of Underwater Instructors and the British Sub-Aqua Club and has spent hundreds of hours studying marine fish under water. Since 1983, he has been a member of the Graduate Faculty of the Florida Institute of Technology. Since March 1983, he has conducted a pioneering research program at Harbor Branch Oceanographic Institution and since October 1987, has been head of the HBOI Fish Culture Department.

John W. Tucker, Jr.
316 13th Avenue
Vero Beach, Florida 32962
jwtuckerfish@hotmail.com

1 INTRODUCTION

Why culture marine fish?

PURPOSES

Cultured marine fish are used mostly for food, stock enhancement, bait, or display in aquariums. They also are used in research to determine their requirements (e.g., physical, chemical, dietary needs; interactions with other organisms) to aid managers of wild or captive fish. Development of eggs and larvae is studied and documented to elucidate evolutionary and taxonomic relationships and to enable identification of captured early life stages. Species such as the sheepshead minnow are used in bioassays to determine the effects of pesticides and other potential pollutants.

RECENT STATUS

In many areas of the world the demand for seafood exceeds the total potential catch from exploited and unexploited resources (e.g. the Caribbean region, Tucker and Jory, 1991). In 1993, 97.8% of the world production of Atlantic salmon came from farms, but only 7.2% of Pacific salmon were cultured (FAO, 1995).

Fish culture has been successful with many species in many countries (Tables 1-1 to 1-5); however, the largest operations are with fish such as carps, tilapias, and salmonids, which are relatively easy to produce in hatcheries. Although in the past marine fish were raised mainly from wild juveniles (e.g. eels, milkfish, yellowtail), hatchery production is increasing. Some examples for which hatchery technology is well established are milkfish and Asian sea bass (barramundi) in southeast Asia, red sea bream in Japan, European sea bass and gilthead sea bream in the Mediterranean, and red drum in the United States. Because of greater complexity, not all aspects of technology are as advanced as for freshwater fish culture, but recent breakthroughs make marine fish culture a promising field.

Total world aquaculture production of fish in 1993 was 11,284,725 mt (metric tonnes), including 85.1% freshwater, 11.4% diadromous, and 3.5% marine species (FAO, 1998). In 1995, the proportions were 86.1%, 10.2%, and 3.7% of 14,823,526 mt, and in 1996, 86.2%, 10.0%, and 3.8% of 16,664,491 mt (Table 1-1). Tables 1-2 through 1-4 give 1996 production data for all fish and countries surveyed by FAO (1998), except freshwater fish raised in fresh water. Table 1-5 lists the major species or groups of farmed diadromous and marine fish with the countries producing the most in 1996.

Tilapias were the main freshwater fish raised in brackish water (Table 1-2). Salmonoids, eels, milkfish, and ayu were the main diadromous fish raised in fresh water (Tables 1-3, 1-5). Milkfish were by far the dominant diadromous fish in brackishwater culture, followed by salmonoids, barramundi, and eels; mainly salmonids were raised in salt water. Of marine fish, mainly mullets and sea breams were raised in fresh water (Tables 1-4, 1-5). Mostly groupers, mullets, sea breams, Asian red snappers, and European sea bass were reared in brackish water. The main marine fish raised in salt water were yellowtail, red sea bream, hirame, puffers, sea basses (serranids), and other jacks in the Far East and gilthead sea bream, sea basses (*Dicentrarchus* spp.), and turbot in Europe and the Mediterranean region.

In 1995, about 95.9% of cultured freshwater fish were produced by 10 countries (Table 1-1): China (74.1%, nearly all carps), India (11.7%, nearly all cyprinids), Indonesia (2.4%, mostly carps and tilapias), Bangladesh (2.4%, a variety), United States (1.6%, nearly all channel catfish), Thailand (1.3%, mostly tilapias, Java barb, clariid catfishes), Viet Nam (0.9%, a variety), Philippines (0.6%, nearly all tilapias), Taiwan (0.5%, mostly tilapias), and Myanmar (0.5%, nearly all rohu).

Table 1-1. Aquaculture production of freshwater, diadromous, and marine fish by country during 1996 (FAO, 1998, with permission from the Food and Agricultural Organization of the United Nations).

Country	Freshwater fish (mt)	Diadromous fish (mt)	Marine fish (mt)
Albania	73^{e}	20^{e}	
Algeria	244^{e}	17^{e}	36^{e}
Argentina	10	1,242	
Armenia	1,630^{e}	1,430^{e}	
Australia	33	10,348	2,013
Austria	810	2,140	
Azerbaijan	488^{e}		
Bangladesh	340,828		
Belarus	4,746^{e}	3^{e}	
Belgium	300	546	
Belize	287		
Bhutan	30^{e}		
Bolivia	80	300	
Brazil	40,000	800^{e}	
Brunei Darussalam	20	72	
Bulgaria	1,900^{e}	585	
Burundi	10^{e}		
Cambodia	9,000^{e}		
Cameroon	55^{e}		
Canada		51,826	
Central African Republic	350^{e}		
Chile		199,085	168
China (PRC)	10,651,754	147,316	240,592
Colombia	20,263	4,506	
Congo Democratic Republic	750^{e}		
Congo Republic	5		
Costa Rica	4,115	136	
Cote d'Ivoire	350^{e}		
Croatia	2,071	314	252
Cuba	27,987		
Cyprus		105	348
Czech Republic	17,378	822	
Denmark		41,428	
Dominica	2^{e}		
Dominican Republic	343	300	20
Ecuador	330	35	
Egypt	53,302		22,535
El Salvador	217		
Estonia	61	211	
Ethiopia	85^{e}		
Faeroe Islands		17,584	
Fiji	122		
Finland	57	17,605	
France	10,143	54,405	2,728
French Polynesia		10	
Gabon	3		
Gambia	4		
Germany	12,168	25,289	
Ghana	550^{e}		
Greece	716	2,520	26,390
Guam	168^{e}	25^{e}	5^{e}
Guatemala	1,990		
Guinea	4		
Guyana	150^{e}		
Honduras	232^{e}		
Hong Kong	3,395	144	4,561
Hungary	8,076	4	
Iceland		3,686	
India	1,680,895^{e}		
Indonesia	343,000	160,120	11,300
Iran	27,914	1,900	
Iraq	3,000^{e}		

Table 1-1. (Continued)

Country	Freshwater fish (mt)	Diadromous fish (mt)	Marine fish (mt)
Ireland		15,875	30
Israel	14,562	1,075	1,931
Italy	6,660	43,982	10,550
Jamaica	3,020[e]		
Japan	15,170	65,142	247,822
Jordan	181		
Kazakhstan	1,154[e]		
Kenya	500	105	
Kiribati	2[e]	50[e]	
D.P. Republic of Korea	10,000[e]	2,000[e]	
Republic of Korea	17,354	4,797	11,384
Kuwait			90
Kyrgyzstan	152[e]	9[e]	
Laos	14,400[e]		
Latvia	379	1	
Lebanon		350	
Lesotho	10[e]	4[e]	
Libya	100[e]		
Lithuania	1,537		
Macedonia	341	570	
Madagascar	2,650		
Malawi	235[e]		
Malaysia	19,206	6,000	2,844
Mali	60[e]		
Malta			1,552
Martinique	13		10
Mauritius	71		16
Mayote	1		
Mexico	22,358	2,706	
Moldova Republic	695[e]		
Morocco	750	153[e]	1,225[e]
Mozambique	37[e]		
Myanmar	68,123[e]		
Namibia	5[e]		
Nepal	10,649		
Netherlands	1,200	3,050	
New Zealand		3,500	
Nicaragua	5		
Niger	21[e]		
Nigeria	17,932	12	
Norway		323,893	635
Pakistan	18,000		
Panama	177		
Papua New Guinea	10[e]	8[e]	
Paraguay	350[e]		
Peru	88	1,148	
Philippines	80,287	148,372	703
Poland	21,900	5,800	
Portugal	2	1,412	867
Puerto Rico	68		
Qatar			1
Romania	13,520[e]	380[e]	
Russian Federation	47,870	3,770	120
Rwanda	50[e]		
Saudi Arabia	2,872		
Senegal	53		
Sierra Leone	30[e]		
Singapore	88	266	378
Slovakia	329	625	
Slovenia	212	532	75
South Africa	110	790	
Spain	160	25,975	6,944
Sri Lanka	3,500[e]		

Table 1-1. (Continued)

Country	Freshwater fish (mt)	Diadromous fish (mt)	Marine fish (mt)
St. Lucia	1		
Sudan	1,000		
Suriname	1 [e]		
Swaziland	88 [e]		
Sweden		6,436	
Switzerland	35	1,126	
Syria	6,355		
Taiwan (ROC)	69,343	92,014	13,505
Tajikistan	93		
Tanzania	200 [e]		
Thailand	188,380	3,750	800
Togo	21		
Trinidad & Tobago	5 [e]		
Tunisia	379	158 [e]	804
Turkey	780	18,703	11,530
Turkmenistan	190		
Uganda	210		
Ukraine	32,195	107	157
United Kingdom	133	100,022	
United States	230,853	41,855	
Uruguay	6	12	
Uzbekistan	5,006		
Venezuela	2,400	235	
Viet Nam	130,000 [e]		
Yugoslavia	1,600 [e]	5	
Zambia	4,769		
Zimbabwe	70 [e]	100 [e]	
Totals	14,365,816	1,673,754	624,921
Total, all fish raised in all types of water in 1996			16,664,491

[e] indicates an estimate, in tables in this chapter.

Table 1-2. World aquaculture production of freshwater fish in brackish and salt water during 1996 (FAO, 1998, with permission from the Food and Agricultural Organization of the United Nations).

Scientific name	Common name	1996 production (mt)
	In brackish water	
Oreochromis mossambicus	Mozambique tilapia	31,600
Oreochromis niloticus	Nile tilapia	28,574
Oreochromis spp.	(tilapias)	7,820
Cyprinus carpio	Common carp	2,175
Micropterus salmoides	Largemouth bass	600
Chrysichthys nigrodigitatus	(bagrid catfish)	155
Oreochromis spilurus		70
Oreochromis aureus	Blue tilapia	40 [93]
Eleotridae	(sleepers)	4
Barbodes gonionotus	Java barb	2 [95]
Osteichthyes	(bony fishes)	45,184
Total, freshwater fish raised in brackish water in 1996		116,182
	In salt water	
Oreochromis spp.	(tilapias)	3
Osteichthyes	(bony fishes)	2
Total, freshwater fish raised in salt water in 1996		5

In this book, when listed in tables, general common names are in parentheses.
In tables in this chapter, a superscript number indicates that the amount is for a year before 1996.

Table 1-3. World aquaculture production of diadromous fish in fresh, brackish, and salt water during 1996 (FAO, 1998, with permission from the Food and Agricultural Organization of the United Nations).

Scientific name	Common name	1996 production (mt)
	In fresh water	
Oncorhynchus mykiss	Rainbow trout	277,439
Anguilla japonica	Japanese eel	177,520
Chanos chanos	Milkfish	39,672
Anguilla spp.	("freshwater" eels)	29,065
Salmo spp.	(salmons, trouts)	25,703
Plecoglossus altivelis	Ayu	10,007
Anguilla anguilla	European eel	7,351
Salmo trutta	Brown trout	7,057
Lates calcarifer	Barramundi/Asian sea bass	4,763
Morone chrysops × *M. saxatilis*	Hybrid striped bass	3,848
Coregonus spp.	(whitefishes, ciscoes)	2,218
Acipenseridae	(sturgeons)	903
Salvelinus fontinalis	Brook trout	568
Salmo salar	Atlantic salmon	302
Oncorhynchus tshawytscha	Chinook salmon	200
Acipenser baerii	Siberian sturgeon	172
Salvelinus spp.	(chars)	100
Coregonus lavaretus	Powan	93
Salvelinus alpinus	Arctic char	90
Anguilla rostrata	American eel	38 [95]
Oncorhynchus kisutch	Coho salmon	27
Lates niloticus	Nile perch	12
Thymallus thymallus	Grayling	4
Acipenser ruthenus	Sterlet	1
Huso huso	Beluga	1
Total, diadromous fish raised in fresh water in 1996		587,116
	In brackish water	
Chanos chanos	Milkfish	324,828
Oncorhynchus mykiss	Rainbow trout	14,935
Lates calcarifer	Barramundi/Asian sea bass	10,233
Salmo salar	Atlantic salmon	7,647
Anguilla anguilla	European eel	665
Salvelinus alpinus	Arctic char	541
Anguilla spp.	("freshwater" eels)	350
Centropomus undecimalis	Common snook	300
Anguilla australis	Shortfin eel	201
Anguilla japonica	Japanese eel	93
Coregonus lavaretus	Powan	89
Macquaria ambigua	Golden perch	8
Maccullochella peelii	Murray perch	1
Total, diadromous fish raised in brackish water in 1996		359,891
	In salt water	
Salmo salar	Atlantic salmon	547,694
Oncorhynchus mykiss	Rainbow trout	87,544
Oncorhynchus kisutch	Coho salmon	76,178
Oncorhynchus tshawytscha	Chinook salmon	12,044
Salmo spp.	(salmons, trouts)	1,330
Lates calcarifer	Barramundi/Asian sea bass	888
Salmo trutta	Brown trout	468
Anguilla anguilla	European eel	401
Salvelinus spp.	(chars)	200
Total, diadromous fish raised in salt water in 1996		726,747
Total, diadromous fish raised in all types of water in 1996		1,673,754

Table 1-4. World aquaculture production of marine fish in fresh, brackish, and salt water during 1996 (FAO, 1998, with permission from the Food and Agricultural Organization of the United Nations).

Scientific name	Common name	1996 production (mt)
	In fresh water	
Mugil cephalus	Striped mullet	4,397
Sparidae	(sea breams, porgies)	516
Dicentrarchus labrax	European sea bass	21
Mugilidae	(mullets)	7
Osteichthyes	(bony fishes)	1,150
Total, marine fish raised in fresh water in 1996		6,091
	In brackish water	
Siniperca chuatsi	Chinese perch	58,437
Mugil cephalus	Striped mullet	24,660
Mugilidae	(mullets)	11,676
Sparus auratus	Gilthead sea bream	8,895
Acanthopagrus schlegeli	Black sea bream	3,037
Epinephelus spp.	(groupers)	2,178
Lutjanus argentimaculatus	Asian red snapper	2,007
Dicentrarchus labrax	European sea bass	1,838
Sparidae	(sea breams, porgies)	1,506
Epinephelus tauvina	Greasy grouper	837
Serranidae	(groupers, sea basses)	559
Pagrus major	Red sea bream	559
Decapturus spp.	(scads)	148
Lutjanidae	(snappers)	121
Osteichthyes	(bony fishes)	1,776
Total, marine fish raised in brackish water in 1996		118,234
	In salt water	
Seriola quinqueradiata	Yellowtail	145,889
Pagrus major	Red sea bream	77,319
Sparus auratus	Gilthead sea bream	23,832
Dicentrarchus labrax	European sea bass	19,231
Paralichthys olivaceus	Hirame	16,553
Tetraodontidae	(puffers)	5,552
Dicentrarchus spp.	(sea basses)	5,382
Trachurus japonicus	Japanese horse mackerel	3,869
Scophthalmus maximus	Turbot	2,588
Trachurus spp.	(jack & horse mackerels)	2,343
Scorpaenidae	(scorpionfishes)	2,036
Thunnus maccoyii	Southern bluefin tuna	2,013
Epinephelus areolatus	Squaretail grouper	750
Lutjanus argentimaculatus	Asian red snapper	690
Epinephelus spp.	(groupers)	407
Sparidae	(sea breams, porgies)	368
Epinephelus tauvina	Greasy grouper	360
Trachinotus blochii	Snubnose pompano	325 [95]
Lutjanus russellii	Russell's snapper	300
Lateolabrax japonicus	Japanese sea bass	266
Rhabdosargus sarba	Silver sea bream	240
Mugilidae	(mullets)	225
Pleuronectiformes	(flatfishes)	218
Gadus morhua	Atlantic cod	198
Cantherines spp.	(filefishes)	148 [94]
Pagrus pagrus	Red porgy	130 [94]
Diplodus sargus	White sea bream	122
Acanthopagrus berda	River bream	90
Mugil cephalus	Striped mullet	82
Lutjanus spp.	(snappers)	80
Thunnus thynnus	Northern bluefin tuna	77

Table 1-4. (Continued)

Scientific name	Common name	1996 production (mt)
Evynnis japonica	Crimson sea bream	52 [94]
Serranidae	(groupers, sea basses)	36
Epinephelus akaara	Redspotted grouper	30 [95]
Solea solea	Common sole	29
Decapterus spp.	(scads)	20
Pleurogrammus azonus	Atka mackerel	19
Acanthopagrus schlegeli	Black sea bream	18
Diplodus spp.	(sea breams)	15
Rachycentron canadum	Cobia	13
Sciaenops ocellatus	Red drum	10
Stephanolepis cirrhifer	Thread-sailfin filefish	7
Sciaenidae	(drums)	2
Siganus spp.	(rabbitfishes)	2
Seriola dumerili	Greater amberjack	1
Dentex dentex	Common dentex	1
Siganus canaliculatus	Pearly spinefoot	1
Osteichthyes	(bony fishes)	189,342
Total, marine fish raised in salt water in 1996		500,596
Total, marine fish raised in all types of water in 1996		624,921

About 95.6% of cultured diadromous fish were produced by 20 countries (Table 1-1): Norway (19.4%, mostly Atlantic salmon, some rainbow trout), Chile (11.9%, salmons, rainbow trout), Indonesia (9.6%, nearly all milkfish), Philippines (8.9%, nearly all milkfish), China (8.8%, eels and others), United Kingdom (6.0%, mostly Atlantic salmon, some rainbow trout), Taiwan (5.5%, milkfish, eels, barramundi, others), Japan (3.9%, mostly eels, coho salmon, rainbow trout, ayu), France (3.2%, mostly rainbow trout), Canada (3.1%, salmons, rainbow trout), Italy (2.6%, rainbow trout, eels, sturgeon), Denmark (2.5%, mainly rainbow trout and eels), United States (2.5%, mostly rainbow trout, Atlantic salmon, hybrid striped bass), Spain (1.6%, mostly rainbow trout, some Atlantic salmon and eels), Germany (1.5%, mostly rainbow trout, some brown trout), Finland (1.0%, nearly all rainbow trout), Faeroe Islands (1.0%, mostly Atlantic salmon, some rainbow trout), Turkey (1.1%, salmonids), Ireland (0.9%, Atlantic salmon and rainbow trout), and Australia (0.6%, Atlantic salmon, rainbow trout, barramundi, eels), with most of the remainder raised in other Asian countries.

About 97.4% of cultured marine fish were produced by 12 countries (Table 1-1): Japan (39.6%, mostly yellowtail, red sea bream, hirame, puffers), China (38.5%, Chinese perch and others), Greece (4.2%, mostly gilthead sea bream, European sea bass, striped mullet), Egypt (3.6%, striped mullet, gilthead sea bream, European sea bass), Taiwan (2.2%, mostly barramundi, black sea bream, striped mullet, groupers, red sea bream, other sea breams), Turkey (1.8%, gilthead sea bream, sea basses), Republic of Korea (1.8%, mostly hirame, scorpionfishes, Japanese sea bass), Indonesia (1.8%, mullets), Italy (1.7%, European sea bass, gilthead sea bream, striped mullet), Spain (1.1%, mostly gilthead sea bream, turbot, European sea bass), Hong Kong (0.7%, mostly striped mullet, groupers, snappers, silver sea bream), and Malaysia (0.4%, mostly eels, barramundi, Asian red snappers, greasy groupers).

More regional information can be obtained from publications such as: Liao (1988), Joseph (1990), and Proceedings of the First, Second, Third, and later Asian Fisheries Forums (Maclean et al., 1986; Hirano and Hanyu, 1990; Chou et al., 1994), Guerrero and Garcia (1991), and Lacanilao et al. (1994) for Asia; Davy (1990) and Yoshimura et al. (1996a) for Japan; Guerrero (1996) and Rice and DeVera (1998) for the Philippines; Al-Thobaity and James (1994) for Saudi Arabia; Kissil (1996) for Israel; Barnabé (1990a) for Europe; Flos et al. (1990) and Larrazabal (1992) for Spain; Aiken (1990) and Needham (1990) for

Table 1-5. Aquaculture production of major diadromous and marine fish by the highest-producing countries during 1996 (FAO, 1998, with permission from the Food and Agricultural Organization of the United Nations). Numbers in parentheses are metric tonnes for the year.

Scientific name	Common name	First	Second	Third	Fourth
		Diadromous			
		Acipenseridae—sturgeons			
Acipenser baeri	Siberian sturgeon	France (160)	Uruguay (12)		
Acipenser ruthenus	Sterlet	Ukraine (1)			
Huso huso	Beluga	Ukraine (1)			
Acipenseridae	(sturgeons)	Russian Federation (600)	Italy (280)	Estonia (17)	Ukraine (5)
		Anguillidae—"freshwater" eels			
Anguilla anguilla	European eel	Italy (3000)	Netherlands (2800)	Denmark (900)	Greece (584)
Anguilla australis	Shortfin eel	Australia (201)			
Anguilla japonica	Japanese eel	China (147,316)	Taiwan (25,063)	Malaysia (3635)	Republic of Korea (1599)
Anguilla rostrata	American eel	Dominican Republic (38)[95]			
Anguilla spp.	("freshwater" eels)	Japan (28,595)	Indonesia (820)		
		Chanidae—milkfish			
Chanos chanos	Milkfish	Indonesia (157,600)	Philippines (148,372)	Taiwan (58,453)	Kiribati (50)[e]
		Osmeridae—smelts			
Plecoglossus altivelis	Ayu	Japan (9775)	Taiwan (232)		
		Salmonidae—salmons, trouts			
Coregonus lavaretus	Powan	Finland (92)	Czech Republic (90)		
Coregonus spp.	(whitefishes, ciscoes)	Armenia (1150)[e]	Russian Federation (970)	Macedonia (70)	Switzerland (26)
Oncorhynchus kisutch	Coho salmon	Chile (66,988)	Japan (8401)	Canada (756)	United States (33)
Oncorhynchus mykiss	Rainbow trout	Chile (54,429)	France (50,625)	Italy (40,702)	Denmark (40,278)
Oncorhynchus tshawytscha	Chinook salmon	Canada (8403)	New Zealand (3500)	Chile (341)	
Salmo salar	Atlantic salmon	Norway (301,426)	United Kingdom (83,344)	Chile (77,327)	Canada (36,352)
Salmo trutta	Brown trout	Germany (2529)	France (2118)	Russian Federation (2200)	United Kingdom (345)
Salmo spp.	(salmons, trouts)	Turkey (18,510)	Japan (4788)	Greece (1927)	Canada (1000)
Salvelinus alpinus	Arctic char	Iceland (541)	France (90)		

Salvelinus fontinalis	Brook trout	France (450)	Czech Republic (82)	Slovakia (25)	Slovenia & U.K. (5)
Salvelinus spp.	(chars)	Norway (200)	Sweden (100)		
Thymallus thymallus	Grayling	France (2)	Slovenia (2)		
		Centropomidae—snooks			
Centropomus undecimalis	Common snook	Dominican Republic (300)			
Lates calcarifer	Barramundi	Taiwan (6981)	Thailand (3750)	Malaysia (2365)	Indonesia (1700)
Lates niloticus	Nile perch	Nigeria (12)			
		Percichthyidae—temperate basses			
Morone chrysops × *M. saxatilis*	Hybrid striped bass	United States (3561)	Israel (287)		
Maccullochella peelii	Murray perch	Australia (1)			
Macquaria ambigua	Golden perch	Australia (8)			
		Marine			
		Gadidae—cods			
Gadus morhua	Atlantic cod	Norway (198)			
		Scorpaenidae—scorpionfishes			
Scorpaenidae	(scorpionfishes)	Republic of Korea (2036)			
		Hexagrammidae—greenlings			
Pleurogrammus azonus	Atka mackerel	Republic of Korea (19)			
		Percichthyidae—temperate basses			
Dicentrarchus labrax	European sea bass	Greece (11,662)	Italy (3800)	France (1997)	Egypt (977)
Dicentrarchus spp.	(sea basses)	Turkey (5210)	Croatia (172)		
Lateolabrax japonicus	Japanese sea bass	Republic of Korea (266)			
		Serranidae—groupers, sea basses			
Epinephelus akaara	Redspotted grouper	Hong Kong (30)[95]			
Epinephelus areolatus	Squaretail grouper	Hong Kong (750)			
Epinephelus tauvina	Greasy grouper	Malaysia (837)	Hong Kong (360)		
Epinephelus spp.	(groupers)	Taiwan (1883)	Thailand (600)	Singapore (93)	Republic of Korea (9)
Siniperca chuatsi	Chinese perch	China (58,437)			
Serranidae	(groupers, sea basses)	Philippines (595)			
		Pomatomidae—bluefishes			
Pomatomus saltatrix	Bluefish	Tunisia (29)[92]			

Table 1-5. (Continued)

Scientific name	Common name	First	Second	Third	Fourth
		Rachycentridae-cobia			
Rachycentron canadum	Cobia	Taiwan (13)			
		Carangidae—jacks			
Decapturus spp.	(scads)	Taiwan (168)			
Seriola dumerili	Greater amberjack	Spain (1)			
Seriola quinqueradiata	Yellowtail	Japan (145,773)	Republic of Korea (116)		
Trachinotus blochii	Snubnose pompano	Hong Kong (325)[95]			
Trachurus japonicus	Japanese horse mackerel	Japan (3869)			
Trachurus spp.	(jack & horse mackerels)	Japan (2343)			
		Lutjanidae—snappers			
Lutjanus argentimaculatus	Asian red snapper	Malaysia (2007)	Hong Kong (690)		
Lutjanus spp.	(snappers)	Singapore (80)			
Lutjanidae	(snappers)	Taiwan (121)			
Lutjanus russellii	Russell's snapper	Hong Kong (300)			
		Sparidae—sea breams, porgies			
Acanthopagrus berda	River bream	Kuwait (90)			
Acanthopagrus schlegeli	Black sea bream	Taiwan (3055)	Japan (80)[94]		
Dentex dentex	Common dentex	Spain (1)			
Diplodus sargus	White sea bream	Greece (122)			
Diplodus spp.	(sea breams)	Cyprus (15)			
Evynnis japonica	Crimson sea bream	Japan (52)[94]			
Pagrus major	Red sea bream	Japan (77,092)	Taiwan (759)	Republic of Korea (27)	
Pagrus pagrus	Red porgy	Greece (130)[94]			
Rhabdosargus sarba	Silver sea bream	Hong Kong (240)			
Sparus auratus	Gilthead sea bream	Greece (13,799)	Turkey (6320)	Spain (3818)	Italy (3650)
Sparidae	(sea breams, porgies)	Taiwan (2363)	Republic of Korea (16)	Cyprus (11)	
		Sciaenidae—drums			
Sciaenops ocellatus	Red drum	Martinique (10)			

Sciaenidae	(drums)	Taiwan (2)			
		Mugilidae—mullets			
Liza ramada	Thinlip mullet	Tunisia (20)[92]			
Mugil cephalus	Striped mullet	Egypt (20,101)	Italy (3100)	Taiwan (2321)	Hong Kong (1561)
Mugilidae	(mullets)	Indonesia (11,300)	Thailand (200)	Ukraine (157)	Spain (125)
		Siganidae—rabbitfishes			
Siganus canaliculatus	Pearly spinefoot	Qatar (1)			
Siganus spp.	(rabbitfishes)	Philippines (2)			
		Scombridae—mackerels			
Thunnus maccoyii	Southern bluefin tuna	Australia (2013)			
Thunnus thynnus	Northern bluefin tuna	Spain (77)			
		Bothidae—lefteye flounders			
Paralichthys olivaceus	Hirame	Republic of Korea (8861)	Japan (7692)		
Scophthalmus maximus	Turbot	Spain (2189)	France (225)	Portugal (144)	Ireland (30)
		Soleidae—soles			
Solea solea	Common sole	Spain (23)	Portugal (6)		
		Pleuronectiformes—flatfishes			
Pleuronectiformes	(flatfishes)	Chile (168)	Russian Federation (50)		
		Balistidae—leatherjackets			
Cantherines spp.	(filefishes)	Japan (148)[94]			
Stephanolepis cirrhifer	Thread-sailfin filefish	Republic of Korea (7)			
		Tetraodontidae—puffers			
Tetraodontidae	(puffers)	Japan (5552)			
		Miscellaneous bony fishes			
Osteichthyes	(bony fishes)	China (182,155)	Japan (5501)	Taiwan (2820)	Hong Kong (660)

[e] estimate. A superscript number indicates the year of the last reported value, which is given.

Canada; Conte (1990) for California; NRC (1992) for the United States; Hargreaves and Alston (1991) and Tucker and Jory (1991) for the Caribbean; Benetti et al. (1994, 1995a) for Ecuador; O'Sullivan (1990, 1992) for Australia; and Hecht et al. (1992) for South Africa. Current news is provided by Aquaculture Europe (magazine), Aquaculture Magazine, Fish Farming International (newspaper), Fish Farming News (newspaper), World Aquaculture (magazine), and others. Scientific information is published mainly in aquaculture journals such as Aquaculture, Aquaculture International, Aquaculture Nutrition, Aquaculture Research, Aquacultural Engineering, International Journal of Aquaculture and Fisheries Technology, Journal of Applied Aquaculture, Journal of Aquaculture in the Tropics, Journal of the World Aquaculture Society, Northern Aquaculture, Progressive Fish-Culturist, Suisanzoshoku (Aquaculture), but also in regional aquaculture journals, fish biology journals (e.g., Asian Fisheries Science, Bulletin of the Japanese Society of Scientific Fisheries—now Fisheries Science), and symposium proceedings.

Development of marine fish culture, at least in the United States, has been slow. In the U.S., only a few freshwater food fish (trout, salmon, channel catfish, and hybrid striped bass), and ornamental fish (e.g., cichlids, goldfish, guppies, platys) are raised in large numbers. Red drum have been raised in small numbers for food, and clownfish, gobies, and dottybacks support a small ornamental marine fish industry. Research has been done on several marine species that have good potential, but actual farming of these seems more likely in other countries. This book describes important aspects of existing technology and identifies sources of additional information.

ORGANIZATION OF THE BOOK

This book is written from the viewpoint of an ex-larval fish taxonomist/physiologist who over the last 27 yr has intensively reared some of the more difficult marine fish (Table 1-6) on research and pilot scales. During hundreds of rearing studies, procedures for spawning, larviculture, and grow-out have been developed.

It is not intended to be a manual for marine fish culture, but a combination reference and textbook. Most of the chapters are literature reviews combined with original data and information gained by experience. Extent of the reviews depends on both the importance and the volume of information available. Fairly extensive reviews are included on reproduction and nutrition, with more limited ones on characteristics of marine fish, water quality, and fish health.

Factors limiting the development of commercial marine fish farming are in five main categories: water quality (chapters 3, 4, and 5), reproduction (chapter 7), nutrition (chapters 8, 9, and 10), health management (chapter 11), and economics (briefly discussed in this chapter). Water quality is mainly an engineering problem and depends on a good source and an appropriate treatment system. Especially during the last 15 yr, much progress has been made in the areas of reproduction and nutrition of marine fish. Fish health depends mostly on water quality and nutrition. Economics is based on supply and demand balances.

Chapter 2 describes characteristics of fish relative to rearing them. Chapter 6 describes containers used for rearing fish. Chapter 12 describes handling and shipping methods for live fish. Chapters 13, 14, and 15 give a synopsis for representative species. Chapter 16 briefly speculates on the future of marine fish culture.

For the sake of completeness and to facilitate comparisons, euryhaline tilapias and diadromous fish such as sturgeons, eels, and salmons are included. Because much already has been written on their culture, their treatment here is brief.

OVERVIEW

Before getting into the details, here are some general comments about biological, engineering, siting, economic, social, environmental, and genetic factors relevant to culture of saltwater fish.

Table 1-6. Fish reared by the author.

Scientific name	Common name	Stages	First year
	Family Anguillidae—"freshwater" eels		
Anguilla rostrata	American eel	Larva to early juvenile	1975
	Family Clupeidae—herrings		
Brevoortia patronus	Gulf menhaden	Egg to adult	1984
	Family Engraulidae—anchovies		
Anchoa mitchilli	Bay anchovy	Egg to mid-stage larva	1978
	Family Cyprinodontidae—killifishes		
Cyprinodon variegatus	Sheepshead minnow	Egg to adult to egg	1971
	Family Poeciliidae—livebearers		
Gambusia affinis	Western mosquitofish	Larva to adult to larva	1971
Poecilia reticulata	Guppy	Larva to adult to larva	1964
Xiphophorus variatus	Variable platyfish	Larva to adult to larva	1984
	Family Centropomidae—snooks		
Centropomus undecimalis	Common snook	Egg to adult	1983
Centropomus pectinatus	Tarpon snook	Larva to adult	1983
Lates calcarifer	Barramundi, Asian sea bass	Egg to late juvenile	1985
	Family Percichthyidae—temperate basses		
Morone chrysops × *M. saxatilis*	White bass × striped bass	Early juvenile to adult	1988
	Family Serranidae—groupers, sea basses		
Centropristis striata	Black sea bass	Egg to early juvenile	1979
Epinephelus fulvus × *E. guttatus*	Coney × red hind	Egg to mid-stage larva	1994
Epinephelus fulvus × *P. furcifer*	Coney × creole-fish	Egg to mid-stage larva	1994
Epinephelus fuscoguttatus	Brownmarbled grouper	Egg to mid-stage larva	1991
Epinephelus guttatus	Red hind	Egg to late juvenile	1994
Epinephelus morio × *C. striata*	Red grouper × black sea bass	Egg to early larva	1994
Epinephelus striatus	Nassau grouper	Egg to adult to egg	1987
Paranthias furcifer	Creole-fish	Egg to mid-stage larva	1994
Plectropomus areolatus	Squaretail coraltrout	Egg to mid-stage larva	1991
Serranus subligarius	Belted sandfish	Egg to adult to egg	1991
	Family Coryphaenidae—dolphins		
Coryphaena hippurus	Dolphin	Egg to juvenile	1994
	Family Lutjanidae—snappers		
Ocyurus chrysurus	Yellowtail snapper	Egg to mid-stage larva	1988
	Family Sparidae—sea breams, porgies		
Archosargus probatocephalus	Sheepshead	Egg to adult	1984
	Family Sciaenidae—drums		
Cynoscion nebulosus	Spotted seatrout	Egg to adult	1985
Leiostomus xanthurus	Spot	Egg to adult	1984
Menticirrhus americanus	Southern kingfish	Egg to early juvenile	1979
Pogonias cromis	Black drum	Early juvenile to adult	1994
Sciaenops ocellatus	Red drum	Early juvenile to adult	1991
	Family Cichlidae—cichlids		
Oreochromis mossambicus	Mozambique tilapia	Egg to adult to egg	1985

Table 1-6. (Continued)

Scientific name	Common name	Stages	First year
	Family Pomacentridae—damselfishes		
Abudefduf saxatilis	Sergeant major	Egg to adult	1994
Stegastes leucostictus	Beaugregory	Egg to mid-stage larva	1993
	Family Mugilidae—mullets		
Mugil cephalus	Striped mullet	Egg to adult	1995
Mugil curema	White mullet	Egg to late juvenile	1972
	Family Labridae—wrasses		
Halichoeres bivittatus	Slippery dick	Egg to early larva	1993
Lachnolaimus maximus	Hogfish	Egg to early larva	1986
Thalassoma bifasciatum	Bluehead wrasse	Egg to early larva	1993
	Family Ephippidae—spadefishes		
Chaetodipterus faber	Atlantic spadefish	Egg to mid-stage juvenile	1980
	Family Bothidae—lefteye flounders		
Scophthalmus aquosus	Windowpane	Egg to adult	1972
	Balistidae—leatherjackets		
Balistes vetula	Queen triggerfish	Egg to late larva	1992

Biological Considerations

The best candidates for aquaculture will grow fast and efficently and will survive well, despite the stresses of captivity. Sufficient seed will be readily available. From a biological point of view, there are four main areas in which available information and knowledge presently limit the potential of marine fish culture: obtaining eggs, maintaining a suitable environment, ensuring good nutrition, and controlling disease. The most common bottleneck is lack of sufficient survival of larvae to the juvenile stage, which would result from good environment, nutrition, and health. Juvenile survival usually is not a problem.

Many marine species can be reared in fresh water after their osmoregulatory capabilities have become adequate. Snook, barramundi, sheepshead, red drum, spotted seatrout, southern flounder, striped mullet, and others can survive and grow well in fresh water at some point before they become juveniles (within 3-5 weeks). But they have to be in salt water for those first few weeks, and salt water adds many more problems for the culturist.

Engineering Considerations

Weather can be especially hazardous to a coastal facility (e.g., hurricanes, northeasters, erosion). Salt brings other problems, notably corrosion of most metals used in construction. Some water quality problems (e.g., higher ammonia toxicity) can be more important in salt water. Maintenance of high water quality often is the most critical factor, but just as often the most difficult and frequently the least understood and most underestimated.

Siting Considerations

Problems can occur in site selection, broodstock acquisition, and security. Coastal land can be expensive. Siting for hatchery and growout facilities would have to minimize both pollution and susceptibility to natural or human disturbances. Away from polluted bays, water quality should not be a problem, but the water must circulate freely to pump intake lines and through cage grow-out areas. Suspended matter should be low. If possible, cages should have at least 4-5 m of water below them. Oceanic water quality is required for eggs and larvae of many species and usually is preferred for other stages; however, for many fish, salinity can be reduced during grow-out, and for some, it should be. Erosion from storms and tides should be anticipated. Power and water supplies must be reliable and have backups. In many areas, fences and possibly guards and dogs would be needed for security. Obstructions such as pipes and barbed wire sometimes are installed in ponds to discourage poachers.

To avoid nearshore conflicts and pollution, cages sometimes are placed offshore, but the risk of storm losses is high. Salmon are raised in large cages in high-energy locations such as the North Sea. "Typhoon-proof" cages have been used with some success off the coast of Japan and similar ones in the Gulf of Mexico.

The number of indigenous candidates is limited by availability. Broodstock, eggs, or juveniles of rare species might have to be imported, and introduction of exotic species is not always the best approach.

Remoteness of farms or markets makes it hard to obtain supplies, keep the product fresh, and get it to market (May, 1978).

Another area of concern is that of possible pollution effects on both fish and consumer health. The impact of past, present, and future industrialization and urbanization should be anticipated in the planning of aquaculture enterprises. Estuarine areas are particularly affected by pollution and urban development. Besides being highly productive nursery grounds for many commercially important animals, clean estuaries or bays are necessary for cage culture of marine fish. Cage farms can become self-polluting. Public health aspects of pollution must be especially considered (Aiken, 1991; Edwards, 1991). The occurrence of human viruses in marine animals from waters lightly to moderately polluted is well documented.

Economic Considerations

To qualify as a commercial aquaculture candidate, a fish must be capable of being produced and sold with an acceptable profit margin. The necessity for a saltwater hatchery and live larval foods increases the cost of raising marine fish. Chapters 13 to 15 give some perspective to species that already are cultured and others that might be considered for culture. Some species that grow slowly in nature will grow more than twice as fast in captivity. Feed conversion and feed cost vary among species and are factors not to be overlooked (Chapter 9). How valuable is the product? Prices from markets in the northwestern and northeastern U.S. can be found in the Fishery Market News Report, published by the National Marine Fisheries Service. INFOFISH Trade News (published by INFOFISH, P.O. Box 10899, 50728 Kuala Lumpur, Malaysia) provides worldwide market information. If fish are sold as fillets, another major factor is the percentage of edible yield.

In nature, survival of marine fish to eating size is usually much less than 1%. When fish are farmed, survival usually is between 10 and 70%. With increased inputs of feed, labor, and energy, people can raise more fish faster in a given space than nature can. But those people want to make a profit when they sell the fish. Further discussion of aquaculture economics can be found in Allen et al. (1984), Meade (1989), and Shang (1990).

Social Considerations

Mariculture ventures have to compete with other interests for coastal resources (e.g, Stephenson, 1990; worldwide coverage by nine articles in World Aquaculture, vol. 23, no. 2, 1992), and usually have to settle for land undesirable for housing and waters not needed for other uses. Geographic and social obstacles to mariculture in the Caribbean region are discussed by Ferlin and Noriega-Curtis (1989). There is a small potential for sales competition with wild fisheries, but if the fish are worth raising, the demand probably is high enough that prices will be stable. Pipelines, cages, pens, etc. should not interfere with navigation. Creation of eyesores must be avoided.

Environmental Considerations
Noxious exotic species can be introduced by aquaculture, unless strict precautions are taken. In many cases, the fish eventually do escape, for example, because of an accident or ponds overflowing during a flood. Some introduced fish have eaten or displaced more desirable indigenous species or otherwise become a nuisance, for example, Nile perch introduced in certain African lakes and grass carp in the U.S. Tilapia are well established in several states of the U.S. (Courtenay et al., 1991). Exotic species should be raised only in closed, recirculating systems or in such a way that the effluent cannot carry exotic fish or exotic pathogens into local habitats where they could survive.

When fish are farmed, space and resources sometimes are taken away from natural populations—e.g., natural water bodies are filled, fenced, or diked; and cages also take up space. In coastal environments, where productivity typically is high, humans cannot compete well with nature. Monoculture practices often result in more of a particular species than would grow naturally in a given area, but, if left alone, overall productivity in nature usually is higher (Turner, 1977). Alternative approaches include strong offshore cages or land-based grow-out using recirculating water systems whenever possible to minimize effluent disposal problems.

Addition and enhanced cycling of nutrients can increase plant biomass and productivity, sometimes resulting in problems such as phytoplankton blooms or ponds clogged with weeds. Physical and chemical effects of solid or dissolved wastes can reduce habitat quality near culture facilities. Excess sediment and nutrients have to be removed from tank or pond wastewater before it is discharged, and lagoon treatment seems the most practical way (Chen et al., 1997).

Introduction of exotic diseases and parasites (e.g., Thompson, 1990), proliferation of rare pre-existing pathogens, and development of resistant or new strains (Aiken, 1991; Edwards, 1991, and others in the same issue) can occur. Crowded aquaculture conditions often enhance growth of viruses, bacteria, and parasites. Changes in composition, abundance, and antibiotic resistance of bacteria have been documented at fish farms.

Further discussion of the environmental impact of aquaculture can be found in publications such as: Herke (1976, 1977, 1978), Gowen and Bradbury (1987), Smith and Pestano-Smith (1985), Stickney (1988), Anonymous (1989), Lam (1990), De Pauw and Joyce (1992), and Pillay (1992).

Genetic Considerations
In fish farming, artificial selection is practiced: because survival is higher than in nature, the ratio of offspring to parents is much higher. In addition, relatively few broodfish are employed. If large numbers of reared fish are released, genetic diversity in the natural population could be reduced (Bartley et al., 1992; Garcia-Vazquez et al., 1995).

SUMMARY
A species good for commercial aquaculture is likely to have the following characteristics: high value, easy to get eggs, larvae, or juveniles of, easy to rear, easy and cheap to feed, can be crowded, grows fast and efficiently.

Some ideas that might contribute to success in rearing marine fish are: Although the basic needs of some species are very well known, marine fish culture is still very much an art. Manuals might help, but there is no substitute for experience and good judgement. Consider well the needs of the fish in choosing the site, in designing the facility, and in operating the facility. Be sure the water quality is suitable before making a major committment. Plan for all possible contingencies. Look closely at rearing conditions and at how the fish develop and behave, especially in response to the conditions or changes in them. Use the past experience of others and yourself at all stages. Evaluate, reevaluate, and revise rearing strategies whenever possible. Keep what is good, but make sure it is repeatable. Eventually, the art will become somewhat more of a science.

2 CHARACTERISTICS OF MARINE FISH

This chapter summarizes some information on events and processes in the life histories of marine fish that can be important and useful in their culture. It also gives an indication of the diversity in life history strategies.

REPRODUCTION

Spawning strategies and methods are very diverse and seem designed to maximize survival of offspring in the natural environment.

Release of Young

There are three general categories. **Viviparous** fish bear few live young that are very advanced (e.g., some elasmobranchs and some teleosts, including members of the scorpionfish, flyingfish, and clinid families; Wourms, 1994). **Ovoviviparous** fish retain their eggs until hatched; larvae are less developed than for viviparous species, and more of those larvae are planktonic (e.g., some elasmobranchs and some teleosts). Species of the scorpionfish, or rockfish, genus *Sebastes* are ovoviviparous. Examples of their development are given by Okiyama (1988) and Nagasawa and Kobayashi (1995). Ovoviviparous species also occur among the topminnows, surf perches, and four-eyed fishes. **Oviparous** fish release small to large numbers of eggs that are fertilized externally (the great majority of species). Only oviparous species will be considered in this chapter.

Egg Handling

Eggs can be **massed** in rocks, weeds, holes, free floating rafts (not many species) or **single** (most fish). **Demersal** (sinking) eggs usually are larger than planktonic eggs and are common in fresh water, estuaries, and nearshore marine environments. **Planktonic** (drifting) eggs also are found in those areas, but are the dominant type in the open ocean. Planktonic eggs sometimes are called "pelagic," which indicates that they usually are found in open water rather than on or attached to the bottom. "Planktonic" is a more specific, functional, term and indicates the type of organism rather than its location.

Parental Care

A few species devote considerable effort to parental care such as guarding, fanning, or carrying eggs. Some cardinalfish and tilapia are mouthbrooders—they carry eggs in the mouth until they hatch and sometimes then carry the larvae for several days after hatching. In the family Cichlidae, members of the genus *Oreochromis* are maternal mouthbrooders, *Saratherodon* spp. are paternal mouthbrooders, and *Tilapia* spp. are nestbuilders. Syngnathidae is the only temperate marine family with brood pouches (Potts, 1984). Queen triggerfish (male and female) and many damselfish (usually the male only) guard their eggs and help them to hatch by blowing water into and fanning the nest. The former usually bury eggs partially in coarse sand, to which they adhere until just before hatching. The latter usually attach them to the underside of a rock. Most fish provide no parental care.

Number of Offspring and Their Survival Ability

Number of offspring and their survival ability often are inversely-related and depend on the life-history strategy of the species. Extreme examples are 50 eggs per clutch in the surf perch, 12 million in the tarpon, and 300 million in the ocean sunfish (Potts, 1984). Chapter 7 gives more information on natural and captive spawning.

Many marine fish produce large numbers (thousands to millions) of single, planktonic eggs that hatch into relatively weak larvae compared to those of freshwater fish. Larvae from drifting eggs undergo wider dispersal than those from stationary eggs, and the higher numbers can help reduce the impact of predation (Economou, 1991). Survival of the young of such fish in nature is controlled more by unpredictable environmental factors, rather than biological factors. Those species are ***r*-life history strategists** as opposed to ***K*-strategists**, which produce fewer, but more precocious young in more predictable, but more limited environments (Pianka, 1970). More eggs increase the probability that enough larvae will encounter conditions and food supplies appropriate for survival. Larger eggs tend to produce larger larvae that individually are more resistant to starvation and predation (Shirota, 1970; Blaxter and Hunter, 1982).

The significance of this in fish farming is that

(1) In some species (*r*), individual females produce large numbers of eggs, providing the opportunity for humans to produce large numbers of fish to stock or eat. In other species (*K*), each female contributes a small number of eggs, so more spawners are needed.

(2) Keeping *r*-type larvae alive and making them grow usually is much harder than for the *K*-type.

Winemiller and Rose (1992) analyzed data for 216 species in 57 families and described three endpoint life-history strategies rather than two:

(1) **Periodic strategy**: delayed maturation, medium or large size at maturation, often migratory, large clutches of small eggs, fast larval and juvenile growth, and short spawning season (e.g., Atlantic sturgeon, striped bass, groupers, cod, cobia, tunas). Many species in this group could also be called *r*-strategists. They produce large numbers of less-competent offspring, infrequently.

(2) **Opportunistic strategy**: early maturation, small size at maturation, small eggs, fast larval growth, long spawning season, and repeat spawning (e.g., anchovies, silversides, killifishes). They produce large numbers of less-competent offspring, but spread out over a long period.

(3) **Equilibrium strategy**: small or medium sized fish, small clutches of large eggs, parental care, slow juvenile and adult growth, and long spawning season (e.g., marine catfishes, tilapias). This is similar to the *K*-strategy, with species adapted to resource- or space-limited habitats. They produce small numbers of highly-competent offspring, frequently.

In this classification, many species are intermediate. For example, menhadens and smelts are opportunistic-periodic (small species, large clutches of small eggs, seasonal spawning, few spawns); gobies, damselfishes, and sculpins are opportunistic-equilibrium (small clutches of large eggs, parental care); and salmon are periodic-equilibrium (small clutches of large eggs, high parental investment).

With a typical periodic-strategist, small numbers of broodstock can produce many offspring for rearing, but effort must be made to extend the spawning season, and larval survival tends to be lower. With a typical opportunistic-strategist, offspring can be obtained frequently, but many more broodfish are needed to produce practical numbers, and larval survival tends to be lower. With a typical equilibrium-strategist, the supply of offspring is more constant and their survival is high, but many broodfish are needed to produce practical numbers.

Natural spawning seasons vary in length from one to 365 days a year. A distinct (but not too short) spawning season synchronized with environmental cues and events helps to ensure that eggs and larvae encounter favorable temperatures, currents, and feeding conditions in nature. In the hatchery, we try to provide all of these.

Survival of wild eggs and larvae varies among species and conditions, but often is in the range 30-90% per day. Sudden changes in temperature or attacks by predators can wipe out entire patches of eggs or larvae. One adult surviving from one thousand eggs would be normal for sockeye salmon in the wild (Foerster, 1968). Survival from eggs to juveniles often is far less than 1% in nature, but under culture conditions survival can range from that level (sometimes) with relatively undeveloped larvae to 100% (sometimes) with precocious larvae.

DEVELOPMENT

Fish eggs and larvae have a wide range of characteristics, again apparently to maximize survival of offspring in the natural environment (Table 2-1; Figs. 2-1 to 2-7).

Egg Stage

The eggs of most finfish desirable for mariculture are small (typically 0.6-2.0 mm diameter) and planktonic. Data in Table 1-1, while not all-inclusive, are representative and indicate size ranges of: 0.8-15 mm for carried eggs, 0.3-6.5 mm for adhesive sinking eggs, 0.6-1.8 mm for adhesive suspended eggs, 1.5-8.2 mm for buried eggs, 1.8-3.6 mm for non-adhesive sinking eggs, and 0.4-4.1 mm for planktonic eggs. Tilapia, salmon, and striped bass hatch from large, sinking eggs into well-developed larvae that are relatively easy to feed and raise. With many species that are periodic strategists, 100,000-300,000 eggs per kilogram female weight can be obtained by chemically inducing ovulation (see chapter 7).

Major features of the teleost egg are the **chorion** (shell), **yolk**, single cell developing into an **embryo**, and sometimes one or more **oil globules**. The **perivitelline space** (containing fluid) separates the chorion from the yolk, oil, and embryo. The **micropyle** is a pore through which a sperm enters to fertilize the egg. The yolk consists mainly of protein and is used for growth. The oil (triacylglycerols) is a concentrated source of energy and **essential fatty acids** (EFA). **Free amino acids** (FAA) also can be used for energy in eggs and larvae (Fyhn, 1993; Ronnestad et al., 1992, 1994; Finn and Fyhn, 1995; Fyhn and Govoni, 1995). After hatching of species that have a significant supply of oil in the egg, the oil might become more important for energy than FAAs. After fertilization of a typical planktonic egg (Fig. 2-1), the egg absorbs water, the perivitelline space forms, the micropyle closes (Yamamoto and Kobayashi, 1992), and the chorion hardens.

Among different species, egg diameter and incubation time both are inversely related to temperature, but directly (and probably linearly) related to each other (Ware, 1975; Duarte and Alcaraz, 1989; Pepin, 1991). Within some species having long spawning seasons, egg diameter is larger when water temperature is lower (Ware, 1975). Warmwater fish with planktonic eggs usually hatch within 10-48 hours after fertilization (haf), but coldwater species with large sinking eggs might take a few months. Sometimes hatching time is expressed in terms of degree-days [e.g., Nassau grouper, (26°C × 26 h)/24 h = 28.2 degree-days], but the relationship between temperature and time is not linear and cannot accurately be extrapolated beyond linear sections of the curve. Hatching success depends especially on thyroid hormone content, temperature, salinity, internal pH, and hatching enzyme production (Brown and Bern, 1988; Helvik et al., 1991; Luberda et al., 1993).

Larval Stage

Larvae from planktonic eggs usually are small and delicate and require live foods for 3-5 weeks (for some, 8 weeks or more). Larvae from adhesive or demersal eggs usually are much more capable at or soon after hatching. This is only a trend; for example, dolphin eggs are planktonic but the larvae are large and effective predators at first feeding. Another trend, with few exceptions, is for greater fragility, often narrower tolerances, and usually greater difficulty in rearing with smaller egg and hatchling size (from beginning to end of planktonic egg section in Table 2-1, especially eggs smaller than 700 μm). The disadvantages of size can be partly compensated for by factors such as larger mouths, larger yolk and oil supplies, and abundance of small prey.

In **preflexion** larvae, the notochord is straight. In **flexion** larvae, the notochord is turning upward and the caudal skeleton is forming. In **postflexion** larvae, the notochord

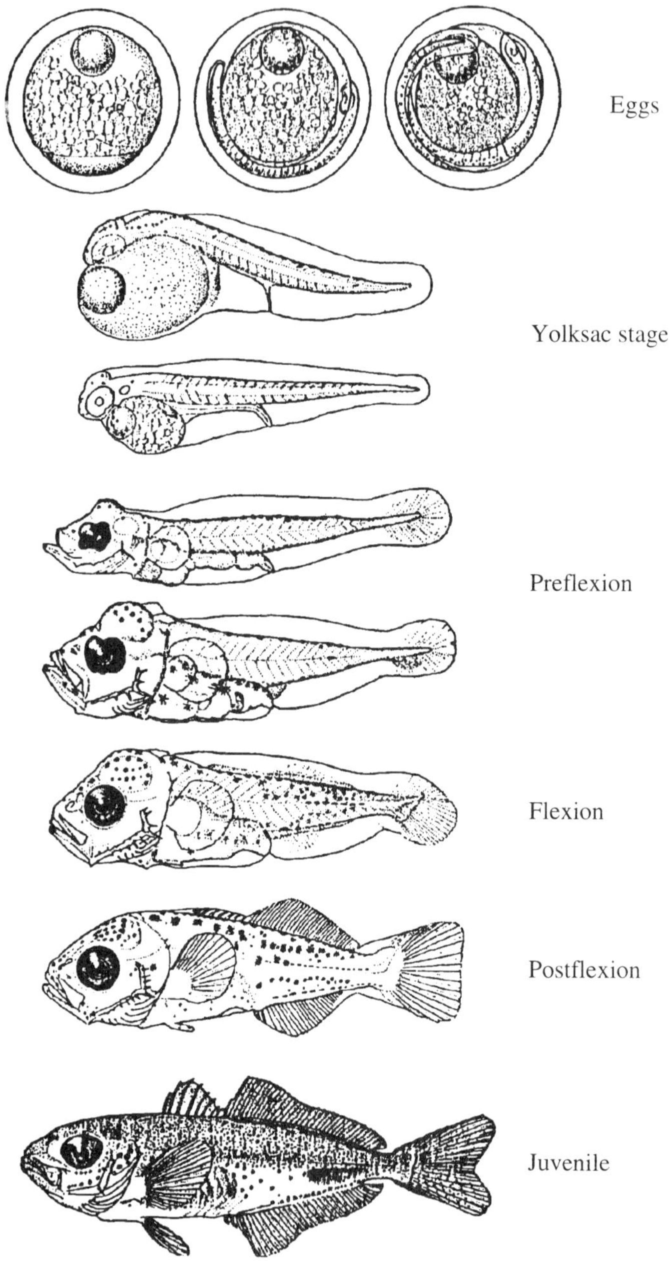

Figure 2-1. Developmental nomenclature for early stages of jack mackerel, a typical marine fish with planktonic eggs: egg diameter 980 μm; larvae 2.0, 2.8, 3.5, 4.9, 7.4, and 10.0 mm BL; juvenile 28 mm BL (modified from Kendall et al., 1984, previously modified from Ahlstrom and Ball, 1954).

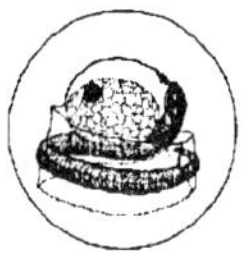

Pacific sardine, 1700 μm
Sardinops sagax
Family Clupeidae
(Matarese and Sandknop, 1984)

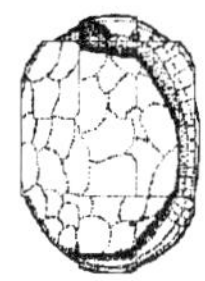

Bay anchovy, 600 x 800 μm
Anchoa mitchilli
Family Engraulidae
(Kuntz, 1914)

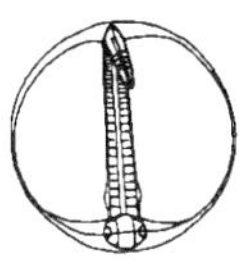

Red cornetfish
Fistularia petimba
Family Fistulariidae
(in Fritzsche, 1984;
from Mito, 1961)

Lined seahorse
Hippocampus erectus
Family Syngnathidae
(Hardy, 1978)

Striped mullet, 780 μm
Mugil cephalus
Family Mugilidae
(Matarese and Sandknop, 1984)

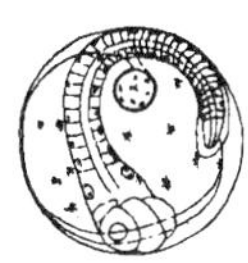

Hirame, 920 μm
Paralichthys olivaceus
Family Bothidae
(in Ahlstrom et al., 1984;
from Mito, 1963)

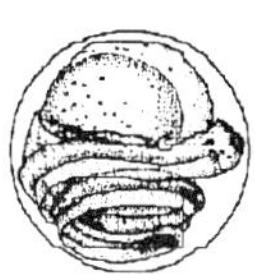

Rex sole, 1900 μm
Errex zachirus
Family Pleuronectidae
(Ahlstrom and Moser, 1980)

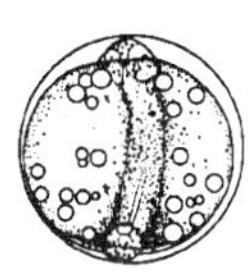

Hogchoker, 760 μm
Trinectes maculatus
Family Soleidae
(Hildebrand and Cable, 1938)

Figure 2-2. Examples of fish eggs.

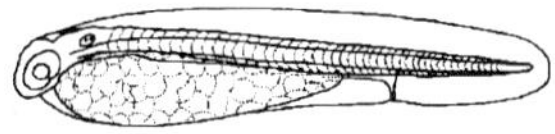

Japanese anchovy, 3.0 mm
Engraulis japonicus
Family Engraulidae
(in McGowan and Berry, 1984;
from Mito, 1961)

Red cornetfish, 7.1 mm
Fistularia petimba
Family Fistulariidae
(in Fritzsche, 1984;
from Mito, 1961)

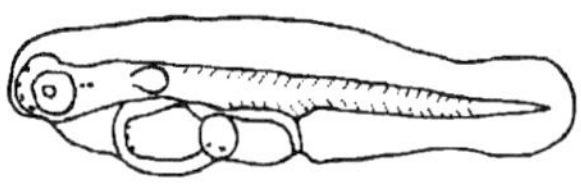

Nassau grouper, 2.6 mm TL
Epinephelus striatus
Family Serranidae
(Powell and Tucker, 1992)

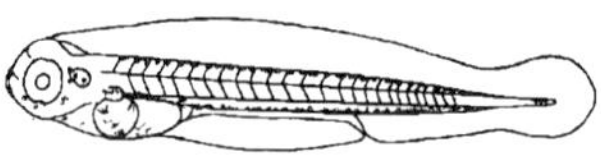

Northern sennet, 3.8 mm SL
Sphyraena borealis
Family Sphyraenidae
(Houde, 1972)

Rough golden puffer
Lagocephalus lunaris, 1.9 mm TL
Family Tetraodontidae
(in Leis, 1984;
after Fujita, 1966)

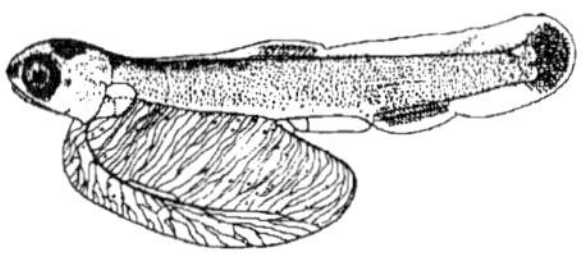

Chinook salmon, 25.0 mm
Oncorhynchus tshawytscha
Family Salmonidae
(Kendall and Behnke, 1984)

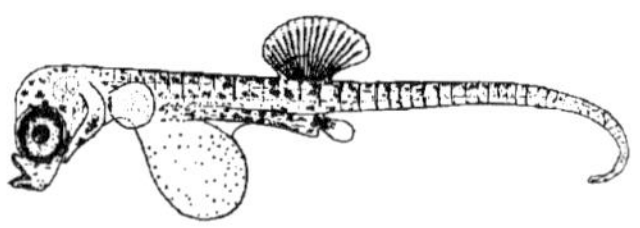

Japanese seahorse, ~6 mm TL
Hippocampus japonicus
Family Syngnathidae
(in Fritzsche, 1984;
from Nakamura, 1937)

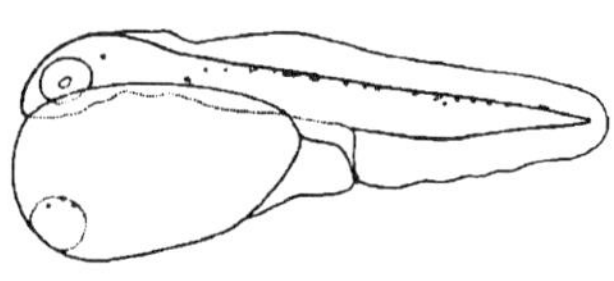

Kelp bass, 2.2 mm
Paralabrax clathratus
Family Serranidae
(Butler et al., 1982)

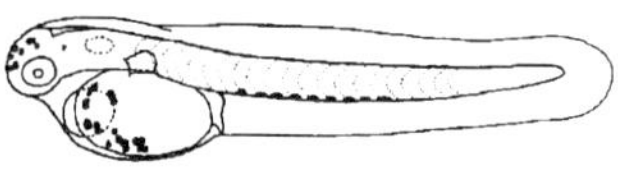

Thread-sailfin filefish
Stephanolepis cirrhifer, 2.1 mm TL
Family Balistidae
(in Aboussouan and Leis, 1984;
after Fujita, 1955)

Figure 2-3. Examples of yolksac larvae.

Eel
Family Anguillidae
(Castle, 1984)

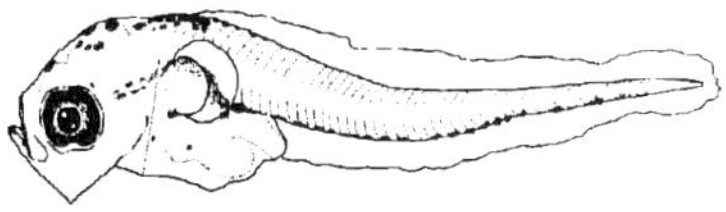

Haddock, 6.1 mm SL
Melanogrammus aeglefinus
Family Gadidae
(Dunn and Matarese, 1984)

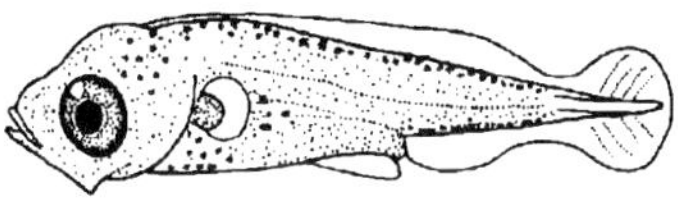

White mullet, 3.7 mm TL
Mugil curema
Family Mugilidae
(Anderson, 1957)

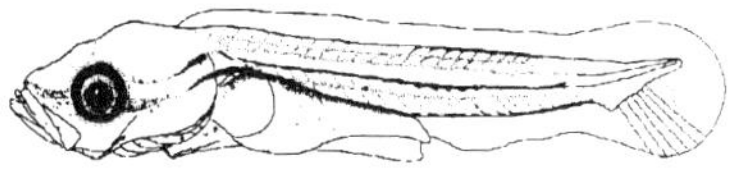

Hogfish, 5.0 mm SL
Lachnolaimus maximus
Family Labridae
(Richards and Leis, 1984)

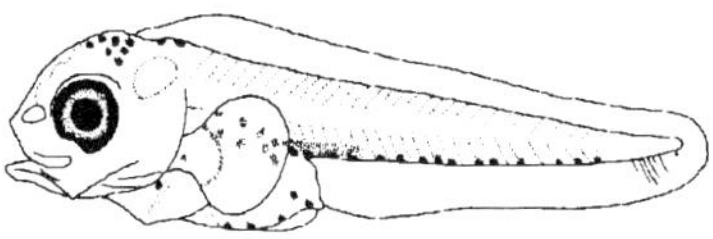

Chub mackerel, 5.0 mm NL
Scomber japonicus
Family Scombridae
(in Collette et al., 1984;
modified after Kramer, 1960)

Slender sole, 5.9 mm
Eopsetta exilis
Family Pleuronectidae
(in Ahlstrom et al., 1984;
from Ahlstrom and Moser, 1975)

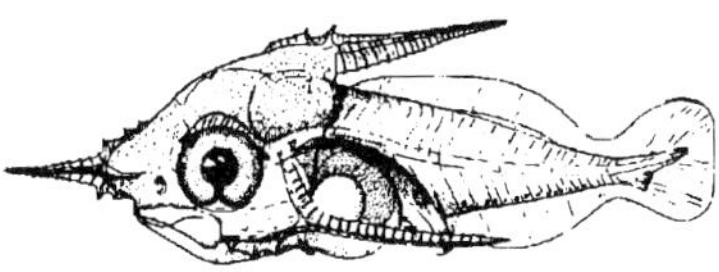

Dusky squirrelfish, 5.0 mm NL
Holocentrus vexillarius
Family Holocentridae
(in Keene and Tighe, 1984;
from McKenney, 1959)

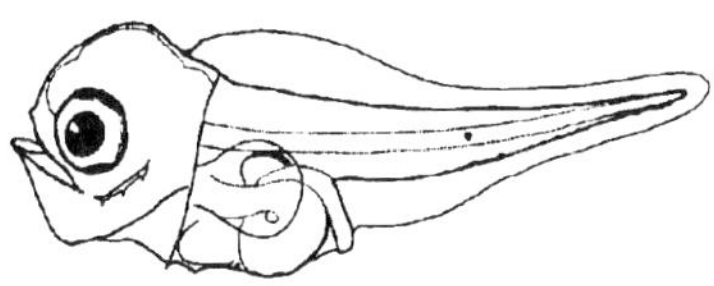

Horned whiff, 2.2 mm NL
Citharichthys cornutus
Family Bothidae
(Tucker, 1982)

Figure 2-4. Examples of early postlarvae.

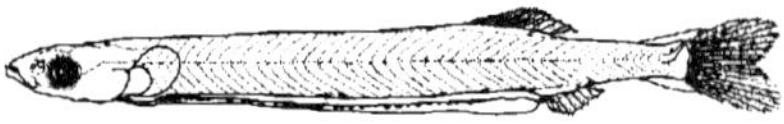

Milkfish, 11.7 mm SL
Chanos chanos
Family Chanidae
(Richards, 1984)

Atlantic cod, 11.0 mm TL
Gadus morhua
Family Gadidae
(Fahay, 1983)

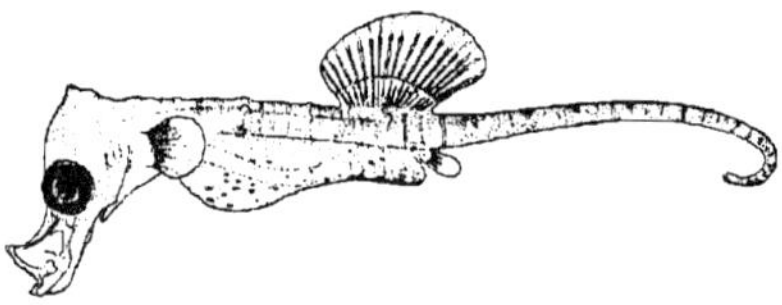

Japanese seahorse, 6.5 mm TL
Hippocampus japonicus
Family Syngnathidae
(in Fritzsche, 1984;
from Nakamura, 1937)

Red cornetfish, 15.6 mm
Fistularia petimba
Family Fistulariidae
(Leis and Rennis, 1983)

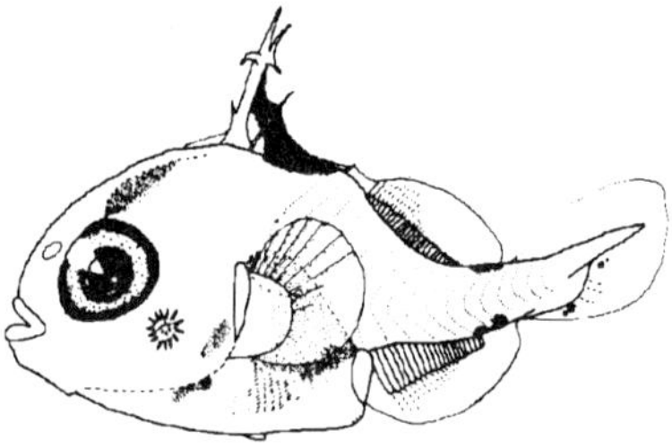

Ocean triggerfish, 3.5 mm
Canthidermis sufflamen
Family Balistidae
(Aboussouan and Leis, 1984)

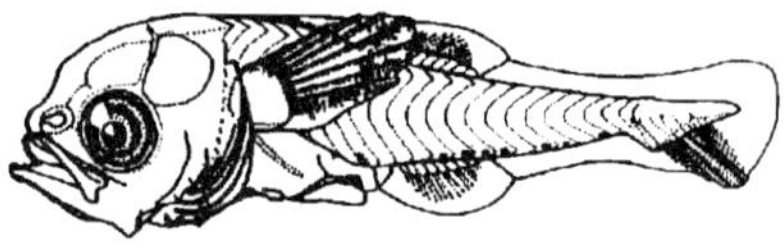

Bay cardinalfish, 4.2 mm SL
Foa fo
Family Apogonidae
(Miller et al., 1979)

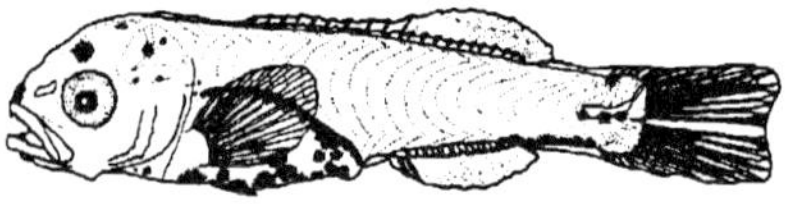

Sobaity, ~10 mm TL
Sparidentex *hasta*
Family Sparidae
(Hussain et al., 1981)

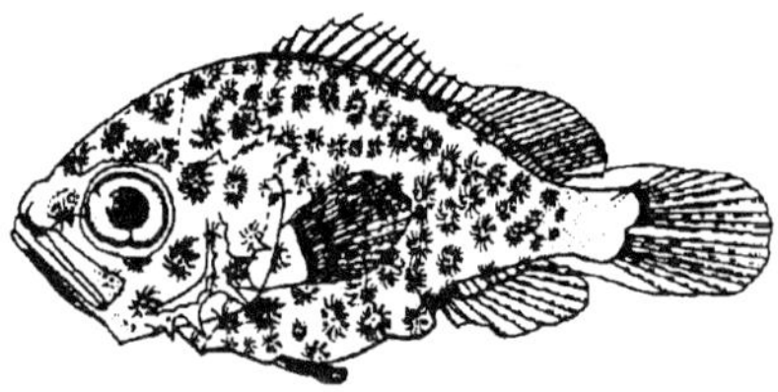

Brown sweetlips, 5.8 mm SL
Pseudopristipoma nigra
Family Haemulidae
(Leis and Rennis, 1983)

Figure 2-4. (Continued)

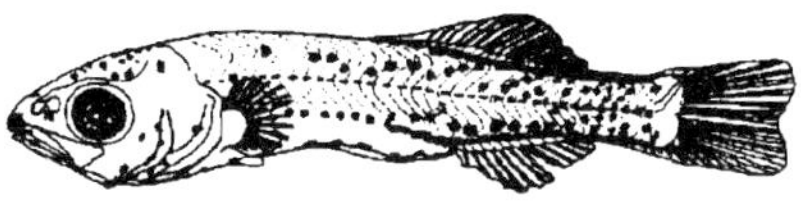

Japanese sea bass, 13.7 mm TL
Lateolabrax japonicus
Family Percichthyidae
(in Johnson, 1984;
from Mito, 1957)

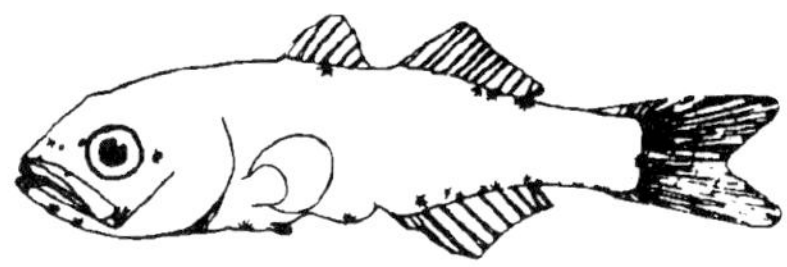

Common snook, 6.3 mm SL
Centropomus undecimalis
Family Centropomidae
(Lau and Shafland, 1982)

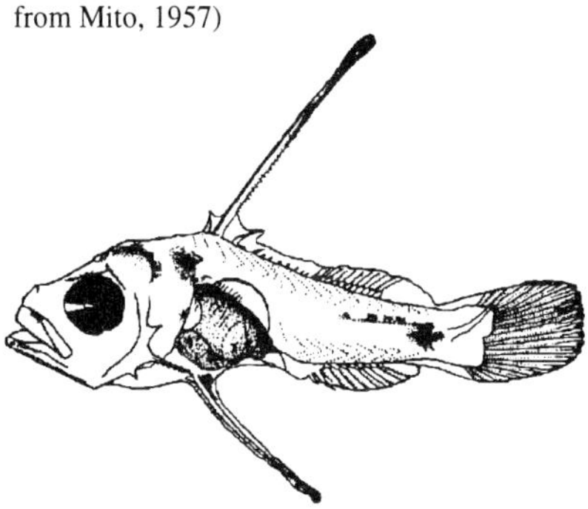

Nassau grouper, 8.1 mm TL
Epinephelus striatus
Family Serranidae
(Powell and Tucker, 1992)

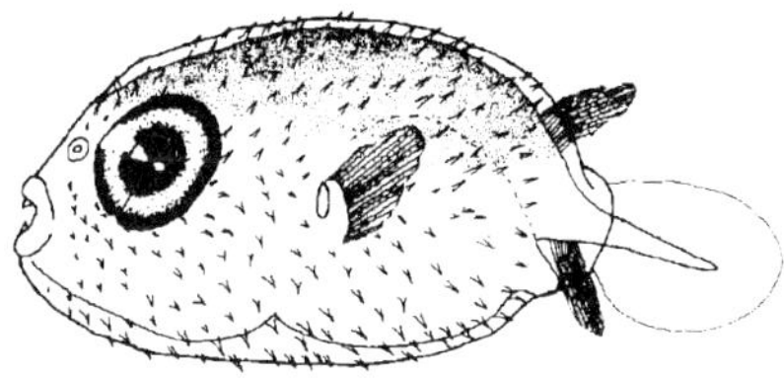

Porcupinefish, 3.6 mm
Canthigaster?
Family Tetraodontidae
(Leis, 1984)

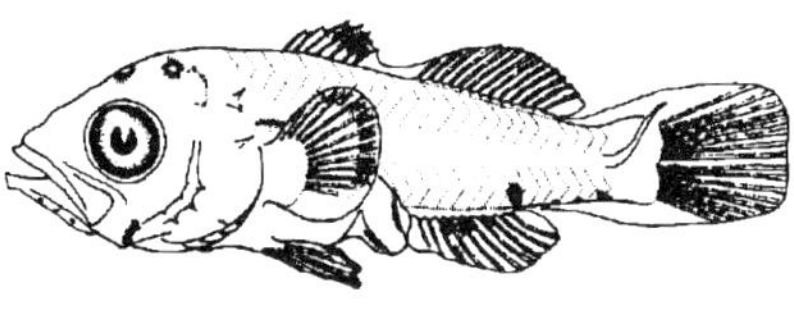

Kelp bass, 7.4 mm
Paralabrax clathratus
Family Serranidae
(Butler et al., 1982)

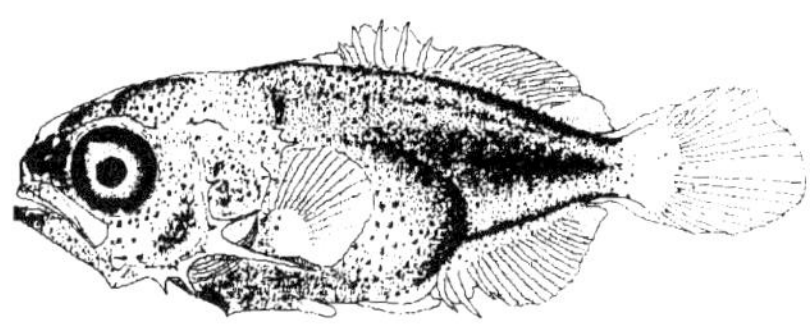

Florida pompano, 5.9 mm
Trachinotus carolinus
Family Carangidae
(Laroche et al., 1984)

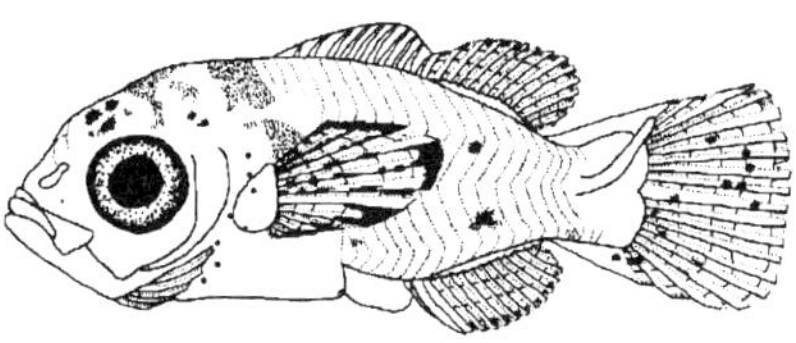

Yellowtail damselfish, 3.7 mm SL
Microspathodon chrysurus
Family Pomacentridae
(Richards and Leis, 1984)

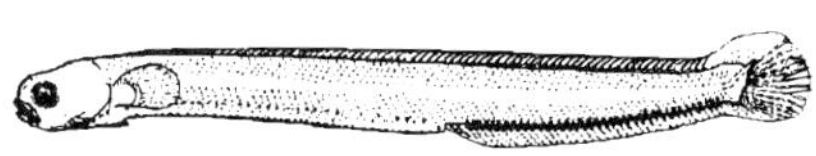

Gunnel, 23.0 mm
Pholis sp.
Family Pholidae
(Matarese et al., 1984)

Figure 2-5. Examples of middle stage postlarvae.

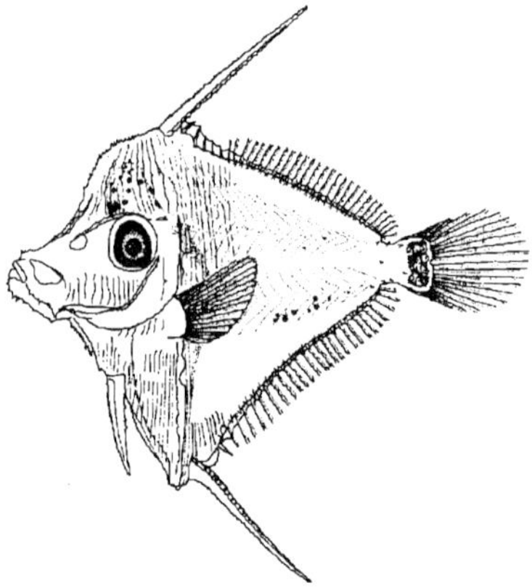

Bluespine unicornfish, 5.9 mm SL
Naso unicornis
Family Acanthuridae
(Leis and Richards, 1984)

Goby, 12.0 mm SL
Luciogobius elongatus
Family Gobiidae
(in Ruple, 1984;
redrawn from Shiogaki and Dotsu, 1972a)

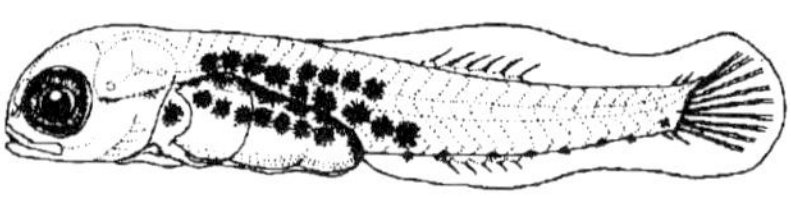

California clingfish, 6.2 mm
Gobiesox rhessodon
Family Gobiesocidae
(Allen, 1984; from Allen, 1979)

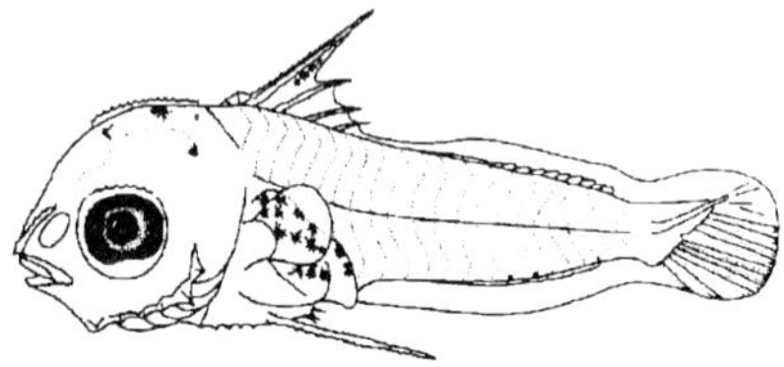

Sandy spinefoot, 5.4 mm TL
Siganus fuscescens
Family Siganidae
(in Leis and Richards, 1984;
modified after Uchida et al., 1958)

Sand lance, 16 mm
Ammodytes marinus
Family Ammodytidae
(in Stevens et al., 1984;
redrawn from Einarsson, 1951)

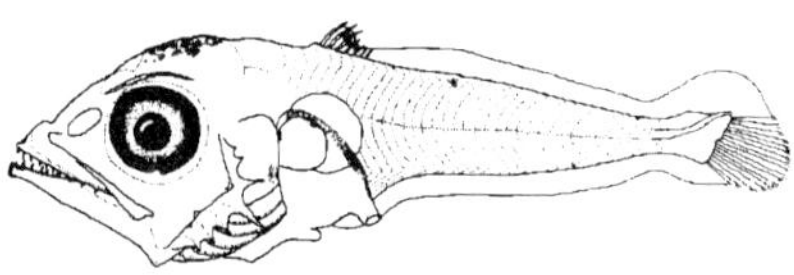

Northern bluefin tuna, 6.0 mm SL
Thunnus thynnus
Family Scombridae
(Collette et al., 1984)

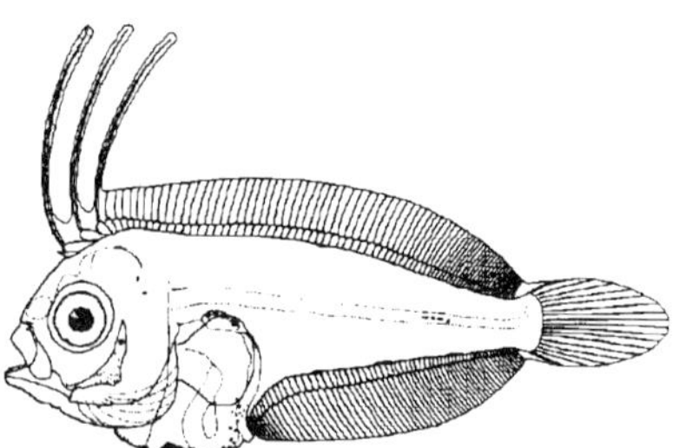

Horned whiff, 8.2 mm SL
Citharichthys cornutus
Family Bothidae
(Tucker, 1982)

Figure 2-5. (Continued)

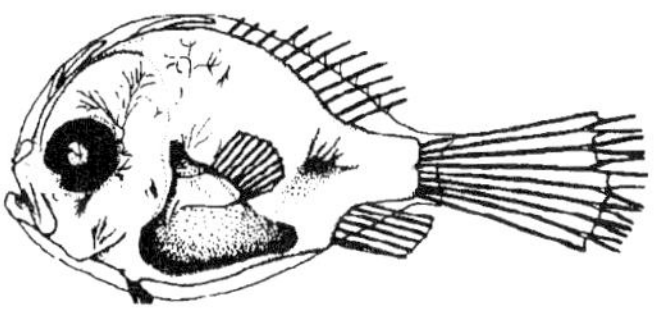

Sargassumfish, 5.3 mm TL
Histrio histrio
Family Antennariidae
(in Pietsch, 1984;
from Adams, 1960)

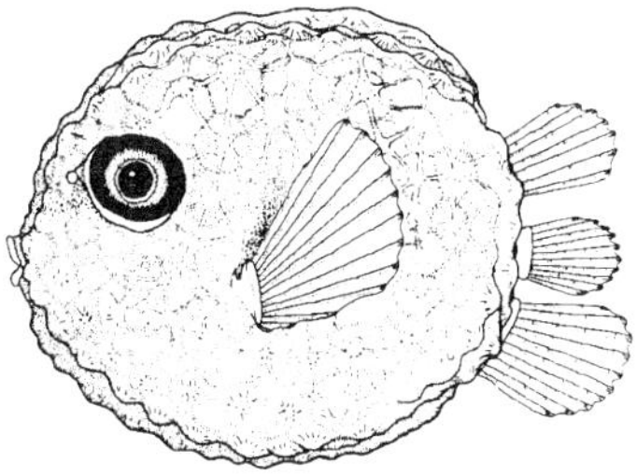

Scrawled cowfish, 6.0 mm SL
Lactophrys quadricornis
Family Ostraciidae
(Aboussouan and Leis, 1984)

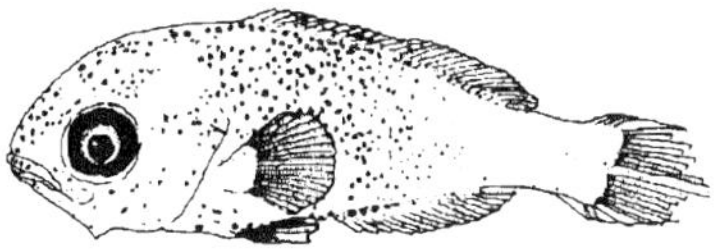

Striped knifejaw, 7.5 mm SL
Oplegnathus fasciatus
Family Oplegnathidae
(Johnson, 1984)

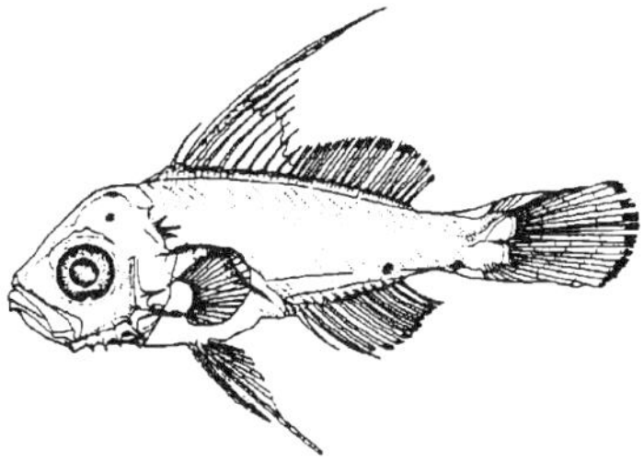

Atlantic red snapper, 7.3 mm SL
Lutjanus campechanus
Family Lutjanidae
(Collins et al., 1980)

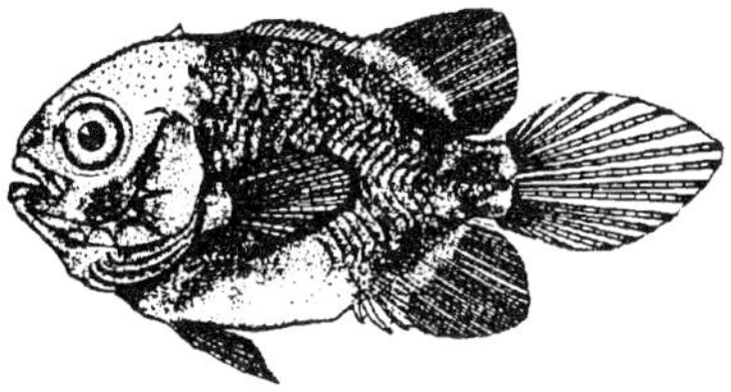

Atlantic spadefish, 9 mm
Chaetodipterus faber
Family Ephippidae
(Hildebrand and Cable, 1938)

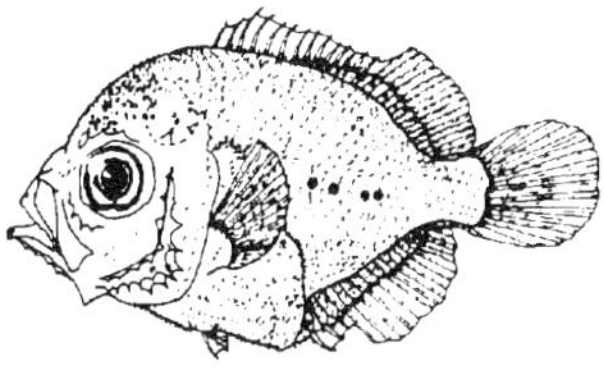

Pygmy angelfish, 4.4 mm SL
Centropyge sp.
Family Pomacanthidae
(Leis and Rennis, 1983)

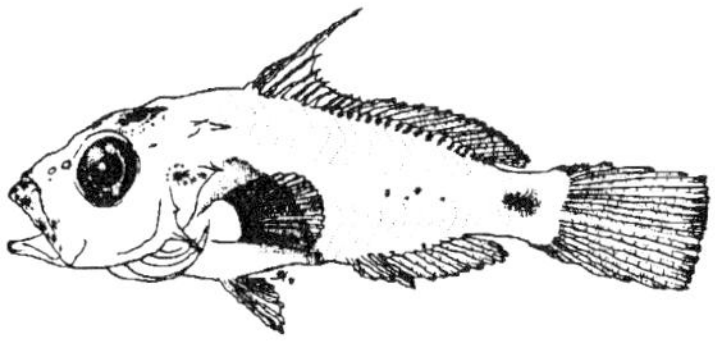

Nassau grouper, 18 mm TL
Epinephelus striatus
Family Serranidae
(Powell and Tucker, 1992)

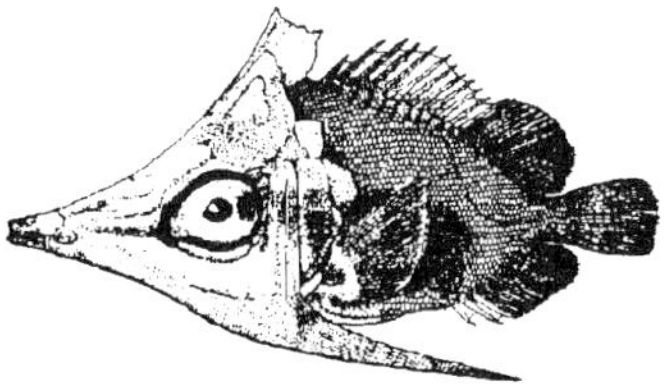

Longnose butterflyfish, 17 mm TL
Forcipiger longirostris
Family Chaetodontidae
(in Johnson, 1984;
from Kendall and Goldsborough, 1911)

Figure 2-6. Examples of late stage postlarvae.

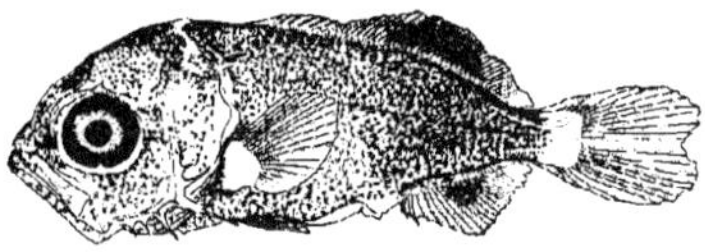

Banded rudderfish, 9.5 mm SL
Seriola zonata
Family Carangidae
(Laroche et al., 1984)

Northern sennet, 14.5 mm SL
Sphyraena borealis
Family Sphyraenidae
(Houde, 1972)

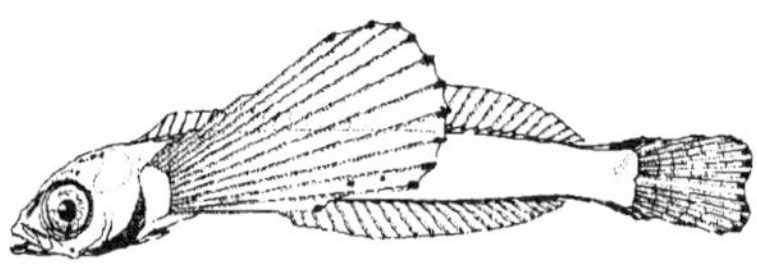

Rockskipper, 11.0 mm SL
Istiblennius zebra
Family Blenniidae
(Miller et al., 1979)

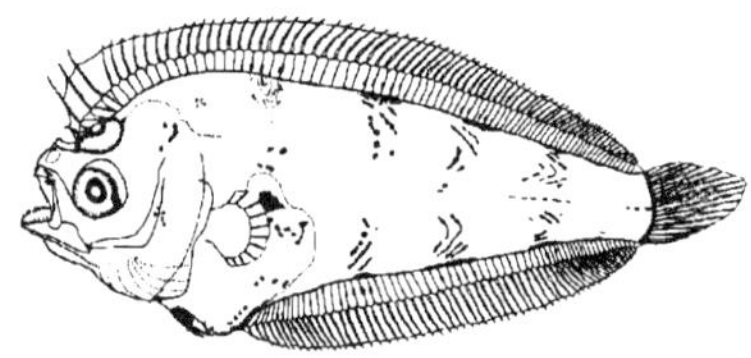

Horned whiff, 17.4 mm SL
Citharichthys cornutus
Family Bothidae
(Tucker, 1982)

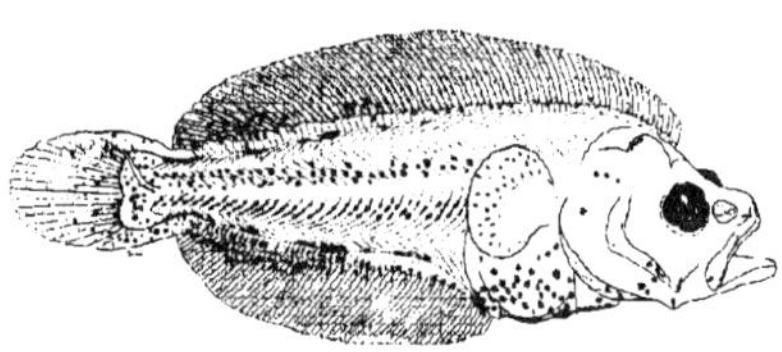

Slender sole, 14.7 mm
Eopsetta exilis
Family Pleuronectidae
(in Ahlstrom and Moser, 1984;
from Ahlstrom and Moser, 1975)

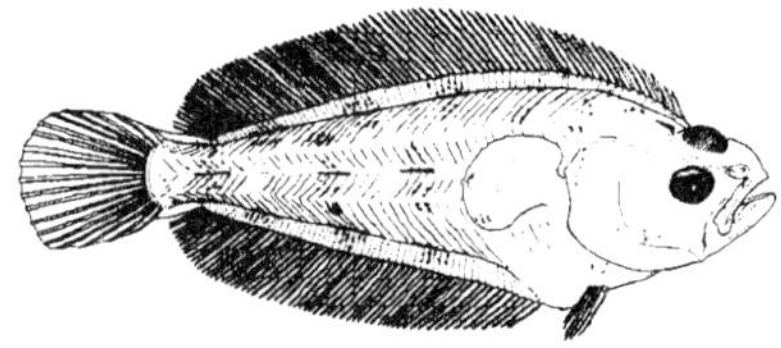

Slender sole, 22.0 mm
Eopsetta exilis
Family Pleuronectidae
(in Ahlstrom et al., 1984;
from Ahlstrom and Moser, 1975)

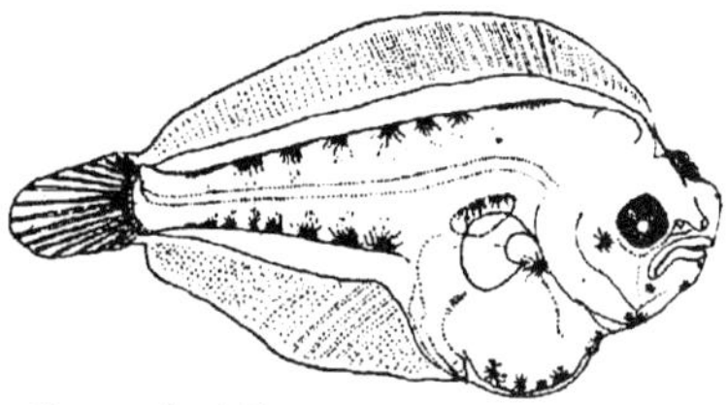

Ovate sole, 4.7 mm
Solea ovata
Family Soleidae
(in Ahlstrom et al., 1984
from Jones and Pantulu, 1958)

Figure 2-6. (Continued)

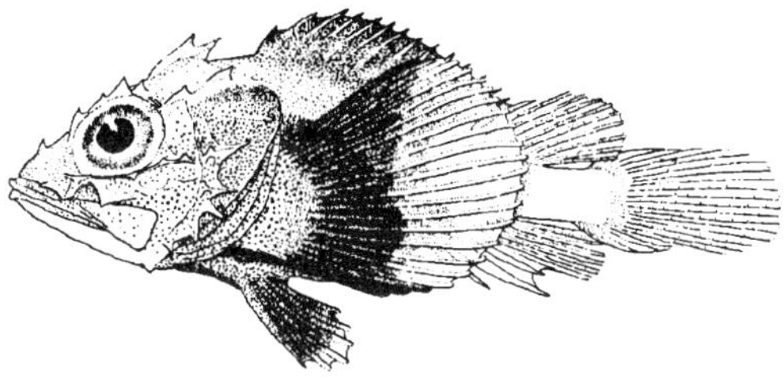

Longspine thornyhead, 26.8 mm pelagic juvenile
Sebastolobus altivelis
Family Scorpaenidae
(Moser et al., 1977)

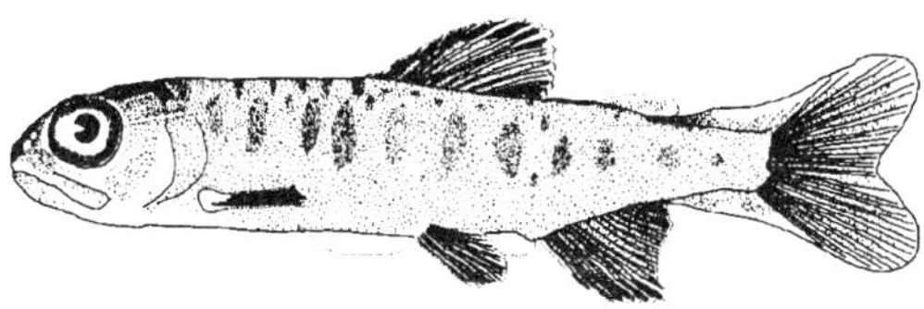

Coho salmon, 37 mm alevin
Oncorhynchus kisutch
Family Salmonidae
(Auer, 1982)

Figure 2-7. Examples of juvenile fish.

has reached its final position and the caudal skeleton is mostly formed. To evaluate fish size, either **wet weight** or **total length** (TL) are appropriate for most purposes. In describing development, **body length** (BL) often is used; this is equivalent to **notochord length** (NL, tip of snout to tip of notochord) from hatching until postflexion and **standard length** (SL, tip of snout to posterior margin of hypural bones at the caudal fin base) thereafter.

At hatching, larvae of most marine species are in the length range 1-5 mm TL. Many are fragile and cannot see, swim well, or catch food for 1-7 d after hatching (dah). Eyes usually become pigmented and functional (**eye pigmentation**, EP) just before **first feeding** (FF)(Blaxter, 1986). The senses of taste and smell develop early—between hatching and 16 dah in several species (Iwai, 1980; Blaxter, 1986). The ability of larvae to detect pressure changes (by neuromasts) in the water around them exists before EP in numerous species, including Atlantic herring (Blaxter et al., 1983), northern anchovy (O'Connell, 1981), Atlantic cod (Morrison, 1993), white seabass (Margulies, 1989), red sea bream (Yamashita, 1982b), Atlantic halibut (Haug, 1990), and others (Iwai, 1967, 1980; Blaxter, 1986).

Good descriptions of **sensory system** development in marine fish larvae are given by O'Connell (1981) and Hunter and Coyne (1982) for northern anchovy, Kawamura and Ishida (1985) for hirame, and Blaxter (1991) for Atlantic herring. The review by Blaxter (1986) is a good summary.

In larvae that hatch at a small size, **gas exchange** occurs through the skin for a while, but within a few to several weeks after hatching, functional gill filaments have formed. This usually is late in larval development or at transformation. Cutaneous gas exchange is not sufficient for larger fish. Northern anchovies have functioning gills at 16-17 mm BL, 32 dah (Blaxter and Hunter, 1982). In Atlantic herring at 20 mm BL, gill filaments are present and by 35-44 mm they are functioning (de Silva, 1974). European plaice at 8 mm BL have gill filaments and secondary lamellae, which begin functioning by 12 mm, at metamorphosis (de Silva, 1974).

Except for some coldwater species, within a few hours to a few days after larvae develop the ability to catch and digest zooplanktonic prey, they have exhausted their yolk and oil and are totally dependent on food. **End of the oil globule** (EOG) often is 1-2 d (or more) later than **end of the yolksac** (EYS). At least in some species, the oil globule lasts longer when larvae are starving. This is partly because the fish are less active when not feeding, and it results in conservation of energy stores and prolongation of life of the fish, sometimes until food becomes available.

The **point of no return** (PNR) is defined as the point beyond which a starving fish that finally encounters sufficient food cannot recover and survive even if it begins feeding (Blaxter and Hempel, 1963; Blaxter and Ehrlich, 1974). PNR usually occurs after EYS, but in some species before and in others after EOG.

The **digestive system** takes time to develop and does not have all the necessary enzymes at first. Most fish have a liver and pancreas at first feeding but no stomach with functional gastric glands until after transformation (Govoni et al., 1986). Some species (e.g., salmon, tilapia) have all of these by the time of first feeding and can eat compound feeds from the beginning. In typical larval fish, three stages of differentiation occur (Govoni et al., 1986). During embryonic development to first feeding, the incipient gut differentiates into a buccopharynx, foregut, midgut, and hindgut. The foregut is mainly for storage, and the last two sections for digestion. During transformation, the foregut becomes the esophagus and stomach, the midgut becomes the anterior intestine, the hindgut becomes the posterior intestine, and adult digestive processes are established.

The gut of a larval barramundi is at first a straight tube, but by 8 dah is completely coiled and has a pouch that will become the stomach (Walford and Lam, 1993). By day 11, the pouch has begun to transform into a stomach. Trypsin-type activity was present in varying amounts during the first few weeks, but dropped to zero by 30 dah (Walford and Lam, 1993). A significant portion of that activity could have been contributed by rotifers in the gut. Pepsin-type activity increased as the stomach developed, going from near zero at 8 dah to a high level on day 23 (approximately at transformation to the juvenile stage).

Most species, including those that are herbivorous as adults, have carnivorous larvae. This is advantageous because (1) animal food is more nutritious, (2) the enzyme cellulase is needed to break down cellulose in algae, and (3) zooplankton contain their own enzymes that will aid the larval fish in digesting them.

Dissolved nutrients and small particles swallowed with water can provide some nutrition. Dissolved glucose was utilized by first-feeding larvae of both European sea bass and gilthead sea bream, but more so by bass (Diaz et al., 1994).

Larval marine fish are known to ingest diatoms, dinoflagellates, flagellates, tintinnids, ciliates, rotifers, cladocerans, copepods, and appendicularians; eggs of bivalves, cladocerans, copepods, chaetognaths, and fish; larvae of polychaetes, lamellibranchs, bivalves, gastropods, barnacles, cladocerans, copepods, decapods, and fish (Blaxter and Hunter, 1982; Kane, 1984; Jenkins, 1987; Fortier and Harris, 1989; Heath et al., 1989; Ozawa et al., 1991; Sánchez-Velasco and Norbis, 1997). If it is the right size, not too prickly, and moves the right way, it is potential food; however, digestibility and nutritional value of these items varies (e.g., diatoms probably are not very useful to larval fish). For many marine fish larvae, copepods are the primary (and possibly the most nutritious) prey; however, smaller or more easily-caught prey often are preferred from first feeding until the fish are more proficient at catching the larger, faster, copepods.

In most cases, the early stages of marine fish larvae do not accept artificial foods and will not survive or grow well on them. However, many species can be weaned on to artificial food just before or during transformation and then will thrive on it. Species, such as summer flounder, that develop a functional stomach during the larval stage (Bisbal and Bengston, 1995a) are more likely to be successfully weaned as larvae than species, such as turbot, that do not have a working stomach until they are juveniles (Segner et al., 1994). Reducing the age at weaning will depend mainly on advances in two areas: feed digestibility and feeding behavior. Fish larvae without fully developed digestive systems probably rely somewhat on free amino acids from their prey and are likely to benefit from the addition of FAAs to compound feeds (Dabrowski and Rusiecki, 1983; Fyhn, 1989). Tricking the larvae into eating non-swimming particles (Barnabé and Guissi, 1994) also is necessary. See chapter 8 for further discussion of weaning.

Walford et al. (1991) reported that the protein membrane enclosing microencapsulated larval feed was broken down in early barramundi larvae only when the larvae had also eaten rotifers. Live prey not only contribute exogenous digestive enzymes, but also tend to stimulate greater secretion of endogenous enzymes than inert particles (see review in Walford and Lam, 1993).

In a typical warmwater species with planktonic eggs, at 2 dah there is less yolk and oil, eyes are pigmented, and the digestive system is functional but simple. At 3-4 dah, yolk is exhausted, a tiny bit of oil might remain, but the larva still has relatively low swimming and feeding ability. At 10 dah, the skeleton is at least partly ossified, fins are forming, and larvae swim faster and feed more efficiently.

Kohno et al. (1994b, 1994c) discussed the length of time after 100% of larvae are feeding until yolk and oil are exhausted as a major factor in **survival ability** and ease of rearing. Those times in five species were: milkfish, 0 h; brownmarbled grouper, 1.5 h; yellowblotch spinefoot, 44.5 h; kawakawa 47 h, and barramundi, 74 h. Milkfish compensate for quick energy utilization by large mouth size (258 μm width), which makes feeding easier, and large body size, which enhances feeding ability and also provides an energy reserve. Barramundi have small mouths and bodies, but feed well and are supported by endogenous nutrition for at least 3 d after feeding is established. Rabbitfish have about 2 d of yolk and oil left after first feeding, but their feeding capabilities are lower than those of barramundi and their small mouths (187 μm width) especially put them at a disadvantage. Grouper larvae are relatively deficient in body size, feeding capability, and endogenous nutrition, so the difficulty of raising them is not surprising. Brownmarbled grouper larvae have larger mouths than others such as greasy grouper, which puts the latter species at an even greater disadvantage, at least in the hatchery (Lim, 1991b).

Juvenile and Adult Stages

Nearly all marine fish larvae do not metamorphose (like butterflies), but pass through a **transformation** phase to become juveniles. Only a few types, such as eels and flatfishes, metamorphose. The larvae of most warmwater and coolwater mariculture candidates transform into juveniles within 20-60 dah; however, some precocious species could be considered juveniles within a few days, coldwater species might take several months, and the American eel as long as a year. In juveniles, the adult complement of meristic elements (bones, fin rays), scales, and well-developed teeth are present, as well as greatly improved ability to escape predators and to feed, plus increased tolerance to handling. Terms like fry and fingerling are vague and unsatisfactory and are avoided by scientists. The adult stage is reached at sexual maturity, which can be after a few months (e.g., tilapia) to 20 yr or more (e.g., possibly giant sea bass).

CRITICAL PERIODS

There are several "critical periods," during which the strength of a wild or captive cohort of larvae can be reduced to zero. In addition to the need for suitable physical conditions and water quality throughout the egg and larval stages, certain other requirements must be met for good survival to result. Some major survival factors related to eight common critical periods or events are listed in chronological order below. For wild larvae, first feeding was the earliest suggested and described critical period, but predation probably can be at least as important in determining cohort survival (Hjort, 1914; Marr, 1956; May, 1974). For cultured larvae, predation from cannibalism can be a major problem in some species. For most species in culture that have small larvae, the period from first feeding until just after yolk and oil exhaustion usually is the most critical.

(1) **At fertilization**
Quality of gametes
Temperature and salinity
Correct timing
Handling

(2) **At hatching**
Strength of embryos
Correct temperature
Dissipation of hatching enzymes
Lack of toxic substances
Correct light intensity (for some species)

(3) **At first feeding**
Strength and development of larvae
Correct size, availability, quality, etc. of food
Appropriate lighting
Correct timing

(4) **At yolk and oil globule exhaustion**
Adequate yolk and oil until food intake is good enough
Correct temperature
Sufficient uptake of nutrients from exogenous sources

(5) **At gas bladder filling**
Strength of larvae
Light not too intense
Appropriate circulation
Lack of surface films

(6) **At a change in diet**
Digestive capability
Correct size, availability, quality, etc. of food
Correct timing
Adequate transition period

(7) **At transition to gill gas exchange**
Overall health of larvae

(8) **At transformation**
Although less vulnerable to physical injury (e.g., protected by fins and scales) juvenile fish in the wild can be more conspicuous to predators than the larvae, which are partly transparent.

GROWTH

Size hierarchy (or growth depensation, wide size range in fish reared from the same spawn) can occur not only because of genetic differences in growth rate but also because of competition for food, social dominance, and lack of predation on small fish (Blaxter and Hunter, 1982). Runts that would be lost to predators in nature often survive in culture, but inhibition of feeding and aggression or cannibalism (in some species) by larger fish might eliminate them anyway.

The simplest measure of growth is the **absolute growth rate** (AGR), which can be used to compare fish starting at the same size but cannot be used to accurately predict intermediate weights because fish growth is not linear. This is simply the average growth per unit time (BW_1 = body weight at beginning, BW_2 = body weight at end).

$$AGR = (BW_2 - BW_1)/days$$

The widely used **specific growth rate** (SGR) is also known as instantaneous, intrinsic, exponential, logarithmic, or compound interest growth rate (Ricker, 1979).

$$SGR = (\ln BW_2 - \ln BW_1)/days$$

The **thermal-unit growth coefficient** (TGC) is considered by Cho (1992) to be more accurate than SGR for estimating weights between BW_1 and BW_2 (for which SGR underestimates) and for extrapolation beyond BW_2 (SGR overestimates).

$$TGC = [(BW_2)^{1/3} - (BW_1)^{1/3}]/\Sigma[(°C)(days)]$$

WILD VS HATCHERY FISH

Factors

Hatchery-reared fish often are different from wild ones, particularly in shape, color, and fitness. Reared fish tend to be shorter and fatter (Blaxter and Hunter, 1982). Two major factors are high feeding rates and sensory deprivation, such as less exposure to extremes of light, temperature, and currents, and to prey and predators. Also, unnatural foods and exposure to chemicals, electric lights, and unnatural surfaces of rearing units can cause anomalies. Factors fall mainly into these categories:

Crowding.—Typical natural density of planktonic marine fish larvae is 1 larva/m^3, with patches as dense as 50-500 larvae/m^3 (Economou, 1991). Typical culture density is ten or more larvae/L, a thousand times more dense than the natural condition.

Different environment.—Planktonic larval fish are not fully prepared to deal with obstacles not present in the ocean, such as walls, screens, pipes, and air streams. With careful management, these barriers can be tolerated. If electric lights are used, the spectrum will be different, but not necessarily harmful. With indoor tanks, the temperature range is likely to be narrower than in the natural habitat, but with outdoor tanks and ponds, the range can be much wider. (see chapter 3)

Different food.—In nature, although the first food might be limited to one or a few species, fish larvae soon grow large enough to partake of a wide variety of zooplankton. In culture, the entire menu might consist of rotifers, *Artemia*, and compound feeds. (see chapters 8 and 9)

More food.—Typical natural density of zooplankton is 15-100 plankters/L (Economou, 1991) but can be higher in patches and in coastal areas (Blaxter and Hunter, 1982). Typical culture density is ten or more plankters/mL (range 0.5-200/mL), about a thousand times more dense than the natural condition. After weaning, cultured fish usually are fed to satiation once to many times a day—good for growth but an unnatural situation. (see chapters 8 and 9)

Metabolic products.—Of special importance are toxic metabolites such as ammonia, urea, and carbon dioxide that can reach harmful concentrations in culture systems but very rarely do so in nature. (see chapters 3-5)

Toxic contaminants.—Plastics, metals, chemicals, etc. used in hatcheries can contaminate culture systems with traces of highly toxic substances that are expensive to measure and that result in disorders or mortality that are not easy to diagnose. (see chapters 3-5)

Inhibitory substances.—These can be metabolites or artificial contaminants that reduce health, growth, food utilization, etc., but are not immediately lethal. (see chapters 3-5)

Quality

In commercial farming the main concerns are for production and marketability. In stock enhancement programs, production and appearance standards must be met, but fitness is of overriding importance. To maximize genetic diversity in released stocks, survival of all fish should be strived for. The runts might be carrying needed genes, and they possibly could have turned out to be the biggest fish, not the smallest, if they had lived under natural conditions. Reared fish might not be as good at avoiding predators, finding food, and surviving in the ocean, but training them in these areas (a seaward-bound program) before release can help (Howell, 1994).

Hatchery and wild fish tend to respond differently to various stresses (Howell, 1994). Differences in nutrition (e.g., fatty acid imbalance), environment (e.g., high nitrite), density (high), behavioral conditioning (e.g., lack of encounters with other organisms and natural structures) could make hatchery fish less likely to survive in a natural habitat. On the other hand, the fish that survived in the hatchery tend to have a genetic structure that made them better able to tolerate hatchery conditions than those that did not survive in the hatchery. If they are later used as broodfish, the survivors' offspring are likely to have higher survival ability than the original group—and so on. The fish that survived well in the hatchery might not have survived as well in nature. Survival traits might not be recognized even by experienced human observers. Therefore, it is very important for stock enhancement hatcheries to avoid selecting against fitness in nature—by using large numbers of broodfish, by replacing them as frequently as possible, and by using as many wild broodfish as possible. Selective breeding is used in food production and might also have potential for stock enhancement (Naevdal, 1994).

SUMMARY

Much variability exists among species in spawning method, number of eggs produced per female, larval competence, and development to the juvenile stage. Some fish can be reared on compound feed from right after hatching. Others must be nursed with live zooplanktonic foods through long and often difficult larval stages. Especially for the latter species, critical periods or events can be times of complete mortality (which can happen in between as well). Survival, growth, appearance, and health of reared fish usually differ from those of wild ones, with the culturist's goal being to improve all of these.

Table 2-1. Some early life history characteristics for about 400 fish species and hybrids, arranged by egg type and approximately in order of decreasing egg size (common names are listed in the species index). Usually, if ranges are wide, they are given, but narrow ranges are represented by mean values. Listed temperatures in most cases are normal for spawning. Optimal rearing temperatures for larvae could be lower or higher. When possible, age and length at transformation refer to the time at which scales have been formed, but in some cases age and size have been given for an early phase of transformation before scales are present. Original sources are not always provided but can be found in the cited references (also see Brownell, 1979, and other compendiums).

Fish	Family	Egg diameter (μm)	Normal temp. (°C)	Time to hatch (d)	Length at hatch (mm)	Age at FF (dah)	Age at EYS (dah)	Age at trans. (dah)	Length at trans. (mm)	References
Eggs Carried										
Galeichthys feliceps	Ariidae	13,190×15,650	22	~75-80	~30	~100	~130	~140	~55	Tilney & Hecht, 1993
Phycodurus eques	Syngnathidae	4000×7000	19	6-7	35	2-3		~14-18		Kuiter, 1988
Apogon conspersus	Apogonidae	4500								Thresher, 1984
Hippocampus erectus	Syngnathidae	?×3500		12-14$^+$	≤3.6				~3.3	Hardy, 1978
Oreochromis mossambicus	Cichlidae	2190×3040	25	4.7	4.1	6-7	11-12	~18	~13	Holden & Bruton, 1992
Pterapogon kauderni	Apogonidae	2500	~26					~20-25	~10-11	Allen, 1996; Marini, 1996; Ounaïs, 1997
Oreochromis urolepis × *O. mossambicus*	Cichlidae	1960×2630	32					~7		Watanabe, p.c., 1993
Bagre marinus	Ariidae	1500×2150			~45		~42-70		~80	Mansueti & Hardy, 1967
Sebastes schlegeli[a]	Scorpaenidae	1250	10	48	6.3			~35	~25	Kusakari, 1991; Yamada & Kusakari, 1991
Assessor macneilli	Plesiopidae	920		16	4	1				Allen & Kuiter, 1976; Thresher, 1984
Syngnathus fuscus	Syngnathidae	875		~10	~3				≤37	Hardy, 1978
Opistognathus aurifrons	Opistognathidae	850	26	8	4	0		9-15$^+$		Thresher, 1984; Walch, 1993
Hippocampus zosterae	Syngnathidae		26	10	8.5	0		0		Breder & Rosen, 1966; Giwojna, 1990
Eggs Adhesive and Sinking										
Macrozoarces americanus	Zoarcidae	8500	~5	75-105	39	0-2	<1		≤73	Fritzche, 1978; Brown et al., 1992
Anarhichas lupus marisalbi	Anarhichadidae	5100	6	116-163	~22	0-6				Pavlov & Novikov, 1987
Opsanus tau	Batrachoididae	5000	19	14-21	7.0		20	~20	~18	Dovel, 1960; Breder & Rosen, 1966
Opsanus beta	Batrachoididae	4100	19							Breder & Rosen, 1966
Acipenser transmontanus	Acipenseridae	3800×4000	15	7.3	8	13	13			Conte et al., 1988
Tylosurus acus	Belonidae	3660	28	~10-12	10.1				~23	Hardy, 1978
Tylosurus acus	Belonidae	3370			14		2			Shirota, 1970
Blepsias cirrhosus	Cottidae	3100	10	~200	11.7					Munehara & Shimazaki, 1991
Sagamia geneionema	Gobiidae	1160×6825	14	17.7	5.5	~18				Shiobara et al., 1990
Cyclopterus lumpus	Cyclopteridae	2600	~10	28.5	5.8	5-8	4		~23	Fritzche, 1978; Brown et al., 1992
Chirolophis japonicus	Stichaeidae	2500	7	~60	12.8	0-1	~14	~39	20-24	Shiogaki, 1983
Acipenser oxyrhynchus	Acipenseridae	2450	18	7	~11		~6		60?	Jones et al., 1978

Table 2-1. (Continued)

Fish	Family	Egg diameter (μm)	Normal temp. (°C)	Time to hatch (d)	Length at hatch (mm)	Age at FF (dah)	Age at EYS (dah)	Age at trans. (dah)	Length at trans. (mm)	References
Opisthocentrus tenuis	Stichaeidae	2300	6	58-64	11.4	0-1	~20	>56	>19	Shiogaki, 1981b
Atherinopsis californiensis	Atherinidae	2250	22	9	8.3	2	2			White et al., 1984; Middaugh et al., 1990
Hemirhamphus sajori	Exocoetidae	2200			7.7		2			Shirota, 1970
Myoxocephalus octodecemspinosus	Cottidae	2100		~90	6.8				~15	Fritzche, 1978
Acanthogobius flavimanus	Gobiidae	900×5500	13	28						Breder & Rosen, 1966
Pterogobius zacalles	Gobiidae	875×4550	10	~27	5.1	0-1	6-9	>31	>16	Shiogaki, 1981a
Opisthocentrus ocellatus	Stichaeidae	1950	5	48	9.5	0-1	23			Shiogaki, 1982
Luciogobius saikaiensis	Gobiidae	1300×3000	18	~10						Breder & Rosen, 1966
Fundulus heteroclitus	Cyprinodontidae	1875	22	11-21	~4-8	0	0-6.5		~25	Hardy, 1978; Marteinsdottir, 1991; Marteinsdottir & Able, 1992
Cheilopogon heterurus	Exocoetidae	1860	21	~14	≤5.8				~19	Hardy, 1978
Pseudoblennius cottoides	Cottidae	1790	16	14	6.7				20-25	Kimura et al., 1987
Cololabis saira	Scomberesocidae	1745			6.2		2			Shirota, 1970
Pseudoblennius percoides	Cottidae	1740	16	13-16	6.9				~20	Kimura et al., 1988
Hypoatherina tsurugae	Atherinidae	1690	~20	13-15	~6.4		2-3		~20	Tsukamoto & Kimura, 1993
Atherinops affinis	Atherinidae	1620	19		5.4	1-2	1-2			White et al., 1984; Middaugh et al., 1990
Gasterosteus aculeatus	Gasterosteidae	1600	17	~6	~3.0		~7		~11	Kuntz & Radcliffe, 1917; Hardy, 1978
Pomacentrus nagasakiensis	Pomacentridae	1100×2300	28	3.5	2.9	~0		15-20	~8	Honda & Imai, 1973; Thresher, 1984
Mogurnda mogurnda	Eleotridae	900×2600	22	13						Breder & Rosen, 1966
Chaeturichthys hexanema	Gobiidae	800×2900	16	~10						Breder & Rosen, 1966
Eutaenichthys gilli	Gobiidae	800×2700	23	8.5						Breder & Rosen, 1966
Pseudomugil mellis	Pseudomugilidae	1465	26	5-8	3.7	~3	~3		~10	Semple, 1991
Anoplarchus purpurescens	Stichaeidae	1450	8	~14						Breder & Rosen, 1966
Clupea pallasii	Clupeidae	1450			6.8		7			Shirota, 1970
Lipophrys pholis	Blenniidae	1420	12	~50						Breder & Rosen, 1966
Amphiprion percula	Pomacentridae	1000×2200	25	8-10	4	0		12		Garnaud, 1951; Breeder's Registry Database, 1995
Amphiprion ocellaris	Pomacentridae	1000×2000	27	7-11	4	0	2-3	~25	~14	Moe, 1973, 1992; Hoff, 1996
Galaxias platei	Galaxiidae	1400	12	20	~5.9					Ortubay and Wegrzyn, 1991
Cyprinodon variegatus	Cyprinodontidae	~1400	24	5-6	~4	2	≥5		~12	Kuntz, 1914; Schimmel & Hansen, 1975; Hardy, 1978; Perschbacher & Strawn, 1988; Breeder's

										Registry Database, 1995
Pholis gunnellus	Pholidae	1400	6	56						Breder & Rosen, 1966
Typhlogobius californiensis	Gobiidae	700×2800	18	~11						Breder & Rosen, 1966
Hirundichthys oxycephalus	Exocoetidae	1370	30	0.5-1	3.4	2	2			Chen, 1978
Alcichthys alcicornis	Cottidae	1350	8	18	4.4				~20	Munehara & Mishima, 1986
Takifugu chrysops	Tetraodontidae	1320	20	6.75	2.9		4	24	9	Fujita & Shinohara, 1986
Luciogobius guttatus	Gobiidae	700×2500	18	~10						Breder & Rosen, 1966
Takifugu porphyreus	Tetraodontidae	1280	~18	5.9	2.5	~4	4	~20	~13	Fujita & Abe, 1992
Gadus macrocephalus	Gadidae	1275	~6	~10	3.6		3-4			Shirota, 1970; Takatsu et al., 1995
Gobiosoma oceanops	Gobiidae	~700×2300	25	6-10	3.5	~0.5	~0.5	18-40	≤8	Valenti, 1972; Colin, 1975; Moe, 1975: Thresher, 1984; Breeder's Registry Database, 1995
Fundulus grandis	Cyprinodontidae	1250	25	14						Tatum et al., 1982b
Pseudochromis fuscus	Pseudochromidae	1250	29	4	2.5	1				Thresher, 1984
Takifugu pardalis	Tetraodontidae	1250	17	21						Breder & Rosen, 1966
Takifugu exascurus	Tetraodontidae	1240	20	6.7	3.0	2	3	16	9.6	Fujita & Honma, 1991
Paraclinus marmoratus	Clinidae	1225	21	10	4.1	~1				Thresher, 1984
Clupea harengus	Clupeidae	1200	8	15	7	4	6	112-168	30-40	Theilacker & Dorsey, 1980; Urho, 1992
Pseudochromis flavivertex	Pseudochromidae	1200	25	4-5	3.9	0-1		35-40	~15	Brons, 1996
Salaria fluviatilis	Blenniidae	1200	20	~14						Breder & Rosen, 1966
Pterogobius zonoleucus	Gobiidae	600×2100	10	15						Breder & Rosen, 1966
Hypoatherina bleekeri	Atherinidae	1114	22	10-11	4.6	0-1	3			Takita & Nakamura, 1986
Etrumeus micropus	Clupeidae	1100			3.8		3			Shirota, 1970
Lovettia sealii	Galaxiidae	1100	13		5	<1				Daintith, 1992
Menidia menidia	Atherinidae	1100	25	6-10	5.2					Bengtson et al., 1987
Microspathodon chrysurus	Pomacentridae	1100		~3		<1		~50	~13	Potthoff et al., 1987
Aspasma minima	Gobiesocidae	925×1300	~21	12-15	3.8	0-1	0-1	~20	~7	Shiogaki & Dotsu, 1971
Acyrtops beryllinus	Gobiesocidae	911×1131	24	10-11	2.6					Gould, 1965
Bostrichthys sinensis	Eleotridae	1015	27		~3.3	4	4	~20		Zhang, 1992
Oxyeleotris marmorata	Eleotridae	660×1560	29	1.9	3.2	4-6	4-5	~40	~13	Tan & Lam, 1973; Senoo et al., 1994a, 1994b
Gramma loreto	Grammatidae	1000	~27	6-7		1		21-30		Moe, 1992; Breeder's Registry Database, 1995
Pseudochromis fridmani	Pseudochromidae	~1000	25	3-5	3.7	0-1		~20-40	≤19	Brons, 1996; Gardner, 1997; Moe, 1997a
Calloplesiops altivelis	Plesiopidae	~1000	26	5-6	3	0	0	~23	~8	Wassink & Brons, 1990
Tripterygion etheostoma	Trypterygiidae	1000	18	16	4.8	~1	~1	~40	~12	Thresher, 1984
Meiacanthus nigrolineatus	Blenniidae	1000	24	10-11		0	0	~20	≤15	Fishelson, 1976
Takifugu niphobles	Tetraodontidae	1000	26	~4.5						Breder & Rosen, 1966
Takifugu rubripes	Tetraodontidae	1000	17	~8.5	2.7		6	~30	9-15	Breder & Rosen, 1966; Shirota, 1970; Kuronuma & Fukusho, 1984; Moksness & Støle, 1997
Tetraodon cutcutia	Tetraodontidae	1000	26	6.5						Breder & Rosen, 1966
Chasmodes bosquianus	Blenniidae	850×970	25	~11	3.6				~12	Fritzche, 1978
Bathygobius soporator	Gobiidae	410×2400	29	~3						Breder & Rosen, 1966

Table 2-1. (Continued)

Fish	Family	Egg diameter (µm)	Normal temp. (°C)	Time to hatch (d)	Length at hatch (mm)	Age at FF (dah)	Age at EYS (dah)	Age at trans. (dah)	Length at trans. (mm)	References
Morone chrysops × *M. saxatilis*	Percichthyidae	980	18	1.8		6		~30		Hodson, 1991
Atherion elymus	Atherinidae	975	21	13-17	4.4		2-3		20	Kimura & Tsukamoto, 1990
Tripterygion etheostoma	Tripterygiidae	965	~17	~16	4.8	~0	~0	~48	~15	Shiogaki & Dotsu, 1973
Plecoglossus altivelis	Osmeridae	960	15	20	6-7	~4-5	4-5	~110	~52	Breder & Rosen, 1966; Azuma, 1964; Tsukamoto & Kajihara, 1984; Tachihara & Kimura, 1991; Iguchi & Yamaguchi, 1994; Lee et al., 1995
Orthonopias triacis	Cottidae	930	13	17.5						Breder & Rosen, 1966
Gobius lidwilli	Gobiidae	700×1200	25	4.5						Breder & Rosen, 1966
Chromis punctipinnis	Pomacentridae	650×1220			3.1					Turner & Ebert, 1963
Abudefduf saxatilis	Pomacentridae	660×1175	26	5-7	3.3	0	3-4	~32	~23	Shaw, 1955; Alshuth et al., 1998
Gobiosoma bosc	Gobiidae	600×1300	27	4	2.0				~15	Kuntz, 1914; Hildebrand & Cable, 1938; Breder & Rosen, 1966; Fritzche, 1978; Fahay, 1983
Parioglossus taeniatus	Eleotridae	600×1300	25	4						Breder & Rosen, 1966
Sphoeroides maculatus	Tetraodontidae	874	19	4.7	~2.4	2-3	5-10		~7	Welsh & Breder, 1922; Martin & Drewry, 1978; Fahay, 1983
Triaenopogon barbatus	Gobiidae	500×1500	25	~4						Breder & Rosen, 1966
Osmerus eperlanus	Osmeridae	850	~10	~11-21	~5.5				≥30	Urho, 1992
Pleuronectes americanus	Pleuronectidae	835	3	21	2.9	5	9	58	7-13	Theilacker & Dorsey, 1980; Fahay, 1983
Salarias enosimae	Blenniidae	835	28	~7	3.6					Mito, 1954
Gobiesox strumosus	Gobiesocidae	710×940	20	5-7	2.9	0^+	0^+		~12	Runyan, 1961; Breder & Rosen, 1966; Martin & Drewry, 1978; Allen, 1985
Brachaluteres ulvarum	Balistidae	820	22	~6	2.1	~3	4-5			Akagawa et al., 1995
Symphodus roissali	Labridae	820								Breder & Rosen, 1966
Abudefduf luridus	Pomacentridae	640×1110			3.1	0-1				Ré, 1980
Chromis dispila	Pomacentridae	610×1000	21	5.5	2.9	3	3		>7	Kingsford, 1985
Chromis sp. (Azorian)	Pomacentridae	650×910			2.8				≤14	Ré & Gomes, 1982
Hypsoblennius hentz	Blenniidae	660×760	26	~11	2.6				~8	Breder & Rosen, 1966; Fritzche, 1978
Abudefduf sordidus	Pomacentridae	550×950								Koh & Kim, 1992
Tridentiger undicerneus	Gobiidae	450×1150	25	~4						Breder & Rosen, 1966
Hypleurochilus geminatus	Blenniidae	700	27	~7						Breder & Rosen, 1966
Pleuronectes yokohamae	Pleuronectidae	700	13	~6.5	3.7			30-35		Kuronuma & Fukusho, 1984; Chen et al., 1991b

Canthigaster valentini	Tetraodontidae	700	24	5.3	1.4	~3	~4			Stroud et al., 1989
Acentrogobius masago	Gobiidae	400×1200	23	4						Breder & Rosen, 1966
Mugilogobius abei	Gobiidae	450×980	24	4	2.1	0-1		44	15-19	Kanabashira et al., 1980
Chromis notata	Pomacentridae	575×760	27	~2.7	2.3		3			Fujita, 1957
Stephanolepis cirrhifer	Balistidae	660	20	2.4						Breder & Rosen, 1966; Thresher, 1984
Lagocephalus lunaris	Tetraodontidae	655	23	3.2	1.9	4	5	26-31	~5-7	Fujita, 1966
Stegastes leucorus	Pomacentridae	450×850	22	5						Breder & Rosen, 1966
Ecsenius bicolor	Blenniidae	500×750	27	9	3.5					Thresher, 1984
Siganus fuscescens	Siganidae	630	24	1.25	2.3	2	2	17	10	Duray, 1990
Canthigaster rivulata	Tetraodontidae	630	22	6.04	1.7		3			Arai & Fujita, 1988
Siganus chrysospilos	Siganidae	620		1.0	2.3					Duray, 1990
Sufflamen fraenatus	Balistidae	600	26		1.6	~3	4			Kawabe, 1984
Dascyllus aruanus	Pomacentridae	475×750	~24	~2.1				35-50		Fishelson, 1964; Danilowicz & Brown, 1992
Dascyllus trimaculatus	Pomacentridae	490×700	24	4-5	2.1					Garnaud, 1957
Trimma grammistes	Gobiidae	475×680	26	2.8	2.0	0-1	~2			Shiobara & Tanaka, 1994
Stegastes leucostictus	Pomacentridae	400×800	29	~5	3	0-1				Brinley, 1939
Balistes vetula	Balistidae	576	25	1-1.5	1.7	2.5	4-5			Tucker, unpub.
Siganus luridus	Siganidae	565	27	1.2		2.0				Popper et al., 1979; Duray, 1990
Siganus rivulatus	Siganidae	~565	27	1.2	2.0	1.8	2.8	~32	~29	Popper et al., 1979; Duray, 1990
Siganus guttatus	Siganidae	560	27	0.79	1.7	2-3	2-5	35-45	~22	Juario et al., 1985; Hara et al., 1986; Duray, 1990; Kohno et al., 1994c
Paramonacanthus japonicus	Balistidae	560		2-3						Nakazano & Kawase, 1993
Siganus vermiculatus	Siganidae	540	30	1.0	1.8			22-25	18-26	Duray, 1990
Siganus canaliculatus	Siganidae	520	27	1.25	1.7	1.6	2	21	20-24	Von Westernhagen & Rosenthal, 1975, 1976; Duray, 1990
Ammodytes tobianus	Ammodytidae	300×800	7	22						Breder & Rosen, 1966
Sicydium japonicum	Gobiidae	450	25	<2						Breder & Rosen, 1966
Eleotris oxycephala	Eleotridae	325×400	27	0.42						Breder & Rosen, 1966
Mallotus villosus	Osmeridae		20	14		0-5	5	~240		Jackson, 1991
Ophiodon elongatus	Hexagrammidae		10	35		5-6		50-55	20-22	Appelbaum et al., 1995
Siganus lineatus	Siganidae		29	1.08	2.5	2		29-35	20-23	Bryan & Madraisau, 1977; Duray, 1990
Siganus virgatus	Siganidae			1.0	2.5					Duray, 1990
Platichthys stellatus	Pleuronectidae		~13	~3.5		5-6	~5	33-90	~9	Policansky & Sieswerda, 1979
Eggs Adhesive and Suspended										
Lophius americanus	Lophiidae	1730×1865			2.5	~5-7	~5-7		≤76	Martin & Drewry, 1978
Scorpaena guttata	Scorpaenidae	1175×1255		2.7	1.9	4-6	4-8		>17	David, 1939; Orton, 1955; Thresher, 1984; Washington et al., 1984
Dendrochirus lunulata	Scorpaenidae	820	28	1.0	1.5	4				Thresher, 1984; Washington et al., 1984
Dendrochirus brachypterus	Scorpaenidae	800	26	1.5	1.1	4			~10-12	Thresher, 1984

Table 2-1. (Continued)

Fish	Family	Egg diameter (μm)	Normal temp. (°C)	Time to hatch (d)	Length at hatch (mm)	Age at FF (dah)	Age at EYS (dah)	Age at trans. (dah)	Length at trans. (mm)	References
Eggs Buried										
Scorpaena miostoma	Scorpaenidae	750×860	22	1.67	1.6	3-4	3			Kimura et al., 1989
Phrynelox scaber	Antennariidae	650×700	24	~3.0	0.9		4-5			Martin & Drewry, 1978
Histrio histrio	Antennariidae	600×700	22	4.5	1.4	~6.5	~6.5		~5	Mosher, 1954; Fahay, 1983; Thresher, 1984
Oncorhynchus keta	Salmonidae	8150	8	70	22	~45	40-50	~61	~55	Pauley et al., 1988; Kaeriyama, 1989; Beacham & Murray, 1990; Groot & Margolis, 1991
Oncorhynchus gorbuscha	Salmonidae	6000	8	75	21.5	~48	78	~106	~60	Kaeriyama, 1989; Beacham & Murray, 1990; Groot & Margolis, 1991
Oncorhynchus masou	Salmonidae	6000					72	~101	~50	Kaeriyama, 1989; Groot & Margolis, 1991
Oncorhynchus nerka	Salmonidae	5700	8	80	19.5	~55			60-100	Pauley et al., 1989; Groot & Margolis, 1991
Salmo salar	Salmonidae	5500	7	74	~20	54				Kane, 1988; Jobling 1995
Oncorhynchus kisutch	Salmonidae	5250	8	60	21	~40	43	~71	~45	Kaeriyama, 1989; Beacham & Murray, 1990; Groot & Margolis, 1991
Salvelinus alpinus	Salmonidae	5200			15				~45	Osse & Boogaart, 1995
Salmo trutta	Salmonidae	~5000	3	28	16.4					Rubin, 1994; Jobling, 1995
Hucho hucho	Salmonidae	4500			13.7				~39	Osse & Boogaart, 1995
Leuresthes tenuis	Atherinidae	1550	18	7.5	7	~0	~4	~40	~17	David, 1939; Hunter, 1981; Vásquez-Yeomans & Sosa-Cordero, 1992
Ammodytes personatus	Ammodytidae	900			3.3		2			Shirota, 1970
Oncorhynchus tshawytscha	Salmonidae		8	65	22	~50				Beacham & Murray, 1990
Eggs Non-adhesive and Sinking										
Morone saxatilis	Percichthyidae	3600	20	1.5-2	3	~4-5	7-14	~37	~30	Breder & Rosen, 1966; Brown, 1987; Houde, 1987; Hill et al., 1989; Webster & Lovell 1990b
Plotosus lineatus	Plotosidae	3310			6.9	~10	~10		~15	Thresher, 1984
Chilomycterus schoepfi	Tetraodontidae	1800								Breder & Rosen, 1966; Fahay, 1983
Eggs Planktonic										
Reinhardtius hipploglossoides	Pleuronectidae	4100	3		~10				~70-85	Crawford, 1992
Hippoglossus hippoglossus	Pleuronectidae	3400	6	13-17	6.5	~30	47	~100	22-34	Russell, 1976; Ahlstrom et al., 1984; Reay, 1984; Haug, 1990; Pittman et al., 1990; Helvik & Walther, 1993

Hippoglossus stenolepis	Pleuronectidae	3400	6	~15	7.5	~40	55			McFarlane et al., 1991
Anguilla anguilla	Anguillidae	2600	17	~1.0	5			~300	~65	Yamamoto, 1981; Castle, 1984; Usui, 1991; Lecomte-Finiger, 1992
Echeneis naucrates	Echeneidae	2600	28	2.60	~6	1-2		17	44	Martin & Drewry, 1978; Nakajima et al., 1987
Hippoglossoides platessoides	Pleuronectidae	2500	4	12.5	~5				18-30	Breder & Rosen, 1966; Fahay, 1983; Ahlstrom et al., 1984
Scomberesox saurus	Scomberesocidae	2420		~14$^+$	≤6				≤25	Hardy, 1978a
Hoplostethus atlanticus	Trachichthyidae	2320	10	9-10						Zeldis et al., 1995
Pleuronectes platessa	Pleuronectidae	2000	9	18	5.8	5	8	40-75	10-14	Theilacker & Dorsey, 1980; Ahlstrom et al., 1984
Verasper variegatus	Pleuronectidae	1970	14	3.8	4.4		6	~32	~15	Cho et al., 1995
Muraenesox cinereus	Muraenesocidae	1950	25	1.5	5.0		8			Umezawa et al., 1991
Fistularia petimba	Fistulariidae	1800	20	7-8	7.1				130$^+$	Hardy, 1978a; Thresher, 1984
Diodon holocanthus	Tetraodontidae	1740	25	4.6	2.6	~2.5		~25	20	Leis, 1978; Sakamoto & Suzuki, 1978; Thresher, 1984
Sardinops sagax	Clupeidae	1700	15	2.8	3.8	4.5		45-50	31-35	Theilacker & Dorsey, 1980
Lactoria fornasini	Ostraciidae	1600×1750	~27	3.8	2.3	~1				Leis & Moyer, 1985
Sardinops melanostictus	Clupeidae	1625	20		3.0	3	3	~50	~40	Shirota, 1970; Nakamura et al., 1991
Xiphias gladius	Xiphiidae	1615								Gordina & Bagnyukova, 1993
Brevoortia tyrannus	Clupeidae	1610	20	~1.7	3.2	3	~4	~65	~27	Hettler, 1981, 1984; Powell, 1993
Scarus croicensis	Scaridae	~700×3400								Walker, 1991
Gadus morhua	Gadidae	1520	6	14	4.5	4-5	6-8	~106	~60	Theilacker & Dorsey, 1980; Fahay, 1983; Solberg & Tilseth, 1984; Huse, 1991
Coryphaena hippurus	Coryphaenidae	1500	26	1.67	4.2	2-3	~3	20$^+$	~16-24$^+$	Johnson, 1978; Kraul et al., 1988; Szyper, 1991; Kim et al., 1993; Ditty et al., 1994b
Theragra chalcogramma	Gadidae	1460								Kendall & Nakatani, 1992
Lactophrys quadricornis	Ostraciidae	1460	28	2.0	6		4.8		~10	Breder & Rosen, 1966; Martin & Drewry, 1978
Melanogrammus aeglefinus	Gadidae	1400	7	~12	3.5		6-7	~106	~60	Hardy, 1978; Hunter, 1981; Fahay, 1983
Lateolabrax japonicus	Percichthyidae	1400	15	3.3	3.9		3-5	~60	~43	Mito, 1957; Breder & Rosen, 1966; Shirota, 1970; Fukuhara & Fushimi, 1982; Huang & Liu, 1992; Zheng et al., 1993
Seriola lalandi	Carangidae	1385	18	4	4.5	~3	4-5	~30	~15	Fujita & Yogata, 1984
Microchirus ocellatus	Soleidae	1375			2.7				7-8$^+$	Ahlstrom et al., 1984; Marinaro, 1991
Dicentrarchus labrax	Percichthyidae	1350	~17	2.2	3.0	4-5	5-8	40-45	~20	Barnabé, 1974; Russell, 1976; Klaoudatos, et al., 1990; Jennings & Pawson, 1991; Marino et al., 1991; Ronzani Cerqueira & Chatain, 1991; Greco et al., 1993a; Person Le Ruyet et al., 1993a, 1993b
Coryphaena equisetis	Coryphaenidae	1350	26	1.58						Ditty et al., 1994b
Scophthalmus rhombus	Bothidae	1350	14	~4.4	3.8		~4-5	61	~17	Jones, 1972
Anguilla japonica	Anguillidae	1300	29	2.0	2.9		~6	95-210	51-69	Yamamoto & Yamauchi, 1974; Tabeta et al., 1987; Tsukamoto et al., 1989; Tzeng, 1990; Usui, 1991; Yu et al., 1991; Ozawa et al., 1992

Table 2-1. (Continued)

Fish	Family	Egg diameter (μm)	Normal temp. (°C)	Time to hatch (d)	Length at hatch (mm)	Age at FF (dah)	Age at EYS (dah)	Age at trans. (dah)	Length at trans. (mm)	References
Brevoortia patronus	Clupeidae	1300	20	1.7	2.8	4	4	~85	31	Hettler, 1984; Powell, 1993
Solea solea	Soleidae	1300	19	~13	3.6	2	~3	42-56	9-10	Breder & Rosen, 1966; Houghton et al., 1985; Tacon, 1988; Person-Le Ruyet et al., 1990a, 1990b; Pepin & Myers, 1991
Atractoscion nobilis	Sciaenidae	1280	16		2.7	4.8	5-6	42-53	16-20	Moser et al., 1983; Orhun, 1989
Rachycentron canadum	Rachycentridae	1250	26		2.8	~5	~5	~30	~29	Hassler & Raineville, 1975; Ditty & Shaw, 1992
Sphyraena borealis	Sphyraenidae	1230	24		2.6	4-5	4-5	~22	~17	Houde, 1972; Fahay, 1983
Cynoglossus abbreviatus	Soleidae	1210	15	3.92	3.3		8	28	14	Fujita et al., 1986
Lachnolaimus maximus	Labridae	1200	26	0.96		4		~34		Colin, 1982
Chanos chanos	Chanidae	1150	29	1.20	3.5	2.5	3	~21	~24	Liao et al., 1979; Taki et al., 1987; Eda et al., 1990b; Liao, 1991
Engraulis japonicus	Engraulidae	1150			2.6		3			Shirota, 1970
Seriola quinqueradiata	Carangidae	1150	20	2.8	3.5	3	4	50-55	45	Breder & Rosen, 1966; Kuronuma & Fukusho, 1984; Fukuhara et al., 1986
Equetus punctatus	Sciaenidae	1150								Moe, 1976
Pseudosciaena polyactis	Sciaenidae	1150	19	1.83	3.2	4	4			Myoung et al., 1994
Eopsetta grigorjewi	Pleuronectidae	1150			3.0		5-6		≥11	Shirota, 1970; Ahlstrom et al., 1984
Scophthalmus aquosus	Bothidae	1140	12	8	~2				~7	Breder & Rosen, 1966; Fahay, 1983; Ahlstrom et al., 1984
Pleuronectes cornutus	Pleuronectidae	1140	18		2.7	~4	~4	51	24	Ahlstrom et al., 1984; Kitajima et al., 1988
Scomber scombrus	Scombridae	1130	14	4.0	~3	4-5	5-6		43^+	Berrien, 1975; Fritzche, 1978
Anguilla rostrata	Anguillidae	1100			<7			365^+	~60	Van Den Avyle, 1984
Macquaria novemaculeata	Percichthyidae	1100	18	2.1	3.1	5-7	8-9	56-70	20-25	Van der wal & Nell, 1986; Battaglene et al., 1989a; Battaglene & Talbot, 1989; Battaglene & Talbot, 1990
Chaetodipterus faber	Ephippidae	1100	27	1.00	2.5	~2.5	~3		~17	Hildebrand & Cable, 1938; Thresher, 1984; Walker, 1991; Ditty et al., 1994a
Scomber japonicus	Scombridae	1100	16	3.6	3.2	2.2	3	25^+	15-25	Fritzche, 1978; Hunter & Kimbrell, 1980; Theilacker & Dorsey, 1980
Tetragonurus atlanticus	Stromateidae	1100			3.0		5		~27	Martin & Drewry, 1978
Seriola dumerili	Carangidae	1090	23	1.6	2.9	3-4	~4	31-36	16-19	Hildebrand & Cable, 1930; Masuma et al., 1990;

										Tachihara et al., 1993
Anchoa hepsetus	Engraulidae	820×1430	20	2	3.6		>1		~45	Hildebrand & Cable, 1930; Jones et al., 1978
Scophthalmus maximus	Bothidae	1080	19	3	3.1	2-3	~7	~30-40	≥18	Jones, 1972; Person-Le Ruyet et al., 1990b, 1991; Tacon, 1988
Kyphosus vaigiensis	Kyphosidae	1050								Thresher, 1984
Prionotus carolinus	Triglidae	1045	22	2.5	~2.6				~27	Kuntz & Radcliffe, 1917; Fritzche, 1978
Glaucosoma hebraicum	Glaucosomatidae	1037	23	~1.33	1.9	3.5	3.3	>45	>10	Pironet & Neira, 1998
Rhabdosargus sarba	Sparidae	1030	21	1.4	2.4	~3-4	3-4	≤45	≤19	Leu, 1994
Girella punctata	Kyphosidae	1030			2.3					Shirota, 1970
Lagodon rhomboides	Sparidae	1020	18	2.0		~3			~25	Hildebrand & Cable, 1938; Cardeilhac, 1976; Johnson, 1978
Tautoga onitis	Labridae	~1020	22	1.8	2.1	≤4	4		≤30	Kuntz & Radcliffe, 1917; Fritzche, 1978
Paralichthys dentatus	Bothidae	1020	~17	3.1	2.6	3-4	4	≥56	25-35	Smith & Fahay, 1970; Smigielski, 1975b; Keefe & Able, 1993; Bisbal & Bengtson, 1995a
Diplodus vulgaris	Sparidae	1010	17	2.5	2.6	5-6	~6			Jug-Dujaković & Glamuzina, 1988
Alosa aestivalis	Clupeidae	~1000	22	~2.1	~3.5		4		≤30	Kuntz & Radcliffe, 1917; Jones et al., 1978
Macrorhamphosus scolopax	Centriscidae	1000			3.0				≥35	Hardy, 1978
Epinephelus amblycephalus × *E. akaara*	Serranidae	~1000	28	0.83	1.7	2-3	2-3			Tseng & Poon, 1983
Pomatomus saltatrix	Pomatomidae	1000	20	2.0	2.1	~4	4-5		≥21	Deuel et al., 1966; Norcross et al., 1974
Equetus lanceolatus	Sciaenidae	~1000	28	0.79				~21	~21	Young, 1980
Centropyge argi	Pomacanthidae	~1000	23	~0.73	1.5					Breeder's Registry Database, 1995
Pomacanthus paru	Pomacanthidae	1000		0.83		2				Moe, 1992
Sparisoma viride	Scaridae	~1000	26	~1.08	~1.5		~3			Koltes, 1993
Scomberomorus maculatus	Scombridae	~1000	26	1.0	2.6	3-4	4		~35	Hildebrand & Cable, 1938; Fritzche, 1978
Thunnus thynnus	Scombridae	1000	~24	1.3	2.8	2-3	~5	≥30	≥13	Breder & Rosen, 1966; Fritzche, 1978; Harada, 1980; Fahay, 1983; Kaji et al., 1996
Katsuwonus pelamis	Scombridae	985	26	1.25	2.4		4			Shirota, 1970; FSFRL, 1973; Kaya et al., 1982
Carangoides malabaricus	Carangidae	980	34		2.4	2-3	3-4			Thangaraja, 1985
Liza haematocheila	Mugilidae	980	18	3.6	3.0	6	7-9	40-76	22-46	Kuronuma & Fukusho, 1984; Ikeda et al., 1991; Yoshimatsu et al. 1992b
Solea senegalensis	Soleidae	980	18		2.7	~2		~30		Dinis, 1992
Stenotomus chrysops	Sparidae	975	22	≤1.7	~2		3		~23	Kuntz & Radcliffe, 1917; Wheatland, 1956; Johnson, 1978
Ocyurus chrysurus	Lutjanidae	960	27	0.96	2.2	3	~3	~62	~20	Riley et al., 1995
Thunnus albacares	Scombridae	960	26	~1.3	2.7	3	3	~58	≥51	Mori et al., 1971; FSFRL, 1973; Fritzche, 1978; Harada et al., 1980; Kuronuma & Fukusho, 1984
Dentex dentex	Sparidae	958	17	3.3	2.2	6-8	7-8			Glamuzina et al., 1989
Dentex gibbosus	Sparidae	956	20	1.46	2.1	~2.5	3			Fernandez-Palacios et al., 1994
Pagrus major (southern)	Sparidae	950	22	1.17	3.1	3-6	~3	25-35	9-14	Battaglene & Talbot, 1992

Table 2-1. (Continued)

Fish	Family	Egg diameter (µm)	Normal temp. (°C)	Time to hatch (d)	Length at hatch (mm)	Age at FF (dah)	Age at EYS (dah)	Age at trans. (dah)	Length at trans. (mm)	References
Sparus auratus	Sparidae	950	18	2.0	2.4	4-5	5	40	~12	Person-le Ruyet & Verillaud, 1980; Tacon, 1988; Ounaïs-Guschemann, 1989; Greco et al., 1993a
Sciaenops ocellatus	Sciaenidae	950	25	1.00	1.7	3	3	35	~25	Henderson-Arzapalo, 1990; Holt, 1990; Holt et al., 1990; Colura et al., 1991
Paranthias furcifer	Serranidae	947	23							Tucker et al., unpub.
Chaetodontoplus duboulayi	Pomacanthidae	945	25	1.02	2.5	~2	2-3			Arai, 1994
Epinephelus fulvus	Serranidae	944	26	1.12	1.7	2.5	5			Tucker et al., unpub.
Epinephelus guttatus	Serranidae	944	26	1.12	1.6	2.5	5	~60	~25	Tucker et al., unpub.
Engraulis mordax	Engraulidae	660×1350	16	2.5	3.0	4	4	50-60	34-40	Lasker et al., 1970; Theilacker & Dorsey, 1980
Pagrus major (northern)	Sparidae	940	18	2.2	2.1	3	4	~30	10-14	Smith & Hataya, 1982; Kuronuma & Fukusho, 1984; Fukuhara, 1985; Foscarini, 1988; Fukusho, 1991
Mugil cephalus	Mugilidae	940	24	1.42	2.8	3	5	28	35-45	Tung, 1973; Kuo et al., 1973; Nash et al., 1974; Nash & Koningsberger, 1981; Eda et al., 1990a; Liao & Chao, 1991
Argyrosomus hololepidotus	Sciaenidae	938	23	1.2	2.2	4	3	~34	15-26	Battaglene & Talbot, 1994a
Haemulon plumieri	Haemulidae	935	24	0.83	2.7	2	~2	~105	~38	Saksena & Richards, 1975
Paralichthys olivaceus	Paralichthyidae	930	18	2.0	2.0	2	3-4	~45-50	10-14⁺	Ahlstrom et al., 1984; Kuronuma & Fukusho, 1984; Yamaha Fishery Journal, 1991
Epinephelus striatus	Serranidae	920	26	1.08	1.9	2.5	~5	46-70	35-50	Tucker et al., 1991; Powell & Tucker, 1992; Tucker & Woodward, 1996
Mycteroperca microlepis	Serranidae	920	21	0.94	2.0			~63	~30	Roberts & Schlieder, 1983; Mullaney & Gale, 1996
Centropristis striata	Serranidae	910	20	2.00	2.1	3.5	5.5	43-52	10-12	Tucker, 1984, 1989
Paralichthys lethostigma	Paralichthyidae	910	18		2.2			>48	>15	Powell & Henley, 1995
Auxis thazard	Scombridae	905	28	~0.96						FSFRL, 1973
Epinephelus tauvina	Serranidae	900	27	1.0	1.7	3-4	4	36-50⁺	≥33	Chen et al., 1977; Lim, 1991b
Trachinotus blochii	Carangidae	900	28	0.83	2.0	2.1				Lin, 1991
Acanthopagrus latus	Sparidae	900	25	1.5	2.0					Mito, 1963; Tsay & Yu, 1991
Pomacanthus arcuatus	Pomacanthidae	~900								Moe, 1976, 1997b
Pomacanthus paru	Pomacanthidae	~900								Moe, 1976, 1997b
Platycephalus indicus	Platycephalidae	895	25	1.0	2.1		4		~16	Breder & Rosen, 1966; Shirota, 1970; Washington

										et al., 1984
Merluccius bilinearis	Gadidae	~890	21	~2.0	~2.8				≤23	Kuntz & Radcliffe, 1917; Hardy, 1978
Cromileptes altivelis	Serranidae	890	29							Chao, p.c., 1993
Mugil liza	Mugilidae	890	24	1.46	2.3	3-4	8	40	~22	Alvarez-Lajonchère et al., 1991
Auxis rochei	Scombridae	890	27	~1.08						FSFRL, 1973
Trachurus japonicus	Carangidae	885	21	1.7	2.4	~3	4	~60	~55	Ochiai et al, 1982
Sillaginodes punctata	Sillaginidae	880	~18	~2.25	2.2	≤8	8		~34	Bruce, 1995
Pleuronectes ferrugineus	Pleuronectidae	880	10	6	2.8	4-5	4-5		12-16	Theilacker & Dorsey, 1980; Fahay, 1983
Plectropomus leopardus	Serranidae	875	26	1.11	1.6	2	~3	~55	~25	Chen et al., 1991a; Masuma et al., 1993
Cynoglossus robustus	Soleidae	875	22	0.71						Breder & Rosen, 1966
Acanthopagrus schlegeli	Sparidae	870	17	2.9	1.9	2-3	2-3	~40	14-18	Mito, 1963; Shirota, 1970; Kuronuma & Fukusho, 1984; Ding et al., 1991; Leu, 1997
Diplodus sargus	Sparidae	870	15	2.0	2.6		5			Mazzola et al., 1983; Greco et al., 1993a
Paralichthys albigutta	Bothidae	870	19		2.1			>45	>12	Powell & Henley, 1995
Liza affinis	Mugilidae	864	22	2.5	2.5	~5	~5	~50	~30	Yoshimatsu et al., 1992a
Epinephelus malabaricus	Serranidae	852	29	0.79	2.3	~3	~3	36-60	~35	Maneewong et al., 1986; Tookwinas, 1990a; L.-T. Lin, p.c., 1992
Paralabrax nebulifer	Serranidae	850	24	~1.18						Gadomski and Caddell, 1996
Oplegnathus fasciatus	Oplegnathidae	850	22	1.5	1.7	3-4	4	~35	~15	Breder & Rosen, 1966; Shirota, 1970; Kuronuma & Fukusho, 1984; Koh & Kim, 1992; Takeuchi et al., 1992b
Archosargus rhomboidalis	Sparidae	850	26	<0.92	2.1	1.5	2.1	33	24	Houde & Potthoff, 1976
Tautogolabrus adspersus	Labridae	850	22	1.7	2.1	2-3	3		~10	Kuntz & Radcliffe, 1917; Fritzche, 1978
Parapercis sexfasciata	Pinguipedidae	850	15	~3						Breder & Rosen, 1966
Paralabrax maculatofasciatus	Serranidae	848	22	0.83	1.8	3	3	35	18	Avíles-Quevedo et al., 1995
Diplodus puntazzo	Sparidae	845	22	1.0	1.7		3.5			Faranda et al., 1985; Greco et al., 1993a
Epinephelus fuscoguttatus	Serranidae	840	28	0.77	1.3	2.5-4	4	35-40	~31	Kohno et al., 1990a, 1990b, 1994c; Lim et al., 1990; Tucker & FitzGerald, 1994
Pterocaesio digramma	Lutjanidae	840	24	1.0	2.2	2-3	3			Yokoyama et al., 1994
Cynoscion regalis	Sciaenidae	840	20	1.5	1.5	2.2	3		32[+]	Wheatland, 1956; Johnson, 1978; Goshorn & Epifanio, 1991
Epinephelus polyphekadion	Serranidae	830		0.88	1.7			~40	~36	Tawada, 1989a
Lutjanus stellatus	Lutjanidae	830	24	1.3	2.5	2-3	4			Hamamoto et al., 1992
Epinephelus akaara	Serranidae	825	24	1.0	1.6	2-4	4	45-50	~30	Ukawa et al., 1966; Tseng & Ho, 1979; Tseng & Ho, 1988; Maruyama et al., 1994
Epinephelus septemfasciatus	Serranidae	820	22	1.4	1.8	3	3-4	~60	~30	Kitajima et al., 1991
Lutjanus campechanus	Lutjanidae	820	24	1.04	2.3	3	3		~22	Collins et al., 1980; Rabalais et al., 1980
Archosargus probatocephalus	Sparidae	820	23	1.17	1.6	3	4	≥39	≤14	Tucker, 1987a; Tucker & Alshuth, 1997
Mugil curema	Mugilidae	820	26-27	~1.67	2.0	1.5	3.5	~30	30-40	Anderson, 1957; Houde et al., 1976
Lutjanus kasmira	Lutjanidae	815	24	0.75	1.8	3	3			Thresher, 1984
Rhabdosargus haffara	Sparidae	810	27	0.95	1.8	~3	~4	35	~12	Chang et al., 1994b

Table 2-1. (Continued)

Fish	Family	Egg diameter (µm)	Normal temp. (°C)	Time to hatch (d)	Length at hatch (mm)	Age at FF (dah)	Age at EYS (dah)	Age at trans. (dah)	Length at trans. (mm)	References
Genyonemus lineatus	Sciaenidae	810	16	~2.00						Gadomski & Caddell, 1996
Epinephelus coioides	Serranidae	807	27	0.92	1.7	3	4	35-40	25-30	Hussain & Higuchi, 1980; Doi et al., 1991; Kawahara et al., 1997
Therapon jarbua	Terapontidae	800	26	0.78	1.8					Breder & Rosen, 1966; Her & Liu, 1991
Lutjanus johnii	Lutjanidae	800	28	0.71	1.6	~3	~3	35	~20	Lim et al., 1985
Leiostomus xanthurus	Sciaenidae	800	20	2.0	1.7	5	5-6		~30	Hildebrand & Cable, 1930; Powell & Gordy, 1980; Powell & Chester, 1985
Genicanthus semifasciatus	Pomacanthidae	800	26	0.65	1.6	2-3	3-4			Suzuki et al., 1979
Paralichthys californicus	Bothidae	800	16	2.08	2.0	4	6	30-52	10-12	Gadomski et al. 1990; Gadomski & Caddell, 1991, 1996; Oda, 1991
Lutjanus argentimaculatus	Lutjanidae	795	28	0.73	1.8	2-3	3-4	~24	~19	Doi & Singhagraiwan, 1993
Sacura margaritacea	Serranidae	790	26	0.64	1.5					Thresher, 1984
Plectorhinchus cinctus	Haemulidae	790	22	1.43	2.2	2-3	3	24	~9	Kobayashi & Iwamoto, 1984 Chang et al., 1994c
Lutjanus analis	Lutjanidae	783	28	~0.71				30-38	~29	Watanabe et al., 1998b
Dactylopterus volitans	Dactylopteridae	760×800		~1	≤1.8	~3	~4		~21	Fritzche, 1978
Gymnocranius griseus	Sparidae	775	21	1.62	1.5					Suzuki & Hioki, 1978
Genicanthus lamarck	Pomacanthidae	775	25	0.74	1.6	3-4	3-4			Suzuki et al., 1979
Cynoscion nebulosus	Sciaenidae	770	28	0.71	1.5	2	2-3	37	~38	Fable et al., 1978; Johnson, 1978; Tucker, 1988
Upeneus bensasi	Mullidae	765			2.2					Thresher, 1984
Cirrhitichthys aureus	Cirrhitidae	765	25	0.86	2.2	3	3			Tanaka & Suzuki, 1991
Xystreurys liolepis	Bothidae	760	20	~1.42						Gadomski & Caddell, 1996
Lutjanus argentiventris	Lutjanidae	758	29	0.64	2.2		2			Avíles-Quevedo et al., 1998
Sphyraena pinguis	Sphyraenidae	755	26	1.0					13-18	Thresher, 1984
Plectropomus areolatus	Serranidae	754	28	0.79		2.5				Tucker & FitzGerald, 1994
Lates calcarifer	Centropomidae	750	28	0.71	1.5	2.5	4-5	25	~17	MacKinnon, 1987; Tucker et al., 1988
Orthopristis chrysoptera	Haemulidae	750	20	1.5	1.5	~2.5	~3		~25	Hildebrand & Cable, 1930
Bairdiella chrysoura	Sciaenidae	750	27	~0.75	~1.5	~2	2		~30	Kuntz, 1913; Lippson & Moran, 1974
Scatophagus argus	Scatophagidae	750			~3				≥5	Johnson, 1984; Barry & Fast, 1992
Peprilus triacanthus	Stromateidae	750	18	2	1.7		3		≥15	Kuntz & Radcliffe, 1917; Martin & Drewry, 1978
Hypoplectrus spp.	Serranidae	~750	~28	1.0		3		21		Domeier, p.c., 1998

Lutjanus griseus	Lutjanidae	745	28	~1	2.3	3-4	~4			Cabrera et al., 1997, Rosas et al. i.p.
Achirus lineatus	Soleidae	735	28	1	1.9	2	3	16?	3-6	Theilacker & Dorsey, 1980; Hunter, 1981; Fahay, 1983; Ahlstrom et al., 1984
Bairdiella icistia	Sciaenidae	725	25	0.83	1.7		2			May, 1975
Malacanthus plumieri	Malacanthidae	720								Thresher, 1984
Chaetodon nippon	Chaetodontidae	720	23	1.21	1.5	~3	~4			Suzuki et al., 1980
Centropomus undecimalis	Centropomidae	700	28	0.71	1.5	2.5	4-5	35	35-40	Tucker, 1987b
Holacanthus ciliaris	Pomacanthidae	~700								Moe, 1976, 1997b
Holacanthus bermudensis	Pomacanthidae	700								Moe, 1976, 1997b
Holacanthus tricolor	Pomacanthidae	700	28	~0.73	~2.5	2-3	2-3			Moe, 1976, 1997b
Callionymus lyra	Callionymidae	700								Breder & Rosen, 1966
Anchoa mitchilli	Engraulidae	600×800	24	1.17	2.0	1.8	2.2	33	~20	Kuntz, 1913; Fahay, 1983; Houde, 1987; Tucker, 1989
Centropyge interruptus	Pomacanthidae	690	25	0.65	1.3	3-4	3-4			Hioki & Suzuki, 1987
Centropyge ferrugatus	Pomacanthidae	685	26	0.68	1.3	3-4	3-4			Hioki et al., 1990
Acanthurus triostegus	Acanthuridae	680	24	1.08	1.7	~4			~20-25	Thresher, 1984
Sparisoma rubripinne	Scaridae	675	26	1.04	1.7	~3	3			Randall & Randall, 1963
Repomucenus beniteguri	Callionymidae	670	24	0.79	1.2	3.3	3-4	~22	~6	Eda et al., 1994
Sillago japonica	Sillaginidae	665			1.3	~4	5	~24	~12	Mito, 1963; Oozeki et al., 1992b
Serranus subligarius	Serranidae	662	27	0.88		2.5	3	~34		Tucker & Woodward, 1995, unpub.
Pseudanthias squamipinnis	Serranidae	660	25	0.66	1.2	~3	~3			Suzuki et al., 1978
Scomberoides tol	Carangidae	660	29		1.6	2-3	4-5			Thangaraja, 1985
Micropogonias undulatus	Sciaenidae	~660	20	~2.0	~1.6			≥35		Middaugh & Yoakum, 1974
Halichoeres poecilopterus	Labridae	660	23	0.8	1.6	~3	3		~21	Kimura & Kiriyama, 1992, 1993
Calotomus japonicus	Scaridae	660	25	1.0						Breder & Rosen, 1966
Selaroides leptolepis	Carangidae	650	28		1.5	3-4	3-4			Thangaraja, 1985
Chaetodon ocellatus	Chaetodontidae	650								Moe, 1976, 1997b
Siganus argenteus	Siganidae	650	28	1.0	2.0	2.0	2-4			Duray, 1990
Alepes djedaba	Carangidae	640	29		1.3					Thangaraja, 1985
Eugerres brasilianus	Gerreidae	640	27	0.8	1.4	~3	~3	~48	~20	Alvarez-Lajonchère et al., 1996
Decapterus russelli	Carangidae	630	30		0.7					Thangaraja, 1985
Centropyge, 6 species	Pomacanthidae	625	22	1.29		~4	4-5			Bauer & Bauer, 1981
Diplectrum formosum	Serranidae	610	26	0.88	1.1		2.0			Manrique, 1987
Callionymus calliste	Callionymidae	600	29	0.46	1.2	2-3	2-3			Takita, 1983
Centropyge loriculus	Pomacanthidae	575	26	0.67	1.6		4			Breeder's Registry Database, 1995
Thalassoma cupido	Labridae	475	23	1.5						Breder & Rosen, 1966
Centropyge resplendens	Pomacanthidae	425	26	0.67	1.0		4			Breeder's Registry Database, 1995
Anoplopoma fimbria	Anoplomatidae		6	12-15	8.7	~21	~41			Whyte et al., 1994
Euthynnus affinis	Scombridae		24		2.8	2-3	4.0			Kohno et al., 1994b
Paralichthys adspersus	Paralichthyidae		~15	~2.5	1.8	4-5	4-5		~17-18	Zuñiga & Acuña, 1992

Table 2-1. (Continued)

Fish	Family	Egg diameter (µm)	Normal temp. (°C)	Time to hatch (d)	Length at hatch (mm)	Age at FF (dah)	Age at EYS (dah)	Age at trans. (dah)	Length at trans. (mm)	References
Paralichthys microps	Paralichthyidae		~15	~2.6	1.6	4-5	4-5		~15-16	Zuñiga & Acuña, 1992
Parapercis cylindrica	Pinguipedidae		27	0.92	1.6		2.6			Thresher, 1984

[a] ovoviviparous.

Credits for reprinted fish illustrations in chapter 2:
Jack mackerel (all), p 20; Pacific sardine, p 21; red cornetfish, p 21; striped mullet, p 21; hirame, p 21; Japanese anchovy, p 22; red cornetfish, p 22; rough golden puffer, p 22; Chinook salmon, p 22; Japanese seahorse, p 22; thread-sailfin filefish, p 22; eel, p 23; chub mackerel, p 23; dusky squirrelfish, p 23; haddock, p 23; hogfish, p 23; slender sole, p 23; milkfish, p 24; Japanese seahorse, p 24; ocean triggerfish, p 24; Japanese sea bass, p 25; yellowtail damselfish, p 25; common snook, p 25; porcupinefish, p 25; Florida pompano, p 25; gunnel, p 25; bluespine unicornfish, p 26; goby, p 26; California clingfish, p 26; sandy rabbitfish, p 26; sand lance, p 26; northern bluefin tuna, p 26; sargassumfish, p 27; striped knifejaw, p 27; scrawled cowfish, p 27; longnose butterflyfish, p 27; banded rudderfish, p 28; slender sole (both), p 28; ovate sole, p 28; with permission from the American Society of Ichthyologists and Herpetologists.
Rex sole, p 21; kelp bass, p 22; kelp bass, p 25; with permission from the California Cooperative Fisheries Investigations.
Bay anchovy, p 21; hogchoker, p 21; northern sennet, p 22; white mullet, p 23; horned whiff, p 23; horned whiff, p 26; Atlantic spadefish, p 27; Atlantic red snapper, p 27; northern sennet, p 28; horned whiff, p 28; longspine thornyhead, p 29; from publications of the U.S. National Marine Fisheries Service.
Lined seahorse, p 21; with permission from the U.S. Fish and Wildlife Service.
Nassau grouper, p 22; Nassau grouper, p 25; Nassau grouper, p 27; with permission from the Bulletin of Marine Science.
Sobaity, p 24; from Aquaculture vol. 22, N. Hussain, S. Akatsu, and C. El-Zahr, Spawning, egg and early development, and growth of *Acanthopagrus cuvier* (Sparidae), p 133, copyright 1981, with permission from Elsevier Science;
Atlantic cod, p 24; with permission from the Northwest Atlantic Fisheries Organization.
Red cornetfish, p 24; brown sweetlips, p 24; pygmy angelfish, p 27; with permission from Jeffrey M. Leis.
Bay cardinalfish, p 24; rockskipper, p 28; with permission from the University of Hawaii Sea Grant College Program.
Coho salmon, p 29; with permission from the Great Lakes Fishery Commission.

3 THE REARING ENVIRONMENT

Although it often seems like larval marine fish are programmed to die, they actually are programmed to live and try very hard to do so. They can succeed if not overwhelmed by individual defects or environmental and nutritional handicaps. In raising marine fish, we hope to simulate the best conditions that nature could provide and sometimes have an opportunity to do better, but more often, compromises must be made because of siting, engineering, or economic constraints.

Many marine fish targeted for aquaculture are not well-adapted for confinement in shallow tanks, but, within limits, they can make adjustments. For example, Atlantic halibut typically spawn at a depth of 300-700 m, and their eggs have been collected at 150-250 m (Helvik and Walther, 1993). Despite the impracticality of simulating these environments, reproduction and larval rearing in captivity have been accomplished under conditions that were close enough.

Tremendous amounts of effort can be required to compensate for apparently minor deficiencies in hatchery facilities. If facilities are truly inadequate, no amount of extra work can make them function well. Grow-out conditions are not as critical, but every effort should be made to provide an optimal habitat for broodfish and larvae. Compromises in hatchery design are likely to be regretted later.

To be useful for aquaculture, fish must survive, resist diseases, develop normally, convert food efficiently, grow reasonably fast, and reproduce. Environmental stress can interfere with all of these functions (Rosenthal and Alderdice, 1976; Adams, 1990; Sumpter, 1993). It can be disruptive during reproduction (Collier et al., 1993), early development (Wedemeyer, 1982; Weis and Weis, 1987; Hall, 1991), or later stages. Although short exposures to contaminants such as methoxychlor or hydrogen cyanide might have no obvious effect on the exposed generation, at least partial reproductive failure can occur in later generations (Handy, 1994).

If water is contaminated with toxins, some fish might survive, but production is likely to be reduced. Even if levels of toxic substances appear to be in acceptable ranges, the additive or synergistic effects might be too much for the fish (Malins and Collier, 1981; Sorensen, 1991). The total effect of all stressors could be complete mortality, even when any single stressor by itself would not be lethal. Disease resistance in fish can be reduced by exposure to toxicants such as phenol, copper, and pesticides (Anderson, 1990). At times, high mortality of fish in hatcheries cannot be explained despite extensive investigation. It could result from factors such as increased body burdens of pollutants (Reish et al., 1988, 1993; Norrgren et al., 1993; Pritchard, 1993).

Gilthead sea bream were reared in floating ponds in what seemed to be a clean area of Algeciras Bay, just east of the Strait of Gilbraltar in Spain (Tarazona et al., 1991). Continuous viral and bacterial diseases, reduced growth, and high mortality occurred in the ponds, but the fish recovered when transferred to laboratory tanks. Analysis of bay water revealed high levels of cadmium, lead, and cyanide and sometimes ammonia, nitrite, phenols, and sulfide (see tables later in this chapter), which probably were major factors.

Because fish gills are very sensitive and can be damaged by salinity extremes, pH extremes, ammonia, chlorine, ozone, metals, pesticides, hydrocarbons, surfactants, and therapeutic compounds (Mallatt, 1985), their condition along with fish behavior can be quick indicators of environmental problems.

In this chapter, tabular information is provided so that the reader is alerted to environmental dangers and limits, and also has an immediate source of representative data plus references for comprehensive data.

Two ideas should be remembered when rearing fish:

1) Either **simulate** or **improve upon** known key conditions to which the species is adapted.
2) Look at the rearing environment from the fishes' point of view to learn more about **what they need.**

Some references for measurement of water quality factors are: Grasshoff et al. (1983), Kopp and McKee (1983), Parsons et al. (1989), Spotte (1992), Wagner et al. (1992), APHA (1995), and manufacturers' manuals.

TEMPERATURE (Table 3-1)

Life processes of fish are controlled by temperature and can be limited by oxygen (Fry, 1971). Temperature is the most important single factor influencing fish growth. Sometimes species are categorized by temperature ranges in which they are healthy and grow best: coldwater (<15°C), coolwater (15-25°C), and warmwater (>25°C). In general, warmwater fish grow faster than cool or coldwater fish.

An increase in temperature reduces oxygen solubility and also can have an indirect effect on water quality by raising the metabolic rates of fish. This could result in reduced dissolved oxygen and increased quantities of metabolic wastes such as carbon dioxide and ammonia, which are toxic to fish.

Departure of temperature from a species's optimal range will reduce growth and increase susceptibility to disease. Extreme temperatures can alter function of the cardiovascular system, nerves, proteins, and enzymes. Some examples of extreme temperature tolerance are a maximum of 41°C for goldfish, minimum of -1°C for Atlantic cod and Atlantic salmon, and a very narrow range at or just below 0°C for Antarctic icefish. Recent long-term exposure to low or high temperature can lower or raise critical temperatures somewhat. Fish respond in different ways to fluctuating and constant temperature. In at least striped bass and some salmonids, the optimal mean fluctuating temperature for growth can be 2°C lower than the optimal constant temperature (Cuenco et al., 1985).

When temperature is higher than optimal, there is a tendency for yolk in fish eggs to be depleted too rapidly and larvae to hatch prematurely. Resulting larvae would have deficient energy reserves and feeding ability. Lower than optimal temperature will lengthen incubation time and increase the opportunity for disease to appear. If it is too low, development of embryos might be retarded enough to reduce viability.

Within the normal range, increasing temperature shortens the time from hatching to transformation. This can reduce exposure time to certain maladies, but also increases the food requirement for survival, maintenance, and growth. Yolk is used faster, but larvae develop faster and feed more actively (Blaxter, 1992), so without the right food, larvae will starve faster. While positive life processes are accelerated, negative ones are, too. Susceptibility to extreme temperature injury depends on temperature history, rate of increase or decrease, duration of exposure, and other interacting factors (Blaxter, 1992).

Some fish have narrow ranges for survival and others have wide ones, but optimal temperature ranges usually are narrow (Table 3-1). At suboptimal temperatures, feeding and growth are reduced. Up to the optimum, increasing temperature enhances growth by increasing food intake and food conversion efficiency. Above that level, increasing metabolic requirements for food processing, locomotion, and other functions bring the law of diminishing returns into effect. Most warmwater fish have an upper limit of 32°C or lower. It is not unusual for the growth optimum to be higher than the feed conversion optimum.

Many of the species under consideration for aquaculture in the Caribbean region have an optimal temperature in the range 26-30°C (Tucker and Jory, 1991). Throughout much of the Caribbean, the annual ocean water range is similar, resulting in a potential year-round

Table 3-1. Temperature limits and optima for diadromous and marine fish. In some cases, ranges could be wider than those reported.

Fish	Spawning (°C)	Best for larvae (°C)	Range for larvae (°C)	Best for juveniles (°C)	Range for juveniles (°C)	References
Acipenser brevirostrum	10-15					Conte et al., 1988
Acipenser fulvescens	12-19	14-17				Conte et al., 1988
Acipenser oxyrhynchus	13-18	14-17				Conte et al., 1988
Acipenser transmontanus	8-22	14-17				Conte et al., 1988
Anguilla anguilla				22-26		Poxton & Allouse, 1982
Anguilla japonica				23-30	12-	Usui, 1991
Brevoortia tyrannus	16-20					Hettler, 1981
Clupea harengus	8-15		(-2-0)1-22(24)			Kelly & Moring, 1986
Anchoa hepsetus					~15-35	Robinette, 1983
Anchoa mitchilli	20-				~5-40	Robinette, 1983
Chanos chanos	25-30		24-33	<34	42	Kelley & Lee, 1986; Villaluz, 1986; Watanabe, 1986
Chanos chanos	25					Kelley & Lee, 1991
Chanos chanos	[26]29-33[34.5]					Chang et al. 1993
Oncorhynchus gorbuscha		~8		15.5	-(24)	Jobling, 1981; Beacham & Murray, 1990
Oncorhynchus keta		~8		13	-(24)	Jobling, 1981; Beacham & Murray, 1990
Oncorhynchus kisutch		~4		15	-(25)	Jobling, 1981; Beacham & Murray, 1990
Oncorhynchus mykiss (fw)	10-12	~6-12		16-18	(<1)6-20(>25)	Jobling, 1981; Billard, 1990
Oncorhynchus mykiss (31 ppt)					-17(>18)	Oorschot & Boon, 1993
Oncorhynchus nerka		~8		15	-(25)	Jobling, 1981; Beacham & Murray, 1990; Petit, 1990
Oncorhynchus tshawytscha		~8		15.5	-(25)	Jobling, 1981; Beacham & Murray, 1990
Salmo salar	4-6	13-16		15-18	-0.8-25(23-28)	Jobling, 1981; Avault, 1985; Saunders, 1986; Laird & Needham, 1988b; Petit, 1990; Poxton, 1991
Salmo trutta				10-15	-(23-26)	Jobling, 1981
Salvelinus alpinus	~4(8)					Jobling et al., 1998
Salvelinus fontinalis				13-16	-(20-25)	Jobling, 1981
Gadus morhua	3-10	~6	2-10(17)	14-16	-1-20(22)	Laurence & Rogers, 1976; Jobling, 1988; Huse, 1991
Melanogrammus aeglefinus	5-10	6	4-10			Laurence & Rogers, 1976; Laurence, 1978
Cyprinodon variegatus	26-		(20)24-35(40)		~11-35	Schimmel & Hanson, 1974
Fundulus grandis	[9]24-28[35]				-1.5-35⁺	Tatum et al., 1982a, 1982b
Fundulus heteroclitus				24-29	-(36)	Jobling, 1981
Hippocampus kuda	19-32	20-31		12-32	8-34	Liang, 1992

Table 3-1. (Continued)

Fish	Spawning (°C)	Best for larvae (°C)	Range for larvae (°C)	Best for juveniles (°C)	Range for juveniles (°C)	References
Hippocampus japonicus	17-33	18-32		9-34	5-36	Liang, 1992
Hippocampus trimaculatus	20-31	21-31		12-32	9-33	Liang, 1992
Sebastes diploproa				13-16	~10-(26-30)	Boehlert, 1981
Sebastes schlegeli	13-17			15-18	(<5)12-20	Jeon, 1995
Centropomus undecimalis		~26-27	~25-29	~27-28	<10-35$^+$	Tucker, 1987b
Lates calcarifer	26-34	27-28	25-35	27-30	(12-16)~21-39(~43)	Maneewong, 1987a; Cheong, 1990; MacKinnon, 1990
Dicentrarchus labrax	10-16	16-20		22-27	2-32	Barnabé, 1990d; Cerqueira, 1991; Coves et al., 1991; Sweetman, 1992; Pelosi et al., 1993
Lateolabrax japonicus	~20			20-25		Huang & Liu, 1992
Macquaria novemaculeata	[11]18	16-20	12-24			Battaglene et al., 1989a; Battaglene & Talbot, 1990
Morone chrysops	18-19					Hodson, 1991
Morone saxatilis	[12]16-20[24]	16-19	12-27	24-25	(<0)10-33	Hill et al., 1989; Nicholson et al., 1990; Hall, 1991; Hodson, 1991
M. saxatilis × *M. chrysops*				25-27	4-32	Hodson, 1991
Morone saxatilis & hybrid lr.		17-18	14-24			Mullis & Smith, 1990
Epinephelus amblycephalus		~24				Tseng & Chan, 1985
Epinephelus malabaricus				26-32		Tookwinas, 1990b
Epinephelus tauvina		[27]29-30[31]				Akatsu et al. 1983; Anon., 1992
Rachycentron canadum					~17-38	Hassler & Rainville, 1975
Pseudocaranx dentex	15-19			24-28	(<8)15-28	Kuronuma & Fukusho, 1984; Kumai, 1995
Seriola dumerili	22-25					Kuronuma & Fukusho, 1984
Seriola lalandi	22-25					Kuronuma & Fukusho, 1984
Seriola mazatlana	25-27					Benetti, 1997
Seriola quinqueradiata	[18]19-21[24.5]	20-23		[18]24-26[27]	7-28(30)	Doumenge, 1990a; Nakada & Murai, 1991; Ikenoue & Kafuku, 1992; Imaizumi, 1993
Trachinotus carolinus				25-30	(8-10)19-31(33-40)	Avault, 1985; Watanabe, 1995
Trachinotus falcatus	24-32	~27-28			(~12)18-33	Lin, 1991
Coryphaena hippurus		25-26	22-28	25-27	~13-28	Szyper, 1991; Ostrowski, 1995
Lutjanus analis					~19-28	Bortone & Williams, 1986

Lutjanus griseus					(11-14)~18-27	Bortone & Williams, 1986
Lutjanus synagris					~15-29	Bortone & Williams, 1986
Ocyurus chrysurus				~24-30	-(34)	
Pomadasys commersonnii				24		Deacon & Hecht, 1995
Acanthopagrus latus	17-21					Kuronuma & Fukusho, 1984
Acanthopagrus schlegeli	17-21				4-36	Kitajima & Tsukashima, 1983; Kuronuma & Fukusho, 1984
Archosargus probatocephalus	~23	~23-28	~22-33	~23-28	~10-35	Tucker, 1987a
Dentex dentex		22-23				Sweetman, 1992
Pagrus major	[11]18-20[25]	16-20	10-	20-28	(4)7-29(32)	Fujiya et al., 1980; Ikenoue & Kafuku, 1992; Foscarini, 1988; Fukusho, 1991; Wu, 1994; Kumai, 1995
Rhabdosargus sarba	17-21			~15	8-	Kitajima & Tsukashima, 1983; Kuronuma & Fukusho, 1984; Woo & Kelley, 1995
Sparidentex hasta		26	22-29			Al-Abdul-Elah et al., 1983a, 1983b
Sparus auratus	18-20	19-20	16-22		5-	Barnabé, 1990d; Chatain & Ounaïs-Guschemann, 1991a; Polo et al., 1991; Sweetman, 1992
Atractoscion nobilis	15-17	~17-23		~23		Orhun, 1989
Bairdiella icistia	24-32	~25	~20-28(31)		10-34	May, 1975
Cynoscion nebulosus	21-35	28	~23-33	~28	~10-35	Taniguchi, 1981; Tucker & Faulkner, 1987
Cynoscion regalis				~28		Lankford & Targett, 1994
Micropogonias undulatus					0-36	Diaz & Onuf, 1985
Sciaenops ocellatus	23-29	27	22-30	~22-25	(2-9)10-33(35)	Henderson-Arzapalo, 1990; Holt, 1990; Holt et al., 1990; Neill, 1990; Roberts, 1990; Colura et al., 1991; Procarione & King, 1993; Whitehurst & Robinette, 1994
Girella nigricans				23-24	-(31)	Jobling, 1981
Girella punctata				~18-23	9-33	Yu & Tung, 1991
Oreochromis aureus				31-37		Arrignon, 1990
Oreochromis mossambicus	19		~24-34	28-34	(4)-(40)	Jobling, 1981; Arrignon, 1990; Subasinghe & Sommerville, 1992
Oreochromis niloticus				31-36	(12)-	Arrignon, 1990
Sarotherodon sp.				28-30		Poxton & Allouse, 1982
Oplegnathus fasciatus	21-26					Kuronuma & Fukusho, 1984
Oplegnathus punctatus	21-26					Kuronuma & Fukusho, 1984
Mugil cephalus	~[12]23[27]	25-26	~20-30	28	(0-3)-	Sylvester & Nash, 1975; Liao, 1981; Nash & Koningsberger, 1981; Poxton & Allouse, 1982; Lee & Kelley, 1991; Tamaru et al., 1993a
Anarhichas lupus		11-14		10-11		McCarthy et al., 1998
Siganus canaliculatus	30-31					McVey, 1972; Bryan et al., 1975; Duray, 1990

Table 3-1. (Continued)

Fish	Spawning (°C)	Best for larvae (°C)	Range for larvae (°C)	Best for juveniles (°C)	Range for juveniles (°C)	References
Siganus guttatus	25-30	~22-26				Hara et al., 1986; Duray, 1990
Thunnus thynnus	22-25			~25-28	(9)14-30	Harada, 1980
Citharichthys stigmaeus	~10.5			~8-13		Rackowski & Pikitch, 1989
Paralichthys californicus	~13-17	~20	~16-24	~24-28	~20-28	Gadomski & Caddell, 1991
Paralichthys dentatus				~18	[3]-	Malloy & Targett, 1991
Paralichthys olivaceus	14-17	16-18		20-23	10-26(>26)	Kuronuma & Fukusho, 1984; Imaizumi, 1993; Iwata et al., 1994; Matsuoka, 1995
Paralichthys woolmani	20-21				-26(>26)	Benetti, 1997
Scophthalmus maximus	[10]12-15[19]	18-20	13-22	[16]18[22]	(2)7-23(30)	Person-Le Ruyet, 1990b; Person-Le Ruyet et al., 1991; Leclercq, 1994; Minkoff & Broadhurst, 1994
Hippoglossus hippoglossus	5-8	~4	~2-7(9)	~13	(<0)2-14^{+}	Bolla & Holmefjord, 1988; Haug, 1990; Holm et al., 1995; Lein et al., 1997b; Ottesen & Bolla, 1998
Hippoglossus stenolepis		6	5-7			McFarlane et al., 1991; Kjorsvik & Holmefjord, 1995
Platichthys flesus				18-20	~2-(26-27)	Fonds et al., 1992
Pleuronectes platessa				18-20	~2-(25-27)	Jobling, 1981; Poxton & Allouse, 1982; Fonds et al., 1992
Pleuronectes vetulus				7-12		Williams & Caldwell, 1978
Rhombosolea tapirina		~15				Hart & Purser, 1995
Solea solea	[8]-[22]	18-20	~10-20	20-25	(3)-(31)	Person-Le Ruyet, 1990b; Howell, 1997
Takifugu rubripes	17-20					Kuronuma & Fukusho, 1984

Lethal, near lethal, or survival-reducing temperatures are in parentheses. Extreme, exceptional, or tentative values are in brackets.

Abbreviations used in this and the following tables are: lr. = larva, jv. = juvenile, ad. = adult, fw = fresh water, bw = brackish water, sw = salt water (occasionally includes bw).

growing season. Outside the tropics, growing seasons usually are limited by winter temperatures, unless heating is used.

Common snook are very sensitive to temperature changes:

<10°C lethal
25°C positive feeding and growth rate inflections
28°C optimal for growth
31°C decrease in growth and growth efficiency
>35°C lethal (somewhere above 35°C)

Important points to remember are:

1) The temperature range for growth is narrower than that for survival.
2) The ranges for spawning, egg survival, and hatching usually are narrower than those for other life history phases.
3) Fish are more susceptible to handling injury and disease at temperatures lower or higher than normal. If fish must be caught or transferred during cold or hot weather, they should be protected from temperature stress and given adequate aeration.
4) When fish are transferred from one body of water to another—if temperature differs by more than 2°C (1°C for eggs and larvae), they need to be acclimated, or tempered first, by gradually changing the temperature of the water they are in to that of their new home.
5) A change of 5-10°C in one day could be too much for juveniles or adults of a sensitive species. Some eggs and larvae could be harmed by a change of more than 2-3°C in a day.
6) If a species suitable for the temperature range of the available water source cannot be chosen, economical temperature control will be necessary.

During the normal spawning season, good ventilation and shading might be enough to keep water temperature in a suitable range for spawning and larval rearing. The water supply/treatment system and pipelines should be protected from weather extremes to avoid wide temperature fluctuations. If water flow rate is slow, pipes exposed to the summer sun can heat up considerably. Out-of-season temperature control might be expensive, but usually is necessary for both broodfish and larvae.

SALINITY (Table 3-2)

Salinity in the appropriate range is important for fish and their prey to maintain salt balance of their body fluids, for facilitating gas exchange, and for maintaining balanced water chemistry. For practical purposes, salinity can be defined as the total grams of solutes dissolved in one kilogram of seawater and is expressed in parts per thousand (ppt). About 99% of the inorganic solutes consist of sodium chloride, magnesium chloride, magnesium sulfate, calcium sulfate, potassium sulfate, calcium carbonate, and potassium or sodium bromide (Moe, 1989). These conservative elements exist in about the same proportions regardless of salinity. The proportions of other elements tend to fluctuate more. At least 97% of the seawater in the world has salinity in the range 33-37 ppt (Moe, 1989; Spotte, 1992). Typical values for the mid-Pacific and mid-Atlantic are 35.5 ppt and 37.0 ppt. Because of evaporation and restricted exchange, areas like the Red Sea have salinity up to 40 ppt or higher. Lagoons have elevated salinities, estuaries have fluctuating salinities, and coastal waters in areas of significant freshwater runoff have slightly (Cape Fear River) or greatly (Rio Grande) reduced salinities.

Ocean-spawning species near the east coast of Florida typically experience salinities of 31-35 ppt offshore and 28-35 ppt nearshore. The range in the Indian River estuary is at least 10-40 ppt, and extremes can interfere with spawning of some species. Reef fish around the Cayman Islands experience a range of 34-37 ppt. Reef fish around Jamaica can encounter much lower salinities—maybe 20 ppt, because freshwater runoff is concentrated in certain coastal areas. In contrast, salinity in the Sargasso Sea south of Bermuda and on Bermudian

Table 3-2. Salinity limits and optima for diadromous and marine fish. In some cases, ranges could be wider than those reported.

Fish	Best for spawning (ppt)	Best for larvae (ppt)	Range for larvae (ppt)	Best for juveniles (ppt)	Range for juveniles (ppt)	References
Acipenser stellatus					0-10	Stickney, 1991
Clupea harengus	25-35	10-20	3-52		[3]5-	Kelly & Moring, 1986
Anchoa hepsetus					~0-44	Robinette, 1983
Anchoa mitchilli					~0-45	Robinette, 1983
Chanos chanos	28-39	27-28	20-40	~3-9	~3-30	Kelley & Lee, 1986, 1991; Villaluz, 1986; Watanabe, 1986; Liao, 1991
Oncorhynchus mykiss			0-8			Morgan et al., 1992
Salmo salar				32-35		Main & Rosenfeld, 1995
Salvelinus alpinus				0-30	≥35	Arnesen et al., 1993
Gadus morhua	28-36	34-36	~25-	~14	(<3)7-35$^+$	Laurence & Rogers, 1976; Huse, 1991; Lambert et al., 1994
Gadus morhua (Europe)	10-15	~15	[2]5-60		(<2)8-	Jobling, 1988; Nissling et al., 1994
Melanogrammus aeglefinus	30-36	34-36	33-			Laurence & Rogers, 1976
Cyprinodon variegatus		15-30	~3-38	0-70	0-125	Schimmel & Hanson, 1974; Nordlie & Haney, 1993
Fundulus grandis	0-		~0-110		~0-145	Tatum et al., 1982a; Perschbacher & Strawn, 1988
Fundulus heteroclitus					0-114	Griffith, 1974; Dunson et al., 1993
Centropomus undecimalis	~28-35	~28-35		~0-11	0-	Tucker, 1987b
Lates calcarifer	28-36	25-31			0-35$^+$	Fortes, 1987; Grey, 1987; Maneewong, 1987b; Ruangpanit, 1987b
Dicentrarchus labrax		25-35	10-37	10-30	0.5-90	Barnabé, 1990d; Coves et al., 1991
Macquaria novemaculeata	~30-35	25-35	10-35			Battaglene et al., 1989a; Battaglene & Talbot, 1990
Morone saxatilis		3-6	0-21		0-35$^+$	Hill et al., 1989; Winger & Lasier, 1994
Morone saxatilis × *M. chrysops*					0-28	Stickney, 1991
Morone saxatilis & hybrids		3-6	~0-16			Winger & Lasier, 1994
Epinephelus akaara ad.					(~3)-	Wu, 1994
Epinephelus amblycephalus		~23				Tseng & Chan, 1985
Epinephelus malabaricus		8-24			20-32	Tookwinas, 1990a; Parado-Estepa, 1991
Epinephelus tauvina		~25	~25-39			Akatsu et al., 1983; Anon., 1992
Rachycentron canadum			~19-35			Hassler & Rainville, 1975
Seriola quinqueradiata	~35	~35		~33	16-	Nakada & Murai, 1991; Ikenoue & Kafuku, 1992
Trachinotus carolinus				~30-35	1-50	Watanabe, 1995

Trachinotus falcatus					5-45	Lin, 1991
Coryphaena hippurus			28-		[5]15-35	Kraul, 1990b; Szyper, 1991
Lutjanus griseus					~1-35	Bortone & Williams, 1986
Lutjanus synagris					19-35	Bortone & Williams, 1986
Acanthopagrus schlegeli					1-	Kitajima & Tsukashima, 1983
Archosargus probatocephalus			[0-36]		0-44	Tucker, 1987a
Pagrus major	31-35	33-35		33-35	(~5)<45(50)	Foscarini, 1988; Iso et al., 1994; Wu, 1994
Rhabdosargus sarba				~15	3-	Kitajima & Tsukashima, 1983; Kraljević et al., 1994; Woo & Kelly, 1995
Sparus auratus	33-35	15-25		~22	2-	Barnabé, 1990d; Stickney, 1991; Castelló-Orvay & Calderer, 1993; Tandler, 1993
Bairdiella icistia	~31-36	~27	5-45		0-58	May, 1975
Cynoscion nebulosus	21-35	~28	(1-6)19-37(43-50)	20-25	0-	Wakeman & Wohlschlag, 1977; Taniguchi, 1981; Mercer, 1984; Tucker & Faulkner, 1987; Banks et al., 1991
Cynoscion regalis				20		Lankford & Targett, 1994
Micropogonias undulatus					0-70	Diaz & Onuf, 1985
Sciaenops ocellatus	~28-35	25-30	10-40	20-35	0-50	Henderson-Arzapalo, 1990; Holt, 1990; Neill, 1990; Colura et al., 1991
Oreochromis aureus	~0-[19]			~0-45	0-53	Watanabe et al., 1985b; Arrignon, 1990
Oreochromis mossambicus	~0-[49]			~0-40	0-120	Watanabe et al., 1985b; Arrignon, 1990
Oreochromis niloticus	0-[29]				0-35	Arrignon, 1990
Oreochromis niloticus				0-2	0-7(10)	Nwadukwe, 1995
Mugil cephalus	30-33	~26	17-36		0-(60-70)	Sylvester et al., 1975; Liao, 1981; Lee & Kelley, 1991; Murashige et al., 1991; Tamaru et al., 1993a
Dormitator maculatus				0-50	0-75(80)	Nordlie & Haney, 1993
Siganus canaliculatus		32	16-32	10	10-32	Duray, 1990
Siganus canaliculatus ad.					(~3)-	Wu, 1994
Siganus guttatus	~32	~24	[2]14-37[55]			Duray, 1990; Young & Dueñas, 1993
Paralichthys lethostigma		~20-33				Daniels et al., 1996
Scophthalmus maximus			[15]20-25[40]	20-27	(3)10-40	Person-Le Ruyet, 1990b; Person-Le Ruyet et al., 1991; Minkoff & Broadhurst, 1994; Gaumet et al., 1995
Hippoglossoides platessoides				≥14	(0)7-	Munro et al., 1994
Hippoglossus hippoglossus	29-34		[~24]-35			Haug, 1990; Lein et al., 1997a
Pleuronectes platessa	~32		~5-60		1-45	Holliday, 1969
Pleuronectes yokohamae					<50(55)	Iso et al., 1994
Solea solea		~30	20-40		(4)≤11-(70)	Poxton & Allouse, 1982; Person-Le Ruyet, 1990b; Champalbert et al., 1994

Lethal, near-lethal, or survival-reducing salinities are in parentheses. Exceptional or tentative values are in brackets.

reefs is very constant at about 36.5 ppt (Jickels et al., 1986).

For many marine species with planktonic eggs, a salinity of at least 30 ppt is necessary for the eggs to float. Even if salinity is lower, they might develop and hatch normally, but resting on the tank bottom could stress the eggs or make them more susceptible to infections. Most marine fish will be stressed by salinities of 40 ppt and higher. Keeping the salinity at 30-34 ppt allows a margin of safety for evaporation. As juveniles and adults, many coastal species can tolerate low salinities (even down to 0 ppt), and some such as common snook (Tucker, 1987b) will grow as well or better in fresh water.

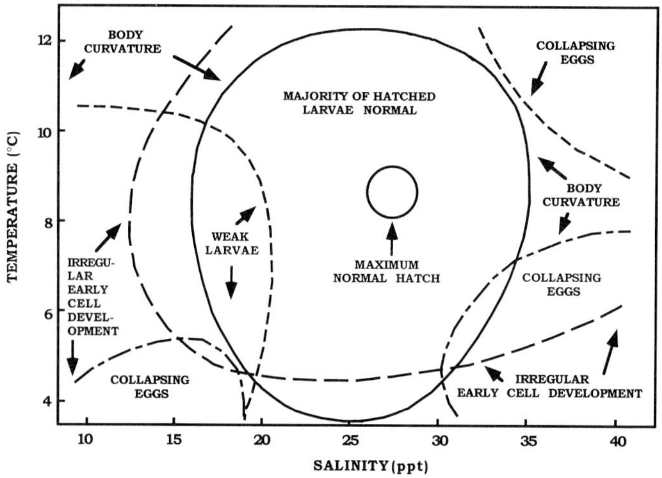

Figure 3-1. Temperature and salinity constraints on early development, hatching, and survival of English sole (from Alderdice and Forrester, 1968; by permission of National Research Council of Canada). Irregular early cell development = usually large differences in cell size (at <20 ppt, 2-12°C; and <6°C, all salinities). Collapsing eggs = collapsing chorion (at >30 ppt, mainly 6°C or 12°C; and <20 ppt, 4°C). Early death (at 2°C, 10-30 ppt). Late death (at 4°C, 15-35 ppt; and 6°C, 10 ppt). Weak larvae = not strong enough to escape chorion (mostly at <20 ppt, 4-10°C). Body curvature (mostly at <20 ppt, 6-12°C; and >30 ppt, 4-8°C). Optimal conditions (at 20-30 ppt, 8-10°C).

Eggs might tolerate wider salinity ranges than larvae, but larval survival is what matters (Holliday, 1969). Fig. 3-1 shows the combined effects of temperature and salinity on eggs of the English sole. The small circle in the middle contains the optimal temperature-salinity combinations. Optimal hatching conditions might be similar to or slightly different from spawning and rearing conditions. Highest hatching rates occurred at 19-25°C and 27-35 ppt for red sea bream (Apostolopoulos, 1976), 19-22°C and 28-32 ppt for silver sea bream (Mihelakakis and Kitajima, 1994a), 24.5°C and 27 ppt for gulf croaker (May, 1975), 23°C and 30 ppt for spotted seatrout (Gray et al., 1991), 25.5°C and 36 ppt for striped mullet (Walsh et al., 1991), and 17°C and 15 ppt for turbot (Kuhlmann and Quantz, 1980). California grunion eggs (which are buried in beaches) hatched best at 22°C (Ehrlich and Muszynski, 1982).

Ionic concentration of internal fluids in most marine fish is similar to that of 10-12 ppt seawater. Theoretically, energy needed for osmoregulation should be minimal in one-third seawater for species that are not stressed at that level. This is true for some. Food conversion tends to be best in isosmotic conditions if other factors are not overriding.

To maintain osmotic balance, marine fish drink much water and have kidneys that excrete much salt but little water. Freshwater fish tend to drink little water and have kidneys that excrete much water but little salt. Stenohaline marine fish have difficulty retaining salt at low salinity, and stenohaline freshwater fish have difficulty preventing overaccumulation of salt (Post, 1987).

Griffith (1974) reviewed the salinity tolerances of most North American species of *Fundulus.* Wu and Woo (1983) reported adult lower lethal limits of 0 ppt for black sea bream; 3 ppt for whitespotted spinefoot, redspotted grouper, yellow grouper, 3-5 ppt for threestripe tigerfish, Russell's snapper; 5 ppt for striped-eel catfish, threeline grunt, silver grunt, red sea bream, silver sea bream, spangled emperor; and >10 ppt for whitefin crevalle.

The original and functional definition of salinity is g salt/kg solution, but indirect determination methods are easier than the complete chemical analysis necessary for direct determination, and they can be nearly as accurate—certainly good enough for aquaculture purposes (APHA, 1995). The most practical means of determining salinity are by refractometers (which measure refractive index of the solution after calibration with distilled water) and inductive salinometers (which measure electrical conductivity of the solution). Grasshoff et al. (1983), Spotte (1992), and APHA (1995) describe the use of refractometers and salinometers. Floating glass hydrometers such as sold in pet stores are easy to use but not always accurate; good ones are expensive and most types break easily. Grasshoff et al. (1983), Parsons et al. (1989), and APHA (1995) describe titration, gravimetric, and densimetric methods, which are not used as often; however, inexpensive titration kits are available. Refractometers are cheaper and more convenient for manual use than salinometers.

SOLIDS

Total solids (TS) include **total suspended solids (TSS)** and **total dissolved solids (TDS)**. By definition, when water is passed through a 2-μm filter, TSS are collected on the filter and TDS pass through (APHA, 1995). **Settleable solids (SS)** are those that settle out after a specified period. Solids are of interest in fish culture mainly for two reasons. 1) High levels of organic solids can reduce oxygen and raise levels of toxicants such as ammonia and hydrogen sulfide. 2) High levels of suspended abrasive solids can injure the gills of fish, leading to hypoxia and infection. Solids are not measured routinely in fish culture but occasionally the need might arise in research or for high density culture. Methods for determining TS, TDS, TSS, SS, and fixed & volatile solids are given by APHA (1995). Moderately expensive meters for measuring TDS are available.

High suspended solids can lower visibility and reduce desired growth of algae. **Turbidity** is a measure of the absorption and scattering of light by living and nonliving particles suspended in water. It is not usually measured for fish culture. APHA (1995) describes the use of nephelometers (the preferred type of turbidimeter).

LIGHT

Choice and management of the spectrum, illuminance (intensity), and duration of light can be critical, especially for early stages.

Spectrum (Table 3-3)

Theoretically, sunlight is the best light source for marine fish. In the ocean, most **ultraviolet (UV)** and **infrared (IR)** radiation is filtered out in the top 1 m of water, and most fish eggs and larvae do not float right at the surface but are suspended from a few meters deep to 100 m or more. Adaptations for protection of eggs and larvae from UV

Table 3-3. Spectral distribution of light from different representative artificial sources (manufacturers' literature; Moe, 1989; Rea, 1995). Wavelengths are given in nm (nanometers). Chlorophyll absorption peaks are at about 400-450 (violet) and 600-650 nm (orange-red). The first five types usually emit less than 2% UV. Cool-white, actinic 03, LPS, plant, and daylight lamps usually emit less than 2% IR, and HPS slightly more.

Source	Ultraviolet 4-400	Violet 380-455	Blue 455-492	Green 492-577	Yellow 577-597	Orange 597-622	Red 622-770	Infrared 700-1,000,000
	Effective and safe for phytoplankton, zooplankton, and larval marine fish							
Cool-white fluorescent	t	+++++++	+++	++++++++	++++++	++++	++++	t
	Have high UV (which could be filtered out), high IR (which possibly could be reflected or tolerated), or unbalanced spectrum (which could be corrected by combining lamps)							
Actinic 03 fluorescent	t	20+	+++++++++	+++				
Low-pressure sodium[h]				t	32+	t	t	
High-pressure sodium[h]	t	+	+++	++++	++++++++++	+++++++++++	++	+
Full-spectrum fluorescent A	t	+++++	++++	+++++++	++	++++	+++++++	+++
Cool-beam incandescent	+	++	+++	++++	++++	+++++	+++++++	++++++
Tungsten incandescent	+	++	++	+++	+++	++++	+++++	++++++++++++
Plant fluorescent	++	++++	+++	+++	++	+++++	+++++++++++++	t
Full-spectrum fluorescent B	++	++	+++++	+++++++++	+++	++++	+++++	++
Mercury vapor[h]	++	++++++	t	++++++++	+++++++++	+++	++	++
Daylight fluorescent	+++	+++++	++++	++++++	+++++++	++++	+++	t
Aquarium fluorescent	+++	++++++	+++++	+++	++	++	++++++++	+++
Quartz-tungsten halogen	+++	+++	+++	++++	+++	++++	+++++	+++++++
Metal halide[h]	+++	++++	+++	+++++++	++++++	+++	+++	+++

[h] high intensity discharge.
t = trace.
Pluses indicate approximate distribution of power among bands, with a total of 32 pluses per row.

damage include shading of vital organs with pigment, vertical migration governed by light intensity, and daily or seasonal restriction of spawning to times when the sun is not high in the sky (Hunter et al., 1981). Shading from natural turbidity and movement from natural turbulence also are protective. Nursery ponds and tanks receiving much light usually have a good bloom of algae.

For practical consideration, ultraviolet radiation can be divided into four bands (Rea, 1995): ozone-producing (180-220 nm), bactericidal (220-300 nm), erythemal (sunburning, 280-320 nm), and black light (300-400 nm). Approximate wavelength ranges for **visible light** are given in Table 3-3. Because the color seen depends on the total effect of the individual colors, light consisting of different spectral combinations can look the same. Infrared radiation can be divided into near (shortwave) infrared (IR-A, 700-1400 nm), middle infrared (IR-B, 1400-3000 nm), and far (longwave) infrared (IR-C, 3000-1,000,000 nm). In larval tanks with low circulation, an excess of infrared can superheat the surface layer or larvae near the surface. Too much infrared also can increase hatchery air cooling costs.

The light spectrum can affect visibility of prey. Early northern anchovy larvae, which have cones but no rods, are most sensitive to peaks at 440 and 600 nm, but older ones, which have cones and rods, have a single peak at 530 nm (Bagarinao and Hunter, 1983). Juveniles and adults of several species have peaks in the green (500-560 nm) range.

In aquaculture, fish are restrained in small containers. Eggs and larvae are closer to the surface and could be burned by UV or overheated by IR. Wavelengths of 280-320 nm (UV) are particularly damaging (Post, 1987), especially to early larvae, which are weaker swimmers and usually have less protective pigment. If artificial lighting is needed only for feeding purposes, lamps approximating a natural spectrum without UV and IR would be useful (Table 3-3). However, if green water is to be maintained, supplemental light in the ranges 400-450 nm and 600-650 nm could be useful for promoting algal growth. For larval rearing, I have used combinations of four lamps for this purpose: cool-white fluorescent for fish and algae, filtered full-spectrum (visible) aquarium fluorescent for fish and algae, actinic 03 (400-470 nm) for algae, and high-pressure sodium (570-630 nm) for algae. The full spectrum lamp can be too intense for early larvae of some species. The HPS lamp should be at a distance of at least 2-3 m or have a good IR filter. Chatain and Ounaïs-Guschemann (1991b) used a 120-watt cold-ray incandescent lamp over each 500-L tank of gilthead sea bream larvae. Quartz-tungsten halogen lamps (300-watt) have been used as nightlights for damselfish larvae (Danilowicz and Brown, 1992). Filters should be used with tungsten halogen lamps, because they can emit dangerous amounts of UV, especially if operating at a color temperature higher than 3100 K (Rea, 1995). Most of the UV produced will be transmitted by fused quartz and standard high-silica glass, but it can be reduced by special high-silica or aluminosilicate glass. Dhert et al. (1994b) reported lower survival up to 22 dah for turbot reared under tungsten halogen (28%) lamps compared with fluorescent lamps (41%). Although spectral peaks were similar (600 nm for halogen, 577 nm for fluorescent), intensity of the tungsten halogen lamps was about three times higher (and UV probably was higher, too). Survival during 23-36 dah was similar. Several of the many available fluorescent lamps are suitable for marine fish larvae, but for many species, cool-white lamps seem to work at least as well as the other choices.

Illuminance (Table 3-4)

Although various measures of light are used in different fields (Rea, 1995), the unit **lux** has most commonly been used to report illuminance (density of radiant power on a surface) in aquaculture. One lux = 1 lumen/m^2 = 0.0929 foot candle. Measurements can be made with inexpensive, fairly accurate light meters. The illuminance of full summer sunlight at noon in the tropics (including Florida) can be more than 130,000 lux (Moe, 1989; pers. obs.). At noon on a clear summer day in Bermuda (15 July 1993), I measured direct

Table 3-4. Light intensities and daylengths for larval rearing.

Species	Feeding threshold (lux)	Suitable range (lux)	Too intense (lux)	Suitable daylength (h/d)	References
Clupea harengus	0.1-0.13				Blaxter, 1980; Bagarinao & Hunter, 1983
Anchoa mitchilli		1400		14^+	Tucker, 1989
Engraulis mordax	~6				Hunter, 1984
Chanos chanos		1000-3000			Chang et al., 1993
Gadus morhua	0-0.5	1-10	87	16-24	Bagarinao & Hunter, 1983; Folkvord & Otterå, 1991; Huse, 1994
Nautichthys oculofasciatus		600		13[t]	Marliave, 1977
Centropomus undecimalis				24	Tucker, unpub.
Lates calcarifer				24	Barlow et al., 1995
Dicentrarchus labrax		800-1000		14-16	Barahona-Fernandes, 1979
Centropristis striata		1400		12^+	Tucker, 1989
Epinephelus coioides		1000-2000		24	Almatar et al., 1997
Sillago ciliata		1360			Battaglene et al., 1994b
Acanthopagrus schlegeli		3000	10,000	13-24	Kiyono & Hirano, 1981
Archosargus rhomboidalis		2500-2800		13-19	Dowd & Houde, 1980
Pagrus major		≤5000	>5000		Watanabe & Kiron, 1995
Sparus auratus		600-1500	10,000	24	Tandler & Mason, 1983, 1984; Tandler & Helps, 1985; Ounaïs-Guschemann, 1989; Chatain & Ounaïs-Guschemann, 1991b
Atractoscion nobilis		2200		14	Orhun, 1989
Abudefduf saxatilis		170-1290		24	Tucker, unpub.
Dascyllus spp.		130-1740		24	Danilowicz & Brown, 1992
Mugil cephalus		500-1400			Nash & Koningsberger, 1981; Lee & Kelley, 1991
Mugil cephalus		400-1800		24	Tucker, unpub.
Mugil curema		2500-2800		13	Houde et al., 1976
Siganus guttatus		1,000		24	Duray & Kohno, 1988
Scophthalmus maximus	~0.1	860-10,800		16-24	Huse, 1994, Minkoff & Broadhurst, 1994
Hippoglossus hippoglossus	≤200	1000-1500			Tilseth et al., 1992
Pleuronectes platessa	0.01-1	860	10,800		Blaxter, 1980; Bagarinao & Hunter, 1983; Huse, 1994
Solea solea				18-24	Fuchs, 1978

[t] Includes tapered twilight periods.

sunlight at 97,000 lux. In winter, especially in temperate regions, the low angle of the sun can cause a large proportion of sunlight to be reflected by greenhouse roofs and water surfaces. Cloud cover can reduce natural illuminance by more than 90%.

Until first feeding, eggs and larvae should be kept well shaded to avoid radiation injury from natural or intense artificial light (Hunter et al., 1982). Stocking of early larvae in unshaded nursery ponds or tanks is best done at night just before the eyes become pigmented. Larvae will be less buoyant than just after hatching and can respond to light appropriately the next morning.

After eyes have developed, light that is too intense will repel larvae. Survival of unfed (Norwegian) Atlantic cod yolksac larvae during 0-20 dah was better in darkness than with 10, 100, or 1000 lux, probably because activity was reduced and energy conserved (Skiftesvik, 1994). Although survival was not affected, the proportion of normal Atlantic halibut produced was higher when yolksac larvae were kept in darkness rather than under 3, 30, or 300 lux during 0-32 dah (Bolla and Holmefjord, 1988).

As they grow stronger, larvae begin to migrate vertically, seeking preferred illuminance levels, which also are likely to be associated with food organisms. Marine fish larvae have been successfully reared at 1-11,000 lux. Microalgae has been cultured with 500-10,000 lux in various places. I have raised larvae in 1-m deep tanks receiving one-third natural light through the greenhouse roof or tank cover. This worked only because the fish were protected by dense phytoplankton.

Up to a point, raising illuminance increases activity and feeding success, but too much light will stress the fish, keep them near the bottom of the tank, hinder feeding, and possibly cause skin diseases. In general, early larvae require at least 10 lux to begin feeding, but some thresholds are lower. Atlantic herring larvae and adults had a feeding threshold at 0.1 lux, but juveniles fed down to 0.01 lux (Blaxter, 1980). Early European plaice larvae also had a threshold at 0.1 lux, but transforming larvae fed with 0.01 lux or less. Atlantic herring larvae (12-14 mm) began feeding with about 0.13 lux, cod (4-5 mm) with more than 0.4 lux, and plaice (6 mm) with at least 1 lux (Bagarinao and Hunter, 1983). Early northern anchovy (6 mm) larvae had a feeding threshold at about 6 lux (Hunter, 1984). Older northern anchovies (10-15 mm) could feed normally at 0.6 lux and even fed some in the dark (10% incidence) when rotifers were at 20-40/mL (Bagarinao and Hunter, 1983). Larvae of other species also can feed to some extent in the dark, possibly using olfaction and vibration to locate prey. With many species, larvae are inactive at very low light levels and feeding is exceptional and inefficient.

Bay anchovy and black sea bass larvae were reared for 12 d in 10-L tanks under two 40-watt cool-white fluorescent bulbs 30 cm above the water, which provided 1400 lux and promoted a heavy algal bloom (Tucker, 1989). Chang et al. (1993) reported that 1000-3000 lux was best for early milkfish larvae. White sea bass larvae were reared under 2200 lux (Orhun, 1989). Feeding of sand whiting during 12-19 dah was better under 1360 lux than with 413 lux (Battaglene et al., 1994). Duray and Kohno (1988) obtained survival up to 40.5% for yellowblotch spinefoot reared at 1000 lux during 3-7 dah.

Barahona-Fernandes (1979) reported that early European sea bass larvae survived better with 150-1000 lux than with 1400-3500 lux (two 20-watt daylight fluorescent bulbs 30 cm above the water, either uncovered or partly blocked). Larvae seemed to function best with 800-1000 lux for the first 7 d and brighter light later.

Ounaïs-Guschemann (1989) reported an optimal light level of about 600 lux (minimum 100-150 lux) for early gilthead sea bream larvae. Growth and survival of gilthead sea bream larvae were better at 1000-1500 lux than at 100-250 or 10,000 lux (Tandler and Mason, 1983, 1984). Growth and survival of gilthead sea bream up to 30 dah were better at 600 and 1300 lux than at 150 and 300 lux (Chatain and Ounaïs-Guschemann, 1991b). Growth and survival of black sea bream larvae during 1-10 dah were better at 3000 lux (58% survival) than at 1000 lux (7%) or 10,000 lux (35%) (Kiyono and Hirano, 1981).

Watanabe and Kiron (1985) recommended shading of outdoor tanks to reduce light to a maximum of 5000 lux for red sea bream larvae.

Sergeant major larvae reared under a daytime range of 170-1720 lux had much more black pigment (melanin) than those reared at 170-410 and 170-1290 lux, which had similar effects (Tucker, unpub.). Increased pigment probably provided shading to protect vital organs. Survival of sergeant major larvae was better with nighttime illumination of 75 and 86 lux than with 0 or 32 lux. Possibly a nightlight allows larvae to orient vertically and avoid settling to the bottom or colliding with the bottom or walls. Feeding activity was not observed with low light. Danilowicz and Brown (1992) used 890-1740 lux for 14 h/d and 130-720 lux for 10 h/d to rear two species of damselfish in the genus *Dascyllus.*

Striped mullet larvae were attracted by 600-1400 lux but repelled by brighter light (Nash and Koningsberger, 1981). A minimum of 500 lux was recommended by Lee and Kelley (1991) for striped mullet. Striped mullet larvae survived and grew better with high density (500,000 cells/mL) phytoplankton, with light attenuation a possible factor (Tamaru et al., 1994). Houde et al. (1976) raised white mullet under 2,500-2,800 lux for 13 h/d and 40-90 lux for 11 h/d.

Huse (1994) reported maximum feeding rates by 5-dah Atlantic cod under 0.1-87 lux and lower rates at 0 and 860 lux. Highest feeding incidence (number of larvae feeding) was at 1-12 lux. A few percent of the larvae fed in the dark, but the dark feeding rate was about 12% of the maximum at 1 lux. Maximum feeding rates by 7-dah European plaice occurred at 12-10,800 lux and a lower rate at 1 lux. Highest feeding incidence was at 87-10,800 lux, and larvae did not feed in the dark. Maximum feeding rates by 4-dah turbot were at 860-10,800 lux and a lower rate at 87 lux; larvae did not feed in the dark. Huse (1994) suggested that cod larvae prefer low illuminance because they feed mainly at dawn and dusk, plaice feed over a wide illuminance range because they migrate farther vertically, and turbot prefer high illuminance because they feed mainly at the surface. Huse (1994) recommended the following ranges to promote feeding: 1-10 lux for cod, at least 100 lux for plaice, and about 1000 lux for turbot. Minkoff and Broadhurst (1994) recommended 2000-4000 lux for larval rearing of turbot.

Some species under certain conditions might do well with lower light intensity. Under normal conditions, Atlantic halibut larvae fed best at 1000-1500 lux and fed very little at <200 lux (Tilseth et al., 1992), but Gulbrandsen (1991a) reported that feeding incidence and rate of first-feeding Atlantic halibut previously held in the dark were better at 0.5 lux than at 5-1000 lux.

We have found that larvae of several Floridian marine fish are healthy with 5% diffuse natural light in the summer and 10-15% in the winter; phytoplankton also blooms well. The range 1000-3000 lux probably is adequate for larvae of most warmwater species. Larvae of high-latitude species and larvae that naturally live under low-light conditions (e.g., well below the surface) seem to need less light. Light over hatchery tanks should be evenly distributed. Bright spots tend to attract and concentrate larvae, sometimes resulting in their death. If natural light is used, larval, juvenile, and adult tanks need to be deep enough so that if the light becomes too intense, the fish can retreat toward the bottom. Older juveniles and adults of some fish like snook become agitated without shade, especially when the sun is overhead.

Daylength (Table 3-4)
The hatching rate of Japanese medaka and zebrafish eggs is higher when they are exposed to light, and hatching of medaka eggs is suppressed by keeping them in constant darkness (Yamagami, 1988). Hatching of sergeant major and triggerfish eggs normally occurs soon after sunset but can be delayed by keeping them in light (pers. obs.). Leaving the lights on will prevent Atlantic halibut eggs from hatching (Helvik and Walther, 1993). Therefore, in some species, secretion of the hatching enzyme probably is controlled by stimulation of

photoreceptors.

Typical early marine fish larvae with functioning eyes are much more active in light than in the dark. They mostly are visual feeders, feeding during the day and drifting/resting at night. Constant bright light sometimes is used to increase feeding time for larvae, but can be stressful to some species, possibly increasing growth but reducing survival (Lubzens et al., 1989). However, survival of even sensitive species might be improved by a low intensity nightlight (10-100 lux) at least to aid orientation.

Juvenile Arctic char had better growth and feed conversion with 12-20 h/d light than with 24 h (Boyer et al., 1994). Survival and growth of juvenile Atlantic cod was similar after 37 d at daylengths of 16 h and 24 h (Folkvord and Otterå, 1991).

Boehlert (1981) found that prejuvenile splitnose rockfish grew faster with 16 h/d light than with 12 h/d. With sailfin sculpin larvae, Marliave (1977) obtained 77% survival with a 13-h day (600 lux) containing two 2-h periods of simulated twilight, followed by 10 lux all night. Survival was 61% with continuous light and 37% with 13 h light:11 h dark. Barlow et al. (1995) found that growth of barramundi larvae improved as daylength increased from 8 to 16 to 24 h/d, but survival did not change; growth and survival of juveniles did not differ at 12, 18, and 24 h/d light. White sea bass larvae were reared with 14 h/d of light (Orhun, 1989). Duray and Kohno (1988) reported better survival and growth for yellowblotch spinefoot larvae raised under continuous light during 3-7 dah than with a normal daylength.

At 1400-3500 lux, European seabass larvae survived best with 12 h/d light, but grew best with 18 h/d (Barahona-Fernandes, 1979). At 800-1000 lux, survival was best with continuous light and growth was best with 14-16 h/d. A compromise of 14-16 h/d was suggested. Continuous light resulted in better growth of surviving European sea bass larvae at 30 dah but lower gas bladder inflation rate and survival than a 9-h daylength (Ronzani Cerqueira and Chatain, 1991). Battaglene and Talbot (1990) found that gas bladder inflation by larval Australian bass was inhibited by continuous light, and even 12 h/d of low light caused some inhibition. Juvenile hybrid striped bass grew faster with daylength increasing from 12 h than when it decreased (Woiwode and Adelman, 1991).

With 1370-3430 lux in white tanks (Tandler and Helps, 1985) and with 600 lux in black tanks (Chatain and Ounaïs-Guschemann, 1991b), growth and survival of gilthead sea bream were better with continuous light than with 12 h/d. Caribbean sea bream larvae survived and grew better up to 16 dah with 13 h/d light than with 7 or 19 h when food density was 100 or 500 wild zooplankters/L but grew best with 19 h light when food was at 25 or 50/L (Dowd and Houde, 1980). Fujiya et al. (1980) suggested that a range of light intensity of 0-2000 lux in hatchery tanks would allow red sea bream larvae to choose the best intensity, depending on stage and individual preferences. Older silver bream juveniles (30-40 mm, ~50 dah) actively fed in the dark, but newly-transformed juveniles (12-15 mm, ~30 dah) did not (Hu, 1991). Growth of black sea bream during 1-12 dah was slightly better but survival was slightly worse with a daylength of 24 h (35% survival) than with 13 h (44%) or 18 h (33%) (Kiyono and Hirano, 1981). For larvae during 13-24 dah, growth continued to be better with 24 h/d light, but survival was not affected by daylength. Neither growth nor survival of juveniles (>24 dah) varied with daylength.

Minkoff and Broadhurst (1994) recommended 16-24 h light/d for larval rearing of turbot. Common sole larvae had similar survival at 12, 18, and 24 h/d light, but grew better with 18 or 24 h (Fuchs, 1978).

Optimal daylength varies among species and has not been determined for many. A good approach for larval rearing is to use either the same (or slightly longer) daylength as during the natural spawning season, a daylength known to be good for related species, or 14-16 h/d as a first approximation.

Because early larvae have short digestion times and small energy reserves, they are susceptible to starvation if they go without food for too many hours. For typical warmwater marine fish, I prefer continuous illumination at 1000-2000 lux for larvae and 18 h at 1000-5000 lux alternating with 6 h dark for early juveniles.

If total or supplemental artificial light is used for grow-out, a daylength equivalent to the annual average or the longest day would be appropriate for frequent feeders to maximize food intake, but less might be better for infrequent feeders to reduce activity and improve feed conversion.

COLOR AND TEXTURE OF HATCHERY TANKS AND FITTINGS

Larvae of many species are attracted to the brightest area of the tank (zooplanktonic prey also). This simple problem can cause high losses of some species, possibly through the factors: crowding stress, abrasion, entrapment in the surface film, localized oxygen depletion, and starvation. I have seen piles of dead larvae appear at the base of white standpipes and around the bottom of glossy (even black) tank walls after the fish spent a day or two trying to swim into the glare. The meniscus can be too bright even with dull black walls, so every precaution should be taken. At times I have placed a dark bucket over the standpipe or created an awning around the inner tank wall to shade the meniscus areas. This seems to help a little. Interior tank surfaces, especially at the waterline, should be non-glossy to minimize distracting reflected light. Dark colors are best, especially for the walls. Black is best for the waterline. Walls can be a little lighter, but not too light. Vertical surfaces of other solid structures (e.g., standpipes, water inlet pipes, surface film skimmers) in the water also should be as dark and non-glossy as possible.

Dark tank color often is said to act as a contrasting background to increase visibility of food to the fish, but considering the limited field of vision of most larvae, the reduction of distracting reflected light might be at least as important. The sighting distance of 5 to 16-mm European pilchard, European plaice, and Atlantic herring larvae was very limited, in the range 1-5 mm (Blaxter, 1980). First-feeding larvae of those three species and northern anchovy first perceived prey at distances shorter than one body length (Hunter, 1980). Volume searched per hour was only 100-200 mL for pilchard, but 200-2000 mL for herring and 100-2200 mL for plaice (Blaxter, 1980).

The background effect is more likely with more-developed fish. For example, Atlantic salmon alevins (8 weeks, 25 mm TL) ate more copepods when their 91-L aquarium was surrounded by red cloth and the copepods were unstained than when the background was blue or green and the prey red or blue (Browman and Marcotte, 1987).

The bottom can be striped or have lighter patches (e.g., blue and black, dull white and black) to increase visibility of fish, their food, algae, and sediment to the culturist. If the bottom is too bright, it can cause prey to migrate downward and be less available to larvae; the larvae themselves might be found groveling on the bottom.

When reared in white-walled tanks, the length increase of surviving gilthead sea bream larvae was one-third more than growth of those reared in black-walled tanks, but survival was only 21% compared to 45% (Chatain and Ounaïs-Guschemann, 1991b). Size-selective mortality might have caused the apparent growth difference. Tank bottom color had no effect.

Dolphin larvae (3-7 dah) survived better in black than in tan 30-L tanks even with less dense prey (Ostrowski, 1989). In tanks that small, the entire volume could be affected by glare. As aeration is increased, reflected light becomes less of a problem. Red drum and some other species can be reared in medium blue or gray tanks. Coloration of some flatfish could be influenced by bottom color.

Naas et al. (1996) found that illumination of clear water in an all-white tank was highest near the walls at the surface, but the bottom also was bright. With black walls and white bottom, light near the walls and bottom was reduced, but the central area was brightest. An entirely black tank had more even and more "natural" illumination. In all three types, green water reduced light overall, but especially near the center.

GREENWATER CULTURE

If not too dense, a healthy growth of phytoplankton in larval rearing water usually is beneficial. It is food for zooplanktonic prey of the fish; can remove ammonia; generates oxygen and keeps pH high; could release anti-bacterial or growth-promoting substances; could promote growth of beneficial bacteria in the tank or in the larval gut; could stimulate behavior, feeding, digestion, or other functions.

Food capture might be enhanced by algal attenuation, diffusion, or dispersion of light, possibly because prey is more visible (Naas et al., 1992; Tamaru et al., 1994) or because larvae can orient better. In clear water, larvae of Nassau grouper, sergeant major, queen triggerfish (pers. obs.), and Atlantic halibut (Naas et al., 1992) congregate at bright spots near the water surface, especially next to the tank walls. In green water, larvae tend to be more active and feed better. This effect might be just as important as increased nutritional value of prey. Feeding incidence, survival, and growth of Atlantic halibut larvae were better in green than in clear water (Naas et al., 1992). While larvae in clear water aggregated near the surface and walls, those in green water spent most of the day searching for prey and eating. The steeper brightness gradient in green water also could help larvae maintain themselves at their preferred illumination level.

STRUCTURED HABITAT

Most marine fish larvae do not require extra habitat, but some species with demersal larvae might benefit from material on the tank bottom. Parker et al. (1990) reported that leaf litter improved survival but not growth of coho salmon from first-feeding to juveniles. The litter certainly provided cover, but it might also promote production of additional food organisms. Juveniles of some flatfish species are healthier if provided with sand, and extra horizontal habitat in the form of shelves could increase carrying capacity for benthic larvae or juveniles.

DISTURBANCES

Handling

Eggs are much tougher than larvae, but some are especially sensitive to handling until gastrulation is complete. Handling of larvae should be avoided. If they have to be moved, lifting larvae in a container of water is the safest way, but this might not work with some groupers and other sensitive species. Nets should be considered for only the toughest larvae (e.g., tilapia). When sensitive larvae have become juveniles and are covered with scales, they can be gently netted. Extra care is needed with juveniles that have relatively small gill surface areas because they can suffocate more easily in air. Some species also simply seem to scare more easily.

Noise

Some larvae can be killed by tapping on the tank, especially if they are floating near the surface during the day after hatching. At that time they are very buoyant and can become trapped in the surface tension and suffocate. Vibrations and noise from pumps and other machinery can be conducted through walls and floors to the tanks. Separating motors from walls and padding their bases can help. Air pumps that seem quiet to humans can transmit very loud sound through airlines into the tanks. An in-line muffler should help.

Aeration and water circulation

Stratification in tanks and ponds can result from heating of the surface water during the day but can be caused by evaporation or by waters of low and high salinity being combined without mixing. Lethal gradients of temperature or salinity and layers with low dissolved oxygen can result (Losordo, 1991b).

Water movement is important to prevent stratification, to insure oxygenation, to disperse live food, to disperse artificial feeds (Backhurst et al., 1989), and (with tank color and light) to keep larvae from aggregating too densely. In some situations, direct aeration has seemed necessary to maintain dissolved oxygen above critical levels even though the fish suffered from the turbulence. Bubbles hitting larvae might stun or injure them. Too much movement might keep larvae from feeding, make them struggle and waste energy, or cause gas bladder overexpansion by rapidly moving larvae upward. Strong circulation (from water jets or wind) can be just as lethal as heavy aeration.

Oxygenation and circulation can be provided without direct aeration in the tank. Culture water should be thoroughly aerated before or as it enters fish tanks. For eggs and larvae, a diffuser consisting of a submerged, unpressurized, vertical water inlet pipe with holes aimed in the direction of desired water flow can be used to create circulation; water is aerated as it enters the top of the pipe. An airstone is placed inside the pipe.

Optimal aeration levels depend on factors such as species, age, tank size and shape, temperature, and water quality. Larvae of some fish (e.g., Percichthyidae, Serranidae, Sparidae) seem less tolerant of turbulence than others (e.g., Percichthyidae, Coryphaenidae, Pomacentridae, Mugilidae), as indicated by reduced growth, survival, or gas bladder inflation rates (Battaglene and Talbot, 1990; pers. obs.). Minimal aeration from airstones during the first 8 d killed walleye larvae reared in a wire basket enclosure by forcing them against the container (Kindschi and MacConnell, 1989). On the other hand, high aeration increased survival of European sea bass larvae (Barahona-Fernandes, 1978). Tougher species can tolerate and sometimes need (e.g., dolphin) strong aeration or circulation right from the beginning.

When direct aeration is used, it should not be so strong that larvae cannot avoid the bubble stream. Light aeration can be used for eggs of fragile species, but newly hatched larvae can be shocked by contact with bubbles. With them, only very gentle circulation of tank water (preferably without aeration) should be used for the first few days. In the beginning, tank covers can be adjusted to allow natural breezes into the tank, or electric fans can be directed, so that a gentle circulation is induced. Very slowly oscillating horizontal paddles or vertical plungers have been used to maintain circulation. Water flow into the tank can help, but should be gentle and diffuse in the beginning.

Later, as biomass increases, maintenance of dissolved oxygen becomes more important; a minimum of 5 mg/L dissolved oxygen is suitable for most larvae. After a few days, when larvae can swim better and are feeding, circulation can be stronger (partly to promote uniform distribution of prey). Circulation can be increased gradually as the fish become more capable swimmers. The right amount of water movement will stimulate the fish to swim more to maintain position or to find food, and a slight increase in exercise could make them healthier.

When air diffusers are used in rearing tanks, the air flow should be gradually increased, but not to the point that larvae are swept helplessly into the stream of bubbles and battered by them or pushed against tank walls. The best valves available should be used for exact control of air flow.

Water exchange

Daily water exchange rate can be increased gradually from about 10% for eggs and early larvae to at least 300% for small juveniles and 500% or more for larger fish, depending on tolerance of the fish, their density, and the need for flushing of wastes.

MICROBIAL ENVIRONMENT

The microbial environment of fish includes mainly viruses, algae, protozoans, bacteria, and fungi. All groups except algae include pathogens, but algae can produce toxins. Aside from disease and toxin producing microorganisms, algae, protozoans, and bacteria are of the most interest in aquaculture. Algae functions as food for zooplankton and also can remove ammonia and nitrate. Protozoans can have a positive effect as food for larval fish or their zooplanktonic prey or a negative effect by competing with fish or zooplankton (e.g., for oxygen, food). Heterotrophic bacteria such as *Escherichia coli*, *Proteus vulgaris*, *Bacillus subtilus*, *Aerobacter cloaci*, *Pseudomonas* spp., and *Flavobacterium* spp. typically obtain their nutrition from dissolved organic matter (Jana, 1994). Nitrifying bacteria such as *Nitrosomonas* spp., *Nitrosococcus* spp., *Nitrobacter* spp., and *Nitrospina* spp. convert ammonia and nitrite into nitrate. Denitrifying bacteria such as *Achromobacter*, *Micrococcus*, *Pseudomonas*, *Bacillus*, and *Vibrio* spp. (Spotte, 1992; Jana, 1994) convert nitrate into nitrogen gas. Probiotic bacteria such as some *Bacillus* spp. can protect aquatic organisms from pathogenic bacteria and/or release growth-promoting substances.

WATER CHEMISTRY

pH (Table 3-5)

pH is the negative logarithm of the hydrogen ion concentration in water and therefore is an index of acidity. Measurements are on a scale of 1 (very acidic) to 14 (very alkaline). At a pH of 7 (neutral, pure water), H^+ (hydrogen) and OH^- (hydroxyl) ions are balanced. For each unit below 7, the ratio of H^+:OH^- increases by a factor of ten. For each unit above 7, the ratio of OH^-:H^+ increases by a factor of ten. Changes in pH can affect fish by altering water chemistry, increasing or decreasing toxicity of dissolved substances, and affecting gas exchange between fish and water. Extreme pH can cause electrolyte or physiological imbalance, impairment of gas exchange, damage to skin, fins, or gills, bone deformity, and mortality. Both low and high pH can influence blood pH and also cause ammonia retention (Randall, 1991).

Many fish seem to tolerate pH in the range 6.5 to 8.4 (Poxton and Allouse, 1982), but pelagic and reef fish (especially as larvae) probably have a narrower range. If at all possible, the pH of full strength seawater should be maintained at 8.0-8.4. The pH of brackish water naturally should be between 7.0 and 8.4, depending on salinity. Seawater has a greater buffering capacity than freshwater, proportional to its salinity, but this capacity tends to decrease with age of the water (Moe, 1989; Spotte, 1992).

Except in certain cases, such as eutrophic ponds in the summer, acid well water, or recirculating systems, pH is not likely to be a problem. When algal growth is heavy, pH rises in the light as carbon dioxide is depleted and oxygen builds up, but both pH and dissolved oxygen drop in the dark when photosynthesis decreases and respiration continues. In recirculating systems, as ammonia is oxidized by nitrifying bacteria, H^+ ions (i.e., acidity) are produced and carbonate alkalinity (i.e., buffering capacity) is reduced (Brownell, 1980b). If pH does become troublesome, several methods can help maintain it in a suitable range. Water exchange is an obvious one. Increased aeration can drive off excess carbon dioxide and add oxygen (raises pH) or drive off excess oxygen (lowers pH). Foam fractionation can remove organic acids, and ozonation can oxidize organic acids, thus raising pH. Sodium bicarbonate can be added to raise pH (Moe, 1989; Spotte, 1992). Bivalve shells, aragonite, calcite, dolomite, marble chips, lime, or calcium carbonate added to closed systems as buffering agents can help prevent the pH from dropping below 7.5 (Spotte, 1992). Before use, such material should be washed, then conditioned by soaking in seawater for at least 24 h.

pH can easily and accurately be measured with a glass electrode and pH meter (Grasshoff et al., 1983; Parsons et al., 1989; Spotte, 1992). Color comparator kits are cheaper but not as accurate.

Acidity

This is the ability of a solution to react with a strong base to reach a specified pH and can be measured by titration as mg/L calcium carbonate equivalent (APHA, 1995). Most fish do not live well in very acidic waters. Acidity can be important in some freshwater or industrially polluted seawater sources, and it can increase in recirculating systems. In the

Table 3-5. pH requirements for diadromous and marine fish. Values are for juveniles, except where noted for larvae.

Fish	Best range	Range for good health	Lethal limits	References
Salmo salar (fw)		5.5-8.5		Laird & Needham, 1988a
Anadromous salmonid parr (fw)		6-9		Wedemeyer et al., 1981
Salmonids (fw)	6.6-8.0	6.0-8.5		Billard, 1990
Morone saxatilis & hybrid lr. (fw)	7.5-8.5	6.5-9.0		Mullis & Smith, 1990
Morone spp. hybrid (fw)	7.0-8.5	6.0-9.5		Hodson, 1991
Freshwater fish		6.7-8.6		Warren et al., 1990
Clupea harengus (sw)	8.2-8.3		-8.45	Schach, 1939
Gaidropsarus capensis lr. (sw)		>5.9,<8.4	4.7-9.2	Brownell, 1980b
Cyprinodon variegatus (sw)			4.2-	Dunson et al., 1993
Fundulus heteroclitus lr. (sw)			3.8-	Dunson et al., 1993
Fundulus heteroclitus jv. (sw)			3.9-	Dunson et al., 1993
Menidia beryllina (sw)			4.6-	Dunson et al., 1993
Dicentrarchus labrax (sw)			~5.2-9.7	Carter, 1964
Acanthopagrus schlegeli (sw)	~7.5-			Hirayama, 1970
Diplodus sargus lr. (sw)		>5.7,<8.5	5.1-9.0	Brownell, 1980b
Pagrus major (sw)	7.6-8.6			Foscarini, 1988
Chelon labrosus (sw)			~4.7-9.9	Carter, 1964
Gobius paganellus (sw)			~4.6-	Carter, 1964
Pleuronectes platessa (sw)			~5.0-9.8	Carter, 1964
Heteromycteris capensis lr. (sw)		>5.9,<8.5	4.5-9.2	Brownell, 1980b
Most fish jv.-ad.		6.5-8.5	5-9	Poxton & Allouse, 1982; Randall, 1991
Marine fish, screening		7.9-8.2		Huguenin & Colt, 1989
Recommended limits lr. (sw)			8.0-8.4	Tucker
Recommended limits jv.-ad. (sw)			7.0-8.5	Tucker

latter case, pH and alkalinity are better indicators.

Alkalinity
This is the ability of a solution to neutralize an acid and usually is expressed in mg/L calcium carbonate equivalent and sometimes as mEq/L (milliequivalents/L) (Moe, 1989; Spotte, 1992). The alkalinity of seawater depends mainly on bicarbonate and carbonate ion concentrations and is high (2.4-2.5 mEq/L, ~125 mg/L $CaCO_3$). Seawater has a high buffering capacity in both directions and tends to maintain its pH. Alkalinity is not usually a consideration in mariculture except in closed systems, where ion concentrations tend to change over time. Alkalinity could be a factor in fresh water and the recommended range for many freshwater fish is 30-200 mg/L. When marine fish are raised in fresh water, if hardness is okay, alkalinity probably is, too.

Nicholson et al. (1990) recommended alkalinity >150 mg/L to stabilize pH for freshwater culture of striped bass and hybrids. Acceptable alkalinity for freshwater salmonids is 10-400 mg/L (Billard, 1990). Acceptable alkalinity for black sea bream is greater than 1.0 mEq/L (Hirayama, 1970).

One acid titration method for total alkalinity is described by APHA (1995), four by Grasshoff et al. (1983), one by Parsons et al. (1989), and two by Spotte (1992). APHA (1995) describes nomographic methods for determining carbonate alkalinity, bicarbonate alkalinity, hydroxide alkalinity, and free CO_2 from the pH, total alkalinity, temperature, and total mineral content of water. It also gives formulas for calculating carbonate alkalinity, bicarbonate alkalinity, hydroxide alkalinity, free CO_2, and total CO_2 from the pH and total alkalinity. Titration and color comparator kits are available.

Hardness
Total hardness could be defined as the amount of minerals dissolved in a liter of water (Moe, 1989). By that definition, seawater would have a total hardness of about 35,000 mg/L. Hardness originally was a measurement of the ability of water to precipitate soap (APHA, 1995). Because calcium and magnesium are the main contributors to this effect, the small contributors (mostly iron, strontium, zinc, and manganese) usually are not evaluated and total hardness is considered equivalent to **carbonate hardness** (APHA, 1995). Carbonate hardness is a measure of calcium and magnesium ions in water and is expressed in terms of mg/L $CaCO_3$ equivalent, mEq/L (= 50 mg/L $CaCO_3$), or dKH (= 17.9 mg/L $CaCO_3$, German). Carbonate hardness can be calculated from the calcium and magnesium content by the folowing formula (APHA, 1995): mg/L $CaCO_3$ equivalent = 2.497(mg/L Ca) + 4.118(mg/L Mg). With 400 mg/L Ca and 1,350 mg/L Mg (Spotte, 1992), typical seawater has a total hardness of 999 + 5559 = 6558 mg/L $CaCO_3$ equivalent. Fresh water with less than 75 mg/L hardness is considered soft. Rainforest water is very soft, sometimes with hardness of 0-5 mg/L. Well water often is hard; artesian water at HBOI has a hardness of 769 mg/L and is very good for culture of both freshwater and euryhaline marine fish.

Recommended hardness for freshwater fish varies from 20 to 400 mg/L. Because the hardness of seawater is very high, it usually is not a consideration in mariculture. However, when marine fish are raised in fresh or nearly fresh water, a certain degree of hardness might be required (because they need magnesium and calcium). However, we have found that barramundi will grow as well in nearly pure water (5 mg/L hardness) as in saline water, if their feed contains enough minerals. Acceptable total hardness for young striped bass and hybrids in fresh water is 200-250 mg/L $CaCO_3$ equivalent (Mullis and Smith, 1990). Acceptable hardness and chlorinity (chlorine content) for juvenile red drum are >100 mg/L $CaCO_3$ equivalent and >150 mg/L Cl^- (Neill, 1990). Pursley and Wolters (1994) found that juvenile red drum in fresh water had better survival, growth rates, and feed conversion at 500 mg/L Cl^- than at 150 mg/L Cl^-. At 500 mg/L Cl^-, survival was improved by raising hardness from 100 to 400 mg/L $CaCO_3$ equivalent. In the absence of specific data, hardness probably should be at least 150 mg/L $CaCO_3$ equivalent to be safe.

APHA (1995) describes a calculation method and an EDTA titration method for hardness. Color comparator kits are available.

Redox potential
The oxidation-reduction potential (ORP or redox potential) is a measurement of the ratio of positive and negative charges on oxidized and reduced molecules in seawater, as millivolts (mv) (Moe, 1989; APHA, 1995). Equal numbers of charges give a redox potential of zero; with more reduced molecules, ORP is negative; with more oxidized molecules, ORP is positive. With low dissolved oxygen, redox potential tends to become negative, but with enough oxygen, it will stay positive. Moderately high redox potential facilitates respiration in cells.

Redox potential is of most interest in recirculating systems. Moe (1989) recommended a range of 200-350 mv for marine aquariums. With less than 200-250 mv, toxic compounds are more likely, and at values higher than 400-450 mv, oxidative injury to the primary inhabitants could occur. Preferred types of algae grow better at higher ORP (i.e., filamentous green at 250-300 mv and large macroalgae >300 mv), while mat-forming blue-green and red species grow better at 200-250 mv. Redox potential over coral reefs is about 350-400 mv. A good target probably is 300 mv.

Increasing temperature or pH will lower redox potential. Increasing organic matter tends to lower it. Ozonation tends to raise it. Aeration, foam fractionation, water exchange, and avoidance of organic overloads are practical means of maintaining redox potential.

Redox potential usually is measured with a platinum indicator electrode and pH meter or voltmeter (Moe, 1989; APHA, 1995).

Biochemical oxygen demand
BOD is an index of the biodegradable organic substances in water (APHA, 1995), which, in fish culture, would be derived mainly from the fish, their food, and sometimes plankton, but also could be present in the water source. BOD is measured as the amount of oxygen required to degrade those substances during a 5-d incubation. Because BOD is not routinely useful and measurements are time-consuming, it is mainly used for effluent monitoring or research. Methods are given by APHA (1995).

WASTE PRODUCTS

Hatching enzymes
These can be toxic to hatchlings if the fish are crowded or are stressed by something else. Water exchange during incubation should be sufficient to flush them out without injuring the eggs or larvae.

Nitrogenous waste (Fig. 3-2; Tables 3-6, 3-7, 3-8)
The main forms of nitrogen in water are: dissolved ammonia gas (NH_3), ammonium (NH_4^+) compounds, ammonium hydroxide (NH_2OH), nitrite (NO_2^-) compounds, nitrate (NO_3^-) compounds, dissolved nitrogen gas (N_2), dissolved organic nitrogen compounds (e.g., amino acids, amines, proteins, humic substances), and particulate organic nitrogen compounds (Jana, 1994). A simplified aquatic nitrogen cycle includes: nitrogen fixation (N_2 to organic forms), ammonification (organic forms to NH_4^+), nitrification (NH_4^+ to NO_2^- to NO_3^-), and denitrification (NO_3^- to N_2).

Chemoautotrophic (obtain carbon from CO_2) bacteria such as *Nitrosomonas europaea* (primarily), *Nitrosococcus oceanus* (ocean), and *Nitrosococcus mobilis* (brackish water) oxidize ammonia into nitrite to obtain energy (Spotte, 1992). Other chemoautotrophs such as *Nitrobacter winogradski*, *Nitrosococcus mobilis*, and *Nitrospina gracilis* oxidize nitrite into nitrate.

The simplified nitrification reactions are:

$$2\ NH_4^+ + 3\ O_2 \rightarrow 2\ NO_2^- + 2\ HOH + 4\ H^+$$

$$2\ NO_2^- + O_2 \rightarrow 2\ NO_3^-$$

In water quality testing, the three forms are measured as µg/L or mg/L of total ammonia nitrogen or TAN (NH_4-N), nitrite nitrogen (NO_2-N), and nitrate nitrogen (NO_3-N). APHA (1995) describes titration (minimum detectable concentration = MDC ~5000 µg/L), ammonia-selective electrode (MDC 30 µg/L), and indophenol blue colorimetric (MDC 20 µg/L) methods for ammonia, a colorimetric (MDC 5 µg/L) method for nitrite, and chromatographic, nitrate electrode (MDC 140 µg/L), cadmium reduction (MDC 10 µg/L), titanous chloride reduction (MDC 10 µg/L), and hydrazine reduction (MDC 10 µg/L) methods for nitrate. Spotte (1992) describes indophenol blue and electrode methods for ammonia, one colorimetric method for nitrite, and cadmium and hydrazine methods for nitrate. Parsons et al. (1989) describe oxidation-colorimetric and indophenol blue methods for ammonia, a colorimetric method for nitrite, and a cadmium method for nitrate. Grasshoff et al. (1983) describe an indophenol blue method for ammonia, a colorimetric method for nitrite, and a cadmium method for nitrate. Color comparator kits for measuring ammonia, nitrite, or nitrate, and ammonia electrodes are readily available.

Ammonia (unionized NH_3, ionized NH_4^+) is the major substance excreted by fish eating protein foods. It is excreted through the gills and kidneys. Ammonia also is formed by decomposition and mineralization of other wastes such as urea, uric acid, amine, creatine, creatinine; uneaten food; and dead plants and animals. The optimal temperature range for ammonification is 30-35°C (Jana, 1994). In aquaculture systems, water exchange, bacterial nitrification, and uptake by algae are the three main routes for ammonia disposal. Ammonia can be strongly adsorbed by colloidal and other particles, especially with high pH and high dissolved humic levels (Jana, 1994).

Increased ammonia in culture water (Table 3-6) can lead to retention of ammonia in the fish and possibly uptake from the water (Moe, 1989). Excess internal ammonia apparently interferes with oxygen transport by hemoglobin. Unionized ammonia (NH_3 = dissolved ammonia gas) is thought to be the form most toxic to fish, because it easily goes through gill membranes. The ammonium ion (NH_4^+) is thought to be relatively non-toxic, but this is not certain. The proportion of unionized ammonia rises with pH (Fig. 3-2) and so does toxicity, but factors other than the NH_3 increase could be contributing (Meade, 1985; Russo, 1985; Spotte, 1992; Handy and Poxton, 1993). In addition, NH_4^+ does seem to exhibit some toxicity. The concentration of unionized ammonia can be calculated from TAN if pH and temperature are known (ionic strength of the solution is a third but minor factor). An increase of one pH unit increases the amount of unionized ammonia by a factor of about ten. Over a pH range of 7.0-8.2, an increase of 10°C doubles the amount of unionized ammonia. Low dissolved oxygen levels further stress the fish and increase the chance of hypoxia (insufficient oxygen) resulting from gills damaged in the presence of ammonia.

In seawater, the availability of sodium ions to exchange for ammonium ions and increased gill membrane permeability (e.g., by calcium) or sodium uptake could make ammonia excretion easier than in fresh water (Soderberg and Meade, 1991). Also, the abundant ions in seawater act to moderate the effects of pH and decrease the NH_3:NH_4^+ ratio. Ammonia tolerance might be increased by previous exposure.

The proportion of NH_3 increases with pH, but a given concentration of NH_3 can be more toxic with lower pH (Russo and Thurston, 1991). Toxicity increases when dissolved oxygen is reduced. Toxicity of ammonia plus copper or zinc was additive. Toxicity of ammonia and hydrogen cyanide combined was more than additive.

Acute ammonia toxicity causes overproduction of gill mucus, rapid escape movements, hyperventilation, irregular ventilation, coughing, equilibrium loss, twisting, spiral swimming,

Table 3-6. Dissolved total ammonia nitrogen (and unionized ammonia nitrogen) in natural and hatchery water.

Purpose	Measured (µg/L)	Considered safe (µg/L)	Considered harmful (µg/L)	Considered lethal (µg/L)	References
Bermuda shelf	2.8				von Bodungen et al., 1982
Sargasso Sea	3.0				von Bodungen et al., 1982
Saronikos Gulf, Greece, western	9.0				Friligos, 1989
Saronikos Gulf, Greece, northeastern	34.9				Friligos, 1989
Algeciras Bay, Spain, eastern	33				Tarazona et al., 1991
Algeciras Bay, Spain, southwestern	1,920				Tarazona et al., 1991
Thailand, coastal aquaculture areas	(10)				Tookwinas, 1990a
Anguilla anguilla jv. (fw, 23°C, pH 7.4-8)		≤6200(120)	>6200(120)	55,900(1079)	Sadler, 1982
Pimephales promelas life (fw)			(150)	(370)	Thurston et al., 1986
Oncorhynchus clarki (fw)				(247-493)	Poxton & Allouse, 1982
Oncorhynchus keta (fw)				(370)	Poxton & Allouse, 1982
Oncorhynchus mykiss (fw)		(25)	(100)	(430)	Poxton & Allouse, 1982
Oncorhynchus mykiss (fw)			(10-70)	43,500(500-800)	Post, 1987
Oncorhynchus mykiss (fw, 7-9°C, pH 8.2)		1240(25)	3629(73)		Hofer et al., 1995
Salmo salar jv. (fw, 1.8°C, pH 6.4)			3000(0.7)	128,500(30)	Knoph, 1992
Salmo salar (fw)		(12.5)			Laird & Needham, 1988a
Salmonids (fw, 19°C, pH 7.0)			1700(10)	8000(45)	Billard, 1990
Anadromous salmonid parrs (fw, 10°C, pH 7.0)[a]		<2680(5)			Wedemeyer et al., 1981
Morone saxatilis & hybrid lr. (fw)		(<16)			Mullis & Smith, 1990
Morone saxatilis (fw, 21°C, pH 8.3)				12,800(1010)	Oppenborn & Goudie, 1993
M. saxatilis × *M. chrysops* (fw, 21°C, pH 8.3)				8100(640)	Oppenborn & Goudie, 1993
Freshwater fish (25°C, pH 7.0)[a]		<3520(20)			Warren et al., 1990
Freshwater fish (25°C, pH 7.0)		4400(25)			Lloyd, 1981
Brachionus plicatilis (23 ppt, 23°C, pH 7.9-8.6)		(2100)		(7800)	Yu & Hirayama, 1986
Brachionus plicatilis (30 ppt, 25°C, pH 7.7)				998,000(14,600)	Snell & Persoone, 1989
Acartia clausi [cop] (sw)			93	312	Buttino, 1994
Eucalanus pileatus[cop] ad. (sw)				12,743(643)	Venkataramiah et al., 1982
Salmo salar jv. (34 ppt, 4-9°C, pH 7.8)		<153(1)	153(1)		Knoph & Olsen, 1994
Gaidropsarus capensis lr. (35 ppt, 15°C, pH 7.8)[a]		1400(20)	4800(70)	29,700(430)	Brownell, 1980a

Cyprinodon variegatus lr. (sw)			90,200(2100)	Miller et al., 1990
Menidia beryllina lr. (sw)	(50)	(74)	(380)	Miller et al., 1990
Dicentrarchus labrax jv. (34 ppt, 18°C, pH 8.2)	~2220(97)		22,200(970)	Person-Le Ruyet et al., 1995
Morone saxatilis (sw, 23°C, pH 8.0)			(35)	Hazel et al., 1971
Epinephelus malabaricus jv. (sw)	(<20)			Tookwinas, 1990a
Epinephelus striatus lr. (34 ppt, 27°C)			100-200(10-20)[e]	Tucker & Woodward, 1996
Epinephelus striatus ad. (36 ppt, 27°C)			100-300(10-30)[e]	Tucker & Woodward, 1996
Coryphaena hippurus (sw)	1000, brief			Szyper, 1991
Diplodus sargus lr. (35 ppt, 15°C, pH 7.8)[a]		2100(30)	20,000(290)	Brownell, 1980a
Lithognathus mormyrus lr. (35 ppt, 15°C, pH 7.8)[a]		4800(70)	23,500(340)	Brownell, 1980a
Pachymetopon blochii lr. (35 ppt, 15°C, pH 7.8)[a]		900(13)	2600(38)	Brownell, 1980a
Pagrus major lr.			(2)	Guillén et al., 1993, 1994a
Sparus auratus jv. (34 ppt, 18°C, pH 8.2)	~5130(222)		51,300(2220)	Person-Le Ruyet et al., 1995
Sparus auratus jv. (40 ppt, 26-27°C, pH 8.1)	1200(64)	8,220(470)	13,000(740)	Wasjbrot et al., 1991, 1993
Cynoscion nebulosus lr. (sw)			(280)	Daniels et al., 1987
Cynoscion nebulosus jv. (sw)			(980)	Daniels et al., 1987
Nibea japonica lr.			(2.7)	Guillén et al., 1994b
Sciaenops ocellatus egg-lr. (~30 ppt, 25°C, pH ~8.2)[a]	<550(45)		3600(260)	Holt & Arnold, 1983; Holt, 1990; Neill, 1990
Sciaenops ocellatus jv. (4 ppt, 20°C, pH 6.8-7.1)[a]	12,500(50)		199,900(800)	Wise et al., 1989
Mugil cephalus jv. (sw)			(12,226)	Venkataramiah et al., 1982
Scophthalmus maximus jv. (34 ppt, 16°C, pH 7.9)[a]	7900(140)		50,600(900)	Alderson, 1979
Scophthalmus maximus jv. (35 ppt, 18°C, pH 8.2)	~3350(147)		33,500(1470)	Person-Le Ruyet et al., 1994, 1995
Scophthalmus maximus jv. (34 ppt, 17°C, pH 7.9-8.0)[a]		4200(90)	37,000(800)	Person-Le Ruyet et al., 1997
Solea solea jv. (34 ppt, 17°C, pH 7.9)[a]	2300(45)		40,100(770)	Alderson, 1979
Monacanthus hispidus jv. (sw)			(690)	Venkataramiah et al., 1982
Research in sw (35 ppt, 28°C, pH 8.3)[a]	10(1)			Kinne, 1976
Marine aquarium fish ad. (35 ppt, 28°C, pH 8.3)[a]	102(10)			Spotte, 1992
Marine fish research (35 ppt, 28°C, pH 8.3)[a]	<10(1)			Huguenin & Colt, 1989
Marine fish production (35 ppt, 28°C, pH 8.3)[a]	<102(10)			Huguenin & Colt, 1989
Marine fish holding (35 ppt, 28°C, pH 8.3)[a]	<408(40)			Huguenin & Colt, 1989
Recommended limit for larval fish in sw	**<10(1)**			Tucker
Recommended limit for older fish in sw	**<50(5)**			Tucker

[a] Total ammonia nitrogen has been estimated for salinity, temperature, and pH. [e] Empirical data. Except for *Brachionus* sp. (rotifers) and *Artemia* sp. (brine shrimp), microalgae and invertebrates are indicated by these abbreviations in this and the following tables: [amp] amphipod, [bac] bacteria, [car] caridean shrimp, [cil] ciliate protozoan, [cld] cladoceran, [clm] clam, [cop] copepod, [crb] crab, [mic] microalgae (see Table 8-1), [mus] mussel, [mys] mysid shrimp, [oys] oyster, [pen] penaeid shrimp, [rot] rotifer.

convulsions, and mortality within minutes to hours (Russo and Thurston, 1991; Knoph, 1992; Spotte, 1992). Dead fish characteristically have gaping mouths, extended gill covers, and rigid u-shaped bodies. With low to moderate poisoning, fish might stay at the surface, either still or swimming slowly and occasionally gulping for air. Chronic sublethal exposure to ammonia could result in reduced feeding, physiological imbalance, disease susceptibility, blood and kidney damage, reduced growth, and reproductive dysfunction. Usually gill injury is present and interferes with gas exchange. This is **environmental gill disease** (Post, 1987). If gills are very swollen, the gill covers might flare outward. Bacteria can invade, causing **bacterial gill disease**, which is worse. In either case, mortality can occur within a few days.

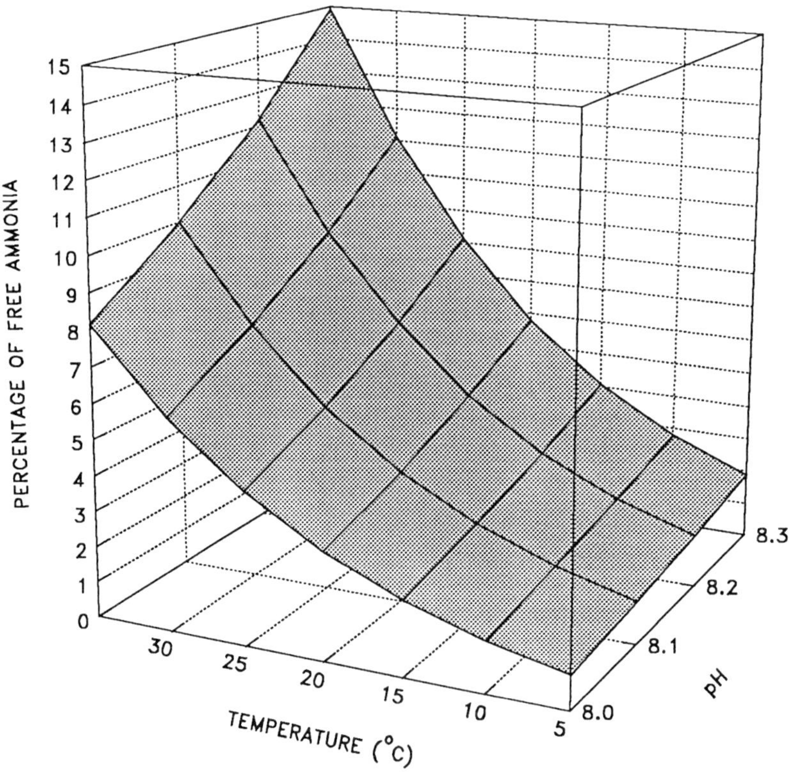

Figure 3-2. Relation between water temperature, pH, and the ratio of free (unionized) ammonia nitrogen (NH_3-N) to total ammonia nitrogen (NH_4-N, or TAN).

Reduced survival or growth of marine fish has occurred at 300 μg/L NH_3-N. But striped bass can be killed by 20 μg/L, and 10-20 μg/L could be lethal to Nassau grouper larvae within an hour or two. In the study by Brownell (1980a), first feeding by larvae of five marine species was affected by NH_3-N levels of 10-100 μg/L.

Ammonia could be more toxic in salt water than in fresh water because the pH is higher. Levels as low as 10 μg/L NH_4-N (1 μg/L NH_3-N) can affect feeding, growth, or survival of sensitive species or stages. Ammonia should be kept near zero, but practical limits are 1 µg/L NH_3-N for larvae and 5 µg/L NH_3-N for older fish.

Nitrite (NO_2^-) also is toxic (Table 3-7) and should not be allowed to accumulate. It binds with hemoglobin to form methemoglobin, thus interfering with oxygen transport to tissues and causing brown-blood disease (additional causes are possible, such as stress, hypoxia, hypercapnia, copper, sulfite, tricaine methanesulfonate; Russo, 1985, Wells et al. 1997). Low dissolved oxygen increases the problem. When affected channel catfish were transferred from water containing nitrite, methemoglobin was reduced to a normal level in 24-48 h; however, in European sea bass, nitrite poisoning caused severe anemia from loss of hemoglobin that persisted for 24 d (Scarano et al., 1984; Lewis and Morris, 1986). Sublethal exposure might lead to some increased tolerance after several days. At least for European sea bass, decreasing temperature can reduce nitrite toxicity (Saroglia et al., 1981). Nitrite is less toxic in salt water than in fresh water because of the other ions present. Mainly chloride and calcium ions provide protection in seawater, but bromide, bicarbonate, sulfate, phosphate, and nitrate also can help (Russo and Thurston, 1991; Tomasso, 1994). Lowered pH tends to increase nitrite toxicity by raising the proportion of nitrous acid (HNO_2). Marine fish kept in fresh water are at increased risk. With a mature biofilter, nitrite should not build up, unless the filter becomes overloaded or the water flow is too slow. Recommended limits for nitrite are 0.1 mg/L for larvae and 1 mg/L for older fish, but zero is better.

Nitrate (NO_3^-) is relatively non-toxic (Table 3-8) but could promote the growth of disease organisms. Sublethal effects could include weakened immune response and haematological and biochemical alterations. Largemouth bass and channel catfish seem to tolerate up to 400 mg/L. Some marine fish can stand 300 mg/L, but 100 mg/L reduced growth and survival of common clownfish larvae and juveniles (Frakes and Hoff, 1982).

Although nitrate is relatively harmless, it should not be allowed to accumulate. It is probably best to keep NO_3-N lower than 20 mg/L for eggs and larvae and lower than 50 mg/L for older fish. Sometimes this is not possible, for instance in fertilized ponds, but in those cases, problems could occur.

Prevention of problems.—The two basic methods depend on how water is used:

1) A flow-through system—water is used once and discarded. A sufficient flow rate will dilute and flush out nitrogenous wastes.
2) If the water supply is limited or if there are problems heating, cooling, conditioning, or disposing of it, then recirculation, or repeated use of the same water, might be necessary. In this case, ammonia must be removed before it builds up to a dangerous level. Aeration won't remove much of the ammonia. The best method is biofiltration. What is necessary is the creation of as much suitable habitat as possible for nitrifying bacteria. *Nitrosomonas* spp. take up to 3-4 weeks to establish, and *Nitrobacter* spp. 3-6 weeks, depending on the conditions. Seeding with bacteria greatly accelerates the process. If the biofilter is the correct size, is fully colonized by bacteria, has enough oxygen, and has sufficient flow of water, both ammonia and nitrite levels should be close to zero. The only practical ways to remove nitrates are by exchanging the water, growing plants in the water and removing them, or using a dentrification system (the latter can be tricky).

Fish are more resistant to ammonia if sufficient dissolved oxygen is present. Care should be taken when transporting fish to prevent ammonia buildup.

Large organic molecules

High levels of some naturally-occurring organic compounds might interfere with growth and reduce disease resistance. Dissolved organic substances in seawater include proteins, amino acids, starches, sugars, vitamins, hormones, pigments, humic acids, phenolic compounds,

Table 3-7. Dissolved nitrite nitrogen in natural and hatchery water.

Purpose	Measured (mg/L)	Considered safe (mg/L)	Considered harmful (mg/L)	Considered lethal (mg/L)	References
Bermuda	0.0004				von Bodungen et al., 1982
Sargasso Sea	0.0007				von Bodungen et al., 1982
Saronikos Gulf, Greece, western	0.0018				Friligos, 1989
Saronikos Gulf, Greece, northeastern	0.0025				Friligos, 1989
Algeciras Bay, Spain, eastern	0				Tarazona et al., 1991
Algeciras Bay, Spain, southwestern	1.77				Tarazona et al., 1991
Thailand, coastal aquaculture areas	0.0039				Tookwinas, 1990a
Chanos chanos jv. (fw)				12	Almendras, 1987
Pimephales promelas (fw)				45	Lewis & Morris, 1986
Ictalurus punctatus (fw)			1	7.1	Lewis & Morris, 1986; Post, 1987
Oncorhynchus mykiss jv. (fw)				3.3	Eddy et al., 1983
Oncorhynchus mykiss (fw)				0.24	Lewis & Morris, 1986
Oncorhynchus mykiss jv. (fw)			0.22		Hofer & Gatumu, 1994
Oncorhynchus tshawytscha jv. (fw)			1.2	≤6	Crawford & Allen, 1977
Salmo salar (fw)		0.0066			Laird & Needham, 1988a
Anadromous salmonid parrs (fw)		30-60			Wedemeyer et al., 1981
Salmonids (fw)		<0.1-0.2	>100		Lloyd, 1981; Billard, 1990
Morone saxatilis & hybrid lr. (fw)		<0.06			Mullis & Smith, 1990
Oreochromis niloticus (fw)				16	Lewis & Morris, 1986
Freshwater fish		<1.0			Warren et al., 1990
Anguilla anguilla elver (sw)				503	Saroglia et al., 1981
Chanos chanos jv. (bw)				675	Almendras, 1987
Oncorhynchus tshawytscha jv. (sw)			≥30	~248	Crawford & Allen, 1977
Salmo salar jv. (sw)				~160	Eddy et al., 1983
Gaidropsarus capensis lr. (sw)			280	1500	Brownell, 1980a
Lates calcarifer jv. (0 ppt, 28°C)			1.45	14.5	Woo & Chiu, 1994

Lates calcarifer jv. (15 ppt, 28°C)			10.5	105	Woo & Chiu, 1994
Lates calcarifer jv. (32 ppt, 28°C)			9.3	93	Woo & Chiu, 1994
Dicentrarchus labrax (sw, 23°C)				220	Saroglia et al., 1981
Dicentrarchus labrax jv. (sw)			13	~95	Scarano et al., 1984
Dicentrarchus labrax		0.1		3	Blanchet-Besson, 1986
Epinephelus striatus lr. (sw)		0.15 [e]	0.30 [e]		Tucker & Woodward, 1996
Diplodus sargus lr. (sw)			150	870	Brownell, 1980a
Lithognathus mormyrus lr. (sw)			60	670	Brownell, 1980a
Cynoscion nebulosus lr. (sw)				980	Daniels et al., 1987
Sciaenops ocellatus egg-lr. (sw)		~10	~100	100-500	Holt & Arnold, 1983; Holt, 1990
Heteromycteris capensis lr. (sw)			190	2170	Brownell, 1980a
Synaptura kleinii lr. (sw)			180	1210	Brownell, 1980a
Research in sw		0.1			Kinne, 1976
Marine fish, screening		<50			Huguenin & Colt, 1989
Marine fish production		100			Huguenin & Colt, 1989
Recommended limit for larval fish in sw		**<0.1**			Tucker
Recommended limit for older fish in sw		**<1**			Tucker

[e] Empirical data.

Table 3-8. Dissolved nitrate nitrogen in natural and hatchery water.

Purpose	Measured (mg/L)	Considered safe (mg/L)	Considered harmful (mg/L)	Considered lethal (mg/L)	References
Bermuda	0.0028				von Bodungen et al., 1982
Sargasso Sea	0.0029				von Bodungen et al., 1982
Saronikos Gulf, Greece, western	0.0073				Friligos, 1989
Saronikos Gulf, Greece, northeastern	0.0203				Friligos, 1989
Raja eglanteria jv. (sw)				>960	Pierce et al., 1993
Ictalurus punctatus (fw)		400			Poxton & Allouse, 1982
Oncorhynchus clarki egg (fw)				20	Kincheloe et al., 1979
Oncorhynchus mykiss egg (fw)				5	Kincheloe et al., 1979
Oncorhynchus mykiss jv. (fw)				6000	Kincheloe et al., 1979

Table 3-8. (Continued)

Purpose	Measured (mg/L)	Considered safe (mg/L)	Considered harmful (mg/L)	Considered lethal (mg/L)	References
Oncorhynchus tshawytscha lr. (fw)				20	Kincheloe et al., 1979
Oncorhynchus tshawytscha jv. (fw)				5800	Kincheloe et al., 1979
Oncorhynchus tshawytscha (fw)				1310	Russo & Thurston, 1991
Salmonids (fw)		0-3			Billard, 1990
Poecilia reticulata (fw)				180-200	Russo & Thurston, 1991
Morone saxatilis & hybrid lr. (fw)		<0.6			Mullis & Smith, 1990
Morone saxatilis jv.-ad. (fw)		≤8.6			Nicholson et al., 1990
Freshwater fish		<1.0			Warren et al., 1990
Gaidropsarus capensis lr. (sw)			360	>4000	Brownell, 1980a
Centropristis striata jv. (sw)				2400	Pierce et al., 1993
Trachinotus carolinus jv. (sw)				1000	Pierce et al., 1993
Diplodus sargus lr. (sw)			270	3100	Brownell, 1980a
Lithognathus mormyrus lr. (sw)			270	3020	Brownell, 1980a
Amphiprion ocellaris jv. (sw)		~20	100		Frakes & Hoff, 1982
Stegastes leucostictus ad. (sw)				>3000	Pierce et al., 1993
Scophthalmus maximus (sw)		342 [e]			Poxton & Allouse, 1982
Heteromycteris capensis lr. (sw)			420	3510	Brownell, 1980a
Monacanthus hispidus jv. (sw)				573	Pierce et al., 1993
Marine aquarium fish ad. (sw)		<50			Spotte, 1992
Research in sw		20			Kinne, 1976
Recommended limit for larval fish in sw		**<20**			Tucker
Recommended limit for older fish in sw		**<50**			Tucker

[e] Empirical data.

hydrocarbons, organic iron, and organic phosphates (Moe, 1989).

A foam fractionator, or protein skimmer, sometimes is used to remove dissolved organic matter. It works by bubbling air, oxygen, or ozone through water so that the large organic molecules collect in a film on the surface, which is automatically skimmed off, usually through a standpipe. Activated carbon sometimes is used to remove these compounds, by adsorption. Special resins can filter some of them out. A final possible option is the use of ozone to oxidize the molecules, but that does not seem to be effective (Spotte, 1992).

GASES

Oxygen (Fig. 3-3; Table 3-9)

Sufficient dissolved oxygen (**DO**) is necessary for fish to survive, feed, grow, and reproduce. Another reason for keeping water oxygenated and preventing buildup of organic matter on tank walls and bottoms is to minimize the activity of anaerobic bacteria. Anaerobic processes such as fermentation tend to produce toxic substances like organic acids, carbon dioxide, ammonia, methane, and hydrogen sulfide. Aerobic processes such as oxidation of ammonia to nitrate usually produce substances that are less toxic.

Sources and sinks.—Oxygen in the water mostly has been dissolved from the atmosphere or produced by aquatic plants. It is reported as mg/L (ppm) or % ASV = air saturation value. The capacity of water to hold oxygen, as for any gas, depends on temperature, salinity, and altitude (an increase lowers the capacity), or pressure (an increase raises the capacity). Natural seawater bodies contain oxygen varying from 0 to 12 mg/L (or more if

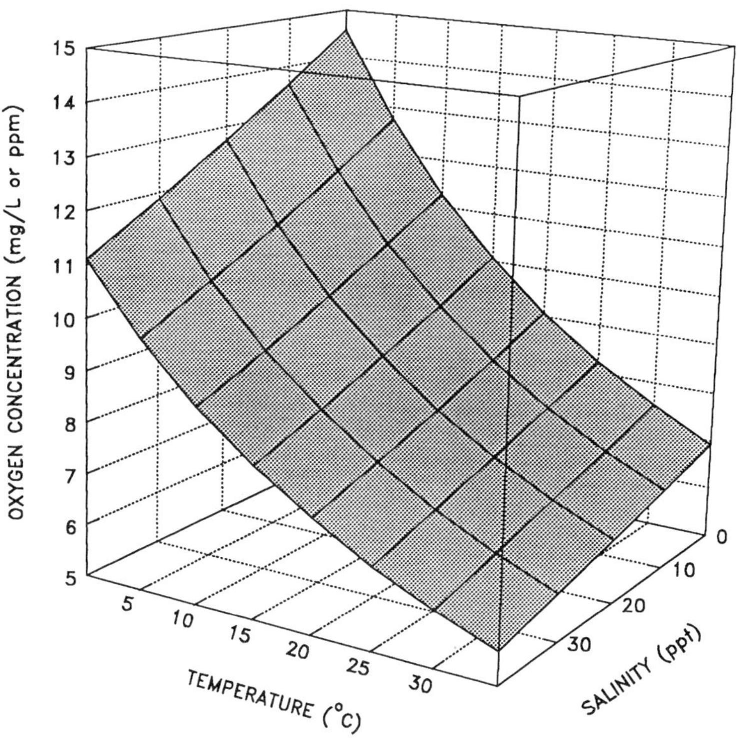

Figure 3-3. Relation between water temperature, salinity, and dissolved oxygen at saturation.

Table 3-9. Dissolved oxygen minima for diadromous and marine fish. Values are for juveniles, except where noted for larvae or adults.

Fish	Best (mg/L)	Minimum for good health (mg/L)	Lethal minimum (mg/L)	References
Anguilla anguilla		4		Poxton & Allouse, 1982
Anguilla japonica		4.5		Yamagata et al., 1983
Anguilla japonica			<1	Usui, 1991
Salmo salar (fw)		6		Laird & Needham, 1988a
Salmonid eggs (fw)	10-12	8		Truelson & Bio, 1994
Salmonid lr. (fw)	>70% saturation	6-7		Billard, 1990
Salmonid jv. (fw)	>70% saturation	5-5.5		Billard, 1990
Salmonid jv. (fw)	7-9	5-6		Truelson & Bio, 1994
Morone saxatilis egg		>5		Rombough, 1988
Morone saxatilis lr.	5-6		2.4	Hill et al., 1989; Hall, 1991
Morone saxatilis & hybrid lr. (fw)	Saturation	4		Mullis & Smith, 1990
Morone saxatilis	6-12		3.0	Hill et al., 1989; Hall, 1991
Morone saxatilis ad.			3	Hill et al., 1989
Morone spp. hybrid (fw)	6	4	<1	Hodson, 1991
Clupea harengus lr. (sw)		>5	3.6	De Silva & Tytler, 1973; Rombough, 1988
Anchoa mitchilli lr. (sw)		≥3.1	≤2.1	Breitburg, 1994
Chanos chanos		>1		Watanabe, 1986
Gadus morhua lr. (sw)		3.5	2.5	Nissling, 1994
Gadus morhua		>70% sat., or 4.3	<1.1	Jobling, 1988; van der Meeren, 1991
Gaidropsarus capensis lr. (sw)		>5.6	2.5	Brownell, 1980b
Fundulus heteroclitus egg		>4.5	2.4-4.5	Rombough, 1988
Fundulus heteroclitus		4	<1-3	Tatum et al., 1982b
Congiopodus spinifer lr. (sw)		>5.3	3.6	Brownell, 1980b
Lates calcarifer (sw)		4	~0.5-1.0	Tookwinas, 1990a; Wu, 1994
Dicentrarchus labrax (sw)		5		Coves et al., 1991
Dicentrarchus labrax (sw)	90% saturation	3-4.5	2	Barnabé, 1990d; Coves et al., 1991

Epinephelus akaara (sw)		>4	~1.0	Wu & Woo, 1984; Wu, 1994
Epinephelus malabaricus (sw)		4		Tookwinas, 1990a
Epinephelus tauvina ad. (sw)			0.5-1.0	Wu, 1994
Seriola quinqueradiata (sw)	>5	4	1	Ikenoue & Kafuku, 1992
Trachinotus carolinus		>3	2.5	Moe et al., 1968
Coryphaena hippurus (sw)	6	5	4-4.5	Kraul, 1990b; Szyper, 1991
Lutjanus russellii ad. (sw)			~1.0	Wu, 1994
Acanthopagrus schlegeli (sw)		>4	0.5-1.0	Wu & Woo, 1984; Wu, 1990
Pachymetopon blochii lr. (sw)		>5.2	3.2	Brownell, 1980b
Pagrus major (sw)	4	3.5	~0.5-1.0	Foscarini, 1988; Wu, 1994
Rhabdosargus sarba (sw)		60% saturation		Chiba, 1983
Pterosmaris axillaris lr. (sw)		>5.7	3.5	Brownell, 1980b
Sciaenops ocellatus lr. (sw)	>5	>3		Henderson-Arzapalo, 1990; Colura et al., 1991
Sciaenops ocellatus (sw)		~5	~2	Neill, 1990; Stahl et al., 1997
Mugil cephalus lr. (sw)		5.0	3.2-4.8	Sylvester et al., 1975
Polydactylus sexfilis	5.8-6.0	4.9		Ostrowski, 1995
Gobiosoma bosc lr. (sw)		≥3.0	≤2.0	Breitburg, 1994
Siganus canaliculatus (sw)			2	Duray, 1990
Siganus canaliculatus ad. (sw)			~1.0	Wu, 1994
Siganus guttatus (sw)			0.7	Duray, 1990
Thunnus thynnus (sw)			1.1	Harada, 1980
Scophthalmus maximus lr. (sw)		6		Person-Le Ruyet et al., 1991
Scophthalmus maximus (sw)	5	3	1	Person-Le Ruyet, 1990b
Pleuronectes platessa lr. (sw)			2.7	De Silva & Tytler, 1973
Heteromycteris capensis lr. (sw)		>4.0	0.94	Brownell, 1980b
Cynoglossus capensis lr. (sw)		>5.4	<2.3	Brownell, 1980b
Salmonid egg-lr. (fw)	9.8	8.8	~6.5	Poxton & Allouse, 1982
Salmonid jv.-ad. (fw)	7.2	5.2	~3.2	Poxton & Allouse, 1982
Marine nonanadromous fish	8.8	6.8	~4.5	Poxton & Allouse, 1982
Marine fish production		>6		Huguenin & Colt, 1989
Recommended limit for marine fish		**5-7**		Tucker

supersaturated).

During the day, production of oxygen by photosynthesis usually exceeds consumption by plant respiration—and dissolved oxygen tends to increase. At night, both plants and animals are using oxygen, but essentially none is produced. High concentrations of aquatic plants, whether macrophytes or phytoplankton, can severely deplete oxygen, especially at night. High concentrations of animals can deplete oxygen day or night. The most dangerous time for fish in a pond or restricted natural water body is early morning, after a night of heavy respiration. During September, in a branch of the Indian River, I have measured 0.4 mg/L DO in the morning and 10.7 mg/L DO (supersaturated) in the afternoon. If the quantity of plants is insignificant, oxygen should not fluctuate as much, unless something else changes (e.g., increase in temperature, accumulation of rotting wasted feed). Intensive cage farming also can deplete oxygen to harmful levels (Lee et al., 1991a, 1991b).

Effects.—As temperature rises, oxygen solubility decreases and fish metabolism and demand for oxygen increase. Activity, feeding, growth, and survival of fish can be limited by DO. Excess DO (environmental hyperoxia) causes the ventilation rate to slow, possibly enough that blood CO_2 is not released and increases to harmful levels (physiological hypercapnia) (Spotte, 1992). This is most likely in sealed oxygenated shipping containers. Low DO (environmental hypoxia) causes ventilation rate and blood pressure to increase and heart rate to decrease in an attempt to get and transport more oxygen (Spotte, 1992). Chronic hypoxia can result in developmental deformities, especially skeletal and circulatory (Rombough, 1988).

Fish react immediately to low oxygen levels, by first trying to avoid such a disagreeable situation. At times, during the warmer months, it is not unusual to see pond fish gasping at the surface or at least staying just under the surface, where DO is highest. If the fish can't escape to better oxygenated waters, slowing down will reduce their oxygen needs.

Two of the most likely causes of sudden fish kills in ponds are low DO, sometimes with high ammonia (mainly in the morning), and high DO with high temperature leading to gas bubble trauma (mainly in the afternoon)(Krom et al., 1985).

Some species have wider tolerances, normally live near the surface, or like the tarpon, can get their oxygen directly from the air by breathing it. Resistance to low oxygen depends on several things:

1) Age and size—older or larger fish need more oxygen.
2) Stage—with larvae, culture water often cannot be aerated as vigorously, dense prey could deplete oxygen more, and some (adult) compensatory mechanisms are not available (Brownell, 1980b).
3) Temperature—higher temperature increases metabolism and the need for oxygen.
4) Activity—more active fish need more oxygen.
5) Feeding level—fish eating need more oxygen.

Prevention of problems.—Aeration is important for keeping DO from becoming too low or too high. There are numerous ways to aerate fish culture water. Small tanks usually are aerated by sprayers, airstones, porous tubing, or injectors. These can be in the culture tank or in a separate supply chamber. Large tanks and ponds also can be aerated by spargers (pipes with drilled holes), various other bubblers, paddlewheels, splashboards, airlifts, boats, and wind.

After an oxygen crisis, fish should not be fed before they are ready, and they should be watched for two weeks for signs of disease.

Limits.—Common carp are healthy with only 3 mg/L DO. Oxygen minima in Table 3-9 are mostly in the range 3-6 mg/L, with a few exceptions. For example, salmon eggs and larvae need 6-9 mg/L, juveniles and adults of the common snook can tolerate 1 mg/L for several hours, and milkfish also survive 1 mg/L. For most marine fish to be healthy, DO should be at least 5 mg/L, and hopefully near saturation (80-100% ASV).

Measurement.—Dissolved oxygen is easily measured (to ±0.1 mg/L or better) electronically with a membrane electrode, and with more effort by Winkler titration (Grasshoff et al., 1983; Parsons et al., 1989; APHA, 1995).

Carbon dioxide (Table 3-10)
Sources and sinks.—Respiration of animals and plants and decomposition of organic matter are major sources of carbon dioxide in culture systems. Photosynthesis uses CO_2. Temperature, atmospheric pressure, salinity, and pH are factors governing the amount of free CO_2. Carbon dioxide buildup is more likely in static or recirculating systems, especially because release of surplus CO_2 from water to the atmosphere is relatively slow (Moe, 1989). Heavy green-water cultures are likely to experience diurnal fluctuations. Extending daylength will reduce the net contribution from plants.

In seawater with a pH higher than 8, most carbon dioxide is converted to bicarbonate (HCO_3), and problems are very unlikely. Even in fresh water, carbon dioxide trouble is rare. But well water sometimes has a dangerous excess of CO_2, because it has been sealed underground like in a bottle. During transport of fish in sealed containers, CO_2 could increase to dangerous levels.

Effect.—High CO_2 in the water (environmental hypercapnia) slows its release from the gills causing it to build up in the blood, reducing blood pH (Spotte, 1992). This decreases the blood's oxygen carrying capacity, leading to physiological hypoxia. Even with DO at 10 mg/L, 140 mg/L dissolved carbon dioxide is lethal to channel catfish. Dissolved carbon dioxide usually is low and can be tolerated up to at least 8 mg/L by most larvae and adults if enough dissolved oxygen is present.

Prevention of problems.—The water source should not contain high CO_2. Aeration will remove it. Hydrated lime will remove an equivalent amount of free CO_2 (~ppm:ppm) from ponds.

Limits.—To be safe, dissolved carbon dioxide should be kept below 8 mg/L. This will be the normal situation in most fish rearing units.

Measurement.—APHA (1995) describes nomographic (from pH, total alkalinity, temperature, and total mineral content) and titration methods for free CO_2 and a calculation method for free and total CO_2 (from pH and total alkalinity). Parsons et al. (1989) describe a method for determining carbonate alkalinity and total carbon dioxide from pH and total alkalinity, and a direct gas analysis (titration) method for total CO_2.

Hydrogen sulfide (Table 3-11)
Sources.—H_2S is formed in low oxygen environments by bacterial reduction of sulfate ions (Grasshoff et al., 1983; Moe, 1989). Likely sites are accumulations of fine sediment, deep unaerated water, and pipelines or filters that have not had oxygenated water flowing through them for a few hours. Problems with H_2S should be less likely in seawater than in fresh water because as pH increases, H_2S dissociates into sulfur and hydrogen ions, which are not toxic (at pH 9, H_2S concentration will be very low). However, anaerobic areas of the rearing system can produce lethal levels even in seawater.

Effects.—Rapid mortality is likely (Post, 1987). Dissolved hydrogen sulfide can be toxic to fish eggs at $\leq$12.0 μg/L and to juveniles at $\leq$2.0 μg/L. It can affect sensitive species at 0.1 μg/L. This is far below the threshold for smelling it. Fish will die in minutes in the fresh artesian water at HBOI if it isn't aerated first to get rid of the H_2S. Hydrogen sulfide stirred up from pond sediments (especially during harvest) easily can result in complete mortality (Krom et al., 1985). Nitrifying bacteria can be inhibited by H_2S as low as 15-30 μg/L (Spotte, 1992).

Prevention of problems.—
1) Oxidize H_2S to non-toxic sulfate SO_4^{2-} by adding air, oxygen, or potassium permanganate (2-6 mg/L for tolerant fish).
2) Remove excess organic matter from rearing units.
3) Raise the pH by aeration or liming.
4) Bottom sediments in ponds should not be stirred up while fish are in them.

Limit.—Zero. Hydrogen sulfide should be excluded from fish culture water.

Measurement.—APHA (1995) describes iodometric titration (minimum detectable concentration, MDC >1000 μg/L), methylene blue colorimetric (MDC 100 μg/L), ion-selective electrode (MDC >30 μg/L), and gas dialysis (MDC 2 μg/L) methods for measuring dissolved sulfide, which can be used with pH and the H_2S ionization constant to calculate unionized hydrogen sulfide. Grasshoff et al. (1983) describe methylene blue and titration methods for H_2S. Parsons et al. (1989) describe a methylene blue method. An Orion Research Corporation manual describes an electrode method for total soluble sulfides (H_2S + HS^- + S^{2-}).

Chlorine (Table 3-12)
Sources and sinks.—Chlorine can come from tapwater, accidental spills, or intentional dumping of waste water from power plants, sewage plants, and paper mills. If algal culture water is chlorinated, highly-toxic chloramines can be present in the algae.

Chlorination of seawater (Spotte, 1992) generates **chlorine-produced oxidants (CPOs)**, including brominated sterilizing agents such as hypobromous acid (HOBr), bromite ion (OBr^-), and bromate ion (BrO_3^-), as well as hypochlorite (OCl^-) and chlorate ions (ClO_3^-). Mutagens and carcinogens such as the trihalomethanes are produced: chloroform ($CHCl_3$), bromoform ($CHBr_3$), bromodichloromethane ($CHBrCl_2$), dibromochloromethane ($CHBr_2Cl$). Bromophenols, formaldehyde, acetaldehyde, and other compounds also occur.

At a pH of 8.5, chlorination of water that contains ammonia will produce monochloramine (NH_2Cl), but at pH 4.5 the main product is dichloramine ($NHCl_2$), which is twice as toxic (White, 1986). It is not much consolation that the less toxic compound is formed in seawater. The sodium thiosulfate (usually at 7 mg/L for each mg/L of free chlorine residuals, Boyd, 1996) used to neutralize chlorine will not affect the chloramines. Over time, a slow hydrolysis will break down chloramines, first forming free chlorine or hypochlorous acid (HOCl), which then reacts with organic nitrogen to produce organic chloramines, which are less toxic than the inorganic ones.

Effects.—Agitation, exhaustion, equilibrium loss, damage to gills and pseudobranchs, deformed larvae, and mortality (Post, 1987); 3 ug/L can be lethal to fish, and lower levels can be harmful. Gill or skin damage by chlorine-based oxidants could predispose fish to vibriosis or other infections (Spotte, 1992).

Prevention of problems.—Chlorination often is used to sterilize water for algae culture, and sometimes is used for fish culture water. If possible, chlorine should not be used in any water that will go into a fish tank, even if it is thought to be removed. Chlorine solutions can be used to sterilize non-porous, non-reactive, fish tanks, if they are thoroughly rinsed and dried. Bleach should not be used because it can contain the stabilizer cyanuric acid, which makes the chlorine difficult to remove. Two precautions to help minimize the chloramine risk are: 1) the use of mature algae cultures, and 2) the removal of as much of the water as possible from the algae (e.g., by sieving, settling, or centrifugation). Sieving can damage cells, highly motile cells do not settle well, and fragile ones do not centrifuge well (all three difficulties apply to *Isochrysis* spp.).

Table 3-10. Dissolved carbon dioxide in natural and hatchery water.

Purpose	Measured (mg/L)	Considered safe (mg/L)	Considered harmful (mg/L)	Considered lethal (mg/L)	References
Normal surface waters	<10				APHA, 1995
Some ground waters	>10				APHA, 1995
Oncorhynchus mykiss (fw)		<9	9-10		Billard, 1990
Salmo salar (fw)		10			Laird & Needham, 1988a
Morone saxatilis & hybrid lr. (fw)		<10			Mullis & Smith, 1990
Freshwater fish		15			Warren et al., 1990
Freshwater fish		<60	90-175		Post, 1987
Clupea pallasii egg (sw)				~2.1-18.5	Kelley, 1946
Salmo salar jv. (sw)		8-12	26		Fivelstad et al., 1998
Heteromycteris capensis lr. (sw)			>7.7		Brownell, 1980b
Freshwater fish		10	>20		Piper et al., 1982
Marine fish production		<10			Huguenin & Colt, 1989
Recommended limit		**<8**			Tucker

Table 3-11. Dissolved hydrogen sulfide in hatchery water.

Purpose	Considered safe (μg/L)	Considered harmful (μg/L)	Considered lethal (μg/L)	References
Gammarus pseudolimnaeus[amp]	2		11	Smith et al., 1976
Ictalurus punctatus (fw)		≤100		Martin, 1985
Oncorhynchus mykiss egg (fw)	3.5		7.3	Smith et al., 1976
Oncorhynchus mykiss lr. (fw)	4.8		7.0	Smith et al., 1976
Oncorhynchus mykiss jv. (fw)	8.5		14.8	Smith et al., 1976
Salmo salar (fw)	2			Laird & Needham, 1988a
Salmonids (fw)	0.1			Billard, 1990
Freshwater fish	<3			Warren et al., 1990
Freshwater fish & invertebrates	2			Smith et al., 1976
Epinephelus malabaricus jv. (sw)	0			Tookwinas, 1990a
Fish	~1		1.8-3.2	Post, 1987
Marine fish production	<1			Huguenin & Colt, 1989
Recommended limit	**0**			Tucker

Table 3-12. Dissolved chlorine-produced oxidants (CPOs) in hatchery water.

Purpose	Considered safe (μg/L)	Considered harmful (μg/L)	Considered lethal (μg/L)	References
Pimephales promelas lr. (fw)			108	Larson et al., 1977
Oncorhynchus gorbuscha jv. (sw)			250[30]	Stober & Hanson, 1974
Oncorhynchus mykiss (fw)			14	Brunel & Fuzeau, 1990
Oncorhynchus tshawytscha jv. (sw)			250[30]	Stober & Hanson, 1974
Salmo salar (fw)	~30			Laird & Needham, 1988a
Salmonids (fw)	2			Billard, 1990
Freshwater fish	<3			Warren et al., 1990
Brachionus plicatilis			180[30]	Capuzzo, 1979a, 1979b
Crassostrea virginica[oys] lr. (bw)			120[30]	Capuzzo, 1979a
Acartia tonsa[cop] (bw)			820[30]	Capuzzo, 1979a
Eucalanus pileatus[cop] ad. (sw)			110	Venkataramiah et al., 1982
Mysidopsis bahia[mys] yg. (bw)	20	44	73	Fisher et al., 1994
Alosa aestivalis egg (bw)		14	150	Morgan & Prince, 1977, 1978
Brevoortia tyrannus jv. (bw)	62		360	Richardson & Burton, 1981
Engraulis mordax egg (sw)			116	Casian, 1991
Menidia beryllina lr. (bw)	87		128	Fisher et al., 1994
Menidia menidia egg (sw)			120	Morgan & Prince, 1977
Menidia menidia			200 (5 min)	Toner & Brooks, 1975
Menidia menidia (sw)			37	Brunel & Fuzeau, 1990
Morone americana egg (bw)		46	150	Morgan & Prince, 1977, 1978
Morone saxatilis egg (bw)		51	190	Morgan & Prince, 1978
Morone saxatilis egg (bw)	<10		10	Middaugh et al., 1977; Hall et al., 1981
Morone saxatilis lr.			40	Hall et al., 1981
Morone saxatilis jv. (bw)			40	Hall, 1991
Cynoscion nebulosus (sw)			10-120	Johnson et al., 1977
Mugil cephalus jv. (sw)			212	Venkataramiah et al., 1982
Gobiosoma bosc (sw)			80	Brunel & Fuzeau, 1990
Pleuronectes platessa egg (sw)			105	Alderson, 1974

Pleuronectes platessa lr. (sw)			25	Alderson, 1974
Pleuronectes platessa jv. (sw)			84	Alderson, 1974
Solea solea lr. (sw)			28	Alderson, 1974
Solea solea jv. (sw)			70	Alderson, 1974
Fish	~<1	40		Post, 1987
Legal limit, surface sw	10			Florida statute, 1993
Marine fish production	<1			Huguenin & Colt, 1989
Recommended limit for larval fish in sw	**0**			Tucker
Recommended limit for older fish in sw	**<<1**			Tucker

[30] 30-min LC50.

Table 3-13. Dissolved ozone-produced oxidants (OPOs) in hatchery water.

Purpose	Considered safe (μg/L)	Considered harmful (μg/L)	Considered lethal (μg/L)	References
Enterococcus seriolicida[bac] (sw)			393	Sugita et al., 1994
Pasteurella piscicida[bac] (sw)			165	Sugita et al., 1994
Vibrio anguillarum[bac] (sw)			229	Sugita et al., 1994
Oncorhynchus mykiss jv. (fw)	2		8	Oakes et al., 1979
Salmo salar (fw)	~5			Laird & Needham, 1988a
Anadromous salmonid parr (fw)	3			Wedemeyer et al., 1981
Brevoortia tyrannus jv. (sw)			360	Richardson & Burton, 1981
Menidia menidia (sw)			80	Toner & Brooks, 1975
Morone saxatilis egg (fw-bw)			10-50	Hall et al., 1981
Morone saxatilis lr. (bw)			80	Hall et al., 1981
Morone saxatilis jv. (bw)			80	Hall et al., 1981
Sillago japonica egg-lr. (sw)			180	Isono et al., 1993
Ocyurus chrysurus ad. (sw)		100		Reid & Arnold, 1994
Sciaenops ocellatus lr. (sw)	~50-100	200-400		Reid & Arnold, 1994
Fish	2	5	30	Post, 1987
Recommended limit for larval fish in sw	**0**			Tucker
Recommended limit for older fish in sw	**<2**			Tucker

Limit.—There should be no chlorine or chlorine-produced oxidants in seawater for larval fish culture. Older fish might tolerate occasional trace amounts.

Measurement.—Iodometric electrode, syringaldazine (MDC 100 µg/L), iodometric titration (MDC 40 µg/L), DPD ferrous titration (MDC 18 µg/L), amperometric titration (MDC 10 µg/L), and DPD colorimetric (MDC 10 µg/L) methods for chlorine analysis are given by APHA (1995). Swimming pool test strips are not sensitive enough (typical detection limit 2000 µg/L), and aquaculture test kits are not much more sensitive (typical detection limit 200 µg/L).

Ozone (Table 3-13)
Source.—Ozone is generated electrically and added to water to sterilize it. Typical treatment levels for marine aquariums would be in the range 13-260 µg/L/h (Moe, 1989). Ozonation of fresh water followed by aeration is relatively safe. However, with bromide present, ozonation of seawater produces the same bromine and chlorine compounds that chlorination does, with similar hazards to marine organisms (Spotte, 1992). The half-lives of these **ozone-produced oxidants (OPOs)**, which are toxic to aquatic life, would be short in fresh water but can be more than 24 h in seawater.

Effects.—Ozone can burn and otherwise damage gills and can cause gill hypertrophy, mineral imbalances, blood cell imbalances, hyperglycemia, and mortality (Post, 1987). Larvae are particularly susceptible. More than 2 µg/L OPOs can be toxic.

Prevention of problems.—Some fish at some stages have been raised in water containing small amounts of ozone. Shrimp routinely are reared in ozonated water. But it should be used only with great caution for fish in seawater. Use of artificial seawater without bromide can reduce the risk from OPOs.

Limit.—There should be no ozone or ozone-produced oxidants in seawater for larval fish culture. Older fish might tolerate occasional trace amounts.

Measurement.—APHA (1995) describes an indigo colorimetric method for measurement of ozone, with an MDC of 2 µg/L.

GAS-RELATED DISORDERS

Gas bladder non-inflation and lordosis (Table 3-14)
Results.—Many fish species have an internal gas bladder for regulating buoyancy. Like those of northern anchovy and sand whiting (Hunter and Sanchez, 1976; Battaglene et al., 1994), larvae of many fish seem to conserve energy during dark periods, when they are not feeding, by filling their gas bladders and hovering at near-neutral buoyancy. During lighted periods, the gas bladders are partly deflated to allow effective feeding and escape maneuvers. If their gas bladders cannot be filled normally, larvae become highly stressed.

In some batches of reared fish, gas bladders of high numbers of larvae do not fill during early development as they should (Kitajima et al., 1994). Because those fish can't maintain neutral buoyancy, they swim abnormally and often develop lordosis or kyphosis (abnormal vertical spinal curvature), which usually becomes obvious in the early juvenile stage (Boglione et al., 1995). After larvae survive the first mortality peak by starting to feed well, they might encounter a second mortality peak from gas bladder non-inflation (Doroshev et al., 1981). Padros and Crespo (1995) reported these defects in gas bladders of turbot larvae and juveniles: inflation disturbances, malformed gas gland and rete mirabile, dystrophic pneumatic duct, and bacterial invasion. Growth and survival rates of such fish nearly always are reduced. (also see pages 513-514)

Causes.—Despite some variation by species and temperature, initial **gas bladder inflation (GBI)** in typical marine fish from planktonic eggs occurs at about the same age, between mouth opening and about 2 weeks after hatching, in several species (Table 3-14). Sometimes the last-chance day is difficult to determine because larvae with hopelessly undeveloped gas bladders might persist for a while, but they are more likely to die.

In **physostomous** fish, the digestive system and gas bladder are connected by a pneumatic duct that can be used for inflation and deflation (in larvae through adults). The time of initial gas bladder filling can be delayed in some physostomous species such as lake whitefish until 2-3 months after hatching (Tait, 1960). **Physoclistous** fish do not have such a duct as juveniles and adults, but often do as larvae. At least some of the latter species are thought to first inflate the gas bladder through this duct, which usually closes later before transformation. Three initial inflation mechanisms seem plausible: (1) gulping gas bubbles in midwater, (2) gulping air from the surface, and (3) secreting gas.

Access to the surface to gulp air seems an important factor for some species. Gilthead sea bream larvae reared in tanks with varying degrees of access had the following gas bladder inflation (GBI) rates: surface film blower, 80-81%; water sprinkler, 36-44%; control, 28%; water jets, 0-13%; mineral oil film, 0% (Chatain and Ounaïs-Guschemann, 1990). Growth was reduced with water jets and mineral oil. Lim (1993) reported that use of a surface oil skimmer improved initial gas bladder filling by greasy grouper from 40% to 80-90%.

Kitajima et al. (1981) demonstrated the need for air gulping and the relationship between non-inflation and lordosis in red sea bream. At 7 dah more than 90% of larvae with surface access had normal gas bladders, but none of those without access had GBI. During approximately 40-120 dah, 90% of the fish without GBI developed obvious lordosis. The pneumatic duct of red sea bream closed at 10-12 dah, but by that time a functioning gas secretion system could change gas bladder volume by a factor of 1.2-2.0, thus facilitating diurnal migration (Kitajima et al. 1985). Thereafter, when light decreased, inflation occurred in about 1 h, and when light increased, deflation required about 2 h.

Northern anchovies remain physostomous, and larvae longer than 10 mm increase gas bladder volume as much as 5 times at night by gulping surface air and make more extensive vertical migrations than red sea bream (Hunter and Sanchez, 1976). A mineral oil film experimentally prevented this. Anchovies and other clupeoids (e.g., Atlantic menhaden, Forward et al., 1993) do not not seem to secrete gas, but rely on air gulping throughout their lives. Possibly many species that become physoclistous use a similar strategy while they are physostomous as larvae—that is they gulp air (or gas bubbles) when they want to be near the surface and release it when they want to descend. Then, by the time the duct permanently closes, they switch to gas secretion.

Surface films in larval rearing tanks usually contain oils, polysaccharides, dead algae, dead zooplankton, and a variety of metabolites. A film containing oil only from dead algae, zooplankton, fish eggs, and fish larvae does not seem to repel larvae. If it is thin, larvae can break through to the air. On the other hand, enrichment oils often do repel larvae and sometimes will suffocate or trap them. Certain combinations of oils and polysaccharides from decaying macroalgae can trap and kill all the larvae in a tank like flypaper, if they are small like groupers. It can be dangerous to allow macroalgae to build up in water treatment systems for early larvae. If eggs are stocked and hatching rate is low, a large amount of oil could be released from the dead eggs (e.g., Boggs and Summerfelt, 1996a, 1996b).

High light intensity or **long daylength** could repel larvae from the surface and interfere with initial filling. Cerqueira and Chatain (1991) reported that more European sea bass larvae reared under continuous (6% GBI) illumination of 500 lux had undeveloped gas bladders than those reared with a 9-h daylength (37% GBI). Only 0.1% of Australian bass reared with continuous fluorescent light (100-200 lux) had GBI (Battaglene and Talbot, 1990); with 12 h/d light at 0.3-50 lux, GBI was 11-18%, and with no light, GBI was 72-74%. Chatain and Ounaïs-Guschemann (1991b) found that neither light intensity (150-1300 lux) nor daylength (12 or 24 h), affected gas bladder inflation in gilthead sea bream.

Too little or too much circulation can reduce the GBI rate. Walleye larvae reared with

Table 3-14. Initial gas bladder inflation in larval fish (peak time in parentheses).

Species	Physostomous larvae	Typical temperature (°C)	Age (dah)	Length (mm)	Limiting factors	References
Coregonus clupeaformis[p]	Yes	4	105		NAS	Tait, 1960
Oncorhynchus mykiss[p]	Yes	9	24		NAS	Tait, 1960
Salmo trutta[p]	Yes	3	119		NAS	Tait, 1960
Gasterosteus aculeatus	?				NAS	Von Ledebur & Wunder, 1938
Hippocampus spp.	Yes	26	0-1	~10		Vincent, 1995c
Lates calcarifer	Yes	26-32	1.5-2^{+}	2		Bagarino & Kungvankij, 1986
Dicentrarchus labrax	Yes	17	6-13	4-7	SF,HS,M,Li	Cerqueira, 1986; Sweetman, 1992; Barnabé & Guissi, 1993; Divanach et al., 1997
Lateolabrax japonicus	Yes		10-18	5-8	SF,HA	Kitajima et al., 1994
Macquaria novemaculeata	Yes	19	5-11	4-5	Li,LS,HA	Battaglene & Talbot, 1990, 1993
Morone saxatilis & hybrids	Yes	18	4-7^{+}		HS,LA,NAS,SW,HT	Doroshev & Cornacchia, 1979; Bulak & Heidinger, 1980; Doroshev et al., 1981; Hadley et al., 1987; Chapman et al., 1988
Epinephelus striatus	Yes	26	6-18	3-7	SF	Tucker, pers. obs.
Epinephelus tauvina	Yes	28	5-7^{+}	2-3	SF	Lim, 1993
Stizostedium vitreum	Yes	21	5(6-9)11	10-12	SF,M,OF,B,GS,HA	Barrows et al., 1988; Kindschi & MacConnell, 1989
Sillago ciliata	Yes?	24	4-25	2-10		Battaglene et al., 1994b
Seriola lalandi	Yes		3-10	5-6	SF	Kitajima et al., 1994
Seriola mazatlana	Yes	25-26	7-8		SF	Benetti, 1997
Eugerres brasilianus	Yes	27	4-5^{+}	~2.5	None	Alvarez-Lajonchère et al., 1996
Acanthopagrus schlegeli	Yes	20			OF	Leu, 1997
Dentex dentex	Yes	22.5	5-12		SF	Sweetman, 1992
Pagrus major	Yes	22	4-12	3-5	SF,HA,M	Kitajima et al., 1981, 1985, 1994; Yamashita, 1982a; Sweetman, 1992
Rhabdosargus sarba	Yes	20			OF	Leu & Liou, 1992
Sparidentex hasta	Yes	21	4-13	3-6	SF,LT,HT,SW	Al-Abdul-Elah et al., 1983a, 1983b
Sparus auratus	Yes	19	6-17	4-5	SF	Ounaïs-Guschemann, 1989; Chatain & Ounaïs-

						Guschemann, 1990, 1991b; Sweetman, 1992; Sarasquete & Yúfera, 1995
Atractoscion nobilis	Yes	17	4-21	4-6		Orhun, 1989
Sciaenops ocellatus	Yes	24	3-4	2-3		Holt et al., 1981
Oreochromis mossambicus	No	25.5	7-9		LO	Doroshev & Cornacchia, 1979
Latris lineata	Yes	17	6-8	5	SF	Goodsell et al., 1996
Lachnolaimus maximus	Yes	26	10-13	7		Colin, 1982
Scophthalmus maximus	Yes	18	2-9	3-5		Cousin & Laurencin, 1985

[P] Physostomous as larvae through adults.
B = bacteria in gas bladder, GS = gas supersaturation, HA = high aeration, HS = high salinity, HT = high temperature, LA = low aeration, Li = light (vs no light), LO = low oxygen, LS = low salinity, LT = low temperature, M = malnutrition, NAS = no access to surface, OF = overfeeding, SF = surface film, SW = still water.

Table 3-15. Dissolved nitrogen or total gas in natural and hatchery water. ΔP = differential pressure in mm Hg; % = % of saturation.

Purpose	Measured ΔP(%)	Considered safe ΔP(%)	Considered harmful ΔP(%)	Considered lethal ΔP(%)	References
Released well water	up to 500(166)				Colt, 1986
1 m shallower at 20°C	73(110)				Colt, 1986
Heated from 10 to 15°C	83(111)				Colt, 1986
DO increase from photosynthesis	300(139)[+]				Colt, 1986
From sea level to 1000 m high	86(113)				Colt, 1986
From sea level to 3000 m high	234(144)				Colt, 1986
Acipenser transmontanus (fw)		131(118)		225(131)	Counihan et al., 1998
Oncorhynchus mykiss lr. (fw)				130(117)	Colt, 1986
Salmo salar (fw)		38(105)			Laird & Needham, 1988a
Salvelinus namaycush jv. (fw)		0(100)	15-30(102-104)	76(110)	Krise, 1993
Salmonids (fw)		<15(102)	15-38(102-105)	106(114)	Colt, 1986; Billard, 1990
Morone saxatilis & hybrid lr. (fw)		<0(100)			Mullis & Smith, 1990
Freshwater fish		<23(103)			Warren et al., 1990
Morone saxatilis lr. (bw)		0(100)	22(103)	42(106)	Cornacchia & Colt, 1984
Coryphaena hippurus (sw)		23(103)		167(122)	Kraul, 1993
Recommended limit		**0(100)**			Tucker

light aeration during the first 8 d had only 8.7% GBI, while those without aeration had 32.7% GBI and 51% better survival (Kindschi and MacConnell, 1989). GBI in Australian bass was better with <50 mL air/min/60 L water than with >1000 mL/min (Battaglene and Talbot, 1993). Gas bladder inflation in red sea bream was increased with moderate aeration (50-100 mL air/min/m^3 water) but decreased with no or high aeration (500 mL air/min/m^3; Iseda et al., 1982). High aeration increased GBI in striped bass (Doroshev & Cornacchia, 1979) and sobaity (Al-Abdul-Elah et al. 1983a). Increasing aeration also tends to increase surface film formation by foam fractionation. Rough handling is a possible factor for red sea bream but seems unlikely (Fujita and Kitajima, 1982).

Divanach et al. (1997) reported that excessive circular current for late larval (2-10 vs 0.1-1 cm/s) to early juvenile (10-30 vs 1.5-2 cm/s) gilthead sea bream caused higher incidence of vertebral deformities (67% vs 7%), apparently a result of a metabolic dysfunction from the forced swimming.

Salinity effects on buoyancy can affect larval abilities. Tandler (1993) reported that reducing the salinity in gilthead sea bream larval tanks from 40 to 25 ppt improved GBI from 65% to 92%. GBI in Australian bass was 69% in 15-35 ppt but only 22% in 10 ppt (Battaglene and Talbot, 1993).

Kindschi and MacConnell (1989) reported that increasing **total gas saturation** to 123% for 8 h during the peak gas bladder inflation period (8 dah) reduced GBI in walleye larvae from 32.7 to 12.5%.

Excessive tank depth could make it more difficult for larvae to reach the surface. Lordosis rate was lower for red sea bream larvae in 18 cm of water than in 87 cm (Kitajima and Tsukashima, 1982).

Low egg quality with resulting weak larvae might contribute. Malnutrition or stress of the larvae (or female) leading to weak swimming and gulping ability could be a factor (Fukusho and Kitajima, 1988; Koven et al., 1990; Barnabé and Guissi, 1993; Kitajima et al., 1994). Barramundi larvae had abnormal gas bladders because of a **structural weakness** induced by malnutrition (Rodgers and Barlow, 1987).

Genetic factors were suggested as a possible cause of non-inflation, lordosis, and kyphosis in gilthead sea bream (Paperna, 1978), but did not seem a likely cause of lordosis in a series of studies with red sea bream (Fukusho and Kitajima, 1988).

Chemical contaminants could cause abnormal development and non-inflation (Wester and Roghair, 1992). **Bacterial infection** might be a factor in, or a result of, non-inflation. Kindschi and MacConnell, 1989) reported that normally inflated gas bladders of walleyes were not infected with bacteria but many uninflated and partly inflated gas bladders were infected and sometimes were filled with bacteria, macrophages, and sloughed epithelium.

Prevention.—To maximize gas bladder inflation (and minimize lordosis), Fukusho and Kitajima (1988) recommended that early red sea bream larvae be reared with moderate current and aeration (50-100 mL air/min/m^3 water), surface film removal, water sprinkled on the surface, low fish density (10-20 larvae/L), and good nutrition.

To maximize GBI, Battaglene and Talbot (1993) recommended that Australian bass larvae be kept in the dark (<1 lux), at a salinity of 15-35 ppt, with low aeration (<50 mL air/min/60 L), and without food during 0-10 dah.

Surface films in small tanks can be skimmed off with a beaker or a peristaltic pump with pasteur pipette attached. Efficiency can be increased by using an air jet skimmer to concentrate the film, unless larvae will be carried into the trap along with the film. Most skimmers consist of a floating collar (rectangular, triangular, spiral) with a small entrance, to contain the concentrated film, and airlines with valve(s) directing a stream of air across the water surface, to drive the film into the collar. Early larvae of many species drift (especially at night), and some seem to rather swim with a current than against it. An air skimmer could not be safely used with larvae of greasy grouper and brownmarbled grouper until 5 dah, when they would swim away from it (Lim, 1993). If the oil is not already very thick, continuously sprinkling water on the surface can disperse it.

Anderson (1993) reported that early transfer of striped bass larvae from the hatchery to ponds promoted better GBI rates than prolonged holding in incubation tanks, indicating that pond conditions were more favorable. For one hatchery in 1989, ages at transfer and GBI rates were: 4-5 dah, 95%; 5 dah, 85%; 5-6 dah, 28-75%; and 6 dah, 33%.

Gas bubble trauma (Table 3-15)

Causes.—All fish are susceptible to gas bubble trauma (GBT) but some more than others (Colt, 1986; Spotte, 1992). GBT is similar to decompression sickness, or the bends, that sometimes occurs in humans diving with compressed air. It affects fish when culture water becomes supersaturated with a gas or gases, which then induces supersaturation of internal fluids. Bubbles form in blood vessels and other tissues, interrupting circulation and transport of oxygen, causing hypoxia and tissue disruption. Bubbles can appear when total gas pressure (TGP) sufficiently exceeds the barometric pressure (BP). The difference between TGP and BP is the differential pressure, ΔP, which can be measured by a membrane-diffusion instrument.

This condition can occur when:

1) Water is heated.
2) Air enters water under high pressure, which is then reduced (e.g., by waterfalls, breaking waves, aeration, air leakage into intake of water pump).
3) Barometric pressure suddenly decreases.
4) Photosynthesis produces an excess of oxygen.
5) Bacteria produces large amounts of hydrogen, oxygen, nitrogen, or methane.

Mainly species, age, ΔP, depth of the fish, temperature, duration of the ΔP increase, and nitrogen:oxygen partial pressure ratio determine susceptibility to GBT. For each 1 m depth of a compressed air outlet, 2.8-3.2% supersaturation results, and this is enough to cause GBT in sensitive fish (Cornacchia and Colt, 1984). Transport of fish in a commercial jet typically would subject them to a TGP reduction from 760 mm Hg to 526 mm Hg. An unpressurized aircraft flying lower than 3000 m would undergo less of a pressure change.

Effects.—Acute GBT can cause agitation; loss of equilibrium; erratic swimming; dark skin; emphysemas and lesions of the mouth, eyes, skin, fins, or organs (especially gills, kidneys, and digestive system); gas embolisms in the circulatory system; overinflated gas bladder; exophthalmia; bacterial infections; and high mortality (usually within 10 d), often from cerebral anoxia and hemorrhages (Colt, 1986; Post, 1987; Spotte, 1992). In severe cases, gas bubbles can form in the skin of fish within minutes. Embolisms can occur almost anywhere in the fish and block blood flow. Signs of chronic GBT include bubbles in the gut and buccal cavity, and overinflated gas bladder, sometimes with low level mortality over weeks to months.

The bubbles usually are mostly nitrogen but could include other gases, especially if the pressure change is rapid. Colorni (1990) reported that juvenile gilthead sea bream reared in ponds with diurnal oxygen fluctuation from supersaturation to undersaturation apparently suffered oxygen bubble embolisms in the bloodstream, resulting in ischemic tissue damage.

The most frequent signs of GBT in larvae are overinflated gas bladder and gut, often causing them to float at the surface. Intestinal bubbles could come from an overfilled gas bladder or could develop in the gut, especially if tiny bubbles that were swallowed or developed from digestion are available to act as nuclei to be enlarged by diffusion of dissolved gas into them. Larvae in supersaturated water might ingest too many small bubbles, which then expand too quickly to be expelled. Ingested microalgae also can produce oxygen bubbles inside the gut.

Gas bladder hyperinflation can be severe enough to make larvae float upside down or on their sides at the surface and die. One day after 14-d-old barramundi from one tank were split into two tanks, nearly all fish in both tanks were found floating on their sides and had gas bladders overinflated to as much as 15 times normal size (Bagarinao and Kungvankij, 1986). Within 9 d, resulting mortality was more than 99%. Possible factors were: handling, high temperature (26-32°C), high salinity (32-34 ppt), high illumination (outdoor tanks), dense diatoms (*Chaetoceros calcitrans*) causing DO fluctuation, and possibly bacterial infection (not likely). High temperature, salinity, and light increased the incidence of gas bladder hyperinflation in European sea bass larvae, but malnutrition, crowding, hypoxia, and toxicants were other possible factors (Johnson and Katavic, 1984; Katavic, 1986). Gas bladder hyperinflation was reduced in striped mullet larvae by increasing aeration—with the intention of keeping larvae from feeding at the surface and gulping too much air (Nash et al., 1977), but keeping them below the surface also reduced light and pressure effects.

Although bubbles form only after a saturation threshold is exceeded, gas bladder hyperinflation caused by diffusion can occur when total gas pressure rises above 100% (Cornacchia and Colt, 1984). A TGP of only 102.9% caused hyperinflation and intestinal bubbles in 10-d-old striped bass larvae.

I have seen gas bladder hyperinflation occur in spotted seatrout larvae in the morning when they migrated vertically from bottom water to warmer surface water in 1-m^3 tanks. It appeared to be a simple case of thermal expansion. Some of the floating larvae were transferred to an unstratified tank, and most fish in both tanks recovered by the afternoon. Perhaps thermal stratification is a common cause. (also see page 508)

Prevention.—More than 100% nitrogen saturation poses a potential threat, because the fish also will be near saturation. More than 110% calls for immediate action—change the water. Water should be degassed before it enters fish tanks (e.g., by sprayer, splashboard, packed column). Leaks must not occur in water intake lines. If the water is well-mixed, swimming downward 1 m will reduce ΔP by about 74.3 mm Hg (Colt, 1986). Increasing the proportion of oxygen decreases the effects of a constant ΔP. Dietary vitamin C gives some protection, but fasting makes fish more susceptible. Reducing the ΔP for at least part of the day lowers the risk. To be extra safe, when transporting eggs or fish by air, an airtight container could be used to maintain the original air pressure, which should be gradually adjusted at the destination, if necessary.

Removal of nitrogen and enrichment of oxygen in a closed (no air contact) culture chamber is a method that has been used to simultaneously prevent GBT and increase holding capacity.

Measurement.—APHA (1995) describes a direct-sensing membrane-diffusion method for determining ΔP.

EFFECTS OF SEAWATER COLLECTION AND USAGE (Poxton and Allouse, 1982)

Changes.—When surface seawater is collected, most protozoans and invertebrates die, thus increasing the amount of organic matter. Then bacteria proliferate and decrease the dissolved oxygen concentration, increase the carbon dioxide concentration, decrease the pH, and, if the water becomes anaerobic, produce toxic substances. For prevention of bacterial buildup, water can be disinfected by ozonation or ultraviolet irradiation. This also will get rid of many disease organisms. But it is important to filter the water as much as is practical. Passing water through a 0.45-µm membrane filter will remove nearly all bacteria, phytoplankton, and larger things, leaving mainly viruses. Membrane filtration is impractical for large quantities of water, which can be filtered through 1-µm cartridges or bags (but these are not perfect and often have bigger holes).

In long-term studies of water chemistry of recirculating aquaria, it was found that:

1) Calcium increases (from egestion).
2) Phosphate increases considerably (from egestion).
3) Potassium increases (from excretion).

4) Magnesium decreases (by precipitation with phosphate).
5) Nitrate increases.
6) Alkalinity decreases because of the loss of magnesium and accumulation of nitrate.
7) Organic substances increase.
8) pH decreases because of accumulating metabolic products.
9) Dissolved oxygen capacity decreases.
10) Carbon dioxide increases.

Storage.—Some guidelines for long-term storage of seawater are (Poxton and Allouse, 1982):
1) Remove the solids with a mechanical filter.
2) Control bacteria by cooling to 4°C, ozonation, or UV treatment.
3) Aerate the water to prevent stagnation.
4) Store it in the dark to prevent growth of algae.
5) Store it in inert containers to prevent contamination.
6) Avoid contact with metal.
7) Have a means of draining and cleaning the system.

TOXIC METALS (Tables 3-16 to 3-34)

Sources and effects.—Burning of fossil fuels and garbage, mining, smelting, manufacturing, landfills, waterfills, and cement production all can contribute to metal contamination of water (Chang and Cockerham, 1994). Burning coal or oil releases arsenic, bismuth, cadmium, chromium, copper, lead, manganese, mercury, nickel, selenium, and zinc. Food and food additives also can contribute significantly to metal intake by fish.

Toxic metals can cause behavioral impairment, avoidance, hypoactivity, hyperactivity; hypo- or hyperventilation; weakness; anorexia; physiological imbalance or deficiency; osmoregulation interference; ion imbalance, metabolic disruption; muscle spasms (tetany); altered blood chemistry, anemia; enzyme interference; decreased stress tolerance, increased disease susceptibility; abnormal pigmentation; deformities, lordosis; exophthalmia; damage to gills, fins, thyroid, liver, heart, blood, hematopoietic tissue, spleen, kidneys, stomach, pancreas, intestines, gonads, eyes, nerves, brain, chemoreceptors, mechanoreceptors, muscles, and bones; reduced growth; tumors; zero or reduced egg production, reduced hatching; suffocation by mucus or precipitates; teratogenicity, mutagenicity, carcinogenicity, and mortality (Sorensen, 1991; reviews by Eisler; Chang and Cockerham, 1994). Mercury, cadmium, and lead seem very unlikely to have beneficial biological functions. Other elements, such as copper, iron, zinc, manganese, selenium, and possibly arsenic are essential nutrients required in small amounts, but usually are obtained from food, not water.

Usually, toxicity is increased with younger fish, smaller fish, higher temperature, lower oxygen, lower water hardness or alkalinity, lower salinity, lack of chelating agents, starvation, and greater parasite load. Increasing pH usually increases toxicity but not always (e.g., with copper). Previous exposure can increase tolerance but not always (e.g., with lead). Toxicity also varies with chemical form, presence of other toxins, and sometimes sex. Some strains or populations of the same species can be more resistant than others.

The surface film of water can contain accumulations of toxicants at higher levels than lower water layers (Langston, 1991). These include cadmium, lead, zinc, silver, copper, tributyltin (TBT), PCBs, and hydrocarbons. This can occur not only in nature but also in aquaculture tanks, possibly with dire consequences to planktonic eggs and larvae.

Some metals can initiate, promote, or sustain carcinogenesis (Chang and Cockerham, 1994). Copper, lead, cadmium, arsenic, and nickel can induce cancers, and zinc, cadmium, nickel, cobalt, and magnesium can enhance the potential of certain carcinogens in humans.

Concentration limits supposed to be safe for individual metals can be misleading. Chronic exposure of Atlantic herring larvae (10 dah) to a combination of metals with each

Table 3-16. Copper (total) measurements from natural seawater, toxicity, and limits. Numbers in parentheses are from samples taken after filtration.

Purpose	Measured (µg/L)	Considered safe (µg/L)	Considered harmful (µg/L)	Considered lethal (µg/L)	References
Oceanic water	0.025-0.64				Langston, 1991
North Pacific Gyre, surface	0.032				Chester, 1990
Mediterranean Sea	<0.04-5.8				Langston, 1991
Sargasso Sea	0.05-0.11				Jickells & Knap, 1984
Gulf Stream	0.06-0.17				Jickells & Knap, 1984
North Atlantic Gyre, surface	0.079				Chester, 1990
Baffin Bay	0.18-0.36				Jickells & Knap, 1984
Bermuda shelf	0.2				Jickells & Knap, 1984
U.S. northeast shelf	0.25				Jickells & Knap, 1984
Baltic Sea	0.31-0.95				Langston, 1991
Danish Sound & Kattegat	0.45				Jickells & Knap, 1984
Scotian shelf	0.56				Jickells & Knap, 1984
North Sea	0.9				Jickells & Knap, 1984
New Jersey, Hackensack, bw	2.3-37				Hall & Pulliam, 1995
U.K., Restronguet Creek	3-176				Langston, 1991
Algeciras Bay, Spain, eastern	67				Tarazona et al., 1991
Algeciras Bay, Spain, southwestern	91				Tarazona et al., 1991
Bermuda, Castle Harbor	<2(0.22)				Original data, 1994
Florida, Indian River, surface sw	<2(<2)				Original data, 1994
Fla., Indian R., 4.5-m sw well ↑ rock	(<10)				Original data, 1993
Fla., Indian R., 4.5-m sw well, ↓ rock	<2(<2)				Original data, 1994
Brachionus calyciflorus (fw)				2.5	Ferrando et al., 1993
Daphnia carinata[cld] (~fw)				28	Mukhopadhyay et al., 1994
Oncorhynchus mykiss egg (fw)		12		19	Hodson et al., 1979
Oncorhynchus tshawytscha egg-lr. (fw)		<21		21	Woltering, 1984
Salmo salar egg (fw)				10	Hodson et al., 1979
Salvelinus fontinalis egg-lr.				5	Van Coillie et al., 1980
Salvelinus fontinalis (fw)		≤15		3000-7000	Post, 1987
Brachionus plicatilis (sw)				100	Hodson et al., 1979
Brachionus plicatilis (30 ppt, 25°C, pH 7.7)				130	Snell & Persoone, 1989
Crassostrea virginica[oys] lr. (bw)				10	Calabrese et al., 1977
Mercenaria mercenaria[clm] lr. (bw)				4.9	Calabrese et al., 1977

Artemia sp. (sw)			1000	Hodson et al., 1979; Bhat, 1993
Acartia spp.^{cop} (sw)		20	~300	Hodson et al., 1979
Acartia tonsa^{cop} (sw)			31	Bhat & Vamsee, 1993
Tigriopus japonicus^{cop} life (sw)		6.4		Langston, 1991
Tisbe battagliai^{cop} life (sw)	6	8	18	Hutchinson et al., 1994
Parhalella natalensis^{amp} (sw)			72	Bhat & Vamsee, 1993
Clupea harengus fertilization (sw)			10	Langston, 1991
Clupea harengus egg-lr. (bw)		10-30		Abbasi et al., 1995
Pimephales promelas ad. (sw)	4.4-10.6		18.4	Sorensen, 1991
Arius felis ad. (sw)			2400	Steele, 1983
Gadus morhua egg & lr. (sw)			≤10	Swedmark & Granmo, 1981
Rivulus marmoratus jv. (sw)			1400	Lin & Dunson, 1993
Cyprinodon variegatus lr. (sw)	120	160	220	Hutchinson et al., 1994
Fundulus heteroclitus jv. (sw)			1700	Lin & Dunson, 1993
Atherinops affinis egg (sw)	55	115		Anderson et al., 1991
Atherinops affinis lr. (sw)	32		56	Anderson et al. 1994
Atherinops affinis lr. (bw)			44	Anderson et al., 1995
Atherinops affinis jv. (sw)	90		235	Anderson et al., 1991
Menidia menidia egg (sw)			0.025 Cu^{2+}	Langston, 1991
Sebastes schlegeli jv. (sw)			180	Choi & Kinae, 1994
Dicentrarchus labrax	~10	50	100	Cosson & Martin, 1981
Morone saxatilis lr. (bw)			24	Hall, 1991
Morone saxatilis jv. (sw)			190	Palawski et al., 1985
Therapon jarbua (sw)			8460	Mukhopadhyay et al., 1994
Trachinotus carolinus jv. (sw)			2000	Birdsong & Avault, 1971
Lagodon rhomboides lr. (sw)			150	Hodson et al., 1979
Leiostomus xanthurus egg (sw)			0.064 Cu^{2+}	Langston, 1991
Oreochromis mossambicus (sw)			34,750	Mukhopadhyay et al., 1994
Chelon labrosus jv. (sw)			1400	Taylor et al., 1985
Liza parsia (sw)			2850	Mukhopadhyay et al., 1994
Mugil cephalus jv. (sw)			2320	Choi & Kinae, 1994
Apocryptes bato (sw)			37	Mukhopadhyay et al., 1994
Paralichthys olivaceus jv. (sw)			16,100	Choi & Kinae, 1994
Pleuronectes limanda jv. (sw)			300	Taylor et al., 1985
Pleuronectes platessa (sw)		10		Langston, 1991
	Recommended Limits			
Salmonids (fw)	9.5			Billard, 1990
Salmo salar (fw)	6			Laird & Needham, 1988a
Anadromous salmonid parr (fw)	<5-30	20-30	25-55	Wedemeyer et al., 1981
Morone spp. hybrid lr. (fw)	<100			Mullis & Smith, 1990

Table 3-16. (Continued)

Purpose	Measured (μg/L)	Considered safe (μg/L)	Considered harmful (μg/L)	Considered lethal (μg/L)	References
Freshwater fish		6-30			Warren et al., 1990
Legal limit, surface sw		2.9			Florida statute, 1993; CFR, 1994
Marine fish production		<3			Huguenin & Colt, 1989
Marine fish larvae		<20			Blaxter, 1981
Marine fish		**<3**			Tucker

↑ rock = well water from above the first rock layer; ↓ rock = well water from below the first rock layer.

Table 3-17. Mercury measurements from natural seawater, toxicity, and limits. Numbers in parentheses are from samples taken after filtration.

Purpose	Measured (μg/L)	Considered safe (μg/L)	Considered harmful (μg/L)	Considered lethal (μg/L)	References
			Total		
Baltic Sea	0.001-0.006				Langston, 1991
Oceanic water	0.004-0.012				Langston, 1991
Mediterranean Sea	0.005-0.08				Langston, 1991
Open ocean	0.0053				Eisler, 1987a
Tasmania, Derwent estuary	<0.1-16				Langston, 1991
Fla., Indian R., surface sw	(<0.2)				Original data, 1994
Fla., Indian R., 4.5-m sw well, ↓ rock	(<0.2)				Original data, 1994
Fundulus heteroclitus (sw)				230	Hawkins & Lloyd, 1981
			Inorganic		
Anguilla rostrata (fw)				140	Sorensen, 1991
Carassius auratus (fw)			3	820	Sorensen, 1991
Pimephales promelas (fw)		<0.26	0.26		Woltering, 1984
Oncorhynchus keta jv.			5		Brown & Parsons, 1978
Oncorhynchus mykiss egg-lr. (fw)				<0.1	Eisler, 1987a
Oncorhynchus mykiss jv. (fw)				155-200	Eisler, 1987a
Morone saxatilis (fw)				90	Sorensen, 1991
Uronema marinum[cil] (sw)				6.0	Eisler, 1987a

Crassostrea virginica[oys] lr. (bw)			3.3	Calabrese et al., 1977
Mercenaria mercenaria[clm] lr. (bw)			4.0	Calabrese et al., 1977
Acartia tonsa[cop] ad. (sw)			10-15	Eisler, 1987a
Melanogrammus aeglefinus lr. (sw)			98	Eisler, 1987a
Fundulus heteroclitus egg (sw)	<<10		10	Langston, 1991; Sorensen, 1991
Sebastes schlegeli jv. (sw)			10	Choi & Kinae, 1994
Morone saxatilis jv. (bw)		5		Dawson et al., 1977
Oreochromis niloticus ad. (fw)		~100	500	Canli & Erdem, 1994
Mugil cephalus jv. (sw)			100	Choi & Kinae, 1994
Paralichthys olivaceus jv. (sw)			280	Choi & Kinae, 1994
	Organic			
Pimephales promelas ad. (fw)	0.07	0.12	0.13	Woltering, 1984; Eisler, 1987a
Oncorhynchus mykiss (fw)		0.04		Eisler, 1987a
Oncorhynchus mykiss jv. (fw)			5-42	Eisler, 1987a
Salvelinus fontinalis egg (fw)		0.88		Eisler, 1987a
Salvelinus fontinalis life (fw)	<0.03	0.29	0.93	McKim et al., 1976
Gammarus duebeni [amp] (sw)			150	Eisler, 1987a
Fundulus heteroclitus egg		30		Weis et al., 1981
	Recommended Limits (total)			
Salmo salar (fw)	2			Laird & Needham, 1988a
Anadromous salmonid parr, ave. (fw)	0.05	1.9		Wedemeyer et al., 1981
Salmonids (fw)	0.05-2			Billard, 1990
Freshwater fish	<200			Warren et al., 1990
Various fish, ave.	0.5		1000	Post, 1987
Legal limit, surface sw	0.025			Florida statute, 1993; CFR, 1994
Marine fish production	<0.1			Huguenin & Colt, 1989
Marine fish	**≤0.01**			Tucker

Table 3-18. Tin measurements from natural seawater, toxicity, and limits. Numbers in parentheses are from samples taken after filtration.

Purpose	Measured (µg/L)	Considered safe (µg/L)	Considered harmful (µg/L)	Considered lethal (µg/L)	References
			Total		
California, San Diego Bay	≤0.07				Eisler, 1989d
Northeastern Mediterranean	≤0.32				Eisler, 1989d
Chesapeake Bay	≤0.46				Eisler, 1989d
France, Arcachon Bay, 1982-1985	5.05-1.00				Eisler, 1989d
			Total Organotin		
France, Arcachon Bay, 1982-1985	0.20-<0.15				Eisler, 1989d
			Triphenyltin chloride (as TPTC)		
Chasmichthys dolichognathus (sw)				19	Shimizu & Kimura, 1991
			Tributyltin (as TBT)		
Bermuda, Hamilton Harbor, 0.5 m deep	0.0009-0.041				Stebbing et al., 1990
California, San Diego Bay	0.01-0.93				Eisler, 1989d
Bermuda, Hamilton Harbor surface microlayer	0.3				Stebbing et al., 1990
U.K., coastal	-0.43				Eisler, 1989d
U.K., marinas	-2.3				Eisler, 1989d
Maryland, Baltimore Harbor surface microlayer	-4.57				Eisler, 1989d
Dockyard effluent	24,400				Bell & Chadwick, 1994
Some neogastropods		<0.001	0.005		Bell & Chadwick, 1994
Phoxinus phoxinus			0.8	~8	Fent, 1992
Pimephales promelas (fw)				45	Eisler, 1989d
Oncorhynchus mykiss yg. (fw)			0.2	5	George, 1991
Oreochromis mossambicus (fw)				28	Eisler, 1989d
Crassostrea virginica[oys] lr. (sw)			0.9		Eisler, 1989d
Acartia tonsa[cop] (sw)			0.4	0.55	Eisler, 1989d
Eurytemora affinis[cop] (sw)		0.0125		0.1	Eisler, 1989d
Mysidopsis bahia[mys] (sw)				8	Eisler, 1989d
Palaemonetes pugio[car] ad. (sw)				18	Mayer, 1987
Brevoortia tyrannus (sw)			15		Eisler, 1989d
Cyprinodon variegatus (sw)				0.96	Eisler, 1989d

Purpose	Measured (µg/L)	Considered safe (µg/L)	Considered harmful (µg/L)	Considered lethal (µg/L)	References
Fundulus heteroclitus (sw)			9	24	Eisler, 1989d
Morone saxatilis lr. (bw)				2	Hall, 1991
Citharichthys stigmaeus (sw)				20	Eisler, 1989d
Solea solea lr. (sw)				2	Eisler, 1989d
Recommended limit, surface water		0.017			Eisler, 1989d
Legal limit, surface sw (U.K.)		0.001			Bell & Chadwick, 1994
Marine fish		<0.0067			Bell & Chadwick, 1994
Marine fish		**0.001**			Tucker

Table 3-19. Lead measurements from natural seawater, toxicity, and limits. Numbers in parentheses are from samples taken after filtration.

Purpose	Measured (µg/L)	Considered safe (µg/L)	Considered harmful (µg/L)	Considered lethal (µg/L)	References
		Total			
Oceanic water	0.006-0.015				Langston, 1991
North Pacific Gyre, surface	0.016				Chester, 1990
Oceans	0.02				Eisler, 1988b
Baltic Sea	0.02-0.11				Langston, 1991
Gulf Stream	0.03				Jickells & Knap, 1984
North Atlantic Gyre, surface	0.035				Chester, 1990
Danish Sound & Kattegat	0.04				Jickells & Knap, 1984
Bermuda shelf	0.15				Jickells & Knap, 1984
New Jersey, Hackensack, bw	0.5-25.2				Hall & Pulliam, 1995
Greece, industrial area, sw	2-5.5				Eisler, 1988b
Tasmania, Derwent estuary	4-16				Langston, 1991
Algeciras Bay, Spain, eastern	442				Tarazona et al., 1991
Algeciras Bay, Spain, southwestern	414				Tarazona et al., 1991
Bermuda, Castle Harbor	<2(<2)				Original data, 1994
Florida, Indian River, surface sw	<2(<2)				Original data, 1994
Fla., Indian R., 4.5-m sw well, ↑ rock	(<10)				Original data, 1993
Fla., Indian R., 4.5-m sw well, ↓ rock	<2(<2)				Original data, 1994
Daphnia magna[cld] (fw)			1	300	Eisler, 1988b
Mysidopsis bahia[mys] life (sw)		17	37		Eisler, 1988b
Fundulus heteroclitus (sw)				315	Eisler, 1988b
		Total Dissolved			
Oncorhynchus mykiss life (fw)		4.1	7.6	1200	Eisler, 1988b

Table 3-19. (Continued)

Purpose	Measured (µg/L)	Considered safe (µg/L)	Considered harmful (µg/L)	Considered lethal (µg/L)	References
		Divalent (Pb^{2+})			
Oncorhynchus mykiss yg. (fw)			13		Eisler, 1988b
Phaeodactylum tricornutum[mic] (sw)				>5000	Eisler, 1988b
Mytilus edulis[mus] lr. (sw)				476	Eisler, 1988b
Chelon labrosus jv. (sw)				>4500	Taylor et al., 1985
Pleuronectes platessa (sw)				180,000	Eisler, 1988b
		Tetraethyl			
Phaeodactylum tricornutum[mic] (sw)				100	Eisler, 1988b
Pleuronectes platessa (sw)				230	Eisler, 1988b
		Tetramethyl			
Oncorhynchus mykiss yg. (fw)				3.5	Eisler, 1988b
Phaeodactylum tricornutum[mic] (sw)				1300	Eisler, 1988b
Pleuronectes platessa (sw)				50	Eisler, 1988b
		Recommended limits (total)			
Salmo salar (fw)		30			Laird & Needham, 1988a
Anadromous salmonid parr (fw)		30			Wedemeyer et al., 1981
Salmonids (fw)		30			Billard, 1990
Freshwater fish		<20			Warren et al., 1990
Various fish		30		1000-400,000	Post, 1987
Legal limit, surface sw		5.6			Florida statute, 1993
Legal limit, surface sw		8.5			CFR, 1994
Marine fish production		<4			Huguenin & Colt, 1989
Marine fish		**<1**			Tucker

at 10% of the U.K. Environmental Quality Standard (in µg/L, EQS = mercury 0.3, cadmium 2.5, copper 5, chromium 15, lead 25, nickel 30, zinc 40) led to increased water and electrolyte uptake, which could have had deleterious effects.

Copper (Table 3-16) is added to the environment mainly by burning of fossil fuels, mining, smelting, manufacturing, pesticides, and erosion (Hodson et al., 1979; Sorensen, 1991; Spotte, 1992). Copper concentrations range from about 0.025 µg/L in the open ocean to about 176 µg/L in polluted coastal waters. Copper is less dangerous in seawater than in freshwater because its toxicity is moderated by other ions and chelators. Behavior can be affected with more than 4 µg/L Cu (Sorensen, 1991). Behavioral signs of toxicity include avoidance, disorientation, hyperactivity, hypoactivity, and abnormal schooling. Copper can alter blood chemistry, enzyme activity, respiratory physiology, and cardiac physiology. It can cause damage to gills, eyes, nerves, brain, kidneys, liver, hematopoietic tissue, mechanoreceptors, and chemoreceptors. Other effects include decreased or increased oxygen uptake, decreased tolerance to handling or salinity changes, reduced disease resistance, decreased growth, and mortality. Strong interference with reproduction can occur at 10-30 µg/L Cu (Hodson et al., 1979; Sorensen, 1991). Egg production and larval survival can be reduced, and larval deformities increased. (At 338 µg/L Cu, fathead minnow eggs hatched prematurely in 6 d instead of the normal 7 d (Scudder et al., 1988).

Sources of **mercury** (Table 3-17) addition to the environment include burning of fossil fuels, mining, smelting, pulp and paper mills, other industries, pesticides, and disposal of thermometers, batteries, switches, and fluorescent lamps (Eisler, 1987a; Sorensen, 1991). Organomercury compounds are **strong nerve and kidney poisons** (Chang and Cockerham, 1994). The toxicity of mercury to fish is roughly 25 times that of cadmium or lead (Sorensen, 1991). Acute toxicity signs are gill cover flaring, hyperventilation, sluggishness, equilibrium loss, and mortality. Chronic effects include erratic behavior, incoordination, weak response to light changes, dark skin, hypoventilation, indigestion, anorexia, altered blood chemistry, enzyme disruption, reduced growth, emaciation, cataracts, brain damage, reduced fecundity and hatching, abnormal larvae, teratogenicity, mutagenicity, carcinogenicity, and mortality. Bioconcentration factors for inorganic mercury were up to 14,360 in *Acartia clausi* and 7000 in rainbow trout. For methylmercury, BCFs ranged to 179,200 in copepods and 36,000 in trout. U.S. federal regulations state that a total mercury concentration of 12 ng/L or higher (in water) occurring more than once in 3 yr necessitates analysis of the edible part of key aquatic species to determine if methylmercury is higher than the action level of 1.0 mg/kg (CFR, 1994). McKim et al. (1976) reported that a 20-38 weeks exposure of (juvenile-adult) brook trout to 0.030 µg/L methylmercury left residues higher than 0.5 mg/kg, which is considered dangerous for human consumption. Obvious toxicity to the fish occurred at 0.93 µg/L. The trout survived with whole-body residues of 2.7 mg/kg or less but not with 5-7 mg/kg. Depuration of methylmercury from fish is not practical because its half-life is long (1-3.3 yr in several species, including 1200 d for North Atlantic flounder.

While **inorganic tin** is not especially toxic, some **organotin** compounds are strong poisons (Eisler, 1989d; Chang and Cockerham, 1994; Davies et al., 1998; Table 3-18). Inorganic tin is used in metal alloys, catalysts, glass, flame retardants, fluorescent lamps, soaps, toothpastes, and more. Organic tin is used in polyvinyl chloride (PVC) stabilizers, catalysts, lubricants, pesticides, disinfectants, wood preservatives, textiles, and antifouling paints (the last very dangerous but now highly restricted). Tributyltin can cause sluggishness, hyperventilation and air gulping, altered rheotaxis, pigmentation changes, reduced feeding and growth, altered enzyme activity, damage to gills, corneas, nerves, and bile duct, increased hemoglobin, erythrocytes, and hematocrit, and possibly teratogenicity. Organotin bioconcentration factors for aquatic organisms ranged up to 30,000. Sheepshead minnows bioconcentrated TBT 2600 times.

Lead (Table 3-19) is added to the environment mainly by mining, manufacturing,

Table 3-20. Zinc (total) measurements from natural seawater, toxicity, and limits. Numbers in parentheses are from samples taken after filtration.

Purpose	Measured (µg/L)	Considered safe (µg/L)	Considered harmful (µg/L)	Considered lethal (µg/L)	References
Oceanic water	0.0007-0.588				Langston, 1991
North Atlantic Gyre, surface	0.0039				Chester, 1990
North Pacific Gyre, surface	0.0046				Chester, 1990
Gulf Stream	0.01-0.02				Jickells & Knap, 1984
Sargasso Sea	0.01-0.07				Jickells & Knap, 1984
Mediterranean Sea	0.02-10.0				Langston, 1991
U.S. northeast shelf	0.16				Jickells & Knap, 1984
Bermuda shelf	0.3				Jickells & Knap, 1984
San Diego coastal to harbor	0.5-2.6				Eisler, 1993
Danish Sound & Kattegat	0.80				Jickells & Knap, 1984
Scotian shelf	1.1				Jickells & Knap, 1984
Baltic Sea	1.3-3.3				Langston, 1991
Western Mediterranean, coastal	1.5-10				Eisler, 1993
Irish Sea	3-7				Eisler, 1993
North Sea	3.9				Jickells & Knap, 1984
United Kingdom estuaries	6-26				Eisler, 1993
U.K., Restronguet Creek	22-20,460				Langston, 1991
Algeciras Bay, Spain, eastern	24				Tarazona et al., 1991
Algeciras Bay, Spain, southwestern	25				Tarazona et al., 1991
Bermuda, Castle Harbor	<2(<2)				Original data, 1994
Florida, Indian River, surface sw	2(2)				Original data, 1994
Fla., Indian R., 4.5-m sw well, ↑ rock	(40)				Original data, 1993
Fla., Indian R. 4.5-m sw well, ↓ rock	<2(<2)				Original data, 1994
Daphnia carinata[cld] (~fw)				123	Mukhopadhyay et al., 1994
Pimephales promelas ad. (fw)			180	800	Eisler, 1993
Oncorhynchus clarki (fw)		~360		670	Eisler, 1993
Oncorhynchus kisutch yg. (fw)				820-1810	Eisler, 1993
Oncorhynchus mykiss lr. (fw)				10	Eisler, 1993
Oncorhynchus mykiss jv. (fw)			81	310	Eisler, 1993
Salmo trutta lr. (fw)				4.9	Eisler, 1993
Jordanella floridae life (fw)			51	85	Spehar, 1976
Jordanella floridae ad. (fw)		26	51		Woltering, 1984

Tilapia zilli ad.			13,000	Eisler, 1993
Isochrysis galbana[mic] (sw)		74		Eisler, 1993
Philodena acutiformis[rot] ad. (sw)			500	Eisler, 1993
Crassostrea gigas[oys] lr. (sw)		10-20	119-310	Eisler, 1993
Crassostrea virginica[oys] lr. (sw)		206		Langston, 1991
Crassostrea virginica[oys] lr. (sw)			340	Eisler, 1993
Mercenaria mercenaria[clm] lr. (bw)			50	Calabrese et al., 1977
Acartia tonsa[cop] (sw)			294	Eisler, 1993
Tisbe holothuriae[cop] life (sw)	7		10	Eisler, 1993
Clupea harengus fertilization (sw)			10	Langston, 1991
Clupea harengus egg-lr. (sw)		50		Eisler, 1993
Gadus morhua lr. (sw)	≤500		500-1000	Swedmark & Granmo, 1981
Rivulus marmoratus jv. (sw)			147,900	Lin & Dunson, 1993
Fundulus heteroclitus ad. (sw)	~43,000		52,000-66,000	Eisler, 1993
Scorpaenichthys marmoratus lr. (sw)			192	Eisler, 1993
Sebastes schlegeli jv. (sw)			6950	Choi & Kinae, 1994
Morone saxatilis lr. (fw)			100-119	Eisler, 1993
Morone saxatilis jv. (sw)			430	Palawski et al., 1985
Morone saxatilis adult			6700	Eisler, 1993
Therapon jarbua (sw)			60,500	Mukhopadhyay et al., 1994
Leiostomus xanthurus (sw)			38,000	Eisler, 1993
Oreochromis mossambicus (sw)			19,090	Mukhopadhyay et al., 1994
Chelon labrosus jv. (sw)			21,500	Taylor et al., 1985
Liza parsia (sw)			19,130	Mukhopadhyay et al., 1994
Mugil cephalus jv. (sw)			13,750	Choi & Kinae, 1994
Paralichthys olivaceus jv. (sw)			5200	Choi & Kinae, 1994
	Recommended Limits			
Salmo salar (fw)	30			Laird & Needham, 1988a
Salmonids (fw)	4			Billard, 1990
Freshwater fish	5			Warren et al., 1990
Various fish	50		870-33,000	Post, 1987
Morone saxatilis & hybrid lr. (fw)	<100			Mullis & Smith, 1990
Legal limit, surface sw	86			Florida statute, 1993; CFR, 1994
Marine fish production	<25			Huguenin & Colt, 1989
Marine fish	**<10**			Tucker

Table 3-21. Iron (total) measurements from natural seawater, toxicity, and limits. Numbers in parentheses are from samples taken after filtration.

Purpose	Measured (μg/L)	Considered safe (μg/L)	Considered harmful (μg/L)	Considered lethal (μg/L)	References
Danish Sound & Kattegat	0.2-7.0				Jickells & Knap, 1984
Baffin Bay	0.5-2.5				Jickells & Knap, 1984
North Sea	1				Jickells & Knap, 1984
Bermuda shelf	1.07				Jickells & Knap, 1984
Scotian Shelf	3.2				Jickells & Knap, 1984
Bermuda, Castle Harbor	18(13)				Original data, 1994
Algeciras Bay, Spain, eastern	249				Tarazona et al., 1991
Algeciras Bay, Spain, southwestern	355				Tarazona et al., 1991
Florida, Indian River, surface sw	220(30)				Original data, 1994
Fla., Indian R., 4.5-m sw well, ↑ rock	(270)				Original data, 1993
Fla., Indian R., 4.5-m sw well, ↓ rock	42(17)				Original data, 1994
Morone saxatilis lr. (fw)				4000	Hall, 1991
Morone saxatilis jv. (fw)				6000	Hall, 1991
		Recommended Limits			
Salmo salar (fw)		1000			Laird & Needham, 1988a
Salmonids (fw)		0-150			Billard, 1990
Salmonids (fw)		0 (ferrous)			Billard, 1990
Freshwater fish		<100			Warren et al., 1990
Various fish		30		100-10,000	Post, 1987
Morone saxatilis & hybrid lr. (fw)		<500			Mullis & Smith, 1990
Legal limit, surface sw		300			Florida statute, 1993
Marine fish production		<100			Huguenin & Colt, 1989
Marine fish		**<30**			Tucker

Table 3-22. Chromium measurements from natural water, toxicity, and limits. Numbers in parentheses are from samples taken after filtration.

Purpose	Measured (μg/L)	Considered safe (μg/L)	Considered harmful (μg/L)	Considered lethal (μg/L)	References
		Total			

Normal seawater	0.05				Spotte, 1992
Oceanic water	0.088-0.55				Langston, 1991
Seawater	0-5				Eisler, 1986a
New Jersey, Hackensack, bw	1.7-11				Hall & Pulliam, 1995
Tanning operation outfall	13,100				Eisler, 1986a
Pigment producer outfall	62,900				Eisler, 1986a
Bermuda, Castle Harbor	<1(<1)				Original data, 1994
Florida, Indian River, surface sw	1(<1)				Original data, 1994
Fla., Indian R., 4.5-m sw well, ↓ rock	1(<1)				Original data, 1994
Sebastes schlegeli jv. (sw)				26,650	Choi & Kinae, 1994
Mugil cephalus jv. (sw)				1140	Choi & Kinae, 1994
Paralichthys olivaceus jv. (sw)				8400	Choi & Kinae, 1994
		Trivalent			
Daphnia magna^{cld} (fw)			44	2000	Eisler, 1986a
Oncorhynchus mykiss yg. (fw)		30	157	495	Eisler, 1986a
Oncorhynchus sp. jv. (fw)		~200			Eisler, 1986a
Acartia clausi^{cop} (sw)		~17,000			Eisler, 1986a
Aldrichetta forsteri (sw)				53,000	Eisler, 1986a
Legal limit, surface sw		673,000			Florida statute, 1993
		Hexavalent			
New Jersey, Hackensack, bw	<10-14				Hall & Pulliam, 1995
Daphnia carinata^{cld} (~fw)				83,150	Mukhopadhyay et al., 1994
Daphnia magna^{cld} (fw)				10	Eisler, 1986a
Oncorhynchus mykiss (fw)		51	105	3400	Eisler, 1986a
Oncorhynchus tshawytscha yg. (fw)		22	44		Woltering, 1984
Oncorhynchus sp. jv. (fw)				200	Eisler, 1986a
Morone saxatilis (fw)				30,400	Eisler, 1986a
Acartia clausi^{cop} (sw)				8830-19,270	Eisler, 1986a
Tisbe battagliai^{cop} life (sw)		320	420	560	Hutchinson et al., 1994
Mysidopsis bahia^{mys} (sw)		88	198		Eisler, 1986a
Cyprinodon variegatus lr. (sw)		<10,000	<10,000	32,000	Hutchinson et al., 1994
Fundulus heteroclitus (sw)				91,000	Eisler, 1986a
Menidia menidia lr. (sw)				12,400-14,300	Eisler, 1986a
Morone saxatilis jv. (sw)				58,000	Palawski et al., 1985
Oreochromis mossambicus (sw)				21,050	Mukhopadhyay et al., 1994
Aldrichetta forsteri (sw)		~17,900		24,000	Eisler, 1986a
Chelon labrosus jv. (sw)				47,200	Taylor et al., 1985
Citharichthys stigmaeus (sw)				30,000-31,000	Eisler, 1986a
Pleuronectes limanda jv. (sw)				47,000	Taylor et al., 1985

Table 3-22. (Continued)

Purpose	Measured (µg/L)	Considered safe (µg/L)	Considered harmful (µg/L)	Considered lethal (µg/L)	References
Legal limit, surface sw		50			Florida statute, 1993; CFR, 1994
		Recommended Limits (total)			
Freshwater fish		30			Warren et al., 1990
Various fish		50		5000-118,000	Post, 1987
California limit, surface sw		<8			Eisler, 1986a
Marine fish production		<25			Huguenin & Colt, 1989
Marine fish		**<25**			Tucker

Table 3-23. Aluminum (total) measurements from natural seawater, toxicity, and limits. Numbers in parentheses are from samples taken after filtration.

Purpose	Measured (µg/L)	Considered safe (µg/L)	Considered harmful (µg/L)	Considered lethal (µg/L)	References
Sargasso Sea	0.8				Jickells et al., 1986
Bermuda shelf	1.4				Jickells et al., 1986
Normal seawater	3				Hawkins & Lloyd, 1981
Normal seawater	10				Spotte, 1992
Bermuda, Castle Harbor	<30(<30)				Original data, 1994
Florida, Indian River, surface sw	300(41)				Original data, 1994
Fla., Indian R., 4.5-m sw well, ↑ rock	(5.2)				Original data, 1993
Fla., Indian R., 4.5-m sw well, ↓ rock	<30(<30)				Original data, 1994
Salmo salar (fw)		100			Laird & Needham, 1988a
Morone saxatilis lr. (fw)				100	Hall, 1991
Stizostedion lucioperca lr. (fw)			180	300	Vuorinen et al., 1993
Freshwater fish		<10			Warren et al., 1990
Various fish		100		300	Post, 1987
Legal limit, surface sw		1500			Florida statute, 1993
Marine fish		**<10**			Tucker

Table 3-24. Cadmium (total) measurements from natural seawater, toxicity, and limits. Numbers in parentheses are from samples taken after filtration.

Purpose	Measured (µg/L)	Considered safe (µg/L)	Considered harmful (µg/L)	Considered lethal (µg/L)	References
North Pacific Gyre, surface	0.00017				Chester, 1990
Sargasso Sea	0.0002-0.004				Jickells & Knap, 1984
Oceanic water	0.0002-0.025				Langston, 1991
North Atlantic Gyre, surface	0.00022				Chester, 1990
Gulf Stream	0.0003-0.017				Jickells & Knap, 1984
Bermuda shelf	0.0047				Jickells & Knap, 1984
Mediterranean Sea	<0.02-0.7				Langston, 1991
Baffin Bay	0.020-0.066				Jickells & Knap, 1984
U.S. northeast shelf	0.022				Jickells & Knap, 1984
Danish Sound & Kattegat	0.025				Jickells & Knap, 1984
Baltic Sea	0.03-0.07				Langston, 1991
Scotian shelf	0.036				Jickells & Knap, 1984
U.K., Poole Harbor	0.3-42				Langston, 1991
Algeciras Bay, Spain, eastern	73				Tarazona et al., 1991
Algeciras Bay, Spain, southwestern	71				Tarazona et al., 1991
Bermuda, Castle Harbor	2(1)				Original data, 1994
Fla., Indian R., 4.5-m sw well, ↑ rock	2(2)				Original data, 1993
Fla., Indian R., 4.5-m sw well, ↓ rock	2(2)				Original data, 1994
Daphnia carinata[cld] (~fw)				509	Mukhopadhyay et al., 1994
Diaptomus forbesi[cop] (fw)				6.31	Das & Kaviraj, 1994
Anguilla rostrata jv. (fw)			150		Gill et al., 1993
Cyprinus carpio yg. (fw)				216	Das & Kaviraj, 1994
Oncorhynchus kisutch egg-lr. (fw)		4		12	Woltering, 1984
Jordanella floridae life (fw)			8.1	16	Spehar, 1976
Morone saxatilis lr. (fw)				5	Hall, 1991
Morone saxatilis jv. (fw)				2	Hall, 1991
Brachionus plicatilis (30 ppt, 25°C, pH 7.7)				56,800	Snell & Persoone, 1989
Crassostrea virginica[oys] (sw)				3800	Hawkins & Lloyd, 1981
Artemia sp.				1000	Bhat & Vamsee, 1993
Acartia tonsa[cop]				90	Bhat & Vamsee, 1993
Eurytemora affinis[cop] lr. (bw)				51.6	Hall et al., 1995
Eurytemora affinis[cop]				500	Bhat & Vamsee, 1993
Tigriopus japonicus[cop] life (sw)			44		Langston, 1991
Tisbe battagliai[cop] life (sw)		18	24	180	Hutchinson et al., 1994
Clupea harengus fertilization (sw)				5	Langston, 1991

Table 3-24. (Continued)

Purpose	Measured (μg/L)	Considered safe (μg/L)	Considered harmful (μg/L)	Considered lethal (μg/L)	References
Rivulus marmoratus lr. (bw)				800	Park et al., 1994
Rivulus marmoratus jv. (sw)				21,100	Lin & Dunson, 1993
Cyprinodon variegatus lr. (bw)				180.3	Hall et al., 1995
Cyprinodon variegatus lr. (sw)		560	750	1000	Hutchinson et al., 1994
Fundulus heteroclitus jv. (sw)				18,200	Lin & Dunson, 1993
Fundulus majalis jv. (bw)				21,000	Park et al., 1994
Menidia menidia ad. (sw)				13,000	Park et al., 1994
Sebastes schlegeli jv. (sw)				28,700	Choi & Kinae, 1994
Morone saxatilis jv. (bw)			0.5		Dawson et al., 1977
Morone saxatilis jv. (sw)				75	Palawski et al., 1985; Hill et al., 1989
Girella punctata (sw)				27,500	Kuroshima & Kimura, 1989
Oreochromis mossambicus (sw)				242,310	Mukhopadhyay et al., 1994
Mugil cephalus jv. (sw)				2,550	Choi & Kinae, 1994
Paralichthys olivaceus jv. (sw)				28,400	Choi & Kinae, 1994
Pleuronectes americanus ad. (sw)			25-50		Pereira et al., 1993
Pleuronectes platessa (sw)			5		Langston, 1991
		Recommended Limits			
Salmo salar (fw)		3-4			Laird & Needham, 1988a
Anadromous salmonid parr (fw)		0.4-3	>4		Wedemeyer et al., 1981
Freshwater fish		0.5-5			Warren et al., 1990
Various fish		1		10-10,000	Post, 1987
Legal limit, surface sw		9.3			Florida statute, 1993; CFR, 1994
Marine fish production		<3			Huguenin & Colt, 1989
Marine fish		**<1**			Tucker

Table 3-25. Manganese (total) measurements from natural seawater, toxicity, and limits. Numbers in parentheses are from samples taken after filtration.

Purpose	Measured (µg/L)	Considered safe (µg/L)	Considered harmful (µg/L)	Considered lethal (µg/L)	References
North Pacific Gyre, surface	0.055				Chester, 1990
Sargasso Sea	0.11-0.13				Jickells & Knap, 1984
North Atlantic Gyre, surface	0.132				Chester, 1990
Baffin Bay	0.14-0.38				Jickells & Knap, 1984
Bermuda shelf	0.19				Jickells & Knap, 1984
North Sea	0.37				Jickells & Knap, 1984
Scotian Shelf	0.47				Jickells & Knap, 1984
U.S. northeast shelf	1.15				Jickells & Knap, 1984
Bermuda, Castle Harbor	<1(3)				Original data, 1994
Florida, Indian River, surface sw	15(6)				Original data, 1994
Fla., Indian R., 4.5-m sw well, ↑ rock	(80)				Original data, 1993
Fla., Indian R., 4.5-m sw well, ↓ rock	29(6)				Original data, 1994
		Recommended Limits			
Salmo salar (fw)		10			Laird & Needham, 1988a
Salmonids (fw)		0-10			Billard, 1990
Freshwater fish		<10			Warren et al., 1990
Various fish				2200-4100	Post, 1987
Marine fish production		<25			Huguenin & Colt, 1989
Marine fish		**<10**			Tucker

Table 3-26. Silver (total) measurements from natural seawater, toxicity, and limits. Numbers in parentheses are from samples taken after filtration.

Purpose	Measured (µg/L)	Considered safe (µg/L)	Considered harmful (µg/L)	Considered lethal (µg/L)	References
Normal seawater	0.3				Hawkins & Lloyd, 1981
Normal seawater	0.3				Spotte, 1992
Bermuda, Castle Harbor	<1(<1)				Original data, 1994
Fla., Indian R., 4.5-m sw well, ↑ rock	<1(<1)				Original data, 1993
Fla., Indian R., 4.5-m sw well, ↓ rock	<1(<1)				Original data, 1994
Pimephales promelas jv. (fw)				9	Sulaiman, 1993
Oncorhynchus mykiss yg. (fw)		<0.1	0.1	5.6	Woltering, 1984; Sulaiman, 1993
Crassostrea virginica^{oys} lr. (bw)				14.2	Calabrese et al., 1977

Table 3-26. (Continued)

Purpose	Measured (µg/L)	Considered safe (µg/L)	Considered harmful (µg/L)	Considered lethal (µg/L)	References
Mercenaria mercenaria^{clm} lr. (bw)				18.6	Calabrese et al., 1977
Fundulus heteroclitus (sw)				40	Hawkins & Lloyd, 1981
Freshwater fish		<3			Warren et al., 1990
Various fish		0.1-0.5		4-200	Post, 1987
Legal limit, surface sw		0.05			Florida statute, 1993
Marine fish		**0.3**			Tucker

Table 3-27. Arsenic measurements from natural water, toxicity, and limits. Numbers in parentheses are from samples taken after filtration.

Purpose	Measured (µg/L)	Considered safe (µg/L)	Considered harmful (µg/L)	Considered lethal (µg/L)	References
			Total		
South Australia (particulate)	<0.0006				Eisler, 1988a
Baltic Sea	0.52-7.4				Langston, 1991
Atlantic Ocean	1.0-1.5				Eisler, 1988a
South Australia	1.3				Eisler, 1988a
Oceanic water	1.35-2.5				Langston, 1991
Pacific Ocean	1.4-1.8				Eisler, 1988a
U.K., Restronguet Creek	3.3-65.4				Langston, 1991
Bermuda, Castle Harbor	3(3)				Original data, 1994
Florida, Indian River, surface sw	3(3)				Original data, 1994
Fla., Indian R., 4.5-m sw well, ↓ rock	<2(<2)				Original data, 1994
Daphnia magna (fw)			1000	2800	Eisler, 1988a
			Trivalent (As^{3+})		
South Australia	0.03				Eisler, 1988a
Daphnia magna^{cld} (fw)		630		960	Eisler, 1988a
Pimephales promelas (fw)		2100	4800	14,100	Eisler, 1988a
Crassostrea gigas^{oys} egg (sw)				330	Eisler, 1988a
Acartia clausi^{cop} (sw)				510	Eisler, 1988a
Mysidopsis bahia^{mys} (sw)		630	1270		Eisler, 1988a

Purpose	Measured (µg/L)	Considered safe (µg/L)	Considered harmful (µg/L)	Considered lethal (µg/L)	References
Oncorhynchus gorbuscha (sw)		~2500		3800	Eisler, 1988a
Chelon labrosus jv. (sw)				27,300	Taylor et al., 1985
Pleuronectes limanda jv. (sw)				28,500	Taylor et al., 1985
Recommended limit, surface sw		36			Eisler, 1988a
Legal limit, surface sw		36			Florida statute, 1993
		Pentavalent (As^{5+})			
South Australia	1.29				Eisler, 1988a
Daphnia magna[cld] (fw)			520	930	Eisler, 1988a
Pimephales promelas (fw)		530	1500	25,600	Eisler, 1988a
Morone saxatilis jv. (fw)				30,500	Palawski et al., 1985
Thalassiosira aestivalis[mic] (sw)			75		Eisler, 1988a
Eurytemora affinis[cop] (sw)		25		100	Eisler, 1988a
		Recommended limits (total)			
Freshwater fish		<50			Warren et al., 1990
Fish		700		1100-2200	Post, 1987
Legal limit, surface sw		50			Florida statute, 1993
Legal limit, surface sw		36			CFR, 1994
Marine fish		**<25**			Tucker

Table 3-28. Selenium measurements from natural water, toxicity, and limits. Numbers in parentheses are from samples taken after filtration.

Purpose	Measured (µg/L)	Considered safe (µg/L)	Considered harmful (µg/L)	Considered lethal (µg/L)	References
		Total			
Worldwide sw	0.009-<6.0				Eisler, 1985c
Japan	0.04-0.08				Eisler, 1985c
California, sw	0.058-0.08				Eisler, 1985c
Oceanic water	0.06-0.12				Langston, 1991
Florida, Tampa Bay, 1976	<0.2-23				Nelson, 1976
U.K., Poole Harbor	0.3-29				Langston, 1991
Florida, Tampa Bay, 1994	35-191				Benson et al., 1994
California, San Luis Drain	350				Hamilton & Buhl, 1990
Bermuda, Castle Harbor	<2(<2)				Original data, 1994
Fla., Indian R., 4.5-m sw well, ↑ rock	<2(<2)				Original data, 1994
Fla., Indian R., 4.5-m sw well, ↓ rock	<2(<2)				Original data, 1994
		Total (or not stated)			
Daphnia magna[cld] (fw)				70-120	Eisler, 1985c

Table 3-28. (Continued)

Purpose	Measured (µg/L)	Considered safe (µg/L)	Considered harmful (µg/L)	Considered lethal (µg/L)	References
Oncorhynchus kisutch yg. (fw)				160	Eisler, 1985c
Oncorhynchus mykiss (fw)				60-130	Eisler, 1985c
Acartia tonsa[cop] (sw)				800	Eisler, 1985c
Mysidopsis bahia[mys] jv. (sw)				600	Eisler, 1985c
Oncorhynchus kisutch yg. (sw)				26,200	Hamilton & Buhl, 1990
Melanogrammus aeglefinus lr. (sw)				600	Eisler, 1985c
Cyprinodon variegatus egg-jv. (sw)				470-970	Eisler, 1985c
Menidia menidia (sw)				9725	Eisler, 1985c
Morone saxatilis jv. (sw)				1550	Hall, 1991
Lagodon rhomboides (sw)				4400	Eisler, 1985c
Paralichthys dentatus lr. (sw)				3500	Eisler, 1985c
		Selenite			
Oncorhynchus kisutch yg. (sw)				13,600	Hamilton & Buhl, 1990
Morone saxatilis jv. (sw)				1550	Palawski, 1985
		Selenate			
Oncorhynchus kisutch yg. (sw)				39,000	Hamilton & Buhl, 1990
		Recommended Limits (total)			
Freshwater fish		<10			Warren et al., 1990
Legal limit, surface sw		71			Florida statute, 1993; CFR, 1994
Marine fish		**<10**			Tucker

Table 3-29. Antimony (total) measurements from natural seawater, toxicity, and limits. Numbers in parentheses are from samples taken after filtration.

Purpose	Measured (µg/L)	Considered safe (µg/L)	Considered harmful (µg/L)	Considered lethal (µg/L)	References
Normal seawater	0.3				Hawkins & Lloyd, 1981
Normal seawater	0.3				Spotte, 1992
Bermuda, Castle Harbor	<2(<2)				Original data, 1994

Purpose	Measured (µg/L)	Considered safe (µg/L)	Considered harmful (µg/L)	Considered lethal (µg/L)	References
Fla., Indian R., 4.5-m sw well, ↑ rock	<2(<2)				Original data, 1994
Fla., Indian R., 4.5-m sw well, ↓ rock	<2(<2)				Original data, 1994
Various fish				12,000-20,000	Post, 1987
Legal limit, surface sw		4300			Florida statute, 1993
Marine fish		**2000**			Tucker

Table 3-30. Nickel (total) measurements from natural water, toxicity, and limits. Numbers in parentheses are from samples taken after filtration.

Purpose	Measured (µg/L)	Considered safe (µg/L)	Considered harmful (µg/L)	Considered lethal (µg/L)	References
Sargasso Sea	0.12-0.16				Jickells & Knap, 1984
North Pacific Gyre, surface	0.123				Chester, 1990
North Atlantic Gyre, surface	0.135				Chester, 1990
Bermuda shelf	0.14				Jickells & Knap, 1984
Baffin Bay	0.17-0.29				Jickells & Knap, 1984
Oceanic water	0.18-0.70				Langston, 1991
Gulf Stream	0.23				Jickells & Knap, 1984
Scotian Shelf	0.25				Jickells & Knap, 1984
U.S. northeast shelf	0.33				Jickells & Knap, 1984
North Sea	0.40				Jickells & Knap, 1984
Baltic Sea	0.41-1.76				Langston, 1991
Danish Sound & Kattegat	0.48				Jickells & Knap, 1984
U.K., Restronguet Creek	1-18				Langston, 1991
Bermuda, Castle Harbor	8(7)				Original data, 1994
Florida, Indian River, surface sw	9(8)				Original data, 1994
Fla., Indian R., 4.5-m sw well, ↓ rock	8(8)				Original data, 1994
Pimephales promelas life (fw)		380		730	Woltering, 1984
Morone saxatilis jv. (fw)				3900	Hall, 1991
Oreochromis niloticus jv. (fw)			1,500	≤26,000	Alkahem, 1994
Crassostrea virginica^{oys} lr. (bw)				30	Calabrese et al., 1977
Mercenaria mercenaria^{clm} lr. (bw)				1100	Calabrese et al., 1977
Morone saxatilis jv. (sw)				21,000	Palawski et al., 1985
Chelon labrosus jv. (sw)				118,300	Taylor et al., 1985
Freshwater fish		<10			Warren et al., 1990
Various fish		30		5000-43,000	Post, 1987
Legal limit, surface sw		8.3			Florida statute, 1993; CFR, 1994
Marine fish production		<5			Huguenin & Colt, 1989
Marine fish		**<5**			Tucker

Table 3-31. Beryllium (total) measurements from natural water, toxicity, and limits. Numbers in parentheses are from samples taken after filtration.

Purpose	Measured (µg/L)	Considered safe (µg/L)	Considered harmful (µg/L)	Considered lethal (µg/L)	References
Normal seawater	0.0006				Spotte, 1992
Eastern Europe, surface fw	0.005-58				Jagoe et al., 1993
Bermuda, Castle Harbor	<1(<1)				Original data, 1994
Fla., Indian R., 4.5-m sw well, ↑ rock	<1(<1)				Original data, 1994
Fla., Indian R., 4.5-m sw well, ↓ rock	<1(<1)				Original data, 1994
Perca fluviatilis yg. (fw, pH 4.5)				10	Jagoe et al., 1993
Perca fluviatilis yg. (fw, pH 5.5)				80	Jagoe et al., 1993
Legal limit, surface sw, ave.		0.13			Florida statute, 1993
Marine fish		**0.13**			Tucker

Table 3-32. Vanadium (total) measurements from natural seawater, toxicity, and limits. Numbers in parentheses are from samples taken after filtration.

Purpose	Measured (µg/L)	Considered safe (µg/L)	Considered harmful (µg/L)	Considered lethal (µg/L)	References
Normal seawater	2.0				Hawkins & Lloyd, 1981
Normal seawater	2				Spotte, 1992
Bermuda, Castle Harbor	1(1)				Original data
Fla., Indian R., surface sw	2(1)				Original data
Fla., Indian R., 4.5-m sw well, ↓ rock	<1(<1)				Original data
Jordanella floridae yg. (fw)		41		170	Woltering, 1984
Pleuronectes limanda jv. (sw)				27,800	Taylor et al., 1985
Freshwater fish		<100			Warren et al., 1990
Marine fish		**<40**			Tucker

Table 3-33. Boron measurements from natural water, toxicity, and limits. Numbers in parentheses are from samples taken after filtration.

Purpose	Measured (µg/L)	Considered safe (µg/L)	Considered harmful (µg/L)	Considered lethal (µg/L)	References
			Total		
Agricultural drainage (fw)	~1100-83,000				Eisler, 1990a
Open ocean	4500				Eisler, 1990a

Coastal	4600				Eisler, 1990a
California, San Luis Drain	13,000				Hamilton & Buhl, 1990
Mine drainage (fw)	16,000-390,000				Eisler, 1990a
Oncorhynchus kisutch (fw)				12,000	Eisler, 1990a
Oncorhynchus mykiss egg-lr. (fw)				100	Black et al., 1993
Oncorhynchus mykiss ad. (fw)				339,000	Eisler, 1990a
Micropterus salmoides egg-lr. (fw)			1,000	10,000	Black et al., 1993
Some marine phytoplankton			~30,000		Eisler, 1990a
Pleuronectes limanda (sw)				74,000	Eisler, 1990a
		Boric acid			
Daphnia magna[cld] life (fw)		6400	13,600	27,000	Eisler, 1990a
Oncorhynchus mykiss egg-lr. (fw)				1-100	Eisler, 1990a
Oncorhynchus kisutch yg. (sw)				600,000	Hamilton & Buhl, 1990
		Borate			
Oncorhynchus mykiss egg-lr. (fw)				70	Eisler, 1990a
Pleuronectes limanda jv. (sw)				74,000	Taylor et al., 1985
		Recommended limit (total)			
Marine fish		**5000**			Tucker

Table 3-34. Molybdenum (total) measurements from natural water, toxicity, and limits. Numbers in parentheses are from samples taken after filtration.

Purpose	Measured (μg/L)	Considered safe (μg/L)	Considered harmful (μg/L)	Considered lethal (μg/L)	References
Worldwide sw	<1-12				Eisler, 1989b
Pacific Ocean, all depths	~10.3				Eisler, 1989b
Runoff from mining areas (fw)	100-100,000				Eisler, 1989b
Australia, Victoria power plant effluent	330				Eisler, 1989b
California, San Luis Drain	750				Hamilton & Buhl, 1990
Bermuda, Castle Harbor	10(9)				Original data, 1994
Fla., Indian R., surface sw	7(7)				Original data, 1994
Fla., Indian R., 4.5-m sw well, ↓ rock	4(5)				Original data, 1994
Pimephales promelas (fw)				70,000-360,000	Eisler, 1989b
Oncorhynchus mykiss egg-lr. (fw)				28-125	Eisler, 1989b
Mytilus edulis[mus] lr. (sw)			147,000		Eisler, 1989b
Calanus marshallae[cop] (sw)			20,000-100,000	560,000	Eisler, 1989b
Mysidopsis bahia[mys] (sw)				1,205,000	Eisler, 1989b
Cyprinodon variegatus (sw)				3,057,000	Eisler, 1989b
Marine fish		**<20**			Tucker

pesticides, leaded gasoline, storage battery disposal, and sewage (Eisler, 1988b; Sorensen, 1991). It also is used in paint pigments, solder, cables, ammunition, motor oil, linoleum, and printer's ink. Lead can inhibit formation of hemoglobin, resulting in **anemia** (Chang and Cockerham, 1994). Other effects include suffocation from excess mucus, dark skin, weakness, altered blood chemistry, metabolic disruption, muscular atrophy, reduced growth, damage to caudal fin, gills, nerves, brain, thyroid, liver, kidneys, hematopoietic systems, and gonads, paralysis, lordosis, teratogenicity, mutagenicity, carcinogenicity, and mortality. Bioconcentration factors ranged up to 582,000 for algae, 25,670 for blue mussels, and 726 for rainbow trout.

Major sources of **zinc** (Table 3-20) contamination include mining, smelting, electroplating, domestic and industrial sewage, burning of fossil fuels and solid wastes, corrosion of zinc alloys and galvanized metal, road surface runoff, and agricultural runoff. Metallic zinc is used in galvanized steel and iron, brass, and other alloys (Eisler, 1993). Zinc sulfate is used in fungicides and fertilizer mixes. Zinc oxide is in white pigments and photocopy paper. Water rarely contains more than 40 µg/L zinc and sediment 200 mg/kg; however, contamination can raise zinc to 99 mg/L in water and 118 g/kg in sediments (Eisler, 1993). Note the difference between shallow (surface layer) and deeper (below a rock layer) well concentrations at the same site (Table 3-20). Reported zinc content in marine organisms ranged up to 20 g/kg (oysters). In laboratory tests with aquatic organisms, zinc toxicity was higher to younger stages, to starved animals, with higher temperatures, with cadmium or mercury, without chelating agents, with lower salinity, hardness, or alkalinity, with lower dissolved oxygen, and when zinc concentration fluctuated (Eisler, 1993). Effects include avoidance, hyperactivity followed by lethargy, surface swimming, incoordination, blood chemistry alteration, enzyme disruption, overproduction of gill and body mucus, hypoxia, gill and fin hemorrhages, scale loss, reproductive failure, teratogenicity, possibly carcinogenicity, and mortality. Marine bioconcentration factors ranged up to 1,500,000 for crustaceans and 6900 for fish. Half-life in fish ranged up to 313 d (European plaice).

Aeration of **iron**-containing, low-pH, well water (Table 3-21) causes an increase in pH and oxidation of the soluble ferrous ion (Fe^{2+}) to the relatively insoluble ferric ion (Fe^{3+}). If this happens before mechanical filtration (e.g., sand filter), the precipitated iron can be removed. If it happens in a rearing unit, the **iron salts can precipitate on fish gills** and interfere with gas exchange. Bioconcentration factors for bivalves ranged up to 260,000.

Chromium (Table 3-22) contamination can arise from electroplating, metal finishing, tanning, coal burning, incineration, oil drilling, cement production, and cooling towers (Eisler, 1986a). It (mainly hexavalent chromium, Cr^{6+}) can cause hyperactivity, osmoregulatory interference, metabolic disruption, disease susceptibility, damage to gills, kidneys, and stomach, reduced growth, teratogenicity, mutagenicity, carcinogenicity, and mortality.

Dissolved **aluminum** (Table 3-23) **interferes with gill function**, leading to disturbances in gas exchange and ion regulation (Poleo, 1995). Signs of aluminum toxicity include excessive gill mucus, gill filament edema and fusion, plasma acidosis, hypoxia, hypercapnia, electrolyte loss (especially Na^+ and Cl^-), raised hematocrit, and decreased plasma osmolality. Aluminum also is a nerve poison (Chang and Cockerham, 1994).

Cadmium (Table 3-24) can damage nerves, kidneys, testes, and skeleton (Eisler, 1985a; Chang and Cockerham, 1994).

Manganese (Table 3-25) is a nerve poison (Chang and Cockerham, 1994).

Silver (Table 3-26) is very toxic to fish, but exposure is unlikely.

Arsenic (Table 3-27) is added to the environment mainly by manufacturing, mining, smelting, power plants, and pesticides (Eisler, 1988a; Sorensen, 1991). It is used in glass, textiles, wood preservatives, and livestock growth stimulants. Levels in seafoods tend to be high. Arsenic is a general cytotoxicant and **can damage most cells and organ systems** (Chang and Cockerham, 1994). Effects include abnormal behavior, metabolic disruption, damage to nerves, liver, kidneys, and heart, disease susceptibility, teratogenicity, mutagenicity, carcinogenicity, and mortality. Arsenic is bioconcentrated but not biomagnified very much. Bioconcentration factors mostly are low in animals but high in plants. Living *Chlorella vulgaris* in a contaminated area had 50 g/kg arsenic by dry weight.

Selenium (Table 3-28) is added to the environment mainly by the burning of coal and petroleum, glass making, metallurgy, and mining (Eisler, 1985c; Sorensen, 1991). It is used in alloys, pigments, rubber, textiles, drugs, and photographic chemicals. Selenium often is abundant in soils of dry regions. Effects include avoidance, lethargy, loss of equilibrium and coordination, tetany, altered blood chemistry, anemia, damage to gills, liver, kidneys, heart, and ovaries, reduced hatching, exophthalmia, reduced growth, mutagenicity, and mortality. Fish bioconcentration factors ranged up to 3300.

Antimony (Table 3-29), **nickel** (Table 3-30), **beryllium** (Table 3-31), and **vanadium** (Table 3-32) are not likely to be problems except in areas of high industrial pollution or if contamination occurs through misuse of metals in water systems.

Boron (Table 3-33) contamination can occur from mining, coal burning, water drainage, and borax detergents (Eisler, 1990a). It also is used in glass, enamels, glazes, and agricultural chemicals. Boron toxicity to fish is low. Salmonids seem more sensitive.

Molybdenum (Table 3-34) contamination occurs mainly from mining, smelting, milling, and fossil fuel burning (Eisler, 1989b). It is used mainly in steel alloys. Molybdenum was not very toxic to aquatic invertebrates and fish except for rainbow trout eggs and larvae.

Toxicity of mixtures of zinc and copper to many aquatic organisms, including oyster larvae, copepods, amphipods, and marine fish, is generally believed to be more than additive (Eisler, 1993). Toxicity of lead/zinc was more than additive to copepods. When zinc concentration was high, lead uptake by marine fish was more rapid. Toxicity of chromium/zinc was more than additive to copepods and more toxic than zinc with copper, lead, nickel, or cadmium. Toxicity of mercury/zinc to oyster larvae was more than additive. Toxicity of nickel/zinc was only additive for copepods and sticklebacks. Toxicity of mercury/copper and mercury/selenium can range from antagonistic to more than additive (Sorensen, 1991). Toxicity of copper, zinc, and nickel mixtures can be additive (Sorensen, 1991).

Fourhorn sculpins grown in water containing 32 μg/L arsenic, 0.5 μg/L cadmium, 0.8 μg/L copper, 0.1 μg/L mercury, 1.2 μg/L lead, and 5.3 μg/L zinc had increased incidence of vertebral deformities (Bengtsson and Larsson, 1986). Individual concentrations were relatively small, but the combination was deleterious.

Toxicity of mercury and cadmium occurs through replacement of copper and zinc in metalloenzymes, which prevents them from functioning correctly (Brown and Parsons, 1978; Hamilton and Mehrle, 1986; Roesijadi, 1992). Metallothionein and similar proteins protect fish from heavy metals by binding them. If the number of binding sites are insufficient for the number of heavy metal ions present, spillover occurs and toxic effects can begin. Low-level exposure can induce greater production of the protective proteins.

Sediments can have very high concentrations of toxicants (Langston, 1991) that can affect cage, pen, and pond culture, and even tank culture if sediments add toxicants to incoming water (e.g., during high winds).

Prevention of problems is best accomplished by carefully choosing and treating the water supply. Some metals such as iron, copper, and manganese are partly removed in sand filters by precipitation on the sand grains. Metals can be reduced by adding chelating agents such as EDTA or precipitating agents such as sodium metasilicate (Willey, 1977). This either has to be done in batches or with a finely regulated constant delivery system. Also the sodium metasilicate precipitates are messy, settle very slowly, and usually must be filtered out before the water can be used. Algae will remove some metals. If the algae does not enter the food chain, but is removed by settling, water exchange, or filtration, metal removal can be one of the benefits of green-water culture. Naturally-occurring substances, such as humic acid, fulvic acid, amino acids, polypeptides, and suspended solids, can remove metals by chelation, and others like clay, sand, and gravel can adsorb metals (Sorensen, 1991).

Limits.—Tables 3-16 to 3-34 give representative concentrations of heavy metals and other elements in natural and aquacultural environments, and tolerance limits for representative species of marine fish, their prey, and a few other animals for comparison. Data on certain forms are also given in Table 3-35.

Overall, copper and mercury seem to be the most dangerous metals in the sea (Langston, 1991). Copper can be toxic at concentrations slightly higher than background levels. Mercury background levels are not that high, but bioaccumulation of organic mercury is a serious concern. Damage from cadmium, zinc, silver, lead, and chromium is most likely at highly contaminated sites and probably would be a combined effect. Selenium and nickel are less toxic. Arsenic also generally is less dangerous but can be toxic to phytoplankton at five times background levels.

Measurement.—Inductively coupled plasma (ICP) atomic emission spectroscopy is one of the best laboratory methods for determining metal concentrations in water (APHA, 1995). Practical MDCs for laboratory analysis range from a few tenths to several tens of µg/L. Other methods include atomic absorption spectroscopy, anodic stripping voltammetry, and photometry (Grasshoff et al., 1983). Color comparator kits are available for at least copper, zinc, lead, iron, and manganese, but their lower detection limits often are close to or higher than toxic levels.

PESTICIDES AND OTHER POLLUTANTS (Table 3-35)

Sources.—Local application, upstream application, runoff, spills. Stable, fat-soluble compounds such as aldrin, benzene hexachloride isomers, DDD, DDE, DDT, dieldrin, endrin, heptachlor, and mirex tend to increase through biomagnification in the food chain. Some of them have been biomagnified to as much as 100,000 times the water concentration (Chambers, 1994). The environmental half-life of DDT is about 10 yr, but one of its metabolites (DDE), lindane, and heptachlor can last for decades (Chambers, 1994). Fortunately, all of these have been banned in the U.S. (Ware, 1994); however, significant residues still exist in the U.S., and other countries still use some organochlorine insecticides. Most usage of mercury, arsenic, and cadmium in pesticides has been discontinued in the U.S. because of dangers to both humans and the environment (Meister, 1994; Ware, 1994).

Effects.—Some effects are suffocation, physiological imbalance or deficiency, weakness, anemia, damage to eyes, gills, nerves, liver, spleen, or gonads, reduced growth, scoliosis, lordosis, other deformities, and mortality (Post, 1987).

Limits.—Table 3-35 gives tolerance limits for representative species of marine fish, their prey, and a few other animals for comparison. Contamination of fish culture water with almost any insecticide can cause problems.

Measurement.—Measurement of most pesticides and organic pollutants dangerous to fish would require laboratory analysis by gas chromatography, high-pressure liquid chromatography, or fluorescence spectroscopy (Grasshoff et al., 1983; Parsons et al., 1989; APHA, 1995). Practical MDCs for most of these contaminants are in the range 0.010-50 µg/L.

Representative pesticides.—Insecticides from four major classes—chlorinated hydrocarbons, organophosphates, carbamates, and pyrethroids—are serious nerve poisons that are relatively dangerous to fish (Huckle and Millburn, 1990). Like DDT, other chlorinated hydrocarbon (CHC) pesticides are especially dangerous because they are lipid soluble and relatively stable and thus are bioconcentrated and biomagnified in the food chain (Berry et al., 1974; Huckle

Table 3-35. Effects of some representative pesticides and other pollutants on fish and their food organisms. Tests were conducted under various conditions and are not exactly comparable. These data are provided to show how dangerous some of these compounds can be to cultured aquatic organisms. Names of substances considered especially dangerous are in bold type.

Substance[a]	Class[b]	Organism	Considered safe[c] (μg/L)	Considered harmful[d] (μg/L)	Considered lethal[e] (μg/L)	References
		Acaricides				
Bromopropylate[r]	CH	*Mugil cephalus* jv. (sw)			310	Mayer, 1987
Chlorobenzilate[r]	CH	*Cyprinodon variegatus* jv. (sw)			1000	Mayer, 1987
Chloropropylate[r]	CH	*Leiostomus xanthurus* jv. (sw)			306	Mayer, 1987
Dicofol	CH	*Pimephales promelas* yg. (fw)	19		39	Woltering, 1984
Dicofol	CH	*Cyprinodon variegatus* jv. (sw)			420	Mayer, 1987
Sulphenone[r]	OS	*Fundulus similis* jv. (sw)			5600	Mayer, 1987
Sulphenone[r]	OS	*Mugil cephalus* jv. (sw)			12,000	Mayer, 1987
Tetradifon[r]	OS	*Cyprinodon variegatus* jv. (sw)			>1000	Mayer, 1987
Tetrasul[r]	OS	*Leiostomus xanthurus* jv. (sw)			>1000	Mayer, 1987
		Fungicides				
Amobam	DC	*Cyprinodon variegatus* jv. (sw)			>1000	Mayer, 1987
Anilazine	TA	*Leiostomus xanthurus* jv. (sw)			7.9	Mayer, 1987
Cadmium chloride (61% Cd)[r]	I	*Menidia peninsulae* lr. (sw)			310 Cd	Mayer, 1987
Captafol[r]	S	*Fundulus similis* jv. (sw)			31	Mayer, 1987
Captan	S	*Pimephales promelas* yg. (fw)	17		40	Woltering, 1984
Captan	S	*Isochrysis galbana*[mic] (sw)		208		Mayer, 1987
Chlorothalonil	CH	*Leiostomus xanthurus* jv. (sw)			32	Mayer, 1987
Dichlofluanid[r]	S	*Leiostomus xanthurus* jv. (sw)			28	Mayer, 1987
Dithianon[r]	Q	*Fundulus similis* jv. (sw)			10	Mayer, 1987
Ferbam	Ca	*Fundulus similis* jv. (sw)			608	Mayer, 1987
Hexachlorobenzene	CH	*Cyprinodon variegatus* ad. (sw)			>13	Mayer, 1987
Maneb	Ca	*Fundulus similis* jv. (sw)			1600	Mayer, 1987
Mercuric chloride (74% Hg)[I, r]	I	*Menidia peninsulae* lr. (sw)			71 Hg	Mayer, 1987
Mercuric chloride (74% Hg)[I, r]	I	*Leiostomus xanthurus* ad. (sw)			36 Hg	Mayer, 1987
Metam-sodium[Ne,H]	Ca	*Cyprinodon variegatus* jv. (sw)			>1000	Mayer, 1987
Niacide-Z[r]	Ca+ThA	*Cyprinodon variegatus* jv. (sw)			>1000	Mayer, 1987
Sodium hypochlorite[Di,Al]	I	*Menidia peninsulae* jv. (sw)			54	Mayer, 1987
Triphenyltin hydroxide[M,Al, r]	OT	*Leiostomus xanthurus* jv. (sw)			46	Mayer, 1987
Zineb[r]	Ca	*Fundulus similis* jv. (sw)			>15,000	Mayer, 1987
Ziram	Ca	*Fundulus similis* jv. (sw)			6400	Mayer, 1987

Table 3-35. (Continued)

Substance[a]	Class[b]	Organism	Considered safe[c] (µg/L)	Considered harmful[d] (µg/L)	Considered lethal[e] (µg/L)	References
		Herbicides				
Acrolein[Al,M, r]	Ald	*Pimephales promelas* yg. (fw)	11		42	Woltering, 1984
Acrolein[Al,M, r]	Ald	*Pimephales promelas* jv. (fw)	11.4		13.7	Sulaiman, 1993; Eisler, 1994a
Acrolein[Al,M, r]	Ald	*Oncorhynchus mykiss* jv. (fw)			16.1	Sulaiman, 1993
Acrolein[Al,M, r]	Ald	*Fundulus similis* jv. (sw)			240	Mayer, 1987
Ametryn[De]	TA	*Isochrysis galbana*[mic] (sw)		10		Mayer, 1987
Ametryn[De]	TA	*Leiostomus xanthurus* jv. (sw)			>1000	Mayer, 1987
Atrazine[r]	TA	*Oncorhynchus mykiss* (fw)			12,600	Pimentel, 1971
Atrazine[r]	TA	*Isochrysis galbana*[mic] (sw)		100		Mayer, 1987
Atrazine[r]	TA	*Nannochloropsis oculata*[mic] (sw)		15		Eisler, 1989a
Atrazine[r]	TA	*Acartia tonsa*[cop] (sw)			94	Eisler, 1989a
Atrazine[r]	TA	*Eurytemora affinis*[cop] lr. (bw)			500	Hall et al., 1994
Atrazine[r]	TA	*Mysidopsis bahia*[mys] (sw)	80	190	1000	Eisler, 1989a
Atrazine[r]	TA	*Salvelinus fontinalis* ad. (sw)	65	120		Woltering, 1984
Atrazine[r]	TA	*Cyprinodon variegatus* lr. (bw)			2000	Hall et al., 1994
Atrazine[r]	TA	*Cyprinodon variegatus* (sw)	1900	3400	>16,000	Eisler, 1989a
Atrazine[r]	TA	*Leiostomus xanthurus* (sw)			8500	Eisler, 1989a
Bensulide	Am	*Leiostomus xanthurus* ad. (sw)			300	Mayer, 1987
Benthiocarb	Ca	*Brachionus calyciflorus*			6500	Fernandez-Casalderry et al., 1992
Benthiocarb	Ca	*Cyprinodon variegatus* ad. (sw)			1370	Mayer, 1987
Benthiocarb	Ca	*Menidia menidia* lr. (sw)			200	Mayer, 1987
Bromacil	U	*Leiostomus xanthurus* jv. (sw)			>950	Mayer, 1987
Cacodylic acid[De]	Ar	*Fundulus similis* jv. (sw)			>39	Mayer, 1987
Chloramben[r]	Be	*Isochrysis galbana*[mic] (sw)		15,000		Mayer, 1987
Chloramben ammonium salt[r]	Be	*Isochrysis galbana*[mic] (sw)		1,500,000		Mayer, 1987
Chloramben methyl ester[r]	Be	*Isochrysis galbana*[mic] (sw)		1,500		Mayer, 1987
4-Chloroaniline	CN	*Daphnia magna*[cld] yg. (fw)		110		Lee et al., 1993
2,4-D	POA	*Cyprinus carpio* jv. (fw)		150,000	200,000	Neskovic et al., 1994
2,4-D	POA	*Pimephales promelas* life (fw)	300		1500	Woltering, 1984
2,4-D	POA	Rec. limit, salmonids (fw)	0.5			Billard, 1990
2,4-D	POA	*Isochrysis galbana*[mic] (sw)		48,500		Mayer, 1987
2,4-D butoxyethanol ester	POA	*Isochrysis galbana*[mic] (sw)		52,500		Mayer, 1987

2,4-D butoxyethanol ester	POA	*Fundulus similis* adult (sw)			3500	Mayer, 1987
Dalapon$^{NI, r}$	CAA	*Oncorhynchus mykiss* (fw)			340,000	Pimentel, 1971
Dalapon$^{NI, r}$	CAA	*Isochrysis galbana*^{mic} (sw)		15,200		Mayer, 1987
Dalapon sodium salt^{r}	CAA	*Fundulus similis* jv. (sw)			>850	Mayer, 1987
DCPA	PhA	*Cyprinodon variegatus* jv. (sw)			>1000	Mayer, 1987
DEF^{De}	OP	*Leiostomus xanthurus* adult (sw)			130	Mayer, 1987
Dicamba	Be	*Leiostomus xanthurus* jv. (sw)			>870	Mayer, 1987
Dichlobenil^{Al}	BN	*Isochrysis galbana*^{mic} (sw)		59,300		Mayer, 1987
Dichlobenil	BN	*Cyprinodon variegatus* jv. (sw)			13,846	Mayer, 1987
Diquat^{De,Al}	BP	*Daphnia magna*^{cld} (fw)			7100	Brown, 1978
Diquat^{De,Al}	BP	*Oncorhynchus mykiss* (fw)			12,300	Brown, 1978
Diquat^{De,Al}	BP	*Morone saxatilis* lr. (fw)			1000	Hall, 1991
Diquat^{De,Al}	BP	*Stizostedion vitreum* lr. (fw)	480		600	Paul et al., 1994
Diquat^{De,Al}	BP	Rec. limit, salmonids (fw)	0.5			Billard, 1990
Diquat^{De,Al}	BP	*Isochrysis galbana*^{mic} (sw)		6,000		Mayer, 1987
Diquat^{De,Al}	BP	*Fundulus similis* jv. (sw)			>16	Mayer, 1987
Disodium methanearsonate (26% As)	Ar	*Fundulus similis* jv. (sw)			>10,400 As	Mayer, 1987
Diuron	PU	*Oncorhynchus mykiss* (fw)			4300	Pimentel, 1971
Diuron	PU	*Morone saxatilis* lr. (fw)			500	Hall, 1991
Diuron	PU	*Isochrysis galbana*^{mic} (sw)		10		Mayer, 1987
Diuron	PU	*Mugil cephalus* jv. (sw)			5985	Mayer, 1987
Endothall^{Al,De}	PhA	*Morone saxatilis* jv. (fw)			610,000	Hall, 1991
Endothall^{Al,De}	PhA	*Stizostedion vitreum* lr. (fw)	5700	11,000	16,000	Paul et al., 1994
Endothall^{Al,De}	PhA	*Isochrysis galbana*^{mic} (sw)		25,000		Mayer, 1987
Endothall Aquathol Plus^{Al}	PhA	*Leiostomus xanthurus* jv. (sw)			>1000	Mayer, 1987
EPTC^{NI}	Ca	*Fundulus similis* jv. (sw)			>19,600	Mayer, 1987
Fenac sodium salt^{r}	Be	*Leiostomus xanthurus* jv. (sw)			>300	Mayer, 1987
Fenuron	PU	*Isochrysis galbana*^{mic} (sw)		750		Mayer, 1987
Fenuron	PU	*Leiostomus xanthurus* jv. (sw)			>1000	Mayer, 1987
Fluridone	Pr	*Stizostedion vitreum* lr. (fw)	780	1,200	1800	Paul et al., 1994
Molinate	Ca	*Leiostomus xanthurus* jv. (sw)			>1000	Mayer, 1987
Monuron$^{NI, r}$	PU	*Isochrysis galbana*^{mic} (sw)		100		Mayer, 1987
Monuron$^{NI, r}$	PU	*Mugil cephalus* jv. (sw)			16,000	Mayer, 1987
Neburon^{r}	PU	*Isochrysis galbana*^{mic} (sw)		20		Mayer, 1987
Neburon^{r}	PU	*Leiostomus xanthurus* jv. (sw)			320	Mayer, 1987
Paraquat$^{De, r}$	BP	*Oncorhynchus mykiss* (fw)			15-32	Eisler, 1990c
Paraquat$^{De, r}$	BP	*Salmo trutta* (fw)			25	Eisler, 1990c
Paraquat$^{De, r}$	BP	*Isochrysis galbana*^{mic} (sw)		2100		Mayer, 1987
Paraquat$^{De, r}$	BP	*Rithropanopeus harrisii*^{crb} lr. (sw)			0.86	Eisler, 1990c
Paraquat$^{De, r}$	BP	*Mugil cephalus* (sw)			1	Eisler, 1990c
Paraquat L$^{De, r}$	BP	*Fundulus similis* jv. (sw)			>212	Mayer, 1987

Table 3-35. (Continued)

Substance[a]	Class[b]	Organism	Considered safe[c] (µg/L)	Considered harmful[d] (µg/L)	Considered lethal[e] (µg/L)	References
Pebulate	Ca	*Mugil cephalus* jv. (sw)			4914	Mayer, 1987
Picloram/2,4-D[r]	PiA	*Isochrysis galbana*[mic] (sw)		50,000		Mayer, 1987
Picloram[r]	PiA	*Oncorhynchus mykiss* (fw)			150,000	Pimentel, 1971
Prometon	TA	*Isochrysis galbana*[mic] (sw)		1000		Mayer, 1987
Prometryn	TA	*Leiostomus xanthurus* jv. (sw)			>990	Mayer, 1987
Rodeo		*Morone saxatilis* jv. (fw)			23,500	Bills et al., 1993
Silvex[Al, r]	POA	*Isochrysis galbana*[mic] (sw)		5000		Mayer, 1987
Silvex[Al, r]	POA	Rec. limit, salmonids (fw)	2			Billard, 1990
Silvex propylene glycol butyl ether ester[r]	POA	*Leiostomus xanthurus* jv. (sw)			248	Mayer, 1987
Simazine[Al]	TA	*Oncorhynchus mykiss* (fw)			5000	Pimentel, 1971
Simazine[Al]	TA	*Morone saxatilis* jv. (fw)			250	Hall, 1991
Simazine[Al]	TA	*Isochrysis galbana*[mic] (sw)		490		Mayer, 1987
2,4,5-T[r]	POA	*Isochrysis galbana*[mic] (sw)		50,000		Mayer, 1987
2,4,5-T[r]	POA	*Mugil cephalus* jv. (sw)			>50,000	Mayer, 1987
2,4,5-T propylene glycol butyl ether ester[r]	POA	*Leiostomus xanthurus* jv. (sw)			221	Mayer, 1987
TCA sodium salt[r]	CAA	*Mugil cephalus* jv. (sw)			>1000	Mayer, 1987
Terbutryn[r]	TA	*Fundulus similis* jv. (sw)			>978	Mayer, 1987
Tetrachlorodioxin (TCDD) (contaminant of 2,4,5-T)[r]	POA	*Oncorhynchus mykiss* yg. (fw)	0.0001		0.001	Woltering, 1984
Triallate	Ca	*Isochrysis galbana*[mic] (sw)		386		Mayer, 1987
Trifluralin[NI]	DN	*Oncorhynchus mykiss* (fw)			210	Pimentel, 1971
Trifluralin[NI]	DN	*Isochrysis galbana*[mic] (sw)		2400		Mayer, 1987
Trifluralin[NI]	DN	*Cyprinodon variegatus* jv. (sw)			154	Mayer, 1987
Vernolate	Ca	*Oncorhynchus mykiss* (fw)			6200	Brown, 1978
Vernolate	Ca	*Leiostomus xanthurus* jv. (sw)			>980	Mayer, 1987
		Insecticides				
Acephate[A,Ne,C, r]	OP	*Lagodon rhomboides* ad. (sw)			85,000	Mayer, 1987
Aldicarb[A,Ne,C, r]	Ca	*Pimephales promelas* yg. (fw)	78		156	Woltering, 1984
Aldicarb[A,Ne,C, r]	Ca	*Cyprinodon variegatus* ad. (sw)			41	Mayer, 1987

Aldicarb[A,Ne,C, r]	Ca	*Lagodon rhomboides* ad. (sw)			80	Mayer, 1987
Aldrin[r]	CH	*Anguilla rostrata*			5	Brown, 1978
Aldrin[r]	CH	*Mugil cephalus* jv. (sw)			1.9	Mayer, 1987
Aldrin[r]	CH	*Thalassoma bifasciatum* (sw)			12	Brown, 1978
Aldrin[r]	CH	Rec. limit, salmonids (fw)	0.003			Billard, 1990
Aldrin[r]	CH	Legal limit, surface sw	1.3			Florida statute, 1993
Aminocarb[r]	Ca	*Cyprinodon variegatus* jv. (sw)			>990	Mayer, 1987
Aspon[r]	OP	*Cyprinodon variegatus* jv. (sw)			>1000	Mayer, 1987
Azinphos-methyl[A,C]	OP	*Pimephales promelas* life (fw)	0.3	0.5		Woltering, 1984
Azinphos-methyl[A,C]	OP	*Mugil cephalus* jv. (sw)			3.1	Mayer, 1987
Azinphos-methyl[A,C]	OP	Legal limit, surface sw	0.01			Florida statute, 1993
Benzene hexachloride[r]	CH	*Mugil cephalus* jv. (sw)			520	Mayer, 1987
Bufencarb[r]	Ca	*Cyprinodon variegatus* jv. (sw)			33	Mayer, 1987
Calcium arsenate (38% As)[F,H, r]	I	*Fundulus similis* jv. (sw)			12	Mayer, 1987
Carbaryl[M]	Ca	*Pimephales promelas* life (fw)	210		680	Woltering, 1984
Carbaryl[M]	Ca	*Morone saxatilis* jv. (fw)			760	Palawski et al., 1985
Carbaryl[M]	Ca	*Fundulus similis* jv. (sw)			1595	Mayer, 1987
Carbaryl[M]	Ca	*Morone saxatilis* jv. (sw)			2300	Palawski et al., 1985
Carbaryl[M]	Ca	*Mugil cephalus* jv. (sw)			2392	Mayer, 1987
Carbofuran[Ne,M]	Ca	*Oncorhynchus kisutch* (fw)			530	Eisler, 1985b
Carbofuran[Ne,M]	Ca	*Oncorhynchus mykiss*			380	Eisler, 1985b
Carbofuran[Ne,M]	Ca	*Salmo trutta* (fw)			560	Eisler, 1985b
Carbofuran[Ne,M]	Ca	*Cancer magister* lr. (sw)		~1.0	~2.5	Eisler, 1985b
Carbofuran[Ne,M]	Ca	*Cyprinodon variegatus* (sw)	15	23	49-386	Eisler, 1985b
Carbophenothion[A,C, r]	OP	*Cyprinodon variegatus* ad. (sw)			2.8	Mayer, 1987
Chlordane[r]	CH	*Palaemonetes pugio* (sw)			4.2	Eisler, 1990b
Chlordane[r]	CH	*Oncorhynchus mykiss* (sw)			8-47	Eisler, 1990b
Chlordane[r]	CH	*Cyprinodon variegatus* life (sw)	0.5		0.8	Eisler, 1990b
Chlordane[r]	CH	*Morone saxatilis* (sw)			11.8	Eisler, 1990b
Chlordane[r]	CH	*Lagodon rhomboides* (sw)			5.4	Eisler, 1990b
Chlordane[r]	CH	*Mugil cephalus* jv. (sw)			3.2	Mayer, 1987
Chlordane[r]	CH	Rec. limit, salmonids (fw)	0.01			Billard, 1990
Chlordecone (kepone)[r]	CH	*Cyprinodon variegatus* life (sw)	0.07	0.12		Woltering, 1984
Chlordecone (kepone)[r]	CH	*Cyprinodon variegatus* ad. (sw)			70	Mayer, 1987
Chlordecone (kepone)[r]	CH	*Leiostomus xanthurus* ad. (sw)			6.6	Mayer, 1987
Chlordecone (kepone)[r]	CH	*Mugil cephalus* jv. (sw)			30	Mayer, 1987
Chlorpyrifos	OP	*Pimephales promelas* (fw)			0.13	Odenkirchen & Eisler, 1988
Chlorpyrifos	OP	*Oncorhynchus mykiss* (fw)			7.1	Odenkirchen & Eisler, 1988
Chlorpyrifos	OP	*Isochrysis galbana*[mic] (sw)		129		Mayer, 1987
Chlorpyrifos	OP	*Mysidopsis bahia*[mys] (sw)			0.035	Odenkirchen & Eisler, 1988
Chlorpyrifos	OP	*Cyprinodon variegatus* (sw)			136	Odenkirchen & Eisler, 1988
Chlorpyrifos	OP	*Fundulus grandis* jv. (sw)			1.8	Mayer, 1987

Table 3-35. (Continued)

Substance[a]	Class[b]	Organism	Considered safe[c] (μg/L)	Considered harmful[d] (μg/L)	Considered lethal[e] (μg/L)	References
Chlorpyrifos	OP	*Menidia peninsulae* lr. (sw)			0.42	Mayer, 1987
Chlorpyrifos	OP	*Menidia peninsulae* (sw)			0.7	Odenkirchen & Eisler, 1988
Chlorpyrifos	OP	*Morone saxatilis* (sw)			0.58	Korn & Earnest, 1974
Chlorpyrifos	OP	*Leiostomus xanthurus* jv. (sw)			6.4	Mayer, 1987
Chlorpyrifos	OP	*Mugil cephalus* jv. (sw)			5.4	Mayer, 1987
Coumaphos	OP	*Cyprinodon variegatus* jv. (sw)			266	Mayer, 1987
Crotoxyphos	OP	*Cyprinodon variegatus* jv. (sw)			>860	Mayer, 1987
Cypermethrin[r]	Py	*Palaemonetes pugio*[car] ad. (sw)			0.015	Mayer, 1987
Cypermethrin[r]	Py	*Cyprinodon variegatus* (sw)			1.0	Clark et al., 1989
Cypermethrin[r]	Py	*Salmo salar*			2.0	Clark et al., 1989
DDD (TDE)[r]	CH	*Leiostomus xanthurus* jv. (sw)			20	Mayer, 1987
DDE (DDT metabolite)[r]	CH	*Leiostomus xanthurus* jv. (sw)			>99	Mayer, 1987
DDT[R, r]	CH	*Pimephales promelas* life (fw)	0.5		2.0	Woltering, 1984
DDT[R, r]	CH	*Fundulus similis* jv. (sw)			2.8	Mayer, 1987
DDT[R, r]	CH	*Morone saxatilis* (sw)			0.53	Korn & Earnest, 1974
DDT[R, r]	CH	*Lagodon rhomboides* jv. (sw)			0.30	Mayer, 1987
DDT[R, r]	CH	*Mugil cephalus* jv. (sw)			0.40	Mayer, 1987
DDT[R, r]	CH	Rec. limit, salmonids (fw)	0.001			Billard, 1990
Demeton[Ac, r]	OP	*Leiostomus xanthurus* jv. (sw)			294	Mayer, 1987
Demeton[Ac, r]	OP	Legal limit, surface sw	0.1			Florida statute, 1993
Diazinon[Ne]	OP	*Brachionus calyciflorus*			29,220	Fernandez-Casalderry et al., 1992
Diazinon[Ne]	OP	*Daphnia magna*[cld] yg. (fw)		0.9		Lee et al., 1993c
Diazinon[Ne]	OP	*Mysidopsis bahia*[mys] (sw)		3.2	4.8	Eisler, 1986b
Diazinon[Ne]	OP	*Cyprinodon variegatus* life (sw)		0.47	1470	Eisler, 1986b
Diazinon[Ne]	OP	*Mugil cephalus* jv. (sw)			143	Mayer, 1987
Diazinon[Ne]	OP	Rec. limit, salmonids (fw)	0.002			Billard, 1990
Dichlorvos[Ac, r]	OP	*Isochrysis galbana*[mic] (sw)		13,720		Mayer, 1987
Dichlorvos[Ac, r]	OP	*Leiostomus xanthurus* jv. (sw)			314	Mayer, 1987
Dichlorvos[Ac, r]	OP	*Liza parsia* jv. (bw)		32	482	Mohapatra & Noble, 1993
Dichlorvos[Ac, r]	OP	*Mugil cephalus* jv. (sw)			314	Mayer, 1987
Dicrotophos	OP	*Fundulus similis* jv. (sw)			>800	Mayer, 1987
Dieldrin[r]	CH	*Anguilla rostrata*			0.9	Brown, 1978

Dieldrin[r]	CH	*Morone saxatilis* lr. (fw)			1	Hall, 1991
Dieldrin[r]	CH	*Morone saxatilis* (sw)			19.7	Korn & Earnest, 1974
Dieldrin[r]	CH	*Leiostomus xanthurus* jv. (sw)			3.2	Mayer, 1987
Dieldrin[r]	CH	*Mugil cephalus* jv. (sw)			3.2	Mayer, 1987
Dieldrin[r]	CH	*Thalassoma bifasciatum* (sw)			6	Brown, 1978
Dieldrin[r]	CH	Legal limit, surface sw	0.0019			Florida statute, 1993
Diflubenzuron[r]	AC	*Acartia tonsa*[cop] (sw)			1	Eisler, 1992a
Diflubenzuron[r]	AC	*Mysidopsis bahia*[mys] (sw)			0.075	Eisler, 1992a
Diflubenzuron[r]	AC	*Fundulus heteroclitus* (sw)	29,800		>33,000	Eisler, 1992a
Dimethoate[Ac]	OP	*Fundulus similis* jv. (sw)			>993	Mayer, 1987
Dimetilan[r]	Ca	*Leiostomus xanthurus* jv. (sw)			>961	Mayer, 1987
Disulfoton[Ac,Ne]	OP	*Cyprinodon variegatus* jv. (sw)			500	Mayer, 1987
Endosulfan[Ac]	CH	*Brachionus calyciflorus*			5150	Fernandez-Casalderry et al., 1992
Endosulfan[Ac]	CH	*Pimephales promelas* life (fw)	0.2		0.4	Woltering, 1984
Endosulfan[Ac]	CH	*Morone saxatilis* (sw)			0.10	Korn & Earnest, 1974
Endosulfan[Ac]	CH	*Lagodon rhomboides* ad. (sw)			0.30	Mayer, 1987
Endosulfan[Ac]	CH	*Leiostomus xanthurus* ad. (sw)			0.090	Mayer, 1987
Endosulfan[Ac]	CH	*Mugil cephalus* jv. (sw)			0.31	Mayer, 1987
Endosulfan[Ac]	CH	Legal limit, surface sw	0.0087			Florida statute, 1993
Endrin[R,Av, r]	CH	*Oncorhynchus tshawytscha* (fw)			1	Brown, 1978
Endrin[R,Av, r]	CH	*Anguilla rostrata*			0.6	Brown, 1978
Endrin[R,Av, r]	CH	*Brevoortia patronus* jv. (sw)			0.78	Mayer, 1987
Endrin[R,Av, r]	CH	*Cyprinodon variegatus* life (sw)	0.1		0.3	Woltering, 1984
Endrin[R,Av, r]	CH	*Fundulus similis* jv. (sw)			0.22	Mayer, 1987
Endrin[R,Av, r]	CH	*Morone saxatilis* (sw)			0.094	Korn & Earnest, 1974
Endrin[R,Av, r]	CH	*Leiostomus xanthurus* jv. (sw)			0.29	Mayer, 1987
Endrin[R,Av, r]	CH	*Mugil cephalus* jv. (sw)			0.40	Mayer, 1987
Endrin[R,Av, r]	CH	*Mugil cephalus* (sw)			0.3	Brown, 1978
Endrin[R,Av, r]	CH	*Thalassoma bifasciatum* (sw)			0.1	Brown, 1978
Endrin[R,Av, r]	CH	Rec. limit, salmonids (fw)	0.004			Billard, 1990
Endrin[R,Av, r]	CH	Legal limit, surface sw	0.0023			Florida statute, 1993
EPN[Ac, r]	OP	*Morone saxatilis* jv. (sw)			600	Hall, 1991
EPN[Ac, r]	OP	*Lagodon rhomboides* ad. (sw)			18	Mayer, 1987
Esfenvalerate	Py	*Melanotaenia fluviatilis* lr. (fw)			0.32	Holdway et al., 1994
Ethion[A,Av, r]	OP	*Leiostomus xanthurus* jv. (sw)			66	Mayer, 1987
Ethylan[r]	CH	*Isochrysis galbana*[mic] (sw)		70		Mayer, 1987
Famphur	OP	*Oncorhynchus mykiss*	2100		4900	Eisler, 1994b
Fenitrothion[r]	OP	*Anguilla anguilla* jv. (fw)		40		Sancho et al., 1996
Fenitrothion[r]	OP	*Cyprinodon variegatus* jv. (sw)			>750	Mayer, 1987
Fenthion[Ac,Av]	OP	*Daphnia* sp.[cld] first instar (fw)			0.80	Johnson & Finley, 1980
Fenthion[Ac,Av]	OP	*Oncorhynchus mykiss* (fw)			930	Brown, 1978
Fenthion[Ac,Av]	OP	*Acartia tonsa*[cop] ad. (sw)			40	Tucker & Thompson, 1987

Table 3-35. (Continued)

Substance[a]	Class[b]	Organism	Considered safe[c] (µg/L)	Considered harmful[d] (µg/L)	Considered lethal[e] (µg/L)	References
Fenthion[Ac,Av]	OP	*Centropomus undecimalis* egg-lr. (sw)			1200	Tucker & Thompson, 1987
Fenthion[Ac,Av]	OP	*Morone saxatilis* jv. (sw)			453	Korn & Earnest, 1974
Fenthion[Ac,Av]	OP	*Lagodon rhomboides* yg. (bw)		~320		Folmar et al., 1993
Fenthion[Ac,Av]	OP	*Cynoscion nebulosus* egg-lr. (sw)			447	Tucker & Thompson, 1987
Fenthion[Ac,Av]	OP	*Leiostomus xanthurus* egg-lr. (sw)		90	1250	Tucker & Thompson, 1987
Fenthion[Ac,Av]	OP	*Leiostomus xanthurus* jv. (sw)			1116	Mayer, 1987
Fenthion[Ac,Av]	OP	*Mugil cephalus* jv. (sw)			1488	Mayer, 1987
Fentrifanil	BA	*Cyprinodon variegatus* jv. (sw)			9.0	Mayer, 1987
Fenvalerate[r]	Py	*Diaptomus oregonensis*[cop]		0.05	~0.12	Eisler, 1992b
Fenvalerate[r]	Py	*Gammarus pseudolimnaeus*[amp]			0.022	Eisler, 1992b
Fenvalerate[r]	Py	*Pimephales promelas* yg. (fw)	0.33		0.43	Woltering, 1984
Fenvalerate[r]	Py	*Oncorhynchus mykiss*	~0.018		0.08	Eisler, 1992b
Fenvalerate[r]	Py	*Melanotaenia fluviatilis* lr. (fw)			4.5	Holdway et al., 1994
Fenvalerate[r]	Py	*Salmo salar*			1.2	Clark et al., 1989
Fenvalerate[r]	Py	*Isochrysis galbana*[mic] (sw)		>1000		Mayer, 1987
Fenvalerate[r]	Py	*Crassostrea virginica*[oys] lr. (sw)		>1000		Eisler, 1992b
Fenvalerate[r]	Py	*Mysidopsis bahia*[mys] (sw)			0.008-0.021	Eisler, 1992b
Fenvalerate[r]	Py	*Palaemonetes pugio*[car] lr. (sw)		0.0016	0.0032	Eisler, 1992b
Fenvalerate[r]	Py	*Penaeus duoarum*[pen] (sw)			0.84	Eisler, 1992b
Fenvalerate[r]	Py	*Cyprinodon variegatus* (sw)	0.56	2.2	5.0	Clark et al., 1989; Eisler, 1992b
Fenvalerate[r]	Py	*Fundulus heteroclitus* (sw)			1.2-1.8	Eisler, 1992b
Fenvalerate[r]	Py	*Leuresthes tenuis* lr. (sw)			0.29	Mayer, 1987
Fenvalerate[r]	Py	*Menidia menidia* (sw)	0.062		0.31-0.69	Eisler, 1992b
Fenvalerate[r]	Py	*Mugil cephalus* jv. (sw)			0.58	Mayer, 1987
Fenvalerate[r]	Py	*Oreochromis mossambicus* (fw)		9	45	Eisler, 1992b
Flucythrinate[r]	Py	*Cyprinodon variegatus* (sw)	0.032		1.1	Clark et al., 1989
Flucythrinate[r]	Py	*Menidia beryllina* jv. (sw)			0.32	Mayer, 1987
Fonofos	OP	*Pimephales promelas* yg. (fw)	16		33	Woltering, 1984
Fonofos	OP	*Leiostomus xanthurus* jv. (sw)			228	Mayer, 1987
Heptachlor[r]	CH	*Cyprinodon variegatus* life (sw)	<2.8		2.8	Woltering, 1984
Heptachlor[r]	CH	*Lagodon rhomboides* ad. (sw)			3.8	Mayer, 1987
Heptachlor[r]	CH	*Leistomus xanthurus* ad. (sw)			0.85	Mayer, 1987

Heptachlor[r]	CH	*Mugil cephalus* jv. (sw)			2.4	Mayer, 1987
Hydrogen cyanide[r]	I	*Salmo salar* egg (fw)	<10		10	Woltering, 1984
Hydrogen cyanide[r]	I	*Jordanella floridae* egg-lr. (fw)		65	65-150	Cheng & Ruby, 1980
Hydrogen cyanide[r]	I	Rec. limit, salmonids (fw)	5			Billard, 1990
Hydrogen cyanide[r]	I	Freshwater fish	<5			Warren et al., 1990
Hydrogen cyanide[r]	I	Legal limit, surface sw	1			Florida statute, 1993
Hydramethylnon	FH	*Isochrysis galbana*[mic] (sw)		2.8		Mayer, 1987
Hydramethylnon	FH	*Lagodon rhomboides* jv. (sw)			192	Mayer, 1987
Isobenzan[r]	TC	*Leiostomus xanthurus* jv. (sw)			0.30	Mayer, 1987
Landrin[r]	Ca	*Lagodon rhomboides* jv. (sw)			1800	Mayer, 1987
Leptophos[r]	OP	*Leiostomus xanthurus* jv. (sw)			4.1	Mayer, 1987
Lethane 384[r]	TC	*Cyprinodon variegatus* jv. (sw)			>547	Mayer, 1987
Lindane[r]	CH	*Anguilla rostrata* jv.			70	Geyer et al., 1993
Lindane[r]	CH	*Pimephales promelas* ad. (fw)	9		24	Woltering, 1984
Lindane[r]	CH	*Morone saxatilis* (sw)			7.3	Korn & Earnest, 1974
Lindane[r]	CH	*Lagodon rhomboides* ad. (sw)			31	Mayer, 1987
Lindane[r]	CH	*Mugil cephalus* jv. (sw)			23	Mayer, 1987
Lindane[r]	CH	Rec. limit, salmonids (fw)	0.001			Billard, 1990
Malathion[Ac]	OP	*Brachionus calyciflorus*			33,720	Fernandez-Casalderry et al., 1992
Malathion[Ac]	OP	*Daphnia* sp.[cld] first instar (fw)			1.8	Johnson & Finley, 1980
Malathion[Ac]	OP	*Oncorhynchus kisutch* (fw)			100	Brown, 1978
Malathion[Ac]	OP	*Oncorhynchus tshawytscha* (fw)			23	Brown, 1978
Malathion[Ac]	OP	*Morone saxatilis* jv. (fw)			24.5	Palawski et al., 1985
Malathion[Ac]	OP	*Anguilla rostrata*			82	Brown, 1978
Malathion[Ac]	OP	*Brachionus plicatilis* (sw)			45,500	Snell & Persoone, 1989
Malathion[Ac]	OP	*Acartia tonsa*[cop] ad. (sw)			115	Tucker & Thompson, 1987
Malathion[Ac]	OP	*Cyprinodon variegatus* (sw)	4		9	Woltering, 1984
Malathion[Ac]	OP	*Fundulus similis* jv. (sw)			142	Mayer, 1987
Malathion[Ac]	OP	*Centropomus undecimalis* egg-lr. (sw)			9	Tucker & Thompson, 1987
Malathion[Ac]	OP	*Morone saxatilis* jv. (sw)			65	Palawski et al., 1985
Malathion[Ac]	OP	*Cynoscion nebulosus* egg-lr. (sw)			19	Tucker & Thompson, 1987
Malathion[Ac]	OP	*Mugil cephalus* jv. (sw)			314	Mayer, 1987
Malathion[Ac]	OP	*Mugil cephalus* (sw)			550	Brown, 1978
Malathion[Ac]	OP	*Thalassoma bifasciatum* (sw)			27	Brown, 1978
Malathion[Ac]	OP	Rec. limit, salmonids (fw)	0.1			Billard, 1990
Malathion[Ac]	OP	Legal limit, surface sw	0.1			Florida statute, 1993
Methidathion[Ac, r]	OP	*Leiostomus xanthurus* jv. (sw)			32	Mayer, 1987
Methiocarb[Ac,M, r]	Ca	*Fundulus similis* jv. (sw)			317	Mayer, 1987
Methoprene	O	*Lepomis macrochirus* (fw)			4620	Meister, 1994
Methoxychlor	CH	*Cyprinodon variegatus* life (sw)	12		23	Woltering, 1984
Methoxychlor	CH	*Morone saxatilis* (sw)			3.3	Korn & Earnest, 1974
Methoxychlor	CH	*Leiostomus xanthurus* jv. (sw)			23	Mayer, 1987

Table 3-35. (Continued)

Substance[a]	Class[b]	Organism	Considered safe[c] (µg/L)	Considered harmful[d] (µg/L)	Considered lethal[e] (µg/L)	References
Methoxychlor	CH	*Mugil cephalus* jv. (sw)			32	Mayer, 1987
Methoxychlor	CH	*Thalassoma bifasciatum* (sw)			13	Brown, 1978
Methoxychlor	CH	Legal limit, surface sw	0.03			Florida statute, 1993
Methyl parathion[Ac, r]	OP	*Brachionus calyciflorus* (fw)	<3,000	5000	29,190	Fernandez-Casalderry et al., 1992, 1993
Methyl parathion[Ac, r]	OP	*Daphnia magna*[cld] (fw)		0.000031	0.00031	Fernandez-Casalderry et al., 1993
Methyl parathion[Ac, r]	OP	*Pimephales promelas* yg. (fw)	310	590		Woltering, 1984
Methyl parathion[Ac, r]	OP	*Acartia tonsa*[cop] ad. (sw)			28	Mayer, 1987
Methyl parathion[Ac, r]	OP	*Cyprinodon variegatus* jv. (sw)			11,880	Mayer, 1987
Methyl parathion[Ac, r]	OP	*Morone saxatilis* jv. (sw)			790	Korn & Earnest, 1974
Methyl parathion[Ac, r]	OP	*Leiostomus xanthurus* ad. (sw)			59	Mayer, 1987
Methyl parathion[Ac, r]	OP	*Mugil cephalus* (sw)			5200	Brown, 1978
Methyl trithion[Ac]	OP	*Mugil cephalus* jv. (sw)			173	Mayer, 1987
Mevinphos[Ac, r]	OP	*Cyprinodon variegatus* jv. (sw)			640	Mayer, 1987
Mexacarbate[Ac,M, r]	Ca	*Cyprinodon variegatus* jv. (sw)			>900	Mayer, 1987
Mirex[r]	CH	*Pimephales promelas* ad. (fw)	7	13		Woltering, 1984
Mirex[r]	CH	*Leiostomus xanthurus* jv. (sw)			>1960	Mayer, 1987
Naled[Ac]	OP	*Morone saxatilis* jv. (sw)			500	Korn & Earnest, 1974
Naled[Ac]	OP	*Leiostomus xanthurus* jv. (sw)			216	Mayer, 1987
Naled[Ac]	OP	*Mugil cephalus* jv. (sw)			288	Mayer, 1987
Naphthalene	PAH	*Oncorhynchus kisutch* yg. (fw)			3200	Eisler, 1987b
Naphthalene	PAH	*Eurytemora affinis*[cop] (sw)			50	Eisler, 1987b
Naphthalene	PAH	*Palaemonetes* sp.[car] (sw)			2400	Eisler, 1987b
Naphthalene	PAH	*Gadus morhua* egg (sw)		1000-2000		Falk-Petersen & Kjorsvik, 1987
Naphthalene	PAH	*Cyprinodon variegatus* (sw)			2400	Eisler, 1987b
Naphthalene	PAH	Recommended limit (sw)	1			Nagpal, 1994
Nicotine sulfate	Alk	*Pimephales promelas* jv. (fw)			19,700	Sulaiman, 1993
Nicotine sulfate	Alk	*Oncorhynchus mykiss* jv. (fw)			7310	Sulaiman, 1993
Parathion[r]	OP	*Morone saxatilis* (sw)			17.8	Korn & Earnest, 1974
Parathion[r]	OP	*Leiostomus xanthurus* jv. (sw)			18	Mayer, 1987
Parathion[r]	OP	*Mugil cephalus* jv. (sw)			100	Mayer, 1987
Parathion[r]	OP	Rec. limit, salmonids (fw)	0.04			Billard, 1990
Parathion[r]	OP	Legal limit, surface sw	0.04			Florida statute, 1993

Permethrin[r]	Py	*Pimephales promelas* yg. (fw)	0.7		1.4	Woltering, 1984
Permethrin[r]	Py	*Salmo salar*			12.0	Clark et al., 1989
Permethrin[r]	Py	*Cyprinodon variegatus* (sw)	10		7.8	Clark et al., 1989
Permethrin[r]	Py	*Menidia menidia* ad. (sw)			2.2	Mayer, 1987
Permethrin[r]	Py	*Mugil cephalus* jv. (sw)			5.5	Mayer, 1987
Phorate[Ac,Ne]	OP	*Fundulus similis* jv. (sw)			0.32	Mayer, 1987
Phosmet[Ac]	OP	*Mugil cephalus* jv. (sw)			30	Mayer, 1987
Phosphamidon[r]	OP	*Mugil cephalus* jv. (sw)			>913	Mayer, 1987
Phoxim[r]	OP	*Leiostomus xanthurus* jv. (sw)			2.5	Mayer, 1987
Propoxur	Ca	*Cyprinodon variegatus* jv. (sw)			>1000	Mayer, 1987
Ronnel[r]	OP	*Leiostomus xanthurus* jv. (sw)			304	Mayer, 1987
Temephos	OP	*Menidia beryllina* jv. (sw)			3362	Mayer, 1987
Temephos	OP	*Morone saxatilis* jv. (bw)			1000	Hall, 1991
Temephos	OP	*Mugil cephalus* jv. (sw)			>4827	Mayer, 1987
Terpene polychlorinates[r]	CH	*Leiostomus xanthurus* jv. (sw)			3.2	Mayer, 1987
Tetrachlorvinphos	OP	*Leiostomus xanthurus* jv. (sw)			>940	Mayer, 1987
Thanite[r]	TC	*Leiostomus xanthurus* jv. (sw)			26	Mayer, 1987
Thiodicarb	Ca	*Cyprinodon variegatus* jv. (sw)			>990	Mayer, 1987
Toxaphene[F,P, r]	CH	*Pimephales promelas* (fw)	0.025	0.054	18.0	Woltering, 1984; Eisler & Jacknow, 1985
Toxaphene[F,P, r]	CH	*Oncorhynchus kisutch* (fw)			8.0	Eisler & Jacknow, 1985
Toxaphene[F,P, r]	CH	*Oncorhynchus mykiss* (fw)			10.6	Eisler & Jacknow, 1985
Toxaphene[F,P, r]	CH	*Salmo trutta* (fw)			3.1	Eisler & Jacknow, 1985
Toxaphene[F,P, r]	CH	*Cyprinodon variegatus* (sw)			1.1	Eisler & Jacknow, 1985
Toxaphene[F,P, r]	CH	*Morone saxatilis* (sw)			4.4	Eisler & Jacknow, 1985
Toxaphene[F,P, r]	CH	*Lagodon rhomboides* ad. (sw)			0.53	Mayer, 1987
Toxaphene[F,P, r]	CH	*Mugil cephalus* jv. (sw)			3.2	Mayer, 1987
Toxaphene[F,P, r]	CH	Rec. limit, salmonids (fw)	0.005			Billard, 1990
Trichlorfon	OP	*Daphnia pulex*[cld] (fw)		0.18		Johnson & Finley, 1980
Trichlorfon	OP	*Morone saxatilis* lr. (fw)			5000	Hall, 1991
Trichlorfon	OP	*Isochrysis galbana*[mic] (sw)		>49,000		Mayer, 1987
Trichlorfon	OP	*Leiostomus xanthurus* jv. (sw)			>980	Mayer, 1987
Trichloronate[r]	OP	*Leiostomus xanthurus* jv. (sw)			304	Mayer, 1987
Zytron[r]	OP	*Leiostomus xanthurus* jv. (sw)			320	Mayer, 1987
		Molluscicides				
Antibiotic 230	An	*Hypophthalmichthys molitrix* (fw)			2.1	Yao et al., 1993
Niclosamide[P]	BA	*Oncorhynchus mykiss* (fw)			340	Johnson & Finley, 1980
Niclosamide[P]	BA	*Morone saxatilis* jv. (fw)			1050	Hall, 1991
Niclosamide[P]	BA	*Fundulus similis* jv. (sw)		1488		Mayer, 1987

Table 3-35. (Continued)

Substance[a]	Class[b]	Organism	Considered safe[c] (µg/L)	Considered harmful[d] (µg/L)	Considered lethal[e] (µg/L)	References
		Nematicides				
Diamidfos[r]	OP	*Leiostomus xanthurus* jv. (sw)			>1000	Mayer, 1987
Ethoprop[I]	OP	*Lagodon rhomboides* ad. (sw)			6.3	Mayer, 1987
Fenamiphos[I]	OP	*Cyprinodon variegatus* jv. (sw)			115	Mayer, 1987
		Piscicides				
Antimycin A[r]	F	*Oncorhynchus mykiss* yg. (fw)			0.012	Johnson & Finley, 1980
Antimycin A[r]	F	*Leiostomus xanthurus* jv. (sw)			0.22	Mayer, 1987
Rotenone[I]	Bo	*Pimephales promelas* jv. (fw)			6.4	Sulaiman, 1993
Rotenone[I]	Bo	*Oncorhynchus mykiss* jv. (fw)			5	Sulaiman, 1993
Rotenone[I]	Bo	*Morone saxatilis* lr. & jv. (fw)			1	Hall, 1991
Rotenone[I]	Bo	*Praunus flexuosus*[mys] (sw)			2000	Næss et al., 1991
Rotenone[I]	Bo	*Fundulus similis* jv. (sw)			36	Mayer, 1987
Rotenone[I]	Bo	*Symphodus melops* (sw)			50	Næss et al., 1991
Rotenone[I]	Bo	*Gobiusculus flavescens* (sw)			100	Næss et al., 1991
		Rodenticides				
Arsenic trioxide (76% As)[r]	I	*Fundulus similis* jv. (sw)			>30,300	Mayer, 1987
Thallium sulfate[r]	I	Legal limit, surface sw	48 Tl			Florida statute, 1993
		Detergent Chemicals				
Ditallow dimethyl ammonium chloride		*Pimephales promelas* yg. (fw)	50		90	Woltering, 1984
Ditallow dimethyl ammonium chloride		*Oryzias latipes* lr.-jv. (fw)	320	560	1800	Wester & Roghair, 1992
Linear alkylate sulfonate		*Sebastes schlegeli* jv. (sw)			1080	Choi & Kinae, 1994
Linear alkylate sulfonate		*Mugil cephalus* jv. (sw)			60	Choi & Kinae, 1994
Linear alkylate sulfonate		*Paralichthys olivaceus* jv. (sw)			790	Choi & Kinae, 1994
Linear alkalate sulfonates		Fish	200			Post, 1987
Linear alkylbenzene sulfonate		*Oncorhynchus mykiss* lr. (fw)		200		Hofer et al., 1995
Linear alkylbenzene sulfonate		*Gadus morhua* egg & lr. (sw)	<5		5-20	Swedmark & Granmo, 1981
Sodium lauryl sulfate		*Cyprinodon variegatus* jv. (sw)			1700	Mayer, 1987

Detergents		Legal limit, surface sw	500			Florida statute, 1993
		Industrial and Laboratory Chemicals				
Acetone	K	*Oncorhynchus mykiss* yg. (fw)			5,540,000	Johnson & Finley, 1980
Acetone	K	*Cyprinodon variegatus* ad. (sw)			45,000,000	Mayer, 1987
Anthracene	PAH	*Pimephales promelas* ♀-lr. (fw)		6	20	Hall & Oris, 1991
Aroclor 1254	PCB	*Anguilla anguilla* jv. (fw)		200		Klyszejko et al., 1993
Aroclor 1254	PCB	*Penaeus duoarum*[pen] (sw)			1.0	Eisler, 1986c
Aroclor 1254	PCB	*Cyprinodon variegatus* yg. (sw)	0.06		0.16	Schimmel et al., 1974
Aroclor 1254	PCB	*Lagodon rhomboides* (sw)			0.5	Eisler, 1986c
Benzene	AH	*Pimephales promelas* lr. (fw)			15,600	Marchini et al., 1993
Benzene	AH	*Oncorhynchus mykiss* yg. (fw)			9200	Johnson & Finley, 1980
Benzene	AH	*Morone saxatilis* jv. (bw)			5.8	Hall, 1991
Benzene, ave.	AH	Legal limit, surface sw	71.28			Florida statute, 1993
Benzo(a)pyrene	PAH	*Psettichthys melanostictus* egg (sw)			0.1	Eisler, 1987b
Benzo(a)pyrene	PAH	Recommended limit (sw)	0.01			Nagpal, 1994
Bromine (free)	I	Legal limit, surface sw	100			Florida statute, 1993
Bromobenzene	AH	*Pimephales promelas* lr. (fw)			5600	Marchini et al., 1993
Bromoform	BH	*Cyprinodon variegatus* yg. (sw)	<24,000		24,000	Woltering, 1984
Bromoform, ave.	BH	Legal limit, surface sw	360			Florida statute, 1993
Carbon tetrachloride, ave.	CH	Legal limit, surface sw	4.42			Florida statute, 1993
Chloramines[Di]	I	*Brachionus plicatilis*			20	Capuzzo, 1979a, 1979b
Chloramines[Di]	I	*Crassostrea virginica*[oys] lr. (bw)			10	Capuzzo, 1979a
Chloramines[Di]	I	*Acartia tonsa*[cop] (bw)			320	Capuzzo, 1979a
Chloramines[Di]	I	*Oncorhynchus kisutch* lr.-jv. (fw)	11	22	47	Larson et al., 1977
4-chlorophenol	CH	*Cyprinodon variegatus* (sw)	3,200		5400	Krijgsheld & van der Gen, 1986
Chlorophenol	CH	Safe range for different species	0.02-0.90			Warrington, 1994
Cobalt chloride	I	*Diaptomus forbesi*[cop] (fw)			3440	Das & Kaviraj, 1994
Cobalt chloride	I	*Cyprinus carpio* yg. (fw)			333,040	Das & Kaviraj, 1994
Crude oil	O	*Calanus hamatus*[cop] (sw)		10		Buttino, 1994
Crude oil	O	*Mugil cephalus* (sw)		~4500		Minchew & Yarbrough, 1977
Crude oil, weathered	O	*Gadus morhua* egg (sw)		1000		Falk-Petersen & Kjorsvik, 1987
Crude oil (water-sol. fraction)	O	*Gadus morhua* egg-jv. (sw)		~92		Goksøyr et al., 1991
2,4-Dichlorophenol	CH	*Oncorhynchus mykiss* egg-lr. (fw)			70	Krijgsheld & van der Gen, 1986
1,3-Dimethylnaphthalene	PAH	*Gadus morhua* egg (sw)		100-300		Falk-Petersen & Kjorsvik, 1987
Ethanol	Alc	*Oncorhynchus mykiss* yg. (fw)			13,000,000	Johnson & Finley, 1980
Ethylene glycol	Alc	*Oncorhynchus mykiss* yg. (fw)			16,000,000	Johnson & Finley, 1980
Fluorene	PAH	*Pimephales promelas* (fw)			>100,000	Eisler, 1987b
Fluorene	PAH	*Oncorhynchus mykiss* (fw)			820	Eisler, 1987b
Fluorene	PAH	*Palaemonetes pugio*[car] (sw)			320	Eisler, 1987b
Fluorene	PAH	*Cyprinodon variegatus* (sw)			1680	Eisler, 1987b
Fluorescein	H	*Scophthalmus maximus* jv. (sw)			997,000	Pouliquen et al., 1995

Table 3-35. (Continued)

Substance[a]	Class[b]	Organism	Considered safe[c] (µg/L)	Considered harmful[d] (µg/L)	Considered lethal[e] (µg/L)	References
Fluoride	I	Freshwater fish	<500			Warren et al., 1990
Fuel oil, no. 2 (water soluble)[I,Ac]	O	*Fundulus grandis* egg (sw)	~<1100		2200	Ernst et al., 1977
Glycerol	Alc	*Oncorhynchus mykiss* yg. (fw)			54,000,000	Johnson & Finley, 1980
Hexachlorobutadiene	CH	*Lagodon rhomboides* jv. (sw)			392	Mayer, 1987
Hexachlorocyclopentadiene	CH	*Lagodon rhomboides* ad. (sw)			48	Mayer, 1987
Hexachlorocyclopentadiene	CH	*Leiostomus xanthurus* jv. (sw)			31	Mayer, 1987
Methanol	Alc	*Oncorhynchus mykiss* yg. (fw)			19,000,000	Johnson & Finley, 1980
Nickel chloride (45% Ni)	I	*Menidia peninsulae* lr. ()			38,000 Ni	Mayer, 1987
Nickel chloride (45% Ni)	I	*Leiostomus xanthurus* ad. (sw)			70,000 Ni	Mayer, 1987
Oil (total AH)	O	Fish eggs & lr.			14.3	Hurlbut et al., 1991
Oil dispersants		*Gadus morhua* egg (sw)		1000-10,000		Falk-Petersen & Kjorsvik, 1987
Pentachlorobenzene	CH	*Cyprinodon variegatus* jv. (sw)			460	Mayer, 1987
Phenanthrene	PAH	*Palaemonetes* sp.[car] (sw)			370	Eisler, 1987b
Phenol[Di]	Alc	*Oncorhynchus mykiss* yg. (fw)	<200	200	10,500	Woltering, 1984; Sulaiman, 1993
Phenol[Di]	Alc	*Acartia clausi*[cop] (sw)		500	3809	Buttino, 1994
Phenol[Di]	Alc	*Gadus morhua* egg (sw)		1000		Falk-Petersen & Kjorsvik, 1987
Phenol[Di]	Alc	*Cyprinodon variegatus* jv. (sw)			>1980	Mayer, 1987
Phenyls	AH	Rec. limit, salmonids (fw)	1			Billard, 1990
Polychlorinated biphenyls	PCB	Rec. limit, salmonids (fw)	0.001			Billard, 1990
Polychlorinated biphenyls	PCB	Freshwater fish	2			Warren et al., 1990
Polychlorinated biphenyls	PCB	Fish	0.002			Post, 1987
Polychlorinated biphenyls	PCB	Legal limit, surface sw	0.03			Florida statute, 1993
Polychlorinated biphenyls, ave.	PCB	Legal limit, surface sw	0.000045			Florida statute, 1993
Polycyclic arom. hydrocarbons, ave.	PAH	Legal limit, surface sw	0.031			Florida statute, 1993
Potassium dichromate (35% Cr)	I	*Menidia peninsulae* lr. ()			22,000 Cr	Mayer, 1987
Potassium dichromate (35% Cr)	I	*Leiostomus xanthurus* ad. (sw)			27,000 Cr	Mayer, 1987
Sodium azide	I	*Atherinops affinis* lr. (bw)			5400 N_3^-	Anderson et al., 1995
Sodium dodecyl sulfate		*Brachionus plicatilis* (sw)			4420	Snell & Persoone, 1989
Sulfur dioxide	I	*Morone saxatilis* lr. (bw)			2000	Hall, 1991
Tetrabromobisphenol-A	BH	*Daphnia magna*[cld] yg. (fw)		0.1		Lee et al., 1993c
Toluene	AH	*Pimephales promelas* lr. (fw)			17,000	Marchini et al., 1993

Toluene	AH	*Oncorhynchus mykiss* yg. (fw)			24,000	Johnson & Finley, 1980
Toluene	AH	*Oncorhynchus kisutch* lr. (sw)			9360	Korn & Rice, 1981
Toluene	AH	*Cyprinodon variegatus* yg. (sw)	<6700	6700		Woltering, 1984
Toluene	AH	*Morone saxatilis* jv. (bw)			7.3	Hall, 1991
Trimethylnaphthalene	PAH	*Eurytemora affinis*[cop] (sw)			320	Eisler, 1987b
Urea		*Oreochromis niloticus* jv. (fw)		14,000,000	16,000,000	Knud-Hansen & Pautong, 1993
Xylene	AH	*Oncorhynchus mykiss* yg. (fw)		8,200		Johnson & Finley, 1980
m-xylene	AH	*Morone saxatilis* jv. (bw)			9200	Hall, 1992
o-xylene	AH	*Oncorhynchus mykiss* jv. (fw)			8050	Sulaiman, 1993
Xylenes	AH	*Gadus morhua* egg (sw)		4000		Falk-Petersen & Kjørsvik, 1987
		Nitrification (by *Nitrosomonas* spp.) Inhibitor				
Nitrapyrin		*Leiostomus xanthurus* jv. (sw)			>940	Mayer, 1987
		Plastics Components				
Butylbenzyl phthalate	PE	*Crassostrea virginica*[oys] lr. (sw)		780		Mayer, 1987
di-n butylphthalate	PE	*Gammarus pulex*[amp]		500		Burridge & Haya, 1995
di-n butylphthalate	PE	*Palaemonetes pugio*[car]			100	Burridge & Haya, 1995
di-n butylphthalate	PE	*Mysidopsis bahia*[mys]	230		750	Burridge & Haya, 1995
di-n butylphthalate	PE	*Oncorhynchus mykiss*			500	Burridge & Haya, 1995
di-n butylphthalate	PE	*Perca flavescens*			350	Burridge & Haya, 1995
Phthalate esters	PE	*Lepomis macrochirus* (fw)			700	Johnson & Finley, 1980
Phthalate esters	PE	Rec. limit, fish	0.3			Post, 1987
Phthalate esters	PE	Legal limit, surface fw	3.0			Florida statute, 1993
Triethylene glycol	Alc	*Cyprinodon variegatus* ad. (sw)			47,952,000	Mayer, 1987
		Wood Preservatives				
Copper nitrate (34% Cu)[I]	I	*Menidia peninsulae* lr.			140	Mayer, 1987
Copper nitrate (34% Cu)[I]	I	*Leiostomus xanthurus* ad. (sw)			280	Mayer, 1987
Coal tar creosote[r]	O	*Cyprinodon variegatus* jv. (sw)			698	Mayer, 1987
Pentachlorophenol[De,M,I,F,B,NI, r]	CH	*Daphnia magna*[cld] yg. (fw)		250		Lee et al., 1993c
Pentachlorophenol[De,M,I,F,B,NI, r]	CH	*Pimephales promelas* life (fw)	45	48	120	Eisler, 1989c
Pentachlorophenol[De,M,I,F,B,NI, r]	CH	*Oncorhynchus mykiss* yg. (fw)	0.44		19	Eisler, 1989c; Watson et al., 1995
Pentachlorophenol[De,M,I,F,B,NI, r]	CH	*Oncorhynchus nerka* (fw)		3.2	63-68	Eisler, 1989c
Pentachlorophenol[De,M,I,F,B,NI, r]	CH	*Thalassiosira pseudonana*[mic] (sw)		179-189		Eisler, 1989c
Pentachlorophenol[De,M,I,F,B,NI, r]	CH	*Crassostrea virginica*[oys] lr. (sw)		40		Eisler, 1989c
Pentachlorophenol[De,M,I,F,B,NI, r]	CH	*Mysidopsis bahia*[mys] (sw)			320	Eisler, 1989c
Pentachlorophenol[De,M,I,F,B,NI, r]	CH	*Cyprinodon variegatus* life (sw)	47		88	Eisler, 1989c
Pentachlorophenol[De,M,I,F,B,NI, r]	CH	*Fundulus heteroclitus* (bw)	<1	5		Roszell & Anderson, 1994
Pentachlorophenol[De,M,I,F,B,NI, r]	CH	*Menidia beryllina* egg (sw)		0.0005	26	Middaugh et al., 1993
Pentachlorophenol[De,M,I,F,B,NI, r]	CH	*Lagodon rhomboides* lr. (sw)			30	Mayer, 1987
Pentachlorophenol[De,M,I,F,B,NI, r]	CH	*Mugil cephalus* (sw)			112	Eisler, 1989c

Table 3-35. (Continued)

Substance[a]	Class[b]	Organism	Considered safe[c] (μg/L)	Considered harmful[d] (μg/L)	Considered lethal[e] (μg/L)	References
Pentachlorophenol[De,M,I,F,B,NI, r]	CH	*Pleuronectes platessa* lr. (sw)			60-140	Eisler, 1989c
Pentachlorophenol[De,M,I,F,B,NI, r]	CH	Legal limit, surface sw	7.9			Florida statute, 1993
Sodium pentachlorophenate[M,F,B,Al, r]	CH	*Brachionus plicatilis* (sw)			1360	Snell & Persoone, 1989
Sodium pentachlorophenate[M,F,B,Al, r]	CH	*Mugil cephalus* jv. (sw)			110	Mayer, 1987
2,3,4,6-Tetrachlorophenol[r]	CH	*Palaemonetes pugio*[car] ad. (sw)			794	Mayer, 1987
2,4,6-Trichlorophenol[F,B]	CH	*Palaemonetes pugio*[car] ad. (sw)			1078	Mayer, 1987
2,4,6-Trichlorophenol, ave.[F,B]	CH	Legal limit, surface sw	6.5			Florida statute, 1993
Zinc sulfate (40% Zn)[r]	I	*Menidia peninsulae* lr. ()			2240 Zn	Mayer, 1987
Zinc sulfate (40% Zn)[r]	I	*Leiostomus xanthurus* ad. (sw)			15,200 Zn	Mayer, 1987
		Therapeutic and Anesthetic Substances				
Acriflavine		*Morone saxatilis* lr. (fw)			5000	Hall, 1991
Acriflavine		*Morone saxatilis* jv. (fw)			16,000	Hall, 1991
Acriflavine		*Trachinotus carolinus* jv.			114,300	Birdsong & Avault, 1971
Benzalkonium chloride	QA	*Ictalurus punctatus* (fw)			1120	Post, 1987
Benzalkonium chloride	QA	*Morone saxatilis* lr. (fw)			500	Hall, 1991
Benzalkonium chloride	QA	*Morone saxatilis* jv. (fw)			1500	Hall, 1991
Benzocaine		*Morone saxatilis* jv. (fw)			20,000	Bills et al., 1993
Chelated copper[A]	I&O	*Sciaenops ocellatus* jv. (sw)	~200		370	Peppard et al., 1991
Chloramine-T	CA	*Morone saxatilis* jv. (fw)			2800	Bills et al., 1993
Colchicine	Alk	*Dicentrarchus labrax* jv. (sw)		6,250	50,000	Roche et al., 1993
Copper sulfate	I	*Morone saxatilis* jv. (fw)			230	Bills et al., 1993
Copper sulfate	I	*Trachinotus carolinus* jv.			1400-2000	Birdsong & Avault, 1971
Cutrine		*Morone saxatilis* jv. (fw)			100	Hall, 1991
Etomidate		*Morone saxatilis* jv. (fw)			280	Bills et al., 1993
Formalin[B,F,Di]	Ald	*Ictalurus punctatus* (fw)			64	Post, 1987
Formalin[B,F,Di]	Ald	*Oncorhynchus mykiss* jv. (fw)	62		129	van Heerden et al., 1995
Formalin[B,F,Di]	Ald	*Morone saxatilis* lr. (fw)			26,000	Hall, 1991
Formalin[B,F,Di]	Ald	*Morone saxatilis* jv. (fw)			38,000	Hall, 1991
Formalin[B,F,Di]	Ald	*Lates calcarifer* jv. (sw)			239,200	Pascual et al., 1994
Formalin[B,F,Di]	Ald	*Trachinotus carolinus* jv.			69,100-74,900	Birdsong & Avault, 1971
Hyamine 3500		*Morone saxatilis* jv. (fw)			1410	Bills et al., 1993

Hydrogen peroxide (45 min)	I	*Salmo trutta* jv. (fw)	250		500	Rach et al., 1997
Hydrogen peroxide (5 min)	I	*Ocyurus chrysurus* egg (sw)	10,000,000			Douillet & Holt, 1994
Hydrogen peroxide (5 min)	I	*Cynoscion nebulosus* egg (sw)	20,000,000			Douillet & Holt, 1994
Hydrogen peroxide (5 min)	I	*Sciaenops ocellatus* egg (sw)	30,000,000			Douillet & Holt, 1994
Malachite green[F]	I	*Oncorhynchus mykiss* jv. (fw)	0.125		0.267	van Heerden et al., 1995
Malachite green[F]	I	*Morone saxatilis* lr. (fw)			50	Hall, 1991
Malachite green[F]	I	*Morone saxatilis* jv. (fw)			200; 138	Hall, 1991; Bills et al., 1993
Malachite green[F]	I	*Lepomis macrochirus* (fw)			35	Post, 1987
Methylene blue		*Morone saxatilis* lr. (fw)			1000	Hall, 1991
Methylene blue		*Morone saxatilis* jv. (fw)			12,000	Hall, 1991
Metomidate		*Morone saxatilis* jv. (fw)			2000	Bills et al., 1993
Nifurpirinol	NF	*Chanos chanos* jv. (sw)	~500		1060	Tamse & Gacutan, 1994
Nifurpirinol	NF	*Fundulus heteroclitus* (sw)		2,000		Post, 1987
Oxytetracycline (6 h)	An	*Morone saxatilis* jv. (sw)			32,000	Bumguardner & King, 1996
Potassium permanganate	I	*Diaptomus forbesi* [cop] (fw)			220	Das & Kaviraj, 1994
Potassium permanganate	I	*Cyprinus carpio* yg. (fw)			3040	Das & Kaviraj, 1994
Potassium permanganate	I	*Morone saxatilis* lr. & jv. (fw)			~990	Reardon & Harrell, 1994
Potassium permanganate	I	*Morone saxatilis* jv. (fw)			2600; 1580	Hall, 1991; Bills et al., 1993
Potassium permanganate	I	*Morone saxatilis* lr. & jv. (bw)			1410-3260	Reardon & Harrell, 1994
Potassium permanganate	I	*Trachinotus carolinus* jv.			1600-2900	Birdsong & Avault, 1971
Quinaldine	Qn	*Morone saxatilis* jv. (fw & bw)			~4750	Hall, 1991
Quinaldine sulfate	Qu	*Morone saxatilis* jv. (fw)			22,400	Bills et al., 1993
Quinine sulfate	Qn	*Micropterus salmoides* (fw)			6800	Post, 1987
Tricaine methanesulfonate	OS	*Morone saxatilis* jv. (fw & bw)			31,500	Hall, 1991
Tricaine methanesulfonate	OS	*Micropterus salmoides* (fw)			63,000	Post, 1987
Tricaine methanesulfonate	OS	*Pagrus major* jv. (sw)			70,000	Oikawa et al., 1994

lr. = larva, jv. = juvenile, yg. = young, ad. = adult.

sw = salt water, fw = fresh water

[a] Other uses indicated by: [Ac] acaricide, [Al] algicide, [An] antibiotic, [Av] avicide, [B] bactericide, [De] defoliant, [Di] disinfectant, [F] fungicide, [I] insecticide, [M] molluscicide, [Ne] nematicide, [NI] nitrification inhibitor, [P] piscicide

[b] Class indicated by: Alc = alcohol, AH = aromatic hydrocarbon, Ald = aldehyde, Alk = alkaloid, Am = amide, An = antibiotic, Ar = arsenical, AU = acylurea, BA = benzeneamine, Be = benzoic, BH = brominated hydrocarbon, BN = benzonitrile, Bo = botanical, BP = bipyridylium, Ca = carbamate, CA = chloramine, CAA = chlorinated aliphatic acid, CH = chlorinated hydrocarbon, CN = chloroaniline, DN = dinitroaniline, F = fungal, FH = fluorinated hydrocarbon, H = hydrocarbon, K = ketone, NF = nitrofuran, OP = organophosphate, OS = organosulfur, OT = organotin, QA = quaternary ammonium, Qn = quinoline, PAH = polycyclic aromatic hydrocarbon, PhA = phthalic acid, PhE = phthalic ester, PiA = picolinic acid, POA = phenoxy acid, PU = phenyl urea, Pr = pyridinone, Py = pyrethroid, S = sulfenimide, TA = triazine, TC = thiocyanate, ThA = thiazole, U = uracil.

[c] Includes no-observable-effect concentrations (NOEC).

[d] Includes effective concentrations (EC).

[e] Includes lethal concentrations (LC).

[r] Discontinued or highly restricted in the U.S.

and Millburn, 1990). They interfere with function of and cause damage to nerves and liver; many are mutagenic and carcinogenic. Organophosphate and carbamate pesticides are less persistent and are not bioconcentrated or biomagnified as much as CHCs. One class of fungicides is the phthalimides, many of which are teratogenic, mutagenic, and carcinogenic. Most herbicides have low toxicity and bioconcentration potential for fish (Perschbacher et al. (1997).

Paraquat is widely used for aquatic (not in U.S.) and terrestrial weed control (for crops and non-crops in U.S.)(Eisler, 1990c). It is nearly insoluble in lipids and is unlikely to bioaccumulate in animals. It is persistent in soils (surface half-life was ~3 weeks; subsurface half-life was years) and sediments but not in water (half-life in fresh and seawater is ~24-36 h). Experimental effects on fish include metabolic disruption and acetylcholinesterase inhibition. Marine fish are unlikely to be seriously affected by typical post-application water concentrations of less than 1 μg/L; however, substantial amounts of paraquat can enter the food chain through bioconcentration in aquatic plants (13,000-1,300,000 μg/kg measured).

Toxaphene, or chlorinated camphene, was one of the most-used insecticides (Eisler and Jacknow, 1985). In 1980, annual world use was about 105,000 mt. Concentrations higher than 0.070 μg/L are toxic to marine animals, and toxaphene was responsible for many fish kills, especially in the late 1970s. In 1986, it was banned in the U.S. Toxaphene is soluble in lipids but relatively insoluble in water. Bioconcentration and biomagnification rates are high in aquatic animals; bioconcentration factors for several fish species ranged from 1270 to 52,000. It is very persistent with a half-life in soils up to 13-14 yr. Trout could not grow in one lake treated with toxaphene as a piscicide at 40 μg/L until 6 yr later when the concentration reached 0.8 μg/L. Other lakes were toxic for 3-4 yr. Over the years, environmental toxaphene residues have diminished; however, similar widescale problems could occur again if new pesticides were not carefully screened before use or if established application restrictions for potentially dangerous ones were exceeded.

Endosulfan, a chlorinated cyclodiene insecticide, is a central nervous system poison used mainly on non-food crops and for wood preservation (Naqvi and Vaishnavi, 1993). It is very toxic to fish and has caused several fish kills in rivers, mainly from spillage. Massive fish kills along 75 miles of the Rhine River in Germany were caused by endosulfan levels up to 5.5 μg/L. Although normal agricultural spraying is not as dangerous as spills, there is danger of toxicity. The half-life is as long as 2 yr in soil and 1-6 months in water.

Carbofuran, a carbamate insecticide, is an acetylcholinesterase inhibitor used in the U.S. mainly against pests of corn, but also alfalfa, peanuts, rice, and sugar cane (Eisler, 1985b). The half-life in water was 1 week at pH 8.0, but 1.9 months at pH 7.0 and 13.3 yr at pH 6.0. It is soluble in water but relatively insoluble in lipids, and significant bioconcentration and biomagnification are unlikely. Some degradation products are carcinogens. Although it is not approved for water application in the U.S., carbofuran has caused fish kills after normal spraying, and aquatic invertebrates are more sensitive than fish. The danger to farmed fish is not great unless spray drifts into water supplies or rearing units.

Fenvalerate is a good example of an insecticide for discussion because it has been widely used on vegetables, fruits, nuts, sugarcane, soybeans, tobacco, cotton, pine trees, poultry, cattle, in buildings, and for mosquitoes (Eisler, 1992b). Like other pyrethroids (more stable synthetic analogs of natural pyrethrins from plants), it is primarily a nerve poison. In 1989, about one-third of the insecticides used in the world were synthetic pyrethroids. One reason for the increasing popularity of fenvalerates and other pyrethroids is their low toxicity to mammals. However, photolysis of fenvalerate by ultraviolet light produced two cyanide compounds that were as much as 250 times more toxic to mice than the parent compound. Fenvalerate is less likely than some insecticides to enter water in runoff because it is quickly adsorbed to soil particles; however, aerial drift and soil erosion could carry it into water bodies. Tests have shown that release from sediments to water can produce concentrations lethal to mysid, palaemonid, and penaeid shrimp. The half-life was 27-34 d in seawater and as long as 6 weeks in estuarine sediments. Fenvalerate is considered supertoxic to aquatic organisms; concentrations in parts per trillion are lethal to many. Crustaceans are the most sensitive, with fish not far behind. Algae, rotifers, and molluscs are less sensitive. Because of its high toxicity, high bioaccumulation, and persistence in sediments, the risks of using fenvalerate in or near wetlands or drainage systems are unacceptable (Eisler, 1992b).

Fenvalerate has been discontinued in the U.S. and has been replaced by newer pyrethroids such as **esfenvalerate** and **cypermethrin**. Although cypermethrin is highly toxic to fish, it has low water solubility and quickly binds to particles, sediment, and vegetation, so is unlikely to be dangerous to pond fish in normal usage (Huckle and Millburn, 1990).

Chitin formation inhibitors such as **diflubenzuron** and **insect growth regulators** can affect other arthropods such as cladocerans and copepods, which sometimes are important larval fish foods (Eisler, 1992a; Chambers, 1994). Diflubenzuron is bioconcentrated by aquatic algae and has often caused irreversible reduction of crustacean and aquatic insect populations at normal application rates. Application of diflubenzuron to water can result in concentrations as high as 16 µg/L, and it can harm crustaceans at 0.1 µg/L.

Other pollutants.—Cyanide in Algeciras Bay, Spain (30-480 µg/L), in combination with other contaminants, probably contributed to disease, growth reduction, and mortality of caged gilthead sea bream (Tarazona et al., 1991).

Hydrocarbon pollution can occur from spills, natural seepage, incomplete combustion, and biogenic processes (e.g., from phytoplankton) (GESAMP, 1993).

Petroleum components can enter fish through the skin, gills, intestines, and other membranes (Samiullah, 1985). Water solubility of crude oil fractions range from <0.00001 mg/L for benzo(a)pyrene to 82 mg/L for phenol (Overton et al., 1994). Toxic effects include abnormal behavior, deformities (e.g., scoliosis), damage to the skin, gills, fins, kidneys, heart, eyes, nerves, muscles, liver, pancreas, and intestines, carcinogenicity, and mortality (Malins and Hodgins, 1981; Samiullah, 1985; Falk-Petersen and Kjorsvik, 1987; Booman et al., 1995). Viability of fish eggs and growth of larvae can be reduced in the range 0.1-1.0 mg/L petroleum hydrocarbons (Samiullah, 1985).

Even **vegetable oils** (e.g., sunflower), if excessive, can be lethal to marine organisms (Mudge et al., 1995).

Many **toxic organic solvents** such as methanol, ethanol, hexane, benzene, toluene, phenols, xylene, ketones, carbon disulfide, ethylene glycol ethers, and chlorinated hydrocarbons adversely affect the central nervous system (CNS), and some are carcinogens (Hume and Ho, 1994). Methyl ethyl ketone (2-butanone) and methyl-*n*-butyl ketone can potentiate neurotoxicity of other solvents. Trichloromethane (chloroform), which can result from chlorination of water containing organic compounds, is embryotoxic and teratogenic, can damage liver, kidneys, and heart, and is thought to be a carcinogen. Tetrachloromethane (carbon tetrachloride), which has been used extensively for dry cleaning and degreasing, can depress the CNS, damage liver and kidneys, and is a carcinogen. Dichloromethane (methylene chloride) is much used as an aerosol propellant, paint remover, and degreaser, and is likely to be a carcinogen. Trichloroethanes, which are used as dry cleaners, degreasers, and propellants, can depress the CNS, damage liver and kidneys, and could be carcinogens. Trichloroethylene, which has been used as a dry cleaner and degreaser and to decaffeinate coffee, can depress the CNS, damage liver and kidneys, and is a carcinogen. Tetrachloroethylene, which is used as a dry cleaner, degreaser, grain fumigant, and veterinary antihelmintic, depresses the CNS and is a carcinogen. Phenols (10-985 µg/L) probably contributed to problems of gilthead sea bream raised in Algeciras Bay (Tarazona et al., 1991).

Polycyclic aromatic hydrocarbons (PAH) are produced by iron, petroleum, and coal processing, power plants, burning of garbage and brush, and internal combustion engines (Eisler, 1987b). In the late 1970s to early 1980s, aquatic environments received about

230,000 mt of PAH each year. Relatively clean surface waters might contain 0.05-0.25 μg/L total PAHs, but the Rhine River in Germany had 1.12 μg/L and the Thames in England had up to 1.33 μg/L. Effects on fish include weakness, reduced growth, deformed larvae, cytotoxicity, mutagenicity, and cancers (possibly of skin, thyroid, liver, gonad). Benzo(a)pyrene (one of the most toxic of PAHs, Table 3-35) reduced hatching and caused larval deformity in sand tonguefish at 0.1 μg/L. The legal limit for seawater in Florida is 0.031 μg/L average PAHs.

Polychlorinated biphenyls (PCBs) are synthetic halogenated aromatic hydrocarbons, numbering 209, which were used mainly for insulating and cooling electrical transformers and capacitors but also were in hydraulic fluids, heat transfer fluids, plasticizers, adhesives, lubricants, copy papers, flame retardants, extenders, and diluents (Eisler, 1986c; Hansen, 1994). The manufacture and nearly all use of PCBs were banned in the U.S. in 1979. About 1.2 million mt have been produced altogether (Hansen, 1994). PCBs are mostly insoluble in water but soluble in lipids, leading to bioconcentration and biomagnification in aquatic food chains. Concentrations as low as 0.006 μg/L can be harmful to aquatic species. Effects on fish include ion imbalance, hyperglycemia, biochemical alterations, anemia, tetany, liver damage, deformed larvae, reduced growth, histopathology (at 0.5 μg/L), scoliosis, lordosis, mutagenicity, carcinogenicity, and mortality. For sensitive fish, residues of 300 μg/kg total PCBs in eggs and 400 μg/kg in the whole body can be toxic. Because fish and copepods eliminate PCBs via feces and eggs, lowered survival of eggs produced after PCB-exposure of the female is not surprising. The legal limits for seawater in Florida are 0.000045 μg/L 24-h-average PCBs and 0.03 μg/L maximum. Eisler (1986c) suggested that 0.014 μg/L maximum might be more appropriate.

Surfactants can be present in laundry detergents, dishwashing detergents, fabric softeners, bar soaps, shampoos, antistatic agents, food emulsifiers, textiles, leather, mining chemicals, oilfield chemicals, and paints (Lewis, 1992). In 1987, more than 2.5 million mt were used in the U.S. and western Europe. Surfactants tend to do most of their harm by interacting with other toxicants, especially oils. Gill damage and reduced survival and growth are common effects.

TOXICANTS ACCIDENTALLY INTRODUCED BY THE CULTURIST OR THE BUILDING (Table 3-35)

In some situations, many of the materials used in fish culture facilities can be toxic to the fish.

Treated wood.—Because of dangers to humans or the environment, many wood preservatives have been restricted or discontinued in the U.S. (Ware, 1994). These include (with effective year) arsenic pentoxide, bendiocarb, carbofuran, chromic acid, coal tar, coal tar creosote (1989), creosote oil, methyl isothiocyanate, pentachlorophenol (1987), sodium pentachlorophenol, potassium pentachlorophenate (1989), sodium arsenate (1989), sodium dichromate (1992), sodium pyroarsenate (1991). Most of this action was taken only recently, and some of these materials are still in use, often near or in water bodies. Treated lumber often is utilized in hatcheries and rearing enclosures. Certain, but undocumented, cases of toxicity to cultured marine fish have been reported. While wood preservatives are very useful, they should not be allowed to contaminate fish culture water.

Pentachlorophenol (PCP) disrupts energy metabolism in fish (Eisler, 1989c). A typical acute toxicity sequence includes rapid swimming, hyperventilation, equilibrium loss, sinking, and mortality. It is teratogenic to animals, and possibly mutagenic and carcinogenic. Half-life in water was 0.15-15 d. PCP can be bioconcentrated by fish up to 10,000 times. Experimental bioconcentration of PCP in rainbow trout was as high as 572 times and the half-life up to 7 d. PCP applications and spills have caused large fish kills in several countries. Fish surviving some spills had higher than background concentrations of PCP for 6-10 months. The high toxicity of technical PCP results mainly from several other CHC impurities (including chlorophenols, dibenzofurans, dioxins, phenoxyphenols, and hexachlorobenzene). PCP has low water solubility but high lipid solubility. Sodium and potassium salts are water soluble. Accumulated PCP is one cause of off-flavor in fish.

Metals.—Zinc can enter water from galvanized steel or iron used in **construction** of buildings or enclosures. A host of toxic metals can be derived from various alloys in pumps, other machinery, pipes, etc. It is best to keep all metals away from fish culture water, except high grade titanium. Even good stainless steel (type 316) contains molybdenum and is not corrosion-proof in seawater (Wheaton, 1977; Hawkins and Lloyd, 1981).

We have found metal particles from **air compressors** migrating through pipelines toward fish tanks. Even if the particles themselves do not reach the culture water, condensation within the pipe can dissolve some of the metal and transport it. Good equipment, good system design, and air filters can prevent problems.

Since the early 1900s, copper **paints** have been used to reduce growth of fouling organisms on boats (Langston, 1991). For a while, until about 1970, mercury, arsenic, and lead also were used in some paints. Beginning in 1970, tributyltin compounds (sometimes with copper) have been used on boats and aquaculture enclosures. TBT also has been used in industrial cooling systems and paper mills (George, 1991). TBT is very toxic to aquatic organisms and is somewhat persistent. Reported inshore concentrations range from a few ng/L away from boats to more than 600 ng/L (as tin) close to marinas (Langston, 1991).

Concrete is very alkaline when fresh and might require soaking for a few months if sensitive species are to be raised in water contacting it.

Polyester resins and **gel coats** used in making fiberglass and wooden tanks contain mainly polyester resin and styrene monomer. If the resins are not fully cured, styrene released into water can be highly toxic to fish and invertebrates. A third contamination source is the mold wax, which makes it possible to remove the hardened tank from the mold and also helps to seal pores in the gel coat. We always soak and scrub new tanks thoroughly and sometimes steam-clean them to remove at least some of the wax; however, removing too much material can allow seepage of water into the tank wall, possibly resulting in subsequent delamination. We also try to leave new tanks in full sunlight for at least a few months before using them for sensitive fish. But even 1-yr-old weathered tanks can still emit styrene. This is obvious when the tanks are covered with a plastic sheet for a few hours to concentrate the fumes enough to be smelled. Gel coat also contains less than 1 mg/L each of arsenic, cadmium, hexavalent chromium, and lead, but these are tightly bound and probably pose no danger.

Other Plastics.—**Polyvinyl chloride (PVC)** can contain small amounts of lead. When received, PVC piping and other system components can be contaminated with rust or grease, mostly on the outside. Surfaces that will contact culture water (especially for larvae) should be clean. Rust can be removed by sanding. Grease can be removed with acetone and paper towels. Ordinary soap should be avoided. Acetone also will remove labels printed on the pipe, thus improving its appearance. Soaking, scrubbing, and rinsing in hot water finishes the process. PVC pipelines should be flushed with water for 2 weeks before using them to supply sensitive organisms.

Gaskets supplied with PVC bulkhead fittings often are made of **neoprene** not approved for contact with food or drinking water, but these can be replaced with **silicone** gaskets.

PVC cement fumes are extremely toxic to some fish and have been measured in water coming from pipes that were glued 10 yr earlier. Wang and Bricker (1979) measured 4.5 mg/L methyl ethyl ketone and 13 mg/L tetrahydrofuran in a laboratory freshwater supply 6 months after the pipes were glued. Standard industry practice is to let glued joints dry for up to 48 h or more before putting water through the pipe. If water contacts the cement before it is completely dry, the skin becomes less permeable and the solvent will take much longer to be released. We sometimes use a hot air gun (set on medium) to speed up drying, but only after several hours of air drying and carefully so that the pipe and glue do not break down and become weak or decompose into more toxic substances. Gluing PVC near tanks of sensitive organisms can stress or kill them, because the fumes can dissolve from air into water in seconds.

Once, while determining hatching rate of grouper eggs using very clean water, I made the mistake of sealing the beakers in a **polystyrene** foam box to protect them from temperature changes. The effect was dramatic. Hatching rate was near 100%, but immediately after hatching, each larva curled up and died. The plastic odor in the box (probably styrene) was very strong. Other larvae from the same batch of eggs, but hatched in other containers, survived well.

Biofilter materials—Generally, the harder the plastic the more inert it probably is. Softer plastics tend to have more **plasticizers**, such as phthalates, in them. We have measured eight different phthalate esters (965 μg/L total) in water from a biofilter containing plastic bags of gravel and four phthalate esters (388 μg/L total) in completely filtered water (same system) as it entered fish tanks. Phthalates can be toxic to fish at only a few ppm. The softer a plastic, the more plasticizers it is likely to have. Vinyl tubing should be avoided as much as possible. Good-quality airline tubing does not seem to be harmful if only moderate amounts are submerged. Some airline tubing has been reported to contain toxic amounts of lead.

Persons tending marine fish larvae should be extremely careful to avoid poisoning the fish with residues of substances such as **soap**, **chlorine**, **oil**, **gasoline**, and **pesticides** clinging to their hands, nets, or other implements. We rinse our hands very well with hot unchlorinated water before beginning the daily routine. Handling of toxic substances is done by personnel not working with larvae or by culturists only after all larviculture tasks are finished. Hands are not put in the rearing water, except during harvest. In some hatcheries, personnel wear surgical gloves and hairnets when working close to the larval tanks.

Because the pH of microalgae could be high from photosynthesis or low from addition of carbon dioxide (its temperature also could be different), addition to culture water should be done gradually and gently (and not on top of the larvae). If circulation is slow, algae should be added all around the tank; if circulation is fast enough to spread the algae, it can be slowly added at one point (e.g., above an airstone). The algae should have grown to a stage at which it has used up the **added nutrients,** which might have a toxic effect on sensitive larvae.

If hot water is not enough to clean system components, **detergents** such as Alconox® that are used for critical cleaning of laboratory glassware might be safe if thoroughly rinsed off.

Chlorine is used in many hatcheries for disinfection, but usually not when living cultured organisms in uncovered tanks are in the vicinity. Because chlorine is heavier than air, it can hang in a layer near the floor, drift to a lower elevation, or travel down one standpipe and up another, thence directly into tanks in use or into the air system. Great caution should be used with chlorine.

Especially in the winter, **condensation** inside a hatchery can generate a large amount of water that falls directly into larval tanks. Because the water can carry toxicants from the roof or ceiling and its impact can cause injury, precautions should be taken to prevent this. The least toxic materials possible should be used above fish tanks, and if necessary a temporary or permanent "umbrella" can be placed above the tanks.

Prevention of problems.—The utmost care should be taken with any materials placed in a larval rearing tank or its water system. Plastics always should be soaked and pipelines flushed as long as is practical before use. Larger fish are more tolerant but should not be abused.

THERAPEUTIC COMPOUNDS (Table 3-35)
Most substances used to treat fish will also be toxic to the fish if too much is used. Overdoses can cause anemia, damage to liver, kidneys, and nerves, deformities, sores, reduced growth, and mortality. For example, 100 ppm **formalin** might be used to treat fish for parasites, but 30,000 ppm formalin is used to preserve fish. A few hundred ppm formalin can easily be lethal. Excessive use of **antibiotics** can give a competitive advantage to resistant strains of bacteria, allowing them to become more abundant. Treatment with **copper** can make fish more susceptible to vibriosis (Spotte, 1992).

CONTAMINANT EFFECTS ON BENEFICIAL BACTERIA
Many substances can harm beneficial microorganisms like probiotic and nitrifying bacteria.

Calcium, magnesium, manganese, molybdenum, and apparently potassium were necessary for optimal functioning of nitrifying bacteria (Lai and Klontz, 1980). Chloride was beneficial, phosphate seemed to be so, and sulfate did not interfere. Copper, cobalt, and apparently zinc (481 µg/L) inhibited the nitrite to nitrate conversion. Ferric iron (10 mg/L) enhanced conversion of ammonia to nitrite but interfered with conversion of nitrite to nitrate.

The metals copper, zinc, mercury, nickel, silver, and chromium interfered with bacterial nitrification (Painter, 1970). The organic compounds thiourea, allyl-thiourea, 8-hydroxyquinoline, salicyladoxime, histidine, amino acids, mercaptobenzothiazole, perchloroethylene, trichloroethylene, abietic acid, and cyanic acid also interfered.

At typical therapeutic concentrations, six of seven chemicals (13.3 mg/L chloramphenicol, 1.2 mg/L cupric sulfate, 5.3 mg/L gentamycin sulfate, 8.0 mg/L methylene blue, 66.7 mg/L neomycin sulfate, and 0.1 mg/L nifurpirinol, but not 12.0 mg/L quinacrine hydrochloride) inhibited bacterial nitrification in a recirculating system (Bower and Turner, 1982). Methylene blue at 5 mg/L (Collins et al., 1975) and erythromycin at 50 mg/L (Collins et al., 1976) had similar effects.

TESTING AND ELECTRONIC MONITORING OF WATER QUALITY
Measurement of routine water quality factors can be made with simple instruments and test kits. Electronic monitoring systems will record data and automatically make telephone calls if factors such as temperature, oxygen, salinity, and water chemistry move beyond specified limits. Agricultural and industrial contaminants require more sophisticated analyses.

PREDATORS
Predaceous **cnidarians** (hydroids, jellyfish, anemones) and **crustaceans** (isopods, copepods) can be introduced with eggs, water, or live food. Hydroids or jellyfish can build up unnoticed, then suddenly appear by the thousands and sting all the larvae to death very quickly. Hydroids growing harmlessly on the tank bottom can be stirred into the water column or rise to the surface when water circulation is increased and destroy a tank full of larvae within minutes. To avoid such catastrophes, eggs should be washed well, water filtered well, and food cultures kept clean (and also sieved and rinsed before feeding).

Many **aquatic insects** can fly into rearing units and are especially attracted to electric lights at night. The water boatmen (family Corixidae) often are abundant around aquaculture facilities. Those that have invaded our hatchery tanks have small mouthparts and non-raptorial appendages. They feed off of solid surfaces in the tanks, apparently on some combination of epifauna, epiflora, and detritus. Typically, they are found clinging to the bottom or sides of the tank. When they swim to the surface to collect a bubble of air for breathing, the fish are briefly disturbed but quickly return to normal activities. Those water boatmen keep bottom sediments stirred, which probably is beneficial. Some water boatmen are said to eat insect and possibly crustacean larvae and could be dangerous to larval fish. Back swimmers (family Notonectidae) suck the blood from small animals, including fish, and will bite humans. They have long oar-like hind legs and swim on their backs. The predaceous diving beetles (Dytiscidae) typically hang at the surface with their heads down. Their larvae, so fierce they are called "water tigers," eat fish and will attack animals larger than themselves. Giant water bugs (family Belostomatidae) suck the blood of small fish and other animals and will bite humans. Water scorpions (family Nepidae) do not swim fast but could catch small larval fish with their raptorial front legs. Whirligig beetles (family Gyrinidae), which are seen continuously gyrating across the surface, are predacous as adults and larvae and potentially dangerous to larval fish. Openings in hatchery walls should be screened to exclude insects. Sufficient drying and liming of ponds will get rid of most insect eggs and larvae. With good timing and luck, enough larval fish stocked in a recently-filled pond might grow up before predaceous insect populations build up. Unfortunately, during peak insect production, insecticides such as diesel oil or fenthion seem necessary, but they can harm both fish and their prey as well as contaminate the environment and should be avoided.

In large numbers, wading and diving **birds** can decimate juveniles in nursery ponds, and skimming birds could consume a fair number of larvae. Steeply sloping pond edges inhibit waders, and bird nets help keep divers and skimmers out. Predatory **reptiles** such as alligators and **mammals** such as otters can be excluded with good fences. Sometimes pyrotechnics are used to scare birds and mammals away.

NOXIOUS MICROORGANISMS

The toxic effects of certain **microalgae** include paralytic shellfish poisoning, diarrhetic shellfish poisoning, amnesic shellfish poisoning, ciguatera, tetrodotoxin poisoning, fish and shellfish kills, and fish and shellfish tainting. More than 300 species of dinoflagellates, chloromonads, chrysophytes, prymnesioids, diatoms, and cyanophytes have been identified as toxic (Corrales and Maclean, 1995).

Toxins produced by **red** (or **brown**) **tide** organisms can cause heavy mortality or contamination of cultured fish and shellfish (Holmes and Lewis, 1994; Greece, Satsmadjis and Friligos, 1983; Scandinavia, Richardson and Kullenberg, 1987; Hong Kong, Wong and Wu, 1987; Texas, DeYoe and Suttle, 1994; Florida, Buck and Pierce, 1989; Maine, Heinig and Campbell, 1992; Australia, etc.). A number of species of dinoflagellate have been shown to cause red tides, with ciliates, cryptophytes, and cyanophytes also appearing. The dinoflagellate *Pfiesteria piscicida* attacks fish directly (Burkholder et al. 1992; Noga et al., 1993; Glasgow et al., 1995; Burkholder and Glasgow, 1997). Proliferation of pathogenic bacteria can occur along with red tides. Nutrient enrichment combined with favorable physical conditions lead to these blooms. Caged fish are the most vulnerable, but fish in ponds and tanks are not completely safe. Preventive measures include site choice, monitoring, deep cages, barriers around cages, pumping clean water into cages, oxygenation of sediment in ponds or below cages, low fish density to reduce stress, and ability to move fish to a safe place (Corrales and Maclean, 1995). Control measures directed against the algae include skimming, ultrasound, flocculation, and algicides. (also see pages 508-509)

Similar toxins produced by benthic dinoflagellates and concentrated in the food chain cause **ciguatera poisoning** (e.g., Holmes et al., 1991; Morton et al., 1992; Lewis et al., 1994; Corrales and Maclean, 1995; Levine, 1995). Barracuda and many other marine fish (including some groupers) in certain areas (e.g., with "copper bottom" in the Caribbean) are known to be toxic. Farming those species is good assurance that they are free from ciguatera, because their diet is not natural.

Off Scotland, a heavy layer of diatoms **smothered** large numbers of herring eggs in April 1990 (Morrison et al., 1991). **Oxygen depletion** by a silicoflagellate bloom was the most likely cause of mass mortality of caged rainbow trout in Denmark during May 1983

(Henriksen et al., 1993). At least since 1961, blooms of diatoms in British Columbia, Washington, and Scotland have been recognized as killing caged salmon and other fish by **damaging fish gills** with their spines (Harrison et al., 1993).

Off-flavor sometimes results from strong-tasting feed components, contaminants such as PCP, or oxidative rancidity of the processed fish (Tucker and Martin, 1991). More frequently, off-flavor is caused by accumulation of metabolites such as geosmin (muddy-musty flavor) and 2-methylisoborneol (MIB, woody flavor), which are produced by actinomycetes and cyanobacteria (van der Ploeg et al., 1992; van der Ploeg and Tucker, 1994). Treatment of ponds with copper sulfate (not recommended) and depuration of fish in clean water before processing are two common methods of combating this type of off-flavor.

ENVIRONMENTAL PROBLEMS CAUSED BY AQUACULTURE

Sources of environmental degradation from aquaculture include **obstruction**, **pathogens**, **uneaten food**, **excretion**, **egestion**, **fertilizers**, **therapeutants**, **disinfectants**, and **other chemicals** (Brown et al., 1987; Gowen and Bradbury, 1987; Beveridge et al., 1991; Iwama, 1991; Phillips et al., 1991; Weston, 1991; Hargrave et al., 1993; Jenson, 1993; Morimura, 1993; Pullin et al., 1993; Landesman, 1994; Levings, 1994; Rosenthal, 1994; Stewart, 1994).

Because cages take up space, they can obstruct not only boat traffic but also water currents and migrating animals. Release of therapeutic compounds can lead to toxic effects on non-target organisms, bioaccumulation, and antibiotic resistance of fish and human pathogens.

On a typical salmonid cage farm, 20% of the feed is uneaten (Gowen and Bradbury, 1987). Consumed carbon is partitioned into 50% for respiration (CO_2), 30% egested, and 20% for growth. Consumed nitrogen is partitioned into 65% excreted, 10% egested, and 25% for growth. Thus, large proportions of the carbon and nitrogen are released into the environment. Representative values for total (ultimate) biochemical oxygen demand (BOD) in water were 660 mg O_2/g shrimp feed and 330-800 mg O_2/g trout feed (Chieng et al., 1989). Smaller-size feeds break down and pollute water more quickly than large ones.

Although phosphorus often is the limiting nutrient for phytoplankton in fresh water and sometimes in low salinity coastal areas, nitrogen is much more important in most coastal marine areas (Gowen and Bradbury, 1987). Nutrient enrichment could lead to phytoplankton blooms (possibly including red or brown tides), with varying degrees of oxygen depletion. Deposition of uneaten feed and feces on the bottom can cause sediment to become anoxic. One result is the generation of toxic gases—in typical percentages of about 70% methane, 30% carbon dioxide, and 2% hydrogen sulfide (Stewart, 1994). When released into the water, these gases (mainly H_2S) can be toxic to cultured and wild fish, but the latter usually can escape. Another effect is the depletion of benthic organisms.

In typical shallow-water situations, the bottom directly below established fish cages has no or very few macrobenthic organisms. A short distance away, benthic diversity is lower than normal, but biomass ranges from lower to higher than normal (because of increased nutrition and reduced competition). Outside the transitional zone, normal benthic communities are found.

Ye et al. (1991) reported this type of zonation around three 600-m^3 (9 x 11 x 6 m) cages of salmonids located in a 12-m deep coastal bay in **Tasmania**. About 79% of the organic carbon in sediments below the cages came from the feed; there, a single capitellid worm species was dominant. At 10 m, 75% of sediment carbon was from the feed and at 30 m, 63% was; between 10 m and 30 m, benthic diversity was intermediate. At 30 m and beyond, benthic communities were normal, but even at 60 m, 42% of sediment carbon was from the feed.

Tsutsumi (1995) studied severe degradation of a 10-12 m deep bay with a 5-7 m deep connection to the ocean in **Japan**. Caged fish farming began in 1973. At least since 1977, oxygen depletion of bottom water, hydrogen sulfide buildup in sediments, and elimination of nearly all benthic fauna in the bay occurred every summer. In 1966, 77-98% of benthic species at seven stations consisted of 16 bivalve and 21 gastropod molluscs. In 1989, 49-96% were polychaetes, and a single mollusc species remained. In Japan, movement of hypoxic bottom water into fish culture zones or other sensitive areas is called a "blue tide."

Wu et al. (1994a) reported low oxygen, high sulfide, and elimination or depletion of macrobenthos within fish culture zones near **Hong Kong**. Problems were more severe in two bays 3-4 m and 3-7 m deep than in two bays 5-8 m and 5-13 m deep. The shallowest site had a history of large fish kills.

Holmer and Kristensen (1992) studied a rainbow trout cage farm in a shallow brackishwater fjord in **Denmark**. Depth was only 5 m, with 1 m clearance between cages and the bottom. Below the cages, sulfide was high and no macrofauna existed, but 30 m away, benthic communities were normal.

At two moderately large cage farms for European sea bass and gilthead sea bream in the Saronikos Gulf (**Greece**) in 15-35 m (200 mt biomass in 11,000 m^3) and 30-70 m depths (350 mt biomass in 20,000 m^3), water quality was acceptable (3.3-8.2 mg/L DO, 0-1.6 μg/L NH_3-N, 0-20 μg/L NO_2-N) and parasite infestation was low (Papoutsoglou et al., 1996). The benthic community below the shallower farm was normal, and abundant wild fish were present beneath the cages, including white sea bream and thousands of mullet, probably attracted by the cover and wasted feed. It was suggested that reduction of caged fish density from the usual 16 kg/m^3 could raise oxygen and reduce ammonia.

It is best to place fish cages in areas with **good circulation**, **good flushing**, and **at least 10-15 m depth** (Gowen and Bradbury, 1987). Adequate circulation will carry wastes from cages and bring in oxygenated water. In well-flushed areas, environmental degradation is milder and more localized than at sites with less movement. In deeper water, sedimenting wastes are at a safe distance from the fish. **Feeds** that are **well utilized** by the fish (high feed efficiency = low feed conversion ratio) should be employed, and **feed wastage** should be **minimized**.

In tanks and ponds, sufficient **circulation**, **aeration**, and **water exchange** should be maintained for health of the fish. Regular drying or removal of pond sediment will reduce hydrogen sulfide, ammonia, and BOD. Any effluent should be treated before release into the environment to eliminate any pathogens and to reduce solids, BOD, and nitrogen.

Levings et al. (1995) reviewed the legal requirements for locating salmon farms in British Columbia, Washington, New Brunswick, Maine, Iceland, Scotland, Ireland, and Norway and suggested improvements for permitting procedures.

SUMMARY

Good eggs, good rearing conditions, and good nutrition are necessary for producing good-quality healthy marine fish in reasonable numbers. Appropriate tank color, lighting, and water circulation are especially important for larvae. Rearing larvae in green water improves the health of many species. For some, the water surface should be kept clear so that larvae can swim, feed, and inflate their gas bladders without interference. For all stages, temperature, salinity, dissolved oxygen, and water chemistry characteristics must be in suitable ranges. Sufficient hardness and chlorinity are important for marine fish raised in fresh water. Accumulation of waste products and introduced toxicants must be minimized.

Water quality usually is more critical at higher temperatures and salinities, particularly because dissolved oxygen capacity is lower and ammonia can be more toxic. The proportion of free ammonia increases with the higher pH of seawater. Deterioration of water quality (caused by overfeeding or contaminants) that might go unnoticed in coolwater fish rearing systems can be catastrophic in warmwater systems, especially seawater ones.

4 WATER SOURCES

CRITERIA

Besides being the appropriate habitat for aquatic organisms, the most important purpose of culture water is to flush wastes from the rearing units before they reach harmful concentrations. If animals are dense and water flow is high, incoming water also helps maintain dissolved oxygen.

If characteristics of the water supply are very different from the required characteristics, the cost of water treatment can be excessive. Because of other considerations (e.g, land availability, distance from suppliers, distance to markets), compromises often are made in selecting the facility site and water source.

Construction and Maintenance Costs

Initial costs can range from minimal for a low-capacity shallow water intake in a protected area to substantial for high-capacity complicated structures built in high-energy marine environments.

Capacity

Enough water should be available for present, future, and unexpected needs. Flow should be sufficient for filling treatment systems and rearing units without unusually long delays.

Temperature

Except in recirculating systems, heating and cooling of large volumes of water can be very expensive. For large flow-through systems to be economical, source water temperature in each season should be close to the required temperature. Usually heating is cheaper and less complicated than cooling.

Quality

Salinity of 30-32 ppt is necessary to float the eggs of many marine species. Good eggs usually hatch well even if they are not floating, but could be more susceptible to hypoxia and bacterial infection near the bottom. Larvae of many marine species tolerate lower salinities (e.g., 20-30 ppt), and juveniles of many can be raised in fresh water. The osmotic concentration of fish blood is similar to that of one-third strength seawater. Theoretically, the energy requirement for osmoregulation should be reduced under isosmotic conditions. Because of other factors, some species, as well as eggs and larvae, usually are not healthy at 12 ppt; however, other species, usually as juveniles and adults, have better growth efficiency (lower feed conversion ratio) at reduced salinity.

If the source salinity is lower than 32 ppt, artificial sea salts could be added to the incubation water. If salinity is higher than optimal for any stage, fresh water could be added, if economical to do so. Reverse osmosis processors can produce brine or fresh water to be used for raising or lowering salinity.

Hardness.—If fresh water is to be used for dilution or rearing, its hardness should be appropriate and it should be free of pollutants and pathogens. I have raised common snook, sunshine bass, sheepshead, and spotted seatrout in fresh artesian water from a 210-m deep well, that had a hardness of >250 mg/L. For quality, usually well water is best, rain water second, and surface water a distant third.

pH tends to be low in salt well water and high in water with an algal bloom. Sufficient degassing and aeration will correct this. Water of low pH, especially well water, tends to contain more dissolved metals, particularly iron and manganese, which should be checked for. At high pH, ammonia might be more toxic.

Oxygen could be very low (e.g., in well water) or high (e.g., from photosynthesis). Sufficient exposure and aeration should at least equilibrate dissolved oxygen with atmospheric oxygen.

Gas supersaturation.—This can occur when pressure is reduced or temperature increased. Three common reasons are: 1) water is taken from a well, 2) saturated water is taken from a depth greater than 1 m to the surface, 3) air is pulled with water into a pump and dissolved under pressure. Photosynthesis can result in oxygen supersaturation without a pressure change. Because supersaturation can easily occur, degasification should be part of any fish culture water treatment system.

Hydrogen sulfide can be oxidized or dissipated in a degassing tower or in a series of tanks with heavy aeration. **Carbon dioxide** and **methane** also can be driven off by aeration.

Ammonia, nitrites, and nitrates should be within acceptable limits for culture purposes or low enough so that they can be effectively and economically reduced. Aeration does not remove much ammonia. Ammonia and nitrites can be removed by nitrifying bacteria, but a large and/or very efficient system might be needed. Nitrates can be removed by seaweeds, but with a danger of metabolites and decomposition products being added to the water.

Heavy metals, **pesticides**, **hydrocarbons**, and other **pollutants** are difficult to remove and should not be present in harmful concentrations. If justified economically, activated carbon can be placed in the system as a precaution, but using contaminated water is dangerous.

Trace nutrients in the water are necessary for plants and some are useful to animals. Artificial seawater and well water might not have optimal concentrations and ratios. Chelators also help plants utilize nutrients.

Suspended solids can irritate or injure fish gills and deplete oxygen. Many fish seem to tolerate suspended solids up to 100 mg/L (Beveridge, 1987). **Dissolved organic matter** exerts a chemical oxygen demand and could contain toxins.

Pathogens.—A wider variety of potentially harmful organisms is present in surface water; however, even ground water can contain bacteria and viruses. Also, ground water held in open tanks can easily become contaminated with pathogens and even predators. Eutrophic areas seem to have more disease outbreaks. If culture water is not treated (e.g., for cages or pens) it is a good idea to check wild fish living there for diseases, especially fish related to the cultured ones.

Harmful algae.—If dense enough, diatoms can injure fish gills; blooms are most likely in temperate to subtropical areas during the cooler seasons (Beveridge, 1987). Several species of dinoflagellate are toxic to fish (red tides); blooms are more likely at higher temperatures (in temperate areas during the summer or in subtropical areas). Low water exchange, high nutrients (especially nitrates but sometimes phosphates), and possibly chelators promote algal blooms.

Fouling organisms attached to the intake structure, pipelines, and treatment systems can decrease both flow and quality of water.

MAJOR TYPES

The lower the elevation of the pump and the shorter the suction line, the less a given pump will be strained. Usually, a dual system is installed so that one can be used while the other is left off or filled with fresh water for several days and then flushed to reduce growth of fouling organisms inside the pipes. For high usage, several duplicate systems would allow more flexibility for cleaning and maintenance schedules. Constant high flow (>2.5 m/s) will inhibit attachment of fouling organisms (Spotte, 1992). Sometimes chlorine is used to kill fouling organisms and hydrochloric acid to dissolve calcareous deposits, but the danger to cultured organisms and the environment could be high.

The possibility of direct or indirect storm damage is a major consideration. In high-energy sites, scour by waves or currents can remove considerable amounts of sediment in a short time. For example, I have documented an increase in water depth (by removal of sand) from 15 m to 22 m in 24 h by swell from a distant hurricane at a study site about 1.5 km off the North Carolina coast.

Six possible methods of obtaining salt water are compared by the main criteria in Table 4-1. Further discussion can be found in Wheaton (1977), Cansdale (1981), Beveridge (1987), Huguenin and Colt (1989), Petit (1990), Bose et al. (1991), and Spotte (1992).

Open Intake in Shallow Water

These should be outside of popular fishing areas and deep enough that they are not hit by boats. A moderate current across the intake is necessary to prevent stagnation. Low sedimentation and fouling rates are helpful. A structure is needed to support and protect the end of the pipeline and to strain the water. Usually, this is some type of box.

Advantages.—Likely to be simple and inexpensive.

Disadvantages.—The intake can become obstructed by debris and fouling and require constant maintenance. It could be vulnerable to wave damage. Filtration at the intake is partial (only large holes can be used). Water temperature and salinity might vary more than with a deeper intake. Fouling and sedimentation usually are problems.

Open Intake in Deep Water

Water of more nearly constant quality might be obtained from a greater depth.

Advantages.—Probably less debris, fouling, and chance of wave damage and pollution than with shallow intakes. Temperature and salinity are likely to be more constant.

Disadvantages.—The pipeline and maybe the pump are likely to be more expensive. The water might be colder. It might contain less oxygen. Depending on depth, access to the intake structure could be difficult. Fouling and sedimentation might still be problems.

Buried Intake

Suitable locations for buried intakes can be easier to find than locations for wells. In this system, the end of the pipe usually is buried near the center of a hole filled with coarse filter sand or fine gravel, and the water travels vertically from above through the sand, which removes solids and could perform some biofiltration of ammonia and nitrite as well. After enough water passes through the sand and adjacent sediments, the intake area becomes aerobic, with a decrease in ammonia, hydrogen sulfide, carbon dioxide, BOD, and heavy metals.

For each m^3/h of water required, a surface area of about 3 m^2 is needed (Cansdale, 1981). Intake points should be spaced at intervals of at least 7-10 m (Huguenin and Colt, 1989).

Hettler (1971) described and illustrated a dual system for drawing water from a shallow North Carolina estuary. Water passed through 5-cm holes into a concrete chamber containing 5 tons of filter sand, then was pumped to shore. Cansdale (1981) described two types of buried intake, one an inverted box and the other a vertical standard well point. Huguenin and Colt (1989) described those two types plus a third with horizontal pipe collectors.

Advantages.—Part or all of the system can be buried, which helps to keep temperature constant and gives more protection from waves. Debris and fouling are minimal. Initial filtration is good. Biofiltration occurs at the intake if aerobic conditions are maintained. The intake is more isolated from and less likely to be contaminated by oil spills.

Disadvantages.—Likely to be expensive and difficult to modify. The site should be protected from waves and strong currents, which might erode the gravel or sand bed. If water is drawn from anaerobic areas, it is likely to contain hydrogen sulfide, ammonia, and/or methane, and possibly heavy metals.

Saltwater Well
With luck, an aquifer of good quality salt water could be found (usually not far from the ocean). A true aquifer lies between two impermeable layers and is replenished horizontally, and the water is filtered much more than that from a buried intake. Well intake structures range from completely buried standard well points (straining devices not much wider than the well pipe) to large collecting chambers open to the surface and drawing from perforated pipes that radiate outward (Spotte, 1992).

Advantages.—Once built and operating, this can be a very convenient water source. Debris and fouling are not present. Wave and current damage are unlikely except during storms. Pathogens are not likely (but could be introduced once the water is exposed).

Disadvantages.—The cost could be lower or higher than the alternatives. The well site has to be chosen carefully to avoid pollution and freshwater intrusion. Hydrogen sulfide, ammonia, and methane are likely and heavy metals possible.

Transport in Containers
Advantages.—This might be the only choice (maybe for special limited uses). The water could be filtered as it is collected, or later at the hatchery.

Disadvantages.— It is inconvenient. Weather can interfere. The supply is not constant. Extra storage space might be necessary.

Artificial Seawater
Several mixtures are available (Huguenin and Colt, 1989), and some are better than others. The one most likely to be suitable for the purpose at hand should be chosen.

Advantages.—If the fresh water for mixing is consistently good, the mixed seawater will have consistent quality. The risk of pathogen introduction is reduced. A seawater intake system is not required, and the facility does not have to be located near a source of natural seawater.

Disadvantages.—Large quantities of artificial sea salts are expensive. They require a dry, protected storage room. Clean fresh water, special containers, and possibly much effort are needed for dissolving them.

SUMMARY

If possible, the following information should be obtained for a potential water source before a large investment is made: seasonal and annual temperature and salinity ranges; ammonia, nitrite, and nitrate levels; hydrogen sulfide and carbon dioxide levels (mainly for ground water); copper, zinc, lead, iron, and manganese levels; possibility of toxic pollutants (e.g., from oil spills); anything that could be a problem and is suspected.

Table 4-1. Typical water sources. Bold type indicates factors likely to be problems that would require specific treatment to correct.

Type	Initial cost	Maint. cost	Flow capacity	Temp. stabil.	Sal. stabil.	Dis. oxygen	Fouling	Susp. solids	Pathogens	Gas supersat.	Ammonia	H_2S,CO_2, CH_4	Heavy metals	Pesticides, etc.	Trace nutrients
Shallow intake	L-H	M-H	H	**L**	L-H	H	**H**	**M-H**	**L-H**	L	L	L	L-M	L-H	G
Deep intake	M-H	L-H	H	M	L-H	M-H	**M-H**	**M-H**	**L-H**	L-M	L	L	L-M	L-H	G
Buried intake	L-H	L-H	M-H	H	M-H	**L-H**	L	L-M	L-M	L-M	**L-M**	**L-M**	**L-M**	L-M	V
Well	M-H	L	L-H	H	H	**L-M**	L	L-M	L	**L-H**	**L-H**	**L-H**	**L-H**	L-M	V
Transport	L-H	L	L-M	L	L-H	H	NA	**M-H**	**L-H**	L-M	L	L	L-M	L-H	G
Artificial salts	H	L	L-M	M-H	H	H	NA	L	L	L	L	L	L	L	V

L = low, M = moderate, H = high, G = good, V = variable, NA = not applicable.

5 WATER TREATMENT

This chapter is a brief overview of water treatment requirements, principles, and methods relevant to marine fish culture. Requirements for sensitive species and (usually) earlier stages are more critical. Publications such as Wheaton (1977), Allen and Kinney (1981), Wickins and Helm (1981), Haywood and Wells (1989), Huguenin and Colt (1989), Moe (1989, 1992), Thiel (1989a, 1989b), Brune and Tomasso (1991), Spotte (1992), Reinertsen et al. (1993), Wang (1993), Losordo and Westerman (1994), Timmons and Losordo (1994, 1997), Timmons (1995), Libey and Timmons (1996), the journal Aquacultural Engineering, and catalogs provide broader and more detailed discussions.

REQUIREMENTS

It is necessary to limit suspended solids, ammonia, nitrites, other toxins, and pathogens, but maintain appropriate temperature, pH, salinity, and dissolved oxygen.

Single Use

As long as flow is sufficient for control of fish wastes, with a single-use (flow-through, open) culture system, it is necessary to be concerned only with harmful substances, organisms, or deficiencies from the water source or system components.

Re-use

With a re-use (recycled, closed) culture system, additional concerns include the possible accumulation of harmful metabolites and pathogens, and the possible depletion of essential trace substances. Re-use ranges from almost 100%, with replacement of fresh water to offset evaporation, to mostly flow-through with some water recycled through a treatment system.

Static

Rotifers, *Artemia*, copepods, and especially algae often are raised in batches, with no (or occasional and partial) water exchange. After zooplankton density builds up to a suitable level, they are completely or partly harvested and the lost volume is replaced with algae and maybe some water. Before conditions deteriorate, the culture is emptied and restarted.

MATERIALS

Care should be taken to avoid toxicity from tanks, pumps, water lines, connectors, filters, filter media, diffusers, ceiling condensation, etc. Polyethylene and polyvinyl chloride release slightly toxic substances when new and should be flushed with seawater for at least 2 weeks before use (Huguenin and Colt, 1989). Hot air from blowers can cause release of volatile toxins from plastic pipe. Latex, vinyl, and fiberglass can be toxic at least to phytoplankton (Huguenin and Colt, 1989). I have found toxic levels (>1 mg/L) of phthalate plasticizers in water passing over polyethylene. Virgin polyethylene is supposed to be less toxic than recycled polyethylene (personal communication from manufacturer).

Butyl rubber stoppers can release copper and zinc, but silicone rubber stoppers do not (Huguenin and Colt, 1989). Standard neoprene rubber contains potentially toxic hydrocarbons and is not approved for food and water use (personal communication from manufacturer). It also has a strong, nasty odor. To be safe, standard bulkhead fitting gaskets made of neoprene should be replaced with a non-toxic material like silicone rubber.

Titanium might be the only safe metal for seawater systems. Stainless steels could release toxic amounts of chromium or nickel (Huguenin and Colt, 1989).

For maximum strength, newly-cemented PVC joints should dry for at least 24-48 h before water is run through them (cement manufacturer's instructions). At 4°C, joints larger than

3.5 inches (8.9 cm) should dry for at least 8 d. Wetting the joint with water before the cement is completely dry causes formation of an outer skin that retards further drying and results in prolonged release of the toxic solvents. In such cases, water analysis has detected solvents as long as 10 yr after gluing was finished.

Even though the resin in fiberglass-reinforced plastic (FRP) tanks has completely hardened and the tank is said to be "cured," toxic vapors often will continue to be released for extended periods (a year or more). The odor becomes very obvious if the tank is covered with a tarpaulin for several hours to concentrate the fumes. Whether noticeable toxicity occurs depends on concentration and sensitivity, but even low exposure could be stressful. This is unfortunate because FRP tanks are the most convenient for raising fish. Precautions for tanks to house sensitive fish include: using food-grade gel coat inside, using plenty of catalyst to harden both the structural resin and gel coat, steam-cleaning the tank at least once after it arrives, aging the tank dry in a warm place for as long as possible (months), and running salt water through the tank (completely full) for several weeks.

TREATMENT OPTIONS

A comprehensive water treatment system for aquaculture could include:

Straining at the Intake (very coarse)
Sedimentation (coarse to medium)
Mechanical Filtration (medium)
Aeration
Biological Filtration
Aeration
Mechanical Filtration (fine)
Temperature Control
Chemical Filtration
Mechanical Filtration (fine)
Disinfection
Aeration

Of course, not all of these would be needed for all applications, and others might have to be added for special needs. A head tank could be placed near the beginning or end to help insure a short-term steady supply if pumps are off or to help maintain constant water pressure.

Straining at the Intake

Debris and other large objects can be excluded from the intake pipeline and pumps by having water pass through straining devices such as slots or holes in a pipe, a coarse screen, or a gravel bed.

Suspended Solids Removal (Table 5-1)

Suspended solids can injure fish gills, clog filters, generate ammonia, and reduce oxygen (Chen et al., 1994). Natural sources include clay, silt, detritus, and plankton. Culture sources include uneaten feed, feces, bacteria, algae, and other organisms. Total suspended solids (TSS) are those >1 μm in diameter, with settleable SS ≥100 μm and non-settleable SS <100 μm (Chen et al., 1994). The smaller particles (<30 μm) probably cause more mechanical injury to fish and are more difficult to remove. A safe minimum concentration probably is in the range 15-40 mg/L TSS, depending on species, stage, and other stress factors.

Table 5-1. Options for removal of suspended solids.

Type	Initial cost	Maintenance cost	Space	Effective range (µm)	Clogging	Head loss	Main disadvantages
				Sedimentation Devices			
Settling tank	M-H	L-H	H	>100	L	L	Large space, large particles
Tube or plate settler	M-H	M	L-M	>75	L	L	Water loss
Hydrocyclone	M-H	M	L	>50	L	H	Complicated, water loss, high flow needed
				Granular Medium Filters			
Slow sand or gravel							
Downflow	M-H	M-H	H	≥10-30	L-H	L-H	Cleaning
Upflow	M-H	M-H	H	≥10-30	L-M	L-M	Limited flow to retain media, water lost in washing
Horizontal flow	M-H	H	H	≥10-30	L-M	L-M	Cleaning
Rapid sand	M	L	L	≥10-30	L-H	L-H	Possible consolidation of sand
Floating bead	M	L	L	≥20-50	L-H	L-H	Frequent backwashing, large particles
				Screen Filters			
Stationary coarse	L-H	L-M	L-M	>100	L-H	L-H	Large particles
Stationary fine	L-H	L-M	L-M	≥60	M-H	M-H	Low capacity, large particles
Triangle	M-H	L-M	L-M	>60	L-M	L-M	Complicated, low capacity, large particles
Rotating	M-H	L-M	L-M	>40	L-M	L-M	Complicated, low capacity, large particles, metal parts?
				Porous Medium Filters			
Diatomaceous earth	M-H	M-H	L	≥0.1	H	H	Maintaining DE coating to avoid leaks
Cartridge	M-H	L	L	≥1	H	H	Low capacity, refill cost
Bag	L	M	L	≥1	M-H	M-H	Low capacity, leakage
				Molecular			
Foam fractionation	L-M	M	L	<30	L	L	Small particles
Ozonation	M-H	M-H	L	<20	L	L	Small particles

L = low, M = medium, H = high.

Settling tanks.—With a given volume of water, the slower and more even the water flow and the longer and shallower the tank, the sooner particles will settle to the bottom. Particles settle faster in laminar than in turbulent flow. Current next to the bottom at <2 cm/s is not likely to resuspend much of the sediment (Chen et al., 1994). Equipping the tank with baffles or tubes helps divert particles downward, and a trap at the bottom collects sediment and makes removal more efficient. A 50-µm particle takes about an hour to sink 1 m, but a 10-µm particle needs about a day (Timmons, 1994). Residence time in a settling tank probably should be at least 15 min (Huguenin and Colt, 1989). Settling tanks also can function as head tanks (Fig. 5-1). Settling areas can be built into rearing tanks beneath false bottoms.

Figure 5-1. Concrete seawater head tanks on a hill at the Guam Aquaculture Development and Training Center.

If bottom currents in a culture tank are slower than 2 cm/s, sediments are likely to accumulate and can be removed by vacuuming; alteratively, circular water flow and a sloping bottom could lead to concentration of sediments close to a bottom drain for removal. If bottom currents are faster than 2 cm/s, solids will tend to flow out with the water, at the bottom or top.

Centrifuges and **hydrocyclones** use centrifugal force to move solids out of the main water flow (Wheaton, 1977; Chen et al., 1994). They are complicated and not often used in aquaculture, except for special applications.

Screen filters can be used if TSS are low, but can be overwhelmed by high TSS. In older style filters, screens are fixed and could require frequent manual backwashing. Triangle screen filters are cleaned by incoming water. Rotating screen filters are automatically backwashed by water jets and sometimes cleaned by scraper bars. Vibrating screens shed accumulated solids by gravity.

Granular medium filters.—Water flow can be downward or upward, and the filters can be pressurized (pumped into), unpressurized (gravity flow), or under partial vacuum (pumped out of). Upflow filters usually are less likely to clog and can be easier to backwash; however, a wider range of flow rates can be used with downflow filters.

Slow sand (or gravel) filters are fed either by gravity or low pressure pumping. For the same flow rate, they are larger than rapid sand filters, but require less energy and also have increased surface area for biofiltration. Upflow filters often use very coarse sand (2-5 mm), which sheds excess biofilm more easily during backwashing (Chen et al., 1994). Aquarium undergravel filters fit into this category (downflow operated by airlift or electric pump, and upflow by electric pump).

High-rate filters are made of different media in layers, with grain size going downward from largest to smallest. Larger particles are trapped higher in the bed, conserving space in the finer layers for smaller particles and thus delaying clogging. A type of downflow dual-medium aquarium filter contains a top layer of crushed anthracite and a bottom layer of finer sand (Spotte, 1992). A type of downflow multi-medium filter contains those two plus a bottom layer of even finer crushed garnet or ilmenite. Each of these materials has a different density, increasing from top to bottom. For all grains in the three-layer filter to have similar settling velocities, the necessary ratios of grain size (coarse to fine) from top to bottom are 6:3:2. If backwashing is done correctly, the layers stay mostly separated.

Rapid sand filters, such as those used for swimming pools, have water under pressure passing downward through a single sand in the fine to coarse range (0.3-4 mm). A typical filter has these ratings: 136 kg sand capacity, 0.29-m^2 filter surface, 3.5 kg/cm^2 maximum operating pressure, and 238 Lpm design flow rate. In such a filter, 0.5 mm sand retains about 95% of particles down to 15 µm, and 0.3 mm sand removes about 95% of particles down to 6 µm. Although pool filters are easy to install and operate, with heavy sediment loads the sand can become clogged and compacted and require manual stirring.

Floating bead filters typically utilize 3-5 mm, buoyant, polyethylene beads that float at the top of the filter vessel, trapping sediment in the lower layers as water flows upward, and also performing some biofiltration. If backwashed frequently, bead filters are less likely to clog under heavy load than sand filters but do not remove small particles as well. In one pass, a bead filter could remove ~50-65% of 10-50 µm particles and ~100% of 50-100 µm particles (manufacturer's data).

Porous medium filters function like screen filters except that they also have depth, which makes them both more effective and quicker to clog. Because they clog so easily, these filters are best used for fine particle removal after most solids (especially the large particles) have been taken out by other devices.

Diatomaceous earth (DE) filters contain a layer of fossil diatoms held on some kind of screen through which the water passes. Thirty-µm DE can filter about 73% of particles in the range 2.5-20 µm and 28% of those 1.5-2.5 µm (Wickins and Helm, 1981). Maintenance effort can be excessive, and DE can leak into culture tanks, possibly becoming a problem for larvae and small juveniles. DE filters seem most useful for maintaining clarity in display tanks, such as aquariums.

Cartridge filters (Fig. 5-2) can be of woven or pleated plastic, ceramic, wound fibers, resin-bound fibers, or pleated paper (but some of the pleated paper ones can dissolve). They are rated down to about 1 µm, but unless cartridges are very well made and seal tightly, leakage can occur.

Bag filters can be kept in housings or fastened to the ends of pipes. They are rated down to 1 µm but have bigger holes, especially if sewn. If the bags clog and are not cleaned or replaced, leakage can occur at the neck or the bag can burst.

Membrane filters are made of plastic (e.g., polyethylene, polysulfone) and are rated down to 0.2 µm. A 0.45-µm membrane will stop coliform bacteria, but viruses and some bacteria will pass through 0.2-µm pores (Wickins and Helm, 1981). Molecular sieves can filter even smaller particles for special applications.

Absolutely-rated filters are supposed to remove <u>all</u> particles larger than the rated size, but nominally-rated filters only remove <u>most</u> particles larger than the rated size. The choice can be crucial when things like parasites need to be excluded.

Chemical Filtration

Granular activated carbon (GAC) chemically attracts and traps organic molecules (adsorption). One kg could have a surface area of 10,000 m^2 or more. Because it can be made from several raw materials, capacity varies. Carbon made from plants is better because

Figure 5-2. Some 9.75-inch filter cartridges: pleated polyester, wound polypropylene yarn, wound polypropylene fibers, combination activated carbon/wound polypropylene yarn, and activated carbon.

carbon made from bones contains much phosphate, which is likely to pollute the water. GAC made from hardwood has a high surface area and is one of the best (Spotte, 1992). Coconut shell carbon, also good, is often used. Acid washing of the carbon during manufacture is a helpful precaution to remove alkaline substances that could raise pH in culture water (Thiel, 1989b). GAC is not normally used for large volumes of water, but mainly for larval rearing of sensitive species or maintenance of small sensitive broodstock. Water should be well filtered before and after (to remove small bits of carbon) passing through carbon, and the carbon should be changed regularly. When the pores are full, substances strongly attracted can displace those less attracted, possibly resulting in a sudden release of toxic material. One possible indication for replacement is an increase in yellow color of culture water. GAC could remove essential trace substances such as chelators (Wickins and Helm, 1981) from culture water, but this might be more of a problem for algae and invertebrates than for fish. Any kind of container that will hold the carbon and allow water to flow through it (e.g., a length of PVC pipe with restricted ends) will work. Spotte (1992) recommended 1 g of GAC for each liter of recycled aquarium water. Amounts needed for flow-through systems can be determined empirically by the time to exhaustion. Effectiveness is governed more by pore surface area than grain size. Grain size is chosen more for convenience of handling and retention in the contact chamber. Diameters of 1-2 mm are suitable. GAC also acts as a biofilter and like sand must not be allowed to become anaerobic.

Foam fractionation occurs when a gas is bubbled through a column of water and very small particles and organic substances stick together to form a skin around each bubble. The bubbles burst, leaving a froth, which is skimmed off. Both dissolved and suspended matter (including bacteria) are removed (Chen et al., 1992, 1993a, 1993b; Weeks et al., 1992). Thiel (1990) found the following in foam from a reef aquarium: microalgae, phenol, phenyls, amino acids, albumin, urea, food particles, organic material from dead microorganisms, live parasites, live bacteria, fish and invertebrate feces, iron, complex nitrates, complex phosphates, organic acids, ammonia, dimethyl sulfate, colloids, and more. Many of these could directly stress fish, reduce oxygen, or promote an increase of disease organisms. Aeration and degasification are additional benefits of foam fractionation. Wickins and Helm (1981) reported that foam fractionation resulted in more consistent survival of Pacific oyster larvae.

By number, most particles removed have a diameter less than 30 µm, with an average near 10 µm, but, by weight, most are about 15-25 µm (Timmons, 1994). Optimum bubble diameter for fractionation is 800 µm (Spotte, 1992). The simplest foam fractionator is the whole fish tank with aeration and a surface drain, which should skim the surface film before noticeable foam accumulates. The next simplest is an airlift tube with foam diverted away from the tank while most of the water is returned. Countercurrent (opposing water and air flows) foam fractionators are slightly more effective than cocurrent ones because contact time is increased and short-circuiting is less likely (Spotte, 1992).

By producing aggregates, foam fractionation can increase the amount of suspended solids, which might call for more mechanical filtration. It removes bacteria directly, but also could promote growth conditions for them (Wickins and Helm, 1981).

Ozonation can split large refractory molecules, facilitating their removal by bacteria, or other treatments such as sedimentation and foam fractionation (Rueter and Johnson, 1995), but this mechanism is not always effective in practice (Spotte, 1992).

Ion exchange resins will remove ammonia and other toxins from fresh water, but they usually are overwhelmed by the ions in seawater and are of little use in mariculture. Some can be used to remove traces of toxic metals from small amounts of seawater (Wickins and Helm, 1981) for special uses such as pollution experiments. Currently available from aquarium suppliers are resins for removal of organic pollutants, organic acids, and toxic gases, plus **ceramic materials** for removal of phosphates and silicates, but these are relatively expensive.

Carmignani and Bennett (1976) reported that 81% **polyurethane foam** can remove dioctyl phthalate and chlorinated hydrocarbons from distilled water and suggested that it could be used to filter such compounds from aquaculture water.

Chelation of metals by disodium EDTA (disodium ethylenediamine tetracetate) followed by precipitation is one way to remove them (Wickins and Helm, 1981). Removal of excess zinc or other metals by **co-precipitation** with sodium metasilicate (at least 2 mg/L silicon), with or without EDTA, is another possibility (Willey, 1977; Wickins and Helm, 1981). Clay minerals, humic acids, and other substances could be used to remove copper (Wickins and Helm, 1981), but might have side effects. Fuller's earth also can remove metals but then itself has to be separated from the water.

Bicarbonates, carbonates, magnesium, and calcium tend to decrease in closed systems but can be replaced (Wickins and Helm, 1981). Phosphates can increase because of excretion, uneaten food, and decay. In seawater, precipitation of phosphate with **calcium** usually keeps phosphorus concentration $\leq$6 mg/L, but eventually, in closed systems, calcium and maybe magnesium are depleted. Excess phosphate could stress sensitive organisms, or along with nitrate could promote unwanted algal blooms.

Each mg of **sodium thiosulfate** removed 1.10-1.13 mg of ozone-produced oxidants (OPOs) in seawater previously aerated for 1 h but only 0.16-0.19 mg from water aerated for 48 h (Hemdal, 1992). The more labile OPOs probably were easy to remove by either aeration or thiosulfate, while the less labile OPOs (0.19 mg) were resistant to aeration but not thiosulfate.

Biological Filtration (Table 5-2)

Aquatic organisms excrete nitrogen mostly as ammonia, but sometimes as amino acids, volatile amines, or urea (Wickins and Helm, 1981).

Table 5-2. Options for biofiltration.

Type	Initial cost	Maintenance cost	Space	SSA[a] (m^2/m^3)	Eff. range (ammonia)	Solids removal	Clogging	Head loss	Main disadvantages
Submerged sand/gravel									
Downflow (coarse sand)	L	L-M	M	2500	L-M	H	M-H	M-H	Space, cleaning
Upflow (coarse sand)	L-M	L-M	M	2500	L-M	M-H	M	L-M	Space
Trickle (1.9-cm plastic)	L-M	L-M	H	490	M-H	L	L	L	Space
Biodrum (1.9-cm plastic)	H	M	H	490	M-H	L	L	L-M	Space, mechanism breakdown, metal parts?
Biodisc	H	M	H	330	M-H	L	L	L	Space, mechanism breakdown, metal parts?
Fluidized sand bed									
Fine (0.1-0.2 mm)	L-M	L	L	6600	L	M	L-M	L-M	Oxygen depletion
Coarse (1-2 mm)	L-M	L	L	2500	L-M	L-M	L	L-M	Oxygen depletion
Floating bead (3-5 mm)	M	M	M	1300	M	M	L-M	L-M	Space, frequent backwashing
Plant	L-M	H	H		L-H	L	M-H	L-M	Space, cleaning, harmful metabolites?

L = low, M = medium, H = high.

[a] Typical specific surface area (Malone and Drennan, 1993; product literature).

Bacterial nitrification is accomplished mainly by species of *Nitrosomonas* and *Nitrosococcus*, which oxidize ammonia to nitrite, and *Nitrobacter*, *Nitrococcus*, and *Nitrospina*, which oxidize nitrite to nitrate (Spotte, 1992). This happens both in the water and on all solid submerged surfaces that are oxygenated. Biofilters contain solid media designed to maximize the available area for bacteria to colonize and at the same time maintain aerobic conditions and minimize clogging. For comparison of different media, specific surface area (SSA) usually is reported as m^2/m^3. There can be as much as 100 times more nitrifying bacteria in an established filter bed than in the water. The top 5 cm of a static sand filter bed can have 10 times more bacteria than deeper parts. At least for aquaria, 2-5 mm is the optimal medium size for a static bed. Smaller grains clog quickly, and larger ones tend to form channels where the water is short-circuited. For a given diameter, angular particles have more surface area. Filters with moderate porosity (open space) of 30-40% economize on space and allow for reasonable cleaning intervals (Wheaton et al., 1994a).

Submerged biofilters (Fig. 5-3) contain a medium like sand, gravel, rocks, and/or plastic bits immersed in a flow of water (up, down, or sideways). **Trickle** filters are similar but are partially exposed to the air. A **biodrum** filter consists of a volume of medium contained in a cylinder that rotates into and out of the water about three times a minute. A **biodisc** filter consists of discs attached at close intervals on an axle, which rotates them into and out of the water about three times a minute. A **fluidized bed** filter contains a bed of medium (usually sand) that is kept in constant motion by upward water jets; Wheaton et al. (1994a) recommended 1-2 mm sand for >1mg/L TAN and 0.1-0.2 mm sand for <0.3 mg/L TAN. Fluidized bed filters expose all sides of the sand grains for bacterial colonization, and because of its motion, the medium is self-cleaning. A packed bed filter works better for solids removal but is less effective for nitrification than a fluidized bed and has to be backwashed periodically. **Floating bead** filters, which were described earlier, are designed for nitrifying bacteria to grow attached to the beads (Malone, 1992). Frequent backwashing is required to eliminate trapped solids and heterotrophic bacteria on and between the beads, and backwashing also reduces the nitrifiers.

Trickle, biodrum, and biodisc filters by design are very well oxygenated but have relatively low SSA. An exception occurs with highly-porous sintered glass rings from Germany, made for trickle and canister filters. They are reported to have a total SSA of 286,000 m^2/m^3 (manufacturer's literature), but at present are fairly expensive. Not only do these rings provide abundant habitat for nitrifiers, but denitrification will occur in deeper pores within the rings.

Submerged, fluidized bed, and bead filters need supplemental oxygenation. The effectiveness of trickle filters is low, but they are cheap, self-aerating, and easy to set up and maintain. Fluidized bed filters are the most effective, but function best after a solids-removal unit. Bead filters can be more economical because one unit removes both solids and ammonia (Malone, 1992; Malone and Drennan, 1993; Malone et al., 1993; de los Reyes and Lawson, 1996; Hargrove et al., 1996), but neither function is performed as well as by separate solids filtration followed by a fluidized bed.

Full establishment of nitrifying bacteria could take from 10 to 60 d, depending mainly on seeding, nutrient supply, and temperature. Seeding of the biofilter of an aquarium holding juvenile red drum with a commercial mixture of *Nitrosomonas* and *Nitrobacter* resulted in almost immediate control of low levels of ammonia (LaBomascus et al., 1987). However, nitrite increased until 12 d, then declined but had not reached intitial levels by 28 d. Also, TAN suddenly rose from non-detectable during 1-20 d to 0.60 mg/L on day 22. Possibly the capacity of the biofilter was exceeded at that point.

The common basis for measurement of ammonia, nitrite, and nitrate is the nitrogen molecule. NH_4^+ (IA) contains 94% NH_3 (UIA) and 78% N. NO_2^- contains 30% N. NO_3^- contains 22% N.

Dissolved oxygen should be at least 2 mg/L throughout the biofilter bed, but 3 mg/L is better. For each mg of NH_4-N oxidized to NO_2-N, 3.0-3.4 mg O_2 is used, and for each mg of NO_2-N oxidized to NO_3-N, 1.0-1.2 mg O_2 is used, for a total of 4.0-4.6 mg O_2 (Wheaton, 1977; Wheaton et al., 1994a, 1994b). For each mg of TAN in the untreated water, at least 4.0-4.6 mg O_2 has to be supplied to the biofilter, and after water leaves the biofilter, it has to be reoxygenated.

Figure 5-3. Five of the ten 66.7-m^3 biofilter tanks for the 21,200-m^3 aquarium at the Living Seas exhibit at Disney World, Orlando, Florida. They contain 2.5-mm sand. System flow rate is about 144 m^3/min.

The rate of nitrification varies with temperature, and optima range from ≤22°C to ≥30°C for different nitrifiers (Wheaton et al., 1994b). Bright light is inhibiting. Steady pH in the range 7.0-9.0 probably is best for marine nitrifying bacteria. Nitrifiers consume alkalinity, which can lead to unsuitably low pH (Bisogni and Timmons, 1994)—a more serious problem in fresh water. Bacterial populations tend to be more stable with constant nourishment and constant salinity.

Toxic and therapeutic compounds can harm nitrifying bacteria (Wheaton et al., 1991), but brief exposure to low concentrations (≤100 mg/L) of formalin can be tolerated in a fluidized bed (Weaver, pers. comm., 1995). Nitrifying bacteria might also resist moderate amounts of copper sulfate and potassium permanganate (Wheaton et al., 1994b). Nitrifying bacteria could be stimulated by sodium, calcium, magnesium, molybdenum, iron, copper, and phosphate but inhibited by chromium, nickel, >560 µg/L copper, sulfide, and some vitamins, alcohols, and amino acids (Spotte, 1992; Wheaton et al., 1994b). Ammonia inhibition of *Nitrosomonas* can occur at 10-150 mg/L and *Nitrobacter* at 0.1-1.0 mg/L UIA (Wheaton et al., 1994b).

Clogging and fouling can reduce efficiency. About 170-210 µg of nitrifying bacterial biomass is produced per mg TAN oxidized (Wheaton et al., 1991; Wheaton et al., 1994a). This contributes a little to clogging, but not as much as heterotrophic bacteria, which tend to produce more biomass. The latter also occur in the filter system and live off of dissolved organic wastes. It is important to minimize addition of suspended solids to biofilters because they not only tend to clog the medium but also could increase nourishment of heterotrophic bacteria and thus help them compete successfully for space with nitrifying bacteria.

Bacterial denitrification is the reduction of nitrate to nitrogen gas by anaerobic bacteria, possibly with intermediates NO_2 (nitrogen dioxide) and N_2O (nitrous oxide). It can occur in accidentally anoxic areas of rearing systems or in purposefully designed containers. This is one way to eliminate nitrates from recirculating systems, but it requires a high level of maintenance and monitoring to avoid production of toxic gases such as hydrogen sulfide. Because carbon is needed as a food source by the denitrifiers, ethanol or methanol sometimes is added very slowly to provide it, introducing the risk of toxicity from the alcohol if the system malfunctions. A denitrification system should be designed to automatically shut off to keep those compounds out of the rearing units (Whitson et al., 1993). Aeration is necessary between dentrification and the rearing units. Additional biofiltration and aeration after denitrification can help remove NO_2 and H_2S.

Algal filtration.—Algae use carbon dioxide and (depending on species) nitrate, ammonia, and/or urea. Algae can be removed as they grow, but algal filters present serious potential problems. They should have continuous illumination because, in the dark, algae consume oxygen and produce carbon dioxide, which also can reduce pH. When the algae are illuminated, there is a danger that photosynthesis will deplete carbon dioxide so much that pH will rise suddenly and dangerously. Algae are more difficult to maintain than bacteria, especially if undesirable competing species invade. Algae filters usually require much space. Some algae produce toxic (Spotte, 1992) or sticky substances (pers. obs.) that can injure or kill fish. Fragments and debris have to be removed regularly. Bacteria are more efficient than algae in eliminating ammonia.

A thin layer of microalgae such as *Nannochloropsis oculata* and *Tetraselmis suecica* or macroalgae such as *Chaetomorpha* sp. on the walls and bottom of a fish tank can help maintain stable water quality as long as algal debris does not build up too thickly on the bottom (pers. obs. in many successful larval rearing trials). *Chaetomorpha* does require more illumination than is initially given to larval fish tanks. Two Florida species have optima of 4000-5000 lux and 7000-8000 lux (Spotte, 1992). A good procedure is to begin with green water (phytoplankton) at 1000-2000 lux (then seed the water slightly with *Chaetomorpha*, if desired) and gradually increase light intensity until a healthy bloom appears. *Chaetomorpha* particularly needs water movement, and most algae will be healthier after circulation within the tank has been increased.

Disinfection and Sterilization

Disinfection is a reduction in reproductive capacity and/or numbers of viruses, bacteria, and fungi. Sterilization is the elimination of "all" life forms (i.e., at least the killing of disease organisms). Sterilization is rarely accomplished in aquaculture; usually something is still alive, but hopefully it is not dangerous.

Chlorination (White, 1986) often is used for disinfection of surfaces, algae culture water, and effluent water, but not usually for plankton or fish culture water. Effectiveness depends on the form of chlorine, concentration, pH, temperature, contact time, and sensitivity and density of target organisms. When less than 1000 mg/L Cl is added to fresh water with a pH higher than 3, most is converted to hypochlorous acid, HOCl, which is in equlibrium with hypochlorite, OCl^- and H^+ (Wheaton, 1977). At 23°C, as pH increases from 6 to 7, 8, and 9, HOCl decreases from 97% to 78%, 26%, and 3%, and OCl^- increases (White, 1986). HOCl is a stronger disinfectant than OCl^-. HOCl was 20 times more effective against *Escherichia coli* than OCl^-, and at 23°C it was 300 times more effective against ameba cysts (White, 1986). An HOCl residual of 1 mg/L after 30 min contact time is considered effective against viruses and bacteria in relatively clean water (pH ≤8.0); however, 4 mg/L free chlorine residual (5-10 mg/L combined chlorine residual) for 2 h might be better for dirtier water or water with a higher pH. In fresh water with the pH of seawater (8.3) at 25°C, about 15% of the chlorine is present as HOCl; however, in seawater bromine takes over this role. Bromide ion (Br^-) substitutes for the chlorine in HOCl to form hypobromous acid (HOBr) and chloride ion (Cl^-). At 25°C and pH 8.3, the reaction is 99% complete in 10 s. So, testing for chlorine residuals in seawater actually is the measurement of bromine residuals. Hypobromite has similar disinfecting ability to that of hypochlorite but probably is more abundant and therefore more effective at the pH of seawater. At 20°C and pH 8, in seawater 83% of the bromine will be present as HOBr, but in fresh water only 28% of the chlorine will be present as HOCl.

Not only these forms of chlorine and bromine, but many of the chloramines, bromamines, chloroform, bromoform, bromodichloromethane, and other organochlorine and organobromine compounds formed by reactions in seawater are toxic to fish, and some are fairly persistent. Chlorine reacts immediately with ammonia in seawater to form chloramines or bromamines. Sodium thiosulfate usually is added to neutralize chlorine. It also breaks down chloramines and bromamines, but this liberates ammonia, which might have to be dealt with. Sodium thiosulfate also can be toxic to cultured animals (1 mg/L for larval penaeid shrimp; Huguenin and Colt, 1989). Chlorine gas is heavier than air, more than ten times as soluble in water as oxygen, and likely to drift into sensitive areas.

In aquaculture, usually a concentrated solution of sodium or calcium hypochlorite, NaOCl or $Ca(OCl)_2$, is diluted and used for disinfection. Under some conditions, $Ca(OCl)_2$ can leave calcium deposits, but NaOCl is not likely to form a residue.

Autoclaving (e.g., 120°C for 15 min, under pressure) is used to disinfect small batches of water, particularly for phytoplankton culture.

Pasteurization (60°C for 30 min or 71°C for 15 s) also is fairly effective against parasites, bacteria, and viruses (Wheaton, 1977).

Ultraviolet irradiation could deactivate 100% of some disease organisms, such as bacteria in the genus *Vibrio* (Wickins and Helm, 1981). It also can oxidize some organic matter. Effectiveness depends mainly on wavelength, intensity, contact time, UV transparency of the water, and target organism sensitivity. The most effective wavelengths are in the range 2500-2650 Å (Wheaton, 1977; White, 1986). For different organisms, effective exposure values range from less than 35,000 (most bacteria and yeasts, many viruses) to more than 156,000 $\mu Ws/cm^2$ (some viruses, molds, and protozoans)(White, 1986; Huguenin and Colt, 1989; Rodriguez and Gregg, 1993). *Paramecium* require 200,000 $\mu Ws/cm^2$ and spores of the mold *Aspergillus niger* 330,000 $\mu Ws/cm^2$. The pathogenic bacteria *Vibrio anguillarum*, *V. salmonicida*, and *Yersinia ruckeri* in fresh, brackish, and seawater were reduced 99.999% by 2700 $\mu Ws/cm^2$ (Liltved et al., 1995). Under the same conditions, the infectious pancreatic necrosis virus (IPNV) required 122,000 $\mu Ws/cm^2$ for 99.9% reduction. Because UV also removes beneficial bacteria, seeding the culture units might be a good procedure. For example, spores of *Bacillus subtilis*, which has probiotic strains, can be deactivated with 22,000 $\mu Ws/cm^2$ UV. Water flow should be adjusted according to the needed exposure.

UV works best in well-filtered, colorless water. Ordinary glass blocks UV radiation, but quartz glass and teflon tubing do not (Huguenin and Colt, 1989). The portion of the outer tube in contact with water should be cleaned periodically. Because transmittance of the bulb glass diminishes with time, bulbs need to be replaced after continuous usage for 6-12 months, depending on how critical their function is. Usually, output is reduced by 40% after 12 months. Old bulbs should be sent for mercury recycling.

Ozonation can disinfect or sterilize water, depending on the concentration (Oakes et al., 1979; Masschelein, 1982; Langlais et al., 1991; Steslow, 1991; Sugita et al., 1994). It can oxidize disease organisms, organic matter, dyes, sulfide, nitrite, iron, and manganese (Wickins and Helm, 1981; White, 1986). The insoluble oxidized iron and manganese will tend to be removed by precipitation and/or filtration.

Generally, an ozone residual of 0.1-1.0 mg/L for 1-2 min can reduce bacteria by 99% (Huguenin and Colt, 1989). Relative sensitivity of bacteria, viruses, and cysts is illustrated by the following concentrations of ozone needed for 99% deactivation in 10 min: *E. coli*, 0.001 mg/L; polio virus, 0.01 mg/L, and ameba cysts, 0.1 mg/L (White, 1986). (There are exceptions to this trend. A similar one exists for chlorine.) The pathogenic bacteria *Vibrio anguillarum*, *V. salmonicida*, and *Yersinia ruckeri* in fresh, brackish, and seawater were reduced 99.99% by 0.15-0.20 mg/L residual ozone for 180 s (Liltved et al., 1995). Under the same conditions, IPNV required 60 s of 0.10-0.20 mg/L residual ozone for 99.99% reduction.

Reid and Arnold (1994) reported that red drum larvae tolerated 0.1 mg/L ozone (but not 0.2-0.4 mg/L) for 24 h, and juveniles tolerated 0.1 mg/L for 22 weeks. A few µg/L are lethal to sensitive species in a few hours. Ozone and hypobromous acid oxidize cell membranes, with damage to fish being most obvious in the gills, lateral line, and skin.

Safe and effective exposure of culture water to ozone is complicated and expensive. It can be at least irritating to humans, highly toxic to cultured species, and corrosive to equipment. Aeration can dissipate some of the ozone, and UV accelerates conversion of O_3 to O_2. However, ozone reacts with seawater very much like chlorine to form persistent toxic compounds (e.g, bromine, bromoform, bromates, hypobromous acid, bromamines, and possible mutagens). In closed systems, such substances can accumulate. Wheaton (1977) discussed evidence indicating that, even after aging, ozonated water should pass through activated carbon to reduce toxicity before being used. Ozone also could upset water chemistry (e.g., removal of chelators and trace minerals).

Oxygenation

Good aeration maintains high dissolved oxygen and normal carbon dioxide, nitrogen, and total gas pressure. Heavy aeration also can remove most NH_3, but this has to be done at a pH higher than 10 and is not practical for fish culture systems (Lucchetti and Gray, 1988). Sufficient oxygen is necessary for survival and good health of not only the fish, but also live foods and nitrifying bacteria. Fish requirements vary with species, stage, and temperature (see chapter 3) but are relatively high. Carbon dioxide can poison fish even if oxygen is sufficient. To maintain healthy nitrifying bacteria throughout the system, dissolved oxygen should be at least 2-3 mg/L. No part of the water system should be allowed to become hypoxic. If water and filters are not kept well oxygenated, hydrogen sulfide, ammonia, methane, and other toxic compounds could form. Therefore, to be safe, the water should be oxygenated at least at the beginning of the treatment system, before and after biological filters, and just before or as it enters the fish tanks.

The proportion of oxygen in the water can be raised somewhat as long as total dissolved gas pressure (TGP) is not increased above saturation. TGP slightly higher than 100% can lead to gas bubble trauma in sensitive fish. Oxygen can be increased by a given volume without changing TGP if the same volume of nitrogen is removed.

Oxygen transfer from air to water increases with partial pressure of O_2 in the air, exposure area, and exposure time, but decreases with temperature. Gravity aeration is accomplished by splash surfaces and packed columns. A packed column is moderately effective at dissolving oxygen and good for removing nitrogen (Watten, 1994). Submerged devices such as airstones, porous pipes, and drilled pipes can be used in open water or inside a vertical pipe (airlift pump) to both circulate the water and increase oxygen to water contact. Ceramic and fused glass airstones are the best for durability and bubble size (see Chen et al., 1992), followed by heat-bonded silica; cemented sand airstones tend to disintegrate. Oxygen transfer occurs not only by diffusion from bubbles but also by movement of surface water. Airlifts cannot be used with fish too weak to avoid entrainment.

For all applications, air should be filtered before the pump to remove particles such as mold spores. For eggs, larvae, algae, and any other sensitive organisms, air should also be filtered between the pump and the culture tanks, especially for particles from the pump. Activated carbon could be included for airborne pollutants. Required pore size is similar to that for ultrafiltration of water (page 159); 0.2 µm is the lowest practical size and is best for phytoplankton cultures. This can be reached in two or three stages (e.g., 35 µm, 3 µm,

0.2 μm). Condensed water inside the air pipe should be prevented or at least removed frequently to minimize buildup of fungus, which could produce toxins. Aeration by diffusers at depths >1 m should be avoided because supersaturation is likely. Spotte (1992) reported that surface film cleaning was more effective with aeration next to the tank walls and a drain at the center of the surface, but oxygenation was more efficient (more oxygen dissolved per unit of energy) with aeration in the center of a circular tank or in parallel rows arranged across the width of a rectangular tank.

Nearly pure oxygen can be added to culture water (usually only when recirculated) if stocking levels are so high that aeration is insufficient to maintain required DO. Downflow bubble contactors (DCBAs) are very efficient and are not likely to supersaturate. A DCBA is a cone with water and oxygen introduced at the narrow top. The mixture moves downward, and as the cone widens, water velocity slows until it is the same as the upward velocity of the bubbles, which then become nearly stationary. Consequently, contact time is increased, and with good design and adjustments, nearly all the oxygen can be dissolved. Other devices include: packed column, spray tower, multi-stage low-head oxygenator, and counter-current injection column. See Colt and Watten (1988), Boyd and Watten (1989), Colt and Orwicz (1991), Dwyer and Peterson (1993), Wagner et al. (1994), and Watten (1994) for further discussion of these and other types.

In my experience, for sensitive larvae, it is important to aerate water before it enters the tank, maintain green water conditions, and aerate tank water only as much as the larvae can tolerate (their tolerance will increase with time).

Figure 5-4. Electric paddlewheel aerator in a pond.

Paddlewheels (Fig. 5-4) are the most effective devices for pond aeration, because they oxygenate by vigorous splashing and they also mix the water. Diffusers can be used, but supersaturation is possible if they are deeper than 1 m. Water can be pumped from the pond and returned against a splashboard or as a spray; but without mixing, the effect will be localized. Horizontal propellor-aspirator pumps move water but are not as effective as paddlewheels. Vertical pump aerators can be useful. Fountains look nice but tend not to aerate as well as the above devices. (Also see Ruttanagosrigit et al., 1991; Practical Aquaculture and Lake Management, March/April 1992).

For hauling fish, Carmichael et al. (1992) suggested a combination of diffusers and surface aerators, plus ventilation of the headspace. A small bilge pump powered by a 12-volt battery is economical and reliable.

Degasification

Degasification is performed mainly to reduce total gas pressure (mostly from nitrogen) to prevent gas bubble trauma, or to remove carbon dioxide and hydrogen sulfide, but it also drives off some ammonia. Supersaturation in mariculture usually occurs when pressure on water is released, temperature is increased, or air is entrained in a pressurized water system. Of commonly-used devices, only packed columns and paddlewheels remove enough carbon dioxide to be practical (Grace and Piedrahita, 1993, 1994). A packed column probably is the best device for dissipating and oxidizing hydrogen sulfide. This is a container above water level packed with a medium of high porosity, through which water is sprinkled. The medium is not submerged. Spotte (1992) suggested a column about 90-120 cm high containing a 2.5-cm medium, with a water flow rate of 1 L/cm^2 cross-sectional area/min. Pumping air into the bottom of the column helps flush out unwanted gases; the volume of air added should be at least five times the volume of water flowing through (Losordo, 1991a).

Heating and Cooling

Methods range from solar heating and evaporative cooling through simple glass-tube aquarium heaters to expensive, high-capacity, in-line heaters and chillers. For information, see Wheaton (1977), Huguenin and Colt (1989), Spotte (1992), and Timmons and Losordo (1994).

SYSTEM DESIGN

General

Very critical components (e.g., aerators, electricity) that are likely to fail should have backup units installed and ready to switch on automatically or manually, as appropriate. Moderately critical components (e.g., water pumps) should have spares available for quick installation.

Electrical components and sometimes the water itself should be grounded. Wiring and junctions should be weatherproof and should be kept as far from water and as high as possible. Ground-fault circuit breakers are important. As much as possible, only low voltage should be used for electrical devices near or in the water (e.g., sensors and feeders). Glass aquarium heaters should be handled gently because they can easily break, especially from leakage or sudden temperature changes.

Some method to allow cleaning of pipelines should be incorporated, at least on the incoming side of the filter system. Duplicate pipelines permit one to be used while the other is filled with fresh water or allowed to become anaerobic for 1-2 weeks. Flexible polyurethane "pigs" can be placed inside pipes and forced through by water pressure to clean the pipes (Huguenin and Colt, 1989).

Elevations should be such that water and air flow in the right directions, without air locks in the water lines. Access space should be allowed for bleeding off air, draining water from lines, and performing maintenance.

Tees and valves positioned strategically can facilitate future additions to the water system. Valves are needed at low points for draining the lines. Wastewater drains should be large enough to function even when mostly clogged with debris and fouling organisms.

Specific

The marine fish hatchery in Fig. 6-25 is designed to have three types of water treatment system: new salt water, recirculated salt water, and new fresh water.

Salt water is drawn from wells. The two upper incoming saltwater treatment systems need to remove hydrogen sulfide, carbon dioxide, ammonia, nitrite, sulfur bacteria, and dissolved organic matter and also adjust water temperature. Each system could consist of:

Packed column
↓
6,000-L sedimentation/reservoir tank (or one larger main one for all)
↓
Possibly a slow gravel filter (or one larger main one for all)[op]
↓
1-hp pump[fs]
↓
Rapid sand filter (136 kg sand, 238 Lpm)
↓
Three 35-µm pleated cartridges (142-284 Lpm)
↓
Packed column[bio]
↓
4,000-L degassing and foam-fractionating reservoir[bio]
↓ ↑
½-hp pump[bio] 4,000-L degassing and foam-fractionating reservoir
↓ ↑ ↓
437-kcal/min heater/chiller Packed column 1-hp pump[fs]
↓ ↑ ↓
Fluidized-bed sand filter (900 kg sand, 189 Lpm)[bio] ↓
Three 5-µm fiber cartridges (142-284 Lpm)
↓
Three activated carbon cartridges (142-284 Lpm)
↓
Three 1-µm fiber cartridges (142-284 Lpm)
↓
Ultraviolet treatment (142-284 Lpm)[op]
↓
Aeration
↓
Culture tanks

Two optional ([op]) components are the slow gravel filter for water with high suspended solids and the ultraviolet treatment for water suspected of containing pathogens; these are more likely to be needed for surface water than for well water. Components for biofiltration ([bio]) can be omitted if the incoming water will not contain enough ammonia to be harmful;

these components probably would be needed for well water but not for surface water. Pumps that draw from reservoirs filled by other pumps operate via float switches ([fs]) to protect them from running dry if the water level drops too far. If gravity feed is used for one side and a pump for the other side of a recirculating system for fish tanks, float switches usually are not needed. The lower incoming saltwater treatment system (spawning tanks and raceways) is the same as the upper two except the heater/chiller is left out because less water is required and room temperature will mostly take care of it.

The four recirculating saltwater treatment systems for spawning tanks need to remove particulate waste, ammonia, nitrite, dissolved organic matter, and possibly pathogens (optional UV). Each could consist of:

Spawning tank
↓
Settling sump
↓
½-hp pump[fs]
↓
Rapid sand filter (136 kg sand, 238 Lpm)
↓
Three 35-μm pleated cartridges (142-284 Lpm)
↓
Packed column[op]
↓
2,000-L degassing and foam-fractionating reservoir
↓ ↑
½-hp pump 2,000-L degassing and foam-fractionating reservoir
↓ ↑ ↓
↓ Packed column ½-hp pump[fs]
↓ ↑ ↓
Fluidized-bed sand filter (300 kg sand, 76 Lpm) ↓
Three 5-μm fiber cartridges (142-284 Lpm)
↓
Three activated carbon cartridges (142-284 Lpm)
↓
Three 1-μm fiber cartridges (142-284 Lpm)
↓
Ultraviolet treatment (142-284 Lpm)[op]
↓
Aeration
↓
Spawning tank

The 1-μm filters could be useful for removing protozoan parasites. The three recirculating saltwater treatment systems for raceways are mostly the same except the reservoir tank volume is reduced to 900 L.

The incoming freshwater treatment system needs to remove hydrogen sulfide, carbon dioxide, sulfur bacteria, and dissolved organic matter, and also adjust water temperature. It could consist of:

Packed column
↓
6,000-L insulated degassing reservoir tank
↓
1-hp pump^{fs}
↓
Rapid sand filter (136 kg sand, 238 Lpm)
↓
Three 35-µm pleated cartridges (142-284 Lpm)
↓
437 kcal/min heater/chiller
↓
Packed column
↓
4,000-L degassing/foam fractionating reservoir
↓
1-hp pump^{fs}
↓
Aeration
↓
Culture tanks

If the salt or fresh water supply is from the surface rather than from wells (as in this example), treatment systems would have to remove heavier particulate loads, possibly some ammonia and nitrite, dissolved organic matter, and disease organisms. Hydrogen sulfide and ammonia probably will be near zero; therefore, initial degassing and biofiltration might not be needed. However, alternative or additional components might be needed to remove the larger amount of particulate matter and possibly to disinfect the water. For surface fresh water, the full series of particulate filtration might be needed; a settling tank and/or slow gravel filter could replace the initial degassing reservoir tank, and ultraviolet disinfection could be added just after the 1-µm filters.

SYSTEM MONITORING AND CONTROL

The degree of monitoring needed for systems and environmental factors depends on hatchery design, plus fish life stage, density, specific tolerances, and value. Functions such as oxygenation, temperature (e.g., Hanson, 1992), light (e.g., Manci et al., 1992), water level, water flow, and feeding can be automatically controlled. Sensors for the most critical factors can be linked to an automatic telephone dialer that, when activated, will continue to dial several numbers in sequence until acknowledged. For more information, see Wheaton (1977), Gibbs (1991), Water Farming Journal (1992a, 1992b), Wang (1993), Ebeling (1994), Lee (1995b), Schulstad (1997), and catalogs.

High Priority (becoming critical in minutes)

Dissolved oxygen in moderately-stocked tanks can be measured routinely with a portable, battery-operated meter. With high stocking densities and particularly when the fish are depending on pure oxygen additions to the culture water, continuous automatic monitoring of dissolved oxygen is important. Air pumps can be monitored with pressure-sensitive sensors/switches. Aeration or oxygenation devices should have automatic backups or multiple units supplying the same system.

Temperature of culture water should be stable if incoming water is near the correct temperature and hatchery air temperature also is close. Problems can occur when large volumes of water must be heated or chilled before use. Inexpensive thermistor probes with digital displays are reasonably accurate and can be checked whenever someone goes near the tank. Stainless steel dial thermometers are fairly accurate and unbreakable except for the glass dial cover (safer than all-glass thermometers), but they can be damaged by heat. If the fish are very sensitive, if temperature control is not reliable, or if continuous records are needed, measurements can be taken automatically via thermistors or thermocouples (or the more expensive resistance detectors or integrated circuit transducers).

Electricity should be monitored either by having someone at the hatchery continuously or by installing an alarm system that will notify key personnel at home by telephone. In either case, a backup generator with automatic transfer switch is necessary for critical systems (e.g., aeration, biofilter water flow).

Water levels in culture tanks usually are checked visually during hatchery operations. Pipelines and standpipes should be secure enough that they are not knocked loose by ordinary human or fish activities. They should have reasonable protection from storms. Water levels could drop in supply tanks that have pumps leading from them because of supply pump failure, line clogging, or power outage. Continuous automatic sensors in these tanks are justified. Simple float switches could be used for individual pump control. Immersion heaters are best kept in tanks that will not drain down accidentally.

Fire and **burglar alarms** should automatically notify the fire and police departments and key hatchery personnel, as appropriate.

Medium Priority (becoming critical in hours)

Water flow monitoring could be automatic if critical (e.g., for biofilters, in-line heaters, densely-stocked fish tanks); otherwise, routine visual inspection should be enough.

Ammonia monitoring by manual means probably is good enough for most hatcheries. Automatic monitoring is expensive and usually not warranted. Spotte (1992) described an indophenol blue (spectrophotometric) method (~100-2000 µg/L detection range) and a specific ion electrode method (100-14,000 mg/L range) for determining TAN. Inexpensive color comparator kits using the salicylate method are reasonably accurate down to 50-100 µg/L TAN.

UV intensity can be monitored via a photocell covered by a filter that transmits only in the UV range (Rodriguez and Gregg, 1993).

Low Priority (becoming critical in days)

pH is unlikely to change suddenly and can easily be measured in sample beakers with a portable battery-operated or laboratory pH meter. Continuous automatic monitoring is possible.

Nitrite can be measured manually with chemical test kits. Spotte (1992) described a chemical (spectrophotometric) method for determining NO_2-N (5-180 µg/L range). Color comparator kits are reasonably accurate down to about 6 µg/L.

Nitrate usually is not a problem, but should be checked in recirculating systems and at least initially in the water source if contamination is possible. Spotte (1992) described two chemical (spectrophotometric) methods for NO_3-N (0.01-1 mg/L and 4-60 mg/L ranges). Color comparator kits are reasonably accurate down to about 60 μg/L.

Salinity usually will not change suddenly unless brine or fresh water is being added to the supply and the mixing mechanism fails. For this situation, continuous automatic monitoring would be helpful. Otherwise, regular manual measurements with a refractometer or battery-operated salinometer should be enough.

Heavy metals and other likely contaminants could be included in initial analyses and again later if suspected.

ENVIRONMENTAL CONSIDERATIONS

For moral and legal reasons, the effluent from a laboratory or fish farm should not damage the environment or natural populations (Billard and Sevrin-Reyssac, 1993). Obviously, toxic chemicals, drugs, and pathogens should not be released. Nutrients might have to be absorbed by a plant filter of some kind. Suspended solids will continue to break down and release more nutrients. An excess of nitrogen and/or phosphorus is the most likely cause for eutrophication in coastal waters, which often leads to habitat degradation. The most environmentally-acceptable feeds are highly digestible and contain just enough phosphorus to satisfy the requirements of the fish. Feed ingredients of low digestibility will add excess suspended solids to the culture water. Settleable and non-settleable solids should be removed soon after effluent leaves the rearing unit and without pumping, to minimize breakdown into smaller, harder to remove, particles (Ismond, 1993). For removal of solids from the primary effluent, settling, slow sand filtration (Drennan et al., 1993), bead filtration, or straining (Eikebrokk and Ulgenes, 1993; Malone et al., 1993) can be used. A combination of sieving and settling can be effective in removing and concentrating solids, which could then be used as fertilizer (Bergheim and Forsberg, 1993; Bergheim et al., 1993).

SUMMARY

To maximize survival, growth, and production of fish, harmful substances must be kept at safe minimum levels—suspended solids mainly by settling and straining; toxins by choice of water source, biological filtration, and chemical filtration; and pathogens mainly by straining and disinfection. At the same time, helpful substances and conditions (mainly salinity, temperature, oxygen) must be kept at optimal levels by water choice and control devices.

6 CULTURE UNITS

CONSIDERATIONS IN CHOOSING CULTURE UNITS

Factors involved in decisions on rearing unit design and size include: temperature, water quality, and feeding requirements of the fish; growth characteristics of the fish and expected production; site characteristics, site and construction costs, labor and energy costs, water cost; access and security requirements; environmental, social, and legal constraints.

CHARACTERISTICS OF CULTURE UNITS

Aquariums

Structure.—Aquariums usually are made of glass or clear plastic (e.g., acrylic) and range in size from about one liter to several hundred liters. Examples of single-piece containers are flasks, beakers, jars, dishes (Fig. 6-1), bowls, and molded plastic aquariums. Larger aquariums usually are assembled from glass or acrylic and concrete or wood. (A non-toxic compound such as silicone sealer should be used to bond glass.) Public aquarium tanks holding more than 21,000 m^3 exist.

Figure 6-1. Set of 3-L dishes for 1-week feeding experiments with early larvae (at HBOI).

Use.—In aquaculture, aquariums are used mainly for research, incubating eggs, or for spawning and rearing of ornamental species like clownfish and neon gobies. Usually, for eggs and larvae, water flows through and is not re-used. For grow-out, an external biofilter is common. For broodfish, external or internal (e.g., sub-gravel) biofilters are used.

Advantages.—Typical aquariums are clear, portable, and easy to clean. Conditions can be closely controlled, and fish can be closely observed. They are relatively inexpensive to operate, and with a small volume of wastewater, should have low environmental impact.

Disadvantages.—Small size of the unit can limit the scope of a project, particularly because fish usually grow more slowly when normal movement is hindered. Zooplanktonic prey could aggregate at the bottom, sides, and corners of rectangular tanks. Larval fish often do not seem able or willing to back out of a corner and are likely to swim into the corner until exhausted, then die. Larvae can be disoriented by reflections. I once tried to raise common snook larvae in 150-L aquariums outside under shadecloth. Once their eyes developed, the fish oriented to the reflected sun at the bottom and swam upside down until they starved.

Circular and Rectangular Tanks

Structure.—Circular tanks (Figs. 6-2 to 6-6) are better than rectangular ones because they require less water, aerate better, have uniform circulation, distribute food and fish better, and self-clean better (e.g., with center drain and circular flow), but rectangular tanks are easier to build. The Swedish pond actually is a square tank with rounded corners, first developed for Atlantic salmon (Piper et al., 1982). The Burrows pond is a rectangular tank, with a central lengthwise wall dividing most of the water so that it flows in a loop; drains are along the central wall.

Figure 6-2. Eight 10-L fiberglass tanks held in a temperature controlled water table, used for 2-week experiments with early larvae (at National Marine Fisheries Service Laboratory, Beaufort, North Carolina).

Figure 6-3. Fiberglass tanks holding 1 m^3, for long-term (months) nutrition experiments with juveniles (at HBOI).

Figure 6-4. Fiberglass tanks holding 3.5 m^3 (left) and 1 m^3 (right), for larval and juvenile rearing (at HBOI).

Figure 6-5. Fiberglass tanks holding 25 m^3, used mainly for juvenile and adult fish (at Oceanic Institute, Hawaii).

Circular and rectangular tanks usually are made of fiberglass, other plastics, concrete, or wood with epoxy coating or plastic liner. Fiberglass has the best qualities: inertness, strength, durability, portability, and ease of adding fittings. Polyethylene tanks are cheap, light, and portable, but scratch easily and are not as durable as fiberglass and not as easy to attach fittings to. They also can release somewhat toxic plasticizers (e.g., phthalates) into the water. Polypropylene tanks are similar but slightly more inert. Small tanks made of polycarbonate, polymethylpentene, or acrylic are clear, less toxic than softer plastics, less fragile than glass (but still breakable), and fairly easy to make connections to; they mostly are used for research, egg incubation, plankton culture, or *Artemia* hatching. Concrete (Figs. 6-6 to 6-10) is very durable but difficult to seal, harder to clean, usually impossible to move, and subject to cracking if the base is not supported well, if reinforcing steel corrodes, or if empty tanks heat and cool drastically. Wooden tanks are cheap and can be built in place but are not especially strong or durable. Epoxy coatings and plastic liners must be non-toxic.

Figure 6-6. Ferrocement tank holding 3 m^3 (more than 20 yr old), for juveniles and broodstock (at Micronesian Mariculture Demonstration Center, Palau).

Light colors provide much contrast and make fish easy to see, but reflections can disorient the fish. Larvae can be so disoriented and dazed by reflections at the meniscus that they refuse to move until the lights are off or very dim. Sometimes they actually swim up the walls and get stuck above water level. This also can happen (to a lesser degree) next to white objects in the tank such as vertical PVC standpipes. Rotifers, copepods, and especially *Artemia* tend to be attracted to the spot in the tank with the most illumination. If aeration or other circulation is not strong, prey patchiness occurs. If the patchiness makes it easier for the larvae to find and catch food, this is good. If it makes the fish become too crowded or the food and fish go to different parts of the tank, this is not good. An overly bright tank bottom can attract prey downward, but larvae of many marine fish have a strong urge to swim upward. The larvae tend to go to the brightest spot **at the surface**. With other things equal, both prey and fish tend to be attracted to glossy, light-colored walls but are relatively indifferent to dull, dark walls. Uniform lighting will help keep the prey and fish more evenly distributed.

Very dark colors are good for the fish but make them hard to see. A dull or matte (not glossy) black is best. Dark blue might be a second choice. Narrow, dull, white stripes can be painted on the bottom to make the fish and their food more visible to the culturist. Total white area should be no more than 50%; 25% might be safer. White paint on the outside of the tank will help prevent overheating in sunlight.

Use.—Short-term experiments with eggs and larvae can be done in small dishes or tanks (Figs. 6-1, 6-2). Hatchery tanks typically have diameters that range from 1 to 10 m. The minimum size for rearing typical marine larvae is 100 L—but not for long—they will get crowded, and water quality will fluctuate unless it is very carefully monitored and controlled. To avoid growth inhibition, juveniles probably need at least 400 L, and 1-kg fish at least 3000 L. Obviously, the more active species need more room. Larvae of some fish survive better in large tanks (e.g., 3-10 m^3), apparently because of the buffering effect of size.

Fiberglass tanks must be fully cured and cleaned before using them for eggs and larvae. If they have not fully cured during manufacture, months of weathering might be needed.

Fiberglass tanks need to have an interior gel coat that would be safe enough for use with human food and water. If the fiberglass and gel coat are fully cured, such a tank is probably safe for sensitive fish. It is still a good idea to scrub a new tank and soak it for several days before using it. Excess mold wax can be removed by steam cleaning before the tank is soaked.

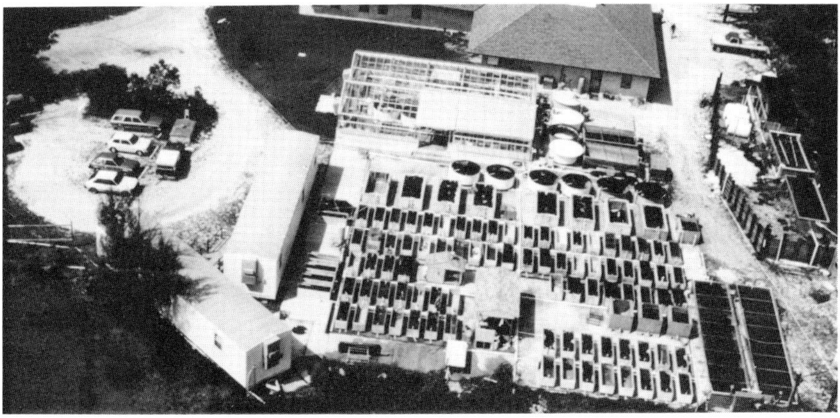

Figure 6-7. Rectangular concrete and aluminum tanks used mostly for raising rotifers, copepods, and seaweeds. The three tanks on the hill to the right are head tanks for the water systems. The six cylindrical fish tanks (above and right of center) are similar to those in Fig. 6-4 (2.4 m in diameter, holding 3.5 m^3). Larvae have been reared inside the greenhouse, but with difficulty because of temperature extremes. Unless temperature control is good, full greenhouses are better for juveniles than for larvae, but buildings that admit smaller amounts of sunlight (2-10%) can be very good for larval rearing (at HBOI).

Figure 6-8. Concrete tanks (or raceways) holding 750 L, good for zooplankton culture, spawning small fish, and raising limited numbers of small juveniles (at HBOI).

Figure 6-9. Concrete raceways holding 37 m^3 (3.0 × 12.2 × 1.0 m deep), used for spawning (e.g., groupers, seatrout) and grow-out (e.g., snook, groupers, sheepshead, seatrout) and 0.014-ha plastic-lined ponds used for minnow production and grow-out of larger species (e.g, snook, sheepshead)(at HBOI).

Figure 6-10. Two of the 37-m^3 raceways in Fig. 6-9 after being covered with a fiberglass roof (at HBOI).

For larvae, the drain can be covered with a screen with small enough mesh so that larvae cannot swim through it and of large enough surface area that larvae are not pulled into it by the current. This screen usually is cylindrical, fits over a normal stand pipe, and is sized for the tank volume and flow rate. Exchange can be improved if either (1) water flows into the tank near the surface and out of the tanks near the bottom (i.e., screened area of the drain is near the bottom), or (2) water flows in near the bottom and out near the top. Option 1 is safer for weak larvae, which tend to stay near the surface and are more likely to be impinged on a surface screen, but option 2 is better for strong larvae because the surface film can be removed with the discharged water. A third and probably the most commonly used option is for water to enter at the surface near or toward the tank wall and flow out from the surface at the center.

An outside standpipe with a bottom valve for complete draining is useful for juveniles and adults. A screen or grate over the drain might be necessary to keep fish from being pulled in. A tank bottom sloping toward the drain allows complete emptying. A horizontal plate fixed just over the drain with a small gap increases horizontal flow near the bottom and improves solid waste removal (Wheaton, 1977). An oval tank with a central divider increases water exchange but can reduce self-cleaning.

As Piper et al. (1982) observed, currents in rearing tanks usually should not be so strong that fish cannot easily maintain their positions. Energy loss, exhaustion, disease, and mortality could result. One exception is dolphin larvae, which need strong water movement to keep them from settling to the bottom.

Flexible plastic bags suspended in larger containers of water or in natural water bodies have been used for larval rearing and for grow-out, mainly in northern Europe (Skaar and Bodvin, 1993; Solaas et al., 1993).

Advantages.—Cylindrical or cylindroconical tanks are the most convenient units for larval rearing and some types of grow-out.

Disadvantages.—Tanks, especially large ones, can be expensive to buy and to operate.

Raceways

Structure and use.—Raceways are long rectangular tanks or ponds (Figs. 6-9, 6-10) usually made of concrete or soil, but occasionally of wood, and sometimes have plastic liners. Water is expected to flow from inlet to drain more effectively than in round or nearly square tanks. Most have dimensions in these ranges: 10-100 m long, 1-10 m wide, and 0.5-15 m deep. Salmon have been raised in plastic pipes in Norway (30 m long × 1.6 m diam.). Silos are another type of enclosed raceway that take little floor space, but a depth of more than 1 m introduces the danger of gas bubble trauma.

Advantages.—If enough water is available, raceways can be stocked at 1000 times the density of a pond (some species at 2203 mt/ha). Because access is easier, observing, feeding, and harvesting of fish are easier in narrow raceways than in wide ponds. The narrowness of raceways increases their potential for efficient water exchange.

Disadvantages.—If fish are densely stocked, much water is needed. Water quality deteriorates toward the outflow end, and fish tend to concentrate at the inflow end. Circulation might not be uniform, with partly stagnant areas resulting. Raceways do not self clean as well as circular tanks unless turbulence is high. More water (more expense) might be needed to flush a raceway. Active fish can have difficulty negotiating corners.

Ponds

Structure.—While tanks are built above ground, ponds are built mostly in the ground. Size might be 0.1 to 15 ha or more. Long narrow ponds can function as raceways. Usually, they should be placed perpendicular to prevailing winds, to reduce erosion from waves.

The bottom of the pond could be in a hole (excavated) or at ground level with built up earthen walls (embankment) or a combination (Wheaton, 1977). The walls can be stabilized by sloping them (Figs. 6-11, 6-14, 6-15) or by building a bulkhead (usually wood or concrete) to hold back the dirt (Fig. 6-16). The bottom should slope to a localized drain that has a basin around it for concentrating the last fish at harvest (Fig. 6-11, 6-12, 6-14). Larger ponds sometimes have lateral ditches leading to the drain.

Earthen ponds (Figs. 6-11 to 6-14) tend to be cheaper to build than lined ponds (Figs. 6-15, 6-16). Water chemistry might be improved by the more natural conditions. More natural foods could be produced. Increased vegetation could provide more habitat. If the soil is too porous, clay can be added as a sealer.

Figure 6-11. Freshwater nursery pond being drained to harvest juvenile barramundi to be stocked in a reservoir (at Brackishwater Fisheries Research Station, Walkamin, Queensland, Australia).

Figure 6-12. Closer view of the concrete harvest structure beyond the far bank in Fig. 6-11. As the pond was drained, the 35- to 100-mm fish first were caught in traps, then dipnetted from the concrete sump inside the pond (Fig. 6-11), and finally concentrated in and dipped from the sump outside the pond (at BFRS).

Pond liners usually are made of plastic. High density polyethylene (40-60 mils) and chlorosulfonated polyethylene (36-45 mils) were recommended by Warren et al. (1990). These are resistant to ultraviolet light and are more durable than PVC. Lined ponds are easier to harvest, easier to clean, and usually easier to maintain (the liner could help prevent erosion) than earthen ones. If lined ponds are left empty for long periods, damage to the liner from sunlight is accelerated and ground water could seep in below the liner and lift it.

Figure 6-13. A 0.1-ha earthen pond used for raising freshwater fish (at Eastern Freshwater Fish Research Hatchery, Grafton, New South Wales, Australia).

Figure 6-14. Closer view of the water control structure (monk = penstock = outlet tower) of the pond in Fig. 6-13 (at EFFRH).

Figure 6-15. Filling of a new plastic-lined pond. (photo by Robert A. Winfree)

Figure 6-16. Excavated ponds with wooden walls sealed with a roofing compound, used for saltwater cage and pond culture of tilapia (at Caribbean Marine Research Center, Lee Stocking Island, Bahamas).

Figure 6-17. Coastal, tidally-fed ponds (or impoundments) used for growing out several marine species (south coast of Taiwan).

Details of pond design, construction, and management are given in Milne (1972), Wheaton (1977), Boyd (1982), Giudice et al. (1982), Strawn et al. (1986), Barnabé (1990b), Chamberlain et al. (1990), Harrell et al. (1990), Manzano and Manzano (1990), Bose et al. (1991), Coche et al. (1997).

The criteria of Warren et al., (1990) for ponds for striped bass and/or hybrids are applicable for other freshwater and marine species. They recommended that nursery ponds be 0.2-0.4 ha in area with a bottom sloping from 0.8 to 1.8 m deep. Tops of berms should be at least 0.6 m above water level and have 2.5:1 to 3:1 inner slopes if the soil is suitable. Both shallow and deep ends should have water inlets, with enough flow capacity to fill the pond in a day. The harvest catch basin or kettle (Figs. 6-12, 6-14) should have a raceway-like configuration with water inlet and outlet and provisions for aeration and lighting at harvest time. There should be an all-weather road at least on the harvest side of each pond. Access is facilitated by a walkway around the pond and stairs to a walkway below the inner part of the drainage structure (Fig. 6-14).

Gravity-fed ponds depend on tidal flow rather than pumping for filling and water exchange (Fig. 6-17). Because tides vary, this type of pond tends to support lower stocking densities; however, supplemental pumping can overcome that disadvantage. If normal pond water level is lower than normal neap high tides and the pond bottom is higher than normal neap low tides, filling or emptying can be started on any day (Milne, 1972).

Use.—If the extra handling effort is justified, ponds rather than tanks can be used for temporary holding of fish. Broodstock can be kept in cages, pens, or open water in ponds for spawning. In nursery ponds, larvae are raised to juveniles. In production ponds, juveniles are grown to market size—sometimes in more than one phase in different-sized ponds. Catch-out or fee-fishing ponds contain fish to be caught by customers.

At harvest, farmed fish should be concentrated from ponds with built-in traps such as kettles (or portable traps) and then removed with nets. Seining of the main part of a pond should be avoided because it usually leads to trouble such as inefficiency, injured fish, and hydrogen sulfide poisoning from bottom sediments.

Advantages.—Water requirements and energy costs are lower than for tanks and raceways.

Disadvantages.—Much space is needed. Stocking density cannot be as high as in flowing water units. If the soil is too permeable and the water table too low, earthen ponds will lose water. Ground water should not be so fresh that it dilutes the culture water unacceptably. Likewise, drainage of rainwater might have to be redirected to prevent pond overdilution. Vegetation could become too dense and crowd the fish, cause extreme dissolved oxygen variation, or die and rot. If fish are to be raised in ponds built in dredge material containment areas (Coleman et al., 1991), sediments should be analyzed for toxic substances first.

Cages

Structure.—Fish cages (Fig. 6-18 to 6-24) are mesh containers that normally float. Walls of the cage need to be strong, light, durable, streamlined, manageable, smooth, resistant to fouling, and inexpensive (Beveridge, 1996). Netting of polyamide (nylon), polyester, polyethylene, or polypropylene is relatively strong and durable and considered good for cage and pen construction (Beveridge, 1996; Anonymous, 1986). Knotless netting is better than knotted netting because it is stronger, smoother, lighter, usually cheaper, less likely to abrade fish, and suffers less from drag and fouling. Nylon is smooth, soft, and flexible, but is expensive and breaks down very quickly (in a few months) in the sunlight of lower latitudes. Polyester is heavy and expensive. Polyethylene is slightly stiffer and can foul more quickly than nylon, but resists sunlight better, is cheap, and was preferred in Singapore (Anonymous, 1986). Polypropylene is expensive and not as durable. Galvanized wire, rigid plastic, and plastic-coated wire sometimes are used, but can be more abrasive to fish and difficult to manipulate. All metals cheap enough to be used for covering cages will corrode with time, first contaminating the local marine environment and later contaminating the disposal site when worn out. Zinc or copper from bare wire can be toxic both to the reared fish and to indigenous organisms. Milne (1972) compared characteristics of several cage wall materials and noted that even stainless steel was attacked and pitted by barnacles.

McCoy (1993) described three ocean cage systems. One was built like a ship with net holds. Another was a large sausage-shaped net with a pipe running through the center. Water or air is added to the pipe to control buoyancy. The cage can be raised halfway out of the water and a boat can enter for servicing the net and grading or harvesting the fish. The cage can be submerged completely for protection from storms. Normally the cage is partly exposed and can be rotated so that exposed sections are dried to control fouling. A third design uses weighted spar buoys and a well-designed spacing system to maintain the cage's shape but keep it flexible and resistant to storms (Figs. 6-23, 6-24). This design and its operation are described more fully in Loverich and Swanson (1993). A more recent design is shown in Fig. 6-25.

More information on cages can be found in the cited references plus Milne (1972), Anonymous (1990), Manzano and Manzano (1989), Barnabé (1990b), Loverich and Croker (1993), Myrseth (1993), Reinertsen et al. (1993), and Huguenin (1997).

Figure 6-18. Small cages for temporary holding of barramundi broodstock (near Weipa, Queensland, Australia).

Figure 6-19. A hapa, or cage for rearing small juveniles (about 1 × 1 × 1 m), particularly barramundi (at Singapore Department of Primary Industry Experimental Fish Farm).

Use.—Small cages of plankton netting sometimes are used to hold eggs or larvae, and larger ones of soft netting (Fig. 6-19) to hold juveniles. Examples of size for grow-out cages are: 1 meter cubes for research or catfish, 5 m × 5 m × 3 m for groupers in Asia, and an octagonal cage for salmon, with 16 m sides. Cages of 50-m diameter or more have been used for open water rearing of salmon and yellowtails. Cage volumes of 2000 m^3 have been used for yellowtail, 10,000 m^3 for salmon, and 50,000 m^3 for bluefin tuna (Huguenin and Ansuini, 1978; Barnabé, 1990a). One of the biggest Norwegian salmon cage farms is 146 m long × 40 m wide. A 20-m octagonal cage 30 m deep (55,000 m^3) could hold 1000 mt of salmon (Gunnarsson, 1993). Larger cages require more complicated handling methods and machinery and increase the potential cost if fish or equipment should be lost.

Cages are best placed in unpolluted waters with good circulation but protected from strong waves and currents. There should be at least 4-5 m of water below the cages at low tide, to reduce the chance of pollution, parasites, and other pathogens from the bottom affecting fish in the cages (Beveridge, 1996; Gowen and Bradbury, 1987; Robinson et al., 1992). Gowen and Bradbury (1987) suggested a minimum clearance of 15 m from the bottom. Such sites can be rare, and cages placed offshore can be quite vulnerable (Figs.

Figure 6-20. The government experimental cage farm (5 × 5 × 3 m units, 40 × 40 m overall) with two commercial farms in the background, where fish such as barramundi, groupers, jacks, and snappers are raised. Groupers also have spawned in some of these cages (at Singapore DPI).

Figure 6-21. Closer view of a cage in Fig. 6-20 (at Singapore DPI).

6-22, 6-24). Cages placed in inshore areas of the Mediterranean are not likely to encounter waves higher than 7 m (Barnabé, 1990b), but waves in the open ocean can be much higher. Salmon and sturgeon were raised in submerged cages at 30 m depth in the Caspian and Black Seas, with automated feeding possible during storms and fish surviving 10-12 m waves (Muravjev et al., 1993).

Barrier nets around the culture nets might be needed to exclude both aquatic and aerial predators, as well as drifting objects that could tear the inner nets.

Figure 6-22. Frame of a 6 × 6 × 3 m cage that was washed ashore on an island west of Taiwan by a typhoon.

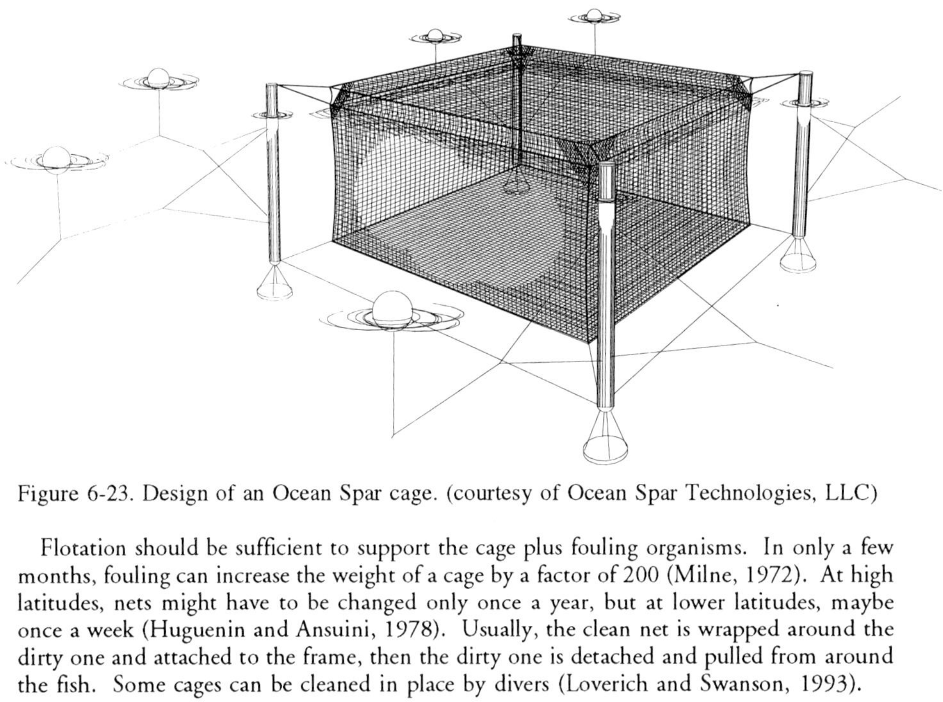

Figure 6-23. Design of an Ocean Spar cage. (courtesy of Ocean Spar Technologies, LLC)

Flotation should be sufficient to support the cage plus fouling organisms. In only a few months, fouling can increase the weight of a cage by a factor of 200 (Milne, 1972). At high latitudes, nets might have to be changed only once a year, but at lower latitudes, maybe once a week (Huguenin and Ansuini, 1978). Usually, the clean net is wrapped around the dirty one and attached to the frame, then the dirty one is detached and pulled from around the fish. Some cages can be cleaned in place by divers (Loverich and Swanson, 1993).

Advantages.—Small cages are inexpensive. Less land is needed than for tanks and ponds. The cost of pumping water is avoided. If an alternative mooring site is available, cages are reasonably portable and often can be towed short distances while fish are in them. This is an advantage if conditions deteriorate at the initial site. If water currents are 10-60 cm/s (0.2-1.2 knots) with only brief slack water periods, exchange should be good (Beveridge, 1996). Waste falls out through the bottom and does not accumulate in the cage. Fish usually can be easily harvested from cages.

Disadvantages.—Boats might be needed for tending cages and feeding fish. Frames and netting are likely to accumulate fouling organisms, and maintenance effort for cleaning and replacing nets can be substantial. In areas with rapid macroalgae growth or settlement of sessile animals, medium-mesh nets might have to be replaced as often as every month and small-meshed ones every week. If the water is shallow, wastes accumulating below cages can reduce water quality. Also, parasites could migrate from the bottom to fish in the cages. Disease organisms cannot be excluded, control can be difficult, and medicated feed might be the only possible treatment method. Cages and their inhabitants can be vulnerable to high winds, high waves, strong currents, pollution, red tides, aquatic predators, drifting debris, theft, and vandalism. Currents stronger than 100 cm/s (2 knots) are too fast for most cages (Beveridge, 1996). Floods and erosion can loosen cages from their moorings. Cages for high energy environments can be especially expensive and difficult to service. Offshore cage farms incur higher costs for cages, boats, grading, and transportation of fish, feed, and workers. Even though the fish might be safe, storms interfere with operations.

Figure 6-24A. Ocean Spar cage installed next to an oil drilling platform in the Gulf of Mexico. (both A & B courtesy of Ocean Spar Technologies, LLC)

Figure 6-24B. The same cage during rough weather.

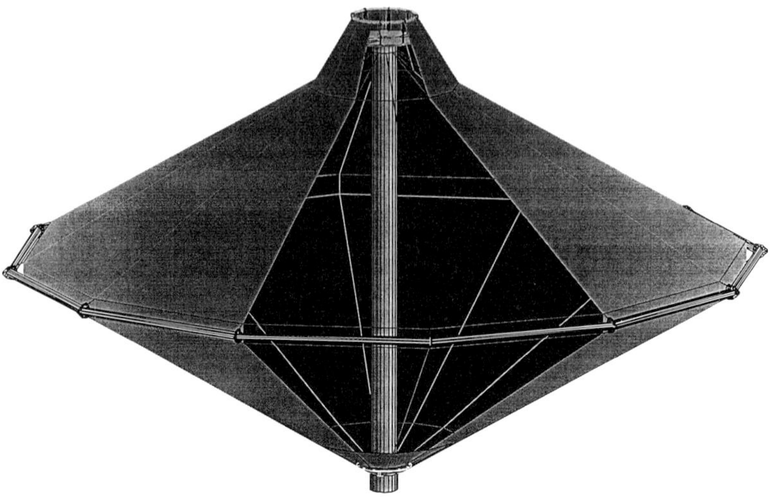

Figure 6-25. Design of the SeaStation™ cage by Ocean Spar Technologies—a fully submersible monospar sea cage design with capacity of about 3000 m^3. (courtesy of Ocean Spar Technologies, LLC)

Pens

Structure.—In a pen, fish are surrounded by mesh walls anchored to the bottom, with or without a top. Typical size might be 20 m square and 2-7 m deep, but pens could be almost any size depending on the site. Beveridge (1996) gives a size range of 0.1 to much more than 1000 ha. Netting similar to that used for cages can be used for pens except replacement is more difficult and rigid netting might be preferred if it is easier to clean *in situ.* Al-Thobaiti et al. (nd) suggested PVC-coated nylon as a durable material for pen netting in the tropics. Even with good circulation, a 1-m minimum depth at the lowest tide is necessary to protect fish from sunburn. For ease of management and harvesting, 2-3 m in many cases would be suitable. Obstructions inside the pen should be minimal, but some fish grow better if they have hiding places. Removable objects such as old tires could be added. Because pens are not easily moved, protection from waves, strong currents, and erosion is important, but sufficient current for flushing of wastes is needed. Building pens perpendicular to the prevailing currents can improve water exchange.

Use.—Al-Thobaiti et al. (nd) suggested initial pen stocking densities of 20/m^3 for herbivores, 15/m^3 for planktivores, and 10/m^3 for carnivores in Saudi Arabian marine waters.

Advantages.—Pens are relatively cheap to build and operate. They can be made in any reasonable shape and size. Natural currents and tides can be used to exchange water.

Disadvantages.—Much space is needed. Pens usually are not portable or easy to disassemble. They and their contents can be vulnerable to high winds, high waves, strong currents, floods, erosion, pollution, red tides, aerial and aquatic predators, drifting debris, theft, and vandalism. Debris is very likely to pile up against pens. The framework and netting is likely to collect fouling organisms very quickly, reducing water exchange and increasing the possibility of storm damage. Supplementary pumping might be necessary to ensure water exchange. Nets might be difficult to change, necessitating their cleaning in place. Circulation is likely to be limited and fish cannot be stocked very densely. Disease organisms cannot be excluded. Waste can build up on the bottom, and parasites that need benthic organisms or sediments to complete their life cycles are more likely to infest fish in pens. It can be difficult to harvest fish from large pens. Pens can be navigational hazards.

Impoundments

Structure.—An impoundment is a diked off portion of a natural water body that functions like a pond or lake or is intermediate between a pond and a pen. The units in Fig. 6-17 could be called ponds or impoundments. Various designs are described in Milne (1972).

Use.—Impoundments are used like ponds. If natural currents or tides are relied upon for water exchange and they vary, stocking density might have to be reduced.

Advantages.—Impoundments could be set up to have better water exchange than ponds.

Disadvantages.—Impounding natural waters at least alters and could eliminate natural habitat. Depending on dimensions and water flow, circulation might not be good. For example, exchange potential would be higher during spring tides than during neap tides. As in a pond, stocking density is limited by water exchange rate. Disease organisms cannot be excluded.

Ocean (ranching)

Mechanism.—Reared fish are released into a natural water body with the expectation that they will be available for capture when they reach harvest size. Ranching works somewhat for non-migratory fish or homing fish like salmon (Thorpe, 1980). Non-migratory fish are hoped to stay where they are put to be harvested. Homing fish are hoped to return to the imprinted "release site" to be harvested. The results of ranching are more predictable in closed bodies of water from which the fish cannot escape and in which they can be found and caught. Red sea bream have been trained to stay in open water areas by a combination of feeding and underwater sound (as a "dinner bell") given three times/d (Foscarini, 1988) and actually have been led to release sites by music played under a boat.

Advantages.—Care of the fish is necessary only during early rearing. The major costs of grow-out feeds and labor are avoided.

Disadvantages.—The fish might be lost to natural causes or fishing. They might not return as expected. Ownership of the fish might be disputed.

APPLICATIONS OF CULTURE UNITS

Holding

Fish temporarily being held for stripping of gametes or other processing can be housed in tanks (Fig. 6-3) or cages (Fig. 6-18) just large enough for them to swim.

Spawning

Voluntarily-spawning marine fish (see Table 7-7) need to be uncrowded and held under simulated natural spawning season conditions, or another combination of factors that will work. Broodstock mostly are kept in tanks (Figs. 6-9, 6-10; Table 6-1), but in a number of cases ponds or cages (Fig. 6-21; Tables 6-2, 6-3) have been used. Traditional Japanese spawning tanks for marine fish are circular, concrete, hold 40-100 m^3, and are outside (Kuronuma and Fukusho, 1984). Gregarious fish like sea bream spawn easily at high densities, but extremely territorial ones like some damselfish can survive and spawn in a moderate-sized tank only as a pair. Simulated natural habitat might improve broodfish health and spawning—for example, sand bottoms for turbot (McEvoy, 1989; Person Le-Ruyet et al., 1991). Spawning substrates might be necessary for the fish or practical for the culturist—for example, removable glass or plastic plates to which damselfish can attach their eggs.

Egg Incubation

Eggs need relatively constant temperature. Higher temperatures require more safeguards against water quality deterioration and bacterial buildup. Attached eggs usually need to be

Table 6-1. Examples of tanks used for spawning (also see Table 7-6).

Species	Shape[a]	Material[b]	Volume (m^3)	Dimensions[c] (m)	Inside[d] color	No. of fish[e]	Age[f] (mo)	Size[f]	References
Engraulidae—anchovies									
Northern anchovy	Cy	Pls	12	4.6D × 0.72H		~1,000			Lasker, 1974
Gadidae—cods									
Atlantic cod	L	Cnc/Gls	10.8	1.3H		7♀,6♂		48-73 cm	Brawn, 1961
	CyHe	PVC-N	175			152♀,116♂	3.6	3.2	Holm & Anderson, 1989
			19.5			1♀,2♂	48	~2.8-6.7 kg	Kjesbu, 1989
Gobiesocidae—clingfishes									
Skilletfish	Re	Acr,Wd	0.113,0.076			1♀,1♂		41,69 mm SL	Martin & Martin, 1971
Atherinidae—silversides									
Topsmelt	Cy		1.4	1.5D × 0.8H		13-21♀,16-20♂		19-65,10-63 g	Middaugh et al., 1992
Centropomidae—snooks									
Barramundi	Re,Cy	Cnc	80-100	≥2H	Gry	18♀,6♂	>36,24-48	≥3.5 kg	Tookwinas, 1990a
Percichthyidae—temperate basses									
European sea bass	Aq		12			6-8♀,12-16♂		0.3-3.6 kg	Zanuy et al., 1986
	Cy	PVC	2		Dark	10♀,10♂	>36,>24	0.8-3,0.3-1.5 kg	Carrillo et al., 1995
Serranidae—groupers, sea basses									
Redspotted grouper	Cy	Cnv	50	8D × 1H		25♀,25♂			Maruyama, p.c., 1992
Orangespotted grouper		Cnc	90			~20♀,4-5♂			Hussain & Higuchi, 1980
Malabar grouper	Re	Cnc	75	4 × 10 × 2H	Gry		36	3-4 kg	Maneewong et al., 1986; Tookwinas, 1990a
Kelp grouper	Sq	Cnc	72	6 × 6 × 2H		2-3♀,1♂			Maruyama, p.c., 1992
Camouflage grouper	Cy		25			2-3 kg/m^3		54-67 cm TL, 2-6 kg	Al-Thobaiti & James, 1997
Nassau grouper	Re	Cnc	37	3.4 × 12.2 × 0.9H	Gry	2-4♀,2♂	≥60	3-8 kg	Tucker & Woodward, 1996
Grammatidae—basslets									
Royal gramma	ReAq	Gls	0.189		Clr	3♀,4♂		25-50 mm	Rosti, 1967
Plesiopidae—longfins									
MacNeill's mini-grouper	ReAq		0.3			1-2♀,1♂[+]		32-38 mm SL	Allen & Kuiter, 1976
Apogonidae—cardinalfishes									

Flamefish	ReAq	Gls	Small		Clr	1♀,1♂$^{+}$		75 mm	Coleman, 1966
Carangidae—jacks									
Bigeye scad	Cy	FRP	9.5	5D × 4.8H		12-72		~124 mm FL, 23 g	Iwai et al., 1996
Mazatlan yellowtail	Re	Cnc	100	5 × 10 × 2H		12		2-8 kg	Benetti, 1997
Yellowtail			90-400			1-1.5♀:1♂	48-72	5-12 kg	Imaizumi, 1993
Coryphaenidae—dolphins									
Dolphin	Do	FRP	10	6.6D		100		1 kg	Kraul, 1990b, 1991
	Do	FRP	~10	6D		2♀,1♂	≥5-7	≥1.5-2.5 kg	Kim et al., 1993
Lutjanidae—snappers									
Atlantic red snapper	Re	FRP	29.9	3.3 x 6 x 1.5H		8		1-3 kg	Arnold et al., 1978
Bluestriped snapper	Aq	Cnc/Gls	4.6	3.3 m^2 x 1.3H	Clr	7♀,6♂,7?$^{+}$		0.2-0.9 kg	Suzuki & Hioki, 1979
Sparidae—sea breams, porgies									
Common dentex			15			50		0.4-2.5 kg	Bibiloni et al., 1993
Pink dentex	Re	FRP	6			11		2.5-3 kg	Fernandez-Palacios et al., 1994
Red sea bream	Re,Cy	Cnc	50-120	1-2H	Gry	90-110♀,90-110♂	36-144	1-3 kg	Fukusho, 1991; Watanabe & Kiron, 1995
			90-400			1-1.5♀:1♂	36-96	2-5 kg	Imaizumi, 1993
	Cy	Cnc	~157	10D × 2H		~157		1-3 kg	Morizane, 1993
	Oc	Cnc	30	5 × 5 × 1.8H		30-50		1-2 kg	Tsujigado & Lee, 1993
Lethrinidae—sweetlips									
Spangled emperor	Re		60	5 × 7.4 × 1.6H		12	~60	2-3 kg	Yonamine, 1989
Sciaenidae—drums									
White seabass	Cy		16.4	4.57D × 1H		23			Orhun, 1989
Spotted seatrout	Re	FRP	~30	3.3 × 6 × 1.5H					Arnold et al., 1976
	Re	Cnc	37	3.4 x 12.2 x 0.9H	Gry	1-7♀,10♂		0.4-2.1 kg	Tucker & Faulkner, 1987
Highhat	SqAq	Gls		0.91 × 0.91 × ?H	Clr	1♀,1♂		13-15 cm	Straughan, 1968
Red drum	Cy	FRP	30	3.3 × 6 × 1.5H		2♀,2♂		10-15 kg	Arnold et al., 1977; Arnold, 1988
	Cy	FRP	5.9	3D × 0.83H		2♀,2♂		3-8 kg	Roberts et al., 1978
	Cy	FRP	13			2♀,2♂		8-18 kg	McCarty, 1990
Chaetodontidae—butterflyfishes									
Blackloin butterflyfish	ReAq	Cnc/Gls	3.6	2.8 m^2 × 1.3H	Clr	24^{+}		44-126 g	Suzuki et al., 1980
Pomacanthidae—angelfishes									
Rusty pygmy angelfish	ReAq	Acr	1.95	1.3 × 2 × 0.75H	Clr	1♀,1♂		66,78 mm	Hioki et al., 1990
Japanese pygmy angelfish	ReAq	Acr	1.95	1.3 × 2 × 0.75H	Clr	3♀,1♂		14-15,17 cm	Hioki & Suzuki, 1987
Cichlidae—cichlids									
Nile tilapia	Cy	Cnc		4D	Gry	50♀,50♂	≥6		Macintosh & Little, 1995
Oreochromis spilurus	Re		9	3 × 3 × 1H		30♀,10♂	≥4	0.2-0.5 kg	Hazza et al., 1989
	CC	Cnc	100		Gry	3♀,1♂/m^2	≥4	~125,225 g	Hazza et al., 1991

Table 6-1. (Continued)

Species	Shape[a]	Material[b]	Volume (m^3)	Dimensions[c] (m)	Inside[d] color	No. of fish[e]	Age[f] (mo)	Size[f]	References
Florida red tilapia			34.2			180♀,60♂	12		Watanabe, 1991
Pomacentridae—damselfishes									
Common clownfish	ReAq	Gls	0.2		Clr	1♀,1♂		80,60 mm	Moe, 1973
Percula clownfish	ReAq	Gls/SB	0.17	0.41 × 1.02 × 0.41H	Clr	1♀,1♂			Mitsch, 1941
	ReAq	Gls	0.050		Clr	1♀,1♂			Garnaud, 1951
Domino damselfish	ReAq	Gls	0.1		Clr	1♀,1♂		≤10,≤15 cm	Garnaud, 1957
Nagasaki damselfish		Acr	0.06			1♀,1♂		67-72,73-75 mm	Honda & Imai, 1973
Whitetail damselfish	ReAq	Gls/SB	1.2	0.91 × 1.22 × 1.07H	Clr	1♀,1♂			Breder & Coates, 1933
Bicolor damselfish	ReAq	Gls	0.8	0.65 × 2.5 × 0.6	Clr	5♀,3♂		4-6,7-8 cm	Myrberg, 1972
Gobiidae—gobies									
Neon goby	ReAq	Gls/SB	0.009		Clr	1♀,1♂			Boyd, 1956
	ReAq	Acr/GB	0.094		Clr	1♀,1♂			Musgrave, 1976
Mugilogobius abei	ReAq	Gls/Slt	0.12	0.3 x 1.14 x 0.43	Clr	5♀,5♂			Kanabashira et al., 1980
Ephippidae—spadefishes									
Atlantic spadefish	UAq		40	2.5 × 9.5 × 2.5H	Clr	5^+		23-30 cm SL	Walker, 1991
Siganidae—rabbitfishes									
Yellowblotch spinefoot		Pls	0.45	0.55H		1♀,1-2♂		0.1-0.7 kg	Hara et al., 1986
Scombridae—mackerels									
Chub mackerel	Cy			7.3D		~200		~0.5-0.7 kg	Lasker, 1974
Atlantic mackerel	Cy	Cnc	54	10D × 0.69H		~1♀:1♂	$2\text{-}10^+$	31-42 cm	Walsh & Johnstone, 1992
Bothidae—lefteye flounders									
Southern flounder	Re	FRP	29.9	3.3 × 6 × 1.5H		6♀,6♂		1-3 kg	Arnold et al., 1977
Chilean flounder	Cy		1.0			3♀,2♂		420-734 g	Silva, 1994
Hirame			90-400			1-1.5♀:1♂	36-72	2-3 kg	Imaizumi, 1993
	Re	Cnc	50	5 × 9 × 1.5H					Tsujigado & Lee, 1993
Ecuadorian flounder	Cy	Cnc	50	6D × 1.77H		6♀,12♂		5-10,1-3 kg	Benetti, 1997
Turbot	Cy	FRP/CB	40	5.1D × 2.7H		4♀,4♂	48-60	2.8-6.3,2.1-3.5 kg	Bromley et al., 1986
		SB	40						McEvoy, 1989
Soleidae—soles									

Common sole	Re,Cy		9-14		1-6/m^2	0.2-1 kg		Baynes et al., 1993
Tetraodontidae—puffers								
Porcupinefish	ReAq	Gls	0.113	Clr	1♀,1♂		15-18 cm	Wolfsheimer, 1957

These notes apply to all tables in this chapter:

[a] Bg = bag, Co = conical, Cy = cylindrical, CyCo = cylindroconical, CyHe = cylindrohemispherical, Do = donut-shaped, Oc = octagonal, Re = rectangular, ReAq = rectangular aquarium, Rsv = reservoir, SC = semicircular, Si = silo, Sq = square, UAq = u-shaped aquarium; CW = with a center wall.

[b] Acr = acrylic, Cnc = concrete, Cnv = canvas, FRP = fiberglass reinforced plastic, Gls = glass, Pls = plastic, PVC-N = PVC-coated nylon fabric; CB = concrete bottom, GB = gravel bottom, SB = sand bottom.

[c] in meters unless other units are given; D = diameter, H = water depth.

[d] Blk = black, Blu = blue, Clr = clear, Gry = gray, Grn = green, Wht = white.

[e] in Tables 6-1 to 6-3, (unless otherwise stated) number of broodfish per container; in Table 6-4, number of eggs or larvae stocked per liter unless other units are given; in Tables 6-5 to 6-8, initial stocking density by number, values in parentheses are final densities by number or weight.

[f] in Tables 6-1 to 6-3, the average or range of ages and weights of broodstock: females, males; in Table 6-4, initial and final ages (an age of 0 means that eggs were stocked and an age of 1 or more means that larvae were stocked); in Tables 6-5 to 6-8, initial and final ages and weights; a number listed in parentheses is duration rather than age; smr = summers.

[g] typical, average, or total, unless otherwise noted; [h] best. [+] other fish species were in the same tank.

[p] parturition.

e = egg; *l* = larva; *j* = juvenile.

Table 6-2. Examples of ponds used for spawning.

Species	Material[b]	Area (ha)	Depth (m)	No. of fish[e]	Age[f] (mo)	Size[f]	References
				Chanidae—milkfish			
Milkfish	Cnc/SB	0.075	1.3	15♀,15♂	11	4.5-7.5 kg	Lin, 1985
				Gadidae—cods			
Atlantic cod	Cnc/Wd	0.011	3	1,600♀,400♂			Dannevig, 1910
				Cyprinodontidae—killifishes			
Gulf killifish	E	0.8	1	500♀,500♂	~36		MacGregor et al., 1983
				Serranidae—groupers, sea basses			
Malabar grouper	Cnc/SB	0.075	1-1.3				Lin, p.c., 1992
				Carangidae—jacks			
Permit	Cnc/SB	0.075	1-1.3				Lin, p.c., 1992
				Cichlidae—cichlids			
Nile tilapia	E	0.17		158♀,158♂		>200 g	Macintosh & Little, 1995
Florida red tilapia	E	0.01	0.5	60♀,20♂		≥30 g	Le Coz et al., 1990

Table 6-3. Examples of cages used for spawning.

Species	Material[b]	Volume (m^3)	Dimensions[e] (m)	No. of fish[e]	Age[f] (mo)	Size[f]	References
				Gadidae—cods			
Atlantic cod	PVC-N	175		63♀,23♂		8.9,6.5 kg	Huse & Jensen, 1983
	PVC-N	175		152♀,116♂		3.6,3.2 kg	Holm & Andersen, 1989
				Centropomidae—snooks			
Barramundi	PE	75	5 × 5 × 3H	<80	>36,>24	≥3 kg	Lim et al., p.c., 1992
				Serranidae—groupers, sea basses			
Brownmarbled grouper	PE	75	5 × 5 × 3H	10♀,10♂		5.4-13.6 kg	Lim et al., 1990; Chao & Lim, 1991
				Scombridae—mackerels			
Northern bluefin tuna		5,655	30D × 8H		~60		Harada, 1980

continually swept with well-oxygenated water and some require jets of water on them just before hatching for assistance. Unattached demersal eggs need good oxygenation and usually at least slight movement to keep them from sticking together or depleting oxygen locally. Salmonid eggs (can take 60-80 d to hatch) usually are incubated in hatching troughs with water flow like in a raceway or in tall or short jars with water upwelling (e.g., MacDonald type). Tabachek et al. (1993) described a 32-jar system. Incubation of white sturgeon eggs (7 d to hatch) in two stages with different upwelling containers was described by Conte et al. (1988). Eggs first were held between horizontal screens in a modified trout egg barrel 20.3 cm in diameter and 51.0 cm high. Then, from neurulation to hatching, the eggs were held in modified MacDonald jars 20 cm in diameter and 50 cm high (13-L). Modified MacDonald jars or round tanks are used for incubating striped bass and hybrid bass eggs (2 d to hatch)(Mullis and Smith, 1990).

Planktonic eggs need good oxygenation and gentle movement to prevent them from sticking together or depleting oxygen. A large surface to volume ratio and carefully controlled circulation helps keep the eggs from piling up. I prefer a 300- to 1000-L round fiberglass tank. In Japan, marine fish eggs sometimes are incubated in fine-mesh cages held in 8- to 10-m^3 tanks (mostly fiberglass, plastic, or wooden) inside the hatchery (Kuronuma and Fukusho, 1984). Temperate marine planktonic eggs can require up to 18 or more days to hatch (typically more than 2 d), but most warmwater planktonic eggs take only 0.5-5 d (typically 2 d or less, Table 2-1). Longer periods often necessitate more elaborate systems and effort for the incubation phase. Devauchelle et al. (1986) and Brownell and Horstman (1987) described laboratory tank systems for maintaining eggs and larvae of marine fish. Person-Le Ruyet et al. (1991) described an elaborate system for incubating turbot eggs, which hatch 3-7 d after fertilization.

Larval Rearing (Table 6-4)

Typical marine fish larvae do not tolerate sudden environmental changes as well as older fish. Constant or slowly changing temperature and salinity usually are best. Containers should not have areas in which larvae could easily be trapped. Larvae and early juveniles usually are reared in tanks made of fiberglass, concrete, etc. and sometimes in nursery ponds. If marine species are to be raised from eggs, a saltwater hatchery is needed during the egg, larval, and early juvenile stages. When fish have been weaned and are strong enough to be handled, they can be moved to cages or grow-out ponds. Intensive rearing in tanks is suitable for most larvae, and extensive rearing in fertilized nursery ponds works for certain species if the weather is stable enough. Recirculating water treatment systems can work, but are complicated for larvae.

Williot (1990) described two methods for rearing sturgeon larvae: (1) cylindroconical concrete tank for the first 3-4 weeks, then a pond for 1 month; (2) cage submerged in a pond for 5-10 d, then larvae free in the pond for 25-40 d. Typical survival was 20-50%.

In Japan, coho salmon larvae and early juveniles have been raised in rectangular raceways, elliptical ponds, and circular concrete or plastic tanks, sometimes in greenhouses (Mahnken, 1991). A typical saltwater acclimation tank for smolts is 5 m × 5 m × 2.5 m deep. In Scandinavia, Atlantic salmon larvae and early juveniles usually are reared in shallow fiberglass tanks with 1-4 m^2 of surface area (Isaakson, 1991).

In Japan, larval rearing tanks usually are circular or rectangular, but sometimes other shapes (e.g., octagonal), are of concrete, hold 0.5-200 m^3, and usually are located outside under shade or inside under skylights (Kuronuma and Fukusho, 1984). Such tanks are common in Asian fish culture.

Sweetman (1992) reported that European larval rearing tanks were mostly in the range 1-30 m^3 for intensive culture and as large as 250 m^3 for semi-intensive culture. Shape ranged from deep cylindroconical to shallow cylindrical. Colors included all black, different shades of gray, and black-sided with a white bottom.

Table 6-4. Rearing units used successfully for larvae of representative species (also see Table 8-4).

Species	Type[a]	Material[b]	Volume (m^3)	Dimensions[c] (m)	Inside color[d]	Density[e] (*e* or *l*/L)	Age[f] (dah)	Survival[g] (%)	References
Acipenseridae—sturgeons									
White sturgeon	Cy	FRP	0.5	1.22D × 0.46H		15-20	1-	50	Conte et al., 1988
Acipenser spp.	CyCo	Cnc		3D			~5-28	55-70	Williot, 1990
Clupeidae—herrings									
Japanese sardine	Sq	FRP	0.7	1.2 × 1.2 × 0.7H		2.1	*e*-80	8	Nakamura et al., 1991
Chanidae—milkfish									
Milkfish	Cy	FRP	0.7	1.25D × 0.6H		2.7	1-21	47[h]	Liao et al., 1979
	Re	Cnc	16	3.5 × 4.5 × 1H	Gry	6.3	1-22	42	Lin, 1985
	Cy	FRP	4	2.5D × 0.8H		12.6	*e*-30	46[h]	Eda et al., 1990b
	Re	Cnc/SB	200-360	200-300 m^2 × 1-1.2H		2-3	*e*-22	~30-50	Chang et al., 1993
Gadidae—cods									
Atlantic cod	Pd	E	30,000-100,000	5-6H			4-50	1-2/m^3	Huse, 1991
	Bg	Pls	113-301	6-8D × 4-6D	Blk	1-3	~1-35	50[h]	Tilseth et al., 1992
		Pls	20.5		Grn	0.2	~1-24	~26	Zhuravleva & Petrov, 1993
	Bg	Pls	5.2		Blk	0.15	1-48	83[h]	van der Meeren et al., 1994
Atherinidae—silversides									
Topsmelt	ReAq	Gls	0.018		Clr	5	1-24	100[h]	Middaugh et al., 1990
California grunion	ReAq	Gls	20	~1.8D × 2.5H	Clr	10	1-40	81[h]	Vásquez-Yeomans & Sosa-Cordero, 1992
Scorpaenidae—scorpionfishes									
Kurosoi	Re	FRP	15	3.5 × 5 × 1.25H	Grn	6.2	p-35	50	Kusakari, 1991
Centropomidae—snooks									
Common snook	RePd	E		0.8 ha		22/m^2	3-32	25	Maciorowski et al., 1987
Barramundi	Cy	FRP	5	2.8D × 0.8H		20-30	1-20	75[h]	Lim et al., 1986
	Re	Cnc	40	3.9 × 6 × 1.75H		10-15	1-20	65[h]	Lim et al., 1986
	CyCo	FRP	0.45		Gry	33	2-25	82[h]	AQUACOP et al., 1990a
			0.002			272	1-22	~100[h]	Rimmer & Reed, 1990
		Cnc	5-25		Gry	50-140	1-25	~76[h]	Tookwinas, 1990a
Percichthyidae—temperate basses									
European sea bass	SC	FRP	1.2		Blk	5-15	1-60	40	Katavic, 1986

	CyCo	FRP	0.5			95	1-30	81^h	Cerqueira & Chatain, 1991
	CyCo	FRP	4		Blk	100	1-45	35-65	Coves et al., 1991
Australian bass		PE	0.040			2.5	~3-21	85^h	van der Wal & Nell, 1986
		PE	0.040			2.5	~21-35	78^h	van der Wal & Nell, 1986
Striped bass	Pd	E				2.6	7-52	72^h	Colura et al., 1976
Striped bass & hybrids	Pd	E	200			~43/m^2	6-40	25-40	Brewer & Rees, 1990
Palmetto bass	Pd	E				12.8/m^2	~2-41	75	Smith, 1989b
				Serranidae—groupers, sea basses					
Redspotted grouper	Sq	Cnc	60			5-28	1-47	34	Maruyama et al., 1994
Blackspotted grouper	Re	FRP	1			12-15	1-40	25^h	Tseng & Chan, 1985
Orangespotted grouper	Cy		5			8	*e*-43	12^h	Doi et al., 1991
	Cy	FRP	4			~10-20	*l*-*j*	13	Almatar et al., 1997
Brownmarbled grouper	CyCo	FRP	0.8	0.12D × 1.2H		40	*e*-12	24	Chao et al., 1993
	Cy	FRP	0.9	0.5D × 1.2H			13-30	91	Chao et al., 1993
	Cy	FRP	5.0	2.5D × 1.2H			31-40	48	Chao et al., 1993
Malabar grouper	Re	Cnc	25	3.8 × 5.8 × 1.2H	Gry	8	1-60	5^h	Maneewong et al., 1986
		Cnc	5-25		Gry	10-60	1-50	1-4	Tookwinas, 1990a
	Re	Cnc,SB	500	20 × 25 × 1H			*e*-40	~30^h	Lin, p.c., 1992
Camouflage grouper			2.8			30	*l*-50	4.7^h	Al-Thobaiti & James, 1997
Nassau grouper	ReAq	Gls	0.11	0.4 × 0.9 × 0.3H	Clr	6.7	*e*-15	25^h	Tucker & Woodward, 1996
	Cy	FRP	3.5	2.4D × 0.75H	Blk	0.3	*e*-98	5^h	Tucker & Woodward, 1996
			34			7	1-62	1.4^h	Watanabe et al., 1995
				Pseudochromidae—dottybacks					
Orchid dottyback	Aq	Gls	0.076			~13	1-36	~35^h	Moe, 1997a
				Carangidae—jacks					
Yellowtail		Cnc	90-150			10	*l*-*j*	5-10	Imaizumi, 1993
				Coryphaenidae—dolphins					
Dolphin	CyCo	FRP	4	2.4D × 1H	Dk	2	2-60	10	Kraul et al., 1988; Kraul, 1991, 1993
	CyCo	FRP	4	2.6D × 0.75H	Blk	10-15	*e*-20	5	Kim et al., 1993
				Lutjanidae—snappers					
Asian red snapper	CyCo	Cnc	3			30	1-55	~16^h	Duray et al., 1996a
Golden snapper	Cy	FRP	5	1.4D × 0.8H		20-25	*e*-35	5^h	Lim et al., 1985
				Gerreidae—mojarras					
Striped patao	Sq	Cnc	5			80	*e*-48	12.5^h	Alvarez-Lajonchère et al., 1996
				Sparidae—sea breams, porgies					
Black sea bream		Cnc	45			10	~1-100	28^h	Leu, 1997
Sheepshead	Cy	FRP	3.5	2.4D × 0.75H	Blk	4	*e*-100	40	Tucker, 1987a
Caribbean sea bream	ReAq	Gls	0.076		Clr/Blk	2	*e*-16	80^h	Dowd & Houde, 1980

Table 6-4. (Continued)

Species	Type[a]	Material[b]	Volume (m^3)	Dimensions[c] (m)	Inside color[d]	Density[e] (*e* or *l*/L)	Age[f] (dah)	Survival[g] (%)	References
Sharpsnout sea bream	Co	FRP	2			~65	1-60	17[h]	Caggiano et al., 1993
Red sea bream	Re,Cy	Cnc	50-100	1-2.5H	Gry	12-72	~1-26	40-50	Fukusho, 1991
		Cnc	90-150			10	*l-j*	30-80	Imaizumi, 1993
	Re/CW	Cnc	60	5 × 9 × 2H		20	*l-j*	10	Tsujigado & Lee, 1993
Silver sea bream	Re	Cnc	5			10	1-45	43[h]	Leu, 1994
Gilthead sea bream	Cy	FRP	3		Wht	~70	1-60	8[h]	Tandler & Helps, 1985
	CyCo	FRP	0.5		Blk	100	1-30	45	Chatain & Ounaïs-Guschemann, 1991b
	CyCo	FRP	0.5		Wht	100	1-30	21	Chatain & Ounaïs-Guschemann, 1991b
	CyCo		0.5				*e*-55	47[h]	Kentouri et al., 1993a
	CyCo	FRP	0.4-1.7			100	*e*-32	20-40	Tandler, 1993
Sciaenidae—drums									
White seabass	ReAq	Gls	0.030	0.25 × 0.8 × 0.4H	Clr	16.7	1-21	60[h]	Orhun, 1989
Spotted seatrout	RePd	E		0.1 ha		100/m^2	2-22	6-11	Porter & Maciorowski, 1984
	Cy	FRP	1	1.5D × 0.6H	Blk	0.2	*e*-50	12.5	Tucker, 1988
	Pd	E				12.9/m^2	2-32	67[h]	Colura et al., 1992
Red drum	Pd	E				75/m^2	~3-28	52	Colura et al., 1991
	CyCo	FRP	~0.15	~0.6D × 0.6H		13	1-32	61	Holt, 1993
Cichlidae—cichlids									
Nile tilapia	Aq	Gls	84		Clr	0.95	6-69	94[h]	Saleh & Eleraky, 1993
Florida red tilapia	CyCo	FRP	0.4				(>25)	62	Le Coz et al., 1990
Pomacentridae—damselfishes									
Common clownfish	ReAq	Gls	0.135		Clr		1-10	60-80	Moe, 1973
	CyCo		0.025	0.25D × 0.65H		4	1-30	42	Alayse, 1984
Percula clownfish	ReAq	Gls/SB	0.17	0.41 × 1.02 × 0.41H	Clr	0.08	1-7	64	Mitsch, 1941
Spot dascyllus	CyCo	FRP	2.5			0.1	*e*-45	41[h]	Danilowicz & Brown, 1992
Mugilidae—mullets									
Striped mullet	Cy	FRP	5			17	1-50	29	Murashige et al., 1991
	Cy	FRP	5			16	1-35	73[h]	Tamaru et al. 1993a
White mullet	ReAq	Gls	0.038		Clr	2.6	1-20	23[h]	Houde et al., 1976
Polynemidae—threadfins									
Pacific threadfin	Cy	FRP	4			20	~*e*-24	~27	Kelley et al., 1995

Anarhichadidae—wolffishes									
Common wolffish	Sq		0.3	1 × 1 × 0.3H	Gry	~0.6	1-108	54^h	Moksness et al., 1989
Gobiidae—gobies									
Neon goby × Genie's cleaner goby	ReAq	Gls	0.12		Clr	(9)	1-26		Colin, 1975
Ephippidae—spadefishes									
Atlantic spadefish			0.5	0.9D × 0.8H			*e*-31		Gaspar, i.p.
Siganidae—rabbitfishes									
Yellowblotch spinefoot	Cy	FRP	0.6			8.3	1-35	37^h	Juario et al., 1985
Bothidae—lefteye flounders									
Southern flounder	ReAq	Gls	0.057		Clr	1	*e*-90	30	Arnold et al., 1977
Hirame		Cnc	90-150			10	*l-j*	25-80	Imaizumi, 1993
			1.5			51	1-85	45	Spedicato et al., 1993a, 1993b
	Re/CW	Cnc	60	5 × 9 × 2H		20		5	Tsujigado & Lee, 1993
Turbot	CyCo,CyHe	FRP	0.15-10			20-40	1-30	40^h	Person-Le Ruyet, 1990b
Pleuronectidae—righteye flounders									
Atlantic halibut			0.25				7-100	27^h	Holmefjord et al., 1993
	Sil		5				10-42	69^h	Harboe et al., 1994
Soleidae—soles									
Common sole	CyCo,CyHe	FRP	0.15-10			60-80	1-30	80^h	Person-Le Ruyet, 1990b
Tetraodontidae—puffers									
Tiger puffer	Re/CW	Cnc	60	5 × 9 × 2H		20		10	Tsujigado & Lee, 1993

Liu and Kelley (nd.b) described larval rearing of striped mullet in 4-m^3 cylindrical fiberglass tanks.

If the environment is favorable, nursery ponds are an alternative for rearing larvae and early juveniles, but these require careful management to ensure adequate natural food and water quality and to limit predators (Colura, 1990b; Colura et al., 1991; Geiger and Turner, 1990; Sturmer, 1990). They also might require more land than is practical. Nursery ponds are used effectively for red drum and hybrid striped bass in the U.S. (Chamberlain et al., 1990; Harrell et al., 1990).

Attempts to rear marine fish larvae in cages and bags have met with varying rates of success (Barnabé, 1990b). Coastal impoundments and 50- to 300-m^3 plastic bags (with natural zooplankton for food) have been used for larval rearing of Atlantic cod, turbot, and Atlantic halibut in Norway (van der Meeren, 1991; Tilseth et al., 1992). Special upwelling tanks are needed for Atlantic halibut larvae during the long yolksac period prior to feeding (Danielsberg et al., 1993; Myhre et al., 1993). Harboe et al. (1994) used silos for yolksac halibut (up to ~30 dah), then plastic bags for feeding larvae (~30-42 dah).

For production of temperate to warmwater marine species, I have used 1000- and 7000-L tanks, but prefer a 3000- to 3500-L (2.4 m diameter × 0.8 m deep) cylindroconical fiberglass tank, black inside with white stripes on the bottom.

Grow-out (Tables 6-5 to 6-8)

Fish being raised to market size need conditions resulting in the best combination of growth rate and feed conversion. The best type of grow-out unit depends on characteristics of the species and the site (advantages and limitations). Ponds (sometimes fenced off into pens) are preferred for bottom feeders such as milkfish, mullet, and killifish (Liao and Chen, 1986; Chamberlain et al., 1990; MacGregor et al., 1983). Small to medium cages often are used for fish that feed from just off the bottom to mid-water like sea bream, sea bass, groupers, and snappers, but ponds or coastal impoundments are used in some areas. Large rounded or many-sided cages or tanks are best for pelagic species such as jacks, tunas, and dolphin, because they swim constantly, often in straight lines, and need to be gently guided in a circle. Also, active species usually cannot be stocked as densely as sedentary ones.

Where land and pumped water are expensive, the most practical containers for growing marine fish out to market size in salt water could be floating cages or blocked (impounded) tidal ponds. Alternatively, after reaching the appropriate stage, euryhaline species could be raised in freshwater ponds farther inland where land is cheaper. The use of tanks or raceways for grow-out could be justified for high value fish where cages cannot be used and land for ponds is scarce, if water is relatively abundant or can be recirculated.

Early juveniles are much stronger and more tolerant than larvae, but still might require more environmental protection and closer control of feeding than larger fish. For these reasons and to achieve maximum stocking density at each phase, grow-out sometimes is broken into two or more phases, with small juveniles being placed in intermediate culture units (Barnabé, 1990a), such as tanks larger than those used for larvae (Person-Le Ruyet et al., 1991) or ponds (Harrell et al., 1990; Hopkins, 1990) or cages (Fig. 6-19; Tookwinas, 1990b; Fukusho, 1991) smaller than those used for final grow-out.

In Scandinavia, mid-sized juvenile Atlantic salmon usually are reared in circular tanks made of fiberglass, concrete, or fiberglass with concrete bottoms, mostly larger than 5 m in diameter (Isaakson, 1991). Final grow-out there often is in large cages, but in Scotland and Iceland, circular concrete tanks and lined ponds 12-25 m in diameter and 2-4 m deep have been used (Isaakson, 1991). Cage volumes of 100-10,000 m^3 have been used in Europe for grow-out; pens and impoundments have met with variable success (Laird and Needham, 1990). In Japan, cages for grow-out of coho salmon typically range from 10 × 10 × 7 m deep to 30 × 30 × 10 m deep (Mahnken, 1991).

Table 6-5. Examples of tanks used for grow-out.

Species	Shape[a]	Material[b]	Volume (m^3)	Dimensions[c] (m)	Inside color[d]	Density[e] ($/m^3$)	Age[f] (mo)	Weight[f] (g)	Survival (%)	References
Centropomidae—snooks										
Common snook	Cy	FRP	3.5	2.4D × 0.75H	Blk	2	~12-14	504-726	100	Tucker, 1987b
Barramundi			1.5				(10)	24-472	90	Fuchs, 1987
	Cy	Cnc	1.6		Gry		6-12	160-566	100	MacKinnon, 1990
Percichthyidae—temperate basses										
European sea bass	Re	Cnc	30	1.8 × 18 × 0.9H	Blk		5-27	1-380		Divanach et al., 1993
Sunshine bass	Re	Cnc	10	2.6 × 3.3 × 1.2H	Gry	20-1.4	1-19	1-1333	64	Tucker et al., 1993a
Serranidae—groupers, sea basses										
Nassau grouper	Re	Cnc	37	3.4 × 12.2 × 0.9H	Gry	1-0.9	~7-19	95-928	100	Tucker & Woodward, 1996
Sparidae—sea breams, porgies										
Sargo	Re	Cnc	30	1.8 × 18 × 0.9H	Blk		(20)	1-150		Divanach et al., 1993
Sharpsnout sea bream	Re	Cnc	30	1.8 × 18 × 0.9H	Blk		5-35	2-330		Divanach et al., 1993
Red porgy	Re	Cnc	30	1.8 × 18 × 0.9H	Blk		(27)	10-848		Divanach et al., 1993
Gilthead sea bream	Re	Cnc	30	1.8 × 18 × 0.9H	Blk		5-23	1-431		Divanach et al., 1993
Sciaenidae—drums										
Spotted seatrout	Cy	FRP	1	1.5D × 0.6H	Blk	4-8	3.4-5.7	24-110	100	Tucker, 1988
Red drum	Cy	FRP	3.5	2.4D × 0.75H	Blk	14-2	2-18	4-2243	100	Tucker et al., 1997a
Cichlidae—cichlids										
Oreochromis spilurus	Cy	Cnc	20			250	4-9	25-250		Hazza et al., 1991
Florida red tilapia			35			(15 kg)	~1-7	1-335	96[h]	Thouard et al., 1990
			23			25	(~6)	1-467	90	Watanabe, 1991
Bothidae—lefteye flounders										
Turbot		Cnc,Pls	20-100	1H		~350	6-36	~0.02-3 kg		Person-Le Ruyet, 1990b

Table 6-6. Examples of ponds used for grow-out.

Species	Material[b]	Area (ha)	Depth (m)	Density[e] (fish/m^2)	Age[f] (mo)	Weight[f] (g)	Survival (%)	References
				Anguillidae—"freshwater" eels				
Japanese eel	E	0.05-0.1	1	≥2.5 kg	~17-23	~100-190	~85	Usui, 1991
				Chanidae—milkfish				
Milkfish	E	1-10	0.3-0.5	0.4-0.5	(~6)	~50-500		Liao & Chen, 1986
	E	1-6	2-3	1.2	(~6)	~50-500		Liao & Chen, 1986
	E	3-5	2-3	1.2	0.7-			Liao, 1991
				Cyprinodontidae—killifishes				
Gulf killifish	E	0.1		33	(3.5)	0.2-2.8	74	Perschbacker & Strawn, 1983
				Centropomidae—snooks				
Common snook	Pls	0.014	1	0.07$^+$	2-20	7-740	100	Tucker, 1987b
Barramundi	E	0.08-2	1-2	0.25-1		20-100		Cheong, 1990
				Percichthyidae—temperate basses				
Sunshine bass				0.68	(11.5)	272-972	96	Smith et al., 1990
Hybrid striped bass				2.15	(8.4)	151-894	72	Jenkins et al., 1993
				Sparidae—sea breams, porgies				
Sheepshead	Pls	0.014	1	0.04	~11-15	314-573	100	Tucker, 1987a
				Sciaenidae—drums				
Red drum		0.1-0.25		0.5	1-21	0.3-1095	97[h]	Hopkins et al., 1987
	E	0.16		~110	1-3	0.4-1	~22	Hopkins, 1990
	E	2		~1.4	3-15	~1-454	90	Hopkins, 1990
	E	8		~0.31	15-27	~454-1362	95	Hopkins, 1990
				Mugilidae—mullets				
Striped mullet	E	8		0.13$^+$	(8)	111-870	100	Shepherd & Bromage, 1988

Table 6-7. Examples of cages used for grow-out.

Species	Material[b]	Volume (m^3)	Dimensions[c] (m)	Density[e] (fish/m^3)	Age[f] (mo)	Weight[f] (g)	Survival (%)	References
				Acipenseridae—sturgeons				
Bester				30-40/m^2	3 smr	4-2500	~68	Williot, 1990
				Chanidae—milkfish				
Milkfish		11-236	3-10D × 1.5-3H					Liao & Chen, 1986
				Centropomidae—snooks				
Barramundi	PE	75	5 × 5 × 3H	44/m^2	(6-8)	90-700	90-95	Anon., 1986
		8		30	6-12	32-677	87	Fuchs, 1987
	PE	6.8	1.5 × 1.5 × 3H	6	(5)	244-879	100	Martosudarmo, 1987
	PE	1.3	1 × 1 × 1.5H	154	4-11	38-551	99[h]	Sakaras, 1987
	PE	300	10 × 10 × 3H	45				Cheong, 1990
				Percichthyidae—temperate basses				
European sea bass		100	5 × 5 × 5	20 kg	5-36	~3-375	~50	Barnabé, 1990d; Coves et al., 1991
				Serranidae—groupers, sea basses				
Greasy grouper	PE	75	5 × 5 × 3H	44/m^2	(6-8)	90-700	90-95	Anon., 1986
Epinephelus spp.		20	2 × 5 × 2H	10-100		9-1300		Tookwinas, 1990a
				Carangidae—jacks				
Greater amberjack		50	10^2 × 5H	14	(15)	73-2959	~24	Porrello et al., 1993
Yellowtail		512	8 × 8 × 8	23	(4)	240-1290	>85	Watanabe et al., 1991a
				Lutjanidae—snappers				
Golden snapper	PE	75	5 × 5 × 3H	44/m^2	(6-8)	90-700	90-95	Anon., 1986
				Sparidae—sea breams, porgies				
Okinawa sea bream				100	(12)	150-500		Shokita et al., 1991
Red sea bream		50-300	4-6H	7-10 kg	~4-34	30-1000		Fukusho, 1991
Gilthead sea bream	Ny	50	5 × 5 × 2H	48	~5-19	7-559	83	Barbato et al., 1993
				Cichlidae—cichlids				
Oreochromis spilurus		125	5 × 5 × 5	9	4-9	25-250		Anon., 1990
Florida red tilapia		1		300	(~3)	9-150	98	Watanabe et al., 1990
		30,60		90-100	70-130	20-80	99	Le Coz et al., 1990
		30		(18 kg)	~1-7	1-348	71	Thouard et al., 1990

Table 6-8. Examples of pens and impoundments used for grow-out.

Species	Type[a]	Area (ha)	Depth (m)	Density[e] (fish/m^2)	Age[f] (mo)	Weight[f] (g)	Survival (%)	References
				Centropomidae—snooks				
Barramundi	Rsv				2-20	~5-5200		MacKinnon, 1990
				Sparidae—sea breams, porgies				
Red sea bream	Pen	5	1-3					Foscarini, 1988
				Cichlidae—cichlids				
Oreochromis spilurus	Pen	0.125	~1-2	28	~1-9	6-270		Al-Thobaiti et al., nd
				Siganidae—rabbitfishes				
Black spinefoot	Pen	0.125	~1-2	15	(7)	25-101	38	Al-Thobaiti et al., nd

In the U.S., striped bass and its hybrids have been grown out mostly in round tanks and raceways, but ponds and cages also are used (Harrell et al., 1990; Carlberg and Massingill, 1995). By computer simulation, Gempesaw et al. (1992) evaluated 0.2 to 12-ha ponds for grow-out of hybrid striped bass and decided the optimal range was 1-4 ha.

Mostly ponds are utilized for all stages of eels in Asia; both ponds and various tanks are used in Europe (Brusle, 1990; Usui, 1991).

In Japan, mostly cages are used for grow-out of marine fish. Pens and impoundments were used more in the past but are rare now (Nakada and Murai, 1991). Cages range in volume from 8 m^3 for less-active fish up to 2000 m^3 for yellowtail and 50,000 m^3 for bluefin tuna (Huguenin and Ansuini, 1978; Kuronuma and Fukusho, 1984). Size of red sea bream cages ranges from 50 to 300 m^3 (Fukusho, 1991).

In southeast Asia, barramundi and groupers are mainly grown out in small to medium cages, but ponds and coastal impoundments also are used (Cheong, 1990; Tookwinas, 1990b).

European sea bass and gilthead sea bream are grown out mainly in medium-sized cages, but raceways and ponds also are used (Barnabé, 1990e; Coves et al., 1991). Turbot have been grown out in tanks, raceways, and cages (Person-Le Ruyet et al., 1991).

Liu and Kelley (nd.b) described two nursery phases for striped mullet in 8- and 20-m^3 tanks.

Cylindrical tanks and raceways are economically feasible only if water is relatively cheap and abundant or can be recirculated. Ponds are more economical for some species in some places.

Plankton Culture (also see chapter 8)
Suitable conditions are necessary to produce sufficient quality and quantity of plankton for hatchery needs. Clean starter cultures of microalgae usually are raised in test tubes (e.g., 25 mL), flasks (e.g., 250-3000 mL), and 19-L glass or plastic carboys. Production cultures are grown in carboys, transparent fiberglass cylinders (normally 87-2700 L, but as large as 35,000 L), opaque fiberglass tanks (1-25 m^3 or more, Fig. 6-5), and concrete tanks holding up to 300 m^3 or more. Sometimes polyethylene bags are used, but care should be taken to avoid even mildly toxic plastics. Algae tanks either have a low surface to volume ratio (deep and narrow) and moderate aeration or a lower profile and high aeration to retard settling.

I have raised rotifers in 3-L flasks, 17-L plastic carboys, 150-L cylindroconical fiberglass tanks, 1000-L cylindrical fiberglass tanks, and 750-L concrete tanks (Fig. 6-8). At 27°C, batch cultures in any of these are good for up to 5 d, at which time numbers peak and water quality will begin to deteriorate. Then, it is best to harvest from, clean, and restart the three smaller units. The two larger ones could be operated with continuous cultures (i.e., with water exchange) until accumulation of algae and debris on the water and walls becomes excessive. In the Far East, mass cultures of rotifers and copepods usually are in outdoor concrete tanks as large as 200-300 m^3 or more (Kuronuma and Fukusho, 1984).

Copepods seem to grow better in containers holding 200 L or more. For convenience, I prefer 1000-L cylindrical tanks (Fig. 6-3). At 27°C, *Acartia tonsa*, *Tigriopus californicus*, and *T. japonicus* batch cultures peak at about 20 d (depending on feeding and harvesting rates) and should be restarted. Continuous cultures can last longer, but increased water exchange and siphoning of the bottom might be necessary.

Brine shrimp (*Artemia*) can be hatched in any container in which vigorous aeration will keep the cysts suspended and moving. A convenient size for moderate numbers is 17 L. Tall conical, cylindroconical, and inverted pyramid shapes are best for hatching. Larger tanks can be used to grow *Artemia* to later stages. The older *Artemia* are kept from settling to the bottom by continuing to use a tall tank or by providing sufficiently strong aeration.

HATCHERY DESIGN

The design for a small research and demonstration hatchery for indigenous marine fish in Florida is shown in Fig. 6-26. Laboratory space included in the hatchery building is mainly for processing eggs, fish, and food. Office, locker room, water chemistry, and workshop facilities should be in separate nearby buildings. This is partly to protect fish from toxic fumes (e.g., cleaning fluids) and office machines from salt air. Some rooms in the hatchery are isolated from others as a sanitation measure. Solvents, fuels, and internal combustion engines are kept at a safe distance. The emergency generator runs on liquid propane rather than diesel oil. Predaceous aquatic insects are prevented from entering the hatchery by keeping doors shut and by screening any other access points.

The hatchery has a wide central hall with windows looking into all sections, and all tanks are arranged for maximum visibility. Broodfish tanks and raceways have one-way windows so the fish can be seen from the hall. Aquariums close to the windows display zooplanktonic foods, small fish, and, when available, novelties like fish eggs. Spawning pairs of belted sandfish or other small species are held in aquariums by a window. The walls have educational displays and videos to illustrate the complete process of raising and stocking marine fish (e.g., pictures and videos of artificial and natural spawning next to the broodfish tanks, enlarged photographs of zooplankton and magnified closed-circuit TV next to the food culture labs, charts of nutritional requirements next to the experimentation lab). For experimental comparisons (e.g, different lighting), tanks in sections 8 and 9 can be separated by curtains but still be visible from the hall. General tours are limited to the hallway, but small special groups and classes can be taken behind the scenes.

In the larval food production sections, standing stock can reach 100 million rotifers and 10 million copepods. Additional tanks would increase food production capacity.

Air temperature can be kept within ±1°C (2°F) of the set temperature. In the top half of the building, it can be adjusted within the range 20-30°C (68-86°F), but usually is kept near 26°C. Each of the five sections at the bottom can be regulated within the range 10-32°C (50-90°F) as necessary to condition broodfish to spawn. Exhaust fans are available in each section.

No bare metal or pressure-treated wood is allowed above the tanks, and use of those materials anywhere is minimized. For example, plastic doors and doorknobs are used wherever possible. Ceiling material cannot shed particles into the tanks. Ceilings are about 3 m high and built so that lights, pipes, and curtains can be suspended from them. Skylights allow about 20% natural light into the tank areas, and this can be reduced with shades. Air blowers are mounted outside at roof level to minimize contamination from dust, exhaust, fumes, etc.

Schematics for water treatment systems are given in chapter 5. Previously aerated and degassed salt well water receives final treatment in three independent filter systems, each of which removes particulate matter, dissolved organic matter, and ammonia. Larval and experimental tanks receive water on a flow-through basis (0-400%/d). Each spawning tank and raceway has a similar treatment system for recirculating water, so that only a 5-10%/d exchange rate is necessary. Much of the wastewater can be diverted to a three-phase sequential polyculture system located between the hatchery and the office building. Three small ponds or raceways can be stocked with (in order) native shrimp, oysters, and seaweeds. This system shows that aquaculture wastewater can not only be cleaned but at the same time can be used to produce multiple crops. Total tank volume is about 91 m^3 in larval and experimental tanks and 220 m^3 in spawning and raceway systems. Peak water usage is about 261 Lpm (379 m^3/d) and routine usage 132 Lpm (191 m^3/d) or less. As much environmentally-friendly technology as possible for water use, recycling, and treatment would be studied, used, and demonstrated in this hatchery.

Another treatment system provides fresh well water for washing and for rearing of euryhaline juvenile and adult marine fish that can live and grow just as well or

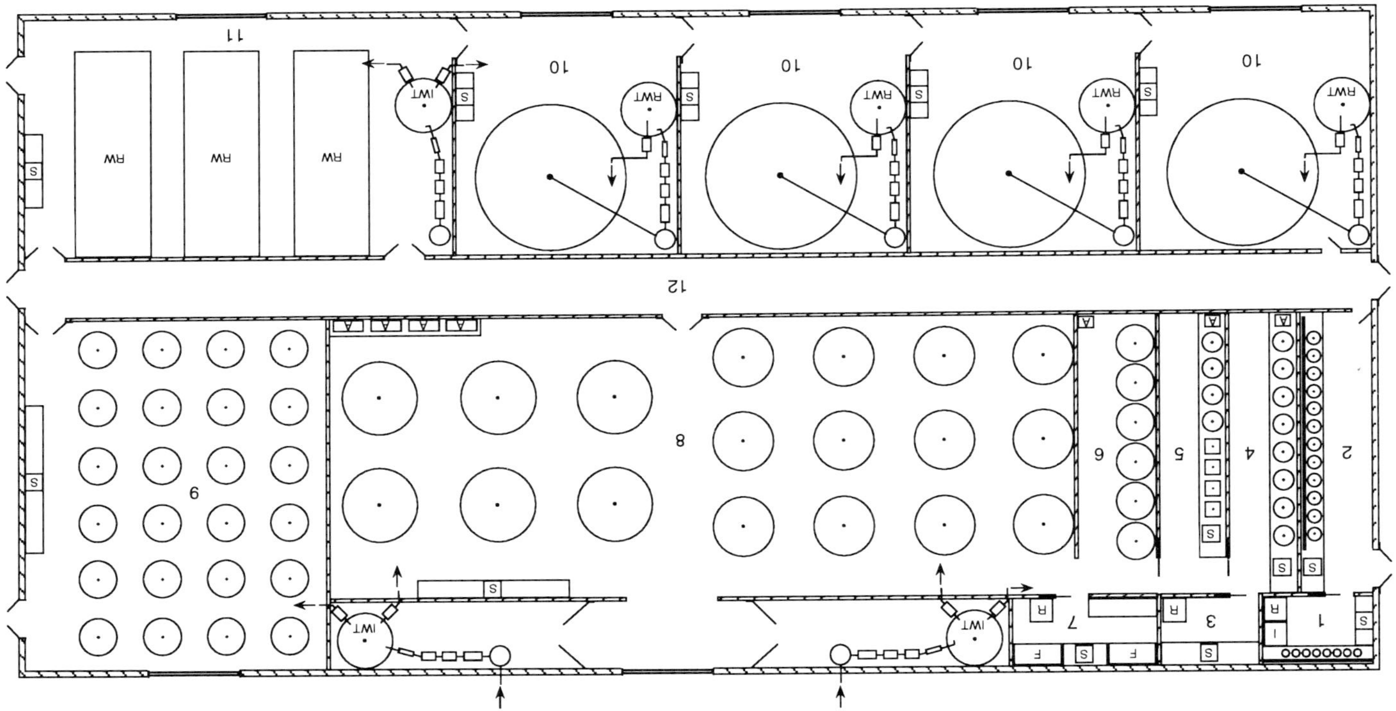

Figure 6-26. Floor plan for an experimental marine fish hatchery (working capacity of tanks is given). 1. Algae Preparation: 17-L polycarbonate carboys. 2. Algae Culture: 80-L clear fiberglass tubes. 3. Zooplankton Preparation. 4. Rotifer Culture: 135-L cylindroconical white fiberglass tanks. 5. Brine Shrimp: 17-L inverted-pyramid acrylic hatching tanks, 135-L cylindroconical white fiberglass culture tanks. 6. Copepod Culture: 1000-L cylindroconical blue fiberglass tanks. 7. Feed Preparation. 8. Larval Rearing: 3500-L cylindroconical and 7000-L cylindrical fiberglass tanks with black walls but black and white bottoms. 9. Experimentation: 1000-L cylindroconical blue fiberglass tanks. 10. Spawning: 45,400-L cylindrical blue fiberglass tanks. 11. Raceways: 22,700-L rectangular blue fiberglass tanks. 12. 2.4-m wide hallway (with displays on walls and windows) looking into all sections. A = aquarium, F = freezer, I = incubator, R = refrigerator, RW = raceway, S = sink. IWT = incoming water treatment, RWT = recirculating water treatment; arrows indicate direction of water flow. The building is 90 × 180 feet (27.4 × 54.9 m).

better in fresh water (e.g., snook, seatrout, red drum, mullet, some snappers and porgies).

Heater/chiller units adjust temperature of incoming water in the two main saltwater systems and the one freshwater system.

SUMMARY

The best rearing units for a given application depend upon several factors, but most importantly: the general requirements of the species, special requirements of each life stage, and advantages and limitations of the site.

7 OBTAINING FISH FOR STOCKING

Two possible sources of fish for growing out are: (A) wild juveniles and (B) domestic juveniles reared in a hatchery.

Five possible sources of eggs for option B are: (1) plankton collections (usually impractical), (2) strip-spawning of running ripe wild adults, (3) strip-spawning of wild or reared adults in which ovulation has been induced by hormone treatment, (4) tank spawning of wild or reared adults previously treated with hormone(s), (5) strip-spawning of untreated captive wild or reared adults, (6) voluntary spawning of untreated captive wild or reared adults. This chapter describes these options, broodfish and egg management, and other considerations for spawning programs.

STOCK FROM THE WILD

Collection of Juveniles

Catching wild juveniles by seining, trawling, trapping, etc. could eliminate the need for a saltwater hatchery, but requires a reliable, available stock of juveniles and is usually impractical, except for very small operations or for unusually available species. In the latter case, enough wild juveniles have been caught for stocking in ponds or cages, but eventually the demand can exceed the supply or wild stocks can be depleted. Examples are milkfish, groupers, and snappers in Asia, eels and yellowtails in Japan, eels in Europe, mullet in the Mediterranean, and southern bluefin tuna in Australia.

Collection of Eggs

Planktonic eggs can be collected with (1) a net towed behind a boat—hard on eggs and not very practical; (2) a net thrown off a structure (e.g., 1000 bay anchovy eggs every 10 minutes by the author during peak of the season in North Carolina)—large numbers are unlikely. Attached eggs can be collected along with their substrate. Atlantic silverside eggs can be found on intertidal vegetation from March through August (Middaugh, 1981)—again numbers are low. Large nests might be worth robbing. Two ornamental species are exemplary. A sergeant major nest can contain more than 25,000 eggs attached to solid objects and a queen triggerfish nest at least 100,000 eggs attached to coarse sand, gravel, shell fragments, etc.

STOCK FROM THE HATCHERY

Usually, a substantial number of juveniles can be reliably obtained only by raising them from eggs taken directly from the parents. Some reviews useful in endocrinology and captive spawning of fish are found in: Donaldson (1973), Fontaine (1976), Lam (1982), Hoar et al. (1983a, 1983b), Lin and Peter (1986), Kah (1986), Crim et al. (1987), Idler et al. (1987), Peter et al. (1988), Watanabe (1988), Schreibman and Margolis-Nunno (1989), Shelton (1989), Zohar (1989, 1991, 1996), Marte (1990), Fenner (1992), Vincent (1992), Yamagami et al. (1992), Harvey and Carolsfeld, 1993; Kime (1993), Kah et al. (1993), Chaudhuri (1994), Wourms, (1994), Bromage (1995), Patiño (1997).

In recent years, a number of marine species have been conditioned to spawn voluntarily in captivity for the first time. Improvements in the use of implants and other hormone treatments have been made. However, routine production of eggs, which is necessary for commercial aquaculture, has not yet been attained for many species, and the technology for each species requires a period of development.

Table 7-1. Egg production by various fish (mostly typical numbers).

Scientific name	Common name	Potential number of spawns/yr	Female size	Type of ovulation	Production (eggs/d, unless noted)	References
Acipenser baeri	Siberian sturgeon		8 kg	Induced	7,000,000	Williot et al., 1991
Lates calcarifer	Barramundi	M	9 kg		7,000,000	Ruangpanit, 1987a
Sciaenops ocellatus	Red drum	M	14 kg	Induced	3,000,000	Colura et al., 1990
Gadus morhua	Atlantic cod	≥19	63 cm, 3.6 kg	Natural	~2,549,090	Kjesbu, 1989; Kjesbu et al., 1991
Epinephelus fuscoguttatus	Brownmarbled grouper	M	6.4 kg	Induced	1,690,000	Tucker & FitzGerald, 1994
Lutjanus argentimaculatus	Asian red snapper	~8	>3.5	Induced	1,300,000	Aqua Farm News, 1992b
Centropomus undecimalis	Common snook	M		Natural	1,027,000	Roberts, 1987
Lutjanus argentimaculatus	Asian red snapper	M		Induced	>1,000,000	Doi & Singharaiwan, 1993
Mugil cephalus	Striped mullet	±M	1 kg	Induced	1,000,000	Lee et al., 1988b
Epinephelus striatus	Nassau grouper	2^+	6.0 kg	Induced	940,000	Tucker & Woodward, 1996
Siganus guttatus	Yellowblotch spinefoot	M	450 g	Natural	900,000	Duray & Juario, 1987
Scatophagus argus	Spotted scat	M	947 g	Induced	807,000	Barry et al., 1993
Epinephelus striatus	Nassau grouper	2^+	6.0 kg	Natural	800,000	Tucker et al., 1997
Dicentrarchus labrax	European sea bass	M	3 kg	Induced	750,000	Barnabé & Barnabé-Quet, 1985
Morone saxatilis	Striped bass		6.2 kg	Induced	736,000	Hodson & Sullivan, 1993
Dicentrarchus labrax	European sea bass	M	2 kg	Induced	700,000	Coves et al., 1991
Cynoscion nebulosus	Spotted seatrout	6^+	~1.4 kg	Natural	>670,000	Tucker & Faulkner, 1987
Sciaenops ocellatus	Red drum	M	11 kg	Natural	636,000	Colura et al., 1991
Chanos chanos	Milkfish		3.0 kg	Natural	600,000	Lacanilao & Marte, 1980; Marte et al., 1984a
Paralichthys californicus	California halibut	13^+	~60 cm	Natural	589,000	Caddell et al., 1990
Centropomus undecimalis	Common snook	M		Induced	500,000	Roberts, 1987
Lutjanus johnii	Golden snapper	4^+	69 cm TL, 5.3 kg	Induced	~475,440	Lim et al., 1985
Dicentrarchus labrax	European sea bass	M	5.4 kg	Induced	439,000?	Harvey et al., 1985
Eugerres brasilianus	Striped patao		300 g	Induced	418,500	Alvarez-Lajonchère et al., 1996
Plectropomus areolatus	Squaretail coraltrout	M	1.5 kg	Induced	408,000	Tucker & FitzGerald, 1994
Chanos chanos	Milkfish		65 cm FL, 4.6 kg	Induced	400,000	Lee et al., 1988a
Trachinotus carolinus	Florida pompano	M	1.4 kg	Induced	400,000	Hoff et al., 1978a
Sillaginodes punctata	Spotted whiting	M	46 cm TL	Natural	396,426	Cockrum & Jones, 1992
Dicentrarchus labrax	European sea bass	M			350,000	Coves et al., 1991
Priacanthus macracanthus	Red bullseye		29 cm FL	Natural	320,000	Liu et al., 1992
Sebastes schelegeli	Kurosoi	M	3 kg	Natural	~300,000 [l]	Jeon, 1995

Heterobranchus longifilis	African catfish	M	3.5 kg	Induced	291,000	Legendre, 1986
Melanogrammus aeglefinus	Haddock	M	41 cm	Natural	280,000	Hardy, 1959
Scophthalmus maximus	Turbot	12	6 kg	Natural	250,000	Person Le-Ruyet, 1990b
Morone chrysops	White bass	~3	810 g	Induced	238,500	Mylonas et al., 1996
Gadus morhua	Atlantic cod	22	6 kg	Natural	227,000	Huse, 1991
Coryphaena hippurus	Dolphin	182	4.5 kg	Natural	200,000	Kraul, 1991
Epinephelus chlorostigma	Brownspotted grouper		40 cm TL	Natural	~178,000	Ghorab et al., 1986
Anoplopoma fimbria	Sablefish			Natural	154,000	Jensen et al., 1992
Sparidentex hasta	Sobaity	≤50	52 cm, 2.7 kg	Natural	~140,500	Al-Marzouk et al., 1994
Phrynelox scaber	Splitlure frogfish			Natural	125,000	Martin & Drewry, 1978
Argyrosomus hololepidotus	Mulloway	M	10 kg	Induced	~100,000	Battaglene & Talbot, 1994
Hippoglossus hippoglossus	Atlantic halibut	6-16		Natural	~100,000	Kjørsvik & Holmefjord, 1995
Rhabdosargus sarba	Silver sea bream	94	902 g	Natural	79,840	Lin et al., 1990
Rhabdosargus sarba	Silver sea bream	≤65	33 cm TL, 1.7 kg	Natural	~69,600	Mihelakakis & Kitajima, 1995
Bairdiella icistia	Gulf croaker	3	75 g	Induced	64,680	Haydock, 1971
Paralichthys lethostigma	Southern flounder			Induced	52,870	Berlinsky et al., 1996
Mallotus villosus	Capelin	1		Natural	50,000	Jackson et al., 1991
Seriola lalandi	Goldstriped amberjack	M	1.5 kg	Natural	~50,000	Kraul, 1985
Pagrus major	Red sea bream	M	1.5 kg	Natural	49,000	Watanabe & Kiron, 1995
Hippoglossus hippoglossus	Atlantic halibut	M		Natural	46,000	Daniel et al., 1993
Acanthopagrus schlegeli	Black sea bream	M	1.6 kg	Natural	45,000	Chou et al., 1993
Scomber scombrus	Atlantic mackerel	20-70	~37 cm	Natural	~32,000	Walsh & Johnstone, 1992
Sillago sihama	Silver whiting	65	201 mm FL	Natural	30,000	Kumai & Nakamura, 1977
Abudefduf saxatilis	Sergeant major	M		Natural	27,000	Prappas et al., 1991
Rhabdosargus sarba	Silver sea bream	~96	~1 kg	Natural	~27,000	Leu, 1994
Tylosurus crocodilus	Houndfish		132 cm	Natural	>25,000	Hardy, 1978
Dascyllus trimaculatus	Domino damselfish	36	100 mm	Natural	25,000	Garnaud, 1957b
Sparus auratus	Gilthead sea bream	120	1 kg	Both	~25,000	Zohar et al., 1995
Coregonus lavaretus	Powan		700 g	Natural	23,882	Gillet, 1991b
Dascyllus albisella	Spot damselfish	~23		Natural	22,701	Danilowicz, 1995
Merlangius merlangus	English whiting	M	315 mm	Natural	20,000	Hislop, 1975
Solea solea	Common sole	8-22	39 cm, 662 g	Natural	~20,000	Houghton et al., 1985
Hippoglossus stenolepis	Pacific halibut	3-11		Natural	17,583	Jensen et al., 1992
Solea solea	Common sole	5	380 g	Natural	14,044	Ramos, 1986b
Solea solea	Common sole	6	449 g	Induced	13,362	Ramos, 1986b
Engraulis mordax	Northern anchovy	20	23 g	Natural	11,000	Hunter & Leong, 1981
Engraulis mordax	Northern anchovy	M	25 g	Induced	~11,000	Leong, 1971
Halichoeres poecilopterus	Multicolorfin wrasse	40		Natural	9500	Kimura & Kiriyama, 1992
Leiostomus xanthurus	Spot	M	250 g	Induced	7500	Hettler & Powell, 1981
Pterogobius zacalles		M	14 cm	Natural	~6270	Shiogaki, 1981a

Table 7-1. (Continued)

Scientific name	Common name	Potential number of spawns/yr	Female size	Type of ovulation	Production (eggs/d, unless noted)	References
Oncorhynchus tshawytscha	Chinook salmon	1	71 cm EHL	Natural	5373	Beacham & Murray, 1993
Stegastes partitus	Bicolor damselfish	M	~60 mm TL	Natural	>5000	Myrberg, 1972
Amblyglyphidodon leucogaster	Whitebelly damselfish	22	12 mm TL	Natural	4009	Goulet, 1997
Anarhichas lupus	Common wolffish	1	1.5 kg	Natural	3669	Pavlov & Moksness, 1996
Oncorhynchus nerka	Sockeye salmon	1	48 cm EHL	Natural	~3600	Beacham & Murray, 1993
Oncorhynchus tshawytscha	Chinook salmon	1	~9 kg	Either	3500	Hunter et al., 1978
Chrysiptera cyanea	Blue devil damselfish	~91	51 mm	Natural	~3500	Breeder's Registry Database, 1994
Paramonacanthus japonicus	Striped filefish		75 mm SL	Natural	3300	Nakazono & Kawase, 1993
Oncorhynchus keta	Chum salmon	1	59 cm EHL	Natural	3034	Beacham & Murray, 1993
Salvelinus alpinus	Arctic char		700 g	Natural	2754	Gillet, 1991a
Oncorhynchus kisutch	Coho salmon	1	51 cm EHL	Natural	2698	Beacham & Murray, 1993
Dascyllus aruanas	Striped damselfish		38 mm TL	Natural	2000	Walker, 1977
Oreochromis niloticus	Nile tilapia	10-12		Natural	2000	Macintosh & Little, 1995
Oncorhynchus gorbuscha	Pink salmon	1	42 cm EHL	Natural	1600	Beacham & Murray, 1993
Amphiprion clarkii	Clark's clownfish	38	100 mm	Natural	≤1500	Breeder's Registry Database, 1994
Cirrhitichthys aureus	Golden curlyfin		10 cm	Natural	~1070	Tanaka & Suzuki, 1991
Anchoa mitchilli	Bay anchovy	M	55 mm FL, 1.6 g	Natural	1069	Zastrow et al., 1991
Chaetodipterus faber	Atlantic spadefish			Natural	>1000	Breeder's Registry Database, 1995
Oreochromis niloticus	Nile tilapia	M	100 g	Natural	1000	Mair et al., 1993
Menidia beryllina	Tidewater silverside	M	Average	Natural	984	Martin & Drewry, 1978
Hippocampus reidi	Longsnout seahorse	M		Natural	900	Vincent, 1995c
Amphiprion percula	Percula clownfish	M	~100 mm	Natural	880	Garnaud, 1951
Amphiprion bicinctus	Twoband clownfish	M		Natural	800	Young, 1991
Oreochromis aureus	Blue tilapia	~27	225 g	Natural	~750	Behrends et al., 1993
Oncorhynchus kisutch	Coho salmon	1	39 cm TL, 727 g	Natural	747	MacQuarrie et al., 1978
Oncorhynchus gorbuscha	Pink salmon	1	40 cm FL, 761 g	Natural	685	MacQuarrie et al., 1979
Amphiprion ocellaris	Common clownfish	26	~80 mm	Natural	650	Squires, 1971; Moe, 1973
Oreochromis spilurus		M		Natural	650	Hazza et al., 1989
Gobiodon citrinus	Lemon goby	~91	62 mm	Natural	~650	Breeder's Registry Database, 1994
Gobiodon okinawae	Yellow clown goby	~91	62 mm	Natural	~650	Breeder's Registry Database, 1994
Gobiesox strumosus	Skilletfish	M	41 mm SL	Natural	645	Martin & Martin, 1971

Gobiosoma oceanops	Neon goby	36	75 mm	Natural	600	Moe, 1975
Amblygobius bynoensis	Bynoe's goby	12	152 mm	Natural	500	Breeder's Registry Database, 1994
Calloplesiops altivelis	Comet			Natural	500	Wassink & Brons, 1990
Menidia menidia	Atlantic silverside	M	Average	Natural	500	Martin & Drewry, 1978
Pseudochromis flavivertex	Sunrise dottyback	26	5 cm TL	Natural	500	Lubbock, 1975
Pomacentrus caeruleus	Caerulean damselfish	~7	32 mm	Natural	~500	Breeder's Registry Database, 1995
Hippocampus erectus	Lined seahorse	M		Natural	450	Vincent, 1995c
Serranus subligarius	Belted sandfish	79	60 mm TL	Natural	411	Tucker & Woodward, 1995
Amphiprion frenatus	Tomato clownfish	M	110 mm	Natural	400	Breeder's Registry Database, 1994
Aspasma minima		M		Natural	330	Shiogaki & Dotsu, 1971
Hippocampus kuda/comes	Spotted seahorse	~12		Natural	325	Schmidt, 1995; Vincent, 1995c
Canthigaster valentini	Sharpnose puffer			Natural	314	Stroud et al., 1989
Meiacanthus grammistes	Striped blenny	M	80 mm	Natural	>300	Breeder's Registry Database, 1994
Chrysiptera cyanea	Blue devil damselfish	~15	70 mm	Natural	300 [+]	Breeder's Registry Database, 1994
Gobiosoma oceanops	Neon goby	M		Natural	300	Walker, 1977
Aidablennius sphynx	Mediterranean blenny	~40		Natural	~300	Kraak & Van Den Berghe, 1992; Kraak, 1996
Syngnathus floridae	Dusky pipefish	M	Average	Natural	~264	Hardy, 1978
Gobiosoma oceanops	Neon goby	36	50 mm	Natural	250	Moe, 1975
Phycodurus eques	Leafy seadragon			Natural	250	Blasiola, 1988
Assessor macneilli	MacNeill's mini-grouper		~35 mm SL	Natural	216	Allen & Kuiter, 1976
Genicanthus melanospilos	Blackspot lyretail angelfish	365	102 mm	Natural	200 [+]	Breeder's Registry Database, 1994
Chrysiptera hemicyanea	Yellowtail blue damselfish	~45		Natural	200	Walker, 1977; Unwin, 1986
Hippocampus histrix	Thorny seahorse	M		Natural	200	Vincent, 1995c
Amphiprion nigripes	Blackfoot clownfish	M	76 mm	Natural	~200	Breeder's Registry Database, 1994
Opsanus tau	Oyster toadfish			Natural	<200	Martin & Drewry, 1978
Centropyge argi	Cherub pygmy angelfish	365		Natural	~175	Breeder's Registry Database, 1994
Centropyge bicolor	Bicolor pygmy angelfish	365		Natural	~175	Breeder's Registry Database, 1994
Centropyge loriculus	Flame pygmy angelfish	365		Natural	~175	Breeder's Registry Database, 1994
Centropyge tibicen	Keyhole angelfish	M		Natural	~175	Breeder's Registry Database, 1994
Amphiprion frenatus	Tomato clownfish	10	89 mm	Natural	150	Breeder's Registry Database, 1994
Centropyge bispinosus	Coral beauty	365	7 g	Natural	150	Bauer & Bauer, 1981
Gasterosteus aculeatus	Threespine stickleback		Average	Natural	150	Hardy, 1978
Hippocampus guttulatus	Continental seahorse	M		Natural	142	Hardy, 1959
Centropyge argi	Cherub pygmy angelfish	365	4.5 g	Natural	119	Bauer & Bauer, 1981
Gramma melacara	Blackcap basslet			Natural	~110	Addison, 1994a, 1994b
Tetraodon cf. *cutcutia*	Ocellated blowfish			Natural	100 [+]	Cohn, 1912
Opistognathus aurifrons	Yellowhead jawfish	~26	85 mm	Natural	100	Walch, 1994b
Gambusia affinis	Mosquitofish	8		Natural	~100 [l]	Hardy, 1978
Oxymonacanthus longirostris	Orangespot filefish		89 mm	Natural	~100	Breeder's Registry Database, 1994
Centropyge fisheri	Fisher's pygmy angelfish	365	5 g	Natural	93	Bauer & Bauer, 1981

Table 7-1. (Continued)

Scientific name	Common name	Potential number of spawns/yr	Female size	Type of ovulation	Production (eggs/d, unless noted)	References
Centropyge acanthops	Orangeback pygmy angelfish	365	4 g	Natural	90	Bauer & Bauer, 1981
Apogon maculatus	Flamefish		~75 mm	Natural	87	Coleman, 1966
Atherinops affinis	Topsmelt	M	15 cm SL, 41 g	Natural	85	Middaugh et al., 1992
Centropyge bispinosus	Coral beauty	M	95 mm	Natural	80	Lamm, 1984
Gramma loreto	Royal gramma	~365		Natural	75	Walch, 1994a
Cyprinodon variegatus	Sheepshead minnow	M		Natural	~70	Hardy, 1978
Centropyge loriculus	Flame pygmy angelfish	365	8.6 g	Natural	60	Bauer & Bauer, 1981
Oreochromis spilurus		M	241 g	Natural	57	Al-Ahmad et al., 1988b
Centropyge flavissimus	Lemonpeel	365	11.5 g	Natural	56	Bauer & Bauer, 1981
Hippocampus fuscus		M		Natural	55	Vincent, 1995c
Fundulus confluentus	Marsh killifish	M		Natural	~48	Hardy, 1978
Arius felis	Hardhead sea catfish		Average	Natural	42	Jones et al., 1978
Acanthemblemaria aspera	Roughhead blenny	M	22 mm SL	Natural	41	Clarke, 1992
Fundulus heteroclitus	Mummichog	M		Natural	30	Hardy, 1978
Hippocampus zosterae	Dwarf seahorse	M		Natural	30	Vincent, 1995c
Lucania parva	Rainwater killifish	M		Natural	24	Hardy, 1978
Acanthemblemaria spinosa	Spinyhead blenny	M	23 mm SL	Natural	21	Clarke, 1992
Fundulus luciae	Spotfin killifish	M		Natural	16	Hardy, 1978
Pseudomugil mellis	Honey blue-eye	~200	20-30 mm SL	Natural	15	Semple, 1991
Thalassoma bifasciatum	Bluehead wrasse	146		Natural		Petersen et al., 1992
Pseudochromis fridmani	Orchid dottyback	42		Natural		Brons, 1996
Gobiosoma oceanops	Neon goby	13-45	37-80 mm TL	Natural		Brown, 1996
Eggs produced per year						
Lates calcarifer	Barramundi	≥2		Natural	8,923,000	Toledo et al., 1991
Rhabdosargus sarba	Silver sea bream	94	902 g	Natural	7,505,000	Lin et al., 1990
Epinephelus akaara	Redspotted grouper	M	1 kg	Natural	>5,000,000	Maruyama et al., 1994
Chanos chanos	Milkfish	M		Natural	4,950,000	Lin & Lin, 1987
Sciaenops ocellatus	Red drum	M	11.4 kg	Natural	4,450,000	Henderson-Arzapalo, 1995
Epinephelus striatus	Nassau grouper	≥2	66 cm TL, 5.8 kg	Natural	≥4,261,000	Tucker et al., 1996
Pagrus major	Red sea bream	M		Natural	~2,500,000	Fukusho, 1991

Sparus auratus	Gilthead sea bream	120	1 kg	Natural	~2,500,000	Zohar et al., 1995
Paralichthys microps	Chilean flounder	≥22	588 g	Natural	2,400,000	Silva, 1994
Gadus morhua	Atlantic cod	M	8.9 kg		2,200,000	Huse & Jensen, 1983
Scophthalmus maximus	Turbot	M	3-5 kg	Natural	1,800,000	Bromley, et al., 1986
Paralichthys olivaceus	Hirame	M	Average	Natural	~400,000	Main & Rosenfeld, 1995
Solea solea	Common sole	8-22	39 cm, 662 g	Natural	318,000	Houghton et al, 1985
Halichoeres poecilopterus	Multicolorfin wrasse	40		Natural	38,000	Kimura & Kiriyama, 1992
Citharichthys stigmaeus	Speckled sandab	M	11 cm TL	Natural	12,100	Rackowski & Pikitch, 1989
Batch fecundity						
Rachycentron canadum	Cobia	M	20 kg		20,000,000	Lotz et al., 1996
Thunnus albacares	Yellowfin tuna	M	50 kg		9,113,000	Suzuki, 1994
Thunnus maccoyii	Southern bluefin tuna	M	19 cm FL		5,000,000	Farley & Davis, 1998
Thunnus albacares	Yellowfin tuna	299	120 cm FL		2,025,000	Schaefer, 1996
Pogonias cromis	Black drum	25-33	8.8 kg		1,360,000	Nieland & Wilson, 1993
Cynoscion nebulosus	Spotted seatrout	24	1.2 kg		507,000	Brown-Peterson et al., 1988
Cynoscion striatus	Striped weakfish	M	53 cm TL		450,000	Macchi, 1998
Euthynnus lineatus	Black skipjack	43-58	3 kg		341,000	Schaefer, 1987
Scomber japonicus	Chub mackerel	~25	1 kg		~108,000	Macewicz & Hunter, 1993
Pagrus major	New Zealand snapper	~12	1 kg		100,000	Scott et al., 1993
Microstomus pacificus	Dover sole	~9	1 kg		~83,000	Hunter et al., 1992
Anoplopoma fimbria	Sablefish	3.37	2.5 kg		82,000	Macewicz & Hunter 1994
Paralabrax clathratus	Kelp bass	~62	300 mm SL, 700 g		~81,000	Oda et al., 1993
Scomber japonicus	Chub mackerel	9^+	443 g		68,356	Dickerson et al., 1992
Trachurus murphyi	Peruvian jack mackerel	7-20			60,956	Lisovenko & Andrianov, 1991
Archosargus probatocephalus	Sheepshead	M			47,000	Render & Wilson, 1992
Sardinops melanostictus	Japanese sardine	M	86 g		40,500	Morimoto, 1998
Dentex tumifrons	Yellow sea bream	M	25 cm FL		35,000	Oki & Tabeta, 1998
Sardinops melanostictus	Japanese sardine	M	~81 g		33,900	Aoki, 1996
Sardinops ocellatus	South African pilchard	M	120 g		~31,000	Le Clus, 1992
Lutjanus vittus	Blackstriped snapper	150	25 cm FL		~27,000	Davis & West, 1993
Scomber scombrus	Atlantic mackerel	30	500 g		26,525	Watson et al., 1992
Gobius niger	Black goby	6-8	82 mm TL		14,614	Joyeaux et al., 1991
Solea solea	Common sole	M	32 cm		8400	Witthames & Walker, 1995
Engraulis mordax	Northern anchovy	M	20 g		7536	Hunter & Macewicz, 1980
Engraulis japonicus	Japanese anchovy	M	125 mm FL		6102	Kim & Kang, 1992
Engraulis encrasicolus ponticus	Black Sea anchovy	≥50			3690	Lisovenko & Andrianov, 1991
Anchoa mitchilli	Bay anchovy	54	59 mm FL		1186	Luo & Musick, 1991
Maurolicus muelleri	Lightfish	M	47 mm, 2 g		364	Melo & Armstrong, 1991
Menidia beryllina	Inland silverside	M	70 mm SL		~250	Middaugh & Hemmer, 1992

Table 7-1. (Continued)

Scientific name	Common name	Potential number of spawns/yr	Female size	Type of ovulation	Production (eggs/d, unless noted)	References
Fecundity						
Mola mola	Ocean sunfish		150 cm TL		300,000,000	Martin & Drewry, 1978
Morone saxatilis	Striped bass		14.5 kg		40,500,000	Hill et al., 1989
Anguilla rostrata	American eel	1	110 cm TL, 2.6 kg		~21,000,000	Barbin & McCleave, 1997
Megalops atlanticus	Tarpon	M	203 cm		12,202,000	Jones et al., 1978; Crabtree et al., 1997
Gadus morhua	Atlantic cod	M	140 cm		12,000,000	Hardy, 1978
Lutjanus campechanus	Atlantic red snapper	M	754 mm FL		9,300,000	Moran, 1988
Plectorhinchus cinctus	Yellowspotted grunt	M	4.6 kg		3,019,000	Chang et al., 1994c
Anguilla rostrata	American eel	1	70 cm TL		2,277,000	Van Den Avyle, 1984
Lophius americanus	Goosefish		90 cm TL		1,500,000	Armstrong et al., 1992
Epinephelus morio	Red grouper		615 mm SL		~1,469,200	Moe, 1969
Pleuronectes americanus	Winter flounder		1 kg		1,389,000	Buckley et al., 1991
Lutjanus analis	Mutton snapper	M	512 mm FL		1,365,975	Bortone & Williams, 1986
Siganus lineatus	Goldline spinefoot	M	31 cm BL		1,193,000	Mobiha, 1991
Scophthalmus maximus	Turbot	M			1,078,000	Jones, 1972
Acipenser oxyrhynchus	Atlantic sturgeon		75 kg		1,030,000	Jones et al., 1978
Mugil cephalus	Striped mullet	±M	40 cm FL		1,004,600	Render et al., 1995
Sebastes schlegeli	Kurosoi	M	60 cm TL		780,000	Haldorson & Love, 1991
Euthynnus affinis	Kawakawa	M	50 cm BL		718,000	Mobiha, 1991
Trachinotus carolinus	Florida pompano		36 cm FL, 576 g		630,000	Finucane, 1969
Sphyraena barracuda	Great barracuda		95 cm FL		600,000	Martin & Drewry, 1978
Lutjanus griseus	Gray snapper	M	315 mm SL		590,000	Bortone & Williams, 1986
Lophius americanus	Goosefish		66 cm		~543,000	Martin & Drewry, 1978
Valamugil seheli	Bluespotted mullet	M	35 cm BL		534,000	Mobiha, 1991
Theragra chalcogramma	Walleye pollock	M	50 cm BL		500,000	Hamatsu et al., 1993
Scophthalmus rhombus	Brill	M			465,000	Jones, 1972
Plectorhinchus celebicus	Celebes sweetlips	M	41 cm BL		428,000	Mobiha, 1991
Alosa sapidissima	American shad	M	Average		412,000	Jones et al., 1978
Solea solea	Common sole	M	600 g		400,000	Millner et al., 1991
Lethrinus hypselopterus	Masked emperor		24 cm BL		369,000	Mobiha, 1991
Merluccius bilinearis	Silver hake	M	30 cm		~367,000	Hardy, 1978

Alosa aestivalis	Blueback herring	M	31 cm	349,700	Jones et al., 1978
Acipenser stellatus	Stellate sturgeon		10.1 kg	285,000	Veshchev, 1991
Lethrinus lentjan	Redspot emperor		34 cm BL	276,000	Mobiha, 1991
Pollachias virens	Pollock	M	Average	~274,000	Hardy, 1978
Lutjanus gibbus	Australian red snapper	M	26 cm BL	267,000	Mobiha, 1991
Alosa pseudoharengus	Alewife	M	29 cm	200,000	Jessop, 1993
Kyphosus cinerascens	Ashen drummer		35 cm BL	198,000	Mobiha, 1991
Sebastes marinus	Ocean perch		63 cm TL	177,800	Haldorson & Love, 1991
Sphoeroides maculatus	Northern puffer		265 mm	176,000	Martin & Drewry, 1978
Alosa aestivalis	Blueback herring	M		158,000	Jessop, 1993
Pleuronectes platessa	European plaice	M		150,000	Jones, 1972
Pleuronectes platessa	European plaice	M	40 cm	146,267	Rijnsdorp, 1991
Clupea harengus	Atlantic herring	M	35 cm	~130,000	Bradford & Stephenson, 1992
Pleuronectes yokohamae	Marbled flounder	1	34 cm BL, 60 g	120,000	Dagang et al., 1991
Scomber japonicus	Chub mackerel	M	500 g	113,500	MacGregor, 1976
Microstomus pacificus	Dover sole	M	45 cm	75,000	Yoklavich & Pikitch, 1989
Microgadus tomcod	Atlantic tomcod	M	35 cm	65,780	Hardy, 1978
Sebastes emphaeus	Puget Sound rockfish	1	156 mm FL	57,787	Beckmann et al., 1998
Syacium papillosum	Dusky flounder	M	168 mm SL	55,200	Martin & Drewry, 1978
Cynoglossus arel	Brown tonguesole		315 mm TL	44,700	Rajaguru, 1992
Cynoglossus lida	Shoulderspot tonguesole		195 mm TL	30,200	Rajaguru, 1992
Trinectes maculatus	Hogchoker	M	~120 mm SL	29,383	Martin & Drewry, 1978
Clupea pallasii	Pacific herring	M	162 g	~27,800	Tanasichuk et al., 1993
Clupea harengus	Atlantic herring	M	188 mm TL	16,026	Parmanne & Kuittinen, 1991
Cheilopogon heterurus	Atlantic flyingfish	M		~9250	Hardy, 1978
Anarhichas lupus marisalbi	White Sea wolffish		Average	~4750	Pavlov & Novikov, 1987
Lampanyctodes hectoris	Lanternfish		1.5 g	969	Prosch, 1991
Maurolicus muelleri	Lightfish	M	1.5 g	501	Prosch, 1991
Fundulus majalis	Striped killifish	M		~500	Hardy, 1978
Fundulus diaphanus	Banded killifish	M		~226	Hardy, 1978
Porichthys plectrodon	Atlantic midshipman		~115 mm SL	140	Martin & Drewry, 1978

M = multiple.
[1] number of larvae released.

Table 7-2. Reported age and size at maturity and spawning season of representative fish.

Scientific name	Common name	Age	Size	Season	References
			Acipenseridae—sturgeons		
Acipenser brevirostrum	Shortnose sturgeon	♂ 4-26, ♀ 6-20 yr	♂ 49-53, ♀ 46 cm TL	~Dec-Sep	Jones et al., 1978; Conte et al., 1988
Acipenser oxyrhynchus	Atlantic sturgeon	♂ 5-34, ♀ 7-30 yr	♂ ~122-165, ♀ 183-244 cm TL	Feb-Jul	Jones et al., 1978; Conte et al., 1988
Acipenser stellatus	Stellate sturgeon	♂ ~7 yr ♀ 9⁺ yr	♂ ~101 cm, ~3 kg ♀ ~116 cm, ~5 kg	~Apr-Aug	Veshchev, 1991
Acipenser transmontanus	White sturgeon	♂ 9-22, ♀ 11-34 yr	♀ 110 cm TL	Feb-Jul	Conte et al., 1988 (wild)
Acipenser transmontanus	White sturgeon	♂ 3-4, ♀ 6-14 yr	♂ 7-14, ♀ 12-61 kg, 151 cm FL		Doroshov et al., 1997 (reared)
Scaphirhynchus albus	Pallid sturgeon	♂ 5-7, ♀ ≤14 yr	♂ ≤60, ♀ ≤85 cm	Mar-Jun	Keenlyne & Jenkins, 1993
			Elopidae-tarpons		
Megalops atlanticus	Tarpon	6.5-7.5 yr	122 cm	Jan-Dec	Jones et al., 1978; Crabtree et al., 1992, 1997
			Albulidae—bonefishes		
Albula vulpes	Bonefish	1 yr	21 cm	Most months	Jones et al., 1978
			Anguillidae—"freshwater" eels		
Anguilla rostrata	American eel	5-18 yr	♂ ~28, ♀ ≤46 cm	poss. Jan-Jul	Hardy, 1978
			Clupeidae—herrings		
Alosa aestivalis	Blueback herring	3-4 yr	≤25 cm	Apr-Jun	Jones et al., 1978
Alosa mediocris	Hickory shad		♂ 29, ♀ 32 cm TL	Apr-Jun	Jones et al., 1978
Alosa pseudoharengus	Alewife	♂ 1-4, ♀ 2-5 yr	♂ 9-15, ♀ 11-17 cm SL	Mar-Aug	Mansueti & Hardy 1967; Jones et al., 1978
Alosa sapidissima	American shad	2-7 yr	♂ ~30-45, ♀ 38-49 cm FL	Apr-Jul	Jones et al., 1978
Brevoortia patronus	Gulf menhaden	1-2 yr	≥15 cm FL	Oct-Mar	Ahrenholz, 1991
Brevoortia tyrannus	Atlantic menhaden	1-3 yr	18-32 cm	Most months	Jones et al., 1978
Clupea harengus	Atlantic herring	2-9 yr	18-31 cm	Mar-Nov	Jones et al., 1978; Rajasilta et al., 1993
Clupea pallasii	Pacific herring			Mar-Apr,Jun-Jul	Tanasichuk et al., 1993
Dorosoma cepedianum	Gizzard shad	1-3 yr	18-28 cm	Mar-Aug	Jones et al., 1978
Dorosoma petenense	Threadfin shad	1-3 yr	49-126 mm	Apr-Jul	Jones et al., 1978
Etrumeus teres	Round herring		10 cm TL	Oct-Jul	Jones et al., 1978
Opisthonema oglinum	Atlantic thread herring		13 cm, ♀ 7 g	Feb-Sep	Jones et al., 1978
Sardinella aurita	Spanish sardine	2-3 yr	12-20 cm	Jan-Dec	Jones et al., 1978
Sardinops melanostictus	Japanese sardine		♀ ~18⁺ cm BL	Jan-Mar	Matsuyama et al., 1991a; Morimoto, 1993
Spratelloides delicatulus	Blue sprat		37 mm SL	Jul-Jun	Milton & Blaber, 1991
Spratelloides lewisi			35 mm SL	Jul-Jun	Milton & Blaber, 1991
			Engraulidae—anchovies		

Anchoa hepsetus	Striped anchovy	~1 yr	~75 mm	Nov-Sep	Jones et al., 1978
Anchoa mitchilli	Bay anchovy	2-10 mo	30-45 mm FL	Jan-Dec	Jones et al., 1978; Luo & Musick, 1991; Zastrow et al., 1991
Encrasicholina devisi			45 mm SL	Mar-Jan	Milton & Blaber, 1991
Encrasicholina heteroloba			45 mm SL	Dec-Oct	Milton & Blaber, 1991
Engraulis encrasicolus	European anchovy	1 yr		Apr-Oct	Palomera, 1992; Ribeiro et al., 1996
Engraulis japonicus	Japanese anchovy		≤9 cm FL	Mar-Sep	Kim & Kang, 1992
Engraulis mordax	Northern anchovy		80-115 mm SL, ~12 g	Jan-Dec	Hunter & Leong, 1981; Hunter & Coyne, 1982
		Chanidae—milkfish			
Chanos chanos	Milkfish	♂ 4, ♀ 5 yr	1-3 kg	Mar-Dec	Kelley & Lee, 1986; Lee, 1986
		Clariidae—airbreathing catfishes			
Heterobranchus longifilis	African catfish	1 yr	1.5 kg	~Jun-Oct	Legendre, 1986
		Ariidae—sea catfishes			
Arius felis	Hardhead sea catfish		♂ 14, ♀ 13 cm	May-Aug	Jones et al., 1978
Bagre marinus	Gafftopsail catfish		♀ 26 cm	May-Aug	Jones et al., 1978
		Plotosidae—marine catfishes			
Cnidoglanis macrocephalus	Cobbler	2-4 yr	42 cm TL	Oct-Jan	Laurenson et al., 1993
		Argentinidae—argentines			
Argentina silus	Greater silversmelt	4-9 yr	27-34 cm TL	Apr-May	Bergstad, 1993
		Osmeridae—smelts			
Mallotus villosus	Capelin	3-4 yr		Apr-Aug	Jackson et al., 1991
Plecoglossus altivelis	Ayu			Aug-Nov	Fenner, 1992
		Salmonidae—salmons, trouts			
Oncorhynchus gorbuscha	Pink salmon	2 yr	~34 cm FL	Jun-Sep	MacQuarrie et al., 1979; Mathisen, 1994
Oncorhynchus keta	Chum salmon	2-7 yr	45 cm FL, 800 g	May-Jan	Pauley et al., 1988; Schreck et al., 1995
Oncorhynchus kisutch	Coho salmon	♂ 2-4, ♀ 3-4 yr		Aug-Apr	Scott & Crossman, 1973; MacQuarrie et al., 1978; Schreck et al., 1995
Oncorhynchus nerka	Sockeye salmon	3-6 yr		Jun-Feb	Pauley et al., 1989; Schreck et al., 1995
		Gonostomatidae—bristlemouths			
Cyclothone alba	Albino anglemouth	♂ 1, ♀ 2 yr	♂ 17, ♀ 21 cm SL	May-Sep	Miya & Nemoto, 1991
Cyclothone atraria	Black bristlemouth	♂ 3, ♀ ≥4 yr	♂ 22, ♀ 40 cm SL	May-Nov	Miya & Nemoto, 1991
Cyclothone pseudopallida	Hiiroonihadaka	♂ 2, ♀ 3 yr	♂ 20, ♀ 31 cm SL	May-Sep	Miya & Nemoto, 1991
Maurolicus muelleri	Lightfish		♂ 24, ♀ 26 mm SL	Mar-Feb	Prosch, 1991
		Myctophidae—lanternfishes			
Benthosema fibulatum	Hokutohadaka		♂ 33-35, ♀ 38-41 mm TL	~Apr-Jun, Sep-Oct	Hussain, 1992
Lampanyctodes hectoris	Lanternfish		36 mm SL	Mar-Feb	Prosch, 1991

Table 7-2. (Continued)

Scientific name	Common name	Age	Size	Season	References
			Gadidae—cods		
Enchelyopus cimbrius	Fourbeard rockling		♀ 15 cm	Jan-Nov	Hardy, 1978
Gadus macrocephalus	Pacific cod			Jan-Mar	Nashida et al., 1991
Gadus morhua	Atlantic cod	2-16 yr	24-100 cm	Jan-Dec	Hardy, 1978
Gadus morhua	Atlantic cod	2-6 yr	4 kg	Feb-Apr	Huse, 1991; Rijnsdorp et al., 1991
Melanogrammus aeglefinus	Haddock	2-6 yr	23-60 cm	Jan-Jul	Hardy, 1978
Merluccius bilinearis	Silver hake	prob. 2 yr	25-30 cm	May-Oct	Hardy, 1978
Microgadus tomcod	Atlantic tomcod	1-4 yr	♀ 17 cm	Nov-Feb	Hardy, 1978
Pollachias virens	Pollack	3-7 yr	34-70 cm	Sep-Jun	Hardy, 1978
Theragra chalcogramma	Walleye pollock	≥3 yr	~25 cm	Dec-Aug	Kim, 1992
Urophycis chuss	Red hake	2 yr	♂ 24-30, ♀ 27-33 cm	May-Sep	Hardy, 1978
Urophycis regia	Spotted hake		♂ 21, ♀ 31 cm	Sep-Mar	Hardy, 1978
Urophycis tenuis	White hake	4 yr	♂ 40, ♀ 48 cm	Most months	Hardy, 1978
			Batrachoididae—toadfishes		
Opsanus tau	Oyster toadfish	3-4 yr	♂ 14-15, ♀ 10-14 cm TL	Apr-Jul	Martin & Drewry, 1978
Porichthys notatus	Midshipman		♂ ≤65, ♀ ≤96 mm SL	May-Aug	Brantley & Bass, 1994
Porichthys plectrodon	Atlantic midshipman	1-2 yr	♂ 80-186, ♀ 92-139 mm SL	Apr-Oct	Martin & Drewry, 1978
			Lophiidae—goosefishes		
Lophius americanus	Goosefish	♂ 3^+, ♀ 4^+ yr	♂ ~37, ♀ ~49 cm TL	Apr-Oct	Martin & Drewry, 1978; Armstrong et al., 1992
			Gobiesocidae—clingfishes		
Acyrtops beryllinus	Emerald clingfish		♂ 17, ♀ 15 mm TL	Oct-Jun	Gould, 1965
Aspasma minima			♂ ≤48, ♀ ≤37 mm TL		Shiogaki & Dotsu, 1971
Gobiesox strumosus	Skilletfish		35-45 mm	Apr-Aug	Martin & Drewry, 1978
			Exocoetidae—flyingfishes		
Cheilopogon heterurus	Atlantic flyingfish	1 yr	20 cm SL	Mar-Jul	Hardy, 1978
Hyporhamphus unifasciatus	Halfbeak		♀ ≤15 cm	Mar-summer	Hardy, 1978
			Belonidae—needlefishes		
Strongylura marina	Atlantic needlefish	poss. 1-2 yr	♀ ≤21 cm	May-Aug	Hardy, 1978
Tylosurus acus	Agujon	1-3 yr	~60-70 cm SL	Mar-Jul	Hardy, 1978
Tylosurus crocodilus	Houndfish	1.5-3 yr	45-80 cm	Feb-Oct	Hardy, 1978

Scomberesocidae—sauries					
Scomberesox saurus	Atlantic saury	2 yr	♂ 23-32, ♀ 23-34 cm	Aug-Jun	Hardy, 1978
Cyprinodontidae—killifishes					
Cyprinodon variegatus	Sheepshead minnow	4-12 mo	♂ 24-25, ♀ 27-28 mm	Jan-Dec	Hardy, 1978
Fundulus confluentus	Marsh killifish		26-40 mm TL	Jan-Dec	Hardy, 1978
Fundulus diaphanus	Banded killifish	2-3 yr	~38-53 mm	May-Aug	Hardy, 1978
Fundulus grandis	Gulf killifish		~75 mm	Feb-Oct	Tatum et al., 1982a
Fundulus heteroclitus	Mummichog	1-2 yr	~32 mm TL	Apr-Sep	Hardy, 1978
Fundulus luciae	Spotfin killifish	≤1 yr	♂ 24-27, ♀ 28-30 mm TL	Apr-Oct	Hardy, 1978
Fundulus majalis	Striped killifish	1-2 yr	♂ ~63, ♀ ~76 mm	Apr-Sep	Hardy, 1978
Lucania parva	Rainwater killifish		~25 mm TL	Jan-Oct	Hardy, 1978
Poeciliidae—livebearers					
Gambusia affinis	Mosquitofish	≤1.5 yr	♂ 18-28, ♀ 22-34 mm	Jan-Dec	Hardy, 1978
Pseudomugilidae—blue-eyes					
Pseudomugil mellis	Honey blue-eye	10-12 mo	20 mm SL	Jan-Dec	Semple, 1991
Atherinidae—silversides					
Atherinomorus lacunosus	Hardyhead silverside	1 yr	85 mm TL	Aug-Dec	Conand, 1993
Atherinops affinis	Topsmelt			May-Aug	Anderson et al., 1991
Atherinosoma microstoma	Smallmouth hardyhead	<1 yr	>30 mm TL	Aug-Dec	Molsher et al., 1994
Hypoatherina bleekeri				May-Aug	Takita & Nakamura, 1986
Membras martinica	Rough silverside		50-63 mm SL	Mar-Sep	Martin & Drewry, 1978
Menidia beryllina	Inland silverside	some ♀ 4-5 mo	some ♀ 48-62 mm SL	Feb-Apr	Middaugh & Hemmer, 1992
Menidia menidia	Atlantic silverside	8-12 mo	50-88 mm	Jan-Sep	Martin & Drewry, 1978; Poole, 1978
Gasterosteidae—sticklebacks					
Apeltes quadracus	Fourspine stickleback	prob. 1 yr	♂ 27, ♀ 33 mm	Apr-Jul	Hardy, 1978
Gasterosteus aculeatus	Threespine stickleback	1-2 yr	20-50 mm	Feb-Sep	Hardy, 1978
Fistulariidae—cornetfishes					
Fistularia tabacaria	Bluespotted cornetfish		poss. 42 cm	~Feb	Hardy, 1978
Syngnathidae—pipefishes, seahorses					
Hippocampus erectus	Lined seahorse		♂ >95 mm	Most months	Hardy, 1978
Hippocampus obtusus	Offshore seahorse		♂ ≥70 mm		Hardy, 1978
Syngnathus floridae	Dusky pipefish		♂ 104, ♀ 75 mm	Feb-Dec	Hardy, 1978
Syngnathus fuscus	Northern pipefish	~1 yr	♂ 83 mm	Mar-Oct	Hardy, 1978
Syngnathus louisianae	Chain pipefish		♂ ≤20 cm	Feb-Sep	Hardy, 1978
Syngnathus pelagicus	Sargassum pipefish		♂ 71 mm SL		Hardy, 1978
Scorpaenidae—scorpionfishes					
Sebastes emphaeus	Puget Sound rockfish	2 yr	11-13 cm FL	Jun-Sep	Beckmann et al., 1998
Sebastes flavidus	Yellowtail rockfish			Jan-Jul	Bowers, 1992

Table 7-2. (Continued)

Scientific name	Common name	Age	Size	Season	References
Sebastes inermis	Mushroom scorpionfish			Dec-Feb	Okiyama, 1988
Sebastes marinus	Ocean perch		♂ 21-26, ♀ 38-41 cm TL		Haldorson & Love, 1991
Sebastes matsubarae	Matsubara's red rockfish			~Dec-Apr	Okiyama, 1988
Sebastes oblongus	Oblong rockfish			Nov-Jan	Okiyama, 1988
Sebastes schlegeli	Kurosoi	♂ 2, ♀ 3 yr	♂ 28, ♀ 35 cm	Mar-Jun	Okiyama, 1988; Jeon, 1995
Sebastes vulpes	Fox jacopever			May-Jun	Okiyama, 1988
Sebasticus marmoratus	Kasago			Nov-May	Okiyama, 1988
			Triglidae—searobins		
Prionotus carolinus	Northern searobin			Jun-Aug	Kuntz & Radcliffe, 1917
			Anoplopomatidae—sablefishes		
Anoplopoma fimbria	Sablefish		♀ 40-65 cm FL	Oct-Apr	Solar et al., 1987; Hunter et al., 1989; Macewicz & Hunter, 1994
			Centropomidae—snooks		
Lates calcarifer	Barramundi	♂ 2 yr ♀ 3 yr	♂ 42 cm TL, 1.1 kg ♀ 50 cm FL, 2.0 kg	Jan-Dec	Duray & Juario, 1987; Toledo et al., 1991
			Percichthyidae—temperate basses		
Dicentrarchus labrax	European sea bass	♂ 2 yr ♀ 3 yr	♂ 25 cm, 200 g ♀ 36 cm, 700 g	Oct-Jun	Carrillo et al., 1993; Carrillo et al., 1995
Morone chrysops	White bass	♂ 2, ♀ 4⁺ yr	♂ 250-280 g, ♀ >2 kg	Mar-Apr	Rees & Harrell, 1990; Berlinsky et al., 1995
Morone saxatilis	Striped bass	♂ 2-3, ♀ 3-6 yr	♂ ~30, ♀ ~50 cm TL	Feb-Jul	Hill et al., 1989; Rees & Harrell, 1990
			Serranidae—groupers, sea basses		
Cromileptes altivelis	Polka dot grouper			Oct-Mar	Chao, p.c., 1993
Epinephelus aeneus	White grouper			Aug	Popper et al., 1994
Epinephelus akaara	Redspotted grouper	4-5 yr	0.5-1 kg	Apr-Sep	Doi et al., 1991
Epinephelus chlorostigma	Brownspotted grouper	~2-4 yr	♂ ~30, ♀ 22-34 cm TL	May-Jul	Ghorab et al., 1986
Epinephelus diacanthus	Sixband grouper	2-6 yr	70-200 g	Apr-May	Doi et al., 1991
Epinephelus fulvus	Coney			Jan-Dec	Robertson, 1991
Epinephelus guttatus	Red hind			Dec-Mar, May-Jul	Robertson, 1991
Epinephelus malabaricus	Malabar grouper			Sep-Nov	Doi et al., 1991
Epinephelus morio	Red grouper	2⁺ yr	♂ ~40, ♀ ~30 cm FL	Jan-May	Brule & Deniel, 1994
Epinephelus polyphekadion	Camouflage rockcod		600 g	Jan-Dec	Tamaru et al., 1996

Epinephelus striatus	Nassau grouper		<50 cm TL	Nov-Sep	Robertson, 1991; Tucker et al., 1993
Epinephelus tauvina	Greasy grouper		7-11 kg	~Aug	Doi et al., 1991
Mycteroperca microlepis	Gag	♂ 5, ♀ 2-6 yr	♂ ~100, ♀ ~35 cm TL	Nov-May	Hood & Schlieder, 1992
Paralabrax clathratus	Kelp bass	♂ 2-4 yr ♀ 2-5 yr	♂ 18-26 cm TL ♀ 18-27 cm TL	~Apr-Oct	Love et al., 1996
Paralabrax nebulifer	Barred sand bass	♂ 2-4 yr ♀ 2-5 yr	♂ 19-26 cm TL ♀ 21-27 cm TL		Love et al., 1996
Plectropomus leopardus	Leopard coraltrout		~20 cm FL	Sep-Nov	Samoilys & Squire, 1994
Serranus subligarius	Belted sandfish	16 mo		Jan-Dec	Tucker & Woodward, 1995
			Pseudochromidae—dottybacks		
Pseudochromis fridmani	Orchid dottyback	4.5 mo			Moe, 1997a
			Priacanthidae—bigeyes		
Priacanthus macracanthus	Red bullseye		17 cm FL	May-Sep	Liu et al., 1992
			Apogonidae—cardinalfishes		
Apogon notatus	Blackspot cardinalfish			May-Sep	Kuwamura, 1983
Archamia zosterophora	Blackbelt cardinalfish		37 mm SL	Jul-Jun	Milton & Blaber, 1991
			Sillaginidae—whitings, sillagos		
Sillaginodes punctata	Spotted whiting	>2 yr	♂ 27-43 ♀ 31-50 cm TL	Mar-Jun	Cockrum & Jones, 1992
Sillaginoides punctata	Spotted whiting	3-5 yr	35-50 cm TL	Jun-Sep	Hyndes et al., 1998
			Malacanthidae—tilefishes		
Malacanthus latovittatus	Blue whiting		22 cm TL	Jun-Sep	Clark & Pohle, 1992
			Pomatomidae—bluefishes		
Pomatomus saltatrix	Bluefish	2 yr	35-40 cm TL	Most months	Haimovici & Krug, 1992; Sabatés & Martin, 1993
			Rachycentridae—cobia		
Rachycentron canadum	Cobia	♂ 1, ♀ 2 yr	♂ 64, ♀ 83 cm FL	Apr-Sep	Biesiot et al., 1994; Lotz et al., 1996
			Echeneidae—remoras		
Echeneis naucrates	Sharksucker		≤41 cm	Feb	Martin & Drewry, 1978
			Carangidae—jacks		
Decapterus macrosoma	Shortfin scad	~2 yr	10-20 cm	Nov-Sep	Widodo, 1991
Selar crumenophthalmus	Bigeye scad	≤1 yr	♂ 19, ♀ 25 cm FL	Apr-Dec	Iwai et al., 1996
Seriola lalandi	Goldstriped amberjack			Apr-May	Fujita & Yogata, 1984
Seriola lalandi	Goldstriped amberjack	~12 mo	1.5 kg	Jan-Dec	Kraul, 1985
Seriola quinqueradiata	Yellowtail			Feb-Jun	Ikenoue & Kafuku, 1992
Trachinotus carolinus	Florida pompano			Apr-Oct	Finucane, 1969, Kloth 1980
Trachinotus falcatus	Permit			Apr-Jun	Finucane, 1969, Kloth, 1980

Table 7-2. (Continued)

Scientific name	Common name	Age	Size	Season	References
Trachinotus mediterraneus	Mediterranean scad	2 yr	16 cm TL	May-Aug	Viette et al., 1997
Trachurus trachurus	Atlantic horse mackerel		~21 cm	Apr-Aug	Macer, 1974; Eltink, 1991
			Coryphaenidae—dolphins		
Coryphaena hippurus	Dolphin	3-6 mo	35-55 cm FL, ~1.5 kg	Jan-Dec	Palko et al., 1982; Uchiyama et al., 1986; Kraul, 1991, Szyper, 1991
			Lutjanidae—snappers		
Lutjanus argentimaculatus	Asian red snapper	2-3 yr	3 kg	~Jan-Nov	Doi & Singharaiwan, 1993
Lutjanus buccanella	Blackfin snapper			Jan-Dec	Robertson, 1991
Lutjanus campechanus	Atlantic red snapper	1-2 yr	375 mm FL	May-Dec	Moran, 1988
Lutjanus campechanus	Atlantic red snapper	~2-3 yr	♂ 50, ♀ 44 cm TL	Jul-Aug	Al-Ogaily et al., 1992
Lutjanus gibbus	Australian red snapper		208 mm BL	Jan-Dec	Mobiha, 1991
Lutjanus griseus	Gray snapper			Jul-Oct	Robertson, 1991
Lutjanus johnii	Golden snapper	♂ 2.5-3 yr ♀ 3-4 yr	♂ 2.3-3.0 kg ♀ 3.5-4.5 kg	Jan-Aug	Lim et al., 1985
Lutjanus synagris	Lane snapper			Mar-Dec	Robertson, 1991
Lutjanus vittus	Blackstriped snapper	1-3 yr	♀ 14-22 cm FL	Jun-Apr	Davis & West, 1993
Lutjanus vivanus	Silk snapper			Jan-Dec	Robertson, 1991
Ocyurus chrysurus	Yellowtail snapper		♂ 26, ♀ 29-31 cm FL	Jan-Dec	Bortone & Williams, 1986; Robertson, 1991
			Gerreidae—mojarras		
Gerres filamentosus	Threadfin silverbelly		11-16 cm SL	Oct-Feb	Kurup & Samuel, 1991b
			Haemulidae—grunts		
Orthopristis chrysoptera	Pigfish	≤2 yr	20 cm TL		Hildebrand & Cable, 1930
Plectorhinchus celebicus	Celebes sweetlips		266 mm BL	Jan-Dec	Mobiha, 1991
Plectorhinchus cinctus	Yellowspotted grunt	~2 yr	♀ ≤1.1 kg	Mar-Apr	Chang et al., 1994c
			Sparidae—sea breams, porgies		
Acanthopagrus berda	River bream		~21 cm FL	May-Aug	Garratt, 1993
Acanthopagrus schlegeli	Black sea bream	♂ 1, ♀ 3 yr		Late winter-spring	Chang et al., 1991
Archosargus probatocephalus	Sheepshead	<2 yr	350-400 g	Feb-Jun	Hildebrand & Cable, 1938; Tucker & Barbera, 1987
Dentex tumifrons	Yellow sea bream	1.5-2 yr	13-15 cm FL	Apr-Jun,Sep-Oct	Oki & Tabeta, 1998
Pagrus major	New Zealand snapper			Sep-Mar	Battaglene & Talbot, 1992; Scott &

					Pankhurst, 1992; Francis, 1994
Pagrus major	Red sea bream	2-5 yr	22-36 cm, 300-750 g	Feb-Jul	Foscarini, 1988; Fukusho, 1991
Lagodon rhomboides	Pinfish			Oct-Mar	Hildebrand & Cable, 1938
Pachymetopon grande	Bronze sea bream	5.5 yr	30 cm FL	Jan-Jun	Buxton & Clarke, 1992
Rhabdosargus sarba	Silver sea bream	2 yr	25 cm, 280 g	Nov-Mar	Lin et al., 1988, 1989
Rhabdosargus sarba	Silver sea bream	≤3 yr	♂ ~592, ♀ ~717 g	Dec-Mar	Leu, 1994
Stenotomus chrysops	Scup			Jun-Aug	Kuntz & Radcliffe, 1917
			Lethrinidae—sweetlips		
Lethrinus hypselopterus	Masked emperor		186 mm BL	Most months	Mobiha, 1991
Lethrinus lentjan	Redspot emperor		184 mm BL	Most months	Mobiha, 1991
			Sciaenidae—drums		
Bairdiella chrysoura	Silver perch	≤2 yr	13 cm TL	Apr-Jul	Hildebrand & Cable, 1930
Cynoscion nebulosus	Spotted seatrout		♂ 36, ♀ 42 cm TL	Apr-Sep	Tucker & Faulkner, 1987
Cynoscion nebulosus	Spotted seatrout		♂ ≤20, ♀ 23-27 cm SL	Apr-Sep	Brown-Peterson et al., 1988
Cynoscion regalis	Weakfish			Apr-Aug	Merriner, 1976; Goshorn & Epifanio, 1991
Cynoscion striatus	Striped weakfish	♀ 3 yr	♀30 cm TL	Oct-Mar	Macchi, 1998
Cynoscion xanthulus	Orangemouth corvina	~2 yr	~62 cm TL	Summer-fall	Thomas et al., 1994
Micropogonias furnieri	Whitemouth croaker			Nov-Mar	Macchi, 1998
Nibea albida			♂ 13-22, ♀ 15-22 cm SL	Jul-Nov	Kurup & Samuel, 1991a
Pogonias cromis	Black drum	♂ 3-5, ♀ 2-5 yr	♂ 55, ♀ 60 cm FL	Jan-May	Nieland & Wilson, 1993; Saucier & Baltz, 1993
Pseudosciaena polyactis	Small yellow croaker	2 yr	♂ 223, ♀ 205 mm TL	Jun	Myoung et al., 1994
Sciaenops ocellatus	Red drum	♂ 1.5-3 yr ♀ 1.5-4 yr	♂ 52-74 cm FL ♀ 73-86 cm FL	Aug-Jan	Matlock, 1990; Arnold, 1991; Ross et al., 1995; Tucker, pers. obs.
			Mullidae—goatfishes		
Mullus surmuletus	Red mullet	0-4 yr	11-27 cm TL	Apr-May	Morales-Nin, 1991
			Kyphosidae—sea chubs		
Kyphosus cinerascens	Ashen drummer		252 mm BL	Most months	Mobiha, 1991
			Pomacanthidae—angelfishes		
Holacanthus ciliaris	Queen angelfish		~125 mm		Straughan, 1956
Holacanthus tricolor	Rock beauty			Feb-Jun	Moe, 1997b
Pomacanthus arcuatus	Gray angelfish			Apr-Sep	Moe, 1997b
Pomacanthus paru	French angelfish		~125 mm		Straughan, 1956
			Cichlidae—cichlids		
Oreochromis niloticus	Nile tilapia	6 mo	<40 g	Jan-Dec	Macintosh & Little, 1995
Oreochromis spilurus		4-5 mo			Hazza et al., 1989
			Pomacentridae—damselfishes		
Abudefduf saxatilis	Sergeant major	≤1 yr		Jan-Dec	Cummings, 1968

Table 7-2. (Continued)

Scientific name	Common name	Age	Size	Season	References
Amphiprion spp.	(clownfishes)	~2 yr			Belanga, 1993
Amphiprion clarkii	Clark's clownfish		≤71 mm SL		Hattori, 1994
Chromis punctipinnis	Blacksmith	2-3 yr	14 cm	~Jun-Aug	Turner & Ebert, 1963
Hypsypops rubicundus	Garibaldi			Apr-Aug	Sikkel, 1994
Pomacentrus amboinensis	Pallid damselfish			Oct-Jan	Meekan et al., 1993
Stegastes leucostictus	Beaugregory			Jan-Dec	Itzkowitz, 1991
			Mugilidae—mullets		
Chelon labrosus	Thicklip mullet			~Mar-Jul	Vidy & Franc, 1992
Liza aurata	Golden mullet			~Oct-May	Vidy & Franc, 1992
Liza ramada	Thinlip mullet			~Nov-May	Vidy & Franc, 1992
Liza saliens	Leaping mullet			~May-Feb	Vidy & Franc, 1992
Mugil cephalus	Striped mullet	♂ 1-7, ♀ 2-8 yr	20-61 cm FL	Jan-Dec	Martin & Drewry, 1978
Mugil cephalus	Striped mullet	3 yr	♂ 20, ♀ 25 cm TL	Oct-Mar	Patillo, 1990
Mugil curema	White mullet	~1 yr	10-30 cm FL	Feb-Nov	Martin & Drewry, 1978
Valamugil seheli	Bluespotted mullet		300 mm BL	Jan-Dec	Mobiha, 1991
			Sphyraenidae—barracudas		
Sphyraena barracuda	Great barracuda	♂ 2, ♀ 3-4 yr	♂ 46, ♀ 66 cm FL	Jan-Dec	Martin & Drewry, 1978
			Labridae—wrasses		
Anampses caeruleopunctatus	Bluespotted tamarin			Mar-May	Colin & Bell, 1991
Gomphosus varius	Birdfish			Mar-May,Nov	Colin & Bell, 1991
Halichoeres marginatus	Speckled rainbowfish			Mar-May	Colin & Bell, 1991
Lachnolaimus maximus	Hogfish			Dec-Apr	Colin, 1982
Tautoga onitis	Tautog	3-4 yr	♂ ~20, ♀ 19 cm	Apr-Sep	Sogard et al., 1992; Perry, 1994
Tautogolabrus adspersus	Cunner			Jun-Aug	Kuntz & Radcliffe, 1917
Thalassoma bifasciatum	Bluehead wrasse		19-26 mm SL	Jan-Dec	Petersen et al., 1992; Shapiro & Rasotto, 1993
Thalassoma cupido	Cupido wrasse		♂ 44, ♀ 42 mm SL		Meyer, 1977
Thalassoma lutescens	Green moon wrasse			Mar-May,Oct-Nov	Colin & Bell, 1991
			Scaridae—parrotfishes		
Cetoscarus bicolor	Bicolor parrotfish			Jan,Mar-May	Colin & Bell, 1991
Scarus globiceps	Violetline parrotfish			Mar-May,Nov	Colin & Bell, 1991
Sparisoma rubripinne	Redfin parrotfish		♂ 194, ♀ 203 mm SL	Jan-Dec	Randall & Randall, 1963

Sparisoma viride	Stoplight parrotfish		♂ 154, ♀ 260 mm SL	Jan-Dec	Koltes, 1993
			Stichaeidae—pricklebacks		
Chirolophis japonicus	Fringed blenny			Nov-Dec	Shiogaki, 1983
Opisthocentrus tenuis		~2 yr	11-15 cm	Dec-Jan	Shiogaki, 1981b
			Anarhichadidae—wolffishes		
Anarhichas lupus marisalbi	White Sea wolffish	prob. 6 yr	♂ 35, ♀ 33 cm; ♂ 400, ♀ 370 g	May-Aug	Pavlov & Novikov, 1987
			Nototheniidae—Antarctic blennies		
Pseudotrematomus eulepidotus			♂ ~18, ♀ ~21 cm TL	Feb-Mar	Duhamel et al., 1993
			Blenniidae—combtooth blennies		
Aidablennius sphynx	Mediterranean blenny		♂ 50, ♀ 33 mm TL	Apr-Aug	Kraak, 1996
Chasmodes bosquianus	Striped blenny			May-Aug	Hildebrand & Cable, 1938
Hypleurochilus geminatus	Crested blenny			May-Sep	Hildebrand & Cable, 1938
Hypsoblennius hentz	Feather blenny			May-Aug	Hildebrand & Cable, 1938
Salaria pavo		≤1 yr	~35 mm TL	Apr-Sep	Ruchon et al., 1993
			Callionymidae—dragonets		
Callionymus lyra	European dragonet		♂ 16, ♀ 17 cm		Holt, 1898
Repomucenus hugenini		≤ 1 yr	45 mm SL	May,Oct-Dec	Gonzales et al., 1997
			Gobiidae—gobies		
Boleophthalmus pectinirostris	Bluespotted mudhopper	♀ 1-3 yr	♀ 90 mm SL	May-Jul	Washio et al., 1993
Gobionellus boleosoma	Darter goby			May-Oct	Hildebrand & Cable, 1938
Gobiosoma bosc	Naked goby			Apr-Sep^{+}	Hildebrand & Cable, 1938
Gobiosoma oceanops	Neon goby	5-6 mo	50 mm	Feb-Apr	Moe, 1975
Gobius niger	Black goby	7-13 mo	♂ 43, ♀ 48 mm TL	Mar-Sep	Joyeaux et al., 1991
Paragobiodon echinocephalus	Pricklehead goby		≤17 mm TL	Apr-Nov	Kuwamura et al., 1993
Pterogobius zacalles				~Apr-May	Shiogaki, 1981a
			Ephippidae—spadefishes		
Chaetodipterus faber	Atlantic spadefish		♀ ≤304 g	Jun-Jul	Hildebrand & Cable, 1938; Gomez, i.p.
			Scatophagidae—scats		
Scatophagus argus	Spotted scat	♀ 7-9 mo	♂ 115, ♀ 140 mm SL	~Jul-Aug	Barry & Fast, 1992
			Siganidae—rabbitfishes		
Siganus canaliculatus	Pearly spinefoot	6-7 mo	4-6 cm, 2-6 g		Von Westernhagen & Rosenthal, 1975
Siganus canaliculatus	Pearly spinefoot	≤1 yr	♂ 110, ♀ 120 mm	Mar-Jun	Tseng & Chan, 1982; Al-Ghais, 1993
Siganus canaliculatus	Pearly spinefoot			~Jan-Oct	Hoque et al., 1998
Siganus guttatus	Yellowblotch spinefoot	♂ 10, ♀ 12 mo	♂ 190, ♀ 215 mm FL	Jan-Dec	Duray & Juario, 1987
Siganus lineatus	Goldline spinefoot		220 mm BL	Jan-Dec	Mobiha, 1991

Table 7-2. (Continued)

Scientific name	Common name	Age	Size	Season	References
			Scombridae—mackerels		
Euthynnus affinis	Kawakawa		42 cm BL	Jan-Dec	Mobiha, 1991
Euthynnus alletteratus	Little tunny		~38 cm FL	Apr-Nov	Anon., 1997b
Euthynnus lineatus	Black skipjack		40-54 cm TL, ~1 kg	Aug-Jun	Schaefer, 1987
Katsuwonus pelamis	Skipjack tuna		~48 cm FL	Jan-Dec	Anon., 1997b
Rastrelliger kanagurta	Indian mackerel		21-25 cm	Apr-May, Sep-Oct	Gopakumar et al., 1991
Sarda sarda	Bonito		40 cm FL	Jun-Jul	Anon., 1997b
Scomber japonicus	Chub mackerel	♀ ~2-4 yr	♀ ≤30 cm FL, ≤350 g	Mar-Oct	Dickerson et al., 1992
Scomberomorus maculatus	Spanish mackerel			Jun-Aug	Hildebrand & Cable, 1938
Thunnus alalunga	Albacore	~5.5 yr	70-90 cm	~Feb-Jul	Wu & Kuo, 1993; Anon., 1997b
Thunnus albacares	Yellowfin tuna	~1-3 yr	♂ ≥53, ♀ ≥57 cm	Jan-Dec	Suzuki, 1994
Thunnus atlanticus	Blackfin tuna			Apr-Nov	Anon., 1997b
Thunnus maccoyii	Southern bluefin tuna		♀ ~1.4 m FL	Aug-May	Farley & Davis, 1998
Thunnus obesus	Bigeye tuna		1-1.3 m FL	Jan-Dec	Anon., 1997b
Thunnus thynnus	Northern bluefin tuna	5-8 yr	1.5-2 m FL, ~40-90 kg	Apr-Jul	Harada, 1980; Anon., 1997b
			Xiphiidae—swordfish		
Xiphias gladius	Swordfish			Jun	Gordina & Bagnyukova, 1993
			Stromateidae—butterfishes		
Peprilus triacanthus	Butterfish	prob. 2 yr	12-18 cm SL	Jun-Sep	Martin & Drewry, 1978
Peprilus triacanthus	Butterfish			Feb-Jul	Rotunno & Cowen, 1997
			Bothidae—lefteye flounders		
Citharichthys arctifrons	Gulf Stream flounder		~50 mm SL	May-Dec	Martin & Drewry, 1978
Citharichthys sordidus	Pacific sandab	♀ 3 yr	♀ ~190 mm TL	~Jul-Oct	Rackowski & Pikitch, 1989
Citharichthys stigmaeus	Speckled sandab	♀ 2 yr	♀ 70-80 mm TL	Apr-Sep	Rackowski & Pikitch, 1989
Etropus crossotus	Fringed flounder		♂ ~91, ♀ ~89 mm	Winter, Mar-Jun	Martin & Drewry, 1978
Etropus microstomus	Smallmouth flounder		50 mm SL	peak Jul-Oct	Martin & Drewry, 1978
Paralichthys dentatus	Summer flounder		♀ 42 cm TL	Sep-Feb	Martin & Drewry, 1978
Paralichthys lethostigma	Southern flounder	♀ 2, ♂ 3 yr	♂ 1.2, ♀ 2.0 kg	Dec-Mar	Daniels et al., 1997
Paralichthys olivaceus	Hirame	~3 yr	~40 cm TL, 750 g	Nov-Jul	Cho & Yang, 1991; Nashida et al., 1991; Saitoh et al., 1991; Subiyanto et al., 1992
Pseudorhombus pentophthalmus	Fivespot flounder		~80 mm BL	~Jun-Oct	Tominaga & Nashida, 1992
Scophthalmus aquosus	Windowpane	3-4 yr	23-25 cm	Apr-Oct	Martin & Drewry, 1978

Scophthalmus maximus	Turbot		♂ 2, ♀ 3.5 kg		Person Le-Ruyet, 1990b
Syacium papillosum	Dusky flounder		82-130 mm SL	Feb-Nov	Martin & Drewry, 1978
		Pleuronectidae—righteye flounders			
Glyptocephalus cynoglossus	Witch flounder	1-10⁺ yr	5-60 cm TL	Mar-Nov	Martin & Drewry, 1978; Burnett et al., 1992
Hippoglossus hippoglossus	Atlantic halibut	♂ 5-17, ♀ 7-12 yr	♂ 70-119, ♀ 90-129 cm	Mar-Sep	Martin & Drewry, 1978
Hippoglossus hippoglossus	Atlantic halibut	♂ 4-17 yr ♀ 6-18 yr	♂ 55 cm, 1.7 kg ♀ 110-115 cm, ~18 kg	Most months	Haug, 1990; Neilson et al., 1993
Microstomus pacificus	Dover sole	♀ 5 yr	♀ 23-44 cm TL	Dec-May	Yoklavich & Pikitch, 1989; Hunter et al., 1992
Pleuronectes americanus	Winter flounder	3-9 yr	♂ ~21, ♀ ~26 cm	Dec-May	Martin & Drewry, 1978
Pleuronectes ferrugineus	Yellowtail flounder	2-6 yr	26-45 cm	Mar-Sep	Martin & Drewry, 1978
Pleuronectes herzensteini	Littlemouth flounder			Feb-May	Nashida et al., 1991
Pleuronectes platessa	European plaice	2-6 yr	20-40 cm	~Jan-Mar	Hardy, 1959; Rijnsdorp et al., 1991
Pleuronectes vetulus	English sole	4 yr	~24 cm	Jan-Mar	Sanborn & Misitano, 1991
Pleuronectes yokohamae	Marbled flounder	♂ 1-2, ♀ 2-4 yr	♂ ≤19, ♀ ≤21 cm BL	Mar-Apr	Dagang et al., 1991
Reinhardtius hippoglossoides	Greenland halibut	10-12 yr	♂ ≤60, ♀ ≤70 cm	~Winter-spring	Crawford, 1992
		Soleidae—soles, tonguefishes			
Cynoglossus arel	Brown tonguesole	~2 yr	♂ 15-26, ♀ 15-29 cm TL	Jun-Mar	Rajaguru, 1992
Cynoglossus lida	Shoulderspot tonguesole	~2 yr	♂ 14-19, ♀ 14-20 cm TL	Feb-Nov	Rajaguru, 1992
Solea bleekeri	Blackhand sole		♂ 55, ♀ 56 mm SL	Sep-Dec	Cyrus, 1991
Solea solea	Common sole	♀ 2-5 yr			Rijnsdorp et al., 1991
Trinectes maculatus	Hogchoker	2-4 yr	♀ 48 mm SL	Jan-Oct	Martin & Drewry, 1978
		Balistidae—leatherjackets			
Monacanthus hispidus	Planehead filefish		♀ ≤81 mm	Most months	Martin & Drewry, 1978
Paramonacanthus japonicus	Striped filefish		♂ ≤76, ♀ 61 mm SL	~May-Oct	Nakazano & Kawase, 1993
		Tetraodontidae—puffers			
Canthigaster rivulata	Sharpnose puffer			Jun-Sep	Arai & Fujita, 1988
Chilomycterus schoepfi	Striped burrfish		♀ ≤19 cm	Spring-fall	Martin & Drewry, 1978
Diodon holocanthus	Balloonfish			~May-Jun	Fujita et al., 1997
Sphoeroides maculatus	Northern puffer		70-100 mm SL	May-Aug	Martin & Drewry 1978
Takifugu exascurus			♀ ≤156 mm TL	Jun-Jul	Fujita & Honma, 1991
Takifugu niphobles	Grass puffer			Jun-Jul	Honma et al., 1980
Takifugu porphyreus	Purple puffer		♂ 26, ♀ 33 cm TL	Jun-Jul	Fujita et al., 1992
Takifugu rubripes	Tiger puffer		40-55 cm, 1.5-2.5 kg	Mar-May	Kuronuma & Fukusho, 1984
Tetraodon cutcutia	Ocellated blowfish	1 yr			Cohn, 1912
		Molidae-molas			
Mola mola	Ocean sunfish		♂ ≤119, ♀ 137 cm TL	Winter	Martin & Drewry, 1978

Egg production by the larger species usually is high, and many smaller species spawn frequently (Table 7-1). Batch fecundity of 40-kg cobia was 190,000,000 eggs (Lotz et al., 1996). Regular large spawns or frequent small ones help make hatchery operations practical. Spawning frequency varies—e.g., Pacific salmon once in a lifetime, some groupers once a year, some tilapias once a month, dolphin every other day, and pygmy angelfish every day.

Some species (Table 7-2) mature quickly (tilapia in a few months) and others slowly (Californian giant sea bass possibly as long as 23 yr; but a mature male about 11 yr old has been captured, Domeier, pers. comm., 1998). Advanced age at maturity can be a handicap for a hatchery. Sometimes mature wild fish can be used until captive broodstock are reared to maturity.

Some species require hormonal manipulation before they will spawn in captivity. Others will spawn voluntarily. Control of reproduction can be achieved by acting at different levels on the brain-pituitary-gonad axis. Final maturation and ovulation of eggs can be chemically-induced or natural, and spawning can be artificial or voluntary. Fish spawn "naturally" in their natural habitat without human interference.

Fish spawn "voluntarily" in any location without hormone or chemical treatment. Fish are "induced" to ovulate by hormone and/or chemical treatment. Fish induced to ovulate sometimes are "strip-spawned" by humans, and sometimes they "tank-spawn," "cage-spawn", or "pond-spawn" on their own. Fish are "conditioned" to ovulate by human control of their environment; sometimes they spawn voluntarily and sometimes they are stripped.

Spermatogenesis (Sperm Formation)

Sperm formation is discussed concisely in Zohar (1989) and at more length in references cited therein. Very briefly, after several mitotic and meiotic divisions, primary spermatogonia are transformed into spermatids, which then differentiate into spermatozoa during spermiogenesis. Finally, sperm are hydrated and diluted during spermiation.

Oogenesis (Egg Formation)

For details, see Yamazaki (1976), Wallace and Selman (1981), Hoar et al. (1983a, 1983b), Bromage and Cumaranatunga (1987), Wallace et al., (1987), Conte et al. (1988), Matsuyama et al. (1988b), Zohar (1989), Berlinsky and Specker (1991), Tamaru et al. (1991a), Thorsen and Fyhn (1991), Kjesbu and Kryvi (1993), Saat (1993), Thorsen et al. (1993), Lotz et al. (1996), Micale et al. (1996). Oogonia which have formed along the margins of folds inside the ovaries go through several mitotic divisions and part of a meiotic division. Then the cells, called primary oocytes, go through several stages to reach maturity.

Previtellogenesis.—Cytoplasm is accumulated and the nucleus grows. In sturgeon, previtellogenesis can last for two or more years (Conte et al., 1988). At the end of this stage, a follicle has formed around the oocyte, with the inner layer called the granulosa and the outer layer the theca (Fig. 7-1).

Vitellogenesis.—During this stage (Figs. 7-1 to 7-4, 7-5a), the oocytes accumulate yolk and oil. Many fish have oocytes that first synthesize cortical alveoli (yolk vesicles) and oil globules internally during "endogenous vitellogenesis." Cortical alveolus formation is controlled by gonadotropin and probably estrogen (Lam, 1982; Wiegand, 1982). During the next phase, exogenous (true) vitellogenesis, yolk globules are formed from vitellogenin taken from the bloodstream. Yolk granule or globule formation is controlled by estrogen and possibly a vitellogenic gonadotropin. The liver is stimulated by estrogen to produce vitellogenin, the precursor of yolk protein (Lazier et al., 1987), and the gonadotropin might be necessary for its incorporation into the oocytes (Abraham et al., 1988; Le Menn et al., 1988; Rodriguez and Le Menn, 1988). In sturgeon, vitellogenesis can take 1-2 yr (Conte et al., 1988). Gonadosomatic index (gonad weight/total weight) can reach at least 16% in male and female Nassau groupers (Tucker et al., 1993b) and 30% in female sturgeon (Conte et al., 1988).

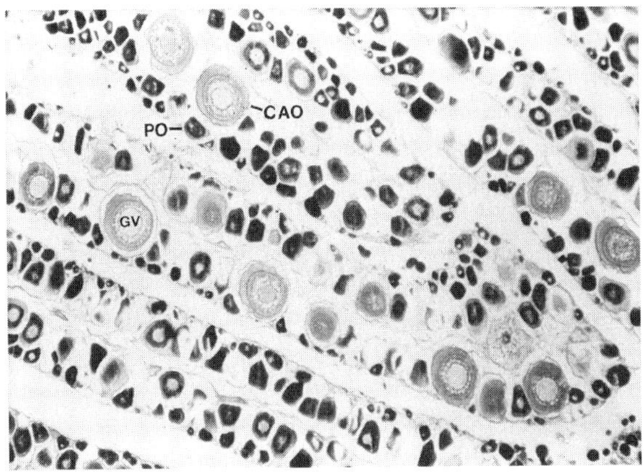

Figure 7-1. Histological section from a common snook ovary in the early stages of reproductive activity. PO = Primary (unyolked) oocyte. CAO = Cortical alveoli stage oocyte. GV = germinal vesicle. (Histology for Figs. 7-1, 7-2, 7-6, and 7-7 by Silas W. Campbell)

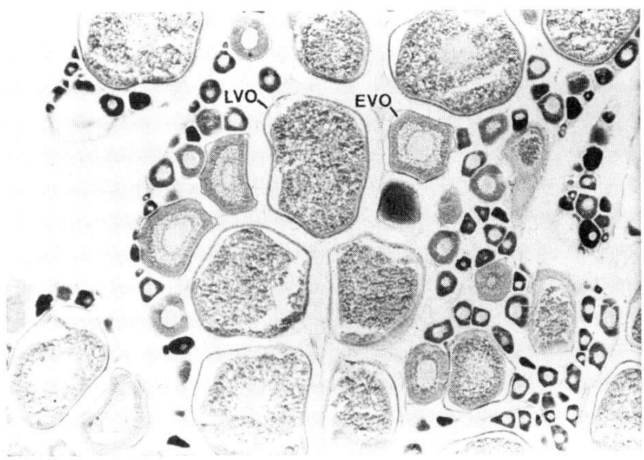

Figure 7-2. Histological section from a nearly ripe common snook ovary. EVO = Early-vitellogenic oocyte, LVO = Late-vitellogenic oocyte.

Oocyte Maturation

This is the period of most interest in the hatchery and will be described in slightly more detail. It usually takes no more than 24 h (Wallace et al., 1987). Sharply rising gonadotropin level causes the follicular envelope to release a maturation-inducing steroid, which stimulates final oocyte maturation (e.g., Trant and Thomas, 1987). At least in some species, a prostaglandin(s) also is released and stimulates ovulation (Goetz et al., 1987). The major occurrences leading up to fertilization are:

Fusion of Yolk and Oil Globules (Fig. 7-5b, 7-5c).—Fusion occurs to different degrees in different species and often results in increased transparency (clearing), especially in planktonic eggs. Yolk partitioning is obvious in some eggs but not in others. Some have their oil dispersed in many small inconspicuous globules and others have most of it in one to several obvious globules.

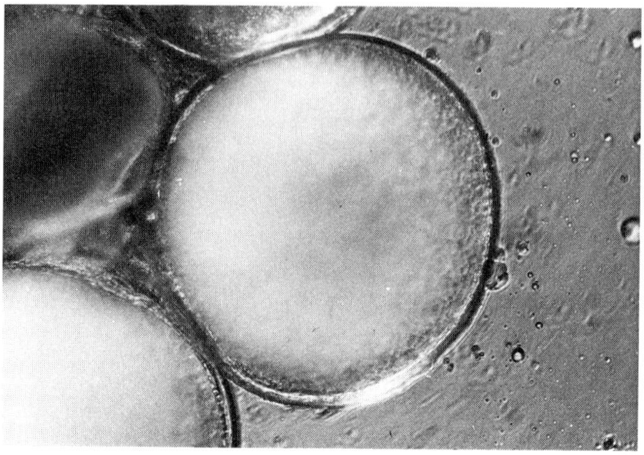

Figure 7-3. Whole late-vitellogenic (fully-yolked) oocytes biopsied from captive spotted seatrout at the end of a spawning season (with overhead illumination). The fish would ovulate if injected with HCG.

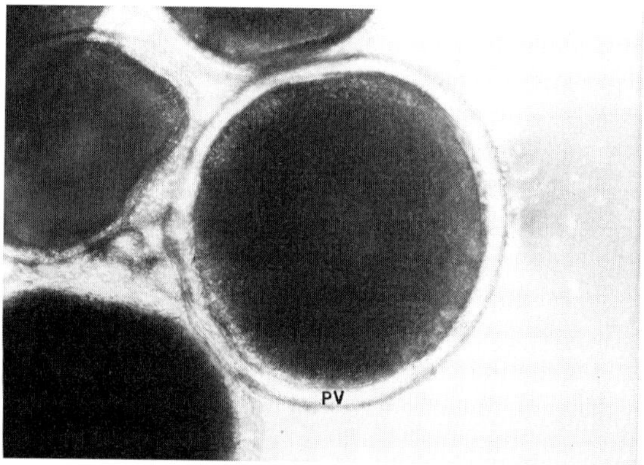

Figure 7-4. Whole late-vitellogenic oocytes biopsied from captive spotted seatrout at the end of a spawning season (the same as in Fig. 7-3 with background illumination). PV = perivitelline space, inside the chorion.

Germinal Vesicle Migration (GVM)(Fig. 7-5c).—The germinal vesicle (nucleus) moves from its central position to the animal pole at the edge of the oocyte next to the micropyle (opening through which the sperm enters).

Peripheral Germinal Vesicle (Fig. 7-5d).—a short transitional phase.

Germinal Vesicle Breakdown (GVBD).—The GV membrane is disrupted and the contents are released into the cytoplasm at the animal pole. Then the delayed first meiotic division is completed. A second meiotic division begins and is arrested. At this point the oocytes have matured into eggs, which can be fertilized; ovulation usually occurs at about the same time (Wallace et al., 1987).

Hydration (Figs. 7-6, 7-5e).—Near the end of maturation, oocytes of some species absorb water and increase noticeably in size.

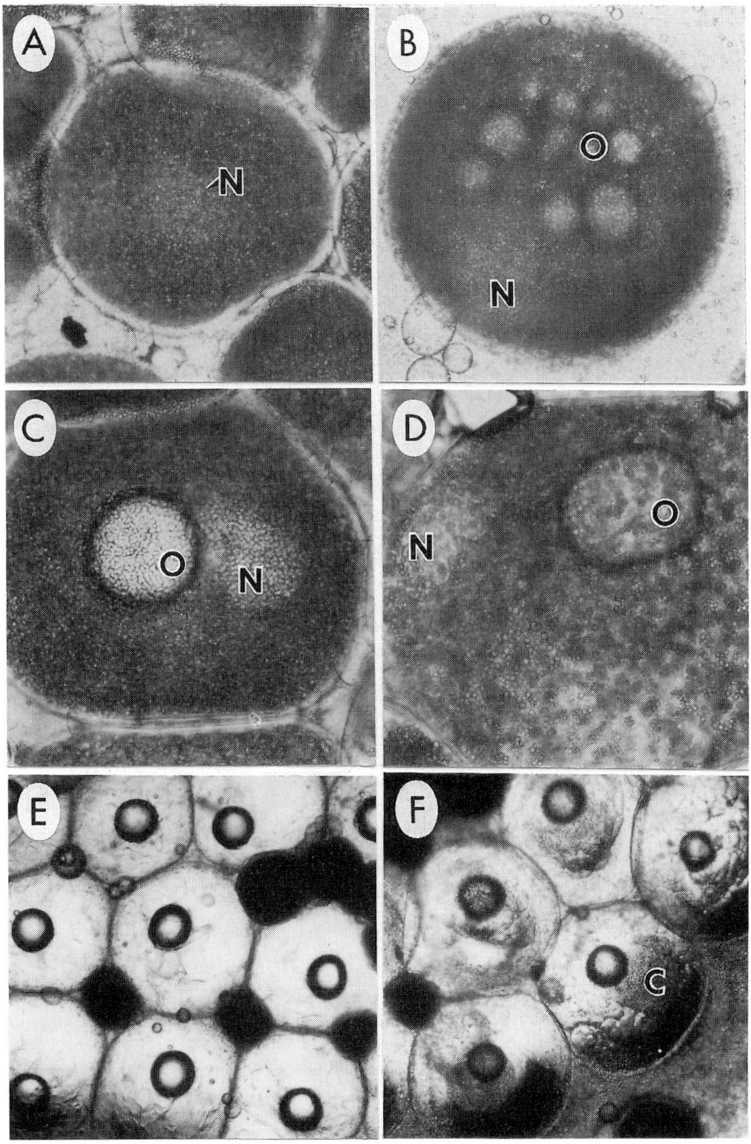

Figure 7-5. Final maturation stages of freshly biopsied whole common snook oocytes, slightly flattened under a coverslip: A) Mature oocyte with central germinal vesicle, cytoplasm full of protein granules but lacking visible oil globules (~400 μm diameter, 100×); B) Oil has accumulated but not yet coalesced into one globule (100×); C) Germinal vesicle migration, single oil globule (100×); D) Germinal vesicle breakdown (100×); E) Hydrated oocytes, protein granules broken down and yolk fused into segments (~700 μm diameter, 40×); F) Ovulated eggs, overripe and dead, with visible degeneration, notably clumped cytoplasm (40×). N = germinal vesicle, or nucleus, O = oil globules, C = clumped cytoplasm. (photos courtesy of Harry J. Grier, Florida Department of Environmental Protection, and Carole L. Neidig, Mote Marine Laboratory).

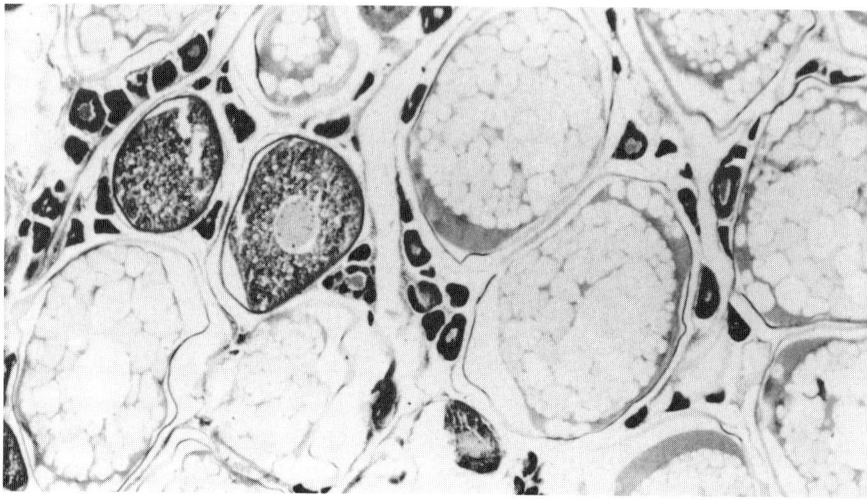

Figure 7-6. Histological section from a common snook ovary just before ovulation. The largest oocytes have undergone germinal vesicle breakdown and have become hydrated.

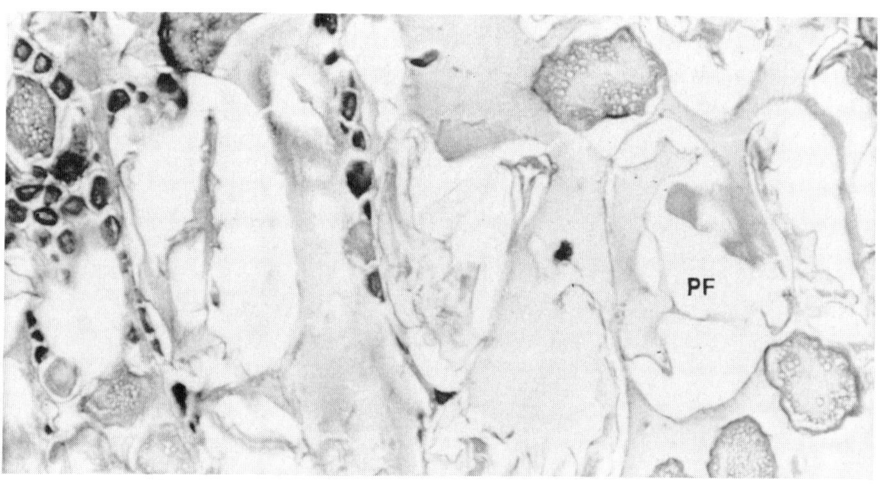

Figure 7-7. Histological section from a common snook ovary just after ovulation. PF = postovulatory follicle.

Ovulation (Fig. 7-7).—Soon after GVBD, the oocyte separates from the follicular envelope and the envelope breaks down, releasing the oocyte into the ovarian lumen or body cavity. Ovulation might take several hours (e.g., sturgeon, Conte et al., 1988) or only a few hours (e.g., Nassau grouper). In species with protracted ovulation, there is a danger that eggs ovulated early could become overripe before the last ones are ovulated. Therefore, eggs are taken from sturgeon at the midpoint, and the late oocytes are close enough that the ovulation process is completed in the fertilization container. McEvoy and McEvoy (1992) suggested that turbot be stripped 4-6 h after predicted ovulation to obtain the most good eggs at 13-14°C.

Fertilization.—A sperm enters the micropyle (which then closes) and fertilizes the egg. Meiosis is finished.

Endocrine Control of Oocyte Maturation and Ovulation

Only the main hormones will be briefly discussed here. Others are discussed in Kah et al. (1993). The two main hormones (**steroids**) secreted by the testes are testosterone and 11-ketotestosterone (both androgens). The four main ones secreted by ovaries are estradiol-17β, estrone (both estrogens), 17α-20β-dihydroprogesterone, and 17α-hydroxyprogesterone (both progestins). Sex pheromones (probably progestins and prostaglandins from the female and androgens from the male) act to synchronize spermiation and spawning behavior with ovulation (Zohar, 1989).

Environmental factors stimulate the brain to signal the hypothalamus to release **gonadotropin-releasing hormone (GnRH)** and possibly decrease the amount of **gonadotropin release-inhibiting factor (GRIF)**, which probably is **dopamine (DA)**. GnRH causes the pituitary to release maturational **gonadotropic hormone (GtH)**. GRIF interferes with stimulation of the pituitary by endogenous GnRH or exogenous substitutes and thus inhibits release of gonadotropin. GtH stimulates the ovaries to release **estradiol**, which causes the liver to release **vitellogenin**, which accumulates in the oocytes. When GtH reaches a critical level, vitellogenic oocytes enter final maturation: germinal vesicle migrates to the periphery of the oocyte; theca and granulosa cells of the follicles release **maturation-inducing steroids (MISs)**; MISs cause GVBD; in some species, MISs influence ions and/or free amino acids to osmotically cause oocytes to absorb water and swell (**hydration**); follicles and surrounding tissue release **prostaglandin**(s); prostaglandin(s) causes follicle rupture = ovulation and probably stimulate the brain to initiate spawning behavior.

A simplified description of final maturation and ovulation is:

Environmental Factors		
stimulate the	Brain	to signal the
	Hypothalamus	to produce & release
Gonadotropin Release Inhibiting Factor (GRIF),		
which inhibits	GnRH	from functioning
&		
Gonadotropin-Releasing Hormone (GnRH),		
which stimulates the	Pituitary (hypophysis)	to produce & release
Gonadotropic Hormones (GtH),		
which stimulate the	Ovaries	to produce & release
Estradiol,		
which stimulates the	Liver	to produce & release
Vitellogenin		
which accumulates in the oocytes.		
GtHs also		
stimulate the	Follicles	to produce & release
Maturation-Inducing Steroids (MIS),		
which stimulate the	Oocytes	to complete maturation
&		
Prostaglandin(s),		
which stimulate the	Follicles	to release oocytes.

Factors acting at each level have been used for obtaining eggs in the hatchery:

Environmental Control	Brain
Gonadotropin-Releasing Hormones (GnRHa ± DA blocker)	Pituitary
Gonadotropins (CP, SP, HCG, PMS)	Ovaries
Maturation-Inducing Steroids (Progesterone)	Oocytes
Prostaglandins	Follicles

Oocyte maturation pattern is related to spawning frequency (Wallace and Selman, 1981; Wallace et al., 1987; Zohar, 1989). Pacific salmon and anguillid eels have synchronous ovaries, in which essentially all oocytes develop and mature together, then the fish spawns and dies. Most teleosts (e.g., rainbow trout, European sea bass, mullet) have group synchronous ovaries, in which at wide intervals (e.g., annually, maybe monthly) one group of oocytes matures and is spawned, while other oocytes remain much less advanced. Group synchronous species might spawn for several consecutive days. Gilthead sea bream (which spawn every day for 2-3 months), and similar fish have asynchronous ovaries, in which oocytes continuously mature during the spawning season. Timing stripping to match the ovulation cycle (typically 3.5 d long) helps to maximize viability of eggs taken from turbot (McEvoy and McEvoy, 1992).

In many species, middle-aged females produce more eggs than young and old ones (Kjørsvik et al., 1990).

In many species, egg size decreases during the spawning season, but this does not necessarily seem to be caused by temperature changes (McEvoy and McEvoy, 1991, 1992). In redspotted grouper, mean diameter of naturally spawned eggs was inversely related to temperature, decreasing from 820 to 710 μm as temperature rose from 24 to 29°C over a two-month period (Hamamoto et al., 1986b). Egg diameter of turbot decreased from 1090 to 980 μm with temperature increase from 11 to 16°C in 1.5 months (Bromley et al., 1986). In some cases, a female turbot can produce twice as many eggs per gram at the end of the season as at the beginning (McEvoy and McEvoy, 1991). In European pilchard and usually in common sole (Baynes et al., 1993), the decrease occurs with rising temperature, but in European plaice and Atlantic halibut, temperature is dropping.

Induced Ovulation

Hundreds of species in dozens of families have been induced to ovulate with hormones (Table 7-3). Fish pituitary extracts, fish gonadotropin, mammalian gonadotropin, gonadotropin releasing hormones and analogs, antiestrogens, maturation-inducing steroids, prostaglandins, catecholamines, and other compounds have been used to induce final maturation and ovulation. Long-term hormone treatment might be needed for certain species (e.g., Japanese eel, milkfish, salmon, winter flounder).

The most commonly used substances are **specific pituitary**, **carp pituitary (CP)**, **salmon pituitary (SP)**, **human chorionic gonadotropin (HCG)**, **pregnant mare serum (PMS)**, and various **gonadotropin releasing hormone-analog**(s) **(GnRHa)**. From the 1930s through the 1960s, ovulation in fish usually was induced by injection of homoplastic (from same species) or heteroplastic (from another) pituitaries. HCG has been used mostly from the early 1960s and fish gonadotropins since the late 1960s. The use of GnRHa on fish began around the early 1970s, and it has become very popular. From the early 1980s to the present, methods for implanted or injected time-release devices have been developed, mainly for GnRHa. Although more and more fish have spawned voluntarily in captivity over the years, reared broodfish of some species (e.g., striped mullet) still require hormone treatment and/or stripping.

Some of the potential problems with induced ovulation were discussed by Fostier and Jalabert (1982). Activity and effectiveness of inducing substances can vary (especially pituitaries). Pituitary extracts contain a mixture of different hormones, some of which might have negative effects. If the oocytes are ovulated prematurely, reduced viability and abnormal development are likely. When the follicles are not completely developed, even if oocyte maturation occurs, ovulation might not.

Pituitary, pituitary homogenate, or pituitary extract from the same species (**conspecific P, PH, PE**), carp (**CP, CPH, CPE**), and salmon (**SP, SPH, SPE**) contain gonadotropins that stimulate the gonads to ripen. The synergistic effect of other hormones in the pituitary could explain why pituitary injection works the best for some species (e.g., striped mullet). At least theoretically, the more closely related the donor and recipient species, the more active the gonadotropins should be. Usually, dried whole pituitaries are ground (with mortar and pestle or tissue grinder) and extracted for about 1 h in chilled 0.6% saline solution, and the supernatant (PE) is injected intramuscularly (IM) or intraperitoneally (IP). Centrifugation can be used to remove the particles first. Dosage of CP ranges at least from 2 to 150 mg/kg body weight.

Fresh pituitaries can be collected from between the brain and the base of the skull of ripe fish. They can be frozen, or dried for storage as follows (Bromage, 1988): Fresh pituitaries are soaked in acetone near 0°C until all are collected. To remove water and fat, the acetone is replaced several times during the next 24 h. Then the pituitaries are dried on clean absorbent material (preferably by vacuum), then sometimes are ground. The whole or powdered glands are placed in sealed containers, and stored at -20°C in a freezer, 4°C in a refrigerator, or at room temperature (in a dessicator if possible), in order of preference. With the best care, potency can last for 15 or more years.

Human chorionic gonadotropin (Fig. 7-8) and horse gonadotropin (pregnant mare serum) will stimulate final maturation and ovulation in many species. They usually are packed as white powder that is first dissolved in saline solution (often provided) and then injected intramuscularly. Dosages of HCG range at least from 100 to 40,000 IU HCG/kg body weight, probably depending on how well HCG mimics endogenous gonadotropin (Lam, 1982) and are typically in the range 200-1500 IU/kg (Table 7-3). For large doses, it is best to mix the HCG with less water so that the volume injected is reduced. Sometimes fish are injected with both pituitary extract and HCG or PMS.

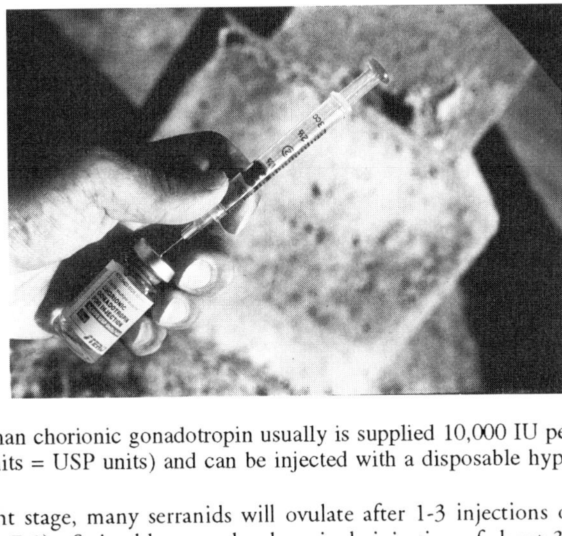

Figure 7-8. Human chorionic gonadotropin usually is supplied 10,000 IU per vial (for HCG, International Units = USP units) and can be injected with a disposable hypodermic syringe.

If at the right stage, many serranids will ovulate after 1-3 injections of about 700 IU HCG/kg (Table 7-3). Striped bass need only a single injection of about 360 IU HCG/kg. Other fish do not always respond as well. Striped mullet require HCG doses 50-100 times higher than striped bass; for them CP and/or GnRHa are more practical.

Table 7-3. Species induced to ovulate by treatment with hormones and other substances: number of treatments(dosage per kg body weight or per fish). Unless stated otherwise, treatments were in the form of intramuscular injections and the fish were strip-spawned. Atypical times are in brackets.

Scientific name	Common name	Location	No. of injections(dosage)	Temp. (°C)	Time to ovul. (h)	Egg di. (μm)	References
			Acipenseridae—sturgeons				
Acipenser baeri	Siberian sturgeon	France	1(4-6 mg CP/kg)	14	40+		Williot et al., 1991
Acipenser desotoi	Gulf of Mexico sturgeon	Florida	1(0.4 mg CP/kg) + 1(3.6 mg CP/kg)	18-25	29-31	2210	Parauka et al., 1991
Acipenser gueldenstaedtii	Russian sturgeon	USSR	1(0.6-0.9 μg GnRHa/kg)				Goncharov et al., 1991b
		USSR	1(60 mg SPE/fish)				Goncharov et al., 1991b
Acipenser ruthenus	Sterlet	USSR	1(4 mg SPE/fish)				Goncharov et al., 1991b
Acipenser stellatus	Starry sturgeon	USSR	1(0.15-0.20 μg GnRHa/kg)				Goncharov et al., 1991b
		USSR	1(25 mg SPE/fish)				Goncharov et al., 1991b
Acipenser transmontanus	White sturgeon	California	1(0.4 mg CP/kg) + 1(3.6 mg CP/kg)	15	32-52		Conte et al., 1988
		California	1(10 μg GnRHa/kg) + 1(90 μg GnRHa/kg)	15	32-52		Conte et al., 1988
Huso huso	Beluga	USSR	1(0.6-1.1 μg GnRHa/kg)				Goncharov et al., 1991b
		USSR	1(250-350 mg SPE/fish)				Goncharov et al., 1991b
			Anguillidae—"freshwater" eels				
Anguilla japonica	Japanese eel	Japan	8-15(4 SP/kg)	17	8-15 wk	1000	Yamamoto, 1980; Yamauchi & Yamamoto, 1982
		Japan	6(G-F + DES + VE) + 5(SP ± VE)	21-22	58 d		Satoh et al., 1992[b]
		Japan	10-13(20 mg SP/fish, IP) + 1(2 mg DHP/kg, IP)	22-23	66-88 d	~1000	Ohta et al., 1996a
			Clupeidae—herrings				
Alosa sapidissima	American shad	Maryland	1(77 μg GnRHa/kg, IM implant)	17-23	2-12		Mylonas et al., 1995
Brevoortia patronus	Gulf menhaden	North Carolina	1(1700 IU HCG/kg, IP) + 1(100 mg CP/kg, IP)	20	34-38		Hettler, 1983[b]
Brevoortia tyrannus	Atlantic menhaden	North Carolina	1(1300 IU HCG/kg, IP) + 1(170 mg CP/kg, IP)	16-20		714	Hettler, 1981[b]
Clupea pallasii	Pacific herring	Br. Columbia	1(10 mg SP/kg)	8-10	3-6 d		Kreiberg et al., 1987
		Br. Columbia	1(100 or 1000 μg SG/kg)	8-10	3-6 d		Kreiberg et al., 1987
		Br. Columbia	1(20 or 200 μg GnRH/kg)	8-10	3-6 d		Kreiberg et al., 1987
		Br. Columbia	1(900 μg GnRHa/kg, IP implant)	9-11	~1-3 wk		Carolsfeld et al., 1988
Sardinops melanostictus	Japanese sardine		1(50 mg GnRHa/fish) + 1(100 IU HCG/fish)				Matsuyama et al., 1991b; Nakamura et al., 1991
			Engraulidae—anchovies				
Engraulis mordax	Northern anchovy	California	1(50 IU HCG/fish) + 1(5 mg SP/fish)	15	<42		Lasker, 1974[b]
		California	1(50 IU HCG/fish) + 1(5 mg CP/fish)	~17	35-42		Leong, 1971, 1989[b]
			Chanidae—milkfish				

Chanos chanos	Milkfish	Taiwan	1(6 mg SG/kg + 400 IU HCG/kg) + 1(6 mg SG/kg + 600 IU HCG/kg)		24	1150	Liao et al., 1979
		Philippines	1(1000 IU HCG/fish)	29-30	16-49		Marte et al., 1988[b]
		Taiwan	(1-1.4 IU HCG/kg)				Lin, 1982, 1984 in Liao, 1991[b]
		Hawaii	1(20-40 μg GnRHa/kg)		12-24	1150	Kelley & Lee, 1991
		Hawaii	1(59 μg GnRHa/kg)		~24	1244	Lee et al., 1986c[b]
		Hawaii	1(1-5 ug GnRHa/kg)				Lee, 1988a
		Hawaii	1(42 μg GnRHa/kg, IP pellet)		~48	1269	Lee et al., 1986c[b]
		Philippines	1(100 μg GnRHa/fish, IP pellet)	29-31	18-36	1157	Marte et al., 1987[b]
		Hawaii	2(200 μg GnRHa/fish, IP pellet + 250 μg MT/fish, IP tube)	24-26	30 d	1269	Lee et al., 1986b, 1986d
			Clariidae—airbreathing catfishes				
Heterobranchus longifilis	African catfish	Ivory coast	1(1000-2500 IU HCG/kg)	27-29	10-14	1540	Legendre, 1986
			Osmeridae—smelts				
Plecoglossus altivelis	Ayu	Japan	1(8200 μg LHRH/ kg, IP)		~48		Hirose & Ishida, 1974
		Japan	1(20.8 mg GnRHa/kg, IP)	~25			Aida, 1983[nf]
		Taiwan	(12,500 IU HCG/kg) or (CP + HCG)	18			Peng et al., 1991
		Taiwan	1(180 μg GnRHa/kg)	~16-20	1-3 d		Chang et al., 1992
			Salmonidae—salmons, trouts				
Oncorhynchus kisutch	Coho salmon	Canada	1(100 μg SG/kg) + 3(100 μg GnRHa/kg)	10	10 d		Donaldson et al., 1981/1982
		Oregon	1(100 μg SG/kg) + 1(62 μg GnRHa/kg)	14	17 d		Sower et al., 1982
		Washington	1(100 μg SG/kg) + 1(0.05 μg GnRHa/kg)	8-12	≤12 d		Sower et al., 1984
		Canada	1(100 μg SG/kg + 20 μg GnRHa/kg)	10	6-14 d		van der Kraak et al., 1985
		Oregon	2(0.5 μg GnRHa/kg)	5-13	7-21 d		Fitzpatrick et al., 1984
		Canada	2(20 μg GnRHa/kg)				Donaldson, 1986
		Washington	1(25 μg GnRHa/fish, IM implant)		9 d		Langer & Zohar, 1992
Oncorhynchus mykiss	Rainbow trout	Newfoundland	1(37-208 μg GnRHa/kg, IP pellet)	0.5-8	14-55 d	4765	Crim et al., 1983
		Maine	1(150 μg GnRHa/fish, IM MS)		9 d		Langer & Zohar, 1992
Oncorhynchus mykiss	Steelhead trout	Washington	1(100 μg SG/kg) + 1 (60 μg GnRHa/kg)	7	≤10 d		Sower et al., 1984
Oncorhynchus nerka	Sockeye salmon	Alaska	2(5 μg GnRHa/kg, IP)/48 h	~11-14	~7-28 d		Slater et al., 1995
Oncorhynchus tshawytscha	Chinook salmon	Canada	1(100 μg SG/kg) + 1(2.5 mg SG/kg)	10-12	3-11 d	8360	Hunter et al., 1978
		Canada	1(100 μg SG/kg) + 1(50 mg SPE/kg)	10-11	3-8 d	8180	Hunter et al., 1978
Salmo salar	Atlantic salmon	Canada	1(125 μg GnRHa/fish, IP pellet)		<9-15 d		Crim & Glebe, 1984
		Maine	1(75 μg GnRHa/fish, IM implant)		12-14 d		Langer & Zohar, 1992
		Norway	2(10 μg GnRHa/kg, IP)/5 d		~7 d		Taranger et al., 1992
Salmo trutta	Brown trout	France	1(5 or 50 μg GnRHa/fish, IP implant + 10 mg pimozide/fish)	4-7	4-8 d		Billard et al., 1984
			Atherinidae—silversides				
Menidia menidia	Atlantic silverside						Bengston et al., 1987
			Anoplopomatidae—sablefishes				
Anoplopoma fimbria	Sablefish	Canada	1(200 μg GnRHa/kg, IP)	7-9	2 wk	1250	Solar et al., 1987
		Canada	1(1 mg SG/kg)	7-9	1-2 wk	1250	Solar et al., 1987

Table 7-3. (Continued)

Scientific name	Common name	Location	No. of injections(dosage)	Temp. (°C)	Time to ovul. (h)	Egg di. (μm)	References
			Centropomidae—snooks				
Centropomus undecimalis	Common snook	Florida	1(500-1000 IU HCG/kg)	28	34	700	Ager et al., 1978; Chapman et al., 1982; Tucker, 1987b
Lates calcarifer	Barramundi	Singapore	1(250 IU HCG/kg)	27-30	0-5 d	820	Lim et al., 1986[b]
		Thailand	1(50-200 IU HCG/kg) + 0-1(100-400 IU HCG/kg)	~28	36		Kungvankij, 1987
		Thailand	1-2(2-3 mg CP/kg + 250-1000 IU HCG/kg)	~28	36		Kungvankij, 1987
		Thailand	(50-500 IU HCG/kg) + (2-4 doses? BP)			800	Maneewong, 1987a
		Singapore	1(75 μg GnRHa/kg)	27-30	0-5 d	820	Lim et al., 1986[b]
		Philippines	2(10 μg GnRHa/kg, IP)		2-4 d		Nacario, 1987[b]
		Philippines	1-4(60-100 μg GnRHa/fish)	26-29	2-5 d		Almendras et al., 1988
		Philippines	1(20-100 μg GnRHa/kg)	27-32	8 h-6 d		Garcia, 1989[b], 1992
		Philippines	1(10-17 μg GnRHa/kg, IP pellet)		2 d		Harvey et al., 1985[b]
		Philippines	1(50-100 μg GnRHa/fish, IP pellet)				Fortes, 1987
		Philippines	1(50-400 μg GnRHa/fish, IP pellet)	26-29	1-4 d		Nacario, 1987[b], Almendras et al., 1988
		Philippines	1(38-200 μg GnRHa/kg, IP pellet)		1-6 d	850	Garcia, 1992
		Philippines	1-7(9 μg GnRHa/fish, osmotic pump)	26-29	1-7 d		Nacario, 1987[b], Almendras et al., 1988
			Percichthyidae—temperate basses				
Dicentrarchus labrax	European sea bass	France	1(800 IU HCG/kg)	14-16	70-75		Barnabé, 1976a
		France	1(1000 IU HCG/kg)		~70		Barnabé & Paris, 1984
		Spain	2(900 IU HCG/kg)/6 h	11-13	48-78		Carrillo et al., 1995[b]
		Spain	1-2(100-200 μg GnRHa/kg, IP) + 0-1(200 IU HCG/kg, IP)	15-19	24-88		Zanuy et al., 1986[b]
		France	1(5-20 μg GnRHa/kg)	14-17	54-83	1187	Barnabé & Barnabé-Quet, 1985
			1(10-20 μg GnRHa/kg)	13-16	24-96		Coves et al., 1991[b]
		Spain	2(10 μg GnRHa/kg)	13-14	32-48		Alvarino et al., 1992
		Spain	1(5 μg GnRHa/kg) + 1(10 μg GnRHa/kg)/5 h		72		Carrillo et al., 1995[b]
Lateolabrax japonicus	Japanese sea bass	Taiwan	(CP + HCG)	20	95-97		Huang & Liu, 1991, 1992
Macquaria novemaculeata	Australian bass	Australia	1(1000 IU HCG/kg)	18	38-40	1100	van der Wal & Nell, 1986[a]; Battaglene et al., 1989a[a]; Battaglene & Talbot, 1990[a]
Morone chrysops	White bass	U.S.	1(1100-2200 IU HCG/kg)	20	16	645	Rees & Harrell, 1990
		Israel	1(62 μg GnRHa/kg, IM implant)	21-23	35-82		Mylonas et al., 1996[a]
Morone saxatilis	Striped bass	South Carolina	1(279 IU HCG/kg)		22-48		Stevens, 1966
		Texas	1(330 IU HCG/kg) + 1(165 IU HCG/kg)	17	45-51		Henderson-Arzapalo & Colura, 1987[b]
		U.S.	1(275-440 IU HCG/kg)	20	16	1200	Rees & Harrell, 1990; Smith & Whitehurst, 1990[b]

		North Carolina	1(300 μg GnRHa/fish, DLS pellet) + 1(330 IU HCG/kg)	19	~56-99		Hodson & Sullivan, 1993
		North Carolina	1(20 μg GnRHa/kg, DLS pellet)				Woods et al., 1995[b]
			Serranidae—groupers, sea basses				
Centropristis striata	Black sea bass	Florida	2(1500 IU HCG/kg)		48-72		Hoff, 1970[a]
		Florida	2(275-550 IU HCG/kg)	17-19	29-48	1054	Roberts et al., 1976
		North Carolina	2-3(560-1600 IU HCG/kg)	18-20	47-63[199]	910	Tucker, 1984
Diplectrum formosum	Sand perch	Venezuela	(500 IU HCG/kg/dose)			610	Manrique, 1987a, 1987b
Epinephelus adscensionis	Rock hind	Florida	(~1000 IU HCG/kg/dose)				Domeier, p.c., 1998
Epinephelus aeneus	White grouper	Israel	1(25, 50, or 75 μg GnRHa/fish, EVAc pellet)	~25	~40-48		Popper et al., 1994; Hassin et al., 1997
Epinephelus akaara	Redspotted grouper	Hong Kong	2(1000-2000 IU HCG/kg)	24-26	38-43	825	Tseng & Ho, 1979
		Hong Kong	2[3](500-1,000 IU HCG/fish) + 0-2(CP/fish)	18-24	40-48	825	Tseng & Ho, 1988
Epinephelus amblycephalus	Blackspotted grouper	Taiwan	3(690 IU HCG/kg + 46 IU PMS/kg)		115	896	Tang et al., 1979
		Hong Kong	1-3(1000 IU HCG/fish)	25-30		1000	Tseng & Poon, 1983
Epinephelus coioides	Orangespotted grouper	Malaysia	2(1000 IU HCG/kg)		36-39		Toledo, p.c., 1992
Epinephelus cruentatus	Graysby	Florida	(~1000 IU HCG/kg/dose)				Domeier, p.c., 1998
Epinephelus fario	Bluespotted grouper	Taiwan	2-3(900-2000 IU HCG/kg)		~48-96	850	Yeh et al., 1987; Kuo et al., 1988[a]
Epinephelus fulvus	Coney	Bahamas	(~1000 IU HCG/kg/dose)				Domeier, p.c., 1998
		Bermuda	2(700 IU HCG/kg)	24-26	42		Tucker et al., unpub.
Epinephelus fuscoguttatus	Brownmarbled grouper	Indonesia	3(150-600 IU HCG/kg)	28-30	62	890	Kohno et al., 1990b[b]; Mayunar et al., 1991[b]
		Palau	1-2(350-700 IU HCG/kg)	28-30	37-39	789	Tucker & FitzGerald, 1994
Epinephelus guttatus	Red hind	Bermuda	2(700 IU HCG/kg)	22-26	39-48	944	Tucker et al., unpub.
Epinephelus malabaricus	Malabar grouper	Thailand	2(400-1000 IU HCG/kg + 2-4 mg CP/kg)	~29-30		900	Chulavitayanukool et al., 1985
		Philippines	2-3(500-1000 IU HCG/kg + 3 mg CP/kg)		36-39		Kungvankij et al., 1986[a]
		Philippines	3(10 μg GnRHa/kg)		36-39		Kungvankij et al., 1986
		Taiwan	2-3(530-1190 IU HCG/kg + 0-1 GP or CP/fish)		48-65	870	Huang et al., 1986, 1987[a]
Epinephelus morio	Red grouper	Florida	2[3](500-1000 IU HCG/kg)	20-23	24-36	975	Koenig, p.c., 1992
Epinephelus polyphekadion	Camouflage grouper	Tahiti	1(500 IU HCG/kg)			~850	AQUACOP et al., 1988[b]
		Tahiti	1(10-20 μg GnRHa/kg)			~850	AQUACOP et al., 1988[b]
		Japan	2(100 mouse units HCG/kg)	26-31			Tawada, 1989b
		Palau	HCG				Madraisau, p.c., 1990
		Palau	1(700-2500 IU HCG/kg) + 1(700-1400 IU HCG/kg)	27	~33-36	798	Tamaru et al., 1996[b]
Epinephelus septemfasciatus	Sevenband grouper	Japan	(500 IU HCG/kg + 7 mg SP/kg)		~48	820	Kitajima et al., 1991
Epinephelus striatus	Nassau grouper	Grand Cayman	1-2(700-1000 IU HCG/kg)	26-28	[29]48-51	920	Tucker et al., 1991
		Bahamas	1-2(700 IU HCG/kg)	23-25	20-31,37-49	917	Tucker & Woodward, 1996[a]
Epinephelus tauvina	Greasy grouper	Singapore	2-4(460-470 IU HCG/kg) + 1(3-6 PP/fish)	24-28	~72-120	900	Chen et al., 1977
		Singapore	1-3(480-780 IU HCG/kg) + 1(0-15 mg SP/fish)	24-28	~36-108	900	Chen et al., 1977
		Singapore	2(500 IU HCG/kg) + 1(20 mg SP/fish)	27	64-72	925	Chen, 1979
Hypoplectrus gemma	Blue hamlet	Florida	1(1000 IU HCG/kg)	~28	~36	~750	Domeier, p.c., 1998; Domeier, 1994

Table 7-3. (Continued)

Scientific name	Common name	Location	No. of injections(dosage)	Temp. (°C)	Time to ovul. (h)	Egg di. (μm)	References
Hypoplectrus puella	Barred hamlet	Florida	1(1000 IU HCG/kg)	~28	~36	~750	Domeier, p.c., 1998; Domeier, 1994
Hypoplectrus unicolor	Butter hamlet	Florida	1(1000 IU HCG/kg)	~28	~36	~750	Domeier, p.c., 1998; Domeier, 1994
Mycteroperca microlepis	Gag	Florida	1(600 IU HCG/kg) + 1(270 IU HCG/kg)	21-23	48	920	Roberts & Schlieder, 1983
		Florida	2[3](500-1000 IU HCG/kg)	20-23	24-36	975	Koenig, p.c., 1992
Mycteroperca phenax	Scamp	Florida	2[3](500-1000 IU HCG/kg)	20-23	24-36	945	Koenig, p.c., 1992
Paralabrax clathratus	Kelp bass	California	1(50 μg GnRHa/kg)	15-21	≤50	880	Oda et al., 1993[a]
Paralabrax maculatofasciatus	Spotted sand bass	Mexico	2(3000 IU HCG/kg)	26	66	813	Perez-Mellado, p.c. 1993
		California	1(50 μg GnRHa/kg)	15-20	≤52	880	Oda et al., 1993[a]
Paralabrax nebulifer	Barred sand bass	California	1(50 μg GnRHa/kg)	17-20	≤53	880	Oda et al., 1993[a]
Paranthias furcifer	Creole-fish	Bermuda	2(700 IU HCG/kg)	23-25	47	947	Tucker et al., unpub.
Plectropomus areolatus	Squaretail coraltrout	Palau	2(350-700 IU HCG/kg)	28-30	34-36	754	Tucker & FitzGerald, 1994
Plectropomus leopardus	Leopard coraltrout	Queensland	2(100-500 IU HCG/kg)		36		Garrett, p.c., 1992
		Queensland	1(25 μg GnRHa/kg)		36		Garrett, p.c., 1992[a]
Serranus tabacarius	Tobaccofish	Florida	1(1000 IU HCG/kg)	~28	~36		Domeier, p.c., 1998
Serranus tigrinus	Harlequin bass	Florida	1(1000 IU HCG/kg)	~28	~36		Domeier, p.c., 1998
			Terapontidae—grunters				
Therapon jarbua	Threestripe tigerfish	Taiwan					Her & Liu, 1991
			Sillaginidae—whitings, sillagos				
Sillago japonica	Japanese whiting						Oozeki et al., 1992b
			Carangidae—jacks				
Seriola dumerili	Greater amberjack	Japan	(670-1000 IU HCG/kg)	23-24			Masuma et al., 1990
		Japan	1(500 IU HCG + 7 mg SP/kg)	23-25	36	1125	Tachihara et al., 1993[b]
Seriola lalandi	Goldstriped amberjack	Japan	CP + HCG			1360	Fujita & Yogata, 1984
Seriola quinqueradiata	Yellowtail	Japan	1(4 RU Syn/kg)		48-72		Kuronuma & Fukusho, 1984
		Japan	1(688 IU HCG/kg)	~20		1188	Mushiake et al., 1993
		Japan	1(600 IU HCG/kg)	18-19		1175	Mushiake et al., 1994[b]; Watanabe et al., 1996a[b]
Trachinotus blochii	Snubnose pompano	Taiwan					Lin, 1991, p.c. 1992
Trachinotus carolinus	Florida pompano	Florida	1(550 IU HCG/kg) + 0-1(275 IU HCG/kg)	~24	29-42	970	Hoff et al., 1978a[a], 1978b; Kloth, 1980
Trachinotus falcatus	Permit	Taiwan		24-32		900	Lin, 1991, p.c. 1992
Trachurus japonicus	Japanese horse mackerel	Japan	(670-1000 IU HCG/kg)				Ochiai et al., 1982
		Japan	1(200 M.U. Gonadotropin/fish)	20-21	35-37		Kuronuma & Fukusho, 1984[b]
		Japan	CP		35-37		Kuronuma & Fukusho, 1984[b]

			Lutjanidae—snappers				
Lutjanus analis	Mutton snapper	Florida	1(1000 IU HCG/kg)				Clarke et al., 1997
		Bahamas	1(500 IU HCG/kg) + 1(1000 IU HCG/kg)	~28	33	783	Watanabe et al., 1998b
Lutjanus argentimaculatus	Asian red snapper	Thailand	1(300-600 IU Gonatropin/fish)		1-11 d	795	Doi & Singhagraiwan, 1993[b]
		Philippines	1(1500 IU HCG/kg)	28-31	27	740	Aqua Farm News, 1992b[b]; Emata et al., 1994
		Philippines	1(100 μg GnRHa/kg)	~26-27			Duray et al., 1996a[b]
Lutjanus griseus	Gray snapper	Venezuela	1(1000-6300 IU HCG/kg)	28	31-70	735	Cabrera et al., 1997; Rosas et al., 1997
		Venezuela	1(30 μg GnRHa/kg)	28	38-46	770	Cabrera et al., 1997; Rosas et al., i.p.
Lutjanus johnii	Golden snapper	Malaysia	2(1000 IU HCG/kg)		36-39		Toledo, p.c., 1992
		Singapore	1(250-500 IU HCG/kg)		1.5-5.5 d	800	Lim et al., 1985[a]
Lutjanus synagris	Lane snapper	Florida	1(500 IU HCG/kg)			725	Clarke et al., 1997
Ocyurus chrysurus	Yellowtail snapper	Florida	1(500 IU HCG/kg)			930	Clarke et al., 1997
			Gerreidae—mojarras				
Eugerres brasilianus	Striped patao	Cuba	Total of 6000-10,000 IU HCG in 2-3 IP injections at 12-h intervals		~1-1.5 d	640	Alvarez-Lajonchère et al., 1996[b]
			Haemulidae—grunts				
Plectorhinchus cinctus	Yellowspotted grunt	Taiwan	1(143 μg GnRHa/kg, IM pellet)		30	780	Chang et al., 1994b
		Taiwan	2(83-225 μg GnRHa/kg)		30-35	780	Chang et al., 1994b
			Sparidae—sea breams, porgies				
Acanthopagrus latus	Yellowfin sea bream	Taiwan	2-3(CP + HCG)				Tsay & Yu, 1991
		Taiwan	1(10 μg GnRHa + 5 μg Dom/kg)		38-59	830	Leu & Chou, 1996[b]
Acanthopagrus schlegeli	Black sea bream	Japan	1(40 RU Syn/fish)		40-50		Kuronuma & Fukusho, 1984
Archosargus probatocephalus	Sheepshead	Florida	1-2(1000 IU HCG/kg)	23-26	75-77	820	Tucker & Barbera 1987[a]
Dentex dentex	Common dentex	Yugoslavia	1(1500 IU HCG/fish)	17-22	117	958	Glamuzina et al., 1989[b]
Diplodus puntazzo	Sharpsnout sea bream	Italy	1(600-1200 IU HCG/kg)	23	47-144	≥845	Faranda et al., 1985[a]
Lagodon rhomboides	Pinfish	Florida	1(4000 IU HCG/kg + steroid mixture + 0-1(20 mg LH/kg)			1020	Cardeilhac, 1976
Pagrus major	New Zealand snapper	Australia	1(1000 IU HCG/kg)	18	24	950	Battaglene & Talbot, 1992
Pagrus major	Red sea bream	Japan	1(100 μg GnRHa/fish, IM pellet)	16-17	18-28		Matsuyama et al., 1993[b]
		Japan	1(200 μg GnRHa/fish, IM pellet)	~16-17	14 d		Matsuyama et al., 1995
		Japan	1 μg GnRHa/d from IP osmotic pump	~16-17	14 d		Matsuyama et al., 1995
		Japan	5 μg GnRHa/d from IP osmotic pump	~16-17	12 d		Matsuyama et al., 1995
Rhabdosargus haffara	Haffara bream	Bahrain	1(500 IU HCG/kg)	20-26	~2-44 d	810	Chang et al., 1994b[b]
Sparus auratus	Gilthead sea bream	Israel	1(20-100 μg GnRHa/kg, IM implant)	~20	48-72	950	Lumare & Villani, 1970 Zohar et al., 1995
		Israel	1-2(100-200 IU HCG/kg, IM or IP)				Gordin & Zohar, 1978[b]
		Israel	1(7.5 μg GnRHa/kg)				Zohar et al., 1987[b]
		Israel	1(100-200 μg GnRHa/kg, MS)		3 d to 4 mo		Langer & Zohar, 1992

Table 7-3. (Continued)

Scientific name	Common name	Location	No. of injections(dosage)	Temp. (°C)	Time to ovul. (h)	Egg di. (µm)	References
			Sciaenidae—drums				
Argyrosomus hololepidotus	Mulloway	Australia	1(1000 IU HCG/kg)	24-26	32	938	Battaglene & Talbot, 1994a
Atractoscion nobilis	White seabass	California					Orhun, 1989
Bairdiella icistia	Gulf croaker	California	1(13 mg SP/kg)	21-23	24-36	750	Haydock, 1971
			1(1000 IU PMS/kg)	21-23	24-36	750	Haydock, 1971
		California	1(100 IU PMS/kg)	22	30-31	725	May, 1972, 1975
Cynoscion nebulosus	Spotted seatrout	Texas	1(1100 IU HCG/kg)	26-28	26-32		Colura, 1974; Arnold et al., 1976
		Texas	1(100 µg GnRHa/kg)	27	30-35		Thomas & Boyd, 1988[b]
Cynoscion regalis	Weakfish		2(200 IU HCG/kg)	23-25	≤48 & ~96		Szedlmayer, 1987[b]
Cynoscion xanthulus	Orangemouth corvina	Texas	1(50 IU HCG/kg)		26-32		Bumguardner et al., 1992
		Texas	1(130 µg GnRHa + 5 mg Pim/kg)	23-27	30		Prentice & Thomas, 1987
		Texas	1(100 µg GnRHa/kg)	24-25	30-35		Thomas & Boyd, 1988[b]
		Texas	1(200 µg GnRHa + 10 mg Pim/fish) + 1(100 µg GnRHa/kg)		>48		Bumguardner et al., 1992
Equetus acuminatus	Highhat						Straughan, 1968
Leiostomus xanthurus	Spot	North Carolina	1(500 IU HCG/kg)	20	30-40		Hettler & Powell, 1981[b]
Micropogonias undulatus	Atlantic croaker	South Carolina	3(125 IU HCG/fish)	19-24	~6-7 d	662	Middaugh & Yoakum, 1974
		Texas	1(100 µg GnRHa/kg)				Gwo et al., 1991
Pogonias cromis	Black drum	Texas	1(5.5 mg CP/kg)	21	42-43		Garza et al., 1978[b]
		Texas	1(275 IU HCG/kg)		1⁺ d		Henderson-Arzapalo et al., 1994
Pseudosciaena crocea	Large yellow croaker	China	1(50-100 µg GnRHa/kg)	17-19	38-60		Lin et al., 1991a
Sciaenops ocellatus	Red drum	Texas	1(500-600 IU HCG/kg)	25	24-30	950	Colura et al., 1990; Colura et al., 1991
		Texas	1(100 µg GnRHa/kg)	25	30-35		Thomas & Boyd, 1988[b]
			Glaucosomatidae—pearl perches				
Glaucosoma hebraicum	West Australia dhufish	Australia	1(100 µg GnRHa/kg, IP implant)			1039	Pironet & Neira, 1998
			Pomacanthidae—angelfishes				
Pomacanthus arcuatus	Gray angelfish	Florida	1(500 IU HCG/kg) + 1(250 IU HCG/kg)	~28	~24	~900	Moe, 1997b
Pomacanthus paru	French beauty	Florida	1(500 IU HCG/kg) + 1(250 IU HCG/kg)	~28	~24	~900	Moe, 1997b
			Cichlidae—cichlids				
Oreochromis niloticus	Nile tilapia	Nigeria	3(1000 IU HCG/kg)		~72		Aluko & Awopetu, 1992
Sarotherodon galilaeus	Mango tilapia	Nigeria	3(1000 IU HCG/kg)		~72		Aluko & Awopetu, 1992

			Oplegnathidae—knifejaws				
Oplegnathus fasciatus	Striped knifejaw	Japan	1(40 RU Syn/kg)		1-3 d		Kuronuma & Fukusho, 1984
			Mugilidae—mullets				
Mugil cephalus	Striped mullet		1(7 mg SG/kg) + 1(14 mg SG/kg)	21	32-36	930	Kuo, 1982, 1995
			1(16,700 IU HCG/kg) + 1 (33,300 IU HCG/kg)	21	32-36	930	Kuo, 1982, 1995
			1(7 mg SG/kg) + 1(33,000 IU HCG/kg)	21	32-36	930	Kuo, 1982, 1995
			1(50 mg CP/kg) + 1(115 mg DOC/kg)	21	32-36	930	Kuo, 1982, 1995
			1(7 mg SG/kg) + 1(115 mg DOC/kg)	21	32-36	930	Kuo, 1982, 1995
		Hawaii	1(20 mg CP/kg) + 1(200⁺ μg GnRHa/kg)	24	35-75	914	Lee et al., 1987; Lee et al., 1988b; Eda et al., 1990a
		Taiwan	2(2.75-5 MP/fish + 20-50 RU Syn/fish)		40-50		Liao & Chao, 1991
		Hawaii	1(20 mg CP/kg) + 1(5 μg GnRHa/kg)	27	36		Lee & Kelley, 1991
		Florida	1(40 mg CP/kg) + 1(300 μg GnRHa/kg)	23	45	968	Tucker, unpub.
Mugil liza	Liza	Cuba	2(3.75 mg MP/kg) + 1(7.5 mg MP/kg)	20-22	72-84	854	Alvarez-Lajonchère et al., 1988[a]
		Cuba	3-4(20 mg MP/kg)	~24	21-46	890	Alvarez Lajonchère et al., 1991
			Callionymidae—dragonets				
Callionymus enneactis	Mangrove dragonet	Japan	1(12.5-25 units Gonatropin/fish)		1-7 d		Takita et al., 1983
			Eleotridae—sleepers				
Oxyeleotris marmorata	Marble goby	Singapore?	1(152 IU HCG/kg)		1 wk	1270	Tan & Lam, 1973
			Gobiidae—gobies				
Acanthogobius flavimanus	Yellowfin goby		1(350 μg LHRH/fish)		3-5 d		Aida et al., 1978
			1(200 IU HCG/fish)		2-5 d		Aida et al., 1978
			1(40 RU Syn/fish)		3-5 d		Aida et al., 1978
			Ephippidae—spadefishes				
Chaetodipterus faber	Atlantic spadefish	Venezuela	1(60-2200 IU HCG/kg)		36-45	950	Gomez, i.p.
		Venezuela	1(12-20 μg GnRHa/kg)		36-40	950	Gomez, i.p.
			Scatophagidae—scats				
Scatophagus argus	Spotted scat	Philippines	1(30-90 μg GnRHa/kg, IM pellet)	27-30	0-16 d	800	Barry et al., 1993
			Siganidae—rabbitfishes				
Siganus argenteus	Streamlined spinefoot		(250 IU HCG/kg)		96	650	Popper et al., 1979
Siganus canaliculatus	Pearly spinefoot	Singapore	5-6(250-300 IU HCG/kg)			470	Soh & Lam, 1973
		Germany	4-5(250-300 IU HCG/kg)	~26-30	6-9 d	532	Von Westernhagen & Rosenthal, 1975, 1976
		Palau	1-2(500 IU HCG/fish)	~30	1-14 d		Bryan et al., 1975
		Kuwait	1(1000 IU HCG/kg)				Higuchi et al., 1980
Siganus guttatus	Yellowblotch spinefoot		2(2 g CC/kg)				Duray, 1990
			4(2000 IU HCG/kg)			460	Palma, 1978
			1-8(2000 IU HCG/kg)	26-32	14-180	550	Juario et al., 1985
		Philippines	1-2(2000 IU HCG/kg)	26-28	22-72	580	Ayson, 1989, 1991
			3-5(1667 IU HCG/kg)		62-112	600	Kausar & Juario, 1991[b]
			(24 mg GnRHa/kg, IP pellet)		8-9 d		Harvey et al., 1985

Table 7-3. (Continued)

Scientific name	Common name	Location	No. of injections(dosage)	Temp. (°C)	Time to ovul. (h)	Egg di. (μm)	References
Siganus luridus	Dusky spinefoot		2(250 IU HCG/kg)	24-25	96	565	Popper et al., 1979
Siganus rivulatus	Black spinefoot		1(250-500 IU HCG/kg)	24-25	36-60		Popper et al., 1979
Siganus vermiculatus	Vermiculated spinefoot		2(2000 IU HCG/kg)		48??	520	Avila, 1984
		Fiji	1(300 IU/kg HCG)	~24-28	~115		Popper & Gundermann, 1976
			Scombridae—mackerels				
Scomber japonicus	Chub mackerel	California	1(1 mg SP/fish) + 1(500 IU PMS/fish)	20	36		Lasker, 1974
			Bothidae—lefteye flounders				
Paralichthys californicus	California halibut	California	1(30-50 μg GnRHa/kg)	~16	41-66		Caddell et al., 1990
Paralichthys dentatus	Summer flounder	Rhode Island	2-6(1.1 or 11 mg CP/kg)	18 or 15	3-14 d	1020	Smigielski, 1975b
		North Carolina	4-18(2 mg CP/kg)	~16	5-19 d		Berlinsky et al., 1997
Paralichthys lethostigma	Southern flounder	South Carolina	1(100 μg GnRHa/fish, SC pellet)	14-20	65-140	~1000	Berlinsky et al., 1996
			Pleuronectidae—righteye flounders				
Hippoglossus stenolepis	Pacific halibut	Br. Columbia	GnRHa				Jensen et al., 1992; McFarlane et al., 1991
Platichthys stellatus	Starry flounder	Massachusetts	7(1 mg CP/kg)	~12	7 d		Policansky & Sieswerda, 1979
Pleuronectes americanus	Winter flounder	Rhode Island	3-6(1.1 mg CP/kg) or 3(11 mg CP/kg)	1-4	3-7 d	800	Smigielski, 1975a
		Rhode Island	2-9(330 IU HCG/kg)	3-5	2-10 d	800	Smigielski, 1975a
		Newfoundland	1(40-120 μg GnRHa/fish)	5	7-40 d		Harmin & Crim, 1992[a]; Harmin et al., 1995
		Canada	1(100 μg GnRHa/fish, IP pellet)	3-7	4-10 wk		Wiegand et al., 1987
Pleuronectes cornutus	Ridged-eye flounder	Japan	CP				Kitajima et al., 1988
Pleuronectes ferrugineus	Yellowtail flounder	Newfoundland	1(~224 μg GnRHa/kg, IM pellet)	5-8	~7 d		Larsson et al., 1997
		Newfoundland	1(75 μg GnRHa/kg, IM MS)	5-8	~7 d		Larsson et al., 1997
Pleuronectes vetulus	English sole	Washington	4-8(150 μg GnRHa/kg)	9-12	13-28		Casillas et al., 1991; Sanborn & Misitano, 1991
Pleuronectes yokohamae	Marbled flounder	Japan	1(1000 μg GnRH/fish)		3-6 d	710	Aida et al., 1978
		Japan	1(3000 IU HCG/fish, IP)	11-13	2-3 d		Hirose et al., 1979
		Japan	1(3 mg SG/fish, IP)	11-13	2-3 d		Hirose et al., 1979
			Soleidae—soles				
Cynoglossus abbreviatus		Japan	(6.5 mg CP/kg)	14-16	2 d	1210	Fujita et al., 1986
Solea solea	Common sole	Spain	1(250-500 IU HCG/kg)		24-96		Ramos, 1986a
		Spain	1(10 μg GnRHa/kg)	17-19	48		Ramos, 1986b
			Tetraodontidae—puffers				
Takifugu exascurus		Japan	1(35 mg CP/kg)	19-21	77	1240	Fujita & Honma, 1991
Takifugu porphyreus	Purple puffer	Japan	1(6 mg CP/kg)	14-16	79-96	1280	Fujita & Abe, 1992

Table 7-3. (Notes).
[a] In some cases, injected fish spawned in the holding unit without stripping.
[b] Injected fish spawned in the holding unit without stripping.
[nf] Eggs were not fertilized?
Substances used were: BP = barramundi pituitary; CC = clomiphene citrate; CP = carp pituitary; DES = diethylstilbestrol-sodium; DHP = 17α,20β-dihydroxy-4-pregnen-3-one); DOC = deoxycorticosterone; Dom = domperidone; GF = Gonagen-Forte (placental gonadotropin + hypophyseal gonadotropin, 10:1); GP = grouper pituitary (number of glands); HCG = human chorionic gonadotropin; LH = mammalian luteinizing hormone; GnRHa = gonadotropin releasing hormone analog; MP = mullet pituitary; MT = methyltestosterone; P = pituitary; Pim = pimozide; PMS = pregnant mare serum (horse gonadotropin); PP = snapper (Pristipomoides spp.) pituitary (number of glands); SG = salmon gonadotropin; SP = salmon pituitary; Syn = Synahorin = human chorionic gonadotropin + mammmalian pituitary extract; VE = vitamin E.
IU = International Unit; MU = Mouse Unit; RU = Rabbit Unit. DLS = in dorsal lymphatic sinus; IM = intramuscular; IP = intraperitoneal; SC = subcutaneous. MS = microspheres (usually for prolonged tank spawning). p.c. = personal communication. i.p. = in press.

HCG does not work well for all fish; however, at least for some members of the families Clupeidae, Centropomidae, Percichthyidae, Serranidae, Carangidae, Lutjanidae, Sparidae, Sciaenidae, Siganidae, and Pleuronectidae, HCG meets at least four of the five requirements set forth by Lin and Peter (1992) for effective hormone-induced spawning. Typically, (1) ovulation rate is high, (2) ovulation is nearly complete, (3) time to ovulation is short, and (4) eggs are fertile and viable. Possible (5) alteration of the spawning cycle (and possible development of resistance) is not important with fish that are caught, held for a few days, spawned, and released. Criterion 5 also does not apply with species that spawn voluntarily in captivity without hormone injection. If hormone-induced ovulation of the same broodfish will be repeated, GnRHa might be a better choice than HCG because it seems to have less lasting impact on the endocrine system. HCG overdoses could be more likely to cause abortion, overhydration, egg retention, and female mortality (Faranda et al., 1985; pers. obs.) or simultaneous hydration and ovulation followed by mortality (Prentice and Thomas, 1987).

GnRHa has been successful with members of several families, particularly Acipenseridae, Chanidae, Salmonidae, Centropomidae, Percichthyidae, Serranidae, Sparidae, Sciaenidae, Mugilidae, and Pleuronectidae (Table 7-3; Crim et al., 1987). After the mammalian decapeptide, luteinizing hormone-releasing hormone (LHRH), was shown to be active in teleost fish, further studies identified GnRH in a wide range of teleosts as [Trp^7-Leu^8]LHRH, with a second uncharacterized form (Kah, 1986; Lin and Peter, 1986; Sherwood, 1987; Zohar, 1989). Different forms of GnRH seem to exist for lampreys, cartilaginous fish (e.g. sharks), and ganoids (e.g., sturgeon)(Sherwood, 1987).

The GnRH molecule is a peptide (chain of amino acids). Mammalian GnRH, salmonid GnRH, and three synthetic analogs have the following molecular formulas (half-life in gilthead sea bream from Zohar et al., 1995):

Mammalian LHRH = LHRH
pGLU-HIS-TRP-SER-TYR-GLY-LEU-ARG-PRO-GLY-NH_2

Salmonid GnRH = [Trp^7-Leu^8]LHRH
pGLU-HIS-TRP-SER-TYR-GLY-TRP-LEU-PRO-GLY-NH_2 (5.5 min)

A superactive GnRH analog = [d-Arg^6, Trp^7-Leu^8, Pro^9-NH-ethylamide]LHRH
pGLU-HIS-TRP-SER-TYR-dARG-TRP-LEU-PRO-NHET (12.5 min)

Another superactive GnRH analog = [d-Trp^6]LHRH
pGLU-HIS-TRP-SER-TYR-dTRP-LEU-ARG-PRO-GLY-NH_2 (17.7 min)

Another superactive GnRH analog = [d-Ala^6, Pro^9-NH-ethylamide]LHRH
pGLU-HIS-TRP-SER-TYR-dALA-LEU-ARG-PRO-NHET (22.1 min)

The first three are decapeptides (ten amino acids in the molecule), and the last two are nonapeptides (nine amino acids). Abbreviations in brackets give the identities and positions of amino acids or other components that differ from those in LHRH. To avoid the question of which gonadotropins are caused to be released when analogs are administered to fish, the trend is to use the term **GnRHa** rather than **LHRHa**. Even though half-lives differ, all three analogs are equally effective for gilthead sea bream females if injected at 2.5 μg/kg or more (Zohar et al., 1995).

LHRH or GnRH and their synthetic analogs, via injection or implantation, stimulate the pituitary to secrete and release endogenous gonadotropin. Some fish in captivity (e.g., gilthead sea bream, Zohar, 1989) will complete vitellogenesis but not ovulate and spawn, probably because endogenous gonadotropin accumulates in the pituitary but is not released. GnRHa can stimulate release of the retained GtH.

In fish, GnRH analogs are much more active than LHRH, [Trp^7-Leu^8]LHRH, or specific GnRH. This is probably because the latter three are recognized and enzymatically degraded quickly, while the body does not recognize the synthetic analog molecules or have as efficient a mechanism for disposing of them (Zohar, 1988, 1989, 1991; Zohar et al., 1995). Zohar (1989, 1991) concluded that d-amino acid substitutions at the sixth and tenth positions are of primary importance in preventing rapid degradation. Injected gilthead sea bream GnRH had a half-life of 5-6 min, but the three GnRH analogs had half-lives of 12-22 min (Zohar, 1991; Zohar et al., 1995). The gonadotropin surge resulting from injection of 10 μg/kg GnRHa was at least four times longer (48 h) than that produced by GnRH (6-12 h). In some fish, GnRHa also exhibits higher affinity for receptor sites in the pituitary. Both factors contribute to the effectiveness of GnRHa (Zohar, 1991). Also, in at least some fish, a GnRH binding protein has higher affinity for analogs and can keep them in the system longer (Zohar et al., 1995).

Because analogs are so potent, only small amounts are needed. GnRHa is effective with some species that are resistant to HCG. It is less likely to cause an immune response after long or repeated treatments, which can result in resistance. GnRHa also seems to cause more prolonged spawning than HCG in some fish. For acute administration, it is dissolved in saline solution and injected intramuscularly at 0.1-300 μg GnRHa/kg body weight with typical doses of 5-30 μg/kg. Usually, two injections of GnRHa work better than one.

Some species, including frequent spawners like European sea bass and gilthead sea bream require a long gonadotropin surge for successful spawning and therefore might have to be injected frequently. To avoid frequent handling and stress to the fish as well as to maintain constant hormone levels, time-release devices for GnRHa have been developed. For chronic administration, GnRHa usually is implanted intraperitoneally in a pellet (usually with cholesterol, cholesterol/cellulose, or polymers; Lee et al., 1986e; Sherwood et al., 1988; Langer and Zohar, 1992) or in microcapsules or microspheres (with polymers; Redding et al., 1984; Zohar, 1991; Langer and Zohar, 1992), containing 5-350 μg GnRHa per dose.

GnRHa implants have worked well in American shad, milkfish, salmonids, barramundi, European sea bass, striped bass, white bass, red sea bream, gilthead sea bream, striped mullet, southern flounder, and winter flounder (Table 7-3).

GnRHa in the fish might last only 30-120 min and the resulting gonadotropin surge could last up to 48 h (Zohar, 1989). Long-term slow-release administration of GnRHa to gilthead sea bream increased the number of females that spawned over long periods. GnRHa injections caused mostly short-term spawning and GnRHa in a silastic pellet had little effect, but GnRHa in a degradable pellet induced mainly long-term spawning (Zohar, 1988). Only 20-25% of gilthead sea bream given GnRHa injections, but 80% of those with GnRHa implants, spawned (Zohar, 1991).

Less invasive means of hormone administration might be possible for some species. Bath treatment was suggested by absorption of GnRH through the gills of goldfish (Sherwood and Harvey, 1986). Two oral doses (2.616 mg/kg BW) of SP 24 h apart induced ovulation in goldfish 27-33 h after the first dose. One oral dose of 50-100 μg des-Gly^{10}[D-Ala^{6}]-LHRH (1-9)-ethylamide and 25-50 μg domperidone/kg BW induced ovulation in Thai carp (Sukumasavin et al., 1992). Thomas and Boyd (1989) reported that spotted seatrout spawned 32-38 h after being fed dead shrimp injected with des-Gly^{10}[D-Ala^{6}]-LHRH (1-9)-ethylamide at 1.0-2.5 mg/kg fish, but not with 0.2 mg/kg. The oral dosage might need to be 10 times higher than the injected requirement (100 μg/kg, Thomas and Boyd, 1988). Oral uptake probably was enhanced by the analog's high resistance to digestion and its hydrophobicity. When 1 mg/kg BW of the same GnRHa was fed to sablefish initially and again 11 d later, ovulation occurred at 14-18 d (Solar et al., 1990). Oral administration is easier on the fish and the culturist, but more testing is needed to determine its efficacy in different species.

A substance that blocks the action, depletes the amount, or slows the synthesis of dopamine sometimes is included with the first GnRHa injection to minimize interference by the GRIF dopamine (Lin and Peter, 1986; Crim et al., 1987; Peter et al., 1987, 1993; Zohar, 1989; Marte, 1990). Examples are pimozide, metaclopramide, haloperidol, spiperone, and domperidone (dopamine receptor antagonists); reserpine, 6-hydroxidopamine, alpha-methyl-para-tyrosine, and carbidopa (dopamine synthesis blockers). Factors such as species sensitivity, oocyte stage, and environment could determine the necessity for the dopamine inhibitor. Milkfish (Marte et al., 1987), Pacific herring (Kreiberg et al., 1987), European sea bass (Barnabé and Barnabé-Quet, 1985), gilthead sea bream (Zohar et al., 1987), Atlantic croaker (Copeland and Thomas, 1989), striped mullet (Lee et al., 1987), sablefish (Solar et al., 1987), and common sole (Ramos, 1986b) are examples of fish that do not require the inhibitor, but cyprinids and clariids seem to need it (Zohar, 1988).

Antiestrogens such as clomiphene citrate and tamoxifen can cause release of gonadotropin, probably by blocking the suppressive effect of endogenous estrogens (Lam, 1982). The best time to use these probably is at the completion of vitellogenesis, when estrogen levels are high. Combining antiestrogens with exogenous gonadotropin seems to enhance the effect of the GtH.

Injected steroids (e.g., progesterone) and prostaglandins have been demonstrated to stimulate oocyte maturation and release from the follicles (Lam, 1982; Scott and Canario, 1987; Ventling and Pankhurst, 1995). MISs seem most useful if gonadotropin level is raised before or at the same time as the steroid injection. Progesterone or deoxycorticosterone (DOC) added to the final injection of HCG could improve results (Davy and Chouinard, 1981). Maturation of striped mullet oocytes was accelerated when testosterone in a silastic capsule was added to a GnRHa pellet treatment, probably because the testosterone was converted to estrogen (Kelley et al., 1987).

MISs alone might cause only maturation and hydration. Apparently, a prostaglandin(s) is necessary for ovulation; it would have to be injected or produced by the follicles and surrounding tissue as a result of indirect or direct stimulation by endogenous or exogenous gonadotropin (Goetz et al., 1987). Injected prostaglandins can stimulate ovulation and possibly spawning behavior in teleost fish (Goetz et al., 1982; Lam, 1982; Zohar, 1989). They could be used to push females with mature oocytes over the edge. Prostaglandins released by females also could act as pheromones on the males.

Corticosteroids seem to promote oocyte maturation, possibly by enhancing the effect of GtH in stimulating MIS production (Zohar, 1989).

Thyroid hormones can stimulate maturation of gonads of immature fish, help reverse oocyte regression, enhance the effect of gonadotropin on development and maintenance of ovaries, and increase growth rate (Lam, 1982, 1994; Cyr et al., 1987; MacKenzie et al., 1987). They might also promote ovulation by making ovaries more sensitive to gonadotropin. Maternal thyroid hormones incorporated in the eggs could contribute toward normal development (e.g., of gas bladder), faster growth, greater environmental tolerance, and better survival (Brown and Bern, 1989; Brown et al., 1989; Dickhoff et al., 1989; Greenblatt et al., 1989; Reddy and Lam, 1991; Lam, 1994; Tachihara et al., 1997). Survival of striped bass (hatchlings to juveniles) from females injected with 20 mg/kg triiodothyronine was four times better after 1 month in a nursery pond (Brown et al., 1989). Thyroid hormones can be given to large fish by injection, to feeding fish in their food, and to eggs and larvae by immersion (Nugegoda and Lam, 1994).

Treating female fish with a combination or series of thyroid hormones, GnRHa, gonadotropins, steroids, and prostaglandins might seem to cover all bases, but becomes exceedingly complex. Effective dosages and timing would be difficult to determine. The simplest treatment that produces enough high quality eggs might be all that is needed. Lack of gonadotropin release by the pituitary seems to be a common barrier to captive reproduction. The two most likely options are to stimulate release of endogenous GtH or substitute exogenous GtH. If broodfish are to be spawned more than once, care should be taken not to permanently upset their endocrine systems.

Figure 7-9. Taking a biopsy sample of oocytes from a squaretail coraltrout with surgical grade polyethylene tubing.

Females are chosen for induced spawning on the basis of oocyte development (diameter, color, structure) as seen in ovarian biopsies (Fig. 7-9, Table 7-4) or determined or inferred by other means. Taking of biopsies should be minimized because the procedure is stressful and sometimes causes infection and reduced egg viability. In at least the groupers with which I have worked, the oviducts do not open to the outside until after ovulation, and taking biopsies earlier injures the fish. (Passing a biopsy needle through the body wall or using an otoscope are two other less practical methods, Lam, 1982). If stage of ovarian development can be predicted with reasonable certainty (e.g., if some of the fish caught at the same time are to be used for food, from internal condition of those dissected), I prefer not to biopsy. With species for which biopsies are not as intrusive and are better tolerated, they can be done sparingly. One of the best materials for taking a biopsy is a short length of surgical polyethylene tubing with inside diameter slightly larger than ovulated egg diameter. If gently passed through the oviduct into the mass of oocytes, it will take a good sample, which can be blown out of the tubing into a Petri dish or onto a slide. Glass tubing should never be used, because it can break inside the fish. With the correct diameter

Table 7-4. Minimum oocyte diameters for successful induced ovulation.

Scientific name	Common name	Initial oocyte diameter (μm)	Egg diameter (μm)	Ratio IOD/ED (%)	References
Epinephelus akaara	Redspotted grouper	250	825	30	Tseng & Ho, 1988
Sparus auratus	Gilthead sea bream	301-400	950	32	Gordin & Zohar, 1978
Centropristis striata	Black sea bass	400-425	982	41	Roberts et al., 1976; Tucker 1984
Chaetodipterus faber	Atlantic spadefish	400	950	42	Gomez, i.p.
Epinephelus malabaricus	Malabar grouper	400	900	44	Kungvankij et al., 1986
Scatophagus argus	Spotted scat	350	800	44	Barry & Fast, 1992; Barry et al., 1993
Paralabrax clathratus	Kelp bass	400	880	45	Oda et al., 1993
Paralabrax maculatofasciatus	Spotted sand bass	400	880	45	Oda et al., 1993
Paralabrax nebulifer	Barred sand bass	400	880	45	Oda et al., 1993
Solea solea	Common sole	600	1250	48	Ramos, 1986a
Dicentrarchus labrax	European sea bass	650-750	1350	48	Zanuy et al., 1986; Coves et al., 1991
Archosargus probatocephalus	Sheepshead	400	820	49	Tucker and Barbera, 1987
Lagodon rhomboides	Pinfish	500	1020	49	Cardeilhac, 1976
Lates calcarifer	Barramundi	400-500	800	50	Lim et al., 1986; Kungvankij, 1987; Nacario, 1987; Garcia, 1989
Paralichthys californicus	California halibut	400	800	50	Caddell et al., 1990
Paralichthys lethostigma	Southern flounder	500	~1000	50	Berlinsky et al., 1996
Plectorhinchus cinctus	Yellowspotted grunt	400	780	51	Chang et al., 1994c
Cynoscion nebulosus	Spotted seatrout	400	770	52	Thomas et al., 1995
Cynoscion xanthulus	Orangemouth corvina	440			Thomas et al., 1995
Sciaenops ocellatus	Red drum	500	950	53	Colura et al., 1991
Micropogonias undulatus	Atlantic croaker	550			Thomas et al., 1995
Epinephelus fuscoguttatus	Brownmarbled grouper	482	890	54	Kohno et al., 1990b
Epinephelus tauvina	Greasy grouper	500	925	54	Chen, 1979
Cynoscion nebulosus	Spotted seatrout	462	840	55	Colura, 1974
Sparus auratus	Gilthead sea bream	530	950	56	Zohar et al., 1995
Chanos chanos	Milkfish	630	1120	56	Marte et al., 1987
Epinephelus fario	Bluespotted grouper	500	850	59	Yeh et al., 1987
Trachinotus carolinus	Florida pompano	580	970	60	Hoff et al., 1978a
Epinephelus septemfasciatus	Sevenband grouper	500	820	61	Kitajima et al., 1991
Siganus guttatus	Yellowblotch spinefoot	350	550	64	Juario et al., 1985
Mugil cephalus	Striped mullet	600	930	64	Kuo, 1982; Lee & Kelley, 1991; Liao & Chao, 1991
Scomber japonicus	Chub mackerel	700	1100	64	Lasker, 1974
Chanos chanos	Milkfish	800	1248	64	Lee et al., 1986c
Chanos chanos	Milkfish	750	1150	65	Tamaru, et al., 1988; Kelley & Lee, 1991
Acanthopagrus berda	River bream	800			Davy & Chouinard, 1981
Acanthopagrus schlegeli	Black sea bream	800			Davy & Chouinard, 1981

of tubing and correct positioning, suction is not likely to be needed. Germinal vesicles of translucent oocytes can be viewed with a microscope if they are flattened slightly by the weight of a cover slip (Rottman et al., 1991D). Opaque oocytes can be cleared by 5-20 min immersion in a mixture of 60% ethanol, 30% formalin, and 10% acetic acid (Stoeckel and Neves, 1992). A simpler but possibly not as effective method is immersion in a series of 5-min baths of 30, 50, 70, 90, and 100% ethanol (Fenner, 1992). Other methods are given by Shelton (1989).

Oocytes of sturgeon and paddlefish sometimes are removed surgically (Conte et al., 1988; Rottman et al., 1991a). A tube containing physiological saline solution is inserted through a 20 to 30-mm incision in the abdominal wall. The saline solution is released into the body cavity and eggs are pulled into the tube by suction. The incision is sutured and treated with an antibiotic. Germinal vesicle status can be determined with a microscope if eggs are first boiled for 5-8 min and cut in half.

Removed oocytes can be tested for maturity by holding them for 12 or more hours at the spawning temperature in a solution containing an appropriate maturation-inducing-steroid. One such solution contains 0.2 mg progesterone dissolved in 0.2 mL absolute ethanol, gently mixed with 20 mL L-15 Leibovitz Medium in physiological saline (Conte et al., 1988; Rottman et al., 1991a). If most of the oocytes proceed to GVBD (usually 80-100%, the fish is eligible for injection. This is practical and useful only with fish like sturgeon in which the final stages are slow.

Striped bass judged to be 9-15 h from ovulation are the best for injecting and stripping (Yeager et al., 1990). Those farther away usually are rejected and those closer might ovulate naturally.

Most success with induced ovulation probably has occurred with fish initially having oocytes in the range from late vitellogenesis (~fully-yolked) to early hydration. Examples of minimum oocyte diameters are given in Table 7-4. Doses at the beginning of the spawning season might have to be higher than late in the season (Lam, 1982). Some possible factors are sensitivity of the fish, ovarian stage, and environment. Number of injections required and intervals between them depends mainly on species, initial degree of ripeness, and temperature. The persistence of injected hormones, etc. in the bloodstream will depend mainly on the substance, species, and temperature (Lam, 1982). In some situations, elevated blood gonadotropin could last for a few days. The usual practice is to space injections at 24-h intervals to help insure constant blood levels. Different situations might call for different intervals. Species that ovulate close to 36 h after an injection might need only one. This is the case in families such as Centropomidae (common snook, Chapman et al., 1982), Percichthyidae (striped bass, Rees and Harrell, 1990), Carangidae (Japanese horse mackerel, Kuronuma and Fukosho, 1984), and Sciaenidae (spotted seatrout, Colura, 1974). If ovulation normally occurs later than 36 h after the first of two injections but oocyte stage is such that ovulation will occur within 36 h, one injection could be enough. An overdose of HCG or a superfluous injection just before ovulation occurs can cause overhydration, possibly resulting in reduction of egg quality, abortion, or loss of the female. With species easily stressed (leading to oocyte regression), females must be injected immediately after capture, but some (e.g., Nassau groupers, red hind) have been injected 1-2 weeks later with excellent results. If fish are kept long enough to adapt well to captivity, even if they do not spawn voluntarily, many are likely to at least complete vitellogenesis and be susceptible to hormone induction.

Hormone injections usually are given intramuscularly (IM) in the thicker back muscles between the dorsal fin and lateral line (Fig. 7-10), but large-scaled fish can be injected under the pectoral fin to avoid dislodging scales (Fig. 7-11). A 26-gauge needle is suitable for most species; finer ones could be used for very small fish, coarser ones for very thick-skinned fish. Pituitary injections and implants usually are placed in the peritoneal cavity. Intraperitoneal (IP) injections should go just inside the abdominal wall. Damage to nerves, large blood vessels, and vital organs should be avoided. The best exact location varies with species. Sensitivity to injections might vary with time of day. The best times to inject striped mullet seem to be 3 and 12 h after daybreak (Lam, 1982). Gilthead sea bream spawned for more days if injected at 2200 rather than 0400, 1000, or 1600 (Zohar, 1988).

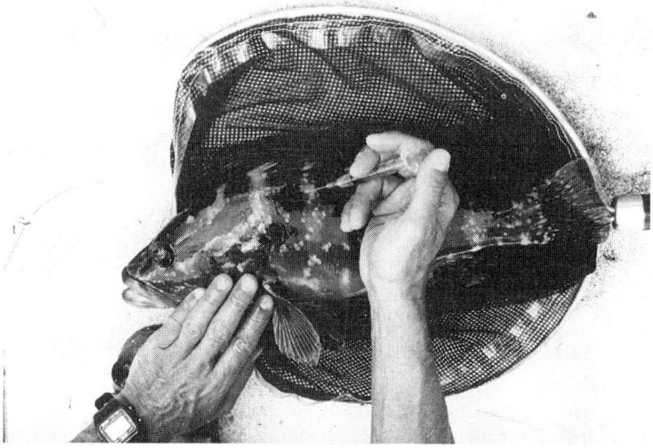

Figure 7-10. Giving an intramuscular injection of HCG to a female Nassau grouper. (Note the soft, knotless nylon net, which is preferred)

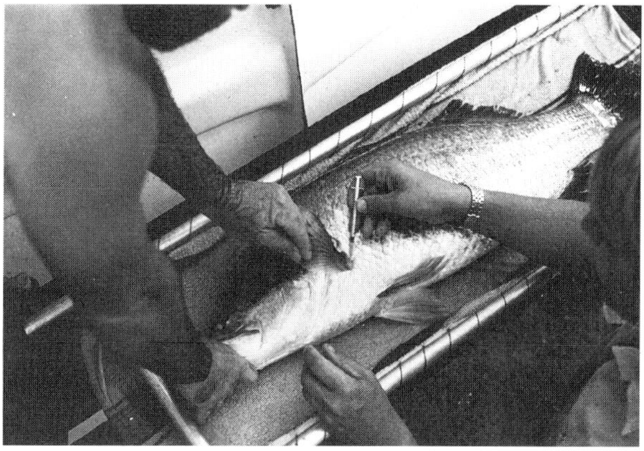

Figure 7-11. Giving an intramuscular injection to a female barramundi behind the pectoral fin to avoid the large scales.

Usually, hormone-treated fish should be held under conditions approximating those during natural spawning, with temperature being most important. Gerking (1982) reported a narrow range of 24-30°C for desert pupfish, outside of which egg viability and production were severely reduced. Lethal limits for the adults are much wider, 2-44°C. When injected broodfish of species such as common snook, Nassau grouper, and red hind are held in water more than 2-3°C higher than optimal, fertilization and hatching rates nearly always are reduced (pers. obs.). Dissolved oxygen should be adequate and disturbance minimal.

Induced spawning attempts with species that ripen simultaneously (e.g., some groupers, Fig. 7-12) should have a high success rate if the fish are caught at the right time. Relatively few broodfish are needed. On the other hand, the window for obtaining spawns is narrow, and hindrances such as bad weather can jeopardize culture programs. Species that do not ripen simultaneously (Fig. 7-13) usually are available for longer periods, and timing is not as critical. However, more broodfish are required to ensure that some are at the right stage, and biopsies are more important to identify those individuals.

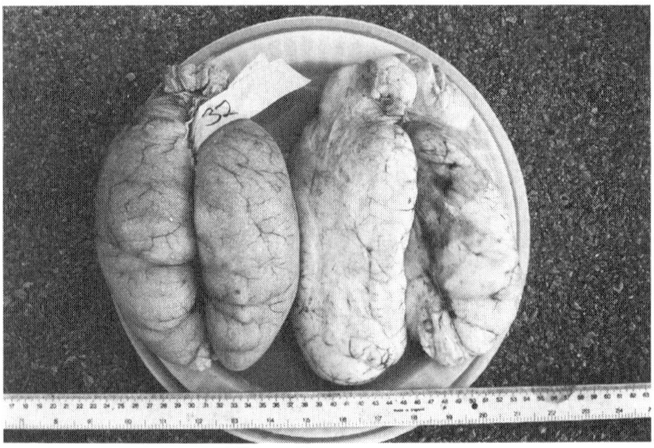

Figure 7-12. Ovaries (left) and testes (right) from ripe Nassau groupers.

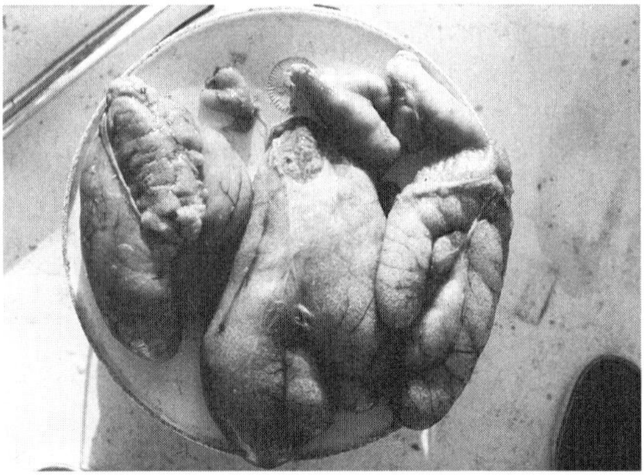

Figure 7-13. Ovaries from gags collected together, covering almost the full range of development from unyolked oocytes (upper left) clockwise to hydrated oocytes on the verge of spawning (center left).

If subjected to spawning conditions, males usually are running ripe already (Fig. 7-14). If not stressed too much, males (e.g, common snook, Nassau grouper) can remain naturally running ripe for months. Sometimes males are injected to dilute and increase the flow of milt, but this usually causes only a temporary increase, followed by premature loss of ripeness. If milt is needed for more than a few days, it is best to avoid injecting the males, unless they can be replaced with fresh ones. At least for warmwater marine fish, one injection 12-24 h before milt is needed usually is sufficient for short-term needs. Sustained-release implants can be used for longer-term milt production. PE, SG, HCG, PMS, LHRH, GnRHa, progestins, methyltestosterone (MT), and other androgens have been used to induce or increase spermiation (Lam, 1982; Lee and Weber, 1986; Lee et al., 1986d,

1992b, 1993; Zohar, 1989; Barry et al., 1991; Garcia, 1990, 1991, 1993; Park et al., 1992; Harmin and Crim, 1993; Pankhurst, 1994; Alvarez-Lajonchère et al., 1996; Ohta et al., 1996b, 1997a, 1997b; Faulkner and Moberg, 1997; Moe, 1997b; Mylonas et al., 1997; and others). Response of males to injections is more consistent than that of females. Usually, males are given one injection when females receive their last one. Usually a lower dose is given to the males (e.g., 50% of the female dose of HCG).

Figure 7-14. A running ripe male squaretail coraltrout releases milt with little or no pressure.

Suquet et al. (1992) reported that turbot sperm number and motility were increased when males were kept in the same tank with females and they were not stripped more often than about once a month.

Timing and Egg Quality

When the time between ovulation and overripening of eggs is brief (Table 7-5), females must be observed carefully for signs of hydration and ovulation. Indicators of ovulation include time after treatment, swollen abdomen, protruding genital papilla, stretching of the membrane holding eggs in (it could become transparent), coloration, and hormone levels. Ultrasound can be used to determine ovulation in some species (Shields et al., 1993). In groupers and some other fish, hydration is indicated by noticeable swelling of the abdomen (Fig. 7-15) and ovulation by protrusion of the egg mass through the genital opening (Fig. 7-16). The female does not swim smoothly, could rest on the bottom more often, and might turn pale, dark, or bicolored. Ultrasonography has been used to measure striped bass gonad diameter to predict time for injection (Blythe et al., 1994). Watson (1992) described a graphical method for predicting ovulation time for striped bass injected with HCG. Normal spawning times for a number of species are given in Table 7-6.

Eggs are released when the protruding membranes are broken by abdominal contractions of the female or manipulation by the male fish or culturist. In six species of *Centropyge*, the male strikes the female's abdomen with his snout to start the egg flow (Bauer and Bauer, 1981). During the beginning of ovulation, a small number of clear (or nearly clear) oocytes sometimes can be obtained by biopsy or forceful stripping; these premature oocytes often have more than the normal final number of oil globules. At an intermediate stage, moderate pressure might produce an impressive number of eggs that are only partially ovulated; clumps of eggs sticking together is typical of this stage. After full normal ovulation, individual, clear, normally-shaped (e.g., spherical, ovoid) eggs with the correct number of oil globules flow easily and quickly (Fig. 7-17). Egg quality indicators are buoyancy at correct salinity, normal transparency, normal shape, normal uniform egg

Table 7-5. Time from ovulation to overripeness (depends mainly on species and temperature).

Scientific name	Common name	Temperature (°C)	Time	References
Oncorhynchus kisutch	Coho salmon		20 d	Fitzpatrick et al., 1987
Clupea pallasii	Pacific herring	9	2 wk	Hay, 1986
Oncorhynchus mykiss	Rainbow trout	10-13	5-7 d	Sakai et al., 1975
Salvelinus alpinus	Arctic char	5	3-9 d	Gillet, 1991a
Coregonus lavaretus	Powan	5	2-4 d	Gillet, 1991b
Pleuronectes yokohamae	Marbled flounder	12	2-3 d	Hirose et al., 1979
Clupea harengus	Atlantic herring	4	2 d	Blaxter, 1955
Salmo trutta	Brown trout		<28 h	Billard & Gillet, 1981
Verasper moseri	Barfin flounder	6	1-3 d	Koya et al., 1994
Plecoglossus altivelis	Ayu		1-2 d	Hirose et al., 1977
Anarhichas lupus marisalbi	White sea wolffish	~2	12 h	Pavlov et al., 1992
Scophthalmus maximus	Turbot	13	10 h	McEvoy, 1984
Gadus morhua	Atlantic cod	5	9 h	Kjørsvik & Lonning, 1983
Anguilla japonica	Japanese eel	22	6-9 h	Ohta et al., 1996a
Hippoglossus hippoglossus	Atlantic halibut	4	6 h	Bromage et al., 1991
Carassius auratus	Goldfish	23	3-12 h	Formacion et al., 1993
Maccullochella peelii	Murray perch		2-3 h	Rowland, 1988
Acipenser transmontanus	White sturgeon	15	~2-3 h	Conte et al., 1988
Pagrus major	New Zealand snapper	~18	≤2 h	Hobby & Pankhurst, 1997
Epinephelus striatus	Nassau grouper	26	1-2 h	Tucker & Woodward, 1996
Bairdiella icistia	Gulf croaker	22	1-2 h	Haydock, 1971
Hippoglossus hippoglossus	Atlantic halibut		1-2 h	Bromage et al., 1990
Morone chrysops	White bass	22	1 h	Mylonas et al., 1996
Morone saxatilis	Striped bass	19	1 h	Stevens, 1966; Rees & Harrell, 1990
Prochilodus platensis	Sabalo		1 h	Fortuny et al., 1988

Table 7-6. Times of voluntary (unless noted otherwise) spawning by representative species.

Scientific name	Common name	Location	Time[a]	References
Gadus morhua	Atlantic cod	Tank	All hours	Kjesbu, 1989
Scomber scombrus	Atlantic mackerel	Nature	All hours	Walsh & Johnstone, 1992
Paralabrax maculatofasciatus	Spotted sand bass	Nature	~0200-0800,1400-2200	Oda et al., 1993
Paralabrax nebulifer	Barred sand bass	Nature	~0200-2000	Oda et al., 1993
Chanos chanos	Milkfish	Pond	0300	Lin, 1985
Chrysiptera cyanea	Blue devil damselfish	Aquarium	Before dawn to ?	Breeder's Registry Database, 1994
Sparus auratus	Gilthead sea bream	Tank	Dawn to +90 min	Barbaro et al., 1997
Abudefduf saxatilis	Sergeant major	Nature	Sunrise to 1830	Foster, 1987
Stegastes dorsopunicans	Orangeback damselfish	Nature	First 2 h of daylight	Petersen & Hess, 1991
Anguilla japonica	Japanese eel	Tank	0500-0900	Satoh et al., 1992[h]
Stegastes partitus	Bicolor damselfish	Nature	0530-0700	Knapp & Warner, 1991
Gomphosus varius	Birdfish	Nature	8-96% of daylight h	Colin & Bell, 1991
Chromis notata	Taillight damselfish	Nature	0600-1700	Nakazono et al., 1979
Scarus globiceps	Violetline parrotfish	Nature	10-98% of daylight h	Colin & Bell, 1991
Siganus lineatus	Goldline spinefoot	Tank	~0630	Bryan & Madraisau, 1977
Brachaluteres ulvarum		Nature	0637	Akagawa et al., 1995
Lophius americanus	Goosefish	Nature	~Early a.m.	Martin & Drewry, 1978
Chrysiptera hemicyanea	Yellowtail blue damselfish	Aquarium	Early a.m.	Unwin, 1986
Meiacanthus grammistes	Striped blenny	Aquarium	Early a.m.	Breeder's Registry Database, 1994
Dascyllus trimaculatus	Domino damselfish	Aquarium	Early a.m.	Garnaud, 1957b
Gobiosoma oceanops	Neon goby	Aquarium	Early a.m., p.m., early night	Moe, 1975
Amphiprion nigripes	Blackfoot clownfish	Aquarium	a.m. to ?	Breeder's Registry Database, 1994
Pseudoblennius cottoides	Asahi-anahaze	Tank	0700-0800,1700-1800	Kimura et al., 1987
Amphiprion frenatus	Tomato clownfish	Aquarium	0700-0830	DiMarco, 1994
Dorosoma petenense	Threadfin shad	Nature	0700,1500	Jones et al., 1978
Menidia menidia	Atlantic silverside	Nature	0700-1935	Middaugh, 1981
Sparisoma viride	Stoplight parrotfish	Nature	0730-0930	van Rooij et al., 1996
Gobiesox strumosus	Skilletfish	Aquarium	0730-1900	Martin & Martin, 1971
Halichoeres poecilopterus	Multicolorfin wrasse	Tank	Mostly 0800-0930	Kimura & Kiriyama, 1992, 1993
Canthigaster valentini	Saddled toby	Nature	0800-1530	Stroud et al., 1989
Paralabrax clathratus	Kelp bass	Nature	0800-1000,1600-0400	Oda et al., 1993
Family Labridae (22 spp.)	(wrasses)	Nature	0800 to sunset	Kuwamura, 1981
Cyprinodon variegatus	Sheepshead minnow	Nature	0800-1840	Hardy, 1978
Thalassoma lutescens	Green moon wrasse	Nature	18-97% of daylight h	Colin & Bell 1991
Thalassoma cupido	Cupido wrasse	Nature	0830-1200	Meyer, 1977
Cetoscarus bicolor	Bicolor parrotfish	Nature	23-93% of daylight h	Colin & Bell, 1991
Tetraodon cf. *cutcutia*	Ocellated blowfish	Aquarium	0900-1200	Cohn, 1912

Table 7-6. (Continued)

Scientific name	Common name	Location	Time[a]	References
Anampses caeruleopunctatus	Bluespotted tamarin	Nature	26-92% of daylight h	Colin & Bell, 1991
Fundulus luciae	Spotfin killifish	Aquarium	1000-1400	Hardy, 1978
Thunnus thynnus	Northern bluefin tuna	Cage	1000, 1700-1900	Harada, 1980
Canthigaster rivulata	Sharpnose puffer	Aquarium	Late morning	Arai & Fujita, 1988
Halichoeres marginatus	Speckled rainbowfish	Nature	40-96% of daylight h	Colin & Bell, 1991
Lutjanus vittus	Blackstriped snapper	Nature	1100-1500	Davis & West, 1993
Paragobiodon echinocephalus	Pricklehead goby	Nature	1100-1700	Kuwamura et al., 1993
Paralichthys olivaceus	Hirame	Tank	1100-1800	Cho & Yang, 1991
Sparisoma rubripinne	Redfin parrotfish	Nature	1125-1740	Randall & Randall, 1963
Apogon notatus	Blackspot cardinalfish	Nature	1135-1447	Kuwamura, 1983
Amphiprion akallopisos	Skunk clownfish	Aquarium	1200	Neugebauer, 1969
Amphiprion ephippium	Red saddleback clownfish	Aquarium	1200	Neugebauer, 1969
Fundulus heteroclitus	Mummichog	Nature	Day	Hardy, 1978
Phrynelox scaber	Splitlure frogfish	Nature	Day	Martin & Drewry, 1978
Amphiprion sebae	Sebae clownfish	Aquarium	Early p.m.	Breeder's Registry Database, 1994
Chrysiptera traceyi	Tracey's damselfish	Aquarium	Early p.m.	Breeder's Registry Database, 1995
Amphiprion percula	Percula clownfish	Aquarium	Day, esp. 1200-1400	Garnaud, 1951
Tautogolabrus adspersus	Cunner	Nature	Mostly 1200-0400	Wicklund, 1970; Ferraro, 1980
Gramma melacara	Blackcap basslet	Aquarium	Midday	Addison, 1994a, 1994b
Dorosoma cepedianum	Gizzard shad	Nature	Midday	Jones et al., 1978
Archosargus probatocephalus	Sheepshead	Nature	Midday	Tucker & Barbera, 1987
Oxymonacanthus longirostris	Orangespot filefish	Aquarium	Midday	Breeder's Registry Database, 1994
Amphiprion ocellaris	Common clownfish	Aquarium	Afternoon	Moe, 1973
Dascyllus aruanus	Threestripe damselfish	Aquarium	~1300-?	Breeder's Registry Database, 1995
Tautoga onitis	Tautog	Aquarium	1330-1600	Olla & Samet, 1977
Fundulus diaphanus	Banded killifish	Nature	1400, 1635	Hardy, 1978
Sparus auratus	Gilthead sea bream	Tank	1400, 0300-0500	Zohar & Gordin, 1979
Paralichthys californicus	California halibut	Tank	1400-2200	Caddell et al., 1990
Repomucenus valenciennei		Aquarium	1425-1720	Takita & Okamoto, 1979
Apogon cyanosoma	Goldstriped cardinalfish	Aquarium	1430-1600	Thresher, 1984
Amphiprion clarkii	Clark's clownfish	Aquarium	1500 to ?	Breeder's Registry Database, 1994; DiMarco, 1994
Amphiprion frenatus	Tomato clownfish	Aquarium	1500 to ?	Breeder's Registry Database, 1994
Lachnolaimus maximus	Hogfish	Nature	Mid-p.m. to 15 min before sunset	Colin, 1982
Tautoga onitis	Tautog	Nature	Mostly 1530-0330	Ferraro, 1980

Euthynnus lineatus	Black skipjack	Nature	1536	Schaefer, 1987
Halichoeres marginatus	Speckled rainbowfish	Nature	1545-1845	Shibuno et al., 1993
Serranus subligarius	Belted sandfish	Nature	1600-1900	Clark, 1959
Pterocaesio digramma	Twin-yellowstriped fusilier	Aquarium	~1600-1900	Yokoyama et al., 1994
Histrio histrio	Sargassumfish	Aquarium	1600-2030	Martin & Drewry, 1978
Trinectes maculatus	Hogchoker	Nature	1630-0030	Ferraro, 1980
Brevoortia tyrannus	Atlantic menhaden	Nature	Mostly 1630-0530	Ferraro, 1980
Fundulus grandis	Gulf killifish	Pond	Mostly 1630-0800	Tatum et al., 1982b
Microgadus tomcod	Atlantic tomcod	Aquarium	~1630-0800	Watson, 1987
Paramonacanthus japonicus	Striped filefish	Nature	1633-1754	Nakazono & Kawase, 1993
Cyclichthys orbicularis	Orbiculate burrfish	Aquarium	1700	Conde, 1986
Amphiprion ephippium	Red saddleback clownfish	Aquarium	1700 to ?	Breeder's Registry Database, 1995
Amphiprion ocellaris	Common clownfish	Aquarium	1700-1830	Breeder's Registry Database, 1995
Parapercis snyderi	Snyder's weever	Nature	1700-1830	Kobayashi et al., 1993
Pagrus major	New Zealand snapper	Aquarium	1700-1900	Scott et al., 1993
Pagrus major	Red sea bream		1700-1900	Matsuyama et al., 1988b
Thunnus thynnus	Northern bluefin tuna	Cage	1700-2000	Takii et al., 1997a
Epinephelus striatus	Nassau grouper	Nature & tank	1700-2100	Tucker et al., 1996
Sparidentex hasta	Sobaity	Tank	1700-2400^{+}	Al-Marzouk et al., 1994
Centropyge interruptus	Japanese pygmy angelfish	Aquarium	1715-1810	Hioki & Suzuki, 1987
Anchoa mitchilli	Bay anchovy	Nature	Mostly 1730-0130	Ferraro, 1980; Luo & Musick, 1991; Zastrow et al., 1991
Paralabrax maculatofasciatus	Spotted sand bass	Tank	1730-2150	Avíles-Quevedo et al., 1995
Cirrhitichthys aureus	Golden curlyfin	Aquarium	1740-1904	Tanaka & Suzuki, 1991
Serranus subligarius	Belted sandfish	Aquarium	Early a.m. to 2400	Tucker & Woodward, 1995
Lachnolaimus maximus	Hogfish	Nature	90-15 min before sunset	Colin, 1982
Centropyge bispinosus	Coral beauty	Aquarium	90-few min before lights off	Lamm, 1984
Centropyge (6 spp.)	(pygmy angelfishes)	Aquarium	1 h before lights off	Bauer & Bauer, 1981
Synchiropus splendidus	Mandarinfish	Aquarium	Just before to just after lights off	Delbeek, 1989
Chaetodon (3 spp.)	(butterflyfishes)	Nature	1 h before to sunset	Lobel, 1978
Nelabrichthys ornatus	Concha wrasse	Nature	Before sunset for 70 min	Andrew et al., 1996
Centropyge potteri	Potter's pygmy angelfish	Nature	1 h before to sunset	Lobel, 1978
Sciaenops ocellatus	Red drum	Tank	Sunset -1 h to +4 h	Roberts et al., 1978
Takifugu niphobles	Grass puffer	Nature	Sunset -50 to +50 min	Honma et al., 1980
Hypoplectrus unicolor	Hamlet	Nature	45 min before to ~sunset	Lobel, 1991
Malacanthus latovittatus	Blue whiting	Nature	32 to 3 min before sunset	Clark & Pohle, 1992
Pomacanthus arcuatus	Gray angelfish	Nature	Sunset -30 to +? min	Moyer et al., 1983
Dendrochirus zebra	Zebra lionfish	Nature	Sunset -26 to +55^{+} min	Moyer & Zaiser, 1981
Cirrhitichthys aprinus	Spotted hawkfish	Nature	Sunset -24 to +34 min	Donaldson, 1990
Centropyge argi	Cherub pygmy angelfish	Nature	18 min before to sunset	Moyer et al., 1983
Chaetodontoplus mesoleucas	Vermiculated angelfish	Nature	Sunset -9 to +7 min	Moyer, 1990
Plectropomus leopardus	Leopard coraltrout	Nature	Sunset -8 to +14 min	Samoilys & Squire, 1994

Table 7-6. (Continued)

Scientific name	Common name	Location	Time[a]	References
Holacanthus tricolor	Rock beauty	Nature	Sunset -6 to +5 min	Moyer et al., 1983
Paracirrhites arcatus	Whiteline hawkfish	Nature	Sunset -2 to +6 min	Donaldson, 1990
Holacanthus ciliaris	Queen angelfish	Nature	~Sunset	Moyer et al., 1983
Pomacanthus paru	French angelfish	Nature	~Sunset	Moyer et al., 1983
Centropyge (6 spp.)	(pygmy angelfishes)	Nature	Sunset	Bauer & Bauer, 1981
Centropyge interruptus	Japanese pygmy angelfish	Nature	Sunset	Moyer & Nakazano, 1978
Paracirrhites forsteri	Freckled hawkfish	Nature	Sunset +1 to +47 min	Donaldson, 1990
Holacanthus passer	King angelfish	Nature	Sunset to +30 min	Moyer et al., 1983
Encrasicholina purpurea	Nehu	Natural	Sunset to +1-3 h	Clarke, 1989
Pagrus major	Red sea bream	Tank	Sunset to night	Foscarini, 1988
Engraulis mordax	Northern anchovy	Nature	Sunset to 0200	Hunter & Macewicz, 1980
Sillago sihama	Silver whiting	Tank	1800-1900	Kumai & Nakamura, 1977
Cynoscion regalis	Weakfish	Nature	1800-2100	Goshorn & Epifanio, 1991
Pogonias cromis	Black drum	Nature	1800-2200	Saucier & Baltz, 1993
Lates calcarifer	Barramundi	Nature	1800-2300	Ruangpanit, 1987a
Rhabdosargus sarba	Silver sea bream	Tank	1800-2300[+]	Mihelakakis & Kitajima, 1995
Engraulis encrasicolus	European anchovy	Natural	1800-0300	Ribeiro et al., 1996
Cynoscion xanthulus	Orangemouth corvina	Tank	Just before dusk	Prentice & Thomas, 1987[h]
Parupeneus multifasciatus	Manybanded goatfish	Nature	Dusk	Lobel, 1978
Amphiprion bicinctus	Twoband clownfish	Aquarium	Dusk	Young, 1991
Centropyge argi	Cherub pygmy angelfish	Aquarium	Dusk to ?	Breeder's Registry Database, 1994
Centropyge bicolor	Bicolor pygmy angelfish	Aquarium	Dusk to ?	Breeder's Registry Database, 1994
Centropyge loriculus	Flame pygmy angelfish	Aquarium	Dusk to ?	Breeder's Registry Database, 1994
Genicanthus melanospilos	Blackspot lyretail angelfish	Aquarium	Dusk to ?	Breeder's Registry Database, 1994
Nemateleotris decora	Decorated firefish	Aquarium	Dusk to 0200	Breeder's Registry Database, 1994
Opistognathus aurifrons	Yellowhead jawfish	Aquarium	Dusk to ?	Breeder's Registry Database, 1994
Alosa mediocris	Hickory shad	Nature	~Dusk to 2400	Jones et al., 1978
Gadus morhua	Atlantic cod	Nature	Dusk to a.m.	Hardy, 1978
Atherinops affinis	Topsmelt	Tank	Dark to 0200	Anderson et al., 1991
Prionotus spp.	(sea robins)	Nature	Mostly 1830-0530	Ferraro, 1980
Chaetodontoplus duboulayi	Scribbled angelfish	Aquarium	1840-2040	Arai, 1994
Centropyge ferrugatus	Rusty pygmy angelfish	Tank	Sunset +60-90 min	Hioki et al., 1990
Selar crumenophthalmus	Bigeye scad	Tank	Sunset +1 h to +14 h	Iwai et al., 1996
Anampses cuvier	Firefish	Aquarium	Just after lights off	DeBernardo, 1975

Chaetodipterus faber	Atlantic spadefish	Tank	~1900	Breeder's Registry Database, 1995
Pomatomus saltatrix	Bluefish	Nature	1900-2100	Norcross et al., 1974
Thunnus obesus	Bigeye tuna	Nature	1900-2400	Nikaido et al., 1991
Atherinops affinis	Topsmelt	Tank	1900-0500	Anderson et al., 1991; Middaugh et al., 1992
Pleuronectes platessa	European plaice	Nature	1900-0600	Simpson, 1971
Sardinella aurita	Spanish sardine	Nature	Peak 1930-2300	Jones et al., 1978
Lutjanus stellatus	Whitespotted snapper	Tank	2000-2300	Hamamoto et al., 1992
Thunnus albacares	Yellowfin tuna	Nature	2000-2400	Suzuki, 1994
Engraulis mordax	Northern anchovy	Tank	2000-2400	Lasker, 1974
Lutjanus kasmira	Bluestriped snapper	Aquarium	2037-2116^{+}	Suzuki & Hioki, 1979
Diodon holocanthus	Balloonfish	Aquarium	2052-2325	Sakamoto & Suzuki, 1978
Cynoscion nebulosus	Spotted seatrout	Tank	2100-2230	Tucker & Faulkner, 1987
Engraulis encrasicolus ponticus	Black Sea anchovy	Nature	2100-2400	Lisovenko & Andrianov, 1991
Sebastes schlegeli	Kurosoi	Tank	2100-0100	Jeon, 1995
Lates calcarifer	Barramundi	Tank	2100-0200	Lim et al., 1986[h]
Sardinops melanostictus	Japanese sardine	Nature	~2100- ?	Morimoto, 1993
Euthynnus affinis	Kawakawa	Tank	2115-2200	Kohno et al., 1994b
Chrysiptera cyanea	Blue devil damselfish	Aquarium	Evening to ?	Breeder's Registry Database, 1994
Acanthopagrus berda	River bream	Nature	Night	Garratt, 1993
Amblygobius bynoensis	Bynoe's goby	Aquarium	Night	Breeder's Registry Database, 1994
Chanos chanos	Milkfish	Nature	Night	Lee, 1986
Epinephelus malabaricus	Malabar grouper	Tank	Night	Maneewong et al., 1986
Etrumeus teres	Round herring	Nature	Night	Jones et al., 1978
Hippoglossus hippoglossus	Atlantic halibut	Aquarium	Night	Martin & Drewry, 1978
Pomacentrus caeruleus	Caerulean damselfish	Aquarium	Night	Breeder's Registry Database, 1995
Scophthalmus maximus	Turbot	Nature & lab	Night	McEvoy & McEvoy, 1992
Solea solea	Common sole	Tanks	Night	Baynes et al., 1993
Syngnathus floridae	Dusky pipefish	Nature	Night to early a.m.	Hardy, 1978
Porichthys notatus	Midshipman	Aquarium	Night to +20 h	Brantley & Bass, 1994
Scorpaena miostoma		Tank	2130-2230	Kimura et al., 1989
Mugil curema	White mullet	Nature	2200-2315	Martin & Drewry, 1978
Rhabdosargus sarba	Silver sea bream	Tank	2200-2400	Leu, 1994
Thunnus albacares	Yellowfin tuna	Nature	2230-0330	Schaefer, 1996
Sprattus sprattus	Sprat	Nature	2200-1200	Simpson, 1971
Sardina pilchardus	European pilchard	Nature	2200-0200	Simpson, 1971
Nemateleotris decora	Decorated firefish	Aquarium	2400-0200	Schiller, 1990
Pleuronectes limanda	Dab	Nature	2400-1200	Simpson, 1971
Mugil cephalus	Striped mullet	Nature	Just after 2400	Arnold & Thompson, 1958
Lutjanus argentimaculatus	Asian red snapper	Tank	0100-0400	Doi & Singharaiwan, 1993

[a] a.m. = morning after sunrise; p.m. = afternoon.
[h] hormone-induced.

diameter, normal size and number of oil globules, normal perivitelline space, good fertilization rate, normal cell division (usually symmetrical at first), lack of inclusions, normal chorion uniformity and strength, normal chromosome pattern, normal osmotic properties, resistance to bacterial infection, and good hatching rate (Yamamoto, 1981; Kjørsvik and Lonning, 1983; McEvoy, 1984; Kjørsvik et al. 1990; Casillas et al., 1991; McEvoy and McEvoy, 1991, 1993; Norberg et al., 1991; Pavlov et al., 1992; Tamaru et al., 1992; Mårstøl et al., 1993; Kjørsvik, 1994; Bromage et al., 1994; Bromage, 1995; Brooks et al., 1997; Shields et al. 1997). Content of protein and amino acids can be used, and possibly pigments, ascorbic acid, and lipid.

Figure 7-15. A female Nassau grouper that has ovulated (HCG-induced) and is about to be stripped.

Figure 7-16. A female Nassau grouper that has ovulated naturally and is about to spawn voluntarily. Note the protruding eggs and the black and white bicoloration, which males assume hours before spawning and females assume just before spawning to indicate readiness. When Nassau groupers are to be strip-spawned, female bicoloration is a signal to the culturist. It usually occurs with both natural and induced ovulation. (photo by Phillippe G. Bush, Cayman Islands Department of the Environment)

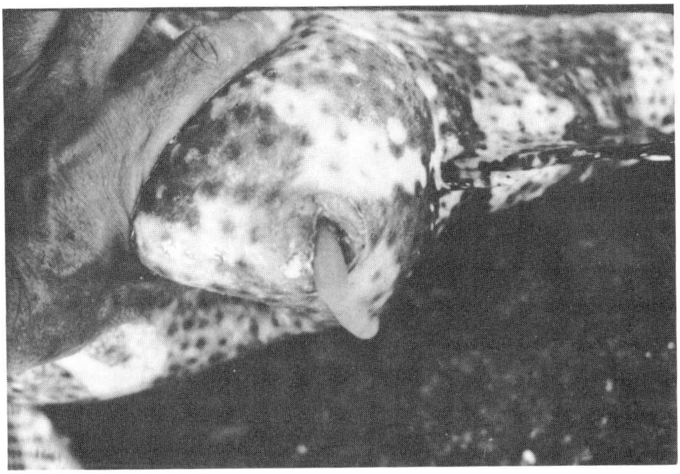

Figure 7-17. Starting the eggs flowing from a brownmarbled grouper. With complete normal ovulation, very little pressure is needed.

Two million or more eggs can be obtained in less than a minute. Masses of clear normal planktonic eggs might appear yellowish or greenish because of oil globules (e.g., especially striped mullet) or could be tinted with other colors, depending on species. Normal salmon eggs are red or red-orange and sturgeon eggs usually are gray or black. Aborted oocytes (infertile ones typically released before full development) usually are less than full size and not normally-shaped; individual oocytes could be yellowish and partly opaque because yolk globules have not finished fusing or yolk has coagulated; often there are more than the normal number of oil globules.

When eggs have been ovulated but then not released and fertilized during the grace period and are losing the ability to be fertilized and develop correctly, they are overripe (overmature; Fig. 7-5f). Overripe eggs could have visible structural irregularities, loss of transparency, and discoloration and might have begun to collapse and lose their normal shape; chorions could be soft and easily torn, cytoplasm "dimpled," and cortical alveoli fused (Kjørsvik et al., 1990; Fauvel et al., 1992; Pavlov et al., 1992). Unless oil globules have been fragmented by handling, overripe eggs typically have the normal number.

If the membrane holding eggs in has been broken too early, the natural process is disrupted to some degree. Two possible results are (1) abortion or (2) loss of a large number of good eggs by leakage. These are good reasons to avoid biopsies.

Artificial Fertilization

Some species will spawn in the holding unit after treatment, but most ordinarily will not (Table 7-3). Artificial fertilization is optional with the first group and usually mandatory with the second. The procedure should be carried out in an area free of sunlight and with constant temperature near that of the water holding the broodfish. I prefer to collect milt from one or more males in hypodermic syringes (without needles) or other clean airtight containers (Fig. 7-18) and set them aside in a safe place where they will not receive sunlight, experience a temperature change, or be dropped. Some hatcheries strip the male(s) at the same time as or just after the females (e.g., striped bass). The female is caught, anesthetized if necessary, then held gently but firmly over a cup, beaker, pan, etc. for receiving the eggs. The container must be very clean, free of toxicants, and for the dry method, dry. I prefer polymethylpentene beakers because they are strong, inert, transparent, graduated, and easy to clean. Fertilization containers can get very dirty from things like oil and milt, which support bacterial growth, so they must be capable of being cleaned. Also, mixing of gametes is more effective in a cylindrical container. A strong plastic bag or soft net can be used to hold the front end of the fish. Dry or rough items like towels will remove the fish's protective mucous and should not be used. If the head covering shields the eyes from light, stress will be reduced and many fish will not struggle as much. Large or unruly fish can be anesthetized; 100 ppm tricaine methanesulfonate is recommended for striped bass (Rees and Harrell, 1990). The anesthetic should be rinsed off before stripping the fish. Light steady pressure on the anterior abdomen should start the eggs flowing. As the flow decreases, pressure is applied further back and on the sides of the abdomen until no more eggs can be easily expressed. If the membrane does not break easily, it might be better to tear it directly rather than apply excessive pressure to the abdomen, which can injure both fish and eggs. Just make sure the timing is right.

Figure 7-18. Collecting milt from a Nassau grouper with a 3-cc hypodermic syringe (without needle) in order not to waste time after eggs are collected.

In the **wet method**, enough clean water to cushion the eggs is added before the eggs. Milt is added immediately on top of the eggs and the mixture is swirled for 5 min. Then the eggs are washed with a large amount of clean water, several times or continuously. Even though the period of sperm motility might be short, if eggs, milt, and water are well mixed within a minute, the wet method should not result in lower fertilization than the dry method.

In the **dry method**, water is added after the eggs and milt (Fig. 7-19). The rest is the same. The dry method facilitates estimation of egg number by measurement of volume; however, it might stress the eggs more than the wet method. Injury to eggs could occur by contact with the container, the weight of eggs above them, or dessication. Rubbing against solid objects could promote bacterial or fungal infection. The natural state for released planktonic eggs is to be suspended in water out of contact with other objects, but if care is taken, the dry method can be just as reliable as the wet one (unless eggs are particularly fragile).

Some species require special procedures for removing their eggs. For example, sturgeon eggs are released into the body cavity, and surgery is the only practical means of removing them. The eggs can be transferred through an 8- to 10- cm incision with a plastic or stainless steel spoon into the fertilization container (Conte et al., 1988).

Figure 7-19. Stripping eggs from a squaretail coraltrout into a dry 3-L plastic beaker (dry method). (Note hard, knotted plastic net, which should not be used if a softer one is available)

When eggs are released into seawater, their fertility begins to decrease, mainly because of micropyle closure from swelling induced by water uptake. Those of some species become infertile in 1 min and others in a few hours. Once activated by water, sperm might last less than half a minute or (rare) a few days. Fertilization of Atlantic halibut eggs decreased within 10-20 min after contact with seawater (Lein, 1991). Striped bass eggs are fertile for 2 min and sperm motile for 30-60 s (Rees and Harrell, 1990). White sturgeon sperm are motile for 2-5 min (Conte et al., 1988). Rainbow trout eggs are fertile for 30 s and sperm motile for 20 s (Brannon, 1991). Turbot sperm are motile for up to 7 min but halibut sperm for hardly more than 1 min (Suquet et al., 1992). The safest approach is to mix the gametes as quickly as possible after stripping. Lower than normal salinity can reduce sperm viability (Lee et al., 1992).

Preliminary fertilization rate can be determined in a little more than an hour with Nassau groupers at 25-27°C (pers. obs.), 2 h with striped bass at 20°C (Rees and Harrell, 1990), and 12-15 h with rainbow trout at 10°C (Ingram, 1988), when the first and second cell divisons are occurring. (Rainbow trout eggs can be cleared by immersion for 10 min in a mixture of equal parts of glacial acetic acid, methanol, and distilled water, Ingram, 1988). A few hours later, the distinction between fertilized and unfertilized eggs will be much more obvious because of further cell division. Eggs that are not transparent can sometimes be cleared so the embryo becomes visible (e.g., white bass, Kerby and Harrell, 1990). Unfertilized eggs often will either sink or float soon after stripping (the opposite of normal eggs) and begin to discolor. Aberrant size or position of cells indicates bad eggs.

Care of eggs

Right after spawning, eggs can be sensitive to handling until they water harden (takes 1-2 h in Atlantic salmon, Isaksson, 1991). After water hardening, resistance might improve. However, the vitelline membrane of at least some species (rainbow trout, Brannon, 1991) can easily be torn during a period from before blastulation begins until blastopore closure. For rainbow trout, this sensitive time is about 2-9 daf. White sturgeon eggs are sensitive to agitation until gastrulation is complete at about 60-78 haf at 15°C. For warmwater fish gastrulation might take only a few hours.

Stripped eggs are relatively clean, but washing is necessary to remove milt, mucus, oil, and debris from them. Tank-spawned eggs need washing to remove even nastier things such as parasites and immature predators. Eggs sometimes also are disinfected. Although few eggs might be lost from individual bacterial infections, sometimes the whole container can become filled with strands of filamentous bacteria, trapping all of the eggs. Hatchlings also can be caught in these webs. Eggs in fresh water tend to get fungal infections, while in salt water, bacterial infections are more likely. For surface disinfection, rainbow trout eggs can be dipped in an aqueous solution of 3333 ppm iodophor with 100 ppm sodium bicarbonate for 5-8 min, then rinsed well (Ingram, 1988). Rainbow trout eggs infected with fungus were successfully treated with 1-1.5 mg/L formalin or 0.5/L hydrogen peroxide for 15 min (Schreier et al., 1996). Salmon eggs routinely have been dipped in iodophor, but it can be toxic (Fowler and Banks, 1990, 1991). Red drum eggs were safely and effectivly dipped in 3% hydrogen peroxide for 5 min (Douillet and Holt, 1994). Five other disinfectants—benzalkonium chloride, formalin, thimerosal, polyvinylpyrrolidone iodine (PVP-I), and sodium hypochlorite—were toxic to red drum eggs at less than effective concentrations. Maximum concentrations of hydrogen peroxide not toxic to yellowtail snapper (1%) and spotted seatrout (2%) were not high enough to eliminate surface bacteria. Atlantic halibut eggs were safely dipped in 400-800 ppm glutaraldehyde at 5°C for 5-10 min, which reduced bacteria by more than 80%; turbot eggs can be dipped in 400-800 ppm glutaraldehyde at 12°C for 2.5 min (Harboe et al., 1994; Salvesen et al., 1997). Salvesen et al. (1991) reported that 10-min baths in 400 mg/L Buffodine, 125 mg/L hypochlorite, 800 mg/L glutaraldehyde, or 2000 mg/L chloramine-T were toxic to European plaice eggs; baths of 400 mg/L glutaraldehyde or 1000 mg/L chloramine-T apparently were safe and reduced bacteria by about 85%. Hydrogen peroxide seems to be the preferred disinfectant for planktonic eggs (Douillet and Holt, 1994).

The stickiness of adhesive eggs usually should be removed to prevent the eggs from clumping together and dying from hypoxia. In most instances, after about 3 min for fertilization, sturgeon eggs are stirred in a suspension of silt and clay for 20-60 min (Conte et al., 1988). Immersion in a solution of urea with sodium chloride or sodium sulfite, followed by a tannic acid rinse sometimes is used. A similar method with urea, sodium chloride, and tannic acid works for white bass eggs (Kerby and Harrell, 1990). White bass eggs also can be treated by aeration in a 150 mg/L tannic acid solution for 10-12 min (Rottman et al., 1988). Eggs deposited in a gelatinous matrix like those of channel catfish might be separated like catfish eggs, as follows. After any initial sensitive stage is finished, about 500 g of eggs are added to 1 L of 1.5 sodium sulfite in hatchery water (pH checked and corrected with HCl if necessary), then are gently kneaded and stirred until the eggs are separate (Rottman et al., 1991b).

Small numbers of eggs can be counted individually with a pipette and large numbers of eggs by measuring their volume in a graduated container or by weighing them.

Eggs should be transferred and incubated under optimal conditions in very clean containers. They should not experience any sudden change in temperature, salinity, or water chemistry. (Water conditions should be relatively constant until the juvenile stage, but if initial conditions are not optimal for growth and development, they can be slowly changed.) Dead eggs should be removed and water should be exchanged at a fast enough rate to prevent oxygen depletion, ammonia buildup, cloudiness (bacteria, fungi), or foul odor. Floating eggs survive better if they are moved to a different container before hatching. If eggs are allowed to hatch in the incubation container, disinfecting them before or during incubation might be especially helpful.

Sinking eggs can be incubated in jars, troughs, or trays (Bromage et al., 1988; Ingram, 1988; Rees and Harrell, 1990; Brannon, 1991; Tabachek et al., 1993). Floating eggs usually are incubated in aquaria, clear acrylic tanks, or fiberglass tanks. If eggs are incubated at high density, the water should be exchanged rapidly. If water exchange is difficult, egg density should be low. Moderate density and moderate exchange are best. These will vary with species and temperature. Planktonic eggs should not be allowed to pile up in layers more than one-egg thick for more than a few minutes. On the other hand, demersal eggs in hatching jars are several layers thick, but oxygenated water moves past them quickly. Hatching rates of Gulf killifish eggs (normally deposited intertidally) incubated in air for 6 or 12 d were better than those exposed for 0 or 18 d, possibly because there was more oxygen and less bacteria (Perschbacher et al., 1995).

To insure quality of stocked larvae, eggs usually are incubated separately and rearing tanks are stocked just before or just after hatching. Fish with fragile larvae such as groupers should be stocked in rearing tanks before hatching. Larvae of striped bass, red drum, damselfish, and many others can be moved by gently dipping with a beaker or siphoning, or they can be allowed to swim into rearing units sometime between hatching and eye pigmentation.

Eggs or larvae of different species have been stocked in rearing tanks at densities of 1-30/L and occasionally higher—mainly coldwater fish (examples in chapters 13-15); the lower the better. Higher stocking densities can promote critical water quality conditions, especially at high water temperatures, but economic factors often force the culturist to stock as many fish as he can get away with. Sometimes the initial density is based on predicted final density (i.e., expected survival from past experience), but survival will vary.

Storage of Gametes

Males of certain species can be difficult to collect in some locations. A possible solution is to cryopreserve sperm when ripe males are available and store it until eggs are obtained. This also opens possibilities of interbreeding different stocks or different species that are not ripe simultaneously. Converting females to males could be another option (see Sex Control, p 296). Cryopreserved gametes usually are stored at less than -130°C (Rana, 1995). Extender solutions (e.g., sugars) keep frozen gametes alive and diluted. Cryoprotectants such as dimethylsulfoxide (DMSO), glycerin, or methanol are necessary to prevent large intracellular ice crystals from forming and to lessen the effect of intracellular salt concentration as water turns to ice (Davy and Chouinard, 1981). Plastic straws, tubes, and bags are convenient containers for milt. Egg stage, extenders, cryoprotectants, cooling and thawing methods and rates are factors affecting success. Unsuccessful cryopreservation largely results from ice crystal damage, especially to membranes (Rana, 1995). Cryoprotectants can be toxic and should be tested for safety and efficacy (Chao et al., 1994; Rana, 1995).

Fish eggs are difficult to cryopreserve (Rana, 1995), but milt is easy. Fifty-seven references up to 1981 on cryopreservation of gametes are listed in Davy and Chouinard (1981). Additional ones on sperm cryopreservation are Mounib et al. (1968) for Atlantic cod; Ott (1975), Scott and Baynes (1980), Stoss and Refstie (1983), Holtz (1993), Lahnsteiner et al. (1996) for salmonids; Hara et al. (1982), Kelley and Lee (1986), Chao and Liao (1987) for milkfish; Ben Amotz and Rosenthal (1981) for Atlantic herring; Stoss and Donaldson (1982), Harvey (1982), Baynes and Scott (1982), Bayrle (1982), Scheerer and Thorgaard (1989), Gallant and McNiven (1991), Wheeler and Thorgaard (1991), Park and Yoon (1992), Gwo et al. (1993), Conget et al. (1996) for rainbow trout; Piironen (1987) for whitefish; Leung (1987), Palmer et al. (1993) for barramundi; Mounib (1978) for European sea bass; Parker et al. (1990) for striped bass; Chao et al. (1992), Gwo (1994) for Malabar groupers; Withler and Lim (1982) for greasy groupers; Gwo (1994) for leopard coraltrout; Caylor et al. (1994) for cobia; Young et al. (1992) for summer whiting; Billard (1978, 1984), Happe and Zohar (1988), Chamberyon and Zohar (1990) for gilthead sea bream; Chao et al. (1986), Chao and Liao (1987) for black sea bream; Kurokura et al. (1986) for red sea bream and crimson sea bream; Thorogood and Blackshaw (1992) for surf sea bream, Gwo et al. (1991) for Atlantic croaker; Chao et al. (1975), Chao and Liao (1978), Chao (1982) for striped mullet; Muiruri (1988) for tilapias; Doi et al. (1982) for northern bluefin tuna; Bolla et al. (1987), Edwardes (1991) for Atlantic halibut; Pullin

(1972) for European plaice; Gwo (1982), Miyaki and Dotsu (1987) for puffers; Gwo et al. (1993) for grass puffer; Lubzens et al. (1997) for common carp. Chao (1991) reviewed research on several species in Taiwan. Later reviews are in Gwo et al. (1991), Billard et al., (1995), and Rana (1995).

The best result for cryopreservation of fish eggs has been with Australian bass (Leung and Jamieson, 1991). Vacuum treatment followed by freezing in a methanol/ dry ice bath successfully preserved up to 21% of the eggs. Reasonable numbers of zebrafish (Harvey, 1983) and rainbow trout blastomeres (Nilsson and Cloud, 1993) survived cryopreservation.

Short-term storage at temperatures near 0°C also works for milt and has been more successful than cryopreservation for eggs. Some references for milt are: Conte et al., (1988) for sturgeon, Ingram (1988) for rainbow trout, and Stoss and Refstie (1983) for Atlantic salmon and brown trout. Rainbow trout milt was viable after storage for 34 d at 0°C with antibiotics and air flow (Rana, 1995). Atlantic halibut milt was partly viable after 60 d storage at -4°C with 5% methanol (Edwardes, 1991). Both eggs and sperm of Nile tilapia were kept in Dulbecco's modified Eagle medium (MEM) for 4 h with negligible loss of fertility (Yeheskel and Avtalion, 1988). Salmon eggs were viable for 20 d when stored in ovarian fluid at -1°C (Rana, 1995). Herring eggs have been stored without water for 36 h and then fertilized, with good results. New Zealand snapper milt was viable for 8 d at 4°C, and eggs held in Leibovitz L15 medium at 17-19°C for 12 h were 45% fertilizable (Hobby and Pankhurst, 1997). Atlantic halibut eggs could not be kept for more than 12 h in ovarian fluid at 4°C without drastic loss of viability (Rana, 1995). Pullin and Bailey (1981) concluded that cryopreservation of European plaice eggs at any stage was unlikely to be successful, but supercooling of fertilized eggs to -1.8°C could be used to delay hatching from 30 d to 43-63 d. Atlantic cod eggs can be stored for 9-10 h, herring eggs for 10-48 h, and Atlantic halibut eggs for 6 h (Kjørsvik et al., 1990).

Natural Ovulation

Many species have spawned voluntarily in captivity (without chemical treatment), mostly during the natural spawning season under conditions of ambient temperature and partial or total natural light (Table 7-7). Some spawned in aquariums outside of the normal season. Small species (usually shorter than 200 mm) usually have spawned in small aquariums, jars, or fiberglass tanks, and large ones in 0.3 to 21,200-m^3 tanks or ponds and 9 to 235-m^3 cages. Many fish will reproduce voluntarily if they are well-nourished and protected from stress—mainly crowding, low water quality, and disturbance. The species that have ovulated naturally and then spawned voluntarily in captivity probably outnumber those that have been induced to ovulate. This approach tends to require more time, facilities, labor, and expense than any other, but egg quality can be the best.

Most medium to large marine fish with planktonic eggs normally spawn sometime from just before sunset until near midnight, but there are exceptions (Table 7-6). Small reef fish often spawn during early to mid-afternoon (e.g., bluehead wrasse).

Broodfish Caught Running Ripe.—This requires very good timing and a fair amount of luck (Table 7-8). Salmon traditionally have been caught during their upstream spawning migration and stripped of gametes. One advantage with salmon is the long time between ovulation and overripening (Table 7-5). With warmwater marine fish, the grace period is much shorter and timing is more critical.

Short-term Captive Broodfish.—Females caught during the spawning season sometimes can be held for hours or days until ovulation occurs naturally. Usually, the fish are then stripped. This has been done with salmon. Chesapeake Bay striped bass judged to be less than 8 h away from ovulation have produced fertilized eggs when paired with males in tanks (Yeager et al., 1990). Similar results have been obtained with Nassau groupers, but stripping is more reliable (Tucker and Woodward, 1996).

Table 7-7. Species that spawned voluntarily in captivity.

Common name	Scientific name	Location	Container	Months	Temp. (°C)	Egg di. (µm)	References
			Petromyzontidae—lampreys				
Petromyzon marinus	Sea lamprey	Michigan	1.1-m^3(2.45 m × 0.75 m × 0.6 m deep) wood Tk	Jun-Jul	16-20		Fredricks & Seelye, 1995
			Muraenidae—moray eels				
Muraenesox cinereus	Pike eel	Japan	50-m^3 Tk	Jul		1950	Umezawa et al., 1991
			Clupeidae—herrings				
Clupea pallasii	Pacific herring	Br. Columbia	7-m^3(3 m diam. × 1 m deep) FRP Tk	Mar-Jun	5-10		Hay, 1986
			Engraulidae—anchovies				
Engraulis mordax	Northern anchovy	California	12-m^3(4.6 m diam. × 0.72 m deep) Pls swimming pool	Apr-Dec	15	~1300	Leong, 1971; Lasker, 1974
			Chanidae—milkfish				
Chanos chanos	Milkfish	Taiwan	860-m^3(750 m^2 × 1-1.3 m deep) Cnc Pd	Apr-Oct	25-33	1170	Lin, 1984, 1985; Lin & Lin, 1987
		Philippines	150 & 200 m^3 Cnc Tks	Apr-Oct			Emata & Marte, 1994
		Philippines	235-m^3(10 m diam. × 3 m deep) cage	Aug	~30-31		Lacanilao & Marte, 1980; Toledo & Gaitan, 1994
		Indonesia	22.6 m^3(6 m diam. × 80 cm deep) round Tk	Jan-Apr		1210	Prijono et al., 1988
			Osmeridae—smelts				
Plecoglossus altivelis	Ayu	Japan	432-L(120 cm × 60 cm × 60 cm deep) Aq	Oct	18-20		Iguchi & Maekawa, 1993
			Salmonidae—salmons, trouts				
Coregonus lavaretus	Powan	France	4 & 12-m^3 Tks	Nov-Apr	4-10		Gillet, 1991b
Oncorhynchus gorbuscha	Pink salmon	Br. Columbia	6.6-m^3(3 m diam. × 0.93 m deep) FRP Tk	Aug,Oct,Nov, Feb,May	10		MacQuarrie et al., 1979
Oncorhynchus kisutch	Coho salmon	Br. Columbia	6.6-m^3(3 m diam. × 0.93 m deep) FRP Tk	Sep,Dec,May	10		MacQuarrie et al., 1978
Oncorhynchus tshawytscha	Chinook salmon	Alaska	171-m^3(6.1 m × 6.1 m × 4.6 m) cage	Aug			Joyce et al., 1993[s]
Salvelinus alpinus	Arctic char	France	4 & 12-m^3 Tks	Oct-Mar	5-8		Gillet, 1991a
			Gadidae—cods				
Gadus macrocephalus	Pacific cod	Japan	5-m^3 round Tk	Dec-Jan	6-8		Sakurai & Hattori, 1996
		Japan	13-m^3 (2.5 m × 5.5 m × 1.2 m deep) RW	Dec-Jan	6-8		Sakurai & Hattori, 1996
Gadus morhua	Atlantic cod	Norway	350-m^3 pond	Feb-	~3-4		Dannevig, 1910
		England?	10.8-m^3(1.3 m deep) L-shaped Aq	Mar	7-10		Brawn, 1961
		Norway	175 m^3 Pls pen	Feb-Apr	3-6	1400	Huse & Jensen, 1983; Huse, 1991
		Norway	20-m^3 Tk	Jan-May	8-9	1400	Kjesbu, 1989; Kjesbu et al., 1991
		Russia	1- 3-m^3 Tks		1-4		Zhuravleva, 1995

Table 7-7. (Continued)

Common name	Scientific name	Location	Container	Months	Temp. (°C)	Egg di. (µm)	References
Merlangius merlangus	English whiting	England	Large Aq	Apr			Cowan, 1938
		Scotland	3.4-m^3(1.5 m × 1.5 m × 1.5 m) Cnc & glass Aq	Feb-May	9-15	1135	Hislop, 1975
Microgadus tomcod	Atlantic tomcod	New York	76-L Aq	~Jan-Apr			Watson, 1987
Pollachius virens	Pollock	England	Large Aq	May			Cowan, 1938
			Batrachoididae—toadfishes				
Porichthys notatus	Midshipman	California	80- to 380-L Aq	~May-Aug		~4500	Brantley & Bass, 1994[b]
			Antennariidae—frogfishes				
Histrio histrio	Sargassumfish	Florida	37-L Aq		27	600 x 700	Mosher, 1954; Paull, 1969
			Gobiesocidae—clingfishes				
Acyrtops beryllinus	Emerald clingfish	Florida	1-L(10 cm × 10 cm × 10 cm) Aq	Nov-May	21-27	1257[L]	Gould, 1965; Jachowski, 1970
Aspasma minima		Japan	Aq			1300[L]	Shiogaki & Dotsu, 1971
Aspasmichthys ciconiae	Yellow clingfish						Shiogaki & Dotsu, 1972
Gobiesox strumosus	Skilletfish	Florida	114-L acrylic Aq & 76-L epoxy-plywood Tk	Sep-Dec	20-28		Martin, 1968; Martin & Martin, 1971
			Cyprinodontidae—killifishes				
Cyprinodon variegatus	Sheepshead minnow	Florida	28-L Aq & 0.14-ha Pls-lined Pd				Tucker, unpub.
Fundulus grandis	Gulf killifish	Alabama	8000-m^3(0.8 ha × 1.0 m deep) Pd	Mar-Oct	20-36		MacGregor et al., 1983
		Texas	0.1-ha Pd				Perschbacker, 1985; Perschbacker & Strawn, 1988
		Alabama	0.08-ha Pd	Feb-Oct	9-35	1250	Tatum et al., 1982b
Fundulus heteroclitus	Mummichog	Georgia	20-L Aq	Mar-Aug			Kneib, 1993
Lucania parva	Rainwater killifish	New York	(36 cm × 14 cm × 14 cm) Aq	Feb	18		Nichols, 1916
			Pseudomugilidae—blue-eyes				
Pseudomugil mellis	Honey blue-eye	Australia	190-L Aq	Jan-Dec	27-30	1465	Semple, 1991
			Atherinidae—silversides				
Atherinops affinis	Topsmelt	California	1.4-m^3(1.5 m diam. × 80 cm deep) round Tk	May-Jan	18	1520	Middaugh et al., 1992
Atherion elymnus	Japanese silverside	Japan	1-m^3 polyethylene Tk	Jun-Jul	~21	975	Kimura & Tsukamoto, 1990
Menidia menidia	Atlantic silverside	Massachusetts	50-m^3(7.4 m diam. × 1.16 m deep) Tk	Jan-Mar	>24		Poole, 1978

			Gasterosteidae—sticklebacks				
Apeltes quadracus	Fourspine stickleback	California	Aq	Jul	23-24		Reisman, 1963
			Syngnathidae—pipefishes, seahorses				
Hippocampus coronatus	Crowned seahorse		Aq				Burgess, 1987
Hippocampus zosterae	Dwarf seahorse		7-L Aq		24		Breeder's Registry Database, 1994
Hippocampus spp.	(seahorses)	France	240-L aquaria				Schmidt, 1995
			Scorpaenidae—scorpionfishes				
Dendrochirus brachypterus	Pygmy lionfish		227-L Aq		23		Brandt, 1976
			250 & 750-L aquaria		23-26		Fishelson, 1978
Scorpaena miostoma		Japan	2-m^3 Cnc Tk	Jun-Jul		860	Kimura et al., 1989
Sebastes schlegeli	Kurosoi	Japan	75-m^3(3.5 m × 5 m × 1.25 m deep) FRP Tk	May-Jul	15	*l*	Kusakari, 1991
		Japan	Tk		12-14	*l*	Takahashi, 1993
		Korea	Tk	Mar-Jun	13-17	*l*	Jeon, 1995
Sebastiscus marmoratus	Kasago	Japan	27 m^3 (3 m × 3 m × 3 m) cage	Jan	13-14	*l*	Kuronuma & Fukusho, 1984
			Anoplopomatidae—sablefishes				
Anoplopoma fimbria	Sablefish	Br. Columbia	13.5-m^3(3.5 m diam. × 1.4 m deep) wood Tk	Feb-Mar	6-7.5		Jensen et al., 1992[s]
			Hexagrammidae—greenlings				
Ophiodon elongatus	Lingcod	Br. Columbia	(3 m diam.) hexagonal Tk		9.5		Appelbaum et al., 1995
			Cottidae—sculpins				
Pseudoblennius cottoides	Asahi-anahaze	Japan	500-L polyethylene Tk	Dec-Feb	16	1790	Kimura et al., 1987
			Cyclopteridae—snailfishes				
Cyclopterus lumpus	Atlantic lumpfish	New Brunswick	400-L Tk	May	5.5		Martin-Robichaud et al., 1994
			Centropomidae—snooks				
Lates calcarifer	Barramundi	Malaysia	100 m^3 (10 m diam. × 1.3 m deep) Cnc Tk	Jan-Dec			Ali, 1987
		Thailand	100 m^3 (10 m diam. × 1.3 m deep) Cnc Tk	Jan-Dec	27-34	800	Kungvankij, 1987; Maneewong, 1987a
		Singapore	75-m^3(5 m × 5 m × 3 m deep) cage				Chou, p.c., 1992
		Philippines	9-m^3 (1.5 m × 3 m × 2 m) cage				Duray & Juario, 1987
		Philippines	48-m^3(4 m × 4 m × 3 m deep) cage	Jun-Oct		~850	Toledo et al., 1991; Garcia, 1992
			Percichthyidae—temperate basses				
Dicentrarchus labrax	European sea bass	Spain	12-m^3 Aq	Oct-May	9-15		Zanuy et al., 1986, Carrillo et al., 1995
		England	3.6-m^3(1.85 m diam.) FRP Tk	Apr	12	1270	Jennings & Pawson, 1991
Lateolabrax japonicus	Japanese sea bass	Japan	75-m^3 round Tk	Dec			Kuronuma & Fukusho, 1984
Morone saxatilis	Striped bass	Maryland	(1.8 m diam.) FRP Tk	May	~14		Woods et al., 1990
Morone saxatilis × *M. chrysops*	Palmetto bass	South Carolina	23-m^3(1.8 m diam. × 0.9 m deep) FRP Tk	Apr			Jenkins & Smith, 1985

Table 7-7. (Continued)

Common name	Scientific name	Location	Container	Months	Temp. (°C)	Egg di. (μm)	References
			Serranidae—groupers, sea basses				
Centropristis striata	Black sea bass	Florida	3.7-m^3 (2.0 m × 2.6 m × 70 cm deep) FRP Tk & a 950-L Tk	Feb-Apr	16-20	928	Cerdà & Wallace, p.c., 1994; Cerdà et al., 1997
Cromileptes altivelis	Polka dot grouper	Singapore	48-m^3 (4 m × 4 m × 3 m deep) cage	Oct-Mar	28	886	Chou, p.c., 1992
		Taiwan					Gomes, p.c., 1992
Epinephelus akaara	Redspotted grouper	Japan	6-m^3 (4.5 m × 2.0 m × 70 cm deep) Pd	Jul-Sep		735	Ukawa et al., 1966
		Japan	7-m^3 (1 m deep) Tk, 50-m^3 (3 m deep) Tk	Jul-Sep	22-29	750	Hamamoto et al., 1986a
		Japan	45-m^3 (10 m × 3 m × 1.5 m deep) Tk	Jun-Aug	21-28	830	Kayano & Oda, 1987
		Hong Kong	1.8-m^3 (1.5 × 1.0 × 1.2 m deep) Cnc Pd	May-Jun		825	Tseng & Ho, 1988
		Japan	50-m^3 (8 m diam. × 1 m deep) Cnc Tk	Jul-Sep	20-27	760	Maruyama et al., 1994
Epinephelus andersoni	Spotted rockcod	South Africa	Public Aq				Ballard, 1971
Epinephelus awoara	Yellow grouper	Taiwan	860-m^3 (750 m^2 × 1-1.3 m deep) Cnc Pd		30		Lin, p.c., 1992
Epinephelus coioides	Orangespotted grouper	Kuwait	90- and 400-m^3 Cnc Tks	Mar-Jul	23-28		Hussain et al. 1975; Higuchi et al., 1980; Hussain and Higuchi,1980; Abdullah et al., 1984
		Philippines	42-m^3 (4.6 m × 4.6 m × 2 m deep) Cnc Tk	Jul-Oct		840	Toledo et al., 1990, 1994; Toledo, p.c., 1992
		Philippines	48-m^3 (4 m × 4 m × 3 m deep) cage	Jul-Oct		840	Toledo et al., 1990, 1994; Toledo, p.c., 1992
		Malaysia	150-m^3 round Tk	Mar-Jul Nov-Jan	28-32	807	Doi et al., 1991
		Philippines	Tk	Jan-Dec			AquaFarm News, 1992a
		Bahrain	120-m^3 Cnc Tk	Feb-Jun	23-27		Kawahara et al., 1997
Epinephelus diacanthus	Sixband grouper	Taiwan	860-m^3 (750 m^2 × 1-1.3 m deep) Cnc Pd		30		Lin, p.c., 1992
Epinephelus fasciatus	Banded grouper	Japan	80-m^3 Tk				Murai et al., 1984
Epinephelus fuscoguttatus	Brownmarbled grouper	Singapore	75-m^3 (5 m × 5 m × 3 m deep) cage	Jun-Jul Sep-Apr	28-29	890	Lim et al., 1990
		Indonesia	5, 10, 30-m^3 Cnc Tks	Jan,Mar-Apr Aug-Nov		890	Mayunar et al., 1991
		Taiwan	860-m^3 (750 m^2 × 1-1.3 m deep) Cnc Pd		30		Lin, p.c., 1992
		Saudi Arabia	25-m^3 (4 m diam. × 2 m deep) FRP Tk	May-Jun	29-32	766	Maneewong & Hazza, 1992; Hassa & Carlos, 1994
Epinephelus malabaricus	Malabar grouper	Japan	240-m^3 (16.2 m diam., 2.9 m wide, 2.0 m deep) donut-shaped Tk			920	Hamamoto et al., 1986b
		Thailand	75-m^3(10 m × 4 m × 2 m deep) Cnc Tk	Nov		852	Maneewong et al., 1986
		Thailand	50-m^3 (5 m × 5 m × 2 m deep) Cnc Tk				Kungvankij, p.c., 1992
		Taiwan	860-m^3 (750 m^2 × 1-1.3 m deep) Cnc Pd		30		Lin, p.c., 1992

Epinephelus moara	Kelp grouper	Japan	240-m^3 (16.2 m diam., 2.9 m wide, 2.0 m deep) donut-shaped Tk	May-Jun	23-27	900	Nakagawa, 1988; Manabe & Kasuga, 1989
		Japan	72-m^3 (6 m × 6 m × 2 m deep) Cnc Tk				Maruyama, p.c., 1992
Epinephelus polyphekadion	Camouflage grouper	Tahiti	10-m^3 round FRP Tk	Nov-Apr		850	AQUACOP et al., 1988
		Japan	74-m^3 (7.4 m × 5.0 m × 2.0 m deep) Tk	May-Sep	25-30		Tawada, 1989b
		Taiwan	860-m^3 (750 m^2 × 1-1.3 m deep) Cnc Pd		30		Lin, p.c., 1992
		Saudi Arabia	25-m^3 round Cnc Tk	Mar-Jun,Aug	29-30	757	Al-Thobaiti & James, 1997; Rasem et al., 1997
Epinephelus striatus	Nassau grouper	Cuba	Public Aq	Apr	25	1024	Manday and Fernandez, 1966
		Grand Cayman	26-m^3 (6.1 m × 2.4 m × 1.8 m deep) cage	Feb	25-27		Tucker et al., 1993b[a]
		Bahamas	1-m^3 (1 m diam. × 1.3 m deep) FRP Tk	Dec-Feb	23-27	1006	Tucker & Woodward, 1996[b]
		Florida	37-m^3 (12.2 m × 3.4 m × 90 cm deep) Cnc Tk	Mar-Apr	24-27	974	Tucker et al., 1996
Epinephelus tukula	Potato bass	Taiwan	860-m^3 (750 m^2 × 1-1.3 m deep) Cnc Pd		30		Lin, p.c., 1992
Hypoplectrus gemma	Blue hamlet	Florida	380-L conical FRP Tk	Apr-Jun	~28	~750	Domeier, 1994; Domeier, p.c., 1998[c]
		Massachusetts	770-m^3(12.2-m diam. × 7.0 m deep) Aq	Apr-Jun	24		Martel, p.c., 1994
Hypoplectrus indigo	Indigo hamlet	Massachusetts	770-m^3(12.2-m diam. × 7.0 m deep) Aq		24		Martel, p.c., 1994
Hypoplectrus nigricans	Black hamlet	Massachusetts	770-m^3(12.2-m diam. × 7.0 m deep) Aq		24		Martel, p.c., 1994
Hypoplectrus puella	Barred hamlet	Massachusetts	770-m^3(12.2-m diam. × 7.0 m deep) Aq		24		Martel, p.c., 1994
Hypoplectrus unicolor	Butter hamlet	Florida	380-L conical FRP Tk		~28	~750	Domeier, 1994; Domeier, p.c., 1998[c]
		Massachusetts	770-m^3(12.2-m diam. × 7.0 m deep) Aq		24		Martel, p.c., 1994
Mycteroperca bonaci	Black grouper	Florida	21,200-m^3 (61.9 m diam., 8.0 m deep) donut-shaped Aq	Feb	26	~800	White, p.c., 1993
Mycteroperca microlepis	Gag	Florida	21,200-m^3 (61.9 m diam., 8.0 m deep) donut-shaped Aq	Feb	26		White, p.c., 1993
Mycteroperca venenosa	Yellowfin grouper	Bahamas	304-m^3 (2.7 m wide, 2.7 m deep) donut-shaped Aq				Miller, p.c., 1992[a]
		Bermuda	5.5-m^3 (2.0 m × 2.3 m × 1.2 m deep) Cnc Aq	Jun-Jul	28-30		Gray, p.c., 1993
Paralabrax maculatofasciatus	Spotted sand bass	Mexico	24-m^3 Tk	All but May,Oct	20-25	848	Avíles-Quevedo et al., 1995
Plectropomus areolatus	Squaretail coraltrout	Saudi Arabia	25-m^3 (4 m diam. × 2 m deep) FRP Tk	May-Jun	29-32	740	Hassa & Carlos, 1994
Plectropomus leopardus	Leopard coraltrout	Japan	200-m^3 (10 m × 10 m × 2 m deep) Tk	May	23-24	875	Masuma, 1989; Teruya et al., 1992; Masuma et al., 1993
		Taiwan	Pd	May-Oct	28-31	790	Chen et al., 1991a
Serranus subligarius	Belted sandfish	Florida	190- & 380-L aquaria				Clark, 1959, 1965
		Florida	28-L Aq	Feb-Jul Sep-Dec	24-27	662	Tucker & Woodward, 1995
Serranus tigrinus	Harlequin bass	Florida	380-L round FRP Tk				Domeier, p.c., 1998[c]
		Texas	300-L (60 cm × 90 cm × 60 cm deep) FRP Tk	Jul	26-27	730	Holt, p.c., 1992

Table 7-7. (Continued)

Common name	Scientific name	Location	Container	Months	Temp. (°C)	Egg di. (μm)	References
			Pseudochromidae—dottybacks				
Pseudochromis flavivertex	Sunrise dottyback	England	0.75-m^2 Aq	Jan-Dec	27		Lubbock, 1975
Pseudochromis olivaceus	Olive dottyback	England	Aq				Lubbock, 1975
Pseudochromis & Ogilbyina spp.	(8 dottybacks)	Puerto Rico	120-L Aq	Jan-Dec			Gardner, 1997
			Grammatidae—basslets				
Gramma loreto	Royal gramma	New York	190-L Aq		26		Rosti, 1967
							Moe, 1992
			Aq				Walch, 1994a
Gramma melacara	Blackcap basslet	Puerto Rico	64-L Aq	Apr-			Addison, 1994a, 1994b
			Plesiopidae—longfins				
Assessor macneilli	MacNeill's mini-grouper	Australia	300-L Aq	Nov-Feb			Allen & Kuiter, 1976
Calloplesiops altivelis	Comet	Netherlands	130-L Aq		26	~1000	Wassink & Brons, 1988
			Apogonidae—cardinalfishes				
Apogon cyanosoma	Goldstriped cardinalfish		150-L Aq		27-28		Thresher, 1984; Burgess, 1987
Apogon maculatus	Flamefish		Small Aq	Apr-May	24		Coleman, 1966
Pterapogon kauderni	Banggai cardinalfish	Indiana	Aq		27		Bates, 1997
			Sillaginidae—whitings, sillagos				
Sillago sihama	Silver whiting	Japan	485-L(99 cm diam. × 63 cm deep) Pls Tk	Jun-Oct	~26-29	686	Kumai & Nakamura, 1977
			Echeneidae—remoras				
Echeneis naucrates	Sharksucker	Japan	4.1-m^3 Aq	Jun-Dec	27-31	2600	Nakajima et al., 1987
			Carangidae—jacks				
Pseudocaranx dentex	Striped jack	Japan	40-100 m^3 Tks	Dec-Jan	15-19		Harada, 1974; Kuronuma & Fukusho, 1984
		Japan	65-m^3 Cnc Tk		22	969	Watanabe et al., 1998a
Selar crumenophthalmus	Bigeye scad	Hawaii	9.5-m^3(5 m diam. × 0.5 m deep) FRP Tk	Jul-Dec	~26?	630	Iwai et al., 1996
Selene vomer	Lookdown	Massachusetts	770-m^3(12.2-m diam. × 7.0 m deep) Aq		24		Martel, p.c., 1994
Seriola dumerili	Greater amberjack	Japan	40-100 m^3 Tks	May-Jun	22-25		Harada, 1974; Kuronuma & Fukusho, 1984

Seriola lalandi	Goldstriped amberjack	Japan	40-100 m^3 Tks	May-Jun	22-25		Harada, 1974; Kuronuma & Fukusho, 1984
		California	Large Aq	Jan-Dec			Kraul, 1985
Seriola mazlatana	Mazatlan yellowtail	Ecuador	100-m^3(5 m × 10 m × 2 m deep) Cnc Tk	Jan-Dec	25-27		Benetti, 1997
Seriola quinqueradiata	Yellowtail	Japan	40-100 m^3 Tks	Mar-Jun	18-20		Harada, 1974; Kuronuma & Fukusho, 1984
Trachinotus falcatus	Permit	Massachusetts	770-m^3(12.2-m diam. × 7.0 m deep) Aq		24		Martel, p.c., 1994
Trachinotus goodei	Palometa	Massachusetts	770-m^3(12.2-m diam. × 7.0 m deep) Aq		24		Martel, p.c., 1994
			Coryphaenidae—dolphins				
Coryphaena hippurus	Dolphin	Hawaii	6.6 m diam. donut-shaped FRP Tk	Jan-Dec	26	1450	Kraul, 1990b, 1991
			Lutjanidae—snappers				
Lutjanus argentimaculatus	Asian red snapper	Thailand	100-m^3(2 m deep) Tk	Oct, Jan			Doi & Singharaiwan, 1993
Lutjanus campechanus	Atlantic red snapper	Texas	30-m^3 round FRP Tk	May-Jun	23-25	800	Arnold et al., 1978
Lutjanus kasmira	Bluestriped snapper	Japan	4.6-m^3(3.3 m^2 × 1.3 m deep) Aq	Aug	22-25	815	Suzuki & Hioki, 1979
Lutjanus stellatus	Whitespotted snapper	Japan	242-m^3(16.2 m diam., 5.8 m wide, 2 m high) donut-shaped Tk	May-Jun	24-26	830	Hamamoto et al., 1992
Pterocaesio digramma	Twin-yellowstriped fusilier	Japan	1134-m^3(27 m × 12 m × 3.5 m deep) Cnc Aq	May-Jul	24-28	840	Yokoyama et al., 1994
			Haemulidae—grunts				
Haemulon sciurus	Bluestriped grunt	Bermuda	5.5-m^3 (2.0 m × 2.3 m × 1.2 m deep) Cnc Aq	May-Jun	25		Gray, p.c., 1994
			Sparidae—sea breams, porgies				
Acanthopagrus latus	Yellowfin sea bream	Japan	40- to 100-m^3 Tks	Oct-Nov	17-21		Harada, 1974; Kuronuma & Fukusho, 1984
		Kuwait	90-m^3 Cnc Tk	Feb-Mar			Higuchi et al., 1980
Acanthopagrus schlegeli	Black sea bream	Japan	40- to 100-m^3 Tks	Mar-Jun	17-21		Harada, 1974; Kuronuma & Fukusho, 1984
		Taiwan	12-m^3(4 m × 2 m × 1.5 m) cement Tk	Feb-Apr	18-20		Chou et al., 1993
Dentex gibbosus	Pink dentex	Canary Islands	6-m^3 rectangular FRP Tk	May-Jun	19-22	956	Fernandez-Palacios et al., 1994
Diplodus vulgaris	Twoband sea bream	Yugoslavia	Aq	Nov	~20	1010	Jug-Dujakovic & Glamuzina, 1988[s]
Pagrus major	New Zealand snapper	New Zealand	Aq				Scott et al., 1993
Pagrus major	Red sea bream	Japan	Cnc Tk		15-22	1000	Ikenoue & Kafuku, 1992
		Japan	50- to 100-m^3(1-2 m deep) rectangular or round Cnc Tk	Mar-Jul	18-20	975	Fukusho, 1991
		Japan	5- to 1000-m^3 Tk (50-m^3 best)	Mar-Jun	17-21		Harada, 1974; Kuronuma & Fukusho, 1984
		Japan	50- to 100-m^3(1-2 m deep) rectangular or round Cnc Tk	Mar-Jul	14-21	845	Watanabe & Kiron, 1995
		Japan	50- to 100-m^3 Tks	Feb-Jun	15-22	845	Foscarini, 1988
Rhabdosargus sarba	Silver sea bream	Japan	40- to 100-m^3 Tks	Mar-Jun	17-21		Harada, 1974; Kuronuma & Fukusho, 1984
		Taiwan	18-m^3(5 m × 3 m × 1.2 m) Cnc Tk	Nov-Mar	14-26	950	Lin et al., 1988, 1989
		Taiwan	100-m^3(7.2 m diam × 2.5 m deep) Cnc Tk	Dec-Mar	14-24	1030	Leu, 1994

Table 7-7. (Continued)

Common name	Scientific name	Location	Container	Months	Temp. (°C)	Egg di. (μm)	References
Rhabdosargus sarba	Silver sea bream	Japan	4.7-m^3(2.0 m diam × 1.5 m deep) Cnc Tk	Apr-Jun	13-21	975	Mihelakakis & Kitajima, 1995
Sparidentex hasta	Sobaity	Kuwait	90-m^3 Cnc Tk	Feb-Mar			Higuchi et al., 1980
		Kuwait	10-m^3(5 m × 2 m × 1.25 m deep) tile-lined Tk	Feb-Apr	20-22		Al-Marzouk et al., 1994
Sparus auratus	Gilthead sea bream	Israel	4- to 20-m^3 Tk	Jan-Dec	~20	950	Zohar et al., 1995
Spondyliosoma sp.			Aq				Wilson, 1958
			Lethrinidae—sweetlips				
Gymnocranius griseus	Wise porgy	Japan	600-m^3 (10 m × 10 m × 6 m deep) Aq	May-Jun	18-26	775	Suzuki & Hioki, 1978
Lethrinus nebulosus	Spangled emperor	Japan	60-m^3(7.4 m × 5 m × 1.6 m deep) Tk	Mar-Apr	21-23	815	Yonamine, 1989
			Sciaenidae—drums				
Atractoscion nobilis	White seabass	California	16.4-m^3(4.6 m diam.) pool	Jan-Jun	15-17	1280	Orhun, 1989
Cynoscion nebulosus	Spotted seatrout	Texas	30-m^3(6 m × 3.3 m × 1.5 m deep) FRP Tk	Jan-Dec	26	770	Arnold et al., 1976; Fable et al., 1978
		Florida	37-m^3 (12.2 m × 3.4 m × 90 cm deep) Cnc Tk	May-Sep	28-32	700	Tucker & Faulkner, 1987
		Florida	12-m^3 round FRP Tk	Jan-Dec	26-28		Capo et al., 1997
Cynoscion xanthulus	Orangemouth corvina	Texas		~Apr-Sep	25-30		Prentice et al., 1989
Equetus acuminatus	Highhat	Florida	91 cm × 91 cm Aq				Straughan, 1968, 1973
Equetus lanceolatus	Jackknifefish	Florida	Aq			~1000	Young, 1980
Pogonias cromis	Black drum	Texas	114-m^3 round Pls Tk	~Mar	21		Garza et al., 1978
Pseudosciaena polyactis	Small yellow croaker	Korea	75-m^3(5 m × 5 m × 5 m) cage	Jun	17-20	1150	Myoung et al., 1994[s]
Sciaenops ocellatus	Red drum	Florida	5.9-m^3(3.0 m diam. × 0.83 m deep) FRP Tk 2.1-m^3(1.82 m diam. × 0.81 m deep) FRP Tk	Jan-Dec	23-25		Roberts, et al., 1978; Roberts, 1990
		Texas	30-m^3(6 m × 3.3 m × 1.5 m deep) FRP Tk 13-m^3 circular FRP Tk	Jan-Dec	24	1000	Arnold et al., 1977; Arnold, 1988; McCarty, 1990; Colura et al., 1991
		Taiwan	265-m^3(15 m diam. × 1.5 m deep) Cnc Pd	Sep-Apr	20-30		Liao et al., 1994
			Chaetodontidae—butterflyfishes				
Chaetodon nippon	Blackloin butterflyfish	Japan	3.6-m^3(2.8 m^2 × 1.3 m deep) Cnc & glass Aq	Mar,Aug-Oct	23-26	720	Suzuki et al., 1980
			Pomacanthidae—angelfishes				

Centropyge acanthops argi, bispinosus, fisheri flavissimus, loriculus	(pygmy angelfishes)	Florida	180-L Aq	Jan-Dec	20-24	625	Bauer & Bauer, 1981
Centropyge argi	Cherub pygmy angelfish	Oklahoma	40-m^3(2.5 m × 9.5 m × 2.5 m deep) U-shaped Aq	Sep	~26	1100	Walker, 1991
			75- to 475-L aquaria		~25		Breeder's Registry Database, 1994
Centropyge bicolor	Bicolor pygmy angelfish	Puerto Rico	75-L Aq		~28		Breeder's Registry Database, 1994
Centropyge bispinosus	Coral beauty	Florida	3.8-m^3 Aq				Bauer & Klaij, 1974 Lobel, 1975
			208-L Aq		25-27	~600	Lamm, 1984
Centropyge ferrugatus	Rusty pygmy angelfish	Japan	1.95-m^3(1.3 m × 2.0 m × 75 cm deep) acrylic Aq	Apr-May	25-27	685	Hioki et al., 1990
Centropyge fisheri	Fisher's pygmy angelfish	South Africa	Aq				Ballard, 1971 Lobel, 1974
Centropyge flavissimus	Lemonpeel						Steene, 1978
Centropyge interruptus	Japanese pygmy angelfish	Japan	1.95-m^3(1.3 m × 2.0 m × 75 cm deep) acrylic Aq	Feb-May	24-26	690	Hioki & Suzuki, 1987
Centropyge loriculus	Flame pygmy angelfish	Maryland	473-L (182 cm × 46 cm × 51 cm) Aq		27	~600	Karanikas, 1989
		Puerto Rico	75-L Aq		~28		Breeder's Registry Database, 1994
Centropyge potteri	Potter's pygmy angelfish	Hawaii	Aq	Jun		<1000	Lobel, 1975
Centropyge tificin (as *tibicen*?)		Maryland	473-L (182 cm × 46 cm × 51 cm) Aq		27		Karanikas, 1989
Chaetodontoplus duboulayi	Scribbled angelfish	Japan	7-m^3(3 m × 1.5 m × 1.5 m) Aq	Mar-May	24-26	945	Arai, 1994
Genicanthus lamarck	Freckletail lyretail angelfish	Japan	5-m^3(3.8 m^2 × 1.3 m deep) Aq	Jul-Nov	21-27	775	Suzuki et al., 1979
Genicanthus melanospilos	Blackspot lyretail angelfish	Japan	800-L(0.9 m^2 × 90 cm deep) Aq	May	24-28	740	Hioki et al., 1982
			266-L Aq		25-27	~1000	Breeder's Registry Database, 1994
Genicanthus semifasciatus	Halfbanded lyretail angelfish	Japan	800-L(0.9 m^2 × 90 cm deep) Aq	Dec-Apr	25-26	800	Suzuki et al., 1979
Pomacanthus arcuatus	Gray angelfish		Small Aq	Feb			Straughan, 1959a, 1959b
Pomacanthus paru	French angelfish	Massachusetts	770-m^3(12.2-m diam. × 7.0 m deep) Aq		24		Martel, p.c., 1994
			Cichlidae—cichlids				
Oreochromis mossambicus	Mozambique tilapia	Malaysia	FRP Tk		27		Subasinghe & Sommerville, 1992
Oreochromis niloticus	Nile tilapia	Thailand	4-m diam. Tk	Jan-Dec	~30		Macintosh & Little, 1995
Oreochromis spilurus		Saudi Arabia	9-m^3 (3 m × 3 m × 1 m deep) wood Tk	Mar-Oct	≥24		Hazza et al., 1989
		Saudi Arabia	9.6-m^3 (3.5 diam × 1 m deep) Cnc Tk	Mar-Oct	≥24		Hazza et al., 1989
		Kuwait	2-m^3(2 m × 2 m × 50 cm deep) Tk	May-Oct	20-31		Al-Ahmad et al., 1988b
Oreochromis urolepis x *O. mossambicus*	Florida red tilapia	Bahamas	34-m^3(7.6 m × 4.5 m × 1.0 m deep) Pls-lined Tk	Jan-Dec	23-31		Ernst et al., 1991; Smith et al., 1991; Watanabe, 1991; Watanabe et al., 1992

Table 7-7. (Continued)

Common name	Scientific name	Location	Container	Months	Temp. (°C)	Egg di. (µm)	References
Pomacentridae—damselfishes							
Abudefduf saxatilis	Sergeant major	Florida	21,200-m^3 (61.9 m diam., 8.0 m deep) donut-shaped Tk	Jan-Dec	24-26	1000^L	Prappas et al., 1991, Andrew, p.c., 1994
		Massachusetts	770-m^3(12.2-m diam. × 7.0 m deep) Aq		24		Martel, p.c., 1994
Amphiprion akallopisos	Skunk clownfish	Germany	350-L epoxy resin Tk	Nov-May	24-26		Neugebauer, 1969
Amphiprion bicinctus	Twoband clownfish						Hackinger, 1967
Amphiprion clarkii	Clark's clownfish						Tanase & Araga, 1975
			200- to 270-L aquaria		25-28	~3500^L	Breeder's Registry Database, 1994
Amphiprion ephippium	Red saddleback clownfish	Batavia		Mar-Jun			Verwey, 1930
		Germany	400-L epoxy resin Tk	Dec-May	24-26		Neugebauer, 1969
			90-L Aq		25-28		Breeder's Registry Database, 1995
Amphiprion frenatus	Tomato clownfish		75- to 1135-L aquaria		24-28		Breeder's Registry Database, 1994, 1995
Amphiprion latezonatus	Lord Howe anemonefish		100-L Aq		24	~2000^L	Breeder's Registry Database, 1994
Amphiprion nigripes	Blackfoot clownfish		114-L Aq		26	~2000^L	Breeder's Registry Database, 1994
Amphiprion ocellaris	Common clownfish	Florida	150-L & 208-L aquaria				Moe, 1973, 1992
			200- to 400-L aquaria		25-28		Breeder's Registry Database, 1994, 1995
Amphiprion percula	Percula clownfish	Batavia		Apr-Nov			Verwey, 1930
			170-L(41 cm × 102 cm × 41 cm deep) Aq		28		Mitsch, 1941
		Monaco	50-L & 800-L aquaria	Feb-Aug	~25	2200^L	Garnaud, 1951
Amphiprion polymnus	Saddleback clownfish	France	450-L Aq	Feb-May	25-26	2000^L	Terver, 1971
Amphiprion sebae	Sebae clownfish		75-L Aq		26		Breeder's Registry Database, 1994
Amphiprion spp.	(clownfishes)						Squires, 1971
Chromis cyanea	Blue chromis		75-L Aq		26		Breeder's Registry Database, 1994
Chromis cyanea	Blue chromis	Massachusetts	770-m^3(12.2-m diam. × 7.0 m deep) Aq		24		Martel, p.c., 1994
Chromis insolata	Sunshinefish	Massachusetts	770-m^3(12.2-m diam. × 7.0 m deep) Aq		24		Martel, p.c., 1994
Chromis multilineata	Brown chromis	Massachusetts	770-m^3(12.2-m diam. × 7.0 m deep) Aq		24		Martel, p.c., 1994
Chromis notata	Taillight damselfish	Japan	Aq	Jun-Jul	20-23	800^L	Nakazono et al., 1979
Chromis scotti	Purple reeffish	Massachusetts	770-m^3(12.2-m diam. × 7.0 m deep) Aq		24		Martel, p.c., 1994
Chrysiptera cyanea	Blue devil damselfish	Puerto Rico	75-L Aq				Breeder's Registry Database, 1994
			200-L Aq		24-27		Breeder's Registry Database, 1994

Chrysiptera hemicyanea	Yellowtail blue damselfish	Texas	98-L Aq	Jan	27		Walker & Herwig, 1976; Walker, 1977
Chrysiptera traceyi	Tracey's damselfish		1500-L Aq		22-24	~1000	Breeder's Registry Database, 1995
Dascyllus aruanas	Threestripe damselfish	Israel	150-L Aq	Jun-Jul	25	750[L]	Fishelson, 1964
			Aq		24-27	<500[L]	Walker, 1977
			600-L Aq		24	<1000[L]	Breeder's Registry Database, 1995
Dascyllus melanurus	Blacktail humbug						Kattman & Kattman, 1974; Kattman, 1977
Dascyllus trimaculatus	Domino damselfish	Monaco	100-L Aq	Sep-Apr	24	700[L]	Garnaud, 1957a, 1957b
			30-L Tk	Dec	28-29		Koenig, 1958
			Aq		24-27	<500[L]	Walker, 1977
Microspathodon chrysurus	Yellowtail damselfish	Massachusetts	770-m^3(12.2-m diam. × 7.0 m deep) Aq		24		Martel, p.c., 1994
Pomacentrus caeruleus	Caerulean damselfish		375-L Aq		29	~1000	Breeder's Registry Database, 1995
Pomacentrus coelestis	Blue damselfish						Matsuoka, 1962
Pomacentrus nagasakiensis	Nagasaki damselfish	Japan	60-L acrylic Aq	Jul-Oct	24-29	2300[L]	Honda & Imai, 1973
Premnas biaculeatus	Spinecheek clownfish	Batavia		Apr-Jun			Verwey, 1930
			208- & 567-L aquaria		24-29		Etherington, 1972
			100-L Aq		24	~2000[L]	Breeder's Registry Database, 1994
Stegastes fuscus	Dusky damselfish	Massachusetts	770-m^3(12.2-m diam. × 7.0 m deep) Aq		24		Martel, p.c., 1994
Stegastes leucorus	Whitetail damselfish	New York	1.2-m^3(122 cm × 91 cm × 107 cm deep) Aq	Spring-Aug		850	Breder & Coates, 1933
Stegastes leucostictus	Beaugregory	Massachusetts	770-m^3(12.2-m diam. × 7.0 m deep) Aq		24		Martel, p.c., 1994
Stegastes partitus	Bicolor damselfish	Florida	800-L(2.5 m x 65 cm × 60 cm deep) Aq	Jan-Dec	25-29		Myrberg, 1972
		Massachusetts	770-m^3(12.2-m diam. × 7.0 m deep) Aq		24		Martel, p.c., 1994
Stegastes planifrons	Threespot damselfish	Massachusetts	770-m^3(12.2-m diam. × 7.0 m deep) Aq		24		Martel, p.c., 1994
Stegastes variabilis	Cocoa damselfish	Massachusetts	770-m^3(12.2-m diam. × 7.0 m deep) Aq		24		Martel, p.c., 1994
			Oplegnathidae—knifejaws				
Oplegnathus fasciatus	Striped knifejaw	Japan	40- to 100-m^3 Tks	May-Jul	21-26		Harada, 1974; Kuronuma & Fukusho, 1984
Oplegnathus punctatus	Spotted knifejaw	Japan	40- to 100-m^3 Tks	May-Jul	21-26		Harada, 1974; Kuronuma & Fukusho, 1984
			Cirrhitidae—hawkfishes				
Cirrhitichthys aureus	Golden curlyfin	Japan	9-L(130 cm × 100 cm × 70 cm) Aq space	Sep-Dec	25	765	Tanaka & Suzuki, 1991
			Labridae—wrasses				
Anampses cuvier	Firefish		76-L Aq		25-26		DeBernardo, 1975
Halichoeres poecilopterus	Multicolorfin wrasse	Japan	10-m^3 Cnc Tk	Jun-Sep	21-30	660	Kimura & Kiriyama, 1992, 1993
Tautoga onitis	Tautog	New Jersey	121-m^3(10.6 m × 4.5 m × 3.0 m) Aq	Mar-May	18-21		Olla & Samet, 1977
		Connecticut	12-m^3(10 m × 1.2 m × 1 m) FRP Tk	Jun-Aug	19-22		Perry et al., 1998
Thalassoma bifasciatum	Bluehead wrasse		1.8-m^3 Aq			~600	Zumpe, 1963; Petersen et al., 1992
Thalassoma cupido	Cupido wrasse	Japan	158-L Aq		28		Meyer, 1977

Table 7-7. (Continued)

Common name	Scientific name	Location	Container	Months	Temp. (°C)	Egg di. (μm)	References
			Scaridae—parrotfishes				
Scarus croicensis	Striped parrotfish	Oklahoma	40-m^3(2.5 m × 9.5 m × 2.5 m deep) U-shaped Aq	Sep	~26	1100	Walker, 1991
			Stichaeidae—pricklebacks				
Chirolophis japonicus	Fringed blenny	Japan	Aq		<10	2500	Shiogaki, 1983
Opisthocentrus ocellatus		Japan	390-L(1 m diam. × 50 cm deep) Aq	Dec-Jan	~5	1940	Shiogaki, 1982
Opisthocentrus tenuis		Japan	60-L(59 cm × 29 cm × 35 cm) Aq	Dec-Jan	~4-9	2300	Shiogaki, 1981b
			Anarhichadidae—wolffishes				
Anarhichas lupus	Common wolffish	Norway	4-m^3(2 m × 2 m × 1 m) Cnc Tk	Oct-Feb		5750	Johannessen et al., 1993
		Norway	620-L(1.5 m × 1.5 m × 0.37 m) Tk	Sep-Jul	4-11	5260	Pavlov & Moksness, 1996[s]
Anarhichas lupus marisalbi	White Sea wolffish	USSR	80 cm × 80 cm × 1.5 m high pyramid-shaped box	Aug	12-14		Pavlov & Novikov, 1987[s]
			Opistognathidae—jawfishes				
Opistognathus aurifrons	Yellowhead jawfish		70- to 76-L aquaria		24-28	~2000	Walch, 1993, 1994b
Opistognathus macrognathus	Banded jawfish		Aq				Thresher, 1984
			Trypterygiidae—threadfin blennies				
Trypterygion etheostoma	Snake blenny	Japan	Aq	Mar-Nov		965	Shiogaki & Dotsu, 1973a, 1973b
			Clinidae—scaled blennies				
Acanthemblemaria sp.	(blenny)		Aq				Thresher, 1984
Chaenopsis sp.	(pikeblenny)		Aq				Thresher, 1984
Emblemaria sp.	(blenny)		Aq				Thresher, 1984
Paraclinus marmoratus	Marbled blenny	New York	Aq	Mar			Breder, 1941
			Blenniidae—combtooth blennies				
Aidablennius sphynx	Mediterranean blenny	France	80-L(80 cm × 40 cm × 25 cm) Aq	~Apr-Aug	~14$^+$		Kraak, 1996
Meiacanthus grammistes	Striped blenny		37-L Aq		26	<1000	Breeder's Registry Database, 1994
Meiacanthus nigrolineatus	Blackline blenny	Israel			24	~1000	Fishelson, 1976
			Callionymidae—dragonets				

Callionymus calliste	Ornate dragonet	Japan	Aq	Jul-Aug		600	Takita, 1983
Callionymus lyra	European dragonet	England	Aq	Feb			Holt, 1898
							Hardy, 1959
							Wilson, 1978
Callionymus maculatus	Spotted dragonet						Hardy, 1959
Callionymus richardsoni	Richardson's dragonet	Japan	1.09-m^3(135 cm diam. × 76 cm deep) Pls Aq & 35.9-m^3(4.8 m × 6.8 m × 1.1 m deep) Cnc Tk	May		700	Takita & Okamoto, 1979; Takita, 1980
Repomucenus valenciennei		Japan	1.09-m^3(135 cm diam. × 76 cm deep) Pls Aq & 35.9-m^3(4.8 m × 6.8 m × 1.1 m deep) Cnc Tk	May		690	Takita & Okamoto, 1979; Takita, 1980
			Eleotridae—sleepers				
Nemateleotris decora	Decorated firefish		151-L(91 cm × 46 cm × 46 cm)	Oct-Nov	26		Schiller, 1990
			150-L Aq		26		Breeder's Registry Database, 1994
			Gobiidae—gobies				
Amblygobius bynoensis	Bynoe's goby		340-L Aq		26	~1000	Breeder's Registry Database, 1994
Chriolepis minutillus	Rubble goby						Kerstitch in Thresher, 1984
Coryphopterus personatus	Masked goby						Brandt, 1974
Gobiodon citrinus	Lemon goby	Puerto Rico	75-L Aq		21-32		Breeder's Registry Database, 1994
Gobiodon okinawae	Yellow clown goby	Puerto Rico	75-L Aq		21-32		Breeder's Registry Database, 1994
Gobiosoma evelynae	Sharknose goby	Florida	Aq		20-23	1850^{L}	Colin, 1975
Gobiosoma genie	Genie's cleaner goby	Florida	Aq		26-27	2500^{L}	Colin, 1975
Gobiosoma horsti	Yellowline goby	Florida	Aq		26-30	2450^{L}	Colin, 1975
Gobiosoma oceanops	Neon goby	Florida	1-L jar & 9-L Aq	Feb-	27-29	1000^{L}	Boyd, 1956
			Aq				Burgess, 1987
							Feddern, 1967
		Florida	40-L Aq	Dec-May	28	700×2300	Valenti, 1972
						3300^{L}	Colin, 1975
		Florida	Aq	Jan-Dec		2000^{L}	Moe, 1975
			95-L acrylic Aq		25-26		Musgrave, 1976
			Aq		21-24		Walker, 1977
Gobiosoma xanthiprora	Yellowstriped cleaner goby	Florida	Aq		24-26	1850^{L}	Colin, 1975
Gobius ruthensparri		England	Large Aq	Spring			Cowan, 1938
Lythrypnus dalli	Bluebanded goby						Turner et al., 1969
							Bauer & Bauer, 1974
Mugilogobius abei		Japan	120-L(114 cm × 30 cm × 43 cm) Aq	May		965	Kanabashira et al., 1980
Pterogobius zacalles		Japan	243-L(120 cm × 45 cm × 45 cm) Aq	Apr-May	13-15	4550^{L}	Shiogaki, 1981a

Table 7-7. (Continued)

Common name	Scientific name	Location	Container	Months	Temp. (°C)	Egg di. (μm)	References
Quisquilius hipoliti	Rusty goby						Bauer & Bauer, 1975
Sagamia geneionema	Hairychin goby	Japan	800-L(130 cm × 70 cm × 90 cm) Aq		14	6825[L]	Shiobara et al., 1990
Signigobius biocellatus	Twospot goby						Hudson, 1977
			Ephippidae—spadefishes				
Chaetodipterus faber	Atlantic spadefish	Massachusetts	770-m^3(12.2-m diam. × 7.0 m deep) Aq		24		Martel, p.c., 1994
		Oklahoma	40-m^3(2.5 m x 9.5 m × 2.5 m deep) U-shaped Aq	Sep	~26	1100	Walker, 1991
			370-m^3 Cnc Tk		23	~1000	Breeder's Registry Database, 1995
			Siganidae—rabbitfishes				
Siganus canaliculatus	Pearly spinefoot	Palau	20-cm deep ferrocement Tk	Jul	31		McVey, 1972
		Palau	500-L ferrocement Tk	Apr,Jun-Aug	~30		Bryan et al., 1975
Siganus guttatus	Yellowblotch spinefoot	Philippines	450-L(1 m diam. × 55 cm deep) Pls tank	Jan-Dec	25-30	564	Hara et al., 1986
		Philippines	500-L FRP Tk				Duray & Juario, 1987
		Philippines	500-L FRP Tk		26-28	580	Ayson, 1991
Siganus lineatus	Goldline spinefoot	Palau	7-m^3(3 m diam. × 1 m deep) Cnc Tk	Mar			Bryan & Madraisau, 1977
Siganus rivulatus	Black spinefoot						Popper et al., 1973
Siganus vermiculatus	Vermiculated spinefoot	Fiji	1.2-m^3 Cnc Tk	Nov	~24-28		Popper & Gundermann, 1976
			Scombridae—mackerels				
Euthynnus affinis	Kawakawa	Japan	2200-m^3 circular Tk	Apr	23		Kohno et al., 1994b
Scomber japonicus	Chub mackerel	California	12-m^3(4.6 m diam. × 0.72 m deep) Pls swimming pool	Jun	20		Lasker, 1974
Scomber scombrus	Atlantic mackerel	Scotland	54-m^3(10 m diam.) Cnc Tk	Apr-Jun	~14-15		Walsh & Johnstone, 1992
Thunnus thynnus	Northern blufin tuna	Japan	Cage	Jun-Jul	22-25		Harada, 1980
		Japan	13,500-m^3(30 m × 30 m × 15 m deep) cage	Jul	~27		Takii et al., 1997a
			Bothidae—lefteye flounders				
Paralichthys californicus	California halibut	California	20-m^3(4.9 m diam. × 2.5 m deep) wood Tk	Feb-Sep	11-19		Caddell et al., 1990
Paralichthys lethostigma	Southern flounder	Texas	30-m^3(6 m × 3.3 m × 1.5 m deep) FRP Tk	Dec-Jan	17		Arnold et al., 1977
Paralichthys microps	Chilean flounder	Chile	1-m^3 round Tk	Aug-Jan	13-17	800	Silva, 1994
Paralichthys olivaceus	Hirame	Japan	8 m^3 (3.7 m diam. × 80 cm deep) round Tk	Feb-Jul	13-16		Harada, 1974; Kuronuma & Fukusho, 1984; Saitoh et al., 1991
		Japan	29-m^3(4.6 m diam. × 1.76 m deep) Tk	Nov-Apr	12-18	919	Cho & Yang, 1991

Paralichthys woolmani	Ecuadorian flounder	Ecuador	50-m^3(6 m diam × 1.77 m deep) Cnc Tk	May-Mar	20-21		Benetti, 1997
Scophthalmus maximus	Turbot	France	5- to 40-m^3 sand-bottom Tks	Jan-Dec	14	1080	Person-Le Ruyet et al., 1991
		England	40-m^3(5.1 m diam. × 2.7 m deep) FRP & Cnc Tk	May-Jul	11-16	1035	Bromley et al., 1986
		Denmark	>40-m^3 Tk	Jan-Dec			Bye & Htun-Han, 1978, 1979; Paulsen, 1987
		France	17-m^3 & 45-m^3 sand-bottom Tks	Jan-Dec	10-15	1059	Devauchelle et al., 1988
		Europe	Various Tks	Dec-Oct	13-14		McEvoy, 1989
			Pleuronectidae—righteye flounders				
Hippoglossus hippoglossus	Atlantic halibut	Norway	Large Tk	~Dec-Mar	5-7	3275	Haug, 1990
Hippoglossus stenolepis	Pacific halibut	Washington	23.6-m^3(5 m diam. × 1.2 m deep) Tk	Feb-Mar	~8-10	3030[+]	Liu et al., 1991a[s]
		Br. Columbia	13.5-m^3(3.5 m diam. × 1.4 m deep) wood Tk	Feb-Apr			Jensen et al., 1992[s]
Pleuronectes americanus	Winter flounder	Rhode Island	1.9-m^3(1.2 m diam. × 0.8 m deep) Aq				Smigielski, 1975a
Pleuronectes cornutus	Ridged-eye flounder	Japan	Tk				Kitajima et al., 1988
Pleuronectes platessa	European plaice	England	Large Aq	Spring			Cowan, 1938
		Russia	1- to 3-m^3 Tk		1-4		Zhuravleva, 1995
Verasper moseri	Barfin flounder	Japan	40-m^3 Cnc Tk	Mar-May	6		Koya et al., 1994
			Soleidae—soles				
Solea solea	Common sole	England	5.3-m^3(3.7 m × 1.8 m × 0.8 m deep) Tk	Apr-Jun	12-17	1300	Houghton et al., 1985
		Spain	300-L Tk	Feb-Mar	17-19		Ramos, 1986b
		France	7- to 17-m^3 circular Tks	Mar-Nov	10	1390	Devauchelle et al., 1987
		Europe	0.15- to 25-m^3 rectangular and circular Tks	Mar-Jul	8-17	~1300	Baynes et al., 1993
			Balistidae—leatherjackets				
Balistes vetula	Queen triggerfish	Florida	$21{,}200\text{-m}^3$ (61.9 m diam., 8.0 m deep) donut-shaped Aq	Mar-Dec	24-26		White, p.c., 1992
		Massachusetts	770-m^3(12.2-m diam. × 7.0 m deep) Aq		24		Martel, p.c., 1994
Brachaluteres ulvarum		Japan	54-L(60 cm × 30 cm × 30 cm) Aq	Jun	19-22	820	Akagawa et al., 1995
Canthidermis sufflamen	Ocean triggerfish	Massachusetts	770-m^3(12.2-m diam. × 7.0 m deep) Aq		24		Martel, p.c., 1994
Melichthys niger	Black durgon	Florida	$21{,}200\text{-m}^3$ (61.9 m diam., 8.0 m deep) donut-shaped Aq	Mar-Dec	24-26		White, p.c., 1992
Oxymonacanthus longirostris	Orangespot filefish		37-L Aq		22-26	~1000	Breeder's Registry Database, 1994
Paramonacanthus japonicus	Striped filefish	Japan	60-L Aq	Aug or Sep		560	Nakazono & Kawase, 1993
Sufflamen fraenatus	Bridled triggerfish	South Africa	Public Aq				Ballard, 1971
		Japan	11-m^3(4.3 m × 1.6 m × 1.6 m) Aq	May-Aug	25-28	~600	Kawabe, 1984
			Tetraodontidae—puffers				
Canthigaster rivulata	Sharpnose puffer	Japan	Aq			630	Arai & Fujita, 1988
Ciclichthys orbicularis	Orbiculate burrfish	France	Aq	Jun			Conde, 1986

Table 7-7. (Continued)

Common name	Scientific name	Location	Container	Months	Temp. (°C)	Egg di. (μm)	References
Diodon holocanthus	Balloonfish	California	113-L Aq	May	26		Wolfsheimer, 1957
		Japan	1.9- to 2.7-m^3 aquaria	May-Aug	24	1740	Sakamoto & Suzuki, 1978
Sphoeroides maculatus	Northern puffer (puffer)						Welsh & Breder, 1922
							Schreitmuller, 1930a, 1930b
Takifugu rubripes	Tiger puffer	Japan	40- to 100-m^3 Tks	Apr-May	17-20		Harada, 1974; Kuronuma & Fukusho, 1984
Tetraodon cf. *cutcutia*	Ocellated blowfish	Germany	(65 cm × 25 cm × 18 cm deep) Aq	May	26	1000	Cohn, 1912

[a] Spawning rush witnessed, but eggs not collected.
[b] Ovulation occurred naturally within a few days after ripe wild fish were transferred to the tanks. Because tanks were small, normal spawning did not occur, but eggs and milt were released voluntarily, resulting in nearly 100% fertilization.
[c] Fertilized eggs were produced within a few hours after ripe adults were placed in the tanks.
[L] Egg length is given.
[s] Fish were strip-spawned.
Aq = aquarium. Cnc = concrete. FRP = fiberglass-reinforced plastic. Pd = pond. Pls = plastic. RW = raceway. Tk = tank. *l* = larvae.
p.c. = personal communication.

Table 7-8. Examples of fish other than salmonids that have been caught near the time of ovulation and strip-spawned or whose eggs were netted.

Scientific name	Common name	Location	Months	References
Anchoa mitchilli	Bay anchovy	North Carolina	Aug	Tucker, unpub.
Chanos chanos	Milkfish			Chaudhuri et al., 1978
Gadus morhua	Atlantic cod	Norway	Mar-Apr	Kjørsvik & Lønning, 1983
Hypoatherina bleekeri		Japan	May-Aug	Takita & Nakamura, 1986
Menidia menidia	Atlantic silverside	Rhode Island	May-Jul	Bengtson et al., 1987
Centropomus undecimalis	Common snook	Florida	~Jun-Sep	Roberts, 1987
Lates calcarifer	Barramundi	Australia	Nov-Mar	Barlow et al., 1996
Lates calcarifer	Barramundi	Thailand	Apr-Sep	Maneewong, 1987a
Epinephelus striatus	Nassau grouper	Bahamas	Dec-Feb	Tucker & Woodward, 1996
Plectropomus leopardus	Leopard coraltrout	Australia	Oct	Rimmer et al., 1994b
Archosargus probatocephalus	Sheepshead	Florida	April	Tucker & Barbera, 1987
Cynoscion regalis	Weakfish	Delaware	May-Aug	Goshorn & Epifanio, 1991
Holacanthus tricolor	Rock beauty	Florida	~Feb-Jun	Moe, 1997b
Liza haematocheila	Redeye mullet	Japan	Apr	Kuronuma & Fukusho, 1984
Mugil cephalus	Striped mullet	Taiwan	Nov-Feb	Liao et al., 1972
Tautoga onitis	Tautog	Connecticut	Jun-Jul	Perry, 1994
Thalassoma bifasciatum	Bluehead wrasse	St. Croix	Jan-Dec	Warner & Swearer, 1991; Petersen et al., 1992[N]
Sparisoma rubripinne	Redfin parrotfish	St. John	Jan-Dec	Randall & Randall, 1963
Siganus rivulatus	Black spinefoot	Israel	Apr-Aug	Popper et al., 1973
Katsuwonus pelamis	Skipjack tuna	California	Oct-Nov	Kaya et al., 1982; Hunter et al., 1986
Thunnus albacares	Yellowfin tuna	Japan	Jul	Kuronuma & Fukusho, 1984
Paralichthys californicus	California halibut	California	Winter-spring	Gadomski & Caddell, 1991
Glyptocephalus cynoglossus	Witch flounder	New Hampshire	Jun-Jul	Bidwell et al., 1997
Pleuronectes americanus	Winter flounder	Rhode Island	Feb-Apr	Buckley et al., 1991
Pleuronectes yokohamae	Marbled flounder	Japan	Jan	Kuronuma & Fukusho, 1984
Canthigaster rivulata	Sharpnose puffer	Japan	Jun-Sep	Arai & Fujita, 1988
Takifugu rubripes	Tiger puffer	Japan	Mar-May	Kuronuma & Fukusho, 1984

[N] net collection.

Medium-term Captive Broodfish.—If caught before or early in the spawning season, handled carefully, and kept under spawning conditions, some species will quickly adapt to captivity and spawn voluntarily within days or weeks. Examples are black sea bass, dolphin, and spotted seatrout. As for induced ovulation, the fish should be kept under conditions similar to natural ones at peak spawning.

Long-term Captive Broodfish.—Some species need more time to adapt to captivity, possibly months to years. Broodfish of some species well-established in aquaculture often are raised from eggs. Usually, barramundi have to be raised in captivity from a small size. Examples of spawning tanks and a cage are shown in Figs. 7-20, 7-21, 6-10, 6-21.

Figure 7-20. Round concrete spawning tanks at the National Institute for Coastal Aquaculture, Songhkla, Thailand. They hold 100-150 m^3 (10 m diam. × 1.3-2 m deep). This size tank seems to work better than narrower ones, maybe because the fish have more space to swim horizontally.

Figure 7-21. Corner of a cage used for rearing and spawning barramundi/sea bass and groupers at the Singapore Department of Primary Industry experimental cage farm (see Figs. 6-20, 6-21). When spawning is expected, the white liner of 500-μm mesh plankton net material is inserted between the cage and broodfish to contain the eggs.

Successful long-term conditioning depends on holding fish under ambient conditions, simulated ambient conditions, or an artificial cycle that works. Voluntary spawning can result from manipulation of environmental factors, especially temperature and daylength (Scott, 1979; Bye, 1984, 1987; Wooton, 1982). Whether increases or decreases in these factors stimulate phases of gonadal maturation and spawning usually is a reflection of natural conditions during those processes. The length of the conditioning procedure might be months or years. Other possible factors include lunar cycle, salinity, water chemistry, water level, current, habitat structure, suitable spawning substrate, sex ratio, and stocking density (Lam and Munro, 1987). Behavioral interaction and pheromones could be important (Liley et al., 1987). Some fish spawn over a wide temperature range and others over a narrow one (Table 7-7). There is somewhat of a trend for temperate species to depend more on daylength as a stimulus and tropical species more on temperature. Whether daylength is increasing or decreasing might be more important than the actual length (Randall et al., 1987). Daylength is a major factor governing the time of salmonid reproduction (Zohar, 1989), but it seems to be a less critical stimulus than temperature for some fish such as striped bass (Smith and Jenkins, 1987), tilapias (Zohar, 1989), and Nassau grouper (Tucker et al., 1996). European sea bass, gilthead sea bream, striped mullet, and red drum are stimulated to reproduce by decreasing temperature and daylength (Zohar, 1989). Manipulations of these two factors can be used to advance, delay, extend, or compress spawning cycles. Rainbow trout held under a compressed seasonal daylength cycle matured and spawned 102 d (20 months old) before the control fish held under a normal cycle (Bromage, 1987). Tidal movements (and thus lunar phase) could influence intertidal spawners (e.g., killifish, silversides, grunion, some puffers), because of the need for placing eggs at the best sites, and reef fish (e.g., groupers), possibly for synchronization of spawners or for rapid dispersal of eggs away from predators (Gibson, 1992).

Voluntary spawning has occurred mostly with well-fed uncrowded fish during the natural spawning season under conditions of ambient temperature and partial or total natural light. Fish are adapted to spawn in response to a standard set of environmental cues appropriate for that species or population (Wooton, 1982). Conditioning fish to spawn in captivity requires that the cues that really matter are present in the right combination, degree, and order. This is not too difficult if natural spawning conditions can be analyzed. Broodfish from different populations might respond differently to environmental cues.

Care of Broodfish

The capture method should be as gentle as possible (Safford and Thomas, 1987). The usual gear includes seines, trammel nets, hand nets, traps, hook and line, and electroshockers. Gill nets and trawls injure fish and should be avoided for most. Tough fish like sturgeon sometimes are caught by gillnetting or snagging. Some, like yellowtail snapper and striped mullet will not tolerate much handling, but others like turbot and Atlantic halibut can be strip-spawned a few times a week without affecting egg production (McEvoy and McEvoy, 1992). Immediately after capture, broodfish should quickly be put in well-aerated or oxygenated water. The water should be filtered or exchanged, the frequency depending on transport time, temperature, fish density, and fish sensitivity. If they are to be kept for more than a few days, the fish should be given plenty of room and low light (shade) and protection from unecessary disturbance.

Broodfish should be given plenty of a complete pelleted diet or a variety of high-quality freshly-frozen natural foods supplemented with vitamins and maybe essential fatty acids (see chapter 9 and Watanabe, 1988; Navas et al., 1995, 1997; Sargent, 1995; Watanabe and Kiron, 1995). More food often results in more eggs, and important nutrients can include essential fatty acids, protein, vitamins C and E, minerals, and pigments (Kjørsvik et al., 1990; Bruce et al., 1993; Chou et al., 1993; McEvoy et al., 1993; Rainuzzo, 1993; Duray et al., 1994; Cerdà et al., 1995; Peleteiro et al., 1995; Zohar et al., 1995; Bell et al., 1996;

Evans et al., 1996; Sheikh-Eldin et al., 1996; Verakunpiriya et al., 1996, 1997b; Fernandez-Palacios et al., 1995, 1997). Good nutrition is especially important during early vitellogenesis (Buddington et al., 1987).

With species that are protogynous hermaphrodites, sex control (Yamazaki, 1983) sometimes is considered necessary to ensure an optimal male:female ratio for maximizing egg production. If males are difficult to collect or do not occur in reared broodstock until years after the fish mature as females, hormone-induced sex-reversal of females is an option (e.g., bluespotted grouper, Kuo et al., 1988; greasy grouper, Chao and Lim, 1991). If too many females change to males, suppression of sex-reversal by administering female hormones such as estradiol or by maintaining enough large dominant males in the tank has been accomplished (e.g., redspotted grouper, Maruyama et al., 1994). With species that are simultaneous hermaphrodites, such as belted sandfish, mature fish of similar size can be paired, unless territoriality and aggression cause excessive conflict (Tucker and Woodward, 1995).

Two important thoughts are: Optimal conditions for reproduction might differ from those for growth. Reproduction is the life function most sensitive to stress (Gerking, 1982; Safford and Thomas, 1987; Sumpter et al., 1987).

Egg Collection

Planktonic eggs can be collected by a surface overflow into a smaller tank with a screened drain. They can be skimmed from the surface with beakers or buckets (after crowding with a net, if necessary). Milkfish and grouper eggs spawned in ponds have been collected in a neuston net placed downstream from a paddlewheel (Lin, 1985, p.c., 1992).

Fish that build nests might require substrate material, such as PVC pipe fittings for damselfish and gravel or crushed shell for triggerfish. With fish that scatter adhesive eggs, if the bottom of the tank is covered with mats, the eggs will be easier to move.

Induced versus Natural Ovulation

Predicting the time of spawning is useful in scheduling hatchery operations. With natural ovulation, the day might not be predictable, but time of day usually is. With induced ovulation, both the day and the time of day often are predictable.

Egg quality could vary with type of spawning, condition of broodstock, and season. Eggs obtained by induced ovulation sometimes are smaller than naturally ovulated ones, but egg diameter also can vary with female size or age, location, physical conditions, and time of the season. Larger eggs might have more energy reserves and therefore higher viability, or they might only have higher moisture content (e.g., wider perivitelline space). Other factors, such as larval culture conditions, might be more important than small differences in egg quality; however, the viability of artificially spawned eggs is more likely to fluctuate. It is best to minimize handicaps, when possible. With captive broodfish, good nutrition, especially proper quality and quantity of fat eaten by the females, is important to help ensure high egg quality (Watanabe, 1985).

Stripped eggs are more easily processed than tank-spawned eggs because they are more concentrated. They are cleaner and do not have to be washed as thoroughly to remove potential pathogens.

Many groupers, jacks, snappers, sea breams, drums, and rabbitfishes will spawn voluntarily, but hormones sometimes are used to control the spawning schedule. Most cultured flatfish species easily spawn without hormones, but some usually need them.

Hormone-induced tank-spawning has been used for striped bass, groupers, snappers, striped mullet, and others. If it works, good eggs are produced with a minimum of problems. The main danger is that some types of fish are more likely to retain ovulated eggs until they become overripe. Sometimes the membrane holding eggs in does not thin in pace with the ovulation process and become fragile enough to break when it should.

BROODFISH ADAPTABILITY

Relatively sedentary species seem to adapt better to captivity and tolerate handling better than more active species. Squaretail coraltrout are easily disturbed by activity near the holding tank and often dash against the tank walls. Although spawning might be straightforward, such broodfish are more subject to injury and juveniles might be more difficult to rear. Dolphins are very active (especially males) and tend to swim in a straight line instead of following the curvature of the tank, sometimes resulting in severe collisions with tank walls. A donut-shaped tank keeps the fish swimming more nearly parallel with the walls and helps prevent collisions. Other constantly-swimming pelagic fish such as tuna and mackerels also would be safer in such tanks. Calmer species rest on or off the bottom and either quietly watch or ignore activities above.

Fish that depend on conservation of food in a guarded territory (sheepshead, Nassau grouper) tend to be more aggressive than others that do not (e.g., common snook). Damselfish tend to be very aggressive, especially when guarding a nest. Schooling species (e.g, spotted seatrout) normally are very unaggressive. Such behavior must be taken into account when setting up broodfish tanks. Groupers will spawn in pairs or groups. Belted sandfish and damselfish usually spawn only in pairs and resent intruders. Spotted seatrout spawn better in groups. Schooling pelagic fish like herrings, anchovies, dolphins, and tunas might spawn in pairs or small groups but do not move far from the school to do so.

Tolerance to handling varies greatly. Nassau groupers are much more docile than the closely related red hinds. Some species, such as Nassau grouper, have a large gill area and can obtain oxygen from the air as long as the gills are moist. Some species, such as Nassau grouper, have very thick, tough skin and tolerate handling well, while others, such as creolefish and yellowtail snapper have very thin soft skin and suffer much from handling. Snook tolerate low water quality (dissolved oxygen <1 mg/L), but lose scales easily when handled. Red drum and spotted seatrout spawn very easily in captivity, but are likely to get skin infections when handled.

EXAMPLES

Wild or reared **sturgeons** usually are induced to ovulate with CP, SP, or GnRHa, and the eggs are surgically removed (Conte et al., 1988; Williot, 1990).

Three- to 4-yr-old reared **Japanese eels** have been induced to mature and ovulate by long-term treatment with various hormones (Table 7-3; Lin et al., 1991b; Chiba et al., 1994; Kagawa et al., 1995, 1997, 1998; Ohta et al., 1997). Because of the complexity of the spawning cycle, a fair amount of work will be needed before routine spawning of eels is possible.

Milkfish have spawned voluntarily in tanks and cages, but not always predictably. They are very sensitive to handling and do not seem to spawn as well in tanks as in cages and ponds, which provide more natural surroundings. A minimum cage diameter of 6 m seems necessary (Harvey and Carolsfeld, 1993). Marte et al. (1984) described rearing and management of milkfish broodstock, including construction of 10-m diameter and 10-m square spawning cages with egg collectors. Liu and Kelley (nd.a) recommended stocking 3-5 kg or larger milkfish at 1 fish/2 m^2 in 100- to 300-m^3 tanks or ponds. Hormones can be used to induce milkfish to spawn at an earlier age (e.g., by 2 yr or more) than they would voluntarily. For best results with GnRHa, initial oocyte diameter should be relatively large (Table 7-4).

Traditionally, captured or reared **salmon** and **trout** have been stripped after natural ovulation occurs, but hormones (e.g., SP, GnRHa) have also been used to accelerate the process. Northern hemisphere salmonids usually spawn in fall or winter when daylength is decreasing or short (Bromage, 1995; Table 7-2). In salmon, a sequence of constant long days abruptly followed by short days can accelerate spawning, and the reverse can delay it, by as much as 3-4 months (Bromage, 1995). This is a simplification of the gradually changing artificial cycles that previously were used to shift spawning times. So far, photoperiod manipulation has been more effective than hormone treatment in altering salmonid spawning cycles (Donaldson, 1986; Fast et al., 1991; Bromage et al., 1992a, 1992b, 1993b). However, hormones are useful for short-term acceleration and synchronization of ovulation and spermiation (Langer and Zohar, 1992; Taranger et al., 1992; Swanson et al., 1997).

As much as 2000 kg of **Atlantic cod** can be stocked in a 175-m^3 plastic spawning pen (Huse, 1991). The natural season typically is 2.5 months long.

In the scorpionfish **kurosoi**, fertilization is internal and the young are born as long as 12 mo after copulation (Jeon, 1995). Pregnant wild females can be caught and held, or reared males and females will reproduce without hormone treatment. Highest numbers of healthy larvae are obtained from 4- to 5-yr-old females. In Korean waters, kurosoi copulate sometime during September-November and deliver well-developed larvae sometime during March-June. In Japan, copulation probably occurs during November-January (Yamada and Kusakari, 1991).

Barramundi are tank- and strip-spawned with and without hormones, but common snook have spawned without stripping only after hormone treatment. Under ambient conditions, voluntary spawning of barramundi occurs near quarter moon phases and can continue for at least 4-5 months (Toledo et al., 1991; Garcia, 1994). They will spawn in cages or tanks; density of the 3-kg or larger fish should be less than 80 fish/75-m^3 cage (Lim et al., 1986). By 5 yr, most will be females. GnRHa pellets can accelerate spawning from July-August to April-May. Females with oocytes of ≥400 μm diameter will spawn after GnRHa injection or implantation. Although barramundi have been conditioned to spawn voluntarily in any month, hormone treatment seems more necessary as distance from the equator increases (Barlow et al., 1996).

European sea bass can be conditioned to spawn any month by expanding or compressing daylength cycles or using long (15L:9D) and short (9L:15D) days (Bye, 1987; Carrillo et al., 1993; Bromage, 1995). Usually, they are kept on a 12-month displaced spawning cycle. Natural spawning season usually is December to March in the Mediterranean region and June in Ireland (Carrillo et al., 1993). Captive broodfish are kept at 8-17°C, but spawning usually occurs at 13-15°C. Fish density can be 3-4 kg/m^3, and tank size probably should be at least 20-m^3 to contain enough broodfish (Barnabé, 1990e). For maturation prior to GnRHa treatment, Coves et al. (1991) recommended a sex ratio of 1 male:2-3 females at a density of 4-13 kg/m^3 in 12-m^3 tanks (males 0.7-2.5 kg, females 1-4 kg). Twenty percent (oldest fish) are replaced every year after the spawning season. Merely reducing temperature to less than 17°C can accelerate spawning by 3 months (Carrillo et al., 1995). Although HCG works reasonably well for European sea bass, better eggs can be obtained by giving females with 650-μm oocytes two intraperitoneal injections of 5 and 10 μg GnRHa/kg 4-6 h apart, which typically results in spawning after 72 h (Carrillo et al., 1995).

Hybrid striped bass usually are produced from strip-spawned eggs because voluntary spawning of the (different-sized) parent species is not always reliable. **Striped bass** have been conditioned to mature their eggs in captivity (then were injected with HCG and strip- or tank-spawned) with the cycle compressed to 6 months, but fertilization rate was variable (Blythe et al., 1994, and contained references). Woods and Sullivan (1993) reported that reared striped bass containing oocytes too immature to respond to HCG could be induced to tank-spawn with GnRHa pellets. Hatchery-reared striped bass and white bass females have been induced to ovulate (Woods et al., 1992; Mylonas et al., 1996; Smith et al., 1997). **Palmetto bass** and **white bass** have been conditioned to mature 2-3 months early (then were injected with HCG and strip-spawned)(Kohler et al., 1994). Palmetto bass have spawned voluntarily in tanks, but the defect rate of hybrid offspring usually is high (Jenkins and Smith, 1985).

Many species of **grouper** and **sea bass** will spawn voluntarily in captivity, and hormone-induced ovulation of ripe wild or captive fish also is reliable (Tucker, 1994). One male and four female wild **Nassau groupers** spawned after 8 months in a 26-m^3 cage (Tucker et al., 1993b), and two males and four females (initial numbers) spawned after 15, 27, and 28 months in a 37-m^3 concrete raceway (Tucker et al., 1996). In both cases, spawning occurred after normal spawning temperatures were reached (for the raceway fish this was 3-4 months later than in their home range). If the interval between capture and attainment of those conditions was shorter, the fish probably would have spawned sooner than 15 months. Nassau groupers reared from eggs were mature at 5-6 yr, but because temperature in their raceway could not be controlled and fluctuated widely, they did not spawn immediately; in April 1998, temperature was more stable and the 8-yr-old groupers spawned. In the Philippines, one female estuary grouper with two males in a 48-m^3 cage spawned 5-10 times a month for 4 months (Toledo et al., 1990, 1994; Toledo, pers. comm., 1992). During December 1989 to October 1990, ten female and ten male brownmarbled groupers in a 75-m^3 cage spawned 2-5 times during each of nine periods of 2-6 d, usually starting between the last quarter moon and new moon (Lim et al., 1990). Forty female and nine male orangespotted groupers in a concrete tank spawned almost continuously for 50 d during April to June 1976 (Hussain and Higuchi, 1980). A pair of belted sandfish in a 28-L aquarium spawned 79 times in 81 d during February to May 1991 (Tucker and Woodward, 1995). Eight female leopard coraltrout in a pond produced eggs 110 times during May to October 1989 (Chen et al., 1991a).

Yellowtails will spawn voluntarily and can be induced to ovulate, but most are reared from wild juveniles because hatchery production is insufficient. Captive **Mazatlan yellowtails** in a 100-m^3 tank spawned for about 21 months (Benetti, 1997).

At 26°C, female **dolphins** will spawn every other night for extended periods, but can eventually become physically depleted (Kim et al., 1993). Broodfish should be kept in a large donut-shaped tank (e.g., 6-m diameter outer wall and 3-m diameter inner wall) to minimize injury from collisions with the tank wall. Usually two females and one male are stocked per tank (aggression among males can lead to injury and death).

At least five species of **snapper** have spawned voluntarily in tanks (Table 7-7), but hormone-induced ovulation also is practiced (Table 7-3). **Golden snappers** (≥2.3 kg) can be stocked at 30-50 fish/75-m^3 cage (Lim et al., 1985). Eight Caribbean red snappers (≥50 cm FL) spawned in a 30-m^3 tank 20 months after capture (Arnold et al., 1978).

Sparids tend to spawn readily in captivity. **Gilthead sea bream** can be conditioned to spawn any month (Bye, 1987). Temperature is kept at 16-20°C, and females ovulate when daylength is quickly shortened to 9 h and then maintained. In Israel, the spawning season for gilthead sea bream can be shifted to any time of the year by manipulation of temperature and daylength (Zohar et al., 1995). Because egg quality was found to suffer when the cycle was compressed, a 12-month cycle is preferred. Typically, 2- to 6-yr-old broodfish (1:1 sex ratio) are kept in 4- to 20-m^3 tanks at a density of 10-15 kg/m^3 and temperature range of 16-21°C. About 3 weeks after the shortest day, they are given intramuscular injections of GnRHa in 3-mm pellets or 50- to 150-μm microspheres (Langer and Zohar, 1992). Beginning 2-3 d later, they spawn every day for up to 4 months. With a 12-month cycle, consistently high-quality eggs can be obtained every day of the year. Progress also has been made in shifting the spawning season of **red sea bream**, which normally spawns every day for up to 2 months (Matsuyama et al., 1993). Copolymer pellets containing 100 μg GnRHa (72 μg/kg) induced spawning for several days 5-6 months before (Nov-Dec) the natural season (Apr-Jun), beginning 18 d after implantation. Voluntary spawning typically is accomplished by moving broodfish from cages to 50- to 100-m^3 tanks 4-6 weeks before spawning is expected (Fukosho, 1991; Watanabe and Kiron, 1995). The same number of 3- to 7-yr-old males and females are stocked at 0.7-1.5 fish/m^3. The natural season lasts for 60-80 d.

Sciaenids spawn easily in captivity. **Red drum** can be put through a compressed year of temperature and photoperiod changes and made to spawn every 3-4 months at a rate of 10-20 times a month (Roberts, 1990; Thomas et al., 1995). Arnold (1988) reported that two females and two males spawned regularly (360 times) for 71 months. They seem to spawn best at 25-26°C with a constant daylength of 12-13 h. Usually two or three males and two or three females (8-18 kg, references in Table 7-7) are stocked in a 6-m^3 or larger tank. Jenkins et al. (1997a) reported that on a cycle of 25°C for 4 d and 21°C for 3 d (13 h light), red drum spawned every week for 14 months. Spawning began the day after temperature increased and ended when temperature dropped. **Spotted seatrout** can begin to spawn within 2 weeks after capture and will spawn year-round (Arnold et al., 1976; Tucker and Faulkner, 1987). Although not necessary for obtaining eggs from captive sciaenids, hormone treatment has been used to induce ovulation in wild fish and to synchronize spawning of captive ones (Thomas and Arnold, 1993).

With high temperatures and long days, **tilapias** will reproduce year-round in almost any reasonable container (Little et al., 1993; Little, 1994; Macintosh and Little, 1995).

Striped mullet have not been reported to spawn voluntarily in captivity, but hormone-induced tank-spawning is standardized (Kuo, 1995). Adult mullet, like milkfish, are very sensitive to handling and can quickly develop external ulcers and bacterial septicemia. The season can be accelerated or delayed so that captive broodfish can be induced to spawn during any month. A 6-h day and 21-22°C were found to be optimal for maturation (Kuo, 1995). At slightly lower (17-20°C) or higher (24°C) temperatures, females can maintain vitellogenic oocytes for as long as 5 months (Kuo, 1995). The best quality eggs can be obtained with the least effort and damage to broodfish by injecting a reared or wild female (having oocyte diameters of 600 μm or more) with CP, then GnRHa, placing it in a tank with one or two reared or acclimatized males and allowing voluntary fertilization to occur (Lee and Tamaru, 1988; pers. obs.).

Female **skipjack tunas** ovulated and released eggs within 65 h after capture, but males did not fertilize them (Kaya et al., 1982). **Northern bluefin tunas** have spawned in a cage (Harada, 1980).

European flatfish tend to spawn in the spring with increasing daylength (Bromage, 1995; Table 7-2). **Turbot** usually are conditioned to ovulate, then are tank- or strip-spawned (McEvoy, 1989; Leclerq, 1994). Turbot in the southern North Sea normally spawn during late May to July (Bye, 1987). Daylength is longer than 15 h and water temperature is in the range 11-16°C. Changes in daylength determine the spawning season. Temperatures below 8°C interfere with gametogenesis and above 16°C reduce egg quality. If the temperature for captive fish rises too quickly, the males could be spent before the females ovulate. So some farmers hold turbot at 10-12°C most of the time and raise them to 15-16°C during the 2 months before spawning. If temperature is constant at 12°C, individuals might not be synchronized. Commercial farms maintain separate turbot broodstocks, usually with each on a 12-month shifted cycle, for year-round egg production. Male (2-7 kg) and female (3.5-8 kg) broodfish typically are stocked in equal numbers at ≤10 kg/m^2 in 5- to 40-m^3 tanks, with 20% (oldest) replaced yearly (Person-Le Ruyet et al., 1991). Normally, a female will spawn 6-8 times over a 2-3 month period. Egg quality in the spring and fall might be a little better. Sometimes tanks next to each other with no barrier have been kept on different light cycles by using submerged lights. Light intensity is relatively low, and dawn or dusk tapering does not seem necessary. Daylength is increased from 6-8 h ("winter") to 16-18 h ("summer"). It probably is best to maintain temperature below 10°C when days are short, increase it slowly to 14°C during the 2-3 months just before spawning, then keep it at 13-14°C during spawning. Turbot usually are strip-spawned when ovulation is detected, especially with off-season broodfish because the artificial conditions often result in variable fertilization with tank-spawning. **Hirame**, the main flatfish species raised in Asia, will spawn voluntarily during most months (Table 7-7). **Atlantic halibut** ovulate in captivity but usually have been strip-spawned (Kjørsvik and

Holmefjord, 1995).

In a review, Baynes et al. (1993) reported that **common sole** broodfish almost always are captured and that they probably do not need sand on the tank bottom. Some eggs can be obtained within a few days after capture, but the first conditioned spawn might be delayed by up to 3 yr while the fish adapt to captivity. Densities are in the range 1-6 fish/m^2 (apparently <5 kg/m^2) and sex ratios 0.5-3 males:1 female. Manipulation of photoperiod and temperature or temperature alone has caused out-of-season spawning; however, the resulting eggs can be of lower quality than in-season ones. Hormone treatment is not needed. Length of the spawning period usually has been in the range 27-111 d. Wild sole populations normally spawn for about 6 weeks during December to May, progressively later going from the English Channel to the North Sea, in a narrow temperature range of 8-12°C (Bye, 1987). Between spawning seasons, a daylength of 13-15 h at a sensitive time might stimulate the beginning of a new cycle, but later, daylength does not seem important. Captive soles have spawned over a wide range of 6.5- to 24-h light/d (decreasing daylength could have an effect). One broodstock kept under a constant 18-h day for 3 yr spawned for more than 6 months. Soles have spawned under artificial light intensity ranging from 20-80 to 1500 lux, with and without tapering. In some hatcheries, soles are kept at 9-16°C; first keeping days short for 4-6 weeks, then increasing them, can result in spawning 22 weeks later.

If kept at the right temperature and photoperiod, some ornamental fish such as **damselfishes**, **gobies**, and **pygmy angelfishes** and baitfish such as **killifishes** will spawn voluntarily year-round in small tanks or aquaria. Damselfish nest building and spawning rituals are fairly predictable (Brinley, 1939; Myrberg et al., 1967; Fishelson, 1970; Jan and Chang, 1984; and references in tables).

GENETIC CONSIDERATIONS

These need to be considered mainly when broodstock are reared from eggs or when stock enhancement is a goal (Wohlfarth and Hulata, 1989; Brannon, 1991; Aquaculture vol. 137, 1995). In either case, inbreeding is to be avoided and genetic diversity is to be maintained. If only wild spawners are used for foodfish production, precautions are not necessary.

Inbreeding occurs when related fish spawn together. It can be used positively in selective breeding programs to develop desired traits, such as deeper color (Bromage, 1988; Chourrout et al., 1988; Wohlfarth, 1990) or faster growth (Murata et al., 1996). Fixation of desirable traits, but not undesirable ones, is the ultimate goal. Fixation is the presence of two identical genes at the same locus. Fish characteristics often result from the effects of several loci, and selective breeding typically attempts to fixate as many alleles as possible to obtain the most pronounced result. On the other hand, inbreeding can increase the incidence of homozygosity of deleterious recessive genes and possibly fixate them as well. This can result in reduced growth, survival, reproductive capacity, or disease resistance, and increased percentage of abnormalities (Tave, 1990a). Ideally, to avoid inbreeding, all spawners must be completely unrelated. The next best situation is pedigreed mating, in which only one son and one daughter per family are chosen as future broodstock. Next best is to choose equal but small numbers of broodstock from each family. Within the second and third cases, the larger the spawning population, the less inbreeding should result. Inbreeding also can be reduced by adding broodstock from other hatcheries (unrelated) or from the wild.

Genetic drift can occur when the number of spawners is small, and it is defined as a change in gene frequencies because of sampling error (Tave, 1990b). When broodfish are chosen from a population of fish, even if they are unrelated, the smaller number selected is likely to have gene frequencies unlike those of the population. Initially, or after further selection and breeding, rarer alleles can be lost. The lost ones might have promoted the ability to survive, grow, resist disease, or reproduce better in the normal environment or under changing conditions, as opposed to relatively constant hatchery conditions.

The idea of genetic engineering to improve fish characteristics—such as growth rate, feed conversion, environmental tolerances, and disease resistance—is growing in popularity (Chourrout et al., 1988; Wohlfarth and Hulata, 1989; Wohlfarth, 1990; Bromage, 1995; Thorgaard, 1995). Caution is required when there is a chance of ecological or genetic damage to natural populations being caused by escaped or released fish. If done carefully to avoid such consequences, this method would have wider application and could be much quicker than breeding programs.

Female salmonids grow faster and mature later than males. Raising all male tilapias will prevent reproduction before harvest, which can lead to stunting. For such purposes, sterile "female milt" (from fish that are genetically female but functionally male) as well as monosex and triploid eggs of trout, salmon, and cyprinids are now produced commercially (Zohar, 1991).

HYBRIDIZATION

Hybridization of marine fish is possible. Resulting offspring could have improved combinations of attributes such as good color, shape, growth rate, feed conversion, and disease resistance (Chevassus, 1983; Chourrout et al., 1988; Wohlfarth and Hulata, 1989; Wohlfarth, 1990; Thorgaard, 1995). Only crosses between closely related species are likely to be viable. Frequently, even when later survival is good, prefeeding survival of hybrids is unusually low because of defects. If egg production is high, this might not be a problem. If and when they reach the feeding stage, survival of hybrids is likely to be intermediate between those of the parent species. Growth rate of hybrids also tends to be intermediate. "Hybrid vigor" actually is more likely when different stocks of the same species are interbred. Still, a faster growth rate from one species could be combined with other qualities from the second to produce a more desirable fish. Some hybrids such as those of striped bass and white bass do exhibit hybrid vigor (Kerby and Harrell, 1990). They survive better, grow faster when young, are more resistant to disease, and generally are hardier than the parent species. They also tolerate higher temperatures than striped bass and can be raised at salinities of 0-30 ppt. Sterile or monosex populations resulting from hybridization could help to optimize growth rates by limiting reproduction in fish like tilapia. The hybrid blue angelfish × queen angelfish has occurred naturally (Feddern, 1968). Some examples of viable artificial hybrids are: angelfish (Moe, 1977), groupers (Tseng and Poon, 1983), and flounders (Kim et al., 1996b).

SEX CONTROL

To obtain the desired male:female ratios for breeding, growing out, or marketing, sometimes the sex of fish is changed by hormone treatments or chromosome manipulation (Johnstone et al., 1978; Yamazaki, 1983; Bye and Lincoln, 1986; Donaldson and Benfey, 1987; Wohlfarth and Hulata, 1989; Zohar, 1989; Pandian and Sheela, 1995; Thorgaard, 1995). Sex control has mostly been applied to carps, salmonids, tilapias, anbantids, topminnows, and killifishes. There are two main classes of sexuality in fish, gonochorism (separate sexes) and hermaphroditism (both sexes in the same individual). Most cultured fish are gonochorists, which fall into two categories: undifferentiated fish have gonads that begin like an ovary, then those of roughly half turn into testes; differentiated fish have gonads that develop directly into ovaries or testes. Hermaphroditism can be simultaneous (female and male), protogynous (female then male), or protandrous (male then female). The sex of differentiated species usually can be controlled only during a certain developmental stage when the sex genes are active, and the result is probably permanent. Sex control of hermaphroditic and undifferentiated fish is more complex because of protracted sex gene activity. For example, groupers changed from female to male by testosterone treatments sometimes revert to female after the treatments stop. Treatment modes include food, immersion, injection, and implantation (Pandian and Sheela, 1995). Immersion is most effective for young fish, but feeding is easiest. At least 34 species (nine families) of hermaphrodites and 47 species (15 families) of gonochores have been sex-reversed (Pandian and Sheela, 1995).

17α-methyltestosterone, 19-norethynyltestosterone, and methylandrosterone are effective in converting females to males or creating all-male cohorts from immature fish (Yamazaki, 1983; Chao and Chow, 1990). Dosage of methyltestosterone has ranged from 0.1 to 100 mg/kg diet. At least 16 androgens have been used for this purpose (Pandian and Sheela, 1995). It might be possible to create all-male cohorts of Japanese eels by feeding elvers ~100 g methyltestosterone/kg feed (Ohta and Takano, 1996).

Mainly 17β-estradiol has been used to convert males to females or create all-female cohorts from immature fish (Yamazaki, 1983; Bromage, 1988). Degani and Kushniro (1992) reported that 70% of European eels fed 60 mg 17β-estradiol/kg diet for 1 yr (16 months after glass eel stage) were phenotypically female, compared to 26% of control eels. By feeding 2- or 7-g juveniles 25 mg 17β-estradiol for 91 d, Andersen et al. (1996) obtained 75-78% females vs 51-54% control females.

To produce all female salmon, an early immersion treatment is necessary. Salmon eggs and alevins have been converted to all females by immersion in 0.5-400 μg 17β-estradiol/L (sometimes with feeding of 10-20 mg/kg diet), and alevins were converted by 20 mg/kg in food alone. Piferrer and Donaldson (1992) produced 100% female chinook salmon by daily 8-h immersion in 400 μg 17β-estradiol/L for 35 d. Happe and Zohar (1988) changed 14-month-old (mean weight 322 g) gilthead sea bream from male to female by feeding them 1 mg 17β-estradiol/kg body weight per day for 60 d. Estrone has been used in the range 10-200 mg/kg diet for several species. Diethylstilbestrol and ethynylestradiol have been used in the range 25-100 mg/kg diet. Feeding the hormones for several weeks might be necessary. A procedure for incorporating hormones into feed using alcohol as a carrier and its use in sex-reversing rainbow trout are described by Ingram (1988). At least 15 estrogens have been used for this purpose (Pandian and Sheela, 1995).

Martin-Robichaud et al. (1994) produced 100% phenotypically female Atlantic lumpfish by giving the larvae *Artemia* containing 231-407 ng 17β-estradiol/mg DW for the first 20 d of feeding.

Chang et al. (1995) reported that feeding 6-month-old striped mullet 1 mg 17β-estradiol/kg feed for 4 months followed by 15 mg/kg for 4 months (2.5 mg eaten per fish) produced 76% females vs 34% in controls.

Social interaction also can help determine sex in fish. In a preliminary study, when European eels were kept in a group for 6 weeks and given an HCG injection after the first 3 weeks, 77% became male, but when other eels were kept solitary, 60% became female (Degani and Kushniro, 1992).

Artificial gynogenesis (female parthenogenesis = total maternal inheritance) caused by chromosome manipulation and by hybridization can be used to produce all female populations, which also happen to be inbred (Thorgaard et al., 1981; Yamazaki, 1983; Chourrout et al., 1988; Wohlfarth and Hulata, 1989; Zohar, 1989; Brannon, 1991). Examples are masu salmon, chum salmon, blue tilapia, European plaice, and North Atlantic flounder. If rainbow trout eggs are fertilized with radiation-inactivated sperm, then heat shocked, gynogenetic diploid females result. These can be masculinized by feeding them 1-3 mg methyltestosterone/kg diet, and when the resulting masculinized females (XX) are used to fertilize eggs from normal females (XX), all female populations result. This prevents low-quality or precocious males from undermining the value of the crop. Androgenesis (total paternal inheritance) also is possible.

Sterile triploid fish have been produced by subjecting fertilized eggs to pressure or temperature shock (Yamazaki, 1983; Bye and Lincoln, 1986; Bromage, 1988; Ingram, 1988; Wohlfarth and Hulata, 1989; Zohar, 1989; Brannon, 1991; Ju, 1994; Felip et al., 1997).

Examples are coho, chinook, pink, and Atlantic salmon, European sea bass, red sea bream, blue tilapia, European plaice, and North Atlantic flounder. Young rainbow trout can be sterilized with 17α-methyltestosterone by giving them four 1-h baths in 300 μg/L or by feeding them 20 mg/kg diet for the first 90 d (Dickison, 1985; Brannon, 1991). Tilapia can be sterilized by feeding them 50 mg MT/kg diet for 60 d (CRC pamphlet).

SUMMARY

In most situations, marine fish culture depends on hatchery-spawned and reared stock. Some species have been caught at or just before the time of ovulation and strip- or tank-spawned, but this requires very good timing and usually some luck. Females of many species with mature or nearly mature oocytes can be induced to ovulate by treatment with a range of substances that act at different levels in the endocrine system. Human chorionic gonadotropin and gonadotropin-releasing hormone analogs are practical agents for inducing ovulation in many species of freshwater and marine fish. Males often do not require treatment to produce milt, but with injections, they usually respond well for a short time, and with implants, longer.

Many species also can be conditioned to spawn voluntarily in captivity if time, facilities (particularly temperature control), water quality, food, and labor are sufficient (especially milkfish, cods, killifishes, serranids, jacks, dolphin, snappers, sea breams, drums, angelfishes, tilapias, damselfishes, gobies, rabbitfishes, lefteye and righteye flounders, common sole, triggerfishes, puffers). In between are some species that have not spawned voluntarily in captivity despite large human efforts, but can be conditioned to a stage at which they are hormonally induced to ovulate (striped mullet). Others will ovulate naturally, but might have to be stripped (sablefish, Pacific halibut). Wild or reared sturgeon usually are induced to ovulate with CP, SP, or GnRHa, and the eggs are surgically removed. Usually, captured or reared salmon and trout are stripped after natural ovulation, but hormones (e.g., SP, GnRHa) have been used to accelerate the process. Barramundi are tank- and strip-spawned with and without hormones, but common snook have tank-spawned only after hormone treatment. Many groupers, jacks, snappers, sea breams, drums, rabbitfishes, and others will spawn voluntarily, but hormones sometimes are used to accelerate, synchronize, or otherwise control the spawning schedule. Some flatfish species easily spawn without hormones, but others usually need them. Development of methods for routine voluntary spawning often requires considerable effort and expense, and even after methods are known, housing and feeding the fish still can be expensive. But unless ripe wild broodfish are readily available when needed, a captive broodstock is the best option.

8 NUTRITION OF LARVAL FISH

This chapter discusses the nutritional requirements of cultured marine fish larvae and some of the techniques used for feeding them. In larvae that have small reserves, the line between well-fed and malnourished (i.e., dead) often is very fine (Figs. 8-1, 8-2).

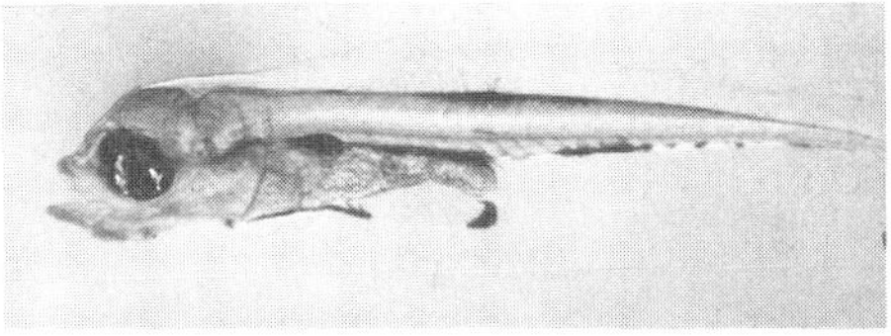

Figure 8-1. Healthy black sea bass larva (11 dah).

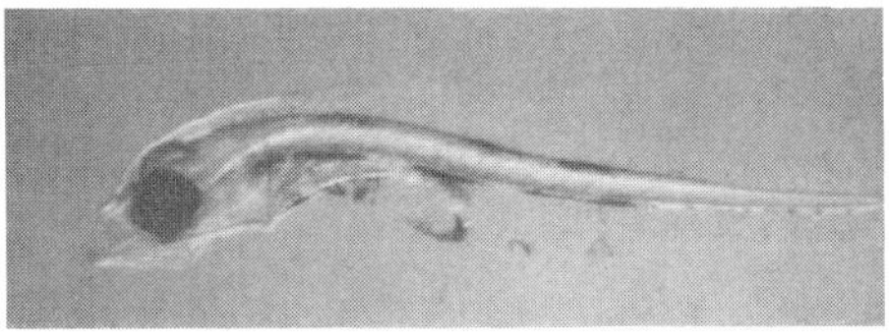

Figure 8-2. Starved black sea bass larva (7 dah).

According to general usage and logic, specific live or lightly processed (e.g., frozen, dried) dietary items will be referred to as "foods," specific formulated dietary items as "feeds," the combination of specific items eaten within a given period (e.g., day, life stage) as the "diet," and foods in general as "foods."

FEEDING CRITERIA

Criteria for acceptable larval fish foods and feeds are: availability, density, appearance, size, possibly smell or taste, nutritional quality, microbial status, digestibility, and stability of the food, plus ability of the fish. Extra care is necessary during first feeding and weaning to avoid high mortality. First the larval fish must locate the food without wasting energy. Then the fish must be attracted to the food, be capable of swallowing it, and be willing to swallow it. The food must contain an appropriate balance of essential nutrients that can be utilized by the fish. The food must not disintegrate or lose an appreciable amount of nutrients before the fish eats it. Larval survival and growth will be enhanced if feeding effort is kept low. Particular attention must be paid to size, fatty acid balance and content, amino acid balance, and nutrient density of foods given to marine fish larvae.

Availability of Food

Healthy rotifers, brine shrimp, and cultured cladocerans, calanoid copepods, and some harpacticoid copepods (e.g., *Euterpina acutifrons*) are mostly planktonic; they will be readily available to planktonic fish larvae, which do not ordinarily encounter solid surfaces in their natural environment. Rotifers stressed by sudden salinity changes (Lubzens et al., 1989) and, except for nauplii, most harpacticoid copepods (e.g., *Tigriopus japonicus, Tisbe furcata*) will adhere to or crawl on surfaces, and unhealthy zooplankters could settle to the bottom, where they would be harder to locate and consume. Rotifers and *Artemia*, especially, are positively phototactic and will tend to aggregate in highly illuminated areas. Tank color also can affect prey distribution; zooplankters tend to be attracted more by brighter colors and glossy surfaces that reflect more light. Keeping prey healthy and keeping light at the correct level will help to insure that food is available to the larvae.

Density of Food

Prey densities used for rearing marine fish larvae range from 0.05 copepod/mL (Houde and Schekter, 1980) to 200 dinoflagellates/mL (*Gymnodinium splendens*, Lasker et al., 1970; Hunter, 1980, 1984). The usual densities are 100 dinoflagellates/mL, 5-20 rotifers/mL, 1-10 copepod nauplii/mL, 0.5-6 *Artemia* nauplii/mL, and 0.5-3 adult copepods/mL; at higher densities, water and prey quality easily deteriorate. High prey density is more of a problem at higher temperatures; both reproduction of prey and metabolism (oxygen consumption and ammonia production) of all living things in the tanks are increased. In aquaculture, average prey densities often are only of theoretical value. Unless the water is very well mixed, zooplankters tend to form patches, with some areas of higher and other areas of lower than average prey density. This is the actual situation in which the larval fish must feed. What matters is that the larvae maximize their net intake of essential nutrients and therefore grow as fast as possible (maximize growth efficiency). Especially with low larval densities, rigid adherence to abitrary average prey density mandates can result in overfeeding.

Prey density could be more of a factor for larvae with less-developed hunting ability. Gulbrandsen (1991b) reported an optimal concentration of 12 rotifers/mL for first-feeding Atlantic halibut larvae, in contrast to an optimum near 1 rotifer/mL for European plaice (Hagen, 1989). Fifteen-day old gilthead sea bream (4-5 mm) needed at least 5 rotifers/mL to maintain a normal ingestion rate; by 22 dah (6-7 mm), that threshold had disappeared (Ounaïs-Guschemann, 1989). Growth and survival of Australian bass larvae in polyethylene (white) tanks were best with prey concentrations of 9-15 rotifers/mL during the first two weeks after yolk absorption and 6 *Artemia* nauplii/mL during the next two weeks (van der Wal and Nell, 1986). Rotifer intake by orangespotted groupers in black tanks at 7 and 14 dah was directly related to rotifer densities of 5, 10, and 20/mL, and survival at 14 dah was best (13.5% vs 1.4 and 1.8%) with 20/mL (Duray et al., 1996b). In black tanks, survival of first feeding dolphin larvae (3-7 dah) was similar (38-53%) with densities of 5, 10, and 20 rotifers/mL (Ostrowski, 1989).

Up to a point, increasing prey density can allow larvae to feed more efficiently. Four-day-old red drum (10 larvae/1-L beaker) had more (~11) oyster trochophores in their guts at 20/mL than at 10/mL and at least as many as at 40/mL (Pickering and Holt, 1993).

Hagen (1993) found that during the first hour of feeding, European plaice consumed more rotifers (4 vs 2) with larvae at 5-10/L and rotifers at 1000-2000/mL than with other combinations including 20 larvae/L and 500 rotifers/mL. The best combination, therefore, was 10 larvae/L and 1000 rotifers/mL.

Initial feeding of turbot at 4 dah was better with 7.5-10 rotifers/mL, but by 5 dah, 2.5 rotifers/mL were sufficient (Olmedo et al., 1995).

Appearance of Food. Tanks with black walls and light-colored bottoms enhance visibility of food from the larvae's viewpoint and visibility of larvae from the culturist's viewpoint. European sea bass grew and survived better in black-walled tanks than in white ones, probably because of better contrast of their prey against the dark background (Ronzani-Cerqueira, 1986). Gilthead sea bream larvae located prey more easily and survived better in black-walled tanks, regardless of whether the bottoms were black or white (Ounaïs-Guschemann, 1989). Survival of first feeding dolphin larvae (3-7 dah) was 67% in black tanks and 29% in tan tanks, probably because higher contrast of rotifers with black tank walls improved feeding efficiency (Ostrowski, 1989). Growth of orangespotted groupers was similar in tan and black tanks before 14 dah, by which time larvae were eating more and growing faster in tan tanks (Duray et al., 1996b). In glass tanks, first feeding common sole larvae ate more *Artemia* that were stained black than those stained four other colors or unstained, probably because the black ones were more visible (Dendrinos et al., 1984). Hinshaw (1985) obtained better feeding (on *Artemia* nauplii), growth, and survival (44.2% vs 4.6% at 14 dah) of yellow perch larvae in black tanks than in white ones. Most of the species in Table 1-1 were reared in black-walled tanks; when white or glass tanks were used, early larvae often became disoriented by the glare. In one trial, common snook larvae in glass tanks did not begin feeding, swam upside down for a few days, and then died (pers. obs.).

Size of Food

Optimal prey size for marine fish larvae seems to be about 25% of the mouth width at first feeding, increasing to 50% within a few days (Hunter, 1984). Shirota (1970) compared sizes of larval mouths of nearly 40 marine species at first feeding with sizes of prey organisms. He demonstrated that species such as skipjack tuna and yellowtail could easily ingest *Artemia* nauplii or large copepods, but others such as tiger puffer and Japanese anchovy would need prey the size of protozoans or young rotifers and copepods at first feeding. Some examples of mouth diameter at first opening are: ~278 μm for common dentex, ~420 for twoband sea bream, and ~456 μm for gilthead sea bream (Glamuzina et al., 1989). Some examples of mouth widths at first feeding are: 187 μm for yellowblotch spinefoot, 200 μm for Asian red snapper, 216 μm for chub mackerel, 224 μm for barramundi, 225 μm for brownmarbled grouper, 258 μm for milkfish, and 291 μm for kawakawa (Hunter and Kimbrell, 1980; Doi and Singhagraiwan, 1993; Kohno et al., 1994b, 1994c). First feeding barramundi selected 40-60 μm microcapsules when offered a range of 15-150 μm (Walford et al., 1991). The optimal width range of prey for most marine fish larvae that have been cultured is in the range 35-100 μm at first feeding. Preferred prey size quickly increases as mouth size and feeding competence increase.

Reared Asian red snapper larvae ate copepod nauplii (100-150 μm TL) at first feeding and copepodites (200-500 μm) 3-4 d later (Doi et al., 1997).

At first feeding (3.5 mm TL), mouth width of gilthead sea bream was about 250 μm and larvae were capable of eating hard microparticles smaller than 100 μm but preferred the range 25-50 μm (Fernández-Diaz et al., 1994). At 4-5 mm, larvae preferred 51-100 μm particles, and beyond 5 mm, their preference changed to 101-150 μm. At 6.5 mm (mouth width ~520 μm), larvae could eat hard particles as large as 200 μm. Size preference for soft particles was higher, with larvae longer than 4.5 mm preferring the range 201-250 μm.

Prey of wild hirame larvae ranged from <100 μm in width and <200 μm in length at a larval size of 2.3-4.0 mm NL (first feeding) to 500 μm x 1210 μm in 7.2-7.9 mm larvae (early metamorphosis)(Ikewaki and Tanaka, 1993). Reared turbot larvae preferred prey about 40% as wide as the mouth and 36% as high; optimal prey width was <140 μm at first feeding (Planas, 1994). Iglesias et al. (1994) reported an optimal rotifer width of <100 μm for turbot at first feeding (4 dah). Mouth width of first-feeding greenback flounder was 215 μm and mean prey width was 96 μm (range 37-201 μm)(Jenkins, 1987).

Apparently minor differences in rotifer size can be important to larvae. Usually, one of two approaches is taken to guarantee a sufficient amount of small rotifers for larvae with small mouths: 1) A small strain of rotifer is cultured. The adult female width range for 13 strains was 92-199 μm (Snell and Carrillo, 1984). 2) Small rotifers are separated from large ones by sieving. Strain characteristics determine the potential width range, but culture conditions determine the actual size obtained. Males are smaller than females and newly hatched rotifers are much smaller (about 50% or less of adult female width). Maintaining optimal conditions for reproduction will maintain a continuous supply of rotifer larvae. Most cultured strains reproduce best at temperatures of 25-30°C and salinity near 20 ppt, and relatively few can be raised in large numbers below 20°C (Lubzens, 1987; Lubzens et al., 1989). At least some strains become smaller with increasing temperature or salinity or with certain diets (Snell and Carrillo, 1984; James, 1993). Eggs of some strains are smaller and hatch faster at higher temperatures (e.g., 0.166-μm^3 eggs in 10.6 h at 30°C vs 0.219-μm^3 eggs in 27.4 h at 20°C, Yúfera, 1987). When rotifers are sieved for food, rapid growth would be a problem only if rotifers are provided in such excess that large numbers escape being eaten. Although adult size might be reached within 36 h after hatching (Snell and Carrillo, 1984), with correct feeding practices most rotifers will be consumed by the fish larvae within 6-12 h, before they can grow much.

Fukusho and Okauchi (1984) observed dominance of a large strain (mean width 177 μm) of rotifer at 22°C in May and a small strain (mean width 96 μm) at 28°C in July in a Japanese eel culture pond. In Florida, the size of a single strain can decrease slightly as water temperature rises in the summer, and 28°C seems near optimal for high reproduction rate and small size.

Fernández-Diaz et al. (1994) used a small strain of rotifer with average adult dimensions of 113 × 140 μm and a large strain with dimensions of 156 × 240 μm.

Artemia nauplii used by Fernández-Diaz et al. (1994) measured 176 × 451 μm. Other reported sizes range from 375 × ~428 μm for a strain from San Francisco Bay to 450 × 517 μm for a strain from Italy (Léger et al., 1986). Length of metanauplii increases from about 500 to 800 μm during the four stages (instars), which occur during about 0.5-5 dah (Bengtson et al., 1991). Orhun (1980) reported that San Francisco Bay nauplii were 428 μm long, instar II metanauplii 578 μm, and instar III metanauplii 670 μm. Pryor and Epifanio (1993) found that San Francisco Bay nauplii were 167 × 449 μm and Great Salt Lake nauplii were 191 × 488 μm. McGeachin et al. (1977) fed *Tetraselmis chuii* to San Francisco Bay *Artemia* and obtained lengths of 380 μm at hatching, 670 μm at 1 dah, 740 μm at 2 dah, 760 μm at 3 dah, and 760 μm at 4 dah. Lavens et al. (1995) gave average lengths of 1.7-2.7 mm for 5-d-old, and 2.1-3.8 mm for 7-d-old *Artemia*, depending on culture density.

Ability of Larvae

Typical marine fish larvae usually begin feeding on small, slow prey (e.g., dinoflagellates, ciliates, mollusc larvae, larval copepods, larval cladocerans) and progress to larger, faster prey (e.g., juvenile and adult copepods and cladocerans, barnacle larvae, larval and juvenile shrimp, amphipods, isopods) as they become more competent (Hunter, 1980, 1981; Kauffman et al., 1981; Heath, 1992; Drake and Arias, 1993). Over the size range 4-23 mm TL, copepods were the major food of wild Pacific cod larvae, with cladocerans contributing a few percent to the diet (Takatsu et al., 1995). Over the size range 10-50 mm SL (transformation at ~63 dah and 25 mm TL), major foods of wild gag groupers shifted from calanoid copepods to gammaridean amphipods, with some mysid shrimp; decapods became more important at 20-50 mm and dominated at 50-125 mm (Mullaney and Gale, 1996). Like many species, wild greenback flounder larvae were relatively opportunistic predators, feeding on a wide variety of zooplankton—especially bivalve veligers, tintinnids, and dinoflagellates for small larvae and cladocerans, copepods, and larvaceans for older larvae; invertebrate eggs also were eaten but not preferred (Jenkins, 1987). Most marine fish larvae do not encounter or eat many rotifers; exceptions can be expected among some brackishwater species such as the small sand goby (Mehner, 1992).

Variations in mouth size, swimming ability and patterns, and chemical and mechanical senses contribute to feeding ability (Shirota, 1970; Houde and Schekter, 1981; Blaxter, 1986; Huse, 1993; Døving and Knutsen, 1993a, 1993b; Kohno et al., 1994c). Milkfish survive well because of their large bodies and mouths, even though first feeding and EYS occur at about the same time (Kohno et al., 1994c). Barramundi survive well despite their smaller bodies and mouths because first feeding occurs about 3 d before EYS and they feed efficiently. Brownmarbled groupers are at a disadvantage because they have small bodies and mouths, first feed 1.5 h before EYS, and do not feed efficiently. Yellowblotch spinefoot larvae have about 2 d between first feeding and EYS, but with smaller mouths than the other three species, can have difficulty at first feeding. Kawakawa larvae first feed 1-2 d before EYS at 4.0 dah (Kohno et al., 1994b).

Typical marine fish larvae do not have functional stomachs or gastric glands until long after first feeding, often not until transformation (e.g., Hjelmeland et al., 1993). Representative times are 3 weeks for red and black sea bream, 6 weeks for milkfish, 6.5 weeks for hirame (Kurokawa and Suzuki, 1997), 3 weeks for striped bass (Huang et al., 1998a), 7 weeks for summer flounder (Huang et al., 1998b), and 14 weeks for ayu (Watanabe and Kiron, 1994). Weight of European sea bass at 63 dah was higher if they were weaned at 39 dah than at 37 or 38 dah (Gennari et al., 1992).

Survival ability of larvae also is related to health and nutrition of the broodstock, particularly because of variation in nutrients incorporated into the eggs (e.g., Lavens et al., 1995).

Chemical Stimuli by Food

This discussion is derived partly from Iwai (1980). Histological evidence supports the hypothesis that the olfactory sense is present in 2-d-old Mozambique tilapia, 3-d-old marble gobies (Senoo et al., 1994a), and 8-d-old red sea bream. Immature taste buds first appear in the freshwater species Mozambique tilapia, 1 dah, pond smelt 2 dah, goldfish 2 dah, and marble goby 6 dah (Senoo et al., 1994a), and in the marine species Japanese sea bass 5 dah, grass puffer 8 dah, stone flounder 14 dah, and red sea bream 16 dah. When clam juice was added to water holding red sea bream larvae 8 dah (smell only), they did not respond, but at 16 dah (smell and taste), clam juice stimulated larvae to greater activity. The mouths of Atlantic halibut larvae do not work until ~26 dah, but they have olfactory organs by 8 dah (Døving and Knutsen, 1993). Thus, the senses of smell and taste are not fully developed in some species during the first days of feeding and are not available for use in detecting potential food or evaluating partially ingested food. These two senses might be of more value in older fish that will encounter a wider range of particles in their search for food. Food size and movement could be the primary feeding stimulants for early larvae. In their natural habitats, larvae are unlikely to encounter anything that is the right size and looks like prey but is not edible. In compound diets (e.g., microparticles), olfactory and gustatory stimulants can be useful and sometimes are essential. Feeding by young yellowtail can be increased by addition of the amino acids proline, alanine, and methionine, with inosine monophosphate. For some fish in which the sense of smell develops first, olfactory stimulants might be effective at earlier stages than gustatory stimulants.

Nutritional Quality of Food

Several forms for designating fatty acids have been used, for example linolenic acid = 18:3ω3, 18:3*n*-3, 18:3(*n*-3), 18:3n3, C18:3ω3, $C_{18:3}$n-3, etc. Before the colon is the number of carbon atoms in the molecule; after the colon is the number of double bonds; after the ω or n- is the location of the last double bond before the terminal methyl group (e.g., ω3 fatty acids have the last double bond between the third and fourth carbon atoms from the end, and ω6 fatty acids have the last double bond between the sixth and seventh carbon atoms from the end).

ω3HUFA = long-chain omega-3 highly-unsaturated fatty acids = fatty acids in the ω3 series (the last double bond is between the third and fourth carbon atoms from the end) with 20 or more carbon atoms and 4-6 double bonds in the molecule (n-3HUFA is synonymous). The forms "18:3ω3" and "ω3HUFA" are used here rather than "18:3n-3" and "n-3HUFA" because of tradition and because "ω" means end, which is a more direct way of designating the position of the double bond than "n" for number of atoms. ADA = arachidonic acid = 20:4ω6; EPA = eicosapentaenoic acid = 20:5ω3; DHA = docosahexaenoic acid = 22:6ω3; fatty acid level = percentage of total lipids or of total fatty acids; fatty acid content = percentage of dry organism. FA = fatty acids; EFA = essential fatty acids; DW = dry weight. ω-yeast is baker's yeast raised in a medium containing marine oil (often cuttlefish liver oil) so that ω3HUFA's are incorporated into the yeast.

Either EPA or DHA or both are essential nutrients for the marine fish whose requirements have been determined. Rotifers can be, and *Artemia* usually are, deficient in at least one of these fatty acids, and possibly other nutrients (Watanabe et al., 1983a). The lipid content and fatty acid composition of feeding rotifers, *Artemia*, *Daphnia pulex*, and *Moina* sp. depends on those of their recent foods, but for newly hatched zooplankters, especially male rotifers (which do not feed) and *Artemia*, fat content and composition depend on the composition of the maternal diet (Kanazawa et al., 1979; Oka et al., 1982; Watanabe et al., 1983a; Lubzens et al., 1985, 1989; Léger et al., 1986; Rainuzzo et al., 1989; Mims et al., 1991). Copepods tend to maintain good contents of fat and ω3HUFA but also can benefit from enrichment (Fukusho et al., 1980). Fatty acid and possibly amino acid deficiencies make *Artemia* nauplii risky to use, but after they begin feeding, enrichment can do much to eliminate the deficiencies. When their diets lack EPA and DHA, 15-d-old *Artemia* can convert small amounts of linoleic (18:2ω6) and linolenic (18:3ω3) acids to EPA and DHA but not enough to supply marine fish larvae (Schauer and Simpson, 1985). At least several species of marine fish have a slight ability to convert linolenic acid to DHA, but apparently not enough to meet their requirements (Schauer and Simpson, 1985). Excess linoleic or linolenic acid can be harmful (see chapter 9).

Robin (1995) concluded that arachidonic acid (20:4ω6) could be just as essential as EPA and DHA for at least some marine fish. Different batches of *Artemia* might or might not contain any. Algae and oil emulsions fed to rotifers and *Artemia* do not necessarily contain it. Ishizaki et al. (1998) reported lower survival and growth for yellowtail larvae given arachidonic acid-enriched *Artemia* compared with DHA, EPA, or oleic acid

Scott and Middleton (1979) reported that growth and survival rates of turbot larvae reared on rotifers with the microalga *Dunaliella tertiolecta* added to the fish tanks were very low compared to those reared with *Phaeodactylum tricornutum*, *Pavlova lutheri*, or *Isochrysis galbana*. They presented evidence that rotifers feeding on *D. tertiolecta* were nutritionally unsuitable as food for turbot larvae because those rotifers lacked EPA and DHA and turbot cannot synthesize ω3HUFA from shorter-chain fatty acids. Experiments with *Artemia* nauplii indicated that turbot larvae could not synthesize DHA from EPA. Scott and Middleton also suggested that certain large, fragile dinoflagellates (typically containing a level of DHA in the range 15-34%) are eaten and can be digested by wild turbot larvae, while algae used in research is tougher and sometimes is ingested but probably cannot be digested by the fish. Moffatt (1981) reported that northern anchovy larvae obtained nutritional benefit from ingested *Chlorella* sp. cells—possibly extra energy and/or trace nutrients. Reitan et al. (1991, 1994a) reported that Atlantic halibut larvae ingested a large number of *Tetraselmis* sp. cells but assimilation efficiency was negligible (1-5%).

Some marine fish require both EPA and DHA, while others seem to need only one or the other. Turbot require both EPA and DHA, but Atlantic herring and European plaice need only EPA (Lubzens et al., 1989). Both EPA and DHA seem to be essential for common sole larvae (Dendrinos and Thorpe, 1987). In addition, growth of both common sole (Dendrinos and Thorpe, 1987) and turbot (Gatesoupe et al., 1977a) can be promoted by linolenic acid.

DHA seems more important than EPA in diets for some species (e.g., larvae of Atlantic herring, Pacific cod, striped jack, yellowtail, red sea bream, striped knifejaw, hirame, turbot, and marbled flounder; Watanabe and Kiron, 1995; Takeuchi et al., 1996; Kanazawa, 1997; Sargent et al., 1997). Watanabe et al. (1989b) reported that rotifers enriched with DHA plus methyl oleate (3.9% EPA, 21.8% DHA of total lipids; 4.0% ω3HUFA and 17.8% total lipid of DW) were superior as food for larval red sea bream (growth and vitality) to those enriched with methyl oleate alone, EPA plus methyl oleate, or fatty acid methyl esters.

DHA was considered more valuable than EPA. Juvenile striped jack juveniles (0.5-3 g) grew and survived as well on a compound feed containing no EPA and 1.2% DHA as they did with 0.7% EPA and 0.6% DHA (Watanabe et al., 1989a). A feed containing 1.4% EPA without DHA was inferior. EPA and DHA were the only ω3HUFA's present in the feeds. Takeuchi et al. (1996) found that raising DHA to 1.6-2.2% (with 0.9-1.1% EPA in 3.3-5.0% ω3HUFA) in *Artemia* increased survival and vitality of larval striped jack, but up to 3.1-8.4% EPA without DHA (in 4.0-8.7% ω3HUFA) did not. Furuita et al. (1996b) reported similar survival of yellowtail fed *Artemia* containing 1.0% EPA + 1.8% DHA (2.9% ω3HUFA) or 1.6% EPA + 3.3% DHA (5.2% ω3HUFA), but better growth and vitality with the higher content. Corneillie et al. (1990) found 16.4% EPA and 16.3% DHA (of FA) in ~4.5-month-old wild European sea bass, 4.0% EPA and 5.5% DHA in normal 60-d-old hatchery fish, and only 2.8% EPA and a trace of DHA in 40-d-old hatchery fish with whirling disease.

DHA is especially high in the eyes and other neural tissues of healthy fish and apparently is of critical importance for normal function leading to good survival and growth (Mourente and Tocher, 1992; Bell et al., 1996; Tocher et al., 1997). An optimal ratio of ADA/EPA/DHA in food might be necessary.

DHA might be more important for larvae than for older fish. Starved gilthead sea bream larvae (18°C) retained relatively more ω3HUFA than other fatty acids (particularly DHA, but not EPA) during starvation than when they were fed rotifers enriched with *I. galbana* (Koven et al., 1989). Conservation of DHA could be related to its importance in that species.

Generally, growth and survival of marine animals seen in different studies has been good with *Artemia* having levels of EPA higher than 4% and bad when EPA was less than 3% (Léger et al., 1986); however, EPA content (of DW) also depends on the percentage of total lipid, and enrichment with DHA usually can compensate for a shortage of EPA.

The EPA range in nauplii of four *Artemia* strains was 1.2-10.4% (of total lipids); no DHA was detected (Webster and Lovell, 1990b). Two batches from San Francisco Bay had very different levels of palmitoleic acid, linolenic acid, and EPA. Nauplii with 9.3 and 10.4% EPA supported better growth and survival of striped bass larvae at 19°C than those with 1.2 and 3.0% EPA.

Growth and survival (92-93%) of European sea bass (37-56 dah) were nearly identical on diets of reference strain *Artemia* nauplii (1.1% EPA, 0% DHA, 1.1% ω3HUFA of DW) and enriched San Pablo Bay metanauplii (0.2% EPA, 0.3% DHA, 1.0% ω3HUFA)(van Ballaer et al., 1985). Growth and survival were reduced with three other types of *Artemia* having ω3HUFA below 0.7%. If fat content was average (20%, Léger et al., 1986), actual ω3HUFA content would have been about 1.9-2.2% for the better foods and less than 1.4% for the inferior ones.

Izquierdo et al. (1989) reported that larval red sea bream needed 3.3-3.5% ω3HUFA (of DW) in rotifers. Of the effective diets tested, the one with the least ω3HUFA had levels of 6.8% EPA and 3.6% DHA (of total lipids), with contents of 1.9% EPA, 1.0% DHA, and 3.3% ω3HUFA (of DW); 3.0% ω3HUFA in *Artemia* was enough (Watanabe and Kiron, 1995).

Striped knifejaw needed >3.0% ω3HUFA (of DW) in rotifers (Watanabe and Kiron, 1995). Tamaru et al. (1991b) suggested minimum contents of about 0.7% EPA and 0.5% DHA in rotifers for larval striped mullet.

Live food requirements for turbot were in the range (0.4%, marginal?)1.2-3.2% ω3HUFA (Gatesoupe et al., 1977b; Minkoff and Broadhurst, 1994; Rainuzzo et al., 1994; Reitan et al., 1994b; Watanabe and Kiron, 1995).

Reported ω3HUFA requirements in *Artemia* were: 1.6-2.1% DHA for Pacific cod (Zheng et al., 1996), 1.3% ω3HUFA (0.5% DHA) for palmetto bass (recommended by Tuncer et al., 1993), 0.9-1.1% EPA + 1.6-2.2% DHA in 3.3-5.0% ω3HUFA for striped jack

(Takeuchi et al., 1996), about 1.2% EPA + 2.6% DHA in 4.1% ω3HUFA for yellowtail (Furuita et al., 1996b), 0.9% EPA + 1.6% DHA in 2.8% ω3HUFA for red sea bream (Furuita et al., 1996a), at least 2.1% EPA + 0.9% DHA in 3.0% ω3HUFA for gilthead sea bream (Koven et al., 1992), >3.0% in *Artemia* for striped knifejaw (Watanabe and Kiron, 1995), and 3.5-4.0% for hirame (Izquierdo et al., 1992).

It seems prudent to provide both EPA and DHA (possibly with some ADA and linolenic acid) in foods for marine fish. A dietary ω3HUFA content of 2-4% of DW, including at least 1.0% EPA and 1.0% DHA, should satisfy or exceed the requirements of most marine fish larvae. Some species will thrive with less, but a slight surplus is better than a deficiency. For *Artemia* to contain this much ω3HUFA, they usually must be enriched. Enrichment of zooplankton is accomplished by feeding them ω3HUFA contained in oil added to yeast, an emulsion, or microfeeds.

Microbial Status of Food

Bacteria and other microbiota can affect the health and production of cultured algae (Munro et al., 1995), rotifers (Skjermo and Vadstein, 1993; Yu et al., 1994; Verdonck et al., 1997), *Artemia* (Gorospe and Nakamura, 1996), copepods (Guérin and Rieper-Kirchner, 1991), and fish (Kusuda et al., 1986; Nicolás et al., 1989; Hansen et al., 1992; Strøm and Ringo, 1993; Gatesoupe, 1994; Grisez et al., 1996). Live foods should be free of pathogens. Ideally, they should contain probiotic (beneficial) bacteria. See chapter 11 for further discussion.

Digestibility and Assimilation of Food

Enzymes necessary for digestion of rotifers are present and assimilation is high in larvae of the marine fish that have been investigated (Lubzens et al., 1989). Digestibility of *Artemia* can vary with age of the *Artemia* and with species and age of the fish. Until about 12 h after hatching (instar II metanauplii), the digestive system of *Artemia* is not open, apparently restricting entry of external digestive enzymes and making them less digestible to early larvae of some marine fish. During the day after hatching, digestive enzymes in *Artemia* increase quickly to about five times the amount at hatching (Pan et al., 1991). Decapsulated *Artemia* cysts are especially resistant to digestion (Dendrinos and Thorpe, 1987).

The prey's own enzymes might aid in their digestion. Munilla-Moran et al. (1990) estimated the contribution of prey to the digestive enzymes present in the gut of first feeding turbot (3 dah) to be: 15-27% of the amylase, 43-60% of the proteases, 79-88% of the exonuclease, and 89-94% of the esterase. The prey of 30-d-old metamorphosed turbot had the potential to provide 100-1000 times more enzymes than the turbot's digestive system, but by then levels in the fish were sufficient to digest foods that lack enzymes (turbot can be weaned to dry feed during 25-35 dah or a little sooner). Variation in production of digestive enzymes among individuals probably results in some larvae being able to digest food better than others (Munilla-Moran et al., 1990) and possibly grow and survive better because of this. Typically, a number of larvae die at first feeding and when major dietary changes are made (especially at weaning), while other larvae easily adapt to the changes. Low enzyme levels might be enough to handle small amounts of food. With high ingestion rates (as sometimes occurs with *Artemia* nauplii), endogenous enzymes might be too diluted to be effective by themselves or they might not be able to penetrate the prey and release its enzymes. Incomplete digestion, such as sometimes occurs with *Artemia*, could result.

Fyhn (1989) and Fyhn et al. (1993) suggested that certain free amino acids are important energy sources not only in marine fish embryos but also in feeding larvae and that they might be essential in larval diets, whether living or formulated. This could be another factor contributing to the superiority of live foods for early larvae.

Diaz et al. (1994) reported that first-feeding European sea bass and (to a lesser degree) gilthead sea bream could utilize dissolved glucose from swallowed water, which delayed starvation effects. This does not indicate a practical feeding method but does show that larvae can benefit from dissolved nutrients.

Stability of Food

The nutritional value of unfed zooplankters will diminish. Except at low temperatures (<20°C), rotifers starved for 6 h have begun to lose much of their nutritional value (Hoff and Snell, 1989; Olsen et al., 1989, 1993a; Ortega et al., 1995). At 24°C, lipid content of starved rotifers previously cultured on *Dunaliella tertiolecta* dropped from 59 to 19 ng/rotifer in 2 d, protein went from 147 to 78 ng, and carbohydrate from 44 to 12 ng (Scott and Baynes, 1978). After 49 h of starvation at 16°C, the fatty acid composition of rotifers (previously fed baker's yeast plus seafood meals and oils) did not change much, but total lipid decreased by 25% (Rainuzzo et al., 1989). Two or more feedings per day will help keep prey nutrient density high for the most sensitive early larval stages, which need every reasonable form of assistance.

Rotifers and other zooplankters can be stored in a refrigerator before or after enrichment to reduce metabolism and prolong their usefulness. When rotifers were held for 10 d at 1°C and 20 ppt salinity without food, survival in four trials ranged from 40 to 83% (Lubzens et al. 1990a). *Artemia* nauplii usually can be stored for at least 48 h at 4°C with minimal mortality or loss of nutritional value (Léger et al., 1986). Increase in size is negligible, and the *Artemia* swim more slowly after being chilled, making them easier to catch so that first feeding on them might be accelerated. Shock-frozen *Artemia* are more nutritious than slow-frozen ones, but both are likely to lose nutrients, especially through oxidation (e.g., lipids) and leaching (e.g., free amino acids). Frozen *Artemia* is best restricted to use as a weaning food or feed attractant. In terms of growth and survival (28%), live *Artemia* nauplii were more nutritious than frozen or freeze-dried nauplii for striped bass from first feeding at 6 dah to 19 dah (Webster and Lovell, 1990a). Live *Artemia* nauplii have been shown to be much more nutritious than dried, brined, or pasted *Artemia*.

Much of the enzymatic activity was retained when zooplankton was frozen or freeze-dried (Grabner et al., 1981/1982); however, contact with liquid water caused rapid deterioration. L-lactate dehydrogenase was 1920 times more stable (half-life 280 d) in freeze-dried samples at room temperature than in thawed frozen ones (half-life 3.5 h). After 10 min in fresh water, about three-fourths of the enzymatic activity of frozen or freeze-dried zooplankton was lost. After 60 min in water, about one-third of the total amino acids and nearly all of the free amino acids were lost.

Compound larval feeds will deteriorate quickly in water, especially at higher temperatures, with microcapsules usually lasting longer than microparticles because their contents are better protected. Soluble components such as vitamins will leach out and lipids will become rancid. Bacteria and fungi will grow on the feed, possibly producing toxins. Non-living feeds (especially larval ones) should be stored in a freezer whenever possible.

CHOICE AND CULTURE OF FOODS

Most species of marine fish that have been cultured can be reared on a sequential diet of: (1) rotifers (*Brachionus plicatilis*), (2) brine shrimp (*Artemia* sp.), and (3) dry feeds. When facilities consist of little more than a shelter, the culturist might be forced to rear the entire food chain from algae to fish in a small space.

Microalgae

Microalgae (Table 8-1) is the customary food given to zooplankton that will be fed to larval fish. Much has been written on the culture of microalgae. Useful references are: Stein (1973), Microalgal Technology Research Group (1986), Richmond (1986), Walsh et al. (1987), Hoff and Snell (1989), Barnabé (1990e), LeBorgne (1990), Myers and Boisvert (1990), Benemann (1992), Carlos et al. (1992), Fukusho and Hirayama (1992), Starr and Zeikus (1993), Brown et al. (1997b), Tsovenis et al. (1997), Chebil and Yamasaki (1998).

Table 8-1. Ranges of total lipid content and percent eicosapentaenoic (EPA) and docosahexaenoic (DHA) acid contents of total fatty acids in microalgae used as food for ciliates, rotifers, copepods, and *Artemia*. Under different culture conditions, wider lipid and fatty acid ranges are possible. Temperatures, given for comparison, are approximate values for typical strains and will vary for different strains.

Species	Best growth (°C)	Total lipid (% DW)	Highest total lipid[a]	EPA (%)	DHA (%)	Highest % ω3HUFA[a]
<u>Class Chlorophyceae</u>						
Chlorella regularis[b]		2		trace	trace	
Chlorella sp.		2		0	0	
Dunaliella tertiolecta	25	7		0-0.1	0-0.1	
Nannochloris atomus		14.7		3.2-19.4	0-0.5	
<u>Class Prasinophyceae</u>						
Tetraselmis chuii		5		4.1	0	
Tetraselmis gracilis	~21	1.7-6.9	EST	~4.6-7.9	~0	LST
Tetraselmis suecica	25	2-18	EXP	0-8.4	0-trace	LST
Tetraselmis tetrathele	25	10		3.1-6.4	0	
<u>Class Bacillariophyceae</u>						
Chaetoceros calcitrans	29	4-17	EST	4.3-22.9	0-1.9	EST
Chaetoceros gracilis	29	4-20		4.1-24.8	0-6.3	
Phaeodactylum tricornutum	23	8-23	EST	6.3-26.4	0.3-5.3	EST
Thalassiosira pseudonana	20	9-14		7.7-23.9	0-3.9	
<u>Class Prymnesiophyceae</u>						
Isochrysis galbana	18	5-31	LST	0-25.2	0-18.9	LST
Isochrysis sp. (Tahitian)	28	19-25		0-2.3	7.9-19.4	
Pavlova lutheri	20	13-32	LST	0-28.3	0-15.5	LST
Pavlova salina	27			25.4-28.2	10.2-11.0	
<u>Class Eustigmatophyceae</u>						
Nannochloropsis oculata	28	23-32	LST	9.7-37.1	0-2.6	EXP
Nannochloropsis salina	28	29-60[c]		14.8	0	
Nannochloropsis sp.				12.1-36.6	0	
Nannochloropsis sp. (MFD-2)	~25	20		23.7	0.4	
<u>Class Cryptophyceae</u>						
Rhodomonas baltica				10.3	9.2	
Rhodomonas sp.		14-32	LST	0-7.0	0-3.1	EXP
Chroomonas salina				7.0-11.9	5.2-5.7	
Schizochytrium sp.		32		0.6	24.0	
<u>Class Euglenophyceae</u>						
Euglena gracilis				2.7-75.7	0-76.0	

[a] EXP = exponential phase, EST = early stationary phase, LST = late stationary phase.
[b] Freshwater species for comparison only.
[c] Ash-free dry weight basis.

Compiled from Epifanio (1979), Scott and Middleton (1979), Epifanio et al. (1981), Webb and Chu (1981), Trotta (1983), Pillsbury (1985), Fabregas et al. (1986), Maruyama et al. (1986), Microalgal Technology Research Group (1986), Ben-Amotz et al. (1987), Helm and Laing (1987), Whyte (1987), Mortensen et al. (1988), Volkman and Hallegraeff (1988), Fernández-Reiriz et al. (1989), James et al. (1989), Volkman et al. (1989), Okauchi et al. (1990), Villegas et al. (1990), Okauchi (1991), Thompson et al. (1990), Volkman et al. (1991), Burgess et al. (1993), Hayashi et al. (1993a); Reitan et al. (1993c), Alonso et al. (1994), Spedicato et al. (1994), Su et al. (1994); Whyte et al. (1994), Montaini et al. (1995), Robin (1995), Støttrup et al. (1995), Barclay and Zeller (1996), Fernández-Reiriz and Labarta (1996), Lourenço et al. (1997), Reitan et al. (1997).

Stock cultures (glass tubes or small flasks) are kept under very clean conditions in an incubator or controlled-environment laboratory and are used or replaced before they deteriorate. The first one or two levels of scaling up usually are done in the laboratory (Fig. 8-3). Final stages can be inside under artificial light, in a greenhouse with partial sunlight (Fig. 8-4), or outside with full sunlight (Fig. 8-5). The type of culture, temperature, nutrients, other conditions, and growth phase all can affect the nutritional value of microalgae to zooplankton and to the fish larvae eating them. Higher lipid content in a *Nannochloropsis* sp. occurred at 32°C than at 18 or 25°C, but maximum growth rate and maximum EPA percentage was at 25°C (Sukenik et al., 1993). Percentage EPA was higher (37.6 vs 15.9% of FA) when the *Nannochloropsis* sp. was grown under 35 as opposed to 290 µmol quanta/m^2/s. When it was grown in unshaded outdoor ponds in Israel, both lipid (39.4% of DW) and EPA (3.8%) were highest in winter.

Figure 8-3. Microalgae cultures in 250-mL flasks and 18-L carboys.

Figure 8-4. Microalgae cultures in fiberglass tubes.

Figure 8-5. Microalgae culture in a concrete tank, which also can be shaded and used for rotifers.

In our experience, medium-sized batch cultures of Tahitian *Isochrysis* mature 3-4 d after inoculation and *Nannochloropsis oculata* in 4-5 d. With good sanitation, chemostat cultures of both can last for 3 months or longer.

Rotifers

Moderate to lengthy discussions of rotifer biology and culture (mainly *Brachionus plicatilis*, Fig. 8-6) appear in Theilacker and McMaster (1971), Fontaine and Revera (1980), Schlüter and Groeneweg (1981, 1985), Walker (1981), James et al. (1983), Fukusho and Okauchi (1984), Opuszynski et al. (1984), Reguera (1984), Snell and Carrillo (1984), Yu and Hirayama (1986), Lubzens (1987), Snell et al. (1987), Yamasaki et al. (1987), Yúfera (1987), Yúfera et al. (1993), Korstad (1988), Lubzens et al. (1989, 1990a, 1990b, 1995a, 1995b), Hoff and Snell (1989), James and Abu-Rezeq (1989a, 1989b), Snell and Persoone (1989), Chamberlain et al. (1990), Coves et al. (1990), Pourriot (1990), Talbot et al. (1990), Fulks and Main (1991), Ramírez-Sevilla et al. (1991), Carlos et al. (1992), Fukusho and Hirayama (1992), Nagata and Whyte (1992), Attramadal and Sandvand (1993); Bennett and Boraas (1993), Bennett et al. (1993), Boraas (1993a, 1993b), Fu et al. (1993), Hino (1993), James (1993); Lubzens et al. (1993, 1995a), Rothhaupt (1993); Skjermo and Vadstein (1993), Walz (1993), Hagiwara (1994), Hagiwara et al. (1994, 1995a), Lavens et al. (1994), Skaramuca (1994), Yu et al. (1994), Aoki et al. (1995), Hirayama and Hagiwara (1995), Korstad et al. (1995), Lebedeva and Orlenko (1995), Aoki and Hino (1996), Rueda-Jasso (1996), Yoshimura et al. (1996a, 1996b), and Reitan et al. (1997). Resting egg formation and use are discussed by Hino and Hirano (1984, 1985, 1993), Hagiwara and Hirayama (1993), Hagiwara et al. (1985, 1988, 1993a, 1993b), Hamada et al. (1993), Lubzens et al. (1993), Yamasaki and Tanaka (1996), Balompapueng et al (1997), Hagiwara et al (1997), and Douillet (1998).

Local strains of *Brachionus plicatilis* (Ramírez-Sevilla, et al. 1991) can be isolated and raised. Freshwater rotifers such as *Brachionus calyciflorus* (Awaiss et al., 1992; Rico-Martinez and Dodson, 1992) are useful for fish larvae raised at low salinity such as those of sunshine bass, which usually are raised in ponds. Depending on strain, food, temperature, and water quality, lifespan can range up to at least 17 d, but with the best conditions, 7-12 d probably

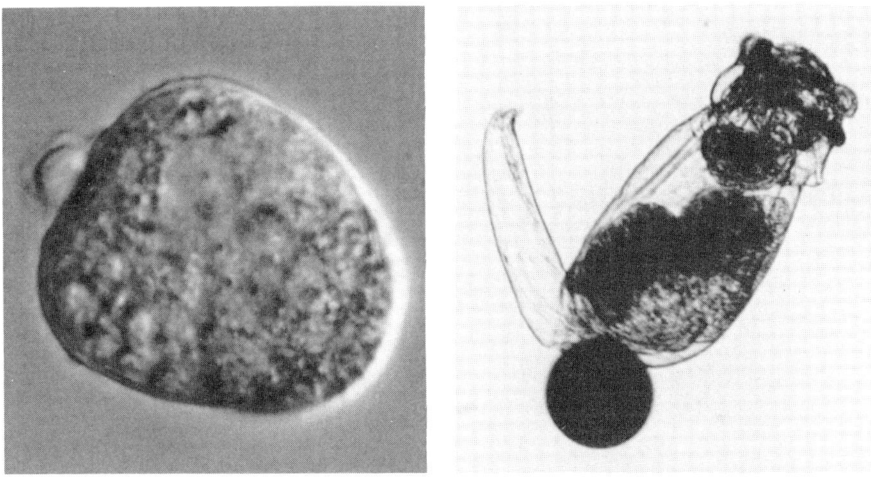

Figure 8-6. Left: Hardshell clam egg, *Mercenaria mercenaria*, ~50 µm (photo by R. LeRoy Creswell). After several hours, the egg hatches into a trochophore larva of about the same shape and size but possessing an apical tuft of sensory cilia (forming already) and two bands of cilia around the middle for swimming and feeding. Right: Female brackishwater rotifer, *Brachionus plicatilis*, which also uses cilia to swim and feed, carrying one egg (egg ~50 µm).

is more typical (Cabrera et al., 1993). Lubzens et al. (1990) reported that when rotifers were held in the dark at 4°C and fed yeast, with water changes every 4-8 d, population replacement occurred in about 12 d. Cultures held at 3°C without food for 3-4 weeks remained viable (Ortega et al., 1995).

Rotifers are reared in many types of containers, both inside (Fig. 8-7) and outside (Figs. 8-8, 8-5). Health and nutritional quality of rotifers depends on several culture factors, including type of culture (e.g., Schmid-Araya, 1992), water quality, temperature, foods (e.g., Carić et al., 1993), rotifer density, and age of the culture (e.g., Carić et al., 1993). James and Abu Rezeq (1989a) fed a local species of *Nannochloropsis* (20 million cells/mL, constantly) and baker's yeast (four times/d) to continuous cultures of *Brachionus plicatilis rotundiformis* in 100-L and 1000-L chemostats. With addition of 144 L algae/d to 100-L chemostats, a maximum production of 319 rotifers/mL/d was reached, with doubling in 1.0 d. With 960 L algae/d added to 1000-L chemostats, maximum production was 261 rotifers/mL/d, with doubling in 0.9 d. With the use of filtration mats, oxygen injection, pH control, defoaming agents, and freshwater *Chlorella* sp. as food, densities of up to 30,000/mL are possible; tank space and operating cost per rotifer are reduced, but additional equipment and attention are needed (Yoshimura et al., 1996a, 1996b). More typical harvest densities are in the range 100-1000/mL.

Rotifers have been reared on many species of microalgae. *Brachionus plicatilis* consume food in the width range 3-17 µm but prefer food smaller than 12-15 µm (Walker, 1981; Hoff and Snell, 1989). All of the microalgal species in Table 8-1 are well within this range except the diatoms (Bacillariophyceae), which are at the upper size limit and could be more difficult to ingest because of their setae or shape. *Chaetoceros calcitrans* and *Thalassiosira pseudonana* rarely are fed to rotifers. The prymnesiophytes and eustigmatophytes listed are mostly less than 8 µm in width and are more suitable for the smaller rotifer strains and for young rotifers. Algae fed to rotifers should contain significant amounts of EPA and DHA because one or both of these are essential fatty acids (EFA) in the diet of marine fish. The ability of rotifers to synthesize ω3HUFA is limited and their diet must include a generous

Figure 8-7. Rotifer cultures in indoor fiberglass tanks.

Figure 8-8. Rotifer cultures in outdoor fiberglass tanks.

portion of these if the requirements of marine fish larvae eating the rotifers are to be met (Lubzens et al., 1985).

Choice of algal species to feed rotifers depends partly on climate or season and whether algae is cultured inside or outside (Table 8-1). Culture conditions and growth phase strongly affect lipid and fatty acid contents. The chlorophytes are low in ω3HUFA and not very useful. The use of *Dunaliella tertiolecta* alone to feed rotifers resulted in low survival and growth of turbot, but good growth occurred when *Isochrysis galbana*, *Phaeodactylum lutheri*, and *Phaeodactylum tricornutum* were used (Scott and Baynes, 1978). The prasinophytes grow fast and are good food for rotifers but also are very low in DHA. Under the conditions of several studies, the bacillariophytes and eustigmatophytes seem to be reliable as EPA sources; their DHA content appears to fluctuate more. The prymnesiophytes are most likely as sources of DHA. *Nannochloropsis oculata* (previously called Japanese marine *Chlorella*) is widely used as a rotifer food because it is easy to grow, resisting both contamination and competition, and it supports rotifer growth very well. With good reasons, many culturists believe it to be the single most important algae species for marine fish larviculture. To avoid the effects of a possible DHA deficiency, *N. oculata* could be supplemented with at least one other species likely to have a good DHA content, such as Tahitian *Isochrysis*. Other species combinations can be used.

Fatty acid variations definitely result from differences in culture conditions. Optimal temperatures for algal growth usually do not produce the highest EPA and DHA levels. Mortensen et al. (1988) found that *C. gracilis* had higher EPA and DHA levels when cultured at 18°C than at 22-32°C, even though growth was best at about 30°C. Theoretically, it might be better to harvest and use algae when ω3HUFA content is highest. However, it might not be practical to use algae when cell density is low and residual nutrients are high. In one study, *N. oculata* had EPA levels 2.5-3.3 times higher during exponential growth than during the stationary phase (Okauchi et al. 1990). *I. galbana* contained zero EPA and DHA during exponential growth, but had levels of 6.9% EPA and 2.4% DHA late in the stationary phase (Fernández-Reiriz et al., 1989). At temperatures much higher or lower than optimal, fatty acid distributions can change drastically (James et al., 1989). *Chlorella saccharophila* had an EPA level 2.6 times higher at 25°C than at 28°C (Teshima et al., 1983). Thompson et al. (1990) observed in *C. calcitrans*, *C. gracilis*, *P. tricornutum*, *P. lutheri*, and *I. galbana* a tendency for DHA levels to increase, and in *C. gracilis*, *P. tricornutum*, and *Pavlova lutheri* a tendency for EPA levels to decrease, with increasing illumination.

The use of frozen concentrated microalgae allows algal culture to be done in advance or at facilities distant from the fish hatchery (Lubzens et al., 1995b). Feeding concentrated *Nannochloropsis* sp. to rotifers permits higher rotifer density and the production of resting eggs. In that study, ω3HUFA uptake was similar for fresh and frozen *Nannochloropsis* sp., but reproductive rate was better with fresh algae. Microalgae also can be cryopreserved (Cañavate and Lubiñn, 1995).

To reduce algae production costs, baker's or marine yeast sometimes is used as the main food or as a supplement (James et al., 1987; Lubzens, 1987). Sanitation is more critical when using yeast. Foods other than live algae and yeast have been less successful for continuous production (Lubzens, 1987). Vitamin B_{12} is likely to be deficient (for rotifers) in baker's yeast and should be supplemented (Hirayama, 1987). Marine yeast has a better content of ω3HUFA but is more expensive than baker's yeast. Commercial dry rotifer feeds have recently become available, but extra sanitation measures are required and enrichment might still be needed (Lavens et al., 1994). Rotifers can grow well with bacteria as food (Ushiro et al., 1980; Yasuda and Taga, 1980), but resulting rotifer compositions have not been well investigated. There is some potential for using ω3HUFA-producing bacteria to fortify rotifers raised on yeast (Watanabe et al., 1992).

Brine Shrimp

Useful references on biology and culture of *Artemia* (Fig. 8-9) are Persoone et al. (1980), Léger et al. (1986), Sorgeloos et al. (1983, 1986, 1993), Verischele et al. (1990), Browne et al. (1991), Talbot and Battaglene (1991), Dhert et al. (1992a), Planko and O'Sullivan (1993), Nimura et al. (1994a, 1994b), Barata et al. (1996), and Fábregas et al. (1996). Other anostracans such as *Thamnocephalus platyurus* are potential larval foods (Mura, 1995).

Hydration of *Artemia* cysts in fresh water for 1-2 h at 25°C causes the slightly shriveled chorion to regain a spherical shape, which facilitates hatching. Usually, hydrated cysts are incubated in salt water in some sort of cone- or v-shaped tank (Fig. 8-10) with moderately high light and turbulence until the nauplii hatch. Then the hatching tank becomes a separatory funnel. Aeration is turned off, heavy debris sinks and is drained off to be discarded, next nauplii are drained off up to the surface layer of floating debris, which is discarded. Nauplii usually will sink quickly to the bottom of the tank, but a strong light can be used to attract them. Planko and O'Sullivan (1993) suggested the following conditions for best hatching: 25-30°C, 20-25 ppt, >2 mg/L DO, and <5 g cysts/L water.

Because rinsing or decapsulation of cysts does not completely disinfect *Artemia* (Dehasque et al. 1991; Verdonck et al., 1991), GomezGil-RS et al. (1994) suggested that *Artemia* be

Figure 8-9. Brine shrimp, *Artemia* sp. (left to right): empty shell (~250 μm), prenauplius inside hatching membrane, newly hatched instar I nauplius (~430 μm TL), instar II metanauplius 12 h after hatching (~580 μm TL); bottom: adult female (~9 mm TL). Rotating the upper left panel 90° clockwise shows the typical natural positions of shell and prenauplius at hatching. (photos courtesy of Artemia Reference Center, University of Ghent, Belgium)

Figure 8-10. Transparent acrylic hatching tank for *Artemia*.

hatched without decapsulation, then rinsed and disinfected with 5 mg/L sodium hypochlorite for 15 min. The nauplii would have to be rinsed very thoroughly before they are fed to fish. Theisen et al. (1998) reported that *Artemia* nauplii survived exposure to ozone that eliminated nearly all bacteria from them; it might be safer than chlorine.

Variations in hatching success, naupliar size, nutritional value, and freedom from contaminants in different batches of cysts (Léger et al., 1986) affect the usefulness of *Artemia* as a larval fish food. The reported dimension range of nauplii is 375-450 µm wide × ~428-517 µm long (Léger et al., 1986), with compressed widths as narrow as 176 µm (Fernández-Diaz et al., 1994). Decapsulated *Artemia* cysts and newly hatched nauplii are not very digestible for marine fish larvae and should not be relied on as a main food. Decapsulated cysts also are not very attractive and not likely to be eaten well by typical marine fish larvae.

Klein-MacPhee et al. (1982) concluded that *Artemia* containing higher levels of EPA were more suitable for feeding to marine fish larvae and reported that the Reference, Australian, and Brazilian strains were the best that had been evaluated at that time. However, fatty acid profiles vary considerably not only by general area of collection but also by time and specific location of collection. The fatty acid content of *Artemia* adults and their cysts are determined by the fatty acids of the adults' food supply, which can vary by location, year, and season (Bengtson et al., 1991). *Artemia* containing linolenic acid but not much EPA have been considered more useful for freshwater fish, and those with enough EPA have been considered suitable for freshwater and marine fish (Navarro et al., 1993). However, the natural EPA content of *Artemia* usually is not optimal for marine fish.

Survival of marine fish larvae fed *Artemia* nauplii is highly unpredictable because of their nutritional variability (Léger et al., 1986; Bengtson et al., 1991). In some studies, *Artemia* nauplii have been starved before feeding them to fish (Léger et al., 1986). The resulting loss of lipid and shifts in amino acid and fatty acid profiles might or might not improve their quality, depending on original composition, culture conditions, and type of predator. *Artemia* could have slight deficiencies in niacin, pyridoxine, riboflavin, and selenium (Bengtson et al., 1991). If they are fed to marine fish larvae (with a few exceptions, such as European plaice) for more than a few days, all *Artemia* from all sources should be enriched with essential fatty acids and possibly vitamins (C especially).

Other Live Foods

Early larvae of some species (e.g., butterflyfish, angelfish, wrasses, parrotfish) have small mouths and might require prey smaller than marine rotifers, which mostly range in width from 35 to 200 µm. The marine dinoflagellate *Gymnodinium splendens* (Lasker et al., 1970; Thomas et al., 1973; Scura and Jerde, 1977), marine ciliates such as *Euplotes* sp. (May, 1970; Reguera, 1984; Hoff and Snell, 1989; Tucker, unpub.), the young of freshwater ciliates such as *Paramecium* sp. (Mitchell, 1991), and larvae of many copepods have widths of 20-50 µm and are suitable food for larvae of some fish. Larval fish will consume ciliates that occasionally contaminate algae or rotifer cultures. Their value, if any, is unknown for most fish species. Ciliates have been raised with microalgae such as *D. tertiolecta*, Tahitian *Isochrysis* sp., and *N. oculata* (Reguera, 1984; Tucker, unpub.). They seem to do better with a heavy organic load, (e.g., when they contaminate older rotifer cultures).

Copepods are the main natural food for marine fish larvae (Poulet and Williams, 1991), usually are nutritionally complete (Corner and O'Hara, 1986), and make useful supplements. Examples of calanoid copepods successfully cultured and used as fish food are: *Acartia clausi* (Zillioux, 1969; Klein Breteler and Gonzalez, 1982; Watanabe et al., 1983a; Buttino, 1994), *Acartia tonsa* (Zillioux, 1969; Parrish and Wilson, 1978; Kuhlmann et al., 1981; Turk et al., 1982; Støttrup et al., 1986; Jónasdóttir, 1994; Lutz et al., 1994; Næss, 1996; Tucker, unpub.), *Eurytemora affinis* (Kuhlmann et al., 1981; Barthel, 1983; Naess et al., 1996), *Oithona* sp. (Watanabe et al., 1983a; Shansudin et al., 1997), *Paracalanus* sp.

(Watanabe et al., 1983a), and *Pseudocalanus* sp. (Klein Breteler and Gonzalez, 1982). Examples of harpacticod copepods are: *Euterpina acutifrons* (Zurlini et al., 1978; Ben-Amotz et al., 1987; Szyper, 1989; Kraul, 1990a), *Tigriopus japonicus* (Fig. 8-11; Iwasaki, 1973; Kitajima, 1973; Watanabe, 1979; Lee and Hu, 1981; Lee et al., 1981; Smith and Hataya, 1982; Watanabe et al., 1983a; Hagiwara et al., 1995b; Tucker, unpub.), *Tigriopus californicus* (Shaw et al., 1994; Tucker, unpub.), *Tigriopus brevicornis* (Vilela, 1992), *Tisbe* spp., usually *T. furcata* or *T. holothuriae* (Johnson and Olson, 1948; Battaglia, 1970; Hunter, 1976; Kahan et al., 1981/1982; Lopez, 1982; Uhlig, 1984; Guerin and Rieper-Kirchner, 1991; Rieper-Kirchner et al., 1991; Zhang and Uhlig, 1991, 1993; Miliou et al., 1992; Heath, 1994; Norsker and Støttrup, 1994; Støttrup and Norsker, 1997; Tucker, unpub.), and *Amphiascoides atopus* (Sun and Fleeger, 1995). Culture techniques are similar to those used for rotifers, but copepods are more expensive to produce because most take 5-10 times longer than rotifers to double and they cannot be crowded as much. Foods given to copepods include: microalgae—*Chlamydomonas* sp., *Dunaliella tertiolecta*, *Nannochloris* sp., *Tetraselmis* sp., *T. maculata*, *T. suecica*, *T. tetrathele*, *Chaetoceros danicus*, *C. gracilis*, *C. muelleri*, *Phaeodactylum tricornutum*, *Isochrysis galbana*, *Pavlova lutheri*, *Nannochloropsis oculata*, *Skeletonema costatum*, *Thalassiosira pseudonana*, *Rhodomonas baltica*, *Rhodomonas lens*, *Chroomonas salina*, *Gymnodinium* sp., *Oxyrrhis marina*; and a variety of others—*Ulva* sp., lettuce, baker's yeast, baker's yeast with cuttlefish liver or other marine oils, marine yeast, rice bran, wheat, alcohol fermentation wastes, soy sauce cake, fish muscle, dried mollusc, tunicates, and powdered or flake fish feed. Copepods also eat microorganisms that feed on the added foods. Smaller foods (4-8 µm diameter) such as *I. galbana* and *N. oculata* seem of more value to early stages of copepods than to adults (Zillioux, 1969; Støttrup et al., 1986). Feeding bristles of *Acartia clausi* are spaced at 4-6.5 µm in stage VI nauplii and 15-25 µm in adults (Zillioux, 1969). *Acartia tonsa* copepodites consume particles in the range 14-59 µm (Raymont, 1983).

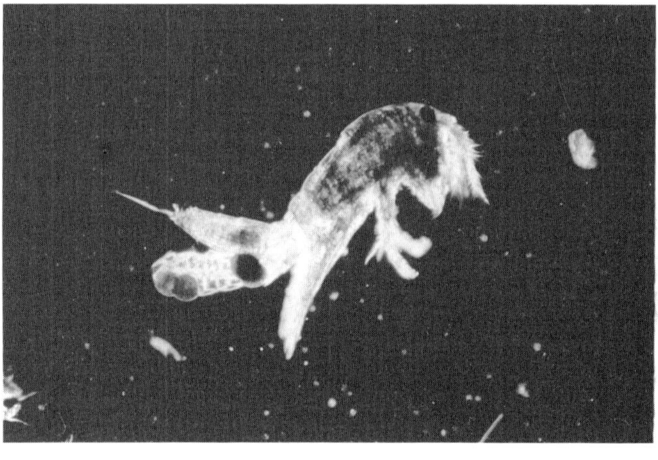

Figure 8-11. A harpacticoid copepod, *Tigriopus japonicus*, female carrying four eggs.

Freshwater cladocerans such as *Moina* sp. and *Daphnia* sp. (Watanabe, 1979; Rees and Oldfather, 1980; Ventura and Enderez, 1980; Oka et al., 1982; Watanabe et al., 1983a; Hoff and Snell, 1989; Shim, 1988; Villegas, 1990; Mims et al., 1991; Sun et al., 1991; Bonou and Saint-Jean, 1998), brackishwater *Diaphanosoma aspinosum* (Segawa and Yang, 1988), and saltwater *Podon* sp. (Kuhlmann et al., 1981) raised on bacteria, microalgae, baker's yeast, rice bran, pig manure, and poultry manure have been used or have potential.

Other copepods and cladocerans as well as larvae of molluscs, polychaetes, barnacles, sea urchins, ascidians, and fish have been tried, but usually these are not as reliable in quantity or quality (May, 1970; Kuronuma and Fukusho, 1984). Some large copepods might attack fish larvae, so care should be taken when untested species are used. Appelbaum et al. (1995) reported that male *Tigriopus californicus* have large pinchers with which they held and killed lingcod larvae, but they have successfully been fed to several other species. Nematodes similar in size to rotifer and copepod larvae (50 µm long) can be cultured and fed to fish (Biedenbach et al., 1989; Rouse et al., 1992).

In the past, oyster larvae were used as a larval food for red sea bream, but were found to be no better nutritionally than rotifers and more difficult to produce (Kuronuma and Fukusho, 1984). Rotifers were of considerably more value to turbot and common sole larvae than mussel trochophores (Howell, 1973). On the other hand, we have found that hard clam (*Mercenaria mercenaria*) trochophores (Fig. 8-6) can be a good supplemental starter food if given with small rotifers and then replaced by rotifers as soon as the fish are ready. Sanitary broodstock holding and spawning methods should be employed to minimize pathogens such as *Vibrio* spp. and *Pseudomonas* spp., which can be associated with molluscs. Clam or oyster trochophores can be cryopreserved and stored for long periods (Chao et al., 1997). When needed, they are thawed within a few minutes; if handled properly, a reasonable proportion can be revived, and they do not develop shells or grow larger. Alternatively, shell development and further growth can be prevented by ultraviolet irradiation of the mollusc sperm to the point that about half of them stop swimming (e.g., expose a 1-mm-thick dilute aqueous suspension 28 cm from a 254-nm, 30-watt, germicidal bulb for 15-180 s, usually 60 s) before the eggs are fertilized (Scarpa, 1997).

Wild plankton can be collected with various nets and traps (Fig. 8-12)(Barnabé, 1990d; Tacon et al., 1991b; Daintith, 1992; Andersson, 1995). Nutritional quality is likely to be very high, and it has been used safely in many studies (e.g., tintinnids, copepods, and others, Houde and Schekter, 1980; copepods and crab megalopa, Tucker, 1987a). However, the appreciable chance of introducing pathogens or pests such as parasitic or predaceous copepods, isopods, medusae, anemones, or worms, or toxic dinoflagellates, plus variation in quantity and quality limit the use of wild plankton. Another alternative is extensive culture of zooplankton in ponds or impoundments (Quin, 1993). Also, young fish in cages can feed on zooplankters that swim through the mesh (Fermin and Seronay, 1997).

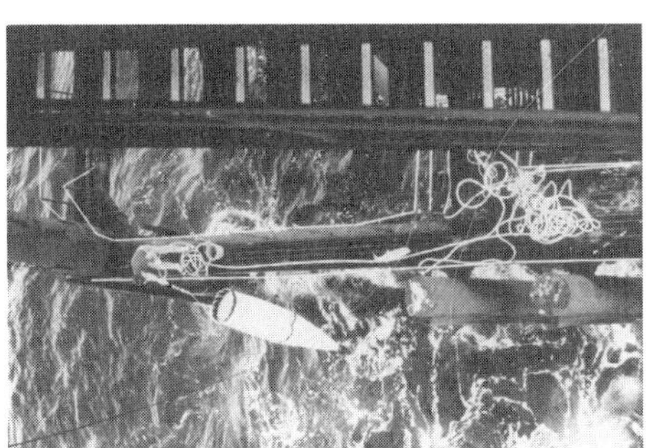

Figure 8-12. Collecting wild plankton from a light tower 40 km off the North Carolina coast.

Enrichment of Live Foods

To reduce uncertainty concerning lipid quality of zooplankton, they can be enriched. Some marine oils are reliable sources of EPA and DHA, and mixtures of purified oils also are used (Table 8-2). Enrichment of rotifers can be easier than the culture manipulation and biochemical analyses that seem necessary if only algae is used. However, enrichment of rotifers with high concentrations of oil often is messy and can be lethal to the rotifers. When the lipid content of rotifers is very high, they do not survive as well, especially in cold water (Olsen et al., 1993b). Therefore, enrichment should be designed to increase mainly essential fatty acids rather than nonessential ones, and still keep all of them balanced well enough so the fish can utilize them normally. *Euglena* enriched with essential fatty acids can be fed to rotifers or *Artemia* (Hayashi et al., 1993a, 1993b), but that requires an extra step.

Because rotifers cultured on baker's yeast alone are deficient in ω3HUFA (Ben-Amotz et al., 1987), they can be enriched for up to 48 h with yeast enriched with oil and sometimes vitamins, or with other materials (Table 8-2; Smith and Hataya, 1982; Watanabe et al., 1983a; Hoff and Snell, 1989; Lubzens et al., 1989). Addition of cuttlefish liver oil to baker's yeast fed to rotifers raised their EPA level from 1-2% to 9-12% and DHA level from zero to 7-9% (Watanabe et al., 1983a).

Léger et al. (1986) and Bengtson et al. (1991) reviewed five types of methods for enriching *Artemia*, which also apply to other types of zooplankton. In the British technique, nauplii are fed microalgae (usually *Isochrysis* sp.) for 24 h or longer. In the Japanese indirect method, either microalgae (usually *N. oculata*) or ω-yeast is fed for at least 24 h. In the Japanese direct method, baker's yeast with emulsified marine oil (usually cuttlefish or pollock liver oil) is fed for up to 24 h. In the French technique, a compound feed containing ω3HUFA is fed for up to 24-48 h. In the Belgian technique, a micronutrient-fortified marine oil emulsion is fed for up to 12-24 h. Retention of DHA can vary some with strain or species of *Artemia* (Evjemo et al., 1997).

Sorgeloos et al. (1988) reported that feeding rotifers or *Artemia* enriched with a Selco® emulsion to larvae of species such as milkfish, barramundi, European sea bass, dolphin, gilthead sea bream, and yellowblotch spinefoot resulted in higher growth and survival rates. Growth and survival of species such as European sea bass and turbot have been enhanced when they were fed rotifers or *Artemia* enriched with Frippak Booster® microcapsules. Rimmer and Reed (1990) reported very good growth and high survival of barramundi larvae fed *Artemia* enriched with Frippak Booster® microcapsules compared with those fed unenriched *Artemia.* Tocher et al. (1997) found that rotifers and *Artemia* enriched with emulsions made from fish heads were of very good quality. Robin et al. (1993) found that, even though EFA levels were highest in rotifers fed an oil emulsion, survival and growth of turbot larvae were better (but not with statistical significance) when they were fed rotifers given microcapsules or microparticles; such results might occur from fatty acid imbalances or from excess oil contributing to surface films.

When enriched with an emulsified mixture of fatty acid methyl esters (FAME) containing 36% EPA and 43% DHA, high ω3HUFA levels were reached in rotifers after 3-6 h at 21-24 °C (59% ω3HUFA in FA and 3.5% in wet rotifers), in *Artemia* after 12 h at 23-24°C (20% ω3HUFA in FA and 0.9% in wet *Artemia*), and in *Moina* after 24 h at 18-24°C (45% ω3HUFA in FA)(Watanabe et al., 1983b). When fat soluble vitamins A and D were added to cuttlefish liver oil used for enrichment of rotifers, incorporated content of vitamin A reached a maximum after 6 h and vitamin E after 24 h.

Takeuchi et al. (1992) found that maximum uptake by *Artemia* of ω3HUFA from triglyceride, methyl ester, and ethyl ester mixtures occurred between 6 and 12 h at 22-24 C and between 12 and 24 h at 24-28°C. A free fatty acid mixture resulted in maximum uptake by 12 h in both temperature ranges. All four mixtures contained 43% ω3HUFA.

Støttrup et al. (1995) reported that the ω3HUFA content of rotifers was strongly altered

Table 8-2. Percentages of palmitic (16:0), palmitoleic (16:1ω7), oleic (18:1ω9), linoleic (18:2ω6), linolenic (18:3ω3), arachidonic (20:4ω6), eicosapentaenoic (20:5ω3), and docosahexaenoic (22:6ω3) acids and total ω3HUFA in some materials used to feed or enrich zooplankton fed to marine fish larvae.

Material	16:0	16:1ω7	18:1ω9	18:2ω6	18:3ω3	20:4ω6	20:5ω3	22:6ω3	DHA/EPA	ω3HUFA	References
				% of total fatty acids							
Soybean oil triglycerides	12.6	0	23.6	53.5	4.4	0	0	0		0	Miyashita et al., 1995
ICES no-HUFA enrichment medium	10.1	0	10.8$^+$	2.4	0	0	0	0		0	Ashraf et al., 1993a
Baker's yeast (*Saccharomyces cerevisiae*)	23.8		43.4$^\#$	6.7	0.9		0	0.2		0.2	James et al., 1987
Marine yeast (*Candida* sp.)	10.6		53.2$^\#$	9.5	1.4		0.8	0.7	0.88	1.5	James et al., 1987
Capelin oil	11.8	8.7	9.0	1.6	0.7	0.4	7.4	3.9	0.53	11.7	Kissil & Koven, 1990
Pollock liver oil	12.6	17.4$^+$	10.7$^+$	1.7	0.8	0.2	10.9	2.2	0.20	13.5	Izquierdo et al., 1989
Atlantic menhaden oil	24.0	9.2	12.4	5.7	0.6		9.1	5.5	0.60	15.7	Davis & Olla, 1992
ICES medium-HUFA enrichment medium	15.6	7.5	30.8$^+$	7.0	3.0	0.6	10.0	5.6	0.56	16.2	Ashraf et al., 1993a
Walleye pollock oil	11.2	7.3	12.2	0.8	0.6	0.3	11.6	7.4	0.64	≥19.0	Ackman & Kean-Howie, 1995
Cuttlefish liver oil	12.8	11.3$^+$	8.6$^+$	2.0	2.0	0.5	14.3	5.1	0.36	20.1	Izquierdo et al., 1989
Menhaden oil (*Brevoortia* spp.)		11.1	14.0		2.0		14.0	8.1	0.58	23.1	Clawson & Lovell, 1992
Japanese sardine oil	22.0	7.9	11.6	0.9	0.8	1.2	13.7	9.3	0.68	≥23.1	Ackman & Kean-Howie, 1995
Tuna eye oil	17.2	6.0	18.1	1.1	0.6	1.5	5.7	21.3	3.74	30.0	Tocher et al., 1997
Squid oil	16.8		19.1	2.0	1.5	0	14.6	16.9	1.16	32.0	Kanazawa, 1993a
Arrow squid liver oil	21.8	5.8	9.6	0.8	0.4	2.9	20.2	11.5	0.57	32.2	Hayashi, 1996
Copepod emulsion (mainly *Calanus finmarchicus*)	13.3	9.6	5.6	1.7	2.8	4.2	14.1	19.3	1.37	34.0	Rainuzzo et al., 1994
Bonito head oil	20.7	7.1	13.2	1.5		1.9	7.1	24.0	3.38	35.8	Ando et al., 1997
Triacylglycerol mixture	10.0	9.4	10.0	2.0	1.1	1.5	25.0	11.1	0.44	39.9	Rodriguez et al., 1996
Several experimental oil mixtures	8.0	9.4	7.5	1.3	1.0	1.2	27.0	12.4	0.46	43.0	Takeuchi et al., 1992a
Atlantic halibut roe emulsion	17.0	3.6	10.1	4.1	0.9	1.1	9.4	37.2	3.96	48.4	Rainuzzo et al., 1994
Protein Selco® (30.3% lipid, DW)	10.7	5.2	9.8	7.3	1.1	1.9	24.2	24.0	0.99	51.7	Fernández-Reiriz et al., 1993
Dry Selco® (56.1% lipid, DW)	11.8	4.3	9.8	4.8	1.0	0	25.3	24.1	0.95	52.7	Fernández-Reiriz et al., 1993
Super Rotifer® (28.1% lipid, DW)	11.7	1.8$^+$	8.6	2.0	2.3		22.9	30.8	1.34	53.7	Estévez & Kanazawa, 1996
Super *Artemia*® (24.5% lipid, DW)	7.4	0.7$^+$	2.2	4.4	0.5		20.1	35.0	1.74	55.1	Estévez & Kanazawa, 1996
ICES high-HUFA enrichment medium	2.3	1.2	12.5$^+$	1.6	3.4	1.6	26.4	28.9	1.09	58.9	Ashraf et al., 1993a
Super Selco® (94.2% lipid, DW)	2.7	0	8.4	1.8	0.8	2.2	29.4	34.8	1.18	69.1	Fernández-Reiriz et al., 1993
Ester-85® HUFA mixture	0.9		1.2	4.7	1.2	0	28.8	41.7	1.45	76.1	Kanazawa, 1993b
				% of dry weight							
Super Selco®	0.19	0.05	0.82	0.28	0.21	0.06	1.62	2.15	1.33		Kraul et al., 1992

$^+$ Analysis included other monoenes. $^\#$ Analysis included 18:0.

after 3 h of feeding on any of four species of algae. Rodríguez et al. (1996) found that maximum uptake (4% of DW) by rotifers of ω3HUFA from a 4% emulsion of a triacylglycerol mixture containing 40% ω3HUFA occurred between 6 and 12 h at 17-19°C. Raising the concentration to 8 or 12% increased rotifer mortality with no appreciable rise in uptake. A methyl ester mixture resulted in less ω3HUFA uptake and might be toxic to rotifers and fish eating them.

Rimmer and Reed (1990) raised EPA levels from 11-12% to 13% (of FA) and DHA from 0% to 3% in rotifers by enriching them with Frippak Booster® microcapsules. Rotifer fatty acid profiles were similar for enrichment periods of 4 and 20 h, and no advantage of the longer enrichment was indicated. Newly hatched *Artemia* contained negligible EPA and DHA. In unfed *Artemia*, 2% EPA was present at 4 hah and 3% EPA at 24 hah. No advantage was indicated for holding *Artemia* for 24 h instead of 12 h before enriching them. Metanauplii enriched for 4 h (9% EPA, 2% DHA) had a better fatty acid profile than those enriched for 20 h (7% EPA, 1% DHA).

For maximum uptake of ω3HUFA, the manufacturer's recommended enrichment regimens for Artemia Systems Super Selco® emulsion at 25-28°C were 100 mg/million rotifers for 3-4 h, twice, and 1 g/million *Artemia* for 12 h, twice. For Frippak Booster® microcapsules at 20°C, the recommendations were 1 g/million rotifers for 4 h, once, and 2 g/million *Artemia* for 8 h, once. From 20°C to 26-30°C, the rate of uptake should more than double, according to Van't Hoff's Law. Also, decomposition of enrichment media is very likely to result in fouling of water and high bacterial growth after 4 h. Apparently, avoidance of fouling during long enrichment with high-lipid emulsions requires nearly sterile water, which can be impractical. During enrichment of *Artemia* at 27°C, oxidative rancidity of oils can occur as early as 19-23 h, and it probably is prudent to limit the enrichment period to 18 h or less (McEvoy et al. 1995). We usually rely on microalgae as the only rotifer food but sometimes enrich rotifers for up to 3-4 h in salt water filtered to 5 μm (24-28°C). We usually enrich *Artemia* for 4-8 h.

Nutritional Quality of Live Foods (Table 8-3)

Protein content (of DW) of rotifers has been reported to be in the range 30-72% (Scott and Baynes, 1978; Watanabe et al., 1983a; Lubzens et al., 1989; Tamaru et al., 1993; Chang and Yi, 1994; Øie et al., 1994; Rimmer et al., 1994a; Whyte et al., 1994; Ortega et al., 1995). Average protein content of *Artemia* nauplii is about 50% (range 37-71%), with preadults and adults containing about 56% (Léger et al., 1986). Dendrinos and Thorpe (1987) reported 42% protein in San Francisco Bay, California (SFB) *Artemia* nauplii. Adult *Artemia* that were fed algae contained 57% protein (Kim et al., 1996). Watanabe et al. (1983a) reported ranges of 63-75% for *Tigriopus japonicus*, 70-71% for *Acartia clausi*, 70% for *Daphnia* sp., and 68-78% for *Moina* sp. *Tisbe holothuriae* contained 71-73% protein (Miliou et al., 1992). *Moina macrocopa* raised by Villegas (1990) contained 56% protein. The nematode *Panagrellus* sp. contained 48% protein (Biedenbach et al., 1989).

Amino acid balance of rotifers, copepods, and cladocerans tends to be consistently good regardless of culture conditions; however, starving zooplankters will be depleted of protein (Watanabe et al., 1983a; Lubzens et al., 1989). *Artemia* nauplii could be deficient in threonine, histidine, methionine, and phenylalanine, but adults have a good balance (Watanabe et al., 1983a), and generally amino acid content of nauplii and metanauplii is good for fish, with methionine being the first-limiting amino acid (Léger et al., 1986; Bengtson et al., 1991).

Total lipid content (of DW) of rotifers has been reported to be in the range 4-37% (Scott and Baynes, 1978; James et al., 1987; Lubzens et al., 1989; Olsen et al., 1989; Rainuzzo et al., 1989; Chang and Yi, 1994; Øie et al., 1994; Ortega et al., 1995; Rodriguez et al., 1996; Table 8-3 and included references). Minkoff and Broadhurst (1994) reported these ω3HUFA rotifer contents (of DW) and DHA/EPA ratios (in parentheses): 3.1 (0.3) after mass culture, 5.1 (0.2) after feeding on *Nannochloris* sp., 4.0 (0.7) after feeding on *Isochrysis galbana*, 14.2 (0.7) after enrichment with Protein Selco,® 19.2 (0.8) after enrichment with Dry Selco,® and 38.7 (0.7) after enrichment with Super Selco.®

Total lipid content (of DW) in *Artemia* decapsulated cysts and nauplii was in the range 9-30% (Watanabe et al., 1983a; Léger et al. 1986; Navarro et al., 1993). Dendrinos and Thorpe (1987) measured 22.6% total lipid in one batch of SFB nauplii. Bengtson et al. (1991) and Sorgeloos et al. (1993) reported zero or very low DHA and the following EPA level (of FA) ranges in nauplii: 0.3-0.4% for the North Arm of Great Salt Lake, Utah (GSL); 0.3-13.3% for SFB; 1.3-15.4% for Bohai Bay, China; 2.7-3.6% for the South Arm of GSL; 3.5-10.6% for Macau, Brazil; and 5.2-9.5% for Chaplin Lake, Canada. Some brands of *Artemia* are sold with analysis values on the can labels, which are very helpful to the user. Adult *Artemia* that were fed algae contained 10.1% lipid (Kim et al., 1996a).

Watanabe et al. (1983a) reported lipid contents (of DW) of 14-26% for *T. japonicus*, 8-12% for *A. clausi*, 13% for *Daphnia* sp., and 12-27% for *Moina* sp. Støttrup (1993) reported 1.2% linolenic acid, 2.0% EPA, and 0.9% DHA (of DW) in *Calanus finmarchicus* and 0.4% linolenic acid, 1.4% EPA, and 1.7% DHA in *Tisbe holothuriae*. *T. holothuriae* contained 9.7-10.8% lipid (Miliou et al., 1992). Lipids of *Euchaeta japonica* ranged from 64% in eggs to a minimum of 12% in copepodite II and 49% in adult males and 52% in adult females (Sargent and Henderson, 1986). In June (good feeding conditions), copepodites and males of *Calanus helgolandicus* had 34% lipid and females 28%, but in January (conditions not as good), copepodites had 16% lipid, males 5%, and females 3% (Sargent and Henderson, 1986). *Moina macrocopa* raised by Villegas (1990) contained 6% lipid. The nematode *Panagrellus* sp. contained 17% lipid (Biedenbach et al., 1989).

Table 8-3 illustrates the effects of different culture and enrichment methods on the fatty acid composition and total lipid content of prey organisms typically used for marine fish larvae. The amount of ω3HUFA in prey can be raised by a combination of increasing ω3HUFA in the prey's food and increasing retention of fat in the prey, but an unhealthy excess of nonessential fatty acids must be avoided.

For typical marine fish larvae, the wild copepods are the most natural representatives in Table 8-3 and have a good balance of essential fatty acids. Wild mixed zooplankton (mainly calanoid copepods) had DHA/EPA ratios of 1.03-1.73. Wild *Acartia clausi* had levels of 16.6-29.2% EPA and 12.3-28.6% DHA (of total lipids); on the average, they contained 2.31% EPA, 2.38% DHA, and 4.82% ω3HUFA (of DW)(Watanabe et al., 1983a). Fujita et al. (1980) reported levels of 18.7% EPA and 20.7% DHA (of total lipid)(0.4% ω3HUFA of WW) in wild *Acartia clausi*. *Tigriopus japonicus* contained 1.87% EPA, 1.48% DHA, and 4.08% ω3HUFA (of DW) when reared under natural conditions and 2.19% EPA, 2.76% DHA, and 5.40% ω3HUFA when enriched with cuttlefish liver oil; the fed *T. japonicus* had a higher DHA/EPA ratio (Watanabe et al., 1983a).

Unless enriched, *Daphnia* could be deficient in at least DHA and *Moina* could be deficient in both EPA and DHA. *Daphnia* sp. had levels of 16.5% EPA and 0.2% DHA (of total lipids)(Watanabe et al., 1983a). Mims et al. (1991) reported that *Daphnia pulex* from a fertilized pond (10°C) had contents of 2.43% EPA, 0.33% DHA, 2.95% ω3HUFA, and 23.6% total lipid (of DW) and those raised in tanks with horse manure (18°C) had 1.19% EPA, 0.33% DHA, 1.53% ω3HUFA, and 20.9% total lipid. *Moina* sp. had levels of 0.2-1.5% EPA (of total lipid) and zero DHA when raised on baker's yeast, 14.5-20.8% EPA plus a trace of DHA on poultry manure, and 7.0% EPA plus 0.3% DHA with both yeast and manure (Watanabe et al., 1983a). Oka et al. (1982) reported that initial contents of 0.17% EPA, 0.03% DHA, 0.44% ω3HUFA, and 10.8% total lipid (of DW) in *Moina* sp. were altered by feeding to 0.46% EPA, 0.03% DHA, 0.84% ω3HUFA, and 14.4% total lipid on baker's yeast; 0.23% EPA, 0.02% DHA, 0.50% ω3HUFA, and 14.7% total lipid on freshwater *Chlorella regularis*; and 4.24% EPA, 1.02% DHA, 6.10% ω3HUFA, and 27.5% total lipid on ω-yeast.

Table 8-3. Percentages of palmitic (16:0), palmitoleic (16:1ω7), oleic (18:1ω9), linoleic (18:2ω6), linolenic (18:3ω3), arachidonic (20:4ω6), eicosapentaenoic (20:5ω3), and docosahexaenoic (22:6ω3) acids, DHA/EPA ratio, total ω3HUFA, total lipids (dry weight), and total ω3HUFA (dry weight)—for nematodes, copepods, cladocerans, rotifers, and brine shrimp (Z = zooplankter, fed nothing or various combinations of bacteria, yeasts, microalgae, oils, squid, microcapsules, and emulsions).

Species	16:0	16:1ω7	18:1ω9	18:2ω6	18:3ω3	20:4ω6	20:5ω3	22:6ω3	DHA/EPA	ω3HUFA	Lipid	ω3HUFA/Z	Treatment and reference
							Nematodes						
<u>Moderate</u>													
Panagrellus redivivus	12.9	10.5		9.9^{H}	9.3	4.6	7.4	3.2^{L}	0.43	10.7	14.9	1.59	Cultured at 20-24°C with wheat flour, yeast, cod liver oil, and chicken egg yolk (Rouse et al., 1992)
							Oyster trochophore larvae						
<u>High</u>													
Ostrea edulis	29.7	4.0	4.0	2.2	2.8	1.6	18.1	10.8	0.60	31.3	23.7	7.42	Early trochophores from wild oysters; all numbers except total lipids are based on neutral lipids (Helm et a., 1991)
Crassostrea gigas	19.8	4.2	2.9	1.6	1.9	0.9	15.6	13.5	0.86	33.6			Cryopreserved trochophores (Jones, p.c., 1998)
							Calanoid copepods						
<u>Moderate</u>													
Euterpina acutifrons	16.5	3.4	1.0^{+}	33.8	4.7		2.7	6.0	2.22	8.7			Cultured at 22°C with *Isochrysis galbana* (Ben-Amotz et al., 1987)
Euterpina acutifrons	24.5	4.7	0	6.5	2.2		11.1	16.1	1.45	27.2			Wild (Ben-Amotz et al., 1987)
Wild zooplankton	12.8	1.4	8.7	2.0	0.2	0.1	15.9	27.6	1.73	43.7	0.94	0.41^{L}	Mainly *Oithona* sp. (Shamsudin, 1994)
<u>High</u>													
Wild zooplankton	20.4	1.1	2.3	2.5	6.4		14.7	21.0	1.43	~36.8	~9.7		Almost entirely *Acartia clausi* & *A. tonsa* (Davis & Olla, 1992)
Acartia clausi	17.2	5.0^{+}	4.4^{+}	0.9	0.9		22.0	22.7	1.03	46.0	10.5	4.82	Wild (Watanabe et al., 1983a)
Wild zooplankton	16.9	1.7	2.9	2.0	1.5	0.8	21.1	32.9	1.56	55.3	13.0	7.19	Mainly *Eurytemora affinis* (Næss et al., 1995b)
							Harpacticoid copepods						
<u>Moderate</u>													

Harpacticus littoralis							57.0	1.9^{L}	0.03	~58.9	5.0	~2.97	Fed a microalgae mixture + detritus (Khanaichenko et al., 1994)
High													
Tigriopus japonicus	14.0	6.3^{+}	20.1^{+}	2.5	8.2		8.2	6.5	0.79	17.9	22.8	4.08	Reared under natural conditions with no food added (Watanabe et al., 1983a)
Tigriopus japonicus	12.1	8.1^{+}	21.3^{+}	1.7	1.3		10.8	13.6	1.26	26.6	20.3	5.40	Fed baker's yeast + cuttlefish liver oil (Watanabe et al., 1983a)
Cladocerans—water fleas													
Low													
Moina sp.	13.8	6.0^{+}	6.3^{+}	21.5^{H}	20.5		1.6	0.2	0.12	3.4	14.7	0.50^{L}	Cultured at 14°C with freshwater *Chlorella regularis* (Oka et al., 1982)
Moina macropa	5.8	18.9	23.0	5.9	3.5	3.7	7.8	0.3^{L}	0.04	8.1			Cultured with chicken manure extract (Villegas, 1990)
Moderate													
Daphnia pulex	14.7	6.7	12.1	5.1	12.2	3.3	10.3	1.4^{L}	0.14	12.5	23.6		Reared in fertilized pond at 10°C with no food added (Mims et al., 1991)
High													
Moina sp.	11.2	15.1^{+}	22.0^{+}	2.5	0.9		15.4	3.7	0.24	22.2	27.5	6.10	Cultured at 14°C with ω-yeast = baker's yeast cultured with cuttlefish liver oil (Oka et al., 1982)
Rotifers													
Low													
Brachionus plicatilis	7.9	0	44.9	15.8	5.9	0	0	0		0		0	Fed yeast (*Saccharomyces cerevisiae*) for 1 d & starved for 24 h at 23°C, then fed bacteria (strain ACAM 456) for 24 h at 18°C (Nichols et al., 1996)
Brachionus plicatilis	7.4	31.3	10.6^{+}	0	34.5^{H}		0	0		0	4.5	0	Cultured at 22°C with baker's yeast (Ben-Amotz et al., 1987)
Brachionus plicatilis	22.5	6.9	17.2	19.1	0	4.2	0.6	0		0.6			Cultured at 24-32°C with *Chlorella* sp. (Villegas et al., 1990)
Brachionus calyciflorus				4.8	0.1		0.5	0.1	0.20	~0.6			Cultured at 25°C with *Dictyosphaerium chlorelloides* (Awaiss et al., 1992)
Brachionus plicatilis	8.5	20.0	22.3	9.2	0.4	0.3	0.6	0.1	0.17	0.7	11.2		Cultured at ~17°C with baker's yeast (Fernández-Reiriz & Labarta, 1996)
Brachionus plicatilis	36.0	0	20.4	14.3	0	2.1	1.2	0		1.2			Cultured at 24-31°C with *Tetraselmis tetrathele* (Villegas et al., 1990)

Table 8-3. (Continued)

Species	16:0	16:1ω7	18:1ω9	18:2ω6	18:3ω3	20:4ω6	20:5ω3	22:6ω3	DHA/EPA	ω3HUFA	Lipid	ω3HUFA/Z	Treatment and reference
Brachionus plicatilis	21.8		27.0^{+}	~9.9	~1.2	0	1.3	0		1.3	~2.4	0.03	Cultured at 25°C with yeast (Su et al., 1994)
Brachionus plicatilis	7.3	26.0	29.9	7.2^{H}	0.5		1.4	0.3	0.21	2.4	19.4	0.46	Cultured with baker's yeast (Watanabe et al., 1983a)
Brachionus plicatilis	7.3	23.6	28.5	2.6	1.6	0	0	0.8		2.5	9.8	0.24	Cultured at 20°C with baker's yeast (Whyte & Nagata, 1990)
Brachionus plicatilis	2.2	2.7	3.0	1.6	0.2	0.2	1.1	1.3	1.18	2.6	22.5		Cultured with baker's yeast, then fed an oil emulsion for 18 h at 18.5°C (Rodríguez et al., 1997)
Brachionus plicatilis	8.7	26.8	24.7	2.2	0	0.7	2.4	0.9	0.38	3.3			At 20°C, fed baker's yeast for 9 h, then capelin oil for 6 h (Kissil & Koven, 1990)
Brachionus plicatilis	25.3	5.9	10.0^{+}	8.9	2.8		0.8	0.5	0.62	3.1	7.4		Cultured at 22°C with *Chlorella stigmatophora* (Ben-Amotz et al., 1987)
Brachionus plicatilis	19.7	7.7	5.1	12.9^{H}	0		0.7	0		3.2	9.8	0.32	Cultured at 20°C with *Chlorella saccharophila* (Whyte & Nagata, 1990)
Brachionus plicatilis	23.9	5.9^{+}	14.7^{+}	~26.5	15.3		3.3	0		3.3		0.10	Cultured at 22°C with *Chlorella regularis* (Takeyama et al., 1996)
Brachionus plicatilis	11.4	17.1	18.8	9.3^{H}	0.9	1.2	3.2	0.8	0.25	4.0	14.0	0.36	Cultured with baker's yeast, then fed *Isochrysis galbana* + *Nannochloropsis gaditans* for 24 h at 17°C (Fernández-Reiriz & Labarta, 1996)
Brachionus plicatilis	9.8		24.0	11.8	1.0	2.2	1.8	2.2	1.22	4.0			Cultured at 28°C with baker's yeast (Tamaru et al., 1993c)
Brachionus plicatilis	11.4	15.3	28.0$^{\#}$	8.6	3.2	~1.9	3.2	0.3	0.94	4.6			Cultured at 28-30°C with baker's yeast (Walford & Lam, 1987)
Brachionus plicatilis	9.6	6.4	10.4	13.5	5.6		2.4	4.8	2.00	~7.4	~6.3		Fed Microfeast L-10 for 24 h (Davis & Olla, 1992)
Brachionus plicatilis	26.9	6.8^{+}	13.1^{+}	~22.7	13.5		9.2	0		9.2		0.42	Cultured at 22°C on *Nannochloropsis oculata* (Takeyama et al., 1996)

Brachionus plicatilis	16.1	8.2⁺	21.2⁺	~7.1	10.0		6.0	4.3	0.72	10.3		0.52	Cultured at 22°C with *Chlorella regularis* for 7 d, then Tahitian *Isochrysis* for 24 h (Takeyama et al. 1996)
Brachionus plicatilis	7.7	0.4	12.7	2.1	1.4	trace	9.4	1.3	0.14	12.2		0.51	Fed yeast (*Saccharomyces cerevisiae*) for 1 d & starved for 24 h, then fed bacteria (strain ACAM 456) for 24 h at 18°C (Nichols et al., 1996)
Brachionus plicatilis	16.0	7.3⁺	25.9⁺	~6.6	5.4		8.3	4.6	0.55	12.9		0.51	Cultured at 22°C with Tahitian *Isochrysis* (Takeyama et al., 1996)
<u>Questionable</u>													
Brachionus plicatilis	9.0		24.7	7.4^{H}	1.1		3.6	0.7^{L}	0.19	5.1	34.9		Cultured at 25-27°C with marine yeast, baker's yeast, and *Chlorella* sp. (James et al., 1987)
Brachionus plicatilis	9.1		27.0$^{\#}$	7.3^{H}	1.3		3.2	1.9^{L}	0.59	5.6	26.1		Cultured at 25-27°C with baker's yeast and *Chlorella* sp. (James et al., 1987)
Brachionus plicatilis	18.3	29.2	8.9⁺	3.8	3.8		2.8	1.0	0.36	7.0	20.1		Cultured at 22°C with *Chaetoceros gracilis* (Ben-Amotz et al., 1987)
Brachionus plicatilis				9.7	11.5	1.5	6.8	0.4^{L}	0.06	8.6	11.4	1.14	Fed *Tetraselmis suecica* (Robin, 1995)
Brachionus plicatilis	18.3	20.7	16.7⁺	7.3	1.8		6.5	0.8	0.12	8.9	16.4		Cultured at 22°C with *Nannochloropsis salina* (Ben-Amotz et al., 1987)
Brachionus plicatilis	12.4	9.6	9.6⁺	14.9	4.2		1.6	4.2	2.62	9.2	11.1		Cultured at 22°C with *Isochrysis galbana* (Ben-Amotz et al., 1987)
Brachionus plicatilis	9.8				6.7	0.6	4.4	5.6	1.27	10.0			Cultured with Tahitian *Isochrysis*, then fed *Rhodomonas baltica* for 3 h at 18°C (Støttrup et al., 1995)
Brachionus plicatilis	18.3	22.5	6.5⁺	4.2	3.6		1.8	0.8	0.44	12.7	8.9		Cultured at 22°C with *Phaeodactylum tricornutum* (Ben-Amotz et al., 1987)
Brachionus plicatilis	13.3	1.8	5.3	9.5^{H}	14.4		4.9	1.4	0.28	13.0	13.1	1.70	Cultured at 20°C with *Tetraselmis suecica* (Whyte and Nagata, 1990)
Brachionus plicatilis	6.3	25.9	22.3	4.9	0.8	0.6	9.1	4.1	0.45	13.9			At 20°C, fed baker's yeast for 9 h, then enriched with HUFA-enhanced capelin oil for 6 h (Kissil & Koven, 1990)
Brachionus plicatilis	10.4	22.2⁺	27.8⁺	~6.3		2.7	11.5	0^{L}		14.8	7.7		Cultured with *Nannochloropsis oculata* (Rimmer et al., 1994a)
Brachionus plicatilis	9.4	15.2⁺	14.1	~3.9	0.2	3.7	8.7	0.3^{L}	0.03	15.1	10.8		From resting eggs from a culture fed *Nannochloropsis oculata*, just hatched (Hagiwara et al., 1993a)

Table 8-3. (Continued)

Species	16:0	16:1ω7	18:1ω9	18:2ω6	18:3ω3	20:4ω6	20:5ω3	22:6ω3	DHA/EPA	ω3HUFA	Lipid	ω3HUFA/Z	Treatment and reference
<u>Moderate</u>													
Brachionus plicatilis	15.6	5.2^{+}	12.9^{+}	6.5^{H}	22.4^{H}	2.4	5.8	0.2^{L}	0.03	12.4	27.8	3	Cultured with *Tetraselmis tetrathele* (Fukusho et al., 1984)
Brachionus plicatilis	11.5	15.2	15.0	5.6	2.3	0.4	7.2	5.7	0.79	14.1	17.9	1.5	Cultured at 20°C with baker's yeast + copepod emulsion (Rainuzzo et al., 1994)
Brachionus plicatilis	13.5	9.5	23.5$^{\#}$	6.7	1.7	~2.7	8.7	5.9	0.68	17.1			Cultured with baker's yeast, then fed Frippak AR 121® microcapsules for 8 h at 28-30°C (Walford & Lam, 1987)
Brachionus plicatilis	16.4		10.3^{+}	~3.8	~7.1	2.3	9.4	7.7	0.82	17.4	~6.5	1.13^{L}	Wild, from a pond (Su et al., 1994)
Brachionus plicatilis	12.3	19.0^{+}	22.3^{+}	~8.9		2.3	13.0	2.5^{L}	0.19	18.0	9.4		Cultured on *Nannochloropsis oculata*, then fed Frippak Booster® microcapsules for 4-22 h at ~29°C (Rimmer et al., 1994a)
Brachionus plicatilis					0		9.7	5.0	0.52	14.7			Cultured at 25°C with baker's yeast + *Nannochloropsis* sp. strain MFD-2 (James & Abu Rezeq, 1989a)
Brachionus plicatilis	10.6	3.4	15.2	7.2^{H}	6.5		2.2	8.5	3.86	18.1	15.8	2.86	Cultured at 20°C with Tahitian *Isochrysis* sp. (Whyte & Nagata, 1990)
Brachionus plicatilis	8.2	5.8	4.9	18.7^{H}	6.8	3.4	5.6	3.0	0.54	19.3	11.0	2.13	Cultured at 15-20°C with *Chroomonas salina* (Whyte et al., 1994)
Brachionus plicatilis	24.8	9.9^{+}	2.3^{+}	8.0	1.4	1.4	0.3^{L}	18.3	61.0	19.8		1.49	Fed *Schizochytrium* sp. for 8 h at 26°C (Barclay & Zeller, 1996)
Brachionus plicatilis	11.3	35.6^{+}	6.3	3.2		3.5	17.9	0.9^{L}	0.05	21.2			Fed frozen *Nannochloropsis* sp. for 12 h at 30°C (Lubzens et al., 1995b)
Brachionus plicatilis	9.0	15.3	18.2	9.5^{H}	1.1	0.7	8.1	10.6	1.31	21.5	14.9	1.5^{L}	Cultured at 20°C with baker's yeast + halibut roe emulsion (Rainuzzo et al., 1994)
Brachionus plicatilis	12.7	17.4	7.3	5.6	0.3	5.1	15.0	0.3^{L}	0.02	22.2	11.3	2.51	Cultured at 15-20°C with *Nannochloropsis oculata* (Whyte et al., 1994)
Brachionus plicatilis	17.4		13.1^{+}	~6.9	~7.2	2.8	19.5	2.7	0.14	22.2	~16.1	3.57	Cultured at 25°C with *Nannochloropsis*

													oculata (Su et al., 1994)
Brachionus plicatilis	8.1	10.7	9.0	10.9^{H}	3.9	1.7	3.2^{L}	5.2^{L}	1.62	23.5	11.6	2.72	Cultured at 15-20°C with Tahitian *Isochrysis* (Whyte et al., 1994)
Brachionus plicatilis	9.9	11.8	18.5	11.9^{H}	0.9	1.3	12.8	8.5	0.66	24.0	17.5	1.57^{L}	Cultured with baker's yeast, then fed Dry Selco® for 6 h at 20°C (Fernández-Reiriz et al., 1993)
Brachionus plicatilis	19.0		6.0	4.4	0.1	6.1	17.9	7.3	0.41	25.2			Cultured at 28°C with *Nannochloropsis oculata* (Tamaru et al., 1993c)
Brachionus plicatilis	9.5	12.6	19.0	10.0^{H}	0.8	1.4	12.5	9.9	0.79	25.4	18.5	1.67^{L}	Cultured with baker's yeast, then fed Protein Selco® for 6 h at 20°C (Fernández-Reiriz et al., 1993)
Brachionus plicatilis	16.9	22.4	10.4	3.8^{H}	0.2		24.9	0.2^{L}	0.01	28.6	29.8	8.75	Cultured with *Nannochloropsis oculata* (Watanabe et al., 1983a)
Brachionus plicatilis	12.8	7.3^{+}	15.9	13.9	2.8		16.0	13.0	0.81	29.0	18.2		Cultured with baker's yeast + *Nannochloropsis oculata*, then fed Super Rotifer® for 6 h at 20-22°C (Estévez & Kanazawa, 1996)
Brachionus plicatilis	12.8	13.2	0	1.2	1.3		20.5	6.0^{L}	0.29	34.2	10.0	3.42	Cultured at 20°C with *Thalassiosira pseudonana* (Whyte & Nagata, 1990)
High													
Brachionus plicatilis	8.5	20.0^{+}	24.1^{+}	7.6^{H}	0.9	0.6	6.8	3.6	0.53	11.6	33.7	3.29	Cultured with baker's yeast, then fed pollock liver oil for 8-10 h at 22-24°C (Izquierdo et al., 1989)
Brachionus plicatilis	8.2	17.4^{+}	25.4^{+}	7.9^{H}	1.4	1.1	10.0	7.6	0.76	19.7	36.3	5.97	Cultured with baker's yeast, then fed cuttlefish liver oil for 8-10 h at 22-24°C (Izquierdo et al., 1989)
Brachionus plicatilis			11.7	2.9	0.9	3.5	14.5	9.0	0.62				Cultured at 26-30°C with *Nannochloropsis oculata*, then fed squid oil for 2 h (Ostrowski & Divakaran, 1990)
Brachionus plicatilis	8.7		4.0^{+}	$\sim 11.1^{H}$	~11	2.1	14.0	17.9	1.28	31.9	~20.5	6.54	Fed *Tetraselmis chuii* + ω3HUFA mixture for 6 h at 25°C (Su et al., 1994)
Brachionus plicatilis	10.0		11.5^{+}	8.7^{H}	~1.1	2.8	16.3	17.8	1.09	34.1	~37.0	12.62	Fed *Nannochloropsis oculata* + ω3HUFA mixture for 6 h at 25°C (Su et al., 1994)
Brachionus plicatilis	5.6	7.8	20.4	8.2^{H}	1.1	1.8	17.7	16.4	0.93	37.8	19.8	2.95	Cultured on baker's yeast, then fed Super Selco® for 6 h at 20°C (Fernández-Reiriz et al., 1993)

Table 8-3. (Continued)

Species	16:0	16:1ω7	18:1ω9	18:2ω6	18:3ω3	20:4ω6	20:5ω3	22:6ω3	DHA/EPA	ω3HUFA	Lipid	ω3HUFA/Z	Treatment and reference
Brachionus plicatilis							12	19	1.58	38	10.5		Cultured at 20°C with baker's yeast + capelin oil, then fed squid meal for 24 h at 7-8°C (Rainuzzo et al., 1989)
Brachionus plicatilis	7.1		15.2^{+}	$\sim 10.1^{H}$	~1.1	1.2	14.3	26.3	1.84	40.6	~14.5	5.89	Fed yeast + ω3HUFA mixture for 6 h at 25°C (Su et al., 1994)
Brachionus plicatilis	6.6	8.7^{+}	14.8^{+}	7.5	0.6	1.7	16.2	21.7	1.34	42.9			Cultured at 20°C on baker's yeast + Super Selco® (Reitan et al., 1993c)
Brachionus plicatilis							15.9	55.5	3.5	77.2			Enriched in 300 µL/L emulsion of oil mixture containing 45.2% DHA and 6.7% EPA for 12 h at 27°C (Dhert et al., 1993)
Anostracans—brine shrimp & relatives													
Low													
Artemia sp. cysts	20.6	4.7	22.8	5.4	11.9		0	0		0	9.2	0	Decapsulated GSL cysts (Navarro et al., 1993)
Artemia sp. cysts & nauplii	5.7	3.1	14.0	1.6	0.4		0.2						Minimum values from Léger et al. (1986)
Artemia sp. nauplii	10.8	5.9^{+}	22.0^{+}	$\sim 10.8^{H}$		0	0.9	0		0.9	16.4		From Colombian cysts, just hatched (Rimmer et al., 1994a)
Artemia sp. nauplii	14.4	9.8	21.8	6.6	20.1		1.7	0		~1.7	~17.1		From SFB cysts, just hatched (Davis & Olla, 1992)
Artemia sp. nauplii	15.5	3.7	27.0	6.3	28.2^{H}	0.5	1.9	0		1.9	15.7	0.30	From EG cysts (Dickey-Collas & Geffen, 1992; Dickey-Collas, 1993)
Artemia sp. metanauplii	10.8	5.6^{+}	22.1^{+}	$\sim 9.9^{H}$		0	2.1	0		2.1	14.3		From Columbian cysts, not fed, 24 hah (Rimmer et al., 1994a)
Artemia sp. metanauplii	8.0	3.3	13.1	4.1	14.1	1.0	3.2	0		3.6	10.8		Fed baker's yeast during 6-24 hah at 27°C (McEvoy et al., 1996)
Artemia sp. nauplii	11.6	4.6	17.4	5.9	22.1	1.1	3.9	0		4.5	12.3		From EG cysts, 6 hah (McEvoy et al., 1996)
Artemia sp. metanauplii	12.6	4.0	22.5	6.8	20.3	2.3	3.6	0.2	0.06	4.5	13.0	0.58	From AF cysts, not fed during 0-24 hah (Næss et al., 1995b)
Artemia sp. cysts	5.7	3.1	14	2.6	2.3	1.6	0.3	0		4.7			Minimum values for decapsulated SFB

													cysts (Mura, 1995)
Artemia sp. nauplii					25.8	1.6	4.1	0.4	0.10	5.0			From Chinese (PRC no. 1188) cysts, just hatched (Dhert et al., 1993)
Artemia sp. metanauplii	12.0	3.4^{+}	35.7^{+}	6.6	36.2	1.7	4.6	1.0	0.22	5.9			Fed squid meal for 24-36 h at 25°C (Reitan et al., 1993c)
Artemia sp. nauplii					48.6	0.7	3.4	0	0	6.0			From GSL (South Arm) cysts, just hatched (Dhert et al., 1993)
Moderate													
Artemia sp. cysts	13.8	12.5	24.7	4.7	2.7	1.2	6.5	0^{L}		6.5	17.9	1.02	Decapsulated cysts from La Mata, Spain (Navarro et al., 1993)
Artemia sp. metanauplii	13.4	3.5	20.5	6.0	25.6^{H}	0.9	5.1	0.2^{L}	0.04	8.3	18.0	1.50	From SFB cysts, fed *Nannochloropsis oculata* during 12-24 hah at ~27°C (Whyte et al., 1994)
Artemia sp. metanauplii	12.7	2.9	19.5	7.9^{H}	25.1^{H}	0.8	4.4	0.5^{L}	0.11	8.9	17.6	1.58	From SFB cysts, fed *Chroomonas salina* during 12-24 hah at ~27°C (Whyte et al., 1994)
Artemia sp. metanauplii	13.0	3.6	20.1	6.2	24.7^{H}	0.8	4.0	0.5^{L}	0.12	9.2	18.8	1.73	From SFB cysts, fed Tahitian *Isochrysis* during 12-24 hah at ~27°C (Whyte et al., 1994)
Artemia sp. metanauplii	9.4	6.1^{+}	20.9^{+}	$\sim 12.4^{H}$		0	7.7	1.4^{L}	0.18	9.8	14.4		From Colombian cysts, fed Frippak Booster® microcapsules during 0-24 hah at ~29°C (Rimmer et al., 1994a)
Artemia sp. metanauplii	18.7	5.0	18.6	6.6	22.9	1.5	7.1	2.3^{L}	0.32	10.2	15.6		Fed liposomes containing phosphatidylcholine fraction of Atlantic herring roe during 6-30 hah at 22°C (McEvoy et al., 1996)
Artemia sp metanauplii		15.5	16.1		7.5		10.5	$<0.1^{L}$	<0.01	10.9			From Chinese cysts, not fed (Clawson & Lovell, 1992)
Artemia sp. metanauplii				8.8	16.0	1.1	8.3	2.7^{L}	0.32	~11.8		2.13	From GSL cysts, fed cod liver oil for 24 h (Robin, 1995)
Artemia sp. metanauplii	17.4	5.5^{+}	13.3^{+}	4.2	26.7	2.7	5.3	7.3	1.38	12.6		1.28^{L}	Fed *Shizochytrium* sp. for 24 h at 30°C (Barclay & Zeller, 1996)
Artemia sp. metanauplii	11.6	3.8^{+}	24.4	5.0	21.3		7.9	5.6	0.71	13.5	18.2		Fed Super *Artemia*® for 6 h at 20-22°C (Estévez & Kanazawa, 1996)
Artemia sp metanauplii	17.6	7.9	15.5	3.9	13.4	1.3	8.8	5.6	0.64	14.6	21.5	1.8	Cultured at 20°C with baker's yeast + copepod emulsion (Rainuzzo et al., 1994)

Table 8-3. (Continued)

Species	16:0	16:1ω7	18:1ω9	18:2ω6	18:3ω3	20:4ω6	20:5ω3	22:6ω3	DHA/EPA	ω3HUFA	Lipid	ω3HUFA/Z	Treatment and reference
Artemia sp. metanauplii		5.6	15.6		18.2		11.3	4.1[L]	0.36	16.0			From GSL cysts, fed menhaden oil for 3-9 h (Clawson & Lovell, 1992)
Artemia sp. metanauplii	9.9	9.5	34.7[#]	5.1	2.5	~2.7	12.2	3.6[L]	0.30	16.9			Fed baker's yeast for 20 h, then Frippak AR 121® microcapsules for 8 h at 28-30°C (Walford & Lam, 1987)
Thamnocephalus platyrus cysts	13.3	0	20.1	0	3.2	1.8	16.2	1.2[L]	0.07	17.7			Decapsulated cysts (Mura, 1995)
Artemia sp. nauplii					2.4	4.3	16.1	0.6	0.04	17.9			From SFB cysts, just hatched (Dhert et al., 1993)
Artemia sp. metanauplii	9.8	4.3	27.5[+]	0	20.7	0	13.7	6.3	0.46	20.0			From GSL cysts, fed ICES high-HUFA emulsion for 19 h at ~25°C (Ashraf et al., 1993a)
Artemia sp. cysts	26.6	30.6	37.5	10	33.6	1.6	13.8	0[L]		20.9			Maximum values for decapsulated SFB cysts (Mura, 1995)
Artemia sp. cysts & nauplii	26.6	30.6	37.5	15.6[H]	33.6[H]		15.4	~0[L]					Maximum values from Léger et al. (1986)
Artemia sp. metanauplii	8.5	2.8		4.1[H]	19.1[H]	0.3	28.6	0[L]		29.4	32.5	8.68	From Utah cysts, fed EPA + egg yolk for 15-22 h at 23-25°C (Furuita et al., 1996a)
<u>High</u>													
Artemia sp. metanauplii	10.6	3.2		4.8[H]	23.2[H]	0.4	4.7	10.8	2.30	16.5	29.6	4.06	From Utah cysts, fed DHA + egg yolk for 15-22 h at 23-25°C (Furuita et al., 1996a)
Artemia sp. metanauplii	11.0	3.9	19.7	6.8[H]	18.4[H]	1.7	9.8	5.0	0.51	16.6	29.0	4.81	From AF cysts, fed Super Selco® during 6-26 hah (Næss et al., 1995b)
Artemia sp. metanauplii	11.0	2.5	21.2	4.9	23.7[H]	0.6	10.1	6.9	0.68	17.7	23.2	4.11	From EG cysts, fed Super Selco® during 12-24 hah (Dickey-Collas & Geffen, 1992; Dickey-Collas, 1993)
Artemia sp. metanauplii	16.0	5.0	19.0	4.9	14.2	1.6	10.5	7.2	0.68	18.4	23.1	2.6	Cultured at 20°C with baker's yeast + halibut roe emulsion (Rainuzzo et al., 1994)

Artemia sp. metanauplii				5.3	17.8	1.3	12.7	7.2	0.57	21.0		3.29	From GSL cysts, fed Super Selco® for 24 h (Robin, 1995)
Artemia sp. metanauplii	10.0	14.4	14.3	4.1[H]	3.1	1.8	17.0	5.4	0.32	23.7	38.0	9.00	From Chinese cysts, fed fatty acid methyl ester mixture (75% ω3HUFA) for 24 h (Izquierdo et al., 1992)
Artemia sp. metanauplii	9.3	3.5	18.1	6.5	15.6	2.3	18.8	7.4	0.39	29.5	16.8		Fed Super Selco® during 6-24 hah at 27°C (McEvoy et al., 1996)
Artemia sp. metanauplii	16.7	5.8	24.4	5.7	14.2	3.0	10.8	19.4	1.80	42.8	19.7		Fed Atlantic herring roe polar lipid + tuna eye oil during 6-24 hah at 27°C (McEvoy et al., 1996)
Artemia sp. metanauplii					39.8	1.0	14.8	35.0	2.36	54.7			From GSL (South Arm) cysts, enriched in 600 μL/L emulsion of oil mixture containing 45.2% DHA and 6.7% EPA for 12 h at 27°C (Dhert et al., 1993)
Artemia sp. metanauplii					3.5	4.3	24.0	28.8	1.20	54.9			From SFB cysts, enriched in 600 μL/L emulsion of oil mixture containing 45.2% DHA and 6.7% EPA for 12 h at 27°C (Dhert et al., 1993)
Artemia sp. metanauplii					18.4	1.8	14.6	72.1	4.94	92.1			From Chinese (PRC no. 1188) cysts, enriched in 600 μL/L emulsion of oil mixture containing 45.2% DHA and 6.7% EPA for 12 h at 27°C (Dhert et al., 1993)

High = Contained at least 2% ω3HUFA, including at least 1.0% EPA and 1.0% DHA, which should be sufficient for most marine fish larvae.
Moderate = Contained at least 1% ω3HUFA. Because EPA and/or DHA content were less than 1%, supplementation might be needed for some species.
Questionable = Because of low EPA and/or DHA (or insufficient data), these seem to be unsuitable for long-term sustenance of typical marine fish larvae.
Low = probably contained insufficient ω3HUFA, EPA, and/or DHA for long-term sustenance of typical marine fish larvae.
[L] Low EPA, DHA, or ω3HUFA (in whole animal, dry weight basis).
[H] Prey contained more than 1% linoleic acid or more than 4% linolenic acid (dry weight basis). Use of those animals as the main food could be a matter of concern for some species.
[+] Analysis included other monoenes.
[#] Analysis included 18:0.
AF = from Artemia Systems. EG = from Artemia Systems. GSL = Great Salt Lake, Utah. SFB = San Francisco Bay, California.

Reported fatty acid contents of rotifers (Table 8-3) ranged from zero EPA, DHA, and ω3HUFA when fed only baker's yeast to 7.42% EPA (of DW) with *N. oculata* (Watanabe et al., 1983a), ~6.59% DHA with *N. oculata* + ω3HUFA mixture (Su et al., 1994), and 12.62% ω3HUFA with *N. oculata* + ω3HUFA mixture (Su et al., 1994).

Koven et al. (1990) enriched rotifers with four combinations of soy, capelin, or squid oil plus dried squid, and with *Nannochloropsis* sp. at 19-23°C The highest ω3HUFA content (0.84% of DW) occurred in those fed squid oil plus ground whole squid. Growth of larval gilthead sea bream at 19-23°C was strongly related to the ω3HUFA content of the five groups of rotifers fed to them.

James and Abu-Rezeq (1989b) found that rotifers of a small Spanish strain (S1) had levels of 2.3-6.2% EPA plus 1.8-3.7% DHA, and a small Japanese strain (S = *B. p. rotundiformis*) had 3.2-6.8% EPA plus 1.7-5.3% DHA, when fed baker's yeast and *Chlorella* sp. at 23-28°C.

When ten diets consisting of various combinations of baker's yeast and seafood meals and oils were fed to rotifers at 7-8°C for 24 h, the highest combined levels of EPA and DHA occurred in those fed squid meal (contents: 1.26% EPA, 2.00% DHA, 3.99% ω3HUFA of DW), followed closely by baker's yeast with cod liver oil and then herring roe meal (Rainuzzo et al., 1989).

Izquierdo et al. (1989) reported that enrichment of rotifers with cuttlefish liver oil resulted in higher ω3HUFA uptake than pollock liver oil, but both rotifer groups produced good growth and survival of red sea bream larvae. Both of the natural oils were more effective in enriching rotifers than fatty acid methyl esters, the use of which resulted in abnormalities in the larvae. DHA/EPA ratios were 0.53 for PLO, 0.76 for CLO, and 0.70-1.35 for FAMEs.

Ostrowski and Divakaran (1990) reported that dolphin larvae grew to about the same weight when they were fed rotifers reared on *Nannochloropsis oculata* that then were not enriched or were enriched for 2 h with olive, soybean, corn, or squid oil. However, survival of larvae fed unenriched rotifers (52%) was less than for those given enriched rotifers (66% for squid oil). EPA content was 12.9% (of FA) with algae only and 14.5% with algae then squid oil. Rotifers fed algae, soybean oil, and corn oil contained no DHA.

As might be expected, wild rotifers from a pond contained about 38 times more ω3HUFA (1.13% of DW) than yeast-fed rotifers and probably would support growth of many fish species (Su et al., 1994). However, rotifers fed one of two algal species or yeast plus an ω3HUFA mixture had very good EPA, DHA, and total ω3HUFA contents.

Fernández-Reiriz et al. (1993) demonstrated that rotifers previously cultured on baker's yeast could have a high ω3HUFA level (2.95% of DW) after 6 h of enrichment with Super Selco.® The same treatment with Dry Selco® or Protein Selco® resulted in 1.57% and 1.67% ω3HUFA.

Rainuzzo et al. (1989) showed that rotifers fed squid meal for 24 h had at least as much ω3HUFA (38% of FA) as those fed baker's yeast, yeast + capelin oil, yeast + cod liver oil, shrimp meal, herring roe meal, fish meal, or three mixed diets.

Reitan et al. (1993) reported that rotifers cultured with baker's yeast and Super Selco® had very high ω3HUFA (42.9% of FA).

Tocher et al. (1997) reported that rotifers enriched with Protein Selco® for 8 h contained 0.9% EPA, 0.4% DHA, and 1.57% ω3HUFA (of DW). Rotifers enriched with tuna eye silage for 8 h contained 1.1% EPA, 2.0% DHA, and 3.46% ω3HUFA, thus indicating an advantage for the silage over the dry enricher.

The highest percent ω3HUFA for rotifers listed in Table 8-3 was for those enriched for 12 h with an oil emulsion having a DHA/EPA ratio of 6.7 (Dhert et al., 1993).

Léger et al. (1986) reported a wide range for linolenic acid (18:3ω3) and narrow, low range for EPA, with oleic acid (18:1ω9) usually dominating lipids of *Artemia* nauplii. Enrichment with marine oils (e.g., Watanabe et al., 1978) improves the pattern of short-chain fatty acids (more like natural foods), but most importantly ensures EPA and DHA content. Usually *Artemia* do not contain measurable DHA, but Schauer et al. (1980) reported low levels of 0.26% DHA (of total lipid) in Australian nauplii and 0.06% DHA in Brazilian nauplii.

Essential fatty acid level ranges reported by Fujita et al. (1980) for unfed *Artemia* (San Francisco, Canadian, South American, Chinese strains) 24 h after hatching were 0.8-12.8% EPA and 0-0.1% DHA (of total lipids). *Artemia* fed for 24 h had levels of 1.7-15.0% EPA and 0-0.3% DHA with *N. oculata*, 0.9-14.3% EPA and 0-0.2% DHA with baker's yeast, and 3.4-10.8 % EPA and 0.4-2.3% DHA with ω-yeast (baker's yeast, *Spirulina*, and cuttlefish liver oil).

Tocher et al. (1997) reported that *Artemia* enriched with Protein Selco® for 16 h contained 0.9% EPA, 0.4% DHA, and 1.31% ω3HUFA (of DW). *Artemia* enriched with tuna eye silage for 16 h contained 0.5% EPA, 0.4% DHA, and 1.0% ω3HUFA, thus indicating no advantage in percentages for the silage over the dry enricher.

EPA and DHA contents in *Artemia* were proportional to enrichment levels of EPA or DHA ethyl esters (Furuita et al., 1996). Enrichment with 99% EPA resulted in no DHA in rotifers, but enrichment with 99% DHA resulted in a 2.3 DHA/EPA ratio.

Naess et al. (1995b) reported that *Artemia* enriched with Super Selco® for 20 h contained eight times more ω3HUFA (4.81% of DW) than unenriched metanauplii (0.58%) and had a DHA/EPA ratio of 3.32.

Dickey-Collas and Geffen (1992) found that *Artemia* enriched with Super Selco® for 12 h had 14 times more (4.11% of DW) ω3HUFA than just-hatched nauplii (0.30%) and had a DHA/EPA ratio of 2.56.

Rainuzzo et al. (1994) reported that *Artemia* cultured with baker's yeast + copepod emulsion contained 1.8% (of DW) ω3HUFA, and those cultured with yeast + halibut roe emulsion contained 2.6% ω3HUFA. DHA/EPA ratios were 0.64 and 0.68.

Robin (1995) observed that *Artemia* enriched with Super Selco® for 24 h had more ω3HUFA (3.29% of DW) and a better DHA:EPA ratio (0.57) than those enriched with cod liver oil (2.13%, 0.32).

Izquierdo et al. (1992) reported that enrichment with a 75% ω3HUFA (methyl ester) mixture for 24 h resulted in 9.00% (of DW) ω3HUFA and a DHA/EPA ratio of 0.32 in *Artemia*.

McEvoy et al. (1996) found that *Artemia* enriched with Super Selco® or herring roe polar lipid + tuna eye oil for 18 h contained very high levels of ω3HUFA (29.5%, 42.8% of FA) and moderate total lipids (16.8%, 19.7%, of DW). DHA/EPA ratios were 0.39 and 1.80.

The highest percent ω3HUFA for *Artemia* listed in Table 8-3 was for those enriched for 24 h with an oil emulsion having a DHA/EPA ratio of 6.7 (Dhert et al., 1993). Chinese *Artemia* held the most, with those from San Francisco Bay and Great Salt Lake both holding 40% less.

Compound Larval Feeds

The three main specialized types are microencapsulated (MED), microbound (MBD), and microcoated (MCD) diets (feeds). See chapter 9 for more details.

Although many early marine fish larvae have difficulty in accepting and digesting microcapsules and microparticles (Dabrowski and Culver, 1991), some success has been had with a few species (e.g., Barnabé, 1976c; Appelbaum, 1985; Person-Le Ruyet, 1990a; Dabrowski and Culver, 1991; Leu and Liou, 1992; Holt, 1993; Mourente and Tocher, 1993). Early weaning was the original goal, but co-feeding of compound feeds with live foods can at least reduce the live food requirement. Microencapsulated feeds provide an alternative way to administer vaccines and therapeutic agents to larvae. Especially important are attractiveness, digestibility, and nutritional quality of the feeds. Near-neutral buoyancy helps to keep them suspended. During early stages, some larvae might have difficulty recognizing inert particles as food (Iwai, 1980). In addition, their digestive systems are simple and do not seem to have the enzymatic activity necessary for breaking down non-living feeds (Dabrowski and Rusiecki, 1983; Dabrowski and Culver, 1991; Infante and Cahu, 1994; Infante et al., 1996). Typical early marine fish larvae probably depend to a great degree on small colloidal proteins in zooplankton because they do not have the enzymes necessary for digesting and absorbing larger protein molecules (Hino, 1993). Older larvae have greater capabilities to make more kinds of enzyme and to adjust enzyme production according to the type of food (Infante and Cahu, 1994). Early nylon-protein-membrane microcapsules were relatively indigestible to first feeding red sea bream (Kanazawa et al., 1982). In a later study (Kanazawa et al., 1989), beginning at 10 dah, artificial microbound (with zein) feeds were partly substituted for live food (rotifers and *Artemia*) of red sea bream and completely substituted for live food of hirame reared at 18-20°C, resulting in reasonably good survival and growth rates. Cross-linked-protein microcapsules for crustacean larvae were developed by Jones et al. (1987). The latter type of microcapsule was tested on barramundi larvae by Walford et al. (1991). With continuous feeding and gentle aeration, larvae easily ingested 40- to 60-μm microcapsules from first feeding (3 dah); however, they could not digest them through 8 dah and probably longer. Gut residence time was 60 min or slightly less in larvae anesthetized with 2-phenoxyethanol. Digestion of microcapsules occurred only when rotifers were eaten along with them and provided the necessary proteolytic enzymes for breaking down their protein wall. Larger microcapsules (80-150 μm) tended to block the gut. Microbound feeds tend to be more digestible than MCDs (Person Le Ruyet et al., 1993a). Yúfera et al. (1995) reported that gilthead sea bream consumed protein-walled microcapsules from first feeding but could not break down the capsules until about 20 dah; soft-walled microcapsules were easier to digest than hard-walled ones, but larvae fed rotifers alone survived and grew better than those given MCD or MCD and rotifers (Yúfera et al., 1996; Fernández-Diáz and Yúfera, 1997). Even microbound (carrageenan) feeds were not well digested by first-feeding white sturgeon larvae (Gawlicka et al., 1996), but moist or semi-moist feeds were utilized. Larval feeds are likely to cause increases in dissolved ammonia and bacteria (Muir and Sutton, 1994), which could necessitate higher water exchange and vigilance for health problems.

Thus, even if highly palatable and relatively digestible larval feeds are developed, larvae of most marine fish might still require live foods for the first few days or weeks (Rosenlund et al., 1997). Larvae of the majority of cultured marine fish will learn to eat compound feed and will thrive on it if they are weaned during transformation.

Theoretically, compound larval feeds that are attractive and digestible and simulate the composition of ideal natural foods would be best. For nutritional purposes, amino acid profiles of marine animals do not differ as much as fatty acids. *Acartia clausi* contained 3.5% isoleucine, 5.5% leucine, 1.5% methionine, 0.8% cystine, 3.7% phenylalanine, 3.6% tyrosine, 4.2% threonine, 1.1% tryptophan, 4.5% valine, 5.4% lysine, 4.3% arginine, 1.9% histidine, 5.4% alanine, 9.0% aspartic acid, 9.5% glutamic acid, 4.6% glycine, 4.6% proline, and 3.3% serine (of crude protein); *Tigriopus japonicus* amino acids were similar (Watanabe et al., 1983a). The amino acid profile of barramundi raised on mixed zooplankton in a pond was similar to those of the two copepods (Rimmer et al., 1994a). If proteins are mainly from marine animals, amino acid content should be good; supplementation for balancing amino acids normally is needed only when large amounts of plant proteins are used. Free amino acids in larval feeds could be more easily utilized by early larvae than water-soluble protein, particularly because of the short residence time of food in the simple gut (Walford et al., 1991; Walford and Lam, 1993). Up to 5-10% crystalline (free) amino acids can be included in feeds for larval red sea bream (López-Alvarado and Kanazawa, 1994, 1995, 1997). Likewise, if lipids are mostly from marine animals, their balance should be good. Phospholipid content should be high (Person Le Ruyet et al., 1993a). An excess of vitamins C and E will help protect lipids from oxidation.

Addition of digestive enzymes or substances that stimulate the fish to secrete enzymes can increase digestibility (Person Le Ruyet et al., 1993a). Such things occur in live foods (Kolkovski et al., 1997a, 1997b). Kolkovski et al. (1993) found that, during 20-32 dah, gilthead sea bream grew best on enriched *Artemia*, very little on a microparticulate feed without added pancreatin, and moderately on the same feed with 0.05% pancreatin. Hormones (e.g., thyroid) and other growth factors occur in some feedstuffs or can be added.

Weaning

For most marine species being reared, weaning from live to dry food can begin a few days before transformation and will be finished by the time the fish are juveniles. Feeding of crumbles or pellets can be done by hand or by machine, but early weaning by hand (Fig. 8-13) helps to minimize feed wastage and water fouling. This process might be finished in 3 d or (rarely) could take as long as 3 weeks. The transition should be gradual enough so that slow learners can survive the process. Size of food particles should be the largest that can easily be swallowed by the fish (25-50% of mouth width for soft items and ~25% for hard ones).

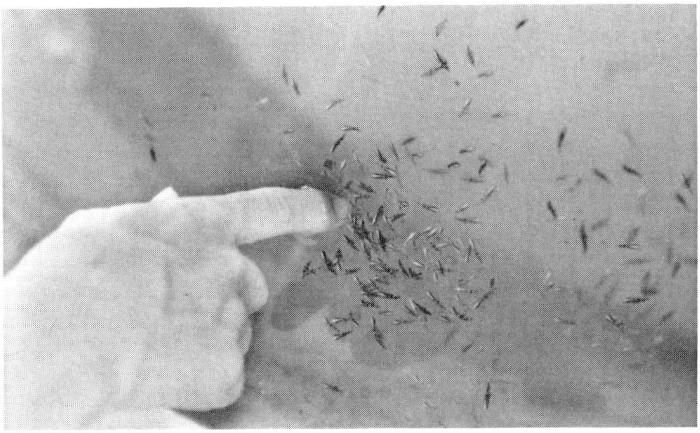

Figure 8-13. At transformation, these barramundi accustomed to being hand-fed were very easy to wean; however, automated feeding does not increase the difficulty very much.

Sometimes larvae can be weaned from live plankton to dry feed, just by gradually or abruptly withdrawing the live food. Examples are: European sea bass 25-30 dah (Person-Le Ruyet et al., 1991); dolphin 25-30 dah (Kim et al., 1993); Asian red snapper 40-50 dah (Duray et al., 1996a); spotted seatrout 33-36 dah (Tucker, 1988); red drum 8 dah (Holt, 1993); common sole (Bromley, 1977), 10-15 dah (Gatesoupe, 1983); turbot, 14-19 dah (Bromley, 1978), 23-29 dah (Bromley and Howell, 1983), and about 36-43 dah with good survival and growth (Kuhlmann et al., 1981).

Weaning fish too early can result in lower production and higher juvenile cost because of reduced growth and survival (Gennari et al., 1992). Attractiveness and digestibility of the compound feeds are critical. Early weaning of some robust species with reasonable survival and growth has been accomplished (e.g., European sea bass at 20 dah, Person Le Ruyet et al., 1993b, Cahu and Infante, 1994, Barnabé and Guissi, 1994; red drum at 8 dah, Holt, 1993; common sole at 10 dah, Gatesoupe, 1983, Appelbaum, 1985). Red sea bream were limited to 10% live food by 10 dah (Kanazawa et al., 1989).

For efficiency, weaning often is initiated by offering highly attractive "appetizer" foods to the larvae before or during presentation (side by side, or mixed) of dry feed. Examples are:

frozen *Artemia* plus 10-30% *Artemia* powder in feed pellets for European sea bass 35-42 dah (Barahona-Fernandes and Girin, 1976); trash fish for greasy grouper (Chen, 1979); freeze-dried *Artemia*, krill, shrimp, or mussel for European sea bass (Barahona-Fernandes, 1982); freeze-dried scallop meal for common snook 35-38 dah (Tucker, 1987b); frozen *Artemia* for barramundi 20-25 dah (Tucker et al., 1988); freeze-dried raw scallop and shrimp meal for sheepshead 27-30 dah (Tucker, 1987a); fresh squid and shrimp for red drum 28-35 dah (Soletchnik et al., 1990), and minced shrimp for Nassau grouper 56-69 dah (Tucker and Woodward, 1996). This takes more effort, but typically helps slow learners and increases survival during weaning.

Late larvae that will become herbivores, omnivores, or bottom fish, such as sparids, mugilids, pleuronectids, and soleids seem to be less particular about food and could be slightly easier to wean than more raptorial fish, such as centropomids, percichthyids, and serranids (Kentouri, 1981; pers. obs.). Among sciaenids, bottom-feeders like red drum are easier to wean than more raptorial species like spotted seatrout (pers. obs.). Cannibalism (Fig. 8-14) tends to decrease after weaning.

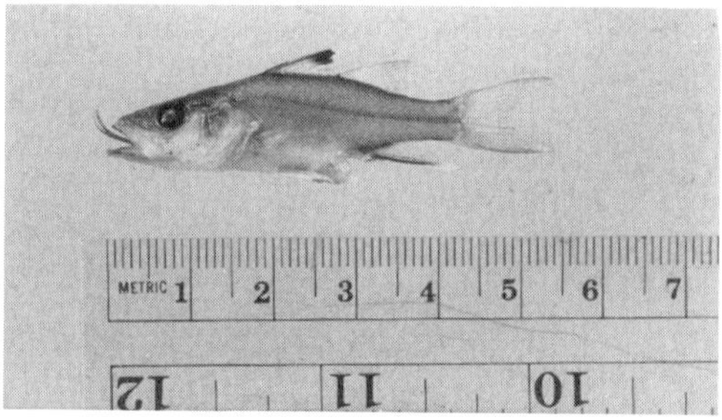

Figure 8-14. Some cannibalism occurs in early juveniles of piscivorous species like common snook even when size differences are small (cannibal 56 mm, victim 40 mm TL). It can become a problem if the fish are not very well fed and graded when necessary. (from Tucker, 1987b; photo by Thomas A. Smoyer)

Certain types of fish are more adaptable than others. Older wild juveniles of some species that have well-established feeding habits (e.g., snook, seatrout) are very difficult to train to eat pellets. Others, even as adults, (e.g., some groupers and sea bream) will easily eat hard pellets within a day after capture.

GENERAL FEEDING PRACTICES

Various methods have been used for feeding larvae of different species (Table 8-4). Efficacy and efficiency can be attained by balancing standardization with flexibility. The culturist should observe larvae very closely (especially when they are feeding) to help determine needs and detect problems. Survival can be enhanced by making the transition from one live food to the next over a period of at least 5 d (pers. obs.).

Overfeeding of rotifers can: (1) dangerously reduce water quality (e.g., reduce oxygen and raise ammonia) and (2) result in too many rotifers losing their nutritive value before they are eaten. Overfeeding of *Artemia* (particularly nauplii) can: (1) dangerously reduce water quality, (2) allow larvae to eat so fast that too little nutrition is obtained from the *Artemia*, (3) cause gut blockages in larvae, (4) result in *Artemia* escaping and growing up to crowd the fish.

Clearwater vs Greenwater Culture

Greenwater culture is a standard method for larvae of some types of fish such as mullet and milkfish and is good for many others. For sensitive species that need high water quality, carefully-controlled low-density green water can be successful, but high algal densities are dangerous. Addition of microalgae to fish rearing tanks can help zooplanktonic prey to conserve their nutritional value longer. Ingested microalgae might provide trace nutrients, enzymes, feeding stimulants, or non-specific immunological stimulants (Lavens et al., 1995). Green water seldom is intentionally used past the early juvenile stage.

High density rotifer cultures require high densities of algae (e.g., 600 rotifers/mL with 20,000,000 algal cells/mL, James and Abu Rezeq, 1989a), but lower densities of rotifers can be maintained with much less algae (e.g., 6 rotifers/mL with 200,000 algal cells/mL). Some examples of algal cell densities used in fish rearing tanks are: 30,000-50,000 cells/mL *I. galbana* alone during 2-3 dah, then combined with 200,000-300,000 cells/mL *Chlorella* sp. or *Tetraselmis* sp. up to 10 dah for rotifers being eaten by yellowblotch spinefoot (Juario et al., 1985); 200,000-500,000 cells/mL *Chlorella virginica* 1-21 dah, supplemented with 20,000-50,000 cells/mL *I. galbana* 2-10 dah and *Tetraselmis chuii* 2-21 dah for rotifers and copepods being eaten by milkfish (Liao, 1991); 500,000 cells/mL *N. oculata* for rotifers being eaten by red sea bream (Smith and Hataya, 1982); 500,000-700,000 cells/mL *N. oculata* for rotifers being eaten by striped mullet (Lee and Kelley, 1991). In copepod production cultures, reproduction rates were high for *E. acutifrons* with algal food densities (*T. suecica*, *Chaetoceros danicus*, *P. tricornutum*, *Gymnodinium* sp.) in the range 10,000-1,000,000 cells/mL (Zurlini et al., 1978), and for *Acartia tonsa* with 20,000 cells/mL *Rhodomonas baltica* (Støttrup et al., 1986). In this context, our upper limit of 50,000-100,000 cells/mL seems reasonable. The greenwater method can be used for sensitive fish such as groupers as long as the water quality is maintained, but this could be difficult. Densities of fish larvae, their prey, and algae must be kept in balance. Unused nutrients from the algal culture could kill larvae outright. Only mature algae that has used up its nutrients should be used. Too much algae will cause fluctuations in dissolved oxygen and pH. A sudden drop in culture water transparency to less than 1 m (because of phytoplankton) is a danger signal. Likewise, too many prey organisms can cause a dangerous decrease in dissolved oxygen and increase in ammonia. Large numbers of rotifers should not be allowed to remain in clear water for long but can be left in green water. Up to 1 starving rotifer/mL in clear water at the next rotifer addition probably can be tolerated occasionally but is not optimal. Greenwater culture provides more constant quality and density of rotifers.

AQUACOP et al. (1990a) found that survival of barramundi larvae raised in tanks tended to be lower, but growth better, in clear water than in green water. Naas et al. (1992) reported that Atlantic halibut larvae fed much better on unenriched *Artemia* metanauplii in green water (50-84 flagellates <5 μm/μL during 33-43 dah) than in clear water, resulting in higher growth and survival at 54 dah. Enhancement of feeding rate by turbidity seemed more important than nutritional effects of the algae. Reitan et al. (1993c) reported that survival, growth, and vigor of turbot larvae were improved by the maintenance of 60,000 cells/mL Tahitian *Isochrysis* or 130,000 cells/mL *Tetraselmis* sp. in the larval rearing tanks. Øie et al. (1997) also reported better survival and growth of turbot larvae with Tahitian *Isochrysis* in the tanks.

Composition and treatment of the water supply are important factors. If all necessary nutrients are not present, algae added to fish culture water will not live for long. Algae also will not survive without enough light. If ammonia is present in the incoming water, algae could help protect larvae by removing at least some of it. For long-term protection, a thin layer of microalgae or macroalgae (e.g., *Chaetomorpha* sp., *Enteromorpha* sp.) and nitrifying bacteria growing on the tank walls could be safer and more stable than suspended phytoplankton for removing ammonia. (In a crisis, phytoplankton could be added cautiously for quick ammonia removal, if macroalgae and bacteria are thin, but this could complicate the problem if the phytoplankton suddenly blooms.) One approach for

Table 8-4. Successful feeding schedules for larvae of several species of marine fish. Unless otherwise noted, the number under container size is tank volume. If reported, the highest survival values are given; otherwise, they are typical or average.

Species	Container size (L)	Stocking density	Food & density (dah provided)	Transformation (dah)	Transformation (mm TL)	Survival (dah)	Survival (%)	References
			Clupeidae—herrings					
Atlantic menhaden *Brevoortia tyrannus*	670	14.9 *l*/L	*Chl* sp. (0); R 10/mL (5-14); Z (11-30); AN (15-30); F (15-30)	~60	~30	*e*-30	~20	Hettler, 1981
Japanese sardine *Sardinops melanostictus*	700	2.1 *e*/L	*Nan* sp. 1,000/μL (0-14); R 5/mL (3-); AN (14-); *Pag maj* E&L (>14-); F (>14-)	~50	~40	*e*-80	7.7	Nakamura et al., 1991
			Engraulidae—anchovies					
Northern anchovy *Engraulis mordax*	400	15 *e*/L	*Gym spl* 100⁺/ml (0-12); *Nns* sp. (5-74); R 118/mL (5-20), 15/mL (21-74); *Tis fur* (6-48); AN 2/mL (48-74); AA (66-74)	≤74	~35	*e*-74	12.5	Hunter, 1976
			Chanidae—milkfish					
Milkfish *Chanos chanos*	200	6 *l*/L	*Chl* sp. 0.5-3.5/μL (1-21); OE&L ≥30/mL (2-7); R ≥10/mL (2-21); Co (14-21); AN (14-21)			1-21	40	Liao et al., 1979
Milkfish	700	2.7 *l*/L	*Chl* sp. 0.5-3.5/μL (1-21); OE&L ≥30/mL (2-7); R ≥10/mL (2-21)			1-21	47	Liao et al., 1979
Milkfish	600	6.5 *l*/L	*Chl vir* 200-500/μL (1-21); *Iso gal* 20-50/μL (2-10); *Tet chu* 20-50/μL (2-21); R 20-30/mL (2-9); *Tis elo* (7-21); R 10-20/mL (10-21); AN (10-21)			1-21	43	Juario et al., 1984; Liao, 1991
Milkfish	2500		*Tet* sp. 20-60/μL, *Chl* sp. 20-30/μL, or *Mon* sp. 20-50/μL; R 10-15/mL; AN 0.5-2⁺/mL (10-21)			1-21	52	Prijono et al., 1988
Milkfish	4000	12.6 *l*/L	*Nan ocu* 700-1,000/μL (2-20); R 10-20/mL (2-30); AN (12-30)			1-30	46	Eda et al., 1990b
Milkfish	4000	20 *e*/L	*Nan ocu* 500/μL (2-25); R 20/mL (3-); EAM 0.01-4/mL (15-25); F (20-25W)					Liu & Kelley, nd.a
			Osmeridae—smelts					
Ayu^{FW} *Plecoglossus altivelis*	45,000	13.1 *l*/L	R (1-44); A (44-50); MPD (7-)			1-50	88	Kyowa, nd.b
			Galaxiidae—troutminnows					

Tasmanian whitebait *Lovettia seallii*	4000		ER (1-30); EAM (20-45); F (~45-)			~1-37	~30	Daintith, 1992
Tasmanian whitebait	4000		Z (1-ad)			~1-37	>50	Daintith, 1992
Gadidae—cods								
Atlantic cod *Gadus morhua*	145-3000		R, Z	30-40				Larsson & Pickova, 1993
Atlantic cod	5200 bag	0.15 *l*/L	Z: mainly *Eur aff* (1-)			1-48	83	van der Meeren, 1994
Atlantic cod	22	55 *l*/L	R (1-22); MPD (15-71)			1-71	~35	Baskerville-Bridges & Kling, 1997
Pseudomugilidae—blue-eyes								
Honey blue-eye[FW] *Pseudomugil mellis*			Co, Cl, LE, FF, LF (~1-); AN (12-)		≥10			Semple, 1991
Atherinidae—silversides								
Topsmelt *Atherinops affinis*	3		AN 5/mL (1-24)			1-24	100	Middaugh et al., 1990
Jacksmelt *Atherinopsis californiensis*	18	5 *l*/L	AN 5/mL (1-24)			1-24	91	Middaugh et al., 1990
California grunion *Leuresthes tenuis*	20	10 *l*/L	*Tet sue* 5/µL (0-); AN 0.5/mL (4-23); AN 1/mL (24-43)	~40	~18	1-40	81	Vásquez-Yeomans & Sosa-Cordero, 1992
Syngnathidae—pipefishes, seahorses								
Seahorses *Hippocampus* spp.	10		R (0-); AN (0-); *Poe ret* H (later)					Schmidt, 1995
Seahorses			C,Z (~1-14); AN (~10-180); G, My, Sh, *Poe ret* H (later)					Vincent, 1995c
Scorpaenidae—scorpionfishes								
Kurosoi *Sebastes schlegeli*	15,000	6.2 *l*/L	*Chl* sp. (0-); R 5/mL (0-10); AN 2/mL (4-25); OA (10-35); MF (15-40); F (20-); *Tig jpn*, MySh, *The cha* E (periodically)	~35	~25	*p*-35	50	Kusakari, 1991
Anoplomatidae—sablefishes								
Sablefish *Anoplopoma fimbria*	4	25 *l*/L	R 25-50/mL (21-); EAM 20-40/mL (65-)			21-65	10	Whyte et al., 1994
Hexagrammidae—greenlings								
Lingcod *Ophiodon elongatus*	5	20 *l*/L	EAM ≥20/mL (11-45)	~50	~20	4-45	~9	Appelbaum et al., 1995

Table 8-4. (Continued)

Species	Container size (L)	Stocking density	Food & density (dah provided)	Transformation		Survival		References
				(dah)	(mm TL)	(dah)	(%)	
			Cyclopteridae—snailfishes					
Atlantic lumpfish *Cyclopterus lumpus*	26	19.2 *l*/L	EAM (5-55); F (30-)			1-100	21	Martin-Robichaud et al., 1994
			Centropomidae—snooks					
Common snook *Centropomus undecimalis*	Pond	22 *l*/m^2	Z: mainly R, *Aca ton* (3-)			3-32	25	Macioroswki et al., 1987
Common snook	3000	4.5 *l*/L	MA; *Mer mer* E&L (2-7); R (2-28); Co (3-30); AN, EAM (25-40); F (35-)	~35	~35	1-55	7.0	Tucker et al., unpub.
Barramundi *Lates calcarifer*	5000	5*e*/L	*Chl* sp. or *Tet* sp. (1-15); R 2-3/mL (2); R 3-5/mL (3-10); AN <0.2/mL (11-12); R 5-10/mL (11-15); AN 0.5-1.0/mL (13-20); *Moi mic* 0.10-0.15/mL (18-20); AN, *Moi mic* (~21-40); MF (~21-)	~25	~11	1-20	75	Lim et al., 1986
Barramundi		30 *l*/L	*Chl* sp. 100-200/μL (1-21); R 10-15/mL (2-15); R 5/mL (16-21); AN 1/mL (15-21); A 2/mL (22-30); F (15-)					Duray & Juario, 1987
Barramundi	30,000	~20 *l*/L	*Chl* sp., *Tet* sp. 5% tank vol./d (0-15); R 5-15/mL (3-7); R 15-20/mL (8-15); AN 1-4/mL (8 to 10-14); AN 4-7/mL (15-20); *Moi mic* (~15-35); AN 6-7/mL (20-25); MF (20-)	~25	~10	1-45	~29	Maneewong, 1987b
Barramundi	1000		R (3-15); AN (10-25); F (23-)	~25	~17			Tucker et al., 1988
Barramundi	200	30 *l*/L	R (2-12); AN (10-21); EAM (11-21) [total A 0.1-1.5/mL]			1-21	48	Dhert et al., 1990
Barramundi		30 *l*/L	*Chl* sp. 100-300/μL (1-); R 15-20/mL (1-12); AN 0.5-2/mL/d (12-14); R <15/mL (13-15); AN 5-10/mL/d (15-17); EAM 5-10/mL/d (18-23); OA, AA, *Moi* sp					Parazo et al., 1990
Barramundi			*Tet* sp. or *Chl* sp. 8-40/μL (0-15); R 10-20/mL (2-15); AN (8-30); *Dap* sp. (20-30); MF (25-)			1-40	~62	Tookwinas, 1990a
Barramundi	3000	30 *l*/L	*Chl* sp. 100-200/μL (0-15); R 15-20/mL (1-15); AN 0.5-3/mL (11-14); AN 5-10/mL (15-17); EAM 5-10/mL (18-25); A, FA, MF (23-)			1-30	~34	Parazo et al., 1991b

Barramundi	Pond	83 l/m^2	Z: mainly C, R, Cl		2-22	~43	Rutledge & Rimmer, 1991
Barramundi	2	275 l/L	R 10-20/mL (2-14); AM, EAM 2-5/mL (8-12); AM, EAM 5/mL (13-30)		1-30	94	Rimmer et al., 1994a
			Percichthyidae—temperate basses				
European sea bass *Dicentrarchus labrax*	150	60 l/L	R ≤6/mL (3-15); AN ≤10/mL (10-); OA ~3/mL (23-)		1-30	56	Barahona-Fernandes, 1978
European sea bass	10,000	20 l/L	*Tet sue* ~75/μL (0); R ~1.5/mL (5-20); AN ~0.5/mL (9-35);		1-35	35	Bedier, 1981
European sea bass			R (2-15); A (10-23); MBD (20-); OA (17-35); F (35-)				Person-Le Ruyet, 1990a
European sea bass	50-250	60-80, then 10-20 l/L	ER (5-14); AN (10-15); EOA (13-19); MPD (20-40); F (41)	~35-40	1-60	35	Person Le Ruyet et al., 1993a
European sea bass	7000 & 15,000	45 l/L	MA (6-12); R (6-12); AN (8-16); EAM (12-); F (27-)		*e*-1 g	15	Sweetman, 1992
European sea bass	500	160 *e*/L	AN (4-30)		*e*-30	38	Barnabé & Guissi, 1993
European sea bass	500	144 l/L	EAM (5-30)		*e*-30	~26	Barnabé & Guissi, 1994
Japanese sea bass *Lateolabrax japonicus*			R, AN, C, MMt	~60	1-60	~65	Zheng et al., 1993
Australian bass *Macquaria novemaculeata*	40	2.5 l/L	R 9/mL (3-21); AN 6/mL (21-35)		1-35	65	van der Wal & Nell, 1986
White bass *Morone chrysops*	16-40	78 l/L	R ≤10/mL (3-11); AN 3/mL (8-24); F (11-24)		3-24	57	Denson & Smith, 1996
Striped bass *Morone saxatilis*			AN (5-30); FF (12-22); F (18-)				Nicholson et al., 1990
Striped bass	Pond	12.6 l/m^2	Z: mainly *Aca ton*		7-52	72	Colura et al., 1976
Striped bass	70	2 l/L	AN (5-9); EAM (10-32)		*l*-32	82	Tuncer & Harrell, 1992
Striped bass	70	20 l/L	EAM 5/mL (7-28)		6-28	30	Ozkizilcik & Chu, 1994
Striped bass	Pond	74 l/m^2	MA, R, Cl, Co, Z (5-43); F (14-42)		5-43	87	Barkoh, 1996
Sunshine bass *Morone chrysops* × *M. saxatilis*	Pond		Z: mainly R, Cl, Co		5-40	59	Ludwig, 1993
Sunshine bass	76	19.6 l/L	R 20/mL (5-27); F (16-)		5-41	19.6	Ludwig, 1994
Palmetto bass *Morone saxatilis* × *M. chrysops*	70	2 l/L	AN (5-9); EAM (10-32)		*l*-32	81	Tuncer & Harrell, 1992

Table 8-4. (Continued)

Species	Container size (L)	Stocking density	Food & density (dah provided)	Transformation (dah)	Transformation (mm TL)	Survival (dah)	Survival (%)	References
Serranidae—groupers, sea basses								
Black sea bass *Centropristis striata*	2000	5 *l*/L	*Dun* sp. or *Nan* sp. (0); R (3-13); Z (7-24); AN (14-24); FF, MK, FA (21-23)			1-24	10	Roberts et al., 1976
Redspotted grouper *Epinephelus akaara*	60,000	5-28 *e*/L	*Nan* sp. 300-500/μL (0-40); R (2-37); A (28-50); F (20-50)	~50	~30	1-47	34	Maruyama et al., 1994
Whitespotted green grouper *Epinephelus amblycephalus*	1000	12-15 *l*/L	*Cra gig* E&L 5/mL (2-8); R 10-15/mL (2-20); *Tig jpn* 5/mL (8-22); AN 0.1-0.3/mL (20-35); MSh (33-40)	~40	~35	1-40	~10	Tseng & Chan, 1985
Orangespotted grouper *Epinephelus coioides*	5000	~7.9 *l*/L	MA; OE&L (2-4); R (2-25); EAM, Co (13-40); F (2-35); MF (25-)	35-40	25-30	*e*-43	12.1	Doi et al., 1991
Orangespotted grouper	3000		*Chl vir* 300/μL (2-31); R 20/mL (2-20); R 10/mL (21-31); AN (21-25); EAM (26-30); OA 1-3/mL (31-50); MF (45-60)	~35-45	~27	1-60	3.2	Duray et al., 1997
Brownmarbled grouper *Epinephelus fuscoguttatus*	900	35.6 *l*/L	*Chl* sp. 200-300/μL (2-12); R 5-10/mL (2-12); EAM 1-5/mL (10-30); EAM 5-10/mL (30-40)	~40	~31	1-40	13.2	Lim et al., 1990
Malabar grouper *Epinephelus malabaricus*	26,000	~14 *l*/L	*Tet* sp. 20-50/μL (2-20); R 10-20/mL (2-25); *Tig jpn* ~0.16/mL (13-30); AN (15-55); *Moi* sp. (30-); AA (35-); MF (40-)	~60	~28	1-30	~10	Maneewong et al., 1986
Malabar grouper	26,000		*Tet* sp. or *Chl* sp. (0-18); R ≥10/mL (3-30); AN (12 to 15-50); MF (45-); F (~50-)	~40	~35	1-50	4.4	Tookwinas, 1990a
Malabar grouper			R (1-24): MED (1-10); AN (12-24); OA (25-34); AA (35-50)			50	~14	Quinitio & Toledo, 1991; AquaFarm News, 1992a
Camouflage grouper *Epinephelus polyphekadion*	2800	30*l*/L	*Nan* sp. 300-500/μL; R; EAM; F (35-)	~50	22-47	1-50	4.7	Al-Thobaiti & James, 1997
Sevenband grouper *Epinephelus septemfasciatus*			*Nan* sp. (1-); *Cra gig* L 15-20/mL (3-6); R 5/mL (3-38); AN (33-63); *Aca ste*, *Tig jpn* (33-63); MK, MF (59-80)	~60	~30	80	1.7	Kitajima et al., 1991
Nassau grouper *Epinephelus striatus*	3500	0.3 *e*/L	MA (0-25); ER (3-49); AN (11-60); EOA (21-60); MSh (56-61, 63,66,69); F (56-61W)	46-70	35-50	*e*-98	5.0	Tucker & Woodward, 1996
Nassau grouper	33,800	7 *l*/L	*Nan ocu* 500/μL (0-20); Pro (0-); C (0-), R 6-18/mL (2-5), 8-27/mL (6-15); 1-7/mL (16-20); AN (15-17);	45-62	~30	1-62	1.4	Watanabe et al., 1996b

			EAM 0.3-0.5/mL (18-62); MPD (20-62)					
Greasy grouper *Epinephelus tauvina*	500-3000		*Chl* sp. (0-28); ML (3-15); R (3-28); Co, AN (11-40); MSh MF (38-)					Mayunar, 1992
Spotted sand bass *Paralabrax maculatofasciatus*	1000	40*e*/L	*Tet chu* 10/μL (3-29); R 5/mL (3-7), 10/mL (8-30); AN 1-1.5/mL (20-35); OA 0.5-1/mL (24-45); C 1/mL (23-38) MF (31-60); F (40-)	35	18	0-30	2.5	Avilés-Quevedo et al., 1995
Leopard coraltrout *Plectropomus leopardus*	60,000		R (2-55); AN (11-70); OA (38-78); F (59-98)	~65	~35	55	3.2	Masuma et al., 1993
			Pseudochromidae—dottybacks					
Orchid dottyback *Pseudochromis fridmani*	76	~13 *l*/L	*Nan ocu* (0-); R (0-23); AN (8-); GrSh (9-); Z (1-33); F (12-)	~20-40	≤19	1-36	~35	Moe, 1997a
			Grammatidae—basslets					
Royal gramma *Gramma loreto*		~5 *l*/L	MA (1-20); R (1-20); AN (~10-); EAM (~20-); F (30-)	~30				Walch, 1994a
Blackcap basslet *Gramma melacara*			*Nan ocu*, *Iso* sp. (0-14); R (0-); AN, Co (9-35)	21-35				Addison, 1994a, 1994b
			Plesiopidae—longfins					
Comet *Calloplesiops altivelis*	33		MA (0-); R (1-); F (3-); AN (7-); MSh, MH (14-)	~23	~8			Wassink & Brons, 1990
			Sillaginidae—whitings, sillagos					
Japanese whiting *Sillago japonica*	30	31.7 *e*/L	OL (3-11); R (6-23); C (18-28)			*e*-28	~30	Lee & Hirano, 1981
Japanese whiting	1000	30 *l*/L	R 5-10/mL (2-27)	≥25	≥12			Oozeki et al., 1992b
			Rachycentridae—cobia					
Cobia *Rachycentron canadum*	38 & 76	0.6-7.9 *l*/L	*Ana* sp. (0-); Z (2-39); *Gam aff* (40-); MSh, CF, F	~30	~29			Hassler & Raineville, 1975
			Carangidae—jacks					
Greater amberjack *Seriola dumerili*	10,000		R; A; FL, Mixed	~30	~10			Masuma et al., 1990
Greater amberjack			*Nan ocu*; R; A, FE	~31	~16			Tachihara et al., 1993
Goldstriped amberjack *Seriola lalandi*			R; Z; *Tig jpn*; *Opl fas* E&L, MF	~30	~15			Fujita & Yogata, 1984
Goldstriped amberjack			*Cha gra* & *Tet chu* ~10/μL (2-); R 1-3/mL (2-23); *Eut acu* 1-3/mL (2-23); F (21-)			30	40	Kraul, 1985

Table 8-4. (Continued)

Species	Container size (L)	Stocking density	Food & density (dah provided)	Transformation (dah)	Transformation (mm TL)	Survival (dah)	Survival (%)	References
Mazatlan yellowtail *Seriola mazatlana*	15,000	20-50 *l*/L	*Tet* sp. 10/μL (3-21); R 5-10/mL (3-7); ER 5-10/mL (7-16); AN 1-3/mL (10-14); EAM 1-3/mL (14-35); Z (14-35); F (27-)			*l-j*	~1	Benetti, 1997
Yellowtail *Seriola quinqueradiata*	~9000		OL (3-7); R (5-11); Co (8-46); *Acan sch* L (24-46); MF (33-46)					Kuronuma & Fukusho, 1984
Florida pompano *Trachinotus carolinus*			*Dun* sp., *Tet* sp. (0); Z, Pr, R, *Oit* sp. (3-13); AN (14-) MSh, MF, (~15-)	~24	~13	3-18	1.7	Hoff et al., 1978b
Florida pompano			R; A; F (~22-)	~22	~15	*e*-22	~9.6	McMaster, 1988
Permit *Trachinotus falcatus*			OE&L, Pr, R (2-); A, FM, F			~20	36	Lin, 1991
			Coryphaenidae—dolphins					
Dolphin *Coryphaena hippurus*		1 *l*/L	*Cha gra*, *Tet chu*, *Chl* sp. 200/μL; R and/or *Eut acu* (2-6); *Eut acu* (4-12); AN (8-20); OA (15-W); MSq, MF, MK (20-30W)					Szyper, 1991
Dolphin	4000	10-15 *l*/L	AN 0.25-0.3/mL (3-4); EAM(16-he) 0.05/mL (5-6); EAM(36-he) 0.05/mL (6-13); EOA (14-30); F (25-)			*e*-20 *e*-45	5 2.5	Kim et al., 1993
Dolphin	4000	2 *l*/L	*Cha gra* 5-20/μL (3-25); R 1-2/mL initially (2-8); *Eut acu* (6-12); EAM (10-25); *Cor hip* H (18-40); MSq or F (~30-)			2-30 2-60	40 10	Kraul, 1993
			Lutjanidae—snappers					
Mutton snapper *Lutjanus analis*	380		T-*Iso* (0-13); R (0-13); Z (0-13); AN (14-)	~28	~20			Clarke et al., 1997
Mutton snapper	30,000	10.5 *e*/L	*Nan ocu* (-7 to 28); Z (-7 to 38); R (-2 to 28); AN (7-9); EAM (10-35); F (24-37)	30-38	~29	2-38	14.3	Watanabe et al., 1998b
Asian red snapper *Lutjanus argentimaculatus*	190,000	~3.6 *l*/L	Co (3-8); R 10-20/mL (6-12); AN (~10-30); MF or MSh (~25-)	~24	~19	1-20	~0.9	Doi & Singhagraiwan, 1993
Asian red snapper	3000	30 *l*/L	*Chl vir* 100-300/μL (2-34); R 20/mL (2-20); R 10/mL (21-28); AN (21-24); EAM (25-30); OA 1-3/mL (23-50); MF (40-)	~35	~17	1-55	16.2	Duray et al., 1996a
Golden snapper	4000	20-25 *e*/L	*Chl* sp. 250-500/μL; *Per vir* L (3-13); R (5-23); AN (12-30);	~35	~20	*e*-35	5	Lim et al., 1985

Lutjanus johnii			Z (20-35); *Moi mic* (27-35); MM, MF, MSh (30-)					
Yellowtail snapper *Ocyurus chrysurus*	300 & 600		Z, R, AN	~62	~20			Riley et al., 1995
Yellowtail snapper	380		T-*Iso* (0-13); R (0-13); Z (0-13); AN (14-)	~31	~17			Clarke et al., 1997
			Gerreidae—mojarras					
Striped patao *Eugerres brasilianus*	5000	80 *e*/L	*Nan ocu* (0-32); ER 3-10/mL (1-32); *Oit* sp. 0.5/mL (7-31); EAM 2-8/mL (13-57); F (35-)	~48	~20	*e*-48	12.5	Alvarez-Lajonchère et al., 1996
			Sparidae—sea breams, porgies					
Yellowfin porgy *Acanthopagrus latus*	45,000	10.2 *l*/L	*Cra gig* L 5-10/mL (2-10); R 10-20/mL (3-20); R 5-10/mL (21-35); AN 1-2/mL (18-43); Co ~5/mL (20-44); MCD (30-)	~38	~12	1-50	28	Leu & Chou, 1996
Black sea bream *Acanthopagrus schlegeli*			R 2/mL (1-2); R 5/mL (3-9); R 10/mL (10-17); R 15/mL (18-28); A 4/mL (19-24); A 6/mL (25-44); M*Tap* sp. (26-39); MSF (27-)					Park, 1991
Black sea bream	45,000	10 *l*/L	R 10-15/mL (~3-30); AN 1-2/mL (20-40); MCD (30-)	~40	~14	~1-40 ~1-100	50 28	Leu, 1997
Sheepshead *Archosargus probatocephalus*	3500	4 *e*/L	*Tet* sp., T-*Iso* (1-24); R 5-10/mL (3-24); R 0-5/mL (25-31); AN (14-30); AA(31-47); *Tig jpn* (18-70); ScM, ShM (24-77); *Aca ton*, CrL (occ 28-45); F (27-)	≥39	≤14	*e*-62 *e*-100	 40	Tucker, 1987a; Tucker & Alshuth, 1997
Common dentex *Dentex dentex*	7000 & 15,000	45 *l*/L	MA 100/μL (3-20); R 2-4/mL (3-12); ER 4-6/mL (13-20); AN (14-18); EAM (19-); F (19-)			*e*-1 g	15	Sweetman, 1992
Sargo *Diplodus sargus*	200	80 *l*/L	MA 230/μL (2-30); R 5-7/mL (2-12); AN 0.5/mL (10-20); EAM 0.5/mL (18-30); F (20-)	~30	~12	1-30 1-94	15 11	Pousão-Ferreira et al., 1997b
Twoband sea bream *Diplodus vulgaris*			R 5-8/mL (5-17); R <5/mL (18-28); AN 1-2/mL (15-28)			1-28	~12	Jug-Dujaković & Glamuzina, 1988
New Zealand snapper *Pagrus major*	5000		R 10/mL (4-21); R 5/mL (22-27); EAM 0.4-1.5/mL (23-40); MF (33-)	~35	~14			Battaglene & Talbot, 1992
Red sea bream *Pagrus major*	90,000	18.1 *l*/L	R (4-37); FR (11-34); AN (24-37); FA (26-30,32-34,36-37); FE (30,32,34-35); MK (29-37); MF (35-37)			1-37	24	Tanaka & Fujiwara, 1983; Foscarini, 1988
Red sea bream	500	~6.7 *l*/L	*Nan ocu* 500/μL (~0-10); R (3-26); Co (21-25); ClJ, ShJ (22-33); MSh, MF (25-60); Mixed (25-32); F (32-)			4-30	50	Kuronuma & Fukusho, 1984
Red sea bream	100,000	15.8 *l*/L	R (4-46); FR (16-41); AN (28-46); FA (30-39); MK (30-46)			1-47	32	Tabaru et al., 1984; Foscarini, 1988
Red sea bream			R 2/mL (1-3); R 5/mL (4-9); R 10/mL (10-18); R 15/mL (19-30); A 4/mL (15-21); A 6/mL (22-40); M*Tap* sp. (24-41); MSF (33-)					Park, 1991

Table 8-4. (Continued)

Species	Container size (L)	Stocking density	Food & density (dah provided)	Transformation		Survival		References
				(dah)	(mm TL)	(dah)	(%)	
Red sea bream	7000 & 15,000	45 *l*/L	MA 100/μL (3-18); R 2-4/mL (3-9); ER 4-6/mL (10-18); AN (10-24); EAM (14-); F (19-)			*e*-1 g	>7	Sweetman, 1992
Red sea bream			R (3-25); MPD (8-); A (26-39)					Kyowa nd.a
Haffara sea bream *Rhabdosargus haffara*	6,000	7.6 *l*/L	*Chl* sp., *Tet* sp. 200-500/μL (3-20); R 5-10/mL (3-37); F (8-33); AN 0.5-1/mL (28-45); CfP (40-70)	~38	~13	1-70	10.9	Chang et al., 1994b
Silver sea bream *Rhabdosargus sarba*	5000	10 *l*/L	*Nan ocu* (0-45); R 5-15/mL (2-30); AN 0.5-1/mL (20-30); AN 1-2/mL (31-40); MCD (30-45)	≤45	≤19	1-45 1-105	43 24	Leu, 1994
Gilthead sea bream *Sparus auratus*			*Pra mar* 20-25/mL (4-6); R (5-16); AN (12-25); AM (22-40); OA (30-45); F (40-)	~40	~12	1-60	11	Person-le Ruyet & Verillaud, 1980
Gilthead sea bream	3000	~70 *l*/L	*Chl* sp. 700/μL; R 8-15/mL (3-29); AN 0.02-?/mL (8-); FF (40-); OA	≤70		1-70	7.8	Tandler & Helps, 1985
Gilthead sea bream	7000 & 15,000	45 *l*/L	MA 100/μL (4-20); R 2-4/mL (4-13); ER 5-6/mL (14-20); AN (9-22); EAM (17-); F (17-25-)			*e*-1 g	7	Sweetman, 1992
Gilthead sea bream	600	100 *e*/L	*Nan* sp. 500-700/μL (3-); R 20/mL (3-); AN (15-17); EAM ≤5/mL (18-); F (30-)					Zohar et al., 1995
			Sciaenidae—drums					
Mulloway *Argyrosomus hololepidotus*	2000	50 *l*/L	R 10/mL (3-15); EAM 0.2-0.6/mL (12-40); AA (33-68); MF (38-); F (39-)	~34	15-26	1-23	7.3	Battaglene & Talbot, 1994
White seabass *Atractoscion nobilis*	30	16.7 *l*/L	R (3-21); AN (~3-); *Tig cal*; MK	~42	~19	1-21	60	Margulies, 1989; Orhun, 1989
Spotted seatrout *Cynoscion nebulosus*		≤5 *l*/L	*Tet chu* (0-); R ≥20/mL (2-5); R (6-8); AN 3-5/mL (5-8); AN (9-); OA (15-)			1-30	30	Arnold et al., 1976; Fable et al., 1978
Spotted seatrout	1000	0.2 *e*/L	MA (2-10); T*I*, *No* (2-10); R (2-22); AN (7-36); F (33-)	33-37	~38	*e*-50	12.5	Tucker, 1988
Spotted seatrout	Pond	12.9 *l*/m^2	MA, Z: mainly R, Co, Pol (2-32)			2-32	67	Colura et al., 1992
Highhat *Equetes acuminatus*			*Nns* sp. 10/mL (2-9); R 5-10/mL (2-13); EAM 3-5/mL (10-45) FF (~45-)	~30				Harris, 1996

Red drum *Sciaenops ocellatus*	Pond	24 *l*/m^2	Z: mainly C			2-32	65	Colura et al., 1976
Red drum	57-74	1-2 *l*/L	R (3-); AN (6-); MSh (15-)			1-30	30	Arnold et al., 1977
Red drum	150		R 3-5/mL (3-10); A 0.5-1/mL (10-19); MSh (15-35); F (15-)	~35	~25			Holt et al., 1990
Red drum	150	13.3 *l*/L	R 3-5/mL (3-12); AN 1-2/mL (13-18)			1-18	63	Holt, 1993
Red drum	150	13.3 *l*/L	R 2-3/mL (3-7); MPD (3-32)	~32	~33	1-18 1-32	84 61	Holt, 1993
			Cichlidae—cichlids					
Nile tilapia[FW] *Oreochromis niloticus*	84	0.95 *l*/L	MA, Cl, worms, insect L (6-27); F (14-69)			6-69	94	Saleh & Eleraky, 1993
			Pomacentridae—damselfishes					
Sergeant major *Abudefduf saxatilis*	1600		MA (0-24); *Cra gig* L (0-9); R (0-24); Co (2-32); AN (11-12); OA (13-32); EAM (13-31); F (32-)					Alshuth et al., 1998; Tucker et al., unpub.
Clownfishes *Amphiprion* spp.	240		*Rha* sp., *Iso* sp. (0-); R (0-); AN (5-26); F (~19-)					Juhl, 1992
Clownfishes	101		R (0-7 to 10); FF (3-10 or 12); AN (5-12 or 15); KM (8-15 or 17); F (8-); Gel (17-)	~25	~14	1-30	89	Hoff, 1996
Twoband clownfish *Amphiprion bicinctus*	1000		R (0-7; AN (~5-17); F (~15-)	≥17				Young, 1991
Tomato clownfish *Amphiprion frenatus*	45		MA (0-); R (0-); AN (3-30); FF (14-); FA (~30-)					DiMarco, 1994
Spot dascyllus *Dascyllus albisella*	2500	0.1 *l*/L	*Tets* >2/μL (0-13); R 1/mL (0-13); Z >0.1/mL (5-23); AN 0.1/mL (17-)	23-45		0-45	41	Danilowicz & Brown, 1992
Threestripe damselfish *Dascyllus aruanus*	2500	0.16 *l*/L	*Tets* >2/μL (0-13); R 1/mL (0-13); Z >0.1/mL (5-23); AN 0.1/mL (17-)	35-50		0-50	~25	Danilowicz & Brown, 1992
			Oplegnathidae—knifejaws					
Striped knifejaw *Oplegnathus fasciatus*			MA (1-7); R (5-27); AN (16-30); F, FE (26-36)	~35	~15			Koh & Kim, 1992
			Mugilidae—mullets					
Redeye mullet *Liza haematocheila*	1000	30-40 *l*/L	R (3-22); EAM (19-29); F (~27-)	25-30	12-14			Yoshimatsu et al., 1992b
Striped mullet *Mugil cephalus*	4000	20.7 *l*/L	*Nan ocu* 500-700/μL (2-30); R 5-10/mL (2-50); AN 0.1-2/mL (20-60); F (~27-)			1-60	52	Lee et al., 1988a; Eda et al., 1990a; Lee & Kelley, 1991

Table 8-4. (Continued)

Species	Container size (L)	Stocking density	Food & density (dah provided)	Transformation (dah)	Transformation (mm TL)	Survival (dah)	Survival (%)	References
Striped mullet			*Nan ocu* or *Chl* sp. & *Nav* sp. 500-700/μL (1-2⁺); OE&L ≤500/mL; (3-13); R 3-10/mL (5-19); Co 2-3/mL (10-45); AN 2-3⁺/mL; (16-45); G 1-2/mL (29-45); RB, Fl (29-45); F (5-)	~28	~12			Liao & Chao, 1991
White mullet *Mugil curema*	38	2.6 *e*/L	*Chl* sp., *Ana* sp.; Co 1-2/mL (1-); F			*e*-20	23	Houde et al., 1976
Liza *Mugil liza*	1800	20 *l*/L	*Chl* sp.; R; AN			1-26	11.1	Alvarez-Lajonchère et al., 1988
Mugil soiuy	4000		*Eup* sp. 20/mL (3-4); R 3-5/mL (4-19); Z (4-29); C 0.5-1/mL (4-29); AN 0.5-1/mL (10-29)					Saifulina, 1991
			Polynemidae—threadfins					
Pacific threadfin *Polydactylus sexfilis*	4000	20 *l*/L	*Nan ocu* 500/μL (2-15); R 20/mL (2-15); AN 0.3-3/mL (7-19); FA, FFE, GrSq, GrF, GrKr, (15-25)			~24	~27	Kelley et al., 1995
Pacific threadfin	4000	6-39 *l*/L	*Nan ocu* 500/μL (3-14); R 20/mL (3-14); EAM (10-26); R ≤20 (15-26)	≥26				Ostrowski et al., 1996b
			Labridae—wrasses					
Multicolorfin wrasse *Halichoeres poecilopterus*	500		*Nan ocu* 200/μL (3-); *Cra gig* L 60/mL (3-); R, AN, Z, F	≥35	~21			Kimura & Kiriyama, 1993
			Anarhichadidae—wolffishes					
Common wolffish *Anarhichas lupus*	300	~0.5 *l*/L	F (16-); AN (24-71)			1-108	54	Moksness et al., 1989
White Sea wolffish *Anarhichas lupus marisalbi*	200		A (0-); MT (10-15); F (10-); OA, MF, ME (15-)					Pavlov & Novikov, 1987
			Opistognathidae—jawfishes					
Yellowhead jawfish *Opistognathus aurifrons*	76	5-7 *l*/L	*Tet* sp., *Nan* sp. (0-14); R 5-20/mL (0-14); AN (10-); EAM (15-); F (~28-)	~21-28				Walch, 1993, 1994b
			Eleotridae—sleepers					

Marble goby[FW] *Oxyeleotris marmorata*	1000	16 *l*/L	*Chl* sp. 500/μL; F (3-7); R (4-24); AN (16-55); *Moi* sp. (26-59); *Chi* sp. L (39-)	~40	~13	5-10 5-70	32 22	Senoo et al., 1994a, 1994b
Gobiidae—gobies								
Mugilogobius abei	30		OE&L (0-15); R (16-43); MF, MgL (44-)	~44	~17			Kanabashira et al., 1980
Ephippidae—spadefishes								
Atlantic spadefish *Chaetodipterus faber*	75		R (2-19); AN (9-); FF, MSh, MF, GD, AA (20-)	~20				Walker, 1991
Atlantic spadefish	509		*Tet* sp. 50/μL (2-31): R 2-20/mL (2-14); AN 2-10/mL (5-24), 10/mL (25-31); MySh (24-31)	~31	24-33			Gomez, i.p.
Siganidae—rabbitfishes								
Yellowblotch spinefoot *Siganus guttatus*	600	50 *l*/L	*Iso gal* 30-50/μL (1-10); *Chl* sp., *Tet* sp. 200-300/μL (3-35); R 10-20/mL (2-18); R 5-10/mL (19-35); AN 1-2/mL (18-35); F (20-)	~35	~18	1-35	37	Juario et al. 1985
Yellowblotch spinefoot	11,000	18.2 *l*/L	*Chl* sp. 500/μL (0-45); R 5-20/mL (1-30); AN 0.2-1/mL (15-38); F (23-)	~35	~19	45	16.6	Hara et al., 1986
Yellowblotch spinefoot			*Chl* sp. 100/μL (1-15); *Chl* sp. 50/μL (16-30); R 10-15/mL (1-15); R 5/mL (16-25); AN 0.3/mL (15-20); A 0.5/mL (21-25); A 1-2/mL (26-35); F (25-)					Duray & Juario, 1987
Goldline spinefoot *Siganus lineatus*	6000	0.83 *l*/L	MA ~1,700/μL (1-20); R 2-48/mL (1-14); *Oit* sp. (12-22, (31-35); AN (15-18, 22-33)	~35	~23	1-35	16	Bryan & Madraisau, 1977
Whitespotted spinefoot *Siganus oramin*	500/3 wk, then cage		R (3-30); AN (12-42); Co (22-50); MSh, F (30-)					Higuchi et al., 1980
Scombridae—mackerels								
Chub mackerel *Scomber japonicus*	200 to 400	15 *e*/L	*Tet* sp. 300-500/μL (2-); R 0-30⁺/mL (2-); *Tis* sp. 0.5-1/mL (5-)	16-24⁺	~15			Hunter & Kimbrell, 1981
Yellowfin tuna *Thunnus albacares*	15,000	0.02 *l*/L	R (2-10); Co (8-20)	~58	~51	1-17 1-21	14 0	FSFRL, 1973
Northern bluefin tuna *Thunnus thynnus*	5000	6 *l*/L	ER (3-30); AN (12-30); *Ple leo* E&L (12-30); OA (24-30); F (26-30)	≥30	≥13	2-35	0.2	Kaji et al., 1996
Bothidae—lefteye flounders								
California halibut *Paralichthys californicus*	300	60-70 *e*/L	R 15-20/mL (4-20); R (21-); AN 5/mL (21-)	~47	~15			Gadomski et al., 1990
Summer flounder *Paralichthys dentatus*	40	12.5 *e*/L	*Chl* sp., R, *Aca cla*, *Cen ham*, *Tem lon* (4-); AN (~14-)	≥56	≥12	~1-56	5	Smigielski, 1975b

Table 8-4. (Continued)

Species	Container size (L)	Stocking density	Food & density (dah provided)	Transformation (dah)	Transformation (mm TL)	Survival (dah)	Survival (%)	References
Summer flounder			R (3-23); A (16-*m*)	35-65				Bengtson, 1997
Summer flounder		30 *l*/L	R (3-); EAM (~14-); F (~35-)			1-35	~33	Nardi, 1997
Southern flounder *Paralichthys lethostigma*	57	1.2 *e*/L	R (6-); AN (14-); MSh (35-)			*e*-90	30	Arnold et al., 1977
Southern flounder			R (5-17); A (17-48)	30-50		~*l*-50	10-15	Daniels et al., 1997
Southern flounder	Pond	28.5/m^2	Z: mainly R 2.2/mL initially (5-); F (35-)	≥34	≥15-22	5-206	6.1	Jenkins et al., 1997b
Chilean flounder *Paralichthys microps*	200		R 5/mL (~4-); AN 10/mL		~15-16			Zuñiga & Acuña, 1992
Hirame *Paralichthys olivaceus*	1500	51 *l*/L	*Tet sue* 50-70/µL (0-25); R 12-18/mL (4-34); EAM 1-2/mL (20-60)	~45	~15	1-85	45	Spedicato et al., 1994
Hirame	50,000	20 *l*/L	R (2-30); A (12-46); F (16-)	~50	~25	50	25	Yamaha Fishery Journal, 1991
Hirame			R (3-25); MPD (8-); A (26-40)					Kyowa nd.a
Ecuadorian flounder *Paralichthys woolmani*	15,000	20-50 *l*/L	*Tet* sp. 10/µL (3-21); R 5-10/mL (3-7); ER 5-10/mL (7-16); AN 1-3/mL (10-14); EAM 1-3/mL (14-35); Z (14-35); F (27-)			*l*-*j*	8	Benetti, 1997
Turbot *Scophthalmus maximus*	90, then 3000	3-22 *l*/L	*Iso gal* ~111/µL (0-); R ~7/mL (0-); AN (6-)	43-55	~30	2-55	~25	Alderson & Ball, 1977
Turbot	2000	~30 *l*/L	ER (2-11); EAM (8-30); F (~30-)	~30-40	≥18	1-30	40	Person-Le Ruyet, 1990b
Turbot	400		R (3-16); *Har lit*, *Aca cla* (3-30); AN (6-30); *Tis fur* (10-30); AM (15-30); *Moi mic* (18-30); OA (20-30)	~30	~16			Khanaichenko et al., 1994
Turbot	200	10 *e*/L	*Tet sue* (1-12); ER 3-5/mL (2-4); ER 5/mL (5-12); EAM 0.5-1.5/mL (13-)			1-27	52	Reitan et al., 1994b
			Pleuronectidae—righteye flounders					
Witch flounder *Glyptocephalus cynoglossus*	200		T-*Iso* ~200/µL (0-); R 3/mL; Z					Bidwell et al., 1997
Atlantic halibut	562	1.3 *l*/L	MA 7-90/µL (33-54); AM 0.2/mL (33-54)	≥90	~47	33-54	30	Haug, 1990; Naas et al.,

Hippoglossus hippoglossus								1992
Atlantic halibut	150	3.3 *l*/L	MA 7-17/µL (33-90); EAM (33-52); Z (53-90)	≥90	~47	33-52	26	Haug, 1990; Næss et al., 1995b
Lemon sole *Microstomus kitt*			R (~8-52); AN (~40-104)	~85	~17	~1-104	~50	Howell, 1974
Starry flounder *Platichthys stellatus*	38	13.1 *e*/L	*Dun pri* (3-); R 5/mL (3-)	33-90	~9	~90	1.8	Policansky & Sieswerda, 1979
Ridged-eye flounder *Pleuronectes cornutus*			R (4-55); AN (20-80); *Tig jpn* (43-80)	~51	~24	75	~17	Kitajima et al., 1988
European plaice *Pleuronectes platessa*	60	15 *l*/L	R, AN (5-7); EAM 5/mL (8-15); AN 5/mL (16-50)	~35	~14	1-50	~39	Dickey-Collas & Geffen, 1992
Greenback flounder *Rhomobosolea tapirina*			ER (-14); AM 10/mL (9-22); AM ≤10/mL (23-42); F (23-)	≥33	≥16			Hart & Purser, 1996
Spotted halibut *Verasper variegatus*		24 /L	*Chl* sp. 500/µL; R 20/mL (~4-)	~32	~15	25	44	Cho et al., 1995
			Soleidae—soles, tonguefishes					
Cynoglossus abbreviatus			R (5-32); AN (22-58); *Tig jpn* (30-75); Co (30-75)	28-35	~14			Fujita et al., 1986
Solea senegalensis	150, then 1500	60 *l*/L	R (2-4); AN (5-40); FA (31-40); F (41-)					Dinis, 1992
Common sole *Solea solea*	2000	~70 *l*/L	EAM (2-31); F (~31-)	~15	~10	1-30	80	Person-Le Ruyet, 1990b
			Tetraodontidae—puffers					
Takifugu chrysops	60	33.3 *l*/L	*Nan ocu*; R (3-30); AN (23-41); *Tig jpn* (30-50); MC (40-90)	~24	~9			Fujita & Shinohara, 1986
Takifugu exascurus	60	33.3 *l*/L	*Nan ocu* (4-21); R (4-21), AN (10-51); MF (33-)	~16	~10			Fujita & Honma, 1991
Tiger puffer *Takifugu rubripes*			R (3-21); F (8-37); *A* (14-37)					Kanazawa, 1991b

FW Freshwater species.
e = fish eggs stocked; *l* = fish larvae stocked; *m* = metamorphosis (of flatfish); *p* = parturition.
E = eggs; H = hatchlings; L = larvae (used for food).
W = weaning period.
he = hours enriched.

Table 8-4. (Notes continued)
Microalgae: *Ana* sp. = *Anacystis* species; *Chl* sp. = *Chlorella* species; *Chl vir* = *Chlorella virginica*; *Dun pri* = *Dunaliella primalecta*; *Iso* sp. = *Isochrysis* species; *Iso gal* = *Isochrysis galbana*; MA = microalgae; *Mon* sp. = *Monochrysis* species; *Nan ocu* = *Nannochloropsis oculata*; *Nan* sp. = *Nannochloropsis* species; *Nav* sp. = *Navicula* species; *Nns* sp. = *Nannochloris* species; *Pra mar* = *Prasinocladus marinus*; *Rho* sp. = *Rhodomonas* species; *Tet chu* = *Tetraselmis chuii*; T*Iso* = Tahitian *Isochrysis*; *Tet* sp. = *Tetraselmis* sp.
Protozoans: *Eup* sp. = *Euplotes* sp; Pro = protozoans.
Molluscs: Cl = clam; *Cra gig* = *Crassostrea gigas* (Pacific oyster); *Mer mer* = *Mercenaria mercenaria* (hard clam); Mu = mussel larvae; Oy = oyster; *Per vir* = *Perna viridis* (green mussel); *Tap* sp. = *Tapes* sp. (short-necked clam).
Rotifers: ER = enriched rotifers; FR = frozen rotifers; R = rotifers.
Artemia: A = *Artemia*; AA = adult *Artemia*; OA = older *Artemia* (i.e., between nauplius and adult); EA = enriched *Artemia*; EAM = enriched *Artemia* metanauplii; FA = frozen *Artemia*.
Cladocerans: Cl = cladocerans; *Moi mic* = *Moina micrura*; *Moi* sp. = *Moina* species.
Copepods: C = copepods.
Calanoid copepods: *Aca cla* = *Acartia clausi*; *Aca ste* = *Acartia steueri*; *Aca ton* = *Acartia tonsa*; *Cen ham* = *Centropages hamatus*; *Eur aff* = *Eurytemora affinis*; *Eut acu* = *Euterpina acutifrons*; *Oit* sp. = *Oithona* sp.; *Tem lon* = *Temora longicornis*.
Harpacticoid copepods: *Har lit* = *Harpacticus littoralis*; *Tig jpn* = *Tigriopus japonicus*; *Tis fur* = *Tisbe furcata*; *Tsb elo* = *Tisbintra elongata*.
Mixed zooplankton: Z = mixed zooplankton (including copepods).
Miscellaneous animals: G = amphipods; MySh = mysid shrimp; Pol = polychaetes.
Live fish: *Acan sch* = *Acanthopagrus schlegeli* (black sea bream); *Cor hip* = *Coryphaena hippurus* (dolphin); FE = fish eggs; FFE = frozen fish eggs; FL = fish larvae; *Gam aff* = *Gambusia affinis* (mosquitofish); *Poe ret* = *Poecilia reticulata* (guppy); *Opl fas* = *Oplegnathus fasciatus* (striped knifejaw); *Pag maj* E&L = *Pagrus major* (red sea bream); *Ple leo* = *Plectropomus leopardus* (leopard coraltrout); *The cha* = *Theragra chalcogramma* (walleye pollock).
Nonliving seafoods: CF = cooked fish; CfP = cuttlefish paste; CJ = clam juice; GrF = ground fish; GrKr = ground krill; GrSh = ground shrimp; KM = krill meal; MB = minced bivalve; MC = minced clam; MF = minced fish; MK = minced krill; MM = minced mussel; MSh = minced shrimp; MSq = minced squid; MxSF = mixed seafoods; ScM = scallop meal; SF = seafoods; ShJ = shrimp juice; ShM = shrimp meal.
Processed plant matter: Fl = flour; RB = rice bran.
Processed animal matter: LE = liver extract; ME = minced earthworms; MH = minced beef heart; MMt = minced meat; MTu = minced tubifex worms.
Compound feeds: GD = gelatin-bound diet; F = compound feed; FF = flake feed; LF = liquid feed; MBD = micro-bound diet; MCD = micro-coated diet; MED = microencapsulated diet; MPD = microparticulate diet.

conditioning water consists of adding small amounts of phytoplankton every day from 1 d before stocking eggs until a healthy layer of macroalgae is present on the tank walls (maybe 5-10 d later). If water exchange is sufficient and light intensity is low, macroalgae will not get out of control.

With high densities of larvae and food, and without algae and enough light for it to grow, a high exchange rate with ammonia-free water is necessary. Toxic levels of ammonia in pipelines from metabolism of fouling organisms and microplankton and decomposition of detritus should be avoided.

Nursery Ponds

Extensive larval rearing in ponds or impoundments can produce larger numbers of juveniles more economically than intensive rearing in tanks, but nursery ponds need to be managed very carefully, and weather changes can cause problems. For methods, see Geiger (1983), Opuszynski et al. (1984), Rutledge (1988), Chamberlain et al. (1990), Colura (1990), Davis (1990b), Geiger and Turner (1990), Naas et al., (1991), Habib and Rahman (1993), Anderson and Tave (1994), and Li et al. (1996).

SPECIFIC FEEDING PRACTICES (Table 8-4)

Length (i.e., ~swimming ability) and mouth size (i.e., range of food choice) at first feeding are key factors in larval survival. Lengths reported in the literature more often are for hatchlings and juveniles rather than for first feeding larvae, and not many mouth widths have been reported. The hatchling lengths given in this section can be viewed as indicators of relative ability at first feeding, which usually occurs soon after hatching. Larvae of species that hatch at 1-2 mm usually are fairly difficult to feed and probably need the smallest possible rotifers or other small foods. Those 2-3 mm long tend to first feed well on small to medium rotifers. Some longer than 3 mm can first feed on *Artemia* nauplii, and some longer than 4 mm might not need live food at all. Mouth size is not necessarily proportional to length. Some species develop a large amount between hatching and first feeding (e.g. salmon, halibut), but most do not. For more information on development, see Table 2-1.

Typical rearing temperatures are given in parentheses. If the water supply was not thoroughly treated, substantial amounts of food (e.g., protozoans, copepods) besides those added (i.e., listed) might have been present.

Acipenseridae

Sturgeons are large at hatching (~10 mm TL), have large yolks, do not require live food, will first feed (8-14 dah for white sturgeon) on moist or semi-moist salmonid feeds, and will eat dry feeds well within 19-41 dah (Conte et al., 1988; Williot, 1990). Oligochaetes, *Artemia*, *Daphnia*, and other live or fresh foods sometimes are used at first feeding or mixed with salmonid feeds for several days, but these might not be necessary or practical and weaning can be difficult.

Siberian sturgeon larvae reared for 29 d on two experimental diets based on beef liver and yeast had survival of 68% and 73% and final wet weight of 379 mg and 392 mg (Charlon and Bergot, 1991). Specific growth rate was 10%/d and FCR was 2. Commercial trout starter was fed next.

Italian sturgeon larvae were fed a commercial feed (Larvit by Hendrix) supplemented with egg yolk, curdled milk, and vitamins for 15 d, and then a combination of commercial sturgeon feed, brown trout starter, fresh fish, and vitamins (Giovannini et al., 1991).

Gulf of Mexico sturgeon should survive well (>50%) when fed *Artemia* nauplii for a week, a dry larval feed for 3 weeks, then semimoist feed (Bardi et al., 1998).

Anguillidae

American eel larvae drift in the oceanic plankton for about 10 months before reaching the coast and becoming elvers (Usui, 1991). **European eels** take 22 months or more. Many attempts to raise leptocephali have ended in starvation. The diet of early eel larvae in nature is not known. An article in the Yamaha Fishery Journal (1992a) stated that before metamorphosis wild eels eat zooplankton, during metamorphosis they fast, and afterwards (as they migrate up rivers) they successively eat detritus, zooplankton, larger crustaceans, aquatic insects, small fish, and other animals. Elvers do not require live food. During the earliest stages, some dissolved nutrients might be absorbed through the skin or the gut with minimal digestive effort (Otake, 1996). In Japanese eels, eyes are pigmented, digestive enzymes are present, the yolk is exhausted by 7 dah, and they should start feeding (Kurokawa et al., 1995). Recent studies showing that leptocephali will eat solid food helps give the study of larval eel nutrition new direction. When Mochioka et al. (1993) presented copepodites, small ctenophores, jack muscle paste, and squid paste to leptocephali of four species from **three eel families** in an aquarium, the squid paste (but none of the other foods) was eaten by leptocephali of a moray and a conger eel, but not by those of a nettastomatid and another moray eel. Tanaka et al. (1995) presented rotifers to Japanese eel larvae starting 8 dah; the next day larvae were seen making striking motions and beginning 13 dah rotifers were found in larval guts. Unfortunately, the eels did not survive to 18 dah. Kurokawa et al. (1996) found that larvae would eat rotifers and absorb proteins from them at 11 dah.

Japanese elvers can be fed by a schedule like the following (Usui, 1991):

Weeks 1-2: minced bivalve or earthworm flesh

Weeks 3-4: minced fish flesh and *Tubifex* sp. or similar worms
Weeks 5-10: minced fish flesh or compound feed
Compound elver feeds have been been developed and are replacing much of the fresh food.

Clupeidae
Atlantic menhaden (20°C) hatch at 3.2 mm TL, first feed 3 dah, have EYS ~4 dah, and transform ~65 dah (~27 mm). **Gulf menhaden** (20°C) hatch at 2.8 mm TL, first feed and have EYS 4 dah, and transform ~85 dah (~31 mm). Both species were reared on similar diet sequences by Hettler (1981; Table 8-4). Tucker (unpub.) also raised gulf menhaden in a 1000-L tank on rotifers, *Artemia* nauplii, and commercial salmon feed.

Japanese sardines (20°C) hatch at 3.0 mm TL, first feed and have EYS 3 dah, and transform ~50 dah (~40 mm). They were raised on rotifers, *Artemia* nauplii, and red sea bream eggs & larvae (Nakamura et al., 1991). Survival from eggs to 80 dah was 7.7%.

Engraulidae
Northern anchovies (16°C) hatch at 3.0 mm TL, first feed and have EYS 4 dah, and transform 50-60 dah (34-40 mm). They were raised on dinoflagellates, rotifers, copepods, and *Artemia*, with 12.5% survival during 0-74 dah (Hunter, 1976). Survival was lower in other trials without copepods, but 6% of larvae fed only rotifers survived to 42 dah. Overdependence on *Artemia* nauplii typically led to high mortality during 20-40 dah.

Chanidae
Milkfish (29°C) hatched at 3.7 mm TL, first fed 77 hah (5.4 mm) with a mouth width of 258 µm, all fed by 125 hah, and also had EYS at about 125 hah (Kohno et al. 1994c). Gastric glands were developed at about 6 weeks (Watanabe and Kiron, 1994).

Milkfish have been shown to grow faster when *Artemia* are enriched with ω3HUFA (Bengtson et al., 1991). Rotifers fed *Tetraselmis tetrathele* or *Isochrysis galbana* supported faster growth than those fed *Chlorella* sp. (Villegas et al., 1990). Tamaru et al. (1993b) reported a greater mean length of milkfish raised through transformation (2-30 dah) on rotifers cultured on baker's yeast with or without *N. oculata* than for *N. oculata* alone, but better survival of 52% with algae-fed rotifers vs 41% (yeast), and 30% (yeast & algae) could have led to more small fish affecting the measurements.

When milkfish larvae were abruptly weaned from rotifers to one of six non-living diets or *Artemia* nauplii in 20-L aquaria, survival (63%) was best with a high-fat artificial plankton larval feed, but growth was best with *Artemia* (Duray and Bagarino, 1984).

Marte et al. (1994) reported increased growth and survival when microencapsulated or microbound feeds were added with *Chlorella* sp., rotifers, and *Artemia.*

Over a 30 d period, wild early juvenile milkfish fed *Moina macrocopa* (cultured with chicken manure extract) grew to a weight 2.78 times greater, 4.7 to 213 mg, than those fed rotifers (cultured on "marine *Chlorella*") (Villegas, 1990).

Osmeridae
Ayu (15°C) hatch at 6-7 mm TL, have EYS 4-5 dah, and transform at ~70 mm. In nature, some ayu spend the whole life cycle in freshwater lakes, but the eggs of river ayu are deposited on pebbles and their larvae drift down to the sea to develop (Yamaha Fishery Journal, 1992b). There, copepods are the main food, with some squid and bivalve larvae. (Cladocerans are the main food of lake ayu larvae.) Later, as they reach coastal waters before ascending rivers, their diet includes zooplankton, midges, and macroalgae. After entering rivers, they mostly eat macroalgae but can eat zooplankton and phytoplankton. Larvae can utilize microparticulate feeds even though gastric glands are not developed until about 6 weeks (Watanabe and Kiron, 1994). Kyowa (nd.b) reported a survival of 88% (1-50 dah) for ayu raised with a microparticulate feed substituted for half the usual number of rotifers. Total foods used to raise 520,000 juveniles (23.7 mm TL, 27.3 mg WW) were 6 trillion rotifers (11,586/fish), 136 million *Artemia* (261/fish), and 6.17 kg feed (11.9

mg/fish).

Galaxiidae

Tasmanian whitebait (13°C) hatch at 5 mm TL and begin feeding within 24 h. Larvae reared in tanks on wild zooplankton had similar growth but better survival than those fed rotifers, *Artemia*, and pellets (Daintith, 1992).

Salmonidae

Salmon (3-8°C) eggs take weeks or months to hatch into large (16-22 mm TL) larvae that are ready to begin feeding several weeks after hatching, when about 80% of their yolk has been exhausted (Needham, 1988a). They do not need live food and usually are given dry (sometimes semi-moist) starter feeds. However, Kim et al. (1996a) found that coho salmon larvae grew almost as well on adult *Artemia* (5.8% body weight gain/d) as on natural zooplankton (6.0%) and better than on a commercial starter feed (4.0%). This probably resulted from higher food intake rather than a nutritional advantage.

Gadidae

Atlantic cod (6°C) hatch at 4.5 mm TL, first feed 4-5 dah, have EYS 6-8 dah, and transform ~106 dah (~60 mm). Production of pepsin increased sharply when cod reached 14 mm SL (Hjelmeland et al., 1995b).

Atlantic cod larvae have been raised mostly in large tanks, plastic bags, or impoundments containing blooms of wild zooplankton—especially tintinnids, rotifers, copepods, and cladocerans (Holm et al., 1991; Huse, 1991; Tilseth et al., 1992; van der Meeren and Naess, 1993; van der Meeren et al., 1994; van der Meeren, 1995). Tintinnids seem to be a preferred first food. Opstad et al. (1989) obtained better growth of larvae in 200-L tanks by enriching rotifers with cod roe meal (for at least 4 h) than when enrichment was done with fish meal.

Tilseth et al. (1992) reported survival to transformation as high as 50% for cod raised in plastic enclosures. Enclosures with volumes greater than 100 m^3 were the most successful. The primary problem has been zooplankton shortage and resulting increases in cannibalism; backup supplies of *Artemia* or frozen zooplankton can help. Fish stocking density has to be kept low, and weaning can be difficult.

Cyprinodontidae

Killifishes (26°C) typically hatch at ~4-8 mm TL and first feed within a day or two. They usually are raised extensively in fertilized ponds on zooplankton supplemented with fine compound feeds (Tatum et al., 1982b; Perschbacher and Strawn, 1983; Strawn et al., 1986).

Atherinidae

Atlantic silversides (25°C) hatch at 5.2 mm TL. **Inland silverside** larvae had similar growth and survival when fed *Artemia* enriched with low-, medium-, and high-HUFA ICES emulsions and, unlike Atlantic silversides, did not seem to require EPA (Ashraf et al., 1993a). Inland silverside larvae fed badly on a microencapsulated feed containing 60% casein but well on a microbound feed containing 56% krill; after 3 weeks (from 10 dah), mean weight was similar but survival lower (100% vs 60 & 73%) with the krill feed than with unenriched *Artemia* nauplii.

Leibovitz et al. (1991) found that inland silverside larvae survived as well but grew less with microencapsulated *Artemia* than with live ones; however, the albumen-alginate coating was better than coatings tried earlier.

California grunion (18°C) hatch at 7 mm TL, first feed within hours, have EYS ~4 dah, and transform ~40 dah (~17 mm). They were easy to raise with microalgae and *Artemia* nauplii (Vásquez-Yeomans and Sosa-Cordero, 1992).

Syngnathidae

Seahorses (25°C) have been raised on rotifers, *Artemia*, and newly hatched fish (such as guppies); survival usually was not high (Schmidt, 1995). Vincent (1995c) used cultured zooplankton and *Artemia* and cautioned that a diet of only *Artemia* could be unbalanced and difficult to digest.

Hexagrammidae

Lingcod (10°C) eggs require a long incubation period (~35 d at 10°C), first feed 5-6 dah, have EYS ~6-7 dah, and transform ~50-55 dah (~20-22 mm TL). Larvae were raised on enriched *Artemia* metanauplii only, but higher dietary EFA probably would have increased survival (Appelbaum et al., 1995).

Cyclopteridae

Atlantic lumpfish (10°C) hatch at 5.8 mm, first feed 4-7 dah, have EYS 4 dah, and transform ≥56 dah (~30 mm). Larvae were raised on enriched *Artemia*, and juveniles on a mixture of chopped fish and bivalves or squid plus a vitamin/mineral premix (Brown et al., 1992). Martin-Robichaud et al. (1994), using enriched *Artemia* metanauplii and different dry feeds for juveniles, obtained up to 21% survival during 1-100 dah. Feeding enriched *Artemia* metanauplii 2-3 times/d resulted in better growth than continuous (drip) feeding (Brown et al., 1997c).

Centropomidae

Barramundi (27°C) hatched at a small size (1.4 mm TL), first fed 54 hah (2.6 mm) with a mouth width of 224 μm, all fed by 71 hah, and had EYS about 74 h after all larvae were feeding (Kohno et al. 1994c). Transformation occurs ~25 dah (~17 mm TL).

Because barramundi can be fiercely cannibalistic, especially if reared at high density and not fed well enough, frequent grading might be necessary once many of the fish reach 18 mm TL (Parazo et al., 1991a; Barlow et al., 1993). Juario et al. (1991) obtained high survival and very little cannibalism when barramundi were raised on enriched *Artemia* to 60 dah, good survival and slightly more cannibalism after weaning to compound feed at ~20 dah, and low survival and high cannibalism after weaning at ~10 and ~15 dah.

The feeding schedules for barramundi in Table 8-2 are similar. Rotifers usually are fed during at least 2-15 dah and *Artemia* are started by 10-12 dah. Although barramundi can be reared with unenriched *Artemia*, if *Artemia* are the main food, they should be enriched. Survival and stress resistance was strongly correlated with ω3HUFA in *Artemia* (Bengtson et al., 1991). Cladocerans such as *Moina macrocopa* can be substituted for *Artemia* beginning at 15-16 dah (Fermin and Bolivar, 1994). Weaning to conventional dry crumbles can be accomplished by 26 dah (Tucker et al., 1988).

A schedule used for barramundi in 5- to 25-m^3 tanks in Thailand included rotifers, *Artemia* nauplii, *Daphnia* sp. (optional), and ground trash fish beginning 25 dah (11 mm TL)(Tookwinas, 1990a).

Parazo et al. (1990) recommended that algae and rotifers be added to larval tanks before first feeding (~36 hah) to establish stable populations. Enrichment of *Artemia* seems especially important during 18-23 dah; however, Parazo et al. (1991b) obtained similar survival and growth of barramundi given *Artemia* that had been fed *Chlorella* sp. as for enriched ones. *Moina* sp. can be partially substituted for *Artemia*, but salinity should be reduced to ~10 ppt so they survive better.

Barramundi were raised in 1000-L tanks with rotifers (fed *N. oculata*) during 3-15 dah, and *Artemia* nauplii 10-25 dah (Tucker et al., 1988). Salmon starter was first offered 20 dah. Transformation occurred during 21-25 dah (17 mm TL). The fish began to eat commercial salmon starter 23 dah and were weaned by 26 dah. A cannibalistic tendency in juveniles was mostly controlled by grading and feeding well.

Rimmer and Reed (1990) fed rotifers to barramundi during 2-14 dah and *Artemia* during

8-22 dah in 2-L chambers. Growth (10 mm TL at 22 dah) and survival (near 100%, 1-22 dah) were significantly higher when the *Artemia* were enriched with Frippak Booster® microcapsules, regardless of whether rotifers were enriched. Rimmer et al. (1994a) stated that rotifer enrichment is of little value for barramundi, but *Artemia* enrichment is necessary to prevent a mortality syndrome, probably caused by ω3HUFA deficiency, that can begin at 18-20 dah.

Survival and growth of barramundi larvae reared in 200-L tanks were similar whether they were fed unenriched or Selco-enriched *Artemia* during 10-21 dah (two treatments each); however, those fed enriched *Artemia* survived a high-salinity stress test better at 21 dah (Dhert et al., 1990). Finally, during 22-27 dah, survival of those fed enriched *Artemia* was 100%, while survival with unenriched *Artemia* was zero.

Although barramundi consumed a microbound (with gelatin) feed during the first five feeding days, they could not digest them and died (Southgate and Lee, 1993). Walford and Lam (1991) obtained slightly better growth but similar survival of barramundi when a microencapsulated feed was added to the basal diet of rotifers.

At 12 dah, ω3HUFA-deficient larvae were less aggressive and less resistant to stress. ω3HUFA-deficient early juveniles were slower, less coordinated, less responsive, less resistant to stress, swam higher in the water, were darker than normal fish, and did not school (Dhert et al., 1992b). Buranapanidgit et al. (1988) reported an ω3HUFA optimum of 1.72% (of DW, in 13% total lipids) for juvenile barramundi.

Rutledge and Rimmer (1991) raised barramundi extensively in a 0.12-ha pond on mixed zooplankton, obtaining ~43% survival during 2-22 dah and a final mean size of 35.5 mm TL. Some cannibalism was witnessed as early as 18 dah.

Common snook (28°C) hatch at a small size of 1.5 mm TL, first feed ~2.5 dah, have EYS 4-5 dah, and transform ~35 dah (35-40 mm TL). Larvae were reared in 1000-L and 3500-L tanks with Tahitian *Isochrysis*, *N. oculata*, rotifers (initially 10/mL) 2-20 dah, and *Artemia* nauplii 12-38 dah (Tucker, 1987b). Weaning to commercial salmon starter was accomplished during transformation, 35-38 dah (40-43 mm TL). Sometimes *Tigriopus japonicus* were used up to 42 dah, and sometimes an appetizer of freeze-dried bay scallop meal was used during weaning. Survival from fertilization through transformation was about 7-10% in some trials. In 1995 and 1996, a simplified diet sequence of rotifers, copepods, and *Artemia* was used (with the addition of clam trochophores, which probably were unnecessary, Table 8-4) and the best result was 946 advanced juveniles (55 dah, 7.0% survival) produced in one 3000-L tank despite losses to cannibalism (Tucker et al., unpub.). Our research has shown that cannibalism is worse in early juveniles than in late larvae and is increased by higher fish density and food shortages. Survival and growth were improved when larval tanks were inoculated with a probiotic *Bacillus* sp. isolated from healthy snook raised in 1995 (Kennedy et al., 1998).

Maciorowski et al. (1987) raised snook extensively in ponds on mixed zooplankton. When two moderately-stocked (initially >217,000 larvae/ha) 0.8-ha ponds were harvested during transformation (32 dah), nearly all the snook died from handling and salinity stress; recovery would have been 17% and 25%. Recovery from two lightly-stocked (39,000 larvae/ha) 0.4-ha ponds was 0.03 and 25.1% at 36 dah (early juveniles). Recovery from two moderately-stocked (239,000 larvae/ha) 0.4-ha ponds was 0.6% and 5.1% at 42 dah. Probably food fluctuations and cannibalism, and possibly other predation were factors.

Percichthyidae

European sea bass (18°C) hatch at 3.0 mm TL, first feed 4-5 dah, have EYS 5 dah, and transform ~45 dah (~20 mm TL). Although larvae are capable of feeding 4 dah, Johnson and Katavic (1986) found that feeding could be delayed until 9 dah at 15°C and 13 or 26 ppt salinity without decreasing growth or survival.

European sea bass secrete trypsin at hatching, chymotrypsin, esterase, and amylase beginning 5 dah, lipase beginning 15 dah, and pepsin beginning sometime after 25 dah; they transform by about 40 dah (Person Le Ruyet et al., 1993a, 1993b). Juvenile weaning feeds have been successfully used at 30 dah or later. Larvae did not eat microparticulate or microbound feeds well until 20-25 dah. Before that age, microfeeds could be substituted only partially for *Artemia*, but at 20 dah, with larval weight 3-5 mg, *Artemia* could abruptly be replaced with MPD, followed by juvenile feeds starting at 41 dah; at 60 dah, weight was ~77 mg and survival 35% (from hatching). Gennari et al. (1992) reported that sea bass reared on enriched *Artemia* then abruptly weaned to dry feed at 39 dah (40 mg) reached a mean weight of 363 mg at 63 dah.

Average commercial-scale survival of 15% from eggs to 1-g juvenile sea bass has been attained (Sweetman, 1992). Reducing dependence on live phytoplankton and zooplankton by improvements in compound feeds has been a high priority.

Some alternative feeding schedules are given in Table 8-4. Strong correlations have been reported between survival and EPA content of *Artemia* and between growth and DHA content (Bengtson et al., 1991).

Gatesoupe and Luquet (1981) raised larvae in 60-L tanks on rotifers (fed freeze-dried *T. suecica* plus a compound feed containing 4% cod liver oil, then enriched with a mixture containing 10% cod liver oil) during 5-14 dah, and *Artemia* nauplii during 10-20 dah. Survival (50%) from hatching to 20 dah was better than for larvae raised on rotifers fed only dried algae (3%) or algae plus the compound feed (27%). Using similar foods (150 and 250-L tanks), Gatesoupe and Robin (1982) obtained 32-87% survival from hatching to 21 dah without additional rotifer enrichment when 10% cod liver oil was included in the compound rotifer feed.

European sea bass have been reared from first feeding to 120 dah with only frozen rotifers, copepods, and *Daphnia* (Kentouri, 1980). Coves et al. (1991) reported that one feeding regimen developed for European sea bass larvae in France was simply *Artemia* during 10-57 dah, with weaning to dry feed during 50-57 dah. Devresse et al. (1991) weaned European sea bass from *Artemia* to microparticulate feeds during 35-56 dah, with ~67% survival and good growth. Person-Le Ruyet (1990a) described the following feeding schedule: rotifers 2-15 dah, *Artemia* nauplii 10-30 dah, and *Artemia* metanauplii beginning at 22 dah, with weaning to microfeeds possible at 20-23 dah. Weaning to conventional dry feed has been accomplished as early as 36-45 dah but sometimes is delayed until 60-80 dah (Barnabé, 1990e; Person-Le Ruyet, 1990a).

Barnabé and Guissi (1994) obtained better growth and survival for larvae fed enriched *Artemia* during 5-30 dah than for those fed rotifers & *Artemia* nauplii, rotifers & microparticulate feeds, *Artemia* nauplii, and *Artemia* nauplii & microparticulate feeds.

To grow to 12 mm TL, a European sea bass would have to consume about 14,720 rotifers, 440 copepods sieved to 185-280 μm, and 230 copepods sieved to 280-407 μm or 20,600 rotifers only (Barnabé, 1990e).

Striped bass (20°C) hatch at 3 mm TL, first feed ~4-5 dah, have EYS 7-14 dah, and transform ~37 dah (~30 mm TL). Striped bass and **palmetto bass** (from striped bass eggs) larvae have large mouths at first feeding and can be raised on *Artemia* alone. They also are raised in ponds on mixed zooplankton (Barkoh, 1996; Buurma et al., 1996). Survival and growth of striped bass and palmetto bass were very good on a simple diet of *Artemia* nauplii and enriched metanauplii (Tuncer and Harrell, 1992). Tuncer et al. (1993) recommended a minimum of 1.26% ω3HUFA (of DW), including 0.45% DHA in enriched *Artemia* fed to larval palmetto bass to increase viability and growth. Ozkizilcik and Chu (1996) designed a microencapsulated feed for striped bass larvae.

White bass (18°C) first feed 6 dah and transform ~30 dah. White bass and **sunshine bass** (from white bass eggs) larvae have small mouths and require rotifers or similar-sized foods for at least a few days before *Artemia* can be eaten. They often are raised in ponds on mixed zooplankton (Ludwig, 1993), but can be raised in tanks with reasonable survival

(Ludwig, 1994). White bass were weaned to salmon starter crumbles by 41 dah (Denson and Smith, 1996). Denson and Smith (1997) concluded that sunshine bass needed rotifers for only 4-7 dah, could eat *Artemia* beginning at 8 dah, and could be weaned to dry feed by 32 dah.

Australian bass (20°C) hatch at 3.1 mm TL, first feed 5-7 dah, have EYS 8-9 dah, and transform 56-70 dah (20-25 mm). Larvae can be fed rotifers 5-19 dah and *Artemia* beginning about 14 dah; transformation occurs during 56-70 dah (20-25 mm TL)(Battaglene et al., 1989a). On a diet of only rotifers and *Artemia* nauplii, survival was 65% (van der Wal and Nell, 1986). The best food densities were 9 rotifers/mL and 6 *Artemia*/mL.

Japanese sea bass (20°C) hatch at 3.9 mm TL, have EYS 3-5 dah, and transform ~60 dah (~43 mm). Larvae survived well (~65%) on a mixed diet (Zheng et al., 1993).

Serranidae

Grouper larvae are small and fragile, and they have relatively small mouths at first feeding. The larval period is longer and they tend to require live food longer than typical marine fish.

Greasy groupers (27°C) hatch at 1.7 mm TL, first feed 3-4 dah, have EYS ~4 dah, and transform 36-50+ dah (≥33 mm). Chen (1979) described feeding methods used for raising larvae in 1000-L tanks. Foods were: boiled chicken egg yolk 3-5 dah, rotifers 3-15 dah, cladocerans (*Diaphanosoma* sp.) from 10 dah until transformation, mysids (*Mesopodopsis* sp.) from a few days before transformation until weaning to shredded fish or dry pellets at 55-60 mm TL. When weaning was attempted with smaller juveniles (33 mm), they had trouble digesting the pellets and did not survive well. The initial weaning food was 80% ground trash fish mixed with 20% pellets, becoming 100% pellets by weaning day 6. Transformation began at 33 dah for large fish, but was not finished until 50-60 dah for small fish.

Malabar groupers (29°) hatch at 2.3 mm TL, first feed and have EYS ~3 dah, and transform 36-60 dah (~35 mm). They accepted minced fish by 40-45 dah and semi-moist compound feed at about 50 dah (Maneewong et al., 1986; Tookwinas, 1990a).

Redspotted groupers (24°C) hatch at 1.6 mm TL, first feed 2-4 dah, have EYS 4 dah, and transform 45-50 dah (~30 mm). They have been successfully raised on a large scale in Japan (Maruyama et al., 1994). In 1989, survival was 16% from eggs to 47 dah and 34% from hatchlings to 47 dah. Mortality from nutritional deficiencies sometimes occurred at ~20 dah. Mortality after 40 dah was mostly from cannibalism. They can be weaned by transformation at ~50 dah.

Brownmarbled groupers (28°) hatch at 1.3 mm TL, first feed ~69 hah (2.8 mm) with a mouth width of 225 µm, all feed by ~92 h, and have EYS ~1.5 h after all are feeding (Kohno et al. 1994c). Lim et al. (1990) reported that brownmarbled grouper hatchlings were slightly larger than those of greasy groupers, transformation occurred at ~40 dah (~31 mm TL), and the former species survived better under similar hatchery conditions. In 1990, the best survival from hatching to 40 dah was 13.2%.

Lim (1991b) stated that the main advantage of brownmarbled groupers was that they could eat small rotifers at first feeding, while greasy groupers could not and had been fed mussel eggs and trochophores at first. Also the larval period of greasy grouper was longer (40-50+ vs 35-40 d). Cannibalism, which became serious at 30 dah in brownmarbled grouper and at 35 dah in greasy grouper, and low stress resistance could have been related to nutritional deficiencies.

Nassau groupers (26°C) hatch at 1.9 mm TL, first feed ~2.5 dah, and have EYS ~5 dah. The larval period is long with a wide time range of 46-70 dah for transformation (35-50 mm TL, Tucker and Woodward, 1996). In 1990, larvae reared in 3500-L tanks received rotifers enriched with Frippak Booster® microcapsules, *Artemia* nauplii, and Frippak Booster-enriched 2- to 5-d-old *Artemia.* Larvae were planktonic until 42 dah, when they began orienting to the tank walls and bottoms. Then, white plastic milk crates were added to provide cover. Transformation of all fish occurred during 46-70 dah, with individuals making the change in less than a week. Weaning to 1.6 mm dry pellets was accomplished by feeding minced penaeid shrimp and crumbled pellets during 56-61 dah (40-60 mm TL), but a small amount of shrimp was given to the fish 63, 66, and 69 dah to enhance growth of small ones. Mortality peaks were at 5 dah (yolk and oil exhausted) and at about 20 dah, when consumption of *Artemia* nauplii was highest. In 1990, survival from fertilization to 98 dah was 5.0% and would have been higher if fewer *Artemia* nauplii had been used. (We later conducted fatty acid analysis of the main algae used for rearing rotifers and *Artemia* during the period in which Nassau groupers and sergeant majors were reared. It contained less than 1% EPA + DHA of FA and could not have been *Nannochloropsis oculata* as was thought, but might have been a *Chlorella* sp. In 1995, the deficient algae was replaced with true *N. oculata,* containing 23.2% EPA and 3.2% DHA.) Transformation might have been accelerated by providing cover or feeding shrimp earlier. Ten of those fish have been raised for 8 yr. In 1994, trochophores, rotifers, *Artemia,* and copepods were used. In one trial, survival from eggs to 15 dah in a 120-L aquarium was 25%. In another, survival during 10-75 dah in a fiberglass tank was 59%.

In 1994, Watanabe et al. (1996b) raised Nassau grouper larvae in 33.8-m^3 rectangular tanks on wild zooplankton, rotifers, and *Artemia,* with highest survival from hatching to 62 dah of 1.4%. Preliminary research showed that survival was better (8.0 vs 2.1%) when rotifers sieved to ≤200 μm were fed during 2-10 dah than when rotifers were not sieved (lorica length range 50-264 μm in the culture population). Survival during 1-10 dah was better with a mixed diet of oyster trochophores and rotifers (15.6%) than with rotifers alone (9.7%) or a sequential diet of trochophores and rotifers (mixed during 5-7 dah, 2.6%); growth was better with the rotifer and combined diets (4.1, 3.7, 3.1 mm NL).

Orangespotted groupers (27°C) hatch at 1.7 mm TL, first feed 3 dah, have EYS 4 dah, and transform 35-40 dah (25-30 mm). At first feeding, the mouth gape was about 180 μm and small rotifers were eaten (Duray, 1994a). By 20 dah (>6 mm TL), they preferred *Artemia* over rotifers. When *Artemia* were given to them at 15 dah, survival decreased, but if *Artemia* feeding was delayed until 21 dah, survival was improved.

Dry feeds were presented to orangespotted groupers during 2-35 dah and minced fish starting at 25 dah (Doi et al., 1991). Larvae were given minced fish only after 40 dah. In one 1990 trial, survival from fertilization to 10 dah was 38%, but by 43 dah, survival had decreased to 12.1% (still good for groupers).

In 1995, survival of orangespotted groupers from hatching to 24 dah was better in 3000-L (20 vs 7%) than in 500-L tanks (Duray et al., 1997). Growth (5.36 vs 4.23 mm TL) and survival (21 vs 11%) were better when rotifers fed during the first 2 weeks were sieved to <90 μm (average lorica dimensions before sieving were 120 μm × 162 μm). Survival was better but growth was worse when only rotifers were fed up to 35 dah (32%, 11.9 mm, 4.6 mg DW) than when *Artemia* alone (6%, 15.1 mm, 8.8 mg) or rotifers + *Artemia* (13.3%, 17.1 mm, 10.7 mg) were provided during 15-35 dah. Survival and growth (starting with 5 larvae/L) during 21-35 dah were better when *Artemia* nauplii were fed at 3/mL (30%, 12.0 mm TL, 4.9 mg DW) than at 1/mL (20%, 10.4 mm, 2.8 mg) or 2/mL (27%, 11.9 mm, 4.4 mg). Under similar conditions, survival was better with 12 *Artemia*/mL/d (50.3%) than with 6/mL/d (39.7%) or 3/mL/d (18.0%).

Pseudochromidae

Dottybacks (25°) hatch at 3-4 mm TL, begin feeding within a day, exhaust their yolk (Brons, 1996; Gardner, 1997; Moe, 1997a). Rotifers probably are sufficient as a first food, but wild zooplankton also has been used.

Grammatidae
Gramma larvae typically are robust and begin feeding within a day after hatching.

Royal grammas (27°C) hatch at 6-7 mm TL, feed ≤1 dah, and transform 21-30 dah. Walch (1994a) raised larvae on rotifers and *Artemia*. After transformation (~30 dah), they can be weaned to dry feed.

Blackcap basslets were raised on rotifers, *Artemia* nauplii, and small copepods (Addison, 1994a, 1994b). After transformation, they can be weaned to various foods.

Plesiopidae
Comets (26°C) hatch at 3 mm TL, have EYS at about the same time, first feed right after hatching, and transform ~23 dah (~8 mm). Wassink and Brons (1990) raised them on rotifers, 300-μm dry feed, and *Artemia* nauplii, with minced meat added at 14 dah.

Carangidae
Jack larvae have been fed oyster larvae, rotifers, copepods, *Artemia*, and fish eggs and larvae. Most probably can be raised on rotifers and *Artemia*, but copepods can increase vigor and survival.

Yellowtails (20°C) hatch at 3.5 mm TL, first feed ~3 dah, have EYS ~4 dah, and transform 50-55 dah (~45 mm). Wild larvae eat mainly planktonic crustaceans, but at ~30 mm, they begin eating fish larvae, too; by 130 mm, fish are the main food. Hatchery methods were developed during 1954-1977 (Yamaha Fishery Journal, 1989), but the demand is so high that yellowtails have been raised mostly from wild mojako (late larvae to juveniles, 15-150 mm). They are fed minced fish (often sand eels), with increasing use of soft-moist or granulated dry feeds (Nakada and Murai, 1991). A possible feeding schedule for larvae was given by Kuronuma and Fukusho (1984). Imaizumi (1993) considered rotifers, *Artemia* nauplii, and older *Artemia* indispensable and other zooplankton (e.g., copepods) and microfeeds necessary for yellowtail larvae; typical survival from larvae to 30-mm juveniles was 5-10% in Japanese hatcheries. Furuita et al. (1996b) reported a requirement of about 1.6% EPA + 3.3% DHA (5.2% ω3HUFA) in *Artemia* for best survival, growth, and vitality.

Mazatlan yellowtail larvae were raised through transformation on rotifers, *Artemia*, wild zooplankton, and compound feeds (Benetti, 1997). Cannibalism was intense during and after transformation (older than 4 weeks).

Striped jacks (23°C) raised through transformation (30 to 60 dah, 8.2 to 38.5 mm TL) on a microparticulate feed containing 3% squid liver oil and 2% ω3HUFA mixture had good survival (≥80%), growth, and resistance to air exposure (Kanazawa, 1993b). Takeuchi et al. (1996) reported a requirement of 0.9-1.1% EPA + 1.6-2.2% DHA (3.3-5.0% ω3HUFA) in *Artemia* for best survival and vitality of larvae.

Coryphaenidae
Dolphins (26°C) hatch at 4.2 mm TL, first feed 2-3 dah, have EYS ~3 dah, and transform at ≥20 dah (~16-24+ mm). Survival of dolphin larvae became more consistent after *Artemia* were enriched with ω3HUFA (Bengtson et al., 1991). During 3-7 dah, survival of larvae (in 30-L tanks) fed rotifers enriched with olive, soybean, corn, or squid oil (mean 68%) was significantly higher that those fed rotifers raised on and enriched with *N. oculata* (52%)(Ostrowski and Divakaran, 1990). Larvae do not appear to be able to convert shorter chain ω3 fatty acids to DHA but tend to conserve both DHA and arachidonic acid if their food does not contain enough or they are deprived of food. When diets are deficient in ω3HUFAs, resulting high mortality does not occur until 8-12 dah, possibly because DHA reserves from the egg have dropped below a critical level by then.

Szyper (1991) described two alternative feeding schedules for dolphin larvae: (1) rotifers 2-8 dah, *Euterpina acutifrons* 6-25 dah, dolphin hatchlings 18-45 dah, chopped squid 30-45 dah; (2) rotifers and/or *E. acutifrons* 2-6 dah, *E. acutifrons* 4-12 dah, *Artemia* nauplii 8-20 dah, fed *Artemia* 15-30 dah, and chopped squid, fish, or krill 20-30 dah. Weaning to dry feeds is accomplished in juveniles 30 dah or older.

Kraul et al. (1988, 1992) and Kraul (1993) summarized some refinements in dolphin larval rearing. Rotifers were the best first food, and their fatty acid content was not especially important because early dolphin larvae had large ω3HUFA reserves. Copepods (*Euterpina acutifrons*) were the best second food (0.09% EPA and 0.10% DHA of DW)—better for survival than enriched *Artemia* that contained the minimum requirements of ~0.17% EPA and 0.40% DHA (of DW). Dolphin yolksac larvae were the best third food, one reason being their content of 0.27% EPA and 0.82% DHA (12 times more DHA than in enriched juvenile *Artemia*). The use of enriched *Artemia* metanauplii and juveniles can offset the need for copepods and dolphin larvae, but also can entail higher mortality and loss of vigor. Survival can be improved by delaying weaning until 30 dah.

Kim et al. (1993) preferred higher stocking densities of 10-15 larvae/L and fed dolphin only different ages of *Artemia* (nauplii to 10 dah) to match growth of the fish.

Lutjanidae

Snapper larvae typically are small, slender, and fragile, and have a short time between first feeding and EYS. Some have small mouths and extended larval periods like groupers.

Golden snappers (28°C) hatch at 1.6 mm TL, first feed and have EYS ~3 dah, and transform ~35 dah (~20 mm). Lim et al. (1985) raised larvae on a diverse diet. Critical nutritional periods occurred during 4-7 dah and 18-28 dah. Cannibalism sometimes was a problem after 25 dah. The best survival from eggs to early juveniles was 5%.

Asian red snappers (28°C) hatch at 1.8 mm, first feed 2-3 dah, have EYS 3-4 dah, and transform ≥35 dah (≥19 mm). Doi and Singhagraiwan (1993) reported rearing results mainly from 1992. Larvae had a mouth width of 145 µm at first opening and 200 µm at first feeding. Although the amount of yolk is similar to that of barramundi, the smaller size of the oil globule could be one reason this snapper is more vulnerable to first feeding difficulties. First feeding by Asian red snappers occurred at 2-3 dah depending on temperature, and yolk and oil were exhausted by about 101 hah. Maximum food length at first feeding was ~100 µm. Small foods such as small or sieved rotifers and/or copepod nauplii (e.g., *Acartia* or *Oithona* sp.), possibly supplemented with oyster eggs and larvae, were necessary. Establishment of copepod populations in the larval rearing tanks could have been a key factor for survival.

Duray et al. (1996a) raised Asian red snapper larvae on rotifers and *Artemia*. Survival from hatching to ~24 dah was better in 3000-L (27%) than in 500-L tanks (3%). The best survival during 25-55 dah was 60%. Growth was similar and survival (5.6% vs 3.2%) better when sieved rotifers (<90 µm) were fed up to 14 dah. Final weight was similar for larvae fed 1, 2, or 3 *Artemia*/mL during 21-35 dah (fish density not given), but length was greater (14.2, 12.6, 12.2 mm TL) and survival lower (54, 65, 81%) with higher *Artemia* density. Under similar conditions, survival was better with 8 *Artemia*/mL/d (63%) than with 4/mL/d (25%) or 2/mL/d (17%); growth was better with the lower feeding level (9.0 and 8.9 vs 16.8 mg), probably because there were fewer fish.

Gerreidae

Striped pataos (29°C) hatch at 1.4 mm TL, first feed 2-3 dah, have EYS 3 dah, and transform ~48 dah (~20 mm). Larvae were raised on rotifers, copepods, and *Artemia* (Alvarez-Lajonchère et al., 1996). Final density of 48-d-old juveniles in 5-m^3 tanks was as high as 10/L. For larval densities of 30-40/L, those authors recommended food densities of 500 *N. oculata* cells/µL, 10-15 rotifers/mL (especially during the first 4 feeding days), and 1-2 copepods/mL, 5-10 *Artemia*/mL, with paste or microfeeds (<250 µm) beginning before 30 dah.

Sparidae

Sea bream and **porgy** larvae have been fed diverse diets, but most can be raised on rotifers and *Artemia*. Although larvae of some species are small at hatching, they become effective feeders. Juveniles and adults of most cultured species are omnivorous, frozen food usually is well accepted by late larvae, and weaning is relatively easy. Some species can become territorial or cannibalistic as juveniles. Between 5 and 7 mm, the feeding mode of red sea bream changes from swallowing to biting (Kohno et al., 1983). Then, larvae can start biting others, and density might need to be reduced soon afterwards (Ikenoue and Kafuku, 1992). Gastric glands are developed in red sea bream and black sea bream at about 3 weeks (Watanabe and Kiron, 1994).

Red sea bream (18°C) hatch at 2.1 mm TL, first feed at 3 dah, have EYS 4 dah, and transform ~30 dah (10-12 mm). Wild larvae and early juveniles eat mainly protozoans and copepods, then amphipods; at ~30 mm TL, they begin eating isopods, shrimp, crabs, and polychaetes, as well as amphipods (Yamaha Fishery Journal, 1990; Kim and Myoung, 1995). Development of red sea bream culture began in the 1950s and 1960s. Kuronuma and Fukusho (1984) reported a variety of larval diets used at different hatcheries in Japan during the 1950s to 1970s. Rotifers were fed from 3 dah, copepods as early as 12 dah, minced clam as early as 15 dah, and *Artemia* as early as 18 dah, with formulated feeds started during 30-40 dah. Typical size and survival in a Korean hatchery were 13.5 mm TL and 25% at 40 dah, then 95 mm and 7.5% at 130 dah (Kim and Myoung, 1995). Imaizumi (1993) considered rotifers and *Artemia* nauplii indispensable and microfeeds necessary for red sea bream larvae; typical survival from larvae to 30-mm juveniles was 30-80% in Japanese hatcheries. Wild copepods are good, and crab larvae can be eaten at >15 mm TL (Sun and Liu, 1992).

In one experiment, red sea bream (7-12 mm) survived and grew better on *Artemia* fed ω-yeast than those fed *N. oculata* or newly hatched nauplii (Fujita et al., 1980). In a second experiment, they (9-24 mm) survived better on *A. clausi* than on newly hatched *Artemia* or *Artemia* fed *Spirulina* or ω-yeast; growth was best on the fed *Artemia*, followed by *A. clausi* and newly hatched *Artemia*.

Izquierdo et al. (1989) estimated a minimum requirement of 3.46% ω3HUFA of DW (0.4% of WW) in rotifers fed to red sea bream larvae. Growth and activity of late larvae were good when they were fed *Artemia* containing 4.9% EPA + 0.9% DHA in 6.4% ω3HUFA (of DW)(Takeuchi et al., 1992). Furuita et al. (1996a) obtained the best early juvenile survival with *Artemia* containing a minimum of 2.06% EPA and no DHA in 2.25% ω3HUFA or 0.85% EPA + 0.95% DHA in 2.05% ω3HUFA (of DW); however, a minimum of 0.89% EPA + 1.62% DHA in 2.79% ω3HUFA was required for the best stress resistance—better than that provided by 8.44% EPA and no DHA in 8.68 ω3HUFA.

Black sea bream (21°C) hatch at 1.8 mm TL, first feed 2-3 dah, have EYS 2-3 dah, and transform ~40 dah (13-18 mm). Larvae were raised on rotifers and *Artemia* nauplii (Leu, 1997). In 1990, survival from hatching to 40 dah was 50%, and survival from hatching to 100 dah was 28%. Weaning was accomplished at 41 dah.

Park (1991) described feeding schedules used for red and black sea bream larvae in Korea as including rotifers and *Artemia* nauplii (sometimes enriched older *Artemia*). Red sea bream were weaned to mixed prepared seafoods by 45 dah and black sea bream by 41 dah.

Gilthead sea bream (18°C) hatch at 2.4 mm TL, first feed 4-5 dah, have EYS 5 dah, and transform ~40 dah (~12 mm). After several rearing trials in 1978, with up to 11% survival from hatching to 60 dah, Person-le Ruyet and Verillaud (1980) recommended the feeding schedule in Table 8-4. Weaning to dry feed was accomplished just after transformation, at 40-45 dah.

Yúfera et al. (1991) found that some gilthead sea bream larger than 40 μg and half of those larger than 80 μg would eat *Artemia* nauplii. They suggested that *Artemia* could be first added to a diet of small rotifers at 12.5 dah or with large rotifers at 15 dah.

Sweetman (1992) and Zohar (1995) described recent feeding schedules for gilthead sea bream, with compound feeds given to larvae. Average commercial scale survival of 7% from eggs to 1-g juveniles was reported by Sweetman (1992). High levels of EPA and DHA in *Artemia* contribute to growth and survival, and high DHA can increase stress resistance (Bengtson et al., 1991).

Koven et al. (1992) obtained the best growth of gilthead sea bream larvae during 22-36 dah with *Artemia* containing the highest ω3HUFA tested (2.06% EPA + 0.92% DHA in 2.98% ω3HUFA, of DW; 37% lipid). Rodríguez et al. (1997) obtained better growth during 3-14 dah with a DHA:EPA ratio in rotifers of 1.2:1 and 1.4:1 than with 0.3:1 and 0.6:1. Salhi et al. (1994) found that during 14-27 dah, gilthead sea bream survived better (29.8 vs 18.6%) on microparticulate feeds containing 0.68% EPA + 1.22% DHA in 2.05% ω3HUFA (of DW)(20.0% lipid) than with 0.26% EPA + 0.51% DHA in 0.82% ω3HUFA (13.6% lipid). Tandler and Kolkovski (1991) recommended that microfeeds be fed with live food (e.g., 80% substitution of rotifers) until more attractive and digestible feeds can be used.

Sheepshead (23°C) hatch at 1.6 mm TL, first feed 3 dah, have EYS and EOG 4 dah, and transform ≥39 dah (≤14 mm). In 1984, sheepshead were reared in 1000-L and 3500-L tanks on a diverse diet, with commercial salmon starter beginning 27 dah and live foods tapering off to small rations during 48-70 dah (Tucker, 1987a). From fertilization to 7 dah (yolk and oil exhausted at 4 dah) survival was 62%, and from 7 dah to 100 dah it was 64% (40% overall), with no further mortality of tank-reared fish for more than 3 yr.

Silver sea bream (21°C) hatch at 2.3 mm TL, first feed ~3 dah, have EYS 3-4 dah, and transform ~42 dah (~17 mm). Larvae were raised on rotifers and *Artemia* (Leu, 1994). Survival from hatching to 45 dah was 43%, and survival from hatching to 105 dah was 24%. Weaning was accomplished at 41 dah, after the juvenile digestive system had developed.

Yellowfin sea bream (22°C) hatch at 2.0 mm TL, first feed 3-4 dah, have EYS 2-3 dah, and transform ~38 dah (~12 mm). Larvae were raised on a diverse diet, with the best survival (from hatching to 50 dah) of 27.5% occurring in 1990-1991 (Leu and Chou (1996). Transformation was finished at ~38 dah and weaning was accomplished by 45 dah.

Sciaenidae

Although not especially large, **drum** larvae typically are robust and easy to raise on rotifers and *Artemia. Cynoscion* spp. are hardy, grow fast, and have high survival potential as long as they are not crowded.

Red drum (25°C) hatch at 1.7 mm TL, first feed and have EYS ~3 dah, and transform ~35 dah (~25 mm). Rearing of larvae was successful in the mid-1970s, with survival from hatching to 30 dah of up to 65% in ponds with wild zooplankton (Colura et al., 1976) and 30% in aquaria with rotifers and *Artemia* (Arnold et al., 1977). As for many other species, the first 3 weeks is the most difficult nutritional period for red drum (Holt, 1992).

Craig et al. (1994) found no difference in growth rates of red drum fed rotifers cultured with yeast, fish emulsion, and wild algae and those further enriched with ICES medium- or high-HUFA emulsions, but growth was lower when rotifers were fed the ICES no-HUFA emulsion.

Pickering and Holt (1993) suggested that cryopreserved oyster trochophores might be used to partially substitute for rotifers and help reduce live food production effort.

Progress has been made in the use of larval feeds for early weaning of red drum (Holt, 1992, 1993). Survival was better, growth similar, and low-salinity stress resistance similar for red drum fed rotifers & a microparticulate feed compared with those fed a control diet of rotifers & *Artemia* nauplii (Table 8-4; Holt, 1993). After red drum were fed rotifers and semipurified microparticulate feeds (50% red drum muscle, 10% albumin, dextrin, and menhaden oil) during 3-7 dah, then feeds only during 8-17 dah, the best growth and survival were with the control commercial feed (24.8% lipid), and the next best growth was with an experimental feed containing intermediate lipid (18.3%)(Brinkmeyer and Holt, 1995).

Spotted seatrout (28°C) hatch at 1.5 mm TL, first feed 2 dah, have EYS 2-3 dah, and transform ~37 dah (~38 mm). Arnold et al. (1976) raised larvae during 1973-1974 on rotifers and *Artemia*. They recommended a stocking density of ≤5 larvae/L for best growth and survival. Only 1-2 larvae/L were stocked in 14,000-L outdoor tanks. High fish density (e.g., 25/L) makes cannibalism more likely than lower densities (e.g., 5/L)(Taniguchi, 1981). In 1985 and 1986, Tucker (1988) raised spotted seatrout in 1000 and 3500-L tanks on rotifers and *Artemia* nauplii. Dry feed was offered begining 19 dah, but was not well accepted until the beginning of transformation at 33 dah. Both transformation and weaning were finished by 37 dah (38 mm TL). The best survival (from a low stocking density of 0.2 egg/L) was 12.5% from fertilization to 50 dah, beyond which mortality was negligible. In 1997, Tucker et al. (unpub.) raised spotted seatrout again in one 3000-L tank on rotifers, copepods, *Artemia* nauplii, and enriched *Artemia* metanauplii. Dry feed was offered from 21 dah on, but was not eaten until 26 dah; live food was discontinued after 32 dah. Survival (from 6.5 hatchlings/L) to 35 dah was at least 10%. Size was very uniform until weaning, probably because larval tanks were inoculated with probiotic *Bacillus* sp. and the larvae were very well fed. This is a highly raptorial species with a strong cannibalistic tendency. Weaning required only 5-7 d, and cannibalism was controlled by weaning as early as possible, feeding often, and grading the fish occasionally.

Colura et al. (1992) reported the results of 52 pond rearing trials with spotted seatrout during 1983-1988. The statistically most important factor influencing variations in survival and production was the density of polychaete larvae, especially during ~18-32 dah (400-1200/L best). Temperature was better in the lower part of the range 24-29°C than the upper. Salinity less than ~17 ppt and minimum dissolved oxygen less than 1 mg/L were deleterious. The best survival of 67% during 2-32 dah was from a low stocking density of 12.9 larvae/m^2. Bumguardner et al. (1992) raised **orangemouth corvina** under similar conditions (19.5 larvae/m^2 stocked) and obtained a best survival of ~96% during 2-29 dah.

Weakfish (20°C) hatch at 1.5 mm TL, first feed 2.2 dah, have EYS 3 dah, and transform at 32^+ mm. Their larvae preferred large rotifers over small rotifers and *Artemia* nauplii during 5-8 dah, but preferred small nauplii over rotifers during 14-17 dah (Pryor and Epifanio, 1993). At 8-17 dah, small (SFB) nauplii always were preferred over large (GSL) nauplii even though the fish could easily eat the larger prey and obtain more nutritional benefit. Possibly a difference in taste or smell led to the preference. Growth and survival up to 9 dah (15 larvae/L stocked in 19-L aquaria) were better with 0.5-10 rotifers/mL than with 0.05-0.10 R/mL (Goshorn and Epifanio, 1991).

White seabass (16°C) hatch at 2.7 mm TL, first feed 4.8 dah, have EYS 5-6 dah and EOG 7-8 dah, and transform 42-53 dah (≥16 mm). Larvae have been reared on rotifers, *Artemia*, and cultured copepods (Margulies, 1989; Orhun, 1989; Dutton 1992). They can eat small *Artemia* nauplii at first feeding. By about 30 dah, cannibalism can start causing high losses.

Mulloways (18°C) hatch at 2.2 mm TL, first feed 3-4 dah, have EYS 3 dah, and transform ~34 dah (15-26 mm). They were raised on rotifers and *Artemia* (Battaglene and Talbot, 1994). Cannibalism was first seen at 22 dah and was a major cause of mortality. At the end of transformation, during 34-40 dah, larvae became bottom feeders. They were weaned later than necessary at 5 weeks after transformation.

Highhat (an ornamental drum) larvae were raised on rotifers and *Artemia*, then weaned 2 weeks after transformation (Harris, 1996).

Pomacanthidae

Pygmy angelfish (27°C) eggs are ~425-1000 µm in diameter. Larvae hatch at ~1.0-1.6 mm

TL, first feed 3-4 dah, and have EYS 3-5 dah. They are considered difficult to rear because early larvae have very small mouths and require small prey. They usually have not survived past 12 dah (Lamm, 1984; Moe, 1992; Breeder's Registry Database, 1994).

Rock beauties, which hatch at about 2.5 mm TL, first feed and exhaust their yolk at 2-3 dah, have larger mouths and have been raised on a mixture of wild and cultured plankton (Moe, 1977, 1992). **Gray** and **French angelfish** also were raised.

Cichlidae

Tilapias (29°C) hatch at ~4 mm, first feed after several days, have EYS after a few more days, and transform within a few weeks. They do not require live food. The young also are easily raised in eutrophic ponds or tanks without added feed (Phillips et al., 1994). Survival after first feeding typically is near 100% (Suresh and Lin, 1992).

Pomacentridae

Damselfishes (especially clownfishes) produce large robust larvae that are ready to feed soon after hatching.

Clownfishes (27°C) hatch at ~4 mm TL, first feed within hours, and transform ~25 dah (~14 mm). The following foods were recommended for producing juvenile clownfish (Hoff and Snell, 1989; Hoff, 1996) rotifers; flake food; *Artemia*; semi-moist salmon starter; krill, shrimp, or *Spirulina* meal; and salmon grower or gelatin-based feeds. Weaning can be accomplished between 13 and 16 dah, several days before transformation is completed. Juveniles of about 14 mm TL can be removed during 25-30 dah, with survival as high as 89%.

Spot dascyllus and **threestripe damselfish** hatched at 2.2 mm TL, first fed soon after, and transformed by 45-50 dah (Danilowicz and Brown, 1992). Larvae were relatively easy to raise on rotifers, wild zooplankton, and *Artemia*.

Sergeant majors (26°C) hatch at 3.3 mm TL, first feed within hours, have EYS 3-4 dah, and transform ~32 dah (~23 mm). They were easily reared for description of egg to juvenile development (Alshuth et al. 1998; Tucker et al., unpub.). Rotifers, cultured copepods, and *Artemia* were the main foods; oyster trochophores were given for several days but were not necessary. Weaning was accomplished quickly during 32-33 dah, and could have been done sooner.

Oplegnathidae

Striped knifejaws (22°C) hatch at 1.7 mm TL, first feed 3-4 dah, have EYS ~4 dah, and transform ~35 dah (~15 mm). Growth and activity of late larvae were good when they were fed *Artemia* containing 4.1% EPA + 0.7% DHA in 5.3% ω3HUFA (of DW)(Takeuchi et al., 1992b).

Survival (71 vs 42%) and resistance to air exposure were better for striped knifejaw larvae (3-23 dah) raised on rotifers and *Tigriopus japonicus* than for those fed rotifers and *Artemia* (Kanazawa, 1993b). DHA was five times higher in copepods than in *Artemia*. EPA was similar but DHA was several times higher in the larvae eating copepods. In another experiment, survival (71%), growth, and resistance to air exposure were better for striped knifejaw raised through transformation (21-42 dah) on a feed containing 1% DHA (of DW) than with 2% DHA, 1% EPA, 2% EPA, and 1% EPA + 1% DHA.

Mugilidae

Mullet larvae are relatively robust and can easily be raised on rotifers and *Artemia*. Growth is best in green water (e.g., Tamaru et al., 1994; Yoshimatsu et al., 1995). Mullet are easily weaned by the time they are juveniles, after which they continue feeding in the water column but also scour the walls of the rearing unit for food (pers. obs.).

Striped mullet (24°C) hatch at 2.8 mm TL, first feed 3 dah, have EYS 5 dah, and transform ~28 dah (35-45 mm). Eda et al. (1990a) reported that striped mullet ate small and large rotifers and a few *Artemia* at first feeding (~3 dah); during 6-10 dah, preference shifted from rotifers to *Artemia* nauplii. Extended feeding of rotifers improved survival. Oozeki et al. (1992a) reported that striped mullet could eat large rotifers well at first feeding, *Artemia* nauplii by 9 dah (4.3 mm TL), and formulated food by about 6.3 mm.

Tamaru et al. (1993c) found that striped mullet survived and grew better during 2-15 dah if given rotifers fed *N. oculata* with or without baker's yeast than for yeast alone; best results occurred when rotifers were fed the mixed diet.

During 15-35 dah, survival of striped mullet was ~98% when they were fed *Artemia* metanauplii enriched overnight with menhaden oil, with or without rotifers fed baker's yeast and *N. oculata* (Ako et al., 1994). Survival was ~24% for fish fed only *Artemia* nauplii and 55% for fish fed rotifers and nauplii.

In the second rearing cycle of striped mullet from a Florida population, larvae were raised in 3000-L tanks on a diet of rotifers, copepods, and *Artemia*; survival was about 30% from hatching to 60 dah (Tucker et al., unpub.). Larvae and juveniles were much more robust than in the first cycle, probably because larval tanks were inoculated with a local probiotic *Bacillus* sp.

Polynemidae

Pacific threadfins (26°C) first fed ~3 dah and transformed ≥26 dah. Larvae were raised on rotifers and *Artemia* (Ostrowski et al., 1996b). They were weaned to pellets at ≥26 dah.

Labridae

Tautogs (22°C) hatch at 2.1 mm TL, first feed ≤4 dah, have EYS 4 dah, and transform at ≤30 mm. Perry et al. (1998) raised larvae in 20-L and 850-L tanks on protozoans up to 4 dah, then enriched rotifers and *Artemia* plus wild zooplankton. *Artemia*, wild zooplankton, and commercial feeds were given to juveniles.

Zoarcidae

Ocean pouts (5°C) hatch at 39 mm TL, have EYS 1 dah, first feed by 2 dah, and could be considered juveniles at hatching. They were raised on enriched *Artemia* and a mixture of chopped fish and bivalves or squid plus a vitamin/mineral premix (Brown et al., 1992). Survival was 95% at 8 weeks and 89% at 1 yr.

Anarhichadidae

Common wolffish preferred enriched *Artemia* metanauplii to dry feed during the first 3 weeks, then ate both until 38 dah, when their preference shifted to dry feed (Brown et al., 1997c).

Opistognathidae

Yellowhead jawfish (26°C) hatch at 4 mm TL, first feed within hours, and transform 9-15 dah. Walch (1993, 1994b) raised larvae on rotifers and *Artemia*. After transformation (~21-28 dah), they can be weaned to prepared foods.

Gobiidae

Neon gobies (25°C) hatch at 3.5-4 mm, first feed and have EYS within 12 hah, and transform 18-40 dah (≥8 mm). Like many other members of this family, they are easy to raise (Moe, 1975).

Yellow prawn gobies hatched at ~2.0 mm TL with some yolk left, fed soon after hatching, settled in about 1 week, and transformed at ~20-25 dah (Brown, 1996). Larvae were first fed enriched rotifers, then starting at 15 dah, copepods and enriched *Artemia*.

Ephippidae

Atlantic spadefish (27°C) hatch at 2.5 mm TL, first feed ~2.5 dah, have EYS ~3 dah, and transform at ~17 mm. Larvae are fairly robust. Walker (1991) raised them on rotifers and *Artemia* nauplii. Gomez (in press) used rotifers, *Artemia*, and mysid shrimp.

Siganidae

Rabbitfish larvae have small mouths at first feeding but can be raised on rotifers and *Artemia*, if enough small rotifers are available during the first few feeding days.

Goldline spinefoots (27°C) hatch at 2.5 mm TL, first feed 2 dah, and transform ~35 dah (~23 mm). Larvae were raised on rotifers, copepods, and *Artemia* nauplii, with survival of 16% from hatching to 5 dah (Bryan and Madraisau, 1977). Three early feeding stages were defined: dark-head carnivorous larvae, brown-head omnivorous transforming larvae, and herbivorous juveniles. The gut was highly coiled in early juveniles and about 1.63 times body length. The authors recommended that *Artemia* be reduced during transformation because if late larvae or early juveniles gorge on *Artemia*, their newly coiled intestines could easily become blocked.

Yellowblotch spinefoots (27°C) hatched at 1.9 mm TL, first fed at 55 hah (3.0 mm) with a mouth width of 187 μm, all fed by 82 h, and had EYS about 44 h after all were feeding (Kohno et al. 1994c). Transformation occurs 35-45 dah (~22 mm). Juario et al. (1985) raised larvae in 600-L tanks with microalgae, rotifers, *Artemia* nauplii, and compound feeds. Sometimes, oyster trochophores were used during 1-3 dah. A critical period occurred during 3-5 dah with peak mortality at 4 dah when yolk and oil were exhausted (cumulative mortality 50-90%). At 15 dah, chasing and apparent fin biting began but did not seem a serious problem. Additional mortality of the smaller fish coincided with the beginning of transformation at 24 dah. The best survival was 37% from hatching to transformation (35 dah). The fish can be weaned by that age (Duray and Juario, 1987). Hara et al. (1986) reported that survival at 21 dah was better for larvae that had been fed rotifers sieved to <80μm during 2-3 dah than for those fed the full size range (lorica length up to >203 μm). High ω3HUFA in *Artemia* increased stress resistance of larvae (Bengtson et al., 1991).

Scombridae

Scombrid larvae are fairly robust at hatching. **Tunas** have large mouths at first feeding and will eat rotifers, *Artemia*, and copepods. Despite this, they have not survived well in captivity (Mori et al., 1971; FSFRL, 1973). Prey quality must be high, and water movement and cannibalism seem to be factors needing control.

Chub mackerels (19°C) hatch at 3.2 mm TL, first feed 2-2.5 dah (mouth width 216 μm), have EYS 3 dah, and transform by 25 dah or a little later (15-25 mm). Hunter and Kimbrell (1980) raised them on rotifers and copepods.

Kawakawas (24°C) hatch at 2.8 mm TL, first feed 2-3 dah, and have EYS 4.0 dah and EOG 4.9 dah (Kohno et al., 1994b).

Northern bluefin tunas (24°C) hatch at 2.8 mm TL, first feed 2-3 dah, have EYS ~5 dah, and transform at ≥17 mm.

Yellowfin tunas (26°C) hatch at 2.7 mm TL, first feed and have EYS ~3 dah, and transform at ~58 dah (~51 mm). Harada et al. (1980) raised this species to the juvenile stage on rotifers, copepods, and fish larvae.

Bothidae

Hirame (18°C) hatch at 2.0 mm TL, first feed 2 dah, have EYS 3-4 dah, and metamorphose and transform 24-50 dah (beginning at 10-14+ mm). Wild larvae mostly ate appendicularians (*Oikopleura* sp.), copepods, and invertebrate eggs (Ikewaki and Tanaka, 1993). By ~30 mm, they begin shifting to a fish diet, and cannibalism can be a problem. The pancreas and gastric glands are not completely developed until transformation (~45 dah), helping to explain why digestion of compound feeds is difficult for larvae (Kurokawa and Suzuki, 1996).

Hirame were first reared in 1965, and commercial-scale hatchery methods were developed by the 1980s (Yamaha Fishery Journal (1991). Before 1987, a variety of foods were used, but since then, mainly rotifers, *Artemia* nauplii, and compound feeds.

Imaizumi (1993) considered rotifers and *Artemia* nauplii indispensable and older *Artemia* and microfeeds necessary for hirame larvae; typical survival from larvae to 30-mm juveniles was 25-80% in Japanese hatcheries.

Izquierdo et al. (1992) reported that weight of juvenile hirame (fed for 32 d) was 11% greater with *Artemia* containing 5.87% ω3HUFA (of DW) than for 3.48% ω3HUFA, but length was similar for 3.48% and 5.87% and greater than for 2.68% ω3HUFA. Survival did not change significantly over the treatment range 1.84-9.00% ω3HUFA. The DHA:EPA ratio was low for all treatments (0.01-0.32) and did not seem correlated with the incidence of albinism, which was lowest (10%) with a ratio of 0.18 and highest (28%) with a ratio of 0.32. Conclusions were that hirame larvae need 3.5-4.0% ω3HUFA in *Artemia* and that DHA might promote growth of hirame but seems not to be essential for survival.

Spedicato et al. (1994) raised hirame on rotifers and enriched *Artemia*, with 45% survival from hatching to 85 dah. In a preliminary trial, enrichment of *Artemia* with more DHA reduced the incidence of juvenile albinism from 80% to 37%.

Hirame larvae that were fed *Artemia* enriched (for ≥18 h) with more than 4 mg vitamin A palmitate/L of enrichment medium during 10-50 dah survived and were pigmented as well as those with lower levels, but grew slower and had more bone deformities (Dedi et al., 1995). The recommended safe level was 2 mg vitamin A palmitate/L of medium, which should produce less than 50 IU vitamin A/g of *Artemia*. Effects on vertebral development were further documented by Dedi et al. (1997). Dedi et al. (1998) reported fusion of caudal skeleton elements in hirame that had been fed *Artemia* with 45 IU vitamin A/g; therefore, enrichment with 2 mg vitamin A palmitate/L probably is too high.

Turbot (19°C) hatch at 3.1 mm TL (0.1-0.2 mg WW), first feed 2-3 dah, have EYS ~7 dah, metamorphose ~30 dah, and begin benthic feeding ~40 dah (≥18 mm)(Person-Le Ruyet, 1990b). At first feeding, larvae obtained about 90% of their energy from the oil globule (Rønnestad et al., 1992). At 15°C, EYS occurred during 7-11 dah and EOG during 9-12 dah in unfed larvae (Segner et al., 1994). Rotifers (or rotifer-sized food) are needed for about the first 10 dah of feeding. Weaning to microfeeds has been accomplished as early as 25-30 dah (50 mg WW). A functional stomach, with gastric glands and pepsin secretion, is not present until metamorphosis (Segner et al., 1994).

Turbot were successfully raised in the 1970s. Alderson and Ball (1977) obtained ~25% survival during 2-55 dah and 85-95% normal pigmentation in juveniles, using rotifers and *Artemia* nauplii.

Kuhlmann et al. (1981) fed turbot larvae rotifers, *Eurytemora affinis*, and *Acartia tonsa* with *Nannochloris* sp. in a 2000-L tank from first feeding at 3 dah through transformation at 30-36 dah (22 mm TL). Weaning of early juveniles to dry pellets required about one week. Survival from fertilization to 36 dah was about 25%. Person-Le Ruyet et al. (1991) described the following typical feeding schedule for turbot: rotifers, 3-13 dah; *Artemia* nauplii, 8-26 dah; enriched metanauplii 18-30 dah, weaning to dry pellets, about 25-30 dah.

Minkoff and Broadhurst (1994) described an acceptable feeding schedule for turbot: rotifers during 2-12 dah, *Artemia* nauplii during 8-25 dah, and metanauplii during 18 dah to weaning. Although ω3HUFA contents of >1.6% (of DW) in rotifers and >1.2% in *Artemia* have been recommended, turbot can be successfully raised on live foods containing 0.4% ω3HUFA, such as rotifers fed *Nannochloris* sp. or *Isochrysis galbana*. A dietary DHA:EPA ratio of about 2:1 was considered best. The highest losses usually occur during 9-16 dah, primarily because of nutritional inadequacy of *Artemia* and/or bacterial infections. The weaning period usually is 10-14 d long and can be started at 30-35 dah (~75 mg WW). Factors possibly leading to juvenile albinism include low DHA/EPA, low ω3HUFA, unenriched *Artemia*, and fast growth.

Planas (1994) described two types of larval rearing for turbot. Intensive systems with small to medium tanks, stocking of 20-40 larvae/L, and enriched rotifers and *Artemia* for food were popular in southern European hatcheries. Extensive systems with large tanks, bags, or impoundments, stocking of <1 larva/L, and wild zooplankton for food were popular in northern Europe.

Urup (1994) described a commercial hatchery system for turbot with computer control of microalgae, calanoid copepod, and larval cultures that was expensive to build and operate but eliminated the need for rotifers and *Artemia*.

Cunha and Planas (1995) proposed an optimal feeding schedule for turbot at 18-20°C: <4 mm SL, rotifers <150 µm; 4-4.5 mm, rotifers 125-200 µm; 4.5-5.5 mm, rotifers 125-200 µm and *Artemia* nauplii (200-250 µm); >5.5 mm, *Artemia* nauplii.

Addition of *Isochrysis galbana* or *Tetraselmis* sp. to larval turbot tanks along with rotifers increased their survival, growth, and resistance to out-of-water stress (Reitan et al., 1993c).

When *Rhodomonas baltica*, Tahitian *Isochrysis*, *Pavlova lutheri*, or *Chlorella* sp. were added to larval turbot tanks along with rotifers, survival during 1-18 dah was in the range 18-42%, but when *Dunaliella tertiolecta* was used, survival was only 1-8% (Støttrup et al., 1995). *D. tertiolecta* and associated rotifers contained negligible ω3HUFA, unlike the other algae and rotifers.

Turbot survived better (21, 28%) to 26 dah when fed rotifers enriched with Dry Selco® and *Artemia* enriched with Super Selco® than with 11 other enrichment combinations, but growth was similar for the top nine treatments (Støttrup and Attramadal, 1992). Individual rotifers contained 0.73 ng EPA and 0.70 ng DHA, and *Artemia* contained 41.76 ng EPA and 21.55 ng DHA.

Turbot survived better on rotifers enriched with experimental or commercial cod roe mixtures than on those fed five other mixtures (Opstad et al., 1993).

Reitan et al. (1994b) concluded that a DHA/EPA ratio of about 1.7 in turbot at 12 dah should maximize (~90%) normal pigmentation at metamorphosis. ω3HUFA levels of 1.5% (of DW) in rotifers and 1.8% in *Artemia* were sufficent for good growth and survival of larvae (Rainuzzo et al., 1994), but optima of 1.3-3.2% ω3HUFA were previously reported. A DHA/EPA ratio of 0.79 in rotifers was correlated with higher incidence of malpigmentation (66%) than for ratios of 1.31 (42%) and 2.78 (32%).

Although a minimum amount of dietary DHA is required by turbot larvae, overfeeding of DHA reduced their resistance to high-salinity stress (Dhert et al., 1994).

To grow to 50-75 mg WW (~30 dah), turbot need to eat about 3200 rotifers, 10,000 *Artemia* nauplii, and 7000 *Artemia* metanauplii (Person-Le Ruyet, 1990b).

Summer flounders (17°C) hatch at 2.6 mm TL, first feed 3-4 dah, have EYS 4 dah, and transform ≥56 dah (25-35 mm). At all stages, wild larvae fed almost exclusively during daylight and ate mostly copepods with some tintinnids (Grover, 1998). Larvae were raised in 1972 in aquaria on rotifers and calanoid copepods, with the best survival from hatching to 56 dah at 5% (Smigielski, 1975b). The juvenile digestive system is complete at about 31 dah (Bisbal and Bengtson, 1995a). At 16 dah, 90% of the larvae fed rotifers from the time of mouth opening (4 dah) had survived, but only 36% of those first fed at 6 dah had survived (Bisbal and Bengtson, 1995b).

Southern flounders (17°C) hatch at 2.2 mm TL, first feed 5-6 dah, and transform >48 dah (>15 mm). Larval rearing was successful in the mid-1970s, with 30% survival from fertilization to 90 dah obtained in aquaria with rotifers and *Artemia* (Arnold et al., 1977). They also have been raised in a pond on mixed zooplankton (Jenkins et al., 1997b).

California halibut (20°C) hatch at 2.0 mm TL, first feed ~4 dah, have EYS ~5 dah, settle ~19 dah, metamorphose ~27 dah, and transform (form scales) at ~47 dah. Larvae were raised on rotifers and *Artemia* nauplii (Gadomski et al., 1990). Weaning capability was not discussed.

Ecuadorian flounder larvae were raised through transformation on rotifers, *Artemia*, wild zooplankton, and compound feeds (Benetti, 1997). A large number of fish did not complete metamorphosis, possibly because of a deficiency in the eggs. The incidence of pseudo-albinism (white stripes or patches) was high in juveniles and possibly resulted from a nutritional deficiency.

Pleuronectidae

Atlantic halibut (6°C) culture research made significant advances during the 1980s, including establishment of captive broodstock (Pittman et al., 1987; Haug, 1990; Kjørsvik and Holmefjord, 1995). Long developmental times increase the difficulty of raising larvae. At 6°C, hatching of 6.5-mm larvae occurs at ~16 dah, the yolksac lasts about 48 d, and distinct first feeding occurs during 33-50 dah. Metamorphosis occurs at ≥90 dah (~47 mm TL). Beginning at ~27 dah, a significant amount of trypsin was produced (Ugelstad et al., 1995). At 30 dah, eyes were fully pigmented and the gall bladder was yellowish-green; peristalsis was first seen at 33 d (Pittman et al., 1987). Digestion of small food items might be possible as early as 23-26 dah. Lein and Holmefjord (1992) found the highest feeding incidence and most full guts at 4.5°C when larvae were first fed during 44-59 dah. Atlantic halibut have been raised in tanks (Naas et al., 1992) and in plastic bags (Næss et al., 1995b). Wild or cultured zooplankton often is used, but halibut can be raised on *Artemia* alone. Larvae can be weaned at ≤70 dah.

Holmefjord and Olsen (1991) stated that the best growth has been obtained in green water with wild zooplankton, but enriched cultured foods supported comparable growth. Naas et al., (1992) obtained better survival and growth of halibut fed unenriched *Artema* metanauplii in green water than in clear water.

Næss et al. (1995a, 1995b) reported that feeding enriched *Artemia* to Atlantic halibut larvae resulted in good growth and survival, but if fed for 57 d (33-90 dah), the resulting juveniles were malpigmented. Larvae fed wild zooplankton from first feeding (33 dah), or after enriched *Artemia* given during 33-51 dah, had no malpigmentation. Thus, for juveniles to be normal, a certain proportion of prey more natural than *Artemia* seems necessary in the larval diet, beginning at about 52 dah.

Lemon soles raised on rotifers and *Artemia* nauplii had ~50% survival (~1-104 dah), but a number of the juveniles were uncoordinated (possibly an EFA deficiency sign)(Howell, 1974).

European plaice (9°C) larvae hatch at 5.8 mm TL, first feed ~5 dah, have EYS 8 dah, metamorphose at ~35 dah, and transform at 40-75 d (10-14 mm). Rotifers are optional during ~5-7 dah. Dickey-Collas and Geffen (1992) obtained good survival (~39%) of plaice larvae from hatching to 50 dah when raised on *Artemia* only. They also found that growth and survival of plaice larvae was not related to dietary DHA, even if the EPA content was low, helping to explain why they are easy to raise on *Artemia* of varying quality. In that study, larvae grew and survived just as well on unenriched nauplii (1.9% EPA of FA and no other ω3HUFA) as on enriched metanauplii (10.1% EPA, 6.9% DHA, 17.7% ω3HUFA). However, Dickey-Collas (1993) later reported that only 29% of metamorphosed plaice reared on unenriched nauplii had normal pigmentation vs 63% of those fed enriched metanauplii. Duration of the diet seemed more important than its timing within the rearing period (40-50 dah).

Greenback flounder (16°C) larvae were raised on rotifers and *Artemia*; metamorphosis occurred at ≥33 dah (≥16 mm TL)(Hart and Purser, 1996). The authors concluded that a 20-d weaning period beginning at 23 dah should result in the best survival and growth, but a 10-d period would be more economical and almost as good.

Winter flounders (15°C) hatch at 5.8 mm TL, first feed 5 dah, have EYS 8 dah, and transform 40-75 dah (10-14 mm). Metamorphosis began at 26 dah and was finished in some fish at 33 dah (Lee and Litvak, 1996a). Weaning was accomplished during 47-52 dah (starting at 12.5 mm, 42 mg WW) with loss of only ~8% of the fish.

Soleidae

Common soles (19°C) are moderately large at hatching (3.6 mm TL, 0.5-0.6 mg WW), first feed 2 dah (~4 mm), have EYS ~3 dah, have EOG ~10 dah, metamorphose and begin benthic feeding at ~15 dah, and transform 42-56 dah (9-10 mm)(Person-Le Ruyet, 1990a, 1990b). Rotifers are not needed. To grow to 50-75 mg WW (~30 dah), common sole need to eat about 8000 *Artemia* nauplii and 11,000 metanauplii. Weaning is possible at 30-40 dah with conventional dry feed (Métailler et al., 1983) and about 10 dah with microfeeds (but with reduced growth).

Gatesoupe (1983) reared common sole on a simple diet of *Artemia* (nauplii, unenriched metanauplii, or enriched metanauplii) during 2-15 dah and semi-moist crumbles beginning 10 dah; the best survival was 61% from hatching to 62 dah.

Adding *Tisbe holothuriae* (10% by number) to the diet of Brazilian *Artemia* nauplii improved growth and pigmentation of common sole after metamorphosis (Heath and Moore (1997).

Senegal soles hatched at 2.7 mm TL, first fed ~2 dah (~2.8 mm), settled ~19 dah, and transformed ~30 dah (Dinis, 1992). Larvae were raised on rotifers, *Artemia*, and rehydratable feeds.

Tetraodontidae
Tiger puffers (17°C) hatch at 2.7 mm TL, have EYS 6 dah, and transform ~30 dah (9-10 mm). Larvae grew and survived better during 23-60 dah on a combination of rotifers, *Artemia*, and microparticulate feeds than on the live foods alone (Kanazawa, 1991b).

GENERAL METHODS USED IN OUR HATCHERY

Twenty-μm mesh sieves are used to strain clumps from algae and enrichment microcapsules; 20-μm sieves for mollusc eggs and trochophores; 35-, 70-, 118-, and 250-μm sieves for rotifers; 35- to 600-μm sieves for copepods; and 118-μm sieves for *Artemia*.

Primary cultures of rotifers (a Russian strain of *Brachionus plicatilis*) are kept in 3-L flasks in a clean room. Larger quantities have been raised in semi-continuous cultures at 24-30°C and 18-35 ppt in 76-L or 152-L white polyethylene trash cans, 150- to 1000-L fiberglass tanks, and 724-L concrete tanks. These sizes are preferred because they are easy to manage. Cultures seem to be healthier in clear and white tanks than in black tanks. Rotifer food consists mainly of *Nannochloropsis oculata* and sometimes Tahitian *Isochrysis* and/or small amounts of commercial powders. Small cultures are restarted every 5-7 d and large ones after several weeks, before water quality and bacterial composition deteriorate. Well-maintained cultures double every 18 h, producing at least 100-200 rotifers/mL/d. Rotifers are enriched only for experimental comparisons, not for routine feeding.

Microalgae is added to the fish tanks (up to 5000 cells/mL), mainly as food for zooplankton but also to remove ammonia. The algae culture should be old (dense) enough so that nearly all of the nutrients have been taken up.

Artemia cysts are not decapsulated but are disinfected (e.g., in a 2% sodium hypochlorite solution) before incubation in 17- to 140-L and larger acrylic tanks. Newly-hatched *Artemia* nauplii are fed sparingly to larval fish. Between 12 and 24 h after *Artemia* hatch, the metanauplii are enriched for 4-8 h, washed, concentrated, and fed to the fish. At lower temperatures, *Artemia* might need to be enriched longer. Depending on the size of larvae, some *Artemia* are raised up to 5 d or more (occasionally to adults) on *N. oculata* and Tahitian *Isochrysis* before being enriched and given to the fish.

Copepods of suitable size can be used as a supplemental food. Care should be taken not to cross-contaminate the food cultures. As long as it is economical, feeding of copepods can be continued until weaning begins or the supply is depleted.

Larvae of most warmwater marine species from planktonic eggs will have pigmented eyes and begin searching for food within 1.5-5.0 d after hatching. If larvae are ready to feed at night, the lights are turned on and the larvae are fed. The fish are watched for signs of distress anytime something is added to the tank. Larvae routinely are observed at close range for 15- to 30-min periods (especially at first feeding and when diets are changing) to determine if they are successfully feeding. When rotifer density is such that larvae have to swim more than a couple of body lengths to find one to eat, more are added, but not so many that the tank becomes swamped. If fish density is low, overstocking of rotifers can easily occur. With low fish density and high rotifer density, rotifer additions should be limited and algal additions maintained. A large rotifer surplus might have to be removed by careful flushing or siphoning—if the fish can survive the disturbance.

Feeding rotifers twice/d should be enough, but high fish density might call for three or four additions. If the tank is fully stocked with larvae (10 or more per liter), initial density of rotifers can be 5-10/mL, but only just before larvae begin feeding. The number and frequency of later additions is determined by consumption. If many rotifers are left from the first feeding of the day, the second feeding might have to be cut back; close observation of the fish will help determine what they need. There should be light for at least 3 h after the last feeding.

When larvae appear large enough (e.g., 7 dah for spotted seatrout; 8-10 dah for barramundi, common snook, Nassau grouper, red drum; 14 dah for sheepshead), about 50 *Artemia* nauplii can be placed in the tank where the fish are densest. The fish are observed for at least 15-30 min to determine if the *Artemia* are eaten. This is done once in the morning and once in the late afternoon until *Artemia* are eaten. During the next few days, small numbers of nauplii can be cautiously added, making sure that they all are eaten by the end of the day. Soon after all the fish are eating nauplii well, rotifers and nauplii can become secondary foods and be reduced to one addition per day, while 1-d-old enriched metanauplii become the primary food (compound feeds might also be included but are likely to be of secondary importance at early stages). Metanauplii are fed two or more times per day, with numbers and sizes cautiously increased, depending on how competent (add more) and how gluttonous (add fewer) the fish are. The larvae should not be allowed to eat *Artemia* so fast that their stomachs look like they might burst (that can kill the fish). Because rotifers and copepods are more efficient packagers of ω3HUFA than *Artemia*, it is good to delay or reduce feeding of *Artemia* and extend feeding of rotifers or copepods as long as practical—maybe until weaning begins.

When the fish start to develop juvenile pigmentation, the weaning process can be started. The incidence of cannibalism (Fig. 8-14) can be increased by crowding, underfeeding, and delayed weaning. We have reduced cannibalism by feeding sufficiently, weaning early, and grading juveniles (e.g., spotted seatrout, Tucker, 1988; barramundi, Tucker et al., 1988; common snook, Tucker, 1987b). High quality ingredients, including animal protein, marine fat, and a full complement of vitamins, should be used in the starter feeds (see chapter 9). Prolonged feeding with fresh or frozen food can lead to nutritional deficiencies and increased cannibalism.

INDUSTRIAL-SCALE LARVAL FOOD PROCESSING IN ITALIAN HATCHERIES

Candreva et al. (1996) described rotifer and *Artemia* production in commercial sea bass and sea bream hatcheries. Rotifers were raised in several stages: small vials, 2- and 10-L containers, 200-L plastic bags (algae and yeast), 3-m^3 fiberglass tanks (Culture Selco® at 330 mg/L the first day and 100-150 mg/L thereafter, for 4 d; 200 to 600 rotifers/mL), and a 10-m^3 tank (Culture Selco® for 4 d, 200 to 600 rotifers/mL). Tank conditions were 27°C and 20 ppt salinity, with pure oxygen added. At harvest, the rotifers were fed *Nannochloropsis* sp. (12,000/µL) and *Isochrysis* sp. (5000/µL) for 10-12 h, then Protein Selco® or DHA-Protein Selco® (250 mg/L in two rations) for 8-10 h. Those not used immediately were held in cold storage until needed (5000/mL with algae).

Artemia were hydrated and decapsulated in a 500-L tank (15 kg cysts), with a solution of 11-12% sodium hypochlorite, 20% sodium hydroxide, and 10% antifoam agent. Then they were rinsed sequentially with water, sodium thiosulfate solution (50 g/kg cysts), and water. Cysts (3 g/L) were hatched in 1.2- to 3.5-m^3 inverted pyramid tanks at 30°C.

Nauplii to be used soon were held in a 650-L insulated milk tank at 7°C and, as needed, were pumped to an enrichment tank (280/mL) and fed Culture Selco® or DC Selco.® Unused cysts were stored in brine (270 ppt) for up to 2-3 d at 15-20° or 2 weeks at 4°C.

SUMMARY

Criteria and methods for feeding marine fish larvae have been reviewed. Food must be available, at the right density, visible and attractive, of the right size, with appropriate chemical stimuli, of high nutritional quality, with good microbiota, digestible, and stable until eaten. Larvae of many marine fish can be reared on a simple diet of rotifers (*Brachionus plicatilis*) and brine shrimp (*Artemia* sp.) metanauplii. Foods given to these prey organisms can vary in quality. Enrichment of rotifers previously raised on good-quality algae or the equivalent is not necessary for larvae of many species, especially if rotifers are not used for long or are combined with other foods. To ensure that *Artemia* contain sufficient essential fatty acids (ω3HUFA, mainly eicosapentaenoic acid and docosahexaenoic acid), they should be enriched before fish eat them. However, a large excess of essential fatty acids can be harmful, possibly because of an imbalance. Unenriched *Artemia* nauplii can be used for short periods, in combination with other high-quality foods, or for species that have low ω3HUFA requirements. Copepods are probably the most nutritious live food for marine fish larvae, but large numbers are more expensive to produce than rotifers and *Artemia*. For most marine fish that have been cultured, weaning from live to dry food is most effective if it begins a few days before transformation starts, and weaning usually can be completed by the time the fish are juveniles.

Much of the mortality that occurs among intensively-reared marine fish larvae is a result of feeding them inappropriate foods, particularly: first foods that are too big, too many less-digestible *Artemia* nauplii, or prey deficient in essential fatty acids, or of presenting the foods incorrectly (e.g., in light-colored tanks, with light too intense). These problems can be solved by choice and (when necessary) enrichment of foods and optimal hatchery design. Larvae should be able to continuously and easily locate and capture nutritious digestible prey without wasting time and energy in lengthy searching. Water movement should not be so high that fish are stressed, cannot feed, or cannot fill their gas bladders. Light should not be so intense that larvae are repelled from the surface. Tank walls should be dark to enhance distribution and visibility of prey. Ideally, prey should be about 25% of mouth width for first feeding larvae and 50% for later stages. For many species, prey measuring 35-100 μm in width are appropriate at first feeding. Rotifer size can be controlled by choice of strain or by sieving, and to some degree by culture conditions. Food size and movement probably are the primary feeding stimulants for early larvae. Nutritional quality of prey is mostly determined by their content of ω3HUFA, which can be regulated by careful choice and enrichment of prey organisms. Published ω3HUFA requirements for young marine fish are in the range 1-5% of the dried food, and some natural foods contain 3-8% ω3HUFA, so the target range is high. Aiming for 2-4% ω3HUFA including at least 1.0% EPA and 1.0% DHA (with other fatty acids in balance) would probably be good insurance for most marine fish. Early larvae typically do best with live foods; possibly because they contain labile components such as attractants, digestive enzymes, stimulants for larval enzymes, and free amino acids. Acceptability and digestibility of compound feeds varies with species and age. Typical cultured marine fish can easily and quickly be weaned to dry feeds during transformation, and progress has been made in early weaning of several species.

9 NUTRITION OF JUVENILE AND ADULT FISH

For marine fish farming to be feasible, foods that produce adequate growth and survival at an acceptable cost are necessary. It is useful to group fish into five feeding categories: pelagic carnivores, demersal carnivores, demersal omnivores, demersal herbivores, and planktivores. With few exceptions (e.g., milkfish, ayu, tilapia, striped mullet), most mariculture candidates fall into the first three groups. Requirements vary broadly among categories and on a finer scale among species within categories. Needs also are affected to a degree by factors such as age, size, growth rate, developmental stage, density, food quality, water quality, temperature, salinity, pH, photoperiod, stress, and health. Availability of feed ingredients varies among regions, and will influence the composition of least-cost formulas. High quality trash fish sometimes is used for grow-out of carnivorous and omnivorous marine fish, but unless it is very fresh, such a diet can carry pathogens and toxins, or be nutritionally deficient (Taniguchi, 1983a, 1983b; Nakagawa et al., 1984). In Japan and Europe, some commercial feeds are available, but in other Far Eastern countries, the Middle East, Australia, and the Western Hemisphere, development of feeds for non-salmonid marine fish is relatively recent. Some of the earlier literature on feeding marine fish was reviewed by New (1986). This chapter reviews research on nutrition of representative cultured marine fish and makes some suggestions on fulfilling their requirements.

REQUIREMENTS AND COMPONENTS

In formulating feeds for carnivorous and omnivorous marine fish, particular attention must be paid to protein, fatty acid, and vitamin contents. Requirements for herbivorous fish are less stringent. Initial formulations can be based on simulation of natural diets and research done on similar species. Further refinements would be based on experience and research with the species of interest. If demand is high enough, shipping costs can be reduced by making feeds locally, possibly with cheaper local ingredients. Reducing feed cost by using higher proportions of lower cost ingredients reaches a limit when lower growth rates, lower survival, or increased feed volume begin to increase overall production costs. In farming of fish, feed formulations that are marginally acceptable on nutritional grounds can be fully evaluated only by an economic analysis including all relevant factors, especially any increase in production time that raises labor and overhead costs.

Nutritional requirements of major species have been well examined. These include freshwater fish such as rainbow trout, channel catfish (Li and Robinson, 1994), and tilapia, anadromous fish like salmon, ayu, and striped bass, and marine fish such as European sea bass, yellowtail, gilthead sea bream, red sea bream, red drum, hirame, and turbot.

Shape, color, flavor, texture, nutritional quality, safety, and processing quality can be affected by the way fish are raised, and especially by their foods (Haard, 1992). Essential fatty acids in processed fish vary according to levels in the foods they ate (Kalogeropoulos et al., 1993). Dietary fats, amino acids, and nucleotides could affect flavor of the fish (Cowey, 1993). Environmental contaminants from the water or fish foods could affect the fish or human consumers of the fish (Clarkson, 1995).

A complete feed is formulated to fulfill all of the species requirements and can be used in culture systems (e.g., tanks, cages) that do not allow significant feeding on natural foods. Incomplete feeds do not satisfy the known requirements and are used to supplement natural foods (e.g., in ponds).

From the fish farmer's viewpoint, a major economic factor is the cost of feed necessary to reach a desired production level. Feed conversion ratio (FCR) is defined as the weight of dry feed required to produce a unit weight of wet fish and is used as an index of how efficiently a fish grows (Table 9-1). Multiplying the unit cost of feed by FCR provides the actual cost of feed to produce a given weight of fish. Gross growth efficiency (GGE) is the reciprocal of FCR except that it is calculated on a dry/dry basis (dry weight gained/unit dry weight of feed eaten) and sometimes wet/wet. Active species tend to use less of their food energy for growth than sedentary species (Brett and Groves, 1979). They often have higher feed conversion ratios (i.e., lower gross growth efficiencies) and therefore higher feed costs. With an appropriate high quality feed, reasonable conversion can be obtained even with very active species such as yellowtail (often because they eat more). Feed efficiency ratio (FER) has been defined as wet weight gained/unit dry weight of feed eaten. FCR is preferred in this book because any efficiency should be calculated with factors having the same basis (i.e., dry/dry, wet/wet) and feed efficiency ratios higher than 1.0 do not seem reasonable. FCR and FER are merely reciprocals of each other. Daily ration (DR) is the weight (usually dry) of feed eaten per day divided by wet weight of the fish.

Table 9-1. Feed conversion ratios (weight of dry feed necessary to produce a unit increase in fish wet weight) for 450-gram fish raised in tanks or cages.

Species	FCR	Reference
Common snook	1.0	Tucker, 1987b
Atlantic halibut	1.0	Tuene & Nortvedt, 1995
Barramundi	1.1	MacKinnon, 1990
Nassau grouper	1.1	Tucker & Woodward, 1996
Dolphin	1.1	Ostrowski, 1995
Turbot	~1.2	Sheehan et al., 1994
White sturgeon	1.3	Cui & Hung, 1995
Sunshine bass	1.3	Tucker et al., 1993a
Yellowtail	1.3	Watanabe et al., 1991a
Red drum	1.3	Tucker et al., 1997a
European sea bass	1.4	Tacon, p.c., 1998
Gilthead sea bream	1.4	Tacon, p.c., 1998
Florida red tilapia	1.6	Watanabe, p.c., 1998
Malabar grouper	1.9	Chua & Teng, 1980
Florida pompano	~2.0	McMaster, 1988

Because only about 3% of the total feed needed to raise a 2-kg turbot is eaten by the time they weigh 100 g, using the best quality starter feeds is not necessarily wasteful or significantly polluting (Nijhof, 1994a). More often than not, higher-quality feeds are less polluting. Performance is a high priority for starter feeds. Unit feed cost and environmental friendliness become more important for the feeds that are used in large amounts (i.e., grower feeds).

Protein

Typical protein contains a total of 5650-5710 cal/g, but only about 4000 cal/g are digestible by fish (e.g., rainbow trout, 4020; common carp, 4020; Nile tilapia, 4520; European eel, 5305 digestible cal/g; Steffens, 1989). Protein in fish feed is used mainly for growth but also for energy. Fish must consume enough protein to satisfy growth requirements. All fish that have been studied require the same ten amino acids needed by most monogastric animals including humans (Millikin, 1982b; NRC, 1983; Lovell, 1989; Steffens, 1989; Akiyama et al., 1997). If amino acids are unbalanced, more protein will be required to get enough of the limiting amino acid. Amino acid deficiencies are not likely in diets containing protein mostly from high-quality animal sources. Some signs of amino acid deficiencies are: cataracts, fin erosion, mineral imbalances, renal calcinosis, lordosis, scoliosis, and mortality (Tacon, 1995). An excess of leucine in blood meal can induce an isoleucine deficiency. Possible signs of excess leucine are: opercular deformities, spongiosis of epidermal cells, scale defects, scale loss, and scoliosis.

Because protein tends to be the most expensive component of a fish feed, protein/energy (P/E) ratios should be considered in formulating feeds (Table 9-2). An optimal P/E ratio

Table 9-2. Protein/energy ratios of feeds used successfully for raising representative fish (not necessarily optimal).

Species	Weight (g)	P/E (mg P/kcal)	P/E (mg P/kJ)	Basis[a]	Reference
Atlantic salmon	500-10,000	79	19	GE	Helland et al., 1991
Atlantic salmon	0.13-50	88	21	GE	Helland et al., 1991
Gilthead sea bream	5-30	92	21.9	GE	Vergara et al., 1996a
Gilthead sea bream	44-97	93	22.3	GE	Marais & Kissil, 1979
Greasy grouper	66-130	94	22.5	GE	El-Dakour & George, 1982
Tiger puffer	4-12	99	23.6	GE	Takii et al., 1995a
Siberian sturgeon	19-45	107	25.6	GE	Kaushik et al., 1991
Gilthead sea bream	0.8-3.3	115	27.4	GE	Vergara et al., 1996b
Tiger puffer	3-59	116	27.6	DE	Han & Furuichi, 1995
Red drum	46-77	120	28.7	~GE	Daniels & Robinson, 1986
Barramundi	1-41	125	29.9	GE	Catacutan & Coloso, 1995
Malabar grouper	11-49	143	34.1	GE	Shiau & Lan, 1996
Black sea bream	8-28	112	26.8	GE-DE	Chuang et al., 1985
Yellowtail[b]	240-1290	104	24.9	DE	Watanabe et al., 1991a
Atlantic halibut	7-556	115	27.4	DE	Aksnes et al., 1996
Redlip mullet	1-10	127	30.4	DE	Yoshimatsu et al., 1992b
Florida pompano	0.5-5	129	30.8	DE	Williams et al., 1985
Common dentex	21-46	130	31.0	DE	Tibaldi et al., 1996
Yellowblotch spinefoot[b,om]	1-8	134	31.9	DE	Parazo, 1990
Sunshine bass[b]	207-862	138	33.0	DE	Tucker et al., 1993a
Atlantic cod[b]	173-318	139	33.2	DE	Lie et al., 1988
Red drum[b]	66-6125	140	33.6	DE	Tucker et al., 1997a
Malabar grouper[b]	58-225	142	33.8	DE	Teng et al., 1978
European eel[b]		144	34.5	DE	Brusle, 1990
Red drum[b]	4-63	144	34.3	DE	Tucker et al., 1997a
Common snook[b]	241-726	145	34.7	DE	Tucker, 1987b
Red sea bream[b]	7-29	145	34.7	DE	Takeuchi et al., 1991
Milkfish[om]	0.04-0.18	146	34.9	DE	Lim et al., 1979
Common snook[b]	16-44	147	35.2	DE	Tucker, 1987b
Barramundi[b]	9-62	147	35.2	DE	Tucker et al., 1988
European sea bass[b]	37-57	147	35.0	DE	Hidalgo & Alliot, 1988
Spotted seatrout[b]	24-110	147	35.2	DE	Tucker, 1988
Yellowtail[b]	88-214	148	35.3	DE	Shimeno et al., 1985
Spotted seatrout[b]	52-80	149	35.6	DE	Tucker, 1988
Yellowtail	2-16	151	36.2	DE	Takeuchi et al., 1992e
Yellowtail[b]	65-	154	36.9	DE	Takeda et al., 1975
Nassau grouper[b]	95-1490	158	37.9	DE	Tucker & Woodward, 1996
Sunshine bass[b]	862-1333	160	38.3	DE	Tucker et al., 1993a
Nassau grouper[b]	7-71	160	38.3	DE	Tucker & Woodward, 1996
Barramundi[b]	19-224	166	39.6	DE	Chou, 1985
Rainbow trout	grower	107	25.5	ME	Alsted et al., 1995
Barramundi	0.4-	108	25.8	ME	Tubongbanua, 1987

[a] GE = gross energy; DE = digestible energy; ME = metabolizable energy.
[b] DE was calculated using the values 4 kcal/g protein, 8 kcal/g fat, 3 kcal/g carbohydrate.
[om] Omnivorous at this stage; the rest are carnivorous.

allows the maximum amount of protein to be available for growth by minimizing the amount used for energy (protein sparing). Typical ranges for birds and mammals are 60-70 and 45-65 mg protein/kcal. Fish seem to need higher P/E ratios. Because most fish are satiated by the number of calories ingested, they might stop eating before consuming enough protein for the best growth if P/E is low. Fast-growing small fish need an abundance of high quality protein and fat. When size increases and relative growth slows, proportionally less protein might be acceptable, but this also depends on other factors such as temperature and constraints on activity. Generally, nutrient density should be high for young fish but can be reduced as fish get larger. If basic nutritional requirements are satisfied by the feed, the P/E ratio usually is acceptable. Feeds developed for carnivorous marine fish by the author have P/E ratios in the upper half of the published range (Table 9-2).

Dietary protein requirements for freshwater fish are well documented (NRC, 1981, 1983; Lovell, 1989; Steffens, 1989). The requirement for channel catfish is in the range 25-36%. For tilapia the requirement decreases from 50% at first feeding to 25-35% in adults. Salmonids need about 50% protein at first feeding but only 35% at 1 yr (NRC, 1981). Carnivorous marine fish need relatively large amounts of protein (Table 9-3). Optimal protein contents for even omnivorous marine species (e.g., striped knifejaw, yellowblotch spinefoot) seem high compared to those of channel catfish. The range for marine fish is narrow, with 50-60% dietary protein being necessary for most species studied. Some species might require higher feed protein content as temperature increases, but in most cases that is not a practical consideration. On the other hand, as temperature rises, so do metabolism and energy requirements to support it. This leads to an increase in feed consumption and a decrease in growth efficiency. Protein requirement generally decreases with increasing size or age. When growth rate slows, less protein is needed, and older fish seem capable of using more carbohydrate.

Except for fresh fish or other meat, fish meal (FM) is one of the most palatable, digestible, and nutritious protein sources for fish. To minimize accumulation of toxic decomposition products, the sooner the fish is processed after capture the better (Pike, 1993; Bimbo and Crowther, 1992). Low temperature (steam-dried) fish meal typically is of higher quality than conventional flame-dried fish meal (McCallum and Higgs, 1989). High-temperature processing can break down amino acids or cause them to combine into less digestible forms (Tacon, 1995). Usually, herring, menhaden, or anchovy meal is used in the U.S., but byproduct meals from salmon, sardine/pilchard, and cod/haddock (white fish) also are available. White fish and sardine meals are used extensively in Japan. Capelin and herring meals and fish protein concentrate are used in Europe. Extrusion of trash fish and fish processing wastes might be a feasible alternative for supplying fish proteins for feeds and at the same time minimizing resource wastage (Choudhury, 1995). Up to 61% cuttlefish meal was successfully used for red sea bream, and addition of cuttlefish meal to broodfish diets resulted in better egg quality than when only white fish meal was used (Watanabe et al., 1984a, 1984b). Ostrowski et al. (1989) included 35% squid meal in dolphin feeds.

Fish meal is expensive, and as much as 20-25% of the world supply could be in demand for use in fish feeds by the year 2000 (Bimbo and Crowther, 1992). Other preferred protein sources (up to about 30% of diet) are poultry byproduct meal (Gallagher and Degani, 1988), meat byproduct meal, and cooked soybean products. Other possible sources, used in smaller amounts (usually 1-15%), are fish protein hydrolysate, shrimp meal, krill meal, crab meal, spray-dried blood meal, hydrolyzed feather meal (Fowler, 1990), brewer's yeast, torula yeast, corn gluten meal, sesame seed meal, sunflower seed meal, peanut (groundnut) meal, chick peas, field peas, and cow peas. Up to 39% canola meal low in erucic acid, glucosinolate, and phytic acids can be used for some species. Cottonseed meal and lupin meal have antinutritional factors that are difficult to remove. Up to 34% cottonseed meal with reduced gossypol or ~10% low-alkaloid lupin meal can be used for

Table 9-3. Protein contents of feeds used successfully for raising representative fish (near optimal).

Species	Weight (g)	Protein (%)	Criteria[a]	Reference
Thicklip mullet	15-42	38	G	Ojaveer et al., 1996
Milkfish[om]	0.04-0.18	40	G,FCR	Lim et al., 1979
Malabar grouper	58-225	40-50	G,FCR	Teng et al., 1978
Atlantic salmon	grower	43		Helland et al.. 1991
Gilthead sea bream	44-97	44	G,FCR	Marais & Kissil, 1979
Red drum	46-77	44	G,FCR	Daniels & Robinson, 1986
European eel		45		Degani, 1986
Japanese eel		45		Degani, 1986
Striped bass	38-130	45	G,FCR	Hung et al., 1993
Redlip mullet	1-10	45	G,FCR	Yoshimatsu et al., 1992b
Yellowblotch spinefoot[om]	1-8	45	G,FCR	Parazo, 1990
Striped knifejaw[om]	6-29	45-55	G	Ikeda et al., 1988
Gilthead sea bream	5-30	46	G,FCR	Vergara et al., 1996a
European sea bass		47-50		Alliot et al., 1974
Greasy grouper	58-130	47-60	G,FCR	El-Dakour & George, 1982
Atlantic salmon	starter	48		Helland et al., 1991
Rainbow trout	grower	48		Alsted et al., 1995
Ayu		48		Kanazawa, 1991a
Malabar grouper	4-13	48	G	Chen & Tsai, 1994
Turbot		~48		Devesa, 1994
Barramundi	1-41	50	G,FCR	Catacutan & Coloso, 1995
European sea bass	31-57	50	G,FCR	Hidalgo & Alliot, 1988
Yellowtail		~50		Takeda et al., 1975
Yellowtail	2-16	50	G,FCR	Takeuchi et al., 1992e
Red sea bream	689	~50	R	Watanabe et al., 1984a
Turbot	~20-1680	~50	G,FCR	Danielssen & Hjertnes, 1993
European plaice	14	50	G	Cowey et al., 1972
Tiger puffer[om]	2	50	G,FCR	Kanazawa et al., 1980
Tiger puffer[om]	~0-120	50	G	Kanazawa, 1991b
Russian sturgeon × beluga	22-	52	G,FCR	Gershanovich & Kiselev, 1993
Barramundi	35-173	~52	G,FCR	AQUACOP et al., 1990a
Red drum	66-215	53	G,FCR	Tucker et al., 1997a
Atlantic cod	173-318	54	G	Lie et al., 1988
Blackspotted grouper		~54		Lin et al., 1981
Chinook salmon	2	55	G	DeLong et al., 1958
Striped bass	2-9	55	G,FCR	Millikin, 1982a
Red sea bream		~55	G,FCR	Yone, 1976
Red sea bream	7-29	55	G,FCR	Takeuchi et al., 1991
Common sole		>55		Metailler, 1990
Common dentex	21-46	56	G,FCR	Tibaldi et al., 1996
Gilthead sea bream	0.8-3.3	56	G,FCR	Vergara et al., 1996b
Yellowtail	88-213	57	G,FCR	Shimeno et al., 1985
European plaice	10-20	57	G	Guillaume et al., 1991
Common sole		57	G	Guillaume et al., 1991
Atlantic halibut	34-314	58	G	Hjertnes et al., 1993
Thicklip mullet	15-38	59	FCR	Ojaveer et al., 1996
Gilthead sea bream	2-21	61	G,FCR	Sabaut & Luquet, 1973
Atlantic halibut	5-677	65	G,FCR	Aksnes et al., 1996
European plaice	3-	70	G	Guillaume et al., 1991

[a] G = growth. FCR = feed conversion ratio. R = reproduction.
[om] Omnivorous at this stage; the rest are carnivorous.

some species. Wheat bran and rice bran contain a small amount of protein. Fish protein concentrate is good but expensive. Fish scrap meal or silage can be used in moist feeds (Satoh et al., 1984; Toyama et al., 1984; Asgard and Austreng, 1985; Yone et al., 1986c; Shahidi, 1995). Upper limits for these are determined by factors such as digestibility, nutrient balance (e.g., amino acid, fatty acid), fat content, ash content, and fiber content. A variety of plant protein sources is possible if digestibility is improved and antinutritional factors are removed or inactivated by processing (Tacon and Jackson, 1985; Tacon, 1992); however, little is known about their digestibility in carnivorous marine fish.

Hardy (1996) summarized the defects in some of the most likely sources to substitute for marine animal protein. Poultry byproduct meal varies in quality and has high ash (also high fat). Feather meal varies in digestibility. Corn gluten meal contains high fiber and can cause yellowing of fillets. Wheat gluten meal costs too much.

Among other uses, marine animal hydrolysates can be included in fish feeds (Heras et al., 1994; Kołodziejska and Sikorski, 1995). Essential amino acids (both peptide and free) are higher in hydrolysed than in freeze-dried Atlantic cod muscle and might be better utilized by Atlantic salmon (Espe et al., 1993). Growth of Atlantic salmon (0.15-12 g) was improved by inclusion of 3.3 or 5.3% fish protein hydrolysate (5 or 8% amino acid substitution, 61.3-66.6% fish meal, 41-43% protein, 20% fat, 15% starch)(Berge and Storebakken, 1996). Fish silage should not be stored too long before being used; Espe et al. (1992) reported better results after 2 months of storage than after 5 months.

Growth and FCR of rainbow trout were as good with 17% herring meal + 23% spray-dried blood powder as with 49% herring meal + no blood powder (Luzier et al., 1995). The substitution caused a 38% reduction in phosphorus, but on the basis of protein content, blood powder cost about 24% more than herring meal. Up to 5% spray-dried blood powder could be used for Atlantic salmon, but growth was slightly reduced with 9.5% (Rasmussen, 1994).

Reduced growth when soybean meal (SBM) is substituted for fish meal has been blamed on amino acid deficiencies, lower digestibility, mineral deficiencies, indigestible oligosaccharides, and antinutrient factors (Rumsey et al., 1995). Soybeans are less palatable than animal feedstuffs and contain too much unsuitable fat and antinutritional factors; however, when most of the oil and antinutrients are removed, soybean meal is very good (Hardy, 1996). Soybean protein concentrate (SPC) was nearly as good as high-quality fish meal for Atlantic salmon when used at 38% of the feed (with 18% fish meal vs 50% in the control), and was better than full-fat (FSBM) or defatted soybean meals (DSBM)(Olli et al., 1994a). Up to 30% textured vegetable (soybean) protein has given us excellent results with marine fish such as common snook, barramundi, and spotted seatrout.

Trypsin inhibitor activity (TIA) of 5 mg/g has been suggested as the upper limit for soybean ingredients in fish feeds (Rumsey et al., 1995). When soybean protein concentrate (TIA 1.5 mg/g) or toasted defatted soybean meal (TIA 0.6 mg/g) were substituted for all (57% SPC or 78% DSBM for 56% FM) or half the fish meal (29% SPC or 39% DSBM for 28% FM) in a rainbow trout feed, only the SPC+FM feed consistently supported growth as well as (or better than) the full fish meal control feed. FCR was best for both of those feeds (1.0 at 3-66 g, 1.3 at 52-162 g). The DSBM+FM feed was third BEST and the non-fish meal feeds were far worse. An inflammatory reaction to soy protein was documented.

Canola refers to Canadian varieties of rapeseed that have less than 2% (vs 25-55% of lipids) erucic acid (toxic) and less than one-tenth of the glucosinolates (antinutrients) of standard rapeseed (Higgs et al., 1995). Other factors such as high fiber, phenolic compounds, and phytic acid can limit the amount of canola meal used in feeds and should be reduced as much as possible. Canola meal's amino acid profile more closely resembles that of herring meal than soybean meal's profile does. Lysine is the most-likely limiting amino acid. Up to at least 16-23% canola meal can be fed to juvenile Pacific salmon, ~13-26% to rainbow trout, and 15-41% to tilapia, but dephytinization raises the limits (39%

canola meal, 59% of protein, for rainbow trout). Alcohol-washed, fiber-reduced, canola meal could be used to provide at least 25% of the dietary protein (31% canola meal, 46% total protein) for chinook salmon feed and 40% (38% canola meal, 40% total protein) for rainbow trout (McCurdy and March, 1992). Rapeseed protein concentrate is good but not easy to obtain (Hardy, 1996).

Cottonseed meal use is restricted by the antinutrient gossypol; cyclopropenoid fatty acids also are present but probably in tolerable amounts (Robinson and Li, 1995). Lysine is the most-likely limiting amino acid. Without supplemental lysine, up to 15-20% cottonseed meal can be used for salmonids, catfish, and tilapia, and as much as 34% has been fed to chinook salmon.

Distillers grains with solubles (DGS) have no identified antinutritional factors and seem palatable to fish, but do have fairly high (9.1%) fiber (Webster et al., 1995). Channel catfish grew at least as well and probably better with 35% DGS or 70% DGS + lysine than with no DGS + no lysine (control) and significantly better than with 70% DGS + no lysine. FCR was similar for all feeds (2.2-2.4) except 70% DGS + no lysine (4.0). Up to 8% DGS have been successfully fed to lake trout and 10% distillers dried solubles to rainbow trout.

For a carnivore, rainbow trout can tolerate relatively large amounts of vegetable protein (Moyano et al., 1992; Watanabe et al., 1993a; Olli and Krogdahl, 1994; Gomes et al., 1995; Smith et al., 1995). Sunflower meal was nearly as good as soybean meal (both at 39%) in replacing 40% of fish-meal protein (59% reduced to 35%)(Sanz et al., 1994). Replacement of 60% (35 & 25%) of the brown-fish-meal protein by defatted soybean meal + malt protein flour, soybean meal + corn gluten meal, or meat meal + malt protein flour did not adversely affect growth or conversion (Yamamoto et al., 1995a). Replacement of 30% of brown-fish-meal protein by defatted soybean meal (29%) and 30% by malt protein flour (25%) did not have a large effect on growth and conversion (Akiyama et al., 1995). Malt protein flour alone was substituted for 60% of brown-fish-meal protein without a large effect on growth or conversion (Yamamoto et al., 1994).

Amino acid supplementation can improve feeds with low animal protein contents (Tudor et al., 1996; Davies and Morris, 1997). Soybean meal has low methionine, lysine, cystine, and threonine. Canola meal, cottonseed meal, and distiller's dried grain have low lysine. Methionine is low in plant proteins, yeast, meat & bone meal, blood meal, and hydrolyzed feather meal; lysine is low in oilseeds, hydrolyzed feather meal, and algae (Tacon, 1995).

Unconventional feed ingredients such as single cell proteins (e.g., from algae, yeasts, bacteria, fungi) might be used more in the future (Mahnken et al., 1980; Tacon and Jackson, 1985; Martin et al., 1993; Perera et al., 1995). Yeast in small amounts (e.g., 1-2%) has been used as a natural source of vitamins and immunostimulants, and larger amounts for protein; however, imbalances could occur with too much. Metailler and Huelvan (1993) reported that European sea bass grew slightly better with 10% yeast included in their feed.

Lipids

Typical fat contains a total of 8660-9510 cal/g, and about 8000 cal/g are digestible by fish (e.g., European eel, 7960; rainbow trout, 8010; common carp, 8010-9010; Nile tilapia, 9010 digestible cal/g; Steffens, 1989). Most lipid in fish and their foods is in the form of triglycerides (neutral fats); a triglyceride molecule contains one glycerol molecule and three fatty acid molecules. Fat is included in fish feeds primarily as a concentrated readily available source of energy, with some lipids being used in the synthesis of biological membranes (phospholipids) and in metabolic processes; lipids also transport other, fat-soluble, nutrients—including vitamins A, D, E, and K (Watanabe, 1982; Steffens, 1989). Although carnivorous fish are adapted to using protein for energy, optimal levels of fat and carbohydrate will provide extra energy for maintenance and activity and minimize the amount of protein needed for energy production. Too much total fat can reduce protein intake or increase fat deposition in viscera, skin, or muscle, which might reduce vigor or food quality of the fish; however, too little total fat or the wrong kind of fat can cause physiological problems involving essential fatty acid (EFA) deficiency (NRC, 1981; Steffens, 1989). Choice of feed ingredients can influence quality and acceptability of the product, especially through variations in fatty acid composition (Johnsen and Dupree, 1991).

With an essential fatty acid (EFA) deficiency, or an imbalance in fatty acids, growth rate, growth efficiency, and vigor can be reduced (NRC, 1981; Sargent et al., 1989; Wanakowat et al., 1993). Pigmentation can be abnormal. Anemia is common and mortality can be high (Roberts and Bullock, 1989). Non-essential fatty acids can accumulate in the liver, giving it a swollen, pale, greasy appearance, and intracellular fat globules can be seen in histological sections (Owen et al., 1972; NRC, 1981; Watanabe, 1982; Post, 1987; Roberts and Bullock, 1989; Sargent et al., 1989). A choline deficiency also can cause fatty liver syndrome (Ogino et al., 1970; Ketola, 1976). EFA-deficiency signs in juvenile barramundi were: reddening and hemorrhaging of mouth, fins, opercula, and body; hyperventilation; swallowing difficulty: anorexia; abnormal swimming; pale liver; and intolerance to disturbance or handling (Wanakowat et al., 1993). Four percent linolenic acid (18:3ω3) in the diet caused fatty degeneration of the liver in red sea bream (Fujii and Yone, 1976; Fujii et al., 1976). EFA deficiency in turbot feeds resulted in pathological changes in fat cells (Cowey et al., 1976) and gill tissue, with high mortality reported (Bell et al., 1985a, 1985b). More than 1% of fatty acids in the linoleic series (ω6) depressed growth of coho salmon (Yu and Sinnhuber, 1979). Optimal fatty acid balance can increase resistance to toxic substances (Lemaire et al., 1992) and other stressors. DHA accumulates in the brain and eyes and is especially important for optimal functioning of the nervous system (Mourente and Tocher, 1993).

Natural freshwater prey of juvenile Atlantic salmon contained substantial amounts of linolenic acid and EPA but very little DHA (Bell et al., 1994). DHA was better than linolenic acid, which in turn was better than linoleic acid as EFA for growth of small (1.6 g) juvenile Arctic char; for good growth, the linolenic acid requirement was ~1.0-2.0% (Yang and Dick, 1993).

Watanabe et al. (1985) found that the linoleic acid level of eggs spawned by red sea bream eating feed containing 10% corn oil was 22.1% of total fatty acids, about eight times higher than normal. The feed contained 5.6% linoleic acid, 0.4% EPA, 0.4% DHA, and 0.9% ω3HUFA. Factors in broodstock nutrition (including EFA) are discussed in a later section.

Lie et al. (1986) reported that the linoleic acid content of livers of Atlantic cod given a feed containing 8.4% peanut oil was 13.4% of total fatty acids, about six times higher than in cod eating feeds containing cod liver oil or Greenland halibut oil. The peanut oil feed contained 1.8% linoleic acid, 0.1% EPA, and 0.2% DHA. The cod liver oil feed contained 0.1% linoleic acid, 0.9% EPA, and 1.4% DHA. The halibut oil feed contained 0.1% linoleic acid, 0.4% EPA, and 0.5% DHA. ω3:ω6 ratios were 0.2, 12.6, and 7.4 for the three feeds. Feed consumption and growth rate were similar with the fish oil feeds but slightly reduced with peanut oil.

With moderate to high dietary fat, growth and conversion of juvenile red drum were considerably better with added fish oil than with added soybean oil (Figs. 9-1 and 9-2; Tucker et al., 1997b). Apparently, low levels of soybean oil can be utilized as long as sufficient HUFAs are present.

Long-chain highly unsaturated fatty acids (ω3HUFA), primarily eicosapentaenoic acid (EPA, 20:5ω3) and/or docosahexaenoic acid (DHA, 22:6ω3) are essential in the diet of Japanese eel, kasago, yellowtail, red sea bream, black sea bream, red drum, opaleye, striped mullet, turbot, European plaice, and tiger puffer (Owen et al., 1972; Gatesoupe et al., 1977; Yone, 1978; Kanazawa et al., 1979; Leger, 1980; Yamada et al., 1980; Steffens, 1989; Lochmann and Gatlin, 1993a) and possibly all marine fish—particularly higher carnivores

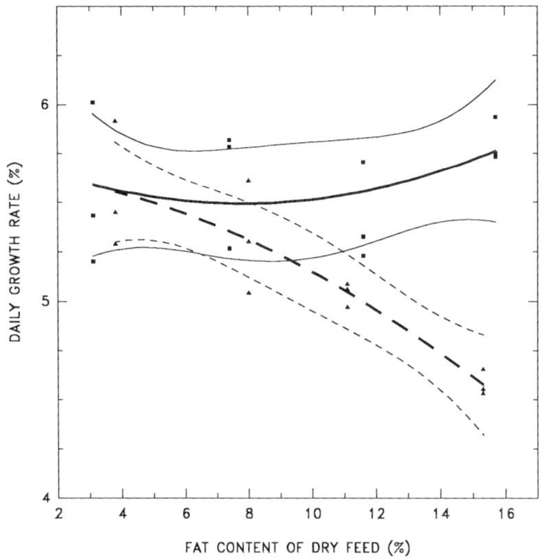

Figure 9-1. Relation between fat content of feed and daily growth rate for red drum over the size range 0.3-19.4 g. Solid lines (squares) represent the second order regression and 95% confidence interval for feeds with added menhaden oil, and dashed lines (triangles) represent data for added soybean oil (three replicates × 1.5%, 5.2%, 9.0%, and 12.7% of each oil). Note that beginning with about 9% total fat (6.8% fish oil added, 7.3% total), growth on the fish oil feeds was significantly better. (Tucker et al., 1997b)

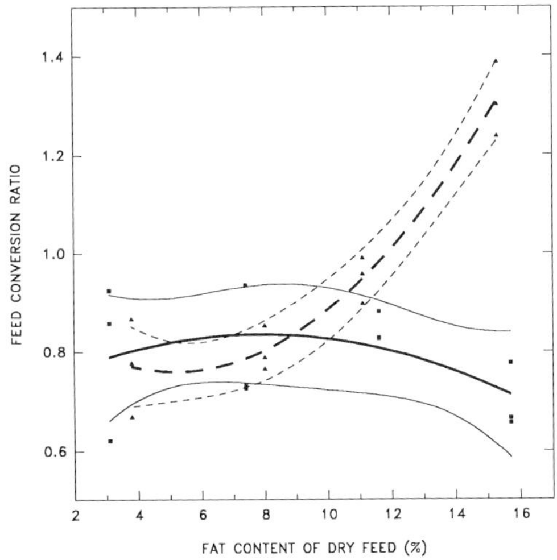

Figure 9-2. Relation between fat content of feed and feed conversion ratio for red drum over the size range 0.3-19.4 g. Solid lines (squares) represent the second order regression and 95% confidence interval for feeds with added menhaden oil, and dashed lines (triangles) represent data for added soybean oil (three replicates × 1.5%, 5.2%, 9.0%, and 12.7% of each oil). Note that beginning with about 10% total fat (7.8% fish oil added, 8.3% total), FCR for the fish oil feeds was significantly better. (Tucker et al., 1997b)

(Cowey and Sargent, 1979; Sargent et al., 1989). Species such as European sea bass have some ability to convert linolenic acid into DHA (Di Bella et al., 1993).

For Japanese and European eels, a 0.5% minimum of both linoleic and linolenic acids was necessary for good growth (Takeuchi et al., 1980; Degani, 1986). Ayu need about 1% linolenic acid and/or EPA (Kanazawa, 1991a). Pacific salmon need 1-2% ω3 fatty acids (Hardy, 1991). Barramundi need 0.5% ω6PUFA and 0.5% ω3PUFA (Feed Development Section, 1994).

The ω3HUFA requirement for juvenile milkfish was ~1-1.5% ω3PUFA (Feed Development Section, 1994), barramundi 1.0-1.7% (Boonyaratpalin, 1991; Wanakowat et al., 1993), yellowtails 2.0-4.6% (Watanabe et al., 1989c; Takeuchi et al., 1992d), red drum ~0.5% or ~7.0% of lipid (Lochmann and Gatlin, 1993a), and red sea bream (3-12 g) 1.0% EPA and 0.5% DHA (Takeuchi et al., 1990) or 20% of dietary lipid (Takeuchi et al., 1992c).

Watanabe et al. (1989c) estimated the minimum dietary ω3HUFA requirement of striped jack (0.2-20 g) to be 1.7%. Nakajima et al. (1995) suggested including 2.3-3.4% ω3HUFA in feed (15-25% of lipid) for juvenile striped jacks (4-40 g); 4.8% (30% in 15% lipid) was excessive.

DHA seems to be more important than EPA, especially for young marine fish (Bell et al., 1985b; Steffens, 1989). Juvenile striped jacks (0.5-3 g) grew and survived as well on a compound feed containing no EPA and 1.2% DHA as they did with 0.7% EPA and 0.6% DHA (Watanabe et al., 1989a). A feed containing 1.4% EPA without DHA was inferior. EPA and DHA were the only ω3HUFAs present in the feeds.

A minimum dietary ω3HUFA content of 2%, including at least 1.0% EPA plus 1.0% DHA for young marine fish and 0.7% EPA plus 0.7% DHA for older fish, might be a little higher than necessary for all species but seems to be a reasonable safeguard (Fujii and Yone, 1976; Watanabe, 1982; NRC, 1983; Bell et al., 1985a; Kanazawa, 1985).

Optimal total fat content for most cultured fish is probably in the range 10-20% (Cowey and Sargent, 1979; NRC, 1981). A wide range of dietary fat levels has been used (Table 9-4). With possible exceptions, sedentary fish might tolerate less dietary fat than more active ones. At least theoretically, flatfish should grow well with low fat, other demersal species with low to intermediate fat, and active pelagic fish with intermediate to high fat. Specific differences in adaptations to natural diets and in energy requirements for activity are important factors. Marine fish are more capable of utilizing oils from marine organisms than from terrestrial sources. In some cases the upper limits for dietary fat might be higher than predicted, especially if inappropriate fats were used to determine those limits. New (1986) listed characteristics of several early commercial feeds for European sea bass, yellowtail, red sea bream, and gilthead sea bream. Protein contents were adequate but fat contents were probably lower than optimal for the first three species.

Canola, soybean, sunflower, linseed, peanut, corn, cottonseed, or coconut oils and pork or beef fat have been used successfully for freshwater fish and anadromous salmonids (NRC, 1981, 1983; Dosanjh et al., 1984; Lovell, 1989); however, those fats are deficient in EPA and DHA and cannot be the main sources for marine fish. The best source is marine fish oil, but marine invertebrate oils can be used successfully if available and economical (Table 8-2). Menhaden (12% EPA, 11% DHA), anchovy (17% EPA, 9% DHA), and cod liver oils (12% EPA, 9% DHA) are the best that are commonly used. Herring (7% EPA, 7% DHA) and sand eel (11% EPA, 5% DHA) oils are reasonably good, but capelin (3% EPA, 1% DHA) oil is deficient (Lovell, 1989; Sargent et al., 1989). Cuttlefish liver (11% EPA, 11% DHA) and krill (12% EPA, 8% DHA) oils also are good sources of ω3HUFAs (Shibata, 1983; Watanabe et al., 1984c; Sargent et al., 1989; Watanabe et al., 1991b). Variations in fatty acid profiles do occur, at least by season and location. NRC (1973) recommended adding at least 5% fish oil to fish feeds to provide essential fatty acids. For marine fish, this is probably the bare minimum. Nine percent menhaden oil is required to

Table 9-4. Fat contents of feeds used successfully for raising representative fish (not necessarily optimal).

Species	Weight (g)	Fat (%)	Reference
Red drum	2-10	7	Williams & Robinson, 1988
Milkfish[hr]	juveniles	8	Lim, 1991a
Florida pompano[a]	0.5-5	8	Williams et al., 1985
Yellowblotch spinefoot[om]	1-8	8	Parazo, 1990
Barramundi	19-224	9	Chou, 1985
Gilthead sea bream	0.8-3.3	9	Vergara et al., 1996b
Gilthead sea bream	44-97	9	Marais & Kissil, 1979
Spotted seatrout[c]	53-79	9	Tucker, 1988
Redlip mullet	7-24	9	Yoshimatsu et al., 1993
Turbot[b]		9	Adron et al., 1976
European eel		10	Brusle, 1990
Barramundi[d]	20-181	10	AQUACOP, et al., 1990b
Sunshine bass	862-1333	10	Tucker et al., 1993a
Nassau grouper	7-1490	10	Tucker & Woodward, 1996
Red sea bream[e,f]	34-45	10	Yone et al., 1971
Striped knifejaw[g,om]	6-29	10	Ikeda et al., 1988
Turbot		<10	Metailler, 1990
Common sole		<10	Metailler, 1990
Malabar grouper	58-225	11	Teng et al., 1978
Tiger puffer[om]	4-12	11	Takii et al., 1995a
European sea bass	37-57	12	Hidalgo & Alliot, 1988
Striped bass	38-130	12	Hung et al., 1993
Common dentex	21-46	12	Tibaldi et al., 1996
European sea bass		13	Coves et al., 1991
Sunshine bass	632-927	13	Fair et al., 1993
Turbot		13	Person-Le Ruyet et al., 1991
Common snook[h]	16-44	13	Tucker, 1987b
Barramundi[i]	9-62	13	Tucker et al., 1988
Red drum	66-6125	13	Tucker et al., 1997a, 1997b
Spotted seatrout[c]	24-60	13	Tucker, 1988
Common snook[j]	241-726	14	Tucker, 1987b
Dolphin	5-325	14	Ostrowski et al., 1989
Sheepshead[k,om]	314-573	14	Tucker, 1987a
Barramundi	1-41	15	Catacutan & Coloso, 1995
Yellowtail	2-16	15	Takeuchi et al., 1992e
Gilthead sea bream	5-30	15	Vergara et al., 1996a
Red sea bream[l]	7-29	15	Takeuchi et al., 1991
Atlantic cod	173-318	16	Lie et al., 1988
Sunshine bass	207-862	16	Tucker et al., 1993a
Yellowtail[m]	88-214	16	Shimeno et al., 1985
Yellowtail	65-	17	Takeda et al., 1975
Russian sturgeon × beluga	22-	20	Gershanovich & Kiselev, 1993
Atlantic salmon	juveniles	26	Helland et al., 1991
Yellowtail[n]	240-1290	29	Watanabe et al., 1991a

[a] Growth and FCR better with 4 or 8% menhaden oil as sole lipid source than with 0 or 12%; 8% apparently, but not statistically, best by both criteria.
[b] Highest growth rate and protein conversion at highest fat content tested.
[c] At 53-79 g, growth and FCR similar with 9.3% and 12.9% fat (3.4%, 7.7% fish oil); 12.9% fat better for smaller fish (24-60 g).
[d] Growth and FCR not significantly different with dietary fat levels of 7-14%, but 10% near optimal.
[e] Growth and FCR better with 10% fat than with 20%.
[f] Growth and FCR better with 9% added fish oil than with 3 or 15% (Yone, 1976).
[g] Growth best with the highest fat level in purified feeds.

Table 9-4. (Notes continued)
[h] FCR best with a feed containing 12.9% fat (highest tested), including 7.7% fish oil.
[i] Growth and FCR improved with increasing fat (10.0 to 16.9% fat, 2.3 to 9.4% fish oil); apparent threshold between 11% and 13% fat (4-5% fish oil), above which both growth and FCR leveled off.
[j] Growth and conversion good.
[k] Growth better on a commercial salmon feed than on catfish, trout, or fresh diets.
[l] Growth and FCR best with 15% fat including 2.2% ω3HUFA.
[m] Growth and conversion better with 9.1% added pollock liver oil (16.2% total fat) than with 0%, 0%, or 16.7% (5.7, 7.4, 22.2% total fat).
[n] A high level of fat from white fish meal, brown fish meal, krill meal, pollock liver oil, and sardine oil resulted in good growth and FCR.
[hr] Herbivorous at this stage. [om] Omnivorous at this stage. The rest are carnivorous.

provide 1.1% EPA and 1.0% DHA, and six percent to provide 0.7% of each. Part of the requirement is satisfied by ω3HUFAs from the oil in fish meal and possibly other ingredients.

The phospholipids in lecithin (phosphatidylcholine, phosphatidylinositol, and maybe phosphatidylethanolamine) appear to be essential in the diet of at least the larvae of marine fish (Kanazawa et al., 1983a, 1983b; Kanazawa, 1985, 1993a; Meyers, 1995, Orthoefer et al., 1995; Coutteau et al., 1997; Craig and Gatlin, 1997). These are important components of biomembranes. Lecithin improves growth and feed conversion of young salmonids (Poston, 1990a, 1990b). Poston (1990a) suggested that phospholipids could aid in digestion of fats by emulsifying them, limit peroxidation of lipids, and improve palatability of feeds. Lecithin might contain unidentified growth factors as well (Meyers, 1985).

Carbohydrate

Typical carbohydrate contains 4100-4210 cal/g, but fish vary widely in their ability to utilize carbohydrate (e.g., rainbow trout: raw starch, 1150-1650, cooked starch 2530, dextrin 3820, glucose 3730; European eel: potato starch 1620; common carp: α-starch + dextrin 3510; Nile tilapia: α-starch + dextrin 4010 digestible cal/g; Steffens, 1989). Carbohydrate apparently is not essential in the diet of carnivorous fish (Steffens, 1989). It is included in fish feeds mainly as a low-cost energy source for protein sparing, and it also acts as a binder. Carnivorous marine fish do not utilize carbohydrates as well as omnivorous or herbivorous species (NRC, 1987; Steffens, 1989). Carbohydrate digestibility decreases as its percentage increases. An excess of carbohydrate interferes with digestion of proteins. Excessive carbohydrates in salmonid feeds can cause hyperglycemia, glycogen deposition in the liver, and enlarged liver (Post, 1983; Steffens, 1989). Simple carbohydrates (sugars vs starches) are often more digestible and metabolizable, but too much can cause unacceptable blood sugar fluctuations, so complex carbohydrates are safer. Digestibility of starch is increased by gelatinization—partly because it becomes easier to hydrolyze and partly because it is retained longer in the digestive system than raw starch (Spannhof and Plantikow, 1983; Sturmbauer and Hofer, 1985). In addition, amylase inhibitors that occur in wheat and several other grains are inactivated by heat. Fiber is indigestible for fish and includes complex carbohydrates like lignin and cellulose that can be digested only by bacteria (Feed Development Section, 1994). A small amount (e.g., 2%) can improve FCR by slowing digestion time.

Carbohydrate digestibility for marine fish is generally lower than for freshwater fish, and carbohydrates are much less digestible than fat and protein for most fish (Steffens, 1989). Common sole can utilize carbohydrates better than turbot (Metailler, 1990). Common carp (freshwater omnivore) tolerated 30% dextrin, red sea bream (saltwater carnivore/omnivore) 20%, and yellowtail (saltwater carnivore) 10% (Furuichi and Yone, 1980).

Livers and intestines of white sturgeon (initial weight 28-68 g) fed 27% maltose, glucose, dextrin, or raw corn starch were normal, but the use of 27% sucrose, lactose, or fructose resulted in abnormally high intestinal water retention and histologically detectable abnormalities in the liver (Hung et al., 1990).

Knösche and Schreckenbach (1995) obtained experimental evidence with hybrid sturgeon, rainbow trout, and common carp to support the hypothesis that metabolic disturbances were related to excess dietary carbohydrate. Sturgeon were less tolerant of carbohydrate than trout and carp.

Growth and feed conversion in Siberian sturgeon (90-400 g) were better with 38% hydrolyzed corn starch than with raw starch (Kaushik et al., 1989). However, the high level of even cooked starch changed morphology, increased size, and diminished functioning of livers.

Growth rate of European eels (45 g, initially) steadily declined (0.90 to 0.58%/d) with increasing carbohydrate (20-50% cooked starch)(Garcia-Gallego et al., 1993). FCR stayed in the range 0.41-0.49, but protein conversion ratio (PCR) improved from 1.16 to 0.73 as protein content decreased from 44 to 20%. Feed intake was 31% less with the high-carbohydrate feed; stimulating the eels to eat more of that feed might have increased their growth rate some.

Hydrolyzed potato starch was not well utilized by Atlantic cod of 292-450 g (Hemre et al., 1989). As starch content increased from 12 to 36%, starch digestibility decreased from 40 to 26%. The energy contribution of carbohydrate appeared to be negligible.

Growth and conversion of Atlantic salmon were as good with experimental feeds containing 20% glucose or dextrin as with a high-quality commercial feed containing 20% soy flour; growth was slower with 20% raw starch or no starch + 8% more herring oil (Fynn-Aikins et al., 1995).

Feed conversion ratio in common snook increased as carbohydrate content of their feed increased (Fig. 9-3).

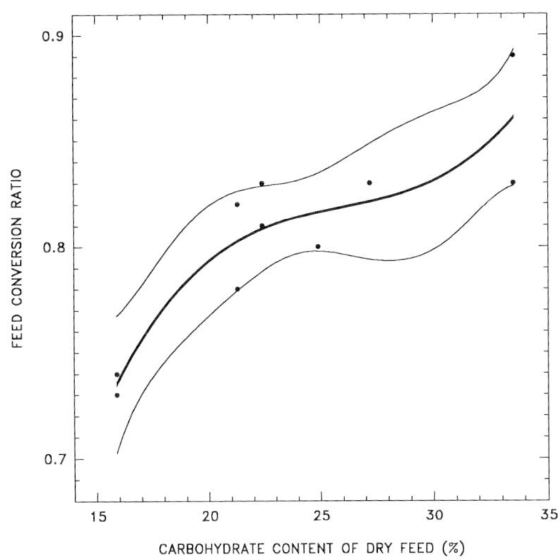

Figure 9-3. Relation between carbohydrate content of feed and feed conversion ratio for common snook over the size range 15.6-44.2 g (means). A third order regression curve and 95% confidence interval are shown for six feeds each tested twice. (Tucker, 1987b)

Digestibility of carbohydrate sources for palmetto bass decreased in about this order (rankings for 6-g and 95-g fish in parentheses): wheat flour (1,1), dextrin (1,2), wheat starch (1,3), corn starch (2,2), wheat middlings (2,3), and potato starch (3,3)(Gallagher, 1997). Digestibility values for 10% and 40% hydrolyzed potato starch fed to yellowtail were 57%

and 39% (Shimeno et al., 1979). Digestibility of 10% and 20% hydrolyzed potato starch for yellowtail (76-148 g) was 56% and 52% (Furuichi et al., 1986); feed efficiency was 70% (FCR 1.43) with 10% starch, but 58% (FCR 1.71) with 20% starch. Glucose was not well utilized because it was quickly absorbed and excreted. Shimeno et al., (1979, 1985) obtained the best growth and feed conversion in yellowtail with a feed containing the most fat (16%) and intermediate carbohydrate (10% as dextrin).

Growth and conversion of red sea bream were reduced when dextrin was increased from 10% to 20% (Yone, 1976). For red sea bream (24-37 g; Furuichi et al., 1987), digestibility was 67% with 15% hydrolyzed starch, 38% with 25% hydrolyzed starch, and 32-38% with 15-35% raw starch (feed efficiency range 32-55%).

When 10% chitin, chitosan, or cellulose were added to the diet of red sea bream, Japanese eel, and yellowtail, chitinase activity was detected in the stomachs of all three species but not chitosanase or cellulase (Kono et al., 1987). With chitin supplementation, growth of all three species was best, and conversion for sea bream and eel was best. With chitosan, growth and conversion were worst for all three, possibly because of interference with digestion or absorption. With cellulose, growth and conversion were slightly better than with the control feed for sea bream and yellowtail, possibly because of increased digestion time.

Practical limits for good-quality, relatively digestible carbohydrates are about 10% for young fish, and in the ranges 10-20% for strict carnivores, 15-35% for omnivores, and 30-50% for herbivores.

Digestible (e.g., hydrolyzed) starch is the preferred type of carbohydrate for marine fish. Good sources include: soybean meal, high gluten wheat flour, wheat middlings, whole wheat flour, Irish potatoes, cassava, sago palm, sorghum, rice, rice bran, and corn meal. The use of carbohydrate is limited not only by species' tolerances, but sometimes by high fiber content. Crushing and steam cooking of carbohydrates improves their digestibility.

Micronutrients

The immune response of fish has been shown to be responsive to levels of vitamins (Table 9-5), minerals (Table 9-6), and other factors that exceed the minimum requirement for good growth (Landolt, 1989). Initial vitamin and mineral levels used in my research were based on information in published reviews (Millikin, 1982b; NRC, 1981, 1983) with safety margins added to compensate for processing and storage losses, to allow for increased requirements for different species and different ages, and to stimulate growth. Prudent overfortification with vitamins should not raise feed costs excessively. Minerals are not overfortified. Levels of both were revised during early feeding studies with practical feeds and slight adjustments were made by reference to later reviews (Cho et al., 1985; Halver, 1985, 1989; Poston, 1986; Lall, 1989; Steffens, 1989). The following discussions are mostly based on these references. Most research on individual micronutrients has been done with rainbow trout, common carp, channel catfish and Japanese eel, with some studies on salmon, yellowtail, red sea bream, red drum, and others. Requirements for vitamins in freshwater fish are well documented, and minerals not so well. Requirements for neither are very well documented for marine fish, with a few exceptions. Values given here are relatively accurate but subject to revision. Comprehensive lists of deficiency and overdose signs from the literature are given in the following sections. The most typical and obvious signs are marked by asterisks. Characteristic external signs should lead the culturist to suspect certain deficiencies and to investigate further (e.g., histologically, hematologically, biochemically), if possible and if necessary.

Care should be exercised in choosing and mixing ingredients because interactions among vitamins, minerals, and other nutrients can cause deterioration of feed quality (Hilton, 1989). Vitamins and minerals usually are added in the form of premixes, including some anti-clumping filler like wheat middlings. These mixtures usually are added at a rate of about 1-2% of the total ingredients.

Vitamins
Fifteen vitamins have been shown to be essential in the diet of fish (Table 9-5; NRC, 1983). Para-aminobenzoic acid (PABA), lipoic acid, citrin, and unidentified growth factors might act as vitamins (Halver, 1989). PABA currently is not considered essential (Steffens, 1989). The safest approach in formulating feeds is to assume that vitamins are not contributed by other ingredients and add the full complement. If liver oils are used, possible overdose of vitamin A and D should be considered and avoided.

Thiamin (B_1) is very important for carbohydrate, lipid, and protein metabolism, for nerve function, and for good appetite; as part of the coenzyme cocarboxylase, it is especially important in decarboxylation. The requirement increases with higher carbohydrate in feeds, higher rearing temperature, and stress. When dry, thiamin is relatively resistant to heat and storage, but moisture, alkalis, sulfites, cooking, and extended storage can degrade it. Thiamin mononitrate is stable in premixes that do not contain choline chloride or trace minerals. Thiamin is destroyed by thiaminase, which can be inactivated by heat. A level of 10-20 mg thiamine hydrochloride/kg diet was recommended for rainbow trout (Steffens, 1989). Yellowtail required 1.2-11.2 mg/kg and should be given 22.4 mg/kg (Aoki, 1995). The minimum requirement for turbot is at least 0.6-2.6 mg/kg (Devesa, 1994). Extrusion of feed at 80°C did not cause loss of thiamin, but 120°C reduced it by one-third (Haaland et al., 1993). To compensate for variations in need and losses in processing and storage, I suggest that 40 mg thiamin/kg be added to feeds for typical marine fish.

Deficiency signs include: anorexia; *hyperirritability; *muscle spasms and cramps; *ataxia; slow, clumsy, spasmodic swimming; *convulsions; whirling; hyperventilation; blindness; *dark coloration; depigmentation; exophthalmia; cataracts; *hemorrhage of skin and fins; detachment of opercles; edema; hyperemia of fins; exudation; erythropoietic disturbance; liver damage; low transketolase activity in erythrocytes and kidney; increased pyruvate in muscles; muscular atrophy; low growth efficiency; low growth rate; and mortality.

Riboflavin (B_2) is essential for normal carbohydrate, lipid, and protein metabolism and eye structure; in the body, it occurs as the enzymes flavin adenine dinucleotide (FAD) and flavin mononucleotide (FMN). The requirement could vary because of caloric density, other feed ingredients, and environmental conditions. Riboflavin is relatively resistant to heat and storage when protected from light and oxygen. About 25% can be lost during the extrusion process, and 40% can be lost after 20 min of leaching in water. A level of 20-35 mg riboflavin/kg diet was needed for maximum liver uptake by rainbow trout (Steffens, 1989). Yellowtail required 2.9-11.0 mg/kg and should be given 22 mg/kg (Aoki, 1995). I suggest that 60 mg riboflavin/kg be added to feeds for typical marine fish.

Deficiency signs include: anorexia; lethargy; incoordination; muscle weakness; *slow, clumsy swimming; *visual disturbances; *corneal vascularization; abnormal pigmentation of iris; *photophobia; adhesion of lens and cornea; *cataracts; *blindness; nervous disturbances; *dark or otherwise altered coloration; skin inflammation and lesions; *hemorrhage of eyes, opercula, fins, heart, liver, and abdomen; necrosis of fins, gills, opercles, and kidneys; clubbed gills, anemia; fatty liver; glycogen deficiency; impaired activity of erythrocyte glutathione reductase and D-amino acid oxidase; renal atrophy; muscular atrophy; spinal curvature; short, dwarfish body form; low growth efficiency; low growth rate; and mortality.

Pyridoxin (B_6) is very important in protein metabolism; as the coenzyme codecarboxylase (pyridoxal phosphate), it is essential for decarboxylation of amino acids, and many other enzymatic processes; it has a role in fat metabolism and synthesis of messenger RNA; it occurs in pyridoxal phosphate, a coenzyme in protein metabolism. The requirement increases with higher protein in feeds, with growth rate, and during seasonal cycles of endocrine activity. Phosphorylated pyridoxin compounds are unstable when exposed to

Table 9-5. Suggested amounts of vitamins to be included in feeds for salmonids and marine fish.

Vitamin	Usual forms for supplementation	Salmon (requirement)[a]	Salmonids (add to feed)[b]	Marine fish (add to feed)[c]
Thiamin (B_1)	Thiamin hydrochloride or mononitrate	10-15	20-40	40 mg/kg
Riboflavin (B_2)	Riboflavin	20-25	40-80	60 mg/kg
Pyridoxine (B_6)	Pyridoxine hydrochloride	15-20	20-40	40 mg/kg
Pantothenic acid (B_5)	D-calcium or sodium pantothenate	40-50	80-160	300 mg/kg
Niacin (B_3)	Niacin, nicotinic acid, niacinamide	150-200	300-600	300 mg/kg
Biotin (H)	Biotin	1-1.5	2-4	2 mg/kg
Folacin (B_9)	Folacin	6-10	10-20	20 mg/kg
Cyanocobalamin (B_{12})	Cyanocobalamin	0.015-0.02	0.08	0.3 mg/kg
Choline	Choline chloride	600-800	3000-6000	6000 mg/kg
Inositol	Myoinositol	300-400		900 mg/kg
Ascorbic acid (C)	L-ascorbyl-2-polyphosphate			400 mg/kg
Ascorbic acid (C)	Uncoated or coated	100-150	400-800	700 mg/kg
Retinol (A)	Retinol acetate, palmitate, or propionate, liver oils	2000-2500	5000-10,000	10,000 IU/kg
Cholecalciferol (D)	Cholecalciferol	2400	3600-7200	3000 IU/kg
Tocopherol (E)	α-tocopherol, α-tocopherol acetate	30-35	60-120	500 IU/kg
Menadione (K)	α-phylloquinone, menadione bisulfite or bisulfate salts	10	8-16	20-50 IU/kg

[a] Halver, 1989; Helland et al., 1991.
[b] Rumsey, pers. comm., 1992; low values for old fish, high values for young fish.
[c] Tucker, developed during 1984-1991.

Table 9-6. Mineral requirements for several species and guidelines for salmonids and marine fish. All values are in mg/kg.

Mineral	Usual forms for supplementation	Several species (requirements)[a]	Salmonids (guidelines)[b]	Marine fish (guidelines)[c]
Required				
Calcium (Ca)	Calcium lactate, monobasic or dibasic calcium phosphate	2000-7000	10,000	6000
Chlorine (Cl)	Sodium chloride		2000	
Chromium (Cr)*	Chromium potassium sulfate			1

Cobalt (Co)*	Cyanocobalamin, cobalt chloride	0.05		0.05
Copper (Cu)*	Copper chloride or sulfate	3-6	5-20	9
Iodine (I)*	Potassium iodide	0.6-45	0.5-5	4
Iron (Fe)*	Ferrous or ferric chloride, ferrous sulfate, iron citrate	30-200	100	200
Magnesium (Mg)	Magnesium sulfate or oxide	200-700	1000	700
Manganese (Mn)*	Manganese sulfate, oxide, or carbonate	2-50	60	60
Available phosphorus (P)	Monobasic sodium or potassium phosphate	4000-9000	8000-20,000	12,000
Potassium (K)	Potassium sulfate	8000	8000	
Selenium (Se)*	Sodium selenite	0.05-0.5	0.20	0.5
Sodium (Na)	Sodium chloride	1000-2200	3000	
Sulfur (S)	Sulfates		2000	
Zinc (Zn)*	Zinc sulfate, oxide, or carbonate	15-200	150	200
Probably required or helpful				
Arsenic (As)				
Boron (B)				
Fluorine (F)	Sodium fluoride	4.5		
Molybdenum (Mb)				
Nickel (Ni)				
Silicon (Si)				
Tin (Sn)				
Vanadium (V)				
Possibly required or helpful				
Aluminum (Al)	Aluminum chloride	15		
Bromine (Br)				
Cadmium (Cd)				
Lead (Pb)				
Strontium (Sr)				
Titanium (Ti)				

[a] From review in Steffens (1989).
[b] Rumsey, pers. comm., 1992; low values for old fish, high values for young fish.
[c] Tucker, developed during 1984-1991 for marine fish reared in salt water and fresh water (total from feedstuffs and supplements).
* Sometimes should be added to feed; the others probably are available from macroingredients and culture water.

ultraviolet radiation. Pyridoxal phosphate can degrade in air. Air, light, and heat combine with moisture to decompose free forms of pyridoxal and pyridoxamine. Pyridoxin hydrochloride is relatively stable, but premixes should not contain trace minerals, which can degrade it. Storage of feeds at 20°C for 6 months caused negligible loss of pyridoxin. A level of 10-20 mg pyridoxin hydrochloride/kg diet was recommended for rainbow trout (Steffens, 1989). Barramundi needed 5 mg/kg to grow normally, but 10 mg/kg helps ensure normal lymphocyte counts (Boonyaratpalin, 1991). Yellowtail required 2.5-11.7 mg/kg and should be given 23.4 mg/kg (Aoki, 1995). For red sea bream, 2-5 mg/kg was sufficient (Yone, 1976). Gilthead sea bream needed 3-5 mg/kg for good growth and enzymatic activity (Kissil, 1991). Red hybrid tilapia needed 3 mg/kg (Lim et al., 1995). Turbot needed at least 1-2.5 mg/kg (Guillaume et al., 1991). I suggest that 40 mg pyridoxin/kg be added to feeds for typical marine fish.

Deficiency signs include: anorexia; lethargy; *hyperirritability; handling intolerance; ataxia; tetany; muscle spasms; *epilectic-like seizures and convulsions; *hyperventilation and gasping; *opercular flexing; *erratic circular, spiral, or surface swimming; loss of distance perception; unsuccessful feeding attempts; nervous disturbances; eye lens lesions; exophthalmia; blindness; scale protrusion; altered pigmentation; skin inflammation; lesions of mouth, snout, liver, and kidneys; widespread hemorrhages; edema; ascites; normocytic, normochromic anemia; microcytic, hypochromic anemia; leucopenia; liver hypertrophy; fat accumulation and glycogen deficiency in liver; histological changes in digestive tract, kidneys, liver, pancreas, and thymus; decreased aminotransferase activity in muscle, liver, pancreas, and erythrocytes; muscular dystrophy; low growth rate; mortality; and *rapid rigor mortis.

A sign of overdosage (20 mg/kg) in channel catfish was microcytic normochromic anemia.

Pantothenic acid (B_5), as part of acetyl coenzyme A, is important in the metabolism of carbohydrate, lipid, and protein and is essential for combustion reactions and energy production; this coenzyme is necessary for many enzymatic processes, development of the central nervous system, and normal physiology. The requirement increases with higher temperature and seems to increase with smaller fish size. Pantothenic acid is very unstable with heat, moisture, acid, or alkali. D-calcium and d-sodium pantothenate are relatively stable, but some can be lost with high heat. Common carp need 30-42 mg d-calcium pantothenate/kg diet, rainbow trout 50-60 mg/kg, and channel catfish 140-250 mg/kg; 280 mg/kg increased complement activity in red sea bream (Steffens, 1989). Barramundi need 15-90 mg/kg (Boonyaratpalin, 1997). Yellowtail required 13.5-35.9 mg/kg and should be given 71.8 mg/kg (Aoki, 1995). I suggest that 300 mg pantothenic acid/kg be added to feeds for typical marine fish.

Deficiency signs include: anorexia; lethargy; opercular flexing; breathing difficulty; ataxia; *slow, clumsy swimming; stress intolerance; metabolic upsets; exophthalmia; blotchy skin; hyperemia of skin; dark coloration; hemorrhage, erosion, and scarring of skin, fins, and gills; mucous membrane lesions; *lesions, adhesions, and puffiness of gills; *clubbed (hyperplasia), exudate covered, fused gill lamellae; swollen opercles; anemia; fatty liver; cellular atrophy; lesions and necrosis of kidney and pancreas; atrophy of pancreatic acinar cells; vacuoles and hyaline bodies in kidney tubules; probably increased susceptibility to infection; low growth efficiency; low growth rate; and mortality.

Niacin (B_3) is necessary for normal metabolism of carbohydrate, lipid, and protein; it is involved in many enzymatic processes, especially as part of the coenzymes nicotinamide adenine dinucleotide (NAD) and nicotinamide adenine dinucleotide phosphate (NADP). Niacin and niacinamide are very stable; however, about 20% of niacinamide can be lost during extrusion. A level of 50-200 mg niacin/kg diet was recommended for rainbow trout (Steffens, 1989). Yellowtail need 12.0 mg/kg nicotinic acid and should be given 96 mg/kg (Aoki, 1995). I suggest that 300 mg niacin/kg be added to feeds for typical marine fish.

Deficiency signs include: anorexia; lethargy, ataxia; feebleness; *muscle weakness, spasms, and cramps; *jerky movements; abnormal swimming, whirling; flexing and detachment of opercles; exophthalmia; *photosensitivity; dark coloration; *hemorrhage and lesions of skin and fins; fin erosion; swollen gills; jaw deformities; *edema of stomach and intestines; intestinal lesions; anemia, impairment of NAD and NADP systems; low growth efficiency; low growth rate; and mortality.

Niacin overdose (10,000 mg/kg) can deplete choline, resulting in fatty liver and growth depression.

Biotin (H) is necessary for carbohydrate, lipid, and protein metabolism, especially through carboxylation and decarboxylation, and for energy production; in various forms, it functions as a coenzyme. The requirement is raised by increased dietary fat. Biotin can be degraded by strong oxidizing agents or conditions. It is inactivated by avidin, an enzyme present in uncooked egg white. Lake trout required 0.5-1.0 mg biotin/kg diet and common carp 1 mg/kg; 1-2 mg/kg was recommended for all species (Steffens, 1989). Yellowtail required 0.22-0.67 mg/kg and should be given 1.34 mg/kg (Aoki, 1995). I suggest that 2 mg biotin/kg be added to feeds for typical marine fish.

Deficiency signs include: anorexia; lethargy; abnormal swimming; *spastic convulsions; dark coloration; depigmentation; *short, clubbed gill lamellae; proliferation of dermal mucous cells; *skin lesions ("blue slime"); intestinal lesions; anemia; lesions and fragmentation of erythrocytes; increased leucocytes and immature erythrocytes; fatty liver; decreased activity of pyruvate-carboxylase, acetyl CoA-carboxylase, citrate synthase, and α-ketoglutarate dehydrogenase in the liver; increased lactate in liver; abnormal synthesis of fatty acids and glycogen; atrophy and necrosis of pancreatic acinar cells; glycogen deposition in kidney tubules and lateral muscle; *muscular atrophy; low growth efficiency; low growth rate; and mortality.

Folacin (folic acid, B_9) is necessary for synthesis of DNA, RNA, and normal red blood cells; it is important for bone marrow formation, blood glucose level maintenance; and cell membrane function; it acts as a coenzyme in one-carbon transfer processes. The requirement is increased when fish are treated with sulfonamides. Sunlight and extended storage will degrade folacin. Processing and storage can reduce folic acid by 5-10% in steam-pelleted crumbles and by 3-7% in extruded crumbles. A level of 5-10 mg folacin/kg diet was recommended for rainbow trout (Steffens, 1989). Yellowtail required 0.8-1.2 mg/kg and should be given 2.4 mg/kg (Aoki, 1995). I suggest that 20 mg folacin/kg be added to feeds for typical marine fish.

Deficiency signs include: anorexia; lethargy; clumsy swimming; exophthalmia; *dark coloration; *pale gills; *deterioration of caudal fin; ascites; *macrocytic, normochromic, megaloblastic anemia; normocytic, normochromic anemia; increased eosinophils, neutrophils, platelets, and immature erythrocytes; decreased lymphocytes, normal mature erythrocytes, hematocrit, and hemoglobin; limited proerythrocytes in anterior kidney; large immature segmented erythrocytes; anisocytosis; poikilocytosis; reduced plasma protein; necrosis of spleen; liver hypertrophy; abdominal bloating; low growth efficiency; low growth rate; and mortality.

Cyanocobalamin (B_{12}) is essential for normal metabolism and growth, DNA synthesis, red blood cell maturation, and nerve function; it acts as a coenzyme. Cyanocobalamin is very stable at moderate temperatures but can degrade with heat and low pH or with extended storage. Growth was improved with 0.012-0.025 mg cyanocobalamin/kg diet for common carp and 0.030-0.040 mg/kg for rainbow trout; 0.05 mg/kg was recommended for salmonids (Steffens, 1989). Yellowtail required 0.053 mg/kg and should be given 0.424 mg/kg (Aoki, 1995). I suggest that 0.2-0.4 mg cyanocobalamin/kg be added to feeds for typical marine fish.

Deficiency signs include: anorexia; *microcytic hypochromic anemia; megaloblastic anemia; decreased hematocrit, hemoglobin, erythrocytes, and plasma protein; erythrocyte fragmentation; increased platelets; decreased neutrophils; low growth efficiency; and low growth rate.

Choline is a phospholipid that combines with acetyl CoA to form the neurotransmitter acetylcholine; it is a methyl donor in several biochemical reactions; it facilitates lipid transport from the liver and prevents fatty liver degeneration; it is important for growth. Choline is contained in lecithin. It should be kept separate from vitamins E and K, which it will react with and probably inactivate. Using an aqueous solution of choline to fortify feeds helps to maintain the separation. A combination of choline chloride and lecithin might be more effective than either alone. The requirement is raised by decreased methionine or increased fat in the diet. For lake trout, 1000 mg choline/kg diet was enough; common carp need 2000-4000 choline chloride/kg; 500 mg choline chloride/kg was required by red sea bream, but 8000 mg/kg was better (Steffens, 1989). Juvenile white sturgeon needed 1700-3200 mg/kg and 4000-6000 was recommended (Hung, 1991b). Yellowtail required 2100-2920 mg/kg and should be given 5840 mg/kg (Aoki, 1995). From research with semipurified feeds, Kanazawa (1993a) reported that striped knifejaw larvae needed 50,000 mg/kg lecithin (16,000 mg/kg choline + inositol) and juveniles needed 30,000 mg/kg lecithin (10,000 mg/kg choline + inositol); juvenile hirame grew better with 70,000 mg/kg lecithin than with 50,000 mg/kg (~22,000 mg/kg choline + inositol). I suggest that 6000 mg choline/kg (10,000 mg choline chloride/kg) be added to feeds for typical marine fish. Lecithin can be used.

Deficiency signs include: anorexia; exophthalmia; pale coloration; anemia; reduced cholinesterase; slow emptying of stomach; *enlarged, pale or yellowish, fatty liver; *hemorrhage of kidneys and intestines; *whitish-gray intestines; abdominal distension; low growth efficiency; low growth rate; and mortality.

Inositol is a phospholipid involved in carbohydrate metabolism and might promote amylase activity and biotin synthesis; it can help prevent fatty liver degeneration; its primary role seems to be in lipositol, a structural element in animal cells; it is important for growth. The requirement might be raised by increased dietary glucose. Inositol is contained in lecithin and is relatively stable. Young common carp need 440 mg inositol/kg diet; 350-500 mg/kg were recommended for rainbow trout (Steffens, 1989). Yellowtail required 190-423 mg/kg and should be given 846 mg/kg (Aoki, 1995). Red sea bream needed 550-900 mg/kg (Yone et al., 1971). I suggest that 900 mg inositol/kg be added to feeds for typical marine fish.

Deficiency signs include: anorexia; lethargy; uncoordinated swimming; dark coloration; skin lesions, especially at base of dorsal fin; *deterioration of fins, especially the caudal fin; edema; *slow emptying of stomach; *distended stomach; *whitish-gray intestines; *increased neutral lipids, cholesterol, and triglycerides in liver; anemia; reduced cholinesterase and transaminase activities; decreased total phospholipid, phosphotidylcholine, phosphotidylethanolamine, and phosphotidylinositol; low growth efficiency; low growth rate; and mortality.

Ascorbic acid (C) is very important in the synthesis of collagen, cartilage, and corticosteroids, in the ossification of bone, and in the healing process; it aids in thyroid function, calcium uptake from water, red blood cell maturation, and hydroxylation reactions; it acts as an antioxidant along with tocopherol and selenium; it reduces the toxic effects of ammonia, nitrite, heavy metals, nitrites, and pesticides and increases resistance to infections and stress. The requirement is increased by toxins, infections, metabolic disorders, stress, high temperature, and smaller size. Stress causes depletion of body stores. Seasonal requirements might vary.

Ascorbic acid is very sensitive to oxygen and heat. Most unprotected ascorbic acid in a fish feed will be lost after 3 months of storage at room temperature. As much as 17-32% ascorbic acid can be lost during extrusion or steam pelleting, 67-83% after feed storage at 20°C for 6 months, and 15-67% by water leaching of feed for 10 s. In the past, protective coatings such as ethylcellulose were used to prolong shelf life of ascorbic acid in feeds. Recently, more stable compounds such as L-ascorbyl-2-polyphosphate (APP) have become popular. There is practically no loss of APP in typical processing and little loss after 6 months of storage (10-40%, Grant et al., 1989). L-ascorbyl-2-sulfate and ascorbyl-6-palmitate also are very stable in feeds and can be used by fish (Soliman et al., 1987; Steffens, 1989). On an equimolar basis, APP biological activity seems to be nearly equivalent to that of ascorbic acid (Grant et al., 1989), but the other two forms appear to be less active (Albrektsen et al., 1988; Sato et al., 1991). Because of its instability, ascorbic acid is not incorporated into premixes, but is added separately to the ingredient mixture just before feeds are pelletized.

Indian major carp needed 670-750 mg ascorbic acid/kg diet at first feeding; 300-3000 mg/kg were required to fight infection in channel catfish; 1000 mg/kg raised the hemagglutinin titre in red sea bream; 1250 mg/kg improved viability of offspring in Mozambique tilapia; 7000-8000 mg/kg helped rainbow trout resist nitrite toxicity; at least 400 mg ascorbic acid/kg diet was recommended for young rainbow trout and 200 mg/kg for older ones (Steffens, 1989). In Atlantic salmon, deficiency signs occurred after 20 weeks on a feed with 50 mg/kg ascorbic acid, but not with 310 mg/kg (White et al., 1993). Immunity enhancement was not detected in Atlantic salmon fed 3170 mg/kg vitamin C compared with the normal amount of 440 mg/kg (Thompson et al., 1993). Barramundi needed 500 mg/kg to grow normally, but 1100 mg/kg helps ensure normal tissue content (Boonyaratpalin, 1991); 30 mg/kg ascorbyl polyphosphate-Mg (12.6 mg/kg ascorbic acid equivalent) was enough to prevent deficiency signs in barramundi (Boonyaratpalin et al., 1994, Boonyaratpalin, 1997, Phromkunthong et al., 1997). Without supplementation, deficiency signs were: bad appetite, equilibrium loss, immobility, fin erosion, scale loss, exophthalmia, low growth rate, short operculum, short body, detached isthmus, scoliosis, lordosis, broken back and associated hemorrhage. APP was better than ascorbyl 2-sulfate for European sea bass (Amerio et al., 1998). Yellowtail required 122 mg/kg and should be given 976-1950 mg/kg (Shimeno, 1991; Aoki, 1995). For normal growth and health of striped knifejaw, 250 mg/kg ascorbic acid was needed, but for tissue saturation, ~500 mg/kg would be necessary (Ishibashi et al., 1992). L-ascorbyl-2-phosphate-Mg at 50 mg/kg feed was enough for good growth, survival, and stress resistance of juvenile red sea bream (5-48 g)(Kosutarak et al., 1994); higher levels are needed to protect against oxidized lipids, and 5 g/kg was not toxic to juveniles (3-30 g)(Kosutarak et al., 1995b). Gilthead sea bream (500 mg initially) had kidney damage after 4 months of eating a feed containing only 50 mg/kg ascorbic acid (Alexis et al., 1997). For hirame, 60-100 mg/kg L-ascorbyl-2-phosphate-Mg was enough for good growth and survival (Teshima et al., 1993). With 2% ω3HUFA in a feed for young hirame (6.1-9.9 g), 10 mg/kg L-ascorbyl-2-phosphate-Mg resulted in better growth (1.71%/d vs 1.50 and 1.56) and FCR (0.58 vs 0.67 and 0.64) than 0 or 1000 mg/kg (Kosutarak et al., 1995a). The minimum requirement for turbot was 50-100 mg/kg (Devesa, 1994). However, Merchie et al. (1996, 1997) determined a minimum of 20 mg/kg ascorbic acid (as APP) for normal survival, growth, and liver histology in early juvenile European sea bass and turbot. Contents up to 150-2750 mg/kg and higher have been shown to increase disease resistance, mainly in channel catfish and salmonids (Lall and Olivier, 1993). I suggest that 400-1000 mg ascorbic acid/kg (mostly as APP) be added to feeds for typical marine fish.

Deficiency signs include: anorexia; lethargy; equilibrium loss; erratic or surface swimming; exophthalmia; iritis; blindness; dark coloration; enhanced or reduced pigmentation; ascites; *abnormal collagen and cartilage; defective support cartilage for and lesions of eyes, head, gills, opercles, jaws, and fins; clubbed gills; hyperplasia of jaw and snout; erosion of lower jaw, fins, and skin; *hemorrhage of eyes, head, skin, fins, liver, kidneys, intestines, and muscles; petechial hemorrhage; fin necrosis; scale loss; muscle degeneration; *lordosis; *scoliosis; spinal fracture; shortened body; liver, spleen, and adrenal hypertrophy; pancreatic and hematopoietic tissue atrophy; microcytic anemia; hypochromic anemia; leucopenia; reduced ascorbic acid in liver and kidneys; renal granuloma; fatty liver; increased liver cholesterol; increased hematocrit; increased plasma cholesterol and triglycerides; decreased serum alkaline phosphatase and serum thyroid hormone (T_3); delayed wound healing; increased susceptibility to infection; reduced egg hatching and larval survival; low growth efficiency; low growth rate; and mortality. After a few months, vitamin C-deficient turbot can die from renal granulomatous syndrome caused by disruption of tyrosine catabolism (Gouillou-Coustans and Guillaume, 1993).

Retinol, or other forms, (A) is essential in producing pigments needed for normal vision (especially in low light); it is important for normal eye and gill structure and function, bone formation, fertility, cold resistance, and maintenance of epithelium and mucous membranes. The requirement increases with stress, infection, and ultraviolet light. Retinol (vitamin A) occurs only in animals and is stored mainly as retinyl palmitate in the liver (Booth et al., 1992). Fish (especially liver) oil is a source. Carotenoids (provitamin A) occur mostly in plants, but also in some animals (crustaceans, egg yolk). Fifty of the 400 identified carotenoids have been shown to have vitamin A activity, with β-carotene being one of the strongest. Retinol is sensitive to oxygen, and some activity is lost during feed storage. One International Unit (IU) of vitamin A is the equivalent of 0.3 μg all-trans retinol (Lovell, 1989). Rainbow trout need about 2500 IU/kg diet for maximum weight gain and 50,000 IU retinol/kg for maximum retinol uptake; common carp need 4000-20,000 IU/kg (Steffens, 1989). Yellowtail required 5.68 mg/kg (18,933 IU) and should be given 6.88 mg/kg (22,933 IU)(Aoki, 1995). I suggest that 10,000-20,000 IU retinol/kg be added to feeds for typical marine fish and suggest that carotenoids be included, too.

Deficiency signs include: anorexia; *opercular flexing or twisting; *exophthalmia; xerophthalmia; night blindness; lens displacement; corneal thinning and expansion; retinal degeneration; *depigmentation; epithelial keratinization; *hemorrhage of eyes, skin, fins, and kidneys; scale loss; fin erosion; abnormal bone formation; nerve degeneration; anemia; increased serum alkaline phosphatase; *liver atrophy; white spots on liver; edema; *ascites; increased susceptibility to infection; low growth rate; and mortality.

Overdose (900,000-2,700,000 IU/kg in rainbow trout, Hilton, 1983) signs include: *lesions of skin and fins; erosion and necrosis of fins; epithelial keratinization; hyperplasia of head cartilage; abnormal bone formation; ankylosis; *spinal deformity; vertebral fusion; anemia; reduced hematocrit; increased serum alkaline phosphatase; *pale liver; liver atrophy; liver and spleen hypertrophy; reduced body fat; low growth efficiency; low growth rate; and mortality.

Cholecalciferol, or other forms, (D) promotes absorption, mobilization, transport, and utilization of calcium and phosphorus and is necessary for normal bone mineralization. Fish (especially liver) oil is a source. One IU of vitamin D is the equivalent of 0.025 μg cholecalciferol (Lovell, 1989). Cholecalciferol stability is high during processing and storage. Channel catfish had maximum growth with 1000-4000 IU cholecalciferol/kg diet; most species probably need 500-3000 IU/kg (Steffens, 1989). I suggest that 3000 IU cholecalciferol/kg be added to feeds for typical marine fish.

Deficiency signs include: *tetany of white skeletal muscles; "droopy tail" syndrome; fatty liver; impaired calcium homeostasis; low body calcium, phosphorus, and ash; possibly increased blood cell counts; increased serum thyroid hormone (T_3); ultrastructural changes in white muscle fibers; low growth efficiency; and low growth rate.

Overdosage (3.75 million IU/kg) signs include: anorexia; lethargy; tetany; dark coloration; fragile bones; enlarged liver; fatty liver and muscle; increased hematocrit and serum calcium;

decreased hemoglobin; and low growth efficiency and rate.

Tocopherols (E) are important in DNA formation, respiratory chain phosphorylation, enzymatic processes, carbohydrate turnover, fat metabolism, reproduction; as antioxidants, they protect lipids and phospholipid-containing membranes. Tocopherols increase resistance to stress, toxins, and diseases. Most vegetable oils that occur in fish feeds are sources. The requirement is raised with lower temperature, decreased amounts of other antioxidants, and increased unsaturated fatty acids in the diet or body; size and growth rate might affect requirements; spawning fish should be given more. Tocopherol is fairly resistant to heat but sensitive to oxidizing agents. Free tocopherol is rapidly oxidized in oxygen and light and is used up by labile substances such as polyunsaturated fish oils. The acetate and phosphate esters are much more stable. One IU of vitamin E is the equivalent of 1 mg D-α-tocopherol (Lovell, 1989). The pathological signs of skeletal muscle degeneration, increased plasma aspartate aminotransferase (ASAT) and creatine kinase (CK) activity, and increased muscle content of thiobarbituric acid reactive substances (TBARS) in European sea bass eating oxidized oil with no vitamin E added were diminished with 40 mg/kg Dl α-tocopherol acetate and were not observed with 300 mg/kg (Stephan et al., 1993). The minimum requirement for turbot is ~30 mg/kg vitamin E (with 6-15% fat)(Devesa, 1994). Usually 100-150 IU tocopherol/kg diet is recommended; however, about 500 IU/kg diet was beneficial to Nile tilapia, common carp, and Atlantic salmon (Steffens, 1989). The minimum requirement for Japanese eels was 200 mg/kg (Arai, 1991). Yellowtail required 119 mg/kg α-tocopherol and should be given 595 mg/kg (Shimeno, 1991). Antibody production in rainbow trout rose when vitamin E was increased from 23 to 60 and 600 mg/kg (Lall and Olivier, 1993). Commercial feeds usually contain 100-400 mg/kg. Higher levels seem justified for marine fish, especially considering the dangers of lipid peroxidation (e.g., jaundice in yellowtail, Sakai et al., 1998). I suggest that 400-600 IU tocopherol/kg be added to feeds for typical marine fish.

Deficiency signs include: anorexia; lethargy; *feebleness; interrupted swimming; *handling intolerance; exophthalmia; xerophthalmia; *depigmentation; *pale gills; gill membrane adhesion; clubbed gills; blistering of skin; hemorrhage of skin and fins; lordosis; *anemia; reduced hematocrit and hemoglobin; impaired erythropoiesis; erythrocyte fragility and fragmentation; increased immature erythrocytes; erythrocyte size and shape variation; anisocytosis; poikilocytosis; increased plasma protein; diminished function of pituitary and islets of Langerhans; epicarditis; lipid peroxidation; liver damage; ceroid deposition in liver, spleen, and blood vessels; *yellow-orange, fatty liver; enlarged gall bladder; hemosiderin deposition in spleen, pancreas, and kidneys; lesions of pancreas and kidney; exudative diathesis; *ascites; decreased body protein and increased moisture, increased or decreased fat; decreased ovarian protein and fat and increased moisture; decreased linoleic acid content of all tissue fat; decreased myosin and actomycin; muscular dystrophy and lesions; impaired reproduction; *low survival of eggs and larvae; increased susceptibility to disease; low growth efficiency; low growth rate; and mortality.

Overdosage (≥2500-5000 mg/kg) signs include: decreased hematocrit; decreased linoleic acid in liver; depressed immunity; fatty liver; liver atrophy; toxic liver reaction; low growth rate, and mortality. Hypervitaminosis could occur with as little as 1000 mg/kg (Lall and Olivier, 1993).

Menadione, or other forms, (K) is essential for the formation of prothrombin and proconvertin, which are necessary for blood clotting; it is involved in electron transport and oxidative phosphorylation. The requirement increases with administration of sulphonamides or antibiotics. Vitamin K is degraded by heat, ultraviolet light, oxygen, moisture, strong oxidizing or reducing conditions, and extended storage. It was much less stable than thiamin, pyridoxin, folacin, and vitamin A during extrusion at 80-120°C (Haaland et al., 1993). A level of 10-20 mg menadione/kg diet was recommended, but a range of 10-40 mg/kg is used in rainbow trout feeds (Steffens, 1989). Growth of Atlantic cod (1.7 g initially) might have been reduced by 21.5 mg/kg menadione sodium bisulphite compared with 9.8 mg/kg (Grahl-Madsen and Lie, 1997). I suggest that 20-50 mg menadione/kg or equivalent be added to feeds for typical marine fish.

Deficiency signs include: *hemorrhage of eyes, skin, fins, gills, vascular tissue, and viscera; *increased blood clotting time; anemia; decreased hematocrit; lipid peroxidation; liver atrophy; lordosis, scoliosis; and muscular dystrophy.

Minerals

At least 27 (including the four main bulk elements carbon, hydrogen, nitrogen, and oxygen) minerals are essential for animals (Table 9-6; Underwood, 1977). Most or all of these are required by fish (NRC, 1981; Watanabe et al., 1988, 1997; Lall, 1989; Steffens, 1989; Lovell, 1989). Fish have some ability to substitute an abundant related mineral for a deficient one, which can handicap tests for essentiality and minimum requirements. Unlike terrestrial animals, fish can obtain some of the necessary minerals from the water in which they live. Because salt water has a much higher mineral content than fresh water, deficiencies are less likely for fish reared in salt water. Practical feeds tend to have a good supply of minerals, but semipurified feeds often need fortification. Ingredients such as white fish meal, which have mineral imbalances, can promote deficiencies. High dietary ash content (especially with high calcium) increases the need for supplementation of trace minerals. Marine fish reared in salt water on a practical feed containing a moderate amount of ash are relatively safe. With practical feeds, most of the mineral requirement will be provided by other feed ingredients and culture water, but some (marked with asterisks) might have to be added to the feed, especially if animal content is low. Biologically available forms are preferred. Marine algae has been proposed as a source of minerals for fish feeds (Fabregas and Herrero, 1986). Algae also is a protein source for some species and might contribute unidentified or unknown growth factors (Nakagawa et al., 1987). Sugiura et al. (1998) found that adding 5% citric acid to a feed for rainbow trout (63% herring meal) increased the apparent availability of calcium, phosphorus, magnesium, iron, manganese, and strontium.

Calcium is a major component of the skeleton (about 37% of bone); it has roles in muscle and nerve activity, in blood clotting, enzyme activation, vitamin metabolism, and membrane permeability; a large amount is stored in the skin and scales. Fish take up calcium from culture water through the gills and digestive tract. Adequate dissolved calcium increases tolerance to stress and higher temperature. It also can reduce the effects of toxic substances such as aluminum; however, dissolved aluminum can reduce uptake and storage of calcium. If dissolved calcium is low (fresh or brackish water), more will be required from the diet. Even when fish are raised in soft fresh water, the calcium requirement usually will be met by a practical feed without supplementation, because ingredients like fish, crustacean, mollusc, and meat byproduct meals are good sources. Rainbow trout need 2 g Ca/kg diet, channel catfish 4.5 g/kg, blue tilapia 7 g/kg, Japanese eels 2.7 g/kg, and red sea bream 3.4 g/kg. Redlip mullet (El-Zibdeh et al., 1996a), tiger puffers (Hossain and Furuichi, 1998), and possibly red sea bream (Sakamoto and Yone, 1973) could not obtain enough calcium from seawater alone, but yellow croaker (29-90 g) seemed to have that capability (El-Zibdeh et al., 1995). Calcium was not available to tiger puffers from tricalcium phosphate but was from calcium lactate (Hossain and Furuichi, 1998).

Deficiency signs include: anorexia; decreased bone ash; low growth efficiency; low growth rate. Overdose signs include growth reduction in young chinook salmon (73 g Ca/kg and 51 g total P/kg) and channel catfish (20 g Ca/kg).

Phosphorus is a major skeletal component (about 16% of bone); it also occurs in nucleic acids, energy-rich phosphates, phospholipids, and many enzymes; it is important in carbohydrate, lipid, and protein metabolism, muscle and nerve function, pH maintenance in body fluids, and essentially all cellular energy production. Only a small amount of phosphorus is obtained from culture water; most comes from the diet. Ingredients like fish, crustacean, mollusc, and meat byproduct meals are good sources. Fish with stomachs can use less soluble forms (e.g., dibasic or tribasic calcium phosphate) better than stomachless fish, which need more soluble forms (e.g., monobasic sodium or potassium phosphate). Stomachless fish have trouble obtaining phosphorus from fish meal and could require additional ingredients or supplementation. Availability of phosphorus from casein, yeast, and monobasic calcium phosphate is in the range 90-97% for both rainbow trout (with stomach) and common carp (stomachless). Because high dietary calcium can interfere with uptake of dietary phosphorus, calcium should be limited, if possible. Excess dietary phosphorus will increase the amount excreted, and could contribute to eutrophication of culture water or discharge sites. Atlantic and chum salmon require 5-6 g available P/kg diet, rainbow trout 6 g/kg, common carp 4-7 g/kg, channel catfish 4.2-4.5 g/kg, Nile tilapia 9 g/kg, Japanese eels 5.8 g/kg, and red sea bream 6.8 g/kg. Striped bass needed 5.8 g/kg for optimal growth and bone mineralization (Dougall et al., 1996). Rodehutscord (1996) recommended 0.25 g available phosphorus/MJ DE for rainbow trout. Dietary mineral ratios could be important; red sea bream grew best with a Ca/P ratio of 1/2 (Yone, 1976).

Deficiency signs include: anorexia; head and bone deformities; possibly reduced hematocrit; decreased inorganic phosphates in blood; possibly increased cholesterol and bilirubin in serum; increased serum alkaline phosphatase and vertebral hydroxyproline; decreased liver glycogen; increased gluconeogenic enzymes in liver; increased fat and decreased water in muscles, vertebrae, liver, and viscera; low bone mineralization; decreased bone ash; decreased phosphorus and calcium in body and vertebrae; curved enlarged spongy vertebrae; low growth efficiency; and low growth rate.

Overdose signs (51 g total P/kg with 73 g Ca/kg for chinook salmon; 19 g available P/kg for common carp) include slow growth rate.

Magnesium is important in phosphorylation reactions, enzyme systems, metabolic processes, and neuromuscular transmission. At last some fish can obtain a large part of their requirement by drinking culture water, especially salt water. In fresh water, rainbow trout need 200-700 mg Mg/kg diet, depending on the water Mg content. Common carp require 400-500 mg/kg, channel catfish at least 400 mg/kg, guppies 540 mg/kg, and Japanese eels 400 mg/kg. Sakamoto and Yone (1979) reported that magnesium supplementation was unecessary for red sea bream if their diet contained at least 120 mg/kg; the fish were equally healthy with 120 or 660 mg/kg. Yellow croakers (28-83) seemed to obtain enough calcium from either their feed or seawater containing 53 mg/L (El-Zibdeh et al., 1996b). Practical feeds should provide enough magnesium. Ordinarily, only purified feeds need supplementation.

Deficiency signs include: anorexia; lethargy; muscle flaccidity; possibly muscle cramps; epileptiform convulsions; cataracts; pathology of muscles, pyloric cecae, and gill lamellae; decreased bone ash; increased calcium content of muscle, kidneys, and skeleton; nephrocalcinosis; reduced magnesium content of whole body, plasma, muscle, and bones; increased sodium content of muscles and skeleton; bone deformities; spinal curvature; low growth efficiency; low growth rate; and mortality.

Potassium is important in the stimulation of the nervous and muscular systems, in osmoregulation, and in maintenance of the acid-base balance; it is the main intracellular cation. Fish take up appreciable amounts of dissolved potassium through the gills but not through the digestive tract. Plant feedstuffs tend to provide more potassium than animal sources. In seawater, red sea bream do not seem to require dietary potassium. Pacific salmon needed 8000 mg/kg (Hardy, 1991). Practical feeds usually contain ample potassium.

Deficiency signs include: anorexia; decreased body potassium; increased body sodium and magnesium; low growth efficiency; and low growth rate.

Sodium is important in osmoregulation and in maintenance of the acid-base balance; it is the main extracellular cation. Animal protein sources supply more sodium than plant sources. Dietary content of 1-2.2 g Na/kg is sufficient for rainbow trout. Practical feeds usually contain enough.

Chlorine is very important in osmoregulation; chloride is the main extracellular anion. Freshwater fish can take up NaCl from hypotonic solutions through the gills, and small amounts are excreted in the urine. Channel catfish did not need a salt supplement. Marine fish continually absorb NaCl by diffusion and excrete it through the gills. Fish reared in fresh water can tolerate more handling and temperature stress if about 10 ppt salt is added to the water. Addition of salt to feeds for tilapia (100 g NaCl/kg diet), salmon, and rainbow trout raised in fresh water improved their survival when they were moved to salt water. Animal protein sources are good suppliers of chlorine. Practical feeds usually contain enough.

For young coho salmon (freshwater stage), an overdose of 30 g/kg dietary NaCl reduced growth rate; however, growth efficiency and growth rate of 2-yr-old Atlantic salmon (saltwater stage) were not appreciably affected by up to 120 g/kg salt. Between 50 and 110 g/kg dietary salt, growth rate of rainbow trout began to decrease.

Sulfur is important as a component of the amino acids methionine and cystine. A very small amount of dissolved sulfate is absorbed through the gills; more is absorbed from food but little is used. Requirements and deficiency signs in fish have not been established. Deficiency is unlikely if the requirement for sulfur-bearing amino acids is met.

Iron* is a component of blood pigments and facilitates oxygen transport; it is involved in oxidation-reduction reactions and electron transport. Fish can take up dissolved iron through the gills, but most is absorbed from food in the intestines. Phytic and tannic acids can reduce absorption of iron from food, but ascorbic acid improves it. Rainbow trout probably need more than 39 mg Fe/kg diet, common carp 200 mg/kg, young channel catfish 30 mg/kg, Japanese eels 170 mg/kg, and red sea bream 150 mg/kg as ferrous or ferric chloride and 200 mg/kg as ferric citrate. The minimum requirement for Atlantic salmon was 73 mg/kg (Helland et al., 1991). Yellow croaker (28-83 g) given a feed with only 26.5 mg/kg iron were anemic, but with 145.2 mg/kg they were normal (El-Zibdeh et al., 1996b). Ukawa et al. (1994) reported that 4 g/kg iron proteinate could prevent sub-normal blood characteristics in yellowtail eating feeds with 45% fish meal and 20% soy protein concentrate. Commercial salmonid feeds usually contain 150-800 mg/kg (Lall and Olivier, 1993). Feeds with reasonable amounts of animal ingredients will provide enough iron; if animal ingredients are low, supplementation might be needed.

Deficiency signs include: *microcytic hypochromic anemia; decreased serum iron, transferrin saturation, hemoglobin, hematocrit, mean corpuscular hemoglobin (MCH), MCH concentration (MCHC), mean corpuscular volume (MCV); yellowish-white liver; low growth efficiency; and low growth rate.

Overdose (1380 mg Fe/kg diet) signs include: anorexia; diarrhea; liver cell damage; low growth efficiency; low growth rate; and mortality.

Zinc* is important in different enzyme systems, in carbohydrate, lipid, and protein metabolism, and in gonad development. Fish can absorb dissolved zinc through the gills and some is useful, but 1 mg/L or less dissolved zinc can be toxic. Food is the main source. Zinc is better absorbed from animal than from plant sources. High ash, calcium (e.g., in white fish meal), or phytic acid (and other plant substances) can interfere with zinc absorption.

Depending on phytic acid content of the feed, rainbow trout and common carp need 15-

100 mg Zn/kg diet and channel catfish up to 200 mg/kg. For maximum growth of channel catfish, addition of zinc to feeds naturally containing 50-59 mg/kg was not necessary, but zinc content of bone was higher with 200 mg/kg than with 0 or 45 mg/kg (Li and Robinson, 1996). First-feeding Atlantic salmon needed 57-97 mg/kg (Maage et al., 1993). Shearer et al. (1992) suggested that a minimum of 100 mg/kg zinc is suitable in Atlantic salmon feeds containing up to 17.5% ash. Japanese eels needed 50-100 mg/kg (Arai, 1991) and red sea bream 24 mg/kg. Growth of turbot was slightly improved by raising dietary zinc from 100 mg/kg to 1050 mg/kg. A level of 60.1 mg/kg was sufficient but 24.5 mg/kg was deficient for optimal bone mineralization in yellow croaker (29-90 g)(El-Zibdeh et al., 1995). Feedstuffs usually do not provide enough zinc, and supplementation is recommended.

Deficiency signs include: anorexia; cataracts; necrosis and erosion of skin and fins; decreased serum alkaline phosphatase; decreased zinc in serum, eyes, intestines, hepatopancreas, kidneys, and vertebrae; increased iron and copper in intestines and hepatopancreas; decreased zinc and calcium in bones; decreased copper and manganese in vertebrae; increased body iron; impaired protein and fat digestion; decreased body protein and fat content; increased body water content; dwarfism; low growth rate; and mortality.

Growth was reduced in rainbow trout by an overdose of 2000 mg Zn/kg diet (with low calcium of 0.1 mg/kg) and in common carp with 294 mg Zn/kg.

Manganese* is a cofactor or component of different enzymes and is necessary for normal carbohydrate, lipid, and protein metabolism and normal growth. Fish obtain it more easily from food than from water. High dietary calcium and phosphorus interfere with manganese absorption. The young of rainbow trout need about 12-13 mg Mn/kg diet, common carp about 12-13 mg/kg, channel catfish 2-3 mg/kg, and red sea bream 18 mg/kg. Salmon and trout broodstock need at least 50 mg/kg. Lorentzen et al. (1996) recommended adding 15 mg/kg to Atlantic salmon parr feeds. A level of 5.0 mg/kg was sufficient but 0.9 mg/kg was deficient for optimal bone mineralization in yellow croaker (29-90 g)(El-Zibdeh et al., 1995). Feedstuffs usually do not provide enough manganese, and supplementation is recommended.

Deficiency signs include: anorexia; equilibrium loss; cataracts; skeletal and tail abnormalities; decreased manganese in liver, gonads, and skeleton; decreased zinc in gonads and vertebrae; decreased calcium in vertebrae; decreased magnesium, potassium, sodium, zinc, copper, and phosphorus in liver; decreased activity of Cu-Zn-superoxide-dismutase in liver; decreased activity of Mn-superoxide-dismutase in liver and heart muscle; decreased egg hatching rate; dwarfism; low growth efficiency; low growth rate; and mortality.

Copper* is necessary for the formation of hemoglobin; it is a component of important enzymes, and plays a role in metabolism and iron absorption. Rainbow trout need 3 mg/kg, common carp 3 mg/kg, channel catfish 5 mg/kg, and red sea bream 5 mg/kg. The minimum requirement for Atlantic salmon was 6 mg/kg (Helland et al., 1991). A level of 4.1 mg/kg was sufficient, but 0.1 mg/kg was deficient for optimal bone mineralization in yellow croaker (29-90 g)(El-Zibdeh et al., 1995). Many feedstuffs have high copper contents, sometimes from contamination during processing. Supplementation usually is not necessary.

Deficiency signs include: anemia; cataracts; decreased copper in liver, vertebrae, and body; decreased Cu-Zn-superoxide dismutase in liver; decreased cytochrome c oxidase activity in heart; and slow growth.

As little as 0.1 mg/kg dissolved copper can be toxic. Rainbow trout tolerated as much as 600 mg Cu/kg diet; only with more than 730 mg/kg did they exhibit overdose signs including: anorexia, low growth efficiency, and low growth rates. However, as little as 16 mg/kg added to channel catfish feeds increased copper in the liver, caused slight anemia, and reduced growth efficiency and growth rate.

Iodine* is necessary for the formation of the metabolism-regulating thyroid hormones thyroxine and triiodothyronine; it is important for cellular oxidation reactions, neuromuscular and endocrine function, circulation, and growth. In salt water fish probably can obtain enough dissolved iodine through the gills, but in fresh water deficiencies can occur. A level of 4.5 mg/kg was necessary to protect Atlantic salmon smolts from bacterial kidney disease. Chinook salmon need 0.6-1.1 mg/kg to reach maximum storage levels. The requirement for rainbow trout probably is in the range 0.6-2.8 mg/kg. Some marine fish meals and seaweeds contain generous amounts of iodine. Other fish meals, animal meals, and plant ingredients contain little iodine. Because of high variability of iodine in feedstuffs, I suggest that 4 mg/kg be included in feeds for typical marine fish.

Deficiency signs are thyroid hyperplasia (goiter), low growth rate, and mortality.

Cobalt* is a necessary component of cyanocobalamin. Fish can obtain it from both water (through the gills) and from food. The requirement for fish seems to be about 0.05 mg Co/kg diet and this can usually be provided by a practical feed, particularly if cyanocobalamin is supplemented.

Deficiency signs are the same as for cyanocobalamin. Overdose (0.1-5 mg Co/kg diet) signs include hemorrhage of the digestive tract and abnormal white blood cells.

Selenium* works together with vitamin E; it is important in glucose and lipid metabolism; as a component of glutathione peroxidase (which destroys hydrogen peroxide and hydroperoxides) it protects against oxidation of cells and membranes; it can reduce mercury and cadmium toxicity. Fish can obtain it from water through the gills and from food. Seawater contains less dissolved selenium than freshwater (about 0.09 μg/L). Atlantic salmon need 0.15 mg Se/kg diet, rainbow trout 0.2-0.4 mg/kg, common carp 0.05-0.1 mg/kg, channel catfish 0.25 mg/kg, and Japanese eels 0.3-0.5 mg/kg. Most fish probably require 0.2-0.5 mg/kg. Selenium content of feedstuffs is variable. If enough good sources such as marine fish meals (except tuna) and byproducts are included in the feed, supplementation is probably unnecessary. If not, supplementation is recommended.

Deficiency signs include: anorexia; lethargy; ataxia; cataracts; hemorrhages; anemia; decreased hematocrit; increased hemolysis; increased pyruvatekinase activity in plasma; exudative diathesis; ascites; pericardial edema; myocardial degeneration; nerve and liver lesions; decreased glutathione peroxidase activity in plasma and liver; increased glutathione transferase activity in liver; decreased selenium in liver, plasma, and other tissues; decreased vitamin E in liver and other tissues; pancreatic abnormalities; muscular dystrophy; low growth efficiency; low growth rate; and mortality.

Dissolved selenium is toxic in the range 40-130 μg/L. Overdose (13 mg Se/kg diet for rainbow trout, 15 mg/kg for channel catfish) signs include: spiral swimming, increased calcium and magnesium in kidneys; nephrocalcinosis; increased magnesium in liver; low growth efficiency; low growth rate; and mortality.

Chromium* (trivalent, Cr^{3+}) activates several enzymes and is necessary for normal carbohydrate and lipid metabolism. Deficiency signs in fish are not yet established. Because of possible variation in feedstuffs, supplementation is recommended.

Fluorine plays a role in bone and scale formation. Deficiency signs in fish have not been well established. A level of 4.5 mg/kg was necessary to protect Atlantic salmon smolts from bacterial kidney disease (Lall and Olivier, 1993). Fish seem to tolerate a wide range of dietary fluorine. Krill meal can contain as much as 2000-3000 mg/kg.

Boron supplementation improved growth and survival of carp. Deficiency signs in fish have not been established.

Molybdenum functions in enzymatic processes. Supplementation improved growth and survival of carp. Deficiency signs in fish have not been established.

Silicon deficiency has been documented in some animals but not fish.

Arsenic, **nickel**, **tin**, and **vanadium** deficiencies have been documented in some animals but not fish. Their functions are not clear. **Bromine** and **strontium** might be needed.

Aluminum has been said to benefit some animals but has not clearly been proven essential (Poston, 1991). An aluminum deficiency was implicated in dwarfism of Japanese eel. Japanese eels were reported to need 15 mg/kg (Arai, 1991). Dietary aluminum in the range 0-2000 mg/kg did not affect survival, growth, or conversion of Atlantic salmon, but 2000 mg/kg caused increased body fat and 1000-2000 mg/kg caused slightly reduced ash (Poston, 1991). Toxicity of dissolved aluminum in acidic fresh water occurred at less than 50 μg/L.

Essentiality of **bromine**, **cadmium**, **lead**, or **strontium** has not been confirmed (Lall, 1989). **Titanium** might be needed by some species (Dawes, 1989).

Pigments

Pigments are necessary in the diet of some fish for full color development, and they have a positive effect on reproductive efficiency and other functions (NRC, 1981, 1983). Carotenoids could protect against ultraviolet radiation, provide vitamin A activity, act as antioxidants, increase resistance to high ammonia and low oxygen, increase growth, enhance communication (through skin coloration), accelerate maturation, increase egg production, and improve egg quality (Meyers, 1994; Torrissen and Christiansen, 1995). Astaxanthin or other carotenoids might be better sources of vitamin A than vitamin A esters for marine fish larvae. Torrissen and Christiansen (1995) recommended including 1% astaxanthin or canthaxanthin in all fish feeds.

Carotenoid sources for use in fish feeds include: synthetic astaxanthin, synthetic canthaxanthin, synthetic lutein, krill, other crustaceans, palm oil, *Spirulina*, marigold meal, red pepper, yellow corn, corn gluten, alfalfa, ipil-ipil leaf, other plants, the yeast *Phaffia rhodozyma*, and possibly the microalga *Haematococcus pluvialis* (Meyers, 1994; Meyers and No, 1995; Wan et al., 1995).

The red carotenoid pigment astaxanthin is largely responsible for the red coloration of the skin and flesh of salmonids (Arai et al., 1987; Gentles and Haard, 1991) and the skin of red sea bream and other species (Fujita et al., 1983b; Ibrahim et al., 1984). Increased hatching rate, larval survival, liver health, vitamin A activity, and possible antioxidant protection for lipids have been cited as benefits of dietary astaxanthin (Sanderson and Jolly, 1994).

The red carotenoid pigment canthaxanthin is effective for providing color in salmon but not in red sea bream (NRC, 1981, 1983; Hardy, 1989). Although some species can convert different pigments into astaxanthin (NRC, 1983), others, including red sea bream, cannot (Fujita et al., 1983b; Ibrahim et al., 1984). For such species, astaxanthin in the diet is necessary for full color development. To promote full development of the preferred red coloration, extra astaxanthin usually is provided to red sea bream in the form of krill or red shrimp during the few months before harvest (Yahama Fishery Journal, 1990). Because exposure to excessive ultraviolet light causes caged red sea bream to darken (with melanin), the cages often are shaded for at least 2-3 months.

The yellow carotenoid pigment tunaxanthin is largely responsible for the yellow and blue-green coloration of the skin of yellowtail and other species (Fujita et al., 1983a). At least some fish (e.g., red sea bream and yellowtail) can convert astaxanthin to tunaxanthin (Fujita et al. 1983a, 1983b; Ibrahim et al., 1984).

Therefore, astaxanthin seems to be the most useful pigment for marine fish feeds. Some sources of astaxanthin that have been used in fish feeds are: crayfish, crabs, various shrimps, krill, mysids, yeast, and algae. At least 20% (wet weight) of crustacean is necessary in salmonid feeds (Hardy, 1989). With red sea bream, 56% fresh krill was more effective than 20% krill meal, possibly because of degradation of the pigments during processing (Ibrahim et al., 1984). Krill oil (5% of diet) and especially astaxanthin diester (0.15% of diet) extracted from krill oil were very effective for red sea bream (Fujita et al., 1983b). Two percent krill oil was nearly as effective as a diet of fresh sand lance for yellowtail (Fujita et al., 1983a). Astaxanthin diester was stable for at least a year when stored properly (Miki et al., 1983). Concentrated carotenoid extracts (3-7% of diet) from crayfish and crabs have been used for salmonids (Hardy, 1989). Consumption of astaxanthin-supplemented feeds for 4-8 weeks was enough for full skin color development in rainbow trout, yellowtail, and red sea bream (Fujita et al., 1983a, 1983b; Ibrahim et al., 1984; Gentles and Haard, 1991). Feeds containing astaxanthin or canthaxanthin might have to be given to Atlantic salmon for a year before harvest to develop full flesh color (Helland et al., 1991). To enhance skin color in farmed ayu, 30 g *Spirulina* (containing zeaxanthin) and 40 mg lutein per kg feed can be included in formulations (Kanazawa, 1991a).

Carotenoids usually are included at 5-10% of the feed for enhancing pigmentation of rainbow trout and Pacific salmon; a minimum muscle content of 0.4-0.5% is sufficient for good color (Meyers, 1994). When rainbow trout (especially) and salmon are growing, more carotenoids tend to be deposited in the skin. During maturation, carotenoids tend to shift from muscle to skin to eggs. Wild salmon eggs contain 0.21-1.01% astaxanthin, which, along with canthaxanthin, is depleted during yolk absorption—possibly being used as provitamin A.

Egg buoyancy, hatching rate, percentage of normal larvae, and hatchery production of red sea bream were improved by the addition of raw krill, krill oil, or 2 mg astaxanthin to the broodstock diet (Watanabe and Miki, 1993). Krill oil was more effective than defatted krill meal. The beneficial components probably were choline in the polar lipids and astaxanthin in the nonpolar lipids.

Dietary pigments are not always beneficial. An excess of xanthophylls from corn gluten and alfalfa meal can cause an undesirable yellow color in trout and catfish flesh (NRC, 1981).

Growth was similar for rainbow trout given feeds containing 10% brewer's yeast or *Phaffia rhodozyma* (which contains 3000 mg/kg astaxanthin), but muscle carotenoid content was near zero with brewer's yeast and more than 5 mg/kg with *Phaffia* (Sanderson and Jolly, 1994).

Algae

Dietary algae is good for the health of carnivores as well as plant-eating fish. Growth was improved in nibbler with 5% *Chlorella, Spirulina,* or *Ulva* meal (Mustafa and Nakagawa, 1995). Growth and conversion were better in red sea bream with 5% *Ascophyllum, Porphyra, Ulva,* or *Spirulina* meal. Recovery from handling and anesthesia were improved for 0.8-g red sea bream by including 2.5% *Ascophyllum nodosum* in their feed (Nakagawa et al., 1997). Dietary algae is a potential source of protein and fat (Wahbeh, 1991) and also is said to improve digestion, liver function, hematological characteristics, carcass quality, and resistance to anesthesia, hypoxia, removal from water, starvation, and disease. The wrong kind, or too much of the right kind, can be detrimental. Growth and FCR were worse for thicklipped mullet (1-9 g) when 16.5% *Porphyra purpurea* was included in their feed, replacing 6 of 68% herring meal, 7.5 of 15% corn starch & dextrin, and 3.4 of 6.8% cellulose (Davies et al., 1997b).

Immunostimulants

Consumption by fish of substances such as yeasts (e.g., *Candida utilis*; baker's yeast, *Saccharomyces cerevisiae*), β-glucans (from yeast cell walls), astaxanthin, alginate, certain synthetic peptides, betaine, and deacylated chitin have been shown to increase non-specific immunity in combating pathogens such as *Aeromonas salmonicida, Vibrio anguillarum,* and *Vibrio salmonicida* (Raa et al., 1992; Siwicki et al., 1994; Christiansen et al., 1995; Nordmo et al., 1995; Verlhac et al., 1996). Feeding β-glucan or baker's yeast to channel catfish enhanced non-specific immunity but did not increase resistance to *Edwardsiella ictaluri*, possibly because the dosage was not high enough or the stimulated defense mechanism was not appropriate (Duncan and Klesius, 1996).

Non-nutritive Components

Fiber is not required by fish, but small amounts might aid digestion slightly by increasing gut residence time (Steffens, 1989). Too much fiber dilutes nutrient density and can obstruct digestive enzyme activity or decrease absorption of nutrients (Leary and Lovell, 1975). Excessive fiber content also increases the volume of feces and has the potential to reduce water quality. Early juveniles probably would be the most affected by excessive fiber. Omnivores and herbivores are less likely to be adversely affected by increased dietary fiber than carnivores. Growth, conversion, and nutrient uptake were enhanced in red sea bream when 5% of the seaweed wakame (*Undaria pinnatifida*) was added to the diet, probably because of a higher fiber level, which increased digestion time (Yone et al., 1986a, 1986b). Protein and lipid digestibility in European sea bass were not diminished with levels of the feed binder sodium alginate as high as 8% (Spyridakis, et al., 1989); however, the sum of cellulose and alginate in the test feeds apparently was constant. All six alginates tested at 5% in moist (48% moisture) feeds for rainbow trout significantly reduced protein and fat digestibility (Storebakken and Austreng, 1987). Small amounts of carboxymethyl cellulose (CMC) have been variously reported to reduce nutrient availability and fish growth in salmonids (2%, NRC, 1981), or to increase digestion time and nutrient absorption and possibly growth and feed efficiency in red sea bream (Furuichi et al., 1983). Although some species can tolerate a fiber content as high as 8-20% (Leary and Lovell, 1975; Steffens, 1989), for carnivorous fish, fiber probably should be kept at 4% or less to minimize inhibitory effects. Low levels of fiber can and must be tolerated in a practical feed, because even high-quality ingredients contain some.

Ash varies mainly with bone or shell content of animal ingredients. Some can be tolerated, but ash reduces the nutritional density of feed. Ash content should be kept low, especially for small fish that need large rations to sustain high growth rates. High ash can interfere with absorption of zinc and other nutrients. It also contributes to phosphorus pollution.

Moisture in feed higher than 11% can contribute to loss of vitamins, oxidative rancidity of lipids, and growth of fungi and bacteria (NRC, 1983; Lall, 1989). The industry standard is 10% maximum moisture in "dry" feeds.

Spoilage of ingredients or feeds by contamination with bacteria, fungi, or yeast can be reduced by sanitary processing and storage procedures (Hardy, 1989). More than 20 antimicrobial compounds have been used in the feed industry to retard growth of bacteria and fungi on semimoist and moist feeds. These include propionates, ascorbates, sorbates, sodium benzoate, ascorbic acid, benzoic acid, sorbic acid, and citric acid. Growth of aerobic microorganisms and oxidation of fats can be reduced by removing oxygen from the feed container (e.g., packing under nitrogen, carbon dioxide, or partial vacuum). Sometimes water activity in semimoist feeds is controlled by addition of substances such as sucrose, sodium chloride, glycerol, and propylene glycol.

Digestive enzymes have been used in feeds for terrestrial livestock and might also be useful for fish. However, in one study (Carter et al., 1992), dietary α-amylase had no significant effect on consumption, growth, or FCR of Atlantic salmon parrs (12-21 g).

Binders stabilize a feed long enough for the fish to consume it. Some examples are starches, gelatin, glutens, collagen, carageenan, alginates, agar, guar gum, chitosan, lignin sulfonate, hemicellulose, carboxymethyl cellulose, sodium bentonite, and calcium bentonite (Hardy, 1989).

Palatability factors in feed for carnivorous fish decrease as the proportion of animal ingredients is reduced, but they can be replaced by a combination of attractants, arrestants, incitants, and/or stimulants (Mackie and Mitchell, 1985). Most marine animal and good-quality land animal feedstuffs are naturally attractive. Pellets containing 18% fish meal and 13% fish oil that were coated with 10% liquid krill hydrolysate were eaten in larger amounts by rainbow trout than uncoated pellets (Oikawa and March, 1997). Notable examples of additives that promote feeding are: free amino acids for Anguillidae, Cobitidae, Salmonidae, Serranidae, Carangidae, Haemulidae, Sparidae, Siganidae, Scombridae; glycine betaine for Carangidae, Siganidae; glycine betaine plus amino acids for Haemulidae, Sparidae, Soleidae, and Tetraodontidae; nucleotides for Carangidae, Haemulidae, Scombridae, Bothidae; lipids for Carangidae; nitrogenous bases for Cobitidae; spices for Cobitidae, Carangidae (Harada, 1985a, 1985b; Mackie and Mitchell, 1985; Ishida and Hidaka, 1987; Harada, 1990; Shimizu et al., 1990). Alkaline and neutral attractants like glycine, proline, taurine, valine, and betaine seem better for carnivores and more acidic attractants like aspartic acid and glutamic acid for herbivores (Goddard, 1996). Daily ration of largemouth bass (4-9 g) was inversely related to soybean meal content (4.8 to 1.3% DR with 0-60% SBM)(Kubitza et al., 1997). Inosine-5-monophosphate at 2800 mg/kg increased feed intake by 23%, but 10% fish meal increased intake by 58%. Ineffective treatments were: 1400-5600 mg/kg inosine, 9351 mg/kg amino acid mixture, 6100 mg/kg betaine, and 9351 mg/kg amino acids + 6100 mg/kg betaine. Although betaine (an organic base from crustaceans) and inosine did not work for largemouth bass, they seem to be effective for some marine fish. Of 36 vegetable extracts tested (Harada et al., 1996), onion was the most attractive for juvenile yellowtail, followed by Japanese radish, cucumber, carrot, and perilla (mint family). Cabbage, leaf lettuce, rakkyo (lily family), turnip, Irish potato, broad bean, soybean, eggplant, celery, and mioga (ginger family) also were attractive. Wasabi was the least attractive.

Bovine, ovine, porcine, and piscine growth hormones as well as thyroid hormones, insulin, and anabolic steroids have been shown to increase growth in fish (Donaldson et al., 1979; Higgs et al., 1982; Matty and Lone, 1985; Weatherley and Gill, 1987b). Weatherley and Gill (1987a) suggested that good quality species that are unpopular as food because of small size or slow growth might be feasible for aquaculture if growth could be increased enough. Growth hormone has been isolated from sharks, sturgeon, eels, carp, salmon, yellowtail, tilapia, and tuna (Kawazoe, et al., 1988). Growth and sex hormones accidentally occur in fish meals, especially low temperature meals (Hardy, 1989; Feist and Schreck, 1990), and this might be one of the reasons for the superiority of fish meal as an ingredient in fish feeds. Chua and Teng (1980) suggested that growth-promoting substances could be added to feeds for the Malabar grouper. This species grew from 54 g to 500 g in 81 d on feed containing nitrovin and in 138 d without nitrovin. The hormone-like compound dimethyl-B-propiothetin occurs in various aquatic organisms, including algae and fish. Dietary DMPT enhanced growth and conversion of yellowtail, hirame, goldfish, common carp, and red sea bream and also increased swimming strength of the last three species (Nakajima, 1991; Nakajima et al., 1990;). Dietary L-carnitine (100-250 mg/fish/d) increased fat utilization, thus reducing fat deposition and increasing protein deposition (growth) in juvenile European sea bass (Santulli and D'Amelio, 1986; Santulli et al., 1988). Dissolved L-carnitine (1.5 g/L) had a similar growth-stimulating effect on larval European sea bass (Santulli et al., 1990). Growth-promoting substances might have an application in fish culture, but only to the extent that the human consumer is not adversely affected.

Antioxidants are necessary. Oxidative rancidity of lipids, even at low levels, can reduce survival and growth of fish by affecting several organ systems (Taniguchi, 1983a; Ketola et al., 1989; Roberts and Bullock, 1989; Steffens, 1989; Sekiya et al., 1991). The natural antioxidants tocopherol and selenium are present in feeds, but they usually are supplemented with synthetic ones. The synthetic antioxidants most commonly added to fish feeds (NRC, 1983, Lovell, 1989; Hardy, 1989) are 150 mg ethoxyquin/kg feed, and 0.02% of the feed fat content of BHA (butylated hydroxyanisole) or BHT (butylated hydroxytoluene).

Carotenoids also help (Okimasu et al., 1992). Tocopherol seems to be the most effective antioxidant in fish feeds (Steffens, 1989).

BROODSTOCK NUTRITION

Factors of particular importance include fatty acid composition, lipid content, protein quality and content, vitamin E, vitamin C, and pigments (Lie and Mangor-Jensen, 1993).

Broodstock feeds for Atlantic cod can influence fat and pigment components in eggs (Lie at al., 1993; Grung et al., 1993), possibly affecting their viability. Cardiac disorders (i.e., muscle necrosis, low stress resistance) has been observed in salmon fed high levels of ω6PUFA even when ω3PUFA were sufficient (Sargent, 1995). Too high an ω6/ω3 ratio in the diet of female broodfish might also affect stress resistance of their larvae. Sargent (1995) also suggested that the ratio of 2/1 DHA/EPA measured in most fish eggs should be appropriate for broodstock feeds.

Broodfish diet effects on offspring also have been investigated for European sea bass (Cerdà et al., 1994a, 1995). Broodfish given a feed containing 34% protein, 14% lipid, and 32% carbohydrate produced eggs with lower buoyancy and hatchability than broodfish given a feed with 51% protein, 13% lipid, and 10% carbohydrate (Cerdà et al., 1994b). Egg viability and hatchability were higher for broodfish fed more ADA, EPA, and DHA (Navas et al., 1995).

Orchid and sunrise dottyback broodfish were heavily fed chopped clams, squid, and krill, plus *Artemia* and dry feeds (Brons, 1996).

Two-year-old reared yellowtails produced twice as many eggs (after HCG injection) when they were fed Oregon-type soft moist pellets, but egg quality was better with soft-dry pellets; both quantity (2% of moist pellet number) and quality were much lower for those given frozen raw fish (Watanabe et al., 1996a). Fertilization rates were 57% for SDP, 34% for MP, and 3% for RF. The SDP pellet contained 55% sardine meal, 10% krill meal, 12% wheat flour, 3% wheat gluten, 3% potato starch, and 13% pollack liver oil. Vitamin E, squid liver oil, and carotenoid oil are considered valuable components of yellowtail broodstock feed (Mushiake et al., 1993). Including 10% krill in feeds for yellowtail broodstock resulted in higher egg quality, probably because of carotenoids, but fatty acid and vitamin E contents of eggs also were related to the feed composition (Verakunpiriya et al., 1996). In contrast, Verakunpiriya et al. (1997a) obtained lower hatching (30 & 11 vs 51%) and fewer normal larvae (33 & 12 vs 50%) from eggs of yellowtails given feeds with sardine meal (47, 37, 67%) replaced by higher levels of krill meal (20, 30, 0%).

Red sea bream produced better eggs when their feeds contain krill or krill lipids, astaxanthin, and sufficient vitamin E (Watanabe and Kiron, 1995). Substitution of 7-10% corn oil for marine oil in broodstock feeds for red sea bream resulted in reduced viability of eggs, and a high percentage of abnormal eggs and larvae (Watanabe, 1984b, 1984c). These effects might have resulted from an excess of linoleic acid, (18:2ω6); levels of 28.2-34.0% in bad feeds vs 0.9-3.0% in good feeds (of total fatty acids) represented actual contents of 3.0-5.6% and 0.1-0.4% linoleic acid. ω3HUFA deficiency certainly was a factor; ω3HUFA levels were 0-5.2% of total fatty acids in bad feeds vs 17.1-29.6% in good feeds. Bad feeds contained 0-0.4% EPA, 0-0.4% DHA, and 0-0.9% ω3HUFA. Good feeds contained 1.2-1.5% EPA, 0.9-1.9% DHA, and 2.5-3.4% ω3HUFA. Later, Watanabe et al. (1991c) reported that including vitamin E or krill oil (with no vegetable oils) in the diet of red sea bream improved egg quality.

Chou et al. (1993) recommended adding 4% fish oil (2.97% HUFA in feed) to broodstock feeds for black sea bream for good egg production and quality.

About 1.60% ω3HUFA in broodstock feed for gilthead sea bream resulted in the best combination of egg production and quality (Fernández-Palacios et al., 1995, 1997). Squid protein probably also had a positive effect.

Although *Centropyge* spp. are mostly herbivorous, captive rusty pygmy angelfish spawned well when fed chopped fish, clams, and shrimp (Hioki et al., 1990).

Tamaru et al. (1992) reported higher fertilization rates for eggs from striped mullet held in outdoor ponds (66, 78%) vs those from indoor tanks (50%). The most likely detected factor in egg composition was lower arachidonic acid in eggs from indoor broodstock.

Yellowblotch spinefoot broodstock given 17.4% dietary lipid produced more and better eggs than those given 9.9 or 13.6% (Duray, 1990, 1994b). Broodfish given feeds with cod liver oil + soybean oil + lecithin or cod liver oil alone (9.9-16.5% lipid) spawned more often that those fed less lipid (0.3-7.4%)(Hara et al., 1986).

NUTRITIONAL DISORDERS

The causes and signs of nutritional disorders are discussed by NRC (1981, 1983), Post (1987), Ghittino (1989), Roberts (1989e), Roberts and Bullock (1989), Tacon (1995), and many research reports such as Laurencin et al. (1990). A summary of the most common ones is given here.

Overdoses

These are most likely with vitamin A, vitamin E, copper, selenium, and fluoride (Lall and Olivier, 1993).

Starvation

This can occur when food is not adequate, access to food is unequal, or fish refuse to eat. Starving fish usually are dark, have soft flesh, slender bodies, little abdominal fat, pale gills, and sometimes more parasites. Amino acid deficiency can result in dark color, reduced growth, cataracts, dorsal fin erosion, and spinal deformity. Essential fatty acid deficiency usually causes enlarged pale fatty livers, with anemia and increased mortality. Carbohydrate excess can lead to liver cell degeneration, high liver glycogen, and high visceral fat.

Antinutritional factors and toxins

Occurrence and effects in fish foods are thoroughly discussed in NRC (1981, 1983), Hendricks and Bailey (1989), Lovell (1989), Roberts (1989a), Roberts and Bullock (1989), Steffens (1989), Tacon (1992), Mitchell et al. (1993), Feed Development Section (1994), and Tacon (1995). Correct choice, preparation, storage, and testing of ingredients and feeds will help to avoid problems with these contaminants.

Feed additives such as antioxidants, binders, drugs, and hormones can have harmful side effects. Nitrites and nitrates are added to fish meal as antioxidants. They cause the formation of nitrosamines and should be limited to less than 200 ppm in fish feeds. BHA, BHT, and ethoxyquin can be toxic. Reversible liver damage and enlargement occurred in Japanese eels (1 g, initially) fed 2 mg/kg BHT for 12 weeks (Hwang et al., 1992). Large amounts of chemically-substituted polycellulose binders in feeds could cause a degenerative hepatorenal syndrome with dark skin, liver and kidney damage, and mortality. When photosensitizing drugs or natural compounds that have been ingested are present in the skin, they can be energized by ultraviolet radiation and cause necrosis and ulcers. Erythromycin overdose has caused kidney damage, and sulfonamide overdose has caused reduced growth, damage to liver and kidneys, and visceral arteriosclerosis.

Manmade or caused contaminants that are most likely include heavy metals, pesticides, and solvent residues. These can be avoided by routine testing and, if necessary, rejection of contaminated materials. Pesticide contamination of feed usually is not bad enough to be acutely toxic, but can cause toxemia, necrosis, and hemorrhage of both liver and kidneys, and mortality. Chronic pesticide or heavy metal toxicity usually interferes with growth or reproduction.

Natural contaminants include toxins from fungi (Williams and Blaney, 1992), bacteria, algae, higher plants, or protozoans, antinutrients, and products of oxidative rancidity. Aflatoxins, produced by the fungi *Aspergillus* spp., can be toxic to fish at 0.01 ppb. Results can be cancer of liver, spleen, kidney, or gills, lymphoma, liver necrosis, branchial edema, and widespread hemorrhaging. Spoilage of ingredients can lead to botulism. Large dietary amounts of proteins from some bacteria have resulted in biliary carcinoma. Even low doses of toxic alkaloids from plants can cause liver and kidney damage. Raw soybean meal contains trypsin inhibitors, toxic phytates, and toxic saponines, which mostly can be removed by cooking (Olli et al., 1994b). Rapeseed meal contains: erucuc acid, which can damage the heart, adrenal glands, and mitochondria; glucosinolates, which interfere with thyroid function; tannins, which interfere with protein utilization; sinapine, which is very bitter; and phytic acid, which interferes with protein and zinc utilization. Canola contains much less erucic acid and glucosinolates; tannins and sinapine can be washed out, and phytic acid also can be removed. Cottonseed meal contains the yellow pigment gossypol and toxic cyclopropenoic acid. Gossypol can cause anorexia, anemia, damage to liver and kidneys, fatty liver, and tumors. It can be avoided by using glandless varieties of cottonseed or by chemically removing it (e.g., with ferrous sulfate); the processing method causes variation in content. Cyclopropenoic acid could still be a problem; toxicity signs include liver damage, enzyme alterations, and carcinogenic synergism with aflatoxins (Tacon, 1995). Thiaminase, an enzyme that destroys thiamin, occurs in many seafoods and can be especially dangerous when trash fish are frequently fed to cultured fish (Figs. 9-4, 9-5). It can be deactivated by cooking. Plant estrogens, which can interfere with reproduction, can be solvent extracted. Tannins, which bind proteins and inhibit trypsin, can be removed with the hulls of legumes. Cyanogens in cassava, which can release hydrocyanic acid, can be soaked out. Ipil-ipil (*Leucena leucocephala*) pods and leaves sometimes included in fish feeds in tropical regions contain varying levels of the toxic amino acid mimosine, which can reduce growth, interfere with liver enzymes, and cause cachexia; it can be soaked out. Some other naturally-occurring toxins are: lead, cadmium, mercury, oxyanions, bacterial toxins, shellfish toxins, avidin, and many compounds from plants, such as cyanoglycosides, oxalates, polyphenol, and enzyme inhibitors. Large amounts of offal (e.g., pig spleen, horse liver) in feeds have been known to induce cataracts in fish. Heating of fish meals containing free histamine and histidine can cause formation of gizzerosine, which can damage the stomach in fish (Tacon, 1995).

Lipid oxidation increases the amounts of free radicals, peroxides, hydroperoxides, aldehydes, and ketones, which are toxic and also degrade proteins and some vitamins. With time, lipoid liver degeneration will occur. Signs are anorexia, cataracts, pale gills, clubbed gills, microcytic anemia, decreased hematocrit and hemoglobin, bronzed rounded heart, enlarged liver with rounded edges, spleen and kidney damage, and eventually ceroid deposition in the liver, spleen, kidney, and choroid. Other signs of feeding on rancid lipids are darkening, exophthalmia, hemorrhages, steatitis, muscular dystrophy, skeletal myopathy, lordosis, and low growth efficiency and rate.

Fatty livers

These are common in reared fish. Cod eating feeds with protein energy/total energy ratios of 0.84 had hepatosomatic indices (HSI) similar to those of wild fish, but usually, the lower the PE/TE ratio, the larger the liver (Jobling, 1988). HSIs in the range 2-6% are normal and 8-9% probably abnormal. Cod eating a wide variety of formulated feeds have enlarged livers. Jobling (1988) believed that when the digestive capacity is overloaded by formulated feeds being released from the stomach faster than natural foods, excess fat is stored wherever possible, and also, because they cannot be stored for long, even amino acids can be converted into lipids. In fish like salmon, fillets become fatty; in fish like cod, livers become fatty. A solution more likely than reducing feed amounts or feeding frequency, is to use feed that is digested more slowly so that the supply of nutrients is more constant. In some cases, giving feeds in combination with or alternately with fresh food are two options.

Although dry livers of some fish like Atlantic cod normally contain 50-60% lipid (Lie et

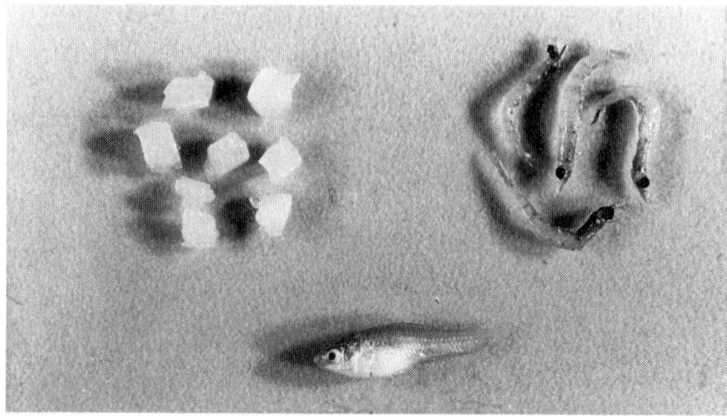

Figure 9-4. Fresh and frozen foods (e.g., chopped fish, whole krill, whole molly) must be used with caution to avoid nutritional imbalances.

Figure 9-5. Probable thiaminase-induced thiamin deficiency in common snook given the foods shown in Fig. 9-4 (40 mm TL). (from Tucker, 1987b; photo by Robert A. Winfree)

al., 1986), livers of wild red drum usually have little or no detectible lipid (Anonymous, 1994); when red drum were reared in ponds with supplemental salmon starter feed, then released to natural waters, their HSI dropped from 2.9 to 1.2%, similar to that of wild fish. Cultured red drum, especially those given formulated feeds, often have increased liver lipid content. When red drum were given semi-purified feeds (containing 6-10% lipid from menhaden or red drum), liver lipid content was 33.3% of wet liver in 20.4-g fish (Craig et al., 1995), 36.8% of wet liver in 34.6-g fish (Lochmann and Gatlin, 1993b), 66.5% of dry liver in 25.6-g fish (Daniels and Robinson, 1986), and 93.2% of dry liver in 29.1-g fish (Serrano et al., 1992). HSI values for cultured 20- to 29-g red drum eating semi-purified feeds to satiation twice a day were in the range 1.99-3.18% (Daniels and Robinson, 1986; Serrano et al., 1992; Craig et al., 1995). Red drum (0.4-175 g) fed one of six commercial or 15 experimental feeds twice a day to satiation had 27.4-75.2% lipid in dry liver (4.4-44.7% wet) and HSIs in the range 1.04-4.95% (Tucker et al., ms.). In a study comparing menhaden and soybean oils (1.5-12.7%) in starter feeds for red drum, Tucker et al. (1997b) found that all fish had some degree of liver fattiness as well as liver histological features differing from those of wild fish. It was concluded that up to a point liver fattiness in red drum is reversible and probably not harmful; it might even be a useful energy storage mechanism. Beyond that point, permanent damage and dysfunction are possible.

Flood et al. (1996) reported that high levels of liver lipid in red drum raised on compound feeds were correlated with the loss of liver cytochrome P-450 and other enzymatic alterations and suggested that those changes could reduce the ability of farm-raised fish to metabolize hydrocarbon pollutants.

At release, hatchery-reared black sea bream (45 mm BL, 3.7 g) had 6.0% liver lipid, but within 22 d, it had dropped to 3.3%, similar to wild levels (Nakagawa et al., 1994). In wild fish, liver lipid rose from 2.1% in August to 18.5% in November (114 d later), illustrating a natural seasonal cycle.

Common Signs of Nutritional Disorders

Tacon (1992) listed possible causes for six pathological conditions related to food. Exopthalmia could be caused by deficiency of pantothenic acid, niacin, folic acid, vitamin A, or vitamin E, or by toxicity of oxidized fish oil. Cataracts could be caused by deficiencies of methionine, tryptophan, riboflavin, vitamin A, magnesium, zinc, manganese, copper, or selenium, or by toxicity of choline or oxidized fish oil. Fin or skin hemorrhage could be caused by deficiency of thiamin, riboflavin, pantothenic acid, niacin, inositol, vitamin C, vitamin A, or vitamin K, or by toxicity of oxidized fish oil. Fin erosion could be caused by deficiency of lysine, tryptophan, riboflavin, niacin, inositol, vitamin C, or zinc, or by toxicity of lead or vitamin A. Fatty liver could be caused by deficiency of choline or essential fatty acids, or by toxicity of oxidized fish oil. Scoliosis or lordosis could be caused by deficiency of tryptophan, vitamin C, phosphorus, magnesium, or essential fatty acids, or by toxicity of lead, cadmium, leucine, vitamin A, or oxidized fish oil.

ENVIRONMENTAL CONSIDERATIONS

Increased suspended solids, chemical oxygen demand, and phosphorus content are three of the major concerns when fish culture water is discharged to the environment and also when it is recirculated. It is important to ensure that feeds are attractive, palatable, and of the right size and that they do not disintegrate before the fish can eat them. Increasing nutrient availability (e.g., high digestibility, best processing), balance (e.g., avoidance of surpluses and deficits, especially with amino acids, fatty acids, and micronutrients; optimal protein/energy ratio), and density (e.g., avoidance of high fiber and ash, or excess carbohydrate) helps minimize waste (Cho et al., 1994; Talbot and Hole, 1994). Most of the phosphorus contamination comes from feed (both eaten and uneaten). Improving feed conversion ratio usually reduces pollution and often reduces the cost of feeding. Avoidance of underfeeding and overfeeding by knowing and conforming to the fishes' natural behavior patterns is one strategy (Smith et al., 1993; Thorpe and Cho, 1995). Reducing total dietary phosphorus by lowering ash content is another. Decreased dietary phosphorus combined with the use of ingredients with higher bioavailability of phosphorus (e.g., blood meal with 81% availability vs herring meal with 52%) can result in more phosphorus being incorporated in the fish and less building up in the water (Luzier et al., 1995). Robert et al. (1993) reported about a two-third reduction of effluent phosphorus in rainbow trout culture water when dietary ash was reduced from 12.3 to 7.6% and phosphorus from 1.9 to 1.0% (0.86 to 0.58% available phosphorus).

Johnsen et al. (1993) reported no difference in growth when Atlantic salmon ate feeds with 30% fat (55% fish meal) instead of 22% (49% fish meal), but FCR improved from 1.07 to 0.96, and reductions of 38% ammonium nitrogen, 22% phosphorus, and 23% organic matter occurred.

FEED STUDIES

A wide range of ingredients has been used in feeds for marine fish. Typical analysis values for common ones are given in Table 9-7. Prices among ingredients vary widely, and for each ingredient prices vary with season and supply-demand pressures. Average U.S. price changes between March 1992 and May 1998 for one metric ton of each of the following feedstuffs were: wheat middlings \$88 to \$75, rice bran \$70 to \$69, wheat bran \$97 to \$82, dehydrated alfalfa pellets (17% protein) \$118 to \$151, canola meal (36% protein) \$166 to \$131, condensed fish solubles \$198 to \$227, peanut meal \$165 to \$214, meat & bone meal \$252 to \$207, soybean meal (48-49% protein) \$220 to \$190, corn gluten meal (60%

Table 9-7. Typical characteristics of some ingredients used in fish feeds. Carbohydrate contents of plant ingredients were estimated by subtraction. Except where indicated, analysis values are from Allen (1991). Because ingredients and analyses vary, some of the more recent values differ slightly from those in NRC (1983). Values in this table and in NRC (1983) are representative. For precise formulations, the particular lots of ingredients to be used should be analyzed.

Ingredient	International feed number	Crude protein (%)	Crude fat (%)	Carbohydrate (difference)[a] (%)	Crude fiber (%)	Ash (%)	Moisture (%)	Digestible energy (kcal/kg)
Alfalfa meal, dehydrated	1-00-024	21.5	3.8	41.9	21.5	11.3	6.9	560[e]
Blood meal, animal, spray-dried[a]	5-00-381	93.0	1.4		1.1	7.1	7.0	3410[e]
Bovine spleen, dried		65.3	18.6			7.3		6503[f]
Brine shrimp meal		63.8	6.5	14.9	1.5	13.3	7.2[g]	
Canola meal	5-08-135	41.7	4.2	34.0	12.2	7.9	8.8	
Casein, dried	5-01-162	88.9	0.6	6.4	0.2	3.9	10.0	4400[h]
Cassava tuber meal	4-09-598	2.7	0.3	84.9	8.7	3.4	12.7	
Cod roe, dried		79.0	4.1			4.5		5246[f]
Copra meal, solvent[a]	5-01-573	23.4	3.9	50.7	15.4	6.6	9.0	
Corn distiller's solubles, dehy.[a]	5-28-237	29.7	9.2	48.3	5.0	7.8	7.0	2440[e]
Corn gluten meal, (60%)	5-09-318	66.7	2.2	26.3	2.8	2.0	10.0	4040[e]
Cottonseed meal		38.4	3.3	45.2	5.4	7.7	12.6[i]	
Crab meal, process residue[a]	5-01-663	34.8	2.1		11.6	44.6	8.0	
Crab protein concentrate[a]	5-16-422	67.1	0.5			6.8	10.0	
Distiller's grains with solubles		29.5	9.2	47.4	9.1	4.8[j]		
Feather meal, poultry, hydrolyzed	5-03-795	91.2	2.7		1.6	4.2	6.8	3670[h]
Fish meal, AAFCO	5-01-977	66.7	6.3		1.1	22.8	11.6	
Fish meal, Atlantic herring	5-02-000	77.4	10.8		1.1	11.2	7.0	4720[e]
Fish meal, menhaden	5-02-009	67.4	11.1		1.1	21.7	8.0	4240[h]
Fish meal, Peruvian anchovy	5-01-985	71.4	11.0		1.1	16.5	9.0	4570[e]
Fish meal, red fish	5-07-973	62.0	8.7		1.1	28.3	8.0	
Fish meal, sardine	5-02-015	70.6	6.0		1.1	17.4	8.0	
Fish meal, tuna	5-02-023	64.5	7.5		1.1	24.7	7.0	
Fish meal, white	5-02-025	67.0	4.4		1.1	26.4	9.0	3490[e]
Fish protein hydrolysate[b]		80.6	12.4		0.5	6.4	7.0	
Fish protein hydrolysate		73.3	21.3			4.9	1.2[k]	
Fish solubles, condensed[a]	5-01-969	65.3	11.2		0.9	19.2	50.0	

Fish solubles, dehydrated[a]	5-01-971	69.2	8.9		1.5	13.5	7.0	3680[e]
Green mussel meal (*Perna viridis*)		63.1	11.3	12.2	3.1	10.4	5.7[g]	
Krill meal	5-16-423	33.3	27.8			27.8	82.0	
Meat & bone meal, 50%	5-09-322	54.0	9.2		3.0	35.6	7.4	3390[e]
Meat meal, 55%	5-09-323	59.3	7.8		2.7	26.9	7.2	
Oat groats (dehulled oats)	4-03-331	17.4	6.5	70.9	2.8	2.4	7.8	
Peanut meal, solvent	5-03-656	52.4	1.6	30.7	7.4	7.9	8.5	
Poultry byproduct meal	5-03-798	61.7	14.9		2.6	17.0	6.0	3720[e]
Poultry protein hydrolysate[b]		82.1	9.5		0.5	6.3	5.0	
Rice bran, unextracted	4-26-201	14.8	6.5	52.3	14.3	12.1	9.0	
Rice bran, with germ[a]	4-03-928	14.1	15.1	45.2	12.8	12.8	9.0	
Rice grain, rough	4-03-939	8.2	1.9	73.6	11.2	5.1	11.0	
Rice middlings[a]	1-03-941	6.8	5.6	39.0	31.5	17.1	8.0	
Rice polishings[a]	4-03-943	13.4	13.9	60.8	3.6	8.3	10.0	
Safflower seed meal, solvent	5-07-959	46.2	1.4	27.2	16.6	8.6	9.2	
Scallop meat (*Chlamys* sp.)		65.2	10.9	8.8	1.4	13.7	7.3[g]	
Shrimp meal, process residue[a]	5-04-226	44.2	4.3		15.6	29.7	10.0	
Sorghum gluten meal	5-04-388	46.7	4.8	42.6	3.9	2.0	10.0	
Soybean meal, dehulled, solvent	5-04-612	54.3	1.1	34.6	3.3	6.7	10.7	
Soybean meal, expeller	5-04-600	47.2	3.9	34.9	7.3	6.7	11.0	
Soybean meal, solvent	5-04-604	49.1	0.6	35.8	7.8	6.7	10.4	2980[e]
Soybean protein concentrate[a]	5-08-038	91.9	0.6	3.6	0.1	3.8	8.0	
Soybeans, full-fat, cooked	5-04-597	42.2	20.0	27.1	5.6	5.1	10.0	4190[e]
Soy textured vegetable protein[c]		53.2	1.6	36.4	2.1	6.7	6.8	
Squid meal (*Loligo* sp.)		76.2	6.4	11.1	0.4	5.9	6.2[g]	
Sunflower meal, expeller	5-04-738	44.1	8.2	26.4	14.0	7.3	7.0	
Sunflower meal, solvent	5-04-739	45.2	2.5	30.0	14.0	8.3	7.0	
Wheat bran	4-05-190	16.6	4.5	60.5	11.2	7.2	11.0	2790[h]
Wheat flour[a]	4-05-199	13.4	1.4	83.2	1.5	0.5	12.0	
Wheat flour, high-gluten		16.1	1.9	72.8	3.2	6.0	12.0	
Wheat germ meal	5-05-218	28.1	7.9	54.1	3.9	6.0	11.0	
Wheat middlings	4-05-205	20.0	4.0	61.9	7.9	6.2	11.0	1800[e]
Yeast, brewer's, dried	7-05-527	48.4	0.4	42.6	1.6	7.0	7.0	2710[e]
Yeast (*Candida utilis*)		50.1	1.3			18.2	82.9[l]	
Yeast (*Phaffia rhodozyma*)		22	23			3	5[m]	
Yeast, torula, dried	7-05-534	52.0	2.1	34.4	2.9	8.6	6.8	

Table 9-7. (Notes)
[a] Also includes cumulative error.
[b] NRC, 1983.
[c] California Spray Dry Co., 1992.
[d] Tucker, unpub.
[e] Value for rainbow trout (Lovell, 1989).
[f] Heinsbroek & Kreuger, 1992.
[g] Feed Development Section, 1994.
[h] Value for channel catfish (Lovell, 1989).
[i] Nengas et al., 1995.
[j] Webster et al., 1995
[k] Berge & Storebakken, 1996
[l] Martin et al., 1993
[m] Sanderson & Jolly, 1994

protein) $310 to $260, poultry byproduct meal $297 to $261, hydrolyzed feather meal $248 to $248, menhaden meal $482 to $603, and dried brewer's yeast $682 to $873; in both August 1997 and May 1998, tuna meal cost $385 and anchovy meal $594 (Anonymous, 1992, 1997a, 1998b, 1998c). When the best materials are not available, substitutions can be made within limits set to maintain acceptable performance.

When more than one or two factors vary in a feed comparison experiment, it is more difficult to interpret results; however, some trends seem to exist. Results of some early studies were confounded by factors such as ω3HUFA deficiencies, excessive fiber, and unpalatable or indigestible ingredients. When the range of treatments did not encompass the optimal treatment, the optimal treatment was not determined. At times large differences in ω3HUFA content among feeds were overlooked. Tacon (1995) recommended ways of improving and standardizing nutrient requirement determinations: 1) practical culture conditions should be simulated, 2) feedstuffs should be fully described, 3) the test fish should be capable of maximum growth rates, 4) the form of feedstuffs should be well-defined and a wide-enough range tested, 5) in pond work, isotope tracing can separate the contribution of natural foods, and 6) economic evaluation should be included and actual farm studies done when possible.

Cho et al. (1985) presented a general marine fish feed formula as follows: 50-60% fish meal; 15% soybean meal; 10% meat & bone meal; 10-20% corn, tapioca, wheat, or rice bran; and 3.5% fish oil.

In my work, evaluations in tanks with 10 experimental feeds (Table 9-8) and seven species (Table 9-9) resulted in good growth and feed conversion. High performance of the feeds probably resulted from the use of high quality macroingredients and increased amounts of vitamins. The only important pathologies observed were: (1) A few common snook had undeveloped gas bladders and lordosis, possible factors being films on the water surface and nutritional deficiencies during the larval stage. (2) Handling stress caused mortality of a few sunshine bass between trials. (3) Acute ammonia toxicity (above 0.1 ppm TAN) from contamination of the water source caused mortality of some Nassau groupers between trials.

Of particular interest are similarities among species in growth rate and FCR, especially for common snook, sunshine bass, and Nassau grouper. Barramundi, red drum, and spotted seatrout grow faster than the other three (pers. obs.). With higher carbohydrate and lower fish oil, JT8520 is the lowest quality feed listed.

Acipenseridae

Sturgeons have mostly been grown out on commercial salmonid feeds containing 10-20% fat (Hung, 1991a, 1991b), but those are not completely appropriate and experimental feeds have produced slightly better results. Some examples are given in Tables 9-10 and 9-11. Both white sturgeon (Hung, 1991a) and Siberian sturgeon (Kaushik et al., 1991) need about 44-50% protein in grower feeds.

White sturgeon juveniles did not need cholesterol or lecithin in their feed but did need 0.17-0.32% choline (Hung, 1991a). With a range of 20-53% protein tested, 40% was enough for good growth. The best protein was a combination of 62% casein, 30% wheat gluten, and 8% egg white. Other proteins tested, in decreasing order of quality (% energy retained), were: casein, defatted shrimp meal, defatted herring meal, soybean concentrate, egg white, gelatin, and defatted zein meal. The best carbohydrates tested were maltose, glucose, dextrin, and raw cornstarch; sucrose, lactose, and fructose were unsuitable. Cui and Hung (1995) presented a feeding and growth table.

Hybrid sturgeon (Russian sturgeon x beluga) given feeds containing all combinations of 45, 52, and 60% protein with 10, 15, and 20% fat, grew best (2.7%/d) and had the best FCR (0.9) with 52% protein and 20% fat (Gershanovich and Kiselev, 1993).

Atlantic sturgeon (2.5-31 g) had better growth (4.17%/d) and FCR (0.54) with a commercial feed containing 59% protein and 16.0% fat than with another commercial and two experimental feeds containing 53-54% protein and 13.1-22.5% fat (Mohler et al., 1996).

Anguillidae

Feed with 45% protein was suitable for **Japanese** and **European eels**; a 0.5% minimum of both linoleic and linolenic acids was needed for good growth (Takeuchi et al., 1980; Degani, 1986).

European eels (<30 cm TL) in nature had a diet of about two-thirds crustaceans and one-third insects (Barak and Mason, 1992). Farmed eels are fed a variety of foods—pastes, meals, granules, pellets (Table 9-12).

In Europe, a preferred food for glass eels has been cod roe, but survival of glass eels to market size was only 25-40% (Heinsbroek and Kreuger, 1992). In northwestern Europe, cod roe was fed for the first 2-4 weeks, then starter crumbles (for trout or eels), then pellets (for trout or eels)(Heinsbroek, 1991). In Italy, paste foods were used more. Heinsbroek (1991) gave analysis values of five foods used for European eels.

European glass eels (300-730 mg) grew better with 10% chicken oil in feed than with 5% chicken oil, 5% soybean oil, or 10% soybean oil (Degani, 1986).

Addition of cod roe extract, bovine spleen extract, or synthetic amino acids to trout starter crumbles increased consumption and growth of European eels (0.3 g initially), but growth was still better on whole roe or spleen (Heinsbroek and Kreuger, 1992).

Japanese eels in nature eat crustaceans, aquatic insects, small fish, and many other animals (Yamaha Fishery Journal, 1992a). In Japan, cultured eels were given compound feed on about 70% of the farms and raw fish on the other 30% (Usui, 1991). Usually frozen sand lance, sardines, anchovies, Pacific saury, mackerels, Atka mackerel (*Pleurogrammus* sp.), or Japanese horse mackerel were used.

The first formula in Table 9-12 is typical for Japan; 5-10 parts oil are mixed with 100 parts of the dry mixture, then 80-100 parts water are added with a blender to make a paste (Arai, 1991). Pastes are popular because eels have difficulty eating pellets and pellet size has to be carefully controlled to match eel size.

Mixed fresh/compound feeds used in Japan during the glass eel stage have resulted in 70-90% survival and growth similar to that obtained with *Tubifex* spp. (Heinsbroek and Kreuger, 1992). Heinsbroek (1991) gave the composition of five foods used for Japanese eels. Some ingredients were: white fish meal, krill meal, skim milk powder, casein, soybean meal, linseed meal, gelatinized potato starch, dextrin, bread crumbs, yeast, fish liver & viscera, liver powder, wheat germ, soy protein, lactose, bile powder, bone extract, animal fat, and blood meal.

Clupeidae

Wild juvenile **scaled sardines** (21-30 mm SL) ate mostly veliger larvae, crab larvae, and copepods (Carr and Adams, 1973). Wild adults (82-122 mm SL) ate mostly copepods, barnacle larvae, cladocerans, and trematodes, and also amphipods, bivalves, and invertebrate eggs (Motta et al., 1995).

Engraulidae

Wild juvenile **striped anchovies** (25-50 mm SL) ate mostly veliger larvae, zoea larvae,

Table 9-8. Composition of 10 experimental feeds (dry pellets) developed by the author. All values except protein:energy are in percent. Except for moisture, proximate analysis values are moisture free.

Component	Grower JT8520	Grower JT8560	Grower JT8630	Grower JT8825	Starter JT9010	Grower JT9020	Grower JT9120	Starter JT9210	Grower JT9220	Broodfish JT9261[d]
Formulation										
Anchovy meal	26.9	26.9	60.0	30.0						
Low-temperature herring meal					60.4	45.0	45.0	50.0	30.0	40.5
Shrimp meal										10.0[d]
Poultry meal								12.5	15.0	8.0
Meat & bone meal								12.5	15.0	8.0
Poultry/meat meal	26.2	28.3	13.5	30.0						
Blood meal					3.5	3.5	3.5			
Feather meal					5.0	10.0	10.0			
Textured vegetable protein	18.3	30.1	11.7	25.0	8.0	13.0	13.0	10.0	20.0	10.0
Corn gluten meal					5.0	5.0	5.0			5.0
High-gluten wheat flour			8.7	8.0	4.9	13.9	15.6	6.2	12.7	12.2[d]
Wheat flour	26.7	8.3								
Wheat middlings			3.5							
Brewer's yeast					1.0	1.0	1.0			1.0
Bone meal					2.0	2.0				
Cod liver oil	1.1	5.5								
Menhaden oil			1.4	2.9	7.5	3.5	3.5	6.0	4.5	2.5[d]
Soybean oil				2.9						
Lecithin					1.0	1.0	1.0	1.0	1.0	1.0
Microingredients[a]	0.8	0.9	1.2	1.2	1.7	2.1	2.4	1.8	1.8	1.8
Proximate analysis (vacuum dry)										
Crude protein (Kjeldahl)	50.9	53.8	51.6	53.9	60.1	59.7	59.9	58.5	53.4	58.6
Crude fat (ether extract)	8.7	12.9	13.7	15.5	16.2	10.3	10.3	18.0	13.3	11.8[d]
Carbohydrate (difference)	24.9	15.8	13.2	16.9	9.6	17.2	18.7	9.7	20.1	16.7
Fiber (acid detergent)	4.0	4.9	5.8	1.1	0.3	0.9	0.9	0.9	0.9	1.7
Ash (600°C for 4 h)	11.5	12.6	15.7	12.6	13.8	11.9	10.2	12.9	12.3	11.2

Total fish oil	3.3	7.7	6.2	5.3	13.4	8.0	8.0	11.7	8.2	6.2 [d]
Total crustacean oil										1.5
EPA[b]	0.5	1.0	1.1	0.8	1.3	0.7	0.7	1.1	0.8	0.8 [d]
DHA[b]	0.3	0.7	0.6	0.5	1.4	0.7	0.7	1.2	0.9	0.8 [d]
Protein/energy (mg P/kcal DE)[c]	146	147	145	138	151	160	158	144	140	155
Protein/energy (mg P/kJ DE)[c]	35.0	35.2	34.7	33.0	36.0	38.3	37.9	34.3	33.6	37.0
Proximate analysis (of air dry weight)										
Moisture	11.1	10.0	8.3	6.2	8.0	7.6	7.8	5.9	6.5	7.9

[a] Microingredients were:
A. Feeds JT8560 and JT8520 contained, in mg/kg feed: U.S. Fish and Wildlife Service vitamin mix no. 30, 4,000; ascorbic acid, 700; choline (chloride), 1,250; USFWS mineral mix no. 2, 1,000; ethoxyquin, 150.
B. Feeds JT8630 and JT8825 contained, in mg/kg feed: iodine (potassium iodide) 4, zinc (oxide) 60, manganese (oxide) 5, chromium (chromium potassium sulfate) 1, thiamin (mononitrate) 40, riboflavin 60, pyridoxine 40, pantothenic acid (d-calcium pantothenate) 200, niacin 300, biotin 2, folacin 20, cyanocobalamin 0.2, choline 4,000, inositol 900, ascorbic acid 700, ethoxyquin 200; in IU/kg feed: retinol 20,000, cholecalciferol 3,000, tocopherol 500, menadione 50.
C. Feeds JT9010 and JT9020 contained: same as B, except (1) iron was added (ferrous sulfate) 50 mg/kg, (2) choline was increased to 4800 mg/kg, (3) substituted for ascorbic acid (L-ascorbyl-2-polyphosphate) 300 mg/kg equivalent, (4) ethoxyquin was decreased to 150 mg/kg.
D. Feed JT9120 contained: same as B, except (1) zinc was increased to 90 mg/kg, (2) manganese was increased to 40 mg/kg, (3) copper was added (sulfate) 3 mg/kg, (4) choline was increased to 4800 mg/kg, (5) substituted for ascorbic acid (L-ascorbyl-2-polyphosphate) 400 mg/kg, (6) ethoxyquin was decreased to 150 mg/kg.
E. Feeds JT9210, JT9220, and JT9261 contained: same as D, except copper was deleted

[b] Contents of eicosapentaenoic and docosahexaenoic acids were calculated using average values for herring and menhaden oils from Lovell (1989) and Sargent et al. (1989).

[c] Digestible energy was calculated using the values 4 kcal/g protein, 8 kcal/g fat, 3 kcal/g carbohydrate.

[d] Feed JT9261 later was revised to JT9262 by replacing 10.0% shrimp meal with 10.0% krill meal, reducing high-gluten wheat flour to 10.2%, and increasing menhaden oil to 4.5%. Analysis was similar, except JT9262 had 13.8% fat, 8.2% fish oil, 1.0% EPA, and 0.9% DHA.

Table 9-9. Performance of eight experimental feeds (common snook, Tucker, 1987b; barramundi, Tucker et al. 1988; sunshine bass, Tucker et al., 1993a; Nassau grouper, Tucker and Woodward, 1996; spotted seatrout, Tucker, 1988; red drum, Tucker et al., 1997a, unpub.; black drum, Tucker et al., unpub.). Feed formulas are given in Table 9-8. Daily growth = specific growth rate. Feed conversion ratio = weight of dry feed that produced a unit weight of wet fish.

Species	Feed	Time (d)	Initial Weight (g)	Final Weight (g)	Mean Growth (g/d)	Daily Growth (%)	Daily Ration (%)	Feed Conv. Ratio
Common snook	JT8560	50	16	44	0.56	2.08	1.52	0.73
	JT8520	80	49	124	0.94	1.15	1.21	1.05
	JT8520	104	95	230	1.30	0.85	0.92	1.08
	JT8630	54	241	376	2.50	0.83	0.74	0.89
	JT8630	54	504	726	4.11	0.67	0.74	1.10
Barramundi	JT8560	75	9	62	0.71	2.49	2.32	0.93
Sunshine bass	JT8825	50	207	348	2.82	1.04	1.09	1.05
	JT8825	182	348	862	2.82	0.50	1.46	2.92
	JT9020	156	862	1333	3.02	0.28	0.88	3.16
Nassau grouper	JT9010[a]	96	0.1	6.1	0.06	4.28		
	JT9020	145	7	71	0.44	1.57	1.41	0.90
	JT9120	227	95	448	1.56	2.01	2.04	1.02
	JT9120	138	210	594	2.78	0.75	0.77	1.02
	JT9120[b]	128	483	928	3.46	0.51		
	JT9120	140	598	906	2.20	0.30	0.39	1.32
	JT9120	410	174	1490	3.21	0.52	0.73	1.41
	JT9261	158	1396	2030	4.01	0.24		
	JT9261	42	1682	1870	4.48	0.25	0.45	1.80
Spotted seatrout	JT8560	30	24	60	1.20	3.10	2.42	0.78
	JT8560	21	53	79	1.24	1.94	2.37	1.22
	JT8560	20	87	110	1.15	1.17	1.81	1.55

Red drum	JT9210	75	4	63	0.79	3.57	2.86	0.80
	JT9220	76	66	215	1.96	1.56	1.73	1.11
	JT9220	107	216	610	3.68	0.97	1.28	1.32
	JT9220	145	626	1581	6.59	0.64	0.98	1.53
	JT9220	89	1640	2243	6.78	0.35	0.56	1.59
	JT9220	150	2025	2953	6.19	0.25	0.66	2.65
	JT9220	182	2953	4740	9.82	0.26	0.46	1.76
	JT9220	168	4740	6125	8.24	0.15	0.77	5.08
Black drum	JT9220	86	705	1099	4.57	0.52	0.81	1.58

[a] Fed rotifers and *Artemia* during 3-60 dah, minced penaeid shrimp during 56-61, 63, 66, 69 dah, and starter feed (JT9010) during 56-98 dah.

[b] Because these fish were to be released, the pellets were supplemented with live goldfish, cyprinid minnows, penaeid shrimp, and fiddler crabs plus frozen fish, queen conch, shrimp, and scallops.

Table 9-10. Composition (%, except as noted) and performance of moist, semi-moist, and dry feeds used to experimentally raise sterlet and its hybrid (with male Siberian sturgeon) in 300-L tanks for 56 d in Hungary (Ronyai et al., 1991).

Component	Feed 1	Feed 2	Feed 3[a]
Formulation			
Fish meal	13	13	+
Minced fresh headless fish		35	
Meat meal	5.8	5.8	
Boiled slaughterhouse blood	35		
Wheat meal	26	26	+
Yeast	19.5	19.5	+
Vitamin premix	0.7	0.7	+
Proximate analysis (dry)			
Crude protein	59.4	54.5	47.2
Moisture	37	31	11
Performance of sterlet			
Fish weight (g)	17-50	17-57	18-54
Daily growth (%)	1.93	2.13	2.00
Performance of hybrid			
Fish weight (g)	28-94	29-113	22-81
Daily growth (%)	2.13	2.43	2.39
Combined			
Feed conversion ratio (kg/kg)	3.6	3.0	2.2

[a] Also contained soybean meal, lysine, starch, and unspecified oil.
+ Amounts not given.

Table 9-11. Composition (%, except as noted) and performance of dry pellets used to experimentally raise Siberian sturgeon in sand-bottomed concrete tanks for 45 d in France (Kaushik et al., 1991). Performance was best with the highest fish meal and protein, but a feed intermediate between D35 and D45 was nearly as good as D45.

Component	D25	D35	D45
Formulation			
Fish meal	27.5	41.5	55.5
Soybean meal	6.1	7.5	9.5
Extruded wheat	15.0	12.5	8.5
Crude corn starch	6.0	11.8	18.0
Gelatinized starch	33.3	16.4	
Fish oil	6.1	4.3	2.5
Vitamin premix[a]	2.0	2.0	2.0
Mineral premix[b]	2.0	2.0	2.0
Binder	1.0	1.0	1.0
Chromic oxide	1.0	1.0	1.0
Proximate analysis (dry)			
Crude protein	29.4	40.1	52.6
Crude fat	7.5	8.7	9.9
Protein/energy (mg P/kcal GE)	62	83	107
Protein/energy (mg P/kJ GE)	14.9	19.9	25.6

Table 9-11. (Continued)

Component	D25	D35	D45
Performance			
Fish weight (g)	22-39	21-40	19-45
Daily growth (%)	1.27	1.41	1.93
Feed conversion ratio (kg/kg)	0.87	0.80	0.66

[a] In mg/kg feed: thiamin 15, riboflavin 30, pyridoxine 15, calcium pantothenate 50, nicotinic acid 175, biotin 2.5, folacin 5, cyanocobalamin 0.05, choline 2000, inositol 1000, ascorbic acid 500, vitamin K 5; in IU/kg feed: vitamin A 15,000, cholecalciferol 3000, tocopherol 60.

[b] In mg/kg feed: dibasic calcium phosphate 10,000, calcium carbonate 4300, magnesium hydroxide 2480, potassium chloride 1800, potassium iodide 8, sodium chloride 800, ferric citrate 400, zinc sulfate 80, copper sulfate 60, manganese sulfate 60, cobalt sulfate 4.

Table 9-12. Composition (%) of feeds used for commercial eel farming: paste (with water and 5-9% oil) for Japanese eels in Japan (Arai, 1989), meal for Japanese eels in Taiwan (Chen and Liao, 1991), and paste or granules for European eels in Europe (Brusle, 1990).

Component	Japanese eel (Japan)	Japanese eel (Taiwan)	European eel (elvers)	European eel (young eels)	European eel
Formulation					
Brown fish meal		21			+
White fish meal	65	50			
Shrimp meal					+
Meat meal					+
Liver meal	2				
Blood meal					+
Soybean meal					+
Soybean flour	5				
Cooked cereals					+
Gelatinized potato starch	22.5				
Tapioca starch		22			
Corn starch		4			
Yeast	3				
Animal fats					+
Vegetable fats					+
Vitamin premix	0.5	1			+[a]
Mineral premix	2	2			+[b]
Proximate analysis (dry)					
Crude protein		50.4	58.4	50.0	51.1
Crude fat		6.4	9.6	7.8	9.8
Carbohydrate			15.6		
Fiber			2.2	3.3	2.2
Ash		16.8	12.4	11.1	13.0
Protein/energy (mg P/kcal DE)[c]			163	144	144
Protein/energy (mgP/kJ DE)[c]			39.1	34.6	34.4
Proximate analysis (of air dry weight)					
Moisture		7.8	11.0	10.0	8

[a] In mg/kg feed: thiamin 5, riboflavin 20, pyridoxine 5, pantothenic acid 50, folacin 3, cyanocobalamin 0.04, ascorbic acid 100, tocopherol 20, vitamin K 2, PP 150; in IU/kg feed: vitamin A 10,000, cholecalciferol 1000.

[b] In mg/kg feed: sodium chloride 1000, iron 120, zinc 27, manganese 25, cobalt 0.2.

[c] Digestible energy was calculated using the values 4 kcal/g protein, 8 kcal/g fat, 3 kcal/g carbohydrate.

Table 9-13. Composition (%, except as noted) and performance of feeds used experimentally for wild milkfish larvae in the Philippines. Data reprinted from Aquaculture vol. 71, V.R. Alava and C. Lim, Artificial diets for milkfish, *Chanos chanos* (Forsskal), fry reared in seawater, p 340-341, copyright 1988, with permission from Elsevier Science.

Component	I	II	IV
Formulation			
Fish meal	30.0	30.0	30.0
Shrimp head meal	16.0	16.0	
Meat & bone meal			13.4
Soybean meal	20.0		20.0
Corn gluten meal		20.4	
Wheat flour	15.0	13.6	16.8
Rice bran	11.5	12.5	13.0
Cod liver oil	3.0	3.0	2.3
Vitamin premix[a]	1.0	1.0	1.0
Mineral premix[b]	3.5	3.5	3.5
Proximate analysis (dry)			
Crude protein	40.9	40.7	40.8
Crude fat	9.2	8.9	8.1
Carbohydrate (NFE)	30.0	30.1	31.7
Crude fiber	4.9	5.8	2.8
Ash	15.0	14.5	16.6
Protein/energy (mg P/kcal DE)[c]	125	126	126
Protein/energy (mg P/kJ DE)[c]	29.9	30.0	30.2
Performance			
Fish weight (mg)	9-202	9-198	9-181
Daily growth (%)	11.11	11.03	10.72
Feed conversion ratio (kg/kg)	0.86	0.94	1.04
Survival (%)	97	92	97

[a] In mg/kg feed: thiamin hydrochloride 20, riboflavin 40, pyridoxine hydrochloride 30, pantothenic acid 60, niacin 200, biotin 2, para-aminobenzoic acid 100, folacin 5, cyanocobalamin 0.1, choline chloride 1000, inositol 600, ascorbic acid 2000, beta-carotene 20, vitamin D 10, tocopherol 100, vitamin K 10, BHT 10.

[b] In mg/kg feed: dibasic calcium phosphate dihydrate 12,938, calcium carbonate 8750, magnesium sulfate heptahydrate 1875, monobasic potassium phosphate 6250, potassium chloride 625, potassium iodide 6, monobasic sodium phosphate 3750, ferric sulfate heptahydrate 313, zinc sulfate heptahydrate 281, copper sulfate pentahydrate 19, manganese sulfate heptahydrate 188, cobalt chloride hexahydrate 1.

[c] Digestible energy was calculated using the values 4 kcal/g protein, 8 kcal/g fat, 3 kcal/g carbohydrate.

Table 9-14. Composition (%, except as noted) and performance of feeds used experimentally for milkfish in Taiwan (Chen and Liao, 1991). Despite differences in formulas, performance was similar. Data reprinted with permission from: J.P. McVey (ed.), Handbook of Mariculture, Vol. II, Finfish Aquaculture, copyright CRC Press, Boca Raton, Florida.

Component	Milkfish 7	Milkfish 9	Milkfish 10
Formulation			
Fish meal			25
Unfermented soybean meal		25	
Fermented soybean meal	25		
Peanut meal	25	25	25
Wheat gluten	10	10	10
Wheat flour	15	15	15
Spirulina	15	15	15
Fish oil	2	2	2
Soybean oil	2	2	2
Vitamin premix	3	3	3
Mineral premix	3	3	3

Table 9-14. (Continued)

Component	Milkfish 7	Milkfish 9	Milkfish 10
Proximate analysis			
Crude protein	38.8	40.4	45.8
Crude fat	8.0	8.3	6.7
Ash	7.2	6.7	9.6
Performance			
Fish weight (g)	9-52	11-56	13-79
Daily growth (%)	1.71	1.64	1.66
Feed conversion ratio (kg/kg)	1.89	1.83	1.77

Table 9-15. Composition (%) of two dry feeds for milkfish (Lim, 1991a). Data reprinted with permission from: R.P. Wilson (ed.), Handbook of Nutrient Requirements of Finfish, copyright CRC Press, Boca Raton, Florida.

Component	Pond grower	Broodfish
Formulation		
Anchovy or menhaden meal	8.0	22.0
Soybean meal (48% protein)	31.5	35.0
Wheat flour		15.0
Corn meal		20.9
Grains or grain byproducts	56.0	
Fish oil		4.6
Vitamin premix	0.5 [a]	1.0 [b]
Mineral premix[c]	0.5	0.5
Dicalcium phosphate	1.5	1.0
Hemicellulose or lignin sulfonate	2.0	
Proximate analysis (dry)		
Crude protein	26	36

[a] Supplemental premix for warmwater fish (NRC, 1977).
[b] Complete premix for warmwater fish (NRC, 1977).
[c] Practical premix for warmwater fish (NRC, 1977).

Table 9-16. Composition (%, except as noted) and performance of two dry feeds used to experimentally raise milkfish (0.6 fish/m^2) in 1-ha ponds for 3 months (Sumagaysay et al., 1990).

Component	Low energy	High energy
Formulation		
Fish meal	5.0	5.0
Soybean meal	20.0	23.5
Copra meal	35.0	35.0
Rice bran	15.0	30.0
Wheat pollard	22.5	
Coconut oil		4.0
Vitamins	0.5	0.5
Tricalcium phosphate	2.0	2.0
Proximate analysis (dry, calculated)		
Crude protein	25.0	24.7
Crude fat	7.7	12.7
Carbohydrate (NFE)	51.8	47.9
Crude fiber	7.3	6.4
Ash	8.2	8.3
Protein/energy (mg P/kcal DE)[a]	79	72
Protein/energy (mg P/kJ DE)[a]	18.8	17.2

Table 9-16. (Continued)

Component	Low energy	High energy
Performance (6000 fish/ha)		
Fish weight (g)	2-143	2-125
Daily growth (%)	4.7	4.6
Feed conversion ratio (kg/kg)	1.6	1.6
Survival (%)	81	91
Production (kg/ha)	695	682
Performance (9000 fish/ha)		
Fish weight (g)	2-125	2-143
Daily growth (%)	4.6	4.7
Feed conversion ratio (kg/kg)	2.2	1.8
Survival (%)	88	85
Production (kg/ha)	990	1094

[a] Digestible energy was calculated using the values 4 kcal/g protein, 8 kcal/g fat, 3 kcal/g carbohydrate.

Table 9-17. Composition (%, except as noted) and performance of natural food and a dry supplemental feed used for milkfish (1.5 fish/m^2) raised in 12-m^2 tanks for 96 d (Sumagaysay, 1993).

Component	Natural	Supplemented
Natural food from fertilized soil	Yes	Yes
Formulation		
Peruvian fish meal		15.0
Soybean meal		35.0
Copra meal		19.0
Rice bran		25.0
Bread flour		2.0
Cod liver oil		1.0
Ascorbic acid		0.05
Vitamins & minerals[a]		2.0
Dicalcium phosphate		0.95
Proximate analysis (dry)		
Crude protein		34.3
Crude fat		7.0
Carbohydrate (NFE)		42.7
Crude fiber		5.5
Ash		10.5
Protein/energy (mg P/kcal GE)		80
Protein/energy (mg P/kJ GE)		19.1
Performance		
Fish weight (g)	5.6-76.6	5.4-116.4
Daily growth (%)	1.16	3.09
Feed conversion ratio (kg/kg)	0.58	1.54

[a] In mg/kg feed: thiamine 0.088, riboflavin 0.176, pyridoxine 0.022, niacin 1.20, folic acid 0.0044, cyanocobalamin 0.88, calcium pantothenate 0.24, choline chloride 8.80, vitamin A 0.1056, vitamin D_3 0.0033, vitamin E 0.154, vitamin K 0.024; potassium iodide 0.088, ferrous sulfate 1.76, calcium (carbonate, phosphate, sulfate) 24.00, cobalt sulfate 0.0088, copper sulfate 0.088, magnesium sulfate 1.32; manganese sulfate 2.40; L-lysine hydrochloride 1.32, methionine 1.76.

copepods, and mysid shrimp (Carr and Adams, 1973). Wild adults (86-100 mm SL) ate mostly copepods, and also bivalves, shrimp, and gastropods (Motta et al., 1995). Wild juvenile **bay anchovies** (15-23 mm SL) ate mostly veligers, copepods, and invertebrate eggs (Carr and Adams, 1973).

Chanidae

Milkfish in ponds grow well on lab-lab, the combination of microbenthic bacteria, algae, protozoans, copepods, ostracods, worms, mollusc larvae, etc., that is the best natural food (Santiago, 1986). Published dietary requirements include: 40% protein for late larvae, 30-40% protein for juveniles, 7-10% lipid, 1-1.5% ω3PUFA, 25% carbohydrate, and 2500-3500 cal/g digestible energy (Feed Development Section, 1994). Feed formulas for larvae, juveniles, and broodfish are given by Feed Development Section (1994). Milkfish are fed meals and dry or moist, floating or sinking, pellets (Lim, 1991a). In ponds, they usually are fed 2-3 times a day. Alava and Lim (1988) fed milkfish larvae six experimental compound feeds for 28 d and found no significant differences in performance; however, diet I appeared to be the best (Table 9-13). Chen and Liao (1991) reported results of a comparison of ten compound feeds with milkfish (Table 9-14). The best feeds (7, 9, 10) contained 25% peanut meal and 25% of another protein source. Lim (1991a) presented practical formulas for pond production of milkfish and for maintenance of broodstock (Table 9-15). For grow-out, milkfish needed at least 40% protein, with animal sources being better utilized than plants in some studies but not others (Emata, 1994). Fish meal or meat & bone meal are preferred to shrimp head meal, and soybean meal is preferred to copra or ipil-ipil. Some commercial feeds are low in protein (e.g., 23-27%) but work reasonably well. Fat should be in the range 7-10%. Successful experimental feeds have contained up to 35% dextrin, and commercial feeds with 45% or more carbohydrate are used.

Clam and squid meals were better protein sources than shrimp head and fish meals for 17-mg milkfish (Carolin and Jose, 1995). Protein conversion was best with clam meal and worst with shrimp meal.

Addition of 1% linolenic acid to a semipurified feed containing 6% lauric acid (12:0) was better than 1% linoleic acid for growth (1.55-3.7 g)(Bautista and de la Cruz, 1988).

Better growth (1.53%/d) and FCR (1.75) for milkfish (8.6 g initially) were obtained with a semipurified feed containing 5% tristearin (18:0), 1.0% linolenic acid, 0.5% EPA, and 0.5% DHA than with simpler combinations (Borlongan et al., 1992).

At a stocking density of 0.6 fish/m^2, growth of milkfish in ponds was slightly better with a low energy feed, but at 0.9 fish/m^2, growth was better with a high energy feed (Table 9-16; Sumagaysay et al., 1990). Production was better for both feeds at the higher stocking level. Supplementary feeding improved growth of milkfish (1.5 fish/m^2) in fertilized earth-bottomed tanks (Table 9-17; Sumagaysay, 1993).

Further information is given by Boonyaratpalin (1997).

Osmeridae

Ayu are mainly herbivorous but can utilize animal foods. Lake ayu often continue to feed on zooplankton as well as phytoplankton (Yamaha Fishery Journal, 1992b). In the 1960s, they were fed fresh minced fish (especially Japanese horse mackerel, sardine, sand lance) mixed with wheat or other flour. Later, farmers began using compound feeds containing fish meal and flour. Wild ayu (eating their preferred foods) have a particular aroma (mostly in the viscera and skin), and *Spirulina* added to feeds could help cultured fish maintain a similar aroma. Compounds causing the cucumber- or watermelon-like aroma probably result from enzymatic lipid oxidation in the fish (Hirano et al., 1992).

Ayu need 40-50% dietary protein (Kanazawa, 1991a). Forty percent protein was sufficient in microparticles fed to larval ayu when the amino acid composition of the feed was similar to that of the fish's body (Table 9-18). Because they can convert linolenic acid into EPA, about 1% of either or both is sufficient in the diet. Dietary phospholipids improved survival, growth, and morphology of young ayu.

Table 9-18. Composition (%, except as noted) and performance of an experimental microparticulate feed used in Japan for ayu during 10-40 dah (Kanazawa, 1991a). Data reprinted with permission from: R.P. Wilson (ed.), Handbook of Nutrient Requirements of Finfish, copyright CRC Press, Boca Raton, Florida.

Component	Amount
Formulation	
White fish meal	21.86
Bonito eggs	17.85
Squid meal	6.77
Krill meal	5.63
Soybean protein	9.67
Gluten meal	4.11
Brewer's yeast	4.70
Pollock liver oil	6.00
Soy lecithin	3.00
Vitamin premix	6.00
Mineral premix	5.00
Cellulose	9.41
Proximate analysis (dry)	
Protein	40.0
Performance	
Fish weight (mg)	1.5-12
Daily growth (%)	7
Survival (%)	72

Salmonidae

Nutritional requirements for **salmons** are much better known than for most marine fish and numerous practical feeds are available (Tables 9-19, 9-20). They are fed dry or semi-moist sinking pellets or (mainly in cages) floating pellets.

Atlantic salmon need 40-45% protein (of metabolizable energy) in grower feeds and 45-50% in starter and broodstock feeds (Helland et al., 1991). Juveniles grow best with 25-27% fat. Salt or betaine sometimes is added to feed given during the transition from fresh to salt water (smoltification) to help the fish maintain ionic and osmotic equilibrium.

Atlantic salmon grew slightly faster (100 to 623 vs 512-605 g) with lower protein and higher lipid (35/32.1 vs 37/29.5, 39/25.5, 42/21.1), but FCR was slightly better with the 39/25.5 feed (0.87 vs 0.88-0.93)(Hillestad and Johnsen, 1994).

Table 9-19. Composition (%) of three Oregon Moist Pellet feeds used commercially for salmonids (from Hardy, 1989, with permission).

Component	Feed OM-3 (Oregon Mash)	Feed OP-4	Feed OP-2 (≥3.2 mm pellet)
Herring meal[a]	49.9	47.5	28.0
Wet fish	20.0	30.0	30.0
Dried whey product or dried whey	8.0	4.0	5.0
Wheat germ meal	10.0	Remainder	Remainder
Cottonseed meal (48.5% protein)			15.0
Corn distiller's dried solids			4.0
Fish oil	10.0	6.5-7.0	6.0-6.75
Vitamin premix[b]	1.5	1.5	1.5
Choline chloride (70% choline)	0.5	0.5	0.5
Mineral premix[b]	0.1	0.1	0.1
Sodium bentonite (binder)		3.0	

[a] Anchovy or hake meal can be substituted for half the herring meal in OP-4 and OP-2.
[b] See Hardy (1991) for vitamins and minerals.

Table 9-20. Composition (%) of three Abernathy dry feeds used commercially for salmon (from Hardy, 1989, with permission).

Component	Feed S8-2 (starter)	Feed 18-2 (crumbles)	Feed 19-2 (pellets)
Herring meal	58	55	50
Condensed fish solubles	3	3	3
Poultry byproduct meal	1.5	1.5	1.5
Blood flour or meal	10	10	10
Dried whey	5	5	5
Wheat germ meal		5	5
Wheat middlings, mill run, or shorts	6.27	7.22	12.22
Fish oil and up to 2% soy lecithin	12	9	9
Vitamin premix[a]	1.5	1.5	1.5
Choline chloride (60%)	0.58	0.58	0.58
Ascorbic acid	0.1	0.1	0.1
Mineral premix[a]	0.05	0.1	0.1
Lignin sulfonate (binder)	2.0	2.0	2.0

[a] See Hardy (1991) for vitamins and minerals.

Pacific salmon need 45-50% protein (Hardy, 1996) and 1-2% ω3 fatty acids (Hardy, 1991). Twenty percent dextrin was optimal for chinook salmon. High-performance feeds might contain 42-46% protein and 30-33% fish oil (Forster, 1995). Hardy (1996) gave this general formula for an extruded salmon feed: 50% fish meal, 8% poultry byproduct meal, 5% soybean meal, 10% ground whole wheat, 7% wheat byproducts, 3% molasses, 14% fish oil, 2% vitamin premix, 0.4% choline chloride (60%), 0.5% ascorbic acid, and 0.1% mineral premix (46% protein, >20% fat, 8% moisture).

Arctic char (5-100 g) had good growth and conversion with 54% protein and 20% lipid, but the cheapest feed per unit gain had 44% protein and 20% lipid (Jobling, 1991).

Gadidae

Atlantic cod are demersal carnivores. Lie et al. (1988) suggested that an optimal practical feed would provide 60% available energy from protein, 25% from fat, and 15% from carbohydrate. A feed containing 52% protein, 11% fat, 17% carbohydrate, and 20% fiber and ash would be suitable. Table 9-21 describes an experimental feed that performed well.

Prolongation of live foods for cod can make weaning more difficult, but maybe more so for slow growers (i.e., slow learners?). When they were weaned late, larger cod (~108 g) ate, survived (90 vs 40%), and grew better than small ones (~16 g)(Otterå et al., 1994). Survival did not vary with water content of the feeds, but growth during the 21-d period was better with ~50% dietary moisture than with ~30 or ~0%.

In a study by dos Santos et al. (1993), growth was best for cod given whole herring, and conversion was best with prawns (Table 9-22). Enriched herring paste produced intermediate growth and the worst conversion.

When Clark et al. (1995) offered the same weight of two feeds and whole capelin to cod in cages, differences in growth probably resulted from differences in consumption (Table 9-23). The cod ate capelin better than pellets and also grew better.

Cyprinodontidae

Killifishes are not very demanding nutritionally and usually are raised in ponds with abundant natural food. Wild adult goldspotted killifish (46-87 mm SL) ate mostly copepods, and also ostracods, invertebrate eggs, gastropods, and amphipods (Motta et al., 1995). Wild adult longnose killifish (73-114 mm SL) ate mostly bivalves, with some trematodes and invertebrate eggs. Various minnow, catfish, trout, and even poultry feeds have been given to killifish (Trimble et al., 1981; pers. obs.). As with freshwater baitfish,

Table 9-21. Composition (%, except as noted) and performance of an experimental feed (semi-moist, bulk) used in Norway for Atlantic cod (Lie et al., 1986, 1988).

Component	Amount
Formulation	
Squid mantle	91.3
Starch	4.3
Capelin oil	4.1
Vitamin premix[a]	0.1
Mineral premix[b]	0.2
Proximate analysis (dry)	
Protein	57
Fat	16
Carbohydrate	18
Ash	5
Protein/energy (mg P/kcal DE)[c]	139
Protein/energy (mg P/kJ DE)[c]	33.2
Performance	
Fish weight (g)	173-318
Daily growth (%)	1.01
Feed conversion ratio (kg/kg)	0.8

[a] In mg/kg premix: thiamin hydrochloride 2.5, riboflavin 5.0, pyridoxine 2.5, calcium pantothenate 10.0, niacin 37.5, folacin 1.25, ascorbic acid 100, Rovimix AD_3 (type 500/100, containing vitamins A and D) 1.0, Rovimix E-50 Adsorbate (containing α-tocopherol acetate) 15.0.
[b] Commercial mineral premix for poultry and swine.
[c] Digestible energy was calculated using the values 4 kcal/g protein, 8 kcal/g fat, 3 kcal/g carbohydrate.

cottonseed meal at 57-225 kg/ha is good for fertilizing ponds, but complete formulated feeds are better for health and growth of the fish (Collins, 1994; Lochmann and Phillips, 1996).

Atherinidae

Wild juvenile **inland silversides** ate: (11-26 mm SL) mostly veliger larvae and detritus, with some copepods and crab larvae; (37-60 mm) mostly veligers, copepods, crab larvae, and mysid shrimp (Carr and Adams, 1973).

Inland silversides, after 7 d of rotifers and 3 d of *Artemia*, were offered a microencapsulated feed containing 56% krill meal and 25% whole egg, with or without 0.5% proteases and 0.5% amylase (42% protein, 15% fat)(Ashraf et al., 1993b). Both krill feeds resulted in growth (during 3 weeks) similar to that with the control diet of *Artemia* (3.2 to 21-22 mg), but survival was lower (73 and 60 vs 100%). Another feed with 60% casein was not eaten well.

Syngnathidae

Pipefishes and **seahorses** stay healthier if fed mostly live and fresh food but can get by on frozen seafoods at least for a while (Giwojna, 1996). The foods should be small and moving. Unless a formulated feed is used, vitamin and EFA enrichment will help. Wild adult **gulf pipefish** (91-149 mm SL) ate mostly amphipods, but also some shrimp and barnacle larvae (Motta et al., 1995). To help seahorses feed continuously (which is best for them), Baldassano (1996) suggested using a cylinder of white netting material as a feeding platform at which *Artemia* will aggregate.

Scorpaenidae

Kurosoi were weaned to minced sand lance after reaching 12 mm TL (Kusakari, 1991). Between 12 and 50 mm TL, minced, then chopped, then whole, sand lances could be eaten.

Table 9-22. Composition (%, except as noted) and performance of three fresh foods used to experimentally raise Atlantic cod in 350-L tanks for 45 d in Norway (dos Santos et al., 1993).

Component	Prawn	Herring	Enriched herring
Formulation			
Prawns (*Pandalus borealis*)	100		
Whole Atlantic herring		100	79
Herring meal			8
Capelin oil			6
Gelatin			2
Water			5
Proximate analysis (wet)			
Protein	11.5	17.2	25.1
Lipid	3.5	12.0	21.1
Ash	2.9	3.1	3.3
Moisture	79.0	63.0	50.8
Protein/energy (mg P/kcal GE)	105	71	78
Protein/energy (mg P/kJ GE)	25.0	17.0	18.7
Performance			
Fish weight (g)	528-666	509-679	466-607
Daily growth (%)	~0.45	~0.57	~0.52
Feed conversion ratio (kg/kg)	1.0	1.25	1.67
Protein conversion ratio (kg/kg)	0.53	0.56	0.77

Table 9-23. Composition (%, except as noted) and performance of three foods used for captive wild Atlantic cod raised experimentally in 145-m^3 cages in Canada (Clark et al., 1995).

Component	Low-protein feed	High-protein feed	Capelin
Formulation			
Whole capelin	22.0	15.0	100.0
Fish silage	10.0	10.0	
Capelin meal	43.0	53.0	
Wheat middlings	4.2	8.0	
Starch	2.0	2.0	
Herring oil	14.8	8.0	
Guar gum	2.0	2.0	
Vitamins	1.0	1.0	
Minerals	1.0	1.0	
Proximate analysis (dry)			
Protein	48.9	61.4	82.9
Lipid	28.9	20.3	38.2
Ash	7.7	7.7	10.6
Gross energy (kcal/g)			
Gross energy (kJ/g)	18.7	18.6	27.6
Protein/energy (mg P/kcal GE)	109	138	126
Protein/energy (mg P/kJ GE)	26.1	33.0	30.0
Proximate analysis (of air dry weight)			
Moisture	27.8	31.1	80.1
Performance			
Fish weight (g)	1.9-2.6	1.9-2.4	1.9-2.9

Commercial crumbles also were fed to fish larger than 14 mm. Kurosoi also are fed moist pellets (Jeon, 1995). In cage culture, they are fed five times a day at first, then twice a day after reaching 100 g (Jeon, 1995). Cannibalism can be a problem when kurosoi are 18-70 mm TL.

Suitable dietary levels are 40% protein (8-300 g fish), 7-10% lipids, up to 25% digestible carbohydrate, and 125 mg protein/kcal DE (36 g, initially)(Jeon, 1995). The ω3HUFA requirement for small juveniles (2 g, initially) was met by 1% EPA or DHA, but DHA was better.

Growth and health of kurosoi (1-121 g) were improved by adding 5% seamustard powder (*Undaria pinnatifida*) to their feed (Yi and Chang, 1994).

Growth was better for kurosoi (138-208 g) given a feed containing 70% white fish meal and 13% dextrin (51% protein, 9.2% fat) than for two feeds with 35% casein and 15% gelatin (54/12.6, supplemented with four nonessential amino acids; 51/11.6, supplemented with eight essential amino acids plus cystine)(Lee et al., 1993a). FCR was similar for the fish-meal (1.02) and essential-AA feeds (1.04), and better than for the nonessential-AA feed (1.18).

Growth and FCR were similar with a feed containing 55% white fish meal and two containing 20% soybean meal, or 10% soybean meal + 10% corn gluten meal + 10% meat & bone meal substituted for part of the fish meal (Table 9-24; Lee and Jeon, 1996). Including 0.5% DL-methionine and 0.5% L-lysine with 40% fish meal and 20% soybean meal did not significantly improve performance.

Growth and FCR were not significantly worsened by inclusion of up to 15% cottonseed meal, but did become worse when more than 5% (10-15%) rapeseed meal was added (Table 9-24; Lee and Yoo, 1996).

Growth and FCR were similar for feeds containing 24-52% white fish meal and 0 or 12.5% meat meal, 0-5% blood meal, 0 or 5% feather meal, 0 or 5% soybean meal, 0-7.5% corn gluten meal, 19-32% wheat flour, but they were worse when 5 or 10% extruded soybean meal and 10 or 20% extruded corn were included (Table 9-24; Lee et al., 1996).

Centropomidae

Barramundi published dietary requirements include: 43% protein for juveniles, 10% lipid, 0.5% ω6PUFA, 0.5% ω3PUFA, and 20-25% carbohydrate (Feed Development Section, 1994). Boonyaratpalin (1991) reported that, for grow-out, barramundi need at least 40-45% protein (with 12% fat) and juveniles need 1.0-1.7% ω3HUFA. Boonyaratpalin and Wanakowat (1993) showed that thiamin, riboflavin, and pantothenic acid are essential for barramundi, but inositol is not as critical. In Asia, the use of sinking soft-moist or dry pellets is increasing, but trash fish is still widely used. A feed formula popular in Thailand is described in Table 9-25. Another formula is given by Feed Development Section (1994).

Simple feeds containing 77% fish meal with moisture adjusted to 7-32% were well utilized (Chou, 1984). Chou (1985) later fed barramundi (19-223 g) three similar feeds (containing 54-55% protein, 7.4-9.4% fat, 12-13% carbohydrate) differing in fish meal source: Norwegian herring meal, Thai fish meal, and Singapore fish meal, plus a fourth diet of deboned fish. The best growth (1.3%/d) and FCR (1.2) occurred with the Singapore fish meal (highest fat) feed. Freshness of the fish meal might have been a factor.

Barramundi (9-566 g) were raised in tank trials with commercial and experimental feeds (MacKinnon et al., 1987; Tucker et al., 1988; MacKinnon, 1990). Growth rate ranged from 2.68 to 0.67%/d and FCR from 0.89 to 1.13; a mean weight of 560 g was reached 12 months after hatching (Table 9-9). Barramundi (15-71 g) were fed a series of six dry feeds containing 20, 40, or 60% anchovy meal and 0.7-4.6% menhaden oil (2.3-9.4% total fish oil, 10.0-16.9% total fat), with 48-52% protein and 8-14% carbohydrate (Tucker et al., 1988). Protein and carbohydrate ranges were narrow and probably not factors. Fish meal content alone did not seem to affect growth or conversion. The feed containing the

Table 9-24. Composition (%, except as noted) and performance of eight feeds used to experimentally raise kurosoi in aquariums in Korea (SBM = soybean meal evaluation for 9 weeks, Lee and Jeon, 1996; CSM = cottonseed meal, RSM = rapeseed meal evaluations for 9 weeks, Lee and Yoo, 1996; SPAP = substitution of plant and animal proteins for 15 weeks, Lee et. al., 1996).

Component	SBM 1	SBM 5	SBM 8	CSM/RSM	CSM 15	RSM 5	SPAP 1	SPAP 5
Formulation								
White fish meal	55.0	41.8	35.0	55.0	47.5	52.5	52.0	24.0
Soybean meal		20.0	10.0					5.0
Meat meal								12.5
Meat & bone meal			10.0					
Blood meal								4.5
Feather meal								5.0
Cottonseed meal					15.0			
Rapeseed meal						5.0		
Wheat flour	28.3	21.0	17.7	28.3	20.4	25.7	29.5	28.8
Corn gluten meal			10.0					0.7
Yeast	3	3	3	3	3	3	3.0	3.0
L-lysine HCl								0.4
L-methionine								0.6
Squid liver oil	3.7	4.2	4.3	3.7	4.1	3.8	4.3	4.3
Enzyme mixture								0.2
Vitamin premix	3 [a]	3 [a]	3 [a]	3	3	3	3.0	3.0
Mineral premix	4 [b]	4 [b]	4 [b]	4	4	4	4.0	4.0
Sodium alginate	4	3	3	3	3	3	4.0	4.0
Proximate analysis (dry)								
Crude protein	49.0	47.3	48.0	49.0	46.7	47.6	44.6	48.7
Crude lipid	7.5	7.5	7.8	7.5	7.7	7.7	8.6	8.1
Carbohydrate (difference)				27.5	26.6	27.9	26.8	26.3
Crude fiber				2.1	5.8	3.1	3.9	4.1
Crude ash	13.9	12.6	16.3	13.9	13.2	13.7	13.1	9.8
Protein/energy (mg P/kcal DE)[c]				145	142	142	136	144
Protein/energy (mg P/kJ DE)[c]				34.6	34.0	33.9	32.5	34.4
Performance								
Fish weight (g)	23-60	23-55	23-56	23-60	23-57	23-55	114-170	114-169
Daily growth (%)	~1.54	~1.40	~1.42	~1.53	~1.44	~1.39	~0.38	~0.37
Feed conversion ratio (kg/kg)	1.20	1.24	1.24	1.20	1.24	1.19	1.79	1.87

Table 9-24. (Notes)

[a] In mg/kg premix: thiamine 2.1, riboflavin 7, pyridoxine 1.4, niacin 27.8, folacin 0.5, cyanocobalamin 0.003, Ca-D-pantothenate 9.7, D-biotin 4.2, choline chloride 278.3, myo-inositol 139.1, PABA 13.9, vitamin C 92.7, vitamin A 0.6, vitamin D_3 0.002, α-tocopheryl acetate 14.5, vitamin K_3 1.4.

[b] In mg/kg premix: $MgSO_4 \cdot 7H_2O$ 80, $NaH_2PO_4 \cdot 2H_2O$ 370, potassium chloride 130, ferric citrate 40, $ZnSO_4 \cdot 7H_2O$ 20, calcium lactate 356.5, copper chloride 0.2, $AlCl_3 \cdot 6H_2O$ 0.15, $Na_2Se_2O_3$ 0.01, $MnSO_4 \cdot H_2O$ 2, $CoCl_2 \cdot 6H_2O$ 1.

[c] Digestible energy was calculated using the values 4 kcal/g protein, 8 kcal/g fat, 3 kcal/g carbohydrate.

Table 9-25. Composition (%) of a moist feed used in Thailand for barramundi (Boonyaratpalin, 1988).

Component	Amount
Formulation	
Fish meal	70.0
Rice bran	12.4
Gelatinized starch	10.0
Marine fish oil	1.5
Vegetable oil	3.0
Vitamin premix	1.0
Ascorbic acid	0.1
Mineral premix	2.0
Combine with	
Water	40-50

most fish meal (60%) and fat (16.9% fat, 9.4% fish oil) produced the best growth (2.68%/d) and FCR (0.89); however a feed containing only 20% fish meal and 13.4% fat (5.3% fish oil) produced similar results (2.64%/d, FCR 1.04). Feeds with 1.4-4.6% added fish oil (5.3-9.4% total fish oil; 13.4-16.9% total fat) produced better growth and feed conversion than feeds with ≤1% added fish oil (≤4.2% total fish oil; ≤11.6% total fat). This suggests that feeds for young barramundi should contain at least 5-6% fish oil (or equivalent). Total fat in the range 13-17% with at least half from fish oil (or equivalent) seems appropriate.

Fuchs (1987) raised barramundi in tanks and cages on a starter feed (56% protein, 16% fat) and a grower feed (55% protein, 12% fat) containing normal fish meal (35, 33%), fish protein concentrate (21, 18.9%), meat & bone meal (5, 4.7%); dried whey (0, 3.7%); soybean meal (8, 10.3%), lactic yeast (9, 9.4%); leaf protein concentrate (1.5, 2.8%); wheat (4.5, 8%); corn (6, 0%); guaranate (2, 2%), red capelin oil (2, 3.7%), and corn oil (2.6, 0%). With the starter feed, fish were raised from 1.1 g to 25.6 g in 95 d (growth 3.28%/d, FCR 0.9). With the grower feed, fish grew from 32 g to 677 g in 187 d (growth 1.62%/d, FCR 1.4). A mean weight of 500 g was reached about 12 months after hatching.

Tubongbanua (1987) raised barramundi (0.36 g, initially) in earthen ponds for 114 d on five feeds containing either fish meal (pelleted or mashed), raw trash fish, pasteurized trash fish, or sun-dried trash fish. Each contained about 40% protein (108 mg protein/kcal). Best growth was with pelleted fish meal or sun-dried trash fish (3.93%/d), but best FCR (4.8, moist) was with pasteurized trash fish.

AQUACOP et al. (1990b) obtained better growth (1.8%/d) and FCR (1.3) with barramundi (35-173 g) on a control feed containing 34% Norseamink, 20% fish concentrate, 4% meat & bone meal, 4% dried whey, 10% soybean concentrate, 8% whole wheat, 10% yeast, 3% alfalfa, and 4% capelin oil (48% protein, 10.6% fat) than on test feeds containing no fish concentrate but more or less protein (34-54%) and fat (8-12%).

Barramundi growth and conversion apparently improved when frozen fish was added to a dry feed, but the only significant difference was that final weight with the dry feed was

Table 9-26. Composition (%, except as noted) and performance of three experimental foods (a dry pellet, frozen fish, and the two combined) used for barramundi and greasy groupers raised in 1-m^3 cages for 155 d in Indonesia (from Tacon et al., 1991a, with permission).

Component	Dry pellet RD-1	Moist pellet	Frozen fish
Formulation			
Frozen splendid pony fish		60	100
Brown fish meal	75.0	30.0	
Wheat middlings	8.6	3.4	
Wheat flour	10.0	4.0	
Fish oil	4.2	1.7	
Lecithin	0.75	0.3	
Choline chloride (50%)	0.40	0.16	
Vitamin premix[a]	0.33	0.13	
Mineral premix[b]	0.037	0.15	
Zeolite	0.75	0.3	
Proximate analysis (dry)			
Crude protein	49.7	55.0	67.2
Crude lipid	10.8	8.6	13.0
Carbohydrate (NFE)	23.2	19.0	—
Crude fiber	5.0	3.9	—
Ash	11.2	13.5	22.2
Protein/energy (mg P/kcal GE)[c]	100	114	134
Protein/energy (mg P/kJ GE)[c]	23.9	27.2	32.1
Proximate analysis (of air dry weight)			
Moisture	10.0	39.0	74.5
Performance for barramundi			
Fish weight (g)	7.2-73.5	8.5-164	8.9-177
Daily growth (g/d)	0.66	1.55	1.68
Daily growth (%)	2.32	2.96	2.99
Feed conversion ratio (kg/kg)	1.69	1.15	0.96
Protein conversion ratio (kg/kg)	0.85	0.64	0.64
Performance for greasy groupers			
Fish weight (g)	23.8-108	25.8-246	23.8-266
Daily growth (g/d)	0.84	2.20	2.42
Daily growth (%)	1.51	2.25	2.41
Feed conversion ratio (kg/kg)	1.36	1.06	0.89
Protein conversion ratio (kg/kg)	0.68	0.58	0.61

[a] In mg/kg feed: thiamine 32, riboflavin 40, pyridoxine 32, nicotinic acid 160, folic acid 8, cyanocobalamin 0.04, pantothenic acid 120, biotin 0.4, inositol 300, vitamin C 800, vitamin E 200, vitamin K_3 8; in IU/kg feed: vitamin A 4000, vitamin D_3 2000.

[b] In mg/kg feed: iodine 3, iron 30, zinc 50, copper 3, manganese 25, trivalent chromium 0.25, selenium 0.1, cobalt 0.5.

[c] Gross energy was calculated by the authors, using the values 5.6 kcal/g protein, 9.5 kcal/g fat, 4.1 kcal/g carbohydrate.

lowest of the three (Table 9-26; Tacon et al., 1991a).

Barramundi (89-601 g) grew faster with trash fish than with two experimental moist pellets (Chaitanawisuti and Piyatiratitivorakul, 1994).

When moist feeds containing 5-33% fish meal, 0-35% trash fish, copra meal, soybean meal, rice bran, and yeast (~35% moisture) were given to barramundi (0.26-1.6 g, 4-14 g), growth was better with the four feeds containing 0-35% unpasteurized trash fish than with another containing 35% pasteurized trash fish (Marasigan, 1994).

Partial substitution of shrimp head meal for a portion of the fish meal in barramundi feed resulted in better growth (Duray, 1994b). Equal amounts of cod liver oil and soybean oil were better for growth and survival than either oil alone.

When barramundi (1.3-41 g) were given feeds with 35, 42.5, or 50% protein and 5, 10, or 15% lipid, growth was best with 50% protein/15% lipid, but not significantly better than with 42.5/10 or 50/10 (Catacutan and Coloso, 1995). FCR was significantly better with the 50/15 feed than with any other. Protein conversion was significantly better with the 35/10 and 35/15 feeds.

Common snook (initial weight 3.2 g) had better growth and conversion on commercial salmon feeds than on trout or catfish feeds (Tucker, 1987b). Trials with experimental feeds were conducted with fish in the size range 16-726 g (Table 9-9). Growth rate ranged from 2.08 to 0.67%/d and FCR from 0.73 to 1.10. Snook in the range 504-726 g gained 4.1 g/d with a feed conversion ratio of 1.10. In 1986, a 500-g snook could be produced in 12 months with a feed cost of US$0.33. For 16-44 gram snook, feed conversion with JT8560 was significantly better than with JT8520, possibly because of a fatty acid deficiency or a carbohydrate excess in JT8520 (Tucker, 1987b). Linear regression of feed conversion ratio (0.74-0.86) vs percent carbohydrate (15.9-33.5%) for six feeds given to juvenile common snook was significant ($r^2 = 0.736$, $p = 0.0004$; Fig. 9-3). Growth and feed conversion were good when more than 50% animal ingredients and less than 25% carbohydrate were included.

Percichthyidae

European sea bass's arginine requirement (2.1 g initially) was 18.1 g/kg (Tibaldi et al., 1994). A number of studies have been done on digestive enzymes in larvae (e.g., Cahu and Infante, 1994; Infante and Cahu, 1994a, 1994b) and juveniles (e.g., Caruso and Genovese, 1995; Lindner et al., 1995).

An experimental extruded weaning feed contained 25% cod powder, 12% egg white albumin, 4% hemoglobin, 11% whey protein concentrate, 11% isolated soybean protein, 3% wheat gluten, 13% native corn starch, 3% α-cellulose, 4% hydrogenated coconut oil, vitamins, minerals, and other additives; varying amounts of fish oil were added (Coutteau et al., 1995, 1996).

Hidalgo and Alliot (1988) obtained the best growth (1.01%/d) and FCR (1.56) with European sea bass (37-57 g) on an intermediate feed containing 66.3% fish meal, 15.0% hydrolyzed corn starch, 7.4% cellulose, 2% guaranate, 2.3% cod liver oil, 4.0% coconut oil, and 3% vitamins and minerals (50% protein, 12% fat, 15% digestible carbohydrate, calculated) out of a range of feeds containing 40-80% fish meal (30-60% protein, 12% fat, 3-39% carbohydrate).

Pérez et al. (1997) concluded that 45% protein, 12-14% lipid, and ≤30% carbohydrate were suitable for sea bass in the range 2.5-18.4 g. Their best feed (2.05%/d growth, 1.34 FCR) contained 41% anchovy meal, 8.5% blood meal, 12% brewer's yeast, 10% wheat shorts, 12.5% precooked starch, 9.1% fish oil, 4.3% cellulose, and 2% binder.

Feeds developed by IFREMER that are suitable for weaning and ongrowing of European sea bass (Coves et al., 1991) are described in Table 9-27.

Striped bass and **hybrid striped bass** feeds were generally discussed by Nicholson et al. (1990). Some of the literature on striped and hybrid bass feeding studies was reviewed by Tucker et al. (1993a). Commercial feeds designated for striped bass are available, but there is room for improvement. Out of commercial feeds used previously, catfish feeds were definitely unsuitable, trout feeds were marginally good, and salmon feeds were the best, but contained too much fat.

When given a daily ration of 1.0-1.5% commercial salmon feed (45% protein, 12.4% fat), growth, conversion, composition, and physiological indices were near optimal for striped bass (38-130 g) at 19°C (Hung et al., 1993).

An unidentified deficiency in commercial trout feed led to ascites and liver, heart, and skeletal muscle necrosis in striped bass (Lemm et al., 1993). Salmon feed resulted in better

Table 9-27. Composition (%) of two IFREMER feeds (extruded, rehydratable, dry pellets) recommended for European sea bass (Coves et al., 1991), turbot (Person-Le Ruyet et al., 1991), and other marine fish. Reprinted with permission from: J.P. McVey (ed.), Handbook of Mariculture, Vol. II, Finfish Aquaculture, copyright CRC Press, Boca Raton, Florida.

Component	Starter	Grower
Formulation		
Norwegian capelin meal	42.0	42.0
Fish protein concentrate	8.0	8.0
Blood meal	4.0	4.0
Hydrolyzed feather meal	3.0	3.0
Corn gluten	7.0	7.0
Wheat meal	4.0	4.0
Wheat germ meal	3.0	3.0
Wheat middlings	4.0	4.0
Brewer's yeast	3.0	3.0
DL-methionine	0.45	0.45
Potato alpha starch	11.0	11.0
Mineral premix[a]	1.0	1.0
Added just before feeding		
Cod liver oil	4.9	6.8
Soy lecithin	1.0	1.0
Vitamin premix[b]	2.0	1.0
Choline chloride (50%)	1.2	0.6
Ascorbic acid	0.28	0.14
Inosine	0.16	
BHT	0.01	0.01
Proximate analysis (dry)		
Crude protein	51.0	51.0
Crude fat	11.3	13.2

[a] In mg/kg premix: dibasic calcium phosphate dihydrate 50.0, calcium carbonate 21.5, magnesium carbonate 12.4, potassium chloride 9.0, potassium iodide 0.004, sodium chloride 4.0, sodium fluoride 0.1, iron sulfate heptahydrate 2.0, zinc sulfate monohydrate 0.4, copper sulfate pentahydrate 0.3, manganese sulfate monohydrate 0.3, cobalt sulfate heptahydrate 0.002.

[b] In mg/kg premix: thiamin 1000, riboflavin 2500, pyridoxine 1000, d-calcium pantothenate 5000, niacin 10,000, biotin 100, folacin 500, cyanocobalamin 6, meso inositol 100,000, tocopherol acetate 4000, vitamin K-3 100; in IU/kg premix: vitamin A acetate 1,000,000, cholecalciferol 100,000.

growth and did not cause pathology, but was not optimal.

Sunshine bass were raised in tanks on a commercial salmon starter feed for 6 months after hatching, followed by experimental feeds (Tucker et al., 1993a). During 6-19 months, growth rate went from 1.04 to 0.28%/d, and FCR from 1.05 to 3.16; average weight gain was 2.8-3.0 g/d and a mean weight of 1333 g was reached (Table 9-9). Growth and conversion were better with experimental feeds than with a commercial trout feed. To grow from 0.85 to 900 g, the average sunshine bass needed 291 g salmon starter plus 2816 g trout grower during 1-19 months (FCR 3.46) or 291 g salmon starter plus 1770 g experimental grower during 1-14.5 months (FCR 2.29). To reach 1333 g, each needed 291 g salmon starter plus 3257 g experimental grower during 1-19 months (FCR 2.66).

Sunshine bass (632-927 g) grew better (0.30%/d) with 8% menhaden oil in dry pellets than with 0, 4, or 12% (Fair et al., 1993; Table 9-28).

Keembiyehetty and Gatlin (1997) reported similar growth (0.5 g/d) and FCR (1.7) for sunshine bass (9-38 g) given feeds with 57% menhaden meal or 14.2% menhaden meal, 55.9% soybean meal, and methionine.

Palmetto bass weaning was facilitated with freeze-dried fish silage (Gallagher, 1993). Bass fed *Artemia* and the silage mix up to 28 dah were larger than those fed only *Artemia*,

Table 9-28. Composition (%, except as noted) of two feeds for sunshine bass raised experimentally in 2-m^3 tanks for 18 weeks in South Carolina. Data reprinted from Aquaculture vol. 116, P.H. Fair, W.P. Williams, and T.I.J. Smith, Effect of dietary menhaden oil on growth and muscle fatty acid composition of hybrid striped bass, *Morone chrysops* × *M. saxatilis*. p 174, 177, copyright 1993, with permission from Elsevier Science.

Component	4% fish oil	8% fish oil
Formulation		
Anchovy meal	28.3	28.3
Liquid fish solubles	2.5	2.5
Blood meal (85% protein)	5.0	5.0
Hydrolyzed feather meal	10.0	10.0
Texturized vegetable protein	10.0	10.0
Whole wheat	18.7	16.6
Gelatinized grain	18.7	16.6
Menhaden oil	4.0	8.0
Lecithin	0.5	0.5
Choline chloride (70%)	0.1	0.1
Ascorbyl polyphosphate	0.3	0.3
Vitamins (USFWS no. 30)	0.2	0.2
Minerals (USFWS no. 3)	0.1	0.1
Ameribond	1.7	1.7
Proximate analysis (dry)		
Crude protein	39.6	39.9
Crude fat	8.6	13.0
Carbohydrate (NFE)	35.1	32.1
Crude ash	7.5	7.8
Protein/energy (mg P/kcal DE)[a]	119	111
Protein/energy (mg P/kJ DE)[a]	28.5	26.5
Proximate analysis (% of air dry weight)		
Moisture	9.2	7.2
Performance		
Weight (g)	616-860	632-927
Daily growth (%)	0.27	0.30
Survival (%)	93	96

[a] Digestible energy was calculated using the values 4 kcal/g protein, 8 kcal/g fat, 3 kcal/g carbohydrate.

but with weaning at 20 dah, growth and transformation were retarded.

Even with 6 or 12% fish oil, semipurified feeds containing casein were unpalatable to palmetto bass unless at least 5-10% fish meal was included (Brown et al., 1993). Growth and survival (5 or 11 g initially) were better with a commercial salmon starter (55% protein/15% lipid) than with experimental and other commercial feeds. Fish from all groups had cell damage and accumulation of glycogen and lipid in their livers.

Final weight (162, 164 g) and FCR (1.9, 1.8) were similar for palmetto bass (12 g initially) given an experimental feed (36% protein/6% lipid) and a commercial hybrid striped bass feed (36/8) and were better than with two other experimental feeds (32/6, 40/6)(Swann et al., 1994).

Gallagher (1994) reported good growth and FCR for palmetto bass (100-150 g initially) given a feed containing 16.5% menhaden meal, 44% soybean meal, 10% dextrin, 10% wheat middlings, and 8% fish oil, but the feed was not as suitable for 5-g fish.

Gallagher and LaDouceur (1995) observed that herring meal protein could be replaced up to 25% by blood meal (9.5% of feed), and 75% by hydrolyzed (34%) or low-ash (36%) poultry meals without significantly worsening growth or FCR of palmetto bass (5-31 g). Control fish meal content was 47%.

Brown et al. (1997a) concluded that palmetto bass could be fed 30% solvent-extracted soybean meal without added minerals and 40% with supplementation (22.6% menhaden meal, 40% soybean meal, 23% corn, 5.8% fish oil).

Sullivan and Reigh (1995) ranked these ingredients for 50-g palmetto bass according to apparent digestibility of dry matter (crude protein, energy): menhaden meal (3,1), blood meal (5,?), meat & bone meal (8,2), cottonseed meal (6,3), wheat middlings (2,4), wheat flour (1,6), soybean meal (7,5), rice bran (9,7), and corn grain (4,8). Digestibility ranges were 27-84% for dry matter, 71-93% for protein, and 40-96% for energy.

A sensory evaluation panel could distinguish (flavor and texture) between palmetto bass raised on feeds containing or not containing soybean meal, but acceptability was similar (Postel et al., 1996). The panel also could distinguish pond- from tank-reared fish, but again acceptability was similar.

Serranidae

Wild **groupers** mainly eat fish, crabs, shrimp, mantis shrimp, lobsters, and molluscs (Brulé and Canché, 1993; Brulé et al., 1994. Red groupers seem to prefer crabs first, then shrimp. In Thailand and other areas, groupers have been fed mainly trash fish (with vitamins & minerals), secondarily moist or semimoist pellets, and rarely high-protein dry pellets (Ruangpanit and Yashiro, 1995). Barramundi feeds work fairly well.

Greasy groupers (197-631 g) grew slightly but not significantly better (2.82 vs 2.34 g/d) with experimental pellets (81.5% fish meal, 2.0% cod liver oil, 12% tapioca α-starch) than with trash fish (Chou and Wong, 1985). FCR was better with trash fish (1.7 vs 2.4, DW).

Growth and conversion of greasy groupers apparently improved when frozen fish was added to a dry feed, but the only significant difference was that final weight with the dry feed was lowest of the three (Tacon et al., 1991a; Table 9-26).

Malabar groupers (58-229 g) were raised on feeds containing sun-dried tuna muscle (23-80%), dextrin (0-20%), alpha-cellulose (10-31%), carboxymethylcellulose (6%), cod liver oil (3%), and corn oil (0-14%), with protein contents of 20-70% and similar energy levels (Teng et al., 1978). The best growth (1.5-1.6%/d) and FCR (1.10) occurred with feeds containing 40% or 50% protein, 13.5% or 9.1% fat, and 12.0% or 11.1% digestible carbohydrate.

Chen and Chen (1986) raised Malabar groupers on a series of feeds based on white fish meal and dextrin. The best growth (2.12%/d) occurred with the feed containing the most protein (54%, moist) and fat (7.8%, moist)

A preliminary feed formula used in Thailand for Malabar groupers consisted of 75% fish meal, rice bran, bananas, about 0.5% soybean oil, vitamins, and minerals (Tookwinas, 1990b).

In semipurified feeds for Malabar groupers, 48% protein was optimal for growth, and 48 or 54% resulted in the best FCR (Chen and Tsai, 1994). Shiau and Lan (1996) suggested that protein can be reduced to 44% if gross energy is 3400-3750 cal/g, because differences in growth and conversion were small.

Nassau groupers (40 mm TL, 1 g) raised in 1990 were weaned from *Artemia* to a starter feed, JT9010 (Tables 9-8, 9-9; Tucker and Woodward, 1996). After the fish reached 6-7 g at ~3 months, grower feeds were used. Under less than optimal conditions (including a 17-33°C temperature range) the groupers reached a mean weight of 1.5 kg at 23 months (FCR rising from 0.9 to 1.3) and 2.0 kg in 28 months (FCR up to 1.8, Table 9-9). When the pelleted feed was supplemented with live and frozen food, growth was slightly better. Improvement of early growth rates will allow production of 600-g groupers within 12 months, 1-kg groupers within 18 months, and 2-kg groupers within 24 months.

Nassau groupers (2.3-14.2 g) raised for 56 d in 1993 grew better with an experimental dolphin (62% protein/14.2% lipid; 3.19%/d) feed and a commercial Japanese carnivorous-fish (56/7.8; 3.05%/d) feed than with commercial salmon (53/15.2; 1.71%/d) and trout (44/5.9; 0.93%/d) growers (Ellis et al., 1996). FCR was 0.94 for the carnivorous-fish feed,

1.32 for the dolphin feed, 1.60 for the salmon grower, and 5.55 for the trout grower.

Nassau groupers (1994) given a commercial starter feed (46% protein, 10% lipid) grew fastest (3.3-12.4 g in 63 d; 2.07%/d) at 31°C and had slightly but not significantly better FCR (1.04) than at 22 and 25° (Ellis et al., 1997).

Belted sandfish spawn readily if they are kept healthy (Tucker and Woodward, 1995). Feeding a prepared feed to broodfish is a good test of its quality (Watanabe, 1985). Experimental feed JT9120 was the sole diet for a pair of belted sandfish (~60 mm TL) during an 8-month period. The pair spawned 79 times in 81 d, producing 65,054 eggs, with a mean combined fertilization & hatching rate of 92%.

Mérou wild juveniles grew faster on commercial pellets (7-25 g in 69 d) than on octopus (7-21 g)(Castelló-Orvay and Vilar, 1993).

Apogonidae

Cardinalfishes naturally eat small fish and free-living invertebrates, mostly crustaceans (Fenner, 1996). In aquariums, livebearers, shrimp, and fresh or frozen meaty foods are suitable. They are not likely to thrive on existing commercial compound feeds.

Carangidae

Striped jacks in nature sift through sand to find patches of prey and then aggregate to feed on them (Masuda et al., 1995). Wild juvenile **Florida pompanos** ate calanoid copepods, polychaetes, coquina clams, mole crabs, amphipods, and penaeid shrimp; the adults' diet included shrimp, crabs, mussels, clams, and fish (Watanabe, 1995). Wild juvenile **permits** ate: (15-20 mm SL) mostly shrimp and fish; (21-35 mm) mostly fish, with some crabs, shrimp, and worms; (36-70 mm) mostly crabs and gastropods (Carr and Adams, 1973). Wild juvenile **leatherjackets** ate: (21-40 mm SL) mostly copepods (free-living and parasitic), mysid shrimp, whole fish, and fish skin, flesh, and scales; (61-101 mm) mysids, other shrimp, and free-living copepods (Carr and Adams, 1973).

Yellowtails, unlike some less active fish, can utilize large amounts of lipid without growth reduction from premature satiation, and this has a useful protein-sparing effect (Shimeno et al., 1996b). However, yellowtails do not tolerate more than 10% carbohydrate well. They require at least 50% protein. In addition to frozen fish, yellowtails in Japan are fed three types of pellets. The Oregon moist pellet (OMP) contains about 10 parts powdered dry feed, 10 parts minced fish, and 1 part fish oil. The single moist pellet contains about 10 parts powdered dry feed, 4 parts water, and 1 part fish oil. Extruded pellets (EP) also are used.

Aoki (1995) reported that dry pellets were commonly used to raise yellowtails to 1 kg, but raw trash fish (e.g., sardines, anchovies, and sand lances) with vitamins added often were fed to larger fish. Yellowtails smaller than 100 g should be fed 2-3 times a day. Acceptable dietary protein and fat ranges are 50-52% & 7-10% at 0.2-25 g, 46-51% & 10-15% at 20-400 g, and 41-45% & 16-22 at 300-2500 g and larger.

Shimeno et al. (1985) obtained the best growth (2.94%/d) and FCR (about 1.3) by yellowtails (88-214 g) with a feed containing 57.1% protein, 16.2% fat, and 9.6% digestible carbohydrate (ranges: 55-67% protein, 5.7-22.2% fat, 2-20% carbohydrate).

With 15% dietary lipid, 2.1-3.1% ω3HUFA (of the feed) was good for small juvenile yellowtails (2-16 g), and with 20% lipid, 3.4-4.6% ω3HUFA was appropriate (14-22% of total lipids)(Takeuchi et al., 1992c). Takeuchi et al. (1992e) suggested that 50% protein and 15-20% lipid (2.1% ω3HUFA) was suitable for young juveniles (3-14 g). Takeda et al. (1975) determined that yellowtails (65 g, initially) grew best with a feed containing 144 mg protein/kcal (34.4 mg protein/kJ) metabolizable energy (55% protein, 17% fat, and 0% carbohydrate). Shimeno et al. (1980) obtained the best results with yellowtail (105 g, initially) raised on a feed with the same P/E ratio.

Watanabe et al. (1991a) obtained good results by feeding yellowtails the soft, dry pellet described in Table 9-29. The fish grew from 240 g to 1290 g over a 4-month period. The

Table 9-29. Composition (%, except as noted) and performance of two experimental feeds (dry pellets) used for yellowtail (Watanabe et al., 1991a; Viyakarn et al., 1992).

Component	Watanabe	Viyakarn
Formulation		
White fish meal	35	
Brown fish meal	20	
Sardine meal		25
Krill meal	10	10
Defatted soybean meal		30
Wheat flour	8	7
Wheat gluten	3	3
Potato starch	3	3
Sardine oil	10	
Pollock liver oil	5	18
Vitamin premix	3	2
Mineral premix	3	2
Proximate analysis (dry)		
Protein	47.9	42.2
Fat	29.2	22.8
Starch		14.9
Ash	11.3	8.5
Protein/energy (mg P/kcal DE)[a]	~104	106
Protein/energy (mg P/kJ DE)[a]	~24.9	25.5
Proximate analysis (of air dry weight)		
Moisture	10.6	7.7
Performance		
Fish weight (g)	240-1290	39-175
Daily growth (%)	1.38	3.57
Feed conversion ratio (kg/kg)	1.5	1.13

[a] Digestible energy was calculated using the values 4 kcal/g protein, 8 kcal/g fat, 3 kcal/g carbohydrate.

aversion of yellowtails to conventional dry pellets had made previous feeds less successful. The best one had been an extruded feed that was soaked in water and oil just before feeding, but it was inconvenient to make and use, and also not of the best quality (Watanabe and Sakamoto, 1992).

Yellowtails (39-179 g) given feeds containing 0, 20, and 30% defatted soybean meal with 55, 35, and 25% sardine meal had better growth (3.6 vs 3.1-3.4%/d) and slightly better FCR (1.0-1.1 vs 1.2-1.3) than feeds containing 40 and 50% DSBM with 15 and 5% sardine meal (Table 9-29; Viyakarn et al., 1992). A combination of 25% DSBM, 15% corn gluten meal, and 15% sardine meal (3.9%/d, 1.1 FCR) was slightly better than 40% DSBM and 15% sardine meal (3.8%/d, 1.2 FCR).

Up to 20% corn gluten meal or 20-30% meat meal can be included in yellowtail (21-87 g) feeds (Shimeno et al., 1993b). Shimeno et al. (1994) found that for young fish (17-69 g) malt protein flour could be substituted for fish meal to as much as 20% of the feed, but 30% dietary MPF decreased growth and conversion.

Feeds containing 30% defatted soybean meal with or without 15% meat meal were not as good for larger juveniles (131-784 g) as one with all protein from fish meal, but they gave reasonably good results (Shimeno et al., 1995a). Fish meal was partly replaced with 10% defatted soybean meal or 20% meat meal for smaller juveniles (20-70 g) without reducing growth or conversion, and the 20%-meat-meal feed probably was better than the control.

Fermented soybean meal (30%) in feed resulted in slightly better growth and conversion for young yellowtails (42-135 g) than unfermented SBM (30%), but soy protein concentrate

(23%) or no soybean ingredient were best (Shimeno et al., 1993a).

Feeds with purified soy protein (16-20%) were nearly as good as the non-soybean control feed for yellowtails in the range 17-69 g (Shimeno et al., 1995c). Full-fat, defatted, and extruded defatted soybean meal (all heated) at 30% gave intermediate results. Raw defatted soybean meal at 30% caused weight loss, probably because of trypsin inhibitors and antigens.

An extruded feed containing 30% full-fat soybean meal resulted in better growth and probably better conversion (13-60 g) than non-extruded feeds containing only fish meal as the protein source or up to 30% soybean meal (Shimeno et al., 1997).

For fish in the range 197-617 g, growth was similar with 10, 15, and 20% (16.4-23.8% total lipid) pollock liver oil, but conversion worsened with increased oil (Shimeno et al., 1995b).

Ruchimat et al. (1997) reported a lysine requirement of 1.78% of the dry diet or 4.13% of dietary protein for juveniles.

To identify potential feeding attractants, Hidaka et al. (1992) investigated taste reactions for yellowtails (69-928 g) to 24 organic compounds and found positive responses to several of them (e.g., L-lactic acid). Harada and Miyasaki (1997) found five of 19 fruits attractive for yellowtail.

Greater amberjacks (103 g) had better feed conversion when fed chopped fish to 90% satiation than with 70 or 100% García et al., 1993). Amberjacks (106-500 g) grew better on Atlantic horse mackerel with vitamins & minerals than on 50% mackerel + 50% chicken livers (Greco et al., 1993b).

Striped jacks have been raised on raw fish and moist or dry pellets; grower feeds for red sea bream and yellowtail have been used. Juveniles (34-67 g) could utilize up to 15% starch if more than half of it was gelatinized (Arakawa et al., 1993).

Florida pompanos in past research and commercial ventures often were fed trout feed, sometimes mixed with raw fish, and usually did not grow fast or have good conversion. Small juveniles (0.5-4.6 g) had better growth (4.64%/d) and FCR (1.86) with 8% menhaden oil (~42% protein, 8.09% total fat) than with 0, 4, or 12% (Williams et al., 1985). The experimental feeds contained 50% defatted menhaden meal, 20% defatted soybean meal, and 8.6% dextrin.

Palometa (85-405 g in 16 weeks) growth (~2.8 g/d) and FCR (~2.15) were similar on an experimental feed and three commercial salmonid feeds containing 41-60% protein and 10.7-16.1% lipid (Cole et al., 1997).

Coryphaenidae

Dolphins were weaned to a feed containing low-temperature white fish meal, 55% protein and 20% lipid (Kim et al., 1993). Growth and FCR have not been as good in studies with experimental and commercial dry pelleted feeds, including turbot feeds (Anonymous, 1991), as with fresh or semi-moist feeds. Although dolphins grew rapidly on fresh squid (1.0 kg at 135 dah), they grew from 5 g to only 325 g (125 dah) on a semi-moist pelleted feed that cost US$1.76/kg (Table 9-30; Ostrowski et al., 1989). Lower feeding rates and abdominal bloating were suggested as two factors causing lower growth rates with the pelleted feed. Szyper and Ako (1990) obtained the best growth with dolphins raised in tanks on a moist feed containing 50% pureed squid (range 41-77%), 45% menhaden meal (range 18-46%), and 3% menhaden oil (range 3-10%)(Table 9-30). Suitable levels for the first 3 months are 55-60% protein, 10-14% fat, ≤12% carbohydrate, and 134-146 mg protein/kcal metabolizable energy (Ostrowski, 1995). Only the highest-quality fish meals processed at low temperature should be used in dolphin feeds (Ostrowski et al., 1996a).

Lutjanidae

Mutton snappers fed a commercial dolphin feed (56% protein, 14% fat) grew from 10.5 to 140.8 g in 168 d, with FCR of 1.2 (Watanabe et al., 1998b).

Table 9-30. Composition (%, except as noted) and performance of two feeds used experimentally for dolphins: a moist pellet (Szyper and Ako, 1990) and a semi-moist pellet (Ostrowski et al., 1989).

Component	Moist diet	Semi-moist pellet
Formulation		
Squid meal		35.0
Frozen squid	50.0	
Anchovy meal		23.0
Menhaden meal	45.3	
Corn flour		5.0
High-gluten wheat flour		15.0
Gelatin		3.0
Menhaden oil	3.0	14.0
Vitamin premix	0.8 [a]	1.7 [b]
Choline	0.1	
Choline chloride		1.3
Ascorbic acid	0.2	0.3 [c]
Mineral premix	0.8 [d]	1.7 [e]
Proximate analysis (dry)		
Crude protein		49.7
Crude fat		14.3
Ash		15.5
Proximate analysis (of air dry weight)		
Moisture	44.5	18.6
Performance		
Fish weight (g)	3-7	5-325
Daily growth (%)	10.1	4.64
Feed conversion ratio (kg/kg)	1.8	1.36

[a] AIN vitamins made to 10x concentration by Nutrius, Inc.
[b] In g/550 g premix: thiamin 1.5, riboflavin 1.5, pyridoxine 1.5, calcium pantothenate 6, niacin 100, biotin 0.15, folacin 0.5, inositol 20, ascorbic acid 12, menadione sodium bisulfite 12, vitamin A acetate (500,000 IU/g) 0.003, cholecalciferol (400,000 IU/g) 0.5, tocopherol (250 IU/g) 20, BHT 1, balance as α-cellulose.
[c] Vitamin C sulfate.
[d] BTm mineral mix with iron fumarate increased to 4%.
[e] In g/kg premix: dibasic calcium phosphate dihydrate 366.046, calcium carbonate 261.714, magnesium sulfate anhydrous 53.1, monobasic potassium phosphate 176.834, potassium chloride 17.683, potassium iodate 0.177, sodium chloride 106.1, sodium molybdate 0.147, sodium selenite 0.004, ferrous sulfate 8.842, zinc carbonate 2.653, copper sulfate 0.531, manganous sulfate 6.189, cobalt chloride 0.030.

Gerreidae

Wild juvenile **silver jennies** ate: (11-30 mm SL) mostly copepods, with increasing numbers of polychaetes; (31-60 mm) nearly all polychaetes (Carr and Adams, 1973). Wild adults (64-89 mm SL) ate mostly polychaetes, bivalves, cumaceans, amphipods, and gastropods, with some copepods (Motta et al., 1995).

Haemulidae

Wild juvenile **white grunts** ate: (21-35 mm SL) mostly copepods with a few shrimp; (36-40 mm) shrimp and copepods (Carr and Adams, 1973). Wild juvenile **pigfish** ate: (16-30 mm SL) mostly copepods and shrimp; (31-80) mostly shrimp and polychaetes, with some copepods, amphipods, crabs, and fish larvae (Carr and Adams, 1973).

Sparidae

Red sea bream in nature are carnivorous and their diet includes shrimp, amphipods, crabs, mantis shrimp, starfish, small finfish, and other bottom-dwelling animals (Yamaha Fishery

Journal, 1990). **Pinfish** utilize animal food better than plants, but eelgrass and other seagrasses are important components of their diet (Montgomery and Targett, 1992). Wild juvenile pinfish ate: (10-25 mm SL) mostly copepods and amphipods; (26-35 mm) amphipods, shrimp, and macroalgae; (36-65) macroalgae, with some shrimp, fish, and crustaceans; (66-110 mm) mostly fish, shrimp, and macroalgae (Carr and Adams, 1973). Wild adult pinfish (123-159 mm SL) ate mostly algae and tunicates, with some seagrass (Motta et al., 1995).

Red sea bream larvae survived best and grew well on an MBD containing 35.1% sardine powder, 9.4% white fish meal, 8.3% scallop powder, 5.3% squid powder, 13.6% bean curd refuse powder, and 4% pollock liver oil (Kanazawa et al., 1989). Kyowa (nd.a) recommended simultaneous feeding of live foods and microparticulate feed beginning at 8 dah, with weaning at about 40 d.

Red sea bream weaned to microbound feeds during transformation grew and survived best with two feeds containing 10% white fish meal, 23 or 17% krill meal, 7 or 9% casein, and 17 or 19% soybean protein concentrate (López-Alvarado and Kanazawa, 1997). Based on research with semipurified feeds and fish weighing 7-30 g, Takeuchi et al. (1992c) recommended 20% ω3HUFA of total dietary lipid (10-20%) for juveniles.

Juveniles can be reared with Japanese commercial starter feeds containing 50% hydrolyzed fish meal and 25% krill meal, plus other high-protein ingredients (55% protein, 11% fat)

Table 9-31. Composition (%, except as noted) and performance of two experimental feeds (dry pellets) used for red sea bream: Wakame diet (Yone et al., 1986a) and diet 8 (Takeuchi et al., 1991). Adding wakame to feed for red sea bream in the earlier study enhanced growth, but performance of diet 8 was better.

Component	Wakame diet	Diet 8
Formulation		
White fish meal	68.5	57
Casein		15
α-starch	9.5	10
Defatted rice bran	9.5	
Wakame powder	4.8	
Pollock liver oil	4.8	10
Vitamin premix	1[a]	2
Tocopherol (50%)		0.1
Choline chloride		0.9
Minerals	1.9[b]	5
Proximate analysis (dry)		
Crude protein	54.5	54.6
Crude fat	9.9	15.4
Carbohydrate		11.5
Ash	18.0	
Protein/energy (mg P/kcal DE)[c]	~156	145
Protein/energy (mg P/kJ DE)[c]	~37.2	34.7
Proximate analysis (of air dry weight)		
Moisture	8.3	4.5
Performance		
Fish weight (g)	12-41	8-29
Daily growth (%)	1.77	3.24
Feed conversion ratio (kg/kg)	1.58	0.98

[a] Halver's (1957) vitamin premix plus α-cellulose.
[b] In mg/kg feed: monobasic calcium phosphate monohydrate 19,000, ferric sulfate 480.
[c] Digestible energy was calculated using the values 4 kcal/g protein, 8 kcal/g fat, 3 kcal/g carbohydrate.

(Kumai, 1995). Commercial grower feeds for red sea bream in Japan usually contain 40-45% protein, 3-6% fat, and <20% carbohydrate (Yamaha Fishery Journal, 1990).

Two feed formulas for red sea bream are given in Table 9-31. In the study by Takeuchi et al. (1991), the best growth rate and FCR were obtained with diet 8.

FCR of juveniles (10-31 g) was better (1.40 vs 1.49-1.65) for a feed with 49% white fish meal + 29% malt protein flour than with combinations of more or less of both (Yamamoto et al., 1996). However, growth decreased slightly with increasing MPF.

Including 5% of the algae *Ascophyllum nodosum, Porphyra yezoensis, Spirulina maxima,* or *Undaria pinnatifida* in feeds for red sea bream increased their growth and conversion (Yone et al., 1986b; Mustafa et al., 1994, 1995).

Growth (3.7-3.8 vs 3.3-3.4%/d) and FCR (1.1 vs 1.2) of juveniles (9-48 g) were better with 1087 or 2088 mg/kg dietary L-carnitine than with 75 or 545 mg/kg (Chatzifotis et al., 1995). However, starved red sea bream did not seem to benefit from prior feeding with carnitine (Chatzifotis and Takeuchi, 1997).

Water current of 3.0 body lengths/s reduced weight gain of juveniles (4.1-4.2 vs 4.4-4.6%/d) but not conversion (0.79-0.94) compared with currents of 0 and 1.5 BL/s (2.5-15 g)(Forster and Ogata, 1996).

Black sea bream (8-28 g) were raised on five experimental feeds containing brown fish meal (43-63%), cassava α-starch (19%), wheat flour (3-19%), carboxymethylcellulose (4%), cellulose (0-5%), tallow (1.5-6%), and fish oil (1.5-4%)(Chuang et al., 1985). Best growth (about 1.16%/d) and FCR (1.4) were obtained with the highest protein content (40%, range 28-40%).

Liu et al. (1995) recommended 50% protein in feeds for juveniles (40-79 g) with about equal amounts from animal and plant sources. Liu et al. (1994) reported that 0.68% phosphorus with a Ca/P ratio of 1/2 were best for growth and conversion of black sea bream (0.5-2.4 g).

Addition of 2.5% *Ulva pertusa* meal to feed (34-36% protein, 6.6-6.9% lipid) for black sea bream (56-106 g) caused a reduction of both hepatosomatic index (1.99 to 1.69% of BW) and intraperitoneal fat index (2.92 to 1.34% of BW)(Nakagawa et al., 1993).

Gilthead sea bream larvae (15-27 dah) grew and survived fairly well when fed rotifers and a microfeed containing 0.69% EPA and 1.31% DHA (of DW)(Salhi et al., 1994).

Both a view of *Artemia* and *Artemia* metabolites (especially betaine, arginine, alanine, glycine) in the water increased feeding on a microfeed by gilthead larvae (Kolkovski et al., 1997a, 1997b).

Vergara and Jauncey (1993) suggested these as suitable feed contents for juvenile gilthead sea bream: 48% digestible crude protein, 15.8% digestible lipid, and 117 mg protein/kcal digestible energy. Vergara et al. (1996a) recommended 46% protein, 15% lipid, 92 mg protein/kcal gross energy in feeds given to gilthead sea bream (5-30 g) in a daily ration of 3% at 21-22°C. For smaller fish, (0.8-3.3 g) 56% protein (8.9% lipid, 115 mg P/kcal GE) was better than more or less (Vergara et al., 1996b). A dietary DHA/EPA ratio of 0.5 was better than 1.0 or 2.0 for growth rate (2.3%/d) and minimizing HSI (1.67%) in gilthead sea bream (11.5-43 g)(Ibeas et al., 1997).

In a study evaluating a synthetic amino acid mixture as a feed attractant for gilthead sea bream, the feed containing 1% of the mixture apparently performed the best (Table 9-32; Tandler et al., 1982). A simple practical formula was described by Millan et al. (1989). More than 40% protein and about 8-12% fat are required for grow-out of gilthead sea bream (Kissil, 1991).

FCR (1.07 vs 1.18-1.42) was better for gilthead sea bream (10-64 g) given a feed containing 65% fish meal and 21% wheat meal (56% protein/11.4% fat) than with feeds containing less FM and more WM (40-50% protein/10.8-12.0% fat)(Table 9-33; Santinha et al., 1996). Growth (2.3%/d) was similar with 48% FM/36% WM, 57% FM/28% WM, and 65% FM/21% WM but lower (2.1%/d) with 40% FM/43% WM.

Table 9-32. Composition (%, except as noted) and performance of two experimental dry pellets used for gilthead sea bream in Israel (Tandler et al., 1982; Millan et al., 1989).

Component	Tandler	Millan
Formulation		
Fish meal	14.8	47.5
Poultry/meat byproduct meal	9.9	
Soybean meal (oil-extracted)	44.5	
Wheat flour	19.8	
Wheat middlings		44.0
Fish oil		6.0
Soybean oil	5	
Simulated *Mytilus edulis* extract	1	
Vitamin premix	1[a]	0.5
Mineral premix	4[b]	
Binder		2.0
Antioxidant		0.015
Performance		
Fish weight (g)	10	
Daily growth (%)	2.47	
Feed conversion ratio (kg/kg)	1.32	

[a] Commercial premix for trout.
[b] In mg/kg feed: dibasic calcium phosphate anhydrous 20,000, calcium chloride 8600, calcium sulfate 0.8, magnesium carbonate 5000, potassium chloride 3600, potassium iodide 1.6, sodium chloride 1600, sodium fluoride 40, ferric sulfate 800, zinc sulfate 160, copper sulfate 120, manganese sulfate 120.

Growth (1.79-1.85%/d) and FCR (1.52-1.58) for gilthead sea bream (4.5-21 g) were not significantly different with feed containing 74% white fish meal and no meat & bone meal compared with three feeds containing 47% WFM and 33-42% of one of three grades of M&BM (Table 9-33; Davies et al., 1993).

Robaina et al. (1997) found that corn gluten meal could be substituted (40-86 g) for up to 30% of total protein, or meat & bone meal for 20% (~15% of feed, with ~45% FM).

Growth of juveniles (37-65 g) was slightly better with 10 or 20% soybean meal or 10, 20, or 30% lupin seed meal substituted for fish meal (Table 9-33; Robaina et al., 1995). FCR did not vary much. Lipid droplet accumulation in the liver occurred with 20% of either meal and was higher at 30%. Hepatocyte alterations occurred with 20% SBM and were more severe with 30% SBM. The safe limit for both meals might be ~20%.

Aksnes et al. (1997) found that gilthead sea bream (67-165 g) grew slightly faster with higher quality fish meal (1.0%/d vs 0.9%/d), but FCR was similar; feeds had 61-63% FM.

Nengas et al. (1995) ranked ingredients for 45-g gilthead sea bream according to apparent digestibility (protein, energy): herring meal (1,2), skim milk (2,1), soybean meal (3,10), corn gluten (4,4), poultry meat meal (5,6), sunflower meal (6,?), poultry byproduct meal (7,3), cottonseed meal (8,11), fullfat soybean meal (9,8), standard meat & bone meal (10,5), hydrolyzed feather/poultry/meat (11,12), corn gluten feed (12,15), flaked maize (13,13), feather meal B (14,7), blood meal (15,9), intermediate-grade meat & bone meal (16,14), low-grade meat & bone meal (17,16), feather meal A (18,18), and tomato pulp meal (19,17). The first 10 had protein digestibility of 72-96%; the next six, 44-66%; and the last three, 20-35%. The first six (energy ranking) had energy digestibility of 67 to ~100%; 7-14, 32-64%; and 15-18, 7-24%.

Sheepshead can grow to a mean weight of 480 g within 12 months after hatching on commercial salmon feed (Tucker 1987a). Red sea bream and gilthead sea bream need a few more months to reach this size. Sheepshead initially weighing 220 g had better growth rates and conversion on salmon feed than on trout or catfish feed or fresh food. Because those

Table 9-33. Composition (%, except as noted) and performance of four dry feeds used experimentally for gilthead sea bream in Portugal (Santinha et al., 1996), Greece (Davies et al., 1993), and the Canary Islands (Robaina et al., 1995).

Component	Santinha	Davies	Davies	Robaina
Formulation				
Fish meal	65.0			61.3
White fish meal		74.0	46.7	
Fish protein concentrate	10.0			
Meat & bone meal			42.5	
Soybean meal				20.2
Wheat meal	20.6			
Fish oil	2.6			7.2
Cod liver oil		3.6	0.6	
Corn starch		4.0	4.0	
Starch				4.1
Dextrin		4.0	4.0	1.4
Choline chloride	0.3	0.5	0.5	
Ascorbic acid		0.1	0.1	
Vitamin premix	0.5 [a]	0.4	0.4	2.0
Mineral premix	1.0 [b]	0.7	0.7	0.4
α-cellulose		12.3		
Carboxymethyl cellulose				0.5
Cellulose				3.0
Guar gum		0.4	0.4	
Proximate analysis (dry)				
Protein	55.9	47.2	49.0	55.4
Fat	11.4	9.2	9.2	13.2
Carbohydrate (NFE)	21.8	7.2	8.3	
Fiber		13.5	1.2	
Ash	10.9	17.0	26.0	12.0
Protein/energy (mg P/kcal GE)	103	105	120	126
Protein/energy (mg P/kJ GE)	24.7	25.0	28.7	30.0
Proximate analysis (of air dry weight)				
Moisture		5.9	6.4	8.7
Performance				
Fish weight (g)	9.9-64	4.5-21	4.7-22	39.4-62
Daily growth (%)	2.3	1.81	1.85	0.76
Feed conversion ratio (kg/kg)	1.07	1.58	1.53	1.64
Protein conversion ratio (kg/kg)	0.60	0.79	0.80	0.83

[a] In mg/kg feed: thiamin 5, riboflavin 5, pyridoxine 5, pantothenic acid 100, niacin 10, biotin 0.5, folacin 2, cyanocobalamin 0.05, choline chloride 500, inositol 500, PABA 50, ascorbic acid 200, vitamin E 20, vitamin K_3 5; in IU/kg feed: vitamin A 10,000, vitamin D_3 2000.
[b] In mg/kg feed: magnesium 100, iodine 0.6, fluorine 1, iron 40, copper 5, manganese 10, cobalt 0.4.

fish had been transferred from 0.014-ha ponds to 2.4-m diameter tanks, aggressive encounters caused feed wastage and elevated activity, leading to inflated FCR for all the feeds.

Caribbean sea bream are mostly herbivorous. Cendejas (1993) reported that wild fish ate 72% plants (mostly algae and turtlegrass), 12% crustaceans (mainly amphipods), and 8% molluscs. Small fish (3-7 cm) tended to be more carnivorous, medium ones (7-11 mm) herbivorous, and large ones (>11 mm) omnivorous.

Silver sea bream were weaned to a microcoated feed containing 5% squid meal, 25% oyster extract, 5% shrimp extract, 15% fish solubles, 15% egg yolk, 5% egg albumin, 5% yeast, 2.5% EPA oil, 5% soybean oil, 5% soy lecithin, 2.5% x-carageenan, 0.05% papain, vitamins, and minerals, coated with 83% cholesterol and 17% egg lecithin (adding 9% to formed feed)(Leu, 1994).

Silver sea bream (86-138 g) growth rate increased (0.51 to 1.11%/d)(1.55 to 1.49 FCR) with daily ration (1.0 to 2.5)(Mihelakakis and Kitajima, 1994b).

Under nearly isosmotic conditions (15 ppt), growth of silver sea bream (150-250 g, initially) increased with increasing dietary protein (30-55% protein, 52-90% white fish meal)(Woo and Kelly, 1995).

Yellowfin sea bream larvae were weaned to a microcoated feed containing the same ingredients as for silver sea bream in Leu (1994) but with egg albumin increased to 10%, papain increased to 0.5%, yeast deleted, and a zein coating instead of cholesterol/lecithin (Leu and Chou, 1996).

Yellowfin sea bream (5-11 g) were given a feed containing 60% fish meal, 2% sheep liver, 6.8% soybean meal, 10% corn meal, 5% wheat flour, 10% wheat bran, 3% alfalfa meal, 3% fish oil, and 0.2% vitamins and minerals (50% protein, 11% fat, 18% carbohydrate, 15% ash, 6% moisture)(Jafri et al., 1981). Growth was about 1.76%/d and FCR 2.61.

Chien and Liao (1994) reported that the optimal dietary content of acetone-insoluble lecithin for early juvenile yellowfin sea bream is more than 2.0%.

Common dentex (21-46 g) had the best growth (1.32%/d) and FCR (1.83-1.89) with two feeds containing 65% fish meal (56% protein, 12.0 or 17.5% lipid), but a feed with 56% fish meal (44% protein, 17.2% lipid) was nearly as good (1.25%/d, 1.92 FCR)(Tibaldi et al., 1996).

Sciaenidae

Wild juvenile **spotted seatrout** (20-130 mm SL) ate mostly fish and shrimp (Carr and Adams, 1973). Wild juvenile **silver perch** ate: (6-35 mm SL) mostly copepods, with some shrimp and fish; (36-160) mostly shrimp, with some fish, amphipods, and polychaetes (Carr and Adams, 1973). **Red drum** juveniles in nature eat a wide variety of animal foods, including crustaceans, fish, polychaetes, molluscs, and echinoderms (Overstreet and Heard, 1978).

Red drum nutritional requirements were reviewed by Robinson (1991) and Gatlin (1995). A summary of requirements compared with two experimental feeds is in Table 9-34. Lochmann and Gatlin (1993a) reported a minimum dietary ω3HUFA requirement of ~0.5% (7% of lipid) for red drum (3-17 g). With less ω3HUFA, fin erosion, bleeding gills, shock syndrome, and low growth efficiency, growth rate, and survival occurrred. With too much (1.5%), growth was reduced.

For red drum (0.8-35 g) given feeds containing red or black drum muscle, the best specific growth rate from each study was in the range 3.7-6.5%/d and FCR was 1.05-1.6 (Moon and Gatlin, 1989, 1991; Gatlin et al., 1991; Craig and Gatlin, 1992; Serrano et al., 1992; Lochmann and Gatlin, 1993a, 1993b; Craig et al., 1995). For red drum (1-34 g) given feeds without drum muscle, best growth was in the range 1.9-3.9%/d and FCR was 1.2-2.7 (Daniels and Robinson, 1986; Davis and Robinson, 1987, Brown et al., 1988; Williams and Robinson, 1988; Ellis and Reigh, 1991, Reigh and Ellis, 1992).

Reigh and Ellis (1992) found that a 34%-protein feed containing a 1/1 soy protein/fish protein ratio supported growth as well as higher fish-protein feeds and was more economical. Ellis and Reigh (1991) reported that red drum (2.9 g initially) had better growth and conversion with 10% (highest tested) dietary lipid and 24% (lowest) carbohydrate than with less lipid or more carbohydrate.

Williams and Robinson (1988) obtained better growth and FCR for red drum (1.9-10 g) given 7.4 or 11.2% dietary lipid (mostly menhaden oil) than with 4.0 or 15.0%. Red drum (3.25-20.4 g) had better growth and FCR with 7 or 14% menhaden oil in their feed than with 7% corn, coconut, or saturated menhaden oil (Craig et al., 1995). Craig and Gatlin (1995) reported that red drum (2.6-15.8 g) had better growth and conversion during

Table 9-34. Comparison of experimental feed composition with published dietary requirements for juvenile red drum.

Component	Published (%)	JT9210 (%)	JT9220 (%)	References
Energy (kcal/g of feed)	3.8	4.1	3.8	Daniels & Robinson, 1986
Protein (% of feed)	35-44	58.5	53.4	Daniels & Robinson, 1986
Protein (% of feed)	40	58.5	53.4	Serrano et al., 1992
Lysine (% of feed)	1.55	4.06	3.43	Craig & Gatlin, 1992
Lysine (% of protein)	4.43	6.73	6.32	Craig & Gatlin, 1992
Lysine (% of protein)	5.7	6.73	6.32	Brown et al., 1988
Methionine + cystine (% of feed)	1.06	2.14	1.80	Moon & Gatlin, 1991
Methionine + cystine (% of protein)	3.03	3.54	3.31	Moon & Gatlin, 1991
Threonine (% of feed)	0.8	2.26	1.98	Boren & Gatlin, 1995
Threonine (% of protein)	2.28	3.86	3.72	Boren & Gatlin, 1995
Lipid (% of feed)	7.4-11.2	18.0	13.3	Williams & Robinson, 1988
EPA + DHA (% of lipid)	7	15.8	13.0	Lochmann & Gatlin, 1993a, 1993b
Phosphorus (% of feed)	0.86	~1.22	~1.07	Davis & Robinson, 1987
Zinc (mg/kg of feed)	20-25	205	192	Gatlin et al., 1991
Ascorbic acid (mg/kg of feed)	60-75	400	400	Collins, 1990b

a 6-week period with feeds containing 4.4% coconut oil + 2.0% menhaden oil or 5.4% beef tallow + 1% ω3HUFA mixture (containing 60% EPA + DHA) than with similar feeds containing 6.4% menhaden oil or tricaprylin + menhaden oil or tricaprylin + ω3HUFA mixture.

When Wurts and Stickney (1993) raised red drum on 32%-protein catfish feed plus crayfish and tilapia in a freshwater pond, the fish grew from 35 to 258 g in 61 d (3.27%/d). Growth was good, but initial and final numbers of fish, survival, and likelihood of cannibalism were not reported. Sandifer et al. (1993) raised red drum in saltwater ponds on a 38%-protein commercial trout feed (and natural foods). The fish grew from 1.9 to 1340 g in 543 d with FCR of 2.6. Early growth was 2.5 to 447 g in 244 d (1.82 g/d, 2.12%/d). Later growth was 496 to 1355 g in 198 d (4.36 g/d, 0.51%/d). It is difficult to compare growth in ponds with growth in tanks, because of natural foods, lower visibility, less control, and other factors in ponds. When Wurts and Stickney (1993) raised red drum on a 48%-protein commercial salmon feed in brackish water in tanks, the fish grew from 18 to 50 g in 28 d (3.65%/d) with an FCR of 0.8.

Feeds containing 40-60% fish meal (with up to 20% soybean meal and/or soy protein isolate) were superior to those with less fish meal and more soybean ingredients, and 30% fish meal (29% soybean) was almost as good as 40% (Davis et al., 1995). Consumption of and growth on a 10% fish meal feed with fish or seafood flavoring was better than for the same feed with or without shrimp flavoring and a commercial feed, but not as good as for a 40% fish meal feed. Feed consumption and growth for red drum (13-117 g) decreased when dietary fish meal content was reduced from 30-40% to 20% and 10% (Meilahn et al., 1996).

In a 56-d trial starting with 0.4-g red drum, Davis et al. (1995) obtained an FCR of 0.96 with a feed containing 60.0% menhaden meal, 5.0% shrimp head meal, 2.1% soybean meal, 17.7% wheat starch, 5.0% wheat gluten, 3.1% menhaden oil, and 0.5% soybean oil (45% protein).

When Tucker et al. (unpub.) raised red drum (0.5-20 g) on nine commercial and experimental feeds, the best performance resulted from three feeds containing 58-60% protein and 11-18% fat. The third best feed, JT9110, was similar to JT9120, but with more fish meal (60%) and added fish oil (7.5%). The best feed, JT9210, resulted in slightly faster growth and was more economical than the second best, a popular commercial freeze-dried starter feed (FryFeedKyowa C-2700,® Kyowa Hakko Kogyo Co., Tokyo).

During 1991-1994, red drum were raised from juveniles averaging 2.5 g to adults averaging 6.1 kg (Tucker et al., 1997a). Growth rates with simple experimental starter and grower feeds were similar to those with one of the best-performing commercial starter feeds for marine fish, which was used as a reference (Tables 9-8, 9-9; Fig. 9-6). Over the size range 4.3-2243 g (Fig. 9-7), experimental feed cost per unit of growth gradually increased from \$0.88/kg to \$1.54/kg live weight (1996 prices). Thereafter, FCR and cost fluctuated, apparently because of temperature variation and fish maturation. Growth from 4.3 to 4740 g was accomplished in 27 months with an average feed cost of \$1.74/kg growth. Daily gain peaked at 9.8 g/d between 2.9 and 4.7 kg. Raising the fish beyond 4.7 kg was more costly because of worsening FCR. Thus, considering feeding costs, an increase in harvest size of red drum from the usual 1.0-2.0 kg range up to 4.0-5.0 kg seems feasible.

Jirsa et al. (1997) obtained results similar to those of trial RD-2 (Tables 9-8, 9-9) with their diet 4 (Table 9-35). They reported the best performance with feeds containing more fish meal. Feed cost to produce a unit weight of red drum on diet 4 would be 63% of that on diet 1. Chemical oxygen demand decreased with increasing feed quality, but total phosphorous in the water increased slightly.

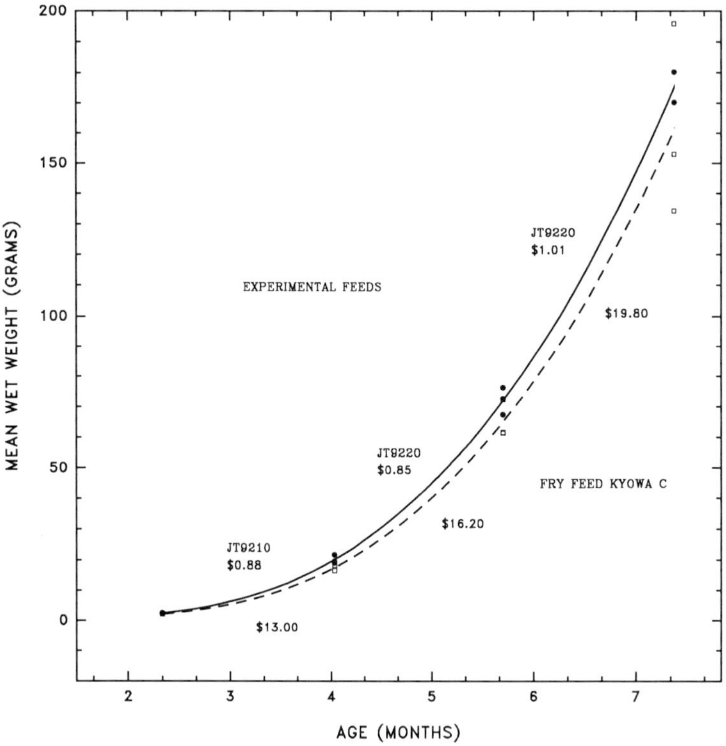

Figure 9-6. Red drum growth and cost (for 1996) of feed per kilogram of growth (Tucker et al., 1997a). The solid line and dots represent two experimental feeds. The dashed line and squares represent FryFeedKyowa C-2700.® Growth (4.1, 2.4, 1.6%/d) and survival (93-100%) were not significantly different. FCR was significantly better for the commercial feed during the first phase (0.65, 0.80) but was only slightly better during the second (0.81, 0.88) and third (0.99, 1.04). The only practical difference was in the cost.

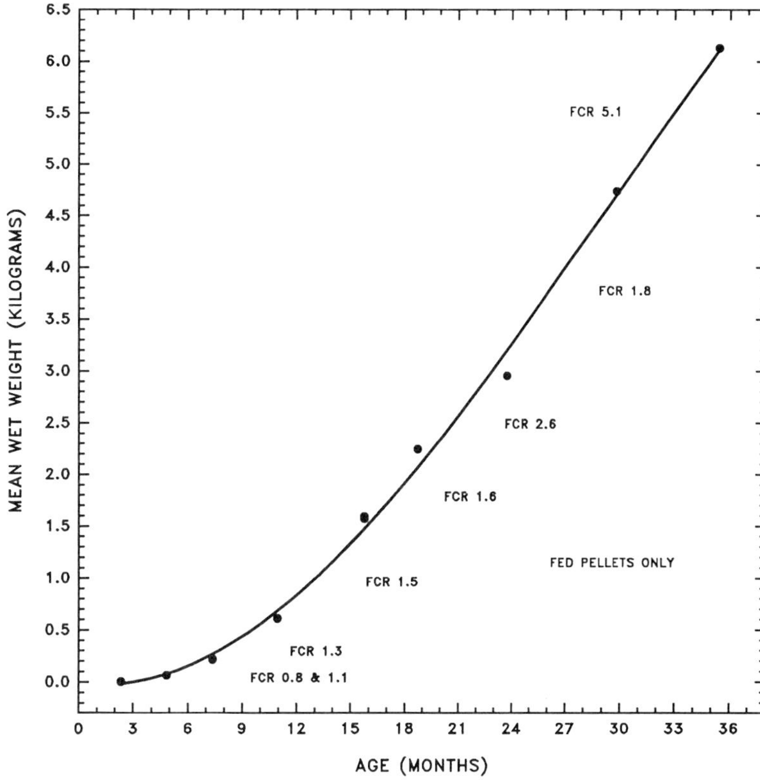

Figure 9-7. Red drum growth and feed conversion ratio (Tucker et al., 1997a). JT9210 was used between the first and second points and JT9220 thereafter.

Gaylord and Gatlin (1996) ranked these ingredients for 150-g red drum according to apparent digestibility of organic matter (crude protein, energy): select menhaden meal (2,1), regular menhaden meal (6,2), poultry byproduct meal (7,4), meat & bone meal (5,3), dehulled soybean meal (3,6), cottonseed meal (4,5), and wheat (1,7). All had a protein digestibility of 77% or more except poultry meal at 48%.

McGoogan and Reigh (1996) ranked commonly used ingredients for 220-g red drum according to apparent digestibility of dry matter (crude protein, energy) as follows: menhaden meal (2,1), blood meal (1,2), sorghum grain (7,5), meat & bone meal (9,4), corn grain (4,3), soybean meal (5,6), cottonseed meal (8,8), wheat middlings (3,7), and rice bran (6,9). All had a protein digestibility of 74% or more.

A feed containing 5.5% anchovy meal and 66% soybean meal was not statistically worse that a control feed (54.7% anchovy meal, no SBM) for 5-g red drum, but 2.7% anchovy meal and 69.6% SBM was not as good unless 2% glycine or 4% fish solubles were added (McGoogan and Gatlin, 1997). L-methionine also was included. Intermediate combinations (6-54% anchovy meal) were not tested.

Spotted seatrout (24-110 g) trials with experimental feeds resulted in daily growth rates from 3.10 to 1.17% and FCRs from 0.78 to 1.55 (Tucker, 1988; Table 9-9). Seatrout in the weight range 22-60 g grew better on feed JT8560 than on five others that had more whole wheat flour (14-37%) and less textured vegetable protein (17-29%). Five months after hatching, the seatrout weighed 110 g.

Table 9-35. Composition (%, except as noted) of feeds used to experimentally raise red drum in 220-L tank systems for 10 weeks in Texas (from Jirsa et al., 1997, with permission).

Component	Diet 1	Diet 2	Diet 3	Diet 4
Formulation				
Menhaden meal	28.6	32.2	35.9	39.6
Fish solubles	1.0	1.0	1.0	1.0
Commercial protein mixture	7.2	8.1	9.0	9.9
Soybean meal	7.2	8.1	9	9.9
Wheat gluten	3.6	4.0	33.1	26.1
Wheat starch	46.8	40.0	33.1	26.1
Menhaden oil	1.0	2.0	3.0	4.0
Lecithin	0.5	0.5	0.5	0.5
Vitamins[a]	3.0	3.0	3.0	3.0
Ascorbyl polyphosphate	0.1	0.1	0.1	0.1
Minerals[b]	0.5	0.5	0.5	0.5
Dipotassium phosphate	0.5	0.5	0.5	0.5
Proximate analysis				
Protein	32.0	36.0	40.0	44.0
Lipid	4.5	5.9	7.3	8.7
Protein/energy (mg P/kcal DE)[c]	100	106	111	116
Protein/energy (mg P/kJ DE)[c]	23.9	25.3	26.6	27.7
Performance				
Fish weight (g)	89-157	89-197	89-222	89-226
Daily growth (g/d)	0.96	1.54	1.90	1.95
Daily growth (%)	0.80	1.13	1.30	1.33
Feed conversion ratio (kg/kg)	2.62	1.88	1.51	1.25
Protein conversion ratio (kg/kg)	4.90	3.87	3.33	3.20

[a] In g/100 g premix: thiamin HCL 0.5, riboflavin 3, pyridoxine HCL 1, Dl-calcium pantothenate 5, nicotinic acid 5, biotin 0.05, folacin 0.18, cyanocobalamin 0.002, choline chloride 100, inositol 5; In IU/kg feed: vitamin A acetate 100,000, dL-tocopherol acetate 2000, menadione 2.
[b] In g/kg premix: cobalt chloride 0.004, cupric sulfate pentahydrate 0.250, ferrous sulfate 4, magnesium sulfate heptahydrate 28.398, manganous sulfate monohydrate 0.650, potassium iodide 0.067, sodium selenite 0.010, zinc sulfate heptahydrate 13.193.
[c] Digestible energy calculated by the authors.

Atlantic croakers had better growth (3.8 to 20.8 g in 7 weeks, ~127 dah) and FCR (1.22) with 45% protein and 8% lipid than with 30-40% protein or 12-16% lipid (Davis and Arnold, 1997).

Chaetodontidae

Butterflyfish have been separated into five feeding groups of varying difficulty for the culturist (Allen, 1985, Giwojna, 1989): 1) Those that are opportunistic omnivores (e.g., **raccoon** and **spotfin butterflyfishes, common bannerfish**) will readily eat fresh, frozen, flake, or freeze-dried foods. A flake feed with 40-50% protein alternating with a high-vegetable flake can be the standard diet, but for good health, a variety of other foods, including algae or other whole plants (e.g., *Ulva*, kelp, Romaine lettuce) should be added. 2) Those that eat copepods, shrimp, pelagic tunicates, various larvae (e.g., **pyramid** and **black pyramid butterflyfishes, schooling bannerfish**) can be fed live zooplankton and *Artemia*, and finely-chopped frozen fish, beef heart, shrimp, squid, clams, and scallops. 3) Those that eat benthic invertebrates such as polychaetes, shrimp, and crabs (e.g., **threadfin, copperbanded**, and **longnose butterflyfishes**) can be given the same foods as the second group plus white worms, tubifex, black worms, bloodworms, glassworms, and mosquito larvae. 4) Those that feed on soft (and sometimes hard) coral polyps (e.g., **lined** and

fourspot butterflyfishes) are not as easy to feed but can be given the same foods as the third group, plus pale anemones (*Aiptasia* sp.) if available. 5) Those that eat only hard coral polyps and mucus (e.g., **ornate**, **chevron**, and **exquisite butterflyfishes**) can be kept alive only by feeding them live coral.

Pomacanthidae

Among wild angelfish, there are three main feeding groups (Allen, 1985). **Pygmy angelfish** and many other small species eat mainly algae. Most of the large angelfish (***Holacanthus*** and ***Pomacanthus*** **spp.**) eat mainly sponges, but also algae and other sessile plants and animals (e.g., tunicates, zoantharians, gorgonians, hydroids, and various eggs). ***Genicanthus*** **spp.** eat mostly pelagic tunicates, but also some bryozoans, polychaetes, and algae. Wrobel (1988) recommended that pygmy angelfish be fed fresh and/or frozen algae (e.g., *Caulerpa* sp.), plus natural foods (e.g., detritus, coral polyps, sponges, anemones, polychaetes, other invertebrates) when possible, *Artemia*, chopped seafoods (e.g., squid, shrimp), and frozen foods, as necessary.

Cichlidae

Tilapias are not very demanding nutritionally, and their requirements are relatively easy to meet (Tables 9-36, 9-37). Wild Nile tilapia ate 63% microalgae and plant detritus, 21% zooplankton, and 10% zoobenthos (Abdelghany, 1993). Luquet (1991) reviewed tilapia nutrition. Published dietary requirements for **Nile tilapia** include: 35% protein for larvae, 25-30% protein for juveniles, 6-10% lipid, 0.5% linoleic acid, <0.9% phosphorus, 25% carbohydrate, and 2500-4300 cal/g digestible energy (Feed Development Section, 1994). Feed formulas for Nile tilapia larvae, juveniles, and broodfish are given by Feed Development Section (1994). Recent examples of literature on their nutrition are: Yakupitiyage (1991), Nour et al. (1992), Ross et al. (1992), Akiyama (1993), Saleh and Eleraky (1993), Fagbenro (1994), Fagbenro et al. (1994), Mazid et al. (1994), Phillips et al. (1994), Wu et al. (1994b, 1995a, 1995b, 1996a, 1996b), Balogun and Fagbenro (1995), Bishop et al. (1995), Guerrero (1995), Macintosh and Little (1995), Reyes-Sosa and Castellanos-Molina (1995), Al-Owafeir and Belal (1996), Diana et al. (1996), Park and Jeong (1996), and Rodríguez-Serna et al. (1996). Some for **blue tilapia** are: Yousif and Alhadhrami (1993), Yousif et al. (1994, 1996); for **Mozambique** tilapia: Gopal et al. (1997); for **hybrid tilapia**: Shiau and Suen (1992), Siddiqui et al. (1994), and Shiau and Liang (1995).

Pomacentridae

Wild adults (~110-135 mm SL) of four species of the damselfish genus ***Chromis*** ate mostly

Table 9-36. Composition (%) of commercial dry feeds used for tilapia (*Oreochromis*) hybrids in Israel (Viola and Arieli, 1983).

Component	Standard	Fish meal
Formulation		
Fish meal	15	25
Soybean meal	20	
Wheat, ground	20	20
Sorghum, ground	45	55
Proximate analysis (dry)		
Protein	25.0	25.0
Fat	3.0	4.0
Ash	5.0	6.0
Calcium	0.6	0.8
Phosphorus	0.7	0.7

Table 9-37. Composition (%) of model dry feeds for tilapia (Lim, 1989, permission to reprint with modifications from Kluwer Academic Publishers).

Component	Pond grower	Raceway grower
Formulation		
Anchovy or menhaden meal	8.0	12.0
Soybean meal (48% protein)	28.8	43.0
Grains or grain byproducts	59.4	38.8
Fat		1.9
Vitamin premix[a]	0.25	0.25
Mineral premix[b]	0.05	0.05
Dicalcium phosphate	1.5	2.0
Binder	2.0	2.0
Proximate analysis (dry)		
Protein	25	32

[a] In mg/kg feed: thiamin 10, riboflavin 15, pyridoxine 10, pantothenic acid 50, niacin 30, folacin 5, choline 500, vitamin C 200; In IU/kg feed: vitamin A 4000, vitamin D 2000, tocopherol 50, vitamin K 10.

[b] In mg/kg feed: iodine 5, iron 44, zinc 100, copper 3, manganese 25, cobalt 0.05, selenium 0.3.

copepods, with some larvaceans, but only a few shrimp larvae, chaeognaths, and polychaete larvae (Tribble and Nishikawa, 1982).

Clownfishes, like other ornamental species, often are fed a combination of flake food, high-quality frozen seafoods, gelatin mixtures, and sometimes live adult *Artemia* or other zooplankton. Astaxanthin is the most important dietary pigment for clownfish coloration; it is converted to zeaxanthin, which becomes β-carotene (Hoff, 1996). Hoff (1996) suggested a gelatin mixture containing beef or chicken heart, fish eggs, fish flesh, fish meal, shrimp, krill, clams, kelp, vitamins, and minerals for both juveniles and broodfish.

Oplegnathidae

Striped knifejaw larvae grew and survived well during 5-38 dah on a microparticulate feed containing fish meal, squid meal, and soybean meal (Watanabe and Kiron, 1994).

Mugilidae

Mullets will eat nearly any organic material that is the right size and does not move away too fast (Marais, 1980; Brusle, 1981; Larson and Shanks, 1996). They feed day and night but more during the day. Once they are juveniles, mullet in tanks scour the tank walls and bottoms, keeping them fairly clean (pers. obs.). Larvae have mouths large enough to easily eat rotifers, but mouth size does not increase proportionately with body size. Therefore, feed size for juveniles cannot be increased as fast as it can for many marine fish. Although not well investigated, the nutritional requirements of mullet are not difficult to meet, and various feeds have been successful (Table 9-38).

Striped mullet have been raised in our hatchery on experimental marine fish feeds and commercial salmon feeds, but they probably would grow nearly as well on lower quality (e.g., trout) feeds.

Redlip mullet up to 1 yr have good growth and conversion with 40-45% protein, 7-11% lipid, and 29-35% carbohydrate (Yoshimatsu et al., 1992c, 1993). After 1 yr, 35% protein and 41% carbohydrate can be used. Based on research with purified feeds, El-Zibdeh et al. (1996a) recommended adding calcium, aluminum, zinc, manganese, copper, iodine, and cobalt to redlip mullet feed.

Thinlip mullet wild juveniles (0.13-1.58 g) had the best growth (3.46%/d) and FCR (1.51) at 36 ppt with a feed containing 60% fish meal, 35% wheat bran, 2% soybean oil, and 3% vitamins & minerals (41% protein, 13.4% lipid)(El-Sayed, 1991). Growth and

Table 9-38. Composition (%) of a dry feed suggested for liza and white mullet in Brazil (Benetti and Fagundes Netto, 1991) and another used as a supplemental food for striped mullet in Taiwan (Liao and Chao, 1991).

Component	Brazil	Taiwan[a]
Formulation		
Fish meal		20
Anchovy meal	29	
Poultry byproduct meal	29	
Cottonseed meal	19	
Rice polishings	19	
Rice bran		30
Wheat flour		10
Yeast		40
Corn oil	1.5	
Saturated fat	1.5	
Vitamin and mineral premix	1	
Proximate analysis (dry, calculated)		
Crude protein	45	
Crude fat	11.3	
Carbohydrate (NFE)	28.2	
Fiber	4.1	
Ash	11.4	
Protein/energy (mg P/kcal DE)[b]	127	
Protein/energy (mg P/kJ DE)[b]	30.3	

[a] Also contained 500 mg/kg vitamin E.
[b] Digestible energy was calculated using the values 4 kcal/g protein, 8 kcal/g fat, 3 kcal/g carbohydrate.

conversion became worse with decreasing salinity.

Thicklip mullet wild juveniles (15-42 g) grew best (1.20%/d, 1.67 FCR) with a feed containing 54% fish meal and 32% corn starch & dextrin (38% protein, 22.8% lipid) but had better FCR (1.09%/d, 1.36 FCR) with a feed containing 85% fish meal and 11.7% starch & dextrin (59% protein, 12.6% lipid)(Ojaveer et al., 1996).

Polynemidae

Pacific threadfins (40-70 dah) grow well with 41-48% protein; 10-14% lipid is suitable and 17% is too much; carbohydrate is well utilized (Ostrowski, 1995).

Stichaeidae

Monkeyface pricklebacks (7-24 g, 4 weeks before the experiment) were raised for 12 weeks on a feed made of seaweeds from their natural diet (10% protein), and the same seaweeds plus casein to reach 30 and 50% protein (Horn et al., 1995). Daily ration was 0.6% with 50% protein, 1.0% with 30% protein, and 1.3% with 10% protein; energy consumption was lower with 50% protein, but nitrogen consumption, assimilation, and excretion rates were similar for all three, suggesting that the protein, rather than energy requirement, was governing appetite. Growth was better with 30% protein (0.41%/d) than with 10 or 50% (0.27, 0.24%/d). FCR was better with 30% protein (1.96) than with 10% (4.54) or 50% (2.44).

Anarhichadidae

Common wolffish had better growth with three dry feeds (22-58 g, 0.24-0.29%/d) than with two moist feeds (24-47 g, 0.13 and 0.16%/d)(Stefanussen et al., 1993). Muscle fat (1.6-4.0%), liver fat (22.2-35.2%), and HSI (4.69% for dry pellets, 3.29% for moist pellets) increased with dietary fat (13.8-26.5%). For good growth, feeds should have >50% protein and <20% carbohydrate.

When given compound feeds from first feeding for 60 d, common wolffish preferred a floating feed to a sinking one and grew better (86-799 g, 3.7%/d vs 86-575 g, 3.2%/d)(Strand et al., 1995).

Blenniidae
Wild juvenile **Florida blennies** (11-40 mm SL) ate mostly amphipods and detritus, with some polychaetes, copepods, bivalves, crabs, fish, and isopods (Carr and Adams, 1973).

Callionymidae
Mandarinfish in aquariums will eat mysid shrimp, amphipods, isopods, benthic copepods, small worms, and protozoans (Delbeek, 1989). They also can be fed frozen mysids, different sizes of *Artemia*, and other finely chopped fresh or frozen foods.

Gobiidae
Wild juvenile **clown gobies** (11-40 mm SL) ate mostly detritus, copepods, and amphipods, with some macroalgae and polychaetes (Carr and Adams, 1973). Wild juvenile **darter gobies** (10-39 mm SL) ate mostly harpacticoid copepods, and also diatoms, ostracods, cyclopoid copepods, and nematodes (Toepfer and Fleeger, 1995).

Siganidae
Rabbitfishes are herbivorous and in nature they consume mainly macroalgae such as *Enteromorpha* sp., *Chaetomorpha* sp., *Gracilaria* sp., *Caulerpa* sp., *Dictyota* sp., and many others (Duray, 1990; Lundberg and Golani, 1995).

Yellowblotch spinefoots became more omnivorous in captivity and would eat a variety of foods (Parazo, 1990). The best growth and conversion were obtained with a feed containing 45% protein (highest tested of six), 7.9% fat, and ~31% carbohydrate (Table 9-39, experimental feed 6), but the same FCR and better protein conversion occurred with a feed containing 36% protein, 8.9% fat, and ~40% carbohydrate. Results with the commercial feed used as a control were slightly better.

Pearly spinefoots grew better on a completely formulated feed (Table 9-40, diet C) than on pure chicken feed or an intermediate mixture. El-Sayed et al. (1995) reported the best growth (0.95 g/d) of captive wild pearly spinefoots (13-141 g) with a density of 30 fish/m^2 (lowest tested) and best FCR (0.92) with a daily ration of 2-3% (lowest tested). However the best production (7.04 kg/m^3) was with 60 fish/m^2 and 3 to >7% daily ration.

Bothidae
Hirame in nature consume anchovies and sand lances especially, and also young horse mackerel, chub mackerel, Japanese whiting, scorpionfish, and righteye flounders (Yamaha Fishery Journal, 1991). Wild juvenile **bay whiffs** (10-49 mm SL) ate mostly *Acartia tonsa* and mysid shrimp, and also polychaetes and a few amphipods (Toepfer and Fleeger, 1995).

Hirame larvae survived best and grew well on a microbound feed containing 24.6% sardine powder, 8.4% white fish meal, 11.5% krill meal, 11.6% scallop powder, 10.7% squid powder, 4.7% bean curd refuse powder, and 6% pollock liver oil (Kanazawa et al., 1989). Hirame older than 20 dah could break down the tripalmitin walls of microencapsulated feeds containing microcrystalline amino acids; however, the walls were up to 95% of the capsule, leaving little room for enclosed nutrients (López-Alavarado et al., 1994). The MCDs were more water-stable than microbound feeds (with alginate, carageenan, or zein). The lipid coating was more stable than protein.

During grow-out, hirame are fed small whole fish, dry pellets, or moist pellets containing minced fish such as sardine, mackerel, or sand lance (Yamaha Fishery Journal, 1991; Matsuoka, 1995). Culturists prefer to use dry pellets from weaning up to 100 mm.

In feeds for juvenile hirame (~1 g initially), hydrolyzed feather meal can be substituted for 20-40% of the white fish meal protein (Kikuchi et al., 1994a), defatted soybean meal for ~50% (Kikuchi et al., 1994b), or meat & bone meal for 20% (Kikuchi et al., 1997a)

Table 9-39. Composition (%, except as noted) and performance of a semi-purified experimental feed (dry crumbles) and a commercial feed (dry pellets) used to experimentally raise yellowblotch spinefoot in 250-L tanks for 8 weeks in the Philippines. Data reprinted from Aquaculture vol. 86, M.M. Parazo, Effect of dietary protein and energy level on growth, protein utilization and carcass composition of rabbitfish, *Siganus guttatus*, p 43, 45-46, copyright 1990, with permission from Elsevier Science.

Component	Experimental feed 6	Commercial feed
Formulation		
Fish meal	10.0	+
Fish solubles		+
Meat & bone meal		+
Casein	28.7	
Gelatin	9.6	
Rice bran		+
Dextrin	19.1	
Corn gluten meal		+
Fats		+
Cholesterol	0.5	
Soy lecithin	2.0	
Cod liver oil	7.0	
Vitamin premix	5.0 [a]	+
Mineral premix	5.0 [b]	+
Limestone powder		+
Sodium chloride		+
Carboxymethyl cellulose	5.0	
Celufil	8.1	
Proximate analysis (dry)		
Crude protein	45.1	47.4
Crude fat	7.9	3.5
Carbohydrate (NFE)	31.4	32.3
Crude fiber	12.1	4.6
Ash	3.5	12.2
Protein/energy (mg P/kcal DE)[c]	134	151
Protein/energy (mg P/kJ DE)[c]	31.9	36.0
Performance		
Fish weight (g)	1-8	1-8
Daily growth (%)	3.77	3.85
Feed conversion ratio (kg/kg)	1.54	1.60

[a] In mg/kg feed: thiamin hydrochloride 100, riboflavin 330, pyridoxine hydrochloride 70, pantothenate 460, niacin 1330, biotin 10, folacin 25, cyanocobalamin 0.1, para-aminobenzoic acid 660, choline chloride 13,330, inositol 6670, ascorbic acid 5000, beta carotene 20, ergocalciferol 10, dl-alpha tocopheryl acetate (50%) 1330, vitamin K 70, BHT 500, Celufil 20,700.

[b] In mg/kg feed: monobasic calcium phosphate dihydrate 670, calcium lactate 1670, magnesium sulfate heptahydrate 670, dibasic potassium phosphate 1160, potassium iodide 30, monobasic sodium phosphate monohydrate 420, ferric citrate 160, zinc sulfate heptahydrate 210, copper sulfate pentahydrate 50, manganese sulfate monohydrate 30, cobalt chloride hexahydrate 30, aluminum chloride hexahydrate 30, Celufil 44,870.

[c] Digestible energy was calculated using the values 4 kcal/g protein, 8 kcal/g fat, 3 kcal/g carbohydrate.

\+ Amounts not given.

(Table 9-41). Up to 20% of white fish meal protein can be replaced by malt protein flour (14-15 g intially, Yamamoto et al., 1995b). Growth was only 6% less and FCR was similar (0.7) for hirame (2.5-30.3 g) given a feed containing 35% white fish meal and 44% defatted soybean meal compared with 70% WFM and no SBM (Kikuchi et al., 1994b). Meat meal also can be substituted without amino acids (Sato and Kikuchi, 1997). When blue mussel

Table 9-40. Composition (%, except as noted) and performance of three dry feeds used experimentally for pearly spinefoot in Thailand (Kungvankij et al., 1990). Diet C was considerably better but probably could still be improved.

Component	Diet A	Diet B	Diet C
Formulation			
Broiler chicken starter feed	100	70	
Fish meal		15	20
Soybean meal			25
Corn meal			10
Rice bran		8	27.5
Coconut oil cake			10
Fish oil		2	2
Vitamin premix[a]			0.5
Seaweed gum (*Eucheuma spinosum*)		5	5
Proximate analysis (dry)			
Crude protein	18.6	24.2	30.5
Crude fat	4.4	6.9	5.9
Carbohydrate	38.6	31.0	10.3
Ash	8.3	13.4	5.1
Calcium	1.3	1.0	1.5
Protein/energy (mg P/kcal DE)[b]	83	99	152
Protein/energy (mg P/kJ DE)[b]	19.8	23.6	36.4
Performance			
Fish weight (g)	5-52	5-62	5-97
Daily growth (%)	2.01	2.16	2.52
Feed conversion ratio (kg/kg)	7.05	6.44	4.05
Survival (%)	96	89	92

[a] In mg/kg feed: thiamin 1, riboflavin 4, pyridoxine 1, calcium D-pantothenate 8, nicotinamide 25, biotin 0.025, folacin 0.15, cyanocobalamin 0.012, choline 300, tocopherol 15, vitamin K_3 3.06, silicic acid 100, "trace treatment" (minerals?) 350, BHT 10; in IU/kg feed: vitamin A 12,000, cholecalciferol 2000.
[b] Digestible energy was calculated using the values 4 kcal/g protein, 8 kcal/g fat, 3 kcal/g carbohydrate.

meal was substituted for up to two-thirds of the white fish meal (0-50% BMM, 29-74% WFM). FCR was similar (0.75-0.8) and growth apparently better (222-272% gain in 6 weeks vs 213%) for hirame weighing 7-28 g (Kikuchi and Sakaguchi, 1997). Over the range 3-23 g, growth and FCR were similar with feeds containing 80% white fish meal (gain 553%, FCR 0.63) and with 60% of the fish meal replaced by meat meal plus amino acids (520%, 0.69)(Kikuchi et al., 1997b).

Addition of 2% *Ulva* sp. to a commercial feed given to hirame (5-120 g) resulted in the best growth (2.2%/d vs 2.0 with 0% *Ulva*, 2.1 with 4%, 1.9 with 8%) and FCR (1.34 vs 1.62 with 0%, 1.56 with 4%, 1.92 with 8%)(Xu et al., 1993).

Turbot can be raised on the IFREMER weaning and ongrowing feeds also used for European sea bass (Table 9-27; Person-Le Ruyet et al., 1991). Weaning feeds for turbot typically contain 50-60% protein and 10-15% fat. Turbot might need more than 60% protein for best growth (Guillaume et al., 1991). Cooked, but not raw, starch is tolerated up to at least 20%. Feed composition and ration are likely to affect body composition during grow-out (Bell et al., 1994) and also egg quality (McEvoy et al., 1993). Nijhof (1994a) gave a comprehensive discussion of practical factors in feed formulation for turbot farming.

Arachidonic acid promoted growth in small turbot (0.9-3.0 g)(Castell et al., 1994). Growth was best with 0.9% ADA/8.4%DHA (of FA), and was better with 0.4/5.7, 0.7/5.5, 5.2/1.1 than with 0.1/6.2 or 2.3/4.3.

Some farms feed dry pellets until fish weigh 500 g, then moist pellets, which are easier

Table 9-41. Composition (%, except as noted) of four experimental feeds used for hirame in Japan (Kikuchi et al., 1994a, 1994b, 1997a; Yamamoto et al., 1995b).

Component	Kikuchi, 1994a	Kikuchi, 1994b	Kikuchi, 1997a	Yamamoto
Formulation				
White fish meal	65	35	56.0	56.0
Meat & bone meal			18.0	
Hydrolyzed feather meal	12			
Malt protein flour				9.5
Amino acid supplement	1.2	4.6	0.6	
Defatted soybean meal		44		11.3
Wheat gluten				5.0
Potato starch	8.8	1.4	12.4	
Gelatinized potato starch				2.1
Pollock liver oil	4	6	6.0	
Pollock viscera oil				4.4
Vitamin premix	4	4	4.2	3.0
Mineral premix	5	5	2.8	3.0
Cellulose				4.7
Chromium oxide				1.0
Proximate analysis (dry)				
Crude protein	53.2	50.5	48.4	55.6
Crude lipid	11.7	9.3	14.3	11.1
Carbohydrate (difference)	14.1	22.5	14.9	
Crude ash	12.0	12.2	19.4	
Protein/energy (mg P/kcal GE)				108
Protein/energy (mg P/kJ GE)				26.0
Protein/energy (mg P/kcal DE)[a]	152	147	137	
Protein/energy (mg P/kJ DE)[a]	36.5	35.1	32.8	
Proximate analysis (of air dry weight)				
Moisture	9.0	5.5	3.0	
Performance				
Fish weight (g)	2.7-26	2.5-28.5	2.6-15	15-40
Daily growth (%)	1.77	4.34	4.16	3.5
Feed conversion ratio (kg/kg)	0.63	0.71	0.83	~0.68
Protein conversion ratio (kg/kg)	0.33	0.36	0.40	

[a] Digestible energy was calculated using the values 4 kcal/g protein, 8 kcal/g fat, 3 kcal/g carbohydrate.

for large fish to handle (Devesa, 1994). Typically, small fish are fed five times a day and large fish once or twice. FCR of moist feeds containing minced trash fish (assuming 40.6% moisture) ranged from 0.62 at 1-50 g to 1.60 at 1000-2000 g, and specific growth rate from 8 to 0.34%/d. Dry commercial turbot feeds contained 45-52% (of DW) protein and 10-20% fat. Protein energy/fat energy ratios of 81/19 to 95/5 were best for 5- to 20-g fish and 70/30 for >20-g fish. Optimal protein content is about 45-50%.

Growth and FCR were better for turbot (480-947 g) with one commercial turbot feed (49% protein, 18.3% fat, 0.51%/d, 1.25 FCR) than with another turbot feed (60% protein, 15.0% fat, 0.25%/d, 2.27 FCR) or a fat-reduced trout feed (53% protein, 4.8% fat, 0.28%/d, 2.81 FCR)(Sheehan et al., 1994).

Bromley (1980) found that dietary water content (0-74%) did not affect feeding and growth of turbot, but instead, energy consumption near satiation was fairly constant at 68-72 cal/g, a 6% daily ration (Table 9-42). Danielssen and Hjertnes (1993) obtained the best growth and FCR for turbot raised for about 675 d with the feed containing the most fish meal and protein (Table 9-42). Nijhof (1993) obtained better growth and FCR for a moist feed with 70% of gross energy from protein and 30% from fat compared with 50/50, 60/40, and 80/20 (Table 9-42).

Table 9-42. Composition (%, except as noted) and performance of three feeds used experimentally for turbot in Great Britain (Bromley, 1980), Norway (Danielssen and Hjertnes, 1993), and The Netherlands (Nijhof, 1993).

Component	Bromley	D & H	Nijhof
Formulation			
Minced fish			56.7
Fish meal		74.0	27.0
Gadoid meal	68.8		
Fish solubles			5.0
Dextrinized wheat		6.4	
Fish oil		15.0	4.8
Cod liver oil	4.8		
Lecithin		0.6	
Vitamin premix	1.6	1.5 [a]	0.5
Mineral premix		0.5 [b]	
Sodium carboxymethyl cellulose	4.8		
Binder		2.0	6.0
Water	20		
Proximate analysis (dry)			
Protein	63.2	54.2	64.9
Lipid	9.2 [c]	22.9	16.7
Carbohydrate (difference)		7.3	
Ash		8.9	
Protein/energy (mg P/kcal GE)	138		124 [d]
Protein/energy (mg P/kJ GE)	33.0		29.6 [d]
Protein/energy (mg P/kcal ME)		130 [d]	
Protein/energy (mg P/kJ ME)		31.1 [d]	
Proximate analysis (of air dry weight)			
Moisture		6.7	47.2
Performance			
Fish weight (g)	15-83	~20-1680	~50-76
Daily growth (%)	3.00	~0.66	~1.3
Feed conversion ratio (kg/kg)	~0.63	0.96	

[a] In mg/kg feed: thiamin 7.5, riboflavin 15, pyridoxine 7.5, niacin 75, folacin 2.5, cyanocobalamin 0.01, pantothenic acid 20, biotin 0.5, inositol 100, ascorbic acid 900, vitamin E 50, menadione bisulfite 20; in IU/kg feed: vitamin A 3750, vitamin D_3 750.
[b] In mg/kg feed: calcium 624, phosphorus 478, iron 100, zinc 50, copper 10, selenium 0.4.
[c] Calculated by the author.
[d] Energy was calculated by the authors.

Cachelou et al. (1989) reported an FCR of 1.7-1.8 for 0.5 to 2-kg turbot, with growth to 1.5-2.5 kg in 3 yr on the Atlantic coast of France. In northwestern Spain they reached 2 kg in 2 yr.

Pleuronectidae

Atlantic halibut (34-314 g) had the best growth (0.76%/d) with the one of six feeds tested having the highest protein (79% fish meal, 15.3% fish oil, 58% protein, 23.8% fat)(Hjertnes et al., 1993). Atlantic halibut (mean 422 g, initially) given commercial turbot feed (59% protein, 21.5% fat, 8.9% carbohydrate, 10.7% ash of DW, and 6.6% moisture) grew 0.5%/d and had an FCR of 0.95 (Tuene and Nortvedt, 1995). Feed consumption, growth, and FCR of individual fish was observed to vary.

Atlantic halibut had the best growth (1.04%/d) and second best FCR (0.73) with a simple feed containing 80% fish meal and 16% fish oil compared with nine other feeds (Table 9-43; Aksnes et al., 1996). The best FCR of 0.67 but slower growth (0.92%/d) was obtained

Table 9-43. Composition (%, except as noted) and performance of three feeds used to experimentally raise Atlantic halibut in 250-L tanks in Norway. Data reprinted from Aquaculture vol. 145, A. Aksnes, T. Hjertnes, and J. Opstvedt, Effect of dietary protein level on growth and carcass composition in Atlantic halibut (*Hippoglossus hippoglossus* L.), p 228-229, copyright 1996, with permission from Elsevier Science.

Component	Diet 1	Diet 3	Diet 5
Formulation			
Fish meal	47.0	80.0	86.1
Extruded wheat	30.0	0	7.8
Fish oil	19.0	16.0	4.4
Vitamins[a]	1.5	1.5	1.0
Minerals[b]	0.5	0.5	0.5
Binder	2.0	2.0	
Proximate analysis (dry)			
Crude protein	41.0	61.8	68.0
Lipid	25.4	24.8	12.7
Carbohydrate (difference)	26.9	3.1	6.6
Ash	6.7	10.2	12.7
Protein/energy (mg P/kcal GE)[c]	70	103	128
Protein/energy (mg P/kJ GE)[c]	16.8	24.7	30.6
Protein/energy (mg P/kcal DE)[d]	92	115	173
protein/energy (mg P/kJ DE)[d]	21.9	27.4	41.3
Performance			
Fish weight (g)	6.7-395	7.2-556	5.3-677
Daily growth (%)	0.97	1.04	0.92
Feed conversion ratio (kg/kg)	1.01	0.73	0.67

[a] In mg/kg feed: thiamin 7.5, riboflavin 15, pyridoxine-HCl 7.5, Ca-pantothenic acid 20, niacin 75, biotin 0.5, folic acid 2.5, cyanocobalamin 0.01, inositol 100, ascorbic acid 100, dl-a-tocopherylacetate 50, menadione bisulfite 20; in IU/kg feed: vitamin A 3750, vitamin D_3 750.
[b] In mg/kg feed: calcium 624, phosphorus 478, iron 100, zinc 50, copper 10, selenium 0.4.
[c] Gross energy was calculated by the authors.
[d] Digestible energy was calculated using the values 4 kcal/g protein, 8 kcal/g fat, 3 kcal/g carbohydrate.

with 86% fish meal. Growth and FCR were acceptable with a feed containing 47% fish meal.

When Atlantic halibut (2.4-5.8 kg) were raised on lean capelin, fat capelin, lean & fat capelin, dry salmon feed, moist locally-made feed, or salmon moist-silage feed, growth was similar (0.11-0.14%/d for immature fish, 0.8-0.10%/d for males)(Bjørnsson et al., 1992). Before males matured, FCR was 0.9-1.1 with whole fish, ~1.4 with dry salmon feed, 1.6 with local moist feed, and 3.0 with silage feed.

Maximum edible pellet size and minimum moisture content contribute to good growth and FCR (Nortvedt and Tuene, 1995).

European plaice (3 g, initially) grew better with a semi-purified feed containing 70% protein than with less protein, and 57% protein in a practical feed was best for growth of 10-20 g fish (Guillaume et al., 1991). Feed containing 10% dextrin and 10% glucose resulted in better growth and FCR of 3-g European plaice than feeds with no carbohydrate.

Winter flounder wild juveniles (1.1-1.8 g) given a commercial feed for 3 months had better growth (0.53 vs 0.04%/d) than those (1.2-1.4 g) given SFB *Artemia* fed Tahitian *Isochrysis*. FCR with pellets improved from 15.9 to 3.4 over the period (Lee and Litvak, 1996b).

When given a feed containing 45% frozen shrimp and 45% herring meal plus water for ~6 months, captive wild winter flounders grew from ~7 to 30 g with an FCR of 1 (Hoornbeek et al., 1982).

English sole captive wild juveniles (4-6 g, intitially) fed Oregon moist pellets for 12 weeks grew best (0.95%/d) with a 16% daily ration at 9.5°C (Williams and Caldwell, 1978). FCR (dry/dry) was best (10.5%) with a daily ration of 8% at 9.5°C.

Soleidae

Wild juvenile **hogchokers** (18-35 mm SL) ate mostly polychaetes and detritus, with a few foraminiferans and amphipods (Carr and Adams, 1973). Wild juvenile **blackcheek tonguefish** ((10-49 mm SL) ate mostly harpacticoid copepods, with a few amphipods and polychaetes (Toepfer and Fleeger, 1995).

Common sole larvae could utilize a microencapsulated feed (containing cod roe, mussel, fungal yeast protein, egg albumin, and more) after 10 d of live food (Appelbaum, 1985). Juveniles grew better with 57-58% dietary protein than with more or less (Guillaume et al., 1991). They feed more on the bottom and need food more often than turbot (Howell, 1997).

Tetraodontidae

Wild juvenile **southern puffers** (6-25 mm SL) ate mostly detritus and fecal pellets, with some polychaetes, gastropods, crabs, and shrimp (Carr and Adams, 1973). Wild adult **striped burrfish** (119-210 mm SL) ate mostly gastropods, and also bivalves, barnacles, and crabs (Motta et al., 1995). In the past, puffers in the Far East usually were fed minced trash fish (sand lance, sardines, mackerel) or krill, but moist and dry pellets are now available (Kanazawa, 1991b).

Tiger puffer juveniles grew fast (3.7-12 g in 20 d), and a dietary fat content of 11.5% was considered suitable (Table 9-44; Takii et al., 1995a). Han and Furuichi (1995) reported the best growth and FCR with 21.6% fat (2.7-59 g in 56 d). Takii et al. (1995b) found that tiger puffers could tolerate ~20% carbohydrate (16% dextrin). A combination of L-serine, L-aspartic acid, glycine, and L-alanine, and betaine in a feed containing 69% casein and 15% pollock oil stimulated feeding by 14- to 40-g tiger puffers (Takaoka et al., 1995). An experimental feed for grow-out is described in the first column of Table 9-44. Kanazawa (1991b) reported that tiger puffers need 50% protein and about 6-10% fat, and can tolerate 30% or more dextrin.

SUGGESTED FEED FORMULAS

Choice of fish feed ingredients will depend on quality, cost, and availability. Least-cost formulas are based on meeting the minimum requirements using the most economical combination of ingredients available at the time. In formulating feeds for carnivorous marine fish, it is necessary to include enough high quality protein, ω3HUFA, vitamins, and minerals; to limit carbohydrate, fiber, ash, and moisture; and to maintain a moderate protein to energy balance in order to obtain good growth but minimize protein wastage.

Suggested constraints (Table 9-45) and sample formulas (Table 9-8) are based on literature review and experience with several marine species. JT9210, JT9220, and JT9262 are simple but good feeds. Up to about 10% krill can be included in JT9210. Moderate variation in percentages of ingredients and substitution of components should not appreciably affect performance. High quality fish meal and oil, poultry and beef byproduct meals, low-fat cooked soybean products, and wheat flour are preferred macroingredients. Suitability of other protein and carbohydrate sources depends partly on availability. Starter feeds should contain about 50-60% high quality protein. As the fish grow, protein content can be reduced at a rate depending on the species. Generally, omnivores and herbivores (at least those larger than 10 g) can get by with less protein (as low as 35%) or fat and more carbohydrate (up to 30% or more) or fiber than carnivores. Because terrestrial vegetable oils might not be as useful as marine oils and apparently contribute to fatty liver syndrome in some marine fish, large amounts should not be added to their feeds. Linolenic and linoleic acid contents are less than 0.3% in the sample formulas. The ω3HUFA requirement will be met if enough fish oil or other marine fat is included. If fish meal is replaced by

Table 9-44. Composition (%) of experimental feeds (dry pellets) used in Japan for tiger puffer (Nakagawa, 1990; Takii et al. 1995a; Han and Furuichi, 1995; Takii et al., 1995b).

Component	Nakagawa	Takii 1995a	H & F [a]	Takii 1995b
Formulation				
Brown fish meal		48	50	49
White fish meal	50			
Casein-Na			10	
Wheat gluten		10		10
α-starch			4	
Dextrin	25	17	10	16
Pollock liver oil	5	7.5	15	5
Vitamin premix	3	3	3	3
Mineral premix	12	8	8	8
Cellulose	4	3.5		6
Carboxymethyl cellulose		3		3
Binder	1			
Proximate analysis (dry)				
Crude protein		43.6	50.3	43.3
Crude fat		11.5	21.6	8.9
Carbohydrate			17.5	
Crude sugar		21.0		20.5
Crude ash		11.2		11.9
Protein/energy (mg P/kcal GE)[b]		99		105
Protein/energy (mg P/kJ GE)[b]		23.6		25.0
Protein/energy (mg P/kcal DE)[b]			116	
Protein/energy (mg P/kJ DE)[b]			27.6	
Performance				
Fish weight (g)		3.7-12	2.7-59	10-27.3
Daily growth (%)		5.77	3.51	3.45
Feed conversion ratio (kg/kg)		0.74	0.70	0.87
Protein conversion ratio (kg/kg)		0.32		0.38

[a] In mg/kg premix: $NaH_2PO_4 \cdot 2H_2O$ 493, KI 0.14, iron citrate 24, $ZnSO_4 \cdot 7H_2O$ 2.9, CuCl 0.09, $MnSO_4 \cdot 4\text{-}6H_2O$ 0.65, $AlCl_3 \cdot 6H_20$ 0.15, $CoCl_2 \cdot 6H_2O$ 0.86, α-cellulose 478.
[b] Energy was calculated by the authors.

ingredients not containing ω3HUFA, fish oil might have to be increased. Poultry and meat meal levels are limited because of their high fat content (e.g., 14% and 7%) and lack of ω3HUFA. The same problem occurs with full-fat (18%) soybean meal. Textured vegetable protein and solvent-extracted dehulled soybean meal contain only 1-2% fat. Although amino acid balance and digestibility of soy protein compare favorably with those of fish meal (Steffens, 1989), the use of soybean products is limited by their high carbohydrate content (e.g., 34%). Carbohydrate levels of feeds should be kept near or below 10% for the youngest fish but probably can be tolerated up to 15-20% by older carnivorous and omnivorous species. High-gluten wheat flour is a good binder.

Because some minerals will be present in the macronutrients, full amounts would not have to be added. Usually, zinc, manganese, iodine, chromium, and sometimes, iron, copper, cobalt, and selenium should be added. With less animal content in the feed, more mineral supplementation might be required. Ethoxyquin is included as an antioxidant.

This is offered as a working guide. Fine tuning awaits the acquisition of more data. With slight modifications, the basic diets should provide enough nutrients for many carnivorous marine perciforms and probably are suitable for other types of fish as well.

MAKING AND STORING FEEDS

Procedures for the formulation and manufacture of fish feeds are described in UNDP/FAO (1980), Cho et al. (1985), McEllhiney (1985), Hardy (1989, 1998), Lovell (1989), Pigott

Table 9-45. Suggested levels (%, except protein/energy) of macronutrients in feeds for carnivorous marine perciform fish (guidelines used in the author's feed development program).

Component	Weight range of fish			
	1-10 g	10-150 g	150-500 g	>500 g
Vacuum dry basis				
Crude protein (minimum)[a]	60	55	50	45
Crude fat (minimum)[b]	16	12	10	9
Carbohydrate (maximum)	13	15	17	20
Fiber (maximum)	2	3	3	4
Ash (maximum)	12	16	20	22
Protein/energy (mg P/kcal DE)[c]	147	152	151	144
Protein/energy (mg P/kJ DE)[c]	35.2	36.3	36.1	34.5
ω3HUFA (minimum)	2.0	2.0	2.0	2.0
Eicosapentaenoic acid (minimum)	1.0	0.8	0.7	0.7
Docosahexanoic acid (minimum)	1.0	0.8	0.7	0.7
Air dry basis				
Moisture (maximum)	10	10	10	10

[a] For 1-10 gram fish, the protein components should consist of at least 50% whole fish meal (or equivalent), up to 30% poultry meal, up to 20% meat meal, up to 30% soybean (or equivalent). For larger fish, protein should be provided by at least 33% whole fish meal (or equivalent), up to 50% poultry and meat meal, up to 40% soybean (or equivalent).

[b] Fat content can be adjusted upward for very active species or downward for very sedentary ones. At least half of the total fat should be fish oil (or equivalent), with a minimum content of 5%. Linoleic acid should be kept well below 1% and linolenic acid well below 4%.

[c] Digestible energy was calculated using the values 4 kcal/g protein, 8 kcal/g fat, 3 kcal/g carbohydrate.

and Tucker (1989), Ensminger et al. (1990), and Feed Development Section (1994). Commercial feeds usually are made for two to five stages—from starter to grower or finisher. Generally, nutrient density ranges from highest in starter feeds to lowest in grower feeds. Nutrient density of broodstock feeds is closer to that of grower feeds, but they have higher quality ingredients, more essential fatty acids, and possibly extra vitamins and pigments.

Manufacture of commercial feeds begins with specification of lower, upper, or fixed limits for each potential ingredient—based on the known or estimated requirements of the fish, feed type, and ingredient quality. Then, within these constraints, choice of ingredients and amounts actually used are determined mainly by availability and cost. Least-cost formulations usually are devised by solving simultaneous equations with a computer program. This permits nutritional requirements to be fulfilled at the lowest cost. Fish feeds are made in several forms.

Compressed Dry Pellets (Fig. 9-8)

Ingredients are finely ground, mixed, cooked briefly (<1 min, often 15-30 sec) with steam under pressure (at 70-95°C and 15-18% moisture), compressed and forced through a die of correct size, cut with a knife, cooled, dried, and top-dressed with oil (if necessary). Usually, the mixture does not expand much as it exits the die, and sinking pellets (density 500-600 g/L) with 9-10% moisture are produced. Ingredients, compression, temperature, and moisture can be varied to modify pellet hardness. Starch is partly gelatinized (<40%) by the heat, and moderate amounts help to bind the pellet, but high levels of fat or fiber make the pellet less stable. Up to 12% oil can be included during pelleting, and 5-6% can be sprayed on (top-dressed) later, but heavier top-dressing tends to waste oil, which runs off

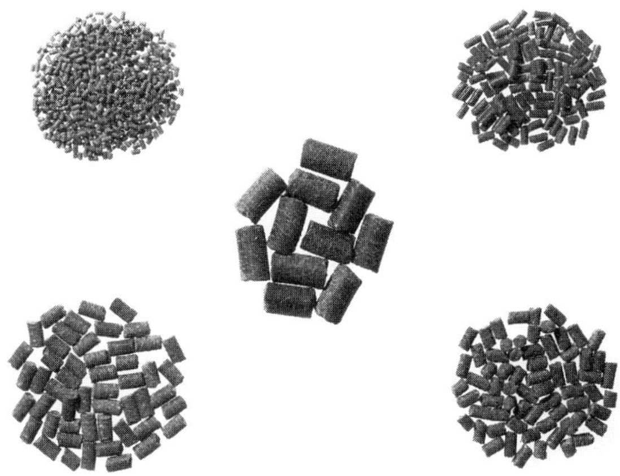

Figure 9-8. Marine fish feeds developed by the author (1.6-9.5 mm diameter).

of the low-porosity pellets. Diameters of pellets for fish customarily are in the range 1.6-9.5 mm. The cost, storage stability, water stability, nutrient loss during manufacture, and starch digestibility are lower for compressed pellets than for extruded pellets.

Extruded Dry Pellets
Ingredients are finely ground, mixed, cooked briefly (2-5 min) with steam under pressure (at 104-180°C and 20-33% moisture), kneaded, forced through a die at high pressure, cut with a knife, cooled, dried (to 9-10% moisture), and top-dressed with oil (if necessary). Expanding steam within the material enlarges it as it leaves the die, possibly resulting in a low-density floating pellet (range 400-550 g/L, neutral buoyancy at ~480 g/L). Lower temperature and higher moisture during processing produces less dense pellets. Sinking pellets can be produced by raising temperature and reducing moisture. Extruded pellets tend to be harder, less dusty, and more stable than compressed dry pellets, but require more time and expense to produce. The increased processing temperature improves starch digestibility (>80% gelatinization) and binding, but vitamin loss can be higher (Haaland et al., 1993). Oil can be top-dressed to a total of about 38%.

In a method intermediate between compression pelleting and extrusion (Hardy, 1998), the ingredients are cooked for 2-3 min with 16-18% moisture and extruded at high speed. Storage and water stabilities are similar to those of extruded pellets. Nutrient loss is similar to that with compression pelleting. Production rate, cost, starch digestibility (60-80% gelatinization), and fat limit (~30%) are intermediate. Density range is 400-600 g/L.

Extruded Semi-moist Pellets
Ingredients are finely ground, mixed, forced through a modified noodle extruder without heat, cut with a knife, and stored without drying (frozen or not frozen). A similar product can be made by compression pelleting without drying.

Moist Feeds
Moist feed can be made in the form of cut or uncut extruded noodles or simply a paste, but the latter is likely to be unstable unless a good binder (e.g., starch) is used. Some eel feeds are kept dry until just before feeding, when they are mixed with water (and maybe oil) and molded into the needed size and shape.

Crumbles
Whole and broken dry pellets are ground between rollers and separated into size ranges by

sieving. Crumbles can be of almost any size but customarily range from less than 0.6 mm (meal) to 2.0-3.4 mm (no. 4).

Meals
These are unconsolidated mixtures mainly used as supplemental food for small bait and ornamental fish raised in ponds, where the fish also obtain natural foods. Meals quickly disintegrate, degrade, and leach nutrients.

Flake Feeds
Very finely ground ingredients are mixed with water and applied to heated, rotating drums, which dry the mixture into a sheet. The sheet is sheared off of the drums by a knife and then crumbled or ground to the desired size.

Larval Feeds (also see chapter 8)
Stabilization, density (buoyancy), appearance, and protection from nutrient leaching and oxidation are primary considerations. There are three main types (Hardy, 1989; Watanabe and Kiron, 1994). **Microencapsulated feeds, or diets,** (MEDs) are enclosed in a membrane formed from a substance such as nylon-protein, gelatin-gum acacia, lipid, chitosan, glycopeptide, egg albumen, or whole chicken egg. **Microbound feeds, or diets,** (MBDs) are made by grinding dried flake feed (bound with agar, zein, gelatin, carageenan, alginates, or others). Microbound feeds coated with materials such as cholesterol & lecithin, modified corn gluten, zein, or nylon-protein are known as **microcoated feeds, or diets,** (MCDs). The coating of MEDs is a large part of the feed. It has to break down easily and not obstruct the gut. Ideally, the coating will provide some nutrition, but at least it should not interfere by causing an imbalance. Because of the small particle size, larval feeds are especially vulnerable to oxidative degradation of fats and micronutrients. They usually are packed in nitrogen and must be stored dry in a freezer after the container is opened.

Specialized Feeds
Uncompressed pellets can be produced by spraying a water mist over dry ingredients rotating on a disc. Ingredients can be mixed with a **gelatin** solution and chilled to produce a high-moisture, semi-solid feed. Feeds can be made by the **surimi** process from fishery byproducts and trash fish. A moist mixture can be **freeze-dried** and then broken up into soft crumbles.

Storage
Ideally, feeds should be stored with as little air as possible in the container in a dry atmosphere below 0°C in the dark. This helps to prevent loss of micronutrients, oxidative rancidity of lipids, and growth of fungi and bacteria. In many cases, large enough ideal storage facilities are not available. Then, feed must be kept as dry (<75% relative humidity) and cool (<20°C) as possible and used quickly, usually within 3 months of manufacture (2 months for unfrozen semimoist feeds). Frozen semimoist feeds should be stored at -20°C.

Semimoist feeds without antimicrobial agents can be expected to support mold growth within 3 d at 22°C, 10-20 d at 1-3°C, and not at all below 0°C (Hardy, 1989).

FEEDING METHODS

Most cultured marine fish naturally feed in the water column or on the bottom and sometimes are reluctant to eat floating pellets. Fish in the families Centropomidae, Percichthyidae, Serranidae, Carangidae, Lutjanidae, Sparidae, Sciaenidae and many others will eat and often prefer slow-sinking dry pellets. Some species are more difficult to train to eat dry pellets; early training (at weaning) takes the least effort. Palatability can be affected if bad-tasting ingredients are present (Mackie and Mitchell, 1985). Size, shape, and texture could affect the amount of feed eaten. Young Atlantic salmon, which are adapted to eating insect larvae, preferred long thin pellets to round or fat ones (Stradmeyer et al.,

1988). The feed particle diameter best for growth of Arctic char ranged from ~1.2 mm at 3 g (73 mm FL, ~21% of mouth width) to ~3.1 mm at 21 g (133 mm, ~32% of mouth width)(Tabachek, 1988). Fish adapted to swallowing relatively large prey (e.g., piscivores) probably also will eat large pellets. Common snook and Nassau groupers of 250 mm total length prefer 9.5 mm pellets to smaller ones (pers. obs.).

Timing, location, and movement are important with larval and weaning feeds (Backhurst et al., 1989).

Young fish, herbivores, species with small or no stomachs, and very active fish need to be fed frequently. In tanks, small milkfish (600 mg) fed eight times a day had better growth and FCR than those fed half as often (Boonyaratpalin, 1997). In 20-L aquaria, gilthead sea bream (250 mg) grew better with continuous feeding for 19 h than with five short meals in 19 h (Goldan et al., 1997). Collins (1981) reported that in nature striped mullet fed mainly from dawn to dusk, with a peak at midday. Ayu accumulated unacceptable amounts of muscle fat if they were fed less than four times a day (Yao et al., 1994).

Once they reach about 100 mm in length, one good meal a day probably is enough for fish with large stomachs that normally eat large infrequent meals (e.g., Japanese eels, Usui, 1991; greasy groupers, Chua and Teng, 1978; European plaice, Jobling, 1982; barramundi and common snook, Tucker et al., 1988) and even others that normally eat smaller, more frequent meals (e.g., Atlantic cod, Lie et al., 1988; whitemouth croaker, Abud, 1990). For high density rearing, more frequent feeding might be necessary to assure that all fish get their share.

Feed Development Section (1994) recommended feeding milkfish in ponds a full ration at >28°C, three-quarter ration at 25-28°, and half ration at 21-24°C, twice a day when natural food is abundant and three times a day if it is not; broodstock in cages or tanks should be fed twice a day. It was recommended to also feed tilapia twice a day when natural food is abundant and three times when it is not. In Taiwan, groupers were fed twice a day until they weighed 200 g, then once a day (Liao et al., 1995). In Japan, hirame were fed 4-6 times a day until they weighed 100 g, 3-4 times until 300 g, and once or twice thereafter (Matsuoka, 1995).

Under production conditions, at ≤10°, growth and conversion were not improved by more than one meal per day for Arctic char at 10°C, one meal for European eels at 15°C, three meals for European sea bass at 20°C, four meals for European eels at 30°C, and probably four meals for salmon at 20°C (Talbot, 1993). At 18.5°C, growth and FCR were better for white sturgeon (8.6 g initially) fed continuously (24 h) than for those fed less frquently (Cui et al., 1997). At 14.5°C, powan juveniles (77-144 g) had similar growth and FCR when fed 24 times during periods of 6, 12, 18, or 24 h/d (Koskela et al., 1997). At 19-22°C, three meals per day were optimal for ~60-g European sea bass (Tsevis et al., 1992). Yellowtails need to be fed often because of their large appetite and high growth rate (Shimeno, 1991).

Many fish compete for food (e.g., gobies) regardless of the supply, while a few others (e.g., topminnows, *Fundulus* spp.) compete only when food is scarce (Cuenco et al., 1985). Competition increases with higher density. For competitive fish, it is likely that the pecking order will be followed at feeding time, with the most aggressive fish satisfied first and the least aggressive last. Feeding the fish until every one refuses to eat more is ideal. With a large number of fish, an even food distribution will allow the smaller or less-aggressive ones to feed without getting in the way of the others.

Manual or automatic feeding (Fig. 9-9) can be employed for early juveniles, with meals given from a few times per day to almost continuously. Older fish usually are fed manually or mechanically (e.g., by tractor-mounted blower, Fig. 9-10). When they are large enough to activate demand feeders (Fig. 9-11), many types of fish reared in tanks or ponds can be trained to use them (e.g., salmonids, percichthyids, sparids, sciaenids). European sea bass can learn to use rod-activated demand feeders well by the time they weigh 65 g or less

(Boujard et al. 1996), and they would bite a quartz activator at ≤178 g (Anras, 1995). European sea bass tended to prefer levers on demand feeders with higher visual contrast (Kentouri et al., 1993c). Electronically-activated demand feeders have been developed and are useful for small fish, which might not have the desire or strength to push a lever. Demand feeders can reduce wastage, stress, and size heterogeneity and increase growth (Paspatis and Boujard, 1996).

Mechanisms for automatically releasing feed include vibrators, conveyer belts, augers, rotating compartments, and compressed air. In cage farming, feed sometimes is moved by water or air flow from a central storage bin through pipes to the cages. Automatic feeders should be adjusted to spread the feed well enough so that even the least aggressive fish can eat well.

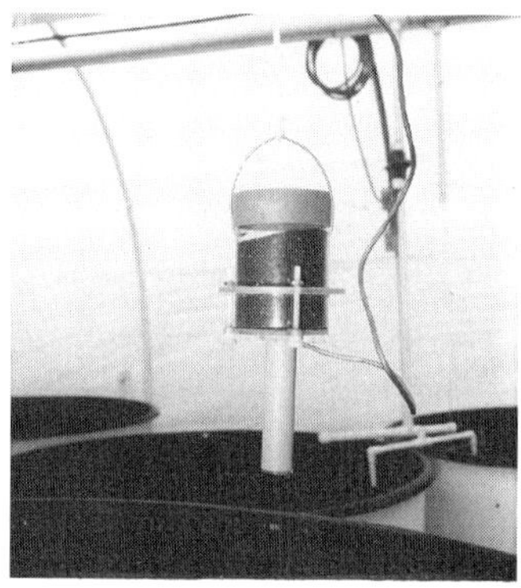

Figure 9-9. Small automatic vibrating feeder for young fish.

Figure 9-10. Tractor-mounted blower-type feeder.

Some feeders protect their contents better than others. Feed should not be exposed to heat, oxygen, or moisture long enough for vitamin loss, rancidity, or fungal growth to occur.

To minimize fouling of water by waste, compound feed is best presented to eels on a perforated tray lowered into the pond and fresh food by hanging from a wire into the pond (Usui, 1991).

Feeding tables can be developed from daily ration, growth, and FCR data from feeding trials with different sizes of fish at the optimal or different temperatures. The culturist can decide to try for maximum growth, minimum FCR, or a compromise.

Figure 9-11. Demand feeder for sheepshead in a raceway.

Juell (1991) and Bjordal et al. (1993) described an echolocation system that continued to release feed for caged salmon until surface swimming activity dropped below a set threshold. Blyth et al. (1993) described an electronic system for regulating feed input for caged salmon according to detected wastage. Kimura et al. (1993) described a system used in openwater ranching.

Gilthead sea bream (4-252 g) had better growth and much better conversion when they used demand feeders than with manual feeding to satiation 3 timesa day or fixed-ration feeding 13 times a day (Kentouri et al., 1993b).

Fish have been known to consume a daily ration from 1% to 50% of body weight, depending on species and size (NRC, 1983). Although predictions can be made, presenting a fixed amount of feed to cultured fish is somewhat unrealistic, because their appetites will vary from day to day, often because of factors not considered in making the predictions. Overfeeding will occur sometimes and underfeeding at other times. By computer modeling, Cuenco et al. (1985) demonstrated that channel catfish would grow faster and ponds would have higher yields when fish were fed to satiation than when they were fed a fixed ration of 1% body weight/d. If fish can be observed, feeding to satiation is preferred to giving them fixed rations. In my experience and that of others, many fish seem to feed best and possibly convert food energy best if they are fed late in the day, just before dark (Chua and Teng, 1978).

SUMMARY

Typical carnivorous and omnivorous marine fish grow best on high-protein, low-carbohydrate feeds, well fortified with vitamins and minerals, and containing sufficient essential fatty acids. The best protein and fat levels (dry basis) for most marine species studied are in the ranges 40-60% and 10-20%. Typical upper limits are 10-20% carbohydrate, 4% fiber, 10-25% ash, and 10% moisture. Herbivores are adapted to less dietary protein and more carbohydrate. Attractive substances can be added to feeds to increase ingestion, growth stimulators to increase growth, and pigments to enhance color. A minimum dietary content of 2% ω3HUFA (long-chain highly-unsaturated fatty acids), including at least 1.0% each of eicosapentaenoic and docosahexanoic acids for young fish and 0.7% of each for older fish seems to be a reasonable safeguard against essential fatty acid deficiency, but some species might need more. Antioxidants should be included to protect against oxidative rancidity. Suggestions for dietary constraints and sample formulas for practical starter, grower, and broodstock feeds have been provided.

10 ENERGETICS

Efficient utilization of food energy is especially important in enhancing the ability of larvae to survive and older fish to grow. In rearing fish, knowledge of energetics is helpful in examining special requirements of the species concerned. Consideration of energy budgets allows the culturist to visualize, and, when data is available, to quantify the effects of major factors such as species, stage, age, food quality, food availability, temperature, daylength, and dissolved oxygen, as well as stressors (e.g., crowding, reduced water quality). The study of energetics is of practical significance in the optimization of feed formulas and feeding practices. In some cases, hatchery design can benefit from initial or later improvements based on energetic considerations. Because the cost of feed tends to be the main cost of producing fish, an understanding of energetics is useful at almost every stage—in choosing species, writing business plans, designing facilities, and raising the fish. This chapter provides an introduction to basic concepts of the energy budget plus some specific examples.

Useful references, most of which include derivation and application of energy budgets for fish, are Winberg (1956), Beamish et al. (1975), Gerking (1978), Elliott (1979), Hoar et al. (1979), Kutty (1981), Blaxter and Hunter (1982), Cho et al. (1982), Kerr (1982), Kitchell (1983), Tytler and Calow (1985), Dabrowski et al. (1988), Hewett and Johnson (1989), Smith (1989a), Steffens (1989), Adams and Breck (1990), Rice (1990), Cho (1992), Houde and Zastrow (1993), Kaushik and Médale (1994), Brett (1995), Cho and Bureau (1995), and Pedersen (1997). Kleiber's (1975) book is a classic general reference.

ENERGY BUDGET COMPONENTS AND INFLUENCING FACTORS

An energy budget (Fig. 10-1) shows the proportions of ingested **gross energy** (**I**) lost through **egestion** (**F**) and **excretion** (**U**) and the proportions used for **metabolism** (**M**) and **growth** (**G**).

In fish farming, the most commonly used concept related to energetics is the feed conversion ratio (FCR), the weight of feed eaten divided by the weight of fish gained. This is a practical number. FCRs allow a farmer to estimate the total amount of feed required to produce a crop. **Gross growth efficiency** (GGE, K_1, **G** in energy budget) is another "more scientific" way of representing the same thing. **G** is the weight (or energy) of fish gained divided by the weight (or energy) of food eaten in a given time, usually on a dry:dry basis. **Protein conversion ratio** (PCR), the weight of protein eaten divided by fish weight gain, and **protein efficiency ratio** (PER), fish weight gain divided by weight of protein eaten, focus on the effect of food protein alone on overall growth. Finally, **productive protein value** (PPV), weight of protein gained by the fish divided by weight of protein eaten, eliminates other components. PCR, PER, and PPV permit closer scrutiny of the effect of varying protein intake on growth. Usually, maximum growth rates occur with feed protein contents higher than those at maximum PER or PPV. In many cases, the most cost-effective feed has a protein content intermediate between those causing the best conversion and best growth; that range can be narrow or wide. With less dietary protein, reduced growth is the overiding factor; with more protein, increased feed cost is the main limitation. When all factors and costs are considered, the highest **G** might not occur at the point of least cost, but is an ideal. **G** can be maximized by providing fish with optimal living conditions and the best quality food in the right amounts at the right times.

Gross energy (**ingestion**, **I**) is the total energy of food consumed in a given time, usually a day (daily ration). **I** and **G** can decrease when temperature drops within the normal range, dissolved oxygen falls below a critical minimum, toxic or inhibitory chemicals build up, or stress increases. **I** is determined by measurement of feeding rate (usually by direct

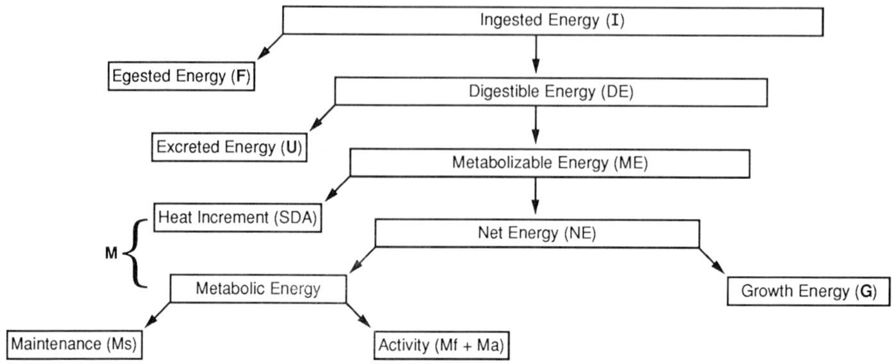

$$I = F + U + M + G \quad \text{or} \quad G = I - F - U - M$$

Figure 10-1. Components of the energy budget.

observation), followed by calculation of energy consumed in the daily ration. Energy content of food and fish often is determined by combustion in a calorimeter. Proximate analysis (% protein, fat, and carbohydrate) followed by conversion with appropriate factors is another, indirect, method for estimating energy content. Standard gross energy equivalents for fish food are 5.65 kcal/g protein, 8.66 kcal/g fat, and 4.10 kcal/g carbohydrate (Brett and Groves, 1979). Energy actually available to fish from these components will be somewhat less depending on efficiency of digestion and assimilation.

Digestible energy (DE) is gross energy minus loss through **egestion** (**F**). **F** depends especially on species, age, type of food, amount of food ingested, and temperature. **F** can increase if diets are unbalanced or feeding rate is too high (thus reducing gut residence time). **F** usually is estimated by difference, but sometimes by collection of feces and measurement of its energy content or measurement of a tracer (this is difficult to do accurately). Rounded off average DE values for fish like salmonids and carp sometimes are used as approximations in food comparisons: 4 kcal/g protein, 8 kcal/g fat, and 3 kcal/g carbohydrate (Brett and Groves, 1979; Steffens, 1989).

Metabolizable energy is digestible energy minus loss through **excretion** (**U**). **U** depends especially on species, age, and type of food. **U** is related to protein content of food and is relatively small and constant unless dietary protein changes drastically. **U** often is estimated along with **F** by difference, sometimes is assumed, and rarely is determined by measurement of ammonia and urea produced by the fish (best done by tracer). Average values (Brett and Groves, 1979) of metabolizable energy for freshwater teleosts are 4.2 kcal/g protein, 8.0 kcal/g fat, and 3.3 kcal/g carbohydrate (cooked starch). Note that the difference between DE and ME is small. The ME equivalents are similar to the DE ones above, which are conservative.

Net energy is metabolizable energy minus loss through the **heat increment,** (HE, also called specific dynamic action, or effect, SDA.) Heat increment is the energy (heat) loss from digestion, transformation, and assimilation of food.

Total metabolism (**M**) includes energy for maintenance, activity, and HE. **M** depends on species, size, age, temperature, activity level, stress level, and ease in locating and capturing prey or other food. **M** can increase when fish grow larger, more effort is required for obtaining food or other activities, temperature rises, daylength increases, stress increases, or departures from optimal water quality ranges occur. On an individual basis (mg oxygen/fish/h), **M** increases with size, but on a weight basis (mg oxygen/kg/h), **M** decreases.

For a variety of fish within normal temperature ranges, a rise of 10°C increases standard metabolism by a factor (Q_{10}) averaging 2.3 (Brett and Groves, 1979). **M** also increases after fish feed because energy is used to process food (HE). **M** usually is estimated by applying an oxycalorific coefficient of 4.63 kcal/ L oxygen to oxygen uptake data obtained by respirometry. This is an average value for a carnivorous fish eating a mixed diet but deriving most of its energy from protein. The correct value for herbivores might be closer to 4.8 kcal/L. Specific values are 4.58 for protein, 4.69 for fat, and 5.04 for carbohydrate. Activity under experimental conditions is not the same as when the fish are in their normal environment, and the data usually are adjusted by a factor to compensate for this. Choosing the factor is critical for the results, but not easy. Metabolism of an active fish can be 8-10 times standard (resting) metabolism, with burst swimming of some species capable of increasing it 1000 times (Knights, 1985). Because routine (≈standard) metabolism is measured in confined unfed fish, the resulting values often are used as an estimate of metabolism for fish in the dark and the values are multiplied by a factor (typically 2) to roughly estimate metabolism for normally swimming and feeding fish under daytime conditions (Brett and Groves, 1979).

Net energy is the energy available for maintenance, activity, and growth (**G**). **G** for an individual or population can be positive, negative, or zero. At first feeding, many species are slightly less efficient than they are a few days later at swimming, capturing food, and digesting food. Overall efficiency in obtaining the requirements of life tends to gradually improve up to the juvenile stage and then improve at a faster rate during that stage. These changes are reflected by an increase in **G** (i.e., the bottom line). Later, as fish grow large and their maintenance requirements increase substantially, growth efficiency decreases. Larger fish, if fed well, also tend to accumulate more fat.

Weight-specific metabolic needs tend to be lower when temperature drops (within the normal range), activity increases, stress increases, or as fish grow larger.

In at least some salmonids and largemouth bass, fluctuations of dissolved oxygen above and below the minimum for good growth will depress growth and can also raise the minimum necessary DO level (Cuenco et al., 1985). It seems that periods of reduced DO cannot be compensated for by periods of increased DO.

NRC (1983) and Smith (1989a) discussed four of the reasons why fish use energy more efficiently than mammals and birds: 1) It is cheaper energetically to excrete ammonia primarily (~85% of excretion by fish) rather than urea as in mammals or uric acid as in birds. 2) The heat increment is much lower (3-5% of ME in rainbow trout vs 30% in some mammals); much of this is attributable to the efficiency of ammonia excretion. 3) Fish do not have to maintain their body temperature in a narrow range like mammals and birds. 4) Fish do not use as much energy for maintaining their positions in the water.

Because fish are energy-efficient, they require less dietary energy and therefore need higher protein/energy ratios in feed to maximize protein assimilation (Smith, 1989). Because of their high energy requirements, cattle need only 10% dietary protein, swine 16%, and chickens 20% for good growth. Suitable P/E ratios (mg P/kcal DE) are: 40 for cattle, 48 for swine, 68 for chickens, 88 for catfish, and 117 for rainbow trout. Corresponding feed conversion ratios are: 8.0 for cattle, 4.0 for swine, 2.5 for chickens, 1.8 for catfish, and 1.5 for trout. Protein conversion ratios are: 0.80 for cattle, 0.64 for swine, 0.50 for chickens, 0.48 for catfish, and 0.52 for trout. Energy input in kilocalories needed to produce a gram of unprocessed protein are about 685 for beef, 149 for chicken, 83-165 for catfish, 30-159 for salmon, 6-53 for carp, and 4-65 for tilapia (Ackefors et al., 1994).

From an economic viewpoint, as much carbohydrate as can be tolerated should be included in feed, then as much fat as possible, and then just enough protein for acceptable growth. As discussed in chapter 9, carbohydrate and fat tolerance vary among species and life stages because of differences in digestive capabilities and assimilation patterns. Also, the best growth usually occurs when dietary protein is higher than what is economically optimal.

When fish consume rations that maximize growth efficiency (i.e, minimize FCR), digestion is efficient and there is a good surplus of energy for growth. With lower rations, a smaller proportion of energy is available for growth after other needs are satisifed. With higher rations, digestion efficiency decreases and a higher proportion of energy is egested and thus lost for growth.

SPECIFIC EXAMPLES

Energy budgets are useful in comparing different stages and species. Budgets based on conditions close enough to hatchery or farm conditions can help predict growth efficiencies and rates. Budgets for a variety of fish are given in Fig. 10-2. Species, age, size, food, measurement methods, and experimental conditions are important factors affecting results. Comparisons between different studies should be done cautiously.

Eggs and Larvae

Variations in energy content of eggs caused by differences between individuals, populations, broodstock nutrition, season, and holding conditions can affect survival ability of larvae (Quattro and Weeks, 1991).

The fact that the oil globule of barramundi (even at normally high temperatures of 26-28°C) lasts until 120-140 hah is a major factor contributing to robustness and high survival of their larvae (Kohno et al., 1986).

Chub mackerel during 3-5 dah (38-85 μg) ate a daily ration of 87% (165-538 rotifers) and had a G of 33% at ~19° (Hunter and Kimbrell, 1980).

Meng (1993) found that striped bass smaller than those represented in Fig. 10-2, O (138 μg at 12-13 dah) grew less efficiently eating *Artemia* nauplii at 17°C: F&U = 18%, M = 50%, and G = 32%. During 15-22 dah, G decreased from 35 to 23% as larvae grew from 204 to 291 μg. During the 19th dah, typical larvae ate 57 nauplii.

Older Fish

From data on consumption and oxygen uptake, Hartman and Brandt (1995) estimated scope for growth at optimal temperatures for three species at two sizes eating bay anchovies and Atlantic menhaden twice/d. At 30 g, striped bass could grow as much as 5.9%/d at 18.5°C, bluefish could grow 6.7%/d at 20°C, and weakfish could grow 7.1%/d at 23.5°C. At 1 kg, all three species could grow about 2.7%/d at 15°C for striped bass, 20°C for bluefish, and 23.5° for weakfish.

Takii et al. (1997b) found that juvenile red sea bream (with stomachs) eating moist pellets had higher digestible, metabolizable, and retention energy than tiger puffers (stomachless) but also higher heat increment and activity needs, thus suggesting that puffers compensate for lower utilization efficiency by conserving energy used for food processing and movement.

Jobling (1990) reported that exercise improved both growth rate and FCR of Arctic char. When fish had to swim against a moderate current, the energy saved via reduced aggression was greater than that lost by constant swimming. Protein deposition probably was higher and lipid deposition lower in exercised fish. Swimming speeds of 1.5-2.0 BL/s for 10- to 100-g fish and 1 BL/s for larger fish probably were optimal for growth and conversion.

Based on chemical oxygen demand of components, Nijhof (1994a) estimated that a 1-kg turbot would convert the material in a feed with CODs of 67% for protein, 29% for lipid, and 4% for carbohydrate into 45% for respiration, 16% for excretion, 12% for egestion, and 27% for growth.

In European plaice and North Atlantic flounder (~100 g at ~18-22°C), M was 29% or more and G was about 37% (Fonds et al., 1992).

Ikeda (1996) estimated that a lightfish, over a 1.8-yr lifetime, would have an average daily ration of 2.9% and would use 9075 cal for metabolism and 1820 cal for growth. Females would invest ~1.3% of assimilated energy in gonads and males ~0.6%.

ENERGY BUDGETS FOR FARMS

In aquaculture, energetics can be useful also at the farm level, at least for examining social and environmental impacts.

Mathews et al. (1976) compared cage farming of Pacific salmon with ocean ranching, and decided that ranching is potentially very efficient. In a similar analysis with Atlantic salmon, Folke (1988) concluded that ocean ranching with delayed release (smolts 3 months older) was more efficient than conventional ranching or cage farming.

Pitcher (1977) constructed an energy budget for a hypothetical 1-hectare rainbow trout pond farm with an annual marketable production of 25 mt. He determined that the cost for feeding was 93% of the total energy input, with the cost of South African fishmeal at 53% of the total, and concluded that use of cheaper fish meal or substitution for part of the fish meal had the greatest potential for improving the farm's economy. Even with the higher costs, production was cheaper than that of U.K. capture fisheries. With the reduction in feed costs, rainbow trout production compared favorably with those of potatoes, bread, and milk.

For a hypothetical rainbow trout cage farm producing 50 mt/yr, energy input was partitioned into 94.5% for feed, 3.7% for fish to stock, 1.6% for equipment, and 0.2% for labor (Ackefors et al., 1994).

In a hypothetical salmon cage farm, with feed containing 44% carbon and 7.7% nitrogen, ~20% of the feed could be wasted and 80% eaten; 30% of C & 10% of N egested and 65% of N excreted; leaving 50% of C for respiration and 20% of C & 25% of N for growth (Gowen and Bradbury, 1987). A farm producing 50 mt fish/yr might use 100 mt feed, with total wastage amounting to 19,360 kg carbon and 6160 kg nitrogen. Thus improvements in feed composition and feeding methods that lower FCR even by small amounts not only will save money but also can reduce pollution substantially.

SUMMARY

Knowledge of energy utilization patterns of individual fish, groups of fish, and whole farms can be useful for predicting and planning for fish growth, food requirements, labor requirements, and costs in aquaculture. In a biological energy budget, subtracting the components for egestion (F), excretion (U), and metabolism (M) from gross food energy (I) provides an estimate of the energy available for growth (G). A farm (or operational) energy budget is like a cost benefit analysis with dollars replaced by energy units.

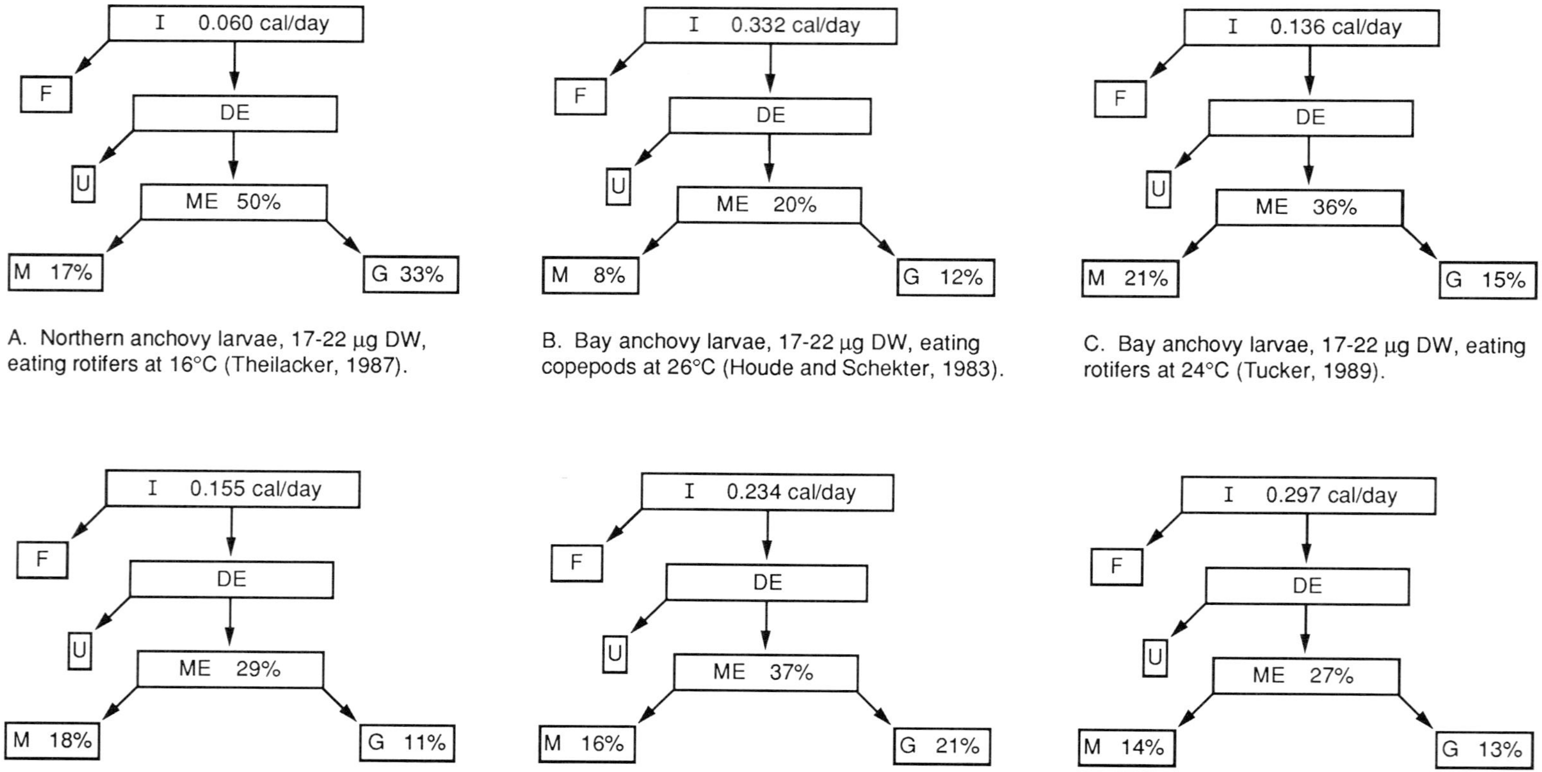

A. Northern anchovy larvae, 17-22 µg DW, eating rotifers at 16°C (Theilacker, 1987).

B. Bay anchovy larvae, 17-22 µg DW, eating copepods at 26°C (Houde and Schekter, 1983).

C. Bay anchovy larvae, 17-22 µg DW, eating rotifers at 24°C (Tucker, 1989).

D. Black sea bass larvae, 17-22 µg DW, eating rotifers at 20°C (Tucker, 1989).

E. Sea bream larvae, 17-22 µg DW, eating copepods at 26°C (Houde and Schekter, 1983).

F. Lined sole larvae, 17-22 µg DW, eating copepods at 28°C (Houde and Shekter, 1983).

Figure 10-2. Examples of energy budgets.

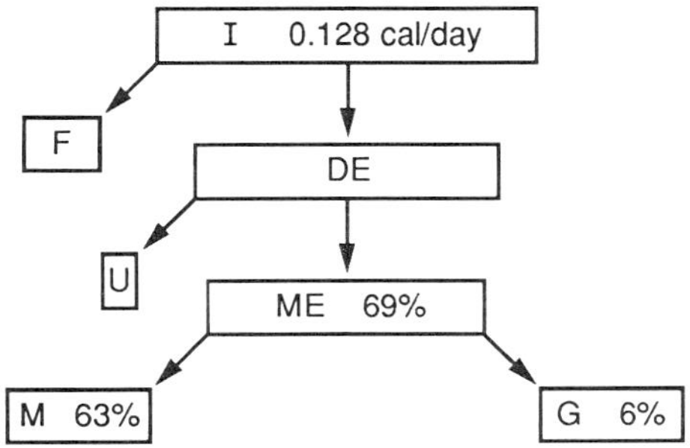

G. Winter flounder larvae, 20 μg DW, eating copepods at 8°C (Laurence, 1977).

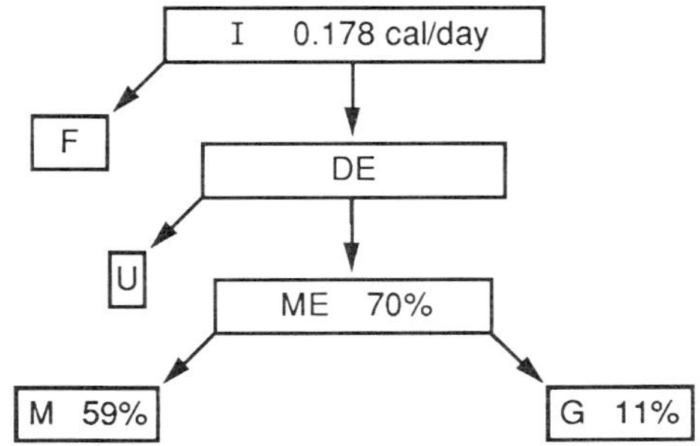

H. Winter flounder larvae, 50 μg DW, eating copepods at 8°C (Laurence, 1977).

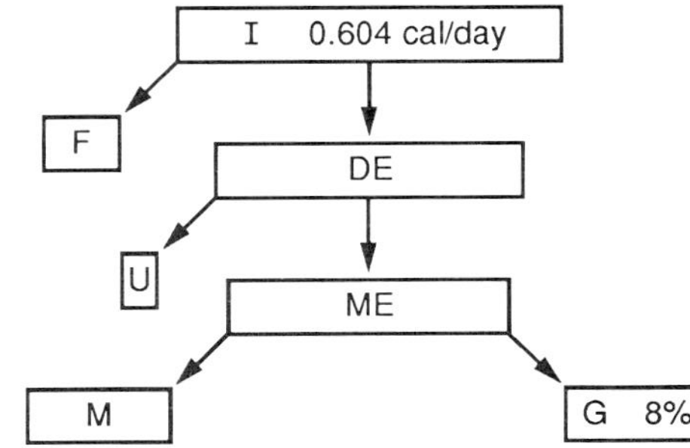

I. Pacific herring larvae, 110-220 μg DW, eating rotifers at 12°C (Eldridge et al., 1977).

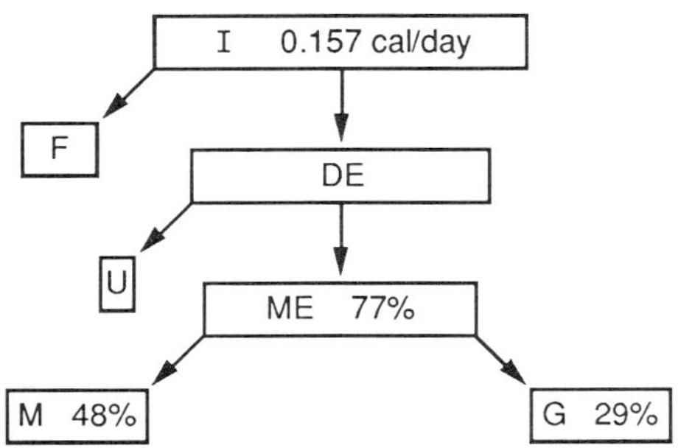

J. Generalized energy budget for first-feeding marine fish larvae at 10°C (Houde, 1989).

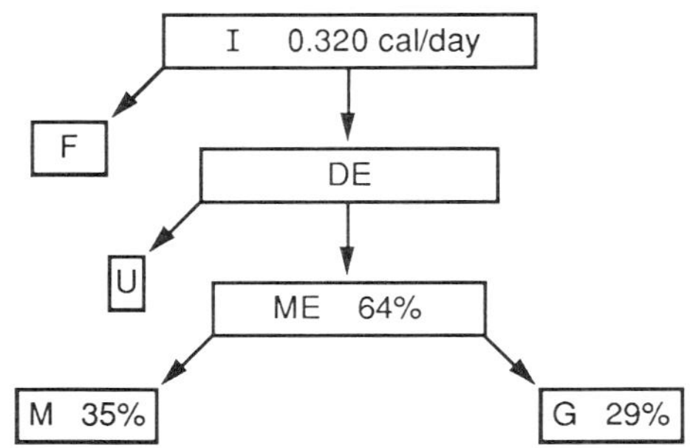

K. Generalized energy budget for first-feeding marine fish larvae at 20°C (Houde, 1989).

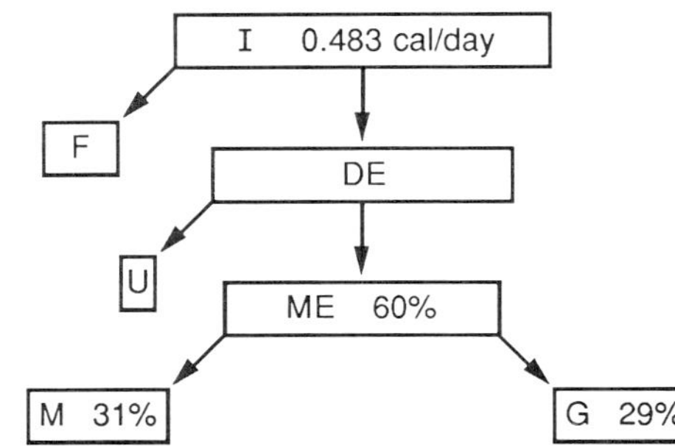

L. Generalized energy budget for first-feeding marine fish larvae at 30°C (Houde, 1989).

Figure 10-2. (Continued)

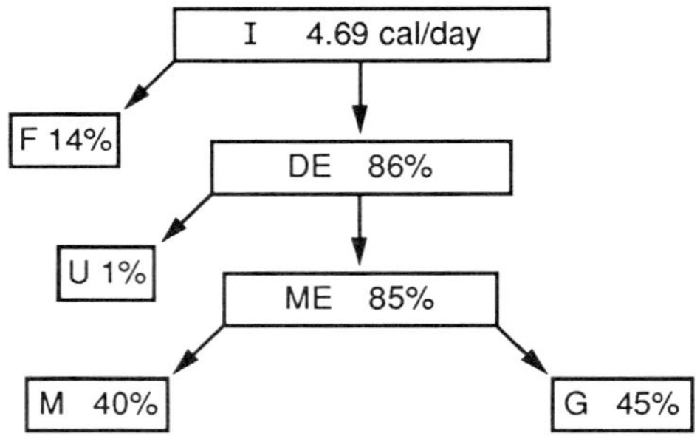

M. Dolphin larvae, 0.304-2.83 mg DW, eating *Artemia* at 26°C (Benetti, 1992).

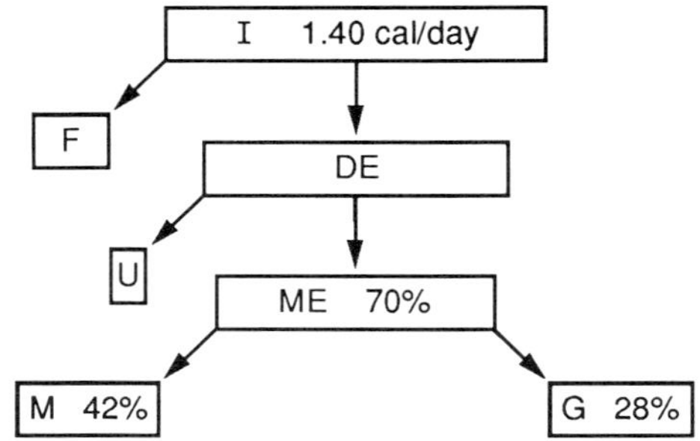

N. Winter flounder larvae, 1 mg DW, eating copepods at 8°C (Laurence, 1977).

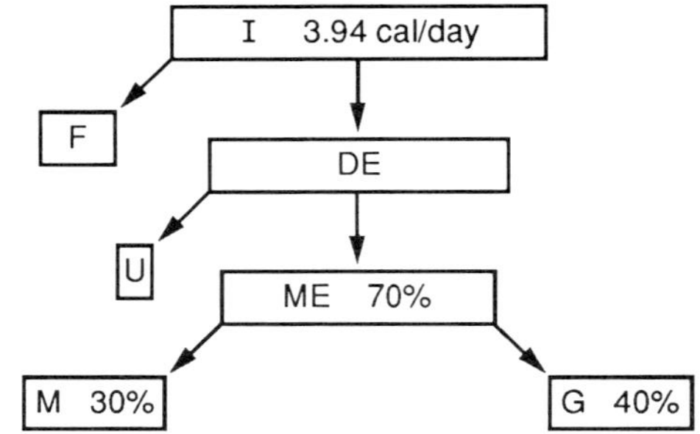

O. Striped bass larvae, 3 mg DW, eating *Artemia* at 18°C (Eldridge et al., 1982).

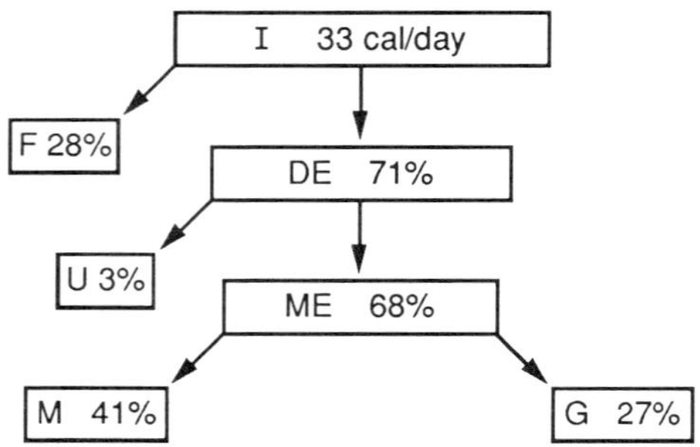

P. Yellow perch juveniles, 32 mg DW, eating *Daphnia* at 19°C (Mills and Forney, 1981).

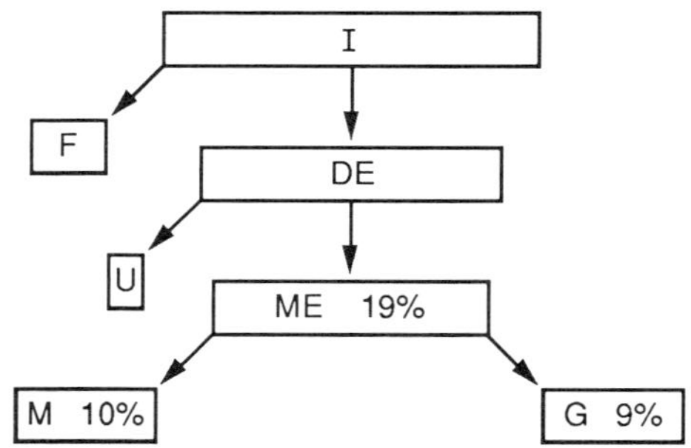

Q. Mozambique tilapia larvae to juveniles, final weight 40 mg DW, eating a variety of animal foods at 25°C (Mironova, 1974).

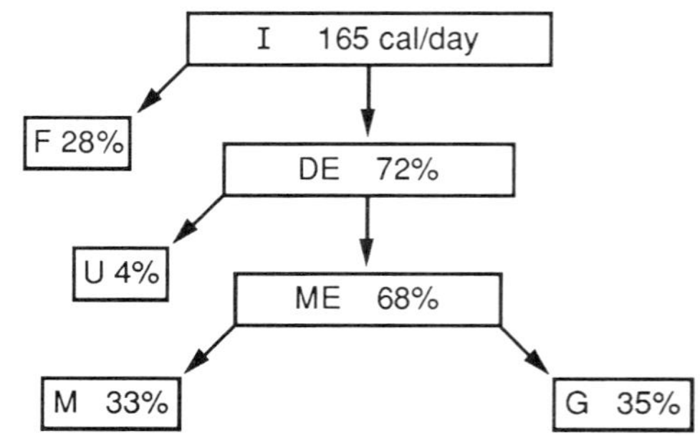

R. Sargassum fish juveniles, 1 g WW, eating sargassum shrimp at 22.5°C (Smith, 1973).

Figure 10-2. (Continued)

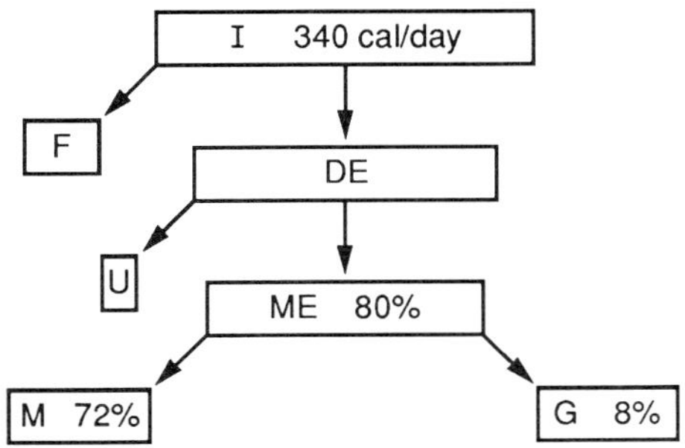

S. Blueback herring juveniles, 2.4 g WW, field and laboratory study at 13-28°C (Burbidge, 1974).

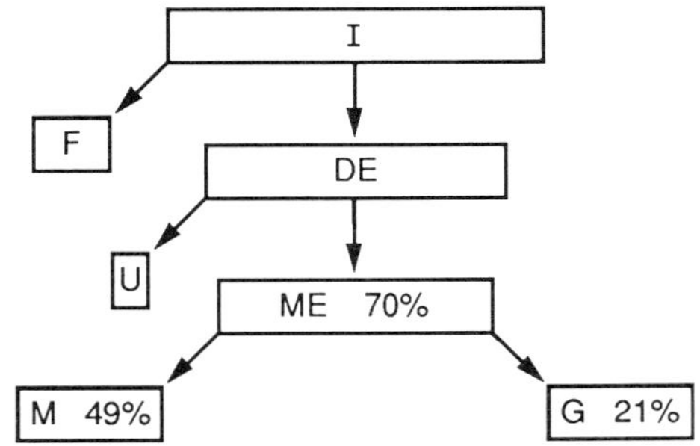

T. Sockeye salmon juveniles, 5.4 g WW (Brett, 1986).

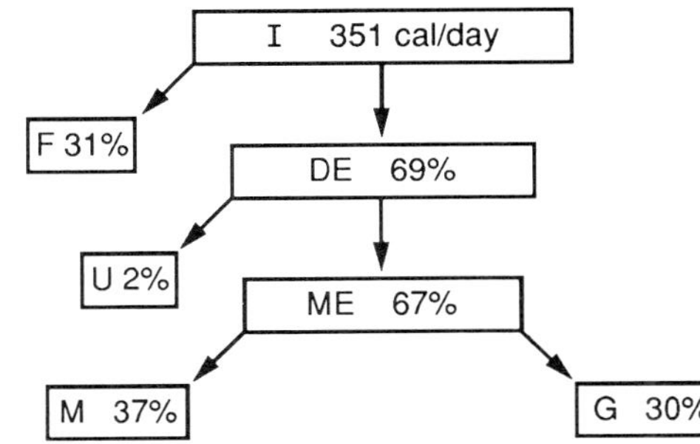

U. White bass juveniles, 1.27 g DW, field and laboratory study (Wissing, 1974).

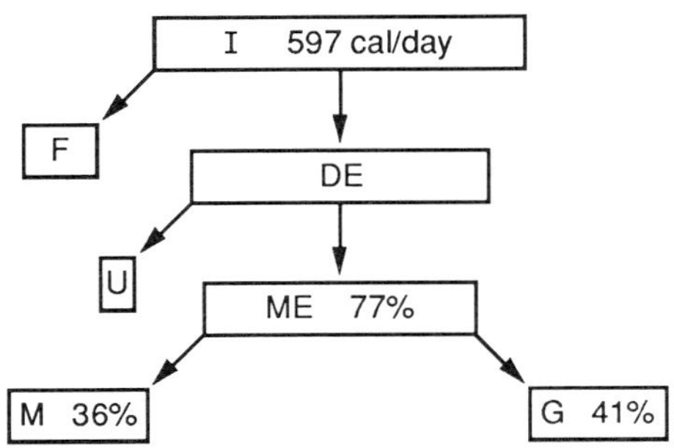

V. Rainbow trout juveniles, 4.1-9.9 g WW, eating experimental trout feed at 10.7-11.7°C (Staples and Nomura, 1976).

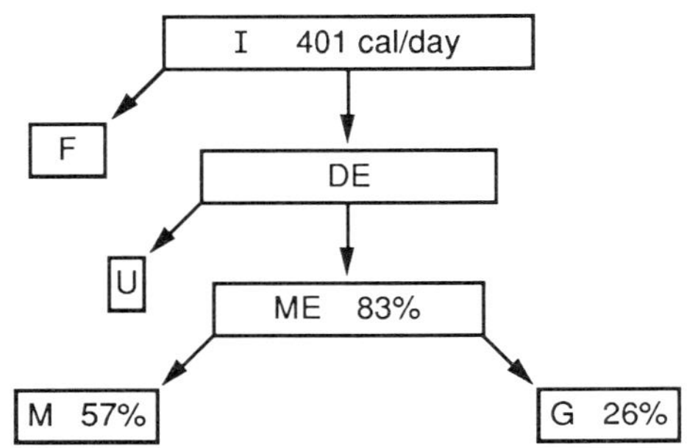

W. Striped bass juveniles, 2.29-16 g WW, eating commercial salmon starter at 23°C (Tuncer et al., 1990).

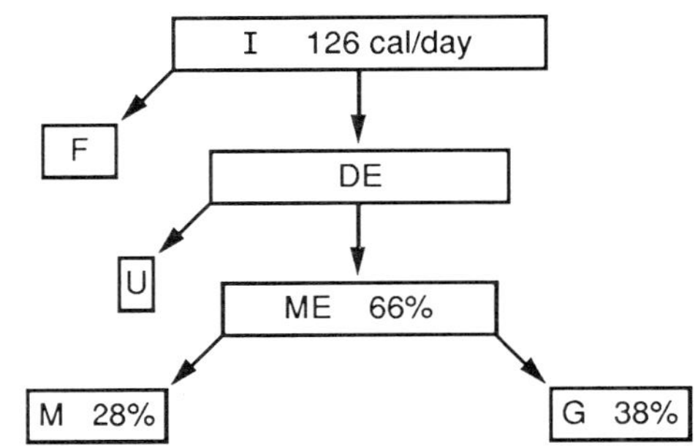

X. Northern anchovy adults, 1.95-2.56 g DW, eating krill at 17°C (Boggs, 1991).

Figure 10-2. (Continued)

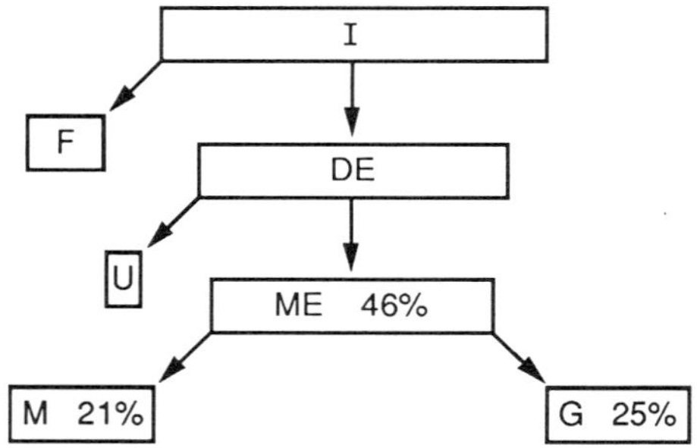

Y. Mozambique tilapia juveniles, 7.6-16.8 g WW, eating a variety of animal foods at 25°C (Mironova, 1974).

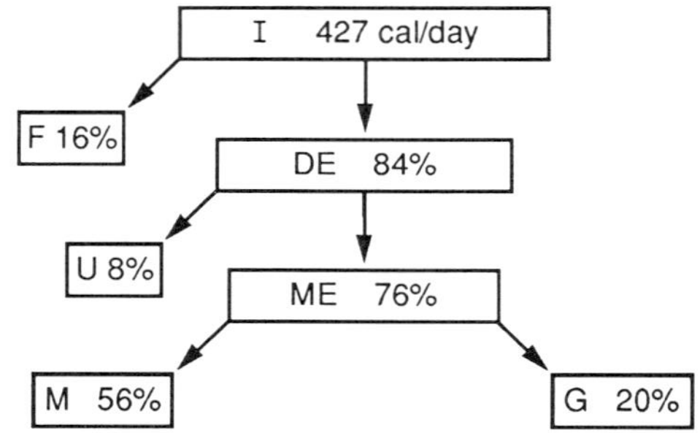

Z. European perch juveniles, 11.4-13.4 g WW, eating amphipods at 14°C (Solomon and Brafield, 1972).

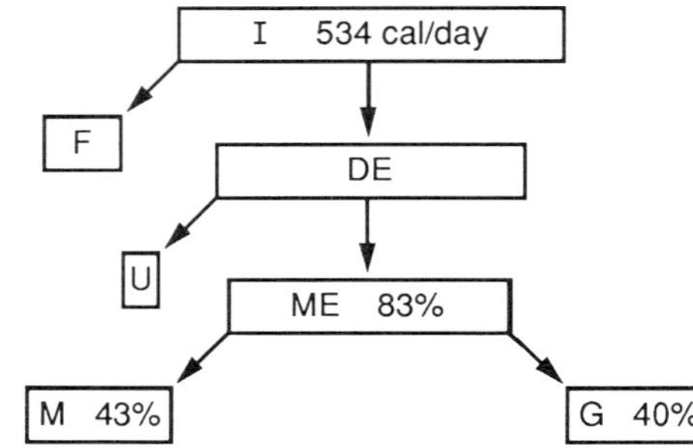

AA. Sunshine bass juveniles, 1.75-25 g WW, eating commercial salmon starter at 23°C (Tuncer et al., 1990).

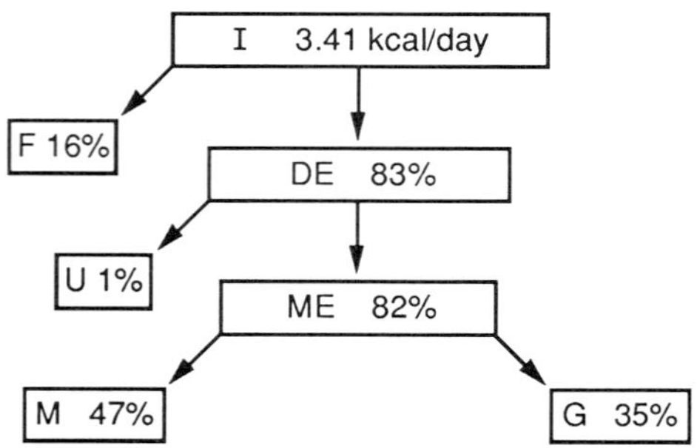

BB. Dolphin juveniles, 23 g WW, eating *Artemia* at 26°C (Benetti, 1992).

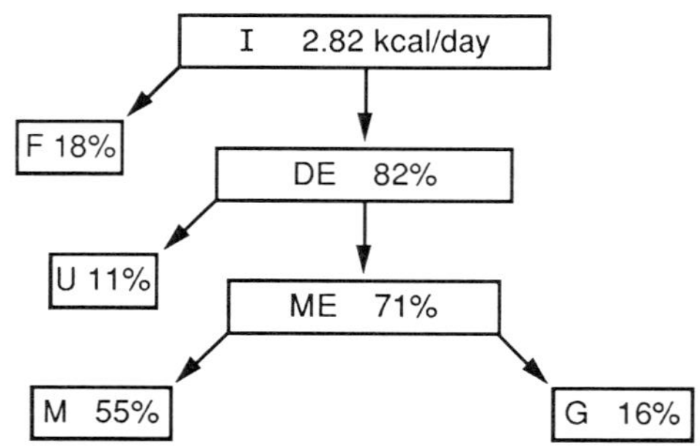

CC. Sargassum fish adults, 28 g WW, eating sargassum shrimp at 22.5°C (Smith, 1973).

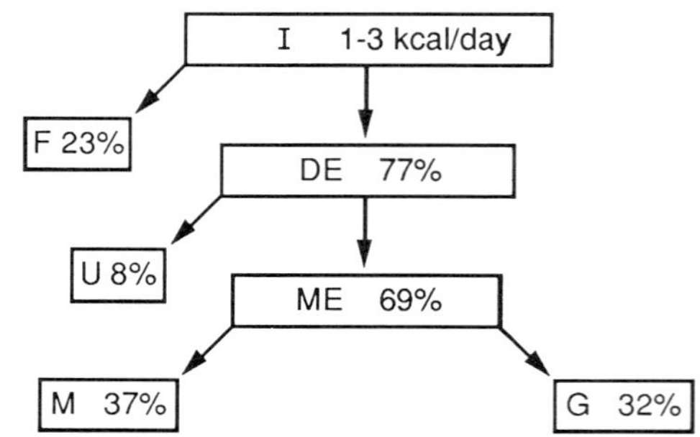

DD. Brown trout juveniles, 50 g WW, eating amphipods at 7-13°C (Elliott, 1976).

Figure 10-2. (Continued)

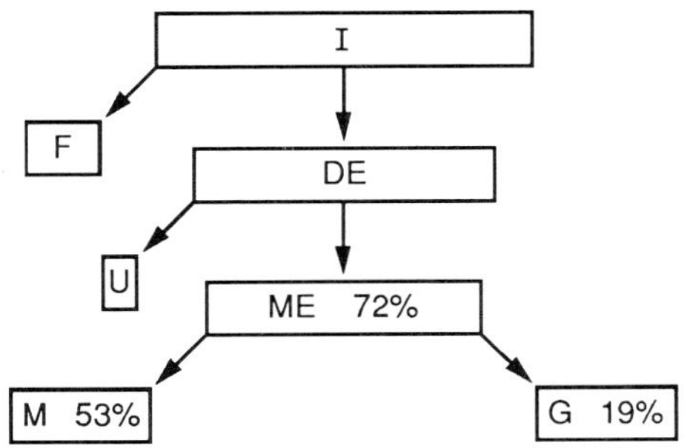

EE. Mozambique tilapia adults, 68.4-76.4 g WW (19.4-21.8 g DW), eating a variety of animal foods at 25°C (Mironova, 1974).

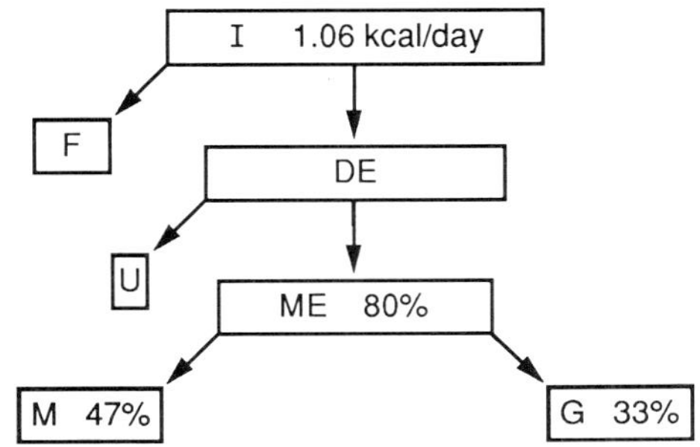

FF. Bluegill sunfish juveniles, 92.1 g WW, eating mayfly nymphs at 20°C (Pierce and Wissing, 1974).

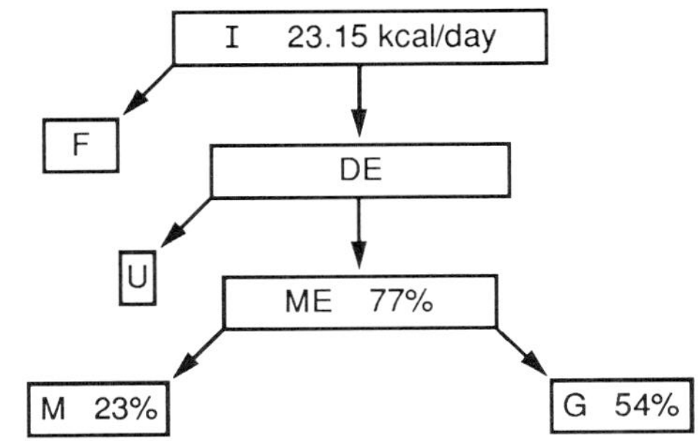

GG. Rainbow trout juveniles, 262-410 g WW, eating experimental trout feed at 10.7-11.7°C (Staples and Nomura, 1976).

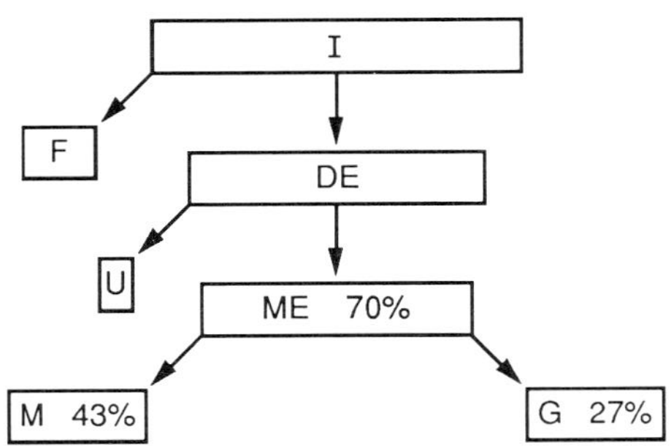

HH. Sockeye salmon juveniles, 420 g WW, (Brett, 1986).

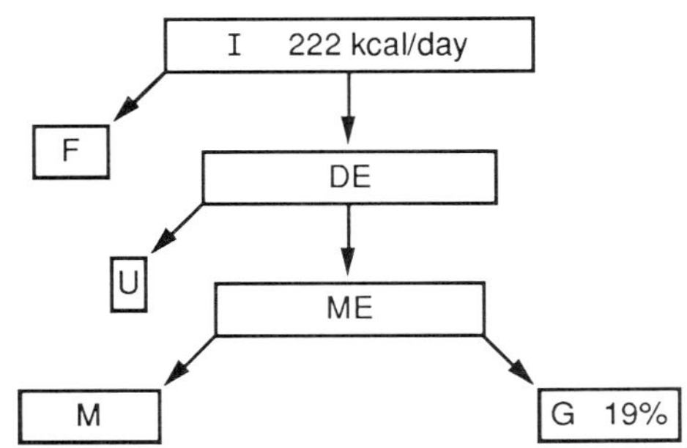

II. Atlantic cod, 487-525 g WW, eating plaice fillets at 16°C (Edwards et al., 1972).

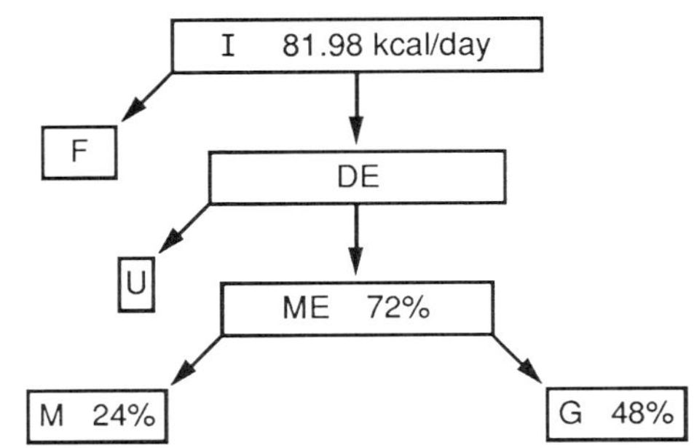

JJ. Rainbow trout adult females, 1603-1982 g WW, eating experimental trout feed at 10.7-11.7°C (Staples and Nomura, 1976).

Figure 10-2. (Continued)

11 MAINTAINING HEALTHY MARINE FISH

At times, disease is a major factor limiting the commercial farming of marine fish, and it can be a source of frustration and disappointment with fish kept for research and ornamental applications.

DISEASE PREVENTION

Fish pathogens often are present in culture systems, but fish usually are resistant to them unless they are stressed too much. Diseases usually result from some form of system deficiency or operator error. They can be avoided by minimizing stress from factors such as malnutrition, crowding, handling, fluctuating temperature, toxicants, chemicals or drugs, low dissolved oxygen, and disturbing light or noise, and by not introducing pathogens to the system (via eggs, water, air, foods, hands, nets, etc.). Choosing resistant species or strains of fish to raise also is a good idea.

The relationship between stress and disease is discussed in numerous publications, including Pickering (1981), Piper (1982), Heming and Paleczny (1987, rainbow trout), Pickering and Pottinger (1987a, 1987b, salmonids), Laird and Needham (1988a, salmonids), Roberts (1989f), Sindermann (1990), Barton and Iwama (1991), Moe (1992, ornamental fish), Iwama et al. (1995, 1997), Iwama and Nakanishi (1996), Wedemeyer (1996), Rotllant et al. (1997), and Mugnier et al. (1998). Healthy fish normally can resist facultative pathogens present in the culture system and latent pathogens present in their bodies, until stress becomes too great. Stress can make fish more susceptible to infectious diseases, partly by increasing secretion of corticosteroids, which suppress the numbers and function of circulating lymphocytes. Stress-induced vitamin depletion also can reduce disease resistance. Enough vitamins should be provided in the feed to allow for this. Under good conditions, salmonids have been raised successfully on feeds containing no ascorbic acid, but with the usual stresses on a farm they could need as much as 200 mg available ascorbic acid/kg diet (Jackson, 1988). Stress can interfere with reproduction by upsetting hormone balances. Stress can decrease growth rates and growth efficiency by reducing food intake or by diverting food energy into activity required to cope with the stressor. Crowding of fish could reduce growth through one or more mechanisms, including: aggressive behavior, alteration of the energy budget, reduction of food intake, and alteration of metabolic hormone levels. Increases in plasma cortisol, glucose, and lactic acid result from stress and can be used as indicators. Hematological, histological, biochemical, and other measurements on healthy fish can be used as a baseline (Buckley et al., 1985; Audet et al., 1993; Canfield et al., 1994; Kakuno and Koyama, 1994; Young et al., 1994).

Some recommendations for minimizing stress are:

- Maintain fish health through good husbandry.
- Maintain near optimal culture conditions.
- Avoid multiple disturbances.
- Allow time for recovery between disturbances.
- Be especially gentle with sensitive stages.
- Use anesthetics when necessary.
- Be aware of the needs of your fish.
- Be aware of their normal behavior and responses to stress.

An observant and experienced fish culturist often will notice the early warning signs of a potential health problem (e.g., unusual behavior, darkening, blanching, scraping against the tank, not eating, sores) and interrupt the process before it escalates.

High water quality must be maintained, because some disease organisms (especially bacteria and protozoans) will proliferate when fish are crowded or water is dirty. If necessary, culture water for fish and zooplankton is best disinfected with ultraviolet radiation (chapter 5). Ozonation is more effective, but can be risky. Unless bromine-free artificial seawater is used and dissipation of oxidants is assured, chlorinating or ozonating fish culture water could be unsafe. Water for algal culture can be chlorinated, but with a small risk of chlorinated hydrocarbons passing through the food chain to the fish. Autoclaving of algal culture water is preferred, at least for small quantities, and UV for larger amounts, but chlorination is the most practical method. Sanitation is essential. Tanks should be disinfected between batches of fish; everything used to handle fish should be disinfected between batches. Quaternary ammonium compounds also are used (e.g., 50% benzalkonium chloride), but we prefer the methods in Table 11-1. Traditional pasteurization (raising temperature to 63°C for 30 min) kills essentially all pathogens (Tortora et al., 1992). Short, high-temperature pasteurization (raising temperature to 72°C for 15 sec) kills essentially all pathogens and reduces total bacteria. Either method could be used to disinfect food, but the first might be easier. Corrales and Maclean (1995) suggested 40-45°C for 30-90 min, 500 ppm chlorine, or 5000 ppm hydrogen peroxide to eliminate harmful microalgae in ballast water of ships to help avoid exotic introductions.

Table 11-1. Methods for disinfection.

<u>Hands</u>
1. Rinse thoroughly with hot fresh water
2. Rinse with 70% isopropyl or ethyl alcohol, or
2. Rinse with 250 ppm iodophors
3. Rinse thoroughly with hot fresh water

<u>Equipment</u>
1. Scrub thoroughly with hot fresh water
2. Soak in 1% formalin (10 µL/L) for 1 h, or
2. Soak in 100 mg chlorine/L for 1 h, or
2. Soak in 200 mg chlorine/L for 30 min
3. Hang in sunlight to dry

<u>Tanks</u>
1. Scrub and wash well with fresh water
2. Add 10 mg chlorine/L for 24 h[a], or
2. Add 100 mg chlorine/L for at least 1 h, or
2. Add 200 mg chlorine/L for at least 30 min[b]
3. Rinse very well with fresh water.
4. Keep completely dry for at least 48 h

<u>Cages</u>
1. Clean with pressurized fresh water
2. Soak in 200 mg chlorine/L for 30 min
3. Leave in sunlight to dry

<u>Ponds</u>
1. Drain
2. Spread up to 380 g calcium oxide/square meter[c] in pond, or
2. Spread 110-280 g calcium hydroxide/square meter in pond
3. Keep as dry as possible for at least 1 week

[a] 25 mg/L or more chlorine might be necessary for some viruses.
[b] Serious viral contamination might require longer treatments with 100-200 mg chlorine/L (Piper et al., 1982).
[c] Up to 1.5 kg lime/square meter has been used after epidemics.

Fish can be scared to death. If removed from the water, very sensitive species like yellowtail snapper, will flip continuously until dead. Others such as Nassau grouper sometimes can survive for 1-2 h completely out of water (Tucker and Woodward, 1996).

Typically, an isolated fish does not grow as well as one in a group, partly because competition for food stimulates feeding. A low density of a territorial species results in much aggressive behavior, but increasing the density usually reduces aggression, probably because there are so many targets that potential aggressors would wear themselves out chasing all of them.

Captive fish sometimes are more vulnerable to parasites and diseases simply because they cannot avoid or escape them. Coral reefs often have cleaner animals (e.g., gobies, shrimp) that feed on parasitic organisms attached to fish. Parasite and bacteria populations can be limited by sunlight (ultraviolet) in shallow water. Outdoor rearing units benefit from this, but indoor ones are more likely to need artificially disinfected water.

All life stages of marine fish are susceptible to viral, bacterial, parasitical, and nutritional diseases, but larvae of most species have weak or no immune systems. For example, kasagos do not have a well developed immune system until they are about 2 months old (~27 mm TL, 250 mg)(Nakanishi, 1986). The culturist can assist larvae in combating diseases in several ways.

Probiotic Organisms

Percent survival during larval and early juvenile periods is negatively correlated with physiological stress and presence of pathogens (Kusuda et al., 1986; Nicolas et al., 1989; Gatesoupe, 1991c; Olsson et al., 1992). Although fish culturists routinely monitor many of the environmental factors that contribute to physiological stress, they seldom monitor the microbial environment, leaving much to chance.

Gut microflora in fish affects nutrition, growth, and susceptibility to disease. Bacteria ingested by larvae can contribute to establishment of the primary intestinal microflora, which subsequently furnishes nutrients and enzymes, or causes pathogenicity. Because of their naive and immature immune systems, larval and juvenile fish are more at risk than adults. The presence of pathogens combined with the stress of handling or long-term crowding can lead to bacterial septicemia with high morbidity and mortality.

Bacterial contamination of eggs usually originates from the parents or from contaminated water. Surface disinfection of eggs should be thought of as only a temporary measure, because exposure to pathogenic bacteria eventually can occur through the water, air, foods, people, later environments, or other sources. *Vibrio anguillarum*, which probably came from rotifers, was dominant in gilthead sea bream larvae that suffered massive mortality (Grisez et al., 1997). Gatesoupe et al. 1997) reported that iron supplementation might not be necessary and could be harmful in microdiets for larvae because it can promote growth of bacteria in rearing systems.

Food organisms should be screened because many pathogenic bacteria can occur on wild and cultured plankton (Verdonck et al., 1994). Some *Pseudomonas* spp. are fish pathogens, but others promote growth of rotifers and *Artemia* (Hagiwara et al. 1995a; Gorospe and Nakamura, 1996; Gorospe et al., 1996). Before first feeding, fish larvae establish their gut flora non-selectively, and appropriate flora in the water can give them a good start. After feeding begins, contaminated food (especially mollusc larvae, rotifers, and copepods, and sometimes *Artemia*) has been identified as the primary infection route for fish. Antibiotics sometimes are used to reduce the bacterial load in food cultures, but the safer strategies of probiotic inoculations and improved environmental management are better choices.

The use of probiotic bacteria is well established in agriculture (Conway, 1989; Maruta et al., 1996) but is mostly experimental in aquaculture. Numerous studies have reported a reduction in number of pathogenic bacteria (especially vibrios) after inoculation of fish or their environment with *Bacillus* spp. or others (Nicolas et al., 1989; Gatesoupe, 1991b,

1991c, 1993, 1994; Strøm and Ringo, 1993; Ringo et al., 1995; Kennedy et al., 1998; Ringø and Gatesoupe, 1998; Fig. 11-1). Some *Bacillus* spp. are known to improve nutritional absorption and decrease numbers of pathogens in the gut of fish and other animals (Conway, 1989; Maruta et al., 1996). Shiranee et al. (1993) reported that the amount of protease produced by and the final weight of pearl spot were positively correlated with the number of gram-positive rods in the gut. de la Banda et al. (1992) reported that larval turbot receiving live lactic bacteria survived six times better than the control group.

We observed increased survival, growth, and size uniformity of common snook, spotted seatrout, and striped mullet that had been reared in water inoculated with a probiotic strain of *Bacillus subtilis* previously isolated from healthy reared snook (Fig. 11-1; Kennedy et al., 1998; Tucker et al., unpub.).

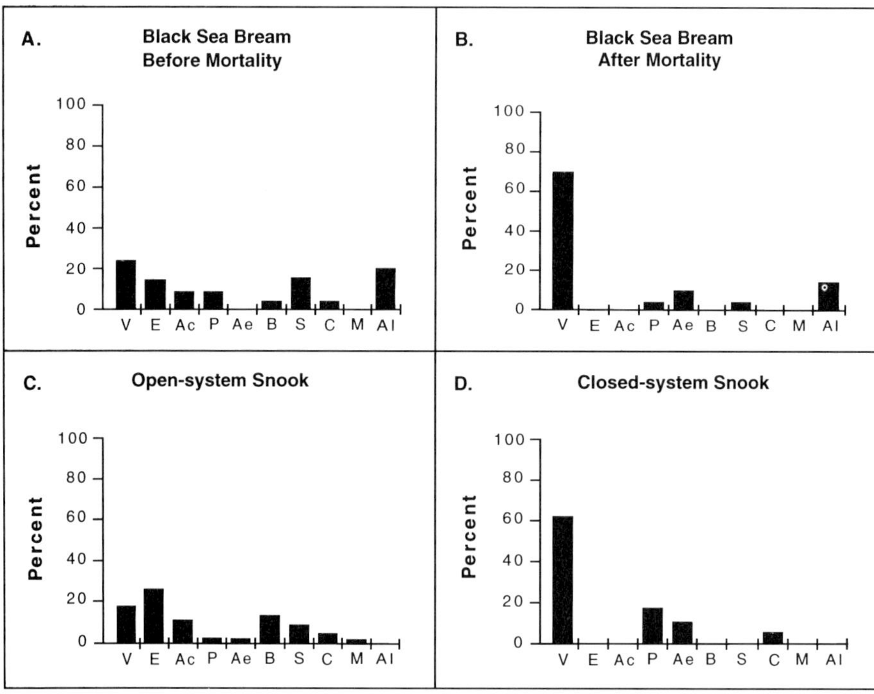

Figure 11-1. Percent biomass of bacterial genera of: A) Healthy black sea bream larvae (6 dah), B) Black sea bream larvae after mass mortality (14 dah), C) Early juvenile common snook raised in an open system (38 dah), D) Early juvenile common snook raised in a closed system (45 dah). (sea bream data from Kusuda et al., 1986; snook data from Kennedy et al., 1998) Probiotic bacteria were not added in any of these trials. V = *Vibrio/Photobacterium* spp., E = gram negative environmental bacteria (e.g., *Necromona/Chromobacterium* spp., *Flavobacterium* spp.), AC = *Moraxella/Acinetobacter* spp., P = *Pseudomonas* spp., Ae = *Aeromonas/Plesiomonas* spp., B = *Bacillus* spp., S = *Staphylococcus* spp., C = *Corynebacterium* spp., M = *Micrococcus* spp., Al = *Alcaligenes* spp. (figure from Kennedy et al., 1998)

Skjermo et al. (1997) obtained better survival of Atlantic halibut yolksac larvae when incubation water was microbially matured before use. Turbot larvae also survived and grew better in aged water, but their growth was better and more consistent when *Tetraselmis* sp.

was added, too. Stabilization of the microflora seemed to inhibit growth of pathogenic opportunistic bacteria.

Immunostimulants

Immunostimulants are compounds or mechanisms that enhance non-specific or specific immunity in fish. The two types of action overlap. Immunostimulants include: killed mycobacteria, synthetic peptides, glucans, alginates, chitin, chitosan, aluminum salts, levamisole, Bacille Calmette Guerin, *Corynebacterium parvum*, bacterial cell wall extracts (LPS antigen), and others (Anderson, 1992; Isayeva and Kozinenko, 1992; Raa et al., 1992; Sakai et al., 1992; Jeney and Anderson, 1993; Jørgensen et al., 1993a, 1993b; Anderson and Siwicki, 1994; Skjermo et al., 1995; Iwama and Nakanishi, 1996; Vadstein, 1997). The term adjuvant is used mostly to indicate an immunostimulant that is mixed and injected with a vaccine to stimulate the immune response to the vaccine, i.e., increase the titer (Anderson, 1992; e.g., Rørstad et al., 1993). Pesticides, some antibiotics, and other toxins tend to depress the immune system.

Vaccines

Vaccines cause the immune system to produce antibodies as if the actual disease was present (Leong and Fryer, 1993; BIOMED, 1994; Brackett, 1994; Mitchell, 1996; Iwama and Nakanishi, 1996). They are given by injection, hyperosmotic immersion (20-sec dip), immersion, bath (30 min to hours), spray, and ingestion—listed in decreasing order of effectiveness because of the type of immunity conferred. Hypersalinity increases permeability of gills and antigen uptake, but can be stressful. Normal immersion is less stressful and can result in sufficient uptake and protection. Oral vaccines have been improved recently by coating the antigen (e.g., as microspheres) to protect it from digestion and have the potential to induce better mucosal immunity than other methods. If perfected, oral vaccines would be not only very effective, but also less stressful and cheaper.

The main types of vaccine are: killed, attenuated, and subunit. Killed vaccines are made by culturing the pathogen to produce enough, then killing, processing, and injecting it into healthy animals (Armstrong, 1997). They are relatively safe because the pathogens cannot be revived. In emergency situations, this type has been made (without full testing) for quick response only on the farm from which the disease was isolated or for fish from that farm. The whole process might take 6 weeks. Because of the limited use, they are called autogenous vaccines. They have been made for *Aeromonas* spp., *Edwardsiella tarda*, *Enterococcus* sp., *Flexibacter psychrophilus*, *Lactobacillus* sp., *Pasteurella piscicida*, *Streptococcus* sp., *Vibrio* spp., *Yersinia ruckeri*, and others. When the need becomes great enough, fully tested and licensed commercial vaccines should be produced and made available to all farmers. Attenuated (modified-live) vaccines contain altered, non-virulent, pathogens. They can be more effective because they simulate the disease and stimulate the immune system better, or more accurately (e.g., with *Edwardsiella*). Two possible dangers are the return of virulence for the treated species or the infection of other fish accidentally exposed. Subunit vaccines contain parts of pathogens rather than whole pathogens. Passive immunity is accomplished by transferring antibodies from other fish or non-fish species (e.g., maternal antibodies passed on to the eggs).

Vaccines for *Aeromonas salmonicida*, *Aeromonas* sp., *Edwardsiella ictaluri*, infectious pancreatic necrosis, *Pasteurella* sp., *Streptococcus* sp., *Vibrio anguillarum*, *Vibrio anguillarum-ordalii*, *Vibrio anguillarum-salmonicida*, *V. parahaemolyticus*, *V. salmonicida*, *Vibrio* sp., viral hemorrhagic septicemia, *Yersinis ruckeri*, and *Yersinia* sp. are licensed (Schnick et al., 1997). Most vaccines have been made for salmonids, carp, and catfish, but some also will work on marine fish and others are being developed (Iwama and Nakanishi, 1996). Two are available for European sea bass and gilthead sea bream: *Vibrio anguillarum* serotype I and *Pasteurella piscicida* (Le Breton, 1996). A *V. anguillarum* vaccine is available for turbot (Toranzo et al., 1994). Vaccinations are not necessarily effective for all fish treated, especially when active immunosuppression is occurring ((e.g., high stress). Most vaccinations have to be repeated (boosters) to ensure long-term effectiveness (>6 months).

In addition, the exposure of larvae to a diverse range of bacteria including non-pathogenic vibrios should enhance immunity of juveniles later released or stocked in cages. Letting larval immune systems encounter low levels of non-virulent bacteria or low-virulence strains of pathogenic bacteria can help them build cross-protective immunity against fully pathogenic bacteria (Sindermann, 1990; Gatesoupe, 1994; Kennedy et al., 1998).

Monitoring

Fish should be checked regularly for signs of disease, especially at feeding time, when abnormal behavior can be the most obvious. Stressful events should be followed by remedial measures and greater vigilance. A veterinarian, fish health specialist, or someone trained in veterinary epidemiology should be on call, and the most-likely therapeutants should be kept in stock or readily available. A health lab with up-to-date diagnosis methods will minimize the time between detection and treatment (Plumb, 1996; Reddington, 1996; Thorburn, 1996; Noga, 1996). If repeated outbreaks are expected, it is wise to have a veterinarian help draw up a disease management plan. In the U.S., a valid veterinarian-client-patient relationship must be established in order to obtain therapeutants that require a prescription. When an aquatic veterinarian is not available, large-animal or production veterinarians can be of great value. Production (cow, chicken, pig) veterinarians specialize in preventive health plans and protocols. Non-veterinary fish health specialists might or might not have this kind of background.

THERAPEUTIC AGENTS

If culture conditions are suboptimal, and sometimes if they are optimal, diseases can develop. Because of possible harm to the fish's immune system, the culture system (e.g., nitrifying bacteria), or outside non-target organisms, the use of chemicals to control fish diseases should be minimized. This section discusses mainly U.S. legal guidelines. European guidelines and maximum residue limits were reviewed by Bates (1996).

Approved Drugs

Only the drugs listed below have been fully approved by the FDA (U.S. Food and Drug Administration) for use on food fish (Federal Joint Subcommittee on Aquaculture, 1994; Schnick, pers. comm., 1998).

Formalin

- In tanks, up to 250 ppm for 1 h daily at <10°C, up to 170 ppm for 1 h daily at >10°C. In earthen ponds, 15-25 ppm indefinitely, as a parasiticide for all finfish.
- 1-2 ppt for 15 min as a fungicide for all finfish. (The maximum treatment will be toxic to eggs of many species.)

Oxytetracycline

- 250 mg/kg fish/d in feed for 4 d for skeletal tagging of Pacific salmon. Fish cannot be released for at least 7 d after treatment.
- 55-82 mg/kg fish in feed for 10 d as a bactericide for salmonids and catfish. Salmonids should not be treated below 9°C and catfish below 16.7°C. Fish cannot be processed for food or released for at least 21 d after treatment.

Romet 30® (sulfadimethoxine and ormetoprim)

- 50 mg/kg fish/d in feed for 5 d as a bactericide (furunculosis) for salmonids. Fish cannot be processed for food or released for 42 d.
- 50 mg/kg fish/d in feed for 5 d as a bactericide (enteric septicemia) for catfish. Fish cannot be processed for food or released for 3 d.

Sulfamerazine

- 220 mg/kg fish/d in feed for up to 14 d as a bactericide (furunculosis) for trout. Fish cannot be processed for food or released for 21 d. Because of economic reasons, sulfamerazine is not currently being sold.

Tricaine methanesulfonate

- 15-330 mg/L as an anesthetic for ictalurids, salmonids, esocids, and percids at >10°C. For other families, use is restricted to hatchery and research purposes. Fish cannot be processed for food or released for 21 d.

Unapproved Drugs of Low Regulatory Priority

The following substances were determined to have low regulatory priority, meaning that objections to their use would be unlikely if: (1) they are used for the indicated purposes, (2) they are used at the prescribed doses, (3) they are used with good management practices, (4) the appropriate grade is used, and (5) adverse effects on the environment are unlikely (Federal Joint Subcommittee on Aquaculture, 1994; also MacMillan, 1995).

Acetic acid

- 1,000-2,000 ppm dip for 1-10 min as a parasiticide.

Calcium chloride

- up to 150 mg/L in fresh water to help freshwater fish maintain osmotic equlibrium during handling or transport.
- enough to raise calcium content to 10-20 mg/L $CaCo_3$ to help eggs harden.

Calcium oxide

- 2,000 mg/L for 5 sec as an external protozoacide.

Carbon dioxide

- dissolved as an anesthetic for cold, cool, and warmwater fish.

Fuller's earth

- to reduce adhesiveness of eggs and improve hatching.

Garlic

- to control helminths and sea lice in marine salmonids.

Hydrogen peroxide

- 250-500 mg/L to control fungi on all life stages of all species.

Ice

- to lower metabolism of transported fish.

Magnesium sulfate

- 30,000 mg/L with 7,000 mg/L sodium chloride for 5-10 min for externally parasitic monogenean trematodes and crustaceans for all life stages of freshwater fish.

Onion

- to treat and repel sea lice infesting external surfaces of fish at all life stages.

Papain

- 0.2% solution to remove the gelatinous matrix from egg masses, improve hatching, and reduce disease.

Potassium chloride

- enough to raise chloride content to 10-20,000 mg/L to reduce stress in freshwater fish by helping them osmoregulate.

Povidone iodine

- 50 ppm for 30 min to disinfect eggs during water hardening.
- 100 ppm for 10 min to disinfect eggs after water hardening.

Sodium bicarbonate
- 142-642 mg/L for 5 min to produce carbon dioxide to anesthetize fish.

Sodium chloride
- 0.5-1% for an indefinite time to aid osmoregulation to relieve stress and prevent shock.
- 3% for 10-30 min as a parasiticide.

Sodium sulfite
- 15% for 5-8 min to improve hatching of eggs (mainly catfish, largemouth bass, smallmouth bass).

Urea and tannic acid
- 3000 mg/L urea plus 4000 mg/L sodium chloride for 6 min, then 150 mg/L tannic acid for 6 min, to reduce adhesiveness of eggs.

Investigational New Animal Drugs
Drugs not yet approved must go through a fairly strict process to become approved by the U.S. Food and Drug Administration (Animal Drug Availability Act of 1996; Schnick and Armstrong, 1997; also see Prater et al., 1996). Investigational new animal drugs (INADs) are undergoing clinical and nonclinical trials to demonstrate their safety and efficacy. New animal drugs (NADs) have been submitted for approval (e.g., erythromycin, Moffitt, 1991). Both researchers and producers can have opportunities to participate in INAD trials, but very good records must be kept. The U.S. Joint Subcommittee on Aquaculture has made this list of priority drugs for the approval process (Schnick and Armstrong, 1997):

Amoxicillin, for bacteria
Benzocaine, anesthetic
Chloramine-T, for bacterial gill disease and *Flexibacter*
Copper sulfate, for fungus, ectoparasites, and surface bacteria
Cutrine-Plus,® for fungus, ectoparasites, and surface bacteria
Diquat, for fungus and surface bacteria
Erythromycin, for bacteria
Formalin, for fungus and ectoparasites
Hydrogen peroxide, for fungus, ectoparasites, and surface bacteria
Neomycin sulfate, for vibriosis
Oxytetracycline, for bacteria and skeletal marking
Potassium permanganate, for ectoparasites and surface bacteria
Praziquantel, for trematodes and cestodes
Trichlorfon, for ectoparasites
Various compounds, for sea lice
Common carp pituitary, for spawning
Human chorionic gonadotropin, for spawning
Gonadotropin-releasing hormone analog, for spawning
17-methyltestosterone, for gender manipulation

Thirteen more drugs have been added to this list (Schnick, pers. comm., 1998):

Aqui-S,® anesthetic
Azimethiphos, for sea lice
Calcein, marking agent
Calcide,® for sea lice
Cypermethrin, for sea lice and freshwater parasites
EarthTec,® for external microbes
Florfenicol, orally for bacteria
Fumagillin, for myxosporidians
Pyceze,® for fungus and external bacteria

Quinine, for parasites and myxosporidians
Salartect 500,® for sea lice
GnRHa, for spawning
Ovaprim,® for spawning

The above three lists (approved, unapproved/low priority, investigational) will change as new treatments are developed and approved. Culturists should remember that in the U.S., drugs must be approved, used under proper extra-label or investigational exemptions, or have low regulatory priority status. Any other use is considered a violation of the Food, Drug, and Cosmetic Act.

Extra-label Use
The policy of "extra-label use" has made drugs labeled for certain fish and for terrestrial food animals available for species of fish not on the label, but this has been allowed only under a veterinarian's prescription when a valid veterinarian-client-patient relationship existed. The Animal Medicinal Drug Use Clarification Act of 1994 prohibited extra-label use of medicated feeds for aquatic animals; however, that provision is under discussion at the FDA (Schnick, pers. comm., 1998).

Unapproved Drugs
Because some compounds are banned for use in farmed fish and the approved lists are revised as necessary, a current list from the CVM (Center for Veterinary Medicine of the FDA) or the equivalent authority in other countries should be consulted before using drugs for fish. Substances known to be dangerous to the health of aquaculturists, consumers, or the environment include: chloramphenicol, nitrofurans (e.g., nitrofurazone, furacin, furanace, furazolidone, furoxone), malachite green, gentian violet, and certain steroid hormones. These drugs should never be used on fish to be consumed by humans, and if determined necessary for non-food fish, they should be used only cautiously and under direct veterinary guidance. Examples of their use on food fish later in this chapter are given only for the sake of thoroughness.

Chloramphenicol is an important antibiotic, previously used on humans for typhoid fever and other infections, that is now banned because it can cause (fatal) aplastic anemia. Its use in humans and all food animals is prohibited in the U.S. and the E.U. Certain other drugs used for humans could be banned or restricted for fish because of the chance of development of resistant human pathogenic bacteria in the fish, culture systems, or water discharge areas (for more on resistance, see Austin and Austin, 1987; Sindermann, 1990).

Drug Criteria
Austin and Austin (1987) listed criteria for choosing a drug to use on fish (modified):

(1) The drug must reach the pathogen.
(2) The drug must kill the pathogen quickly.
(3) The drug must not make the disease worse.
(4) The drug and its breakdown products must not harm the fish or other non-target organisms.
(5) The drug must not leave long-term residues in food fish.
(6) The drug must be stable under normal storage and use.
(7) The drug should not be used if suspected to cause resistant strain development.
(8) Drugs used for treatment of humans should be avoided if a chance of inducing resistant strains of human pathogens exists.

Treatment Methods
Dip—fish are held in a net and briefly immersed in the treatment solution.
Flush—the drug is added to water flowing through the container (tank).
Bath—fish are held in a constant concentration of the drug for a specified period (tank

or pond).
Oral administration—drug is added to the food (only if fish are eating).
Topical application—drug is applied directly to the lesion (valuable fish).
Injection—drug is given by intramuscular (IM), intraperitoneal (IP), or (rarely) subcutaneous injection (for valuable fish).

The first three methods result in a quantity of water that has to be disposed of without infecting or injuring non-target organisms.

Quarantine Procedures
When mixing fish from different stocks (especially if adding wild fish to captive populations), the new fish should be held separately and observed for a period of 1-4 weeks. For immunostimulation, extra amounts of vitamins A, C, and E plus 5 g/kg beta glucan can be included in the diet (Kennedy et al., 1998). During this time, appropriate treatments for suspected pathogens should be employed to protect both old and new groups of fish. Some treatments that have been used on marine fish during quarantine (Kennedy, pers. comm., 1996; Montgomery and Brock, 1997; Schmidt et al., 1997) are listed below. These are not necessarily approved by the CVM or equivalent authority, which should be consulted for currently approved procedures.

For bacteria
- 25 mg/kg BW oxytetracycline, IM or IP injection (10-d effect)

For external parasites (e.g., protozoans, monogeneans, copepods, isopods)
- Freshwater bath for up to 1 h (if fish can take it)
  or
- 20-30 mL/L hydrogen peroxide dip for 3-5 min
  or
- 150-250 mg/L hydrogen peroxide bath for 30 min

For ectoparasitic protozoans (e.g., *Amyloodinium*, *Cryptocaryon*)
- Freshwater dip before and after one of the following
- 10 g/L chloroquine diphosphate continuous bath for 21-45 d (treatment period determined by life cycle of parasite)
  or
- 5 g/kg chloroquine diphosphate in feed for 21-45 d

For monogeneans, and internal protozoans, digenean trematodes, nematodes
- 50 mg/kg BW fenbendazole (or other drug from the benzimidazole family), orally
  or
- 1-2 mg/L fenbendazole bath 6 h/d for 3 d, repeat after 3 weeks

For internal nematodes (and immunostimulation, especially for immunosupressed fish)
- 11 mg/kg BW levamisole, IM injection or orally, twice at 1 week interval

Gallet de St. Aurin et al. (1990) used a quarantine process primarily to eradicate *Cryptocaryon irritans* from tropical marine fish. Water was filtered to 5 μm. Days 1, 5, 8, 11: (1) 200 mg formalin/L bath for 30 min, (2) freshwater bath for 5 min, (3) either furaltodone chlorhydrate or quaternary ammonia bath for 30 min. Day 13: transfer fish to rearing tank. Day 14: step (3) antiseptic bath one last time.

Treatment during Transport
This is mainly to protect fish from infections caused by handling but also could help prevent disease introductions. Methods will vary with species and expected dangers. Fish to be shipped more than a short distance should have a health inspection before shipment and/or just after arrival, and then be quarantined. Fish diseases of all kinds have been spread around the world by shipments. Current veterinary guidelines should be followed

(de Kinkelin and Hedrick, 1991; Arthur and Ogawa, 1996; Daelman, 1996).

BIOLOGICAL DISEASE AGENTS AND THEIR CONTROL

Comprehensive discussions of marine fish diseases and defects are provided by Paperna (1983), Kinne (1984), Kabata (1985); Austin and Austin (1987, 1989), Post (1987), Roberts (1989a), Sindermann (1977, 1990), Grabda (1991), Meidi and Zhifeng (1991), Gratzek and Matthews (1992), Leong, (1992), Plumb, 1992, Woo (1992), Stoskopf (1993), Bruno and Poppe (1996), Divanach et al. (1996), Main and Rosenfeld (1996), Noga (1996), Subasinghe et al. (1996) and articles in the series: Annual Review of Fish Diseases, Diseases in Asian Aquaculture, and other symposium proceedings. Gratzek and Matthews (1992), Moe (1992), and Spotte (1992) describe major diseases of aquarium fish. In addition, there is an extensive array of reports and reviews covering one to several diseases or hosts (e.g., Paperna and Overstreet, 1981; Paperna, 1984c; Chong and Chao, 1986a; Tung and Kou, 1986). This chapter gives only a brief overview of marine fish diseases and examples of treatment methods. Common names and causative agents are listed in Table 11-2.

Laboratory methods for diagnosis and transportation methods for specimens are discussed by Piper (1982), Austin and Austin (1987), Post (1987), Austin and Austin (1989), Bullock (1989), Tortora et al. (1992), Noga (1996), and others. If fish specimens need to be sent to a diagnostic laboratory, the order of preference for their condition is: (1) live, (2) on ice, (3) frozen, or (4) fixed in 10% formalin.

This chapter provides examples of preventive measures and treatments used for significant diseases of marine, anadromous, and catadromous fish. Treatments in some of the references are of historical interest only, because the chemicals are too dangerous. No guarantees are made or responsibility assumed for accidents or overdosages. Special care must be taken because fish sensitivity varies among different species, ages, and conditions (Chapter 3; Bills et al., 1993). The culturist should avoid dangerous compounds, read relevant references, and read label directions and cautions before proceeding with chemical treatment.

Neoplasias

(Post, 1987; Roberts, 1989c; Sindermann, 1990) Excessive, uncoordinated production of cells, which persists after the initiating stimulus is gone, results in tumors and other disorders such as leukemia. These are rare in fish and not of much economic importance. Thorough discussions are available in the references given and others. Some possible factors are age, heredity, immunology, carcinogens, irritants, trauma, and viruses (Post, 1987). Aflatoxins, chlorinated hydrocarbons, and a large number of other natural and artificial substances are carcinogenic. Treatment of neoplasia in fish is not really practical, and prevention is based on good husbandry and avoidance of carcinogens. What look like tumors sometimes can actually be swollen inflamed tissue caused by parasites (Hard et al., 1979). Histopathology can provide a definitive diagnosis.

Viruses

(Austin and Austin, 1989; Post, 1987; Hetrick, 1989; Smail and Munro, 1989; Sindermann, 1990; Hedrick et al., 1991; Kimura and Yoshimizu, 1991; Ahne, 1994) The many viral diseases of fish produce dozens of external and internal signs. Several cause serious problems, including systemic infections, ulcers, hemorrhages, nerve disorders, necrosis of internal organs, hyperplasias, tumors, and high mortality. Apparently, many pathogenic viruses are present in healthy fish but do not cause disease until the fish are stressed. Also, most of them are lethal only to young fish and are latent in adults. Rhabdoviruses and herpesviruses are responsible for most known viral diseases of marine fish. Diagnosis is mainly by monolayer tissue culture, identification of viral proteins or nucleic acids, and sometimes electron or light microscopy or determination of cytopathic effects (Austin and Austin, 1989; Bullock, 1989; Tortora et al., 1992). Developing methods include enzyme

Table 11-2. Major disorders of cultured marine fish.

Common name	Causative agent
Anchor worm	*Lernaea cyprinacea*
Athiaminosis	Feeding raw seafood containing thiaminase
Augenkrankheit (eye disease, Atlantic cod)	*Vibrio* sp.
Bacterial gill disease	Myxobacteria
Bacterial kidney disease	*Renibacterium salmoninarum*
Bacterial septicemia	*Aeromonas* sp., *Edwardsiella* sp., *Pseudomonas* sp., *Vibrio* sp., *Yersinia* sp.
Black ich	*Paravortex* sp.
Black patch necrosis (common sole)	*Flexibacter maritimus*
Black spot disease	Digenean trematode (e.g., *Cryptocotyle lingua*)
Black spot disease (yellowtail)	*Lymphocystis* sp.
Blood flukes	*Sanguinicola davisi*
Blue-sac disease (salmonid larvae)	Environmental
Boil disease	Sporozoans
Botulism	*Clostridium botulinum*
Branchionephritis (Japanese eel)	Unknown
Brown blood disease (=methemoglobinemia)	Nitrites
Cauliflower disease	*Lymphocystis* sp.
Codworm	*Pseudoterranova* (=*Phocanema*) *decipiens*
Coldwater disease (salmonids)	Myxobacteria (e.g., *Cytophaga psychrophila*)
Coldwater vibriosis (salmon)	*Vibrio salmonicida*
Columnaris disease	*Cytophaga columnaris*
Coral fish disease	*Amyloodinium ocellatum*
Costiasis	*Ichthyobodo* (=*Costia*) spp.
Crippled body disease (Japanese eel)	*Pleistophora anguillarum*
Cryptocaryoniasis	*Cryptocaryon irritans*
Distended gut syndrome (gilthead sea bream)	Virus?
Edwardsiellosis	*Edwardsiella* sp.
Eel viral kidney disease	Birnavirus
Egtved disease (=viral hemorrhagic septicemia)	Rhabdovirus
Encephalitis	Virus
Enteric redmouth (salmonids)	*Yersinia ruckeri*
Enterococcosis (yellowtail)	*Enterococcus seriolicida*
Epithelial necrosis (Japanese cultured fish)	Herpesvirus or Paramyxovirus
Epitheliocystis	Chlamydial bacteria (probably)
Epizootic ulcerative syndrome (barramundi)	Virus, bacteria, and/or fungi
Eubacterial meningitis	*Eubacterium tarantellus*
Exocrine pancreas disease (Atlantic salmon)	Unknown (possibly metabolic or nutritional)
Eye disease (Pacific salmon, cod, etc.)	*Vibrio* sp.
Eye fluke	*Diplostomum flexicaudum*, *D. spathaceum*
Fin rot	Bacteria (e.g., *Aeromonas hydrophila*, *Pseudomonas* sp.), maybe protozoans
Fish louse (fresh water)	*Argulus* sp.
Furunculosis	*Aeromonas salmonicida*
Gas bladder fibrosarcoma (Atlantic salmon)	Leucovirus (probably)
Gas bubble disease	Nitrogen bubbles
Gill fluke	Monogenean flatworm
Gill rot	*Branchiomyces* sp.
Gill rot (Japanese eel)	*Aeromonas* sp. and *Saprolegnia* sp.
Gliding bacterial disease	*Flexibacter maritimus*
Golden-eye disease (groupers)	Astro-like virus
Granulomatous hypertyrosinemia	Ascorbic acid deficiency
Head and lateral line disease	Nutritional deficiency or virus
Hexamitiasis (salmonids)	*Hexamita* (=*Octomitus*) sp.
Hirame virus disease	*Rhabdovirus olivaceus*
Hitra disease (salmonids)	*Vibrio* sp.

Table 11-2. (Continued)

Common name	Causative agent
Hole in the head (lateral line erosion)	Ascorbic acid deficiency (probably)
Ich	*Ichthyophthirius multifiliis*
Infectious dropsy (?)	Virus
Infectious hematopoietic necrosis (salmonids)	Rhabdovirus
Infectious pancreatic necrosis (salmonids, etc)	Birnavirus
Jellied fish (bluefin tuna)	*Kudoa clupeidae*
Lateral line disease	Nutritional deficiency or virus
Lordosis	Undeveloped gas bladder, nutritional deficiency
Lymphocystis disease	*Lymphocystis* sp. (iridovirus)
Milkiness (salmon, flatfish, etc.)	*Kudoa thyrsites*
Milky eye disease (milkfish)	*Vibrio* sp., possibly also a fungus
Motile aeromonad septicemia	*Aeromonas* sp., *Pseudomonas* sp.
Mushiness (halibut)	*Kudoa* sp.
Mycobacteriosis	*Mycobacterium* sp.
Nervous suffering disease (groupers)	Rancid dietary lipids (probably)
Nocardiosis	*Nocardia* sp.
Oral papilloma (gilthead sea bream)	Leucovirus (probably)
Papillomatosis (Atlantic salmon)	Virus-like particles
Paracolo disease (Japanese eel)	*Edwardsiella tarda*
Pasteurellosis	*Pasteurella piscicida*
Peduncle disease (=coldwater disease)	Myxobacteria (e.g., *Flexibacter psychrophila*)
Proliferative kidney disease (salmonids)	Unknown (myxosporidian parasite?)
Pseudotuberculosis (yellowtail)	*Pasteurella piscicida*
Red boil	*Vibrio anguillarum*
Red disease (European eel)	*Vibrio anguillarum*
Red disease (Japanese eel)	*Aeromonas hydrophila*
Red fin disease (Japanese eel)	*Edwardsiella anguillimortiferum* and *Aeromonas hydrophila*
Red pest (eels)	*Vibrio anguillarum*
Red plague	*Vibrio anguillarum*
Red sore	*Epistylis* sp. and bacteria
Red sickness (European eel)	Unknown
Red spot disease (Japanese eel)	*Pseudomonas anguilliseptica*
Red spot disease (milkfish)	*Vibrio anguillarum*
Red tide	Dinoflagellate toxins (e.g., *Ptychodiscus brevis*, *Gonyaulax monilata*)
Roundworm	Nematode
Salmon louse (salmonids, and others)	*Caligus elongatus*, *Lepeophtheirus salmonis*
Salmonid blood spot	*Yersinia ruckeri*
Saltwater furunculosis	*Vibrio anguillarum*
Saltwater columnaris (salmonids)	*Sporocytophaga* sp.
Saltwater ich	*Cryptocaryon irritans*
Saprolegniasis	*Saprolegnia* sp.
Scale disease (milkfish)	*Vibrio parahemolyticus*
Scoliosis	Nutritional deficiency
Seabass virus (European sea bass)	Birnavirus
Seahorse disease	*Glugea* sp.
Sea lice (Atlantic salmon)	*Caligus elongatus*, *Lepophtheirus salmonis*
Skin fluke	Monogenean trematode
Sleepy grouper disease	Iridovirus (probably)
Spaghetti worms	*Poecilancistrium caryophyllum* larvae
Spinning disease (Atlantic menhaden)	Birnavirus
Spinning grouper disease	Picorna-like virus
Spiny-headed worm	Acanthocephalan
Stomatopapilloma (European eel)	Orthomyxovirus (possibly)
Streptomycosis (salmonids)	*Streptoverticillum salmonis*

Table 11-2. (Continued)

Common name	Causative agent
Sure disease (Japanese eel)	*Saprolegnia* sp.
Swinging disease	*Ichthyophonus hoferi*
Systemic granuloma (sparids)	Nutritional (probably)
Tapeworm	Cestode
Tapioca disease (Pacific salmon)	*Henneguya salminicola*
Trichodiniasis	*Trichodina* sp.
"Tuberculosis" (=mycobacteriosis)	*Mycobacterium* sp. (e.g., *M. marinum*)
Ulcerative dermal necrosis (salmonids)	Sunlight, virus, and/or *Saprolegnia* sp.
Ulcus syndrome (Atlantic cod)	Rhabdovirus & iridovirus (probably)
Undeveloped gas bladder	Environmental or nutritional deficiency
Velvet disease	*Amyloodinium ocellatum*
Vibriosis	*Vibrio* sp.
Viral deformity (yellowtail)	Virus
Viral erythrocytic necrosis	Virus (probably iridovirus)
Viral hemorrhagic septicemia	Rhabdovirus
Viral nervous necrosis	Virus
Whirling disease	Virus or protozoan (e.g., *Myxosoma cerebralis*)
White grub	*Posthodiplostomum minimum*
White head disease (European sea bass)	Probably bacteria
White mouth disease (puffers)	Virus
White spot disease (=ich)	*Ichthyophthirius multifiliis*
White spot disease	*Amyloodinium ocellatum*
White spot disease	*Cryptocaryon irritans*
White spot disease	*Glugea* sp.
White-spot disease (salmonid larvae)	Environmental
Winter disease (gilthead sea bream)	*Pseudomonas anguilliseptica* (possibly)
Worminess (halibut)	*Unicapsula muscularis*
Viral nerve necrosis (striped knifejaw)	Picornavirus (possibly)
Yamame tumor virus (salmon)	Herpesvirus
Yellow grub	*Clinostomum complanatum*
Yellowtail ascites	Virus
Yersiniosis	*Yersinia* sp.

immunoassays and DNA hybridization (Sanz and Coll, 1992). Immunization or selection of resistant strains of fish should be practiced whenever possible. Generally, viral diseases cannot be treated, and the only option is to isolate infected individuals and maintain good conditions and nutrition. Prevention through exclusion of known pathogens, vaccination (when possible), sanitation, and good husbandry are very important when potential exists for significant viral diseases. Vaccines and chemical treatments for most fish viruses have not been fully developed (Iwama and Nakanishi, 1996). Most viruses can be inactivated by 56°C for 1-2 h, ionizing or ultraviolet radiation, pH above 11 or below 3, chlorine, iodine, ozone, formaldehyde, glutaraldehyde, ammonia, and some detergents (Smail and Munro, 1989). Naturally occurring substances such as one extracted from bacteria (Myouga et al., 1995) might be used in the future. Serious viral diseases include infectious hematopoietic necrosis, viral hemorrhagic septicemia, and eel viruses.

Several viruses have been isolated from eels (Sindermann, 1990; Kimura and Yoshimizu, 1991). EVE is the most dangerous to Japanese eels. EVA and EVX are more pathogenic to rainbow trout than to Japanese eels. ICDV can cause discoloration, fin congestion, increased mucus, and high mortality of Japanese eels.

Herpesvirus infections in salmon can cause anorexia, dark coloration, sluggishness, attraction to water inlets, petechial hemorrhages of the skin, tumors near the mouth, and mortality (Piper et al., 1982; Post, 1987; Kimura and Yoshimizu, 1991). One or more herpesviruses probably were responsible for a condition characterized by epidermal hyperplasia or necrosis in scorpionfish and young and old hirame.

Infectious hematopoietic necrosis is caused by a rhabdovirus (IHNV, Piper et al., 1982; Post, 1987; Shepherd, 1988; Sindermann, 1990; Kimura and Yoshimizu, 1991; Winton, 1991). It has been found in salmonids and ayu. It results in necrosis of hematopoietic tissue in the kidneys and other organs. Other signs include V-shaped muscular hemorrhages, pethechial hemorrhages on body cavity wall and in fat deposits, and mortality. Eggs can be disinfected with iodine, and an experimental vaccine has been shown to be effective for prevention.

Infectious pancreatic necrosis is caused by a birnavirus (IPNV, Piper et al., 1982; Post, 1987; Shepherd, 1988; Sindermann, 1990; Kimura and Yoshimizu, 1991). The disease has occurred in many marine fish, causing significant mortality, and the virus has been found in molluscs and crustaceans. It probably causes spinning disease of Atlantic menhaden. Signs include anorexia, dark coloration, exophthalmia, abdominal distention, pale internal organs, and mortality.

Lymphocystis commonly infects marine fish and causes what probably is their most visible viral disease (Piper et al., 1982; Chao, 1984, 1988; Post, 1987; Smail and Munro, 1989; Johnson, 1990; Sindermann, 1990; Moe, 1992). Infection is mainly from water through the gills or skin abrasions. This iridovirus causes smooth, irregular, gray or white growths usually less than 3 mm in diameter that give the fish a warty or granular appearance. Bacteria could enter through the growths, and growths in the mouth could interfere with feeding, but usually mortality does not occur and recovery is spontaneous. Internal infections also occur (Sindermann, 1990).

Pancreatic-hepatic necrosis can cause anemia, ascites, hemorrhage of liver, petechiae of pyloric caeca, and mortality of young yellowtails (Kimura and Yoshimizu, 1991).

Possibly picornaviruses caused nerve necrosis and mortality in striped knifejaws (Kimura and Yoshimizu, 1991) and larval barramundi (Glazebrook and Heasman, 1992).

Encephalitis in barramundi, European sea bass, striped knifejaw, striped jack, and turbot is caused by viruses similar to those in the family Nodaviridae (Comps et al., 1994).

Rhabodovirus olivaceus (hirame Rhabdovirus = HRV) was found first in hirame and later in other marine species. Signs include hemorrhage of muscles and fins, ascites, congestion and hemorrhage of gonads, kidney and spleen damage, and mortality.

Viral erythrocytic necrosis (VHS) is caused by an iridovirus (Post, 1987; Sindermann, 1990; Kimura and Yoshimizu, 1991). The disease has been found in at least 17 fish families. It can cause anemia and decreased resistance, but is not known to result in high mortality.

Viral hemorrhagic septicemia (VHS) is caused by a rhabdovirus (Piper et al., 1982; Post, 1987; Shepherd, 1988; Sindermann, 1990; Meier et al., 1994). It is known mainly from salmonids but also European sea bass and turbot. The main signs are intramuscular hemorrhage, necrosis of kidneys and spleen, and high mortality.

Bacteria

(Tung and Kou, 1986; Austin and Austin, 1987, 1989; Post, 1987; Frerichs and Roberts, 1989; Shotts and Teska, 1989; Sindermann, 1990; Moe, 1992; Hauck, 1995; Moriarty, 1997). Pathogenic bacteria are identified mainly by morphological evaluation, culturing, histology (e.g., differential staining), flow cytometry, serology, phage typing, nucleic acid hybridization, genetic recombination, and biochemical methods (Austin and Austin, 1989; Bullock, 1989; Tortora et al., 1992). Bacteria can be divided into two groups by the way they respond to Gram staining (Tortora et al., 1992). Those that retain a purple color after washing with ethanol or ethanol-acetone are gram-positive, and those that lose the color are gram-negative. Overall shape is another basis for identification (e.g., rods, cocci). The more serious bacterial fish diseases are caused by gram-negative bacteria and, economically, the most serious by *Vibrio* spp. In fact, most fish pathogens are gram-negative bacteria of three families: Enterobacteriaceae, Pseudomonadaceae, and Vibrionaceae and are in the shape of short rods (Frerichs and Roberts, 1989). In the recent past, neomycin and chloramphenicol have been two of the most effective antibiotics against gram-negative bacteria, with erythromycin, tetracycline, and potentiated sulfonamides being almost as effective, and penicillin, streptomycin, bacitracin, and sulfa drugs ineffective. The effectiveness of antibiotics depends to a large degree on how much they have been used previously. The more they are used, the less effective they become because of increased bacterial resistance. Some plant extracts and related chemicals could be used to control bacteria; in vitro (fresh water), tannic acid, propyl gallate, and methyl gallate inhibited growth of *Aeromonas hydrophila, A. sobria, Edwardsiella ictaluri, E. tarda, Pseudomonas fluorescens*, and *Escherichia coli* (Chung et al., 1995). Growth of *P. fluorescens* was completely inhibited by 50 µg/L of any of the three compounds. Naturally occurring substances such as those extracted from crustaceans (Ueda et al., 1996) might be used in the future. Beneficial nitrifying bacteria in biofilters (*Nitrosomonas* spp., *Nitrobacter* spp.) also are gram-negative and can be adversely affected by antibiotics.

Clinical signs of gram-negative bacterial infection include erosion, hemorrhages, and ulcers of skin on sides of body and fins and around eyes and mouth, cloudy eyes, blindness; hemorrhage of eyes, mouth, gill cavities, and abdominal organs; increased ventilation rate, lethargy, erratic swimming, anemia, abdominal edema, and mortality.

Bacteria gain entry to the body most often through mucous membranes —the digestive system, nasal passages, reproductive system, gills, or lateral line pores—but sometimes through breaks in the skin (Iwama and Nakanishi, 1996; Moe, 1992). Bacterial gill disease (Piper et al., 1982; Post, 1987) often occurs after chronic exposure to sublethal but damaging ammonia concentrations. *Flavobacterium* sp. (gram-negative) has been associated with bacterial gill disease in salmonids (Roberts, 1982; Austin and Austin, 1987; Sindermann, 1990; Egusa, 1992).

Bacterial fin rot usually occurs after fins are injured by handling or from bacterial septicemia with the bacteria settling within the small (almost dead-end) capillaries of the fins (Piper et al., 1982; Egusa, 1992). *Vibrio* and *Aeromonas* spp. can cause finrot, and *Flexibacter columnaris* has been found in lesions of fin rot of milkfish reared in fresh and brackish water (Tung and Kou, 1986). Other bacteria are responsible in some cases.

Some bacterial vaccines are available and a myriad of treatments have been tried. Vaccination can be by injection, spraying, immersion, or in feed, with effectiveness tending to decrease in that order (Austin and Austin, 1987; Post, 1987; Ellis, 1989). Bath treatments are reasonably good for external infections, and medicated feed is good for both external and internal infections. Moe (1992) suggested some antibiotic treatments for ornamental fish. The particular bacterial strain might be resistant to some antibiotics, so if the first does not work, others might be needed. Antibacterial chemicals (e.g., nitrofurazone) can be effective also, but some are not legal or might present health hazards. Tung and Kou (1986) reported that when 22 treatments were tried against *Vibrio anguillarum* in sensitivity tests in Taiwan, chloramphenicol, furadoine, furazolidone, tetracycline, and trimethoprim + sulfamethoxazole (potentiated sulfonamide) were the most effective, with chlortetracycline, doxycycline, and oxytetracycline moderately effective, kanamycin and novobiocin least effective, and the rest mostly not effective. Before any antibiotics are used, bacterial cultures should be taken for identification and sensitivity testing. After cultures are taken, treatment can begin with a broad-spectrum antibiotic. If the signs do not improve in 2-3 d, the identification and sensitivity results can be used to select more effective antibiotics. Indiscriminant use and switching of antibiotics without these tests is not only bad science and against all current therapeutic protocols, but is responsible for the buildup of resistant strains. With so few antibiotics approved for fish, it is imperative that good decisions are made so that some will be available for future use

(Roberts, 1982; Austin and Austin, 1987; Sindermann, 1990; Stoskopf, 1993). Antibiotic sensitivities vary with location. Keeping a record of sensitivities will help the culturist make future treatment decisions and monitor developing resistance. Although pasteurization will kill most pathogenic bacteria, freezing of fresh food will not. Vibrios, like many other bacteria, are very resistant to low temperature, even to -30°C (Wong et al., 1995).

Aeromonas spp. infections are more likely in fresh water and *Vibrio* and *Pseudomonas* spp. infections in salt water.

Aeromonas hydrophila (gram-negative rod) is a major cause of hemorrhagic septicemia in pond-cultured freshwater fish (Piper et al., 1982; Austin and Austin, 1987; Post, 1987; Frerichs and Roberts, 1989). Rising temperature and stress contribute to outbreaks. It occasionally affects marine fish. Signs are dark coloration, large irregular red hemorrhages on the body and fins, ascites; congestion, hemorrhage, and necrosis of internal organs; and mortality. Signs are very similar to those of *Pseudomonas*, but lesions are usually shallower than those caused by *Vibrio* spp. An immersion vaccine has been developed (Austin and Austin, 1987; Ellis, 1989). Treatment is through reduction of organic pollutants and temperature. Antibiotics such as potentiated sulfonamides are given in the feed, but valuable anorexic fish might need injections.

Aeromonas salmonicida (gram-negative rod) causes furunculosis, which is especially bad for Atlantic salmon (Piper et al., 1982; Austin and Austin, 1987; Post, 1987; Shepherd, 1988; Frerichs and Roberts, 1989; Isaksson, 1991). Contributing factors are high temperature, low oxygen, and crowding. Signs include anorexia, dark coloration, internal and external red boils (the external ones with clear red exudate), hemorrhage in gills and fin bases, cardiac necrosis, gathering of fish near drains, and mortality. A partially effective immersion vaccine has been developed (Austin and Austin, 1987; Ellis, 1989). The usual treatment consists of removing fish with furuncles, improving culture conditions, and feeding antibiotics such as potentiated sulfonamides.

Chlamydia-like organisms (gram-negative coccus) cause epitheliocystis in freshwater and marine fish, (Paperna, 1977; Frerichs and Roberts, 1989; Smail and Munro, 1989; Colorni, 1990; Fryer and Lannan, 1994; Venizelos and Benetti, 1996). Nearly all cultured species probably are susceptible. Signs are white miliary lesions on gills and skin, swollen epidermal cells, and hyperplasia of gills. No treatment is known.

Edwardsiella tarda (gram-negative rod) causes Edwardsiellosis in freshwater and marine fish (Austin and Austin, 1987: Post, 1987; Shepherd, 1988; Frerichs and Roberts, 1989). Lesions extend through the skin into the muscles; fibrinous peritonitis and necrosis of liver and kidney occur, and the liver is covered with a fibrinous exudate. An immersion vaccine for *Edwardsiella* sp. has been developed (Austin and Austin, 1987; Ellis, 1989). An experimental vaccine for hirame worked better when injected than when given by immersion (Iida and Wakabayashi, 1992). Improving water quality and reducing fish density are important measures, and sulfonamide or tetracycline treatment can help reduce mortality. When yolk antibody induced in chickens' eggs was extracted and fed to Japanese eels (400 mg/fish), they were protected from intestinal inoculation with *Edwardsiella tarda* (Gutierrez et al., 1993, 1994).

Cytophaga columnaris (gram-negative gliding bacteria) causes columnaris, or saddleback, disease which can affect most freshwater and anadromous, or euryhaline, fish (Piper et al., 1982; Austin and Austin, 1987; Post, 1987; Shepherd, 1988; Frerichs and Roberts, 1989; Sindermann, 1990). It usually occurs at temperatures of 18-20°C or higher. Hard water, high pH, and high organic levels are contributing factors. Typically, whitish lesions with reddish edges appear on the head, back, and gills, with ulceration, hemorrhage, and necrosis, and sloughing of skin occurring later. Mortality can be drawn out or sudden (i.e., in systemic infections). Treatments consist of improving water quality, decreasing temperature, and giving antibiotic or chemical therapy.

Flexibacter psychrophila (gram-negative gliding bacteria) causes coldwater, or peduncle, disease in salmonids (Piper et al., 1982; Austin and Austin, 1987; Post, 1987; Shepherd, 1988; Frerichs and Roberts, 1989). Low temperature (4-12°C), high fish density, and high organic levels promote it. It results in thick lesions on the fins and tail with many mucus cells, which slough off, allowing secondary bacterial infection. Increasing water quality and temperature are the most effective measures, but drugs might help.

Myxobacteria affect many marine species, especially as young fish injured by handling or abraded by cages (Austin and Austin, 1987; Sindermann, 1990). Black patch necrosis is caused in common sole by *Flexibacter maritimus*, a myxobacterium similar to *C. columnaris*.

Pasteurella piscicida (gram-negative rod) causes pasteurellosis (or pseudotuberculosis) in freshwater and marine fish (Hawke and Minton, 1985; Austin and Austin, 1987; Post, 1987; Frerichs and Roberts, 1989). Signs are anorexia, dark coloration, granulomas in liver and spleen, generalized hemorrhagic septicemia, and mortality. Because of widespread use of antibiotics in Japan, resistant strains are common. Effective vaccines have recently been developed (Austin and Austin, 1987). An experimental vaccine worked for yellowtail (Kawakami et al., 1997).

Pseudomonas anguilliseptica (gram-negative rod) causes red spot disease in Japanese eels and also infects European eels, but not as badly (Austin and Austin, 1987; Frerichs and Roberts, 1989). Other fish infected include salmonids, European sea bass, black sea bream, gilthead sea bream, and turbot (Doménech et al., 1997). Signs are anorexia, dark coloration, one or more hemorrhagic ulcers from below the epidermis to the hypodermis, visceral and muscular hemorrhages, a swollen bright red spleen, necrosis of liver, heart, spleen, and kidney, and mortality. Treatments include water quality improvement, temperature reduction, and antibiotics.

Pseudomonas fluorescens (gram-negative rod) is one cause of hemorrhagic septicemia of freshwater and marine fish (Austin and Austin, 1987; Post, 1987; Frerichs and Roberts, 1989). High temperature and crowding are contributing factors. It sometimes is secondary to chronic viral infection or traumatic injury. Lesions begin in blood vessels of the skin and soon become large hemorrhagic ulcers extending into the muscles. Signs also include ascites, abdominal hemorrhage, fibrinous peritonitis, and mortality. Reduction of fish density and improvement of water quality could be essential for recovery. Injection (IP) of kanamycin, and oxytetracycline given in the feed have produced good results.

Vibrio spp., especially *V. anguillarum*, (gram-negative rod; Burke and Rodgers, 1981; Piper et al., 1982; Austin and Austin, 1987; Post, 1987; Shepherd, 1988; Frerichs and Roberts, 1989; Isaksson, 1991, Sindermann, 1990) are the usual causes of serious bacterial septicemias in marine fish, while *Pseudomonas* spp. are the most likely causes of external infections. *V. anguillarum* has a nearly worldwide distribution, but at least some strains are host-specific. It has caused mortality in larvae of gilthead sea bream (probably), turbot, and Atlantic halibut (Grisez et al., 1997). A *Vibrio* sp. causes infectious necrosis of flounder larvae (INFL) in hirame (Muroga et al., 1990). *V. alginolyticus* might be beneficial for larvae of some fish (e.g., European sea bass, gilthead sea bream, turbot) but pathogenic for others (e.g., black sea bream, red sea bream)(Grisez et al., 1997).

Vibrio spp. mostly occur in marine and estuarine areas, but some are in fresh water (Austin and Austin, 1987; Frerichs and Roberts, 1989). Vibriosis has been found in many marine teleosts. It is most common and most dangerous when temperatures are highest in late summer. An exception is *V. salmonicida*, which causes cold-water vibriosis in European salmon (Isaksson, 1991). Vibriosis occurs in salmonids and turbot at 10-11°C and higher and in pleuronectids and eels at 15-16°C and higher. It is more likely after a temperature increase, salinity increase, or rough handling, and can originate from the gut.

With floating cages, *Vibrio* can be a major problem for both cold and warmwater fish, but commercial vaccines are available (Austin and Austin, 1987; Ellis, 1989). At least one brand immunized Florida pompanos for a year (Jory, pers. comm., 1991). Coldwater vibriosis can be controlled in salmonids by vaccination, but is still a critical factor in their culture (Sindermann, 1990). An immersion vaccine developed for salmonids protected 2-g palmetto bass from *Vibrio anguillarum* challenge tests 35 d later (Rogers and Xu, 1992).

Signs of vibriosis include anorexia, dark coloration, clouding and ulceration of eyes, periorbital and abdominal dropsy, anemia, external and internal hemorrhaging, swollen deep dark necrotic skin ulcers exuding reddish fluid, ascites, fibrinous visceral adhesions, enlargement and liquefaction of spleen, and sudden death. Fuzzy, whitish-gray, head and body lesions can give the illusion of fungal infection. Genetic selection and vaccination are two means of prevention. The usual drugs given via the feed have been oxytetracycline, potentiated sulfonamides, and nitrofurans. Flumequine or oxolinic acid (orally or as baths) can be used for Atlantic halibut (Samuelsen and Lunestad, 1996).

Yersinia ruckeri (gram-negative rod) causes enteric redmouth disease of salmonids (Piper et al., 1982; Austin and Austin, 1987; Post, 1987; Frerichs and Roberts, 1989). It usually occurs at temperatures higher than 10°C after carrier fish have been stressed. Signs include congestion of veins and capillaries (especially in eyes, brain, and mouth), dark coloration, hemorrhaging ulcers of mouth (but not always), generalized hemorrhagic septicemia, and mortality. Removal of infected fish, improvement of culture conditions, and (usually) antibiotic treatment are necessary. An immersion vaccine has been developed (Austin and Austin, 1987; Ellis, 1989).

Anaerobic bacteria include *Clostridium botulinum*, the cause of botulism poisoning which occasionally occurs in fish, and *Eubacterium tarantellus* (gram-positive rod), which causes a type of whirling disease (Austin and Austin, 1987; Sindermann, 1990). *E. taranıellus* has been found in at least 12 marine species (Sindermann, 1990). Signs are disorientation, possibly whirling, emaciation, and mortality.

Mycobacterium marinum (gram-positive rod; acid-fast) causes fish "tuberculosis" in tropical freshwater and marine fish but not temperate ones (Austin and Austin, 1987; Post, 1987; Frerichs and Roberts, 1989; Sindermann, 1990; Dailloux et al., 1992; Moe, 1992; Smith, 1996; Colorni et al., 1998). It occurs mainly in salt water, especially when fish are crowded and water quality is low. It is somewhat contagious between live fish and can be transmitted by using raw infected fish for food. Grayish granulomas form on the internal organs, which degenerate. Early signs are dark coloration, anorexia, listlessness, secretiveness, and increased ventilation rate. Later signs include cloudy eyes, exophthalmia, whitish blotches, raised scales, ragged fins, swollen or sunken abdomen, miliary tubercles in liver, spleen, kidney, and other organs, lying on the side, and mortality. Humans also can be infected (Engbæk et al., 1980). Because no effective treatment is available, infected food fish stock should be destroyed and facilities disinfected (Frerichs and Roberts, 1989). Moe (1992) suggested isolating infected aquarium fish and treating them with streptomycin + isoniazid baths and rifamopin in the feed. *Mycobacterium chelonei* causes a similar disease of salmonids in fresh and salt water, and there are a few other species.

Nocardia kampachi (acid-fast) causes a chronic tubercular condition in yellowtails (Austin and Austin, 1987; Post, 1987; Frerichs and Roberts, 1989; Sindermann, 1990). Outbreaks occur mainly in late summer to early fall. Signs are emaciation, skin discoloration, lumps on the skin turning into yellow ulcers, white gill lesions, hemorrhages on skin and around mouth, caseation of muscle, dermal, spleen, kidney, and gas bladder tissue, and mortality. Streptomycin and sulfasoxizole seem to be active against this pathogen. Chen and Wang (1993) reported in-vitro sensitivities of *Nocardia asteroides* to 24 therapeutants.

Renibacterium salmoninarum (gram-positive rod) causes bacterial kidney disease (BKD) in salmonids (Austin and Austin, 1987; Piper et al., 1982; Post, 1987; Shepherd, 1988; Frerichs and Roberts, 1989; Isaksson, 1991; Sindermann, 1990). Changing temperature (mainly rising, in spring to early summer) contributes to outbreaks. Infection could occur through water, abrasions, food, parasites, or eggs (maternally). Signs include dark coloration, exophthalmia, lethargy, listlessness, small hemorrhages at the bases of pectoral fins, grayish-white abcesses in the kidneys and other internal organs, spleen and liver damage, and gradual mortality. A vaccine has been developed (Austin and Austin, 1987). Injecting female broodstock with erythromycin can help prevent vertical transmission to the eggs, but egg bath treatments are not very effective because the bacteria can be in the yolk (Elliott et al., 1989). Prevention and isolation are important, because treatment with drugs is not very effective. Raising dietary iodine and fluorine could increase host resistance.

Streptococcus spp. (gram-positive coccus) infections occur in freshwater and marine fish, but are a special problem in yellowtail (Austin and Austin, 1987; Post, 1987; Frerichs and Roberts, 1989; Sindermann, 1990). Signs in yellowtail are dark coloration, ulcers, ascites, enteritis, peritonitis, abdominal hemorrhages, abdominal distension, adhesion of internal organs, liver damage, and mortality. A vaccine recently has been shown to be effective in yellowtail. Treatment with erythromycin is effective. Signs reported for other species include whirling, external hemorrhages, ascites, hemorrhagic enteritis, and mortality. Recently, in the U.S. an antibiotic-resistant *Streptococcus* has been responsible for mass mortality of tilapia and hybrid striped bass, and has infected humans who contacted the fish.

Fungi

(Post, 1987; Hatai, 1989; Roberts, 1989d; Sindermann, 1990; Grabda, 1991; Moe, 1992) One of the most significant fungal diseases of marine fish is a systemic one caused by *Ichthyophonus* spp. Several others cause smaller-scale problems. Factors contributing to external fungal infections include traumatic injury, crowding, low water quality, and viral or bacterial infection (Roberts, 1989d). Identification is mainly through the steps: isolation, culture, histology, and morphological evaluation (Austin and Austin, 1989; Bullock, 1989; Tortora et al., 1992).

Achlya spp. mainly affect fish in fresh water.

Aspergillus spp. (usually *A. flavus* and *A. niger*) not only produce aflatoxins, but also can infect fish internally (Roberts, 1989b). So far, this has been documented only in tilapia that were fed contaminated feed. Signs are dark coloration, lethargy, abdominal distension, liver necrosis, and mortality. Internal organs contain hyphae. To prevent aspergillomycosis, feed quality must be maintained.

Branchiomyces sanguinis and *B. demigrans* can cause gill rot, mainly in freshwater ponds (Post, 1987; Shepherd, 1988; Roberts, 1989d). Contributing factors are eutrophication, high organic content of water, algal blooms, and temperature higher than 20°C. The result is infarctive necrosis of the gill, with mortality possible in 48 h. Removal of dead fish, reduced feeding, and improvement of water quality can aid recovery of infected fish. Draining and liming ponds can reduce populations of the pathogen.

Dermocystidium spp. cause a gill disease resembling epitheliocystis in salmon in fresh water (Roberts, 1989d; Sindermann, 1990). At temperatures less than 15°C it seems to be more serious. Signs are many cysts (about 1 mm diameter) in gills, hyperplasia of gill lamellae, lesions in spleen, and mortality usually from inability to ventilate. A 2 mg/L trichlorfon bath for 1 h is a suggested treatment for salmon.

Exophiala sp. causes a systemic infection in salmon and marine fish (Sindermann, 1990). Signs are mycobacterial-like granulomas, visceral lesions, kidney enlargement, and mortality.

Ichthyophonus spp. (usually called *I. hoferi*) often are present in healthy freshwater and marine fish, but if the fish are weakened enough by stress, the fungus can flourish, causing degeneration of internal organs and high mortality (Post, 1987; Roberts, 1989d; Sindermann, 1990; Moe, 1992). Host-specificity of strains could occur. It causes a systemic granulomatosis. Infection begins in the digestive tract and spreads via the bloodstream. Consumption of infected raw seafoods (including live copepods) can cause infection. Signs of advanced infection include anorexia, listlessness, dark (black) subcutaneous nodules about 1 mm in diameter (dermal granulomata), raised scales, necrotic skin abcesses or ulcers, exophthalmia, white granulomas on nearly any internal organ, ataxia, circular swimming, upside-down swimming, ascites, and mortality. The granulomas contain spores and other stages. They cause the skin to have a sandpaper-like texture, called the "sandpaper effect." There is no effective treatment. Seafoods must be pasteurized to eliminate the fungus. Infected fish should be isolated. Water quality and nutrition of all fish possibly infected should receive special attention. Some fish can recover after several stress-free days.

Phialophora sp. has caused a systemic mycosis of Atlantic salmon parrs (Roberts, 1989d; Sindermann, 1990). Low temperature was a factor. Signs were petechial hemorrhages at fin bases and ventral surface, inflammatory abdominal edema, swollen kidneys, gas bladder deflation, and fungal growths on internal organs.

Phoma herbarum in the feed has caused a granulatomous condition in salmonids and ayu (Roberts, 1989d; Sindermann, 1990). Mainly the kidney is affected and mortality is low. Other signs are ataxia, swollen vents, hemorrhagic lesions of skin and fins, and fungal growths in internal organs. A *Phoma* sp. has caused high mortality of young ayu.

Saprolegnia spp. cause the typical external fungal infections seen in freshwater and marine fish (Tung and Kou, 1986; Post, 1987; Shepherd, 1988; Roberts, 1989d; Johnson, 1990; Sindermann, 1990; Moe, 1992), especially those stressed by sudden decreases in temperature. Signs include grayish-white patches or cotton-like tufts near the mouth and on gills, body, and fins, milky-looking eyes, a thin white coating on sides, and skin erosion, with secondary bacterial infections likely, and mortality (possibly within 36 h in salmonids). Lesions are nearly circular until they grow large and merge. Occurrence of fungal hyphae in lesions helps confirm the diagnosis. Raising water temperature higher than 18°C stopped saprolegniasis in Japanese eels, but excessively high temperatures might promote it in some species. Fish often recover without treatment. High water and food quality and lack of crowding are important. Chemicals used for bath treatment include formalin, copper sulfate, malachite green, potassium permanganate (fresh water only), and salinity change. Swabbing lesions with iodine can help, but might not be necessary. Extracts from eggs of ayu and *Tribolodon hakonensis* had either antifungal or fungicidal effects on *Saprolegnia parasitica* in vitro (Kudo and Teshima, 1991). Internal saprolegniasis has been documented in a few species, and *Saprolegnia* growing on dead fish eggs often spreads to adjacent live ones.

Scolecobasidium humicola has caused a systemic mycosis in coho salmon (Roberts, 1989d; Sindermann, 1990). Signs were necrotic skin lesions with hyphae, enlarged abdomen, ascites, and adhesions of internal organs.

Protozoans

(Wootten, 1989; Sindermann, 1990; Grabda, 1991; MacMillan, 1991; Paperna, 1991; Moe, 1992; Heckmann, 1996a, 1996b) Fish less than 1 yr old seem particularly susceptible to protozoan diseases. Many protozoans that parasitize fish seem to have a direct life cycle with no intermediate host, and they feed on the skin and gills. Hemoflagellates and hemogregarines that live in the blood of fish need intermediate hosts, most likely leeches. Some of the internal coccidian, microsporidian, and myxosporidian parasites also might need intermediate hosts.

Flagellates. Hemoflagellates in the genera *Trypanosoma*, *Trypanoplasma*, and *Cryptobia* are found in the blood of marine and freshwater fish (Wootten, 1989; Sindermann, 1990; Grabda, 1991). Signs of infestation include anemia, vascular lesions, inflammatory lesions, and mortality. *Trypanoplasma bullocki* caused anemia, splenomegaly, ascites, and mortality in young summer flounder. *Trypanosoma murmanensis* reduced hematocrit, hemoglobin, and plasma proteins in Atlantic cod. *Cryptobia salmositica* has been documented as causing mortality of Pacific salmon.

Infestation by the dinoflagellate *Amyloodinium ocellatum* (Lom and Lawler, 1973; Paperna 1980, 1983, 1984a, 1984b, 1991; Baticados and Quinitio, 1984; Aiello and D'Alba, 1986; Bower, 1987; Wootten, 1989; Gallet de St. Aurin et al. 1990; Johnson, 1990; Sindermann,

1990; Grabda, 1991; MacMillan, 1991; Moe, 1992; Landsberg et al., 1994; Montgomery and Brock, 1997; Vermeer et al., submitted) can be a severe problem for both ornamental and food fish, especially at higher temperatures. This parasite is relatively non-host-specific and attacks many cultured marine fish. Invertebrates might carry it. The life cycle is simple. A mature cyst, called a tomont, releases about 256 free-swimming flagellated dinospores, which search for hosts. After attaching (preferably to gill filaments, but also to skin), they become immature cysts, or trophonts. After feeding for several days, trophonts detach and become tomonts. At detachment, trophont length is from 85 μm to somewhat more than 100 μm. Depending on temperature (19-24°C), the trophont stage lasts for 2-7 d, dinospore development within the tomont takes 3-6 d, and the entire life cycle can be completed within 6-12 d. Depending on the strain, dinospores can survive for 2-30 d before finding a host. Because of more rapid reproduction, *Amyloodinium* is a bigger problem in warmwater culture systems. Reproduction by tomonts is most effective in the ranges 18-30°C and 12-50 ppt. Growth, reproduction, and infestation are so slow below 18°C, that problems are unlikely. At less than 16°C, tomonts do not divide, and below 10°C they die. Young fish, sciaenids (especially red drum), and clownfish are particularly susceptible. Larval densities higher than 5/L and low water quality promote infestation. The main pathological effect is through interference with gas exchange by hyperplasia and fusion of gill lamellae and overproduction of mucus, with total mortality possible in a few days. Signs are: increased ventilation rate, irritability, and aggressiveness, scraping against solid objects (flashing), white powdery appearance of skin, and faded colors. At first, fish often will swim aimlessly at the surface, but in later stages they are likely to lie on their sides on the bottom. Repeated or long exposure to fresh water, chelated copper, and formalin have provided some control. Mostly, a continuous copper bath at about 0.2 mg/L Cu^{2+} for 3 weeks has been used (Reed and Francis-Floyd, 1994). Copper can be toxic to fish and nitrifying bacteria and could precipitate quickly. One of the best treatments for red drum and clownfish is 10 mg/L chloroquine diphosphate in a continuous bath for 21-45 d (Vermeer et al., submitted; Kennedy, unpub.). Freshwater treatment causes detachment of the parasites but does not kill them right away; cysts can remain viable in fresh water for as long as 6 weeks. It is important to dispose of bath water so that it will not lead to infestation of the same or other fish. After recovery from *Amyloodinium*, some fish seem to have acquired a degree of resistance to the parasite.

Ichthyobodo necator is a freshwater flagellate that causes costiasis in freshwater fish (Piper et al., 1982; Post, 1987; Shepherd, 1988; Wootten, 1989; Grabda, 1991; Isaksson, 1991; MacMillan, 1991). This is a serious problem for young salmon. The skin of infested fish turns bluish white or gray and eventually sloughs off, causing osmoregulatory failure and death. Formalin baths can eliminate *Ichthyobodo*. *Hexamita* sp. could cause enteritis in salmonids in fresh water (Piper et al., 1982; Post, 1987; Wootten, 1989).

Coccidians. These can cause damage to internal organs (especially gonads, liver, intestines, and gas bladder), reduced fecundity, emaciation, and mortality in freshwater and marine fish (Post, 1987; Wootten, 1989; Sindermann, 1990; Paperna, 1991). *Eimeria* spp. damage the gonads of clupeids. *Calyptospora* (=*Eimeria*) *funduli* parasitizes the liver and pancreas of killifish and can cause mortality. *Goussia gadi* infestation of the gas bladder can be fatal to gadoids.

Cryptosporidians. (related to coccidians) caused malabsorption and emaciation of surgeonfish (Wootten, 1989), but did not seem to harm young tilapia and gilthead sea bream (Paperna, 1991).

Hemogregarines. These commonly parasitize the blood of marine fish and sometimes can cause a myeloid leucosis, hypertrophy and degeneration of erythrocytes, visceral lesions, intramuscular cysts, and reduced growth (Wootten, 1989; Sindermann, 1990).

Haemogregarina sachai caused tumor-like lesions of somatic muscles, gills, esophagus, stomach, and ovaries of juvenile turbot reared in heated water. One *Haemogregarina* sp. caused visceral lesions in gilthead sea bream and rabbitfish, and another caused intramuscular cysts in mullet.

Microsporidians. These intracellular parasites cause hypertrophy of host cells (Piper et al., 1982; Post, 1987; Wootten, 1989; Sindermann, 1990; Grabda, 1991; Moe, 1992). The skin, muscles, digestive system, and nerves are most affected. Signs include hypertrophy and degeneration of abdominal organs, intestinal occlusion, necrosis of muscles, weakness, reduced growth, and mortality. *Glugea stephani* crowded out the submucosal cells of the intestine in plaice, causing dysfunction and starvation. *Glugea* sp. causes seahorse disease, with dense white patches on the body surface. Fish covered with scutes are more likely to have this (e.g., seahorses, pipefishes, boxfishes). A disease causing white spots in the yolk of clownfish eggs about 5 d after spawning might be the result of a microsporidian (Moe, 1992). Infected embryos were stunted and did not hatch. *Loma* sp. has caused losses in a chinook salmon hatchery. *Pleistophora* spp. can cause mortality in fish, but also often form large cysts in the flesh that make fish unmarketable. *Glugea stephani* isolated from North Atlantic flounder survived -30°C for 23 months and 60°C for 30 min (Amigó et al., 1996). It could be transmitted from other fish fed to farmed fish and possibly from them to human consumers. Sulfathiazole, quinine, quinacrine hydrochloride, or formalin treatment might be of use against some microsporidians.

Myxosporidians. These parasitize both freshwater and marine fish (Piper et al., 1982; Post, 1987; Wootten, 1989; Sindermann, 1990; Grabda, 1991; Paperna, 1991; Sövényi, 1992; Sitjà-Bobadilla and Alvarez-Pellitero, 1992, 1993). Myxosporidians can form large cysts in the flesh and affect its texture, making fish unmarketable. Species of *Ceratomyxa*, *Chloromyxum*, *Myxidium*, and *Leptotheca* live in gallbladders and urinary bladders of teleosts, sometimes causing damage to that organ and the liver, and emaciation. *Ceratomyxa shasta* has caused high mortality in Pacific salmon hatcheries. Signs include ascites, abdominal distension, muscular and visceral nodular lesions, and intestinal hemorrhage. *Henneguya* spp. can cause cysts, heart damage, emaciation, and mortality in salmon and marine fish. *Kudoa* spp. infest many marine fish. Signs include anemia, cysts, degeneration and necrosis of muscles, enlarged kidneys, and mortality, with rapid liquefaction of flesh before or after death (even in harvested fish). The latter condition has been described as mushy, jellied, or milky. *Kudoa cerebralis* causes cysts in the cranial nerves of striped bass, resulting in distortion but not necrosis. Certain species of *Myxobolus* and *Myxostoma* inhabit the cranial cartilages and brains of fish. Signs include erosion or hypertrophy of cartilage, scoliosis, nerve damage, and mortality. *Myxobolus exiguus* cysts caused gill damage, hemorrhage, and mortality of striped mullet. *Myxostoma cerebralis* causes whirling disease and skeletal abnormalities in salmonids, a problem mainly for rainbow trout. Mud (in ponds) is required for it to complete its life cycle. Proliferative kidney disease in freshwater salmonids might be caused by a myxosporidian.

Ciliates. (Cheung et al., 1979; Huff and Burns, 1981; Piper et al., 1982; Colorni, 1985, 1987; Post, 1987; Wootten, 1989; Gallet de St. Aurin, 1990; Johnson, 1990; Grabda, 1991; Paperna, 1991; Moe, 1992; Nilsen, 1995) *Cryptocaryon irritans* also is non-host-specific and causes gill and skin damage, but usually not as severely as *Amyloodinium*. Because the biology and effects of *Cryptocaryon* are so similar to those of the freshwater *Ichthyophthirius multifiliis*, the former often is called saltwater ich. It can be carried by invertebrates, and its life cycle is simple. A mature tomont releases 200 or more free-swimming ciliated tomites, which must find a host within 24-48 h. The trophont stage feeds for 3-7 d, then detaches and matures into a tomont, which can be as large as 450 μm by 350 μm. Tomite development within the tomont takes 4-12 d depending on temperature (20-30°C).

Optimal temperature range for development and reproduction is 25-30°C, and both stop below 10°C. Most infestations occur above 20°C. The first sign of cryptocaryoniasis usually is the appearance of white or grayish spots (pustules) on the body surface and fins. These spots are more distinct than those seen with *Amyloodinium*. Eyes often are attacked. Increased ventilation rate, restlessness, flashing, and color loss are typical. If the disease progresses, the epidermis can slough, and tomonts, mucus, and tissue debris around the gills can block gas exchange and lead to mortality. Lesions often are infected by bacteria. *Cryptocaryon* is more difficult to eradicate than *Amyloodinium* because the trophonts burrow more deeply. Antimalarial drugs are effective but can be expensive for large numbers of fish. A freshwater bath will kill exposed trophonts and tomonts. A combination of freshwater, formalin, and possibly chelated copper baths (e.g., as suggested by Moe, 1992) might be as good as the antimalarial drugs. Tareen and Hadi (1981) reported that 20 mg/L formalin was successful with orangespotted groupers. Transferring fish and cleaning tanks also can help (Colorni, 1987). The situation will dictate which treatment is more practical.

Other parasitic marine ciliates include *Brooklynella hostilis*, *Miamiensis avidus*, and *Caliperia* sp. (Gallet de St. Aurin et al., 1990; Moe, 1992). Typically, small white spots appear on body and fins, then enlarge until epidermal sloughing occurs, resulting in raw ulcers with loose scales, and mortality. Treatment is the same as for *Cryptocaryon*. Some ciliates such as *Ambiphyra* sp. and *Glossatella* sp. are mostly ectocommensal but could cause irritation of the skin. *Chilodonella* spp. are the freshwater equivalents of *Brooklynella*. The most serious effect of *Heteropolaria* (=*Epistylis*) *lwoffi*, which infests salmonids, and *Heteropolaria colisarum*, which infests warmwater fish in fresh water, is to promote secondary bacterial infections. *Trichodina* sp. and *Trichodinella* sp. are serious parasites of freshwater and marine fish. Formalin baths will eliminate *Trichodina*. The freshwater *Trichophyra* sp. does not seem to cause much damage.

Coelenterates

(Grabda, 1991; Raikova, 1994) *Polypodium hydriforme* larvae live inside developing oocytes of sturgeon, (at least 12 species, a hybrid, and paddlefish, too), feeding on and killing them. Just before spawning the parasites exit from the oocytes so that they can go through two (non-parasitic) reproductive stages outside the fish. Sturgeon egg production can be significantly reduced.

Free-living Flatworms

(Anonymous, 1987; Moe, 1992) The turbellarian flatworm *Paravortex* sp. feeds on the skin and gills of the host fish. It does not require an intermediate host and can build up quickly. They can kill at least ornamental species in aquariums. The life cycle is simple. A mature worm on the tank bottom splits and releases about 160 free-swimming young worms. Those that find hosts feed for 5-6 d, then drop to the bottom to mature, which takes a few days. Signs include anorexia, listlessness, flashing, and tiny dark spots (about 0.5 mm) usually just behind the gill openings but sometimes on the fins. Moe (1992) recommended formalin baths and cleaning of tank bottoms for treatment.

Parasitic Flatworms

(Post, 1987; Wootten, 1989, Sindermann, 1990; Grabda, 1991; Paperna, 1991; Schmahl, 1991)

Monogeneans. These are a problem mainly in cages or tanks in which the parasites can build up and the fish are difficult to treat. If older fish are crowded and water flow is restricted, as in tanks, gill flukes can reach epizootic levels. They do not need an intermediate host and feed on blood, tissue, and mucus. Most lay eggs that hatch into ciliated oncomiracidia, which have only a few hours to find a host. Gyrodactylids are viviparous and do not have a larval stage. Damage is caused not only by feeding, but also by the hooks, clamps, and suckers used for attachment. Signs include skin and gill damage, anemia, emaciation, and mortality. On close observation, the flatworms can be found on the fish or in freshwater baths after being made to drop off. Some treatments are: 50 mg/L levamisole for 2 h, 0.1 mg/L niclosamide for 2 h, 10 mg/L praziquantel for 3 h, 100 mg/L mebendazole for 10 min, and 10-20 mg/L toltrazuril for 2-4 h (Schmahl, 1991). Many other simple chemicals, dyes, and antimalarial compounds have been used.

Benedenia seriolae are parasites of yellowtail; they mature 14-18 dah at 20-25°C and can produce about 27 eggs/h (Kearn et al., 1992a, 1992b).

Entobdella hippoglossi are parasites of Atlantic halibut that are resistant to the usual treatments for monogeneans (Svendsen and Haug, 1991). A freshwater bath for 50 min or 5 ppt for 155 min killed adult worms, but 10 ppt for 160 min was not effective. In vitro hypersaline treatment at 40 ppt for 70 min or 50 ppt for 25 min weakened adult worms so that they could not reattach to glass slides. Benzocaine at 80 ppm for 25 min caused temporary detachment, but after 30 min recovery, most had reattached. Formalin at 200 ppm for 24 h was not effective. Soaking eggs in fresh water for 2 h reduced hatching rate by ~90%. Heating eggs in freshwater to 50°C for 5 min reduced hatching by 81-91%. Freshwater or hypersaline treatments seem best to eliminate adult worms, but tolerance of the halibut must be considered, too. The tested treatments were only partly effective against eggs, and the most effective one (heating) could not be used with live halibut. The similar but smaller *Entobdella soleae* infests common sole and turbot.

Neobenedenia melleni are transparent monogeneans about 3-13 mm long that feed on the skin and gills of marine fish, causing damage and allowing bacterial infections to start (Gallet de St. Aurin et al., 1990; Thoney and Hargis, 1991; Paperna, 1991; Cowell et al., 1992; Moe, 1992; Mueller et al., 1994). This species has been a major problem for a wide range of fish in marine aquariums. The life cycle lasts 3-6 d at 29°C. Signs include frequent flashing, scraping, and jumping, plus anorexia, many small bumps on the gills and skin, excessive mucus, corneal thickening and cloudiness, hemorrhaging ulcers on the body and fins, and mortality. If fish are given freshwater baths, the worms will drop off and can be found on the container bottom. Freshwater, formalin, and trichlorfon baths are effective. If cleaner fish can be kept in the tank, they will remove the worms. Older fish are less susceptible, and an immune factor has been found in the mucus of survivors of infestations.

Robinson et al. (1992) reported that caged hybrid red tilapia suffered heavy infestations of *Neobenedenia melleni* when cages were placed in water less than 3 m deep, apparently because the parasite's life cycle could be completed between the cages and bay bottom, resulting in reinfestation. The problem was alleviated by moving cages to deeper water. Reduction of tilapia density in cages and addition of trichlorfon to the feed also have been helpful (Gallet de St. Aurin et al., 1990). Reducing salinity to 15 ppt for 5 d was successful against eggs, juveniles, and adults on Florida red tilapia (Ellis and Watanabe, 1993).

Dactylogyroids parasitize many freshwater and marine fish (Piper et al., 1982; Post, 1987; Wootten, 1989; Thoney and Hargis, 1991; Paperna, 1991). *Pseudodactylogyrus* spp. infest eels, causing hemorrhage and destruction of skin and gills.

Gyrodactyloids parasitize many freshwater and marine fish (Piper et al., 1982; Post, 1987; Wootten, 1989; Thoney and Hargis, 1991; Paperna, 1991). Signs include epidermal hyperplasia, petechial lesions, skin sloughing, gill damage, anemia, and emaciation. *Gyrodactylus salaris* is the most common monogenean infesting cultured Atlantic salmon, mainly juveniles (Isaksson, 1991). Signs are gill damage, frayed fins, and skin ulcers. Formalin or saltwater baths are effective.

To eliminate monogeneans, Thoney and Hargis (1991) suggested using praziquantel or mebendazole baths if possible, with formalin baths the last resort for small culture systems and trichlorfon (continuously) the last resort for large systems.

Digenean Trematodes. Digenean trematodes need one or more intermediate hosts, and fish could act as final or intermediate hosts. The first intermediate host usually is a gastropod or bivalve mollusc. The migrating cercariae and encysted metacercariae can severely damage gills, liver, heart, other internal organs, and muscles, leading to emaciation and sometimes mortality. Treatment is difficult, and prevention is accomplished by eliminating intermediate hosts.

Diplostomum spp. infest the eyes of many European and North American freshwater species, sometimes causing cataracts and blindness (Post, 1987; Wootten, 1989).

Cestodes
(Post, 1987; Wootten, 1989; Sindermann, 1990; Grabda, 1991; Paperna, 1991) Tapeworms require at least one intermediate host, and fish could be either final or intermediate hosts. They can be very damaging to internal organs, partly because of their large size, but mainly because larvae migrate through the organs. Infested fish often have noticeably swollen abdomens. Adult tapeworms usually are obvious upon dissection of the infested fish. Signs of tapeworms include inflammation, skin lesions, bone melanization and erosion, liver and intestinal damage, intestinal obstruction, reduced growth, emaciation, and sometimes mortality.

Nematodes
(Post, 1987; Anderson, 1989; Wootten, 1989; Grabda, 1991; Paperna, 1991; Geets et al., 1992) Nematodes that parasitize fish need at least one more host (usually an arthropod) to complete the life cycle. Fish can be final or intermediate hosts. Adults usually live in the digestive tract and do little harm. Migrating and feeding larvae (and occasionally adults) can cause severe damage to muscles and internal organs (especially the liver, mesenteries, intestines, gas bladder, and gonads), edema, granulomatosis, visceral adhesions, peritonitis, reduced growth, emaciation, and mortality. Treatment is difficult, but freezing of seafoods before giving them to the fish is mostly effective for prevention. Some nematodes such as the codworm (*Pseudoterranova decipiens*) and herring worm (*Anisakis simplex*) not only can render fish unmarketable because of their size, but also can temporarily infest humans if the fish has not been frozen or cooked well.

Acanthocephalans
(Post, 1987; Wootten, 1989; Grabda, 1991; Sindermann, 1990) Larval spiny-headed worms encyst within visceral organs of their intermediate hosts, and adults reside in the intestines of their final hosts. Fish can act as either. Larvae can cause fibrosis of liver and spleen. Adults can cause granulomatosis and severe local damage to the intestine (necrotic hemorrhaging ulcers) including perforation, but do not often reduce growth or survival. They require an invertebrate intermediate host, usually an arthropod.

Crustaceans
Branchiurans. (Post, 1987; Shepherd, 1988; Wootten, 1989; Grabda, 1991; Johnson, 1990; Paperna, 1991; Moe, 1992) Branchiuran eggs are relased into the water and hatch into free-swimming juveniles that have only 2-3 d to find a host. They do not use intermediate hosts. *Argulus* spp. (fish lice) mostly occur in fresh water but sometimes infest marine fish, feeding on their skin and subdermal tissues after injecting digestive and toxic fluids through a piercing proboscis. They attach by means of hooks and suckers. *Argulus bicolor* is reported to infest red drum. Freshwater, formalin, and trichlorfon baths are effective against *Argulus.*

Copepods. (Post, 1987; Shepherd, 1988; Wootten, 1989; Johnson, 1990; Sindermann, 1990; Grabda, 1991; Paperna, 1991; Moe, 1992) Some copepods parasitic on fish have intermediate fish hosts and others do not. Males mostly are not parasitic. Some parasitic copepods crawl around on the body surface feeding lightly on skin or gills, while most are firmly attached by embedded anchors and sometimes feed on deep tissues. The first type usually will respond to freshwater, formalin or trichlorfon, but the second type might be more resistant. Anchored copepods can be removed from a small number of fish with forceps. Most copepods parasitic on red drum attach to the gills, gill cavity, or mouth cavity. Signs in heavily infested fish include anorexia, dark coloration, slow surface swimming; damage, ulcers, and erosion of skin, eyes, gills, and fins; tumors, hematomas, heart damage, jumping, darting, anemia, emaciation, and mortality. Secondary bacterial infection is common.

Salmon lice, *Caligus elongatus* and *Lepeophtheirus salmonis*, (caligid copepods) are the most important parasites of Atlantic salmon in seawater, especially in cages with restricted water exchange (Wootten, 1989; Sindermann, 1990; Isaksson, 1991; Costello, 1993; Kvenseth, 1993; Thomassen, 1993; Bruno and Raynard, 1994; Hogans, 1995; Piasecki and MacKinnon, 1995; Avault, 1996; Roth et al., 1996; Tully et al., 1996; ICES J. Mar. Sci. 54(6):1093-1147, 1997; MacKinnon, 1997; ICES J. Mar. Sci. 55(2):151-200, 1998.). They cause skin erosion, hemorrhage, osmotic stress, and secondary infections. No cure has been perfected yet. If possible, towing cages into fresh water or adding wrasses to the cages can provide good control. Pesticides like azamethiphos, dichlorvos, and permethrin are toxic to non-target organisms, and development of resistance to dichlorvos has been reported for *L. salmonis* (Johnson et al., 1993). Hydrogen peroxide at 1.5 ppt for less than 30 min at 14°C causes lice to drop off, but they might recover; it also is expensive and can be toxic to fish (MacKinnon, 1997). Ivermectin in the feed at 0.05 mg/kg of fish every third day has been used, but that dosage or slight overdosage could have slight toxic effects on Atlantic salmon (Johnson et al., 1993) or other non-target organisms (e.g., polychaetes, Black et al., 1997), but not others (Davies et al., 1997a). Lice probably will develop resistance to ivermectin (Costello, 1993). Growth inhibitors such as diflubenzuron can injure wild arthropods, and withdrawal time is 3 months (MacKinnon, 1997). Simply keeping the cages in water deeper than 10 m greatly reduces the chance of infestation. Other precautions include spacing farms geographically (far enough apart to limit transfer of lice) and fish temporally (avoiding stocking new salmon next to those previously stocked). Another possible solution is a vaccine producing antibodies in salmon that attack the gut of lice beginning to feed on them (Raynard et al. 1994). *Caligus longipedis* is a parasite of farmed striped jack in Japan (Ogawa, 1992), and *Caligus epidemicus* infests several species in Australia (Hallett and Roubal, 1995).

Lernaeid copepods infest both freshwater and marine fish, and lernaeocerid copepods only marine fish. Ergasilid copepods parasitize the gills of many freshwater and marine species.

Isopods. (Sindermann, 1990; Grabda, 1991) Isopod parasites occur as larvae or adults on the gills, skin, and inside of the mouth. Signs include gill and skin damage, reduced growth, and mortality. Secondary microbial infections can occur. Freshwater, formalin, and trichlorfon baths are effective against isopods.

Leeches

(Post, 1987; Wootten, 1989; Grabda, 1991) Leeches lay cocoons of eggs and do not have a larval stage. They feed on the blood of fish, with anemia and bacterial infections possible. Leeches also transmit some blood parasites. Trichlorfon is effective against leeches, and freshwater baths can make marine leeches drop off.

Lampreys

(Wootten, 1989; Grabda, 1991) Lampreys are freshwater or anadromous, jawless fish resembling eels. They spawn in fresh water, and the ammocoete larvae can live for years in muddy bottoms before maturing. Adults of some species migrate to sea. They attach to the skin of fish and rasp the flesh off.

OTHER PATHOLOGICAL CONDITIONS

Genetic Disorders

Abnormalities in fish are not uncommon, especially in aquaculture, partly because predation on weaker individuals is not as likely. Blunt snouts, overshot jaw, undershot jaw, abnormal opercular bones, pugheadedness, missing fins, enlarged fins, malformed fins, fused vertebrae, short tails, malignant melanoma, Siamese twinning, pseudo-albinism, larval die-offs, liver on the left side, and cystic conditions are examples that might have genetic origins (Post, 1987; Roberts, 1989b; Sindermann, 1990).

Environmentally-induced Disorders (also see chapter 3)

Bad rearing conditions. Presence of toxic metals, ammonia, silt, and insufficient gravel in hatching trays can lead to blue-sac, white-spot, or yolksac deformity diseases and mortality of larval salmonids (Roberts, 1989b). Unfavorable rearing conditions such as crowding, intense light, toxins, and thermal, osmotic, or mechanical shock can contribute to developmental abnormalities in fish (Sindermann, 1990). Examples are: distorted jawbones, shortened opercle, lordosis, incomplete fins, and shortened vertebrae in ayu; pugheadedness, crossbite, jaw bone reduction ("sucker mouth"), opercular reduction or malformation, and spinal abnormality in European sea bass; abnormal vertebrae in red sea bream; opercular reduction, hunched back, curved caudal peduncle, disoriented fin rays, and gas bladder hyperplasia in gilthead sea bream. Ambicoloration in reared summer flounder can be controlled by changing the substrate; abnormal pigmentation in flatfish also might be related to nutrition, trauma, parasites, light, or pollution. Among fish in general, three pigmentation anomalies are: xanthochromism (golden), melanism (black), and albinism (no pigment). In many cases of abnormal pigmentation, both heredity and environment seem to be involved. Hargis (1991) reviewed disorders of fish eyes, many of which are related to environmental degradation.

Gas bubble trauma. Supersaturation of nitrogen or oxygen in water can occur when water temperature increases or pressure decreases. Cavitation (drawing in and pressurizing air) by water pumps also can cause this. Gas bubbles (mainly nitrogen and oxygen) can form in fluids or tissues within the fish, interfering with circulation and displacing organs. Chronic exposure to only 102% nitrogen can produce this in rainbow trout alevins (Brannon, 1991). Less sensitive fish might not be affected until a value of 110% is reached (Shepherd, 1988). Rapid mortality will occur at levels near 140% (Post, 1987). Exophthalmia is the most obvious sign. Other signs are bubbles in the eyes, skin, gills, and mouth (in yolk of larvae), edema and degeneration of gill lamellae and intestinal mucosa, and kidney damage (Roberts, 1989b; Spotte, 1992). Some remedies are: aeration (Atlantic salmon, Isaaksson, 1991), water exchange (eels, Usui, 1990), and puncturing bubbles with a needle (seahorses, Giwojna, 1990). Additional references are: Tung and Kou, 1986 (milkfish), Roberts, 1989b (Pacific salmon), and Colorni, 1990 (gilthead sea bream). (also see pages 95-96)

Red tides. About 20 species of dinoflagellates in six genera are known to produce substances toxic to fish. Some examples are *Gonyaulax monilata* and *Ptychodiscus brevis* in the Gulf of Mexico (Johnson, 1990), *Ceratium fusus*, *Gymnodinium* sp., and *Heterosigma akashiwo* along the northwest U.S. coast (Stickney, 1991b), and *Gyrodinium aureoleum* in Norway (Roberts, 1989e; Isaksson, 1991). Blooms of the diatom *Chaetoceros consulutus* also are toxic to salmon in the northwestern U.S. (Stickney, 1991b). The haptophycean flagellate *Chrysochromulina polylepsis* has caused large losses of caged salmon and trout in Norway (Isaksson, 1991). The haptophycean flagellate *Prymnesium parvum* has caused high losses on brackishwater farms in the Mediterranean region (Roberts, 1989d). Blooms of the latter develop during 3-5 d, with fish first concentrating and jumping out of the water, then finally becoming sluggish and dying. Copper and ammonium preparations have been used to control *Prymnesium*. Fish sensitivity to the toxins varies. Hypoxia and gill clogging can occur, too. Blooms occur when temperature, salinity, nutrient, and light conditions are favorable. Near Hong Kong and Florida, this is usually in early spring. They usually last about 5 d. During 1980-1984, red tides at cage fish farms near Hong Kong caused 11 fish kills with a loss of 86 mt. Human control of red tides by short-term actions is a fairly hopeless possibility. Chong and Chao (1986a) suggested the following precautions and remedies. Avoid dangerous areas such as bays with restricted circulation and muddy bottoms, especially if the sediment has a high organic load. Place skirts around cages to exclude surface plankton. Increase circulation to flush plankton away, if possible. (also see page 146)

Pollution. Polluted waters not only contribute to acute or chronic diseases or produce toxic reactions in fish inhabiting them, but also can cause or stimulate a wide array of other problems and defects (Post, 1987; Sindermann, 1990). Just a few possible examples are infections, external lesions, tumors, immunosuppression, chromosome damage, muscular atrophy; abnormal jaws, gills, head, fins, vertebrae; reduced growth, and mortality. Domestic and industrial pollution can deplete oxygen. Toxic metals and organic compounds can cause developmental, morphological, physiological, and reproductive abnormalities, and quick or gradual mortality. Small amounts of chlorine, hydrogen sulfide, and ozone can kill fish within seconds. Small amounts of ammonia can damage gills and kidneys, leading to bacterial gill disease and mortality; higher amounts cause rapid mortality. High carbon dioxide can interfere with oxygen uptake, causing anesthesia, anoxia, and mortality. Pesticides can cause damage to nerves or internal organs, weakness, paralysis, toxemia, and rapid or slow mortality.

Oil. (Chong and Chao, 1986a; Post, 1987) Spilled material could range from crude petroleum to refined compounds. Volatile, dissolved, emulsified, insoluble floating, and insoluble sinking substances can result from spilled oil. Acute or chronic toxicity can result from exposure. The aromatic fractions (e.g., benzene) seem to be the most toxic. Damage and hemorrhages can occur in gills, eyes, and internal organs. Fish gills can be coated, preventing gas exchange. Oxygen can be depleted. Prey populations can be reduced. Fish will avoid oil, if they can. Towing cages to a clean area might not be possible. Bacteria will break down lighter components within about a week, but heavy insoluble materials (e.g., tar) could linger for many months. In the meantime, a skirt could be suspended around fish cages, and the fish could be fed through a funnel or cylinder to avoid contamination from the surface.

Diesel oil sometimes is used to control predatory insects and crustaceans in fish ponds at 38 L/ha every 3 d. In a 1-m deep pond, a concentration of 3800 ppb would result. This is much higher than concentrations that are dangerous to young marine fish and their crustacean prey. Exposure of spotted seatrout hatchlings for 48 h to 10 ppb of the water soluble fraction of no. 2 diesel oil did not cause mortality but did reduce growth and prevent some larvae from developing eye pigment (Johnson et al., 1979). Spotted seatrout exposed to 800 ppb diesel oil for 12 h during hatching grew significantly less than unexposed larvae during the next 4 d (Tucker and Thompson, 1987). The 48hLC50 for spotted seatrout eggs to larvae exposed to a mixture of diesel oil and mineral oil (65:35) was 432 ppb (Thompson and Tucker, 1989). Adult copepods (*Acartia tonsa*) had a $24hLC_{50}$ of 580 ppb and $24hLC_{10}$ of 105 ppb for the same mixture (Tucker and Thompson, 1987). In a fish pond, such oils would be relatively persistent.

Irritating particles. These can be natural dust and silt, bone and shell fragments from feed, or building materials such as cement (Roberts, 1989b). They cause mechanical damage to the gills and interfere with gas exchange.

Sunburn. The burning portion of ultraviolet radiation can penetrate almost 1 m in clear water, and fish can be sunburned at depths greater than 0.5 m (Roberts, 1989b). As expected, lesions usually are on the head, back, and tops of fins. In severe cases, ulceration, then invasion by bacteria, fungi, and/or parasites occur. Photosensitization by natural (e.g., porphyrins in food) or artificial agents (e.g., for salmonids, the antihelmintic phenothiazine) makes fish much more vulnerable to sunburn.

Ulcerative dermal necrosis. This condition occurs in adult salmon and trout, usually as they enter fresh water from the sea (Roberts, 1989b). One of the first signs is the presence of characteristic small gray eroded areas (often distributed uniformly) on the head or adipose fin. Mortality usually occurs after secondary fungal or bacterial invasion. Photosensitizers and viruses have been suggested as possible causes. If secondary infections are prevented, fish can recover from this disease.

Scale loss-blindness-melanism syndrome. This occurred in yellowtail snappers and European sea bass raised in shallow cages at Martinique (Gallet de St. Aurin et al., 1990). Signs were fish at the surface, dark coloration, slow movement, emaciation, retinal degeneration, loss of patches of scales, ulcerated lower lip, whitish-green feces, and mortality (mostly from secondary bacterial infection). The primary cause probably was overexposure to intense sunlight, with high temperature and deficiency of vitamins E and C as possible contributing factors.

Overexposure to Therapeutic Chemicals

(Chong and Chao, 1986a; Post, 1987) Sometimes the safety margin between concentrations or doses that are therapeutic and those that are toxic is narrow.

Antibiotics can cause anemia, abnormal development, skin sores, liver hemorrhages, and kidney damage. Injection of 18.7 mg chloramphenicol/kg BW resulted in sores on Atlantic salmon but not rainbow trout or brown trout. Salmonids given 10-20 mg kanamycin/kg BW/d in the feed for 3-5 weeks suffered liver and kidney damage and mortality. Rainbow trout given 100 mg erythromycin/kg BW/d for 2-3 weeks also had liver and kidney damage. Cutthroat trout given 220 mg sulfamerazine/kg BW/d for 14 d had kidney damage. Post (1987) suggested using 264 mg/kg sulfamerazine for 3 d, then 154 mg/kg for 11 d to reduce side effects.

Furanace overexposure can lead to ulcerative dermatitis. Three to four days in a 0.5 mg/L bath resulted in erosion of skin and muscles in channel catfish. A 2.0 mg/L bath led to developmental abnormalities and mortality in mummichogs.

Alimentary parasiticides. Overdosage of the alimentary parasiticide enheptin (200 mg/kg BW/d in the feed for 10 d, but not 125 mg/kg for 6 d) caused liver and kidney damage and mortality in rainbow trout; residues persisted for nearly 140 d. Normal treatment is 1 g/kg feed for 3 d for intestinal flagellated protozoans. Thiabendazole also can cause liver and kidney damage.

Quaternary ammonium compounds. Although the 48-h LC_{50} for channel catfish exposed to Roccal® was 1.12 mg/L, and for salmonids up to 4.0 mg/L, treatment baths of 4.0 mg/L for 1 h sometimes are employed.

Formalin overexposure can cause severe gill damage. The 96-h LC_{50} for channel catfish was about 173 µL/L, but for green sunfish it was about 468 µL/L. Every 5 parts of formalin can eliminate 1 part of oxygen (i.e., 5 ppm:1 ppm). It is dangerous and should not be used in ponds when water is warmer than 27°C, there is a dense algal bloom, or dissolved oxygen is less than 5 mg/L. Formalin that has been exposed to temperatures less than 5°C should not be used because it could contain paraformaldehyde, which is very toxic. Formalin should not be used if it contains a white precipitate, which probably is paraformaldehyde.

Potassium permanganate in salt water forms toxic manganese dioxide. It probably should be used only in fresh water.

Malachite green can inflame the epithelium of skin, gills, and digestive system and block digestive enzymes. The 96-h LC_{50} for bluegills was 0.035 mg/L, but for coho salmon it was 0.382 mg/L.

Copper. Copper sulfate can cause damage to hematopoietic tissue and kidneys, increased liver fat, and digestive enzyme inhibition. The 48-h LC_{50} for bluegills increased with alkalinity from 3.1 to 44 mg/L. With Cutrine,® the 96-h LC_{50} for largemouth bass was in the range 2.2-70 µL/L (0.198-6.4 mg/L copper).

Egg disinfectants. Iodophors often are used to disinfect eggs of freshwater fish (e.g., Goldes and Mead, 1995). If eggs are treated before water hardening, deformities and mortality can be high. After water hardening, at pH 6.0-8.0, eggs can tolerate 25-100 mg/L iodine from iodophors. A 100 mg/L iodine bath for 15 min is commonly used. Iodophors should not be used for sensitive species or within 3 d of hatching, which precludes most cultured marine fish. Acriflavine has traditionally been used for egg disinfection, but not as much in recent years. Its toxicity to fish has not been characterized, but a 10 mg/L bath for 1 h was highly irritating to adult Nassau groupers (pers. obs.). Thimerosal (merthiolate) is toxic to red drum eggs (Douillet and Holt, 1994) and should not be used anyway because it contains mercury. Hatching rate and survival of red drum eggs were not affected by 0.005% benzalkonium chloride, 0.74% formalin, 3% hydrogen peroxide, 0.01% sodium hypochlorite, or 0.3% polyvinylpyrrolidone iodine (Douillet and Holt, 1994). Five-minute no observable effect concentrations of H_2O_2 were 3% for red drum, 2% for spotted seatrout, and 1% for yellowtail snapper eggs, but 3% was required for good disinfection.

Organophosphorous compounds not only can cause acute toxicity and mortality but repeated exposure to low concentrations can lead to chronic acetylcholinesterase depletion, weakness, reduced growth, and lowered resistance to stressors. With trichlorfon, the 96-h LC_{50} for fathead minnows is about 50 mg/L, which is below therapeutic concentrations.

Toxicity values for two organophosphorous pesticides used to control predatory insects and crustaceans on fish farms are given in Table 11-3. The crustaceans tested were more sensitive to fenthion than malathion, and the fish to malathion, with variations among taxa. Both pesticides are toxic to some crustaceans and fish at concentrations below those commonly recommended for pest control (250-500 ppb fenthion, 250 ppb malathion). Because small amounts can affect survival and growth of cultured and wild animals, extreme care should be exercised in the application of OP pesticides on or near water.

Anesthetics. With quinaldine sulfate, the 96-h LC_{50} for largemouth bass was 6.8 mg/L, and for common carp it was 72.5 mg/L. With tricaine methanesulfonate (TMS, MS-222), the 24-h LC_{50} for lake trout was 33.8 mg/L, and for largemouth bass it was 63.0 mg/L. Toxicity of a mixture (1 part quinaldine sulfate:4 parts TMS) seems to be greater than the sum of the separate toxicities. The 96-h LC_{50} for lake trout was 4.2-16.9 mg/L, and for common carp it was 8.6-34.5 mg/L. Free carbon dioxide at 600 mg carbonic acid/L is tolerated by many fish for up to 15 min.

Nutritional Disorders (also see chapter 9)

Common problems include vitamin or essential fatty acid deficiencies, lipidosis from rancid feed (sign = yellow or pale liver, Figs. 11-2, 11-3), thiamin deficiency from thiaminase in uncooked foods, and mineral imbalances (e.g., from excess whitefish meal in feed). Five types are mentioned here.

Whirling diseases could be caused by viruses, other pathogens, toxins, or environmental stress. One type in European sea bass larvae probably was caused by EFA deficiency (Corneillie et al., 1990).

Table 11-3. Toxicity of two organophosphorous pesticides to crustaceans and fish (after Tucker and Thompson, 1987). Acute toxicity tests determined concentrations at which 50%, 10%, or 0% of the animals died after exposure to the toxicant for a specified time. The last four entries give concentrations for which delayed harmful effects were seen in larvae after an initial 12-h exposure.

Animal and stage	Fenthion (ppb)	Malathion (ppb)
48hLC50		
Daphnia first instar	0.80	1.8
Copepod adult	210	620
Common snook egg to larva	1370	34
Spotted seatrout egg to larva	1333	198
Spot egg to larva	1501	
Common snook juvenile	1015	280
Spotted seatrout juvenile	1100	150
Channel catfish juvenile		14325
48hLC10		
Copepod adult	40	115
Common snook egg to larva	1200	9
Spot egg to larva	1250	
Spotted seatrout egg to larva	447	19
Spotted seatrout juvenile	900	40
48hLC0		
Channel catfish juvenile		4000
96hLC50		
Glass shrimp adult	10	90
Sheepshead minnow juvenile		51
Striped bass juvenile	453	14
Bluegill juvenile	1380	103
12-h exposure during hatching, then reared 3 d past first feeding		
Common snook egg to larva (survival)		18
Spotted seatrout egg to larva (growth)	518	
Spot egg to larva (growth)	90	
Spotted seatrout egg to larva (swimming)		358

Head and Lateral Line Erosion (HLLE). Surgeonfish in aquariums are among the most susceptible. Leakage of slight electrical currents into tank water and malnutrition could be factors. Water systems should always be grounded, and fish should be given the best food that can be afforded. HLLE of palette tangs was relieved in 10 d when green algae and vitamin C (700 mg/kg) were added to their diet (Blasiola, 1988). When broccoli and peas were included in the diet of blue tangs, HLLE did not occur as usual (Collins, 1995). HLLE in angelfish could be caused by a virus (Varner and Lewis, 1991).

Broken back syndrome. Vertebral deformities reported in red sea bream and many other species (Matsuoka, 1982) usually become most noticeable in early juveniles. Broken back syndrome (fractured vertebral column) can occur as a result of severe scoliosis or lordosis, which could be caused by an ascorbic acid deficiency (Gallet de Saint Aurin et al., 1990) or excessive circulation (Divanach et al., 1997).

Granulomatous hypertyrosinemia in turbot probably is caused by ascorbic acid and pyridoxin deficiency and is treated by increasing feed content of ascorbic acid to 60 mg/kg

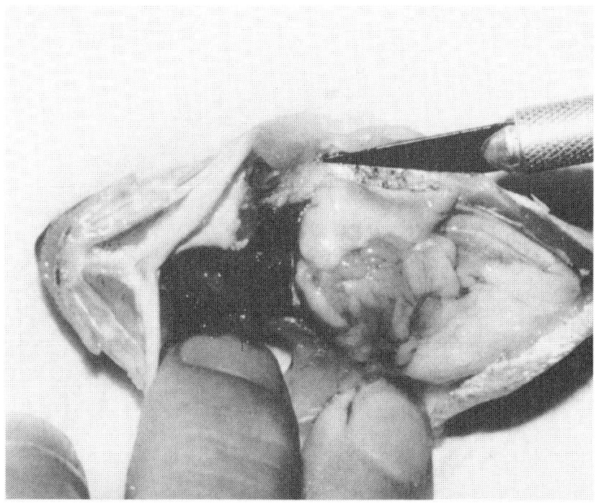

Figure 11-2. Necropsy of a juvenile red drum. The three larger masses below the scalpel blade are: abnormally-pale liver, pyloric caeca, and mesenteric fat bodies.

Figure 11-3. Dissected liver from the red drum in Fig. 11-2, with gall bladder trailing to the right.

BW for 5 d per month (Baudin-Laurencin, 1990). A systemic granuloma disorder in gilthead sea bream that affects the kidneys, spleen, liver, and eyes could be caused by spoiled, deficient, or unbalanced diets (Colorni, 1990).

Anaphylactic Shock
Sometimes certain fish, especially omnivores (e.g. tilapias, carps), if fed high protein feed, can develop a food allergy (Roberts, 1989b). Many fish can become rigid, twitching, and dark immediately after feeding. They usually recover, but chlorpromazine given before feeding has been used as a preventative.

Gas Bladder Disorders
Undeveloped Gas Bladder. Formation of the gas bladder in larvae is a delicate process. During the first two weeks of life, at least a few species must swallow gas bubbles to begin the process. Several species reared in tanks have been found to have gas bladders that never filled or that were filled with fluid instead of gas. Apparently, the larvae did not have access to gas bubbles because of one or more factors, such as: viscous or solid surface film, inappropriate aeration, high light intensity, or weakness (e.g., from malnutrition). The inability of affected fish to regulate buoyancy often causes them to swim with the head tilted upward to maintain position. If the fish survive long enough, the tail muscles and spinal

Figure 11-4. Top: Common snook with undeveloped gas bladder and kypho-lordosis (71 mm FL). Bottom: Normal common snook (76 mm FL). (from Tucker, 1987b; photo by Thomas A. Smoyer, HBOI)

column can become distorted, resulting in lordosis and/or kyphosis, with the tail twisted downward (Fig. 11-4). (also see pages 90-95)

Overinflated Gas Bladder. This is a mysterious malady that results in fish helplessly floating at the surface (Chong and Chao, 1986a). If the gas bladders are deflated to normal volume with a 19-gauge hypodermic needle, many fish will recover.

Urinary Bladder Calculi (stones)
These occur in larvae of several species (e.g., common snook, barramundi, European sea bass, Australian bass, red sea bream, gilthead sea bream, mulloway) and could cause mortality (Nowak and Battaglene, 1996). Food quality, salinity, and stress could be factors.

Incomplete or Twisted Operculum
These developmental defects occur in many species, affecting appearance and sometimes ventilation efficiency and survival (e.g., milkfish, Hilomen-Garcia, 1997; gilthead sea bream, Koumoundouros et al., 1997). Vitamin C deficiency and environmental (e.g., gas supersaturation) or behavioral (e.g., from strong current) stress are possible factors.

Susceptibility
Acipenseridae. Sturgeon are susceptible to gram-negative bacteria (*Vibrio anguillarum*, myxobacteria, *Flexibacter columnaris*, *Yersinia ruckeri*), fungi (*Saprolegnia* sp., and others), and ectoparasitic protozoans (*Ichthyobodo necator*) (Conte et al., 1988; Brun et al., 1991).

Anguillidae. Some of the more important disease agents of cultured eels are viruses (eel virus American, eel virus European), fungi (*Saprolegnia* sp.), gram-negative bacteria (*Vibrio* sp., *Pseudomonas* sp., *Aeromonas* sp., *Cytophaga columnaris*, *Edwardsiella* spp.), monogeneans (*Pseudodactylogyrus* sp.), nematodes, microsporidians (*Pleistophora* sp.), and ectoparasitic protozoans (*Ichthyophthirius multifiliis*) and crustaceans (copepods)(Landsberg, 1989; Gousset, 1990; Usui, 1991; Sprengel and Lüchtenberg, 1991; Ikenoue and Kafuku, 1992).

Chanidae. For milkfish, gram-negative bacteria (*Vibrio* spp., *Aeromonas hydrophila*, *Flexibacter columnaris*), coccidians (*Eimeria* sp.), and ectoparasitic protozoans, copepods

(*Caligus* spp., *Lernaea cyprinacea*), branchiurans (*Argulus* sp.), and isopods (*Rocinella typicus*, *Ichthyoxenous* sp.) are the most significant pathogens (Tung and Kou, 1986; Liao, 1991; Brock et al., 1993). Nineteen different parasites were reported by Regidor and Arthur (1992). Problems also could arise from pollution and malnutrition.

Osmeridae. Ayu pathogens include gram-negative bacteria (*Vibrio anguillarum*, (*Flavobacterium* sp.), gram-positive bacteria (*Streptococcus* sp.), and microsporidians (*Glugea plecoglossi*)(Ikenoue and Kafuku, 1992).

Salmonidae. In seawater cage culture of salmonids in Europe, North America, and Japan, the most common and significant diseases are vibriosis (*Vibrio anguillarum*), coldwater vibriosis (*Vibrio salmonicida*), and furunculosis (*Aeromonas salmonicida*), for which vaccines are available, and bacterial kidney disease (Bruno et al., 1986; Sindermann, 1990; Mahnken, 1991; Torrissen, 1995). Sea lice, infectious hematopoietic necrosis, and myxosporidians are more localized problems. Proliferative kidney disease and costiasis are less important. In Japan, bacterial kidney disease is considered a major problem and furunculosis a moderate one; losses from vibriosis, costiasis, and gill amebae are not significant. In 1992, the greatest disease losses for cultured coho salmon in Japan were from erythrocytic inclusion body syndrome (39%), vibriosis (27%), and bacterial kidney disease (22%)(Sako, 1996). Some of the main salmonid hatchery diseases are furunculosis, bacterial kidney disease, and viral hemorrhagic septicemia. Viral hemorrhagic septicemia and infectious hematopoietic necrosis mainly affect rainbow trout and not salmon (Isaksson, 1991). For Atlantic salmon, infectious pancreatic necrosis is mostly a problem of early juveniles. The freshwater monogenean *Gyrodactylus salaris* is a serious parasite that has spread to many rivers in Norway, probably from infected fish stocked in cages (Hansen and Bakke, 1989). Effective autogenous vaccines have been used for *Aeromonas salmonicida*, *Vibrio anguillarum*, *V. ordalli*, and *V. salmonicida* (Armstrong, 1997).

Gadidae. A protistan endoparasite was found infesting the yolk of cod eggs, but the level of pathology was not determined (Pedersen et al., 1993). Some nematodes infest crustaceans, cod, then marine mammals to complete their life cycles.

Scorpaenidae. Kurosoi in Korea are susceptible to lymphocystis, vibriosis, streptococcosis, and edwardsiellosis (Park and Sohn, 1996).

Centropomidae. For barramundi, gram-negative bacteria (*Vibrio* spp., *Aeromonas* spp., *Pasteurella* spp.), epitheliocystis, ectoparasitic protozoans (*Amyloodinium ocellatus*, *Cryptocaryon irritans*, *Brooklynella* spp., *Trichodina* spp., *Chilodonella* spp.), and monogeneans (*Diplectanum* sp. and others), and lymphocystis are probably the most significant (Chong and Chao, 1986a; Lim et al., 1986; Chonchuenchob et al., 1987; Glazebrook and Campbell, 1987; Humphrey and Langdon, 1987; Anderson and Prior, 1992; Shelley, 1993; Balasuriya and Leong, 1994; Rimmer, 1995).

When reared in fresh well water, juvenile and adult common snook had no disease problems (Tucker, 1987b). Thiaminase-induced athiaminosis could occur in juveniles given fresh or frozen foods (Figs. 9-4, 9-5).

Percichthyidae. European sea bass are susceptible to viruses (infectious pancreatic necrosis, viral hemorrhagic septicemia, viral erythrocytic necrosis, viral nervous necrosis, sea bass birnavirus, possibly a "whirling disease" virus), gram-negative bacteria (*Vibrio* spp., *Cytophaga*-like bacteria, *Flexibacter maritimus*, myxobacteria, *Pasteurella piscicida*, *Pseudomonas* spp., *Rickettsia*-like bacteria), ectoparasitic protozoans (*Amyloodinium* sp., *Cryptocaryon* sp., *Trichodina* sp., *Ichthyobodo necator*), microsporidians, myxosporidians, monogeneans (*Diplectanum aequans*), copepods (*Caligus* sp., *Ergasilus* sp.) and possibly nutritional problems (Breuil and Haffner, 1990; Coves et al., 1991; Varriale et al., 1993; Abela and Le Breton,

1996b; Le Breton, 1996). Mycobacteriosis can be very contagious and cause heavy losses (Colorni and Diamant, 1992). Myxosporidians (*Sphaerospora* spp.) infect wild and farmed sea bass (Sitja-Bobadilla and Alvarez-Pellitero, 1993).

Striped bass are susceptible to lymphocystis, infectious pancreatic necrosis virus, gram-negative bacteria (*Vibrio* spp., *Aeromonas* spp., *Cytophaga columnaris*, *Edwardsiella tarda*, *Pasteurella piscicida*), gram-positive bacteria (*Streptococcus* sp. *Enterococcus* sp.), mycobacteriosis (*Mycobacterium* sp., *Nocardia* sp.), fungi (*Branchiomyces* spp., probably *Achlya* spp. and *Saprolegnia* spp.), ectoparasitic protozoans (*Amyloodinium ocellatum*, *Ichthyophthirius multifiliis*, and others), and ectoparasitic crustaceans (*Argulus* sp., *Ergasilus* sp., and others)(Harrell et al., 1990; Plumb, 1991; Baya et al., 1996).

Serranidae, Lutjanidae, etc. For groupers, snappers, and similar warmwater fish, gram-negative bacteria (*Vibrio* spp., *Aeromonas* spp., *Pseudomonas* spp., *Pasteurella* spp.), *Streptococcus*, mycobacteriosis, ectoparasitic protozoans (*Amyloodinium ocellatus*, sporozoans, *Cryptocaryon irritans*, *Brooklynella* spp.; *Ichthyophthirius* sp.), and monogeneans (*Neobenedenia melleni*, *Diplectanum* spp.) are among the most important (Chong and Chao, 1986a; Lim et al., 1986; Tseng and Ho, 1988; Gallet de Saint Aurin et al., 1990; Dyer et al., 1992; Lee, 1995a). Sleepy grouper disease (lethal) in Singapore probably was caused by a virus introduced with wild juvenile groupers imported for cage farming (Chua et al., 1994). Other viral pathogens and diseases include golden eye disease, red grouper reovirus, spinning grouper disease, and viral nervous necrosis. Nervous suffering disease of groupers (signs could include gill, blood, gas bladder, liver, heart, brain, and nerve damage) probably was caused by rancid dietary lipids (Hua et al., 1994a, 1994b). In Japan, pasteurellosis was a major disease of young redspotted groupers (Sako, 1996). Some snappers can be overwhelmed by parasitic copepods (e.g., *Caligus* sp. on mutton snappers, Kennedy, pers. comm., 1997).

Rachycentridae. Cobia parasites include monogeneans (*Dionchus rachycentris*), digenean trematodes (*Stephanostomum pseudoditrematis*), cestodes, nematodes (*Iheringascaris inguies*), acanthocephalans, and copepods (*Lernaeenicus longiventris*)(Shaffer and Nakamura, 1989).

Carangidae. *Enterococcus seriolicida* (enterococcosis) and *Pasteurella piscicida* (pseudotuberculosis) are economically the first and second most important pathogens of cultured yellowtail in Japan, with vibriosis a distant third (Matsusato, 1975; Doumenge, 1990a; Sindermann, 1990; Kusuda, 1992; Sakata and Kawazu, 1992; Arthur and Ogawa, 1996). These three accounted for 96% (63%, 28%, 5%) of disease losses in 1978. Losses in 1991 were 68% to enterococcosis and 18% to pseudotuberculosis; in 1992, enterococcosis/streptococcosis (58%), pseudotuberculosis (20%), and jaundice (6%)(Sako, 1996). *Nocardia kampachi*, *Mycobacterium* sp., and *Pseudomonas fluorescens* also are important. Other bacterial diseases include hemolytic jaundice, gliding bacterial disease, and staphylococcosis. Red sea bream iridovirus caused large losses during the early 1990s. Yellowtail ascites virus (which could be transmitted vertically from broodfish to larvae; Isshiki et al., 1992), lymphocystis, and viral deformity virus also occur. In Japan, viral epidermal hyperplasia and viral ascites are major diseases of young yellowtail (Sako, 1996). The monogeneans *Benedenia seriolae* and *Heteraxine heterocerca* affect yellowtails (Kearn et al., 1992a, 1992b). Other parasites include microsporidians, myxosporidians, nematodes, and copepods. They also get edwardsiellosis (Park and Sohn, 1996).

In 1992, the greatest disease losses for cultured Japanese horse mackerel in Japan were from vibriosis (81%), and streptococcosis (12%)(Sako, 1996).

Coryphaenidae. Dolphin are susceptible to gram-negative bacteria (*Vibrio* spp., *Aeromonas* sp., *Pasteurella* sp.), occasionally protozoans (*Amyloodinium*, *Cryptocaryon*) (Brock et al., 1993), and parasitic copepods and isopods.

Sparidae. For red sea bream, the disfiguring virus *Lymphocystis* sp., gram-negative bacteria (*Vibrio* spp., *Edwardsiella* spp., *Neisseria* sp., gliding bacteria), monogeneans (*Bivagina* sp.), nematodes (*Philometroides* sp.), and *Longicollum pagrosomi* are major pathogens (Foscarini, 1988; Fukusho, 1991; Kusuda, 1992). Myxobacteriosis has caused large losses of young red sea bream in Japan (Sindermann, 1990). *Cytophaga maritimus*, *Amyloodinium ocellatus*, and *Ichthyophthirius* sp. are important. Occasional malnutrition, "yellow fat" disease, lordosis, vertebral fusion, and urinary calculi occur. In 1992, the greatest disease losses for cultured red sea bream in Japan were from an iridovirus (25%), vibriosis (22%), cryptocaryoniasis (22%), and edwardsiellosis (9%)(Sako, 1996). In Japan, viral epithelial necrosis, gliding bacterial disease, and vibriosis are major diseases of young sea bream (Sako, 1996).

For gilthead sea bream, viruses (*Lymphocystis* sp., aquareovirus), gram-negative bacteria (*Vibrio* spp., *Aeromonas* sp., *Cytophaga*-like bacteria, *Enterobacteria* sp., *Pseudomonas anguilliseptica*), mycobacteriosis, ectoparasitic protozoans (*Amyloodinium ocellatum*, *Cryptocaryon irritans*, *Ichthyobodo* sp., *Trichodina* sp., *Trichodinella* sp.), microsporidians, myxosporidians, epitheliocystis, monogeneans (*Gyrodactylus* sp., *Furnestinia echeneis*, *Microcotyle* sp.), and copepods (*Anilocra physodes*) are the major pathogens, with distended gut syndrome and systemic granuloma also causing problems (Colorni, 1990; Abela and Le Breton, 1996a; Le Breton, 1996). Epitheliocystis causes high mortality of late larvae, and lymphocystis is an occasional problem for older fish (Colorni and Diamant, 1992).

Sciaenidae. For red drum, gram-negative bacteria (*Vibrio* spp., *Aeromonas* spp., *Pseudomonas* spp.), gram-positive bacteria (*Streptococcus* sp., *Catenabacterium* sp.), ectoparasitic protozoans (*Amyloodinium ocellatum*, *Cryptocaryon irritans*, and others), a sporozoan (*Henneguya* sp.), fungi (*Saprolegnia* sp.), and marketability-reducing cestodes (*Poecilancistrium* spp.) probably are the major pathogens (Johnson, 1990; Plumb, 1991; Vermeer et al., submitted). Mycobacteriosis also occurs. *Amyloodinium* has caused major losses (e.g., Sandifer et al., 1993), easily killing all the fish in a unit overnight.

Cichlidae. The monogenean *Neobenedenia melleni* is one of the main health threats to Florida red tilapia raised in salt water in the Caribbean region (Watanabe, 1991; Robinson et al., 1992). In Hawaii, pathogens include ectoparasitic protozoans, and coccidians (*Eimeria* sp.)(Brock et al., 1993).

Mugilidae. Mullets are susceptible to gram-negative bacteria (*Edwardsiella tarda*, *Pseudomonas* spp., *Vibrio* spp., *Aeromonas* sp., *Chromobacterium* sp.), and ectoparasitic protozoans (*Amyloodinium ocellatum*, *Cryptobia* sp., *Cryptocaryon* sp., *Ambiphrya* sp., *Scyphidia* sp., *Trichodina* sp., *Tripartiella* sp.), coccidians (*Eimeria* sp.), monogeneans (*Benedenia monticelli*, *Metamicrocotyla* sp.), digenean trematodes (*Heterophyes* spp.), branchiurans (*Argulus* spp.), and copepods (*Ergasilus lizae*, *Lernaea cyprinacea*)(Landsberg, 1989; Paperna, 1991; Brock et al., 1993; Liu and Kelley, nd.b).

Siganidae. Rabbitfishes are susceptible to *Pseudomonas* spp., *Streptococcus* spp., sporozoans, monogeneans, nematodes, acanthocephalans, and copepods (Duray, 1990).

Bothidae. Hirame larvae can have a herpesvirus (viral epidermal hyperplasia), juveniles a birnavirus, and adults a rhabdovirus (HRV); lymphocystis also occurs (Nguyen et al., 1994; Park and Sohn, 1996). In 1992, the greatest disease losses in Japan were from edwardsiellosis (29%) streptoccosis (22%), benedeniosis (12%), and gliding bacterial disease (5%)(Sako, 1996). *Flexibacter maritimus* can be important (Kusuda, 1992). Vibriosis and pseudotuberculosis also occur. In Japan, birnaviral disease, rhabdoviral disease, vibriosis, bacterial enteritis, edwardsiellosis, and scuticociliatidosis are major diseases of young hirame (Sako, 1995). In Hawaii, pathogens have included bacteria (*Vibrio* spp., *Pasteurella* sp., *Flavobacterium* sp., *Pseudomonas* sp.) and flagellates (*Ichthyobodo* sp.)(Brock et al., 1993).

The most troublesome diseases of turbot are vibriosis and external parasitism, but these can be controlled; viruses constitute the most serious threat to farmed turbot (Person-Le Ruyet et al., 1991). An example is infectious pancreatic necrosis virus, for which *Artemia* and wild crustaceans could be vectors (Mortensen et al., 1993). Turbot pathogens include other viruses (*Lymphocystis* sp., a birnavirus, herpesviruses, a paramyxovirus, a rotavirus, viral hemorrhagic septicemia virus), bacteria (*Vibrio* spp., *Aeromonas* spp., *Flexibacter maritimus*, *Pseudomonas* spp., *Streptococcus* spp.), fungi (*Ichthyophonus* sp.), flagellates (*Ichthyobodo necator*), microsporidians (*Tetramicra brevifilum*, *Glugea stephani*), ciliates (*Trichodina* sp. commonly, *Cryptocaryon* sp. and *Scyphidia* sp. sporadically), cestodes (*Bothriocephalus scorpii*), and leeches (*Hemibdella* sp.)(Nevo, 1983; Toranzo et al., 1994). Larval pathogens include a picornavirus, *Vibrio* spp., *Aeromonas* spp., and *Pseudomonas*-like bacteria (Blanch et al., 1991; Toranzo et al., 1994). *Vibrio* spp. and *Pseudomonas* spp. cause 5-10%, *Tetramicra brevifilum* 3-5%, and filamentous bacteria (myxobacteria) 1-2% of total losses during grow-out (Toranzo et al., 1994).

Pleuronectidae. Atlantic halibut are susceptible to vibriosis, myxosporidians (*Ceratomyxa drepanopsettae*), monogeneans (*Entobdella hippoglossi*), digenean trematodes, cestodes (*Grillotia erinaceus*, *Scolex pleuronectis*), nematodes (*Contracaecum aduncum*, *Anasakis* spp.), acanthocephalans (*Echinorhynchus gadi*), and copepods (*Lepeophtheirus hippoglossi*, *Hatschekia hippoglossi*)(Haug, 1990; Trumble et al., 1993). Egg and larval pathogens include *Vibrio anguillarum*, *Vibrio* spp., and *Flexibacter* sp. (Bergh et al., 1991).

Pacific halibut parasites include myxosporidians (*Kudoa thrysites*), digenean trematodes, and nematodes (*Anisakis simplex*, *Pseudoterranova decipiens*)(Trumble et al., 1993).

Soleidae. Wild broodstock of common sole have leeches and monogeneans frequently and copepods sometimes (Baynes et al., 1993). Black patch necrosis (*Flexibacter maritimus*) at times has been widespread, but it might have been linked with a diet of cooked and stored mussels; soles fed fresh mussels were resistant (Howell, 1997).

Tetraodontidae. Tiger puffer health problems are mainly an iridovirus, viral nervous necrosis, white mouth disease; *Flexibacter maritimus*, *Vibrio anguillarum*; myxosporidians; cryptocaryoniasis, trichodiniasis; monogeneans; and copepods (Arthur and Ogawa, 1996). In 1992, the greatest disease losses for cultured tiger puffers in Japan were from kuchijirosho (40%), heterobothriosis (21%), and cryptocaryoniasis (15%)(Sako, 1996). *Amyloodinium* was a major pathogen of young tiger puffers.

Ornamental fish. Partly because of the way they are confined, ornamental marine fish are susceptible to several disease agents, roughly in this order of decreasing frequency and importance (Sindermann, 1990; Moe, 1992): *Amyloodinium ocellatum*, *Cryptocaryon irritans*, *Brooklynella hostilis*, *Ichthyophonus hoferi*, *Mycobacterium marinum*, *Vibrio anguillarum* (and relatives), *Lymphocystis* sp., *Neobenedenia melleni* (and relatives), and fish lice (branchiurans and copepods).

Larval fish could be especially susceptible to viral, bacterial, and protozoan diseases. Diagnosis is difficult and high losses are possible, especially with small fast-developing warmwater larvae. Effective water filtration and disinfection are especially important. Probiotic bacteria and chemotherapeutants can be administered by feeding them to rotifers or *Artemia* first (Gatesoupe et al., 1989; Chair et al., 1995, 1996; Gapasin et al., 1996; Touraki et al., 1996).

ENVIRONMENTAL AND HUMAN HEALTH CONSIDERATIONS

Numerous diseases have been spread by the transport of marine animals, sediments, water, etc. In many cases, local species have become infected, sometimes with high losses. If exotic fish introductions are judged to be worth the risks and are conducted, great care has to be taken to quarantine the introduced individuals, eliminate any significant diseases in later generations, and prevent escape of the fish and any pathogens. An example was given by Spedicato et al. (1993a), who brought hirame eggs from Japan to Italy, but were careful not to introduce *Rhabdovirus olivaceus*, which could have severe effects on farmed trout.

Treatment of fish with antibiotics is dangerous because bacterial strains resistant to those antibiotics are given an advantage and can build up (e.g., *Pasteurella piscicida*, *Vibrio* spp.; Austin and Austin, 1987). Resistance is becoming a serious problem in human medicine (Service, 1995). Ervik et al. (1994) described a cage design with a funnel at the bottom for collecting sediment, particularly to reduce the amount of antibiotics released. Another danger is the release of nonvirulent live vaccines, which could become virulent again and infect wild fish (Austin, 1993).

Oxytetracycline (OTC) might not be a very good drug for fish in seawater, especially in cages or ponds, because it is not absorbed well (Lunestad and Goksøyr, 1990) and it accumulates in sediments (Pouliquen et al., 1993; Coyne et al., 1994). Oral availability in Atlantic salmon was only 2%, with most of the rest being egested to the environment of the fish (Elema et al., 1996). OTC also has a long residence time in fish, is reported to be immunosuppressive, and can cause liver and spinal damage (Elema et al., 1996; Toften and Jobling, 1996).

When fish are treated with antibiotics, their gut floral composition shifts to resistant bacteria and sometimes total bacteria actually increases (Austin, 1993). The effluent from fish tanks being treated with an antibiotic typically has a higher proportion of bacteria that are resistant to the antibiotic. These could be strains of the target bacteria or other species. Resistance can be transferred between closely related bacteria, possibly from a fish pathogen to a human pathogen. Antibiotics released into the environment from several sources (not just aquaculture) have had this effect. In some coastal areas, even far from aquaculture facilities, resistant strains are common. Many strains of *Aeromonas* spp. were resistant to ampicillin, tetracycline, Romet-30, erythromycin, and others, but not to ceftiofur sodium (Dixon and Issvoran, 1992). Many strains of *Pasteurella piscicida* were resistant to ampicillin, chloramphenicol, kanamycin, nalidixic acid, sulfamonomethoxine, tetracycline, and trimethoprim; florfenicol was the most effective (Aoki and Kitao, 1985; Kim and Aoki, 1993). Strains of *Vibrio parahaemolyticus* varied in resistance to sulfadiazine, and tetracycline, were most resistant to penicillin, erythromycin, kanamycin, streptomycin, and ampicillin, and were least resistant to chloramphenicol and gentamycin (Sanjeev and Stephen, 1992). New compounds such as fluroquinolones and third-generation cephalosporins, for which resistance is less likely to develop, should be safer (Dixon, 1994).

The value of antibiotics important for humans can be negated if they are used, and released, too widely (e.g., in agriculture or aquaculture), then promote development of resistant strains of human pathogens. This is why certain antibiotics are not allowed for non-human applications. In a study by Twiddy (1995), antibiotic-resistant strains of the human pathogens *Salmonella* sp., *Aeromonas hydrophila*, and *Plesiomonas shigelloides* were found in fish from integrated fish farms (chickens with tilapia, carp, and/or catfish) in Southeast Asia.

Drugs such as trimethoprim, oxytetracycline, and sulfadiazine can persist in rainbow trout for as long as 9-11 weeks (Austin, 1993; Namdari et al., 1996). Effects on consumers could include loss of beneficial bacteria, immunosupression, predisposition to *Salmonella*, and buildup of resistant bacteria. In Canada, the required OTC withdrawal times of 60 d at ≤10°C and 40 d at >10°C were confirmed to be realistic by Namdari et al. (1995). In the U.S., required withdrawal time for OTC is only 21 d. For the muscle of palmetto bass (>15°C), elimination time for OTC from a single IP injection of 25 mg/kg was 24 d, but for 50 mg/kg it was 32 d (Xu and Rogers, 1993).

For these and other reasons, the availability, storage, and application of potentially harmful drugs is regulated (Austin, 1993). They should not be released into natural waters. Overuse of a single drug should be avoided; if more than one drug is effective, alternate use can help minimize development of resistant strains. Withdrawal times should be realistic. Possibly, antibiotics that do not promote resistance will be developed for humans and fish (Service, 1995).

SUMMARY

To avoid diseases in cultured marine fish, the culturist must be concerned with:

Sanitation
All containers and equipment that fish contact must be cleaned and disinfected before they are used for a new batch of fish.

Immunization
If available and warranted (e.g., for *Vibrio* in salmonids), vaccination should be used to protect cultured fish, which are more susceptible because of crowding and other stresses.

Water Treatment
The water supply must be filtered and disinfected according to its initial condition and the needs of the fish.

Isolation
The chance of cross-contamination between or within cohorts can be minimized by separating tanks into isolated blocks. When closed systems are used, low tank-to-filter ratios will help minimize potential loss. Broodfish, larval rearing, grow-out, food culture, and food preparation areas should always be separated.

Culture Conditions
Stress must be minimized. Optimal conditions should prevail. Keep water clean, oxygen high, temperature correct, salinity correct, and disturbances few. A beneficial bacterial flora can be promoted by good environmental management and inoculation with probiotic strain(s).

Nutrition
The fish should be given appropriate foods that are fresh and uncontaminated. Immune system function can be enhanced by dietary additives (e.g., vitamins, beta glucan).

Proactive Health Care
Important components include effective: managment and monitoring of tanks, water, bacteria, fish, and foods; established procedures for sanitation of systems, isolation and treatment of sick fish; and education of personnel.

Quarantine
When fish from different stocks are to be mixed, especially by adding wild fish to a captive population, the newcomers should go through a period of treatment and observation (up to 4 weeks). The degree and duration of treatment depends on the value of the fish and the risk of disease transmission.

Isolation
If not all fish are affected, sick ones should be removed from the general population, especially if the disease is contagious.

Chemotherapy
Chemical treatment for disease usually is a last-ditch effort to correct a situation that should not have been allowed to develop. Before proceeding, the problem must be correctly diagnosed. If possible, the gentlest treatments (e.g., salinity or temperature control) should be tried before the more stressful ones.

12 HANDLING AND TRANSPORTING MARINE FISH

SENSITIVITY OF DIFFERENT STAGES

Sensitivity to handling for many species increases in the following order:

Eggs
Juveniles
Adults
Larvae

For most species, eggs are the easiest stage to handle and transport. Because larvae are very fragile, survival of most species is much better if transfers are done before hatching, but not too early or too late. Water-hardening of newly-fertilized eggs might take 1-2 h and eggs seem to be more sensitive during this time, so excessive handling should be avoided at first. Some species might be more sensitive during gastrulation than during later stages. However, late stages require much more oxygen (therefore should not remain crowded), and the eggs might hatch before they can be transported and processed. Handling eggs just before hatching is likely to accelerate hatching and, depending on the degree of shock, might injure some embryos. It is best not to handle eggs within 2-4 h before hatching, if possible. Bromage et al. (1988) stated that salmonid eggs are most resistant to handling as eyed eggs, followed by unfertilized eggs, and then newly-fertilized eggs; transport as eyed eggs is preferred. At 28°C, milkfish eggs began hatching at 25 haf; they were less sensitive to mechanical shock between the end of gastrulation (~10-12 haf) and the tail-free stage (~18-20 haf)(Hilomen-Garcia, 1998).

Eggs can be transferred with small-bore pipets, beakers, or crowding nets with bottoms that hold water. If done very gently but quickly, eggs also can be moved in soft fine nets (e.g., brine shrimp) without water. Between hatching and transformation, typical marine fish (except very tough species) cannot be netted or removed from the water without harming them. Larvae can be transferred with large-bore pipets, beakers, buckets, or larger nets with container bottoms. Eggs and larvae can be concentrated before transfer by careful use of crowding nets, but larvae should not be touched by the net. The robust larvae of damselfish, dolphin, red drum, and others tolerate moving better than those of snook and groupers.

When fish become juveniles, they have scales and relatively thick skin to protect them from abrasion and dessication. Juvenile fish usually tolerate capture, measuring, weighing, transport, etc. better than adults (less mechanical stress, lower oxygen requirement).

Within reasonable limits, for some species, the more often juvenile and adult fish are handled, the tamer and easier to handle they become. Other very shy fish, like snook, could be stressed and not eat for 2 weeks after brief handling.

HANDLING MARINE FISH

Moving Fish

To minimize stress and injury to fish when they are handled (Chong and Chao, 1986a):

Don't feed fish just before handling.
Don't alarm fish excessively (e.g., by chasing them).
Don't drop fish from a height.
Be gentle, but deliberate.
Use non-abrasive handling materials (e.g., knotless nets, smooth plastic bags).

Polyethylene and polypropylene nets are abrasive. One of the gentlest materials with which to handle fish is soft knotless small-meshed nylon, such as good seines are made of. Some commercial dipnets are made of such material. Carrying slings and holding cages can be made of it. Slings, or stretchers, also can be made of smooth, reinforced nylon sheeting. A soft tube-shaped net that completely contains the fish is convenient for transferring larger ones. Tough polyethylene bags are convenient disposable carriers. When fish are removed from water, they and anything they contact should be kept very wet to minimize abrasion and loss of mucus (a towel, even when wet, is very effective at removing mucus and killing fish). Smooth polyethylene food containers and dishpans are made in a wide range of sizes and are very convenient for weighing individuals or small groups. They are non-abrasive and keep the fish from sliding away. Handling materials should be washed and disinfected before they are used with a different batch of fish.

Piling fish on top of each other can injure them, infect them, and reduce their attractiveness to buyers. Crowding in nets should be minimized.

Sturgeon broodfish are large (e.g., 30 kg) and strong and must be handled carefully to protect them and their handlers. Conte et al. (1988) described a tube net and stretcher for handling them. The 2.4-m long fine-mesh net was open at both ends and had handles at both ends. The 2.4-m stretcher was made of smooth reinforced nylon and had a hood at one end to cover the sturgeon's head and hold a water tube in its mouth for irrigating the gills. Although sturgeon can survive for up to 10 min without the water tube, 3 min was the recommended limit.

Most of the common methods for broodfish capture have been used for striped bass: electroshocking, gill nets, bow nets, trap nets, hook and line (Yeager et al., 1990). Red drum broodstock have been captured mainly by hook and line, longline, and beach seine (Tillman and Surovik, 1990).

Some fish at certain sizes (e.g., juvenile salmon and trout) can be safely transported through pipes. If this is attempted with any fish, the pipes must be very smooth inside with no sharp bends and the flow should not be excessively fast. Simply opening a valve wide to allow a torrent of fish and water to rush down the hillside through a pipe or hose is not only a final insult to the fish but probably will bruise and abrade them and possibly break their backs. A certain degree of gentleness is required.

When tanks are drained, fish will become increasingly agitated as the water level drops, especially adult fish, which sense that they might become trapped or stranded. Handling and transport stress might be increased for schooling species if they are separated into groups too small to form a school (Backman and Ross, 1990).

Fish should be especially protected from temperature fluctuations and oxygen depletion during handling. If air temperature is higher than optimal during the day, fish might be best handled early in the morning; however, pond fish could be stressed by the oxygen minimum, which tends to occur then. Juveniles and adults of some species can live out of water for suprisingly long periods if their gills stay wet (e.g., Nassau grouper adults for 2 h or more). Exposure to direct sunlight can kill eggs or larvae near the water surface and can cause stress and sunburn leading to other disorders in juveniles and adults held in shallow water. Also especially vulnerable to sunburn are broodfish caught at depth that are floating upside down in hauling tanks because of expanded or burst gas bladders. They should be kept well shaded and should be deflated as soon as possible.

The gas bladders of physoclistous (without a duct to the outside) fish caught at depth will expand as they are brought to the surface and can easily rupture, filling the body cavity with expanded gas. In mild cases, the fish can swim to the bottom of a tank and stay there until the gas has been absorbed by the body and released. If the fish is floating helplessly, the gas bladder probably has burst. In such severe cases, the pressure on internal organs is high and prompt deflation is necessary to save the fish. Returning the fish in a cage to the depth at which they were caught and gradually bringing them up has been suggested, but usually is impractical (it would take hours). Plastic decompression chambers for ornamental fish are available through diving equipment dealers. Small fish are placed in them by divers before ascending. Then at the surface, the pressure is slowly released. If decompression takes very long, water quality problems could develop. When deflation is necessary, we use a hypodermic syringe with a 20-gauge needle attached and the plunger removed. The fish is held gently but securely upside down just under the water surface. The needle is pushed through the fish's skin at a likely spot between the pectoral fins and anus just until it reaches the body cavity. The exact location is not critical as long as the needle reaches the gas without piercing internal organs. The syringe is held to prevent it from wobbling, and both fish and syringe are pushed below the water. If needle position is correct and the needle is not clogged, bubbles will issue from the open end of the syringe. Repositioning and unclogging of the needle might be necessary before deflation is sufficient. The fish should be deflated only enough so that it can stay on the bottom of the tank without swimming downward. Complete deflation is harmful and can kill the fish. When we caught black sea bass at depths of 18 m or less, they usually could recover overnight in a 1 m deep tank without deflation. If caught deeper than 18 m, manual deflation was necessary. Nassau groupers collected at depths of 18-27 m nearly always recovered if deflated within 1-3 h after capture. If they were caught near 36 m, the expanded gas nearly always inverted the stomach into the mouth and the fish rarely survived. During one collecting trip at a 36-m depth, each time a grouper was hooked, I slid down the fishing line to 10 m depth using scuba, put the fish in a mesh bag, took the hook out, and deflated the fish with a hypodermic needle as I swam to the surface. This prevented the doubling of gas volume that would have occurred (from 2 atm to 1 atm absolute pressure) in the last 10 m of ascent. These fish survived, but the procedure is inconvenient to employ routinely.

Anesthesia

Four levels of anesthesia are important to the fish culturist (McFarland, 1960; Argent, 1987):

Sedation—reactivity to visual and vibrational stimuli is reduced; opercular movement is slightly decreased (useful for transport).

Total loss of equilibrium—swimming stops; fish turns upside down; pressure on caudal fin or peduncle stimulates swimming movements or fin extension; opercular movement is slower (useful for moving fish a short distance).

Total loss of reflex—pressure on caudal fin or peduncle does not stimulate movement; opercular movement is very slow and erratic (useful for weighing, measuring, surgery).

Medullary collapse—opercular movement stops completely (fish might die).

Anesthetics used to facilitate handling of fish include: benzocaine (50 mg/L), metomidate (0.2-0.8 mg/L), 2-phenoxyethanol (0.3 mL/L), quinaldine (2-50 mg/L), tricaine methanesulfonate—also known as MS-222 or TMS (10-1000 mg/L, usually 15-150), urethane (100 mg/L), carbon dioxide, and sodium bicarbonate (to produce carbon dioxide). More information on these and other anesthetics is available in Stickney (1979) and Bromage et al. (1990). Metomidate, phenoxyethanol, and carbon dioxide irritated juvenile chinook salmon more than TMS (Iwama et al., 1989). Only carbon dioxide, sodium bicarbonate, and TMS have been approved for use on food fish in the U.S. Quinaldine and TMS are probably the most practical. Besides not being approved for food fish, quinaldine has other disadvantages. Acetone is necessary to dissolve it, and the safety margin between anesthesia and injury could be smaller than with TMS.

TMS is most commonly used (Table 12-1). In the U.S., it is approved for both non-food and food fish, but, because residues might persist, fish cannot be eaten or processed for food for at least 21 d (at >10°C) after exposure. Efficacy of TMS decreases with decreasing temperature, decreasing water hardness, and increasing fish size (Argent, 1987). Relatively low exposure to TMS can be dangerous. LC_{50} values (lethal concentrations for 50% of the fish) for rainbow trout during exposures of 15-60 min were only 1.8-2.0 times the EC_{50}

Table 12-1. Ranges of concentrations of tricaine methanesulfonate used to anesthetize or sedate fish (Argent, 1987).

Fish	Temperature (°C)	Concentration (mg/L)	Maximum exposure (min)
Rapid anesthesia (1-5 min initial exposure for brief handling)			
Salmon, trout, chars	7-17	80-135	4-12
Channel catfish	7-27	140-270	4-11
Bluegill	10-27	260-330	3-5
Moderately rapid anesthesia (10-20 min initial exposure for longer handling)			
Salmon, trout, chars	7-17	50-60	30+
Channel catfish	7-27	70	30+
Sedation (continuous exposure for transport)			
Salmon, trout, chars	7-17	15-30	360
Channel catfish	7-27	20-40	360
Bluegill		25	480-780

values (effective concentrations for 50% of the fish), and for channel catfish the ratio was 2.4-3.0 (Argent, 1987). When quick anesthesia is the goal rather than long-term sedation, the higher concentrations require the fish to be quickly processed and returned to TMS-free water. TMS can reduce pH of the water. Although unlikely in salt water, if this is suspected, the pH can be checked and the solution buffered if necessary.

Approved standard concentrations based mostly on use with freshwater fish (Schnick et al., 1986) are: 200-400 ppm carbon dioxide for 4 min (anesthetic), 142-642 mg/L sodium bicarbonate for 5 min (anesthetic), 50-330 mg/L TMS for 1-40 min (anesthetic), and 15-66 mg/L TMS for 6-48 h (sedative).

Rees and Harrell (1990) used 100 mg/L TMS to completely knock out striped bass broodfish (sometimes with the same amount of sodium bicarbonate to buffer pH in fresh water). The concentration of TMS required for deep anesthesia of red sea bream within 3-5 min increased from 50 mg/L for 220-µg hatchlings to 100 mg/L for 320-g fish at 595 dah (Oikawa et al., 1994). Roberts (1990) used 122 mg/L TMS for red drum broodstock.

European sea bass broodstock can be anesthetized with 0.3 mL/L propylene glycol (Barnabé, 1990a). Ostrowski et al. (1989) used 350 mL/L 2-phenoxyethanol for 30-45 seconds to anesthetize 25-40 cm SL dolphin. Liu and Kelley (nd.b) used a bath of 15-330 mg/L TMS or 0.2 mL/L phenoxyethanol for 1-2 min to examine striped mullet broodstock.

Ross et al. (1993) found that 1.0 mg/L metomidate was safe for American shad (100 mm TL) for at least 1 h in fresh water at 13-16°C. Falls et al. (1988) reported that 1.6 mg/L etomidate for 2-4 min was suitable for initial sedation and 0.4 mg/L was suitable for maintaining sedation for as long as 23 min in adult red drum.

Because conditions and species vary, we usually determine the necessary concentration each time TMS is used for weighing and measuring, by response of the fish. A fish is placed in a container of water deep enough so it can stay off the bottom, then TMS is added slowly until the fish turns upside down. The concentration is further adjusted for subsequent fish by addition of TMS or water so that loss of reflex occurs in 2 minutes. We have anesthetized and weighed thousands of experimental fish of several species this way, without a single loss. Anesthetic requirements should be determined for the species and size being handled and the current conditions.

The osmolality of teleost blood is about the same as one-third strength seawater (10-12 ppt). Fish raised in fresh water, such as largemouth bass, catfish, and striped bass, seem to tolerate handling and transport better if salinity is raised to 10 ppt (isotonic) with sodium chloride, probably because less energy is required to maintain a salt balance (McCraren, 1990). Even if marine fish have been raised at low salinity, they should not be handled for long unless the salinity is at least 10 ppt.

Wallin and Van Den Avyle (1995) reported that routinely tagged and released striped bass juveniles that were placed in cages in brackish water were less stressed and survived better than those in fresh water. Fish >50 mm TL that were handled more gently survived as well in fresh and brackish water. Fish <50mm survived better in brackish water, even with minimal handling. Addition of 5 g/L sodium chloride and 0.1 g/L Ca^{+2} (as calcium carbonate) to transport water for the larger fish did not improve survival after the transfer. Other euryhaline marine fish like spotted seatrout suffer less from handling if they are kept in brackish water rather than fresh water immediately afterwards (pers. obs.).

Reubush and Heath (1997) found that 5 g/L sodium chloride was suitable for recovery of handled 4-g sunshine bass and that 3 d without food before handling was too long for fish that size because handling stress was greater than for fed fish.

When ~170-g gilthead sea bream were removed from groups of nine fish in 200-L tanks, the remaining fish were stressed for up to 2-3 d (Molinero et al., 1997).

Fish should not be allowed to dessicate or asphyxiate. For brief handling (1-5 min), fish can be dipped in the anesthetic solution, then removed, rinsed in clean water, processed, and returned to clean water. For longer handling periods, a system for irrigating the gills with the correct sedative solution after the anesthetic dip will maintain immobility of the fish and enable gas exchange to keep it alive and healthy.

Handling time and frequency should be minimized. Characteristics of the fish being handled should be known, because some tolerate much more (e.g., serranids) than others (e.g., sciaenids). If necessary to handle fish several times, it might be better to do it all on one day and then leave them alone for a few days rather than spread the handling out over more than one day. If infections result from handling, they usually arise on the second day, and handling the fish then would aggravate the problem. Also, the stress reaction would be prolonged.

Plasma glucose of turbot (600-800 g) was not elevated after a single 9-min underwater confinement in a dipnet, but was elevated after two confinements within 4 h or 24 h (Waring et al., 1997). It was recommended that at least a 24-h recovery period be allowed between handling events. Waring et al. (1997) found similar physiological responses in Atlantic salmon and North Atlantic flounder after net confinement, but the effects were greater for salmon.

Prevention of injury by careful handling is important. Appropriate compounds can be used during or after handling to help reduce stress and the chance of infection (see chapter 11). Intermediate salinity (5-25 ppt, depending on species) lowers the effort required for osmoregulation and thus reduces stress.

If fish to be artificially-spawned are anesthetized, they should be washed well in clean water before being stripped because the anesthetic could affect sperm motility.

GRADING MARINE FISH

Even under good conditions, individual aggressiveness and growth rate can vary. This becomes a problem when a significant number of fish are so intimidated from eating that their growth is further reduced. Stunting, disease, or cannibalism could result. Grading, or separating the fish into size classes (usually 2 or 3, sometimes 4 or 5), is necessary for some species but not others (e.g., turbot, Sunde et al., 1998). It can be done by hand-picking fish, by manually or mechanically pushing a grader (Fig. 12-1) through the tank, or by allowing them to swim through a grader. Grading can range from voluntary to forced. The less the force, the lower the stress and chance of injury. Crowding increases the speed at which fish will swim through a grader. Color or contrast of the grader bars can influence voluntary passage. See Bromage et al. (1988) and Fjæra and Skogesal (1993).

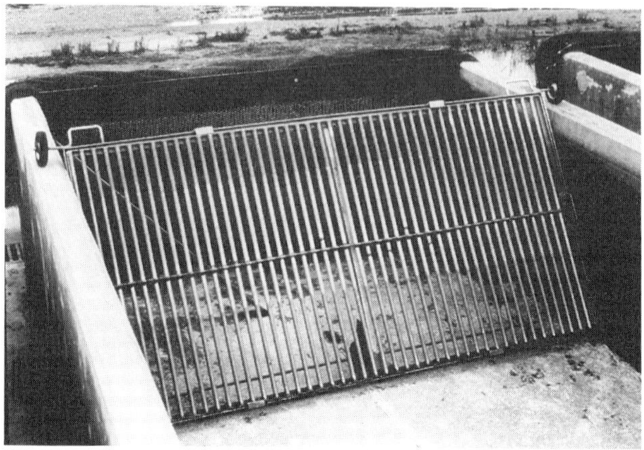

Figure 12-1. A bar-grader for separating fish into different size groups in a rectangular raceway. Small fish swim through, but large fish cannot. (photo by Robert A. Winfree)

TRANSPORTING MARINE FISH

Major Methods

Pipeline. Pumping fish through pipelines tends to be stressful. Alternatively, they can be induced to swim through pipes or canals. Tanks used for different stages can be connected temporarily or for longer by pipelines between hatches built into the tank walls. To minimize stress, the first fish can be induced by a current and the slow ones crowded through. Lekang and Fjæra (1993) found that fish tend to swim against a current of less than 3-4 fish lengths per second but to go with a faster current. Atlantic salmon moved quickest through large hatches in the middle of the wall with a fairly strong current coming out (Lekang et al., 1996). L-leucine also was used to attract fish from one tank to another.

Damp (no water). This is more often done with molluscs and crustaceans, but in some cases fish gametes, juveniles, or adults are successfully transported in chilled, humid chambers. Salmonid milt or eggs in ovarian fluid can be held and transported under oxygen in polyethylene bags for up to 12-24 h (Bromage, 1988; Billard, 1990). Japanese eels (10 kg) sent to market packed in polyethylene bags containing oxygen (about 30 L) and ice can survive for 30 h (Usui, 1991).

Tank (water aerated by agitators, blowers, or compressed air, or oxygenated with bottled oxygen). These usually are carried on a truck or trailer. For moderate-sized operations, a trailer-mounted tank that can be towed by a pickup truck is a more flexible arrangement than a tank permanently mounted on a truck. Small numbers of fish can be carried in tanks or ice chests in the back of a pickup truck or van. Large numbers might require a custom-made tank truck (Fig. 12-2).

From a survey of fish transportation methods in the U.S., Carmichael and Tomasso (1988) reported that 68% of fish-hauling tanks were commercially-manufactured, and 70% were mounted on trucks and 30% on trailers. Most vehicles had only one or two tanks, but most tanks were compartmentalized. Volumes ranged from 60 gal (227 L) to 5200 gal (19,682 L). Most tanks were rectangular or square, but some were oval or round. Tanks mostly were made of fiberglass, aluminum, wood and fiberglass, or stainless steel (marine fish culturists mostly use fiberglass or plastic tanks, but corrosion-resistant steel or aluminum might be better for large tanks). Most tanks did not have a false bottom, but these can be used to facilitate removal of fish. Most were insulated (74%); urethane foam was the most popular insulating material. Most were not refrigerated, but ice often was used for cooling

Figure 12-2. Government fish-hauling truck in Jamaica.

the water. About half of the respondents ventilated the air space of their tanks to help remove carbon dioxide. Oxygenation was used by 83%, agitation by 75%, and both by 58%. Oxygen was delivered mainly by carbon stones or microporous tubing. Circulating pumps were used by 15% and compressed air by only 2%.

Internal baffles designed to reduce shifting and sloshing of water in large tanks will keep the fish calmer and make driving easier. It is safer to operate pumps and agitators off an auxiliary battery charged by the truck's alternator instead of the truck's primary battery. With salt water, circulating pumps might be a better choice than agitators, because the latter tend to corrode. Back-up systems should be available. During long hauls, water replacement or filtration might be necessary to remove wastes.

Yeager et al. (1990) reported that round tanks are less stressful for transporting striped bass broodstock than tanks with corners and that circulating the water helps fish to maintain balance and orientation. They recommended that tank diameter or width should be at least twice the length of the longest fish and depth 0.7-1.0 m. For short-term transport (≤8 h), less sensitive fish like snook, barramundi, and groupers can be held in containers slightly longer than the fish as long as they have room to breathe (pers. obs.).

Although tanks are better for moving fish, broodfish can be held in cages if they are made of soft knotless material and the fish do not have to struggle against current flowing through the cage. The cages can be towed if water flow through them is very restricted at that time. Strong currents can batter recently caught, disoriented fish against the side of a large-meshed cage, leading to injury and infection. Unless it has a complete frame, a small-meshed cage can be collapsed by strong current, trapping the fish in its folds. We have held Nassau grouper broodfish for several days in a salt pond in cages made of nylon seine bags tied inside floating PVC pipe frames (122- and 183-cm cubes). The fish were handled frequently but remained in good condition until they were stripped and released.

Bag (usually with oxygen, sometimes with air). Food-grade polyethylene bags (Fig. 12-3) inside boxes, or boxes alone (Fig. 12-4) are used for shipping fish as cargo and sometimes for moving valuable smaller fish by road. Bags must be at least 0.1 mm thick (4 mil = 0.004 inch), and two or three should be used together to retard water and oxygen loss if leaks occur. Large spiny fish might require extra tough bags. Besides the fish, properly packed bags will contain 20-40% water and 60-80% oxygen. Enough water should be included to at least cover the fish when they are resting upright on the bottom. Water is added first, then the fish, then all air is squeezed from the bag. Oxygen is introduced with airline tubing. With larvae and early (i.e., fragile) juveniles, oxygen is added above the water; with eggs, robust juveniles, or adults, oxygen can be bubbled into the water with an oxygen stone. If oxygen is not available, a lower density of eggs or fish can be shipped under air, but 100% oxygen is far better than 21%. The tops of the bags are tied off with strong rubber bands and the bags are packed in a styrofoam box (or equivalent, e.g., ice chest) in such a way that they will remain upright and not shift. One standard size shipping box has inside measurements of 37 cm × 37 cm × 18 cm high and has a total volume of about 25 L. Weaker styrofoam boxes should be put inside tough cardboard boxes to protect them. Sturdy styrofoam boxes that normally do not need the extra protection are available. A range of sizes and thicknesses is available, and even thicker boxes might be useful (Froese, 1998). With the hope that cargo handlers will be careful, the boxes should be labeled: "LIVE FISH" "KEEP AT ROOM TEMPERATURE" "THIS SIDE UP." Box contents should be clearly identified, especially if they have to go through customs inspection.

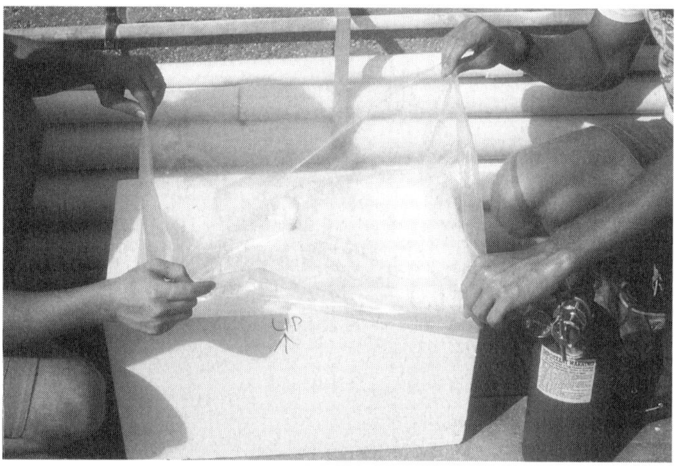

Figure 12-3. Styrofoam box containing a reared 900-g Nassau grouper in triple bags, nearly ready to be driven to the Miami airport and shipped via commercial airline to St. Thomas to be released. The cylinder contains oxygen.

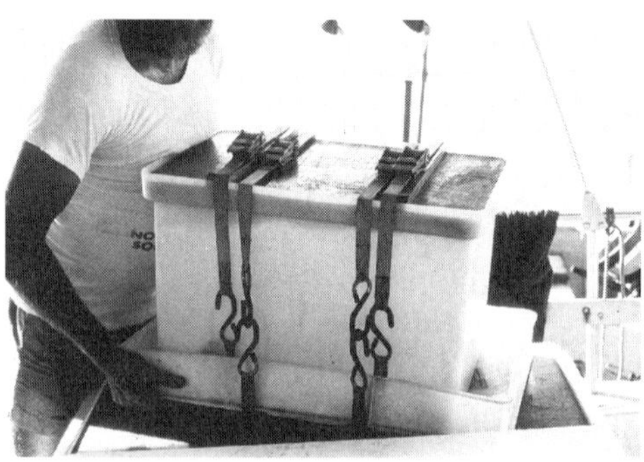

Figure 12-4. Polyethylene battery box containing a wild 50-cm TL male barramundi to be shipped by charter plane from Weipa to the government hatchery in Cairns, Australia.

Precautions

To avoid interruption of the shipment, the shipper should notify everyone who might control it and solve potential problems in advance. He should contact customs offices at both ends and get any required permits. He should contact fisheries offices at both ends and get any required permits. Except for routine shipments, airline cargo offices at both ends should be contacted to make sure they expect the shipment. The shipment should reach the outgoing office on time.

Delays must be avoided. Possible traffic or flight delays should be anticipated and avoided by choosing the most efficient route and time. Boxes should not sit still for long; gas exchange will be retarded. As much as possible, boxes should not be handled roughly.

Collins (1990a) recommended not feeding market-sized fish for 3 d before transport in warm water and 5 d in cold water. Small juveniles might need only 1-2 d of fasting. This allows the digestive tract to empty, thus reducing ammonia and carbon dioxide production, and it also lowers metabolism of the fish, thus reducing oxygen demand. The appropriate fasting time will depend on species, size, and temperature.

During transport, the combination of high carbon dioxide (Table 3-10) and ammonia (Table 3-6) in the water and high lactic acid in the blood decreases the ability of fish to obtain and use oxygen (McCraren, 1990). Dissolved oxygen should be maintained at 5-7 mg/L or higher (very important). More is better because it helps fish to tolerate the increased ammonia and carbon dioxide. Carbon dioxide usually is not a problem, but sometimes it exceeds 30 mg/L, which could be harmful.

Ammonia will build up and cannot be easily removed from salt water. Increasing pH could raise the concentration of unionized ammonia; it should remain in a normal range (8.0-8.4 for salt water). Zeolite can be used for ammonia removal in fresh water but does not work in salt water. Over a 40-h period, 10 g/L clinoptilolite reduced ammonia and 0.04 M Tris-HCl reduced carbon dioxide and helped maintain pH during simulated shipping of mollies in fresh water, thus increasing survival (Kwan et al., 1994).

The fish shipper should anticipate differences between source water and receiving water and avoid sudden large changes in temperature, salinity, and pH (also chloride content and hardness in fresh water). Young freshwater fish usually tolerate a sudden 1°C change in water temperature, and older ones 5°C, with negligible stress (Collins, 1990a). Tolerances of marine fish are similar, but vary among species. Some stenothermal species might be stressed by a 5°C change, and any fish will suffer if the change takes it outside its normal temperature range. Salinity tolerance depends on the species. Many euryhaline marine fish such as common snook and red drum can be moved from 0 to 35 ppt in one step. Going the other direction (35 to 0 ppt) probably should be done more slowly (e.g., overnight). Red drum at 50 mm have been moved directly from 25-28 ppt to fresh water, but to be safe the receiving water should contain at least 125-150 ppm chloride, and a transition period of at least 2-4 h would be even safer (McCraren, 1990). Salinity changes should be considered carefully when stenohaline species are moved.

Because stressed fish are more susceptible to pathological bacteria and protozoans, transport water should be as free of particles and extraneous organisms as possible. Transport tanks and equipment should be disinfected between use (e.g., with 200 mg/L chlorine solution for 1 h). The fish should be handled gently. Physical injury is stressful and can lead to osmotic problems and infections.

Preliminary trials with small numbers of fish could help to anticipate and solve problems before committing large numbers to untested conditions. Maximum stocking density for shipping varies mostly with species, size, temperature, time, and water quality (Tables 12-2, 12-3).

Table 12-2. Examples of stocking densities successfully used for transporting fish in tanks with oxygenation or aeration. In some cases (e.g., low water quality) densities might have to be reduced and in others (e.g., high water quality) they could be increased.

Fish	Stage	Size	Salinity (ppt)	Temperature (°C)	Number/L	Weight/L	Time (h)	References
Channel catfish	Larva	45 mg	0	19	530	24 g	12	McCraren and Millard, 1978
European sea bass	Larva	200 mg	28	17	35	7 g	24	Barnabé, 1990d
European eel	Elver	300 mg	17	8	200	59 g	15	Usui, 1991
Cyprinid bait fish	Adult	25 mm	0	16		119 g	12+	Parker, 1988
European sea bass	Juvenile	4 g	28	17	6.25	25 g	24	Barnabé, 1990d
Red drum	Juvenile	4 g	12	21	7.5	30 g	8+	McCraren, 1990
European eel	Juvenile	200 g	0		5	1000 g	96	Usui, 1991
Channel catfish	Juvenile	454 g	0	19	1.5	681 g	12	McCraren and Millard, 1978
Rainbow trout	Juvenile	120 mm	0	10		56 g	12	Bromage et al., 1988
Milkfish	Adult	2.2 kg	8		0.006	14 g	28	Kelley and Lee, 1986
Striped bass	Adult		10	18		194 g	3+	Yeager et al., 1990
Common snook	Adult	4.5 kg	34	28	0.03	130 g	5	Tucker, pers. obs.
Nassau grouper	Adult[n]	8.6 kg	35	26	0.01	86 g	8	Tucker, pers. obs.
White sturgeon	Adult	30 kg	0	16	0.005	150 g	48	Conte et al., 1988

[n] no aeration, only occasional sloshing.

Table 12-3. Examples of stocking densities successfully used for shipping fish in plastic bags (generally containing 25-40% water and 60-75% oxygen). In some cases (e.g., low water quality) densities might have to be reduced and in others (e.g., high water quality) they could be increased.

Fish	Stage	Size	Salinity (ppt)	Temperature (°C)	Number or weight/L	Time (h)	References
Marine fish	Egg	1 mm	34	27	4,000	12	Tucker, pers. obs.
Grouper	Egg	1 mm			16,000	3	Maneewong et al., 1986
Milkfish	Egg	1.3 mm	34	28	6,000	9	Hilomen-Garcia, 1998
Grouper	Larva	2 mm	33	27	16,000	2+	Quinitio et al., 1991
Turbot	Larva	3 mm		14	4,000	36	Dhert et al., 1998
Milkfish	Larva	3-4 mm	10	28	450	3	Liao, 1991
Striped bass	Larva	3-4 mm	0	18	10,000	8	Rees & Harrell, 1990
Striped bass	Larva	3-4 mm	0	18	5,000	48	Rees & Harrell, 1990
Cyprinid bait fish	Adult	25 mm	0	16	417 g	24	Parker, 1988
Sheepshead	Juvenile	25 mm	34	27	10	12	Tucker, pers. obs.
Milkfish	Juvenile	40 mm	10	28	150	3	Liao, 1991
Milkfish	Juvenile	125 mm	10	28	15	3	Liao, 1991
Tilapia	Juvenile	10 g	~40	20	270 g	24	Hazza et al. 1991
Grouper, barramundi	Juvenile	varies	varies	24	120 g	12	Chong and Chao, 1986a, 1986b
Nassau grouper	Adult	900 g	32	23	90 g	12	Tucker, pers. obs.

Four possible methods to calm fish for transport are:

(1) Anesthetize the fish, then ship in plain water (for short trips).
(2) Sedate the fish for a few hours, then ship in plain water (for short trips).
(3) Pack fish in sedative solution and ship (for longer trips).
(4) Sedate the fish for hours, then ship in water with sedative (for longer trips or excitable fish).

For sedating striped bass transported at 13-18°C, Yeager et al. (1990) suggested using concentrations of 5-8 mg/L TMS for fish smaller than 76 cm and 3-5 mg/L TMS for larger fish. They warned that residual effects might interfere with the spawning cycle. Collins (1990) reported that for transporting freshwater fish the usual anesthetic ranges were 15-40 mg/L TMS or 15-30 mg/L quinaldine (these could be too high for some marine species). For sedation of juvenile red drum during transport, 5-20 mg/L TMS or 0.25-2.0 mg/L quinaldine have been used, with TMS preferred (McCraren, 1990).

Wedemeyer and Wood (1974) recommended that coho salmon and steelhead trout (from fresh water) be handled and transported in water containing a minimum of 1-3 ppt sodium chloride, with calcium chloride added to raise hardness to at least 50 mg/L $CaCO_3$. Parker (1984) reported that adding 10 mg/L sodium chloride to the 25 mg/L TMS solution for transporting striped bass (from fresh water) increased posthauling survival from 75 to 100%.

McCraren (1990) reported that in the U.S. either 1-2 mg/L acriflavine, 5-15 mg/L nitrofurazone, or Terramycin were used for transport of fish, but the legal use of these substances is highly restricted. For groupers, the Singapore Primary Production Department (Anonymous, 1986) suggested a 10 mg/L acriflavine bath for 30 min before shipping. Addition of 10 mg/L nitrofurazone to the shipping water also was suggested. A 100 mg/L formalin dip before the fish are packed will eliminate some external parasites.

Temperature can be lowered 3-4°C below ambient to reduce metabolism (i.e., increased generation of ammonia and carbon dioxide, demand for oxygen). Sometimes ice is placed in the tank or shipping box outside the bag to cool the water. A safer and more accurate approach is to separately chill both the fish and the transport water before putting them in the tank or bag. Anti-foaming agents could be used to keep the surface clear and facilitate gas exchange. During long trips, fish might have to be repacked in clean water at an intermediate location.

Before fish are released at their destination, temperature (no faster than 1°C/10 min) and salinity in tanks or bags should be adjusted to conform with the receiving water to avoid shocking fish (known as tempering). If possible, this should be done during the last part of the trip, so that release of the fish will not be delayed. If possible, the fish should be separated from the transport water so that only fish are added to the rearing unit. Better still, to reduce the chance of introducing parasites or other pathogens, the fish could be held separately and treated before going into the rearing units (e.g., freshwater bath). If conditions in the bags have become critical and keeping fish in the bags any longer will jeopardize their health, they should be released immediately. Temporary holding in a smaller tank might be needed (for recovery of fish and observation for potential problems) before fish go into a large rearing unit. They should be released as gently as possible, even though some fish have been dropped from airplanes.

The effects of overexertion and fatigue might last for a day or more after release. Although the fish might seem normal and feed well the day after transport, biochemical and behavioral fluctuations can occur for several days, and disease can occur up to 1-2 weeks later. McCraren (1990) gives an interesting review of experience in fish transport, with emphasis on red drum.

SUMMARY

Fish being moved should be handled as gently as possible, and precautions should be taken to minimize stress and the chance of infections.

13 CULTURE OF ESTABLISHED AND POTENTIAL SPECIES—FOOD FISH

STATUS

Most marine fish culture is still experimental, but in the last decade or so, many new marine and diadromous food fish species have reached the pilot or commercial stage. Established and new species include: sturgeons, anguillid eels, milkfish, salmonids, barramundi, European sea bass, hybrid striped bass, a few groupers, yellowtail, dolphin, red sea bream, black sea bream, gilthead sea bream, red drum, tilapias, mullets, rabbitfishes, turbot, Atlantic halibut, and tiger puffer.

FAMILY ACCOUNTS

Starry Sturgeon
Acipenser stellatus

Acipenseridae—Sturgeons

Sturgeons can be anadromous, semi-anadromous, or freshwater, but all spawn in fresh water (Conte et al., 1988; Williot, 1990, 1991; Hung, 1991; Struffenegger, 1992; Dettlaff et al., 1993; Fitzgerald, 1995; Logan et al., 1995; Arefyev, 1997). In the U.S., the **green** (Pacific coast) and **Atlantic sturgeons** are the best anadromous examples. Since the early 1960s, sturgeon have been raised in the former Soviet Union. The **bester** (hybrid of beluga × sterlet) can grow to ~3 kg within 3 yr. Usually, eggs are obtained from captive broodfish induced to ovulate or ripe wild fish. Sturgeon have been grown out in ponds, cages, and tanks. The best temperature range for white, lake, and Atlantic sturgeon is 14-17°C.

White sturgeon live mainly in estuaries of large rivers, and those that are ready swim upstream once/yr to areas of fast flow to spawn large adhesive eggs (Conte et al., 1988). Typical female broodfish weigh 36 kg and yield 5000 eggs/kg BW, with hatching rate of 40% (Logan et al., 1995; Doroshov et al., 1997). After they hatch in jars, the young fish are held in a series of increasingly larger tanks as they grow. They are grown out mainly in tanks, raceways, or ponds, with a suitable density of 40-60 kg/m^3. White sturgeon can grow to 7 kg in 36 months (Fitzgerald, 1995) and 63 kg in 11.5 yr, but typical sizes are 1.4 kg in 18.5 months, 4.5 kg in 36 months, and 32 kg in 10.5 yr (Logan et al., 1995). Maximum size is more than 100 kg. White sturgeon can economically be harvested for meat at 18.5 months, but caviar production could require raising them as long as 10 yr.

Elopidae—Tarpons

Tarpon is a very bony fish, but it traditionally has been consumed in some areas, such as northern South America. In northeastern Colombia, wild juvenile tarpon have been grown to market size in earthen ponds on trash fish. This apparently began as an offshoot of the rapidly growing penaeid shrimp farming industry.

Anguillidae—"Freshwater" Eels

Freshwater eels are raised from wild glass eels and elvers caught as they migrate from the sea to freshwater areas, where they remain until maturity, then migrate to the sea to spawn. The difficulty of spawning eels (chapter 7) and the long larval period (chapters 2 and 8) make sustained hatchery production a long-term goal. Larvae were raised from 2.9 mm TL at hatching to 19 mm at 25 dah by Yu et al. (1991). Different rearing systems for juveniles

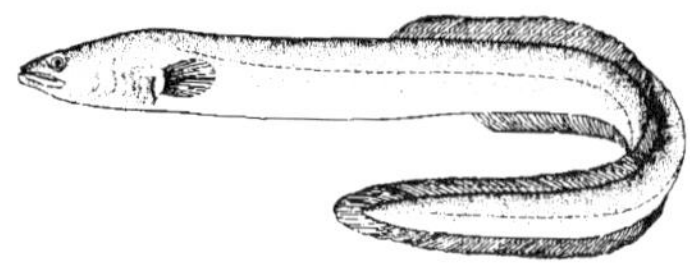

American Eel
Anguilla rostrata

have been used, split mainly into Japanese and European types (Chen, 1976; Brusle, 1990; Gouset, 1990; Arai, 1991; Heinsbroek, 1991; Liu et al., 1991; Usui, 1991; Ikenoue and Kafuku, 1992; Yamaha Fishery Journal, 1992a; Anonymous, 1998a). Typically in Japan, a series of nursery ponds enclosed in greenhouses is used for the early stages, and the final stages are in either greenhouse ponds or outdoor ponds (Heinsbroek, 1991). Tubifex worms, ground beef liver, fish or clams, paste feeds, and dry feeds have been used for starting, and paste or pelleted feeds for growing eels. Initial stocking is during December-April. Good nursery survival is 70-90%, and grow-out survival is 90-95%. In 6 months, 10-20% of the eels have grown to 180-200 g and can be sold; the rest will reach this size in 12-18 months. Harvest weight can be 600-900 times the initial stocking weight. Maximum weight of females (larger) is 1 kg. Early eel culture in Europe was by vallicoltura, blocking of coastal lagoons to trap wild larvae of various species. Recently in Europe, eels have been raised in tanks, ponds, and raceways as large as 500 m^2. Better temperature control is needed more in the north. Cod roe, cow spleen, paste feeds, and dry feeds are used. Both Japanese and European eels grow well at 25-28°C.

Clupeidae—Herrings
Pacific herring culture is well established in Japan (Furuya, 1995).

Atlantic menhaden were induced to ovulate, and larvae were reared at ~15-20°C in a 670-L tank on rotifers, wild zooplankton, *Artemia*, and trout feed in green water, with survival of ~20% from eggs to 30 dah (Table 8-4; Hettler, 1981). Dark tank walls were considered best and rotifers not essential.

Gulf menhaden and **yellowfin menhaden** were spawned and raised with similar methods (Hettler, 1968, 1970; Hettler and Powell, 1981).

Atlantic herring eggs were stripped from ripe wild fish, and larvae stocked at 120/m^3 were raised at 11-12°C in 310-m^3 plastic bags on wild zooplankton (Gamble et al., 1981; Blaxter and Hunter, 1982). Survival was 18% at 1 dah and 1.6% at 50 dah (9.0% for 2-50 dah). The high mortality of hatchlings probably resulted from the combination of their high buoyancy and water turbulence. Atlantic herring larvae from stripped eggs were stocked 7 dah at 2.3 larvae/m^3 in a 4400-m^3 concrete basin and raised at 7-11°C on wild zooplankton (Øiestad and Moksness, 1981). Survival was ~70% at 130 dah.

Scaled sardine larvae from wild eggs were raised at 25-29°C in 75-L aquariums on wild zooplankton and *Artemia* through transformation, then compound feed was added for juveniles (Houde and Palko, 1970). They grew to 76 mm TL at 100 dah, with survival of ~3% from eggs.

Japanese sardine larvae from wild eggs were raised at ~20°C in a 700-L tank on rotifers and *Artemia* in green water, then fish eggs and larvae and compound feed (Table 8-4; Nakamura et al., 1991). Survival was 7.7% from eggs to 80 dah (87 mm TL, 4.5 g).

Engraulidae—Anchovies
Northern anchovy hatchlings are 3.0 TL; larvae were raised at ~16°C in 400-L tanks on dinoflagellates, rotifers, copepods, and *Artemia* in green water, with survival of 12% from eggs to 74 dah (Table 8-4; Hunter, 1976)(Scura and Jerde, 1977; Blaxter and Hunter, 1982; Hunter and Coyne, 1982).

Bay anchovies, which have smaller eggs and hatchlings (2.0 mm), were raised on wild zooplankton in green water (Houde and Schekter, 1983), and on rotifers and *Artemia* in green water (Tucker, 1989).

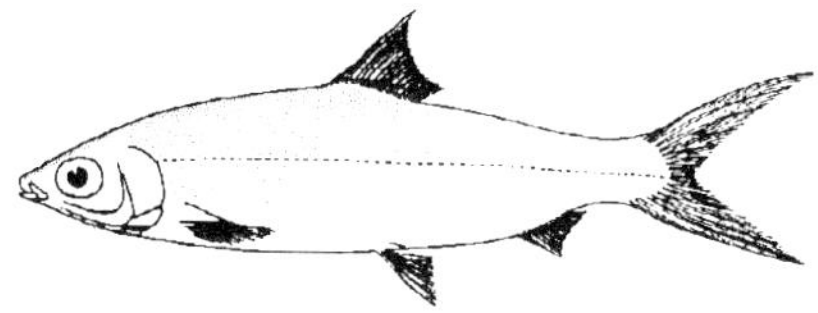

Milkfish
Chanos chanos

Chanidae—Milkfish

Milkfish are very popular in Asia but are hard to catch. They are fairly easy to spawn and raise (Chen, 1976; Liao et al., 1979; Liao, 1991, 1993; Hibler, 1983; Lee and Liao, 1985; Lin, 1985; Taki et al., 1987; Lee et al., 1986a, 1986b, 1988a; Eda et al., 1990b; Chen and Liao, 1991; Kelley and Lee, 1991; Lim, 1991a; Chang et al., 1993; Tamaru et al., 1993b, 1995; Emata, 1994; Kohno et al., 1994c; Liu and Kelley, nd.a). In Asia, milkfish have been raised for at least 700 yr from wild juveniles, but in the last several years, much progress has been made in spawning and larval rearing. The natural spawning season is long (mainly April-September in Taiwan). Wild or captive ripe fish can be induced to ovulate. They also will spawn voluntarily in ponds or large tanks, but broodfish do not tolerate handling well. Larvae are raised in ponds on mixed zooplankton or in tanks on rotifers and *Artemia* (sometimes trochophores and/or copepods, too) in green water (Table 8-4). Hatchery production can be as high as 10/L or more (to 21-30 dah). Total lengths of 60 mm at 30 dah and 150 mm at 70 dah are possible. Juveniles usually are grown out in ponds or outdoor tanks, where they feed on benthic algal mats, but cages and pens have been used, too. Supplemental feeds also are used, and nutritional requirements are not high. Typical stocking densities are 2000-2500/ha at 10-100 g for shallow ponds (1-2 m) and 4-5 times more for deep ponds (2-3 m). Juveniles have wide environmental tolerances (15-40°C, 0-70[158] ppt, low oxygen) and are resistant to diseases. Bait size (80-120 g) can be reached in 4-5 months and food size (200-500 g) in 8-12 months. Maximum weight is ~18 kg. Two or three crops/yr can be produced from the same set of ponds, with annual production of 2000 kg/ha. Milkfish grow well at 26-30°C. Problems include the high costs of maintaining broodstock and spawning oysters, plus high pond temperatures and heavy rains during the hatchery season. Projected production costs for 30-d-old juveniles (without egg cost) were US$0.0067/fish in ponds (2 million/month) and US$0.027/fish in tanks (1 million/month)(Lee et al., 1997).

Ariidae—Sea Catfishes

In Latin America, juveniles of several catfish species are abundant in fresh and brackish water (Tucker and Jory, 1991). They are very resistant to handling and confinement and will mature in captivity. In Colombia, experimental grow-out of wild catfish on trash fish in earthen ponds has produced good results. **New Granada sea catfish** stocked at 7 g reached 186 g in 13 months; stocked at 14 g, they reached 308 g in 16 months. **Crucifix sea catfish** stocked at 9 g reached 455 g in 13 months; stocked at 18 g, they reached 773 g in 16 months. In Venezuela, wild **Pemecou sea catfish** stocked at 4 g in concrete tanks reached 250 g in 19 months. The crucifix catfish grows better and might have more technical potential than the other two, but pressure for aquaculture development is low because the wild stocks are large.

Osmeridae—Smelts

Ayu live for only 1 yr, spawn, and die. They have been induced to ovulate and have spawned voluntarily in captivity (Peng et al., 1991). Natural production can be improved by maintaining a pebble bottom on the spawning grounds (Doumenge, 1990a; Ikenoue and Kafuku, 1992; Yamaha Fishery Journal, 1992b). From Lake Biwa (Japan), adult ayu swim up two artificial rivers to spawn adhesive eggs attached to gravel. Larvae drift back to the lake and grow. At lengths of 20-90 cm, they are harvested and transported to grow-out facilities. The 7 billion juveniles produced in 1992 represented about 70% of the total seedstock for Japan that year. Some juveniles also were raised in hatchery ponds or 30- to 60-m^3 tanks, and some were caught in other lakes or bays. Ayu reared on rotifers, *Artemia*, and an MPD had a survival of 88% from hatching to 50 dah; MPDs can be given to them as early as 7 dah (Table 8-4; Kyowa, nd.b). They are grown out in outdoor and indoor (e.g., greenhouse-covered) concrete ponds with good water flow and exchange (50-100%/h). From 100-300 juveniles/m^2 stocked, 10-30 kg/m^2 can be harvested. Ayu reach at least 75-90 g in 10-11 months, with a maximum size of 200 g. They can be farmed in the range 13-30°C, but growth is best at 20-25°C. Artificially prolonging a summer daylength, starting by August, has been used to keep ayu alive and light-colored for a few more months.

Galaxiidae-Troutminnows
Tasmanian whitebaits spawned voluntarily, and larvae were raised at ~12-14°C in 4-m^3 tanks on rotifers, *Artemia*, and salmon starter crumbles, with survival of ~30% from hatching to 37 dah (Table 8-4; Daintith, 1992).

Salmonidae—Salmons, Trouts
This family contains anadromous and freshwater species. Culture technology for several is well developed (Beacham and Murray, 1990; Billard, 1990; Laird and Needham, 1988a, 1990; Hardy, 1991; Stickney, 1991a; Ikenoue and Kafuku, 1992; Forster, 1995). A few examples are briefly discussed here. Coho and chinook salmon are the main farmed species in the Pacific, and Atlantic salmon in the Atlantic. Both coho and chinook salmon can be raised to 250-400 g 18 months after spawning, and put on weight faster after that (Needham, 1988b). Arctic char also seems to have good potential (Tabachek and March, 1991; Jobling et al., 1998).

Coho salmon spawn mostly during September-January in North America. At 9°C, eggs take ~48 d to hatch (Stickney, 1991c). The young usually become smolts and migrate to the sea during their second year, but some wild fish and one strain reared at higher temperatures (9-13°C) can smoltify at 6 months. Smolts stocked at 40-80 g can grow to 1.5 kg in 8 months (Stickney, 1991b). Stocked at 150-200 g, they can grow to 3 kg in 9 months. Typical cage harvest density is 15 kg/m^3 and size 2-5 kg.

Chinook salmon spawn at different times of the year, depending on the stock (Laird and Needham, 1988b; Fedorenko and Cross, 1991; Stickney, 1991c). At 10°C, eggs take ~90 d to hatch. They remain in freshwater for as little as 4 months or as much as 2 yr after hatching. Most stay at sea for 1-5 yr; they can reach 9-10 kg in 3 yr and 15-18 kg in 5 yr. Maximum weight is >57 kg.

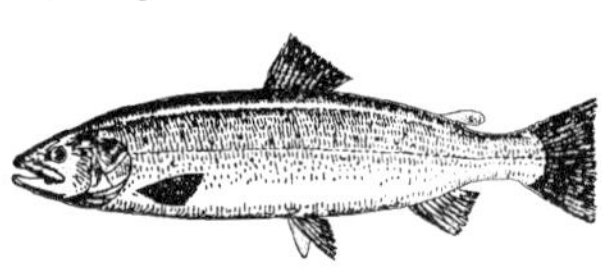

Atlantic Salmon
Salmo salar

Atlantic salmon are farmed in Europe, Iceland, and North America (Kane, 1988; Cook, 1990; Laird and Needham, 1988a, 1990; Michelson, 1991; Jungalwalla, 1991; Poxton, 1991; Stickney, 1991c; Redmayne, 1992; Blakstad, 1993; Forsberg et al., 1993a, 1993b; Hovgaard, 1993; Kjartansson, 1993; Duston and Saunders, 1994; Gueguen and Prouzet, 1994; Holm et al., 1995; Torrissen et al., 1995). In northern Europe, Atlantic salmon spawn mainly during September-November (Isaksson, 1991). At 8°C, eggs take ~62 d to hatch. Typically, a year later (August to March), 30-g smolts are ready to move from fresh to salt water. Then after a year at sea, they can grow to 3 kg, but in culture, 2 kg is more usual. However, 2 yr of grow-out can produce 6- to 8-kg fish (~3 yr old). So, grow-out usually is for 1.5-2 yr. Maximum weight is ~27 kg (122 cm). In Norway, where most salmon have been raised, the industry has been shaken by the combined problems of increasing production, decreasing prices, self-pollution, weather damage, diseases, and toxic algae (Rudi and Dragsund, 1993). To avoid pollution, *Aeromonas*, *Vibrio*, and sea lice epidemics, a shift from cage to tank farming is occurring (Tvinnereim, 1993).

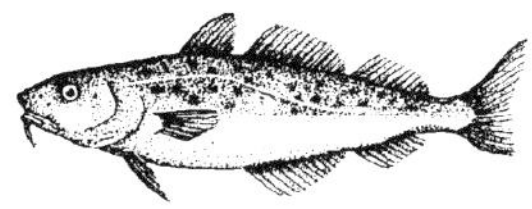

Atlantic Cod
Gadus morhua

Gadidae—Cods

Atlantic cod eggs were first released in Norway for stock enhancement in about 1884 (Holm et al., 1991). Larvae were first reared in 1886, and many eggs and larvae subsequently were released. Cod spawn voluntarily in tanks or cages, possibly 15-20 times during the natural season, which lasts about 2.5 months in February-April in Norway (Kjørsvik and Holmefjord, 1995). A large amount of work has been done on quality and requirements of eggs and larvae (Laurence, 1978; Ellertsen et al., 1981; Thompson and Riley, 1981; Solberg and Tilseth, 1984; Huse, 1991, 1994; Skiftesvik, 1992, 1994; Buckley et al., 1993; van der Meeren and Næss, 1993; Waller et al., 1993; Thorisson, 1994; van der Meeren et al., 1994; Vallin and Nissling, 1994; Baskerville-Bridges and Kling, 1997). Around 1980, concerted work was begun in Norway to develop modern culture methods (Holm et al., 1991). Eggs usually come from voluntary spawning in tanks or bags. At 6°C, they hatch in ~14 d. The larval stage is long, with transformation occurring by ~106 dah (60 mm TL). Larvae often are raised in plastic bag enclosures and impoundments on wild zooplankton but also in circular tanks on wild zooplankton and rotifers, usually in green water (Table 8-4; Holm et al., 1991; van der Meeren, 1991; Tilseth et al., 1992; Larsson and Pickova, 1993; Rosenlund et al., 1993; van der Meeren et al., 1994). Eggs and yolksac larvae will live at 4-10°C but do best at 6-7°C; older larvae and juveniles can tolerate and grow well up to 15°C. *Artemia* sometimes are used, mainly as a supplement. Cod also have been raised on rotifers and *Artemia* without other zooplankton (Rosenlund et al., 1993). Zooplankton population maintenance and weaning sometimes are difficult in bags. Cannibalism among mid-stage larvae to early juveniles can be serious and limits the possible rearing density, especially for bags and impoundments. Survival of larvae reared in a plastic bag on wild zooplankton was 83% from hatching to 48 dah, but the fish were small (~180 mg) compared to warmwater species (van der Meeren et al., 1994). Juveniles sometimes are vaccinated against vibriosis. In nature, they can reach 4-5 kg in 3-4 yr, and a maximum size of ~120 cm in 15 yr. In culture, cod have grown to 2 kg in 20 months, at which time they mature and slow down, reaching only 3 kg in 36 months. With improved methods, 3 kg at 2 yr might be possible. Moksness and Støle (1997) reported a production cost (1994-1995) of US$0.54 for a 50-mm juvenile.

Haddock larvae were raised at ~7°C in aquariums and tanks (Laurence, 1974; Waterman, 1995; Downing and Litvak, 1997). Because they are sensitive to net abrasion, broodfish are best caught by hook and line. Rotifers, *Artemia*, and wild zooplankton have been used as larval foods. **Atlantic tomcod** also have been raised at 3-10°C in 76-L aquariums (Watson, 1987).

Exocoetidae—Flyingfishes

Pointed flyingfish larvae were raised at 27-30°C on *Artemia* to 16 mm TL at 18 dah (Chen, 1978).

Atherinidae—Silversides

The atherinid larvae that have been raised are robust and survive well.

Jacksmelt were raised from collected wild eggs in 20-L aquariums on *Artemia*, with 91% survival from hatching to 24 dah (Table 8-4; Middaugh et al., 1990). **Topsmelt** eggs were obtained from ripe wild females and larvae were raised at 20-22°C in a 3-L beaker on *Artemia*; all survived from hatching to 24 dah (Table 8-4). Temperature-controlled tank-spawning also was accomplished with topsmelt (Middaugh et al., 1992).

Inland silversides spawned voluntarily in captivity, and larvae were raised at 24-26°C in a 37-L aquarium on rotifers and *Artemia* (Ashraf et al., 1993b).

Atlantic silverside eggs were obtained from ripe wild females, and larvae were reared at 17-23°C in small (0.4-0.9 m diam.) tanks on *Artemia* (Barkman et al., 1981). Reared fish grew more slowly than wild ones (~30 vs ~80 mm TL at 80 dah).

California grunion eggs were obtained from ripe wild females, and larvae were raised at 18-19°C in 20-L aquariums on *Artemia* in green water, with survival of 81% from hatching to 40 dah (Table 8-4; Vásquez-Yeomans and Sosa-Cordero, 1992).

Japanese silversides spawned voluntarily in a tank, and larvae were raised at ~20-22°C in a 500-L tank on rotifers, *Artemia*, wild zooplankton, and compound feeds in green water up to 26 mm TL (Kimura and Tsukamoto, 1990). ***Hypoatherina tsurugae*** also spawned voluntarily in a tank, and larvae were reared like the Japanese silversides (Tsukamoto and Kimura, 1993).

Scorpaenidae—Scorpionfishes

Kurosoi are ovoviviparous, with a gestation period of about 48-52 d (Yamada and Kusakari, 1991; Jeon, 1995). Larvae from wild pregnant females were reared in a 15-m^3 tank on rotifers, *Artemia*, minced sand lance, *Tigriopus*, mysids, and fish eggs in green water (Table 8-4; Kusakari, 1991). Survival was 50% from parturition to 35-d old juveniles (25 mm TL). Juveniles were raised on sand lances and compound feed in 10-m^3 concrete tanks and 125-m^3 cages. They grew to 100 mm at 120 d, with overall survival of 45%. Two problems encountered were overfeeding on *Artemia* and parasitic flagellates on the gills. The best temperature for juvenile growth is 15-18°C. They can reach 1 kg within 2 yr. Culture is well established in Japan and the Republic of Korea (Furuya, 1995).

Kasago captive adults spawned voluntarily, and larvae were reared at 13-14°C in 200-L tanks on rotifers and some compound feed in green water, with survival up to 26% at 22 dah and growth from 3.7 mm TL at hatching to ~6.5 mm (Kuronuma and Fukusho, 1984).

Devil stinger culture is well established in Japan (Fujita and Nakuhara, 1955; Furuya, 1995).

Anoplomatidae—Sablefishes

Sablefish eggs were stripped from ripe wild fish, and larvae were reared at 6-7°C in 4-L beakers on rotifers, with survival of 10% from first feeding (21 dah) to 65 dah, then *Artemia* were added to the diet (Table 8-4; Whyte et al., 1994). Wild sablefish grew to 256 mm at 1 yr, 398 mm at 2 yr, and 450 mm at 3 yr (Rutecki and Varosi, 1997).

Hexagrammidae—Greenlings

Lingcod larvae were raised at 10°C from wild eggs and eggs spawned voluntarily in captivity in 5-L buckets on *Artemia*, with survival of ~9% during 4-45 dah (Table 8-4; Appelbaum et al., 1995). Captive wild adults spawned voluntarily, and larvae from the egg masses were raised on *Artemia* (to ~59 dah) and Pacific herring larvae (from 30 dah) in green water (Marliave et al., 1997). Mortality peaks occurred during 2-3 dah and 7-10 dah. Settling began 90 dah.

Cottidae—Sculpins

Elkhorn sculpin eggs were obtained from wild females, and larvae were reared to fully-developed juveniles of 52 mm BL (Munehara and Mishima, 1986).

Asahi-anahaze eggs were obtained from wild females, and larvae were reared at 15-17°C in a 500-L tank on rotifers, *Artemia*, wild zooplankton, fish larvae, and minced fish to fully-developed juveniles of >37 mm TL (Kimura et al., 1987). **Anahaze** were reared at ~16°C in the same manner to fully-developed juveniles of >44 mm (Kimura et al., 1988).

Cyclopteridae—Snailfishes

Atlantic lumpfish are valuable for their roe, and wild catches are variable (Martin-Robichaud, 1991; Brown et al., 1992; Martin-Robichaud et al., 1994). They will spawn voluntarily in tanks, or wild egg masses can be collected. At 9°C, eggs take ~28 d to hatch. Egg masses (which are guarded by the male) were collected, and larvae were raised on *Artemia*, and juveniles on chopped fish and molluscs, with survival of 21% at 1 yr (Brown et al., 1992). Larvae were raised in 26-L tanks on *Artemia* and compound feeds, with survival of 21% from hatching to 100 dah (Table 8-4; Martin-Robichaud et al., 1994). Dietary estradiol treatment (in *Artemia*) can be used to increase the percentage of females.

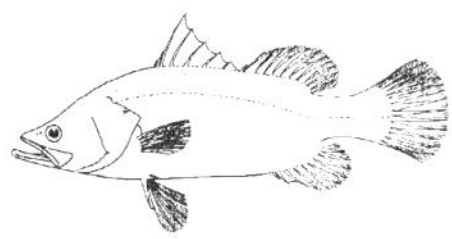

Barramundi
Lates calcarifer

Centropomidae—Snooks

Barramundi (Fig. 13-1) are raised commercially in southeast Asia and northeast Australia. Much work has been done on this important species: feeding abilities and habits of larvae and juveniles (Barlow et al., 1993; Kohno et al., 1994a), biology and culture (Copland and Grey, 1987), hatchery procedures (Lim et al., 1986; Copland and Grey, 1987; Duray and Juario, 1987; AQUACOP et al., 1990a, 1991; MacKinnon, 1990; Parazo et al., 1990, 1991b; Boonyaratpalin, 1991; Rutledge and Rimmer, 1991; Dhert et al., 1992b; Shelley, 1993; Duray, 1994b; Barlow et al., 1996), feeds (see chapter 9), grow-out (Anonymous, 1986; Copland and Grey, 1987; Cheong, 1989, 1990; Nacario, 1991; Singh, 1991; Cash, 1994; Tookwinas et al., 1994; Chou et al., 1995; Liao et al., 1995; Rimmer, 1995).

Figure 13-1. Barramundi juvenile (75 dah, 70 mm TL).

Natural spawning occurs during the warmer months (up to ~6) Sometimes newly-caught ripe fish are stripped after they ovulate without treatment. HCG and GnRHa are effective ovulation inducers, and barramundi have tank-spawned, with and without hormones. They have been conditioned to produce eggs during any month. Induced ovulation can generate at least a quarter million eggs per kg body weight (e.g., 6.8 million/16.5 kg).

Larvae have been raised in tanks with laboratory-cultured, pond-cultured, or wild zooplankton (Table 8-4). They also have been raised in fertilized ponds (Rutledge and Rimmer, 1991; Shelley, 1993). They can be transferred from salt to fresh water at 15 dah. In the nursery phase, barramundi are weaned and grown in tanks, ponds, or cages from ~20 mm to 75-100 mm, the minimum size for grow-out. Cannibalism occurs mainly between 20 mm (weaning) and 120 mm. They have been grown out mainly in cages and ponds, and sometimes in tanks. A pond harvest density of 2.2 kg/m^2 for 500 g fish was reported (Ruangpanit and Yashiro, 1995). Barramundi could be polycultured in ponds with tilapia at ratios of 1/20 to 1/25 (Tesorero et al., 1994). Market size usually is in the range 250-

600 g, but it is economical to raise them to 3 kg or more. Sizes of wild barramundi have decreased as the stocks have, but 30-kg and larger fish occasionally are caught.

Feeding and growth are best at 27-30°C and drop sharply at <25°C. In Australia, at suboptimal temperatures (22-27°C), we raised barramundi in tanks to an average weight of 566 g in 369 d, with FCRs ranging from 0.93 to 1.13. Other studies have indicated a potential for 800-g fish in a year at higher temperatures. Barramundi stocked in a freshwater reservoir reached 5.2 kg at 20 months.

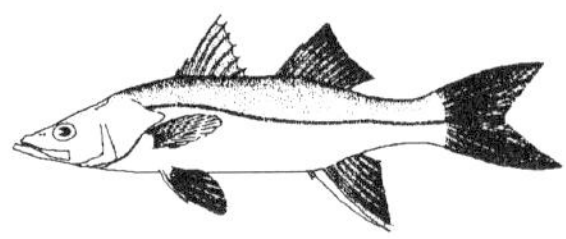

Common Snook
Centropomus undecimalis

Common snook (Fig. 13-2) aquaculture research has been conducted in Florida, Texas, and South America (Shafland and Koehl, 1980; Chapman et al., 1982; Lau and Shafland, 1982; Edwards and Henderson, 1987; Maciorowski et al., 1987; Roberts, 1987; Tucker, 1987b; Kennedy et al., 1998). Although not well known in most of the U.S., they are popular from Florida southward, and other members of the family are becoming better known to consumers internationally (barramundi and Nile perch). This is the largest and fastest-growing member of the family in the western Atlantic (Fig. 13-3). At least in Brazil and Venezuela, wild juvenile common and fat snook have been grown out, mainly on trash fish in earthen ponds, and some were sold locally (Tucker and Jory, 1991). Food quality and acceptance of snook are high (Tucker et al., 1985). They have been raised on experimental pelleted feeds in tanks and ponds to 3 kg (Tucker, 1987b, unpub.). Snook are gregarious and tolerate crowding. Optimal temperature is 27-28°C, but they can survive in the range 10-35°C. Snook can survive transfer to fresh water 15 d after hatching, and juvenile growth is at least as fast as in salt water. They can survive very low water quality (e.g., dissolved oxygen down to 0.9-1.0 mg/L, high turbidity). Common snook were raised from eggs to 725 g mean weight in 15 months (Fig. 13-4) and to larger sizes (Fig. 13-5). FCR was good, 0.7-1.1 for 16- to 725-g fish. Snook also were raised with tilapia in ponds with the idea of reducing overproduction of small fish and providing a second crop; results were promising.

Induced spawning of newly caught or captive wild snook is routine (Chapman et al., 1982; Roberts, 1987); fish usually are stripped but some fertilization has occurred when eggs were released in the holding tank. Egg quality seems to be high, but heavy mortality sometimes is associated with yolk exhaustion. This probably is related to food deficiencies (e.g., prey too large) or bacterial infection. At about 10-12 dah, mortality from EFA deficiency is possible. Then, during the first few days of eating *Artemia*, other digestive or nutritional problems can arise. The main obstacle to snook farming is the need for refining larval culture techniques to increase survival, especially during the first week. We have found that small rotifers and cultured copepod nauplii were sufficient for first feeding. Clam trochophores were tried but found not to be essential if enough small rotifers and copepods were present. In 1995-1996, we began using probiotic bacteria (*Bacillus* no. 48)

Figure 13-2. Common snook late larva (30 dah, 35 mm TL). (photo by Thomas A. Smoyer)

Figure 13-3. Tarpon snook and common snook (1 yr).

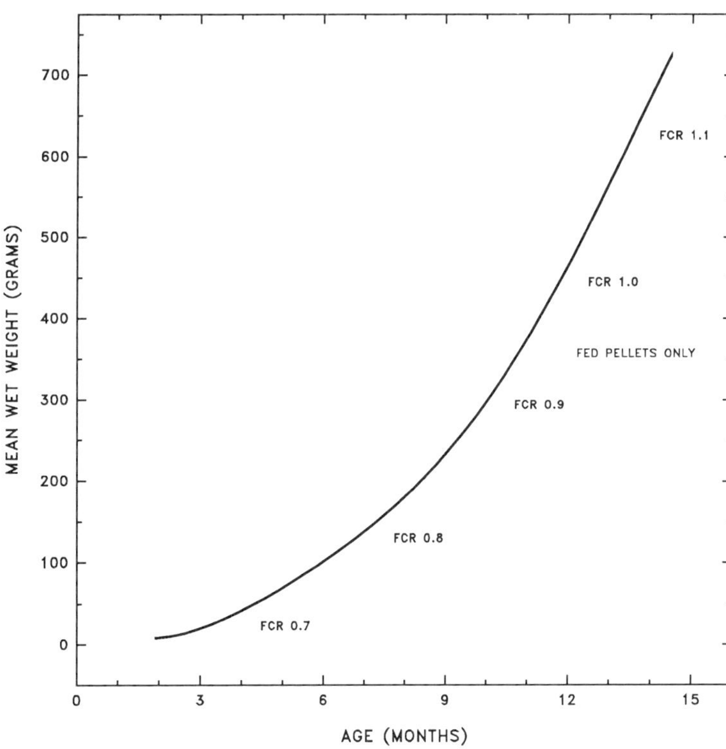

Figure 13-4. Common snook growth and feed conversion ratio (Tucker, 1987b).

Figure 13-5. Common snook (2 yr).

isolated from healthy hatchery-reared snook (Kennedy et al., 1998). The *Bacillus* was kept in culture and used to inoculate egg-incubation and larval-rearing water. Survival of snook from hatching to 50-d-old juveniles was increased from the usual 1-2% to 7.0% (Table 8-4). It is expected that effective control of cannibalism will increase survival further to at least 10-15%. With consistent egg quality and management, this level of survival should be repeatable.

Although common snook culture research has been conducted for two decades, those efforts have been sporadic and less intense than those spent on barramundi. Thus far, barramundi have been found easier to spawn (voluntary or induced tank spawning vs stripping), faster developing (transform at 25 vs 35 dah), and faster growing (454 g within 9-11 vs 11-12 months). FCR appears to be slightly better for snook (1.0 vs 1.1 at 454 g). After the first 2 weeks in saltwater, both species can tolerate transfer to fresh water. Optimal temperature is near 28°C for both. Improved hatchery management and nutrition will increase future survival and growth rates of common snook.

Japanese snook wild juveniles stocked in a 20-m^3 concrete tank and fed raw fish and shrimp at 13-27°C grew from 147 to 346 g in an average of 288 d, then from 358 to 767 g in 211 d (Tashiro and Iwatsuki, 1995). The fish were kept for at least 3 yr altogether, during which all survived and were found to be hardy. They fed well at >25° but not at all at <16°C.

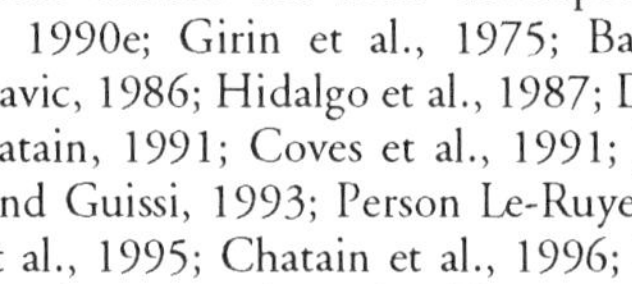

European Sea Bass
Dicentrarchus labrax

Percichthyidae—Temperate Basses

European sea bass culture has been developed since the first rearing study in 1969 (Barnabé, 1972, 1990e; Girin et al., 1975; Barahona-Fernandes, 1978; Bedier, 1981; Johnson and Katavic, 1986; Hidalgo et al., 1987; Divanach and Kentouri, 1990; Klaoudatos et al., 1990; Chatain, 1991; Coves et al., 1991; Jennings and Pawson, 1991; Sweetman, 1992; Barnabé and Guissi, 1993; Person Le-Ruyet et al., 1993b; Planes, 1993; Le Breton, 1993; Carrillo et al., 1995; Chatain et al., 1996; Nehr et al., 1996). Usually, females are induced to ovulate, but spawn in tanks. The natural spawning season is January-March, but environmental control has extended the spawning season for captive broodstock to as early as October and as late as May. Methods for mass-rearing of larvae are well developed, and 40% survival to 1-2 g is possible. Hatchlings can be stocked at 100/L or more. Larvae are raised on rotifers, *Artemia*, and dry feed, and survival is high (Table 8-4). Weaning mostly has been done at ~40 dah, but can be accelerated. Size variation can encourage cannibalism, so juveniles usually are graded, if necessary. They have been grown out in cages, ponds, tanks, and impoundments. Sea bass weigh ~500 mg at 70-80 dah, 3-4 g at 140 dah, and 20-80 g at 9 months. At 25°C, they can grow to a market size of 350-400 g in 15 months. They have wide temperature (2-32°C, but do not feed at <7°C) and salinity (~0-90 ppt) tolerance ranges. The best range for grow-out is 22-27°C. In the early 1980s, rearing studies were done in Martinique, but they were discontinued in favor of western Atlantic species (mainly red drum), partly because of temperature stress and disease problems (Barnabé 1987a, 1987b; Barnabé and Le Coz, 1987; Gallet de St. Aurin, 1987).

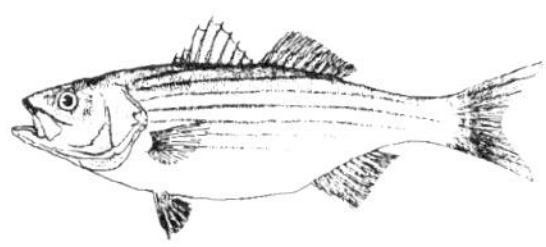

Striped Bass
Morone saxatilis

Striped bass have been stocked for more than a century, and hybrid striped bass support a significant aquaculture industry in the U.S. (Doroshev, 1970; Colura et al., 1976; Wiles, 1981; Geiger, 1983; Houde and Lubbers, 1986; Harrell et al., 1990; Hodson, 1991; Peterson, 1991; Hogans, 1994; Barkoh, 1996). Striped bass larvae were first released in 1884. Methods for hormone-induced ovulation and pond production of juveniles had been developed by the late 1960s. This species is anadromous, and natural spawning of the moderately large, demersal, non-adhesive eggs occurs in freshwater rivers. Ripe broodfish have been collected mostly from wild populations, but much effort has been directed toward establishing captive or reared broodstock (see chapter 7). Striped bass larvae and hybrid larvae from striped bass eggs have large enough mouths to begin feeding on *Artemia* nauplii (Table 8-4). Larvae from white bass eggs need a smaller first food, such as rotifers or copepods. Methods for pond culture of larvae are well developed; supplemental feeding could be required by 21 dah. Transformation of striped bass occurs ~37 dah (~30 mm TL). Striped bass can reach at least 600 g in 12 months, 1 kg in 18 months, 2.6 kg in 3 yr, and 6.5 kg in 5 yr (Parker et al., 1990a). Culture of striped bass and hybrids can be divided into three phases in which: I) 5-d-old larvae stocked at 500,000/ha grow to 25-60 mm in 30-60 d (survival 25-80%), II) phase I fish stocked at 25,000/ha grow to 75-250 mm in 5-9 months (survival ~80%), III) phase II fish stocked at 1000-4000/ha grow to market size in ~9 months (survival 90-100%), resulting in 18-month-old hybrids weighing at least 700 g. Ponds, tanks, and cages have been used for grow-out, but caged striped bass could be susceptible to parasites and temperature changes. Although several hybrids have been produced from members of this family, palmetto bass and sunshine bass have received the most attention. Optimal temperatures are about 24-25°C for striped bass and 25-27°C for palmetto (and probably sunshine) bass. Juvenile striped bass grow well in fresh or salt water, and hybrids grow best at low to intermediate salinity. Market size mostly is in the range 450-1400 g.

White bass reared broodstock have been induced to ovulate; this species can reach at least 650 g in 18 months (Parker et al., 1990; Denson and Smith, 1996; Smith et al., 1997). A temperature of 24-25°C was suitable for larval rearing.

Palmetto bass larvae are robust and survive well, but it takes more effort to obtain female striped bass for producing them (Woods et al., 1985; Harrell et al., 1990; Woiwode and Adelman, 1991; Gempesaw et al., 1992; Buurma et al., 1996; Wolters and DeMay, 1996; Bosworth et al., 1997). As many as 8323 kg/ha (220-755 g in 229 d) have been harvested from grow-out ponds with 93% survival and FCR of 2.3 (Smith et al., 1990).

Sunshine bass eggs are adhesive, and larvae are less robust, harder to feed, and tend to have lower survival, but female white bass for producing them are easier to obtain and easier to spawn (Harrell et al., 1990; Tucker et al., 1993; Ludwig, 1994; Swann et al., 1994). Sunshine bass also grow slightly faster. As many as 9699 kg/ha (272-863 g in 370 d) have been harvested from grow-out ponds with 94% survival and FCR of 2.3 (Smith et al.,

1990). Sunshine bass were raised at 24-26°C in a tank to 1333 g at 19 months old (gaining up to 3.0 g/d) with overall FCR of 2.7, but then slowed to reach only 1710 g at 30 months, partly because they do not grow much larger and partly because they were maturing and diverting food energy into egg production (Tucker et al., 1993a; Figs. 13-6, 13-7).

Sunshine bass are 4-6 mm TL at first feeding (5 dah) and need small zooplankton (Ludwig, 1997). Striped bass and palmetto bass are 6-9 mm at first feeding (5 dah). At

Figure 13-6. Sunshine bass (186 dah, 235 mm TL, 180 g).

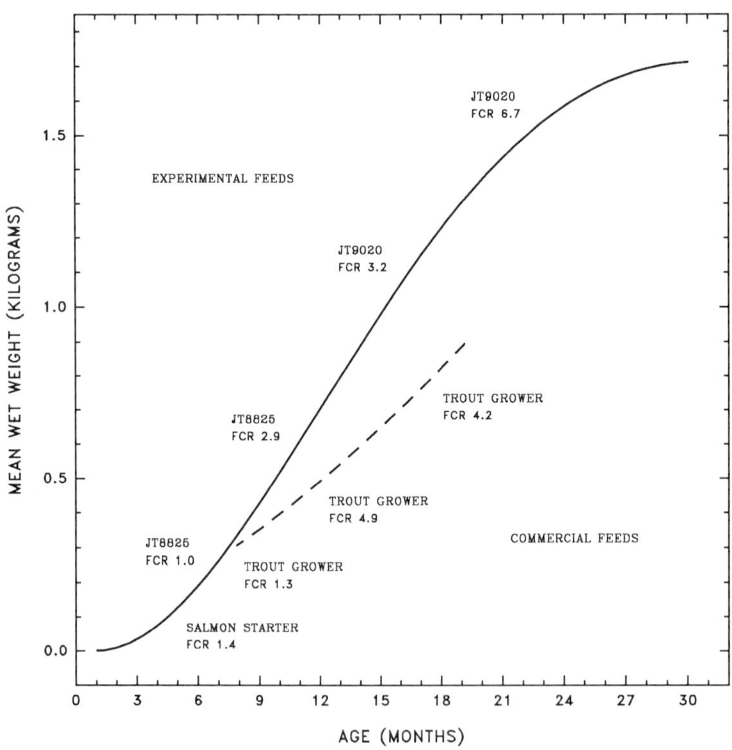

Figure 13-7. Sunshine bass growth and feed conversion ratio (Tucker et al., 1993a).

18-24°C, rotifers (appropriate prey for sunshine bass) will be abundant 5 d after pond filling. Medium-sized zooplankton (appropriate prey for striped and palmetto bass) will be abundant after 2-3 weeks. In 2-4 weeks, predaceous aquatic insects such as backswimmers can build to dangerous numbers (thousands in a 0.4-ha pond). Sunshine bass larvae are more vulnerable to them than the larger larvae. Steps to rear sunshine bass larvae in ponds are: fill ponds at the time of spawning, stock larvae at 5 dah, maintain zooplankton blooms by periodic fertilization (about once a week), begin weaning at 2-3 weeks, and harvest fully-weaned juveniles at ~5-6 weeks.

Australian bass are catadromous, spending most of their lives in fresh water but spawning in brackish water (van der Wal and Nell, 1986; Battaglene et al., 1989a, 1989b; Battaglene and Talbot, 1989, 1993; Battaglene and Allan, 1990). They have been induced to ovulate. Larvae have been raised in tanks on rotifers and *Artemia* and in ponds on mixed plankton. Survival in ponds sometimes was difficult, but the larvae tended to be healthier and cheaper than those from tanks. Eggs and larvae do best at 16-20°C. Australian bass in farm ponds grew to 276 mm TL at 2 yr and 345 mm at 3 yr. They are raised commercially.

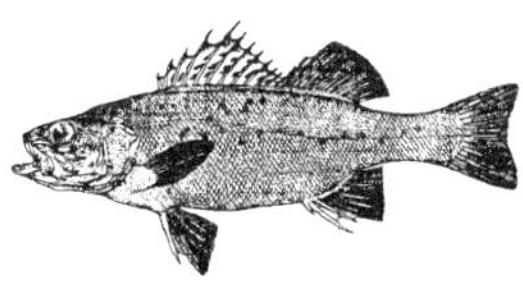

Japanese Sea Bass
Lateolabrax japonicus

Japanese sea bass live in salt and fresh water and have been raised from wild juveniles and from eggs. They have been induced to ovulate and will spawn voluntarily in tanks. Larvae were raised on rotifers, *Artemia*, *Daphnia*, and eel feed, with survival of 2.5% to 90-dah juveniles (40 mm TL)(Huang and Liu, 1991). Larvae reared in 500-L tanks on rotifers and copepods in green water had a survival as high as 54% at 48 dah and growth from 3.9 to 14.5 mm TL (Kuronuma and Fukusho, 1984). Larvae grew and survived well for Zheng et al. (1993). This species can reach a market size of 600 g in 11-13 months. Culture is well established in Japan and Taiwan (Furuya, 1995; Liao et al., 1995). A suitable temperature range is 20-25°C.

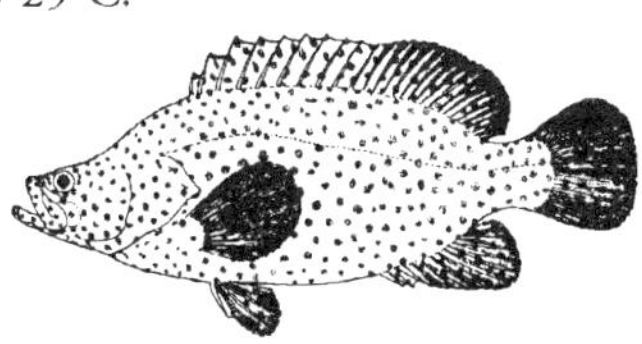

Polka Dot Grouper
Cromileptes altivelis

Serranidae—Groupers, Sea Basses

In the Indo-Pacific and Middle Eastern regions, several species of grouper are farmed in cages, ponds, and tanks, but mostly from 75- to 150-mm wild juveniles and mostly fed trash fish (Anonymous, 1986; Kohno et al., 1988; AQUACOP et al., 1990c; Tookwinas, 1990c; Chen and Liao, 1991; Quinitio and Toledo, 1991; Li, 1995; Aqua Farm News, 1992a; Tookwinas et al., 1994; Liao et al., 1995). They sometimes are fed small tilapia and sometimes polycultured with them. The usual market size is 500-1000 g, which can be reached in 6-8 months of grow-out. The minimum size to begin grow-out, 75-100 mm, can be obtained in nursery tanks, cages, or ponds. They are stocked up to 60 fish/m^3 (<1kg/m^3) in cages. In Taiwan, a pond farm typically has stocked 60,000-80,000 groupers/ha and harvested 80% of them for a production of 30,000-40,000 kg/ha; the fish grow from 46 mm to 600 g in 12 months and 2 kg in 19 months. Most cultured grouper species will spawn voluntarily in tanks, ponds, or cages, but inducing ovulation also is straightforward. Commercial-scale hatcheries for redspotted grouper and kelp grouper in Japan (Maruyama, pers. comm., 1992) and Malabar grouper in Taiwan (Lin, pers. comm., 1992) and Thailand (Kungvankij, pers. comm., 1992) have raised large batches of juveniles, with survival as high as 30%. The larval period is longer than for most cultured fish. Some groupers need small rotifers, trochophores, or copepods at first feeding. Early grouper larvae, especially when stressed, sometimes exude a lot of mucus, which can cause them to stick to each other, to the surface film, or to solid objects. Turbulence in larval tanks has to be just right. With too little, the water stratifies, and zooplankton and fish can aggregate dangerously. With too much, the fish are battered. Gorging on *Artemia* is another source of mortality, and cannibalism among early juveniles can be a problem. Larvae are fragile, and survival from eggs to juveniles often has been only 0-1%, but juveniles and adults are among the hardiest of fish.

Redspotted grouper is a major foodfish farmed in Japan and nearby countries (Ukawa et al., 1966; Tseng and Ho, 1979, 1988; Maruyama et al., 1994; Furuya, 1995). They spawn voluntarily in tanks or ponds and can be induced to ovulate. Larvae can be raised on rotifers, *Artemia*, and dry feed, in green water (Table 8-4; Maruyama et al., 1994). They grow to 25-30 mm TL by 40 dah. In 1989, 403,000 juveniles (29 mm TL) were produced at a Japanese hatchery. The highest survival reported was 34% during 1-47 dah (16% from eggs to 47 dah), the best I know of for groupers. Malnutrition after day 20, cannibalism, and a viral disease were causes of mortality.

Orangespotted grouper is a major foodfish farmed in Malaysia, Taiwan, nearby countries, and the Middle East (Higuchi et al., 1980; Abdullah et al., 1984; Doi et al., 1991; Toledo et al., 1993; Liao et al., 1995; Duray et al., 1996b, 1997; Almatar et al., 1997; Ordonio-Aguilar et al., 1997). They spawn voluntarily in tanks or cages and can be induced to ovulate. Larvae were raised at 28-32°C on oyster eggs & larvae, rotifers, copepods, *Artemia*, dry feed, and minced fish, in green water, with survival of 12% from eggs to 43 dah (Table 8-4; Doi et al., 1991); better nutrition probably was the reason for the increase in larval survival compared with earlier trials. Later, larvae were raised on only rotifers and *Artemia*, then minced fish was given to juveniles; survival was 3.2% from hatching to 60 dah (Duray et al., 1997). Overdependence on *Artemia* was a cause of mortality. Use of larger tanks (3-m^3), starting with small rotifers, prolonging the use of rotifers (up to 35 dah), and splitting *Artemia* into four meals a day contributed to better survival. Almatar et al., (1997) reported larval survival and harvest densities as high as 13.3% and 5.6/L.

Brownmarbled grouper is a good candidate for farming in Southeast Asia (Lim et al., 1990; Supriatna and Kohno, 1990; Chao and Lim, 1991; Chao et al., 1993; Kohno et al., 1994a, 1994c; Tucker and FitzGerald, 1994). They spawn voluntarily in tanks, cages, or ponds, potentially year-round, and can be induced to ovulate. They seem able to eat larger prey at first feeding than some other species, have a shorter larval period, and also might grow faster. Larvae were raised at ~28-30°C on rotifers and *Artemia*, in green water, with survival of 13% from hatching to 40 dah (Table 8-4; Lim et al., 1990). When fed trash fish in cages, they can grow from 90 to 600-800 g in 6-8 months (Anonymous, 1986) or 30 to 600 g within 10 months (Chao et al., 1993).

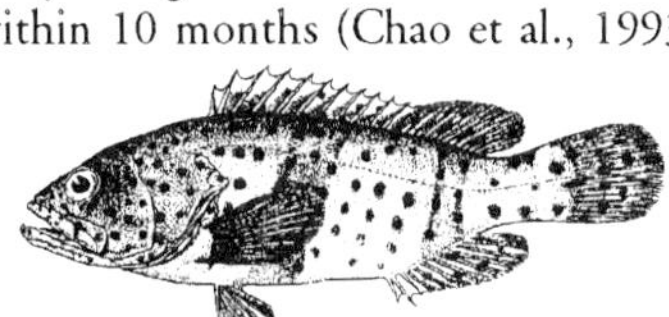

Malabar Grouper
Epinephelus malabaricus

Malabar groupers are farmed in Thailand, Taiwan, the Philippines, and nearby countries (Chua and Teng, 1980; Chulavitayanukool et al., 1985; Huang et al., 1986; Kungvankij et al., 1986; Maneewong et al., 1986; Tookwinas, 1990a, 1990b; Quinitio and Toledo, 1991; Ogburn and Ogburn, 1994; Liao et al., 1995). They spawn voluntarily in tanks or ponds and can be induced to ovulate. Larvae were raised on rotifers, MED, and *Artemia* to 50 dah, with survival of ~14% (Table 8-4; Quinitio and Toledo, 1991). Wild Malabar groupers stocked at 100/m^3 in cages and given compound feed grew from 27 to 587 g in 250 d. In the range 215-385 g, wild Malabar groupers stocked in a 0.1-ha pond grew 2.43 g/d with FCR of 2.4 (dry) on a moist locally-made feed (Ogburn and Ogburn, 1994). A pond harvest density of 1.4 kg/m^2 for 1.25-kg fish was reported (Ruangpanit and Yashiro, 1995). Polyculture with tilapia (1 grouper/15 tilapia) has been successful.

Camouflage groupers spawn voluntarily in tanks or ponds and can be induced to ovulate. Wild adults were induced to ovulate, and larvae were raised to 20 dah (AQUACOP et al., 1988). Larvae from voluntary spawning were raised at 28-29°C in 2.8-m^3 tanks on rotifers, *Artemia*, and compound feed, with survival as high as 4.7% (Table 8-4; Al-Thobaiti and James, 1997). When given a formulated feed in tanks for 12 months, juveniles stocked at 5/L grew from ~57 to ~572 g (~2.8 kg/m^3) and those stocked at 45/L grew to ~478 g (~18.2 kg/m^3).

Greasy groupers are raised in Singapore, Taiwan, and nearby countries (Chen, 1979; Anonymous, 1986; Lim et al., 1990; Chao and Lim, 1991; Lim, 1991b; Liao et al., 1995). They will spawn voluntarily in tanks, but mostly are induced to ovulate. Because wild males rarely are caught, it is not easy to set up captive broodstock of this species. Hormone-induced sex-reversal of females has been practiced. Greasy grouper larvae need smaller first food (e.g., trochophores) than brownmarbled groupers. When fed trash fish in cages, they can grow from 90 g to 600-800 g in 6-8 months (Anonymous, 1986). In the Philippines, greasy groupers stocked at 45 per m^3-cage and fed trash fish grew from 102 to 609 g in 5 months (Aqua Farm News, 1992a).

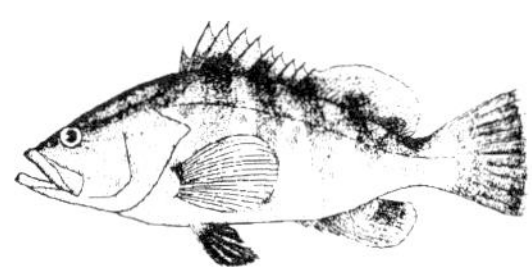

Nassau Grouper
Epinephelus striatus

Nassau grouper induced-spawning and larviculture methods were developed during 1987-1990 in the Cayman Islands and Florida (Tucker et al., 1991; Powell and Tucker, 1992; Tucker and Woodward, 1996). Juvenile and adult Nassau groupers are among the hardiest of fish. A ripe female that sat dry on a boat deck in the sun for 2 h was induced to ovulate 2 d later and was released in fairly good health.

Ovulation was first induced during January 1987. With fresh milt and good water, fertilization rate was 85-86%. Survival from fertilization to first feeding for six spawns was 73-94%. In later work, we found that hormone injection of wild fish was not necessary because newly captured wild female groupers that were about to ovulate could be identifed during the early afternoon and held in 1-m^3 tanks until eggs were ready for release, usually at or just after sunset. If held with males, those fish usually produced 100%-fertilized eggs in the holding tanks. In March 1994 and March-April 1995, captive wild Nassau groupers spawned voluntarily in a 37-m^3 concrete raceway (Table 7-6; Tucker et al., 1996). In April 1998, 8-yr-old reared groupers spawned in a raceway (Tucker, unpub.). A night watchman had witnessed Nassau groupers unexpectedly spawning in the Havana Aquarium more than 30 yr ago (Guitart and Juarez, 1966). Phil Bush also had watched caged Nassau groupers spawn unexpectedly at Grand Cayman in February 1991 (Tucker et al., 1993b). They can be conditioned to spawn during any month, mainly by temperature control.

In the first complete rearing study, 129 juveniles were produced from one spawn in 1990 (Table 8-4, Fig. 13-8). From fertilization to 98 dah, the best survival was 5.0%; during 98-243 dah, overall survival was 97%. Some of those fish were tagged and released (Roberts et al., 1995), and several were raised for more than 8 yr. The best temperature for larvae is ~24-27°C. Feed intake by juveniles and adults was minimal at <18°C, but feeding and growth were good at 22-30°C. The best growth should occur near 30°C. The minimum market size of 2 kg can be reached within 24 months (better than in Fig. 13-9), with FCR

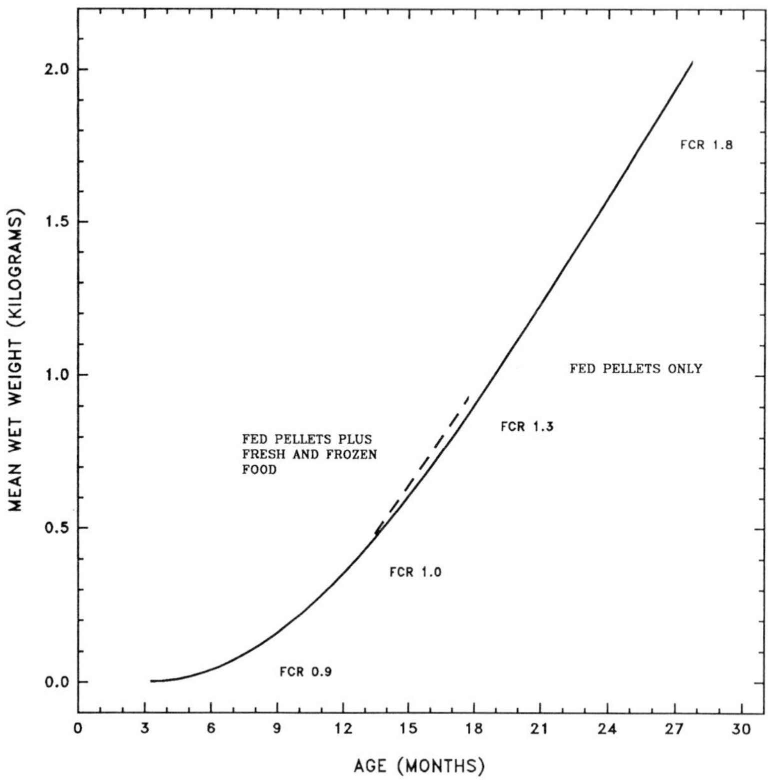

Figure 13-8. Nassau grouper growth and feed conversion ratio (Tucker and Woodward, 1996).

rising from 0.9 to 1.8. Maximum size of wild fish is more than 25 kg.

In 1994, Watanabe et al. (1996b) raised 5651 Nassau groupers to 62 dah in two 34-m^3 tanks in a 70%-shade greenhouse at 24-26°C. Foods were protozoans, rotifers, copepods, and *Artemia* in green water. The best survival from hatching to 62 dah was 1.4%.

Jewfish wild juveniles fed fresh fish in floating cages grew from 787 g to 1788 g in 90 d (Cervigon, 1983). In circular tanks, 3- to 6-kg jewfish reached 12-13 kg in 480 d. Similar studies have been done with black groupers (Alfonso et al., 1983).

Chinese perch are raised in the PRC (Liu et al., 1991b). They are induced to ovulate, and larvae are raised in tanks. Grow-out is in ponds. They can grow to 50-100 g during the first year, ~500 g the second, and 1000-1500 g the third. Maximum size is >5 kg.

Black sea bass, a very high quality serranid that reaches about 3.5 kg, has been spawned and raised experimentally (Hoff, 1970; Roberts et al., 1976; Tucker 1984, 1989), but it does not grow very fast, in nature to 44 g the first year, 126 g the second year, and 220 g (198 mm SL) the third year (Cupka et al., 1973; Mercer, 1989). Possibly, growth rate can be doubled or tripled in culture (Petrocci, 1995).

Spotted sand bass were raised from voluntarily spawned eggs and grown out in cages to adults, which were selectively bred (Avilés-Quevedo et al., 1995).

Studies have been done on spawning and larval rearing of dozens of serranids; some other examples are: **Blackspotted groupers** were induced to ovulate, and larvae were raised to transformation (~40 dah, ~35 mm TL), with survival during 1-40 dah of ~10% (Tseng and

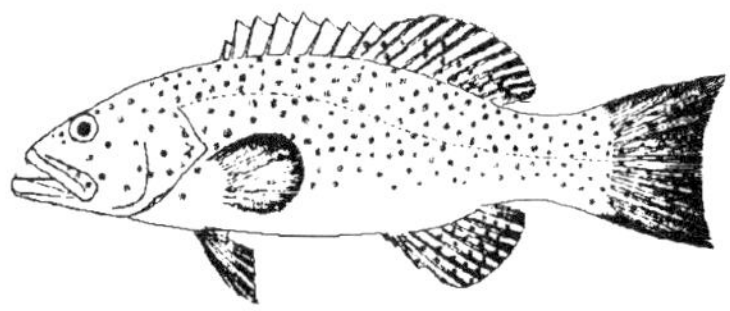

Leopard Coraltrout
Plectropomus leopardus

Chan, 1985). **Gags** were induced to ovulate and eggs and early larvae were described (Roberts et al., 1983). **Sevenband groupers** were induced to ovulate, and larvae were raised to transformation (~60 dah, ~30 mm TL) and beyond (Kitajima et al., 1991). **Leopard coraltrout** have been induce to ovulate and spawned voluntarily, and they were raised to at least 152 dah (Chen et al., 1991a; Masuma et al., 1993); culture in saltwater cages is being developed in Taiwan (Liao et al., 1995). Induced ovulation of ripe **squaretail coraltrout** is routine (Tucker and FitzGerald, 1994). **White grouper** maturation and growth in captivity were studied (Popper et al., 1994). **Barred sand bass** spawned voluntarily (Gadomski and Caddell, 1996). **Red groupers** were induced to ovulate, larvae were raised, and six juveniles raised on pinfish in aquaria during 66-557 dah reached ~168 mm SL (Colin et al., 1996). A growth trial was conducted with ***Epinephelus marginatus*** juveniles (Castelló-Orvay and Gracia-López, 1997). **Sand perch** were induced to ovulate with HCG, and larval development was described (Manrique, 1987a, 1987b).

Terapontidae—Grunters
Threestripe tigerfish have been induced to ovulate and have spawned voluntarily in captivity; culture in brackish and saltwater ponds is being developed in Taiwan (Her and Liu, 1991; Liao, 1993; Liao et al., 1995).

Sillaginidae—Whitings, Sillagos
Japanese whiting captive wild females have been stripped and have spawned voluntarily; they also were induced to ovulate (Lee and Hirano, 1981; Khan et al., 1991; Oozeki et al., 1992b). Larvae were raised at 24-26°C in a 1-m^3 tank on rotifers alone to 27 dah (15 mm TL, 16 mg) and in a 30-L tank on oyster larvae, rotifers, and copepods (Table 8-4).

Sand whitings were induced to ovulate, and larvae were raised in 2-m^3 tanks on rotifers to 25 dah (9.5 mm TL)(Battaglene et al., 1994).

Rachycentridae—Cobia
Cobia larvae from wild eggs were raised at 24-29°C in 38-L and 76-L aquariums on wild zooplankton in green water; mosquitofish, minced shrimp, cooked fish, and trout feed were given to juveniles (Table 8-4; Hassler and Raineville, 1975). At 131 dah, one specimen weighed 80 g (231 mm). Wild female cobia can reach 0.4 kg in 1 yr, 2.4 kg in 2 yr, 6.0 kg in 3 yr, and 31 kg in 7 yr, but males only 0.3, 1.5, 3.7, and 15.0 kg. Culture in saltwater cages is being developed in Taiwan (Liao et al., 1995). Maximum size could be as much as 68 kg.

Echeneidae—Remoras
Sharksuckers spawned in a large aquarium and larvae were raised at 27-31°C in a 100-L tank on wild zooplankton, cultured copepods, *Artemia*, and minced shrimp (Nakajima et al., 1987). Juveniles grew to 55 mm SL by 35 dah.

Carangidae—Jacks
Yellowtail is a high-value species that previously, and still is to a large degree, raised from wild juveniles, mainly in Japan and the Republic of Korea (Kuronuma and Fukusho, 1984; Yamaha Fishery Journal, 1989; Doumenge, 1990a; Nakada and Murai, 1991; Shimeno, 1991; Watanabe et al., 1991a; Ikenoue and Kafuku, 1992; Imaizumi, 1993; Aoki, 1995; Furuya, 1995; Kim and Myoung, 1995). Previously, wild juveniles (~100-150 mm TL) were the only source, but hatchery production is increasing. Usually, juveniles are raised

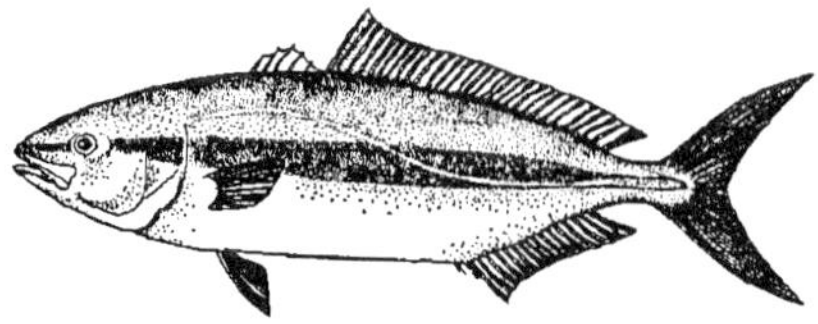

Yellowtail
Seriola quinqueradiata

in cages until mature, then are moved to spawning tanks, where spawning is voluntary or induced. The planktonic eggs are ~1.3 mm in diameter and larvae are fairly long, ~3.5 mm TL. Larvae can be raised on (oyster larvae), rotifers, copepods, and fish larvae, or rotifers, *Artemia*, other zooplankton, and microfeeds (Table 8-4). Hatchery production of 300-500 20-mm fish/m^3 (5-10% survival) is possible. Yellowtails are grown out in cages with volumes up to hundreds of cubic meters. Previously, they were fed trash fish, but by the early 1990s, good pellets had been developed. Because of cannibalism, grading is necessary. They have been grown out mainly in cages, but also in pens and impoundments. Pollution of coastal areas has led to an increase in offshore cage farming. Typical cage rearing density is 7-10 kg/m^3. Larvae do well at 20-22°C. Juveniles can tolerate 7-28°C, feed well at 18-27°C, grow best at 24-26°C, and can tolerate 16 ppt, but high salinity (~33 ppt) is best. They can reach 1.5-2 kg within a year, 4-7 kg in 2 yr, and a maximum of ~13 kg.

Greater amberjack (Masuma et al., 1990; García et al., 1993; Greco et al., 1993b; Porrello et al., 1993; Tachihara et al., 1993). They have been induced to ovulate and have spawned voluntarily in tanks. Larvae have been reared in tanks on rotifers, *Artemia*, and fish eggs or larvae, with or without green water (Table 8-4). At sea, they reached 103 g at 90-120 dah, and in cages on a trash fish diet, 858 g at ~7 months and 3.0 kg at ~18 months. Average FCR with trash fish was 4.4. Maximum size is >50 kg. Culture is well established in Japan (Furuya, 1995). Culture in saltwater cages is being developed in Taiwan (Liao et al., 1995).

Goldstriped amberjack and **banded trevally** will spawn voluntarily in tanks, and their culture is well established in Japan (Furuya, 1995). Goldstriped amberjacks also have been induced to ovulate, and their larvae were reared on rotifers, mixed zooplankton, copepods, fish larvae, and minced fish (Table 8-4; Fujita and Yogata, 1984). At 77 dah, the juveniles were as large as ~195 mm TL (~120 g).

Mazatlan yellowtail is similar to the Japanese yellowtail and closely related to the goldstriped amberjack. Adults spawned voluntarily in tanks, and larvae were raised at 25-26°C in 15-m^3 tanks on wild zooplankton, rotifers, and *Artemia* in green water (Benetti et al., 1994, 1995a, 1997). Juveniles were stocked at only 42/m^3 in 12-m^3 tanks to help reduce cannibalism. They can reach 0.5-1 kg in 4 months. Two of the main problems were epitheliocystis infections during weaning (28-42 dah) and cannibalism.

Florida Pompano
Trachinotus carolinus

Florida pompano has received the most attention in the western Atlantic as a mariculture candidate, especially during 1958 to the mid-1970s (Moe et al., 1968; Finucane, 1970; Smith, 1973b; Wagstaff, 1975; Hoff et al. 1978a, 1978b; Jory et al., 1985; McMaster, 1988; Tucker and Jory, 1991; Watanabe, 1995). Members of the genus *Trachinotus* are considered among the best food fishes in many areas. Research in Florida and Alabama led to several business-sponsored attempts to commercially culture pompanos in Florida and the Dominican Republic. One of the major reasons for failure was the use of inappropriate feeds. These early farming attempts did seem to demonstrate biological feasibility.

There has been recurring interest in farming of pompano and two related species, the permit and palometa, which share many of the same characteristics, but appear more resistant to diseases (Gomez, 1984a). Several factors justify the interest, especially the consistent high market value (pompano is second only to bluefin tuna in Florida). In many places, pompano is preferred by consumers, but the other two species are just as good. Early farming attempts in Florida were hampered by cold shock (<10°C), but pompanos can survive a range of 12-34°C. They tolerate high turbidity, rapid pH changes, and dissolved oxygen down to 3 mg/L. Although optimal salinity is about 30-35 ppt, they tolerate salinity from just above zero to ~45 ppt. Pompanos are moderately resistant to handling and crowding. Wild juveniles will feed a few hours after being collected and stocked. Pompanos can be induced to ovulate year-round. Larvae can be raised on rotifers and *Artemia* in green water (Table 8-4). They have been raised through two generations. Most workers have reported decreased growth rate (at ~350 g) and growth efficiency during intermediate grow-out. Up to 150-200 g, FCR was 3.5, then it jumped to 6.

In Venezuela, spawning, rearing, and nutrition work has been done with all three species (Gomez and Cervigon, 1979, 1984, 1987; Gomez and Scelzo, 1982).

Permits have spawned voluntarily in a large aquarium. Broodstock reared from wild juveniles were induced to spawn and larvae were raised on oyster eggs and larvae, protozoans, rotifers, *Artemia*, fish meal, and dry feed (Table 8-4; Lin, 1991). They reached 20 mm at 20 dah and 500-600 g at 7-10 months. FCR was ~2.0. In Florida, wild permits were raised from 27 to 501 g in 11 months (Jory et al., 1985).

Palometas have spawned voluntarily in a large aquarium. Wild juveniles in cages grew from 15 to 260 g in 5 months on trash fish and to 300 g in 6 months on pellets (Thouard et al., 1990). Palometas were mature by 18 months, were induced to ovulate, and spawned once voluntarily. Wild juveniles fed trash fish in cages grew from 17 to 400 g in 218 d with 91% survival and a harvest density of 2.9 kg/m^3 (Soletchnik et al., 1988). Wild juveniles fed trash fish in tanks grew from 5 to 469 g in 400 d, but survival was only 50% (Gomez, 1977). Within 12 months, palometas can reach 500 g and permits 900 g; permits can reach 1200 g in 18 months (Soletchnik et al., 1987, 1988).

Much valuable information has been accumulated on these three candidates. Spawning and larviculture methods for Florida pompano are fairly reliable, but growth and FCR need improvement. Of high priority is determination of nutritional requirements for this group of fish, which swim continuously and have high metabolic energy demands. A practical feed is a major missing link. Other research needs include closing the life cycle of permit and palometa on a consistent basis, and determining their optimal environmental requirements.

Snubnose pompanos have been induced to ovulate, and larvae were raised (Lin, 1991). Juveniles fed pellets and/or trash fish in cages can grow from 10-15 to 250-600 g in 10 months (Chou et al., 1995). Culture in brackish and saltwater ponds is now well established in Taiwan (Liao et al., 1995). A market size of 400-600 g can be reached at 8-13 months.

Striped jacks spawn voluntarily in tanks and are raised mainly in cages from both wild and hatchery juveniles (Kumai, 1995). For growth, 24-28°C is best. They can reach 400-650 g at 1 yr, 700-1000 g at 2 yr, and 2-2.5 kg at 3 yr. Culture is well established in Japan (Furuya, 1995).

Japanese horse mackerels were induced to ovulate, and larvae were raised at 16-23°C in a 40-m^3 outdoor tank on rotifers, copepods, *Artemia*, fish eggs, compound feed, and minced fish, with survival of 6.3% at 47 dah and growth from 2.5 mm at hatching to 28 mm (Kuronuma and Fukusho, 1984).

Jack mackerel larvae were reared to ~55-mm juveniles in 60 dah, but they grow to only ~150 g in 2 yr (Ochiai et al., 1982; Yamaha Fishery Journal, 1989).

Rainbow runner wild juveniles raised in 12-m^3 tanks on pellets had a growth potential for 1.7 kg at 1 yr (Saint-Félix and Blouin, 1997).

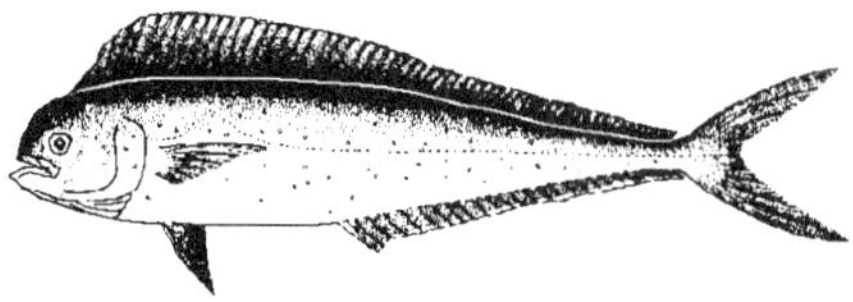

Dolphin
Coryphaena hippurus

Coryphaenidae—Dolphins

Dolphin have some very good characteristics, but also some handicaps (Hagood, 1984; Uchiyama et al. 1986; Kraul et al., 1988, 1992; Ostrowski et al., 1989; AQUACOP et al., 1990c; Kraul, 1990b, 1991, 1993; Szyper, 1991; Kim et al., 1993; Benetti et al., 1995b; Nel, 1995; Ostrowski, 1995, 1997). At 26°C, with good nutrition, dolphin will spawn in captivity throughout the year. A single female can produce 200,000-400,000 eggs every other day. Dolphin grow very fast—45-d-old juveniles weigh ~5 g and can grow on pellets to as much as 1.7 kg at 6 months with FCR of 1.2, and 5 kg at 9 months with FCR of 1.6. With fresh foods, they can reach 10 kg at 1 yr. On the other hand, dolphin are not the easiest fish to rear. They need very clean water and cannot be crowded. Dissolved oxygen should be >6 mg/L for larvae and >5 mg/L for juveniles. Juveniles survive at temperatures of 13-29°C and higher, but the best growth is at ~29°C. A salinity range of 15-35 ppt and possibly lower is acceptable. Larvae are particularly susceptible to bacterial infections, adults also are susceptible to diseases, and adults are easily injured by handling. Larvae can be raised on *Artemia*, but survive better when copepods are included, and rotifers for the first week probably help (Table 8-4). Dolphin yolksac larvae are good food for early juveniles. A very high quality feed is needed for grow-out. Aggression, biting, and possibly cannibalism can be high during 20-45 dah (mainly at 20-50 mm SL) but diminish after complete weaning; water current of 3-4 fish lengths/s helps limit aggression. Because older juveniles and adults normally swim fast and straight, they can injure or kill themselves by ramming tank walls. Donut shaped tanks have been used to protect dolphin broodstock and padded walls for juveniles by the time they are 120 d old. Shade can induce dolphin to swim more slowly as they approach tank or cage walls (possibly like they would slow down under a weedline at sea). The first grow-out phase could start with 50 fish/m^3 at 5 g and end with 42/m^3 at 250 g (90 dah), and the second phase could start with 15/m^3 at 250 g and end with 9/m^3 at 1.7 kg (180 dah) with overall FCR of 1.2 (Ostrowski, 1995, 1997). Stock density up to 18 kg/m^3 has been safely reached. Maximum size is ~40 kg.

Lutjanidae—Snappers

Snappers are very popular for food, but larvae of most are not easy to rear. We raised wild yellowtail and gray snapper juveniles and also induced ovulation in yellowtail snappers and raised larvae for 2 weeks in a 37-L aquarium for specimens. Juvenile and adult yellowtail snappers (midwater fish) have thinner skin and scales and are much more sensitive to handling than gray snappers (bottom fish). Gray snappers also will live in fresh water. The more hardy species might be better choices for farming.

Asian red snappers are farmed in Thailand, the Philippines, Taiwan, and nearby countries (Aqua Farm News, 1992b; Doi and Singhagraiwan, 1993; Emata et al., 1994; Liao et al., 1995; Duray et al., 1996a). They have been induced to ovulate and have spawned voluntarily in tanks. The planktonic eggs (795 µm) and hatchlings (1.75 mm TL) are small (Doi and Singhagraiwan, 1993). At first feeding, larvae had relatively small oil globules, and with a mouth width of ~200 µm, they could eat prey of 60-70 µm width and up to 100 µm length (e.g., oyster eggs and larvae, *Acartia* and *Oithona* nauplii). Larvae reared at 25-29°C on rotifers and *Artemia* in green water, followed by minced fish, had a survival of 16% during 1-55 dah (Table 8-4; Duray et al., 1996a). The best reported growth in cages was from 34 to 963 g at 10 months and 2303 g at 22 months, with 6 fish/m^2 fed trash fish twice/d (FCR 10.8). Growth was better for 50 fish/m^2 than with 110 or 152 fish/m^2.

With 60 fish/m^2, two meals a day were better than one. Fish stocked at 100/m^2 and fed once/d can grow from 20 g to a market size of 500 g in 9-10 months. Fish released at 10 g can grow to 500-600 g after 14-16 months.

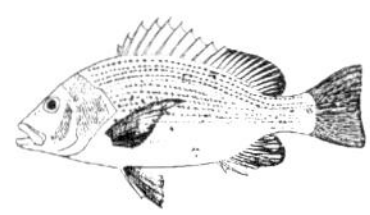

Golden Snapper
Lutjanus johnii

Golden snappers have been induced to ovulate and were stripped or allowed to tank spawn (Lim et al., 1985; Anonymous, 1986; Aqua Farm News, 1992b). Larvae raised at 28-29°C in a 4-m^3 tank on mussel larvae, rotifers, *Artemia*, wild zooplankton, cladocerans, and minced foods had a survival of 5% from eggs to 35 dah (Table 8-4; Lim et al., 1985). When fed trash fish in cages, they can grow from 90 g to 600-800 g in 6-8 months.

Atlantic red snappers have been induced to spawn and have spawned voluntarily at 23-25°C in a tank (Arnold et al., 1978; Collins et al., 1980; Rabalais et al., 1980). Larvae were raised through transformation on mixed zooplankton in green water in 1.2-m^3 outdoor tanks (Anonymous, 1996; Hearn, 1996).

Gray snappers have been induced to ovulate (Damas et al., 1978; Gonzalez et al., 1979; Rosas et al., 1997) and larvae have been raised on wild zooplankton at 26-28°C in 75-L aquariums (Richards and Saksena, 1980). Larvae were fed rotifers and copepods (*Oithona ovalis*) in green water at 28°C; juveniles fed adult *Artemia*, mysid shrimp, and bait fish reached 14.8 g (9.5 mm TL) at 120 dah (Cabrera et al., 1997; Rosas et al., 1997, in press). Wild juveniles grew from 44 to 233 g in 210 d on a diet of fresh fish (León et al., 1996).

Mutton snappers have been induced to ovulate. Larvae were raised on rotifers, wild zooplankton, and *Artemia* (Clarke et al., 1997). Larvae were raised on wild zooplankton, rotifers, *Artemia*, and dry feed in green water (Watanabe et al., 1998b); survival during 2-38 dah was ~14.3%; juveniles reached 141 g at 265 dah (range 30-300 g, 69% weighed 100-180 g).

Lane snappers were induced to ovulate (Borrero et al., 1978; Damas et al., 1979; Milares et al., 1979), and larvae were raised on rotifers, wild zooplankton, and *Artemia* (Clarke et al., 1997).

Yellowtail snappers have been induced to ovulate and have spawned voluntarily; larvae have been raised at ~27-28°C on mixed zooplankton, rotifers, and *Artemia* (Soletchnik et al., 1987, 1989; Riley et al., 1995; Lazo et al., 1997). Juveniles reached ~300 g at 525 dah and needed at last 35% dietary protein.

Wild juveniles of gray, lane, mutton, schoolmaster, and yellowtail snappers were raised in cages, but growth was slow (Elizabeth-Mesnager, 1984; Thouard et al., 1990). Mutton snappers grew to 350 g during 12 months in captivity, and the other four grew much less. Wild mutton snappers fed fresh sardines in concrete tanks grew from 44 to 209 g in 153 d (Gomez and Cervigon, 1987). Most species seem to grow slowly in nature and in culture.

Gerreidae—Mojarras

Wild **mojarras** do not adapt well to confinement, behave aggressively, grow slowly, and do not easily accept pellets, but reared ones might do better (Gomez and Cervigon, 1987).

Striped pataos were induced to ovulate with HCG, and larvae were reared to 73 d with growth rates of 0.3 mm/d and 2.45 mg/d (Borrero and Millares, 1979; Millares et al., 1979). Wild juveniles reared in concrete tanks grew from 4 to 24 g in 300 d (Gomez and Cervigon, 1987). Striped patao eggs (640 µm) and hatchlings (1.43 mm TL) are small (Alvarez-Lajonchère et al., 1996); larvae from induced spawns were raised on rotifers, copepods, and *Artemia* in green water, then dry feed, with a survival of 12% from eggs to 48 dah (Table 8-4); in two seasons, 132,000 juveniles were produced at 5-10/L.

Haemulidae—Grunts

Threeline grunt culture is well established in Japan (Furuya, 1995).

Silver grunt farming in brackish or saltwater ponds is well established in Taiwan (Liao et al., 1995).

White grunt larvae from wild eggs were raised at 24-28°C in a 38-L aquarium on wild zooplankton and *Artemia* (Saksena and Richards, 1975). The last two larvae grew to 46 mm TL at 107 dah and 34 mm at 142 dah.

Porkfish larvae from wild eggs were raised at 25-29°C in 946-L tanks on wild zooplankton, rotifers, copepods, and *Artemia*, with flake feed and ground shrimp added after 3 weeks, through most of the larval period (30 dah, ~13 mm TL)(Potthoff et al., 1984).

Yellowspotted grunts were induced to ovulate, and larvae were reared in 500-L tanks for 9 d on rotifers, *Artemia*, and compound feed in green water, then 10,000 were transferred to a 100-m^2 shaded outdoor pond stocked with algae, rotifers, and copepods (Chang et al., 1994b). A whirling disease occurred while larvae were in tanks eating *Artemia* but not when they were in the pond with copepods. About 18% of stocked fish were harvested at ~27 dah (~17 mm TL). The pond-reared fish were healthier and cheaper to produce. At 2 yr, pond-reared fish weighed 1.2-1.6 kg and were induced to ovulate. Maximum reported size was ~4.6 kg. Culture in brackish and saltwater ponds and cages is being developed in Taiwan (Liao et al., 1995).

Sparidae—Sea Breams, Porgies

At least six sparids are established aquaculture species: the red sea bream, black sea bream, yellowfin sea bream, and silver sea bream in Asia and the gilthead sea bream and common dentex in the Mediterranean region; many others might have potential (Girin, 1983; Liao, 1993).

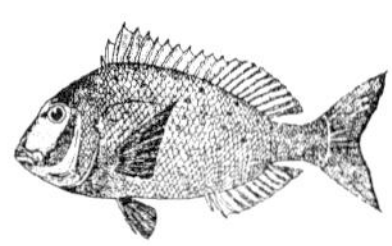

Red Sea Bream
Pagrus major

Red sea bream (Fig. 13-9), is a major food fish farmed in Japan, ROK, Taiwan, PRC, and nearby countries (Kuronuma and Fukusho, 1984; Tabaru et al., 1984; Foscarini, 1988; Fukusho and Kitajima, 1988; Doumenge, 1990a; Yamaha Fishery Journal, 1990; Fukusho, 1991; Ikenoue and Kafuku, 1992; Imaizumi, 1993; Umezawa, 1994; Kim and Myoung, 1995; Kumai, 1995; Liao et al., 1995; Watanabe and Kiron, 1995; Murata et al., 1996; Kyowa nd.a). They were first farmed (from wild juveniles) in 1965. Broodfish (mostly reared) usually are kept in cages until about a month before the natural spawning season (February-July); then they are moved to tanks, where they spawn voluntarily. They also have been induced to ovulate. Temperature control and feeding of frozen krill have allowed extension of the season from the usual 50 d to 6 months. The planktonic eggs (940 µm) and hatchlings (2.1 mm TL) are moderately small. Methods for mass-rearing of larvae are well developed. In the first stage, they usually are reared in tanks (12-72 hatchlings/L) on rotifers and *Artemia* in green water, to transformation (~30 dah, 10-14 mm TL). Larvae reared on rotifers, *Artemia*, and krill had a survival of 32% from hatching to 47 dah (Table 8-4). MPDs are fed as early as 8 dah. Larvae do well at 15-20°C, and juveniles at 20-28°C. Hatchery production of one to five thousand 20-mm juveniles/m^3 (30-80% survival) is possible. Cannibalism can begin as early as 20 dah (~7 mm). In the second stage, juveniles (10-50 mm TL) in cages or tanks are weaned to a mixture of minced fish with clams and/or krill and then are raised on minced fish or compound feeds to 60-80 mm or larger. Good survival to stocking size is ~25%, and most of those should continue to survive. Red sea bream usually are grown out in cages (~300/m^3, initially), pens, or impoundments. About 7 kg/m^3 is an appropriate harvest density for cages. In culture, at ≥14°C, they grow to 600-700 g within 18 months, 1 kg (the usual market size) within 2 yr, and 1.2-1.5 kg after 30 months. In the wild, released red sea bream grow fastest during the third year and then slow down. They reached 770 g during the third year, but some older recovered fish weighed as much as 10 kg. Maximum size is ~13 kg. Under the right conditions, ranching of red sea bream has improved catches in Japan. Recovery of released fish has been as high as 14%. To avoid using an enclosure, it is possible to ranch red sea bream conditioned to assemble at an underwater speaker for feeding and harvest, but some migration and predation will occur. Moksness and Støle (1997) reported a production cost (1993) of US$0.20 for a 70-mm juvenile.

Figure 13-9. Red sea bream (Tokyo fish market).

New Zealand snapper is conspecific with red sea bream and has similar characteristics, but lives in the southern hemisphere in temperate areas around New Zealand and Australia (Henry, 1988; Battaglene and Talbot, 1992; Battaglene and Allan, 1994; Hecht et al., 1996). The natural spawning season is up to 12-15 weeks long. They have been induced to ovulate. Larvae were raised to early juveniles (40 dah, ~18 mm TL) on rotifers, *Artemia*, and minced pilchards. The three improvements needed most were increases in: egg quality, gas bladder inflation success, and survival during weaning. Maximum size is ~20 kg.

Black Sea Bream
Acanthopagrus schlegeli

Black sea bream is another important food fish in the Far East (Kuronuma and Fukusho, 1984; Chen and Liao, 1991; Leu, 1997). They have been induced to ovulate but easily will spawn in tanks. Methods for mass-rearing of larvae are well developed. When they were raised at 17-24°C on rotifers, *Artemia*, and MCD, mean survival was 50% (as high as 75%) from hatching to transformation (Table 8-4; Leu, 1997). Survival to 100 dah was 28%. They can reach at least 190 g at 1 yr, and market size is 200-1000 g. Culture of this species and southern black sea bream in brackishwater ponds is well established in Japan and Taiwan (Furuya, 1995; Liao et al., 1995). Moksness and Støle (1997) reported a production cost (1993) of US$0.28 for a 70-mm juvenile.

Silver sea bream have spawned voluntarily in tanks, and larvae were reared at 19-22°C on rotifers, *Artemia*, and an MCD in green water, with survival of 43% from hatching to 45 dah (Table 8-4; Leu, 1994; Leu and Liou, 1992; Mihelakakis and Kitajima, 1994a, 1995). Silver sea bream also have been induced to ovulate in Taiwan; farming in brackish and saltwater ponds is well established there (Liao, 1993; Liao et al., 1995).

Silver sea bream × black sea bream hybrids and silver sea bream grew faster than black sea bream but were less tolerant to low temperature and salinity (Kitajima and Tsukashima, 1983). The hybrids grew from ~55 to ~340 g in 14 months (~19-months-old).

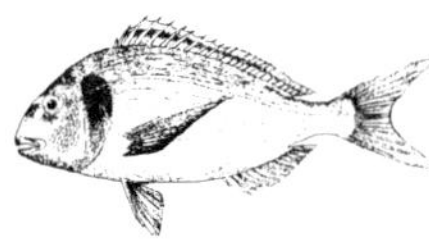

Gilthead Sea Bream
Sparus auratus

Gilthead sea bream is a major food fish in the Mediterranean region (Lumare and Villani, 1970; René, 1974; Person-le Ruyet and Verillaud, 1980; Tandler and Helps, 1985; Porter et al., 1986; Ounaïs-Guschemann, 1989; Divanach and Kentouri, 1990; Chatain, 1991; Kissil, 1991; Tandler, 1991; Shapiro, 1992; Barbato et al., 1993; Le Breton, 1993; Zohar et al., 1995; Chatain et al., 1996; Dhert et al., 1998). It is a protandrous hermaphrodite, with a natural spawning season falling between December and May. Some broodstocks spawn voluntarily and others seem to need hormone treatment, which also can be used to synchronize individuals and set the schedule for egg production. Methods for mass-rearing of larvae are well developed. When they were reared on rotifers and *Artemia* in green water, then dry feed, survival was 11% from hatching to 60 dah (Table 8-4; Person-le Ruyet and Verillaud, 1980). The range 18-26°C is suitable for larvae, and 30°C is best for juveniles. With recent improvements, production of up to 5-12 larvae/L at 32 dah (15 mg) and 40% survival to 1-2 g are possible. Giltheads can reach ~450 g in 14-15 months.

Sobaity will spawn voluntarily in tanks (Higuchi et al., 1980; Al-Abdul-Elah et al., 1983a; 1983b; Almatar et al., 1997). Hatchling to fingerling survival was as high as 28%. As many as 2.8 juveniles/L (35-40 dah) were produced in 4-m^3 tanks. Suitable densities were 200-300/m^3 at 1-60 g and 60/m^3 for grow-out.

Yellowfin sea bream will spawn voluntarily in tanks, but also have been induced to ovulate (Higuchi et al., 1980; Tsay and Yu, 1991; Leu and Chou, 1996). When larvae were raised at 20-24°C on oyster larvae, rotifers, *Artemia*, copepods, and an MCD in clear water, survival was 28% from hatching to 50 dah (Table 8-4; Leu and Chou, 1996). Culture is well established in Japan (Furuya, 1995) and in fresh and brackishwater ponds in Taiwan (Liao et al., 1995).

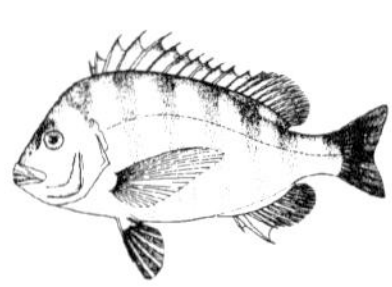

Sheepshead
Archosargus probatocephalus

Sheepshead (Figs. 13-10 to 13-12) is an excellent food fish ranging from Latin America to the northeastern U.S., but is not a targeted species and has limited commercial availability. It often has been substituted for red snapper in certain restaurants. We have induced ovulation with HCG, followed by stripping or voluntary fertilization (Tucker and Barbera, 1987). Considering their adaptability and the ease of inducing spawning, sheepshead probably will spawn voluntarily. In the first complete rearing study, survival from fertilization to adults was 40% (Table 8-4; Tucker, 1987a). More than 6000 juveniles were produced from one spawn. Some were raised in large tanks and ponds on commercial feeds up to 1.0 kg. A few weighed 450 g after 9 months. Juveniles and adults can be

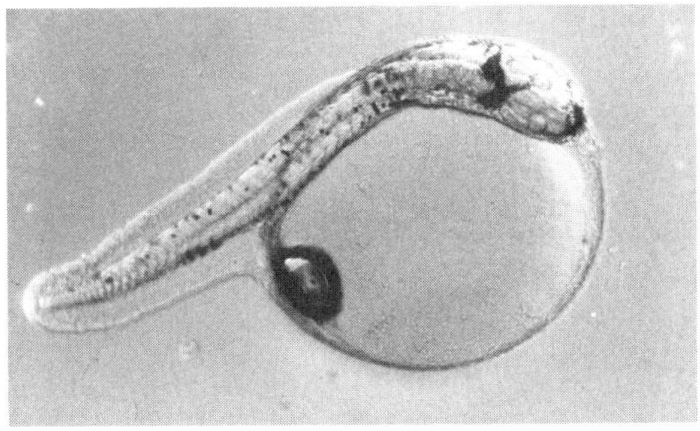

Figure 13-10. Sheepshead hatchling (28 haf, 2.1 mm TL). Note large yolk sac.

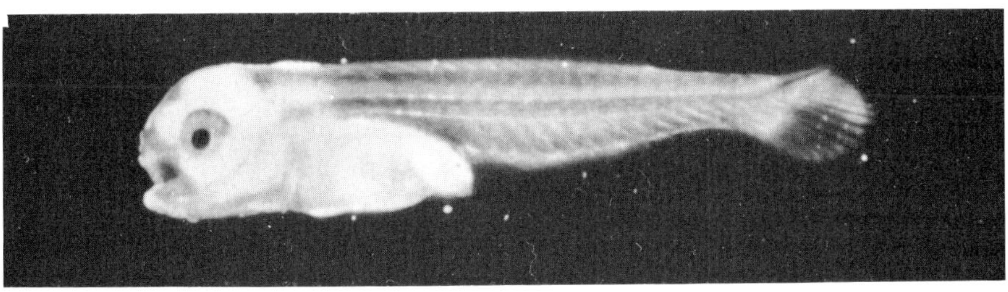

Figure 13-11. Sheepshead larva (14 dah, 6.0 mm TL).

Figure 13-12. Sheepshead juvenile (50 dah, 50 mm TL). (photo by Thomas A. Smoyer)

reared in fresh water. As with some other territorial omnivorous fish, aggression among juveniles is a potential problem that can be controlled by proper management. At 23-30°C, sheepshead will grow to 500 g in 9-11 months.

Caribbean sea bream have been reared on wild zooplankton in green water, but the species is too small to be of much use as food (Houde and Potthoff, 1976; Dowd and Houde, 1980).

Twoband sea bream were stripped after naturally ovulating in captivity, and larvae were reared at ~17°C on rotifers and *Artemia*, with survival of ~12% from hatching to 28 dah (Table 8-4; Jug-Dujaković and Glamuzina, 1988). Mortality was highest during 7-12 dah.

Common dentex have been induced to ovulate, and larvae were reared at ~22-23°C on rotifers and *Artemia* in green water, with typical survival of 15% from egg to 1 g (Glamuzina et al., 1989; Table 8-4; Sweetman, 1992).

Haffara sea bream have been induced to ovulate, and larvae were reared at 18-22°C on rotifers and compound feed in green water, followed by *Artemia* and cuttlefish paste, with survival of 11% from hatching to 70 dah (Table 8-4; Chang et al., 1994a).

Sobaity captive wild adults spawned voluntarily in tanks, and larvae were raised to juveniles in 4-m^3 tanks, then moved to cages and other tanks for grow-out, with a final suitable density of 60/m^3 (Almatar et al., 1997).

White sea bream captive females were induced to ovulate, and larvae were raised at 18-20°C in 200-L tanks on rotifers, *Artemia*, and inert feed, with 15% survival to transformation (Table 8-4; Pousão-Ferreira et al., 1997b). The 12-mm juveniles then grew on inert feeds to 51 mm at 94 dah, with 75% survival.

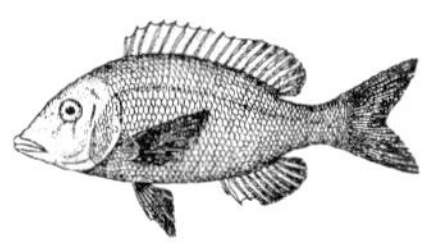

Bluestreak Emperor
Lethrinus choerorhynchus

Lethrinidae—Sweetlips

Spangled emperors have been induced to ovulate and have spawned voluntarily in captivity. Culture is well established in Japan (Furuya, 1995). Farming of them in brackish and saltwater ponds in Taiwan is being developed (Liao, 1993; Liao et al., 1995).

Sciaenidae—Drums

Spawning of sciaenids is relatively easy, and most species used for food have good growth rates.

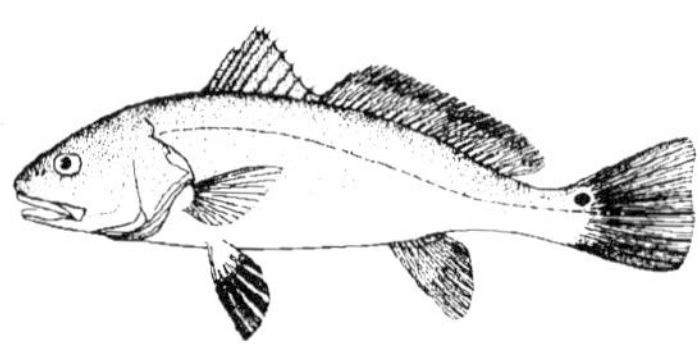

Red Drum
Sciaenops ocellatus

Red drum (Fig. 13-13) has been the subject of much aquaculture research in the southeastern U.S. and in the past decade has been raised in other warm regions (Colura et al., 1976, 1991; Theiling and Loyacano, 1976; Arnold et al., 1977; Roberts et al., 1978; Trimble, 1979; Crocker et al., 1981; Hysmith et al., 1983; Chamberlain, 1985; McCarty et al., 1986; Hopkins et al., 1986; Chamberlain et al., 1990; Davis, 1990a, 1990b; Soletchnik et al., 1990; Thouard et al., 1990; Hesskew, 1992; Holt, 1992, 1993; Sandifer et al., 1993; Thacker and Griffin, 1994; Henderson-Arzapalo, 1995; Liao et al., 1995; McEachron and Daniels, 1995; Thomas et al., 1995; Falguière and Blouin, 1997; Falguière and Paquotte, 1997; Falguière et al., 1997a; Tucker et al., 1997a). More than 200 million have been produced for stock enhancement. Red drum can be conditioned to spawn voluntarily during two or three cycles a year (natural spawning occurs in September-October), or nearly continuously, and can be induced to ovulate, if necessary. Conditioning cycles can be set up so that eggs are produced every month. Methods for mass-rearing of larvae (especially in ponds) are well developed. Pond survival from 5 dah to transformation (~1 month) has been as high as 90%, but usually is 30-70%. Larvae have been raised on rotifers and an MPD only, with weaning at 8 dah and survival of 61% from hatching to 32 dah (Table 8-4; Holt, 1993). Juveniles and adults can be raised in fresh water if hardness is 250 mg/L or higher. When red drum were grown out in 0.1-ha ponds on commercial trout grower feed for 543-547 d, those stocked at 7500/ha grew from 2.5 to

Figure 13-13. Red drum juvenile (122 dah, 126 mm TL, 21 g) without the typical caudal peduncle spot. Reared red drum sometimes have no spot or 2-3 spots.

1355 g (FCR 2.56, survival 88%), and those stocked at 22,500/ha grew from 4.3 to 1200 g (FCR 2.45, survival 91%). During grow-out, red drum in ponds need more than 2.2 mg/L oxygen; with high stocking densities, aeration is necessary. Juvenile temperature range is 8-33°C, but they can survive 4°C for 1-2 d and 0°C for a few hours if the change is not too abrupt. Red drum feeding and growth decrease at <20°C and are negligible at <10°C. Optimal temperature for growth is about 28°C (25-30°C), but *Amyloodinium* infestation (serious external parasite) and *Vibrio* spp. infection (not as serious) are more likely than in the low 20s. Red drum are susceptible to diseases induced by too much handling or crowding. Fatty liver with reduced viability can be caused by inappropriate feeds. The most common market size is 1.4-1.8 kg, which could require 2 yr to reach. Under ambient conditions (15-32°C) in Florida, red drum grew to 1.6 kg at 16 months, 2.2 kg at 19 months, and 4.7 kg at 30 months, but then growth slowed, FCR increased sharply, and the fish reached only 6.1 kg at 36 months (Tucker et al., 1997a, Fig. 9-6). With natural foods and constant good temperature (≥24°C), they can reach 3 kg at 18-20 months (Arnold, 1991; Falguière et al., 1997a). Considering growth rate and FCR, an increase in harvest size from the usual 1.0-2.0 kg range up to 4.0-5.0 kg seems feasible. Falguière and Paquotte (1997) reported that harvest densities of 30 kg/m^3 at 350 g and 28 kg/m^3 at 800 g were possible in cages. Maximum size is at least 43 kg.

Black drum have good potential for farming (Henderson-Arzapalo et al., 1994). They have been induced to ovulate and have spawned voluntarily in tanks. Black drum induced to ovulate also have been hybridized with red drum. Larvae (stocked at 2-3 dah) of all three were reared at 15-26°C in 0.1-ha ponds, and at 39-43 d after stocking, survival was close to 100%. Hybrid juveniles raised at 19-28°C in a pond on salmon and trout feeds grew faster that those of the parent species. After 27 weeks, hybrids weighed 190 g (survival 72%), black drum 144 g (94%), and red drum 142 g (63%). A taste-test panel judged all three to be similar. Juvenile black drum raised in a freshwater pond on catfish starter feed grew from ~30 mm TL to 127 mm (37 g) in 80 d (128 dah)(Garza et al., 1978).

Japanese croaker culture is well established in Japan (Furuya, 1995). Farming in brackish and saltwater ponds is well established in Taiwan (Liao et al., 1995).

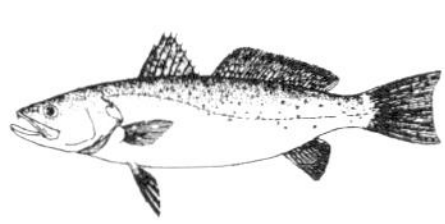

Spotted Seatrout
Cynoscion nebulosus

Spotted seatrout (Figs. 13-14 to 13-16) have been raised for stock enhancement in the U.S. This and other *Cynoscion* spp. might be good candidates for farming (Colura et al., 1976, 1992; Fable et al., 1978; Taniguchi, 1981; Mercer, 1984; Porter and Maciorowski, 1984; Gray et al., 1991). Spotted seatrout can easily be induced to ovulate, and under good conditions, they will spawn voluntarily for many months (Colura, 1974; Arnold et al., 1976;

Tucker and Faulkner, 1987; Capo et al., 1997). In a review of 52 pond culture trials (to ~1 month), Colura et al. (1992) reported that waiting 4 weeks after pond filling before stocking larvae allowed buildup of polychaete populations, which were judged important for survival of larvae (Table 8-4). In one study, spotted seatrout grew from eggs to 110 g in 5 months, with FCR in the range 0.8-1.5 (Tucker, 1988). With fresh foods, growth was

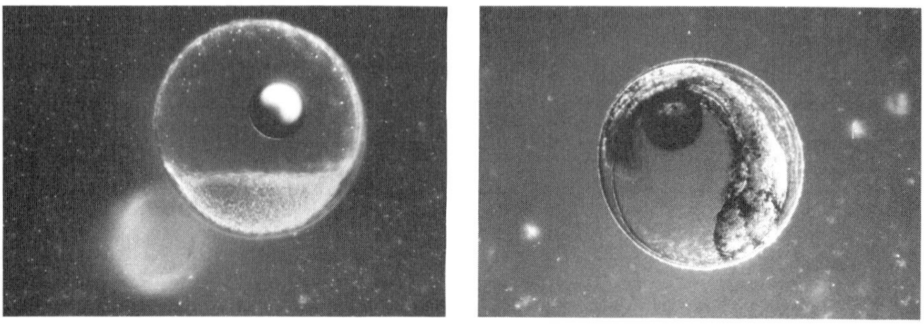

Figure 13-14. Spotted seatrout eggs (770 μm). Left: 2 h after fertilization; Right: 17 h after fertilization, just before hatching.

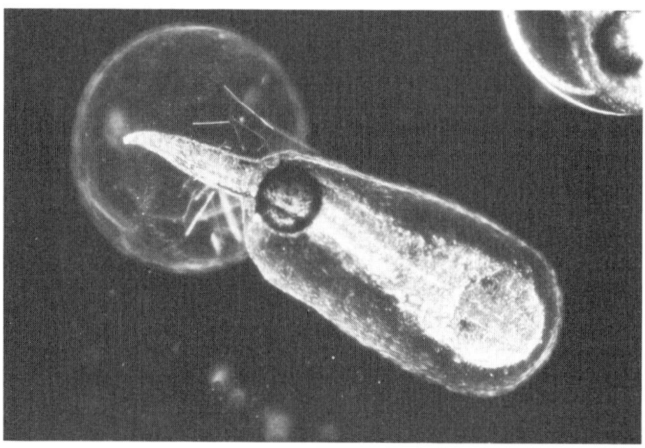

Figure 13-15. Spotted seatrout egg hatching (18 h after fertilization, 1.5 mm TL).

Figure 13-16. Spotted seatrout (53 dah, 73 mm TL, 5 g). (photo by Thomas A. Smoyer)

faster. In another study, they grew from eggs to 500 g in 10-11 months (Tucker et al., unpub). Cannibalism seems to be the main problem (especially in tanks), but can be minimized if the fish are weaned early (by 35-40 dah or sooner), fed well, and graded. For juveniles, 27-30°C probably is best. We have raised juveniles and adults in fresh water with a carbonate hardness of 769 mg/L, but handling tolerance is better if salinity is at least 5-10 ppt.

Orangemouth corvina, spotted seatrout, and hybrid (both crosses) larvae raised at 26-30°C in ponds on mixed zooplankton survived very well to ~1 month, and production was very good: ≤100% & 68,000, ≤96% & 835,900, and ≤100% & 78,000 juveniles (Bumguardner et al., 1992). The orangemouth corvina has good potential and reaches a maximum size of at least 18 kg (Van Olst and Carlberg, 1987). They spawn voluntarily in tanks and can be induced to ovulate.

Weakfish have been induced to ovulate. Eggs were stripped from ripe wild females and voluntary spawning occurred in tanks; larvae were raised at 20-21°C on rotifers and *Artemia* (Epifanio et al., 1988; Goshorn and Epifanio, 1991; Pryor and Epifanio, 1993).

White seabass have been raised for stock enhancement in California (Moser et al., 1983; Margulies, 1989; Orhun, 1989). Tens of thousands have been released. They spawn voluntarily in tanks and can be induced to ovulate. Larvae are ~4 mm TL at first feeding and can be raised on rotifers, *Artemia*, copepods, and minced krill with survival of 60% from hatching to 21 dah (Table 8-4). The larval period is long, and transformation occurs 42-52 dah (~22-35 mm TL). Juveniles grew from 4.5 (60 mm SL) to 76 g (164 mm) during 100-190 dah. Optimal temperature is ~19°C.

Mulloways were induced to ovulate, and larvae raised in tanks on rotifers and *Artemia*, pilchards, and compound feed grew to 21 g (121 mm TL) at 180 dah (Table 8-4; Battaglene and Talbot, 1994). Cannibalism began at ~22 dah. Settling occurred during 34-40 dah.

Small yellow croakers have been induced to ovulate. Eggs were stripped from ripe captive wild females, and larvae raised in a 500-L tank on rotifers, *Artemia*, compound feed, and minced fish grew from 3.2 mm TL at hatching to 65 mm at 2 months (Myoung et al., 1994). Caged yearlings grew from 48 g (169 mm TL) to 290 g (295 mm) in 21 months.

Some other species that have been used for food but do not get large or grow fast have been reared from eggs. **Spot** were induced to ovulate, and larvae were raised at ~20°C on rotifers, wild zooplankton, and *Artemia* through transformation (Powell and Gordy, 1980). **Atlantic croakers** have been induced to ovulate, and they have been raised to juveniles in tanks by methods similar to those for red drum (Hettler and Powell, 1981; Gwo et al., 1991). Early growth (21 g at ~127 dah; Davis and Arnold, 1997) is similar to that of red drum raised under ambient conditions (21 g at 122 dah; Fig. 13-21), but later growth is slower. **Whitemouth croakers** were induced to ovulate (Garcia, 1979).

Glaucosomatidae—Pearl perches

West Australia dhufish were induced to ovulate, and larvae were raised on rotifers and *Artemia* (Pironet and Neira, 1998).

Cichlidae—Cichlids

Tilapia culture in fresh water is widespread in tropical regions (Pullin and Lowe-McConnell, 1982). Experimental culture in salt water has been conducted in areas lacking abundant fresh water, such as the Middle East and Bahama Islands (Suresh and Lin, 1992; Macintosh and Little, 1995). Reproductive success, handling tolerance, low temperature tolerance, and disease resistance can diminish with increased salinity. Some tilapia grow faster in brackish water than in fresh. Larvae can be raised on phytoplankton/zooplankton blooms and older fish on larger pond plants and animals, but supplementary feeding improves growth.

Red hybrid tilapias from crosses between **Mozambique tilapia** and **Wami** or **Nile tilapia** are the most popular for salt water culture in the Caribbean region, and others have been tried. (Ernst et al., 1989, 1991; Watanabe et al., 1989, 1990, 1992, 1993b; Le Coz et al., 1990; Thouard et al., 1990; Watanabe, 1991; Bailey et al., 1992; Robinson et al., 1992; Smith et al., 1991; Ellis et al., 1993; Head et al., 1996; Falguière et al., 1997b). Daily egg and larval production by Florida red tilapia were better at 5 ppt (11.8/female) than at 18 ppt (7.2/female), but either rate could support a hatchery (Ernst et al., 1991). Larvae can be cannibalistic, especially if size varies. Florida red tilapia stocked at 25/m^3 in pools (22-30°C, 36-39 ppt) grew from 1.3 to 467 g in 170 d, with 90% survival and FCR of 1.6 (Watanabe et al., 1990).

Nile tilapia red hybrids were raised from 0.8 to 437 g in 239 d with 42% survival in a recirculating saltwater tank system (Cheong et al., 1987).

Oreochromis spilurus are raised in salt water in the Middle East (Al-Ahmad et al., 1988a, 1988b; Hazza et al., 1989, 1991; Anonymous, 1990; Al-Thobaiti et al., nd; Cruz and Ridha, 1994). Hatching (43 vs 21%), yolksac larval survival (92 vs 66%), and juvenile survival (46-68 vs 25-38%) rates were better in groundwater at 3-4 ppt than in seawater at 38-41 ppt (Al-Ahmad et al., 1988b). However, if larvae are produced in fresh water and acclimated to salt water at 1-2 weeks of age, survival to juveniles can be 70% (Hazza et al., 1989). When stocked at 28 fish/m^2 in saltwater pens, they grew from 6.3 to 270 g in 8 months (Al-Thobaiti et al., nd). In raceways, they can grow to 250 g in ~9 months (Hazza et al., 1991).

Striped Knifejaw
Oplegnathus fasciatus

Oplegnathidae—Knifejaws

Striped knifejaws have spawned voluntarily in tanks. Wild adults were induced to ovulate, and larvae raised at 21-27°C in a 100-m^3 outdoor tank on rotifers, copepods, compound feed, and fish paste grew from 3 mm at hatching to 27 mm at 40 dah (Kuronuma and Fukusho, 1984). Larvae were reared at 21-26°C in a tank on rotifers in green water, then *Artemia*, compound feed, and fish eggs (Table 8-4; Koh and Kim, 1992). Culture is well established in Japan (Furuya, 1995).

Mugilidae—Mullets

Mullets have been farmed in more than 50 countries (Oren, 1981). A few species have been induced to ovulate and are raised from eggs routinely, but most have been reared from wild juveniles. Mullet are hardy and adaptable and grow well in confinement. In extensive culture, without feeding, annual productions of 230-300 kg/ha have been obtained (Noriega-Curtis and Rivas, 1989). Larval culture is best in green water. Larvae can be reared in salt or brackish water, and juveniles grow well in fresh water, but mullet mostly have been grown out in brackishwater ponds. They can be polycultured with shrimp and non-piscivorous fish. Because they jump really high, net covers are needed for tank culture.

Striped Mullet
Mugil cephalus

Striped mullet is the main species raised (Tang, 1964; Liao et al., 1969, 1972; Shehadeh and Ellis, 1970; Nash et al., 1974; Sylvester et al., 1975; Chen, 1976; Anonymous, 1977; Lee and Tamaru, 1988; Eda et al., 1990a; Lee and Kelley, 1991; Liao and Chao, 1991; Liu et al., 1991b; Murashige et al., 1991; Walsh et al., 1991; Leung et al., 1993; Liao, 1993;

Tamaru et al., 1993a, 1994; Ako et al., 1994; Liu and Kelley, nd.b). Wild or captive broodstock routinely are induced to ovulate good eggs (e.g., with carp pituitary and GnRHa), but there have not yet been reports of captive voluntary spawning. Methods for mass-rearing of larvae in tanks and ponds are well developed. They are robust and grow moderately fast on rotifers and *Artemia*, but also can be fed cultured copepods or wild zooplankton (Table 8-4). By 40-45 dah, juveniles will eat detritus and almost anything organic, and grow well on complete or supplementary feeds with low to medium protein and low to medium fat. Juvenile temperature (3-?°C) and salinity (~0-70 ppt) tolerance ranges are wide. Although they transform at ~28 dah, striped mullet do not tolerate handling well until 40-50 dah. Survival of 30-50% from hatching to 60 dah is standard. Striped mullet can grow more than twice as fast in culture as in nature and are capable of reaching 450 g in 10 months, 800 g in 12 months, and 1.5 kg in 15 months (Sarig, 1981; pers. obs.). Striped mullet mature and can be harvested for their gonads at about 3 yr, but the best quality is available for only a few days just before ovulation.

White mullet were reared at 26-27°C in 38-L aquariums on wild copepods in green water to 36 mm TL at 36 dah, with survival at 20 dah as high as 23% (Table 8-4; Houde et al., 1976).

Lizas were raised from 2 to 800 g in 9 months (Tucker and Jory, 1991). Alvarez-Lajonchère et al. (1988) reported 11% survival from hatching to 26 dah for larvae that were fed rotifers and *Artemia* in green water at 23-24°C in a 1.8-m^3 tank (Table 8-4).

Redeye mullet eggs were stripped from ripe wild females, and larvae raised at 19-25°C in a 12-m^3 tank on oyster larvae, rotifers, *Artemia*, copepods, and minced fish settled at 35-40 dah and grew to ~47 mm at 70 dah (Kuronuma and Fukusho, 1984). Larvae were raised in 1-m^3 tanks at 15-20°C on rotifers and *Artemia* to 30-d-old juveniles (~14 mm TL), then at 19-28°C on dry feeds to ~82 dah (~53 mm)(Table 8-4; Yoshimatsu et al., 1992b). They can be induced to ovulate and larvae have been raised in ponds (Liu et al., 1991b).

Largescale mullet culture in fresh and brackishwater ponds is well established in Taiwan (Liao et al., 1995).

Sphyraenidae—Barracudas

Northern sennet larvae were raised from wild eggs at 23-25°C in a 55-L aquarium on wild zooplankton, *Artemia*, and fish eggs (Houde, 1972).

Polynemidae—Threadfins

Pacific threadfins, which are protandrous hermaphrodites, have been induced to ovulate and will spawn voluntarily in tanks and ponds (Szyper et al., 1991; Kelley et al., 1995; Ostrowski, 1995; Ostrowski et al., 1996b). Methods for mass-rearing of larvae are fairly well worked out. They were reared in 4-m^3 tanks on rotifers and *Artemia* in green water, with survival as high as ~27% from egg to 23-26 dah (Table 8-4). A suitable temperature range for larvae is 26-27°C. Cannibalism can be high during 26-40 dah (weaning). They can grow to 500 g at 8 months with FCR of ~1.6 and harvest density as high as 25/m^3.

Striped threadfin farming in brackish and saltwater ponds is being developed in Taiwan (Liao et al., 1995).

Labridae—Wrasses

Tautog larvae were raised on protozoans, rotifers, and *Artemia* (Perry et al., 1998). Juveniles fed *Artemia* and a commercial feed reached 30-500 g at 21 months.

Anarhichadidae—Wolffishes

Wolffish are good for food, but few are caught. Larvae are large and easy to feed, but grow slowly.

White Sea wolffish spawned voluntarily in boxes (Pavlov and Novikov, 1987). Incubation required 4-10 months, depending on temperature. About 5°C was suitable for egg incubation. Larvae were raised at 7-9°C in a 200-L trough on *Artemia*, minced tubifex, salmon starter, minced fish, and minced earthworms (Table 8-4). The fish grew to 34-45 mm at 196 dah.

Common wolffish wild eggs were collected, and larvae were raised at 4-7°C in 300-L tanks on compound feed and *Artemia*, with survival of 54% from hatching to 108 dah (Table 8-4; Moksness et al., 1989). Larvae were ~20 mm TL at first feeding, and juveniles became benthic by the time they were 100 mm. During 80-325 dah, protozoan ectoparasites caused 81% mortality. They grew to ~38 g at 325 dah.

Trichodontidae—Sandfishes
Sailfin sandfish culture is well established in Japan (Furuya, 1995).

Gobiidae—Gobies
Bluespotted mudhopper farming in brackishwater ponds is well established in Taiwan (Liao et al., 1995).

Flathead gobies have been induced to ovulate and have spawned voluntarily in captivity; farming of them in fresh and brackishwater ponds is being developed in Taiwan (Liao, 1993; Liao et al., 1995).

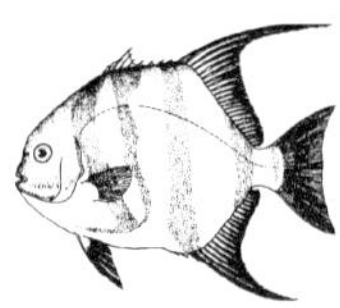

Atlantic Spadefish
Chaetodipterus faber

Ephippidae—Spadefishes
Atlantic spadefish are excellent foodfish but also are popular for large aquariums (also see chapter 15). They have been induced to ovulate with HCG or GnRHa and also have spawned voluntarily in tanks and a large aquarium. In one study, spadefish fed fresh sardines in cages grew from 0.2 to 327 g in a year (Gomez, 1984b). In another, survival to juveniles was 18-22% in aquaria and 7-10% in tanks; 2452 were produced (Manrique et al., 1988). Some of them (1 g) were raised on trash fish in cages to 282 g at 1 yr. Gomez (in press) reported results of successful induced ovulation and 24 larval rearing trials with rotifers, *Artemia* nauplii, and mysid shrimp.

Scatophagidae—Scats
Spotted scats might not grow fast enough to be farmed for food, but have potential as ornamental fish (see chapter 15).

Siganidae—Rabbitfishes
Rabbitfishes are raised mainly from wild juveniles in cages in the Indo-Pacific and Eastern Mediterranean regions (Duray, 1990). Larvae can first feed on rotifers. Juveniles and adults are mainly herbivorous and feed and grow well on a wide variety of algae. Cages, pens, and ponds are used for grow-out. Rabbitfish tolerate handling and crowding, but grow slowly.

Yellowblotch spinefoot is the main farmed species in Southeast Asia (Juario et al., 1985; Hara et al., 1986; Duray and Juario, 1987; Duray and Kohno, 1988; Juario, 1991; Kausar and Juario, 1991; Emata, 1994; Kohno et al., 1994c; Liao et al., 1995). They have been induced to ovulate and also spawn voluntarily in tanks. Larvae are ~2 mm TL at first feeding and can be raised at 26-30°C on rotifers, *Artemia*, and compound feed in green water, with survival from hatching to transformation (35 dah, ~18 mm) as high as 37% (Table 8-4; Juario et al., 1985). Within 12 dah, larvae begin nibbling on tank wall growths. In a pond, they grew from 150 mg to 13.8 g in 52 d. In a cage, juveniles grew from 12 to 75 g in 3 months. From 55 mm, they can reach a market size of 200-300 g in 5-8 months. Females matured at 215 mm FL in 12 months. This is one of the largest species, with a maximum size of more than 1 kg.

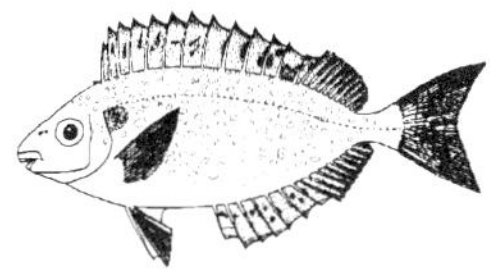

Pearly Spinefoot
Siganus canaliculatus

Pearly spinefoots have been induced to ovulate and will spawn voluntarily in tanks. Culture is well established in Japan (Furuya, 1995). Larvae raised at 28-32°C in 5.5-m^3 tanks on rotifers, copepods, and *Artemia* in green water (flagellates and unarmored dinoflagellates) grew from ~2 mm TL at hatching to juveniles (17-33 mm) by 30 dah (May et al., 1974). Trout feed was eaten as early as 20 dah and *Enteromorpha* as early as 23 dah. Pearly spinefoots were induced to ovulate and the young were raised on rotifers, *Artemia*, copepods, shrimp, and compound feed to ~100 mm TL at 90 dah (Table 8-4; Higuchi et al., 1980). Mortality peaks occurred from difficulties in first feeding and biting during 15-20 dah. In a tank, juveniles at 26-30°C grew from 2-6 g to ~40- to 55-g adults in 4-6 months (von Westernhagen and Rosenthal, 1975). In a cage, 10-g juveniles grew to 114 g in 4 months (Duray, 1990).

Goldline spinefoots spawned voluntarily in a tank, and larvae raised at 25-32°C on rotifers, copepods, and *Artemia* in green water grew from ~2.5 mm TL at hatching to ~23-mm juveniles at 35 dah, with survival of 16% (Table 8-4; Bryan and Madraisau, 1977).

Black spinefoots have been induced to ovulate and will spawn voluntarily in tanks (Popper et al., 1979; Duray, 1990). Larvae raised in a 1-m^3 tank (25-30°C, mostly 26-29°C) on rotifers, *Artemia*, and flake feed in green water began schooling 26 dah, reached 22-36 mm SL at 32 dah, and then grew to 9 g (92 mm) at 96 dah. In a cage, 3-g juveniles grew to 102 g in 6 months (Duray, 1990).

Vermiculated spinefoots have been induced to ovulate and will spawn voluntarily in tanks (Popper and Gundermann, 1976; Duray, 1990). Larvae were raised at 24-28°C on rotifers, copepods, and mosquito larvae to transformation (23 dah). In a pond, juveniles grew from 2 to 175 g in 6 months (Duray, 1990).

Sandy spinefoot farming in brackish and saltwater ponds is well established in Taiwan (Liao et al., 1995).

Scombridae—Mackerels

Several **scombrids** have been spawned in captivity, but mainly by stripping ripe wild females. Larvae have been raised from those eggs and from collected wild eggs, but usually not for long. Problems in rearing tunas include malnutrition, cannibalism, collisions with enclosure walls, high sensitivity to handling (especially when >1 kg), bacterial infections, and parasites (Harada, 1980; Margulies, 1993). Tunas reared from eggs often have not survived past 2-3 months. FSFRL (1973) considered oyster larvae preferable as a first food, rotifers satisfactory, and copepods indispensible as intermediate food for tuna larvae.

Little tunny larvae from wild eggs were raised at ~26°C in aquariums with green water and a 530-L fiberglass tank without green water on wild zooplankton, for up to 18 d (from 2.6 to 8.2 mm TL), but they were difficult to feed (Houde and Richards, 1969). By 10-12 dah, feeding was not good. In later work, little tunnies were raised to >20 mm.

Bullet mackerel eggs were obtained from ripe wild females, and larvae reared in tanks (3- and 30-m^3) grew from 2.5 mm TL at hatching to 15.5 mm at ~50 dah (FSFRL, 1973; Harada et al., 1973). Larvae raised in another hatchery on rotifers, copepods, fish larvae, chopped bivalves, and chopped fish survived to 67 dah (120 mm TL).

Frigate mackerel eggs were taken from ripe wild females, and larvae raised in tanks (3- and 70-m^3) on rotifers, copepods, and ground fish meat grew from 2.5 mm TL at hatching to 120 mm at 33 dah (FSFRL, 1973).

Chub mackerel spawned voluntarily and were induced to ovulate (Lasker, 1974). Larvae were raised at 17-22°C in 400-L tanks on rotifers and copepods in green water (Table 8-4; Hunter and Kimbrell, 1981).

Striped bonito larvae were raised to 99 dah (Harada et al., 1974). Wild juveniles grew from 750 to 1700 g in ~1 yr (FSFRL, 1973).

Skipjack tuna eggs were obtained from ripe wild females, but larvae survived for less than 4.5 d; hatchlings were 2.5 mm TL (FSFRL, 1973).

Yellowfin tuna eggs were stripped from ripe wild females, and larvae were raised at 26-29°C on rotifers, oyster larvae, and copepods, but lived for only 20 d (from ~2.5 to ~9 mm TL)(Table 8-4; Harada et al., 1971; Mori et al., 1971; FSFRL, 1973). Later, larvae were raised to juveniles at ~38 dah (~51 mm)(Harada et al., 1980). Wild yellowfin tunas can grow to 64 cm FL at 1 yr, 105 cm at 2 yr, 135 cm at 3 yr, 156 cm at 4 yr, 169 cm at 5 yr, and 177 cm at 6 yr (Suzuki, 1994). Maximum size is ~120 kg.

Northern Bluefin Tuna
Thunnus thynnus

Northern bluefin tuna spawned at 22-25°C in a large cage, and larvae were raised in (Harada, 1980). They grew from 2.8 mm TL at hatching to ~30 mm at 1 month and ~60 mm at 6 weeks. They can grow to 2-3.5 kg in 1 yr, 7-15 kg in 2 yr, 15-45 kg in 3 yr, 30-70 kg in 4 yr, and 40-90 kg in 5 yr. Oceanic-quality water seems necessary for rearing bluefin tuna. Although they can survive 12°C, exposure to <15°C can cause a reduction in growth later (FSFRL, 1973).

Southern bluefin tuna wild juveniles are grown out commercially in cages at several farms in Australia (Battaglene, 1996).

Black skipjack wild juveniles raised at 24-29°C in tanks on wild zooplankton, molly larvae, and chopped fish grew from ~50 mg to 320 g in ~90 d (Olson and Scholey, 1990).

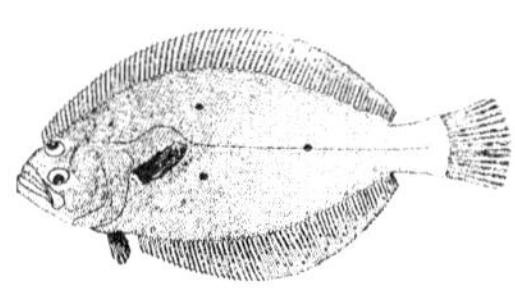

Hirame
Paralichthys olivaceus

Bothidae—Lefteye Flounders

Hirame are raised mainly in Japan and the Republic of Korea (Kuronuma and Fukusho, 1984; Yamaha Fishery Journal, 1991; Ikenoue and Kafuku, 1992; Imaizumi, 1993; Jeon et al., 1993; Spedicato et al., 1993a, 1993b, 1994; Furuya, 1995; Matsuoka, 1995; Min, 1995; Furuta et al., 1997). They spawn voluntarily in tanks and can be induced to ovulate. Methods for mass-rearing of larvae are well developed. In the past, a variety of foods were used, but recently the diet has been simplified to rotifers, *Artemia*, and dry feed. MPDs can be given to them as early as 8 dah (Kyowa, nd.a). They have been raised on rotifers and *Artemia* in green water, with survival of 45% from hatching to 85 dah (Table 8-4). Larvae do well at 16-18°C. Hatchery production of 1500-5000/m^3 at 20 mm (25-80% survival) is possible. Early juveniles tend to be cannibalistic. Grow-out is either in tanks or cages, but the cage bottoms need to be smooth and the waves not large. Stocking density in tanks can be as high as 200/m^2 at 10-g and 38/m^2 at 800 g. Trash fish, dry pellets, or moist pellets are used. Growth is best at 20-24°C. Furuta et al. (1997) reported growth in a tank from 3.5 to 480 g in 337 d, with overall FCR of 1.0 and harvest density of 23 kg/m^3. Because females mature later than males and grow faster until then, producing all-female stocks is desirable (Kato, 1994). Hirame can reach 250-500 g within 1 yr, 700-1000 g in

2 yr, and 1.4 kg or more in 3 yr. Maximum size is ~5 kg. The usual market sizes are 400-1500 g. Moksness and Støle (1997) reported a production cost (1993) of US$0.61 for a 70-mm juvenile.

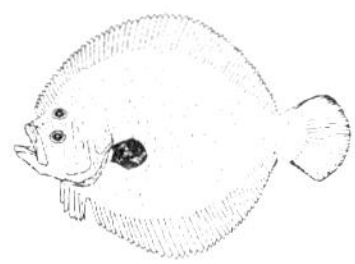

Turbot
Scophthalmus maximus

Turbot are farmed mainly in northern Spain and western France (Jones, 1972; Alderson and Bromley, 1973; Girin, 1974; Jones et al., 1974, 1981; Alderson and Ball, 1977; Nevo, 1983; Benavente and Gatesoupe, 1988; Cachelou et al., 1989; Person-Le Ruyet, 1990b; Person-Le Ruyet et al., 1991; van der Meeren, 1991; Støttrup and Attramadal, 1992; Danielssen and Hjertnes, 1993; Stoss and Røer, 1993; Riaza and Hall, 1993; Støttrup, 1993; Lavens and Remmerswaal, 1994; Reitan et al., 1994b; Cunha and Planas, 1995; Støttrup et al., 1995; Dhert et al., 1998). Rearing turbot is not difficult, but gaining control over spawning has taken some effort. They ripen in captivity without hormones, but voluntary spawning is uncommon and strip-spawning is standard. The natural spawning season is for 2-3 months in the summer, with each female spawning 6-8 times. By control of photoperiod, temperature, and nutrition, turbot now can be spawned year-round. At first feeding, larvae are ~4 mm TL and optimal prey width is <150 µm. They have been raised on rotifers and *Artemia*, with survival of 40% from hatching to metamorphosis (Table 8-4). Green water rearing also has been successful. Methods for mass-rearing larvae in tanks (20-40 larvae/L) with rotifers and *Artemia* (sometimes copepods) and in plastic bags (usually <1 larva/L) with wild zooplankton are well developed. Larvae can be stocked at relatively high density (30/L). Malpigmentation is rare when natural zooplankton is used, but sometimes is common without it. Weaning can be accomplished by ~30 dah. By ~40 dah, the fish become benthic and can be moved to larger tanks. Turbot are relatively fragile until then. They usually are grown-out in tanks or lined ponds, but cages have been used. Possible stocking densities are 1000/m^2 at 2 g, 100/m^2 at 125 g, and 25-120 kg/m^3 (usually ≤50 kg/m^2) for larger fish. Juveniles can survive and grow in the range ~8-25°C, but grow best at ~18°C. Turbot can reach 30 g at 6 months, 450 g within 15 months, 1 kg within 20 months, 2 kg within 28 months, and 3 kg within 3 yr. Market size is in the range 1-5 kg. Maximum weight is ~10 kg.

Summer flounders are being brought into commercial-scale production in the northeastern U.S. (Smigielski, 1975b; Bisbal and Bengtson, 1995a, 1995b; Nardi, 1996, 1997; Bengtson, 1997; Watanabe et al., 1998c). Hormone-induced ovulation of captive broodstock is possible year-round. Summer flounders were induced to ovulate, and larvae were raised at 15-18°C on rotifers, copepods, and *Artemia* in green water, with 5% survival from ~hatching to 56 dah (Table 8-4; Smigielski, 1975b). Larvae raised in 6-L pans on wild zooplankton and rotifers had a survival of 37% from hatching to 30 dah (Klein-MacPhee, 1981). In a commercial operation, captive broodstock were stripped, and at 1 dah, 30 larvae/L were stocked in 3-m rounded square tanks (Nardi, 1997). They were fed rotifers and *Artemia* in green water at 16-20°C, then weaned to compound feed after metamorphosis (~35 dah) and settlement, with 33% survival. Mean weight was 8 g at 5 months. Cannibalism among juveniles can be bad. They grow well on hybrid striped bass feeds and can reach 1 kg in ~2 yr.

Southern flounders have been induced to ovulate and have spawned voluntarily in captivity; larvae have been raised in tanks and ponds (Arnold et al., 1977; Berlinsky et al., 1996; Daniels et al., 1996, 1997; Jenkins et al., 1997b). In one study, adults spawned every second day for 2-week periods in the winter. Larvae from voluntarily-spawned eggs were reared in 57-L aquariums on rotifers, *Artemia*, and minced shrimp, with survival of 30% from eggs to 90 dah (Table 8-4; Arnold et al., 1977). Larvae from induced ovulation were raised in a 0.1-ha lined pond on wild zooplankton including abundant rotifers; growth was good until 3 months (~77 mm TL), but fish reached only 96 mm at 206 dah, probably because live food had been depleted and the juveniles did not eat much of the salmon and trout feeds given to them (Jenkins et al., 1997b). Potential growth has been estimated at 1 kg in 2 yr.

Gulf flounder and southern flounder wild females were induced to ovulate, and larvae were raised (16-19°C, 14-24°C) to about metamorphosis in 100-L tanks on rotifers and *Artemia* (Powell and Henley, 1995).

California halibut have been induced to ovulate and have spawned voluntarily in tanks, and larvae have been raised through transformation at 16-20°C in 300-L and smaller tanks on rotifers and *Artemia* (Table 8-4; Caddell et al., 1990; Gadomski et al., 1990; Gadomski and Caddell, 1991). California halibut and **fantail sole** captive wild females were induced to ovulate, and larvae were raised in 4-L containers (Oda, 1991).

Chilean flounder and ***Paralichthys adspersus*** eggs were obtained from ripe wild females, and larvae were raised at 13-17°C through metamorphosis in 200-L tanks on rotifers and *Artemia* (Table 8-4; Zuñiga and Acuña, 1992). Metamorphosis occurs by ~60 dah, and survival of 5-10% to 90-dah juveniles (30-40 mm) is possible, at least for *P. adspersus* (Alvial and Trujillo, 1993).

Ecuadorian flounders spawned voluntarily in tanks, and larvae were raised at 25-26°C in 15-m^3 tanks on wild zooplankton, rotifers, and *Artemia* in green water (Benetti, 1997). The larvae were moved to 200-L tanks before the beginning of metamorphosis (12-14 dah). When juveniles were 10-15 mm long, they were moved (2000/tank) to 1.5 m diam. × 30 cm high tanks, weaned in 2 weeks, then moved to grow-out tanks. The main problems were lack of metamorphosis, bacterial enteritis, and malpigmentation. Wild juveniles stocked at 30/m^2 in a 400-L tank section and given a 52%-protein semipurified feed at 25-26°C grew from 35 to 73 g in 8 weeks with FCR of 1.58 (Pedrazzoli et al., 1997). Wild juveniles in tanks grew to 1 kg in 1 yr with FCR of 3.9 (Benetti, 1995).

Pleuronectidae—Righteye Flounders

Atlantic halibut, a coldwater species, is the main flatfish farmed in northern Europe (Pittman et al., 1987, 1989, 1990; Haug, 1990; Rabben, 1991; van der Meeren, 1991; Naas et al., 1992; Tilseth et al., 1992; Adoff et al., 1993; Helvik and Walther, 1993; Lein and Holmefjord, 1993; Opstad and Bergh, 1993; Reinertsen et al., 1993; Reitan et al., 1993a, 1993b; Trumble et al., 1993; Björnsson, 1994; Holm et al., 1995; Kjørsvik and Holmefjord, 1995; Næss et al., 1995b; Næss, 1996; Torrissen et al., 1995; Björnsson and Tryggvadóttir, 1996; Pittman, 1996; Berg, 1997; Sutherland, 1997). Some research on eggs and larvae was done in the 1930s, but aquaculture research did not began until about 1974 and large numbers of juveniles (100,000s) were not produced until the early 1990s. Females might spawn 6-16 times during the natural spawning season of January-April. They ripen in captivity without hormones, and some voluntary spawning has occurred, but strip-spawning is regularly employed; egg quality varies. The diameter (3400 μm) and yolk volume are very large for planktonic eggs, and hatching time is long (13-17 d). It is helpful to disinfect eggs 11 daf. Light or turbulence can inhibit hatching. Because yolksac larvae are sensitive and need to be incubated for a long time before first feeding (~230-300 degree-days), incubator tanks should provide gentle upwelling to keep the larvae off the bottom without battering them. Slightly increasing salinity to 35.5 ppt is another strategy. Yolksac larvae are very susceptible to mechanical injury and bacterial infection, which can cause permanent damage and dysfunction, especially to the mouth. Light also can be harmful. In some bag experiments, survival during the yolksac period has been as high as 45%. Larvae are transferred from incubators to rearing tanks before first feeding. The total larval period is long (≥90 d, ~47 mm TL). Mortality peaks can occur especially during 15-35 dah and several days after first feeding. Mortality at hatching could result from defects, bacterial infection, or oxygen concentration (Pittman et al., 1990). Larval mortality during ~10-14 dah is likely a result of defects, bacterial infection, or water flow. Mortality ~25-35 dah could result from energy depletion, behavioral problems, light, or water quality. Mortality ~45-60 dah could result from starvation. Greenwater culture seems best, and wild zooplankton can boost survival and growth. Larvae were reared on *Artemia* only, in green water, from first feeding (33 dah, ~18 mm) to 54 dah, with survival of 30% (Table 8-4). Survival from egg to juvenile might reach 35%. Rotifers, wild zooplankton, and copepods sometimes are used. Cannibalism among juveniles is possible. For the best grow-out production, stocking density should be 1-2 layers of fish on the bottom. Incubation temperature in the lower part of the natural range, 4°C, resulted in better growth, yolk utilization, and development than 6-9°C. For yolksac larvae, 5-7°C is suitable. For older fish, the best temperature for growth decreases with size from 14°C at 10-60 g, to 10°C at 3-5 kg, but for growth efficiency the range is wider, 14 to 6°C. During the first few years, halibut can grow twice as fast in culture as in nature. Atlantic halibut can grow to 40 g in 7 months, 1 kg in ~1 yr, and 3-5 kg in ~3 yr. Females grow faster than males. Wild females fed capelin, herring, squid, and shrimp in tanks grew from 2 to 11 kg in 3 yr, and males from 2 to 6 kg. Market size is 2-5 kg or larger. At 28 yr, wild females weighed 138 kg and males 55 kg. Maximum female size is about 333 kg.

Pacific halibut ripened in captivity and were stripped and also have been induced to ovulate; larvae were raised at ~8°C in a 100-L tank (Liu et al., 1991a, 1993; Jensen et al., 1992).

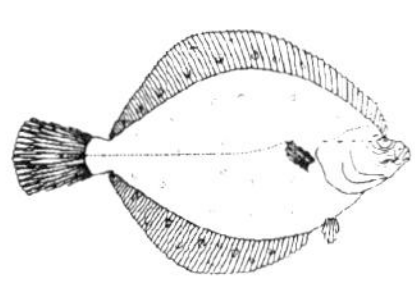

European Plaice
Pleuronectes platessa

European plaice, a coldwater species with a short natural spawning season, spawned voluntarily in outdoor and indoor tanks (Shelbourne, 1974, 1975). Larvae were raised through metamorphosis in 36-L (60 × 30 × 20 cm) aquariums on *Artemia* nauplii, and juveniles were fed adult *Artemia* and white worms (*Enchytraeus albidus*). For larvae, 12°C is good. In one experiment, survival from eggs (27.7/L) to 15-mm fish at 85 dah (103 daf) was 66%, with a harvest density of 3650/m^2. At that stage thinning was needed to avoid fierce aggression and mortality. With 83 eggs/L stocked, a higher juvenile density (6066/m^2) resulted, but survival was only 36%. Larval survival could routinely average 25-30%. Larvae raised in 60-L tanks on rotifers and *Artemia* had a survival of ~39% from hatching to 50 dah (Table 8-4; Dickey-Collas and Geffen, 1992). Moksness and Støle (1997) reported a production cost (1994-1995) of US$0.54 for a 50-mm juvenile.

Winter flounders have been induced to ovulate and have spawned voluntarily in a large aquarium; larvae have been raised to adults that spawned voluntarily (Smigielski and Arnold, 1972; Laurence, 1975, 1977; Smigielski, 1975a; Klein-MacPhee, 1983; Litvak, 1994; Lee and Litvak, 1996a, 1996b). Adhesiveness of the eggs can be removed with diatomaceous earth. Larvae have been raised on wild zooplankton or rotifers and *Artemia*, in clear and green water, in a variety of tanks. Survival of larvae fed either rotifers, copepods, or wild zooplankton in green water was 24-36% at 25 dah (Klein-MacPhee, 1983). For larvae, ~8-10°C is good. Growth and survival (22%, 18%) of larvae fed rotifers or wild zooplankton were similar, but green water enhanced growth with both foods (King et al., 1997). The best temperatures for juveniles are ~15-18°C for northern populations and ~20-23°C for southern populations.

Marbled flounder culture is well established in Japan (Furuya, 1995). Wild females were induced to ovulate, and larvae from the adhesive eggs were raised at 12-14°C in a 150-m^3 outdoor tank on rotifers, copepods, *Artemia*, and minced sandlance in green water, with 70% survival at 54 dah and growth from 3.7 mm at hatching to 18 mm (Kuronuma and Fukusho, 1984).

Barfin flounder culture is well established in Japan (Furuya, 1995).

Starry flounder wild females were induced to ovulate, and larvae were raised through transformation at 12-15°C in 38-L aquariums on rotifers in green water (Table 8-4; Policansky and Sieswerda, 1979).

Ridged-eye flounder wild females were induced to ovulate and captive fish spawned voluntarily; larvae were reared through transformation on rotifers, *Artemia*, and copepods (Table 8-4; Kitajima et al., 1988).

Witch flounder eggs were stripped from ripe wild females, and larvae were raised at ~13°C through transformation in 200-L tanks on wild zooplankton and rotifers in green water (Table 8-4; Bidwell et al., 1997).

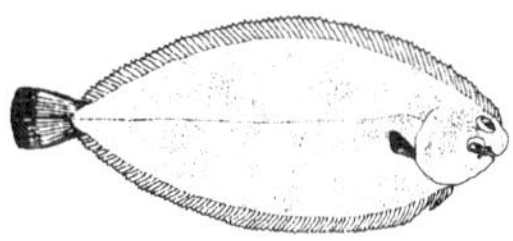

Common Sole
Solea solea

Soleidae—Soles

Common sole larvae are easy to raise, and they survive well, but juveniles grow very slowly (Shelbourne, 1975; Person-Le Ruyet, 1990b). They spawn voluntarily in tanks and can be induced to ovulate. Methods for mass-rearing of larvae are well developed (Table 8-4). Larvae can be stocked at high density (70/L), and early juveniles can be crowded (≤4000/m^2) more than plaice and turbot. In early work (Shelbourne, 1975), larvae were reared through transformation (30-35 dah) in 36-L tanks on *Artemia* nauplii; survival was 80% from hatching to ~30 dah, with a harvest density of 4438/m^2. Larvae in 6.1-m^3 tanks were fed *Artemia* nauplii, then at 10 weeks enchytraeid worms (*Lumbricillus rivalis*) were added to the juveniles' diet; survival was 42% from eggs to 95 daf (~85 dah), with a harvest density of 5487/m^2. They metamorphose and become benthic by ~15 dah but transform 2-3 weeks later. Weaning can be accomplished by 1 month, at which time (or sooner) they can be moved to larger tanks. Normal temperatures for eggs are 10-19°C, depending on the region. Juveniles grow best at ~20-22°C or a little higher. Common sole can reach 20 g at 6 months and 250 g in ~2 yr. They can grow from 5 g to the minimum market size of 125 g (24 cm) in 300 d (Howell, 1997). Although large numbers of common sole and Senegal sole have been produced in hatcheries, weaning difficulties and sensitivity to intensive culture conditions adversely affect survival and growth (Flos et al., 1997).

Senegal sole captive wild adults spawned voluntarily, and larvae were raised through transformation at 18°C in 150-L tanks on rotifers, *Artemia* and compound feed, reaching as much as 28 mm TL and 182 mg (31% survival) at ~130 dah (Dinis, 1992).

Cynoglossus abbreviatus wild females were induced to ovulate, and larvae were raised (starting at 14-16°C) through transformation on rotifers, *Artemia*, and copepods (Table 8-4; Fujita et al., 1986).

Balistidae—Leatherjackets

Balistid larvae have small mouths and are not easy to feed. Juveniles and adults tend to be fiercely aggressive in tanks. This is a difficult family, but some ornamental species might have potential (see chapter 15).

Tetraodontidae—Puffers

Tiger puffers are sold to fugu restaurants (Doumenge, 1990a; Kanazawa, 1991b). Culture is well established in Japan (Furuya, 1995). They have spawned voluntarily in tanks and have been induced to ovulate. The adhesive eggs have been stripped from ripe females. Larvae were raised in a ~45-m^3 tank on barnacle larvae, rotifers, *Artemia*, wild zooplankton, copepods, and minced fish or shrimp in green water, with survival to 40-50 dah (~25 mm) of usually ~10%, and rarely ~20% (Kuronuma and Fukusho, 1984). There were mortality peaks at 12-15 dah (large) and 20-30 dah (small). Larvae can be raised on rotifers, MPDs, and *Artemia* and can grow to 45-50 mm at ~60 dah. For juveniles, 25-26°C is suitable. Market size is 1 kg.

Half-smooth golden puffer larvae were raised on boiled egg yolk and *Artemia* from 1.9 mm TL at hatching to 6.6-mm juveniles at 30 dah (Fujita, 1966).

Balloonfish spawned voluntarily at 24-26°C in a large aquarium, and larvae raised on rotifers, *Tigriopus*, and *Artemia* grew from 2.6 mm TL at hatching to 20-mm juveniles at ~25 dah and 46 mm at 66 dah (Sakamoto and Suzuki, 1978).

Takifugu chrysops eggs were stripped from ripe wild females, and larvae raised in a 60-L tank on rotifers, *Artemia*, *Tigriopus*, and clam meat grew from 2.9 mm TL to 9-mm juveniles at 24 dah and 63 mm at 146 dah (Table 8-4; Fujita and Shinohara, 1986).

Takifugu exascurus were induced to ovulate, and larvae raised on rotifers, *Artemia*, and minced fish grew from 3.0 mm at hatching to 10-mm juveniles at 16 dah and 99 mm at 332 dah (Table 8-4; Fujita and Honma, 1991).

Purple puffers were induced to ovulate, and larvae (20-25°C) were raised on rotifers, *Artemia*, and fish meat to 13-mm juveniles at 20 dah and 101 mm (16-28°C) at 216 dah (Fujita and Abe, 1992). Early juveniles were very aggressive and bit the caudal fins of others.

ENVIRONMENTAL CONSIDERATIONS

The dangers of exotic fish and disease introductions, chemical contamination, eutrophication, and usage conflicts range from high for cage farms to moderate for pond farms to very low for recirculating tank farms. Increased control, protection from environmental changes, and other factors add to the list of advantages for tank farming (see chapter 3).

SUMMARY

Besides salmonids, relatively few coldwater marine fish species seem to have good potential for farming. Nearer the equator, more candidates appear. Table 13-1 compares natural growth of some warmwater species from the western North Atlantic. In culture, growth

Table 13-1. Estimated average number of months required for wild fish to reach 454 g (data from numerous published and unpublished sources, including Manooch and Mason, 1984, 1987; Mason and Manooch, 1985; Matheson et al., 1986; Manooch and Drennon, 1987; Mercer, 1989; Potts and Manooch, 1995; Manooch and Potts, 1997a, 1997b; Manooch, pers. comm., 1998; Potts et al., in press).

Wahoo	3	Mutton snapper	27
Dolphin	6	Queen triggerfish	30
King mackerel	9	Rock hind (grouper)	31
Greater amberjack	10	Blueline tilefish	34
Skipjack, yellowfin, blackfin, bluefin tunas	<12	Atlantic red snapper	34
Warsaw grouper	12	Knobbed porgy	39
Hogfish (wrasse)	13	Yellowtail snapper	44
Black grouper	17	Gray snapper	48
Gag (grouper)	18	Black sea bass	51
Scamp (grouper)	20	White grunt	57
Red hind (grouper)	23	Vermilion snapper	61
Red grouper	27	Lane snapper	79

Table 13-2. Characteristics of representative foodfish raised commercially (C) and experimentally (E).

Fish	Type of culture	Locations	Type of spawning[a]	Egg diameter (μm)	Larval duration (d)	Level of larval rearing[b]	FCR at 454 g[c]	Time to 454 g[c] (months)	Fillet yield at 454 g[d] (%)
Anguillidae									
Japanese eel	C	Far East	HT,HS	2600	~300	10		≥12	L
Chanidae									
Milkfish	C	Indo-Pacific	V,HT,HS,NS	1150	~21	3	~2	~12	L
Gadidae									
Atlantic cod	C	Europe, NE U.S.	V,NS	1520	~106	4			M
Scorpaenidae									
Kurosoi	C	E Asia, Japan, R.O.K.	V		~35	3			M
Centropomidae									
Barramundi	C	S Asia, Australia	V,HT,HS,NS	750	~25	4	1.1	11	H
Common snook	E-C	Florida, Caribbean	HT,HS,NS	700	~35	6	1.0	12	57
Percichthyidae									
European sea bass	C	Europe	V,HT,HS	1350	~45	4	1.4	~16	M
Sunshine bass	C	U.S.	HT,HS,NS[e]	980	~30	4	1.3	8	40
Serranidae									
Brownmarbled grouper	C	SE Asia	V,HT,HS,NS	840	~40	7			M
Greasy grouper	C	SE Asia	V,HS	900	≥36	8			M
Malabar grouper	C	SE Asia	V,HT,HS	852	~50	7	1.9		M
Orangespotted grouper	C	SE Asia, Middle East	V,HT,HS	807	~40	7			M
Redspotted grouper	C	Japan	V,HS	825	~50	7			M
Nassau grouper	E	Caribbean	V,HT,HS,NS	920	~58	7	1.1	12	M
Carangidae									
Yellowtail	C	SE Asia	V,HT,HS	1150	~53	5	1.3	6	H
Florida pompano	E	E U.S.	HT,HS	970	≥23	5	~2.0	11	30
Permit	E-C	SE Asia	HT	900		5		8	M
Coryphaenidae									
Dolphin	E-C	Hawaii, Florida	V	1500	≥20	6	1.1	4	67

Lutjanidae									
Asian red snapper	C	SE Asia	V,HT,HS	795	~35	6		10	M
Sparidae									
Gilthead sea bream	C	Mediterranean	V,HT,HS	950	~40	4	1.4	~14	L
Red sea bream	C	E Asia	V,HT,HS	975	≥30	4	~2.0	~16	L
Black sea bream	C	E Asia	V,HS		~40	4			L
Sheepshead	E	SE U.S.	HT,HS,NS	820	~39	4		11	L
Sciaenidae									
Red drum	C	SE U.S.	V,HT,HS,NS	950	~35	3	1.3	8	31
Spotted seatrout	E	SE U.S.	V,HT,HS,NS	700	~37	3		10	H
Cichlidae									
Florida red tilapia	C	Florida, Caribbean	V,HT	2630 × 1960	≥7	1	1.6	7	33
Mugilidae									
Striped mullet	C	Most warm regions	HT,HS,NS	930	~28	3		10	25
Polynemidae									
Pacific threadfin	E-C	Hawaii	V,HS		≥26	4	≥1.6	≤8	M
Scombridae									
Yellowfin tuna	E	Japan	NS	960	~58	9			H
Bothidae									
Hirame	C	Japan, R.O.K.	V	930	≥27	5	>1.0	12	M
Turbot	C	Europe	V,NS	1080	~40			~15	M
Summer flounder	E-C	NE U.S.	HT,HS	1020	~56	5			M
Southern flounder	E	SE U.S.	V,HS	910	~50	5		~14	M
Pleuronectidae									
Atlantic halibut	C	Northern Europe	V,NS	3400	~100	6		~10	M
Tetraodontidae									
Tiger puffer	C	Far East	V,HS,NS	1000	~30	7			L

[a] Spawning type: V = voluntary spawning; HT = hormone-induced tank spawning; HS = hormone-induced ovulation followed by strip spawning; NS = natural ovulation followed by strip spawning.
[b] Rearing level: 1 = easy; 3 = moderately easy; 5 = moderately difficult; 7 = difficult; 9 = very difficult; 10 = nearly impossible.
[c] with compound feeds.
[d] Yield: L = low; M = medium; H = high.
[e] to produce the hybrid eggs.

rates often can be at least doubled or tripled because of temperature control and greater food availability. Pelagic species typically grow faster than demersal species. Food fish that grow slowly probably are hopeless for aquaculture, but slow-growing bait and ornamental fish might be feasible.

Table 13-2 compares culture characteristics of 34 representatives from 17 families, including some species raised commercially and others that have been spawned and reared on at least a research scale and show good potential. Feed conversion ratio with dry pellets is often near 1.0 for 454-gram fish. For good non-pelagic species, less than 12 months are required to produce 454-gram fish, which are gaining at least 3-5 grams/d and can then be raised to at least 1 kg within 6 more months. Most marine foodfish candidates can be raised from eggs to at least 2 kg in 2 yr. A sophisticated hatchery is required for most of these—a saltwater hatchery for all but a few such as sturgeons, salmonids, ayu, and hybrid striped bass, which need fresh water. Hybrid bass also should be grown out in fresh or slightly brackish water. Barramundi, snooks, striped bass, some snappers, several sea breams, several drums, and mullets can be grown out in fresh or salt water. Groupers can be grown out in brackish water (≥15 ppt for many). Typical pelagic species (e.g., dolphin, tuna) require salt water for all stages.

Most can be grown out to market size in tanks, but more active species tend to require larger tanks. All can grow well in properly-designed cages, but milkfish and mullet are better off in ponds, and pelagic species need large cages. The non-pelagic species do well in ponds. Fish in ponds and cages are more vulnerable to temperature and oxygen extremes and to diseases.

Species that are piscivorous are likely to have cannibalistic tendencies. Species that are very territorial (mainly to guard feeding or nesting areas) are likely to become very aggressive sometime between transformation and maturity. Some active pelagic species grow fast but suffer from confinement more than less active species, might need specially designed enclosures, and might not tolerate high stocking density.

Illustration credits for chapter 13: Japanese sea bass, p 545; Malabar grouper, p 546; reprinted from Katayama (1960), with permission from Tokyo News Service, Ltd. Hirame, p 566; reprinted from Amaoka (1969), with permission from Kunio Amaoka. Black sea bream, p 555; bluestreak emperor, p 558; striped knifejaw 562; reprinted from Lindberg and Krasyukova (1971), with permission from the U.S. National Technical Information Service. Starry sturgeon, p 533; turbot, p 567; common sole, p 570; reprinted from Fischer (1973), with permission from the Food and Agriculture Organization of the United Nations (FAO). Milkfish, p 535; barramundi, p 539; polka dot grouper, p 545; leopard coraltrout, p 549; golden snapper, p 553; pearly spinefoot, p 565; reprinted from Fischer et al. (1974), with permission from FAO. Atlantic salmon, p 536; reprinted from Greenwood, 1975, J.R. Norman: A History of Fishes, with permission. American eel, p 534; common snook, p 540; striped bass, p 543; Nassau grouper, p 547; Florida pompano, p 550; dolphin, p 552; sheepshead, p 556; red drum, p 558; spotted seatrout, p 559; striped mullet, p 562; Atlantic spadefish, p 564; northern bluefin tuna, p 566; reprinted from Fischer (1978), with permission from FAO. European sea bass, p 542; reprinted from Fischer et al. (1981), with permission from FAO. Gilthead sea bream, p 556; European plaice, p 569; reprinted from Whitehead et al. (1984), with permission from the United Nations Educational, Scientific and Cultural Organization (UNESCO). Atlantic cod, p 537; reprinted from Dipper (1987), with permission from Robert Irving. Yellowtail, p 550; red sea bream, p 554; reprinted from: H. Ikenoue and T. Kafuku, Modern Methods of Aquaculture in Japan, copyright 1992, p 131 and 119, with permission from Elsevier Science.

14 CULTURE OF ESTABLISHED AND POTENTIAL SPECIES—BAIT FISH

Some references on freshwater baitfish culture are: Brown and Gratzek (1980), Giudice et al. (1982), Huner and Dupree (1984), Stickney (1986), Engle et al. (1988), Gray (1988), Pounds et al. (1992). Some specifically on golden shiners (the most valuable freshwater baitfish in the U.S.) are: Ludwig (1989), Anonymous (1995a), and Lazur and Zimet (1995).

FAMILY ACCOUNTS

Aplocheilidae—Rivulins

Mangrove rivulus, a hardy easy-to-raise member of this family (related to killifish), can be used for mosquito control and laboratory research (Taylor et al., 1992).

Cyprinodontidae—Killifishes

At least a few members will readily spawn in tanks or ponds. Larvae are large, robust, and capable. Juveniles grow well and are hardy.

Gulf killifish (also called bull minnows) are popular for bait and have been raised experimentally from Texas to Florida (Tatum et al., 1979; Trimble et al., 1981; Tatum et al., 1982a, 1982b; MacGregor et al., 1983; Perschbacher and Strawn, 1983; Wass et al., 1983; Strawn et al., 1986). They reach a market size of 50-75 mm in 4-5 months and a maximum size of ~150 mm.

Gulf killifish can be raised efficiently with a three phase system of breeding ponds, hatching/nursery ponds, and grow-out ponds. Broodfish are stocked at up to 30,000/ha, with two females per male, and fed a daily ration of 3-5%. Spawning can occur in any month when temperature is 20°C or higher. Spawning mats of Spanish moss sandwiched between plastic-coated wire mesh are placed around the windward edge of breeding ponds. Every 7-13 d, the mats are transferred to hatching/nursing ponds (0.2 ha or smaller), with up to 3.75 million eggs/ha stocked. Hatching occurs in about 3 weeks at 20°C and 2 weeks at 25°C. After hatching, powdered minnow meal is provided several times a day. When the young fish reach 0.3-0.8 g, they are transferred to grow-out ponds, with stocking densities up to 400,000 fish/ha. Survival is very high.

Killifish grow well on zooplankton, worms, and insects, but for better growth, especially at high densities, supplemental feeding is needed. Almost any kind of fish feed will suffice and is preferred to single foods (e.g., cottonseed meal, wheat byproducts). A daily ration of ~10% split into two meals is suitable. At temperatures less than 20°C, Gulf killifish grow slowly, and two or three meals a week should be enough. About 45 kg/ha of 12-12-12 fertilizer every week was used to maintain a plankton bloom in clay-bottomed ponds for Gulf killifish. Simpler systems using only one or two pond types or raceways also can be used. The two-pond system uses breeding ponds and hatching/nursing/grow-out ponds, initially stocked with up to 400,000 eggs or hatchlings/ha.

Killifish are very hardy and have wide temperature (0-35°C and higher) and salinity (0-100^{+} ppt) tolerances. I have found mummichog escapees that survived dry under a car seat in the summer for more than an hour. Potential culture problems include broodfish predation on eggs, silt or algal smothering of eggs, fungal infections of eggs, contamination of ponds with competing or predacous species, and somtimes protozoan infestations of the gills or bacterial infections in older fish. As with other pond-cultured fish, the danger from predaceous insects and birds should be considered and avoided.

Sheepshead minnows (Perschbacher and Strawn, 1988) are eagerly eaten by game fish, but Gulf killifish, **mummichogs**, and **striped killifish** are preferred for bait because they stay on a hook better. Sheepshead minnows are extremely hardy and will grow and reproduce in almost any container of salt or brackish water. They have been used in many toxicological studies. They can be a nuisance in ponds of more desirable species such as Gulf killifish. Sheepshead minnow males clear an area of the bottom for egg placement. In 1971, they regularly spawned, and larvae were reared to adults in 30-L aquariums on cultured (collected) zooplankton, flake feed, and minced seafoods (Tucker, unpub.). Mummichogs like to deposit their eggs intertidally in containers such as oyster shells.

Gerreidae—Mojarras
Live mojarras usually are very silvery and are attractive baits for game fish. (See earlier chapters.)

Haemulidae—Grunts
Live grunts are good bait for certain game fish. For example, live pigfish are a preferred natural food and bait for spotted seatrout in Florida. The pigfish's grunting sounds attract the predator. (See earlier chapters.)

Sparidae—Sea Breams, Porgies
Live sparids are good bait for certain game fish such as snook. Sometimes the sharp dorsal spines are trimmed with scissors. (See earlier chapters.)

Mugilidae—Mullets
Small dead whole mullet (75-200 mm TL, sometimes called finger mullet) are popular as bait in many regions; and chopped sections of larger mullet, also. Live mullet sometimes are used, but they do not survive handling and confinement in bait containers as well as killifish. (See earlier chapters.)

Other Families
Depending on regional preferences, other fish can be chosen.

ENVIRONMENTAL CONSIDERATIONS

Anonymous (1995b) discussed some environmental concerns with bait fish culture. If equipment such as nets, waders, buckets, and boats are used both in the culture water and in natural waters outside the farm, there is potential for transferring disease and pest organisms in either direction. Then if the bait fish are sent to another area, the unwanted organisms (e.g., zebra mussels) can be transferred with them and thus introduced to a clean area. Baitfish often are intermediate hosts for parasites. In addition, the bait fish themselves could be an exotic species or from an exotic strain; if they are released, local fish communities could be affected structurally or genetically.

SUMMARY

Because of the relatively low value of bait fish, rearing costs must be kept down. When the fish are shipped alive, care must be taken in handling, sorting, and counting them, which makes the product more expensive. With correct choice of site and species plus good management and marketing, bait fish farming can be feasible. Extensive culture should be considered as an option for bait fish. Ponds provide a more natural environment and a good supply of food. If constructed well, they can be managed efficiently. Tank systems are easier to manage in some ways and alleviate predator and environmental problems, but do not provide natural food and could require more effort to manage water quality. For most species and purposes, the grow-out period will be in the range 2-8 months. On food fish farms, culled or surplus slow-growing fish could be sold as bait, if worth the effort.

15 CULTURE OF ESTABLISHED AND POTENTIAL SPECIES—ORNAMENTAL FISH

Healthy cultured aquarium fish are preferred over wild fish that might have diseases, parasites, or injuries from being captured or drugged. If well nourished, they can be just as colorful as wild fish. Time from spawning to marketing is shorter than for food fish, and price per kilogram is much higher. Some very rare species can sell for thousands of dollars each. The bulk of the U.S. freshwater ornamental fish industry and several wholesalers are based in Florida. Stocks in most traditional collecting areas have diminished, and legal restrictions are rapidly curbing the indiscriminate collecting methods of the past.

Commercial ornamental fish culture has been limited mainly to clownfish and gobies, the easiest and most reliable groups. Dottybacks are becoming commercial. Other damselfish, basslets, comets, jawfishes, blennies, and seahorses have been raised in reasonable numbers, but not yet routinely for sale. Past and present farms in Florida, the Bahamas, Brazil, and the Mediterranean, several public aquariums (e.g., Conde, 1986; Kuiter, 1988; Cooper, 1991; Ounaïs, 1997; Reid, 1997), experimental projects in several countries, and hobbyists have put much effort into developing new species, especially during the 1990s.

Thresher (1984) reviewed much of the literature on spawning and rearing of marine ornamental fish, and should be consulted for basic information on most families. Economic analyses for ornamental freshwater fish farming are given by Teichman (1994). This chapter summarizes rearing information for families and representative species.

FAMILY ACCOUNTS

Eel Families

Garden eels and moray eels have been observed spawning in the field (Thresher, 1984). The difficulty of feeding leptocephali and the long larval period has made rearing of eels difficult (see earlier chapters). The planktonic period can be 3-5 months for spaghetti eels, 6-10 months for morays, ~10 months for some conger eels, and possibly 2 or 3 yr for *Anguilla* spp.

Plotosidae—Eeltail Catfishes

Striped eel-catfish have fairly large non-adhesive demersal eggs (~3.3 mm), large hatchings (6.9 mm), and a yolksac period of ~10 d before they begin feeding. At first feeding, they are well-developed and can eat *Artemia*. Females can grow to 150 mm in a year. Thresher (1984) suggested that plotosid catfishes should be easy to rear beause of their size.

Synodontidae—Lizardfishes

Lizardfish eggs are moderately sized and hatch within ~2-3.5 d into long thin larvae (Thresher, 1984). Settlement occurs at about 30-35 mm TL.

Antennariidae—Frogfishes

These fish have the unusual habit of gluing their eggs into gelatinous rafts, but males of at least one species carry the eggs glued to their sides (Thresher, 1984).

Sargassumfish eggs hatch in 2-5 d, depending on temperature, and the yolksac period is 5-6 d. Another unusual characteristic is the envelope, or bag that surrounds most of the larvae for most of the postlarval period. Transformation is complete at ≥9 mm, with full coloration by 15 mm.

Gobiesocidae—Clingfishes

Clingfish attach their eggs to the underside of something solid like a shell or rock (Thresher,

1984). In some species, the male tends them until hatching. Hatchlings are fairly large (~2.6-6.1 mm TL), and most species probably would be easy to raise.

Yellow clingfish were raised by Shiogaki and Dotsu (1972) from an egg mass deposited on the underside of a roof tile by wild fish in the lower intertidal zone. Larvae fed soon after hatching (5.3 mm TL). Larvae were raised in a 30-L plastic container on *Artemia* nauplii and *Tigriopus japonicus*. They were planktonic for 10 dah, then settled at 9 mm TL. A length of 20.1 mm was reached 55 dah.

Pseudomugilidae—Blue-eyes

Honey blue-eye captive wild adults spawned voluntarily, and larvae were raised in an aquarium by Semple (1991). The adhesive eggs (~1465 µm) were attached to vegetation and hatched in 5-8 d into large larvae (3.7 mm SL). At 25-27°C, larvae were fed copepods, liver extract, dry and liquid feeds, and *Artemia* (Table 8-4). Reared fish spawned 95 dah (~20 mm). Maximum length is ~30 mm.

Fistulariidae—Cornetfishes

Red cornetfish eggs (1.8-2.1 mm) hatch in ~7 d into long thin larvae (7.1 mm)(Thresher, 1984). Because of their small mouths, rearing larvae would require a good supply of small zooplankton.

European Seahorse
Hippocampus hippocampus

Syngnathidae—Pipefishes, Seahorses

Several species of **seahorses** have been raised, with second generations produced for some (Giwojna, 1990; Schmidt, 1995; Vincent 1995a, 1995b, 1995c). Seahorses not only are kept as pets, but also are in demand for traditional Chinese medicine. Both young and adult seahorses are more delicate than they might appear and have to be handled and fed carefully. After a courtship of ~3 d, eggs are deposited and fertilized in the brood pouch of the male. Apparently, the young receive some nourishment from fluid released into the pouch. After several weeks, they are released. Young seahorses do not need much space and have been raised in 10-L aquariums. Dark tank walls and minimal aeration are important for early rearing. Young of several species are released at a length of 6-13 mm, which can triple in 3-4 weeks. Live food is necessary for survival of the young (Table 8-4) and apparently also for the health of older fish, which can be given non-living foods on a supplementary basis. Eventually, a good compound feed might be developed for them. All stages should have objects to hold on to with their tails, and hiding places, if possible.

Azzarello (1991) found no evidence that the male **gulf pipefish** provides nutrition to young held in the brood pouch. Protection and transportation seem to be the most obvious functions of the pouch, but it also could aid the young in osmoregulation during the earliest stages.

Leafy seadragons and **common seadragons** have been raised in small aquariums (Blasiola, 1988). Leafy seadragons were 35 mm TL when released by the brooding male, and common seadragons were 25 mm. They began feeding after 2 d. First food was live mysid shrimp, followed by frozen mysids, then other crustaceans. Leafy seadragons measured 85 mm at 21 d and 140 mm at 3.5 months. Common seadragons were 75 mm long at 21 d.

Scorpaenidae—Scorpionfishes

Chapter 13 discusses culture of scorpaenids used for food. Lionfish have spawned in aquariums, and eggs and larvae were described (Thresher, 1984). They form their eggs into gelatinous balls.

Pygmy lionfish eggs (800 µm) and hatchlings (~1.1 mm) are small. First feeding is at ~4 d, and first foods probably need to be small (Table 2-1). Settlement of wild fish occurs at 10-12 mm.

Serranidae—Groupers, Sea Basses

Chapter 13 discusses culture of serranids used for food. Food species mostly are in the subfamily Epinephelinae, with a few in Serraninae and Anthiinae. Ornamental species are mostly in the latter two subfamilies and Liopropomatinae.

Shy, **butter**, **blue**, and **butter × blue hamlets** (Serraninae) were raised by Domeier (1994) through settlement, and all except the blue hamlets until juvenile colors were lost. Eggs were obtained from voluntary and strip-spawns. At 28°C, larvae were fed wild plankton, rotifers, and *Artemia* in green water, followed by dry feed after settlement. They were grown out in aquariums or large fiberglass tanks.

Belted sandfish (Serraninae), which are simultaneous hermaphrodites, can easily be conditioned to spawn high-quality eggs in small tanks over long periods (Clark, 1959, 1965; Tucker and Woodward, 1995). Eggs (662 µm) and hatchlings (~1.5 mm) are small. Larval rearing difficulty seems to be intermediate between black sea bass (fairly easy) and Nassau groupers (difficult). We raised larvae only once, with two (62, 71 mm TL) reaching maturity and spawning at 16 months after being paired with wild fish. Larvae were fed rotifers and *Artemia*, and juveniles and adults were fed experimental pellets. Transformation and settlement were complete by ~34 dah. More colorful members of the genus *Serranus*, such as harlequin bass and tobaccofish, probably have similar rearing requirements. Wild individuals of all three are sold in pet stores in the U.S.

Some anthiines are popular aquarium fish because they are small, colorful, attractive, and not likely to eat other fish (Thresher, 1984). Because eggs (~600-800 µm), hatchlings (~1.2-1.5 mm TL), and larval mouths are small, larval rearing is not likely to be easy.

Swissguard basslets (Liopropomatinae) have spawned in captivity, but rearing was not reported (Wilkerson, 1996).

Pseudochromidae—Dottybacks

Several species readily spawn in aquariums (Thresher, 1984). **Neon dottybacks** have been reared through at least two generations, and **sunrise** and **orchid dottybacks** also have been raised (Brons, 1995, 1996; Wilkerson, 1996; Moe, 1997a). The adhesive eggs are formed into a ball, which the male tends inside the spawning cave until hatching after ~4.5 d (~1 h after lights are off). Orchid dottyback hatchlings are 3.7 mm TL and sunrise dottybacks 3.9 mm. Rotifers are suitable as the first food. At ~25°C, sunrise dottybacks began settling 25 dah and orchid dottybacks by 26-30 dah. Both become fully-colored juveniles by 35-40 dah (~15 mm). Close to 35% of larvae raised in a 76-L aquarium on wild zooplankton, rotifers, *Artemia*, ground shrimp, and ground flake feed survived to become juveniles (Table 8-4; Moe, 1997a). Gardner (1997) reported that eight species of dottybacks had been reared: **Springer's, neon, sunrise, striped, olive, orchid, yellow,** and **Australian**. The standard method was: green water in a 1000-L tank, static for the first week; rotifers (1-15 dah), enriched *Artemia* (10-? dah), and dry feeds (16-? dah). As they transform, dottybacks become territorial and aggressive, and require separate hiding places (Brons, 1996).

Grammatidae—Basslets

Royal grammas and **blackcap basslets** have been reared (Moe, 1992; Addison, 1994a, 1994b; Walch, 1994a; Wilkerson, 1996). Both species deposit adhesive eggs (~1 mm), preferably in a small cave, where the male (usually larger than the female) guards and tends them until hatching. Typically, 5-7 d later, larvae hatch within an hour after sunset. They were raised in tanks in green water (Table 8-4), with settlement of grammas occurring at ~30 dah and transformation of blackcap basslets during 21-35 dah.

Plesiopidae—Longfins

Comets have been reared (Wassink and Brons, 1988, 1990; Brons, 1995; Wilkerson, 1996). Lack of regular spawning has been the main bottleneck. Adhesive eggs are deposited, preferably in a small cave, where one fish (probably the male) guards them until hatching at 5-6 d. First feeding occurs just after hatching. Wassink and Brons raised larvae at ~26°C in 33-L tanks on rotifers and *Artemia* in green water (Table 8-4). Settling began about 16 dah. They grew to ~15 mm in 2 months, 40 mm in 8 months, and 120 mm in 3 yr.

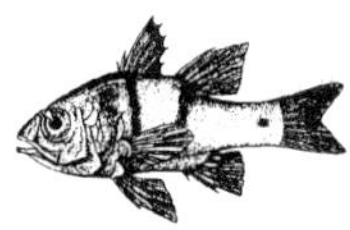

Twobelt Cardinalfish
Apogon taeniatus

Apogonidae—Cardinalfishes

Male cardinalfish brood the ball of eggs in their mouths, and many release the larvae at hatching, about 7-9 d (Thresher, 1984; Fenner, 1996). Egg size range is wide, 0.24-4.5 mm, with those of tropical species tending to be smaller. Some larvae spend a few weeks or more in the plankton. Settlement by some species occurs much sooner and others ~2 months after spawning (10-25 mm). Although larvae are easy to raise, small clutch size limits the commercial potential for cardinalfish.

Banggai cardinalfish captive wild broodstock spawn voluntarily in aquariums (Allen, 1996; Hammer, 1996; Bates, 1997; Marini, 1996). Males continued to hold the larvae after hatching for a total incubation period of 21-24 d (Marini, 1996). The larvae can eat *Artemia* nauplii right after release. Different combinations of rotifers and/or *Artemia* were used to raise them in breeding nets in a 246-L aquarium. From one batch of 28 larvae, 20 survived for 30 d and 13 for 60 d. At a public aquarium, spawning was voluntary, and egg diameter was 2.5 mm (Ounaïs, 1997). Litter size was 10-42, and at release (20-25 daf), the young fish were very advanced (11 mm TL). They were raised on enriched *Artemia.* Bates (1997) considered newly released fish to be juveniles and found them as easy to raise as the adults.

Sciaenidae—Drums

Larvae of most drum species are robust and easy to raise on rotifers and *Artemia* or copepods (see chapter 13). The greatest danger in the hatchery seems to be *Amyloodinium* infestation of early juveniles, which can kill some drums overnight.

Jackknife fish were raised at ~28°C on wild and cultured zooplankton, switching to chopped seafoods and dry feeds after 3 weeks when they were nearing transformation (Young, 1980).

Highhats were induced to ovulate, and larvae were raised in a 760-L system on rotifers, *Artemia*, and flake feed in green water (Table 8-4; Harris, 1996).

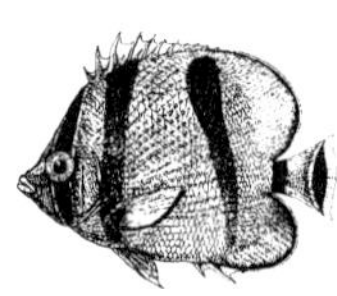

Hoefler's Butterflyfish
Chaetodon hoefleri

Chaetodontidae—Butterflyfishes

The wide range in juvenile and adult feeding habits will be a major factor in rearing butterflyfishes (Allen, 1985; Giwojna 1989). The small planktonic eggs hatch in about a day and soon become armored larvae called tholichthys, which could be planktonic for weeks to months (Allen, 1985). They seem difficult to raise. Suzuki et al. (1980) reared

blackloin butterflyfish (eggs 720 μm, hatchlings 1.5 mm) to 8 dah on oyster trochophores, rotifers, and sea urchin larvae. **Foureye butterflyfish** were 11-15 mm SL at settlement in nature (Robertson, 1992).

Pomacanthidae—Angelfishes

Angelfish spawn in pairs, and at least **pygmy angelfish** and ***Genicanthus* spp.** are protogynous hermaphrodites (Thresher, 1984). Moenich (1990) reviewed methods for keeping ***Holacanthus* spp.** in aquariums. **French angelfish** were 8.5-10.8 mm SL at settlement in nature (Robertson, 1992). **Queen angelfish** were 13.5-16.7 mm SL at settlement in nature (Robertson, 1992). Pygmy angelfish often live in harems, and some of the larger species sometimes do. Eggs are small, with diameters mostly in the range 500-900 μm. Pygmy angelfish have the smallest eggs and larvae.

French Angelfish
Pomacanthus paru

Gray and **French angelfish** eggs were obtained by stripping wild adults, with and without hormone induction; the parent species and their hybrid were raised by Moe (1976, 1977, 1997b, 1997c), who ranked angelfish larvae in the most difficult category. Several hundred 40-d-old juvenile hybrids were produced. None of those or the other angelfish species tried would eat rotifers but did eat copepods well. Eggs (~900 μm) hatch in about 18-30 h. Larger tanks ≥3.8-m^3 were better than small ones for larval rearing. Outdoor 4.5-m^3 concrete tanks with black epoxy paint and 80% shadecloth covers worked well. Typically, in early trials, larvae fed and grew well on copepods, but suddenly died during 5-10 dah. In later trials, survival was improved by treatment with 16 mg/L streptomycin or by continuous filtration of rearing water through activated carbon. Moe (1997) suggested that the main problem had been a dissolved toxin, possibly from *Vibrio anguillarum.*

Pygmy angelfish pairing can be accomplished by putting a large fish with one or more small fish (Thresher, 1984). They usually spawn very readily in captivity (Bauer and Klaij, 1974; Bauer and Bauer, 1981; Lamm, 1984; Wrobel, 1988; Karanikas, 1989; Hioki et al., 1990; Brown, 1994), and natural hybrids occur (Takeshita, 1976). However, most rearing attempts have ended with complete mortality at 9-12 dah. Hatchlings are small (~1.0 mm TL) and possibly fragile. Stability of water characteristics (Wilkerson, 1996) and appropriate foods (small, attractive, and nutritious) probably are the main factors limiting their culture. Larval tanks were inappropriate or too small in some of the attempts. The small number of eggs per spawn make it hard to justify using larger tanks for these fish, but larvae might need the extra space, not only for water quality stability but maybe also to minimize contact with tank walls.

Brown (1994) tried to raise **flame pygmy angelfish** at ~25°C on rotifers, *Acartia* sp., and *Tisbe* sp., but had only two fish survive to 11 dah.

Pomacentridae—Damselfishes

Although **damselfishes** can be aggressive, many species are attractive and popular for aquarium use. Thresher (1984) discussed the life histories of several species. Larvae have been raised in aquariums, but opaque fiberglass or plastic tanks are more practical. Six species of ***Stegastes*** were 8.0-15.1 mm SL at settlement in nature (Robertson, 1992).

Clownfishes are very easy to spawn and rear (Schreiner, 1972; Moe, 1973; Sieswerda, 1977; Anonymous, 1984a, 1984b; DiMarco, 1984; Thresher, 1984; Brosseau, 1991; Young, 1991; Frakes, 1992; Juhl, 1992; Moe, 1992; Shapiro, 1992; Hoff, 1996; Wilkerson, 1992a,

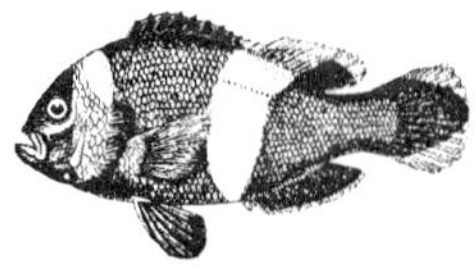

Saddleback Clownfish
Amphiprion polymnus

1992b, 1998). Moe (1976) placed them at level one on a scale of one to ten. At least 17 species have been raised and sold: **skunk**, **orangefin**, **barrier reef**, **twoband**, **Clark's**, **red saddleback**, **tomato**, **cinnamon**, **common**, **percula**, **pink**, **saddleback**, **Australian**, **orange**, **sebae**, **threeband**, and **maroon**. Because they are protandrous hermaphrodites, clownfish are easy to pair. If a small fish is stocked with a large fish, the small one probably is a male and the large one either is or probably will become a female. Another method is simply to put several small fish in a large tank and wait until a pair is formed. On the last day of courtship, the ovipositor of the female becomes obvious. The adhesive eggs are attached to a smooth surface (ceramic tiles or flower pots in most hatcheries). The female defends the territory, and the male tends the eggs until hatching, which usually occurs within a few hours after lights are off after 5-8 d, depending on species and temperature. Larvae can be removed from the breeding tank after hatching, or nests can be removed and artificially incubated with a water or air stream flushing the eggs. Hoff (1996) suggested dipping the nest in 260 ml/L formalin for 1-2 min during the transfer. Because eggs normally hatch within 2 h after lights are off, light control can be used to accelerate hatching to occur in the afternoon rather than at night. Because benthic larvae are adapted to being near solid surfaces and are not likely to become trapped in corners like planktonic larvae, aquariums can be used for raising clownfish (Hoff, 1996). Larvae are large (~4 mm TL) at hatching, begin feeding almost right away, and quickly become strong swimmers. Schooling begins at ~10 dah. Transformation occurs during ~25-35 dah (~14$^+$ mm TL). Market size of 38-50 mm can be reached in 3-4 months. About 24-27°C is suitable for rearing. Rotifers, *Artemia*, dry feeds, and sometimes gelatin feeds, crustacean meal, and algae are used (Table 8-4). Green water sometimes is used. Thyroid hormones could be administered to accelerate development (Doimi et al. 1997). The first 3 d of feeding (0-3 dah), the second day of feeding on *Artemia* (~8 dah), and early transformation (9-11 dah) are critical periods for Clarkii clownfish (Wilkerson, 1992a, 1992b). During the first few days, they can be confused by light reflections. To help prevent choking, *Artemia* should be withheld until the larvae can comfortably swallow them, and shells should be excluded. Gorging on *Artemia* can cause mortality. Until juveniles freely feed at the surface, the fright response is strong enough to result in mortality. Despite the potential problems, overall survival of clownfish can reach 100%.

Sergeant majors spawn easily in captivity and are easy to raise (Alshuth et al., 1998; Tucker, unpub.). In nature, the adhesive eggs are deposited on a rock the male has cleaned, which can be facing up, down, or sideways. The male tends the eggs until hatching, which takes 5-7 d. We raised them from tank-spawned and ocean-spawned eggs. Foods were rotifers, copepods, and *Artemia*, sometimes with oyster trochophores during the first several days (Table 8-4). Transformation occurred mainly during 28-32 dah (17-25 mm TL). In 1 month, they grew to 23 mm TL at 27-29°C, probably faster than wild fish. Sergeant majors were 9.5-14.4 mm SL at settlement in nature (Robertson, 1992).

Spot dascyllus (27°C) and **threestripe damselfish** (24.5°C) from wild eggs were raised in 2500-L tanks on rotifers, wild plankton, and *Artemia*, with *Tetraselmis* sp. up to 13 dah (Table 8-4; Danilowicz and Brown, 1992). Threestripe damselfish schooled for about a week during 20-34 dah, but the other species did not. Spot dascyllus settled during 23-45 dah, and threestripe damselfish during 35-50 dah, both at ~9 mm TL or larger. Volkart (1991) discussed aquarium behavior of these and other *Dascyllus* spp.

Yellowtail damselfish larvae were raised at 27°C on rotifers, flake feed, and *Artemia* in green water (Unwin, 1986). Larvae from wild eggs were reared in 946-L tanks on wild plankton, rotifers, copepods, and *Artemia*, with flake feed and ground shrimp added after 3 weeks, to well developed juveniles (79 dah, 25 mm TL)(Moe and Young, 1981; Potthoff et al., 1987). **Azure damselfish** have been reared (Brons, 1995). **Orangetail damselfish** have been reared; maintaining copepods in the rearing tanks enhanced development of the larvae (Wilkerson, 1996).

Sphyraenidae—Barracudas

Northern sennets were raised from wild eggs on wild zooplankton, including fish larvae (Houde, 1972). Thresher (1984) discussed some early life characteristics of barracudas. There could be some interest in raising small species for aquarium use; however, even the larvae are voracious and they are likely to be very cannibalistic.

Labridae—Wrasses

Some **wrasses** are used as pets and others as food. Most tropical wrasses are protogynous hermaphrodites; some species can change from female to functioning male within 14 d (Thresher, 1984). Males usually are larger than females, and in many tropical species that live and spawn in harems, the dominant males are much more colorful than females and subordinate males. Some of the many types of spawning behavior were reviewed by Thresher (1984). Eggs mostly are planktonic and small. A few species have been raised, mostly for use in larval descriptions. Larvae of some species have small mouths and cannot easily ingest rotifers at first feeding. First food in the range 20-60 μm seems necessary. **Bluehead wrasses** were 10.2-13.7 mm SL at settlement in nature (Robertson, 1992).

Hogfish larvae from wild eggs were raised at 25-27°C in 80-L aquariums on wild zooplankton and *Artemia* in green water (Colin, 1982). By 17 dah, larvae preferred to stay under floating debris; settlement occurred by 34 dah.

Multicolorfin wrasse larvae were raised at ~22-25°C in a 500-L tank on oyster trochophores, rotifers, *Artemia*, and wild zooplankton in green water (Table 8-4; Kimura and Kiriyama, 1993).

Scaridae—Parrotfishes

Parrotfishes are protogynous hermaphrodites and spawn in pairs, groups, harems, and variations thereof, with similarities to wrasses (Thresher, 1984). Larvae are slender and have small mouths, and probably require small prey at first feeding.

Stichaeidae—Pricklebacks

Opisthocentrus tenuis eggs (~2300 μm) stick to each other and hatch in 58-64 d (Shiogaki, 1981b). Captive wild broodfish spawned voluntarily, and larvae were raised on *Artemia* only. At 56 dah, transformation had not occurred, and the largest fish was 19 mm TL. A 32-mm wild juvenile was collected. They reach a maximum size of ~190 mm in 3 yr.

Opistognathidae—Jawfishes

Yellowhead jawfish (*Opistognathus aurifrons, O. gilberti*) have been reared (Thresher, 1984; Walch, 1993, 1994b; Brons, 1995; Wilkerson, 1996; Gardner, 1997). Male jawfish mouthbrood the eggs in or near their burrows until hatching (5-9 d), putting them down occasionally so the male can feed (Hess, 1993). Just after hatching, yellowhead jawfish are almost 4 mm TL, can see, and are well-developed. Larvae were raised in a 76-L aquarium on rotifers, *Artemia*, and dry feeds in green water (Table 8-4). They settle at ~15 dah (~10-15 mm).

Trypterygiidae—Triplefin Blennies

Snake blennies were raised from eggs laid in an aquarium by captive wild broodfish (Shiogaki and Dotsu, 1973a, 1973b). The adhesive eggs were not large (965 μm), but hatchlings were long (~4.8 mm TL) and fed soon after hatching. Larvae were raised in a

30-L aquarium on barnacle nauplii, rotifers, and *Artemia* nauplii. Settling occurred at ~40 dah (~12 mm TL).

Clinidae—Scaled Blennies
Horned blennies have been raised (Wilkerson, 1996).

Blenniidae—Combtooth Blennies
Most blennies spawn large demersal eggs in a small cave, where the male tends them (Thresher, 1984).

Blackline blennies deposit their adhesive eggs (~1.0 mm) on a solid substrate and the male tends them until hatching, after 10-11 d (Fishelson, 1975, 1976). Hatchlings swim well and begin feeding right away. Captive wild broodfish spawned voluntarily, and larvae were reared at 23-25°C in 20-L aquariums on ciliates, rotifers, and *Artemia* nauplii. At 1 dah they are 3.7 mm TL. Transformation occurs by about 20 dah (~15 mm TL) and settling by about 30 dah.

Callionymidae—Dragonets
Dragonets spawn small (~550-750 μm) planktonic eggs (Thresher, 1984).

Beniteguri captive wild adults spawned voluntarily, and larvae were reared at 18-24°C on rotifers, *Artemia*, and minced krill (Eda et al., 1994). Eggs (670 μm) and larvae (1.2 mm TL at H, 2.1 mm at FF) were small. Settling occurred during 22-32 dah (6-13 mm TL).

Mandarinfish (*Synchiropus splendidus*) also have been reared (Brons, 1995; Wilkerson, 1996).

Gobiidae—Gobies
Several species have been raised from eggs collected or spawned in captivity. The adhesive eggs are attached to the substrate. Egg width range is ~0.5-1.0 mm and length ~1.1-3.3 mm (Thresher, 1984).

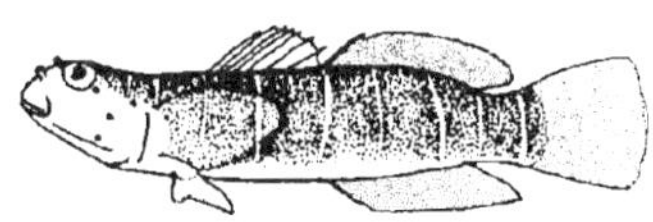

Nineline Goby
Ginsburgellus novemlineatus

Neon, **yellow**, and **citron** gobies have been reared through at least two generations; **Rainford's goby** and other species also have been raised (Moe, 1975, 1992; Brown, 1996; Wilkerson, 1996). At least those, plus **West Indian cleaner, Genie's cleaner, yellow, greenbanded,** and **masked gobies** have been reared and sold (Gardner, 1997).

Neon gobies are very tolerant of captivity and spawn easily in small tanks (Brown, 1996). The adhesive eggs are deposited on the underside of objects like shells and rocks. The male, which usually is larger than the female, tends the eggs until hatching occurs after 6-8 d. Hatchlings are large (~4 mm TL) and begin feeding within half a day. Settlement occurs at ~18-20 dah (~8 mm TL) or could be delayed if conditions are not suitable.

Mugilogobius abei captive wild adults deposited their adhesive eggs (450 × 980 μm) on slate plates and the male tended them until hatching, after at least 4 d (Kanabashira et al., 1980). The 2.1-mm TL hatchlings soon began feeding. Larvae were raised in a 30-L tank on oyster eggs & trochophores and rotifers (Table 8-4). Transformation and settling occurred ~44 dah (15-19.5 mm TL). The fastest growing fish reached ~40 mm in 9 months.

Ephippidae—Spadefishes
Atlantic spadefish (an ornamental and food species) have been reared (Walker, 1991), and some work work has been done toward raising batfish. In 1980, spadefish larvae from wild eggs were raised to mid-stage juveniles in a 70-L aquarium on cultured mixed zooplankton,

Longfin Batfish
Platax pinnatus

flake feed, and minced seafoods (Tucker, unpub.). Spadefish have spawned in tanks voluntarily and by hormone-induction. Eggs (1100 μm) and larvae are planktonic. Larvae were raised in a 75-L aquarium on rotifers, *Artemia*, and mixed foods (Table 8-4). They reached ~60 mm at 4 months.

Scatophagidae—Scats
Spotted scat wild juveniles raised in ponds (daily average range 28-36°C) grew from 0.4 to 15 g in 18 weeks, not fast enough to justify farming them for food (Fast et al., 1989). However, it might be practical to rear wild juveniles to an ornamental-fish market size of 20-50 mm (1-11 g), because fish in that range are difficult to find and catch, but the smaller ones (6-12 mm) are abundant and easy to catch. Culture in brackishwater ponds is being developed in Taiwan (Liao et al., 1995).

Acanthuridae—Surgeonfishes
Because **surgeonfish** eggs are small (~575-700 μm), larvae have very small mouths, and the larval period is long (e.g., 2.5 months), rearing them will be difficult. Three species of *Acanthurus* were 23-33 mm SL at settlement in nature (Robertson, 1992).

Bothidae—Lefteye Flounders
Windowpane eggs were accidentally collected by bucket in 1972, and larvae were raised to adults in a 280-L community aquarium on cultured mixed zooplankton, flake feed, and minced seafoods (Tucker, unpub.).

Balistidae—Leatherjackets
Triggerfish eggs are only ~500 μm in diameter and some filefish eggs a little larger (Thresher, 1984). Eggs of most species are demersal and are tended and guarded ferociously by the female. Most hatch just after dark the day after spawning, into small larvae. Small first food is needed.

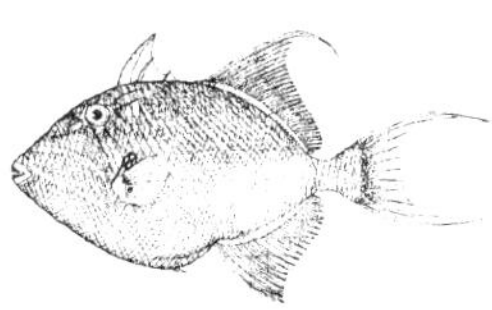

Queen Triggerfish
Balistes vetula

Queen triggerfish larvae had very small mouths but were raised on clam trochophores, rotifers, copepods, and *Artemia* to 23 dah (Tucker et al., unpub.). By that time, the tanks had become overwhelmed by coelenterates and many detached, killing the remaining larvae.

Ostraciidae—Boxfishes
These fish spawn reasonably-sized eggs of about 1.4-2 mm diameter, with hatchlings of at least some species 2.2-2.4 mm long. Larvae become mostly covered with armor plating within a few days (Thresher, 1984).

Scrawled cowfish were raised in a small aquarium (18 or 37 L) from wild eggs collected in a plankton net (Palko and Richards, 1969). *Artemia* nauplii were the primary food for larvae, but wild plankton also can be used; frozen and adult (for fish >6 mm) *Artemia* were used for older fish. Continuous light is good for the first week. They grew to 49 mm TL in ~5 months.

Tetraodontidae—Puffers

Some non-spiny puffers attach adhesive eggs to a rock or coarse sediment, and the males sometimes tend them, but at least one species buries its eggs in beaches at high tide (Thresher, 1984). Depending on species and temperature, hatching might take between 4 and 40 d. (Also see chapter 13.) **Rough golden puffers** were raised on boiled egg yolk and *Artemia* nauplii (Fujita, 1966). By 30 dah, they were 6.6 mm, had good pigment, and could inflate, and they soon settled. **Northern puffers** were raised on rotifers and *Artemia* nauplii (Valenti, 1975). **Sharpnose puffers** were 10.2-23.0 mm SL at settlement in nature (Robertson, 1992).

Some spiny puffers have non-adhesive demersal eggs (striped burrfish) and others have planktonic eggs (e.g., balloonfish). **Balloonfish** hatched at 2.6 mm and the shell covering the body soon thickened (Sakamoto and Suzuki, 1978). They were raised on rotifers, *Artemia* nauplii, and adult *Tigriopus japonicus.* Transformation occurred at ~25 d (~20 mm TL). **Porcupinefish** have been raised up to a month after hatching, on *Artemia* (Conde, 1986).

ENVIRONMENTAL CONSIDERATIONS

Farmed ornamental fish usually are exotic. Preventing introduction of exotic diseases and fish species should be of the highest priority.

SUMMARY

Clownfish are among the easiest of marine fish to rear. Butterflyfish and angelfish are among the hardest. Butterflyfish, pygmy angelfish, wrasse, and triggerfish larvae need small first food(s).

Although larvae of many ornamental species have been raised in aquariums and small tanks, volumes of at least 500-1000 L will maintain more stable conditions and are preferable. Several species of ornamental marine fish have been raised in glass aquaria, but fiberglass tanks without sharp corners probably are best for all but the easiest species. Even clownfish larvae can get trapped in crevices.

Individual females of most ornamental species do not produce large numbers of eggs. If hatchling production is not consistently high enough to fully stock rearing tanks, extra care is needed to avoid overfeeding, because rotifers can overpopulate, reducing water quality and their own nutritional quality, and an excess of *Artemia* often leads to gorging.

Greenwater culture probably will benefit the difficult species with small larvae more than the easy ones with large larvae. For larvae that do not eat added prey within a short time, greenwater culture will help maintain prey quality.

The future of the marine ornamental fish industry lies in intensive culture, and recirculating systems have some obvious advantages. Although grow-out time is shorter (mostly 3-6 months), water quality, nutrition, and health management are at least as critical as for food fish.

Illustration credits for chapter 15: Saddleback clownfish, p 582; longfin batfish, p 585; reprinted from Lindberg and Krasyukova (1971), with permission from the U.S. National Technical Information Service. Queen triggerfish, p 585; French angelfish, p 581; nineline goby, p 584; reprinted from Fischer (1978), with permission from the Food and Agriculture Organization of the United Nations. European seahorse, p 578; twobelt cardinalfish, p 580; Hoefler's butterflyfish, p 580; reprinted from Whitehead et al. (1984), with permission from the United Nations Educational, Scientific and Cultural Organization.

16 THE FUTURE OF MARINE FISH CULTURE

This book has reviewed the culture methods for a number of species. Considering value and demand, many of them seem to be candidates for commercial farming, but harsh biological and engineering realities limit the actual list. The best candidates spawn easily and do not have unusual dietary or environmental requirements.

Criteria for choosing a marine fish for aquaculture are the same as for other organisms. The best candidates are valuable, easy to get seed for, easy to rear, easy to feed, can be crowded, and grow fast and efficiently. Hundreds of species look fairly good by most criteria, but most of those need more work and evaluation. Some have outstanding characteristics that are cancelled out by their deficiencies. Some are mediocre. None are excellent in all respects. However, current research trends are encouraging.

Two phases of development are needed: (1) experimental technology, which is concerned with things like biological, chemical, and nutritional factors; (2) commercial technology, which is concerned especially with engineering and economics. For many fish, the first has been done, but going to the second step can require a large leap.

Factors limiting the development of commercial marine fish farming are in five main categories: (1) reproduction, (2) environmental quality, (3) nutrition, (4) health management, and (5) economics. Water quality needs are mainly engineering problems. Disease control depends mostly on water quality, sanitation, nutrition, and monitoring. Only with well-designed hatcheries and grow-out facilities can requirements in the first four categories be fulfilled. During the last 15-20 yr, much progress has been made in these areas, especially spawning and nutrition. Economic considerations are determined by technology costs at the rearing level as well as supply and demand balances at the marketing level.

Obtaining Fish for Stocking

To maximize operational efficiency and overall production:

Spawning and rearing cycles should occur throughout the year, either for one species or a sequence of more than one. Environmentally-controlled spawning is preferred, but some hormonal control will continue to be needed for at least a while for species like mullet that do not adapt well or for which the appropriate environmental cues are not yet known, or possible to provide. Behavioral and nutritional studies plus domestication of broodstock will lead to easier, more reliable, spawning and better egg quality. Cryopreservation of gametes can make out-of-season egg production possible. Breeding, hybridization, and genetic engineering (e.g., triploidy) can produce faster growing fish.

Environment, Water Treatment, Culture Units

To maximize production per unit volume:

Increase effectiveness and efficiency of filter systems. Make tanks self-clean better. Optimize intensive tank grow-out systems for high stocking density, low water exchange, and low maintenance (e.g., Tvinnereim, 1993). Optimize offshore cage systems for shape retention, fish retention, and predator exclusion; resistance to fouling, wear, and damage from storms and drifting objects; ease of stocking, feeding, cleaning, harvesting, and repairing; waste retention and disposal (e.g., Loverich and Swanson, 1993).

Nutrition

To maximize survival and growth and minimize feeding costs:

Improve broodstock nutrition to maximize quality and numbers of eggs produced.

Optimize nutritional quality of live foods for larvae through choice, culture, and enrichment. Reduce dependence on live foods by increasing digestibility and nutritional value of larval feeds. Increase the efficiency of live food production systems (e.g., Yoshimura et al. 1996a). Improve methods for storage of live foods (e.g., freezing microalgae). Automate feeding of larval foods as much as possible.

Optimize nutritional quality and efficiency of compound feeds. Reduce dependence on fish meal by improving the quality of, and determining effective combinations of, alternative protein sources (In some cases, feeding attractants will be needed.). Expand the capabilities of demand feeders and increase the efficiency of automatic feeders to minimize wastage and pollution while optimizing growth and conversion.

Health

To reduce disease losses and obtain optimal performance from the fish:

When possible, identify and breed for disease resistance. When possible, develop specific-pathogen-free stocks. Gain a better understanding of disease-causing mechanisms and then avoid them. Standardize methods for gaining and maintaining control of microflora in hatchery tanks through good tank and water management, sanitation, and use of probiotic organisms so that treatment with antibiotics and toxic compounds can be avoided (e.g., Vadstein et al., 1993; Kennedy et al., 1998; Ringø and Gatesoupe, 1998). Optimize treatments for enhancing immunity of larvae as they develop immune systems (e.g., immununostimulants, controlled exposure to bacteria). Develop vaccines (preferably by dipping) for significant diseases of marine fish. Develop rapid diagnostic methods for significant diseases. Develop drug treatments with a minimum of side effects. Treat the causes, not the signs, of disease.

Increased value resulting from rising demand and diminishing fisheries will help to widen the profit margin and make additional species feasible for commercial farming. With difficult species, improvements in hatchery design and procedures will be especially important for reducing costs. Despite higher demand and value, some species with unusual requirements or slow growth will not be economical to raise for sale in the foreseeable future. An aquaculture business is most likely to be successful if it puts enough effort into raising at least one proven species and does not risk too much on unproven ones. However, it is more likely to stay successful if it conducts, or has access to, a research and development program designed to improve methods and to permit addition of other species in the future.

One viewpoint is that the extra requirements of marine fish will make it difficult for them to sustain a long-term farming industry. A second viewpoint is that eventually only the few foodfish species that are both in demand and easy to raise will become the "tilapia, trout, catfish, and carp" or the "chickens, ducks, pigs, and cattle" of the sea, and farming of other species will be minimal. A third is that with so many species of marine fish and similarities in requirements, many will prove feasible for farming. The current trend indicates a path between viewpoints two and three.

Aquaculture is not an easy occupation, and marine fish farming is one of the more demanding types.

Knowledge + Experience + Patience + Perseverance → Blue Thumb

LITERATURE CITED

Abbasi, A.R., S.E. Shackley & P.E. King. 1995. Effects of copper on the ultrastructure of muscle cells of herring, *Clupea harengus* L. Pakistan J. Zool. 27:83-87.

Abdelghany, A.E. 1993. Food and feeding habits of Nile tilapia from the Nile River at Cairo, Egypt. In: H. Reinertsen, L.A. Dahle, L. Jørgensen, & K. Tvinnereim (eds.), Fish Farming Technology. Balkema, Rotterdam, p 447-454.

Abdullah, M.A.S., S. Akatsu, K. Al-Abdul-Elah & S.K. Teng. 1984. Refinement of spawning and larval rearing techniques in hamoor (*Epinephelus tauvina*). Kuwait Inst. Sci. Res., Ann. Res. Rep. (1983):55-57.

Abela, M., & A. Le Breton. 1996a. Parasitofauna of cultured Sparidae. In: B. Chatain, M. Saroglia, J. Sweetman & P. Lavens (eds.), Seabass and Seabream Culture: Problems and Prospects. Eur. Aquac. Soc., p 219-222.

Abela, M., & A. Le Breton. 1996b. Parasitofauna of sea bass *Dicentarchus labrax* (L.). In: B. Chatain, M. Saroglia, J. Sweetman & P. Lavens (eds.), Seabass and Seabream Culture: Problems and Prospects. Eur. Aquac. Soc., p 215-218.

Able, K.W. 1984. Cyprinodontiformes: development. Am. Soc. Ichthyol. Herpetol. Spec. Pub. (1):362-368.

Aboussouan, A., & J.M. Leis. 1984. Balistoidei: development. Am. Soc. Ichthyol. Herpetol. Spec. Pub. (1):450-459.

Abraham, M., V. Hilge, K. Akert & C. Sandri. 1988. The teleost oocytes and the follicle cells in relation to transport and incorporation of vitellogenin. In: Y. Zohar & B. Breton (eds.), Reproduction in Fish. Basic and Applied Aspects in Endocrinology and Genetics. Colloq. INRA (44):39-43.

Abud, E.O.A. 1990. Effect of feeding frequency in juvenile croaker, *Micropogonias furnieri* (Desmarest)(Pisces: Sciaenidae). J. Fish Biol. 37:987-988.

Ackefors, H., J.V. Huner & M. Konikoff. 1994. Introduction to the General Principles of Aquaculture. Food Products Press, New York, 172 p.

Ackman, R.G., & J. Kean-Howie. 1995. Fatty acids in aquaculture: are ω3 fatty acids always important? In: C. Lim & D.J. Sessa (eds.), Nutrition and Utilization Technology in Aquaculture. AOSC Press, Champaign, Illinois, p 82-104.

Adams, J.A. 1960. A contribution to the biology and postlarval development of the sargassum fish, *Histrio histrio* (Linnaeus), with a discussion of the *Sargassum* complex. Bull. Mar. Sci. 10:55-82.

Adams, S.M. 1990. Status and use of biological indicators for evaluating the effects of stress on fish. Am. Fish. Soc. Symp. (8):1-8.

Adams, S.M., & J.E. Breck. 1990. Bioenergetics. In: C.B. Schreck & P.B. Moyle (eds.), Methods for Fish Biology. Am. Fish. Soc., Bethesda, Maryland, p 389-415.

Addison, W.M. 1994a. Spawning and rearing the black cap basslet. Breeder's Registry 2(3):1,4.

Addison, B. 1994b. Spawning and rearing the blackcap basslet (*Gramma melacara*). SeaScope, Aquarium Systems, Mentor, Ohio, 11(fall):1,3.

Adoff, G.R., T. Andersson, R. Engelsen & R. Kvalsund. 1993. Land-based farm for ongrowing of halibut. In: H. Reinertsen, L.A. Dahle, L. Jørgensen & K. Tvinnereim (eds.), Fish Farming Technology. Balkema, Rotterdam, p 329-331.

Adron, J. W., A. Blair, C.B. Cowey & A.M. Shanks. 1976. Effects of dietary energy level and dietary energy source on growth, feed conversion and body composition of turbot (*Scophthalmus maximus* L.). Aquaculture 7:125-132.

Ager, L.A., D.E. Hammond & F. Ware. 1978. Artificial spawning of snook. Proc. Ann. Conf. SE Assoc. Game Fish Comm. 30:158-166.

Ahlstrom, E.H., & O.P. Ball. 1954. Description of eggs and larvae of jack mackerel (*Trachurus symmetricus*) and distribution and abundance of larvae in 1950 and 1951. Fish. Bull., U.S. 56:209-245.

Ahlstrom, E.H., & H.G. Moser. 1975. Distributional Atlas of Fish Larvae in the California Current Region: Flatfishes, 1955 through 1960. Cal. Coop. Oceanic Fish. Invest. Atlas No. 23, 207 p.

Ahlstrom, E.H., & H.G. Moser. 1980. Characters useful in identification of pelagic marine fish eggs. Cal. Coop. Oceanic Fish. Invest. Rep. (21):121-131.

Ahlstrom, E.H., K. Amaoka, D.A. Hensley, H.G. Moser & B.Y. Sumida. 1984. Pleuronectiformes: development. Am. Soc. Ichthyol. Herpetol. Spec. Pub. (1):640-670.

Ahne, W. 1994. Viral infections of aquatic animals with special reference to Asian aquaculture. Ann. Rev. Fish Dis. 4:375-388.

Ahrenholz, D.W. 1991. Population biology and life history of the North American menhadens, *Brevoortia* spp. Mar. Fish. Rev. 53(4):3-19.

Aida, K. 1983. Effect of LH-releasing hormone on gonadal development in a salmonoid fish, the ayu. Bull. Jpn. Soc. Sci. Fish. 49:711-718. (in Jpn., Engl. abs.)

Aida, K. R.S. Izumo, H. Satoh & T. Hibiya. 1978. Induction of ovulation in plaice and goby with synthetic LH-releasing hormone. Bull. Jpn. Soc. Sci. Fish. 44:445-450. (in Jpn., Engl. abs.)

Aiello, P., & A. D'Alba. 1986. *Amyloodinium ocellatum* infestation in yellow-tail, *Seriola dumerili*, intensively reared in Sicily, Italy. Bull. Eur. Assoc. Fish Pathol. 6:110-111.

Aiken, D. 1990. Commercial aquaculture in Canada and a glance at some prospects for the future. World Aquac. 21(2)66-75.

Aiken, D. 1991. Editorial: Aquaculture and the influenza virus. World Aquac. 22(1):2.

Akagawa, I., Y. Tsukamoto & M. Okiyama. 1995. Sexual dimorphism and pair spawning into a sponge by the filefish, *Brachaluteres ulvarum*, with a description of the eggs and larvae. Jpn. J. Ichthyol. 41:397-407.

Akatsu, S., K.M. Al-Abdul-Elah & S.K. Teng. 1983. Effects of salinity and water temperature on the survival and growth of brown-spotted grouper larvae (*Epinephelus tauvina*, Serranidae). J. World Maric. Soc. 14:624-635.

Akiyama, T., T. Unuma, T. Yamamoto, P. Marcouli & S. Kishi. 1995. Combinational use of malt protein flour and soybean meal as alternative protein sources of fish meal in fingerling rainbow trout diets. Fish. Sci. 61:828-832.

Akiyama, T., I. Oohara & T. Yamamoto. 1997. Comparison of essential amino acid requirements with A/E ratio among fish species (review paper). Fish. Sci. 63:963-970.

Ako, H., C.S. Tamaru, P. Bass & C.-S. Lee. 1994. Enhancing the resistance to physical stress in larvae of *Mugil cephalus* by the feeding of enriched *Artemia* nauplii. Aquaculture 122:81-90.

Aksnes, A., T. Hjertnes & J. Opstvedt. 1996. Effect of dietary protein level on growth and carcass composition in Atlantic halibut (*Hippoglossus hippoglossus* L). Aquaculture 145:225-233.

Aksnes, A., M.S. Izquierdo, L. Robaina, J.M. Vergara & D. Montero. 1997. Influence of fish meal quality and feed pellet on growth, feed efficiency and muscle composition in gilthead sea bream (*Sparus aurata*). Aquaculture 153:251-261.

Al-Abdul-Elah, K., S. Akatsu & S.K. Teng. 1983a. Effects of water temperature and water turbulence on the growth, survival and occurrence of initial swimbladder inflation in sobaity (*Acanthopagrus cuvieri*) larvae. Ann. Res. Rep. Kuwait Inst. Sci. Res. 8:46-48.

Al-Abdul-Elah, K., S. Akatsu, N. Downing, A.-A. Al-Ameeri, A. Al-Marzouk & C. El-Zahr. 1983b. Factors influencing swim bladder inflation in early larvae. Ann. Res. Rep. Kuwait Inst. Sci. Res. 7:61-64.

Al-Ahmad, T.A., M. Ridha & A.A. Al-Ahmed. 1988a. Production and feed ration of the tilapia *Oreochromis spilurus* in seawater. Aquaculture 73:111-118.

Al-Ahmad, T.A., M. Ridha & A.A. Al-Ahmed. 1988b. Reproductive performance of the tilapia *Oreochromis spilurus* in seawater and brackish groundwater. Aquaculture 73:323-332.

Alava, V.R., & C. Lim. 1988. Artificial diets for milkfish, *Chanos chanos* (Forsskal), fry reared in seawater. Aquaculture 71:339-346.

Alayse, J.-P. 1984. Utilisation dans l'elevage de *Amphiprion ocellaris* Cuvier de techniques destinees a l'aquaculture de poissons marins temperes. Oceanis 10:505-519. (in Fr., Engl. sum.)

Albrektsen, S., O. Lie & K. Sandnes. 1988. Ascorbyl palmitate as a dietary vitamin C source for rainbow trout (*Salmo gairdneri*). Aquaculture 71:359-368.

Alderdice, D.F., & C.R. Forrester. 1968. Some effects of salinity and temperature on early development and survival of the English sole (*Parophrys vetulus*). J. Fish. Res. Bd. Can. 25:495-521.

Alderson, R. 1974. Sea-water chlorination and the survival and growth of the early developmental stages of plaice, *Pleuronectes platessa* L., and Dover sole, *Solea solea* (L.). Aquaculture 4:41-53.

Alderson, R. 1979. The effect of ammonia on the growth of juvenile Dover sole, *Solea solea* (L.) and turbot, *Scophthalmus maximus* (L.). Aquaculture 17:291-309.

Alderson, R., & G.T. Ball. 1977. Towards the development of a rearing technique for the larvae of the turbot *Scophthalmus maximus* L. Min. Agric. Fish. Food, Fish. Lab., Port Erin, Isle of Man, U.K. 37 p. (mimeo)

Alderson, R., & P.J. Bromley. 1973. A method for rearing larvae of the turbot, *Scophthalmus maximus* L. to metamorphosis. ICES Fish. Improvement Committee, C.M. 1973/E:20, 11 p.

Alexis, M.N., K.K. Karanikolas & R.H. Richards. 1997. Pathological findings owing to the lack of ascorbic acid in cultured gilthead bream (*Sparus aurata* L.). Aquaculture 151:209-218.

Alfonso, A., C. Fermin & G. Robaina. 1983. Aspectos biologicos del genero Mycteroperca y su potencial de cultivo. En: Memorias V Simposio, Asociacion Latinoamericana de Acuicultura, Septiembre 5-9, 1983. Valdivia, Chile, 1:2.

Al-Ghais, S.M. 1993. Some aspects of the biology of *Siganus canaliculatus* in the southern Arabian Gulf. Bull. Mar. Sci. 52:886-897.

Alheit, J., & W. Scheibel. 1982. Benthic harpacticoids as a food source for fish. Mar. Biol. 70:141-147.

Ali, H.M. 1987. Sea bass (*Lates calcarifer*) spawning in tanks in Malaysia. Austral. Ctr. Internat. Agric. Res. Proc. (20):129-131.

Alkahem, H.F. 1994. The toxicity of nickel and the effects of sublethal levels on haematological parameters and behaviour of the fish, *Oreochromis niloticus*. J. Univ. Kuwait (Sci.): 21:243-251.

Allen, G.R. 1985. Butterfly and Angelfishes of the World. Aquarium Systems, Mentor, Ohio, 352 p.

Allen, G.R. 1996. The king of the cardinalfishes. Trop. Fish. Hob., 45(5):33-34,36-38,40,44-45.

Allen, G.R., & R.H. Kuiter. 1976. A review of the plesiopid fish genus *Assessor*, with descriptions of two new species. Rec. West. Aust. Mus. 4:201-215.

Allen, L.G. 1979. Larval development of *Gobiesox rhessodon* (Gobiesocidae) with notes on the larva of *Rimicola muscarum*. Fish. Bull., U.S. 77:300-304.

Allen, L.J., & E.C. Kinney. 1981. Proceedings of the Bio-Engineering Symposium for Fish Culture. Fish Cult. Sec., Am. Fish. Soc., Bethesda, Maryland, 307 p.

Allen, P.G., L.W. Botsford, A.M. Schuur & W.E. Johnston. 1984. Bioeconomics of Aquaculture. Dev. Aquac. Fish. Sci. 13, 351 p.

Allen, R.M.D. 1991. Ingredient analysis table: 1991 edition. Feedstuffs 63:24-31.

Alliot, E., A. Febvre, R. Metailler & A. Pastoureaud. 1974. Besoins nutritifs du bar *Dicentrarchus labrax*. Etude du taux de proteines et du taux de lipides dans le regime. Actes Colloq. CNEXO, 1:215-228.

Al-Marzouk, A., K.P. Lone & S.K. Teng. 1994. Photoperiod and temperature effects on the spawning time, fecundity and hatching success of a protandrous teleost, *Sparidentex hasta* Valenciennes. Pakistan J. Zool. 26:321-326.

Almatar, S., T. Al-Ahmed, S.-K. Teng & K. Al-Abdul-Elah. 1997. Advances in research on aquaculture of the blue-fin porgy (*Sparidentex hasta* Valenciennes) and orange-spotted grouper (*Epinephelus coioides* Hamilton 1822) in Kuwait. Island Aquac. Trop. Aquac. Ann. Mtg., Eur. Aquac. Soc., p 15-17.

Almendras, J.M.E. 1987. Acute nitrite toxicity and methemoglobinemia in juvenile milkfish (*Chanos chanos* Forsskal). Aquaculture 61:33-40.

Almendras, J.M., C. Duenas, J. Nacario, N.M. Sherwood & L.W. Crim. 1988. Sustained hormone release. III. Use of gonadotropin releasing hormone analogues to induce multiple spawnings in sea bass, *Lates calcarifer*. Aquaculture 74:97-111.

Al-Ogaily, S.M., N.E.D. Hanafi & A. Hussain. 1992. Some aspects of the biology of red snapper *Lutjanus campechanus* (Rivas, 1966) from the Red Sea. Asian Fish. Sci. 5:327-339.

Alonso, D.L., C.I.S. del Castillo, J.L.G. Sánchez, J.A.S. Pérez & F.G. Camacho. 1994. Quantitative genetics of fatty acid variation in *Isochrysis galbana* (Prymnesiophyceae) and *Phaeodactylum tricornutum* (Bacillariophyceae). J. Phycol. 30:553-558.

Al-Owafeir, M.A., & I.E.H. Belal. 1996. Replacing palm oil for soybean oil in tilapia, *Oreochromis niloticus* (L.), feed. Aquac. Res. 27:221-224.

Alshuth, S.R., J.W. Tucker, Jr. & J. Hateley. 1998. Egg and larval development of laboratory-reared sergeant major, *Abudefduf saxatilis* (Pisces: Pomacentridae). Bull. Mar. Sci. 62: .

Alsted, N., T. Due, N. Hjermitsley & A. Andreasen. 1995. Practical experience with high energy diets, FCR, growth and quality. J. Appl. Ichthyol. 11:329-335.

Al-Thobaiti, S.A., & C.M. James. 1997. Breeding and growout production of the grouper *Epinephelus polyphekadion* (Bleeker) in Saudi Arabia. Island Aquac. Trop. Aquac. Ann. Mtg., Eur. Aquac. Soc., p 20-21.

Al-Thobaiti, S., P.G. White, F.A. Bukhari & P.E. Hofilena. nd. Operational Manual: Culture of Fish in Pens in the Kingdom of Saudi Arabia. Fish Farming Ctr., Jeddah, 50 p.

Al-Thobaity, S.A., & C.M. James. 1994. Fish Farming Center: prospects for aquaculture in Saudi Arabia. Aquac. Mag. 20(6):28-40.

Aluko, P.O., & J.I. Awopetu. 1992. A preliminary study on the induced spawning of three cichlid fishes. J. Aquac. Trop. 7:63-68.

Alvarez-Lajonchère, L., J. Berdayes Arritola, O. Laiz Averhoff & S. Diaz Bellido. 1988. Positive results of induced spawning and larval rearing experiments with *Mugil liza* Val., a grey mullet from Cuban waters. Aquaculture 73:349-355.

Alvarez-Lajonchère, L., O.G. Hernández Molejón & L. Pérez Sánchez. 1991. Production of juveniles of the mullet *Mugil liza* Valenciennes, 1836, by controlled reproduction in Cuba. Ciencias Marinas 17:47-56.

Alvarez-Lajonchère, L., L. Pérez Sánchez, O.G. Hernández Molejón & E. Torres Gómez. 1996. Mass production of striped patao *Eugerres brasilianus* juveniles in Cuba. J. World Aquac. Soc. 27:347-352.

Alvarino, J.M.R., S. Zanuy, F. Prat, M. Carrillo & E. Mananos. 1992. Stimulation of ovulation and steroid secretion by LHRHa injection in the sea bass (*Dicentrarchus labrax*): effect of time of day. Aquaculture 102:177-186.

Alvial, A., & A. Trujillo. 1993. Current status of finfish hatcheries in Chile. In: C.-S. Lee, M.-S. Su & I C. Liao (eds.), Finfish Hatchery in Asia. Tungkang Mar. Lab. Conf. Proc. (3):117-132.

Amaoka, K. 1969. Studies on the Sinistral Flounders Found in the Waters around Japan: Taxonomy, Anatomy and Phylogeny. J. Shimonoseki Univ. Fish. 18(2):65-340.

Amerio, M., C. Ruggi, R.M. Rovelli & L. Völker. 1998. Ascorbic acid availability from ascorbyl 2-polyphosphate and ascorbyl 2-sulfate in sea bass (*Dicentrarchus labrax*). Aquaculture 159:233-237.

Amigó, J.M., M.-P. Gracia, M. Rius, H. Salvadó, P.A. Maillo & C.P. Vivarés. 1996. Longevity and effects of temperature on the viability and polar-tube extrusion of spores of *Glugea stephani*, a microsporidian parasite of commercial flatfish. Parasitol. Res. 82:211-214.

Andersen, D., I. Boetius, L.O. Larsen & P.H. Seidler. 1996. Effects of oestradiol-enriched diet and of feeding with porcine testicular tissue on macroscopic gonadal sex in European eels. J. Fish Biol.

48:484-492.
Anderson, B.S., D.P. Middaugh, J.W. Hunt & S.L. Turpen. 1991. Copper toxicity to sperm, embryos and larvae of topsmelt *Atherinops affinis*, with notes on induced spawning. Mar. Env. Res. 31:17-35.
Anderson, B.S., J.W. Hunt, H.R. McNulty, S.L. Turpen & M. Martin. 1994. Off-season spawning and factors influencing toxicity test development with topsmelt *Atherinops affinis*. Env. Toxicol. Chem. 13:479-485.
Anderson, B.S., J.W. Hunt, W.J. Piekarski, B.M. Phillips, M.A. Englund, R.S. Tjeerdema & J.D. Goetzl. 1995. Influence of salinity on copper and azide toxicity to larval topsmelt *Atherinops affinis* (Ayres). Arch. Env. Contam. Toxicol. 29:366-372.
Anderson, D.P. 1990. Immunological indicators: effects of environmental stress on immune protection and disease outbreaks. Am. Fish. Soc. Symp. (8):38-50.
Anderson, D.P. 1992. Immunostimulants, adjuvants, and vaccine carriers in fish: applications to aquaculture. Ann. Rev. Fish Dis. 281-307.
Anderson, D.P., & A.K. Siwicki. 1994. Duration of protection against *Aeromonas salmonicida* in brook trout immunostimulated with glucan or chitosan by injection or immersion. Prog. Fish-Cult. 56:258-261.
Anderson, I.G., & H.C. Prior. 1992. Subclinical epitheliocystis in barramundi, *Lates calcarifer*, reared in sea cages. Austral. Vet. Assoc. 69(9):226-227.
Anderson, J.A. 1989. Worm infestations in fish. SeaScope, Aquarium Systems, Mentor, Ohio, 6(summer):2-3.
Anderson, R.O. 1993. Apparent problems and potential solutions for production of fingerling striped bass, *Morone saxatilis*. J. Appl. Aquac. 2:101-118.
Anderson, R.O., & D. Tave (eds.). 1993. Strategies and Tactics for Management of Fertilized Hatchery Ponds. Haworth Press, Binghamton, N.Y., 260 p.
Anderson, W.W. 1957. Early development, spawning, growth, and occurrence of the silver mullet (*Mugil curema*) along the south Atlantic coast of the United States. Fish. Bull., U.S. 57:397-414.
Andersson, J. 1995. Two systems for collection of zooplankton for the culture of larval and juvenile fish. Aquac. Internat. 3:336-354.
Ando, Y., M. Kotake & T. Ota. 1997. Lipids and fatty acids in *Artemia* nauplii enriched with fish oil triacylglycerols containing docosahexaenoic acid in different positional distribution patterns. Fish. Sci. 63:605-609.
Andrew, Michael. 1994. The Living Seas, Epcot Center, Disney World, Orlando, Florida, pers. comm.
Andrew, T.G., C.D. Buxton & T. Hecht. 1996. Aspects of the reproductive biology of the concha wrasse, *Nelabrichthys ornatus*, at Tristan da Cunha. Env. Biol. Fishes 46:139-149.
Anonymous. 1977. The success of complete culture of the mullet. Sinorama 2(7):30-35.
Anonymous. 1984a. Conditioning and spawning of marine fishes, part I: adult pairs. SeaScope, Aquarium Systems, Mentor, Ohio, 1(spring):1-3.
Anonymous. 1984b. Larval rearing, part III: marine fish culture. SeaScope, Aquarium Systems, Mentor, Ohio, 1(fall):1,3.
Anonymous. 1986. Manual on Floating Netcage Fish Farming in Singapore's Coastal Waters. Prim. Prod. Dept., Singapore, 17 p.
Anonymous. 1987. Black spot disease—tang turbellaria in review. SeaScope, Aquarium Systems, Mentor, Ohio, 4(summer):3.
Anonymous. 1989. Impact is limited, say Norwegians. Fish Farming Internat. 16(8):21.
Anonymous. 1990. Operation Manual for Sea Cage Culture in the Kingdom of Saudi Arabia. Fish Farming Ctr., Jeddah, 36 p.
Anonymous. 1991. Pelleted feeds tested for mahimahi. Newsline, Oceanic Inst., Honolulu, Hawaii, 4:3.
Anonymous. 1992. Ingredient Market. Feedstuffs 64(13):28.
Anonymous. 1994. Marine Fish Stock Enhancement and Hatchery Legislative Report. Fla. Dept. Env. Prot., St. Petersburg, Fla., 98 p.
Anonymous, 1995a. Baitfish projects. North Florida Aquac. Bull. 4:1-2.
Anonymous. 1995b. Management practices for baitfish production. Aquac. Mag. 21(3):13-15.
Anonymous. 1996. Researchers take critical step in protecting, producing red snappers. Aquac. News:4(12):13.
Anonymous. 1997a. Ingredient Market. Feedstuffs 69(35):20.
Anonymous. 1997b. Tunas in the western Atlantic. Virginia Mar. Resource Bull. 29(1&2):20-22.
Anonymous. 1998a. Eel farming in the Netherlands. Aquac. Mag. 24(4):52,56-62.
Anonymous. 1998b. Ingredient Market. Feedstuffs 70(21):28.
Anonymous. 1998c. Ingredient Market. Feedstuffs 70(23):24.
Anras, M.-L.B. 1995. Demand-feeding behaviour of sea bass kept in ponds: diel and seasonal patterns, and influences of environmental factors. Aquac. Internat. 3:186-195.
Aoki, H. 1995. A review of the nursery and growout culture techniques for yellowtail (*Seriola quinqueradiata*) in Japan. In: K.L. Main & C. Rosenfeld (eds.), Culture of High-value Marine Fishes in Asia and the United States. Oceanic Inst., Honolulu, Hawaii, p 47-55.
Aoki, I. 1996. Batch fecundity of Japanese sardine *Sardinops melanostictus* off southern Kyushu,

southwestern Japan in the peak spawning season, 1991. Fish. Sci. 62:488-489.
Aoki, S., & A. Hino. 1996. Nitrogen flow in a chemostat culture of the rotifer *Brachionus plicatilis*. Fish. Sci. 62:8-14.
Aoki, S., J. Kanda & A. Hino. 1995. Measurements of the nitrogen budget in the rotifer *Brachionus plicatilis* by using ^{15}N. Fish. Sci. 61:406-410.
Aoki, T., & T. Kitao. 1985. Detection of transferable R plasmids in strains of the fish-pathogenic bacterium, *Pasteurella piscicida*. J. Fish Dis. 8:345-350.
APHA (Am. Public Health Assoc.). 1995. Standard Methods for the Examination of Water and Wastewater. APHA, Washington, D.C.
Apostolopoulos, J.S. 1976. Combined effect of temperature and salinity on the hatching rate, hatching time and total body length of the newly hatched larvae of the Japanese red sea bream *Pagrus major*. La Mer 14:23-30.
Appelbaum, S. 1985. Rearing of the Dover sole, *Solea solea* (L), through its larval stages using artificial diets. Aquaculture 49:209-221.
Appelbaum, S., W.C. Clarke, J.E. Shelbourne, J.O.T. Jensen, J.N.C. Whyte & G.K. Iwama. 1995. Studies on rearing of lingcod *Ophiodon elongatus*. Aquaculture 135:219-227.
Aqua Farm News. 1992a. May-Jun, 10(3):20 p.
Aqua Farm News. 1992b. Nov-Dec, 10(6):20 p.
AQUACOP, J. Fuchs, L. Debas, B. Jalabert & E. Morize. 1988. Reproduction of grouper *Epinephelus microdon* in French Polynesia. Natural cycle characteristics and first results in captivity. J. World Aquac. Soc. 19:14A.
AQUACOP, J. Fuchs & G. Nedelec. 1990a. Larval rearing and weaning of seabass, *Lates calcarifer* (Bloch), on experimental compounded diets. Advances in Tropical Aquaculture. AQUACOP, IFREMER, Actes Colloq. (9):677-697.
AQUACOP, G. Cuzon, R. Chou & J. Fuchs. 1990b. Nutrition of the seabass *Lates calcarifer*. Advances in Tropical Aquaculture. AQUACOP, IFREMER, Actes Colloq. (9):757-763.
AQUACOP, J. Fuchs, G. Nedelec & E. Gasset. 1990c. Selection of finfish species as candidates for aquaculture in French Polynesia. Advances in Tropical Aquaculture. AQUACOP, IFREMER, Actes Colloq. (9):461-484.
AQUACOP, E. Thouard, G. Nedelec, A. Bennett & P. Soletchnik. 1991. Status of tropical finfish mariculture at IFREMER. Advances in Finfish and Shellfish Mariculture. Phil. Council Aquat. Mar. Res. Dev., Book Ser. (12):37-49.
Aquaculture. 1995. Special Issue: Genetics in Aquaculture V. v 137.
Arai, H. 1994. Spawning behavior and early ontogeny of a pomacanthid fish, *Chaetodontoplus duboulayi*, in an aquarium. Jpn. J. Ichthyol. 41:181-187. (in Jpn., Engl. abs.)
Arai, H., & S. Fujita. 1988. Spawning behavior and early life history of the sharpnose puffer, *Canthigaster rivulata*, in the aquarium. Jpn. J. Ichthyol. 35:194-202.
Arai, S. 1989. Practical feeding—eels. In: T. Lovell, Nutrition and Feeding of Fish. Van Nostrand Reinhold, New York, p 223-230.
Arai, S. 1991. Eel, *Anguilla* spp. In: R.P. Wilson (ed.), Handbook of Nutrient Requirements of Finfish. CRC Press, Boca Raton, Florida, p 69-75.
Arai, S., T. Mori, W. Miki, K. Yamaguchi, S. Konosu, M. Satake & T. Fujita. 1987. Pigmentation of juvenile coho salmon with carotenoid oil extracted from Antarctic krill. Aquaculture 66:255-264.
Arakawa, T., T. Takeuchi & T. Watanabe. 1993. Suitable levels of starch in diet for juvenile striped jack. Nippon Suisan Gakkaishi 59:1945-1949. (in Jpn., Engl. abs.)
Argent. 1987. Finquel (MS-222). Brochure. Argent Chemical Laboratories, Redmond, Washington, 4 p.
Armstrong, M.P., J.A. Musick & J.A. Colvocoresses. 1992. Age, growth, and reproduction of the goosefish *Lophius americanus* (Pisces:Lophiiformes). Fish. Bull., U.S. 90:217-230.
Armstrong, R. (preparer). 1997. The Salmon Health Report, North. Aquac. Suppl., 30 p.
Arefyev, V.A. 1997. Sturgeon hybrids: natural reality and practical prospects. Aquac. Mag. 23(3):52-58.
Arnesen, A.M., E.H. Jorgensen & M. Jobling. 1993. Feed intake, growth and osmoregulation in Arctic charr, *Salvelinus alpinus* (L.), following abrupt transfer from freshwater to more saline water. Aquaculture 114:327-338.
Arnold, C.R. 1988. Controlled year-round spawning of red drum *Sciaenops ocellatus* in captivity. Contr. Mar. Sci. 30(suppl.):65-70.
Arnold, C.R. 1991. Precocious spawning of red drum. Prog. Fish-Cult. 53:50-51.
Arnold, C.R., J.L. Lasswell, W.H. Bailey, T.D. Williams & W.A. Fable, Jr. 1976. Methods and techniques for spawning and rearing spotted seatrout in the laboratory. Proc. Ann. Mtg. SE Assoc. Fish Wildl. Ag. 30:167-178.
Arnold, C.R., W.H. Bailey, T.D. Williams, A. Johnson & J.L. Lasswell. 1977. Laboratory spawning and larval rearing of red drum and southern flounder. Proc. Ann. Mtg. SE Assoc. Fish Wildl. Ag. 31:437-440.
Arnold, C.R., J.M. Wakeman, T.D. Williams & G.D. Treece. 1978. Spawning of red snapper (*Lutjanus campechanus*) in captivity. Aquaculture 15:301-302.
Arnold, E.L., Jr. & J.R. Thompson. 1958. Offshore spawning of the striped mullet, *Mugil cephalus*,

in the Gulf of Mexico. Copeia 1958:130-132.

Arrignon, J. 1990. Aquaculture in Africa. In: G. Barnabé (ed.), Aquaculture, Vol. 2. Ellis Horwood, New York, p 1035-1069.

Arthur, J.R., & K. Ogawa. 1996. A brief overview of disease problems in the culture of marine finfishes in East and Southest Asia. In: K.L. Main & C. Rosenfeld (eds.), Aquaculture Health Management Strategies for Marine Fishes. Oceanic. Inst., Honolulu, p 9-31.

Asgard, T., & E. Austreng. 1985. Dogfish offal, ensiled or frozen, as feed for salmonids. Aquaculture 49:289-305.

Ashraf, M., D.A. Bengtson & K.L. Simpson. 1993a. Effects of dietary fatty acid enrichment on survival, growth, and salinity-stress-test performance of inland silversides. Prog. Fish-Cult. 55:280-283.

Ashraf, M., K.L. Simpson & D.A. Bengtson. 1993b. Rearing of inland silverside, *Menidia beryllina* larvae on artificial diets. Pakistan J. Zool. 25:273-277.

Attramadal, Y.G., & M. Sandvand. 1993. A rotifer cultivating method resulting in high production and stability. In: H. Reinertsen, L.A. Dahle, L. Jørgensen & K. Tvinnereim (eds.), Fish Farming Technology. Balkema, Rotterdam, p 135-139.

Audet, C., M. Besner, J. Munro & J.-D. Dutil. 1993. Seasonal and diel variations of various blood parameters in Atlantic cod (*Gadus morhua*) and American plaice (*Hippoglossoides platessoides*). Can. J. Zool. 71:611-618.

Auditore, P.J., R.G. Lough & E.A. Broughton. 1994. A review of the comparative development of Atlantic cod (*Gadus morhua* L.) and haddock (*Melanogrammus aeglefinus* L.) based on an illustrated series of larvae and juveniles from Georges Bank. NAFO Sci. Council Stud. 20:7-18.

Auer, N.A. (ed.) 1982. Identification of Larval Fishes of the Great Lake Basin with Emphasis on the Lake Michigan Drainage. Great Lakes Fish. Comm. Spec. Pub. 82-3, 744 p.

Austin, B. 1993. Environmental issues in the control of bacterial diseases of farmed fish. In: R.S.V. Pullin, H. Rosenthal & J.L. Maclean (eds.), Environment and Aquaculture in Developing Countries. ICLARM Conf. Proc. (31):237-251.

Austin, B., & D.A. Austin. 1987. Bacterial Fish Pathogens: Disease in Farmed and Wild Fish. Ellis Horwood, Chichester, England, 364 p.

Austin, B., & D.A. Austin. 1989. Methods for the Microbiological Examination of Fish and Shellfish. Ellis Horwood, Chichester, England, 317 p.

Avault, J.W., Jr. 1985. The importance of water temperature. Aquac. Mag. 11(3):52,54.

Avault, J.W., Jr. 1996. Fungi and crustaceans cause diseases in aquaculture. Aquac. Mag. 22(2):85-87.

Avila, E.M. 1984. Hormone-induced spawning and embryonic development of the rabbitfish, *Siganus vermiculatus* (Pisces: Siganidae). Philip. Sci. 21:75-108.

Avilés-Quevedo, A., U. McGregor-Pardo, R. Rodríguez-Ramos, O. Hirales-Cosio, M.A. Huerta-Bello & M. Iizawa. 1995. Biología y cultivo de la cabrilla arenera. Centro Regional Invest. Pesq. La Paz, Mexico, 80 p.

Avilés-Quevedo, A., R. Rodríguez-Ramos, U. McGregor-Pardo & M. Iizawa. 1998. Descripion del huevo y larva temprana del pargo Amarillo, *Lutjanus argentiventris*. Poster pap., Seminario Interamericano Reproduccion y Cultivo de Peces Marinos y Dulceacuicolas, Inst. Nac. Pesca, Mexico City, March.

Awaiss, A., P. Kestemont & J.-C. Micha. 1992. Nutritional suitability of the rotifer, *Brachionus calyciflorus* Pallas for rearing freshwater fish larvae. J. Appl. Ichthyol. 8:263-270.

Ayson, F.G. 1989. The effect of stress on spawning of brood fish and survival of larvae of the rabbitfish, *Siganus guttatus* (Bloch). Aquaculture 80:241-246.

Ayson, F.G. 1991. Induced spawning of rabbitfish, *Siganus guttatus* (Bloch) using human chorionic gonadotropin (HCG). Aquaculture 95:133-137.

Azuma, M. 1964. The life history of the landlocked ayu-fish (*Plecoglossus altivelis*) in the Lake Biwa, with special references to their developmental stages. Physiol. Ecol. Jpn. 12:55-71.

Azzarello, M.Y. 1991. Some questions concerning the Syngnathidae brood pouch. Bull. Mar. Sci. 49:741-747.

Backhurst, J.R., O. Chaalal & J.H. Harker. 1989. Optimum aeration rates for the suspension of larval feeds. Aquacultural Eng. 8:217-221.

Backman, T.W.H., & R.M. Ross. 1990. Comparison of three techniques for the capture and transport of impounded subyearling American shad. Prog. Fish-Cult. 52:246-252.

Bagarinao, T., & J.R. Hunter. 1983. The visual feeding threshold and action spectrum of northern anchovy (*Engraulis mordax*) larvae. Cal. Coop. Oceanic Fish. Invest. Rep. 24:245-254.

Bagarinao, T., & P. Kungvankij. 1986. An incidence of swimbladder stress syndrome in hatchery-reared sea bass (*Lates calcarifer*) larvae. Aquaculture 51:181-188.

Bailey, D.S., J.A. Hargreaves & J.E. Rakocy. 1992. Enterprise budget analysis of three stocking densities of caged Florida red tilapia. Proc. Gulf Carib. Fish. Inst. 42:455-463.

Balasuriya, L.K.S.W., & T.S. Leong. 1994. Effect of stocking density of seabass *Lates calcarifer* (Bloch) cultured in the floating cages on gill monogenean population. Third Asian Fisheries Forum, Asian Fish. Soc., Manila, p 349-352.

Baldassano, P. 1996. A new feeding strategy for *Hippocampus* sp., and other fishes. J. MaquaCulture

4(3):11,18-19.
Ballard, J. 1971. Aquarium briefs—mating in captivity. Bull. S. Afr. Assoc. Mar. Biol. Res. (9):31-33.
Balogun, A.M., & O.A. Fagbenro. 1995. Use of macadamia presscake as a protein feedstuff in practical diets for tilapia, *Oreochromis niloticus* (L.). Aquac. Res. 26:371-377.
Balompapueng, M.D., A. Hagiwara, A. Nishi, K. Imaizumi & K. Hirayama. 1997. Resting egg formation of the rotifer *Brachionus plicatilis* using a semi-continuous culture method. Fish. Sci. 63:236-241.
Banks, M.A., G.J. Holt & J.M. Wakeman. 1991. Age-linked changes in salinity tolerance of larval spotted seatrout (*Cynoscion nebulosus*, Cuvier). J. Fish Biol. 39:505-514.
Barahona-Fernandes, M.H. 1978. Effect of aeration on the survival and growth of sea bass (*Dicentrarchus labrax* L.) larvae: a preliminary study. Aquaculture 14:67-74.
Barahona-Fernandes, M.H. 1979. Some effects of light intensity and photoperiod on the sea bass larvae (*Dicentrarchus labrax* (L.)) reared at the Centre Oceanologique de Bretagne. Aquaculture 17:311-321.
Barahona-Fernandes, M.H. 1982. Problemes lies au sevrage du bar (*Dicentrarchus labrax*). 3. Influence de l'incorporation d'attractants dans le granule. Pisciculture Francaise 65:10-11.
Barahona-Fernandes, M.-H., & M. Girin. 1976. Preliminary tests on the optimal pellet-adaptation age for sea bass larvae (Pisces, *Dicentrarchus labrax* L. 1758). Aquaculture 8:283-290.
Barak, N.A.-E., & C.F. Mason. 1992. Population density, growth and diet of eels, *Anguilla anguilla* L., in two rivers in eastern England. Aquac. Fish. Mgt. 23:59-70.
Barata, C., F. Hontoria & F. Amat. 1996. Estimation of the biomass production of *Artemia* with regard to its use in aquaculture: temperature and strain effects. Aquaculture 142:171-189.
Barbaro, A., A. Francescon, G. Bozzato, A. Merlin, P. Belvedere & L. Colombo. 1997. Induction of spawning in gilthead seabream, *Sparus aurata* L., by a long-acting GnRH agonist and its effects on egg quality and daily timing of spawning. Aquaculture 154:349-359.
Barbato, F., A. Fanari, F. Meloni & R. Savarino. 1993. Cage culture of gilthead sea bream (*Sparus aurata*) in two Sardinian coastal lagoons. Eur. Aquac. Soc. Spec. Pub. (18):95-104.
Barbin, G.P., & J.D. McCleave. 1997. Fecundity of the American eel *Anguilla rostrata* at 45°N in Maine, U.S.A. J. Fish Biol. 51:840-847.
Barclay, W., & S. Zeller. 1996. Nutritional enhancement of n-3 and n-6 fatty acids in rotifers and *Artemia* nauplii by feeding spray-dried *Schizochytrium* sp. J. World Aquac. Soc. 27:314-322.
Bardi, R.W., Jr., F.A. Chapman & F.T. Barrows. 1998. Feeding trials with hatchery-produced Gulf of Mexico sturgeon larvae. Prog. Fish-Cult. 60:25-31.
Barkman, R.C., D.A. Bengtson & A.D. Beck. 1981. Daily growth of the juvenile fish (*Menidia menidia*) in the natural habitat compared with juveniles reared in the laboratory. Rapp. P.-v. Réun. Cons. Int. Explor. Mer 178:324-326.
Barkoh, A. 1996. Effects of three fertilization treatments on water quality, zooplankton, and striped bass fingerling production in plastic-lined ponds. Prog. Fish-Cult. 58:237-247.
Barlow, C.G., L.J. Rodgers, P.J. Palmer & C.J. Longhurst. 1993. Feeding habits of hatchery-reared barramundi *Lates calcarifer* (Bloch) fry. Aquaculture 109:131-144.
Barlow, C.G., M.G. Pearce, L.J. Rodgers & P. Clayton. 1995. Effects of photoperiod on growth, survival and feeding periodicity of larval and juvenile barramundi *Lates calcarifer* (Bloch). Aquaculture 138:159-168.
Barlow, C., K. Williams & M. Rimmer. 1996. Sea bass culture in Australia. INFOFISH Internat. (Feb):26-29,31-33.
Barnabé, G. 1972. Contribution à l'Étude de la Biologie du Loup (*Dicentrarchus labrax* L.) de la Région de Sète. These 3e cycle, Univ. Sci. Techniques, Languedoc, Montpellier, 160 p.
Barnabé, G. 1974. Mass rearing of the bass *Dicentrarchus labrax* L. In: J.H.S. Blaxter (ed.) The Early Life History of Fish. Springer-Verlag, New York, p 749-753.
Barnabé, G. 1976a. Contribution à la Connaissance de la Biologie du Loup *Dicentrarchus labrax* L. (Poisson Serranidae) de la Région de Sète. Doctoral Thesis. Univ. Sci. Techniques, Languedoc, Montpellier, 426 p.
Barnabé, G. 1976b. Elevage larvaire du loup (*Dicentrarchus labrax* (L.); Pisces, Serranidae) a l'aide d'aliment sec compose. Aquaculture 9:237-252.
Barnabé, G. 1990a. Aquaculture, Vol. 1 & 2. Ellis Horwood, New York, 1112 p.
Barnabé, G. 1990b. Cages or ponds? reality and perspectives for the Mediterranean sea. In: R. Flos, L. Tort & P. Torres (eds.), Mediterranean Aquaculture. Ellis Horwood, New York, p 51-68.
Barnabé, G. 1990c. Harvesting micro-algae. In: G. Barnabé (ed.), Aquaculture, Vol. 1. Ellis Horwood, New York, p 207-212.
Barnabé, G. 1990d. Harvesting zooplankton. In: G. Barnabé (ed.), Aquaculture, Vol. 1. Ellis Horwood, New York, p 264-272.
Barnabé, G. 1990e. Rearing bass and gilthead bream. In: G. Barnabé (ed.), Aquaculture, Vol. 2. Ellis Horwood, New York, p 647-686.
Barnabé, G., & R. Barnabé-Quet. 1985. Avancement et amelioration de la ponte induite chez le loup *Dicentrarchus labrax* (L.) a l'aide d'un analogue de LHRH injecte. Aquaculture 49:125-132.
Barnabé, G., & A. Guissi. 1993. Combined effects of diet and salinity on European sea bass larvae

Dicentrarchus labrax. J. World Aquac. Soc. 24:439-450.

Barnabé, G., & A. Guissi. 1994. Adaptations of the feeding behviour of larvae of the sea bass, *Dicentrarchus labrax* (L.), to an alternating live-food/compound-food feeding regime. Aquac. Fish. Mgt. 25:537-546.

Barnabé. G., & J. Paris. 1984. Ponte avancée et ponte normale du loup *Dicentrarchus labrax* (L.) à la Station de Biologie Marine et Lagunaire de Sète. In: G. Barnabé & R. Billard (eds.), L'Aquaculture du Bar et des Sparidés. INRA, Paris, p 63-72.

Barrows, F.T., W.A. Lellis & J.G. Nickum. 1988. Intensive culture of larval walleyes with dry or formulated feed: note on swim bladder inflation. Prog. Fish-Cult. 50:160-166.

Barry, T.P., & A.W. Fast. 1992. Biology of the spotted scat (*Scatophagus argus*) in the Philippines. Asian Fish. Sci. 5:163-179.

Barry, T.P., M.T. Castanos & A.W. Fast. 1991. Induced spermiation in the male spotted scat (*Scatophagus argus*) by long-term administration of 17 α-methyltestosterone followed by LHRHa. Asian Fish. Sci. 4:137-145.

Barry, T.P., M.T. Castanos, M.P.S.C. Macahilig & A.W. Fast. 1993. Spawning induction in female spotted scat (*Scatophagus argus*). J. Aquac. Trop. 8:121-129.

Barthel, K.-G. 1983. Food uptake and growth efficiency of *Eurytemora affinis* (Copepoda: Calanoida). Mar. Biol. 74:269-274.

Bartley, D., M. Bagley, G. Gall & B. Bentley. 1992. Use of linkage disequilibrium data to estimate effective size of hatchery and natural fish populations. Conserv. Biol. 6:365-375.

Barton, B.A., & G.K. Iwama. 1991. Physiological changes in fish from stress in aquaculture with emphasis on the response and effects of corticosteroids. Ann. Rev. Fish Dis. 1:3-26.

Baskerville-Bridges, B., & L.J. Kling. 1997. Early weaning of Atlantic cod (*Gadus morhua*) larvae onto a microparticulate diet. World Aquac. '97 Book of Abs., p 36-37.

Bates, D. 1996. Drug usage and legislation. In: B. Chatain, M. Saroglia, J. Sweetman & P. Lavens (eds.), Seabass and Seabream culture: Problems and Prospects. Eur. Aquac. Soc., p 119-134.

Bates, J. 1997. The reproduction and early development of the Banggai cardinalfish (*Pterapogon kauderni*). J. MaquaCulture 5(1):16-20.

Baticados, M.C.L., & G.F. Quinitio. 1984. Occurrence and pathology of an *Amyloodinium*-like protozoan parasite on gills of grey mullet, *Mugil cephalus*. Helgolander Meeresuntersuchungen. 37:595-601.

Baudin-Laurencin, F. 1990. Sole and turbot culture: disease. In: G. Barnabé (ed.), Aquaculture, Vol. 2. Ellis Horwood, New York, p 719-723.

Battaglene, S.C. 1996. ILFC symposium. Aquaculture: introduction. Mar. Freshw. Res. 47:209-210.

Battaglene, S.C., & G.L. Allan. 1990. Prawn farmers could turn to bass in the winter. Austral. Fish. 49(1):31-32.

Battaglene, S.C., & G.L. Allan. 1994. Development of hatchery techniques for snapper, *Pagrus auratus*, in Australia. Third Asian Fisheries Forum, Asian Fish. Soc., Manila, p 108-111.

Battaglene, S., & B. Talbot. 1989. Australian bass breeding. New South Wales Dept. Agric. Newsl. 3 p.

Battaglene, S.C., & R.B. Talbot. 1990. Initial swim bladder inflation in intensively reared Australian bass larvae, *Macquaria novemaculeata* (Steindachner)(Perciformes: Percichthyidae). Aquaculture 86:431-442.

Battaglene, S.C., & R.B. Talbot. 1992. Induced spawning and larval rearing of snapper, *Pagrus auratus* (Pisces: Sparidae), from Australian waters. N.Z. J. Mar. Freshw. Res. 26:179-185.

Battaglene, S.C., & R.B. Talbot. 1993. Effects of salinity and aeration on survival of and initial swim bladder inflation in larval Australian bass. Prog. Fish-Cult. 55:35-39.

Battaglene, S.C., & R.B. Talbot. 1994. Hormone induction and larval rearing of mulloway, *Argyrosomus hololepidotus* (Pisces:Sciaenidae). Aquaculture 126:73-81.

Battaglene, S.C., P.J. Beevers & R.B. Talbot. 1989a. A review of research into the artificial propagation of Australian bass (*Macquaria novemaculeata*) at the Brackish Water Fish Culture Research Station, Salamander Bay, 1979 to 1986. New South Wales Agric. Fish., Fish. Bull. 3, 11 p.

Battaglene, S., B. Talbot & P. Beevers. 1989b. Australian bass culture—recent advances. Austral. Fish. 48(7):28-30.

Battaglene, S.C., S. McBride & R.B. Talbot. 1994. Swim bladder inflation in larvae of cultured sand whiting, *Sillago ciliata* Cuvier (Sillaginidae). Aquaculture 128:177-192.

Battaglia, B. 1970. Cultivation of marine copepods for genetic and evolutionary research. Helgoländer wiss. Meeresunters 20:385-392.

Bauer, J., & S. Bauer. 1975. First report of spawning rusty gobies. Octopus 2(5):25.

Bauer, J.A., Jr. & S.E. Bauer. 1981. Reproductive biology of pigmy angelfishes of the genus *Centropyge* (Pomacanthidae). Bull. Mar. Sci. 31:495-513.

Bauer, J.A., Jr. & G. Klaij. 1974. Pigmy angels spawn. Octopus 1(5):7-18.

Bauer, S., & J. Bauer. 1974. Spawning the Catalina goby. Octopus 1(8):6-11.

Bautista, M.N., & M.C. De la Cruz. 1988. Linoleic (ω6) and linolenic (ω3) acids in the diet of fingerling milkfish (*Chanos chanos* Forsskal). Aquaculture 71:347-358.

Baya, A.M., R.B. Navarro & E. Kotopoulis. 1996. Streptococcal infections of hybrid striped bass and tilapia. In: G.S. Libey & M.B. Timmons (eds.), Aquacultural Eng. Soc. Proc. II: Successes and Failures in Commercial Recirculating Aquaculture, p 32-40.

Baynes, S.M., & A.P. Scott. 1982. Cryopreservation of rainbow trout spermatozoa: variation in membrane composition may influence spermatozoan survival. Proc. Internat. Symp. Repr. Physiol. Fish, Wageningen, Netherlands, p 128.

Baynes, S.M., B.R. Howell & T.W. Beard. 1993. A review of egg production by captive sole, *Solea solea* (L.). Aquac. Fish. Mgt. 24:171-180.

Bayrle, H. 1982. Cryopreservation of rainbow trout sperm: effect of equilibration. Proc. Internat. Symp. Repr. Physiol. Fish, Wageningen, Netherlands, p 129-130.

Beacham, T.D., & C.B. Murray. 1990. Temperature, egg size, and development of embryos and alevins of five species of Pacific salmon: a comparative analysis. Trans. Am. Fish. Soc. 119:927-945.

Beacham, T.D., & C.B. Murray. 1993. Fecundity and egg size variation in North American Pacific salmon (*Oncorhynchus*). J. Fish Biol. 42:485-508.

Beamish, F.W.H., A.J. Niimi & P.F.K.P. Lett. 1975. Bioenergetics of teleost fishes: environmental influences. In: L. Bolis, H.P. Maddrell & K. Schmidt-Nielsen (eds.), Comparative Physiology—Functional Aspects of Structural Materials. North-Holland Publishing, Amsterdam, p 187-209.

Beckmann, A.T., D.R. Gunderson, B.S. Miller, R.M. Buckley & B. Goetz. 1998. Reproductive biology, growth, and natural mortality of Puget Sound rockfish, *Sebastes emphaeus* (Starks, 1911). Fish. Bull. U.S. 96:352-356.

Bedier, E. 1981. Pilot scale production of sea bass (*Dicentrarchus labrax*) fry. Rapp. P.-v. Réun. Int. Explor. Mer 178:530-532.

Behrends, L.L., J.B. Kingsley & A.H. Price III. 1993. Hatchery production of blue tilapia, *Oreochromis aureus* (Steindachner), in small suspended hapa nets. Aquac. Fish. Mgt. 24:237-243.

Belanga, S.D. 1993. Clownfish broodstock: selection and pairing. Breeder's Registry 1(2):1,3-4.

Bell, G.M., & J. Chadwick. 1994. Regulatory controls on biocides in the United Kingdom and restrictions on the use of triorganotin-containing antifouling products. Internat. Biodeter. Biodegrad. 34:375-386.

Bell, J.G., C. Ghioni & J.R. Sargent. 1994. Fatty acid compositions of 10 freshwater invertebrates which are natural food organisms of Atlantic salmon parr (*Salmo salar*): a comparison with commercial diets. Aquaculture 128:301-313.

Bell, M.V., R.J Henderson, B.J.S. Pirie & J.R. Sargent. 1985a. Effects of dietary polyunsaturated fatty acid deficiencies on mortality, growth and gill structure in the turbot, *Scophthalmus maximus*. J. Fish Biol. 26:181-191.

Bell, M.V., R.J. Henderson, B.J.S. Pirie & J.R. Sargent. 1985b. Growth, gill structure and fatty acid composition of phospholipids in the turbot (*Scophthalmus maximus*) in relation to dietary polyunsaturated fatty acid deficiencies. In: C.B. Cowey, A.M. Mackie & J.G. Bell (eds.), Nutrition and Feeding in Fish. Academic Press, San Diego, California, p 365-369.

Bell, M.V., J.R. Dick, M. Thrush & J.C. Navarro. 1996a. Decreased 20:4*n*-6/20:5*n*-3 ratio in sperm from cultured sea bass, *Dicentrarchus labrax*, broodstock compared with wild fish. Aquaculture 144:189-199.

Bell, M.V., L.A. McEvoy & J.C. Navarro. 1996b. Deficit of didocosahexaenoyl phospholipid in the eyes of larval sea bass fed an essential fatty acid deficient diet. J. Fish. Biol. 49:941-952.

Ben-Amotz, A., & H. Rosenthal. 1981. Cryopreservation of marine unicellular algae and early life stages of fish for use in mariculture. Eur. Aquac. Soc. Spec. Pub. (6):149-162.

Ben-Amotz, A., R. Fishler & A. Schneller. 1987. Chemical composition of dietary species of marine unicellular algae and rotifers with emphasis on fatty acids. Mar. Biol. 95:31-36.

Benavente, G.P., & F.-J. Gatesoupe. 1988. The continuous distribution of rotifers increases the essential fatty acid reserve of turbot larvae, *Scophthalmus maximus*. Aquaculture 72:109-114.

Benemann, J.R. 1992. Microalgae aquaculture feeds. J. Appl. Phycol. 4:233-245.

Benetti, D.D. 1992. Bioenergetics and growth of dolphin, *Coryphaena hippurus*. Ph.D. Diss., Univ. Miami, 197 p.

Benetti, D.D. 1997. Spawning and larval husbandry of flounder (*Paralichthys woolmani*) and Pacific yellowtail (*Seriola mazatlana*), new candidate species for aquaculture. Aquaculture 155:307-318.

Benetti, D.D., & E.B. Fagundes Netto. 1991. Preliminary results on growth of mullets (*Mugil liza* and *M. curema*) fed artificial diets. World Aquac. 22:55-57.

Benetti, D.D., A. Venizelos & C. Acosta. 1994. Finfish aquaculture development in Ecuador. World Aquac. 25(2):18-25.

Benetti, D.D., C.A. Acosta & J.C. Ayala. 1995a. Cage and pond aquaculture of marine finfish in Ecuador. World Aquac. 26(4):7-13.

Benetti, D.D., E.S. Iversen & A.C. Ostrowski. 1995b. Growth rates of captive dolphin, *Coryphaena hippurus*, in Hawaii. Fish. Bull., U.S. 93:152-157.

Bengtson, D.A. 1997. Paper presented on culture of summer flounder at World Aquac. Soc. Ann. Mtg., Las Vegas.

Bengtson, D.A., R.C. Barkman & W.J. Berry. 1987. Relationships between maternal size, egg diameter, time of spawning season, temperature, and length at hatch of Atlantic silverside, *Menidia menidia*. J. Fish Biol. 31:697-704.

Bengtson, D.A., P. Léger & P. Sorgeloos. 1991. Use of *Artemia* as a food source for aquaculture. In: R.A. Brown, P. Sorgeloos & C.N.A. Trotman (eds.), Artemia Biology. CRC Press, Boca Raton, Florida, p 255-285.

Bengtsson, B.-E., & A. Larsson. 1986. Vertebral deformities and physiological effects in fourhorn sculpin (*Myoxocephalus quadricornis*) after long-term exposure to a simulated heavy metal-containing effluent. Aquat. Toxicol. 9:215-229.

Bennett, W.N., & M.E. Boraas. 1993. Rotifer culture in the turbidostat. Ecological Studies: Analysis and Synthesis 98:30-38.

Bennett, W.N., M.E. Boraas & D.B. Seale. 1993. Turbidostat culture of *Brachionus calyciflorus*: an experimental system to assess biological limits on population growth. Ecological Studies: Analysis and Synthesis 98:77-86.

Benson, R.F., J.A. Saraceno, D.R. Lowell & D.W. Brett. 1994. Selenium levels in the Palm River and McKay Bay region of Tampa Bay. Fla. Sci. 57:177-185.

Berge, G.M., & T. Storebakken. 1996. Fish protein hydrolyzate in starter diets for Atlantic salmon (*Salmo salar*) fry. Aquaculture 145:205-212.

Bergh, Ø., G.H. Hansen, A. Jelmert, A.B. Skiftesvik & R.E. Taxt. 1991. Bacterial diseases of eggs and yolk sac larvae of halibut (*Hippoglossus hippoglossus* L.). Eur. Aquac. Soc. Spec. Pub. (15):389-391.

Bergheim, A., & O.I. Forsberg. 1993. Attempts to reduce effluent loadings from salmon farms by varying feeding frequencies and mechnical effluent treatment. Eur. Aquac. Soc. Spec. Pub. (18):115-124.

Bergheim, A., R. Kristiansen & L. Kelly. 1993. Treatment and utilization of sludge from landbased farms for salmon. In: J.-K. Wang (ed.), Techniques for Modern Aquaculture. Am. Soc. Agric. Eng., St. Joseph, Michigan, p 486-495.

Bergstad, O.A. 1993. Distribution, population structure, growth, and reproduction of the greater silver smelt, *Argentina silus* (Pisces, Argentinidae), of the Skagerrak and the north-eastern North Sea. ICES J. Mar. Sci. 50:129-143.

Berlinsky, D.L., & J.L. Specker. 1991. Changes in gonadal hormones during oocyte development in the striped bass, *Morone saxatilis*. Fish Physiol. Biochem. 9:51-62.

Berlinsky, D.L., L.F. Jackson, T.I.J. Smith & C.V. Sullivan. 1995. The annual reproductive cycle of the white bass *Morone chrysops*. J. World Aquac. Soc. 26:252-260.

Berlinsky, D.L., W. King V, T.I.J. Smith, R.D. Hamilton II, J. Holloway, Jr. & C.V. Sullivan. 1996. Induced ovulation of southern flounder *Paralichthys lethostigma* using gonadotropin releasing hormone analogue implants. J. World Aquac. Soc. 27:143-152.

Berlinsky, D.L., W. King V, R.G. Hodson & C.V. Sullivan. 1997. Hormone induced spawning of summer flounder *Paralichthys dentatus*. J. World Aquac. Soc. 28:79-86.

Berrien, P.L. 1975. A description of Atlantic mackerel, *Scomber scombrus*, eggs and early larvae. Fish. Bull., U.S. 73:186-192.

Berry, J.W., D.W. Osgood & P.A. St. John. 1974. Chemical Villains: A Biology of Pollution. Mosby, Saint Louis, 189 p.

Beveridge, M. 1996. Cage Aquaculture. Fishing News Books, Oxford, 346 p.

Beveridge, M.C.M., M.J. Phillips & R.M. Clarke. 1991. A quantitative and qualitative assessment of wastes from aquatic animal production. Adv. World Aquac. 3:506-533.

Bhat, U.G., & K. Vamsee. 1993. Toxicity of heavy metals Cu, Cd and Hg to the gammarid amphipod *Parhalella natalensis* (Stebbing). Sci. Total Env. Suppl. 1993:887-897.

Bibiloni, G., A. Cladera, M.A. Gomila, S. Massuti & I. Zaratiegui. 1993. A small-scale experiment on rearing of *Dentex dentex*. Eur. Aquac. Soc. Spec. Pub. (19):317.

Bidwell, D., W.H. Howell, N. King, E. Fairchild & A. Tomlinson. 1997. Witch flounder (*Glyptocephalus cynoglossus*) aquaculture. World Aquac. '97 Book of Abs., p 43-44.

Biedenbach, J.M., L.L. Smith, T.K. Thomsen & A.L. Lawrence. 1989. Use of the nematode *Panagrellus redivivus* as an *Artemia* replacement in a larval penaeid diet. J. World Aquac. Soc. 20:61-71.

Biesiot, P.M., R.E. Caylor & J.S. Franks. 1994. Biochemical and histological changes during ovarian development of cobia, *Rachycentron canadum*, from the northern Gulf of Mexico. Fish. Bull., U.S. 92:686-696.

Billard, R. 1978. Some data on gametes preservation and artifical insemination in teleost fish. Actes Coll. CNEXO 8:59-73.

Billard, R. 1984. The control of fish reproduction in aquaculture. Proc. Internat. Conf. Aquac., Venice, 43 p.

Billard, R. 1990. Culture of salmonids in fresh water. In: G. Barnabé (ed.), Aquaculture, Vol. 2. Ellis Horwood, New York, p 549-592.

Billard, R., & C. Gillet. 1981. Vieillisement des ovules et potentialisation par la temperature des effets des micropollutants du milieu aqueux sur les gametes chez la truite. Cah. Lab. Hydrobiologie

Monterau 12:35-42.
Billard, R., & J. Sevrin-Reyssac. 1993. Les impacts negatifs et positifs de la pisciculture d'etang sur l'environnement. Eur. Aquac. Soc. Spec. Pub. (18):17-29.
Billard, R., P. Reinaud, M.G. Hollebecq & B. Breton. 1984. Advancement and synchronisation of spawning in *Salmo gairdneri* and *S. trutta* following administration of LRH-A combined or not with pimozide. Aquaculture 43:57-66.
Billard, R., J. Cosson, L.W. Crim & M. Suquet. 1995. Sperm physiology and quality. In: N.R. Bromage & R.J. Roberts (eds.), Broodstock Management and Egg and Larval Quality. Blackwell, Oxford, p 25-52.
Bills, T.D., L.L. Marking & G.E. Howe. 1993. Sensitivity of juvenile striped bass to chemicals used in aquaculture. U.S. Fish Wildl. Serv., Resource Pub. 192, 11 p.
Bimbo, A.P., & J.B. Crowther. 1992. Fish meal and oil: current uses. J. Am. Oil Chemists Soc. 69:221-227.
BIOMED. 1994. Autogenous vaccines. Biomed Inc., Bellevue, Washington, Tech. Bull. No. 14, 2 p.
Birdsong, C.L., & J.W. Avault, Jr. 1971. Toxicity of certain chemicals to juvenile pompano. Prog. Fish-Cult. 33:76-80.
Bisbal, G.A., & D.A. Bengston. 1995a. Development of the digestive tract in larval summer flounder. J. Fish Biol. 47:277-291.
Bisbal, G.A., & D.A. Bengtson. 1995b. Effects of delayed feeding on survival and growth of summer flounder *Paralichthys dentatus* larvae. Mar. Ecol. Prog. Ser. 121:301-306.
Bishop, C.D., R.A Angus & S.A. Watts. 1995. The use of feather meal as a replacement for fish meal in the diet of *Oreochromis niloticus* fry. Bioresource Technol. 54:291-295.
Bisogni, J.J., Jr. & M.B. Timmons. 1994. Chapter 8: Control of pH in closed cycle aquaculture systems. In: M.B. Timmons & T.M. Losordo (eds.), Aquaculture Water Reuse Systems: Engineering Design and Management. Elsevier, Amsterdam, p 235-245.
Bjordal, Å., J.E. Juell, T. Lindem & A. Fernö. 1993. Hydoacoustic monitoring and feeding control in cage rearing of Atlantic salmon (*Salmo salar* L.). In: H. Reinertsen, L.A. Dahle, L. Jørgensen & K. Tvinnereim (editors), Fish Farming Technology. Balkema, Rotterdam, p 203-208.
Björnsson, B. 1994. Effects of stocking density on growth rate of halibut (*Hippoglossus hippoglossus* L.) reared in large circular tanks for three years. Aquaculture 123:259-270.
Björnsson, B., & S.V. Tryggvadóttir. 1996. Effects of size on optimal temperature for growth and growth efficiency of immature Atlantic halibut (*Hippoglossus hippoglossus* L.). Aquaculture 142:33-42.
Björnsson, B., G. Sigurthorsson, G.-I. Hemre & Ø. Lie. 1992. Growth rate and feed conversion factor of young halibut (*Hippoglossus hippoglossus* L.) fed six different diets. Fisk. Dir. Skr., Ser. Ernæring 5:25-35.
Black, J.A., J.B. Barnum & W.J. Birge. 1993. An integrated assessment of the biological effects of boron to the rainbow trout. Chemosphere 26:1383-1413
Black, K.D., S. Fleming, T.D. Nickell & P.M.F. Pereira. 1997. The effects of ivermectin, used to control sea lice on caged farmed salmonids, on infaunal polychaetes. ICES J. Mar. Sci. 54:276-279.
Blakstad, F. 1993. The economic aspects of land based versus ocean based Atlantic salmon farming. In: H. Reinertsen, L.A. Dahle, L. Jørgensen & K. Tvinnereim (eds.), Fish Farming Technology, Balkema, Rotterdam, p 409-413.
Blanch, R., M. Simón, J.T. Jofre & G. Minkoff. 1991. Bacteria associated with hatchery cultivated turbot:are they implicated in rearing success? Eur. Aquac. Soc. Spec. Pub. (15):392-394.
Blanchet-Besson, C. 1986. Les parametres physico-chimiques en elevage larvaire intensif et la toxicite aigue de l'ammoniaque et des nitrites vis-a-vis du loup. Ph.D. Thesis, Univ. Sci. Tech. Languedoc, Montpellier, 308 p.
Blasiola, G.C., II. 1988. Report on head and lateral line erosion. SeaScope, Aquarium Systems, Mentor, Ohio, 5(summer):1,3.
Blasiola, G.C.. 1992. Diseases of ornamental marine fishes. In: J.B. Gratzek & J.R. Matthews (eds.), Aquariology: The Science of Fish Health Management. Tetra Press, Morris Plains, New Jersey, 275-300.
Blaxter, J.H.S. 1955. Herring rearing—I. The storage of herring gametes. Mar. Res. 3:1-2.
Blaxter, J.H.S. 1980. Vision and the feeding of fishes. In: J.E. Bardach, J.J. Magnuson, R.C. May & J.M. Reinhart (eds.), Fish Behavior and Its Use in the Capture and Culture of Fishes. ICLARM Conf. Proc. (5):32-56.
Blaxter, J.H.S. 1981. The rearing of larval fish. In: A.D. Hawkins (ed.), Aquarium Systems, Academic Press, London, p 303-323.
Blaxter, J.H.S. 1986. Development of sense organs and behaviour of teleost larvae with special reference to feeding and predator avoidance. Trans. Am. Fish. Soc. 115:98-114.
Blaxter, J.H.S. 1991. Sensory systems and behaviour of larval fish. In: J. Mauchline & T. Nemoto (eds.), Marine Biology: Its Accomplishment and Future Prospect. Elsevier, Amsterdam, p 15-38.
Blaxter, J.H.S. 1992. The effect of temperature on larval fishes. Neth. J. Zool. 42:336-357.
Blaxter, J.H.S., & K.F. Ehrlich. 1974. Changes in behaviour during starvation of herring and plaice

larvae. In: J.H.S. Blaxter (ed.) The Early Life History of Fish. Springer-Verlag, New York, p 575-588.

Blaxter, J.H.S., & G. Hempel. 1963. The influence of egg size on herring larvae. J. Cons. 28:211-240.

Blaxter, J.H.S., & J.R. Hunter. 1982. The biology of the clupeoid fishes. Adv. Mar. Biol. 20:1-223.

Blaxter, J.H.S., J.A.B. Gray & A.C.G. Best. 1983. Structure and development of the free neuromasts and lateral line system of the herring. J. Mar. Biol. Assoc. U.K. 63:247-260.

Blyth, P.J., G.J. Purser & J.F. Russell. 1993. Detection of feeding rhythms in seacaged Atlantic salmon using new feeder technology. In: H. Reinertsen, L.A. Dahle, L. Jørgensen & K. Tvinnereim (eds.), Fish Farming Technology. Balkema, Rotterdam, p 209-216.

Blythe, W.G., L.A. Helfrich & G. Libey. 1994. Induced maturation of striped bass *Morone saxatilis* exposed to 6, 9, and 12 month photothermal regimes. J. World Aquac. Soc. 25:183-192.

Boehlert, G.W. 1981. The effects of photoperiod and temperature on laboratory growth of juvenile *Sebastes diploproa* and a comparison with growth in the field. Fish. Bull., U.S. 79:789-794.

Boglione, C., G. Marino, A. Fusari, F. Ferreri, M.G. Finoia & S. Cataudella. 1995. Skeletal anomalies in *Dicentrarchus labrax* juveniles selected for functional swimbladder. ICES Mar. Sci. Symp. 201:163-169.

Boggs, C.H. 1991. Bioenergetics and growth of northern anchovy *Engraulis mordax*. Fish. Bull., U.S. 89:555-566.

Boggs, C.T., & R.C. Summerfelt. 1996a. A method for sampling and estimating the thickness of oil surface microlayers in larviculture tanks. In: G.S. Libey & M.B. Timmons (eds.), Aquacultural Eng. Soc. Proc. II: Successes and Failures in Commercial Recirculating Aquaculture, p 616-622.

Boggs, C.T., & R.C. Summerfelt. 1996b. Origin of the oil film on the water surface of tanks used for the mass culture of larval walleye. In: G.S. Libey & M.B. Timmons (eds.), Aquacultural Eng. Soc. Proc. II: Successes and Failures in Commercial Recirculating Aquaculture, p 623-630.

Bolla, S., & I. Holmefjord. 1988. Effect of temperature and light on development of Atlantic halibut larvae. Aquaculture 74:355-358.

Bolla, S., I. Holmefjord & T. Refstie. 1987. Cryogenic preservation of Atlantic halibut sperm. Aquaculture 65:371-374.

Booman, C., F. Midtøy, A.T. Smith, K. Westrheim & L. Føyn. 1995. Effects of oil on marine organisms with particular reference to first feeding of fish larvae. Fisken og Havet No. 9, 142 p. (in Norw., Engl. sum.)

Boonyaratpalin, M. 1988. Seabass feed. In: Manual for Training Provincial Fisheries Officers. Prajuabkhirikhan Coastal Aquac. Res. Dev. Ctr., Thailand, 21 p. (in Thai)

Boonyaratpalin, M. 1991. Asian seabass, *Lates calcarifer*. In: R.P. Wilson (ed.), Handbook of Nutrient Requirements of Finfish. CRC Press, Boca Raton, Florida, p 5-11.

Boonyaratpalin, M. 1997. Nutrient requirements of marine food fish cultured in Southeast Asia. Aquaculture 151:283-313.

Boonyaratpalin, M., & J. Wanakowat. 1993. Effect of thiamin, riboflavin, pantothenic acid and inositol on growth, feed efficiency and mortality of juvenile seabass. Fish Nutrition in Practice. INRA, Paris, p 819-828.

Boonyaratpalin, M., S. Boonyaratpalin & K. Supamataya. 1994. Ascorbyl-phosphate-Mg as a dietary vitamin C source for seabass (*Lates calcarifer*). Third Asian Fisheries Forum, Asian Fish. Soc., Manila, p 725-728.

Booth, S.L., T. Johns & H.V. Kuhnlein. 1992. Natural food sources of vitamin A and provitamin A. Food. Nutr. Bull. 14:6-19.

Boraas, M.E. 1993a. Semicontinuous culture methods. Ecological Studies: Analysis and Synthesis 98:13-20.

Boraas, M.E. 1993b. The growth of *Brachionus rubens* in semicontinuous culture. Ecological Studies: Analysis and Synthesis 98:41-50.

Boren, R.S., & D.M. Gatlin III. 1995. Dietary threonine requirement of juvenile red drum *Sciaenops ocellatus*. J. World Aquac. Soc. 26:279-283.

Borlongan, I.G. 1992. The essential fatty acid requirement of milkfish (*Chanos chanos* Forsskal). Fish Physiol. Biochem. 9:401-407.

Borrero, M., & N. Millares. 1979. Cultivo de larvas y juveniles del patao *Eugerres plumieri* (Pisces, Gerridae) en condiciones de laboratorio. Revista Cubana de Investigaciones Pesqueras 4:21-42.

Borrero, M., E. Gonzalez, N. Millares & T. Damas. 1978. Desarrollo embrionario y prelarval de la biajaiba (*Lutjanus synagris* Linne, 1758). Revista Cubana de Investigaciones Pesqueras 3:1-28.

Bortone, S.A., & J.L. Williams. 1986. Species profiles: life histories and environmental requirements of coastal fishes and invertebrates (south Florida). gray, lane, mutton, and yellowtail snappers. U.S. Fish Wildl. Serv. Biol. Rep. 82(11.52), 18 p.

Bose, A.N., S.N. Ghosh, C.T. Yang & A. Mitra. 1991. Coastal Aquaculture Engineering. Edward Arnold, London, 365 p.

Bosworth, B.G., G.S. Libey & D.R. Notter. 1997. Egg, larval, and fingerling traits of crosses among striped bass (*Morone saxatilis*), white bass (*M. chrysops*), and their F1 hybrids. Aquaculture 154:201-217.

Boujard, T., M. Jourdan, M. Kentouri & P. Divanach. 1996. Diel feeding activity and the effect of time-restricted self-feeding on growth and feed conversion in European sea bass. Aquaculture 139:117-127.

Bounou, C.A., & L. Saint-Jean. 1998. The regulation mechanisms and yield of brackish water populations of *Moina micrura* reared in tanks. Aquaculture 160:69-79.

Bower, C.E. 1987. Update on *Amyloodinium ocellatum*. SeaScope, Aquarium Systems, Mentor, Ohio, 4(fall):1,3-4.

Bower, C.E., & D.T. Turner. 1982. Effects of seven chemotherapeutic agents on nitrification in closed seawater culture systems. Aquaculture 29:331-345.

Bowers, M.J. 1992. Annual reproductive cycle of oocytes and embryos of yellowtail rockfish *Sebastes flavidus* (family Scorpaenidae). Fish. Bull., U.S. 90:231-242.

Boyd, C.E. 1982. Water Quality Management for Pond Fish Culture. Elsevier, New York.

Boyd, C.E. 1996. Chlorination and water quality in aquaculture ponds. World Aquac. 27(3):41-45.

Boyd, C.E., & B.J. Watten. 1989. Aeration systems in aquaculture. CRC Crit. Rev. Aq. Sci. 1:425-472.

Boyd, D. 1956. Spawning the neon goby *Elacatinus oceanops*. The Aquarium 25:390-396.

Boyer, J.N., W. Van Toever & M.E. Jansen. 1994. Effect of photoperiod on growth of arctic char under commercial production conditions. Prog. Fish-Cult. 56:44-46.

Brackett, J.B. 1994. Trends in aquatic disease control. Third Asian Fisheries Forum., Asian Fish. Soc., Manila, p 394-396.

Bradford, R.G., & R.L. Stephenson. 1992. Egg weight, fecundity, and gonad weight variability among northwest Atlantic herring (*Clupea harengus*) populations. Can. J. Fish. Aquat. Sci. 49:2045-2054

Brandt, C. 1974. First spawning of *Coryphopterus personatus*. Octopus 1(5):21-23.

Brandt, C. 1976. Sea spawnings—dwarf lionfishes. Octopus 3:4-7.

Brannon, E.L. 1991. Rainbow trout culture. In: R.R. Stickney (ed.) Culture of Salmonid Fishes. CRC Press, Boca Raton, Florida, p 21-55.

Brantley, R.K., & A.H. Bass. 1994. Alternative male spawning tactics and acoustic signals in the plainfin midshipman fish *Porichthys notatus* Girard (Teleostei, Batrachoididae). Ethology 96:213-232.

Brawn, V.M. 1961. Reproductive behaviour of the cod (*Gadus callarias* L.). Behaviour 18:177-198.

Breder, C.M., Jr. 1941. On the reproductive behavior of the sponge blenny, *Paraclinus marmoratus* (Steindachner). Zoologica 26:233-235.

Breder, C.M., Jr. & C.W. Coates. 1933. Reproduction and eggs of *Pomacentrus leucoris* Gilbert. Amer. Mus. Nov. No. 612, 6 p.

Breder, C.M., Jr. & D.E. Rosen. 1966. Modes of Reproduction in Fishes. TFH, Jersey City, New Jersey, 941 p.

Breeder's Registry Database. 1994. Sacramento, California.

Breeder's Registry Database. 1995. Sacramento, California.

Breitburg, D.L. 1994. Behavioral response of fish larvae to low dissolved oxygen concentrations in a stratified water column. Mar. Biol. 120:615-625.

Brett, J.R. 1986. Production energetics of a population of sockeye salmon, *Oncorhynchus nerka*. Can. J. Zool. 64:555-564.

Brett, J.R. 1995. Energetics. In: L. Margolis, W.C. Clarke & C. Groot (eds.), Physiological Ecology of Pacific Salmon. UBC Press, Vancouver, p 3-68.

Brett, J.R., & T.D.D. Groves. 1979. Physiological energetics. In: W.S. Hoar, D.J. Randall & J.R. Brett (eds.), Fish Physiology, Volume 8: Bioenergetics and Growth. Academic Press, Orlando, Florida, p 279-352.

Breuil, G., & P. Haffner. 1990. A field report on vibrio disease of seabass (*Dicentrarchus labrax*) in the South of France. In: Advances in Tropical Aquaculture. AQUACOP, IFREMER, Actes Colloq. (9):161-169.

Brewer, D.L., & R.A. Rees. 1990. Pond culture of phase I striped bass fingerlings. In: R.M. Harrell, J.H. Kerby & R.V. Minton (eds.), Culture and Propagation of Striped Bass and Its Hybrids. Am. Fish. Soc., Bethesda, Maryland, p 99-120.

Brinkmeyer, R.L., & G.J. Holt. 1995. Response of red drum larvae to graded levels of menhaden oil in semipurified microparticulate diets. Prog. Fish-Cult. 57:30-36.

Brinley, F.J. 1939. Spawning habits and development of beau-gregory (*Pomacentrus leucostictus*). Copeia 1939:185-188.

Brock, J.A., B.R. LeaMaster & C.-S. Lee. 1993. An overview of pathogens and diseases in marine finfish hatcheries in Hawaii with comments on strategies for health management and disease prevention. In: C.-S. Lee, M.-S. Su & I C. Liao (eds.), Finfish Hatchery in Asia. Tungkang Mar. Lab. Conf. Proc. (3):211-238.

Bromage, N. 1987. The advancement of puberty or time of first-spawning in female rainbow trout (*Salmo gairdneri*) maintained on altered-seasonal light cycles. Proc. Third Internat. Symp. Repr. Physiol. Fish, Memorial Univ. Newfoundland, St. John's, p 303.

Bromage, N.R. 1988. Propagation and stock improvement. In: C.J. Shepherd & N.R. Bromage (eds.), Intensive Fish Farming. Blackwell, Oxford, p 103-153.

Bromage, N. 1995. Broodstock management and seed quality—general considerations. In: N.R. Bromage & R.J. Roberts (eds.), Broodstock Management and Egg and Larval Quality. Blackwell, Oxford, p

1-24.

Bromage, N., & R. Cumaranatunga. 1987. Oocyte development in the rainbow trout with special reference to vitellogenesis and atresia. Proc. Third Internat. Symp. Repr. Physiol. Fish, Memorial Univ. Newfoundland, St. John's, p 194.

Bromage, N., J. Shepard & J. Roberts. 1988. Farming systems and husbandry practice. In: C.J. Shepard & N. Bromage (eds.), Intensive Fish Farming. Blackwell, Oxford, p 50-102.

Bromage, N., J. Jones & C. Randall. 1990. Broodstock care and the effects of hormonal and environmental factors in the induction of spawning. In: R. Flos, L. Tort & P. Torres (eds.), Mediterranean Aquaculture. Ellis Horwood, New York, p 88-101.

Bromage, N., R. Shields, C. Young, M. Bruce, N. Basavaraja, J. Dye, P. Smith, M. Gillespie, J. Gamble & K. Rana. 1991. Egg quality determinants in finfish with special reference to the timing of stripping and methods of fertilization in the Atlantic halibut (*Hippoglossus hippoglossus*). Eur. Aquac. Soc. Spec. Pub. (15):201-202.

Bromage, N., J. Jones, C. Randall, M. Thrush, B. Davies, J. Springate, J. Duston & G. Barker. 1992a. Broodstock management, fecundity, egg quality and the timing of egg production in the rainbow trout (*Oncorhynchus mykiss*). Aquaculture 100:141-166.

Bromage, N., C. Randall, B. Davies & B. McAndrew. 1992b. The control of reproduction in salmonid fish. Icel. Agr. Sci. 6:11-23.

Bromage, N., E.M. Donaldson, S. Zanuy, M. Carrillo & J. Planas (eds.). 1993a. Applications of comparative endocrinology to fish culture. Rec. Adv. Aquac. IV:1-192.

Bromage, N. C. Randall, J. Duston, M. Thrush & J. Jones. 1993b. Environmental control of reproduction in salmonids. Rec. Adv. Aquac. IV:55-65.

Bromage, N., M. Bruce, N. Basavaraja, K. Rana, R. Shields, C. Young, J. Dye, P. Smith, M. Gillespie & J. Gamble. 1994. Egg quality determinants in finfish: the role of overripening with special reference to the timing of stripping in the Atlantic halibut *Hippoglossus hippoglossus*. J. World Aquac. Soc. 25:13-21.

Bromley, P.J. 1977. Methods of weaning juvenile hatchery reared sole (*Solea solea* (L.)) from live food to prepared diets. Aquaculture 12:337-347.

Bromley, P.J. 1978. The weaning of hatchery reared turbot larvae (*Scophthalmus maximus*) on a dry diet. Aquaculture 13:339-345.

Bromley, P.J. 1980. The effect of dietary water content and feeding rate on the growth and food conversion efficiency of turbot (*Scophthalmus maximus* L.). Aquaculture 20:91-99.

Bromley, P.J., & B.R. Howell. 1983. Factors influencing the survival and growth of turbot larvae, *Scophthalmus maximus* L., during the change from live to compound feeds. Aquaculture 31: 31-40.

Bromley, P.J., P.A. Sykes & B.R. Howell. 1986. Egg production of turbot (*Scophthalmus maximus* L.) spawning in tank conditions. Aquaculture 53:287-293.

Brons, R. 1995. Red Sea Fish pHarm. Breeder's Registry 3(1):1-6.

Brons, R. 1996. Reproduction and captive breeding of two Red Sea dottybacks: *Pseudochromis fridmani* and *P. flavivertex*. Freshw. Mar. Aquarium 19(6):48-50,52-62.

Brooks, S., C.R. Tyler & J.P. Sumpter. 1997. Egg quality in fish: what makes a good egg? Rev. Fish Biol. Fish. 7:387-416.

Brosseau, R. 1991. Yet another clownfish breeder. Freshw. Mar. Aquarium 14(5):112,115-116,118,120.

Browman, H.I., & B.M. Marcotte. 1987. Effects of prey color and background color on feeding by Atlantic salmon alevins. Prog. Fish-Cult. 49:140-143.

Brown, A.W.A. 1978. Ecology of Pesticides. Wiley, New York, 525 p.

Brown, C.L., & H.A. Bern. 1989. Thyroid hormones in early development, with special reference to teleost fishes. In: M.P. Schreibman & C.G. Scanes (eds.), Development, Maturation, and Senescence of Neuroendocrine Systems: A Comparative Approach. Academic Press, San Diego, p 289-306.

Brown, C.L., C.V. Sullivan, H.A. Bern & W.W. Dickhoff. 1987. Occurrence of thyroid hormones in early developmental stages of teleost fish. Am. Fish. Soc. Symp. (2):144-150.

Brown, C.L., S.I. Doroshov, M.D. Cochran & H.A. Bern. 1989. Enhanced survival in striped bass fingerlings after maternal triiodothyronine treatment. Fish Physiol. Biochem. 7:295-299.

Brown, D.A., & T.R. Parsons. 1978. Relationship between cytoplasmic distribution of mercury and toxic effects to zooplankton and chum salmon (*Oncorhynchus keta*) exposed to mercury in a controlled ecosystem. J. Fish. Res. Bd. Can. 35:880-884.

Brown, E.E., & J.B. Gratzek. 1980. Fish Farming Handbook. AVI, Westport, Connecticut, 391 p.

Brown, J.A., D.C. Somerton, D.A. Methven & J.R. Watkins. 1992. Recent advances in lumpfish *Cyclopterus lumpus* and ocean pout *Macrozoarces americanus* larviculture. J. World Aquac. Soc. 23:271-276.

Brown, J.A., D. Wiseman & P. Kean. 1997c. The use of behavioural observations in the larviculture of cold-water marine fish. Aquaculture 155:297-306.

Brown, J.R., R.J. Gowen & D.S. McLusky. 1987. The effect of salmon farming on the benthos of a Scottish sea loch. J. Exp. Mar. Biol. Ecol. 109:39-51.

Brown, M.R., S.W. Jeffrey, J.K. Volkman & G.A. Dunstan. 1997b. Nutritional properties of microalgae for mariculture. Aquaculture 151:315-331.

Brown, P.B., D.A. Davis & E.H. Robinson. 1988. An estimate of the dietary lysine requirement of juvenile red drum *Sciaenops ocellatus*. J. World Aquac. Soc. 19:109-112.
Brown, P.B., M.E. Griffin & M.R. White. 1993. Experimental and practical diet evaluations with juvenile hybrid striped bass. J. World Aquac. Soc. 24:80-89.
Brown, P.B., R. Twibell, Y. Jonker & K.A. Wilson. 1997a. Evaluation of three soybean products in diets fed to juvenile hybrid striped bass *Morone saxatilis* x *M. chrysops*. J. World Aquac. Soc. 28:215-223.
Brown, S.D. 1994. Profiling *Centropyge*. Breeder's Registry 2(2):1,5-6.
Brown, S.D. 1996. Gobies. Breeder's Registry 4(1):1,4-5.
Browne, R.A., P. Sorgeloos & C.N.A. Trotman (eds.). 1991. Artemia Biology. CRC Press, Boca Raton, Florida, 374 p.
Brownell, C.L. 1979. Stages in the early development of 40 marine fish species with pelagic eggs from the Cape of Good Hope. J.L.B. Smith Inst. Ichthyol., Ichthyol. Bull. No. 40, 85 p.
Brownell, C.L. 1980a. Water quality requirements for first feeding in marine fish larvae. I. Ammonia, nitrite, and nitrate. J. Exp. Mar. Biol. Ecol. 44:269-283.
Brownell, C.L. 1980b. Water quality requirements for first feeding in marine fish larvae. II. pH, oxygen, and carbon dioxide. J. Exp. Mar. Biol. Ecol. 44:285-298.
Brownell, C.L., & D.A. Horstman. 1987. Two low-maintenance culture systems for small pelagic marine animals. Aquaculture 65:375-383.
Brown-Peterson, N., P. Thomas & C.R. Arnold. 1988. Reproductive biology of the spotted seatrout, *Cynoscion nebulosus*, in south Texas. Fish. Bull., U.S. 86:373-388.
Bruce, B.D. 1995. Larval development of King George whiting, *Sillaginodes punctata*, school whiting, *Sillago bassensis*, and yellow fin whiting, *Sillago schomburgkii* (Percoidei: Sillaginidae), from south Australian waters. Fish. Bull., U.S. 93:27-43.
Bruce, M.P., R.J. Shields, M.V. Bell & N.R. Bromage. 1993. Lipid class and fatty acid composition of eggs of Atlantic halibut, *Hippoglossus hippoglossus* (L.), in relation to egg quality in captive broodstock. Aquac. Fish. Mgt. 24:417-422.
Brulé, T., & L.G.R. Canché, 1993. Food habits of juvenile red groupers, *Epinephelus morio* (Valenciennes, 1828), from Campeche Bank, Yucatan, Mexico. Bull. Mar. Sci. 52:772-779.
Brulé, T., & C. Deniel. 1994. Expose Synoptique des donnees Biologiques sur le Merou Rouge *Epinephelus morio* (Valenciennes, 1828) du Golfe du Mexique. FAO Synopsis sur les Peches No. 155, 39 p. (in Fr.)
Brulé, T., D.O. Avila, M.S. Crespo & C. Déniel. 1994. Seasonal and diel changes in diet composition of juvenile red grouper (*Epinephelus morio*) from Campeche Bank. Bull. Mar. Sci. 55:255-262.
Brun, R., P. Nougayrede, P. Chene, A. Vuillaume & F. Crespeau. 1991. Bilan sanitaire de 2 ans d'elevage d'*Acipenser baeri* en piscicultures intensives. In: P. Williot (managing ed.), Acipenser. CEMAGREF, Bordeaux, p 429-437.
Brune, D.E., & J.R. Tomasso (eds.). 1991. Aquaculture and Water Quality. Adv. World Aquac., Vol. 3, 606 p.
Brunel, G., & P. Fuzeau. 1990. Aquaculture in heated water. In: G. Barnabé (ed.), Aquaculture, Vol. 2. Ellis Horwood, New York, p 801-819.
Bruno, D.W., & T.T. Poppe. 1996. A Color Atlas of Salmonid Diseases. Academic Press, Orlando, Florida, 208 p.
Bruno, D.W., & R.S. Raynard. 1994. Studies on the use of hydrogen peroxide as a method for the control of sea lice on Atlantic salmon. Aquac. Internat. 2:10-18.
Bruno, D.W., T.S. Hastings & A.E. Ellis. 1986. Histopathology, bacteriology and experimental transmission of cold-water vibriosis in Atlantic salmon *Salmo salar*. Dis. Aquatic Org. 1:163-168.
Brusle, J. 1981. Food and feeding in grey mullet. In: O.H. Oren (ed.), Aquaculture of Grey Mullets. IBP 26, Cambridge Univ. Press, Cambridge, U.K., p 185-217.
Brusle, J. 1990. Eels and eel farming. In: G. Barnabé (ed.), Aquaculture: Vol. 2. Ellis Horwood, New York, p 756-785.
Bryan, P.G., & B.B. Madraisau. 1977. Larval rearing and development of *Siganus lineatus* (Pisces: Siganidae) from hatching through metamorphosis. Aquaculture 10:243-252.
Bryan, P.G., B.B. Madraisau & J.P. McVey. 1975. Hormone induced and natural spawning of captive *Siganus canaliculatus* (Pisces: Siganidae) year round. Micronesica 11:199-204.
Buchmann, K., & J. Bjerregaard. 1990. Comparative efficacies of commercially available benzimidazoles against *Pseudodactylogyrus* infestations in eels. Dis. Aquatic Org. 9:117-120.
Buck, J.D., & R.H. Pierce. 1989. Bacteriological aspects of Florida red tides: a revsit and newer observations. Est. Coastal Shelf Sci. 29:317-326.
Buckley, L.J., T.A. Halavik, G.C. Layurence, S.J. Hamilton & P. Yevich. 1985. Comparative swimming stamina, biochemical composition, backbone mechanical properties, and histopathology of juvenile striped bass from rivers and hatcheries of the eastern United States. Trans. Am. Fish. Soc. 114:114-124.
Buckley, L.J., A.S. Smigielski, T.A. Halavik, E.M. Caldarone, B.R. Burns & G.C. Laurence. 1991. Winter flounder *Pseudopleuronectes americanus* reproductive success. II. Effects of spawning time

and female size on size, composition and viability of eggs and larvae. Mar. Ecol. Prog. Ser. 74:125-135.

Buckley, L.J., A.S. Smigielski, T.A. Halavik, B.R. Burns & G.C. Laurence. 1993. Growth and survival of the larvae of three temperate marine fish species at discrete prey densities. II. Cod (*Gadus morhua*), winter flounder (*Pseudopleuronectes americanus*), and silver hake (*Merluccius bilinearis*). In: B.T. Walther & H.J. Fyhn (eds.), Physiological and Biochemical Aspects of Fish Development. Univ. Bergen, Norway, p 183-195.

Buddington, R.K., S.I. Doroshov & G.A.E. Gall. 1987. Intestinal nutrient uptake during the reproductive cycle of the rainbow trout: preliminary information. Proc. Third Internat. Symp. Repr. Physiol. Fish, Memorial Univ. Newfoundland, St. John's, p 195.

Bulak, J.S., & R.C. Heidinger. 1980. Developmental anatomy and inflation of the gas bladder in striped bass, *Morone saxatilis*. Fish. Bull., U.S. 77:1000-1003.

Bullock, A.M. 1989. Laboratory methods. In R.J. Roberts (ed.), Fish Pathology. Bailliere Tindall, London, p 374-405.

Bumguardner, B.W., & T.L. King. 1996. Toxicity of oxytetracycline and calcein to juvenile striped bass. Trans. Am. Fish. Soc. 125:143-145.

Bumguardner, B.W., R.L. Colura, A. Henderson-Arzapalo & A.F. Maciorowski. 1992. Culture of spotted seatrout, orangemouth corvina, and their hybrids. Cal. Fish Game 78:105-115.

Buranapanidgit, J., M. Boonyaratpalin, T. Watanabe, T. Pechmanee & R. Yashiro. 1988. Essential fatty acid requirement of juvenile seabass *Lates calcarifer*. Nat. Inst. Coastal Aquac., Songkhla, Thailand, Tech. Pap. No. 3, 21 p.

Burbidge, R.G. 1974. Distribution, growth, selective feeding, and energy transformations of young-of-the-year blueback herring, *Alosa aestivalis* (Mitchill), in the James River, Virginia. Trans. Am. Fish. Soc. 103:297-311.

Burgess, J.G., K. Iwamoto, Y. Miura, H. Takano & T. Matsunaga. 1993. An optical fibre photobioreactor for enhanced production of the marine unicellular alga *Isochrysis* aff. *galbana* T-Iso (UTEX LB 2307) rich in docosahexaenoic acid. Appl. Microbiol. Biotechnol. 39:456-459.

Burgess, W.E. 1987. A Complete Introduction to Marine Aquariums. T.F.H., Neptune City, New Jersey, 128 p.

Burke, J., & L. Rodgers. 1981. Identification of pathogenic bacteria associated with the occurrence of "red spot" in sea mullet, *Mugil cephalus* L. in southeastern Queensland. J. Fish. Disease 4:153-160.

Burkholder, J.M., & H.B. Glasgow, Jr. 1997. *Pfiesteria piscicida* and other *Pfiesteria*-like dinoflagellates: behavior, impacts, and environmental controls. Limnol. Oceanogr. 42:1052-1075.

Burkholder, J.M., E.J. Noga, C.H. Hobbs & H.B. Glasgow, Jr. 1992. New 'phantom' dinoflagellate is the causative agent of major estuarine fish kills. Nature 358:407-410.

Burnett, J., M.R. Ross & S.H. Clark. 1992. Several biological aspects of the witch flounder (*Glyptocephalus cynoglossus* (L.)) in the Gulf of Maine-Georges Bank region. J. Northw. Atl. Fish. Sci. 12:15-25.

Burridge, L.E., & K. Haya. 1995. A review of di-n-butylphthalate in the aquatic environment: concerns regarding its use in salmonid aquaculture. J. World Aquac. Soc. 26:1-13.

Butler, J.L., H.G. Moser, G.S. Hageman & L.L. Nordgren. 1982. Developmental stages of three California sea basses (*Paralabrax*, Pisces, Serranidae). Cal. Coop. Oceanic Fish. Invest. Rep. 23:252-268.

Buttino, I. 1994. The effect of low concentrations of phenol and ammonia on egg production rates, fecal pellet production and egg viability of the calanoid copepod *Acartia clausi*. Mar. Biol. 119:629-634.

Buurma, B.J., A. Barkoh & G.A. Alexander. 1996. Effects of cottonseed and alfalfa meals as fertilizers on production of palmetto bass (*Morone saxatilis* x *M. chrysops*) fingerlings in plastic-lined ponds. J. Appl. Aquac. 6:59-70.

Buxton, C.D., & J.R. Clarke. 1992. The biology of the bronze bream, *Pachymetopon grande* (Teleostei: Sparidae) from the south-east Cape coast, South Africa. S. Afr. J. Zool. 27:21-32.

Bye, V.J. 1984. The role of environmental factors in the timing of reproductive cycles. In: G.W. Potts & R.J. Wootton (eds.), Fish Reproduction: Strategies and Tactics. Academic Press, London, p 187-205.

Bye, V.J. 1987. Environmental management of marine fish reproduction in Europe. Proc. Third Internat. Symp. Repr. Physiol. Fish, Memorial Univ. Newfoundland, St. John's, p 289-298.

Bye, V.J., & M. Htun-Han. 1978. Daylength control, key to flexible production. Fish Farmer 1(4):10-14.

Bye, V.J., & M. Htun-Han. 1979. Light and temperature are key to controlled spawning. Fish Farmer 2(5):27-28.

Bye, V.J., & R.F. Lincoln. 1986. Commercial methods for the control of sexual maturation in rainbow trout (*Salmo gairdneri* R.). Aquaculture 57:299-309.

Cabrera, T., S.B. Hur & H.J. Kim. 1993. Lifespan and fecundity of three types of rotifer, *Brachionus plicatilis* by an individual culture. Bull. Kor. Fish. Soc. 26:511-518.

Cabrera B., T., J. Rosas C. & J. Millán Q. 1997. Reproducción y desarrollo larvario del pargo dientón (*Lutjanus griseus* L. 1758)(Pisces: Lutjanidae) cultivado en cautiverio. Carib. J. Sci. 33:239-245.

Cachelou, F., D. Duche & A. Jones. 1989. Current status of turbot farming in Europe. World Aquac.

20(3):50-53.
Caddell, S.M., D.M. Gadomski & L.R. Abbott. 1990. Induced spawning of the California halibut, *Paralichthys californicus*, (Pisces: Paralichthyidae) under artificial and natural conditions. Cal. Dept. Fish Game, Fish Bull. 174:175-198.
Caggiano, M., S. Canese, A. Lupo & A. Cirillo. 1993. Experiences of artificial reproduction and larval rearing of sheepshead bream (*Diplodus puntazzo*) in the south of Italy. Eur. Aquac. Soc. Spec. Pub. (19):326.
Cahu, C.L., & J.L.Z. Infante. 1994. Early weaning of sea bass (*Dicentrarchus labrax*) larvae with a compound diet: effect on digestive enzymes. Comp. Biochem. Physiol. 109A:213-222.
Calabrese, A., J.R. MacInnes, D.A. Nelson & J.E. Miller. 1977. Survival and growth of bivalve larvae under heavy-metal stress. Mar. Biol. 41:179-184.
California Spray Dry Co. 1992. Product literature. Stockton, California.
Cañavate, J.P., & L.M. Lubiñn. 1995. Some aspects on the cryopreservation of microalgae used as food for marine species. Aquaculture 136:277-290.
Candreva, P., P. Dhert, A. Novelli & D. Brissi, 1996. Potential gains through alimentation/nutrition improvements in the hatchery. In: B. Chatain, M. Saroglia, J. Sweetman & P. Lavens (eds.), Seabass and Seabream culture: Problems and Prospects. Eur. Aquac. Soc., p 145-159.
Canfield, P.J., N. Quartararo, D.L. Griffin, G.N. Tsoukalas & S.E. Cocaro. 1994. Haematological and biochemical reference values for captive Australian snapper, *Pagrus auratus*. J. Fish Biol. 44:849-856.
Canli, M., & C. Erdem. 1994. Mercury toxicity and effects of exposure concentration and period on mercury accumulation in tissues of a tropical fish *Tilapia nilotica* (L.). Turkish. J. Zool. 18:233-239.
Cansdale, G. 1981. Sea water abstraction. In: A.D. Hawkins (ed.), Aquarium Systems, Academic Press, London, p 47-62.
Capo, T.R., B.B. Wisner, J.E. Serafy & A.E. Boyd. 1997. Extended spawning of captive spotted seatrout (*Cynoscion nebulosus*) under photo-thermal regulation. Island Aquac. Trop. Aquac. Ann. Mtg., Eur. Aquac. Soc., p 62-64.
Capuzzo, J.M. 1979a. The effect of temperature on the toxicity of chlorinated cooling waters to marine animals—a preliminary review. Mar. Poll. Bull. 10:45-47.
Capuzzo, J.M. 1979b. The effects of halogen toxicants on survival, feeding and egg production of the rotifer *Brachionus plicatilis*. Est. Coastal Mar. Sci. 8:307-316.
Cardeilhac, P.T. 1976. Induced maturation and development of pinfish eggs. Aquaculture 8:389-393.
Carić, M., J. Sanko-Njire & B. Skaramuca. 1993. Dietary effects of different feeds on the biochemical composition of the rotifer (*Brachionus plicatilis* Müller). Aquaculture 110:141-150.
Carlberg, J.M., & M.J. Massingill. 1995. Foodfish production 11 - commercial flow-through tank growout systems. Aquac. '95 Book of Abs., p 79-80
Carlos, M.H., M.Y. Seat & S.A. Al-Thobaiti. 1992. Manual for Natural Food Production. Ministry Agric. Water, Fish Farming Centre, Jeddah, Saudi Arabia, UTFN/SAU/010/SAU, 39 p.
Carmichael, G.J., & J.R. Tomasso. 1988. Survey of fish transportation equipment and techniques. Prog. Fish-Cult. 50:155-159.
Carmichael, G.J., R.M. Jones & J.C. Morrow. 1992. Comparative efficacy of oxygen diffusers in a fish-hauling tank. Prog. Fish-Cult. 54:35-40.
Carmignani, G.M., & J. Bennett. 1976. Filter media for the removal of phthalate esters in water of closed aquaculture systems. Aquaculture 8:291-294.
Carolin, E., & S. Jose. 1995. Two novel protein sources of marine origin for the nursery rearing of *Chanos chanos* fry. Fishery Technol. 32:14-18.
Carolsfeld, J., N.M. Sherwood, H. Kreiberg & S.A. Sower. 1988. Induced sexual maturation of herring using GnRH 'quick-release' cholesterol pellets. Aquaculture 70:169-181.
Carr, W.E.S., & C.A. Adams. 1973. Food habits of juvenile marine fishes occupying seagrass beds in the estuarine zone near Crystal River, Florida. Trans. Am. Fish. Soc. 102:511-540.
Carrillo, M., S. Zanuy, F. Prat, R. Serrano & N. Bromage. 1993. Environmental and hormonal control of reproduction in sea bass. Rec. Adv. Aquac. IV:43-54.
Carrillo, M., S. Zanuy, F. Prat, J. Cerda, J. Ramos, E. Mananos & N. Bromage. 1995. Sea bass (*Dicentrarchus labrax*). In: N.R. Bromage & R.J. Roberts (eds.), Broodstock Management and Egg and Larval Quality, Blackwell, Oxford, England, p 138-168.
Carter, C.G., D.F. Houlihan & I.D. McCarthy. 1992. Feed utilization efficiencies of Atlantic salmon (*Salmo salar* L.) parr: effect of a single supplementary enzyme. Comp. Biochem. Physiol. 101A:369-374.
Carter, L. 1964. Effects of acidic and alkaline effluents on fish in seawater. Eff. Wat. Treat. J. 4:484-486.
Caruso, G., & L. Genovese. 1995. Preliminary data on lysozyme in sea-bass (*Dicentrarchus labrax*) juveniles fed diets containing varying levels of vitamin C. Anim. Biol. 4:161-163.
Cash, G. 1994. Growing barramundi in recirculating tanks. Austasia Aquac. 8(4):39-42.
Casian, J.A.R. 1991. Seawater chlorination and the survival and growth of early life stages of northern anchovy (*Engraulis mordax* Girard), with reference to power plant cooling systems. Ciencias Marinas 17:99-117.
**Casillas, E., D. Misitano, L.L. Johnson, L.D. Rhodes, T.K. Collier, J.E. Stein, B.B. McCain & U.

Varanasi. 1991. Inducibility of spawning and reproductive success of female English sole (*Parophrys vetulus*) from urban and nonurban areas of Puget Sound, Washington. Mar. Env. Res. 31:99-122.

Castell, J.D., J.G. Bell, D.R. Tocher & J.R. Sargent. 1994. Effects of purified diets containing different combinations of arachidonic and docosahexaenoic acid on survival, growth and fatty acid composition of juvenile turbot (*Scophthalmus maximus*). Aquaculture 128:315-333.

Castelló-Orvay, F., & A. Calderer. 1993. Growth of gilthead sea bream (*Sparus aurata* L.) under different culture conditions. Eur. Aquac. Soc. Spec. Pub. (18):227-233

Castelló Orvay, F., & V. Gracia López. 1997. Effect of the salinity on the growth and feeding efficiency of the grouper, *Epinephelus marginatus*. Island Aquac. Trop. Aquac. Ann. Mtg., Eur. Aquac. Soc., p 74-75.

Castelló-Orvay, F., & A.F. Vilar. 1993. Effets de différents types d'alimentation sur la croissance du mérou (*Epinephelus guaza* L., 1758) en captivité. Eur. Aquac. Soc. Spec. Pub. (18):235-245. (in Fr., Engl. abs.)

Castle, P.H.J. 1984. Notacanthiformes and Anguilliformes: development. Am. Soc. Ichthyol. Herpetol. Spec. Pub. (1):62-93.

Catacutan, M.R., & R.M. Coloso. 1995. Effect of dietary protein to energy ratios on growth, survival, and body composition of juvenile Asian seabass, *Lates calcarifer*. Aquaculture 131:125-133.

Caylor, R.E., P.M. Biesiot & J.S. Franks. 1994. Culture of cobia (*Rachycentron canadum*): cryopreservation of sperm and induced spawning. Aquaculture 125:81-92.

Cendejas, M.E.V. 1993. Estudio ecotrofico sobre el sargo, *Archosargus rhomboidalis* ¿especie omnivora o herbivora? Revista Invest. Mar. 14:70-85. (in Sp., Engl. abs.)

Cerdà, Joan, & Robin A. Wallace. 1994. Whitney Laboratory, University of Florida, St. Augustine, pers. comm.

Cerdà, J., M. Carrillo, S. Zanuy & J. Ramos. 1994a. Effect of food ration on estrogen and vitellogenin plasma levels, fecundity and larval survival in captive sea bass, *Dicentrarchus labrax*: preliminary observations. Aquat. Living Resour. 7:255-266.

Cerdà, J., M. Carrillo, S. Zanuy, J. Ramos & M. de la Higuera. 1994b. Influence of nutritional composition of diet on sea bass, *Dicentrarchus labrax* L., reproductive performance and egg and larval quality. Aquaculture 128:345-361.

Cerdà, J., S. Zanuy, M. Carrillo, J. Ramos & R. Serrano. 1995. Short- and long-term dietary effects on female sea bass (*Dicentrarchus labrax*): seasonal changes in plasma profiles of lipids and sex steroids in relation to reproduction. Comp. Biochem. Physiol. 111C:83-91.

Cerdà, J., K. Selman, S.-M. Hsiao & R.A. Wallace. 1997. Evidence for the differential regulation of ovarian follicle responsiveness to human chorionic gonadotropin in vitro in a serranid teleost, *Centropristis striata*. Aquaculture 159:143-157.

Cerqueira, V.R., & B. Chatain. 1991. Photoperiodic effects on the growth and feeding rhythm of European seabass, *Dicentrarchus labrax*, larvae in intensive rearing. Eur. Aquac. Soc. Spec. Pub. (15):304-306.

Cervigon, F. (ed.). 1983. La Acuicultura en Venezuela...estado actual y perspectivas. Editorial Arte., Caracas, 121 p.

CFR (Code of Federal Regulations). 1994. Protection of Environment. U.S. Env. Prot. Ag., Title 40, Part 131, Sec. 36, p 352-369.

Chair, M., M. Dehasque, P. Sorgeloos, H. Nelis & A.P. De Leenheer. 1995. Live food mediated drug delivery as a tool for disease treatment in larviculture: a case study with turbot *Scophthalmus maximus*. J. World Aquac. Soc. 26:217-219.

Chair, M., H.J. Nelis, P. Leger, P. Sorgeloos & A.P. De Leenheer. 1996. Accumulation of trimethoprim, sulfamethoxazole, and *N*-acetylsulfamethoxazole in fish and shrimp fed medicated *Artemia franciscana*. Antimicrobial Agents Chemotherapy 40:1649-1652.

Chaitanawisuti, N., & S. Piyatiratitivorakul. 1994. Studies on growth and production of juvenile seabass (*Lates calcarifer*, Bloch) fed exclusively with moist pelleted diets in floating net cages. J. Aquac. Trop. 9:201-208.

Chamberlain, G.W., & G. McCarty. 1985. Why choose redfish? Aquac. Mag. 11:35-42.

Chamberlain, G.W., R.J. Miget & M.G. Haby (compilers). 1990. Red Drum Aquaculture. Texas A&M Sea Grant TAMU-SG-90-603, 236 p.

Chambers, J.E. 1994. Toxicity of pesticides. In: L.G. Cockerham & B.S. Shane (eds.), Basic Environmental Toxicology. CRC Press, Boca Raton, Florida, p 185-198.

Chambeyron, F., & Y. Zohar. 1990. A diluent for sperm cryopreservation of gilthead seabream, *Sparus aurata*. Aquaculture 90:345-352.

Champalbert, G., J. Marchand & J. Le Campion. 1994. Rheotaxis in juvenile sole *Solea solea* (L.): influence of salinity and light conditions. Neth. J. Sea Res. 32:309-319.

Chang, C.-F., W.-S. Yueh & M.-F. Lee. 1991. Effects of LHRH-A and HCG on the steroid profiles of bisexual and mature male and female protandrous black porgy, *Acanthopagrus schlegeli*. Aquaculture 92:83-92.

Chang, C.-F., H.-J. Hu, H.-C. Tang & L.-T. Sun. 1992. Stimulation of ovulation in ayu, *Plecoglossus altivelis*, by treatment with antiestrogens and luteinizing hormone-releasing hormone analog.

Aquaculture 101:329-336.
Chang, C.-F., S.-C. Lan & B.S. Pan. 1995. Feed administration of estradiol-17β stimulates female differentiation in juvenile grey mullet *Mugil cephalus*. Zool. Stud. (Taipei) 34:257-264.
Chang, G.E.G., & C.Y. Yi. 1994. The nutritional value of rotifer (*Brachionus plicatilis*) under different temperatures and diets. Third Asian Fisheries Forum, Asian Fish. Soc., Manila, p 749-751.
Chang, L.W., & L.G. Cockerham. 1994. Toxic metals in the environment. In: L.G. Cockerham & B.S. Shane (eds.), Basic Environmental Toxicology. CRC Press, Boca Raton, Florida, p 109-132.
Chang, R.-M., A.J. Shams, J.S. Al-Aradi & A.A. Al-Ansari. 1994a. The spawning and larval rearing of the Haffara bream *Rhabdosargus haffara* (Forsskal 1775). Third Asian Fisheries Forum, Asian Fish. Soc., Manila, p 116-119.
Chang, S.-L., M.-S. Su & I C. Liao. 1993. Milkfish fry production in Taiwan. In: C.-S. Lee, M.-S. Su, and I C. Liao (editors), Finfish Hatchery in Asia. Tungkang Mar. Lab. Conf. Proc. (3):157-171.
Chang, S.-L., M.-S. Su & I.C. Liao. 1994b. Preliminary report on the propagation of the yellow spotted grunt (*Plectorhinchus cinctus*). Third Asian Fisheries Forum, Asian Fish. Soc., Manila, p 203-206.
Chao, N.-H. 1982. New approaches for cryopreservation of sperm of grey mullet, *Mugil cephalus*. Proc. Internat. Symp. Repr. Physiol. Fish, Wageningen, Netherlands, p 132-133.
Chao, N.-H. 1991. Fish sperm cryopreservation in Taiwan: technology advancement and extension efforts. Bull. Inst. Zool., Acad. Sin., Monograph 16:263-283.
Chao, N.-H., & I.-C. Liao. 1978. Cryogenic preservation of gray mullet *Mugil cephalus* sperm. Cryobiology 15:686.
Chao, N.H., & I.C. Liao 1987. Application of honey in cryopreservation of sperm of milkfish (*Chanos chanos*) and black porgy (*Acanthopagrus schlegeli*). Proc. Third Internat. Symp. Repr. Physiol. Fish, Memorial Univ. Newfoundland, St. John's, p 94-95.
Chao, N.H., H.P. Chen & I.C. Liao. 1975. Study on cryopreservation of grey mullet sperm. Aquaculture 5:389-406.
Chao, N.H., W.C. Chao, K.C. Liu & I.C. Liao. 1986. The biological properties of black porgy (*Acanthopagrus schlegeli*) sperm and its cryopreservation. Proc. Natl. Sci. Council B. Taiwan 10:145-149.
Chao, N.H., H.P. Tsai & I.C. Liao. 1992. Short- and long-term cryopreservation of sperm and sperm suspension of the grouper, *Epinephelus malabaricus* (Bloch and Schneider). Asian Fish. Sci. 5:103-116.
Chao, N.-H., H.-W. Hsu, C.-P. Chiang, H.-P. Tsai, H.-J. Hong & I C. Liao. 1994. Cryoprotectant tolerance in cyprinid loach (*Misgurnus anguillicaudatus*) and Pacific oyster (*Crassostrea gigas*) embryos. Third Asian Fisheries Forum, Asian Fish. Soc., Manila, p 78-81.
Chao, N.-H., T.-T. Lin, Y.-J. Chen, H.-W. Hsu & I-C. Liao. 1997. Cryopreservation of late embryos and early larvae in the oyster and hard clam. Aquaculture 155:31-44.
Chao, T.M. 1984. Studies on the transmissibility of lymphocystis disease occurring in seabass (*Lates calcarifer* Bloch). Sing. J. Pri. Ind. 12:11-16.
Chao, T.M. 1988. Ultrastructure of lymphocystis disease virus in seabass, *Lates calcarifer* (Bloch). Sing. J. Pri. Ind. 16:86-90.
Chao, T.M., & M. Chow. 1990. Effect of methyltestosterone on gonadal development of *Epinephelus tauvina* (Forskal). Sing. J. Pri. Ind. 18:1-14.
Chao, T.M., & L.C. Lim. 1991. Recent developments in the breeding of grouper (*Epinephelus* spp.) in Singapore. Sing. J. Pri. Ind. 19:78-93.
Chao, T.M., L.C. Lim & L.T. Khoo. 1993. Studies on the breeding of brown-marbled grouper (*Epinephelus fuscoguttatus* Forsskal) in Singapore. In: C.-S. Lee, M.-S. Su & I C. Liao (eds.), Finfish Hatchery in Asia. Tungkang Mar. Lab. Conf. Proc. (3):143-156.
Chapman, D.C., W.A. Hubert & U.T. Jackson. 1988. Influence of access to air and of salinity on gas bladder inflation in striped bass. Prog. Fish-Cult. 50:23-27.
Chapman, P., F. Cross, W. Fish & K. Jones. 1982. Final report for sportfish introductions project. Study 1: artificial culture of snook. Fla. Game Fresh Water Fish Comm., Tallahassee, 36 p.
Charlon, N., & P. Bergot. 1991. Alimentation artificielle des larves de l'esturgeon Siberien (*Acipenser baeri* Brandt). In: P. Williot (ed.), Acipenser. CEMAGREF, Bordeaux, p 405-415.
Chatain, B. 1991. Latest improvements in intensive seabass and seabream fry production: the French technique. Eur. Aquac. Soc. Spec. Pub. (15):382.
Chatain, B., & N. Ounaïs-Guschemann. 1990. Improved rate of initial swim bladder inflation in intensively reared *Sparus auratus*. Aquaculture 84:345-353.
Chatain, B., & N. Ounaïs-Guschemann. 1991a. Determination of larval mass rearing standards in the sea bream *Sparus aurata*. Eur. Aquac. Soc. Spec. Pub. (15):314-315.
Chatain, B., & N. Ounaïs-Guschemann. 1991b. The relationships between light and larvae of *Sparus aurata*. Eur. Aquac. Soc. Spec. Pub. (15):310-313.
Chatain, B., M. Saroglia, J. Sweetman & P. Lavens (eds.). 1996. Seabass and Seabream Culture: Problems and Prospects. Eur. Aquac. Soc. 366 p.
Chatzifotis, S., & T. Takeuchi. 1997. Effect of supplemental carnitine on body weight loss, proximate and lipid compositions and carnitine content of red sea bream (*Pagrus major*) during starvation.

Aquaculture 158:129-140.

Chatzifotis, S., T. Takeuchi & T. Seikai. 1995. The effect of dietary L-carnitine on growth performance and lipid composition in red sea bream fingerlings. Fish. Sci. 61:1004-1008.

Chaudhuri, H. 1994. History of induced breeding in fishes and its application to aquaculture. Proc. Zool. Soc. Calcutta 47:1-31.

Chaudhuri, H., J.V. Juario, J.H. Primavera, R. Samson & R. Mateo. 1978. Observations on artificial fertilization of eggs and the embryonic and larval development of milkfish, *Chanos chanos* (Forsskal). Aquaculture 13:95-113.

Chebil, L., & S. Yamasaki. 1998. Improvement of a rotifer ecosystem culture to promote recycling marine microalga, *Nannochloropsis* sp. Aquacultural Eng. 17:1-10.

Chen, C.-H., C.-L. Yen, K.-J. Lin, Y.-K. Fang & C.-L. Chen. 1991a. Propagation of coral trout (*Plectropomus leopardus*). Abs. in program book, Finfish Hatchery in Asia '91. Tungkang Mar. Lab., Tungkang, Taiwan, p 33.

Chen, D., C. Liu & G. Dong. 1991b. Study on fishery biology features of spawning group of *Pseudopleuronectes yokohamae* (Gunther) in the Sheath Bay. Acta Oceanol. Sin. 10:471-476.

Chen, F.Y. 1979. Progress and problems of netcage culture of grouper (*Epinephelus tauvina* F) in Singapore. Proc. World Maric. Soc. 10:260-271.

Chen, F.Y., M. Chow, T.M. Chao & R. Lim. 1977. Artificial spawning and larval rearing of the grouper, *Epinephelus tauvina* (Forskal) in Singapore. Sing. J. Pri. Ind. 5:1-21.

Chen, H.-Y., & I-C. Liao. 1991. Marine finfish nutrition studies in Taiwan. In: J.P. McVey (ed.), Handbook of Mariculture, Vol. II, Finfish Aquaculture. CRC Press, Boca Raton, Florida, p 133-145.

Chen, H.-Y., & J.-C. Tsai. 1994. Optimal dietary protein level for the growth of juvenile grouper, *Epinephelus malabaricus*, fed semipurified diets. Aquaculture 119:265-271.

Chen, S., M.B. Timmons, D.J. Aneshansley & J.J. Bisogni, Jr. 1992. Bubble size distribution in a bubble column applied to aquaculture systems. Aquacultural Eng. 11:267-280.

Chen, S., M.B. Timmons, J.J. Bisogni, Jr. & D.J. Aneshansley. 1993a. Protein and its removal by foam fractionation. Prog. Fish-Cult. 55:76-82.

Chen, S., M.B. Timmons, J.J. Bisogni, Jr. & D.J. Aneshansley. 1993b. Suspended-solids removal by foam fractionation. Prog. Fish-Cult. 55:69-75.

Chen, S., D. Stechey & R.F. Malone. 1994. Suspended solids control in recirculating aquaculture systems. In: M.B. Timmons & T.M. Losordo (eds.), Aquaculture Water Reuse Systems: Engineering Design and Management. Elsevier, Amsterdam, p 61-100.

Chen, S., D.E. Coffin & R.F. Malone. 1997. Sludge production and management for recirculating aquacultural systems. J. World Aquac. Soc. 28:303-315.

Chen, S.-C. 1978. The development of reared larvae of a flying fish, *Hirundichthys oxycephalus*, in the laboratory. Bull. Taiwan Fish. Res. Inst. (30):301-307. (in Ch., Engl. abs.)

Chen, S.-C., & P.-C. Wang. 1993. *In vitro* activity of antimicrobial agents against *Nocardia asteroides*. J. Fish Dis. 16:269-272.

Chen, T.F., & L.L. Chen. 1986. The experiment for the development of artificial diet for the grouper *Epinephelus salmonides*. In: J.L. Chuang & S.Y. Shiau (eds.), Research and Development of Aquatic Animal Feeds in Taiwan, Vol. 1. Fish. Soc. Taiwan, Keelung, Taiwan, p 95.

Chen, T.P. 1976. Aquaculture Practices in Taiwan. Fishing News Books, Farnham, England, 163 p.

Cheng, S.K., & S.M. Ruby. 1980. Effects of sublethal pulse cyanide exposure on several life stages of American flagfish *Jordanella floridae*. Can. Tech. Rep. Fish. Aquat. Sci. (975):55.

Cheong, L. 1990. Status of knowledge on farming of seabass (*Lates calcarifer*) in South East Asia. Advances in Tropical Aquaculture, AQUACOP, IFREMER, Actes Colloq. (9):421-428.

Cheong, L., F.K. Chan, F.J. Wong & R. Chou. 1987. Observations on the culture of red tilapia (*Oreochromis niloticus* hybrid) in seawater under intensive tank condition using a biodrum. Sing. J. Pri. Ind. 15:42-56.

Chester, R. 1990. Marine Geochemistry. Unwin Hyman, London, 698 p.

Cheung, P.J., R.F. Nigrelli & G.D. Ruggieri. 1979. Studies on cryptocaryoniasis in marine fish: effect of temperature and salinity on the reproductive cycle of *Cryptocaryon irritans* Brown 1951. J. Fish Dis. 2:93-97.

Chevassus, B. 1983. Hybridization in fish. Aquaculture 33:245-262.

Chew-Lim, M., R. Singh, T.M. Chao & L. Cheong. 1986. Bacteria and viruses of marine fish and prawns farmed in Singapore. Sing. Vet. J. 10:25-35.

Chiba, H., K. Iwatsuki, K. Hayami, A. Hara & K. Yamauchi. 1994. Changes in serum steroid hormones and vitellogenin levels in cultured female European eels *Anguilla anguilla* during artificially induced ovarian development. J. World Aquac. Soc. 25:553-560.

Chiba, K. 1983. The effect of dissolved oxygen on the growth of young silver bream. Bull. Jpn. Soc. Sci. Fish. 49:601-610.

Chien, Y.-H., & H.-H. Liao. 1994. Effects of lecithin in feed on the survival and growth of yellowfin porgy (*Acanthopagrus latus*) fry. In: The Third Asian Fisheries Forum. Asian Fish. Soc., Manila, p 681-684.

Chieng, C., A. Garcia III & D. Brune. 1989. Oxidation requirements of a formulated micropulverized

feed. J. World Aquac. Soc. 20:24-29.
Cho, C.Y. 1992. Feeding systems for rainbow trout and other salmonids with reference to current estimates of energy and protein requirements. Aquaculture 100:107-123.
Cho, C.Y., & D.P. Bureau. 1995. Determination of the energy requirements of fish with particular reference to salmonids. J. Appl. Ichthyol. 11:141-163.
Cho, C.Y., S.J. Slinger & H.S. Bayley. 1982. Bioenergetics of salmonid fishes: energy intake, expenditure and productivity. Comp. Biochem. Physiol. 73B:25-41.
Cho, C.Y., C.B. Cowey & T. Watanabe. 1985. Methodological approaches to research and development. In: C.Y. Cho, C.B. Cowey & T. Watanabe (eds.), Finfish Nutrition in Asia: Methodological Approaches to Research and Development. Internat. Dev. Res. Cen., Ottawa, Ontario, p 9-80.
Cho, C.Y., J.D. Hynes, K.R. Wood & H.K Yoshida. 1994. Development of high-nutrient-dense, low-pollution diets and prediction of aquaculture wastes using biological approaches. Aquaculture 124:293-305.
Cho, K.C., J.H. Kim, C.S. Go, Y. Kim & K.-K. Kim. 1995. A study on seedling production of the spotted flounder, *Verasper variegatus*. Bull. Nat. Fish. Res. Dev. Ag. 50:41-57. (in Kor., Engl. abs.)
Cho, Y.C., & S.G. Yang. 1991. Study on the early spawning inducement of bastard halibut, *Paralichthys olivaceus*. Bull. Nat. Fish. Res. Dev. Ag. 45:183-195. (in Kor., Engl. abs.)
Choi, M.-S., & N. Kinae. 1994. Toxic effect of micropollutants on coastal organisms. I. Toxicity on some marine fishes. Bull. Kor. Fish. Soc. 27:529-534.
Chonchuenchob, P., S. Sumpawapol & A. Mearoh. 1987. Diseases of cage-cultured sea bass (*Lates calcarifer*) in southwestern Thailand. Austral. Ctr. Internat. Agric. Res. Proc. 20:194-197.
Chong, Y.C., & T.M. Chao. 1986a. Common Diseases of Marine Foodfish. Sing. Pri. Prod. Dept., 34 p.
Chong, Y.C., & T.M. Chao. 1986b. Some problems of air transported estuarine grouper fingerlings. Sing. Vet. J. 10:36-41.
Chou, L.M., et al. (eds.). 1994. The Third Asian Fisheries Forum. Asian Fish. Soc., Manila, 1135 p.
Chou, R. 1984. The effect of dietary water content on the feed intake, food conversion efficiency and growth of young seabass (*Lates calcarifer* Bloch). Sing. J. Pri. Ind. 12:120-127.
Chou, R. 1985. Performance of various fish-meal diets in young sea bass (*Lates calcarifer* Bloch). In: C.Y. Cho, C.B. Cowey & T. Watanabe (eds.), Finfish Nutrition in Asia: Methodological Approaches to Research and Development. Internat. Dev. Res. Ctr., Ottawa, p 82-89.
Chou, Renee. 1992. Marine Aquaculture Section, Singapore Primary Production Department, pers. comm.
Chou, R., & F.J. Wong. 1985. Preliminary observations on the growth and dietary performance of grouper, *Epinephelus tauvina* (Forskal) in floating netcages and fed dry pelleted diet from autofeeders. Sing. J. Pri. Ind. 13:84-91.
Chou, R., H.B. Lee & H.S. Lim. 1995. Fish farming in Singapore: a review of seabass (*Lates calcarifer*), mangrove snapper (*Lutjanus argentimaculatus*) and snub-nose pompano (*Trachinotus blochii*). In: K.L Main & C. Rosenfeld (eds.), Culture of High-value Marine Fishes in Asia and the United States. Oceanic Inst., Honolulu, p 57-65.
Chou, Y.-H., C.-H. Liou, S.-C. Lin & C.-H. Wu. 1993. Effects of highly unsaturated fatty acids in broodstock diets on spawning and egg quality of black porgy (*Acanthopagrus schlegeli*). J. Fish. Soc. Taiwan 20:167-176.
Choudhury, G.S. 1995. Application of extrusion technology to process fish muscle. In: C. Lim & D.J. Sessa (eds.), Nutrition and Utilization Technology in Aquaculture. AOSC Press, Champaign, Ilinois, p 233-245.
Chourrout, D., B. Chevassus, E. Quillet, J.M. Blanc, R. Guyomard, F. Krieg & M. Dorson. 1988. Improvement of salmonids: selective breeding or genetic manipulations? In: Y. Zohar & B. Breton (eds.), Reproduction in Fish. Basic and Applied Aspects in Endocrinology and Genetics. Colloq. INRA (44):183-197.
Christiansen, R., J. Glette, Ø. Lie, O.J. Torrissen & R. Waagbø. 1995. Antioxidant status and immunity in Atlantic salmon, *Salmo salar* L., fed semi-purified diets with and without astaxanthin supplementation. J. Fish Dis. 18:317-328.
Chua, F.H.C., M.L. Ng, K.L. Ng, J.J. Loo & J.Y. Wee. 1994. Investigation of outbreaks of a novel disease, 'sleepy grouper disease', affecting the brown-spotted grouper, *Epinephelus tauvina* Forskal. J. Fish Dis. 17:417-427.
Chua, T.-E., & S.-K. Teng. 1978. Effects of feeding frequency on the growth of young estuary grouper, *Epinephelus tauvina* (Forskal), cultured in floating net-cages. Aquaculture 14:31-47.
Chua, T.-E., & S.-K. Teng. 1980. Economic production of estuary grouper, *Epinephelus salmoides* Maxwell, reared in floating net cages. Aquaculture 20:187-228.
Chuang, J.L., J.L. Lee, Y.Y. Ting & W.T. Lin. 1985. Artificial feeds for black porgy (*Acanthopagrus schlegeli*): a study on protein requirement and utilization by the fish. Proc. AAAP Animal Sci. Congress 2:1252.
Chulavitayanukool, P., C. Puthinuawarat, & N. Sutemechaikul. 1985. Study on the artificial propagation of grouper, *Epinephelus malabaricus* (Bloch and Schneider). Proc. 3rd Seminar Coastal Aquaculture, 22-24 May 1985. Brackishw. Fish. Div., Dept. Fish., Thailand, p 9-21. (in Thai, Engl.

abs.)

Chung, K.-T., G. Zhao, E. Stevens, Jr., B.A. Simco & C.I. Wei. 1995. Growth inhibition of selected aquatic bacteria by tannic acid and related compounds. J. Aquatic Anim. Health 7:46-49.

Clark, D.S., J.A. Brown, S.J. Goddard & J. Moir. 1995. Activity and feeding behaviour of Atlantic cod (*Gadus morhua*) in sea pens. Aquaculture 131:49-57.

Clark, E. 1959. Functional hermaphroditism and self-fertilization in a serranid fish. Science 129:215-216.

Clark, E. 1965. Mating of groupers. Nat. Hist. 74:22-25.

Clark, E., & J.F. Pohle. 1992. Monogamy in tilefish. National Geogr. Res. Explor. 8(3):276-295.

Clark, J.R., L.R. Goodman, P.W. Borthwick, J.M. Patrick, Jr., G.M. Cripe, P.M. Moody, J.C. Moore & E.M. Lores. 1989. Toxicity of pyrethroids to marine invertebrates and fish: a literature review and test results with sediment-sorbed chemicals. Env. Toxicol. Chem. 8:393-401.

Clarke, M.E., M.L. Domeier & W.A. Laroche. 1997. Development of larvae and juveniles of the mutton snapper (*Lutjanus analis*), lane snapper (*Lutjanus synagris*) and yellowtail snapper (*Lutjanus chrysurus*). Bull. Mar. Sci. 61:511-537.

Clarke, R.D. 1992. Effects of microhabitat and metabolic rate on food intake, growth and fecundity of two competing coral reef fishes. Coral Reefs 11:199-205.

Clarke, T.A. 1989. Seasonal differences in spawning, egg size, and early development time of the Hawaiian anchovy or nehu, *Encrasicholina purpurea*. Fish. Bull. 87:593-600.

Clarkson, T.W. 1995. Environmental contaminants in the food chain. Am. J. Clin. Nutr. 61(suppl.):682S-686S.

Clawson, J.A., & R.T. Lovell. 1992. Improvement of nutritional value of *Artemia* for hybrid striped bass/white bass (*Morone saxatilis* x *M. chrysops*) larvae by *n*-3 HUFA enrichment of nauplii with menhaden oil. Aquaculture 108:125-134.

Coche, A.G., J.F. Muir & T. Laughlin. 1997. Management for Freshwater Fish Culture: Ponds and Water Practices. FAO Training Ser. No. 21/1, 233 p.

Cockrum, K.S., & G.K. Jones. 1992. The reproductive biology and fecundity of King George whiting (*Sillaginodes punctata*) in South Australian waters, 1953-1988. Fish. Res. Pap. Dept. Fish. (S. Australia) No. 25, 18 p.

Cohn, F. 1912. *Tetrodon* spec. (cf. *cutcutia*), seine Pflege und Zucht. Blatter fur Aquarien und Terrarien-Kunde 23:582-585. (in Ger.)

Cole, W.M., J.E. Rakocy, K.A. Shultz & J.A. Hargreaves. 1997. Effects of feeding four formulated diets on growth of juvenile palometa, *Trachinotus goodei*. J. Appl. Aquac. 7:51-60.

Coleman, R.E., D. Dugger, J. Homziak & M. Konikoff. 1991. Containment area aquaculture. Aquac. Mag. 17:40-49.

Coleman, R.L. 1966. Spawning red cardinals. Salt Water Aquarium. 2(6):144-148.

Colin, P.L. 1975. Neon Gobies. T.F.H., Neptune City, New Jersey, 304 p.

Colin, P.L. 1982. Spawning and larval development of the hogfish, *Lachnolaimus maximus* (Pisces: Labridae). Fish. Bull., U.S. 80:853-862.

Colin, P.L., & L.J. Bell. 1991. Aspects of the spawning of labrid and scarid fishes (Pisces:Labroidei) at Enewetak Atoll, Marshall Islands with notes on other families. Env. Biol. Fishes 31:229-260.

Collette, B.B., T. Potthoff, W.J. Richards, S. Ueyanagi, J.L. Russo & Y. Nishikawa. 1984. Scombroidei: development and relationships. Am. Soc. Ichthyol. Herpetol. Spec. Pub. (1):591-620.

Collier, T.K., J.E. Stein, H.R. Sanborn, T. Hom, M.S. Myers & U. Varanasi. 1993. A field study of the relationship between bioindicators of maternal contaminant exposure and egg and larval viability of English sole (*Parophrys vetulus*). Mar. Env. Res. 35:171-175.

Collins, C.B. 1990a. Live-hauling warmwater fish. Aquac. Mag. 16:70-76.

Collins, C.B. 1994. Tips on feeds and feeding for catfish and baitfish. Aquac. Mag. 20(4):68-71.

Collins, J. 1990b. The Vitamin C Requirement of Red Drum (*Sciaenops ocellatus*). M.S. Thesis, Miss. State Univ. 46 p.

Collins, L.A., J.H. Finucane & L.E. Barger. 1980. Description of larval and juvenile red snapper, *Lutjanus campechanus*. Fish. Bull., U.S. 77:965-974.

Collins, M.R., 1981. The feeding periodicity of striped mullet, *Mugil cephalus* L., in two Florida habitats. J. Fish. Biol. 19:307-315.

Collins, M.T., J.B. Gratzek, D.L. Dawe & T.G. Nemetz. 1975. Effects of parasiticides on nitrification. J. Fish. Res. Bd. Can. 32:2033-2037.

Collins, M.T., J.B. Gratzek, D.L. Dawe & T.G. Nemetz. 1976. Effects of antibacterial agents on nitrification in an aquatic recirculating system. J. Fish. Res. Bd. Can. 33:215-218.

Collins, S. 1995. Dietary control of HLLE in blue tangs. SeaScope, Aquarium Systems, mentor, Ohio, 12(summer):3.

Colorni, A. 1985. Aspects of the biology of *Cryptocaryon irritans*, and hyposalinity as a control measure in cultured gilt-head sea bream *Sparus aurata*. Dis. Aquat. Org. 1:19-22.

Colorni, A. 1987. Biology of *Cryptocaryon irritans* and strategies for its control. Aquaculture 67:236-237.

Colorni, A. 1990. Pathology of marine warmwater finfish in Israel: problems and research. Advances in Tropical Aquaculture. AQUACOP, IFREMER, Actes Colloq. (9):133-142.

Colorni, A., & A. Diamant. 1992. Pathology in mariculture: specific problems and research in Israel.

Bamidgeh 44:140-141.
Colorni, A., R. Avtalion, W. Knibb, E. Berger, B. Colorni & B. Timan. 1998. Histopathology of sea bass (*Dicentrarchus labrax*) experimentally infected with *Mycobacterium marinum* and treated with streptomycin and garlic (*Allium sativum*) extract. Aquaculture 160:1-17.
Colt, J. 1986. Gas supersaturation—impact on the design and operation of aquatic systems. Aquacultural Eng. 5:49-85.
Colt, J., & C. Orwicz. 1991. Aeration in intensive culture. Adv. World Aquac. 3:198-271.
Colt, J., & B.J. Watten. 1988. Applications of pure oxygen in fish culture. Aquacultural Eng. 7:397-441.
Colura, R.L. 1974. Induced spawning of the spotted seatrout, *Cynoscion nebulosus* (Cuvier). Proc. World Maric. Soc. 5:319-330.
Colura, R.L. 1990. Saltwater pond fertilization. Texas A&M Sea Grant TAMU-SG-90-603, p 78-79.
Colura, R.L., B.T. Hysmith & R.E. Stevens. 1976. Fingerling production of striped bass (*Morone saxatilis*), spotted seatrout (*Cynoscion nebulosus*), and red drum (*Sciaenops ocellatus*), in saltwater ponds. Proc. World Maric. Soc. 7:79-92.
Colura, R.L., A.F. Maciorowski & A. Henderson-Arzapalo. 1990. Induced spawning of spotted seatrout with selected hormone preparations. Prog. Fish-Cult. 52:205-207.
Colura, R.L., A. Henderson-Arzapalo & A.F. Maciorowski. 1991. Culture of red drum. In: J.P. McVey (ed.), Handbook of Mariculture, Vol. II, Finfish Aquaculture. CRC Press, Boca Raton, Florida, p 149-166.
Colura, R.L., T.L. King, J.D. Gray & B.W. Bumguardner. 1992. Analyses of six years of spotted seatrout (*Cynoscion nebulosus*) pond culture trials. Aquaculture 107:313-332.
Comps, M., J.F. Pépin & J.R. Bonami. 1994. Purification and characterization of two fish encephalitis viruses (FEV) infecting *Lates calcarifer* and *Dicentrarchus labrax*. Aquaculture 123:1-10.
Conand, F. 1993. Life history of the silverside *Atherinomorus lacunosus* (Atherinidae) in New Caledonia. J. Fish Biol. 42:851-863.
Conde, B. 1986. Longevity of marine tropicals at the Nancy Aquarium. SeaScope, Aquarium Systems, Mentor, Ohio, 3(summer):1-3.
Conget, P., M. Fernandez, G. Herrara & J.J. Minguell. 1996. Cryopreservation of rainbow trout (*Oncorhynchus mykiss*) spermatozoa using programmable freezing. Aquaculture 143:319-329.
Conte, F.S. 1990. California aquaculture: growth keyed to diversity and markets. World Aquac. 21(3):33-44.
Conte, F.S., S.I. Doroshov, P.B. Lutes & E.M. Strange. 1988. Hatchery Manual for the White Sturgeon *Acipenser transmontanus* Richardson with Application to Other North American Acipenseridae. Div. Agric. Natur. Resour., Univ. Cal., Oakland, 104 p.
Conway, P. 1989. Lactobacilli: fact and fiction. The Regulatory and Protective Role of the Normal Microflora. Proc. Fifth Bengt E. Gustaffson Symp. Macmillan, New York, p 263-281.
Cook, R.H. 1990. Salmon farming in the Bay of Fundy—the challenge of the future. World Aquac. 21(2)46-48,50,52-53.
Cooper, E. 1991. An introduction to rearing larval marine fishes. Freshw. Mar. Aquarium 14(2):120-121,126,128.
Copeland, P.A., & P. Thomas. 1989. Control of gonadotropin release in the Atlantic croaker (*Micropogonias undulatus*): evidence for lack of dopaminergic inhibition. Gen. Comp. Endocrinol. 74:474-483.
Copland, J.W., & D.L. Grey. 1987. Management of Wild and Cultured Sea Bass/Barramundi (*Lates calcarifer*). Austral. Ctr. Internat. Agric. Res., Proc. No. 20, 210 p.
Cornacchia, J.W., & J.E. Colt. 1984. The effects of dissolved gas supersaturation on larval striped bass, *Morone saxatilis* (Walbaum). J. Fish Dis. 7:15-27.
Corneillie, S., C. Agius & F. Ollevier. 1990. Comparison of the fatty acid profile of wild caught fingerlings and yolk sac sea bass (*Dicentrarchus labrax*) larvae with cultured healthy larvae and larvae suffering from whirling disease. Belgian J. Zool. 120:157-164.
Corner, E.D.S., & S.C.M. O'Hara (eds.). 1986. The Biological Chemistry of Marine Copepods. Clarendon Press, Oxford, 349 p.
Corrales, R.A., & J.L. Maclean. 1995. Impacts of harmful algae on seafarming in the Asia-Pacific areas. J. Appl. Phycol. 7:151-162.
Cosson, R.P., & J.-L.M. Martin. 1981. The effects of copper on the embryonic development, larvae, alevins, and juveniles of *Dicentrarchus labrax* (L.). Rapp. P.-v. Réun. Cons. Int. Explor. Mer 178:71-75.
Costello, M.J. 1993. Controlling sea-lice infestations on farmed salmon in northern Europe: options and the use of cleaner fish. World Aquac. 24(1):49-55.
Counihan, T.D., A.I. Miller, M.G. Mesa & M.J. Parsley. 1998. The effects of dissolved gas supersaturation on white sturgeon larvae. Trans. Am. Fish. Soc. 127:316-322.
Courtenay, W.R., Jr., D.P. Jennings & J.D. Williams. 1991. Appendix 2, exotic fishes, In: C.R. Robins (chairman), Common and Scientific Names of Fishes from the United States and Canada. Am. Fish. Soc., Spec. Pub. (20):97-107.
Cousin, J.C.B., & F.B. Laurencin. 1985. Morphogenese de l'appareil digestif et de la vessie gazeuse

du turbot, *Scophthalmus maximus* L. Aquaculture 47:305-319.

Coutteau, P., G. Van Stappen & P. Sorgeloos. 1995. A standard experimental diet for the study of fatty acid requirements of weaning and first ongrowing stages of European sea bass (*Dicentrarchus labrax* L.): selection of the basal diet. ICES Mar. Sci Symp. 201:130-137.

Coutteau, P., G. Van Stappen & P. Sorgeloos. 1996. A standard experimental diet for the study of fatty acid requirements of weaning and first ongrowing stages of the European sea bass *Dicentrarchus labrax* L.: comparison of extruded and extruded/coated diets. Arch. Anim. Nutr. 49:49-59.

Coutteau, P., I. Geurden, M.R. Camara, P. Bergot & P. Sorgeloos. 1997. Review on the dietary effects of phospholipids in fish and crustacean larviculture. Aquaculture 155:149-164.

Coves, D., P. Audineau & J.-L. Nicolas. 1990. Rotifers—rearing technology. In: G. Barnabé (ed.), Aquaculture, Vol. 1. Ellis Horwood, New York, p 232-245.

Coves, D., G. Dewavrin, G. Breuil & N. Devauchelle. 1991. Culture of sea bass (*Dicentrarchus labrax* L.). In: J.P. McVey (ed.), Handbook of Mariculture, Vol. II, Finfish Aquaculture. CRC Press, Boca Raton, Florida, p 3-20.

Cowan, D. 1938. Tank life of fish in the aquarium of the Dove Marine Laboratory, Cullercoats. Rep. Dove. Mar. Lab., Third Ser. (5):36-39.

Cowell, L.E., W.O. Watanabe, W.D. Head, J.M. Shenker & J.J. Grover. 1992. Use of tropical cleaner fish to control the ectoparasite *Neobenedenia melleni* (Monogenea: Capsalidae) on sea-water cultured Florida red tilapia. Abs. in program book, Aquac. '92, p 71.

Cowey, C.B. 1993. Some effects of nutrition on flesh quality of cultured fish. Fish Nutrition in Practice. INRA, Paris, p 227-236.

Cowey, C.B., & J.R. Sargent. 1979. Nutrition. In: W.S. Hoar, D.J. Randall & J.R. Brett (eds.), Fish Physiology, Vol. 8: Bioenergetics and Growth. Academic Press, Orlando, Florida, p 1-69.

Cowey, C.B., J.A. Pope, J.W. Adron & A. Blair. 1972. Studies on the nutrition of marine flatfish. The protein requirement of plaice (*Pleuronectes platessa*). Br. J. Nutr. 28:447-456.

Cowey, C.B, J.W. Adron, J.M. Owen & R.J. Roberts. 1976. The effect of different dietary oils on tissue fatty acids and tissue pathology in turbot *Scophthalmus maximus*. Comp. Biochem. Physiol. 53B:399-403.

Coyne, R., M. Hiney, B. O'Connor, J. Kerry, D. Cazabon & P. Smith. 1994. Concentration and persistence of oxytetracycline in sediments under a marine salmon farm. Aquaculture 123:31-42.

Crabtree, R.E., E.C. Cyr, R.E. Bishop, L.M. Falkenstein & J.M. Dean. 1992. Age and growth of tarpon, *Megalops atlanticus*, larvae in the eastern Gulf of Mexico, with notes on relative abundance and probable spawning areas. Env. Biol. Fishes 35:361-370.

Crabtree, R.E., E.C. Cyr, D.C. Chaverri, W.O. McLarney & J.M. Dean. 1997. Reproduction of tarpon, *Megalops atlanticus*, from Florida and Costa Rican waters and notes on their age and growth. Bull. Mar. Sci. 61:271-285.

Craig, S.R., & D.M. Gatlin III. 1992. Dietary lysine requirement of juvenile red drum *Sciaenops ocellatus*. J. World Aquac. Soc. 23:133-137.

Craig, S.R., & D.M. Gatlin III. 1995. Coconut oil and beef tallow, but not tricaprylin, can replace menhaden oil in the diet of red drum (*Sciaenops ocellatus*) without adversely affecting growth or fatty acid composition. J. Nutr. 125:3041-3048.

Craig, S.R., & D.M. Gatlin III. 1997. Growth and body composition of juvenile red drum (*Sciaenops ocellatus*) fed diets containing lecithin and supplemental choline. Aquaculture 151:259-267.

Craig, S.R., C.R. Arnold & G.J. Holt. 1994. The effects of enriching live foods with highly unsaturated fatty acids on the growth and fatty acid composition of larval red drum *Sciaenops ocellatus*. J. World Aquac. Soc. 25:424-431.

Craig, S.R., W.H. Neill & D.M. Gatlin III. 1995. Effects of dietary lipid and environmental salinity on growth, body composition, and cold tolerance of juvenile red drum (*Sciaenops ocellatus*). Fish Physiol. Biochem. 14:49-61.

Crawford, R.E. 1992. Life history of the Davis Strait Greenland halibut, with reference to the Cumberland Sound fishery. Can. Manuscr. Rep. Fish. Aquat. Sci. 2130, 19 p.

Crawford, R.E., & G.H. Allen. 1977. Seawater inhibition of nitrite toxicity to chinook salmon. Trans. Am. Fish. Soc. 106:105-109.

Crim, L.W., & B.D. Glebe. 1984. Advancement and synchrony of ovulation in Atlantic salmon with pelleted LHRH analog. Aquaculture 43:47-56.

Crim, L.W., A.M. Sutterlin, D.M. Evans & C. Weil. 1983. Accelerated ovulation by pelleted LHRH analogue treatment of spring-spawning rainbow trout (*Salmo gairdneri*) held at low temperature. Aquaculture 35:299-307.

Crim, L.W., R.E. Peter & G.V. Van Der Kraak. 1987. The use of LHRH analogues in aquaculture. In: B. H. Vickery & J. J. Nestor, Jr. (eds.), LHRH and Its Analog: Contraceptive and Therapeutic Applications, Part 2. MTP Press, Boston, p 489-498.

Crocker, P.A., C.R. Arnold, J.A. DeBoer & J.D. Holt. 1981. Preliminary evaluation of survival and growth of juvenile red drum (*Sciaenops ocellata*) in fresh and salt water. J. World Maric. Soc. 12:122-134.

Cruz, E.M., & M.T. Ridha. 1994. Overwintering of tilapia (*Oreochromis spilurus* Gunther) fingerlings using warm underground seawater. Third Asian Fisheries Forum. Asian Fish. Soc., Manila, p 35-38.

Cuenco, M.L., R.R. Stickney & W.E. Grant. 1985. Fish bioenergetics and growth in aquaculture ponds: III. effects of intraspecific competition, stocking rate, stocking size and feeding rate on fish productivity. Ecol. Model. 28:73-95.

Cui, Y., & S.S.O. Hung. 1995. A prototype feeding-growth table for white sturgeon. J. Appl. Aquac. 5:25-34.

Cui, Y., S.S.O. Hung, D.-F. Deng & Y. Yang. 1997. Growth performance of juvenile white sturgeon as affected by feeding regimen. Prog. Fish-Cult. 59:31-35.

Cummings, W.C. 1968. Reproductive Habits of the Sergeant Major, *Abudefduf saxatilis*, (Pisces, Pomacentridae) with Comparative Notes on Four Other Damselfishes in the Bahama Islands. Ph.D. Diss., Univ. Miami, Coral Gables, Florida, 173 p.

Cunha, I., & M. Planas. 1995. Ingestion rates of turbot larvae (*Scophthalmus maximus*) using different-sized live prey. ICES Mar. Sci. Symp. 201:16-20.

Cupka, D.M., R.K. Dias & J. Tucker. 1973. Biology of the black sea bass, *Centropristis striata* (Pisces: Serranidae), from South Carolina waters. S.C. Mar. Resour. Res. Inst., PL-88-309, Project 2-138-R, 97 p.

Cyr, D.G., J. Duston, N.R. Bromage & J.G. Eales. 1987. Annual cycles of serum thyroid hormones and ovarian sex steroids in rainbow trout. Proc. Third Internat. Symp. Repr. Physiol. Fish, Memorial Univ. Newfoundland, St. John's, p 197.

Cyrus, D.P. 1991. The reproductive biology of *Solea bleekeri* (Teleostei) in Lake St Lucia on the south-east coast of Africa. S. Afr. J. Mar. Sci. 10:45-51.

Dabrowski, K., & D. Culver. 1991. The physiology of larval fish digestive tract and formulation of starter diets. Aquac. Mag. 17:49-61.

Dabrowski, K., & M. Rusiecki. 1983. Content of total and free amino acids in zooplanktonic food of fish larvae. Aquaculture 30:31-42.

Dabrowski, K., F. Takashima & Y.K. Law. 1988. Bioenergetic model of planktivorous fish feeding, growth and metabolism: theoretical optimum swimming speed of fish larvae. J. Fish Biol. 32:443-458.

Daelman, W. 1996. Live fish movement and sanitary control: a review of sanitary legislation. In: B. Chatain, M. Saroglia, J. Sweetman & P. Lavens (eds.), Seabass and Seabream culture: Problems and Prospects. Eur. Aquac. Soc., p 111-118.

Dagang, C., L. Chang'an & D. Guangjun. 1991. Study on fishery biology features of spawning group of *Pseudopleuronectes yokamae* (Gunther) in the Sheath Bay. Acta Oceanol. Sin. 10:471-476.

Dailloux, M., C. Henry & D. Terver. 1992. Observation et étude expérimentale de mycobactéries atypiques en aquariums d'eau douce et d'eau de mer. Rev. Sci. L'Eau 5:69-82. (in Fr., Engl. sum.)

Daintith, M. 1992. Comparison of intensive and extensive culture of the Tasmanian whitebait *Lovettia sealii* (Johnston). In: Larval Biology. Austral. Soc. Fish Biol. Workshop, Bur. Rural Resour. Proc. (15):89-94.

Damas, T., M. Borrero, N. Millare & E. Gonzalez. 1978. Desarrollo embrionario y prelarval del caballerote (*Lutjanus griseus* Linne, 1758). Revista Cubana de Investigaciones Pesqueras 3:1-10.

Damas, T., N. Millares, M. Borrero & E. Gonzalez. 1979. Fecundidad en biajaiba (*Lutjanus synagris* Linne, 1758). Revista Cubana de Investigaciones Pesqueras 4:18-30.

Daniel, E.S., C.C. Parrish, D.C. Somerton & J.A. Brown. 1993. Lipids in eggs from first-time and repeat spawning Atlantic halibut, *Hippoglossus hippoglossus* (L.). Aquac. Fish. Mgt. 24:187-191.

Daniels, H.V., C.E. Boyd & R.V. Minton. 1987. Acute toxicity of ammonia and nitrite to spotted seatrout. Prog. Fish-Cult. 49:260-263.

Daniels, H.V., D.L. Berlinsky, R.G. Hodson & C.V. Sullivan. 1996. Effects of stocking density, salinity, and light intensity on growth and survival of southern flounder *Paralichthys lethostigma* larvae. J. World Aquac. Soc. 27:153-159.

Daniels, H.V., R.G. Hodson, C.V. Sullivan & T.I.J. Smith. 1997. Current status of southern flounder (*Paralichthys lethostigma*) culture. World Aquac. '97 Book of Abs., p 104.

Daniels, W.H., & E.H. Robinson. 1986. Protein and energy requirements of juvenile red drum (*Sciaenops ocellatus*). Aquaculture 53:243-252.

Danielsberg, A., A. Berg & T. Lunde. 1993. Design of inlets for incubation of yolk sac larvae of Atlantic halibut (*Hippoglossus hippoglossus* L.). In: H. Reinertsen, L.A. Dahle, L. Jørgensen, and K. Tvinnereim (editors), Fish Farming Technology. Balkema, Rotterdam, p 425-427.

Danielssen, D.S., & T. Hjertnes. 1993. Effect of dietary protein levels in diets for turbot (*Scophthalmus maximus* L.) to market size. Fish Nutrition in Practice. INRA, Paris, p 89-96.

Danilowicz, B.S. 1995. Spatial patterns of spawning in the coral-reef damselfish *Dascyllus albisella*. Mar. Biol. 122:145-155.

Danilowicz, B.S., & C.L. Brown. 1992. Rearing methods for two damselfish species: *Dascyllus albisella* (Gill) and *D. aruanus* (L.). Aquaculture 106:141-149.

Dannevig, G.M. 1910. Apparatus and methods employed at the marine fish hatchery at Flodevig, Norway. Bull. U.S. Bur. Fish. 28:801-810.

Das, B.K., & A. Kaviraj. 1994. Individual and interactive lethal toxicity of cadmium, potassium permanganate and cobalt chloride to fish, worm and plankton. Geobios 21:223-227.

David, L.R. 1939. Embryonic and early larval stages of the grunion, *Leuresthes tenuis*, and of the sculpin, *Scorpaena guttata.* Copeia 1939:75-81.

Davies, I.M., J.G. McHenery & G.H. Rae. 1997a. Environmental risk from dissolved ivermectin to marine organisms. Aquaculture 158:263-275.

Davies, I.M., S.K. Bailey & M.J.C. Harding. 1998. Tributyltin inputs to the North Sea from shipping activities, and potential risk of biological effects. ICES J. Mar. Sci. 55:34-43.

Davies, S.J., & P.C. Morris. 1997. Influence of multiple amino acid supplementation on the performance of rainbow trout, *Oncorhynchus mykiss* (Walbaum), fed soya based diets. Aquac. Res. 28:65-74.

Davies, S.J., I. Nengas & M. Alexis. 1993. Partial substitution of fish meal with different meat meal products in diets for sea bream (*Sparus aurata*). Fish Nutrition in Practice, INRA, p 907-911.

Davies, S.J., M.T. Brown & M. Camilleri. 1997b. Preliminary assessment of the seaweed *Porphyra purpurea* in artificial diets for thick-lipped grey mullet (*Chelon labrosus*). Aquaculture 152:249-258.

Davis, D.A., & C.R. Arnold. 1997. Response of Atlantic croaker fingerlings to practical diet formulations with varying protein and energy contents. J. World Aquac. Soc. 28:241-248.

Davis, D.A., & E.H. Robinson. 1987. Dietary phosphorus requirement of juvenile red drum *Sciaenops ocellatus*. J. World Aquac. Soc. 18:129-136.

Davis, D.A., D. Jirsa & C.R. Arnold. 1995. Evaluation of soybean proteins as replacements for menhaden fish meal in practical diets for the red drum *Sciaenops ocellatus*. J. World Aquac. Soc. 26:48-58.

Davis, J.T. 1990a. Red drum: brood stock and hatchery production. Inst. Food Agric. Sci., Univ. Florida, Gainesville, 4 p.

Davis, J.T. 1990b. Red drum: production of fingerlings and stockers. Inst. Food Agric. Sci., Univ. Florida, Gainesville, 4 p.

Davis, T.L.O., & G.J. West. 1993. Maturation, reproductive seasonality, fecundity, and spawning frequency in *Lutjanus vittus* (Quoy and Gaimard) from the north west shelf of Australia. Fish. Bull., U.S. 91:224-236.

Davy, F.B. 1990. Mariculture in Japan—development of an industry: the first in a three part series. World Aquac. 21(4):36-47.

Davy, F.B., & A. Chouinard (es.). 1981. Induced Fish Breeding in Southeast Asia. Report of a Workshop Held in Singapore, 25-28 November 1980. Internat. Dev. Res. Ctr. Rep. IDRC-178e, 48 p.

Dawes, J.A. 1989. The salt connection. SeaScope, Aquarium Systems, Mentor, Ohio, 7(spring):1,4.

Dawson, M.A., E. Gould, F.P. Thurberg & A. Calabrese. 1977. Physiological response of juvenile striped bass, *Morone saxatilis*, to low levels of cadmium and mercury. Chesapeake Sci. 18:353-359.

Deacon, N., & T. Hecht. 1995. Observations on the thermoregulatory behaviour of juvenile spotted grunter, *Pomadasys commersonnii* (Haemulidae: Pisces). J. Appl. Ichthyol. 11:100-110.

DeBernardo, J.P. 1975. *Anampses cuvieri*: the firefish from Hawaii. Trop. Fish Hob. 24(2):60,62,64-65.

Dedi, J., T. Takeuchi, T. Seikai & T. Watanabe. 1995. Hypervitaminosis and safe levels of vitamin A for larval flounder (*Paralichthys olivaceus*) fed *Artemia* nauplii. Aquaculture 133:135-146.

Dedi, J., T. Takeuchi, T. Seikai, T. Watanabe & K. Hosoya. 1997. Hypervitaminosis A during vertebral morphogenesis in larval Japanese flounder. Fish. Sci. 63:466-473.

Dedi, J., T. Takeuchi, K. Hosoya, T. Watanabe & T. Seikai. 1998. Effect of vitamin A levels in *Artemia* nauplii on the caudal skeleton formation of Japanese flounder *Paralichthys olivaceus*. Fish. Sci. 64:344-345.

Degani, G. 1986. Dietary effects of lipid source, lipid level and temperature on growth of glass eel (*Anguilla anguilla*). Aquaculture 56:207-214.

Degani, G., & D. Kushnirov. 1992. Effects of 17B-estradiol and grouping on sex determination of European eels. Prog. Fish-Cult. 54:88-91.

Dehasque, M., L. Verdonck, P. Sorgeloos, J. Swings, P. Léger & K. Kersters. 1991. Determination of the bacterial contamination in live food production systems in marine fish hatcheries in southern Europe. Eur. Aquac. Soc. Spec. Pub. (15):399-402.

de Kinkelin, P., & R.P. Hedrick. 1991. International veterinary guidelines for the transport of live fish and fish eggs. Ann. Rev. Fish Dis. 1:27-40.

Delbeek, J.C. 1989. The mandarin fish: *Synchiropus splendidus* (Herre). SeaScope, Aquarium Systems, Mentor, Ohio, 6(fall):1,3.

DeLong, D.C., J.E. Halver & E.T. Mertz. 1958. Nutrition of salmonid fishes. VI. Protein requirements of chinook salmon at two water temperatures. J. Nutr. 65:589-599.

delos Reyes, A.A., Jr. & T.B. Lawson. 1996. Combination of a bead filter and rotating biological contactor in a recirculating fish culture system. Aquacultural Eng. 15:27-39.

Dendrinos, P., & J.P. Thorpe. 1987. Experiments on the artificial regulation of the amino acid and fatty acid contents of food organisms to meet the assessed nutritional requirements of larval, post-larval and juvenile Dover sole [*Solea solea* (L.)]. Aquaculture 61:121-154.

Dendrinos, P., S. Dewan & J.P. Thorpe. 1984. Improvement in the feeding efficiency of larval, post larval and juvenile Dover sole (*Solea solea* L.) by the use of staining to improve the visibility of *Artemia* used as food. Aquaculture 38:137-144.

Denson, M.R., & T.I.J. Smith. 1996. Larval rearing and weaning techniques for white bass *Morone chrysops*. J. World Aquac. Soc. 27:194-201.

Denson, M.R., & T.I.J. Smith. 1997a. Effects of larval diets and light intensity on growth, survival and pigmentation of southern flounder. World Aquac. '97 Book of Abs., p 115.

Denson, M.R., & T.I.J. Smith. 1997b. Tank culture of larval sunshine bass. Prog. Fish-Cult. 59:59-63.

De Pauw, N., & J. Joyce. (eds.) 1992. Aquaculture and the Environment. Eur. Aquac. Soc. Spec. Pub. No. 16, 536 p.

de Silva, C. 1974. Development of the respiratory system in herring and plaice larvae. In: J.H.S. Blaxter (ed.) The Early Life History of Fish. Springer-Verlag, New York, p 465-485.

De Silva, C.D., & P. Tytler. 1973. The influence of reduced environmental oxygen on the metabolism and survival of herring and plaice larvae. Neth. J. Sea Res. 7:345-362.

Dettlaff, T.A., A.S. Ginsburg & O.I. Schmalhausen. 1993. Sturgeon Fishes: Developmental Biology and Aquaculture. Springer-Verlag, Berlin, 300 p.

Deuel, D.G., J.R. Clark & A.J. Mansueti. 1966. Description of embryonic and early larval stages of bluefish, *Pomatomus saltatrix*. Trans. Am. Fish. Soc. 95:264-271.

Devauchelle, N., Y. Letty & M. Quere. 1986. Experimental units for incubation and larval rearing with special reference to four marine fish species. Aquaculture 58:297-304.

Devauchelle, N., J.C. Alexandre, N. Le Corre & Y. Letty. 1987. Spawning of sole (*Solea solea*) in captivity. Aquaculture 66:125-147.

Devauchelle, N., J.C. Alexandre, N. Le Corre & Y. Letty. 1988. Spawning of turbot (*Scophthalmus maximus*) in captivity. Aquaculture 69:159-184.

Devesa, S. 1994. Nutrition and feeding of cultured turbot (*Scophthalmus maximus* L.). Eur. Aquac. Soc. Spec. Pub. (22):81-92.

Devresse, B., P. Candreva, P. Léger & P. Sorgeloos. 1991. A new artificial diet for the early weaning of seabass (*Dicentrarchus labrax*) larvae. Eur. Aquac. Soc. Spec. Pub. (15):178-182.

DeYoe, H.R., & C.A. Suttle. 1994. The inability of the Texas "brown tide" alga to use nitrate and the role of nitrogen in the initiation of a persistent bloom of this organism. J. Phycol. 30:800-806.

Dhert, P., P. Lavens, M. Duray & P. Sorgeloos. 1990. Improved larval survival at metamorphosis of Asian seabass (*Lates calcarifer*) using ω3-HUFA-enriched live food. Aquaculture 90:63-74.

Dhert, P., R.B. Bombeo, P. Lavens & P. Sorgeloos. 1992a. A simple semi flow-through culture technique for the controlled super-intensive production of *Artemia* juveniles and adults. Aquacultural Eng. 11:107-119.

Dhert, P., P. Lavens & P. Sorgeloos. 1992b. State of the art of Asian seabass *Lates calcarifer* larviculture. J. World Aquac. Soc. 23:317-329.

Dhert, P., P. Sorgeloos & B. Devresse. 1993. Contributions towards a specific DHA enrichment in the live food *Brachionus plicatilis* and *Artemia* sp. In: H. Reinertsen, L.A. Dahle, L. Jørgensen & K. Tvinnereim (eds.), Fish Farming Technology, Balkema, Rotterdam, p 109-115.

Dhert, P., P. Lavens, M. Dehasque & P. Sorgeloos. 1994. Improvements in the larviculture of turbot *Scophthalmus maximus*: zootechnical and nutritional aspects, possibility for disease control. Eur. Aquac. Soc. Spec. Pub. (22):32-46.

Dhert, P., P. Divanach, M. Kentouri & P. Sorgeloos. 1998. Rearing techniques for difficult marine fish larvae. World Aquac. 29(1):48-55.

Diana, J.S., C.K. Lin & Y. Yi. 1996. Timing of supplemental feeding for tilapia production. J. World Aquac. Soc. 27:410-419.

Diaz, J.P., L. Mani-Ponset, E. Guyot & R. Connes. 1994. Assimilation of dissolved glucose from sea water by the sea bass *Dicentrarchus labrax* and the sea bream *Sparus aurata*. Mar. Biol. 120:181-186.

Diaz, R.J., & C.P. Onuf. 1985. Habitat suitability index models: juvenile Atlantic croaker (rev.). U.S. Fish Wildl. Serv. Biol. Rep. 82(10.98), 23 p.

Di Bella, G., L. Genovese, G. Dugo & M. Amerio. 1993. Fatty acid composition of the liver of the reared European sea bass, *Dicentrarchus labrax* L., in relation to three different diets and to the temperature. Oebalia 19:57-66.

Dickerson, T.L., B.J. Macewicz & J.R. Hunter. 1992. Spawning frequency and batch fecundity of chub mackerel, *Scomber japonicus*, during 1985. Cal. Coop. Oceanic Fish. Invest. Rep. 33:130-140.

Dickey-Collas, M. 1993. The occurrence of juvenile pigmntation abnormalities in plaice (*Pleuronectes platessa*) larvae fed on enriched and unenriched *Artemia salina* nauplii. J. Fish Biol. 42:787-795.

Dickey-Collas, M., & A.J. Geffen. 1992. Importance of the fatty acids 20:5ω3 and 22:6ω3 in the diet of plaice (*Pleuronectes platessa*) larvae. Mar. Biol. 113:463-468.

Dickhoff, W.W., L. Yan, E.M. Plisetskaya, C.V. Sullivan, P. Swanson, A. Hara & M.G. Bernard. 1989. Relationship between metabolic and reproductive hormones in salmonid fish. Fish Physiol. Biochem. 7:147-155.

Dickison, J.A. 1985. Some Effects of 17 alpha-Methyltestosterone on Rainbow Trout, *Salmo gairdneri*, Sex Differentiation and Growth. M.S. Thesis, Univ. Washington, Seattle, 54 p.

DiMarco, R. 1994. Clownfish breeding. Breeder's Registry 2(4):1,3-4.

Ding, Y., J. Li & Y. Ou. 1991. Early scale development in the black porgy *Sparus macrocephalus* (Basilewsky). Tropic Oceanol. 10:16-20.

Dinis, M.T. 1992. Aspects of the potential of *Solea senegalensis* Kaup for aquaculture: larval rearing and weaning to an artificial diet. Aquac. Fish. Mgt. 23:515-520.

Dipper, F. 1987. British Sea Fishes. Underwater World Publications, London, 194 p.

Ditty, J.G., & R.F. Shaw. 1992. Larval development, distribution, and ecology of cobia *Rachycentron canadum* (Family: Rachycentridae) in the northern Gulf of Mexico. Fish. Bull., U.S. 90:668-677.

Ditty, J.G., R.F. Shaw & J.S. Cope. 1994a. A re-description of Atlantic spadefish larvae, *Chaetodipterus faber* (family: Ephippidae), and their distribution, abundance, and seasonal occurrence in the northern Gulf of Mexico. Fish. Bull., U.S. 92:262-274.

Ditty, J.G., R.F. Shaw, C.B. Grimes & J.S. Cope. 1994b. Larval development, distribution, and abundance of common dolphin, *Coryphaena hippurus*, and pompano dolphin, *C. equiselis* (family: Coryphaenidae), in the northern Gulf of Mexico. Fish. Bull., U.S. 92:275-291.

Divanach, P., & M. Kentouri. 1990. Larval rearing in extensive conditions. In: G. Barnabé (ed.) Aquaculture: Vol. 2. Ellis Horwood Limited, Chicester, England, p 820-832.

Divanach, P., M. Kentouri, G. Charalambakis, F. Pouget & A. Sterioti. 1993. Comparison of growth performance of six Mediterranean fish species reared under intensive farming conditions in Crete (Greece), in raceways with the use of self feeders. Eur. Aquac. Soc. Spec. Pub. (18):285-297.

Divanach, P., C. Boglione, B. Menu, G. Koumoudouros, M. Kentouri, S. Cataudella & P. Divanach. 1996. Abnormalities in finfish mariculture: an overview of the problem, causes and solutions. In: B. Chatain, M. Saroglia, J. Sweetman & P. Lavens (eds.), Seabass and Seabream culture: Problems and Prospects. Eur. Aquac. Soc., p 45-66.

Divanach, P., N. Papandroulakis, P. Anastasiadis, G. Koumoundouros & M. Kentouri. 1997. Effect of water currents on the development of skeletal deformities in sea bass (*Dicentrarchus labrax* L.) with functional swimbladder during postlarval and nursery phase. Aquaculture 156:145-155.

Dixon, B.A. 1990. Drug resistance in fish pathogens. Aquarium Fish 2(6):20-24.

Dixon, B.A. 1994. Antibiotic resistance of bacterial fish pathogens. J. World. Aquac. Soc. 25:60-63.

Dixon, B.A., & G.S. Issvoran. 1992. The activity of ceftiofur sodium for *Aeromonas* spp. isolated from ornamental fish. J. Wildl. Dis. 28:453-456.

Doi, M., & T. Singhagraiwan. 1993. Biology and Culture of the Red Snapper, *Lutjanus argentimaculatus*. Res. Proj. Fishery Resource Dev. Kingdom of Thailand, 51 p.

Doi, M., T. Hoshino, Y. Taki & Y. Ogasawara. 1982. Activity of the sperm of the bluefin tuna *Thunnus thynnus* under fresh and preserved conditions. Bull. Jpn. Soc. Sci. Fish. 48:495-498.

Doi, M., M.b.H.M. Nawi, N.R.b.N. Lah & Z.b. Talib. 1991. Artificial propagation of the grouper, *Epinephelus suillus* at the Marine Finfish Hatchery in Tanjong Demong, Terengganu, Malaysia. Dept. Fish., Kuala Lumpur, Malaysia, 41 p.

Doi, M., A. Ohno, H. Kohno, Y. Taki & T. Singhagraiwan. 1997. Development of feeding ability in red snapper *Lutjanus argentimaculatus* early larvae. Fish. Sci. 63:845-853.

Doimi, M., G. Canetti, M. Libralto, L. Dalla Valle, A. Polzonetti-Magni, P. Belvedere & L. Colombo. 1997. Reproduction and hormonal acceleration of metamorphosis in the clown fish, *Amphiprion ocellaris*, an ornamental tropical marine teleost. Island Aquac. Trop. Aquac. Ann. Mtg., Eur. Aquac. Soc., p 116-117.

Domeier, M.L. 1994. Speciation in the serranid fish *Hypoplectrus*. Bull. Mar. Sci. 54:103-141.

Domeier, Michael L. 1998. Pfleger Institute of Environmental Research, Fallbrook, California. pers. comm.

Doménech, A., J.F. Fernández-Garayzábal, P. Lawson, J.A. García, M.T. Cutuli, M. Blanco, A. Gibello, M.A. Moreno, M.D. Collins & L. Domínguez. 1997. Winter disease outbreak in sea-bream (*Sparus aurata*) associated with *Pseudomonas anguilliseptica* infection. Aquaculture 156:317-326.

Donaldson, E.M. 1973. Reproductive endocrinology of fishes. Amer. Zool. 13:909-927.

Donaldson, E.M. 1986. The integrated development and application of controlled reproduction techniques in Pacific salmonid aquaculture. Fish Physiol. Biochem. 2:9-24.

Donaldson, E.M., & T.J. Benfey. 1987. Current status of induced sex manipulation. Proc. Third Internat. Symp. Repr. Physiol. Fish, Memorial Univ. Newfoundland, St. John's, p 108-119.

Donaldson, E.M., U.H.M. Fagerlund, D.A. Higgs & J.R. McBride. 1979. Hormonal enhancement of growth. In: W.S. Hoar, D.J. Randall & J.R. Brett (eds.), Fish Physiology, Vol. 8: Bioenergetics and Growth. Academic Press, Orlando, Florida, p 455-597.

Donaldson, E.M., G.A. Hunter & H.M. Dye. 1981/1982. Induced ovulation in coho salmon (*Oncorhynchus kisutch*). II. Preliminary study of the use of LH-RH and two high potency LH-RH analogues. Aquaculture 26:129-141.

Donaldson, T.J. 1990. Reproductive behavior and social organization of some Pacific hawkfishes (Cirrhitidae). Jpn. J. Ichthyol. 36:439-458.

Doroshev, S.I. 1970. Biological features of the eggs, larvae and young of the striped bass [*Roccus saxatilis* (Walbaum)] in connection with the problem of its acclimatization in the USSR. J. Ichthyol. 10:235-248.

Doroshev, S.I., & J.W. Cornacchia. 1979. Initial swim bladder inflation in the larvae of *Tilapia mossambica* (Peters) and *Morone saxatilis* (Walbaum). Aquaculture 16:57-66.

Doroshev, S.I., J.W. Cornacchia & K. Hogan. 1981. Initial swim bladder inflation in the larvae of physoclistous fishes and its importance for larval culture. Rapp. P.-v. Réun. Cons. Int. Explor. Mer 178:495-500.

Doroshov, S.I., G.P. Moberg & J.P. Van Eenennaam. 1997. Observations on the reproductive cycle of cultured white sturgeon, *Acipenser transmontanus*. Env. Biol. Fishes 48:265-278.

Dosanjh, B.S., D.A. Higgs, M.D. Plotnikoff, J.R. McBride, J.R. Markert & J.T. Buckley. 1984. Efficacy of canola oil, pork lard and marine oil singly and in combination as supplemental dietary lipid sources for juvenile coho salmon (*Oncorhynchus kisutch*) Aquaculture 36:333-345.

dos Santos, J., I.C. Burkow & M. Jobling. 1993. Patterns of growth and lipid deposition in cod (*Gadus morhua* L.) fed natural prey and fish-based feeds. Aquaculture 110:173-189.

Douillet, P. 1998. Disinfection of rotifer cysts leading to bacteria-free populations. J. Exp. Mar. Biol. Ecol. 224:183-192.

Douillet, P.A., & G.J. Holt. 1994. Surface disinfection of red drum (*Sciaenops ocellatus* Linnaeus) eggs leading to bacteria-free larvae. J. Exp. Mar. Biol. Ecol. 179:253-266.

Doumenge, F. 1990a. Aquaculture in Japan. In: G. Barnabé (ed.), Aquaculture, Vol. 2. Ellis Horwood, New York, p 849-945.

Doumenge, F. 1990b. Aquaculture in other Far Eastern countries. In: G. Barnabé (ed.), Aquaculture, Vol. 2. Ellis Horwood, New York, p 968-987.

Dovel, W.L. 1960. Larval development of the oyster toadfish, *Opsanus tau*. Chesapeake Sci. 1:187-195.

Døving, K.B., & J.A. Knutsen. 1993a. Chemokinensis in marine fish larvae. In: B.T. Walther & H.J. Fyhn (eds.), Physiological and Biochemical Aspects of Fish Development. Univ. Bergen, Norway, p 139-145.

Døving, K.B., & J.A. Knutsen. 1993b. Feeding responses and chemotaxis in marine fish larvae. Fish Nutrition in Practice. INRA, Paris, p 579-587.

Dowd, C.E., & E.D. Houde. 1980. Combined effects of prey concentration and photoperiod on survival and growth of larval sea bream, *Archosargus rhomboidalis* (Sparidae). Mar. Ecol. Prog. Ser. 3:181-185.

Downing, G., & M.K. Litvak. 1997. From start to finfish: shedding light on the culture of larval haddock, *Melanogrammus aeglefinus*. World Aquac. '97 Book of Abs., p 129.

Drake, P., & A.M. Arias. 1993. Density and biomass of chironomid larvae in a salt-marsh fish pond system of Cadiz Bay (Spain). Bol. Soc. Port. Entomol. (suppl. 3):483-491.

Drennan II, D.G., M.M. Rahman & R.F. Malone. 1993. Slow sand filtration technologies for the containment of non-indigenous species in recirculating aquaculture effluents. In: J.-K. Wang (ed.), Techniques for Modern Aquaculture. Am. Soc. Agric. Eng., St. Joseph, Michigan, p 335-344.

Duarte, C.M., & M. Alcaraz. 1989. To produce many small or few large eggs: a size-independent reproductive tactic of fish. Oecologia 80:401-404.

Duhamel, G., K.-H. Kock, E. Balguerias & J.-C. Hureau. 1993. Reproduction in fish of the Weddell Sea. Polar Biol. 13:193-200.

Duncan, P.L., & P.H. Klesius. 1996. Dietary immunostimulants enhance nonspecific immune responses in channel catfish but not resistance to *Edwardsiella ictaluri*. J. Aquat. Anim. Health 8:241-248.

Dunn, J.R., & A.C. Matarese. 1984. Gadidae: development and relationships. Am. Soc. Ichthyol. Herpetol. Spec. Pub. (1):283-299.

Dunson, W.A., P. Fricano & W.J. Sadinski. 1993. Variation in tolerance to abiotic stresses among sympatric salt marsh fish. Wetlands 13:16-24.

Duray, M.N. 1990. Biology and Culture of Siganids. SEAFDEC, Tigbauan, Iloilo, Philippines, 47 p.

Duray, M.N. 1994a. Daily rates of ingestion of rotifers and *Artemia* nauplii by laboratory-reared grouper larvae of *Epinephelus suillus*. Phil. Sci. 31:32-41.

Duray, M.N. 1994b. Sea bass, groupers, and snappers. In: F. Lacanilao, R.M. Coloso, G.F. Quinitio (eds.), Proc. of the Seminar-Workshop on Aquaculture Development in Southeast Asia and Prospects for Seafarming and Searanching; 19-23 August 1991; Iloilo City, Philippines. SEAFDEC, Iloilo, p 3-12.

Duray, M., & T. Bagarino. 1984. Weaning of hatchery-bred milkfish larvae from live food to artificial diets. Aquaculture 41:325-332.

Duray, M.N., & J.V. Juario. 1987. Broodstock management and seed production of the rabbitfish *Siganus guttatus* (Bloch) and the sea bass *Lates calcarifer* (Bloch). Perspectives in Aquaculture Development in Southeast Asia and Japan: Proc. Sept. 8-12, 1987, p 195-210.

Duray, M., & H. Kohno. 1988. Effects of continuous lighting on growth and survival of first-feeding larval rabbitfish, *Siganus guttatus*. Aquaculture 72:73-79.

Duray, M., H. Kohno & F. Pascual. 1988. The effect of lipid-enriched broodstock diets on spawning and on egg and larval quality of hatchery-bred rabbitfish (*Siganus guttatus*). Phil. Sci. 31:42-57.

Duray, M., H. Kohno & F. Pascual. 1994. The effect of lipid-enriched broodstock diets on spawning and on egg and larval quality of hatchery-bred rabbitfish (*Siganus guttatus*). Phil. Sci. 31:42-57.

Duray, M.N., L.G. Alpasan & C.B. Estudillo. 1996a. Improved hatchery rearing of mangrove red snapper, *Lutjanus argentimaculatus*, in large tanks with small rotifer (*Brachionus plicatilis*) and *Artemia*. Bamidgeh 48:123-132.

Duray, M.N., C.B. Estudillo & L.G. Alpasan. 1996b. The effect of background color and rotifer density on rotifer intake, growth and survival of the grouper (*Epinephelus suillus*) larvae. Aquaculture 146:217-224.

Duray, M.N., C.B. Estudillo & L.G. Alpasan. 1997. Larval rearing of the grouper *Epinephelus suillus* under laboratory conditions. Aquaculture 150:63-76.

Duston, J., & R.L. Saunders. 1994. Production of underyearling Atlantic salmon smolts, and long-term performance in a sea cage. In: D.D. MacKinlay (ed.), High Performance Fish., Fish Physiol. Assn., Vancouver, p 105-109.

Dutton, P. 1992. Effects of experience on feeding success by larval white seabass, *Atractoscion nobilis*. J. Fish Biol. 41:765-773.

Dwyer, W.P., & J.E. Peterson. 1993. Evaluation of a low head oxygenator at Giant Springs State Fish Hatchery, Great Falls, Montana. Prog. Fish-Cult. 55:121-124.

Dyer, W.G., E.H. Williams, Jr. & L. Bunkley-Wiliams. 1992. *Neobenedenia pargueraensis* n. sp. (Monogenea: Capsalidae) from the red hind, *Epinephelus guttatus*, and comments about *Neobenedenia melleni*. J. Parasitol. 73:399-401.

Ebeling, J.M. 1994. Monitoring and control. In: M.B. Timmons & T.M. Losordo (eds.), Aquaculture Water Reuse Systems: Engineering Design and Management. Elsevier, Amsterdam, p 307-323.

Economou, A.N. 1991. Is dispersal of fish eggs, embryos and larvae an insurance against density dependence? Env. Biol. Fishes 31:313-321.

Eda, H., R. Murashige, Y. Oozeki, A. Hagiwara, B. Eastham, P. Bass, C.S. Tamaru & C.-S. Lee. 1990a. Factors affecting intensive larval rearing of striped mullet, *Mugil cephalus*. Aquaculture 91:281-294.

Eda, H., R. Murashige, B. Eastham, L. Wallace, P. Bass, C.S. Tamaru & C.-S. Lee. 1990b. Survival and growth of milkfish (*Chanos chanos*) larvae in the hatchery. I. Feeding. Aquaculture 89:233-244.

Eda, H., T. Fujiwara & T. Takita. 1994. Embryonic, larval and juvenile development in laboratory-reared dragonets, *Repomucenus beniteguri*. Jpn. J. Ichthyol. 40:465-473.

Eddy, F.B., P.A. Kunzlik & R.N. Bath. 1983. Uptake and loss of nitrite from the blood of rainbow trout, *Salmo gairdneri* Richardson, and Atlantic salmon, *Salmo salar* L. in fresh water and in dilute sea water. J. Fish Biol. 23:105-116.

Edwardes, S. 1991. A Preliminary Investigation into the Seasonal Variation and Short Term Storage of Milt from the Atlantic Halibut *Hippoglossus hippoglossus*. M.S. Thesis, Univ. Stirling, 69 p.

Edwards, P. 1991. Letters: Fish-fowl-pig farming uncommon. World Aquac. 22(3):2-3,6.

Edwards, R., & B.D. Henderson. 1987. An experimental hatchery project: studies of propagation, culture and biology of snook (*Centropomus undecimalis*). Proc. Gulf Carib. Fish. Inst. 38:211-221.

Edwards, R.R.C., D.M. Finlayson & J.H. Steele. 1972. An experimental study of the oxygen consumption, growth, and metabolism of the cod (*Gadus morhua* L.). J. Exp. Mar. Biol. Ecol. 8:299-309.

Ehrlich, K.F., & G. Muszynski. 1982. Effects of temperature on interactions of physiological and behavioural capacities of larval California grunion: adaptations to the planktonic environment. J. Exp. Mar. Biol. Ecol. 60:223-244.

Eikebrokk, B., & Y. Ulgenes. 1993. Characterization of treated and untreated effluents from landbased fish farms. In: H. Reinertsen, L.A. Dahle, L. Jørgensen & K. Tvinnereim (eds.), Fish Farming Technology. Balkema, Rotterdam, p 361-366.

Einarsson, H. 1951. The post-larval stages of sandeels in Faroe, Iceland, and W-Greenland waters. Acta Nat. Isl. 1(7):5-75.

Eisler, R. 1985a. Cadmium hazards to fish, wildlife, and invertebrates: a synoptic review. U.S. Fish Wildl. Serv., Biol. Rep. 85(1.2), 46 p.

Eisler, R. 1985b. Carbofuran hazards to fish, wildlife, and invertebrates: a synoptic review. U.S. Fish Wildl. Serv., Biol. Rep. 85(1.3), 36 p.

Eisler, R. 1985c. Selenium hazards to fish, wildlife, and invertebrates: a synoptic review. U.S. Fish Wildl. Serv., Biol. Rep. 85(1.5), 57 p.

Eisler, R. 1986a. Chromium hazards to fish, wildlife, and invertebrates: a synoptic review. U.S. Fish Wildl. Serv., Biol. Rep. 85(1.6), 60 p.

Eisler, R. 1986b. Diazinon hazards to fish, wildlife, and invertebrates: a synoptic review. U.S. Fish Wildl. Serv., Biol. Rep. 85(1.9), 37 p.

Eisler, R. 1986c. Polychlorinated biphenyl hazards to fish, wildlife, and invertebrates: a synoptic review. U.S. Fish Wildl. Serv., Biol. Rep. 85(1.7), 72 p.

Eisler, R. 1987a. Mercury hazards to fish, wildlife, and invertebrates: a synoptic review. U.S. Fish Wildl. Serv., Biol. Rep. 85(1.10), 90 p.

Eisler, R. 1987b. Polycyclic aromatic hydrocarbon hazards to fish, wildlife, and invertebrates: a synoptic review. U.S. Fish Wildl. Serv., Biol. Rep. 85(1.11), 81 p.

Eisler, R. 1988a. Arsenic hazards to fish, wildlife, and invertebrates: a synoptic review. U.S. Fish Wildl. Serv., Biol. Rep. 85(1.12), 92 p.

Eisler, R. 1988b. Lead hazards to fish, wildlife, and invertebrates: a synoptic review. U.S. Fish Wildl. Serv., Biol. Rep. 85(1.14), 134 p.

Eisler, R. 1989a. Atrazine hazards to fish, wildlife, and invertebrates: a synoptic review. U.S. Fish Wildl.

Serv., Biol. Rep. 85(1.18), 53 p.
Eisler, R. 1989b. Molybdenum hazards to fish, wildlife, and invertebrates: a synoptic review. U.S. Fish Wildl. Serv., Biol. Rep. 85(1.19), 61 p.
Eisler, R. 1989c. Pentachlorophenol hazards to fish, wildlife, and invertebrates: a synoptic review. U.S. Fish Wildl. Serv., Biol. Rep. 85(1.17), 72 p.
Eisler, R. 1989d. Tin hazards to fish, wildlife, and invertebrates: a synoptic review. U.S. Fish Wildl. Serv., Biol. Rep. 85(1.15), 83 p.
Eisler, R. 1990a. Boron hazards to fish, wildlife, and invertebrates: a synoptic review. U.S. Fish Wildl. Serv., Biol. Rep. 85(1.20), 32 p.
Eisler, R. 1990b. Chlordane hazards to fish, wildlife, and invertebrates: a synoptic review. U.S. Fish Wildl. Serv., Biol. Rep. 85(1.21), 49 p.
Eisler, R. 1990c. Paraquat hazards to fish, wildlife, and invertebrates: a synoptic review. U.S. Fish Wildl. Serv., Biol. Rep. 85(1.22), 28 p.
Eisler, R. 1992a. Diflubenzuron hazards to fish, wildlife, and invertebrates: a synoptic review. U.S. Fish Wildl. Serv., Biol. Rep. 4, 36 p.
Eisler, R. 1992b. Fenvalerate hazards to fish, wildlife, and invertebrates: a synoptic review. U.S. Fish Wildl. Serv., Biol. Rep. 2, 43 p.
Eisler, R. 1993. Zinc hazards to fish, wildlife, and invertebrates: a synoptic review. U.S. Fish Wildl. Serv., Biol. Rep. 10, 106 p.
Eisler, R. 1994a. Acrolein hazards to fish, wildlife, and invertebrates: a synoptic review. U.S. Nat. Biol. Survey, Biol. Rep. 23, 29 p.
Eisler, R. 1994b. Famphur hazards to fish, wildlife, and invertebrates: a synoptic review. U.S. Nat. Biol. Survey, Biol. Rep. 20, 23 p.
Eisler, R., & J. Jacknow. 1985. Toxaphene hazards to fish, wildlife, and invertebrates: a synoptic review. U.S. Fish Wildl. Serv., Biol. Rep. 85(1.4), 26 p.
El-Dakour, S., & K.A. George. 1982. Growth of hamoor (*Epinephelus tauvina*) fed on different protein: energy ratios. Kuwait Inst. Sci. Res., Ann. Res. Rep. (1981):75-77.
Eldridge, M.B., T. Echeverria & J.A. Whipple. 1977. Energetics of Pacific herring (*Clupea harengus pallasi*) embryos and larvae exposed to low concentrtaions of benzene, a monoaromatic component of crude oil. Trans. Am. Fish. Soc. 106:452-461.
Eldridge, M.B., J.A. Whipple & M.J. Bowers. 1982. Bioenergetics and growth of striped bass, *Morone saxatilis*, embryos and larvae. Fish. Bull., U.S. 80:461-474.
Elema, M.O., K.A. Hoff & H.G. Kristensen. 1996. Bioavailability of oxytetracycline from medicated feed administered to Atlantic salmon (*Salmo salar* L.) in seawater. Aquaculture 143:7-14.
Elizabeth-Mesnager, D. 1984. Fish, prawn, and algal aquaculture in Martinique. In: T.D. Sleeter (ed.), Assessment of the potential for aquaculture in Bermuda. Bermuda Biol. Sta. Spec. Pub. (27):169-177.
Ellertsen, B., E. Moksness, P. Solemdal, S. Tilseth, T. Westgård & V. Øiestad. 1981. Growth and survival of cod larvae in an enclosure. Experiments and a mathematical model. Rapp. P.-v. Réun. Cons. Int. Explor. Mer 178:45-57.
Elliott, D.G., R.J. Pascho & G.L. Bullock. 1989. Developments in the control of bacterial kidney disease of salmonid fishes. Dis. Aquat. Org. 6:201-215.
Elliott, J.M. 1976. The energetics of feeding, metabolism and growth of brown trout (*Salmo trutta* L.) in relation to body weight, water temperature and ration size. J. Anim. Ecol. 45:923-948.
Elliott, J.M. 1979. Energetics of freshwater teleosts. Symp. Zool. Soc. London 44:29-61.
Ellis, A.E. 1989. The immunology of teleosts. In R.J. Roberts (ed.), Fish Pathology. Bailliere Tindall, London, p 135-152.
Ellis, E.P., & W.O. Watanabe. 1993. The effects of hyposalinity on eggs, juveniles and adults of the marine monogenean, *Neobenedenia melleni*: treatment of ecto-parasitosis in seawater-cultured tilapia. Aquaculture 117:15-27.
Ellis, S.C., & R.C. Reigh. 1991. Effects of dietary lipid and carbohydrate levels on growth and body composition of juvenile red drum, *Sciaenops ocellatus*. Aquaculture 97:383-394.
Ellis, S.C., W.O. Watanabe & W.D. Head. 1993. Effect of initial age variation on production of Florida red tilapia fry under intensive, brackishwter tank culture. Aquac. Fish. Mgt. 24:465-471.
Ellis, S., G. Viala & W.O. Watanabe. 1996. Growth and feed utilization of hatchery-reared juvenile Nassau grouper fed four practical diets. Prog. Fish.-Cult. 58:167-172.
Ellis, S.C., W.O. Watanabe & E.P. Ellis. 1997. Temperature effects on feed utilization and growth of postsettlement stage Nassau grouper. Trans. Am. Fish. Soc. 126:309-315.
El-Sayed, A.-F.M. 1991. Protein requirements for optimum growth of *Liza ramada* fry (Mugilidae) at different water salinities. Aquat. Living Resour. 4:117-123.
El-Sayed, A.-F.M., K.A. Mostafa, J.S. Al-Mohammadi, A.-A. El-Dehaimi & M. Kayid. 1995. Effects of stocking density and feeding levels on growth rates and feed utilization of rabbitfish *Siganus canaliculatus*. J. World Aquac. Soc. 26:212-216.
Eltink, A. 1991. Horse mackerel egg production and spawning stock size in the North Sea in 1990. ICES Rep. C.M.1991/H:27, 14 p.
El-Zibdeh, M., K. Ide, T. Yoshimatsu, S. Matsui & M. Furuichi. 1995. Effects of the deletion of Ca

or trace elements from semi-purified diet on growth and feed utilization of yellow croaker, *Nibea albiflora*. J. Fac. Agric. Kyushu Univ. 40:157-166.

El-Zibdeh, M., T. Yoshimatsu, S. Matsui & M. Furuichi. 1996a. Effects of the deletion of Ca or trace elements from purified diet on growth and efficiency of feed utilization of redlip mullet. J. Fac. Agric. Kyushu Univ. 40:375-382.

El-Zibdeh, M., K. Ide & M. Furuichi. 1996b. Effects of the deletion of Mg or Fe from semi purified diets on growth and efficiency of feed utilization of yellow croaker *Nibea albiflora*. J. Fac. Agric. Kyushu Univ. 40:391-397.

Emata, A.C. 1994. Milkfish, rabbitfish, and mullet. In: F. Lacanilao, R.M. Coloso & G.F. Quinitio (eds.). Proc. of the Seminar-Workshop on Aquaculture Development in Southeast Asia and Prospects for Seafarming and Searanching; 19-23 August 1991; Iloilo City, Philippines. SEAFDEC Aquac. Dept. Iloilo, p 13-23.

Emata, A.C., & C.L. Marte. 1992. Reproductive performance of milkfish broodstock in concrete tanks. Abs. in program, Third Asian Fisheries Forum, Asian Fish. Soc., Manila, p 11.

Emata, A.C., C.L. Marte & L.M.B. Garcia. 1992. Management of Milkfish Broodstock. SEAFDEC Aquac. Ext. Man. No. 20, 22 p.

Emata, A.C., B. Eullaran & T.U. Bagarinao. 1994. Induced spawning and early life description of the mangrove red snapper, *Lutjanus argentimaculatus*. Aquaculture 121:381-387.

Engbæk, H.C., J. Thormann & B. Vergmann. 1980. Aquarium-borne *Mycobacterium marinum* granulomas. Scand. J. Infect. Dis. 12:74-78.

Engle, C.R., L.W. Dorman & D.L. Gray. 1988. Baitfish production: enterprise budget. Southern Reg. Aquac. Ctr., IFAS, Univ. Florida, Gainesville, Pub. No. 122, 4 p.

Ensminger, M.E., J.E. Oldfield & W.W. Heinemann. 1990. Feeds and Nutrition. Ensminger Publishing Co., Clovis, California, 1544 p.

Epifanio, C.E. 1979. Growth in bivalve molluscs: nutritional effects of two or more species of algae in diets fed to the American oyster *Crassostrea virginica* (Gmelin) and the hard clam *Mercenaria mercenaria* (L.). Aquaculture 18:1-12.

Epifanio, C.E., C.C. Valenti & C.L. Turk. 1981. A comparison of *Phaeodactylum tricornutum* and *Thalassiosira pseudonana* as foods for the oyster, *Crassostrea virginica*. Aquaculture 23:347-353.

Epifanio, C.E., D. Goshorn & T.E. Targett. 1988. Induction of spawning in the weakfish, *Cynoscion regalis*. Fish. Bull., U.S. 86:168-171.

Ernst, D.H., L.J. Ellingson, B.L. Olla, R.I. Wicklund, W.O. Watanabe & J.J. Grover. 1989. Production of Florida red tilapia in seawater pools: nursery rearing with chicken manure and growout with prepared feed. Aquaculture 80:247-260.

Ernst, D.H., W.O. Watanabe, L.J. Ellingson, R.I. Wicklund & B.L. Olla. 1991. Commercial-scale production of Florida red tilapia seed in low- and brackish-salinity tanks. J. World Aquac. Soc. 22:36-44.

Ernst, V.V., J.M. Neff & J.W. Anderson. 1977. The effects of the water-soluble fractions of no. 2 fuel oil on the early development of the estuarine fish, *Fundulus grandis* Baird and Girard. Env. Poll. 14:25-35.

Ervik, A., O.B. Samuelsen, J.E. Juell & H. Sveier. 1994. Reduced environmental impact of antibacterial agents applied in fish farms using the LiftUp feed collector system or a hydroacoustic feed detector. Dis. Aquat. Org. 19:101-104.

Espe, M., H. Haaland & L.R. Njaa. 1992. Autolysed fish silage as a feed ingredient for Atlantic salmon (*Salmo salar*). Comp. Biochem. Physiol. 103A:369-372.

Espe, M., E. Lied & K.R. Torrissen. 1993. Changes in plasma and muscle free amino acids in Atlantic salmon (*Salmo salar*) during absorption of diets containing different amounts of hydrolysed cod muscle protein. Comp. Biochem. Physiol. 105A:555-562.

Estévez, A., & A. Kanazawa. 1996. Fatty acid composition of neural tissues of normally pigmented and unpigmented juveniles of Japanese flounder using rotifer and *Artemia* enriched in n-3 HUFA. Fish. Sci. 62:88-93.

Etherington, W. 1972. Mating behavior of the Fiji Damsel. Mar. Aquarist 3(1):32-34.

Evans, R.P., C.C. Parrish, J.A. Brown & P.J. Davis. 1996. Biochemical composition of eggs from repeat and first-time spawning captive Atlantic halibut (*Hippoglossus hippoglossus*). Aquaculture 139:139-149.

Evjemo, J.O., P. Coutteau, Y. Olsen & P. Sorgeloos. 1997. The stability of docosahexaenoic acid in two *Artemia* species following enrichment and subsequent starvation. Aquaculture 155:135-148.

Fable, W.A., Jr., T.D. Williams & C.R. Arnold. 1978. Description of reared eggs and young larvae of the spotted seatrout, *Cynoscion nebulosus*. Fish. Bull., U.S. 76:65-71.

Fabregas, J., & C. Herrero. 1986. Marine microalgae as a potential source of minerals in fish diets. Aquaculture 51:237-243.

Fabregas, J., C. Herrero, B. Cabezas & J. Abalde. 1986. Biomass production and biochemical composition in mass cultures of the marine microalga *Isochrysis galbana* Parke at varying nutrient concentrations. Aquaculture 53:101-113.

Fábregas, J. Otero, E. Morales, B. Cordero & M. Patiño. 1996. *Tetraselmis suecica* cultured in different nutrient concentrations varies in nutritional value to *Artemia*. Aquaculture 143:197-204.

Fagbenro, O.A. 1994. Dried fermented fish silage in diets for *Oreochromis niloticus*. Bamidgeh 46:140-147.

Fagbenro, O., K. Jauncey & G. Haylor. 1994. Nutritive value of diets containing dried lactic acid fermented fish silage and soybean meal for juvenile *Oreochromis niloticus* and *Clarias gariepinus*. Aquat. Living Resour. 7:79-85.

Fahay, M.P. 1983. Guide to the early stages of marine fishes occurring in the western North Atlantic Ocean, Cape Hatteras to the southern Scotian shelf. J. Northw. Atl. Fish. Sci. 4:423 p.

Fair, P.H., W.P. Williams & T.I.J. Smith. 1993. Effect of dietary menhaden oil on growth and muscle fatty acid composition of hybrid striped bass, *Morone chrysops* × *M. saxatilis*. Aquaculture 116:171-189.

Falguière, J.-C., & F. Blouin. 1997. Maturation and spawning of red drum *Sciaenops ocellata* in captivity under tropical condition in Martinique (FWI). Island Aquac. Trop. Aquac. Ann. Mtg., Eur. Aquac. Soc., p 123-124.

Falguière, J.-C., & P. Paquotte. 1997. Technical and financial analysis of red drum (*Sciaenops ocellata*) farming in floating cages in Martinique. Island Aquac. Trop. Aquac. Ann. Mtg., Eur. Aquac. Soc., p 121-122.

Falguière, J.-C., B. Noguerra & P. Dalla-Torre. 1997a. Interaction between sexual maturity and growth in cage culture of red drum *Sciaenops ocellata* under tropical condition in Martinique (FWI). Island Aquaculture and Tropical Aquaculture. Ann. Mtg., Eur. Aquac. Soc., p 125-126.

Falguière, J.-C., O. Denis & V. Vianas. 1997b. Seawater performance in red tilapia (hybrid *O. mossambicus, O. hornorum, O. niloticus,* and *O. aureus*) in Martinique Island (FWI): zootechnical aspects. Island Aquaculture and Tropical Aquaculture. Ann. Mtg., Eur. Aquac. Soc., p 127-128.

Falk-Petersen, I.-B., & E. Kjørsvik. 1987. Acute toxicity tests of the effects of oils and dispersants on marine fish embryos and larvae—a review. Sarsia 72:411-413.

Falls, W.W., G.K. Vermeer & C.W. Dennis. 1988. Evaluation of etomidate as an anesthetic for red drum, *Sciaenops ocellatus*. Contr. Mar. Sci. 30(suppl.):37-42.

FAO, 1995. Aquaculture production statistics 1984-1993. FAO Fish. Circ. No. 815, Rev. 7, 186 p.

FAO, 1998. Aquaculture production statistics 1987-1996. FAO Fish. Circ. No. 815, Rev. 10, 197 p.

Faranda, F., A. Cavaliere, G. Lo Paro, A. Manganaro & A. Mazzola. 1985. Preliminary studies on reproduction of *Puntazzo puntazzo* (Gmelin 1789)(Pisces, Sparidae) under controlled conditions. Aquaculture 49:111-123.

Farley, J.H., & T.L.O. Davis. 1998. Reproductive dynamics of southern bluefin tuna, *Thunnus maccoyii*. Fish. Bull., U.S. 96:223-236.

Fast, A.W., H.D. Biona, T.P. Barry, R. Tabanda & R. Bayogos. 1989. Brackishwater pond culture of the spotted scat (*Scatophagus argus*) in the Philippines. J. Aquat. Trop. 4:37-49.

Fast, A.W., S.T. Katase, E.G. Grau & D.K. Barclay. 1991. Seawater maturation and spawning of rainbow trout in Hawaii. Prog. Fish-Cult. 53:47-49.

Faulkner, I.N., & G.P. Moberg. 1997. Effects of short term management stress on the ability of GnRHa to induce gonadotropin secretion in male white sturgeon, *Acipenser transmontanus*. Aquaculture 159:159-168.

Fauvel, C., M.H. Omnes, M. Suquet & Y. Normant. 1992. Enhancement of the production of turbot, *Scopthalmus maximus* (L.), larvae by controlling overripening in mature females. Aquac. Fish. Mgt. 23:209-216.

Feddern, H.A. 1967. Larval development of the neon goby, *Elacatinus oceanops*, in Florida. Bull. Mar. Sci. 17:367-375.

Feddern, H.A. 1968. Hybridization between the western Atlantic angelfishes, *Holacanthus isabelita* and *H. ciliaris*. Bull. Mar. Sci. 18:351-382.

Federal Joint Subcommittee on Aquaculture. 1994. Guide to Drug, Vaccine, and Pesticide Use in Aquaculture. Texas Agric. Ext. Serv., Texas A&M Univ., College Station, 68 p.

Fedorenko, A.Y., & C.L. Cross. 1991. Seapen rearing to maturity of Squamish River chinook salmon. Can. Manuscr. Rep. Fish. Aquat. Sci. No. 2096, 35 p.

Feed Development Section. 1994. Feeds and Feeding of Milkfish, Nile Tilapia, Asian Sea Bass, and Tiger Shrimp. SEAFDEC Aquac. Dept., Tigbauan, Iloilo, Philippines, 97 p.

Feist, G., & C.B. Schreck. 1990. Hormonal content of commercial fish diets and of young coho salmon (*Oncorhynchus kisutch*) fed these diets. Aquaculture 86:63-75.

Felip, A., S. Zanuy, M. Carrillo, G. Martínez, J. Ramos & F. Piferrer. 1997. Optimal conditions for the induction of triploidy in the sea bass (*Dicentrarchus labrax* L.). Aquaculture 152:287-298.

Fenner, B. 1992. A review of literature on hormonal manipulation of fishes as an aquacultural technique. Freshw. Mar. Aquarium 15(6):148-160.

Fenner, R. 1996. A diversity of aquatic life: cardinalfish, family Apogonidae. SeaScope, Aquarium Systems, Mentor, Ohio, 13(spring):3.

Fent, K. 1992. Embryotoxic effects of tributyltin on the minnow *Phoxinus phoxinus*. Env. Poll. 76:187-194.

Ferlin, P., & P. Noriega-Curtis. 1989. A regional survey of the aquaculture sector in the Caribbean.

FAO/UNDP/ADCP/REP/89/40, 66 p.
Fermin, A.C., & M.E.C. Bolivar. 1994. Feeding live or frozen *Moina macrocopa* (Strauss) to Asian sea bass, *Lates calcarifer* (Bloch), larvae. Bamidgeh 46:132-139.
Fermin, A.C., & G.A. Seronay. 1997. Effects of different illumination levels on zooplankton abundance, feeding periodicity, growth and survival of the Asian sea bass, *Lates calcarifer* (Bloch), fry in illuminated floating nursery cages. Aquaculture 157:227-237.
Fernandez-Casalderry, A., M.D. Ferrando & E. Andreu-Moliner. 1992. Acute toxicity of several pesticicides to rotifer (*Brachionus calyciflorus*). Bull. Env. Contam. Toxicol. 48:14-17.
Fernandez-Casalderry, A., M.D. Ferrando & E. Andreu-Moliner. 1993. Effect of the insecticide methylparathion on filtration and ingestion rates of *Brachionus calyciflorus* and *Daphnia magna*. Sci. Total Env. Suppl. 1993:867-876.
Fernández-Diaz, C., & M. Yúfera. 1997. Detecting growth in gilthead seabream, *Sparus aurata* L., larvae fed microcapsules. Aquaculture 153:93-102.
Fernández-Diaz, C., E. Pascual & M. Yúfera. 1994. Feeding behaviour and prey size selection of gilthead seabream, *Sparus aurata*, larvae fed on inert and live food. Mar. Biol. 118:323-328.
Fernandez-Palacios, H., D. Montero, J. Socorro, M.S. Izquierdo & J.M. Vergara. 1994. First studies on spawning, embryonic and larval development of *Dentex gibbosus* (Rafinesque, 1810) (Osteichthyes, Sparidae) under controlled conditions. Aquaculture 122:63-73.
Fernández-Palacios, H., M.S. Izquierdo, L. Robaina, A. Valencia, M. Salhi & J.M. Vergara. 1995. Effect of *n*-3 HUFA level in broodstock diets on egg quality of gilthead sea bream (*Sparus aurata* L.). Aquaculture 132:325-337.
Fernández-Palacios, H., M. Izquierdo, L. Robaina, A. Valencia, M. Salhi & D. Montero. 1997. The effect of dietary protein and lipid from squid and fish meals on egg quality of broodstock for gilthead seabream (*Sparus aurata*). Aquaculture 148:233-246.
Fernández-Reiriz, M.J., & U. Labarta. 1996. Lipid classes and fatty acid composition of rotifers (*Brachionus plicatilis*) fed two algal diets. Hydrobiologia 330:73-79.
Fernández-Reiriz, M.J., A. Perez-Camacho, M.J. Ferreiro, J. Blanco, M. Planas, M.J. Campos & U. Labarta. 1989. Biomass production and variation in the biochemical profile (total protein, carbohydrates, RNA, lipids and fatty acids) of seven species of marine microalgae. Aquaculture 83:17-37.
Fernández-Reiriz, M.J., U. Labarta & M.J. Ferreiro. 1993. Effects of commercial enrichment diets on the nutritional value of the rotifer (*Brachionus plicatilis*). Aquaculture 112:195-206.
Ferrando, M.D., C. Janssen, E. Andreu & G. Persoone. 1993. Ecotoxicological studies with the freshwater rotifer *Brachionus calyciflorus*. Resource competition between rotifers and daphnids under toxic stress. Sci. Total Env. Suppl. 1993:1059-1069.
Ferraro, S.P. 1980. Daily time of spawning of 12 fishes in the Peconic Bays, New York. Fish. Bull., U.S. 78:455-464.
Finn, R.N., & H.J. Fyhn. 1995. Metabolic fuels in developing cod and turbot embryos and larvae. ICES Mar. Sci. Symp. 201:70-73.
Finucane, J.H. 1969. Ecology of the pompano (*Trachinotus carolinus*) and the permit (*T. falcatus*) in Florida. Trans. Am. Fish. Soc. 98:478-486
Finucane, J.H. 1970. Pompano mariculture in Florida. Proc., Food-Drugs from the Sea Conf., Univ. Rhode Island, p 135-143.
Fischer, W. (ed.). 1973. FAO Species Identification Sheets for Fishery Purposes: Mediterranean and Black Sea (Fishing Area 37). Food Agric. Org. U.N., Rome, 2 vol.
Fischer, W. (ed.). 1977. FAO Species Identification Sheets for Fishery Purposes: Western Central Atlantic: Fishing Area 31. Food Agric. Org. U.N., Rome, 7 vol.
Fischer, W., & P.J.P. Whitehead (eds.). 1974. FAO Species Identification Sheets for Fishery Purposes: Eastern Indian Ocean (Fishing Area 57) and Western Central Pacific (Fishing Area 71). Food Agric. Org. U.N., Rome, 4 vol.
Fischer, W., G. Bianchi & W.B. Scott (eds.). 1981. FAO Species Identification Sheets for Fishery Purposes: Eastern Central Atlantic: Fishing Areas 34, 47 (in part). Food Agric. Org. U.N., Rome, 7 vol.
Fishelson, L. 1964. Observations on the biology and behaviour of Red Sea coral fishes. Bull. Sea Fish. Res. Sta. Haifa 37:11-26.
Fishelson, L. 1970. Behaviour and ecology of a population of *Abudefduf saxatilis* (Pomacentridae, Teleostei) at Eilat (Red Sea). Anim. Behav. 18:225-237.
Fishelson, L. 1975. Observations on behaviour of the fish *Meiacanthus nigrolineatus* Smith-Vaniz (Blenniidae) in nature (Red Sea) and in captivity. Austral. J. Mar. Freshw. Res. 26:329-341.
Fishelson, L. 1976. Spawning and larval development of the blennid fish, *Meiacanthus nigrolineatus* from the Red Sea. Copeia 1976:798-800.
Fishelson, L. 1978. Oogenesis and spawn-formation in the pygmy lion fish *Dendrochirus brachypterus* (Pteroidae). Mar. Biol. 46:341-348.
Fisher, D.J., D.T. Burton, L.T. Yonkos, S.D. Turley, B.S. Turley, G.P. Ziegler & E.J. Zillioux. 1994. Acute and short-term chronic effects of continuous and intermittent chlorination on *Mysidopsis bahia*

and *Menidia beryllina*. Env. Toxicol. Chem. 13:1525-1534.

Fitzgerald, R. 1995. Caviar dreams: California sturgeon farmers look for a roe-sy future. Seafood Leader 14:40-44.

Fitzpatrick, M.S., B.K. Suzumoto, C.B. Schreck & D. Oberbillig. 1984. Luteinizing hormone-releasing hormone analogue induces precocious ovulation in adult coho salmon (*Oncorhynchus kisutch*). Aquaculture 43:67-73.

Fitzpatrick, M.S., C.B. Schreck, F. Ratti & R. Chitwood. 1987. Viabilities of eggs stripped from coho salmon at various times after ovulation. Prog. Fish-Cult. 49:177-180.

Fivelstad, S., H. Haavik, G. Løvik & A.B. Olsen. 1997. Sublethal effects and safe levels of carbon dioxide in seawater for Atlantic salmon postsmolts (*Salmo salar* L.): ion regulation and growth. Aquaculture 160:305-316.

Fjæra, S.O., & G. Skogesal. 1993. Sub surface size-grading of fish. In: H. Reinertsen, L.A. Dahle, L. Jørgensen & K. Tvinnereim (eds.), Fish Farming Technology. Balkema, Rotterdam, p 395-399.

Fleming, I.A., & M.R. Gross. 1993. Breeding success of hatchery and wild coho salmon (*Oncorhynchus kisutch*) in competition. Ecol. Appl. 3:230-245.

Flood, L.P., M.J. Carvan III, L. Jaeger, D.L. Busbee, D.M. Gatlin III & W.H. Neill. 1996. Reduction in hepatic microsomal P-450 and related catalytic activity in farm-raised red drum. J. Aquat. Anim. Health 8:13-21.

Florida statute. 1993. Criteria for surface water quality classifications. Vol. 9, Rule 17-302.530, p 618-17 to 618-26E.

Flos, R., L. Tort & P. Torres. 1990. Mediterranean Aquaculture. Ellis Horwood, New York, 213 p.

Flos, R., L. Reig, P. Ambrosio & J.V. Fernández. 1997. Are exogenous proteolytic enzymes useful in feeds for sole (*Solea sengalensis*)? (abs.) Island Aquac. Trop. Aquac. Ann. Mtg., Eur. Aquac. Soc., p 130-131.

Foerster, R.E. 1968. The sockeye salmon, *Oncorhynchus nerka*. Fish. Res. Bd. Can. Bull. 162, 422 p.

Folke, C. 1988. Energy economy of salmon aquaculture in the Baltic Sea. Env. Management 12:525-537.

Folkvord, A., & H. Otterå. 1991. Effects of size distribution, feeding and light regime on growth and survival of cod juveniles in tanks. Eur. Aquac. Soc. Spec. Pub. (15):299-300.

Folmar, L.C., S. Bonomelli, T. Moody & J. Gibson. 1993. The effect of short-term exposure to three chemicals on the blood chemistry of the pinfish (*Lagodon rhomboides*). Arch. Env. Contam. Toxicol. 24:83-86.

Fonds, M., R. Cronie, A.D. Vethaak & P. van der Puyl. 1992. Metabolism, food consumption and growth of plaice (*Pleuronectes platessa*) and flounder (*Platichthys flesus*) in relation to fish size and temperature. Neth. J. Sea Res. 29:127-143.

Fontaine, C.T., & D.B. Revera. 1980. The mass culture of the rotifer, *Brachionus plicatilis*, for use as foodstuff in aquaculture. Proc. World Maric. Soc. 11:211-218.

Fontaine, M. 1976. Hormones and the control of reproduction in aquaculture. J. Fish. Res. Bd. Can. 33:922-939.

Formacion, M.J., R. Hori & T.J. Lam. 1993. Overripening of ovulated eggs in goldfish I. Morphological changes. Aquaculture 114:155-168.

Forsberg, O.I., S. Sanni & A. Bergheim. 1993a. Biological basis for landbased farming of Atlantic salmon: growth. In: H. Reinertsen, L.A. Dahle, L. Jørgensen & K. Tvinnereim (eds.), Fish Farming Technology. Balkema, Rotterdam, p 463-464.

Forsberg, O.I., S. Sanni & A. Bergheim. 1993b. Operational model for optimized landbased farming of Atlantic salmon. In: H. Reinertsen, L.A. Dahle, L. Jørgensen & K. Tvinnereim (eds.), Fish Farming Technology. Balkema, Rotterdam, p 401-407.

Forster, I.P., & H. Ogata. 1996. Growth and whole-body lipid content of juvenile red sea bream reared under different conditions of exercise training and dietary lipid. Fish. Sci. 62:404-409.

Forster, J. 1995. Salmon aquaculture: meeting the value expectations of a mass market. In: K.L Main & C. Rosenfeld (eds.), Culture of High-value Marine Fishes in Asia and the United States. Oceanic Inst., Honolulu, p 209-218.

Fortes, R.D. 1987. Induced spawning of sea bass (*Lates calcarifer*) in the Philippines. Austral. Ctr. Internat. Agric. Res. Proc. (20):123-125.

Fortier, L., & R.P. Harris. 1989. Optimal foraging and density-dependent competition in marine fish larvae. Mar. Ecol. Prog. Ser. 51:19-33.

Fortuny, A., A. Espinach Ros & V.G. Amutio. 1988. Hormonal induction of final maturation and ovulation in the sabalo, *Prochilodus platensis* Holmberg: treatments, latency and incubation times and viability of ovules retained in the ovary after ovulation. Aquaculture 73:373-381.

Forward, R.B., L.M. McKelvey, W.F. Hettler & D.E. Hoss. 1993. Swimbladder inflation of the Atlantic menhaden *Brevoortia tyrannus*. Fish. Bull., U.S. 91:254-259.

Foscarini, R. 1988. A review: intensive farming procedure for red sea bream (*Pagrus major*) in Japan. Aquaculture 72:191-246.

Foster, S.A. 1987. Diel and lunar patterns of reproduction in the Caribbean and Pacific sergeant major damselfishes *Abudefduf saxatilis* and *A. troschelii*. Mar. Biol. 95:333-343.

Fostier, A., & B. Jalabert 1982. Physiological basis of practical means to induce ovulation in fish. Proc.

Internat. Symp. Repr. Physiol. Fish, Wageningen, Netherlands, p 164-173.

Fowler, L.G. 1990. Feather meal as a dietary protein source during parr-smolt transformation in fall chinook salmon. Aquaculture 89:301-314.

Fowler, L.G., & J.L. Banks. 1990. Iodophor toxicity to eggs and fry of fall Chinook salmon during water hardening. Prog. Fish-Cult. 52:176-178.

Fowler, L.G., & J.L. Banks. 1991. A safe level of iodophor for treating eggs of fall Chinook salmon during water hardening. Prog. Fish-Cult. 53:250-251.

Frakes, T. 1992. Notes on breeding clownfish. SeaScope, Aquarium Systems, Mentor, Ohio, 9(winter):2.

Frakes, T., & F.H. Hoff, Jr. 1982. Effect of high nitrate-N on the growth and survival of juvenile and larval anemonefish, *Amphiprion ocellaris*. Aquaculture 29:155-158.

Francis, M.P. 1994. Duration of larval and spawning periods in *Pagrus auratus* (Sparidae) determined from otolith daily increments. Env. Biol. Fishes 39:137-152.

Fredricks, K.T., & J.G. Seelye. 1995. Flowing recirculated-water system for inducing laboratory spawning of sea lampreys. Prog. Fish-Cult. 57:297-301.

Frerichs, G.N., & R.J. Roberts. 1989. The bacteriology of teleosts. In: R.J. Roberts (editor), Fish Pathology. Bailliere Tindall, London, p 289-319.

Friligos, N. 1989. Hydrographic and plankton variability in the Saronikos Gulf. Toxicol. Env. Chem. 24:155-170.

Fritzsche, R.A. 1978. Development of Fishes of the Mid-Atlantic Bight: An Atlas of Egg, Larval and Juvenile Stages, Vol. 5. U.S. Fish Wildl. Serv., 340 p.

Fritzsche, R.A. 1984. Gasterosteiformes: development and relationships. Am. Soc. Ichthyol. Herpetol. Spec. Pub. (1):398-405.

Froese, R. 1998. Insulating properties of styrofoam boxes used for transporting live fish. Aquaculture 159:283-292.

Fry, F.E.J. 1971. The effect of environmental factors on the physiology of fish. In: W.S. Hoar and D.J. Randall (eds.), Fish Physiology, Vol. 6. Academic Press, New York, p 1-98.

Fryer, J.L., & C.N. Lannan. 1994. Rickettsial and chlamydial infections of freshwater and marine fishes, bivalves, and crustaceans. Zool. Stud. 33:95-107.

FSFRL. 1973. Report on experiments on the development of tuna culturing techniques (April, 1970~March, 1973). S Ser. Far Seas Fish. Res. Lab. No. 8, 165 p.

Fu, Y., A. Hagiwara & K. Hirayama. 1993. Crossing between seven strains of the rotifer *Brachionus plicatilis*. Nippon Suisan Gakkaishi 59:2009-2016.

Fuchs, J. 1978. Influence de la photoperiode sur la croissance et la survie de la larve et du juvenile de sole (*Solea solea*) en elevage. Aquaculture 15:63-74.

Fuchs, J. 1987. Growth of introduced larvae and fingerlings of sea bass (*Lates calcarifer*) in Tahiti. Austral. Ctr. Internat. Agric. Res. Proc. (20):189-192.

Fujii, M., & Y. Yone. 1976. Studies on nutrition of red sea bream—XIII. Effect of dietary linolenic acid and w3 polyunsaturated fatty acids on growth and feed efficiency. Bull. Jpn. Soc. Sci. Fish. 42:583-588.

Fujii, M., H. Nakayama & Y. Yone. 1976. Effect of w3 fatty acids on growth, feed efficiency and fatty acid composition of red sea bream (*Chrysophrys major*). Rep. Fish. Lab. Kyushu Univ., No. 3, p 65-86.

Fujita, S. 1955. On the development and prelarval stages of the filefish, *Monacanthus cirrhifer* Temminck et Schlegel. Sci. Bull. Fac. Agric. Kyushu Univ. 15(2):229-234. (in Jpn., Engl. sum.)

Fujita, S. 1957. On the egg development and prelarval stage of a damselfish, *Chromis notatus* (Temminck et Schlegel). Jpn. J. Ichthyol. 6:87-90. (in Jpn., Engl. abs.)

Fujita, S. 1966. Egg development, larval stages and rearing of the puffer, *Lagocephalus lunaris spadiceus* (Richardson). Jpn. J. Ichthyol. 13:162-168. (in Jpn., Engl. sum.)

Fujita, S., & T. Abe. 1992. Induction of ovarian maturation and development of eggs, larvae, and juveniles of the purple puffer *Takifugu porphyreus* reared in the laboratory. Bull. Jpn. Soc. Sci. Fish. 58:1627-1632. (in Jpn. Engl. abs.)

Fujita, S., & Y. Honma. 1991. Induction of ovarian maturation and devleopment of eggs, larvae and juveniles of the puffer, *Takifugu exascurus*, reared in the laboratory. Jpn. J. Ichthyol. 38:211-218. (in Jpn., Engl. abs.)

Fujita, S., & C. Kitajima. 1982. Occurrence of the lordotic deformity in larvae and juveniles of red sea bream. Kaiyou-Kagaku (Mar. Sci. Monthly) 107:721-727. (in Jpn.)

Fujita, S., & K. Nakahara. 1955. On the development of the egg and the prelarval stages of a scorpaenid fish, *Inimicus japonicus* (Cuvier et Valenciennes). Sci. Bull. Fac. Agric. Kyushu Univ. 15:223-228. (in Jpn., Engl. sum.)

Fujita, S., & M. Shinohara. 1986. Development of eggs, larvae and juveniles of the puffer, *Takifugu chrysops*, reared in the laboratory. Jpn. J. Ichthyol. 33:186-194.

Fujita, S., & T. Yogata. 1984. Induction of ovarian maturation, embryonic development and larvae and juveniles of the amberjack, *Seriola aureovittata*. Japan. J. Ichthyol. 30:426-434. (in Jpn., Engl. abst.)

Fujita, S., T. Watanabe & C. Kitajima. 1980. Nutritional quality of *Artemia* from different localities as a living feed for marine fish from the viewpoint of essential fatty acids. In: G. Persoone, P.

Sorgeloos, O. Roels & E. Jaspers (eds.), The Brine Shrimp *Artemia*, Vol. 3. Ecology, Culturing, Use in Aquaculture. Universa Press, Wetteren, Belgium, p 277-290.

Fujita, S., C. Kitajima & G. Hayashida. 1986. Induction of ovarian maturation and development of eggs, larvae and juveniles of the tonguefish, *Cynoglossus abbreviatus*, reared in the laboratory. Jpn. J. Ichthyol. 33:304-315.

Fujita, S., T. Abe & Y. Dotsu. 1992. Spawning season and spawners of the purple puffer *Takifugu porphyreus* off the coast of Shozanbetsu in northwestern Hokkaido. Bull. Jpn. Soc. Sci. Fish. 58:1621-1625. (in Jpn., Engl. abs.)

Fujita, T., M. Satake, S. Hikichi, M. Takeda, S. Shimeno, H. Kuwabara, W. Miki, K. Yamaguchi & S. Konosu. 1983a. Pigmentation of cultured yellowtail with krill oil. Bull. Jpn. Soc. Sci. Fish. 49:1595-1600.

Fujita, T., M. Satake, T. Watanabe, C. Kitajima, W. Miki, K. Yamaguchi & S. Konosu. 1983b. Pigmentation of cultured red sea bream with astaxanthin diester purified from krill oil. Bull. Jpn. Soc. Sci. Fish. 49:1855-1861.

Fujita T., W. Hamaura, A. Takemura & K. Takano. 1997. Histological observations of annual reproductive cycle and tidal spawning rhythm in the female porcupine fish *Diodon holocanthus*. Fish. Sci. 63:715-720.

Fujiya, M., S. Sakaguchi & O. Fukuhara. 1980. Training of fishes applied to ranching of red sea bream in Japan. In: J.E. Bardach, J.J. Magnuson, R.C. May & J.M. Reinhart (eds.) Fish Behavior and Its Use in the Capture and Culture of Fishes. ICLARM Conf. Proc. (5):200-209.

Fukuhara, O. 1985. Functional morphology and behavior of early life stages of red sea bream. Bull. Jpn. Soc. Sci. Fish. 51:731-743.

Fukuhara, O., & T. Fushimi. 1982. Development of fins and squamation in the percichthyid fish, *Lateolabrax japonicus*. Jpn. J. Ichthyol. 29:173-178.

Fukuhara, O., T. Nakagawa & T. Fukunaga. 1986. Larval and juvenile development of yellowtail reared in the laboratory. Bull. Jpn. Soc. Sci. Fish. 52:2091-2098.

Fukunaga, K., K. Nogami, Y. Yoshida, K. Hamazaki & K. Maruyama. 1990. Recent increase of *E. akaara* fry production amount and its problems at the Tamano Branch Station of the Japan Sea-Farming Association. Saibai Giken. 19(1):33-40. (in Jpn.)

Fukusho, K. 1991. Red sea bream culture in Japan. In: J.P. McVey (ed.), Handbook of Mariculture, Vol. II, Finfish Aquaculture. CRC Press, Boca Raton, Florida, p 73-87.

Fukusho, K., & K. Hirayama (eds.). 1992. The first live feed—*Brachionus plicatilis*. (orig. pub. in Jpn. by Koseisha Koseikaku, Tokyo, 1989), trans. by Asian Interchange Program, Oceanic Inst., Honolulu, Hawaii, 144 p.

Fukusho, K., & C. Kitajima. 1988. Environmental management of larval rearing of marine fishes—a short history of research to prevent lordosis in red sea bream, *Pagrus major*. Environmental Quality and Aquaculture Systems, NOAA Tech. Rep. NMFS 69, p 9-13.

Fukusho, K., & M. Okauchi. 1984. Seasonal isolation between two strains of rotifer *Brachionus plicatilis* in an eel-culture pond. Bull. Jpn. Soc. Sci. Fish. 50:909.

Fukusho, K., T. Arakawa & T. Watanabe. 1980. Food value of a copepod, *Tigriopus japonicus*, cultured with ω-yeast for larvae and juveniles of mud dab, *Limanda yokohamae*. Bull. Jpn. Soc. Sci. Fish. 46:499-503.

Fukusho, K., M. Okauchi, S. Nuraini, A. Tsujigado & T. Watanabe. 1984. Food value of a rotifer *Brachionus plicatilis*, cultured with *Tetraselmis tetrathele* for larvae of red seabream *Pagrus major*. Bull. Jpn. Soc. Sci. Fish. 50:1439-1444.

Fulks, W., & K.L. Main (eds.). 1991. Rotifer and Microalgae Culture Systems. Proc. of a U.S.—Asia Workshop. Oceanic Inst., Honolulu, 364 p.

Furuichi, M., & Y. Yone. 1980. Effect of dietary dextrin levels on the growth and feed efficiency, the chemical composition of liver and dorsal muscle, and the absorption of dietary protein and dextrin in fishes. Bull. Jpn. Soc. Sci. Fish. 46:225-229.

Furuichi,, M., K. Morita & Y. Yone. 1983. Effect of carboxymethylcellulose supplement on the absorption of dietary nutrients, and on the levels of blood sugar and plasma amino nitrogen. Bull. Jpn. Soc. Sci. Fish. 49:1367-1370.

Furuichi, M., H. Taira & Y. Yone. 1986. Availability of carbohydrate in nutrition of yellowtail. Bull. Jpn. Soc. Sci. Fish. 52:99-102.

Furuichi, M., G. Ito & Y. Yone. 1987. Effect of B-starch on growth, feed efficiency, and chemical components of liver, muscle, and blood of carp and red sea bream. Nippon Suisan Gakkaishi 53:1437-1442.

Furuita, H., T. Takeuchi, M. Toyota & T. Watanabe. 1996a. EPA and DHA requirements in early juvenile red sea bream using HUFA enriched *Artemia* nauplii. Fish. Sci. 62:246-251.

Furuita, H., T. Takeuchi, T. Watanabe, H. Fujimoto, S. Sekiya & K. Imaizumi. 1996b. Requirements of larval yellowtail for eicosapentaenoic acid, docosahexaenoic acid, and n-3 highly unsaturated fatty acid. Fish. Sci. 62:372-379.

Furuta, T., H. Honda, K. Kikuchi, N. Iwata & I. Sakaguchi. 1997. Intensive production of Japanese flounder, *Paralichthys olivaceus*, with a closed seawater recirculating culture system. World Aquac.

'97 Book of Abs., p 167.
Furuya, K. 1995. Current condition and subject of marine fish culture in Japan. In: K.L. Main & C. Rosenfeld (eds.), Culture of High-value Marine Fishes in Asia and the United States. Oceanic Inst., Honolulu, p 219-230.
Fyhn, H.J. 1989. First feeding of marine fish larvae: are free amino acids the source of energy? Aquaculture 80:111-120.
Fyhn, H.J. 1993. Multiple functions of free amino acids during embryogenesis in marine fishes. In: B.T. Walther & H.J. Fyhn (eds.), Physiological and Biochemical Aspects of Fish Development. Univ. Bergen, Norway, p 299-308.
Fyhn, H.J., & J.J. Govoni. 1995. Endogenous nutrient mobilization during egg and larval developement in two marine fishes—Atlantic menhaden and spot. ICES Mar. Sci. Symp. 201:64-69.
Fyhn, H.J., R.N. Finn, S. Helland, I. Rønnestad & E.R. Lømsland. 1993. Nutritional value of phyto- and zooplankton as live food for marine fish. In: H. Reinertsen, L.A. Dahle, L. Jørgensen & K. Tvinnereim (eds.), Fish Farming Technology. Balkema, Rotterdam, p 121-126.
Fynn-Aikins, K., S.G. Hughes & G.W. Vandenberg. 1995. Protein retention and liver aminotransferase activities in Atlantic salmon fed diets containing different energy sources. Comp. Biochem. Physiol. 111A:163-170.
Gadomski, D.M., & S.M. Caddell. 1991. Effects of temperature on early-life-history stages of California halibut *Paralichthys californicus*. Fish. Bull., U.S. 89:567-576.
Gadomski, D.M., & S.M. Caddell. 1996. Effects of temperature on the development and survival of eggs of four coastal California fishes. Fish. Bull., U.S. 94:41-48.
Gadomski, D.M., S.M. Caddell, L.R. Abbott & T.C. Caro. 1990. Growth and development of larval and juvenile California halibut, *Paralichthys californicus*, reared in the laboratory. Cal. Dept. Fish Game, Fish. Bull. 174:85-98.
Gaignon, J.-L. 1994. L'élevage intensif des juvéniles en eau douce. In: J.C. Guegin and P. Prouzet (eds.), Le Saumon Atlantique: Biologie et gestion de la Ressource. IFREMER, p 219-237. (in Fr.)
Gallagher, M.L. 1993. Preliminary evaluation of fish silage as a weaning diet for hybrid striped bass. Bull. Aquac. Assn. Can. 93-4:55-58.
Gallagher, M.L. 1994. The use of soybean meal as a replacement for fish meal in diets for hybrid striped bass (*Morone saxatilis* × *M. chrysops*). Aquaculture 126:119-127.
Gallagher, M.L. 1997. Apparent digestibility coefficients for some carbohydrates in diets for hybrid striped bass *Morone saxatilis* × *M. chrysops*. J. World Aquac. Soc. 28:429-431.
Gallagher, M.L., & G. Degani. 1988. Poultry meal and poultry oil as sources of protein and lipid in the diet of European eels (*Anguilla anguilla*). Aquaculture 73:177-187.
Gallagher, M.L., & M. LaDouceur. 1995. The use of blood meal and poultry products as partial replacements for fish meal in diets for juvenile palmetto bass (*Morone saxatilis* × *M. chrysops*). J. Appl. Aquac. 5:57-65.
Gallant, R.K., & M.A. McNiven. 1991. Cryopreservation of rainbow trout spermatozoa. Bull. Aquac. Assn. Can. 91:25-27.
Gallet de Saint Aurin, D., J.C. Raymond & V. Vianas. 1990. Marine finfish pathology: specific problems and research in the French West Indies. Advances in Tropical Aquaculture, AQUACOP, IFREMER, Actes Colloq. (9):143-160.
Gamble, J.C., P. MacLachlan, N.T. Nicoll & I.G. Baxter. 1981. Growth and feeding of Atlantic herring larvae reared in large plastic enclosures. Rapp. P.-v. Réun. Cons. Int. Explor. Mer 178:121-134.
Gapasin, R.S.J., H.J. Nelis, M. Chair & P. Sorgeloos. 1996. Drug assimilation in the tissue of European sea bass (*Dicentrarchus labrax*) fry delivered orally through bioencapsulation. J. Appl. Ichthyol. 12:39-42.
Garciá, B.G., J. Moreno & M.J. Rosique. 1993. Rate of feeding, growth and gross efficiency in juveniles of yellowtail (*Seriola dumerilii*). Eur. Aquac. Soc. Spec. Pub. (18):255-260.
Garcia, L.M.B. 1989. Spawning response of mature female sea bass, *Lates calcarifer* (Bloch), to a single injection of luteinizing hormone-releasing hormone analogue: effect of dose and initial oocyte size. J. Appl Ichthyol. 5:177-184.
Garcia, L.M.B. 1990. Advancement of sexual maturation and spawning of sea bass, *Lates calcarifer* (Bloch), using pelleted luteinizing hormone-releasing hormone analogue and 17alpha-methyltestosterone. Aquaculture 86:333-345.
Garcia, L.M.B. 1991. Spermiation response of mature rabbitfish, *Siganus guttatus* Bloch, to luteinizing hormone-releasing hormone analogue (LHRHa) injection. Aquaculture 97:291-299.
Garcia, L.M.B. 1992. Lunar synchronization of spawning in sea bass, *Lates calcarifer* (Bloch): effect of luteinizing hormone-releasing hormone analogue (LHRHa) treatment. J. Fish Biol. 40:359-370.
Garcia, L.M.B. 1993. Sustained production of milt in rabbitfish, *Siganus guttatus* Bloch, by weekly injection of luteinizing hormone-releasing hormone analogue (LHRHa). Aquaculture 113:261-267.
Garcia, L.M.B. 1994. Sea bass (*Lates calcarifer* Bloch) reproduction in captivity: a research update. Third Asian Fisheries Forum. Asian Fish. Soc., Manila, p 822-825.
Garcia, T. 1979. Desarrollo embrionario y primeros estadios larvales de *Micropogonias furnieri* obtenidos mediante tratamiento hormonal. Ciencias, Ser. 8, Investigaciones Marinas No. 46, 18 p.

García de la Banda, O. Chereguini & I. Rasines. 1992. Influencia de la adición de bacterias lácticas en el cultivo larvario del rodaballo (*Scophthalmus maximus* L.). Bol. Inst. Esp. Oceanogr. 8:247-254.
Garcia-Gallego, M., J. Bazoco, A. Sanz & M.D. Suarez. 1993. A comparative study of the nutritive utilization of dietary carbohydrates by eel and trout. Fish Nutrition in Practice, INRA, Paris, p 939-943.
Garcia-Vazquez, E., P. Moran, A.M. Pendas, J.I. Izquierdo & A.R. Linde. 1995. Effect of parental numbers on chromosome patterns found in artificially produced Atlantic salmon stocks. Trans. Am. Fish. Soc. 124:939-942.
Gardner, T.R. 1997. Commercial breeding of the dottybacks. SeaScope, Aquarium Systems, Mentor, Ohio, 14(summer):1-2.
Garnaud, J. 1951. Nouvelles donnees sur l'ethologie d'un Pomacentride: *Amphiprion percula* Lacepede. Bull. Inst. Oceanogr., Monaco, No. 998, 12 p. (in Fr., Engl. abs.)
Garnaud, J. 1957a. Breeding the black damselfish. The Aquarium 26(7):211-213.
Garnaud, J. 1957b. Ethologie de *Dascyllus trimaculatus* (Ruppell). Bull. Inst. Oceanogr., Monaco, No. 1096, 10 p. (in Fr., Engl. abs.)
Garratt, P.A. 1993. Spawning of riverbream, *Acanthopagrus berda*, in Kosi Estuary. S. Afr. Tydskr. Dierk. 28:26-31.
Garrett, Rod N. 1992. Northern Fisheries Research Centre, Cairns, Australia, pers. comm.
Garza, G., W.H. Bailey & J.L. Lasswell. 1978. Rearing of black drum in fresh water. Prog. Fish-Cult. 40:170.
Gatesoupe, F.-J. 1983. Weaning of sole, *Solea solea*, before metamorphosis, achieved with high growth and survival rates. Aquaculture 32:401-404.
Gatesoupe, F.J. 1994. Lactic acid bacteria increase the resistance of turbot larvae, *Scophthalmus maximus*, against pathogenic vibrio. Aquat. Living Resour. 7:277-282.
Gatesoupe, F.J., & P. Luquet. 1981. Practical diet for mass culture of the rotifer *Brachionus plicatilis*: application to larval rearing of sea bass, *Dicentrarchus labrax*. Aquaculture 22:149-163.
Gatesoupe, F.-J., & J.H. Robin. 1982. The dietary value for sea-bass larvae (*Dicentrarchus labrax*) of the rotifer *Brachionus plicatilis* fed with or without a laboratory-cultured alga. Aquaculture 27:121-127.
Gatesoupe, F.J., C. Léger, M. Boudon, R. Metailler & P. Luquet. 1977a. Alimentation lipidique du Turbot (*Scophthalmus maximus* L.). II. Influence de la supplementation en esters methyliques de l'acide linolenique et de la complementation en acides gras de la serie *n*3 sur la croissance. Ann. Hydrobiol. 8:247-254.
Gatesoupe, F.J., C. Léger, R. Metailler & P. Luquet. 1977b. Alimentation lipidique du Turbot (*Scophthalmus maximus* L.). I. Influence de la longueur de chaine des acides gras de la series ω3. Ann. Hydrobiol. 8:89-97.
Gatesoupe, F.-J., T. Arakawa & T. Watanabe. 1989. The effect of bacterial additives on the production rate and dietary value of rotifers as food for Japanese flounder, *Paralichthys olivaceus*. Aquaculture 83:39-44.
Gatesoupe, F.-J., J.-L. Zambonino Infante, C. Cahu & P. Quazuguel. 1997. Early weaning of seabass larvae, *Dicentrarchus labrax*: the effect on microbiota, with particular attention to iron supply and exozymes. Aquaculture 158:117-127.
Gatlin, D.M., III. 1995. Review of red drum nutrition. In: C. Lim & D.J. Sessa (eds.), Nutrition and Utilization Technology in Aquaculture. AOSC Press, Champaign, Ilinois, p 41-49.
Gatlin, D.M., III, J.P. O'Connell & J. Scarpa. 1991. Dietary zinc requirement of the red drum, *Sciaenops ocellatus*. Aquaculture 92:259-265.
Gaumet, F., G. Boeuf, A. Severe, A. Le Roux & N. Mayer-Gostan. 1995. Effects of salinity on the ionic balance and growth of juvenile turbot. J. Fish Biol. 47:865-876.
Gawlicka, A., L. McLaughlin, S.S.O. Hung & J. de la Noüe. 1996. Limitations of carrageenan microbound diets for feeding white sturgeon, *Acipenser transmontanus*, larvae. Aquaculture 141:245-265.
Gaylord, T.G., & D.M. Gatlin III. 1996. Determination of digestibility coefficients of various feedstuffs for red drum (*Sciaenops ocellatus*). Aquaculture 139:303-314.
Geets, A., E.W. Liewes & F. Ollevier. 1992. Efficacy of some antihelmintics against the swimbladder nematode *Anguillicola crassus* of eel *Anguilla anguilla* under saltwater conditions. Dis. Aquat. Org. 13:123-128.
Geiger, J.G. 1983. A review of pond zooplankton production and fertilization for the culture of larval and fingerling striped bass. Aquaculture 35:353-369.
Geiger, J.G., & C.J. Turner. 1990. Pond fertilization and zooplankton management techniques for production of fingerling striped bass and hybrid striped bass. In: R.M. Harrell, J.H. Kerby & R.V. Minton (eds.), Culture and Propagation of Striped Bass and Its Hybrids. Am. Fish. Soc., Bethesda, Maryland, p 79-98.
Gempesaw, C.M., II, J.R. Bacon & F.F. Wirth. 1992. Economies of pond size for hybrid striped bass growout. J. World Aquac. Soc. 23:38-48.
Gennari, L., A. Roncarati, P. Melotti & O. Mordenti. 1992. Effects of early weaning on growth and

survival of sea bass. Rivista Ital. Acquac. 27:133-143.

Gentles, A., & N.F. Haard. 1991. Pigmentation of rainbow trout with enzyme-treated and spray-dried *Phaffia rhodozyma*. Prog. Fish.-Cult. 53:1-6.

George, S.G. 1991. Biochemical and cytological assessments of metal toxicity in marine animals. In: R.W. Furness & P.S. Rainbow (eds.), Heavy Metals in the Marine Environment. CRC Press, Boca Raton, Florida, p 123-142.

Gerking, S.D. (ed.). 1978. Ecology of Freshwater Fish Production. Wiley, New York.

Gerking, S.D. 1982. The sensitivity of reproduction in fishes to stressful environmental conditions. Proc. Internat. Symp. Repr. Physiol. Fish, Wageningen, Netherlands, p 224-228.

Gershanovich, A.D., & G.A. Kiselev. 1993. Growth and haematological response of sturgeon hybrids Russian sturgeon (*Acipenser guldenstadti brandt*) × beluga (*Huso huso* L.) to protein and lipid contents in the diet. Comp. Biochem. Physiol. 106A:581-586.

GESAMP (IMO/FAO/UNESCO/WMO/WHO/IAEA/UN/UNEP Joint Group of Experts on the Scientific Aspects of Marine Pollution). 1993. Impact of Oil and Related Chemicals and Wastes on the Marine Environment. Reports and Studies GESAMP No. 50, 180 p.

Geyer, H.J., C.E. Steinberg, I. Scheunert, R. Bruggemann, W. Schutz, A. Kettrup & K. Rozman. 1993. A review of the relationship between acute toxicity (LC_{50}) of γ-hexachlorocyclohexane (γ-HCH, Lindane) and total lipid content of different fish species. Toxicology 83:169-179.

Ghittino, P. 1989. Nutrition and fish diseases. In: J.E. Halver (ed.), Fish Nutrition. Academic Press, San Diego, p 605-651.

Ghorab, H.M., A.R. Bayoumi, M.I. Bebars & A.A. Hassan. 1986. The reproductive biology of the grouper *Epinephelus chlorostigma* (Pisces, Serranidae) from the Red Sea. Bull. Inst. Oceanogr. Fish. ARE 12:13-33.

Gibbs, C.R. 1991. Advances in on-line water quality monitoring. Adv. World Aquac. 3:304-321.

Gibson, R.N. 1992. Tidally-synchronised behaviour in marine fishes. In: M.A. Ali (ed.), Rhythms in Fishes. Plenum Press, New York, p 63-81.

Gill, T.S., G. Leitner, S. Porta & A. Epple. 1993. Response of plasma cortisol to environmental cadmium in the eel, *Anguilla rostrata* Lesueur. Comp. Biochem. Physiol. 104C:489-495.

Gillet, C. 1991a. Egg production in an Arctic charr (*Salvelinus alpinus* L.) brood stock: effects of temperature on the timing of spawning and the quality of eggs. Aquat. Living Resour. 4:109-116.

Gillet, C. 1991b. Egg production in a whitefish (*Coregonus shinzi palea*) brood stock: effects of photoperiod on the timing of spawning and the quality of eggs. Aquat. Living Resour. 4:33-39.

Giovannini, G., L. Columbo, P. Bronzi & G. Arlati. 1991. Growth of hatchery produced juveniles of Italian sturgeon, *Acipenser naccarii* Bonaparte, reared intensively in fresh water. In: P. Williot (ed.), Acipenser. CEMAGREF, Bordeaux, p 401-404.

Girin, M. 1974. Nutrition de la larve de turbot [*Scophthalmus maximus* (L.)] avant la métamorphose. In: J.H.S. Blaxter (ed.), The Early Life History of Fish. Springer-Verlag, New York, p 739-746.

Girin, M. 1983. The Sparidae: a warmwater finfish family with world-wide mariculture potential. World Maric. Soc. Spec. Pub. (30 3-14.

Girin, M., M.H. Barahona-Fernandes & A. Le Roux. 1975. Larval rearing of sea bass (*Dicentrarchus labrax* (L.) with a high survival. ICES Paper C.M. 1975/G:14, 7 p.

Giudice, J.J., D.L. Gray & J.M. Martin. 1982. Manual for Bait Fish Culture in the South. Univ. Arkansas, Little Rock, 50 p.

Giwojna, P. 1989. Feeding butterflyfish. SeaScope, Aquarium Systems, Mentor, Ohio, 6(fall):2-3.

Giwojna, P. 1990. A Step-by-step Book about Seahorses. T.F.H., Neptune City, New Jersey, 64 p.

Giwojna, P. 1996. Seahorse nutrition. Freshw. Mar. Aquarium 19(11):72-76,78.

Glamuzina, B., J. Jug-Dujakovic & I. Katavic. 1989. Preliminary studies on reproduction and larval rearing of common dentex, *Dentex dentex* (Linnaeus 1758). Aquaculture 77:75-84.

Glasgow, H.B., Jr., J.M. Burkholder, D.E. Schmechel, P.A. Tester & P.A. Rublee. 1995. Insidiuous effects of a toxic estuarine dinoflagellate on fish survival and human health. J. Toxicol. Env. Health 46:501-522.

Glazebrook, J.S., & R.S.F. Campbell. 1987. Diseases of barramundi (*Lates calcarifer*) in Australia: a review. Austral. Ctr. Internat. Agric. Res. Proc. (20):204-206.

Glazebrook, J.S., & M.P. Heasman. 1992. Diagnosis and control of picorna-like virus infections in larval barramundi, *Lates calcarifer* Bloch. In: Diseases in Asian Aquaculture I. Asian Fish. Soc., Manila, p 267-272.

Goddard, S. 1996. Feed Management in Intensive Aquaculture. Chapman & Hall, New York, 194 p.

Goetz, F.W., & P. Thomas (eds.). 1995. Proc., Fifth International Symposium Reproductive Physiology of Fish. Univ. Texas, Austin, 389 p.

Goetz, F.W., D.C. Smith & S.P. Krickl. 1982. The effects of prostaglandins, phosphodiesterase inhibitors and cyclic AMP on ovulation of brook trout (*Salvelinus fontinalis*) oocytes. Gen. Comp. Endocrinol. 48:154-160.

Goetz, F.W., M. Ranjan, A.K. Berndtson & P. Duman. 1987. The mechanism and hormonal regulation of ovulation: the role of prostaglandins in teleost ovulation. Proc. Third Internat. Symp. Repr. Physiol. Fish, Memorial Univ. Newfoundland, St. John's, p 235-238.

Goldan, O., D. Popper & I. Karplus. 1997. Management of size variation in juvenile gilthead sea bream (*Sparus aurata*). I. Particle size and frequency of feeding dry and live food. Aquaculture 152:181-190.

Goldes, S.A., & S.L. Mead. 1995. Efficacy of iodophor disinfection against egg surface-associated infectious hematopoietic necrosis virus. Prog. Fish-Cult. 57:26-29.

Gomes, E.F., P. Rema & S.J. Kaushik. 1995. Replacement of fish meal by plant proteins in the diet of rainbow trout (*Oncorhynchus mykiss*): digestibility and growth performance. Aquaculture 130:177-186.

Gomes, Luiz A. 1992. Rua Dona Mariana, Rio de Janeiro, Brazil, pers. comm.

Gomez, A. 1977. Crecimiento, conversion, eficacia y mortalidad del pompano *Trachinotus goodei* confinado en estanques de concreto. Bol. Inst. Oceanogr. Univ. Oriente, 16:141-152.

Gomez, A. 1984a. Crecimiento del pampano *Trachinotus carolinus* vacunados contra la vibriosis en la Isla de Margarita, Venezuela. Memorias de la Asociacion Latinoamericana de Acuicultura 5:703-708.

Gomez, A. 1984b. Induccion del desove, desarrollo embrionario y larval de *Chaetodipterus faber* (Broussonet) (Pisces, Ephippidae) en la Isla de Margarita, Venezuela. Anales del Instituto de Investigaciones Marinas, Punta de Betin (14):85-104.

Gomez, A. In press. Induced spawning and larval rearing of spadefish *Chaetodipterus faber* at Margarita Island, Venezuela. Proc. Gulf Carib. Fish. Inst. 47.

Gomez, A., & F. Cervigon. 1979. Perspectivas del cultivo de peces marinos en el Caribe Sur y Nordeste de Sudamerica. Memorias del curso-seminario sobre los recursos del Mar Caribe, CIFCA-Colciencias, Cartagena, Colombia, Noviembre de 1979, 29 p.

Gomez, A., & F. Cervigon. 1984. Crecimiento de *Trachinotus carolinus*, *T. goodei* y *T. falcatus* (Pisces: Carangidae) en jaulas flotantes de 100 m^3 en la Isla de Margarita, Venezuela. Fundacion Cientifica Los Roques, Contribucion, No.16, 42 p.

Gomez, A., & F. Cervigon. 1987. Perspectivas del cultivo de peces marinos en el Caribe Sur y Nordeste de Sudamerica. Revista Latinoamericana de Acuicultura 34:40-52.

Gomez, A., & M. Scelzo. 1982. Polyculture experiments of pompano, *Trachinotus carolinus*, and spotted red shrimp, *Penaeus brasiliensis*, in concrete ponds, Margarita Island, Venezuela. J. World Maric. Soc. 13:146-153.

GomezGil-RS, B., F.A. Abreu-Grobois, J. Romero-Jarero & M. de los Herrera-Vega. 1994. Chemical disinfection of *Artemia* nauplii. J. World Aquac. Soc. 25:579-583.

Goksoyr, A., T.S. Solberg & B. Serigstad. 1991. Immunochemical detection of cytochrome P450IA1 induction in cod larvae and juveniles exposed to a water soluble fraction of North Sea crude oil. Mar. Poll. Bull. 22:122-127.

Goncharov, B.F., L.V. Igumnova, I.S. Polupan & E.A. Savelyeva. 1991a. Comparison of the effects of synthetic gonadotropin releasing hormone and sturgeon pituitaries on gamete maturation in sturgeons. Nauka 22:514-524. (in Rus., Engl. sum.)

Goncharov, B.F., L.V. Igumnova, I.S. Polupan & E.A. Savelieva. 1991b. Induced oocyte maturation, ovulation and spermiation in sturgeons (Acipenseridae) using synthetic analogue of gonadotropin-releasing hormone. In: P. Williot (ed.), *Acipenser*. CEMAGREF, Bordeaux, p 351-364.

Gonzalez, A.A. 1994. Fisheries and aquaculture: the case of turbot in Galicia. Eur. Aquac. Soc. Spec. Pub. (22):187-189.

Gonzales, B.J., N. Taniguchi & O. Okamura. 1997. Spawning cycle of two dragonet species, *Calliurichthys japonicus* and *Repomucenus huguenini*, in Tosa Bay, southern Japan. Fish. Sci. 63:15-21.

Gonzalez, E., T. Damas, N. Millares & M. Borrero. 1979. Desove inducido en el caballerote (*Lutjanus griseus* Linne, 1758) en condiciones de laboratorio. Revista Cubana de Investigaciones Pesqueras 4:43-64.

Goodsell, A., D. Wikeley & L. Searle. 1996. Histological investigation of swim-bladder morphology and inflation in cultured larval striped trumpeter (*Latris lineata*)(Teleostei, Latridae). Mar. Freshw. Res. 47:251-254.

Gopakumar, G., N.G. Pillai & T.A. Omana. 1991. The fishery characteristics and biology of mackerel at Vizhinjam. J. Mar. Biol. Assoc. India 33:107-114.

Gopal, V., S. Prabakaran & P.R. Balasubramanian. 1997. Effcet of a biogas-plant effluent-based pelleted diet on the growth of *Oreochromis mossambicus* fingerlings. Bioresource Technol. 58:315-317.

Gordin, H., & Y. Zohar. 1978. Induced spawning of *Sparus aurata* (L.) by means of hormonal treatments. Ann. Biol. Anim. Biochem. Biophys. 18:985-990.

Gordina, A.D., & T.I. Bagnyukova. 1993. On the spawning swordfish, *Xiphias gladius*, in the Black Sea. J. Ichthyol. 33:144.

Gorospe, J., & K. Nakamura. 1996. Associated bacterial microflora in *Artemia*-rice bran culture. Bamidgeh 48:99-107.

Gorospe, J.N., K. Nakamura, M. Abe & S. Higashi. 1996. Nutritional contribution of *Pseudomonas* sp. in *Artemia* culture. Fish. Sci. 62:914-918.

Goshorn, D.M., & C.E. Epifanio. 1991. Development, survival, and growth of larval weakfish at different prey abundances. Trans. Am. Fish. Soc. 120:693-700.

Gouillou-Coustans, M.F., & J. Guillaume. 1993. Effect of a non specific stressor on the symptoms of ascorbic acid deficiency in turbot (*Scophthalmus maximus*). Fish Nutrition in Practice. INRA, Paris, p 209-213.

Gould, W.R. 1965. The biology and morphology of *Acyrtops beryllinus*, the emerald clingfish. Bull. Mar. Sci. 15:165-188.

Goulet, D. 1997. Reproductive behavior and spawning success of female *Amblyglyphidodon leucogaster* (Pisces: Pomacentridae) from the Red Sea. Env. Biol. Fishes 50:49-60.

Gousset, B. 1990. European eel (*Anguilla anguilla* L.) farming technologies in Europe and in Japan: application of a comparative analysis. Aquaculture 87:209-235.

Govoni, J.J., G.W. Boehlert & Y. Watanabe. 1986. The physiology of digestion in fish larvae. Env. Biol. Fishes 16:59-77.

Gowen, R.J., & N.B. Bradbury. 1987. The ecological impact of salmonid farming in coastal waters: a review. Oceanogr. Mar. Biol. Ann. Rev. 25:563-575.

Grabda, J. 1991. Marine Fish Parasitology. VCH, Weinheim, Fed. Rep. Germany, 306 p.

Grabner, M., W. Wieser & R. Lackner. 1981/1982. The suitability of frozen and freeze-dried zooplankton as food for fish larvae: a biochemical test program. Aquaculture 26:85-94.

Grace, G.R., & R.H. Piedrahita. 1993. Carbon dioxide control with a packed column aerator. In: J.-K. Wang (ed.), Techniques for Modern Aquaculture. Am. Soc. Agric. Eng., St. Joseph, Michigan, p 496-505.

Grace, G.R., & R.H. Piedrahita. 1994. Carbon dioxide control. In: M.B. Timmons & T.M. Losordo (eds.), Aquaculture Water Reuse Systems: Engineering Design and Management. Elsevier, Amsterdam, p 209-234.

Grahl-Madsen, E., & Ø. Lie. 1997. Effects of different levels of vitamin K in diets for cod (*Gadus morhua*). Aquaculture 151:269-274.

Grant, B.F., P.A. Seib, M.-L. Liao & K.E. Corpron. 1989. Phosphorylated L-ascorbic acid: a stable form of vitamin C for aquaculture feeds. J. World Aquac. Soc. 20:143-157.

Grasshoff, K., M. Ehrhardt & K. Kremling. 1983. Methods of Seawater Analysis. Verlag Chemie, Weinheim, Germany, 419 p.

Gratzek, J.B., & G.C. Blasiola. Checklists, quarantine procedures, and calculations of particular use in fish health management. In: J.B. Gratzek & J.R. Matthews (eds.). Aquariology: The Science of Fish Health Management. Tetra Press, Morris Plains, New Jersey, p 301-315.

Gratzek, J.B., & J.R. Matthews (eds.). 1992. Aquariology: the Science of Fish Health Management. Tetra Press, Morris Plains, N.J., 330 p.

Gray, D.L. 1988. Baitfish: feeds and feeding practices. Southern Reg. Aquac. Ctr., IFAS, Univ. Florida, Gainesville, Pub. No. 121, 4 p.

Gray, J.D., T.L. King & R.L. Colura. 1991. Effect of temperature and hypersalinity on hatching success of spotted seatrout eggs. Prog. Fish-Cult. 53:81-84.

Gray, Jennifer. 1993. Bermuda Government Aquarium, Museum, and Zoo, Flatts, pers. comm.

Greco, S., G. Lo Paro, D. Caridi, F. Perdichizzi, S. Cammaroto, V. Micale & L. Genovese. 1993a. Controlled spawning and larval development in the sharpsnout sea bream (*Diplodus puntazzo*, Sparidae). Eur. Aquac. Soc. Spec. Pub. (18):185-188.

Greco, S., D. Caridi, S. Cammaroto & L. Genovese. 1993b. Preliminary studies on artificial feeding of amberjack fingerlings. Eur. Aquac. Soc. Spec. Pub. (18):247-254.

Greenblatt, M., C.L. Brown, M. Lee, S. Dauder & H.A. Bern. 1989. Changes in thyroid hormone levels in eggs and larvae and in iodide uptake by eggs of coho and chinook salmon, *Oncorhynchus kisutch* and *O. tshawytscha*. Fish Physiol. Biochem. 6:261-278.

Greenwood, P.H. 1975. J.R. Norman: A History of Fishes, Third Ed. Wiley, New York, 467 p.

Grey, D.L. 1987. An overview of *Lates calcarifer* in Australia and Asia. In: J.W. Copland & D.L. Grey (eds.), Management of Wild and Cultured Sea Bass/Barramundi (*Lates calcarifer*). Australian Ctr. Internat. Agric. Res. (20):15-21.

Griffith, R.W. 1974. Environment and salinity tolerance in the genus *Fundulus*. Copeia 1974:319-331.

Grisez, L., M. Chair, P. Sorgeloos & F. Ollevier. 1996. Mode of infection and spread of *Vibrio anguillarum* in turbot *Scophthalmus maximus* larvae after oral challenge through live feed. Dis. Aquat. Org. 26:181-187.

Grisez, L., J. Reyniers, L. Verdonck, J. Swings & F. Ollevier. 1997. Dominant intestinal microflora of sea bream and sea bass larvae, from two hatcheries, during larval development. Aquaculture 155:387-399.

Groot, C., & L. Margolis (eds.). 1991. Pacific Salmon Life Histories. UBC Press, Vancouver, 564 p.

Grover, J.J. 1998. Feeding habits of pelagic summer flounder, *Paralichthys dentatus*, larvae in oceanic and estuarine habitats. Fish. Bull., U.S. 96:248-257.

Grung, M., Y.S. Svendsen & S. Liaaen-Jensen. 1993. The carotenoids of eggs of wild and farmed cod (*Gadus morhua*). Comp. Biochem. Physiol. 106B:237-242.

Gueguen, J.C., & P. Prouzet. 1994. Le Saumon Atlantique: Biologie et Gestion de la Ressource; (Atlantic Salmon: Biology and Resource Management). IFREMER, Plouzane, 330 p.

Guérin, J.-P., & M. Rieper-Kirchner. 1991. Influence of three bacteria strains on the population dynamics of *Tisbe holothuriae* (Copepoda, Harpacticoida). Helgoländer Meeresunters 45:493-511.

Guerrero, R.D., III. 1996. Aquaculture in the Philippines. World Aquac. 27(1):7-13.

Guerrero, R.D., III. 1995. Evaluation of home-made feeds used for commercial tilapia production in the Philippines. Proc. 7th Biennial Conf. Internat. Inst. Fish. Econ. Trade, Taipei, Taiwan, R.O.C., July 18-21, 1994, p 254-257.

Guerrero, R.D., III & M.P. Garcia, Jr. 1991. Advances in Finfish and Shellfish Mariculture. Phil. Council Aquat. Mar. Res. Dev., Book Ser. No. 12, 172 p.

Guillaume, J., Coustans, M.-F., R. Métailler, J. Person-Le Ruyet & J. Robin. 1991. Flatfish, turbot, sole, and plaice. In: R.P. Wilson (ed.), Handbook of Nutrient Requirements of Finfish. CRC Press, Boca Raton, Florida, p 77-82.

Guillén, J.L., M. Endo, J.F. Turnbull, H. Kawatsu, R.H. Richards & T. Aoki. 1993. Depressed growth rate and damage to the cartilage of red sea bream larvae associated with exposure to ammonia. Nippon Suisan Gakkaishi 59:1231-1234.

Guillén, J.L., M. Endo & H. Kawatsu. 1994a. Ammonia toxicity to marine fish larvae for aquaculture purposes in southern Japan. International Symposium on Aquatic Animal Health, Univ. California, Davis, p W-15.6.

Guillén, J.L., M. Endo, J.F. Turnbull, H. Kawatsu, R.H. Richards & T. Aoki. 1994b. Skin responses and mortalities in the larvae of Japanese croaker exposed to ammonia. Fish. Sci. 60:547-550.

Guitart M., D., & M. Juarez F. 1966. Desarrollo embrionario y primeros estadios larvales de la cherna criolla, *Epinephelus striatus* (Bloch) (Perciformes: Serranidae). Academia de Ciencias de Cuba, Estudios de Institute Oceanologia, La Habana 1:35-45 (in Sp.).

Gulbrandsen, J. 1991a. Effect of light intensity on feeding success of Atlantic halibut larvae. Eur. Aquac. Soc. Spec. Pub. (15):336-338.

Gulbrandsen, J. 1991b. Functional response of Atlantic halibut larvae related to prey density and distribution. Aquaculture 94:89-98.

Gunnarsson, J. 1993. Bridgestone Hi-Seas fish cage: design and documentation. In: H. Reinertsen, L.A. Dahle, L. Jørgensen & K. Tvinnereim (eds.), Fish Farming Technology. Balkema, Rotterdam, p 199-201.

Gutierrez, M.A., T. Miyazaki, H. Hatta & M. Kim. 1993. Protective properties of egg yolk IgY containing anti-*Edwardsiella tarda* antibody against paracolo disease in the Japanese eel, *Anguilla japonica* Temminck & Schlegel. J. Fish Dis. 16:113-122.

Gutierrez, M.A., T. Miyazaki, K. Mabe, H. Hatta, M. Kim & S. Akashi. 1994. Field tests of passive immunization with anti-*Edwardsiella tarda* egg yolk antibody in the Japanese eel. Third Asian Fisheries Forum. Asian Fish. Soc., Manila, p 317-319.

Gwo, J.-C. 1982. Studies on the Freezing-thawing Damages of Fish semen, Especially on the Morphological Changes and Major Cation Flux of Spermatozoa. M.S. Thesis, Univ. Tokyo, Japan, 74 p. (in Jpn.)

Gwo, J.-C. 1994. Some factors affecting the cryopreservation of the spermatozoa of *Epinephelus malabaricus* and *Plectropomus leopardus* (Teleostei, Serranidae). Third Asian Fisheries Forum. Asian Fish. Soc., Manila, p 524-527.

Gwo, J.-C., K. Strawn, M.T. Longnecker & C.R. Arnold. 1991a. Cryopreservation of Atlantic croaker spermatozoa. Aquaculture 94:355-375.

Gwo, J.-C., K. Strawn & C.R. Arnold. 1991b. Induced spawning of Atlantic croaker (Family: Sciaenidae) with a single injection of luteinizing hormone-releasing hormone analog. Abs. in program book, Finfish Hatchery in Asia '91. Tungkang Mar. Lab., Tungkang, Taiwan, p 34.

Gwo, J.-C., H. Kurokura & R. Hirano. 1993. Cryopreservation of spermatozoa from rainbow trout, common carp, and marine puffer. Bull. Jpn. Soc. Sci. Fish. 59:777-782.

Haaland, H., K. Ladstein, G. Rosenlund & M.A. Oliveira. 1993. Stability of vitamins and pigment in extruded fish feed. In: H. Reinertsen, L.A. Dahle, L. Jørgensen & K. Tvinnereim (eds.), Fish Farming Technology. Balkema, Rotterdam, p 477-479.

Haard, N.F. 1992. Control of chemical composition and food quality attributes of cultured fish. Food Res. Internat. 25:289-307.

Habib, M.A.B., & M.S. Rahman. 1993. Culture of rotifers with reference to some physico-chemical properties of water in nursery pond. Hydrobiologia 255/256:177-184.

Hackinger, A. 1967. Anemonenfische—im Aquarium gezüchtet. Die Aufzucht von *Amphiprion bicinctus* im Aquarium. Aquarien Mag. (Apr):137-141.

Hadley, C.G., M.B. Rust, J.P. Van Eenennaam & S.I. Doroshov. 1987. Factors influencing initial swim bladder inflation by striped bass. Am. Fish. Soc. Symp. (2):164-169.

Hagen, N. 1989. Startfôring av rødspette, *Pleuronectes platessa*. Thesis, Dept. Zoology, Univ. Trondheim, Norway, 64 p.

Hagen, N. 1993. Effect of different prey- and larval densities on the gut content of plaice (*Pleuronectes platessa* L.) at initial feeding. In: B.T. Walther & H.J. Fyhn (eds.), Physiological and Biochemical Aspects of Fish Development. Univ. Bergen, Norway, p 180-182.

Hagiwara, A. 1994. Practical use of rotifer cysts. Bamidgeh 46:13-21.

Hagiwara, A., & K. Hirayama. 1993. Preservation of rotifers and its application in the finfish hatchery. In: C.-S. Lee, M.-S. Su & I C. Liao (eds.), Finfish Hatchery in Asia. Tungkang Mar. Lab. Conf. Proc.

(3):61-71.
Hagiwara, A., A. Hino & R. Hirano. 1985. Combined effects of environmental conditions on the hatching of fertilized eggs of the rotifer *Brachionus plicatilis* collected from an outdoor pond. Bull. Jpn. Soc. Sci. Fish. 51:755-758.
Hagiwara, A., A. Hino & R. Hirano. 1988. Effects of temperature and chlorinity on resting egg formation in the rotifer *Brachionus plicatilis*. Nippon Suisan Gakkaishi 54:569-575.
Hagiwara, A., K. Hamada, A. Nishi, K. Imaizumi & K. Hirayama. 1993a. Dietary value of neonates from rotifer *Brachionus plicatilis* resting eggs for red sea bream larvae. Nippon Suisan Gakkaishi 59:99-104.
Hagiwara, A., K. Hamada, A. Nishi, K. Imaizumi & K. Hirayama. 1993b. Mass production of rotifer *Brachionus plicatilis* resting eggs in 50 m^3 tanks. Nippon Suisan Gakkaishi 59:93-98.
Hagiwara, A., K. Hamada, S. Hori & K. Hirayama. 1994. Increased sexual reproduction in *Brachionus plicatilis* (Rotifera) with the addition of bacteria and rotifer extracts. J. Exp. Mar. Biol. Ecol. 181:1-8.
Hagiwara, A., M.-M. Jung, T. Sato & K. Hirayama. 1995a. Interspecific relations between marine rotifer *Brachionus rotundiformis* and zooplankton species contaminating in the rotifer mass culture tank. Fish. Sci. 61:623-627.
Hagiwara, A., C.-S. Lee & D.J. Shiraishi. 1995b. Some reproductive characteristics of the broods of the harpacticoid copepod *Tigriopus japonicus* cultured in different salinities. Fish. Sci. 61:618-622.
Hagiwara, A., M.D. Balompapueng, N. Munuswamy & K. Hirayama. 1997. Mass production and preservation of the resting eggs of the euryhaline rotifer *Brachionus plicatilis* and *B. rotundiformis*. Aquaculture 155:223-230.
Hagood, R.W. 1984. *Coryphaena* (dolphin-fish) aquaculture. in: T.D. Sleeter (ed.), Assessment of the potential for aquaculture in Bermuda. Bermuda Biol. Sta. Spec. Pub. (27):143-150.
Haimovici, M., & L.C. Krug. 1992. Alimentacao e reproducao da enchova *Pomatomus saltatrix* no litoral sul do Brasil. Rev. Brasil. Biol. 52:503-513. (in Por., Eng. abs.)
Haldorson, L., & M. Love. 1991. Maturity and fecundity in the rockfishes, *Sebastes* spp., a review. Mar. Fish. Rev. 53:25-31
potential for aquaculture in Bermuda. Bermuda Biol. Sta. Spec. Pub. (27):143-150.
Hall, A.T., & J.T. Oris. 1991. Anthracene reduces reproductive potential and is maternally transferred during long-term exposure in fathead minnows. Aquat. Toxicol. 19:249-264.
Hall, L.W., Jr. 1991. A synthesis of water quality and contaminants data on early life stages of striped bass, *Morone saxatilis*. Rev. Aquat. Sci. 4:261-288.
Hall, L.W., Jr., D.T. Burton & L.B. Richardson. 1981. Comparison of ozone and chlorine toxicity to the developmental stages of striped bass, *Morone saxatilis*. Can. J. Fish. Aquat. Sci. 38:752-757.
Hall, L.W., Jr., M.C. Ziegenfuss, S.J. Bushong, J.A. Sullivan & M.A. Unger. 1992. In-situ striped bass (*Morone saxatilis*) contaminant and water-quality studies in the Potomac River and upper Chesapeake Bay in 1989. Aquat. Toxicol. 22:181-221.
Hall, L.W., Jr., M.C. Ziegenfuss, R.D. Anderson, T.D. Spittler & H.C. Leichtweis. 1994. Influence of salinity on atrazine toxicity to a Chesapeake Bay copepod (*Eurytemora affinis*) and fish (*Cyprinodon variegatus*). Estuaries 17:181-186.
Hall, L.W., Jr., M.C. Ziegenfuss, R.D. Anderson & B.L. Lewis. 1995. The effect of salinity on the acute toxicity of total and free cadmium to a Chesapeake Bay copepod and fish. Mar. Poll. Bull. 30:376-384.
Hall, W.S., & G.W. Pulliam. 1995. An assessment of metals in an estuarine wetlands ecosystem. Arch. Env. Contam. Toxicol. 29:164-173.
Hallet, S.L., & F.R. Roubal. 1995. Experiments on the infection dynamics of *Caligus epidemicus* (Copepoda: Caligidae) on the small marine fish, *Ambassis marianus* (Günther). J. Fish Dis. 18:59-66.
Halver, J.E. 1957. Nutrition of salmonoid fishes. III. Water soluble vitamin requirements of chinook salmon. J. Nutr. 62:225-243.
Halver, J.E. 1985. Recent advances in vitamin nutrition and metabolism in fish. In: C.B. Cowey, A.M. Mackie & J.G. Bell (eds.), Nutrition and Feeding in Fish. Academic Press, San Diego, California, p 415-429.
Halver, J.E. (ed.). 1989. Fish Nutrition. Academic Press, San Diego, 798 p.
Hamada, K., A. Hagiwara & K. Hirayama. 1993. Use of preserved diet for rotifer *Brachionus plicatilis* resting egg formation. Nippon Suisan Gakkaishi 59:85-91.
Hamamoto, S., S. Kumagai, K. Nosaka, S. Manabe, A. Kasuga & Y. Iwatsuki. 1992. Reproductive behavior, eggs and larvae of a lutjanid fish, *Lutjanus stellatus*, observed in an aquarium. Jpn. J. Ichthyol. 39:219-228.
Hamamoto, T., K. Yokogawa & M. Tochino. 1986a. Several problems on cultivating the parent fish of red spotted grouper, *Epinephelus akaara* (Temminck et Schlegel), and judging the qualities of the eggs obtained from them. Bull. Kagawa Pref. Fish. Exp. Sta. (2):13-22. (in Jpn., Eng. abs.)
Hamamoto, T., S. Manabe, K. Kasuga & K. Nozaka. 1986b. Spawning and larval rearing of grouper, *Epinephelus salmonoides* (Lacepede). Sabai Giken 15:143-155. (in Jpn.)
Hamatsu, T., K. Yabuki & K. Watanabe. 1993. Maturity and fecundity of walleye pollock *Theragra chalcogramma* in the Pacific coast of southeastern Hokkaido. Bull. Hokkaido Natl. Fish. Res. Inst.

(57):43-51. (in Jpn., Eng. abs.)
Hamilton, S.J., & K.J Buhl. 1990. Acute toxicity of boron, molybdenum, and selenium to fry of chinook salmon and coho salmon. Arch. Env. Contam. Toxicol. 19:366-373.
Hamilton, S.J., & P.M. Mehrle. 1986. Metallothionein in fish: review of its importance in assessing stress from metal contaminants. Trans. Am. Fish. Soc. 115:596-609.
Hammer, H.M. 1996. The Banggai cardinalfish. Freshw. Mar. Aquarium 19(12):112,115-116,118-119.
Han, K.-n., & M. Furuichi. 1995. Effect of dietary lipid level on the growth, feed efficiency and body composition of young puffer fish, *Takifugu rubripes*. Sci. Bull. Fac. Agr. Kyushu Univ. 50:25-29. (in Jpn., Eng. sum.)
Handy, R.D. 1994. Intermittent exposure to aquatic pollutants: assessment, toxicity and sublethal responses in fish and invertebrates. Comp. Biochem. Physiol. 107C:171-184.
Handy, R.D., & M.G. Poxton. 1993. Nitrogen pollution in mariculture: toxicity and excretion of nitrogenous compounds by marine fish. Rev. Fish Biol. Fish. 3:205-241.
Hansen, G.H., E. Strøm & J.A. Olafsen. 1992. Effect of different holding regimens on the intestinal microflora of herring (*Clupea harengus*) larvae. Appl. Env. Microbiol. 58:461-470.
Hansen, L.G. 1994. Halogenated aromatic compounds. In: L.G. Cockerham & B.S. Shane (eds.), Basic Environmental Toxicology. CRC Press, Boca Raton, Florida, p 199-230.
Hansen, L.P., & T.A. Bakke. 1989. Flukes, genetics and escapees. Atlantic Salmon J. 38:26-29.
Hanson, H.M. 1992. A flexible temperature and environmental controller. Prog. Fish-Cult. 54:130-131.
Happe, A., & Y. Zohar. 1988. Self-fertilization in the protandrous hermaphrodite *Sparus aurata*: development of the technology. In: Y. Zohar & B. Breton (eds.), Reproduction in Fish. Basic and Applied Aspects in Endocrinology and Genetics. Colloq. INRA (44):177-180.
Hara, S., J.T. Canto, Jr. & J.M.E. Almendras. 1982. A comparative study of various extenders for milkfish *Chanos chanos* (Forsskal), sperm preservation. Aquaculture 28:339-346.
Hara, S., M.N. Duray, M. Parazo & Y. Taki. 1986. Year-round spawning and seed production of the rabbitfish, *Siganus guttatus*. Aquaculture 59:259-272.
Harada, K. 1985a. Feeding attraction activities of amino acids and lipids for juvenile yellowtail. Bull. Jpn. Soc. Sci. Fish. 51:453-459.
Harada, K. 1985b. Feeding attraction activities of amino acids and nitrogenous bases for oriental weatherfish. Bull. Jpn. Soc. Sci. Fish. 51:461-466.
Harada, K. 1990. Attraction activities of spices for oriental weatherfish and yellowtail. Nippon Suisan Gakkaishi 56:2029-2033.
Harada, K., & T. Miyasaki. 1997. Attractivity of exotic fruit fleshes for the abalone, the oriental weatherfish, and the yellowtail. Fish. Sci. 63:671-675.
Harada, K., T. Miyasaki & A. Karimata. 1996. Attraction activities of terrestrial vegetable extracts for aquatic animals. Fish. Sci. 62:675-682.
Harada, T. 1974. Environmental parameters on maturation and spawning. In: Jpn. Soc. Sci. Fish. (eds.), Maturation and Spawning of Fish - Basis and Its Application. Koseisha Kosekaku 7:66-75.
Harada, T. 1980. Progress and future prospects in tuna culturing studies. Japan Tuna Research Conference, Shimizu, Japan, February 6-7, 1980. Fish. Agency Japan, Far Seas Fish. Res. Lab., p 50-58. (trans. from Jpn. to Eng. by Tamio Otsu, Southwest Fish. Ctr., NMFS, NOAA, Honolulu, Hawaii, Oct 1980)
Harada, T., K. Mizuno, O. Murata, S. Miyashita & H. Hurutani. 1971. On the artificial fertilization and rearing of larvae in yellowfin tuna. Mem. Fac. Agric. Kinki Univ. (4):145-151.
Harada, T., O. Murata & S. Miyashita. 1973. On the artificial fertilization and rearing of larvae in hirasoda, *Auxis thazard*. Mem. Fac. Agric. Kinki Univ. (6):109-112.
Harada, T., O. Murata & S. Miyashita. 1974. On the artificial fertilization and rearing of larvae in bonito. Mem. Fac. Agric. Kinki Univ. (7):1-4.
Harada, T., S. Miyashita & H. Yoneshima. 1980a. Effect of water temperature on yellowfin tuna hatching. Mem. Fac. Agric., Kinki Univ. (13):29-32.
Harada, T., O. Murata & S. Oda. 1980b. Rearing of and morphological changes in larvae and juveniles of yellowfin tuna. Mem. Fac. Agric. Kinki Univ. (13):33-36, pl 1-2. (in Jpn., Eng. abs.)
Harboe, T., I. Huse & G. Øie. 1994. Effects of egg disinfection on yolk sac and first feeding stages of halibut (*Hippoglossus hippoglossus* L.) larvae. Aquaculture 119:157-165.
Hard, G.C., R. Williams & J. Lee. 1979. Histopathology of superficial fish 'tumours' found during a cancer survey of demersal species in Port Philip Bay, Victoria, Australia. J. Fish Dis. 2:455-467.
Hardy, A. 1959. The Open Sea: It's Natural History. Part II. Fish and Fisheries. Houghton Mifflin Co., Boston, 322 p.
Hardy, J.D., Jr. 1978. Development of Fishes of the Mid-Atlantic Bight: An Atlas of Egg, Larval and Juvenile Stages, Vol. 2. U.S. Fish Wildl. Serv., 458 p.
Hardy, R.W. 1989. Diet preparation. In: J.E. Halver (ed.), Fish Nutrition. Academic Press, San Diego, p 475-548.
Hardy, R.W. 1991. Pacific salmon, *Oncorhyncus* spp. In: R.P. Wilson (ed.), Handbook of Nutrient Requirements of Finfish. CRC Press, Boca Raton, Florida, p 105-121.
Hardy, R.W. 1996. Alternate protein sources for salmon and trout diets. Anim. Feed Sci. Technol. 59:71-80.
Hardy, R.W. 1998. Fish nutrition: increasing aquaculture production. Aquaculture Mag. 24(3):86-90.
Hargis, W.J., Jr. 1991. Disorders of the eye in finfish. Ann. Rev. Fish Dis. 1:95-117.

Hargrave, B.T., D.E. Duplisea, E. Pfeiffer & D.J. Wildish. 1993. Seasonal changes in benthic fluxes of dissolved oxygen and ammonium associated with marine cultured Atlantic salmon. Mar. Ecol. Prog. Ser. 96:249-257.

Hargreaves, J.A., & D.E. Alston (eds.). 1991. Status and Potential of Aquaculture in the Caribbean. Adv. World Aquac., Vol. 5, 274 p.

Hargrove, L.L., P.W. Westerman & T.M. Losordo. 1996. Nitrification in three-stage and single-stage floating bead biofilters in a laboratory-scale recirculating aquaculture system. Aquacultural Eng. 15:67-80.

Harmin, S.A., & L.W. Crim. 1992. Gonadotropic hormone-releasing hormone analog (GnRH-A) induced ovulation and spawning in female winter flounder, *Pseudopleuronectes americanus* (Walbaum). Aquaculture 104:375-390.

Harmin, S.A., & L.W. Crim. 1993. Influence of gonadotropic hormone-releasing hormone analog (GnRH-A) on plasma sex steroid profiles and milt production in male winter flounder, *Pseudopleuronectes americanus* (Walbaum). Fish Physiol. Biochem. 10:399-407.

Harmin, S.A., L.W. Crim & M.D. Wiegand. 1995. Manipulation of the seasonal reproductive cycle in winter flounder, *Pleuronectes americanus*, using a gonadotropic hormone releasing hormone. Mar. Biol. 121:611-619.

Harrell, R.M., J.H. Kerby & R.V. Minton (eds.). 1990. Culture and Propagation of Striped Bass and Its Hybrids. Am. Fish. Soc., Bethesda, Maryland, 323 p.

Harris, M.-L.C. 1996. The tropical marine fish breeding program at the National Aquarium in Baltimore. Today's Aquarist 6(8):5-7.

Harrison, P.J., P.A. Thompson, M. Guo & F.J.R. Taylor. 1993. Effects of light, temperature and salinity on the growth rate of harmful marine diatoms, *Chaetoceros convolutus* and *C. concavicornis* that kill netpen salmon. J. Appl. Phycol. 5:259-265.

Hart, P.R., & G.J. Purser. 1995. Effects of salinity and temperature on eggs and yolk sac larvae of the greenback flounder (*Rhombosolea tapirina* Gunther, 1862). Aquaculture 136:221-230.

Hart, P.R., & G.J. Purser. 1996. Weaning of hatchery-reared greenback flounder (*Rhombosolea tapirina* Günther) from live to artificial diets: effects of age and duration of the changeover period. Aquaculture 145:171-181.

Hartman, K.J., & S.B. Brandt. 1995. Comparative energetics and the development of bioenergetics models for sympatric estuarine piscivores. Can. J. Fish. Aquat. Sci. 52:1647-1666.

Harvey, B. 1982. Cryobiology and the storage of teleost gametes. Proc. Internat. Symp. Repr. Physiol. Fish, Wageningen, the Netherlands, p 123-127.

Harvey, B. 1983. Cooling of embryonic cells, isolated blastoderms, and intact embryos of the zebra fish *Brachydanio rerio* to -196 degree C. Cryobiology 20:440-447.

Harvey, B., & J. Carolsfeld. 1993. Induced Breeding in Tropical Fish Culture. Internat. Dev. Res. Ctr., Ottawa, 144 p.

Harvey, B.J., & W.S. Hoar. 1979. The Theory and Practice of Induced Breeding in Fish. Internat. Dev. Res. Ctr. Rep. IDRC-TS21e, 49 p.

Harvey, B., J. Nacario, L.W. Crim, J.V. Juario & C.L. Marte. 1985. Induced spawning of sea bass, *Lates calcarifer*, and rabbitfish, *Siganus guttatus*, after implantation of pelleted LHRH analogue. Aquaculture 47:53-59.

Hassa, M.S., & M.H. Carlos. 1994. Maturation, spawning, and egg hatching of the groupers *Epinephelus fuscoguttatus* (Forskal) and *Plectropomus areolatus* (Ruppel) from the Red Sea. In: S.A. Al-Thobaiti, H.M. Al-Hinty, A.Q. Siddiqui & G. Hussain (eds.), Proc., First Internat. Symp. Aquac. Technol. Investment Opportunities. Min. Agric. Water, Saudi Arabia, p 219-229.

Hassin, S., D. de Monbrison, Y. Hanin, A. Elizur, Y. Zohar & D.M. Popper. 1997. Domestication of the white grouper, *Epinephelus aenaeus*: 1. Growth and reproduction. Aquaculture 156:305-316.

Hassler, W.W., & W.T. Hogarth. 1977. The growth and culture of dolphinfish (*Coryphaena hippurus*), in North Carolina. Aquaculture 12:115-122.

Hassler, W.W., & R.P. Raineville. 1975. Techniques for hatching and rearing cobia, *Rachycentron canadum*, through larval and juvenile stages. Univ. North Carolina Sea Grant UNC-SG-75-30, 26 p.

Hatai, K. 1989. Fungal pathogens/parasites of aquatic animals. In: B. Austin & D.A. Austin (eds.), Methods for the Microbiological Examination of Fish and Shellfish. Ellis Horwood, Chichester, England, p 240-272.

Hattori, A. 1994. Inter-group movement and mate acquisition tactics of the protandrous anemonefish, *Amphiprion clarkii*, on a coral reef, Okinawa. Jpn. J. Ichthyol. 41:159-165.

Hauck, A.K. 1995. Bacterial diseases of commercially important fishes. Aquac. Mag. 21(4):42,44-53.

Haug, T. 1990. Biology of the Atlantic halibut, *Hippoglossus hippoglossus* (L., 1758). Adv. Mar. Biol. 26:1-70.

Hawke, J.P., & R.V. Minton. 1985. Fish pasteurellosis in cultured striped bass (*Morone saxatilis*) in coastal Alabama. Am. Fish. Soc., Fish Health Sec. Newsl. 13:7.

Hawkins, A.D., & R. Lloyd. 1981. Materials for the aquarium. In: A.D. Hawkins (ed.), Aquarium Systems. Academic Press, London, p 171-196.

Hay, D.E. 1986. Effects of delayed spawning on viability of eggs and larvae of Pacific herring. Trans. Am. Fish. Soc. 115:155-161.

Hayashi, K. 1996. Composition and distribution of lipids in different tissues of the arrow squid *Loligo bleekeri*. Fish. Sci. 62:84-87.
Hayashi, M., K. Toda, T. Yoneji, O. Sato & S. Kitaoka. 1993a. Dietary value of rotifers and *Artemia* enriched with *Euglena gracilis* for red sea bream. Nippon Suisan Gakkaishi 59:1051-1058.
Hayashi, M., K. Toda & S. Kitaoka. 1993b. Enriching *Euglena* with unsaturated fatty acids. Biosci. Biotech. Biochem. 57:352-353.
Haydock, I. 1971. Gonad maturation and hormone-induced spawning of the gulf croaker, *Bairdiella icistia*. Fish. Bull., U.S. 69:157-180.
Haywood, M., & S. Wells. 1989. The Manual of Marine Invertebrates. Tetra Press, Morris Plains, New Jersey, 208 p.
Hazel, C.R., W. Thompson & S.J. Meith. 1971. Sensitivity of striped bass and stickleback to ammonia in relation to temperature and salinity. Cal. Fish Game 57:138-153.
Hazza, M.S., S.A. Al-Thobaiti & E.M. Venturanza. 1989. Tilapia Fry Production Manual. Fish Farming Ctr., Jeddah, Saudi Arabia, 27 p.
Hazza, M.S., S.A. Al-Thobaiti & E.V. Ventruanza. 1991. Operational Manual for Tilapia Production Using Baobab Systems. Fish Farming Ctr., Jeddah, Saudi Arabia, 34 p.
Head, W.D., A. Zerbi & W.O. Watanabe. 1996. Economic evaluation of commercial-scale, saltwater pond production of Florida red tilapia in Puerto Rico. J. World Aquac. Soc. 27:275-289.
Hearn, R. 1996. Researchers take critical step in protecting, producing red snappers. Aquac. Mag. 22(6):6-9.
Heath, M., M. Leaver, A. Matthews & N. Nicoll. 1989. Dispersion and feeding of larval herring (*Clupea harengus* L.) in the Moray Firth during September 1985. Est. Coastal Shelf Sci. 28:549-566.
Heath, M.R. 1992. Field investigations of the early life stages of marine fish. Adv. Mar. Biol. (28):1-174.
Heath, P.L. 1994. The Development of *Tisbe holothuriae* as a Live Diet for Larval Flatfish rearing. Ph.D. Thesis, Heriot-Watt Univ., Edinburgh, 265 p.
Heath, P.L., & C.G. Moore. 1997. Rearing Dover sole larvae on *Tisbe* and *Artemia* diets. Aquac. Internat. 5:29-39.
Hecht, T. P.J. Britz & W. Uys. 1992. Aquaculture in South Africa. World Aquac. 23(1):6-19.
Hecht, T., S. Battaglene & B. Talbot. 1996. Effect of larval density and food availability on the behaviour of pre-metamorphosis snapper, *Pagrus auratus* (Sparidae). Mar. Freshw. Res. 47:223-231.
Heckmann, R. 1996a. Protozoan parasites of fish, part 1. Aquac. Mag. 22(3):44-57.
Heckmann, R. 1996b. Protozoan parasites of fish, part 2. Aquac. Mag. 22(4):56-66.
Hedrick, R.P., J.M. Groff, T.S. McDowell & W.H. Wingfield. 1991. Virus infections of cultured white sturgeon (*Acipenser transmontanus*). In: P. Williot (managing ed.), *Acipenser*. CEMAGREF, Bordeaux, p 439-444.
Heinig, C.S., & D.E. Campbell. 1992. The environmental context of a *Gyrodinium aureolum* bloom and shellfish kill in Maquoit Bay, Maine, September 1988. J. Shellfish Res. 11:111-122.
Heinsbroek, L.T.N. 1991. A review of eel culture in Japan and Europe. Aquac. Fish. Mgt. 22:57-72.
Heinsbroek, L.T.N., & J.G. Kreuger. 1992. Feeding and growth of glass eels, *Anguilla anguilla* L.: the effect of feeding stimulants on feed intake, energy metabolism and growth. Aquac. Fish. Mgt. 23:327-336.
Helland, S., T. Storebakken & B. Grisdale-Helland. 1991. Atlantic salmon, *Salmo salar*. In: R.P. Wilson (ed.), Handbook of Nutrient Requirements of Finfish. CRC Press, Boca Raton, Florida, p 13-22.
Helm, M.M., & I. Laing. 1987. Preliminary observations on the nutritional value of 'Tahiti *Isochrysis*' to bivalve larvae. Aquaculture 62:281-288.
Helm, M.M., D.L. Holland, S.D. Utting & J. East. 1991. Fatty acid composition of early non-feeding larvae of the European flat oyster, *Ostrea edulis*. J. Mar. Biol. Assn. U.K. 71:691-705.
Helvik, J.V., & B.T. Walther. 1993. Environmental parameters affecting induction of hatching in halibut (*Hippoglossus hippoglossus*) embryos. Mar. Biol. 116:39-45.
Helvik, J.V., D.O. Oppen-Berntsen & B.T. Walther. 1991. The hatching mechanism in Atlantic halibut (*Hippoglossus hippoglossus*). Int. J. Dev. Biol. 35:9-16.
Hemdal J.F. 1992. Reduction of ozone oxidants in synthetic seawater by use of sodium thiosulfate. Prog. Fish-Cult. 54:54-56.
Heming, T.A., & E.J. Paleczny. 1987. Compositional changes in skin mucus and blood serum during starvation of trout. Aquaculture 66:265-273.
Hemre, G.-I., O. Lie, E. Lied & G. Lambertsen. 1989. Starch as an energy source in feed for cod (*Gadus morhua*): digestibility and retention. Aquaculture 80:261-270.
Henderson-Arzapalo, A. 1990. Red drum egg and larval incubation. Texas A&M Univ. Sea Grant TAMU-SG-90-603, p 51-52.
Henderson-Arzapalo, A. 1995. Review of nursery and growout culture techniques for red drum (*Sciaenops ocellatus*). In: K.L. Main & C. Rosenfeld (eds.), Culture of High-value Marine Fishes in Asia and the United States. Oceanic Inst., Honolulu, p 67-79.
Henderson-Arzapalo, A., & R.L. Colura. 1987. Laboratory maturation and induced spawning of striped bass. Prog. Fish-Cult. 49:60-63.
Henderson-Arzapalo, A., R.L. Colura & A.F. Maciorowski. 1994. A comparison of black drum, red drum, and their hybrid in saltwater pond culture. J. World Aquac. Soc. 25:289-296.
Hendricks, J.D., & G.S. Bailey. 1989. Adventitious toxins. In: J.E. Halver (ed.), Fish Nutrition. Academic

Press, San Diego, p 605-651.

Henriksen, P., F. Knipschildt, O. Moestrup & H.A. Thomsen. 1993. Autecology, life history and toxicology of the silicoflagellate *Dictyocha speculum* (Silicoflagellata, Dictyochophyceae). Phycologia 32:29-39.

Henry, G. 1988. Snapper. New South Wales Agric. Fish., Agfact No. F1.0.3, 6 p.

Her, Y.-S., & D.-C. Liu. 1991. Development and survival of cultured tigerfish (*Therapon jarbua*) larvae. Abs. in program book, Finfish Hatchery in Asia '91. Tungkang Marine Laboratory, Tungkang, Taiwan, p 35.

Heras, H., C.A. McLeod & R.G. Ackman. 1994. Atlantic dogfish silage vs. herring silage in diets for Atlantic salmon (*Salmo salar*): growth and sensory evaluation of fillets. Aquaculture 125:93-106.

Herke, W.H. 1976. Marsh scientist warns against widespread mariculture. National Fisherman 57(5):6B, 14B, 16B, 38C.

Herke, W.H. 1977. Dangers of mariculture in wetlands are further documented by Herke. National Fisherman 58(8):34A.

Herke, W.H. 1978. Letters to the editor: Is coastal aquaculture in the public interest? Mar. Policy, (Jan):78-79.

Herwig, N. 1978. Notes on the treatment of *Cryptocaryon*. Drum and Croaker 18:6-12.

Hess, H.C. 1993. Male mouthbrooding in jawfishes (Opistognathidae): constraints on polygyny. Bull. Mar. Sci. 52:806-818.

Hesskew, J. 1992. Star of Texas:redfish. Seafood Leader 11(1):94-98.

Hetrick, F.M. 1989. Fish viruses. In: B. Austin & D.A. Austin (eds.), Methods for the Microbiological Examination of Fish and Shellfish. Ellis Horwood, Chichester, England, p 216-239.

Hettler, W.F. 1968. Artificial fertilization among yellowfin and Gulf menhaden (*Brevoortia*) and their hybrid. Trans. Am. Fish. Soc. 97:119-123.

Hettler, W.F. 1970. Rearing larvae of yellowfin menhaden, *Brevoortia smithi*. Copeia 1970:775-776.

Hettler, W.F. 1981. Spawning and rearing Atlantic menhaden. Prog. Fish-Cult. 43:80-84.

Hettler, W.F. 1983. Transporting adult and larval Gulf menhaden and techniques for spawning in the laboratory. Prog. Fish-Cult. 45:45-48.

Hettler, W.F. 1984. Description of eggs, larvae, and early juveniles of gulf menhaden, *Brevoortia patronus*, and comparisons with Atlantic menhaden, *B. tyrannus*, and yellowfin menhaden, *B. smithi*. Fish. Bull., U.S. 82:85-95.

Hettler, W.F., & A.B. Powell. 1981. Egg and larval fish production at the NMFS Beaufort Laboratory, Beaufort, N.C. USA. Rapp. P.-v. Réun. Int. Explor. Mer 178:501-503.

Hettler, W.F., Jr., R.W. Lichtenheld & H.R. Gordy. 1971. Open seawater system with controlled temperature and salinity. Prog. Fish-Cult. 33:3-11.

Hewett, S.W., & B.L. Johnson, 1989. A general bioenergetics model for fishes. Am. Fish. Soc. Symp. (6):206-208.

Hibler, M. 1983. A silver-blue revolution: milkfish farming in Asia. Aquac. Mag. 9(3):21-23.

Hidaka, I., C. Zeng & J. Kohbara. 1992. Gustatory responses to organic acids in the yellowtail *Seriola quinqueradiata*. Nippon Suisan Gakkaishi 58:1179-1187.

Hidalgo, F., & E. Alliot. 1988. Influence of water temperature on protein requirement and protein utilization in juvenile sea bass, *Dicentrarchus labrax*. Aquaculture 72:115-129.

Hidalgo, F., E. Alliot, & H. Thebault. 1987. Influence of water temperature on food intake, food efficiency and gross composition of juvenile sea bass, *Dicentrarchus labrax*. Aquaculture 64:199-207.

Higgs, D.A., U.H.M. Fagerlund, J.G. Eales & J.R. McBride. 1982. Application of thyroid and steroid hormones as anabolic agents in fish culture. Comp. Biochem. Physiol. 73B:143-176.

Higgs, D.A., B.S. Dosanjh, A.F. Prendergast, R.M. Beames, R.W. Hardy, W. Riley & G. Deacon. 1995. Use of rapeseed/canola protein products in finfish diets. In: C. Lim & D.J. Sessa (eds.), Nutrition and Utilization Technology in Aquaculture. AOSC Press, Champaign, Ilinois, p 130-156.

Higuchi, M., N. Hussain, S. Akatsu, C. El-Zahr, T. Al-Ahmad & K. Al-Ghemlas. 1980. Studies on the development of large-scale fingerling production techniques of selected Kuwait fishes. Kuwait Inst. Sci. Res., Ann. Res. Rep. (1979):39-40.

Hildebrand, S.F., & L.E. Cable. 1930. Development and life history of fourteen teleostean fishes at Beaufort, N.C. Bull. U.S. Bur. Fish. 46:382-488.

Hildebrand, S.F., & L.E. Cable. 1938. Further notes on the development and life history of some teleosts at Beaufort, N.C. Bull. U.S. Bur. Fish. 48:505-642.

Hill, J., J.W. Evans & M.J. Van Den Avyle. 1989. Species profiles: life histories and environmental requirements of coastal fishes and invertebrates (South Atlantic)—striped bass. U.S. Fish Wildl. Serv. Biol. Rep. 82(11.118) U.S. Army Corps Engineers TR EL-82-4, 35 p.

Hillestad, M., & F. Johnsen. 1994. High-energy/low-protein diets for Atlantic salmon: effects on growth, nutrient retention and slaughter quality. Aquaculture 124:109-116.

Hilomen-Garcia, G.V. 1997. Morphological abnormalities in hatchery-bred milkfish (*Chanos chanos* Forsskal) fry and juveniles. Aquaculture 152:155-166.

Hilomen-Garcia, G.V. 1998. Sensitivity of fertilized milkfish (*Chanos chanos* Forsskal) eggs to mechanical shock and simulated transport. Aquaculture 159:239-247.

Hilton, J.W. 1983. Hypervitaminosis A in rainbow trout (*Salmo gairdneri*): toxicity signs and maximum

tolerable level. J. Nutr. 113:1737-1745.
Hilton, J.W. 1989. The interaction of vitamins, minerals and diet composition in the diet of fish. Aquaculture 79:223-244.
Hino, A. 1993. Present culture systems of the rotifer (*Brachionus plicatilis*) and the function of micro-organisms. In: C.-S. Lee, M.-S. Su & I C. Liao (eds.), Finfish Hatchery in Asia. Tungkang Mar. Lab. Conf. Proc. (3):51-59.
Hino, A., & R. Hirano. 1984. Relationship between water temperature and bisexual reproduction rate in the rotifer *Brachionus plicatilis*. Bull. Jpn. Soc. Sci. Fish. 50:1481-1485.
Hino, A., & R. Hirano. 1985. Relationship between the temperature given at the time of fertilized egg formation and bisexual reproduction pattern in the deriving strain of the rotifer *Brachionus plicatilis*. Bull. Jpn. Soc. Sci. Fish. 51:511-514.
Hinshaw, J.M. 1985. Effects of illumination and prey contrast on survival and growth of larval yellow perch *Perca flavescens*. Trans. Am. Fish. Soc. 114:540-545.
Hioki, S., & K. Suzuki. 1987. Reproduction and early development of the angelfish, *Centropyge interruptus*, in an aquarium. J. Fac. Mar. Sci. Technol. Tokai Univ. (24):133-140.
Hioki, S., K. Suzuki & Y. Tanaka. 1982. Spawning behavior, egg and larval development, and sex succession of the hermaphroditic pomacanthine, *Genicanthus melanospilos*, in the aquarium. J. Fac. Mar. Sci. Technol. Tokai Univ. 15:359-366. (in Jpn., Eng. abs.)
Hioki, S., K. Suzuki & Y. Tanaka. 1990. Development of eggs and larvae in the angelfish, *Centropyge ferrugatus*. Jpn. J. Ichthyol. 37:34-38.
Hirano, R., & I. Hanyu (eds.). 1990. The Second Asian Fisheries Forum: Proc. of the Second Asian Fisheries Forum, Tokyo, Japan, 17-22 April 1989. Asian Fish. Soc., Manila, 991 p.
Hirano, T., C.-H. Zhang, A. Morishita, T. Suzuki & T. Shirai. 1992. Identification of volatile compounds in ayu fish and its feeds. Nippon Suisan Gakkaishi 58:547-557.
Hirayama, K. 1970. Studies on water control by filtration through sand bed in a marine aquarium with closed circulating system—VI. Acidification of aquarium water. Bull. Jpn. Soc. Sci. Fish. 36:26-34.
Hirayama, K. 1987. A consideration of why mass culture of the rotifer *Brachionus plicatilis* with baker's yeast is unstable. Hydrobiologia 147:269-270.
Hirayama, K., & A. Hagiwara. 1995. Recent advances in biological aspects of mass culture of rotifers (*Brachionus plicatilis*) in Japan. ICES Mar. Sci. Symp. 201:153-158.
Hirose, K., & R. Ishida. 1974. Induction of ovulation in the ayu, *Plecoglossus altivelis*, with LH-releasing hormone (LH-RH). Bull. Jpn. Soc. Sci. Fish. 40:1235-1239.
Hirose, K., R. Ishida & K. Sakai. 1977. Induced ovulation of ayu using human chorionic gonadotropin (HCG), with special reference to changes in several characteristics of eggs retained in the body cavity after ovulation. Bull. Jpn. Soc. Sci. Fish. 43:409-416.
Hirose, K., Y. Machida & E.M. Donaldson. 1979. Induced ovulation of Japanese flounder (*Limanda yokohamae*) with human chorionic gonadotropin and salmon gonadotropin, with special reference to changes in quality of eggs retained in the ovarian cavity after ovulation. Bull. Jpn. Soc. Sci. Fish. 45:31-36.
Hislop, J.R.G. 1975. The breeding and growth of whiting, *Merlangius merlangus* in captivity. J. Cons. 36:119-127.
Hjelmeland, K., I. Ugelstad & Y. Olsen. 1993. Proteolytic activity and post mortem autolysis in prey for marine fish larvae. In: Physiology and Biochemical Aspects of Fish Development. Univ. Bergen, Norway, p 229-232.
Hjelmeland, K., I. Ugelstad, J.M. Homme, I. Lein & T. Pedersen. 1995a. Development of enzymatic digestion in marine fish larvae. In: H. Reinertsen, L.A. Dahle, L. Jørgensen & K. Tvinnereim (eds.), Fish Farming Technology. Balkema, Rotterdam, p 127.
Hjelmeland, K., I. Ugelstad & T. Pedersen. 1995b. Ontogenesis of the digestive system in cod (*Gadus morhua* L.): measurement of pepsin as an indicator of metamorphosis. ICES Mar. Sci. Symp. 201:197.
Hjertnes, T., K.E. Gulbrandsen, F. Johnsen & J. Opstvedt. 1993. Effect of dietary protein, carbohydrate and fat levels in dry feed for juvenile halibut (*Hippoglossus hippoglossus* L.). In: Fish Nutrition in Practice, INRA, Paris, p 493-496.
Hjort, J. 1914. Fluctuations in the great fisheries of northern Europe viewed in the light of biological research. Rapp. P.-v. Réun. Cons. Int. Explor. Mer 20:1-228.
Hoar, W.S., D.J. Randall & J.R. Brett (eds.). 1979. Fish Physiology, Vol. 8: Bioenergetics and Growth. Academic Press, New York, 786 p.
Hoar, W.S., D.J. Randall & E.M. Donaldson (eds.). 1983a. Fish Physiology, Vol. 9: Reproduction, Part A, Endocrine Tissues and Hormones. Academic Press, Orlando, 483 p.
Hoar, W.S., D.J. Randall & E.M. Donaldson (editors). 1983b. Fish Physiology, Vol. 9: Reproduction, Part B, Behavior and Fertility Control. Academic Press, Orlando, 477 p.
Hobby, A.C., & N.W. Pankhurst. 1997. Post-ovulatory egg viability in the snapper *Pagrus auratus* (Sparidae). Mar. Freshw. Res. 48:385-389.
Hodson, P.V., U. Borgmann & H. Shear. 1979. Toxicity of copper to aquatic biota. In: J.O. Nriagu (ed.), Copper in the Environment. Wiley, New York, p 307-372.
Hodson, R.G. 1991. Hybrid striped bass culture in ponds. In: J.P. McVey (ed.), Handbook of Mariculture, Vol. II, Finfish Aquaculture. CRC Press, Boca Raton, Florida, p 167-191.

Hodson, R.G., & C.V. Sullivan. 1993. Induced maturation and spawning of domestic and wild striped bass, *Morone saxatilis* (Walbaum), broodstock with implanted GnRH analogue and injected hCG. Aquac. Fish. Mgt. 24:389-398.

Hofer, R., & E. Gatumu. 1994. Necrosis of trout retina (*Oncorynchus mykiss*) after sublethal exposure to nitrite. Arch. Env. Contam. Toxicol. 26:119-123.

Hofer, R., Z. Jeney & F. Bucher. 1995. Chronic effects of linear alkylbenzene sulfonate (LAS) and ammonia on rainbow trout (*Oncorhynchus mykiss*) fry at water criteria limits. Wat. Res. 29:2725-2729.

Hoff, F.H., Jr. 1970. Artificial spawning of black sea bass, *Centropristes striatus melanus* Ginsburg, aided by chorionic gonadotropic hormones. Fla. Dept. Natural Resour. Mar. Res. Lab., Spec. Sci. Rep. No. 25, 17 p.

Hoff, F.H., Jr. 1996. Conditioning, Spawning and Rearing of Fish with Emphasis on Marine Clownfish. Aquac. Consultants, Dade City, Fla., 213 p.

Hoff, F.H., & T.W. Snell. 1989. Plankton Culture Manual. Florida Aqua Farms, Dade City, Florida, 126 p.

Hoff, F., C. Rowell & T. Pulver. 1972. Artificially induced spawning of the Florida pompano under controlled conditions. Proc. World Maric. Soc. 3:53-64.

Hoff, F.H. T. Pulver & J. Mountain. 1978a. Conditioning Florida pompano (*Trachinotus carolinus*) for continuous spawning. Proc. World Maric. Soc. 9:299-309.

Hoff, F.H., J. Mountain, T. Frakes & K. Halcott. 1978b. Spawning, oocyte development and larvae rearing of the Florida pompano (*Trachinotus carolinus*). Proc. World Maric. Soc. 9:279-297.

Hogans, W.E. 1994. Cage culture of striped bass in marine waters of the lower Bay of Fundy. Prog. Fish-Cult. 56:255-257.

Hogans, W.E. 1995. Infection dynamics of sea lice, *Lepeophtheirus salmonis* (Copepoda:Caligidae) parasitic on Atlantic salmon (*Salmo salar*) cultured in marine waters of the lower Bay of Fundy. Can. Tech. Rep. Fish. Aquat. Sci. 2067, 10 p.

Holden, K.K., & M.N. Bruton. 1992. A life-history approach to the early ontogeny of the Mozambique tilapia *Oreochromis mossambicus* (Pisces, Cichlidae). S. Afr. J. Zool. 27:173-191.

Holdway, D.A., M.J. Barry, D.C. Logan, D. Robertson, V. Young & J.T. Ahokas. 1994. Toxicity of pulse-exposed fenvalerate and esfenvalerate to larval Australian crimson-spotted rainbow fish (*Melanotaenia fluviatilis*). Aquat. Toxicol. 28:169-187.

Holliday, F.G.T. 1969. The effects of salinity on the eggs and larvae of teleosts. In: W.S. Hoar & D.J. Randall (eds.), Fish Physiology, Vol. I. Academic Press, New York, p 293-311.

Holm, J.C., & E. Andersen. 1989. Improved spawning pen for Atlantic cod. World Aquac. 20(4):107.

Holm, J.C., T. Svaasand & V. Wennevik (eds.). 1991. Manual for Cod Farming: Breeding Stock and Fry Production. Ocean Research Inst., Bergen, Norway. Eng. vers. trans. by T. Gjesdal & ed. by L. Park. 1993. BRE, Inc., Cape Broyle, Newfoundland, Canada, 125 p.

Holm, J.C., A. Mangor-Jensen & T.J. Hansen. 1995. Recent improvements of farming techniques of Atlantic salmon (*Salmo salar*) and Atlantic halibut (*Hippoglossus hippoglossus*) in Norway. In: K.L. Main & C. Rosenfeld (eds.), Culture of High-value Marine Fishes in Asia and the United States. Oceanic Inst., Honolulu, p 231-239.

Holmefjord, I., & Y. Olsen. 1991. An intensive approach to Atlantic halibut fry production. Eur. Aquac. Soc. Spec. Pub. (15):331.

Holmefjord, I., J. Gulbrandsen, I. Lein, T. Refstie, P. Leger, T. Harboe, I. Huse, P. Sorgeloos, S. Bolla, Y. Olsen, K.I. Reitan, O. Vadstein, G. Øie & A. Danielsberg. 1993. An intensive approach to Atlantic halibut fry production. J. World Aquac. Soc. 24:275-284.

Holmer, M., & E. Kristensen. 1992. Impact of marine fish cage farming on metabolism and sulfate reduction of underlying sediments. Mar. Ecol. Prog. Ser. 80:191-201.

Holmes, M.J., & R.J. Lewis. 1994. The origin of ciguatera. Mem. Queensland Mus., 34:497-504.

Holmes, M.J., R.J. Lewis, M.A. Poli & N.C. Gillespie. 1991. Strain dependent production of ciguatoxin precursors (gambiertoxins) by *Gambierdiscus toxicus* (Dinophyceae) in culture. Toxicon 29:761-775.

Holt, E.W.L. 1898. On the breeding of the dragonet (*Callionymus lyra*) in the Marine Biological Association's Aquarium at Plymouth. Proc. Zool. Soc. London, p 281-315.

Holt, G.J. 1990. Growth and development of red drum eggs and larvae. In: G.W Chamberlain, R.J. Miget & M.G. Haby (eds.), Red Drum Aquaculture. Texas A&M Univ. Sea Grant TAMU-SG-90-603, p 46-50.

Holt, G. Joan. 1992. Marine Science Institute, University of Texas, Port Aransas, pers. comm.

Holt, G.J. 1992. Experimental studies of feeding in larval red drum. J. World Aquac. Soc. 23:265-270.

Holt, G.J. 1993. Feeding larval red drum on microparticulate diets in a closed recirculating water system. J. World Aquac. Soc. 24:225-230.

Holt, G.J., & C.R. Arnold. 1983. Effects of ammonia and nitrite on growth and survival of red drum eggs and larvae. Trans. Am. Fish. Soc. 112:314-318.

Holt, J., A.G. Johnson, C.R. Arnold, W.A. Fable, Jr. & T.D. Williams. 1981. Description of eggs and larvae of laboratory reared red drum, *Sciaenops ocellata*. Copeia 1981:751-756.

Holt, G.J., C.R. Arnold & C.M. Riley. 1990. Intensive culture of larval and post larval red drum. Texas A&M Univ. Sea Grant TAMU-SG-90-603, p 53-56.

Holtz, W. 1993. Cryopreservation of rainbow trout (*Oncorhynchus mykiss*) sperm: practical recommendations. Aquaculture 110:97-100.

Honda, S., & S. Imai. 1973. Breeding and early development of a pomacentrid, *Pomacentrus nagasakiensis* Tanaka in the aquarium. Mem. Fac. Fish. Kagoshima Univ. 22:95-106. (in Jpn., Eng. abs.)

Honma, Y., T. Ozawa & A. Chiba. 1980. Maturation and spawning behavior of the puffer, *Fugu niphobles*, occurring on the coast of Sado Island in the Sea of Japan (a preliminary report). Jpn. J. Ichthyol. 27:129-138.

Hood, P.B., & R.A. Schlieder. 1992. Age, growth, and reproduction of gag, *Mycteroperca microlepis* (Pisces: Serranidae), in the eastern Gulf of Mexico. Bull. Mar. Sci. 51:337-352.

Hoornbeek, F.K., P.J. Sawyer & E.S. Sawyer. 1982. Growth of winter flounder (*Pseudopleuronectes americanus*) and smooth flounder (*Liopsetta putnami*) in heated and unheated water. Aquaculture 28:363-373.

Hopkins, J.S. 1990. Semi-intensive grow out in saltwater. Texas A&M Univ. Sea Grant TAMU-SG-90-603, p 170-177.

Hopkins, J.S., A.D. Stokes, P.A. Sandifer, R.A. Smiley & T.R. McTeer. 1986. Spawning and grow-out trials with red drum in South Carolina. Proc. Ann. Conf. SE Assoc. Fish Wildl. Ag. 40:178-186.

Hoque, M.M., A. Takemura & K. Takano. 1998. Annual changes in oocyte development and serum vitellogenin level in the rabbitfish *Siganus cnaliculatus* (Park) in Okinawa, southern Japan. Fish. Sci. 64:44-51.

Horn, M.H., K.F. Mailhiot, M.B. Fris & L.L. McClanahan. 1995. Growth, consumption, assimilation and excretion in the marine herbivorous fish *Cebidichthys violaceus* (Girard) fed natural and high protein diets. J. Exp. Mar. Biol. Ecol. 190:97-108.

Hossain, M.A., & M. Furuichi. 1998. Availability of environmental and dietary calcium in tiger puffer. Aquac. Internat. 6:121-132.

Houde, E.D. 1972. Development and early life history of the northern sennet, *Sphyraena borealis* DeKay (Pisces: Sphyraenidae) reared in the laboratory. Fish. Bull., U.S. 70:185-195.

Houde, E.D. 1987. Fish early life dynamics and recruitment variability. Am. Fish. Soc. Symp. (2):17-29.

Houde, E.D. 1989. Comparative growth, mortality, and energetics of marine fish larvae: temperature and implied latitudinal effects. Fish. Bull., U.S. 87:471-495.

Houde, E.D., & L. Lubbers III. 1986. Survival and growth of striped bass, *Morone saxatilis*, and *Morone* hybrid larvae: laboratory and pond enclosure experiments. Fish. Bull., U.S. 84:905-914.

Houde, E.D., & B.J. Palko. 1970. Laboratory rearing of the clupeid fish *Harengula pensacolae* from fertilized eggs. Mar. Biol. 5:354-358.

Houde, E.D., & T. Potthoff. 1976. Egg and larval development of the sea bream *Archosargus rhomboidalis* (Linnaeus) (Pisces, Sparidae). Bull. Mar. Sci. 26:506-529.

Houde, E.D., & W.J. Richards. 1969. Rearing larval tunas in the laboratory. Comm. Fish. Rev. 31(12):32-34.

Houde, E.D., & R.C. Schekter. 1980. Feeding by marine fish larvae: developmental and functional responses. Env. Biol. Fish. 5:315-334.

Houde, E.D., & R.C. Schekter. 1981. Growth rates, rations and cohort consumption of marine fish larvae in relation to prey concentrations. Rapp. P.-v. Réun. Cons. Int. Explor. Mer 178:441-453.

Houde, E.D., & R.C. Schekter. 1983. Oxygen uptake and comparative energetics among eggs and larvae of three subtropical marine fishes. Mar. Biol. 72:283-293.

Houde, E.D., & C.E. Zastrow. 1993. Ecosystem- and taxon-specific dynamic and energetics properties of larval fish assemblages. Bull. Mar. Sci. 53:290-335.

Houde, E.D., S.A. Berkeley, J.J. Klinovsky & R.C. Schekter. 1976. Culture of larvae of the white mullet, *Mugil curema* Valenciennes. Aquaculture 8:365-370.

Houghton, R.G., J.M. Last & P.J. Bromley. 1985. Fecundity and egg size of sole (*Solea solea* (L.)) spawning in captivity. J. Cons. 42:162-165.

Hovgaard, P. 1993. A combined approach for better economy in landbased salmon farming. In: H. Reinertsen, L.A. Dahle, L. Jørgensen & K. Tvinnereim (eds.), Fish Farming Technology. Balkema, Rotterdam, p 480.

Howell, B.R. 1973. Marine fish culture in Britain VIII. A marine rotifer, *Brachionus plicatilis* Muller, and the larvae of the mussel, *Mytilus edulis* L., as foods for larval flatfish. J. Cons. 35:1-6.

Howell, B.R. 1974. Problems associated with the feeding of certain flatfish larvae. Informes Tecnicos Inst. Invest. Cient. (14):109-116.

Howell, B.R. 1994. Fitness of hatchery-reared fish for survival in the sea. Aquac. Fish. Mgt. 25(suppl. 1):3-17.

Howell, B.R. 1997. A re-appraisal of the potential of the sole, *Solea solea* (L.), for commercial cultivation. Aquaculture 155:355-365.

Hu, C.S., B.Q. Huang & W.J. Yiea. 1991. Prey capture ability of juvenile silver bream (*Sparus sarba*) under dark-adaptation. Eur. Aquac. Soc. Spec. Pub. (15):307-309.

Hua, D.-K., M.-L. Cai & Z.-Y. Zhang. 1994a. Studies on the grouper's syndrome I. etiology and histopathology. Third Asian Fisheries Forum. Asian Fish. Soc., Manila, p 353-356.

Hua, D.-K., M.-L. Cai & Z.-Y. Zhang. 1994b. Studies on the grouper's syndrome II. serum biochemical determination. Third Asian Fisheries Forum. Asian Fish. Soc., Manila, p 357-360.

Huang, C.-F., & F.-G. Liu. 1991. Artificial propagation and larval rearing of Japanese sea bass (*Lateolabrax japonicus*). Abs. in program book, Finfish Hatchery in Asia '91. Tungkang Mar. Lab.,

Tungkang, Taiwan, p 38.

Huang, L., S. Miwa, D.A. Bengtson & J.L. Specker. 1998a. Effect of triiodothyronine on stomach formation and pigmentation in larval striped bass (*Morone saxatilis*). J. Exp. Zool. 280:231-237.

Huang, L., A.M. Schreiber, B. Soffientino, D.A. Bengtson & J.L. Specker. 1998b. Metamorphosis of summer flounder (*Paralichthys dentatus*): thyroid status and the timing of gastric gland formation. J. Exp. Zool. 280:413-420.

Huang, T.-S., K.-J. Lin, C.-L. Yen, C.-Y. Lin & C.-L. Chen. 1986. Experiments on the artificial propagation of black spotted grouper, *Epinephelus salmonoides* (Lacepede)—I Hormone treatment, ovulation of spawners and embryonic development. Bull. Taiwan Fish. Res. Inst. (40):241-258 (in Ch., Eng. abs.).

Huang, T.-S., C.-L. Yen & C.-Y. Liu. 1987. Experiments on the artificial propagation of salmon-like grouper, *Epinephelus salmonoides* (Lacepede)—III Brood stock nursing induced maturation and spawning. Bull. Taiwan Fish. Res. Inst. (42):317-335 (in Ch., Eng. abs.).

Huckle, K.R., & P. Millburn. 1990. Metabolism, bioconcentration and toxicity of pesticides in fish. In: D.H. Hutson & T.R. Roberts (eds.), Environmental Fate of Pesticides. Wiley, London, p 175-243.

Hudson, R.C.L. 1977. Preliminary observations on the behaviour of the gobiid fish, *Signigobius biocellatus* Hoese and Allen, with particular reference to its burrowing behaviour. Z. Tierpsychol. 43:214-220.

Huff, J.A., & C.D. Burns. 1981. Hypersaline and chemical control of *Cryptocaryon irritans* in red snapper, *Lutjanus campechanus*, monoculture. Aquaculture 22:181-184.

Huguenin, J.E. 1998. The design, operations and economics of cage culture systems. Aquac. Eng. 16:167-203.

Huguenin, J.E., & F.J. Ansuini. 1978. A review of the technology and economics of marine fish cage systems. Aquaculture 15:151-170.

Huguenin, J.E., & J. Colt. 1989. Design and Operating Guide for Aquaculture Seawater Systems. Elsevier, Amsterdam, Dev. Aquac. Fish. Sci. Vol. 20, 264 p.

Hume, A.S., & I.K. Ho. 1994. Toxicity of solvents. In: L.G. Cockerham & B.S. Shane (eds.), Basic Environmental Toxicology. CRC Press, Boca Raton, Florida, p 157-184.

Humphrey, J.D., & J.S. Langdon. 1987. Pathological anatomy and diseases of barramundi (*Lates calcarifer*). Austral. Ctr. Internat. Agric. Res. Proc. (20):198-203.

Huner, J.V., & H.K. Dupree. 1984. Production methods for baitfish: golden shiners and fathead minnows. In: H.K. Dupree & J.V. Huner (eds.), Third Report to the Fish Farmers. U.S. Fish. Wildl. Serv., p 83-96.

Hung, S.S.O. 1991a. Nutrition and feeding of hatchery-produced juvenile white sturgeon (*Acipenser transmontanus*): an overview. In: P. Williot (managing ed.), Acipenser. CEMAGREF, Bordeaux, p 65-77.

Hung, S.S.O. 1991b. Sturgeon, *Acipenser* spp. In: R.P. Wilson (ed.), Handbook of Nutrient Requirements of Finfish. CRC Press, Boca Raton, Florida, p 153-160.

Hung, S.S.O., J.M. Groff, P.B. Lutes & F.K. Fynn-Aikins. 1990. Hepatic and intestinal histology of juvenile white sturgeon fed different carbohydrates. Aquaculture 87:349-360.

Hung, S.S.O., F.S. Conte & E.F. Hallen. 1993. Effects of feeding rates on growth, body composition and nutrient metabolism in striped bass (*Morone saxatilis*) fingerlings. Aquaculture 112:349-361.

Hunter, G.A., E.M. Donaldson, E.T. Stone & H.M. Dye. 1978. Induced ovulation of female chinook salmon (*Oncorhynchus tshawytscha*) at a production hatchery. Aquaculture 15:99-112.

Hunter, J.R. 1976. Culture and growth of northern anchovy, *Engraulis mordax*, larvae. Fish. Bull., U.S. 74:81-88.

Hunter, J.R. 1980. The feeding behavior and ecology of marine fish larvae. In J.E. Bardach, J.J. Magnuson, R.C. May & J.M. Reinhart (eds.), Fish Behavior and Its Use in the Capture and Culture of Fishes. ICLARM Conf. Proc. (5):287-330.

Hunter, J.R. 1981. Feeding ecology and predation of marine fish larvae. In: R. Lasker (ed.), Marine Fish Larvae. Univ. Washington Sea Grant, Seattle, p 33-77.

Hunter, J.R. 1984. Synopsis of culture methods for marine fish larvae. Am. Soc. Ichthyol. Herpetol. Spec. Pub. (1):24-27.

Hunter, J.R., & K.M. Coyne. 1982. The onset of schooling in northern anchovy larvae *Engraulis mordax*. Cal. Coop. Oceanic Fish. Invest. Rep. 23:246-251.

Hunter, J.R., & C.A. Kimbrell. 1980. Early life history of Pacific mackerel, *Scomber japonicus*. Fish. Bull., U.S. 78:89-101.

Hunter, J.R., & C. Kimbrell. 1981. Some aspects of the life history of laboratory-reared Pacific mackerel larvae (*Scomber japonicus*). Rapp. P.-v. Réun. Cons. Int. Explor. Mer 178:344.

Hunter, J.R., & R. Leong. 1981. The spawning energetics of female northern anchovy, *Engraulis mordax*. Fish. Bull., U.S. 79:215-230.

Hunter, J.R., & B.J. Macewicz. 1980. Sexual maturity, batch fecundity, spawning frequency, and temporal pattern of spawning for the northern anchovy, *Engraulis mordax*, during the 1979 spawning season. Cal. Coop. Oceanic Fish. Invest. Rep. 21:139-149.

Hunter, J.R., & C. Sanchez. 1976. Diel changes in swim bladder inflation of the larvae of the northern anchovy, *Engraulis mordax*. Fish. Bull., U.S. 74:847-855.

Hunter. J.R., S.E. Kaupp & J.H. Taylor. 1981. Effects of solar and artificial ultraviolet-B radiation on larval northern anchovy, *Engraulis mordax*. Photochem. Photobiol. 34:477-486.

Hunter, J.R., S.E. Kaupp & J.T. Taylor. 1982. Assessment of effects of UV radiation on marine fish larvae. In: J. Calkins (ed.), The Role of Solar Ultraviolet Radiation in Marine Ecosystems. NATO Conf. Ser. 4: Mar. Sci., Plenum Press, New York, p 459-497.

Hunter, J.R., B.J. Macewicz & J.R. Sibert. 1986. The spawning frequency of skipjack tuna, *Katsuwonus pelamis*, from the South Pacific. Fish. Bull., U.S. 84:895-903.

Hunter, J.R., B.J. Macewicz & C.A. Kimbrell. 1989. Fecundity and other aspects of the reproduction of sablefish, *Anoplopoma fimbria*, in central California waters. Cal. Coop. Oceanic Fish. Invest. Rep. 30:61-72.

Hunter, J.R., B.J. Macewicz, N.C. Lo & C.A. Kimbrell. 1992. Fecundity, spawning, and maturity of female Dover sole *Microstomus pacificus*, with an evaluation of assumptions and precision. Fish. Bull., U.S. 90:101-128.

Hurlbut, S.E., D.P. French & B.J. Taylor. 1991. Evaluation of the potential effects of major oil spills on Grand Banks commercial fish species as a result of impacts on eggs and larvae. Env. Studies Res. Funds, Calgary, Rep. No. 110, 53 p.

Huse, I. 1991. Culturing of cod (*Gadus morhua* L.). In: J.P. McVey (ed.), Handbook of Mariculture, Vol. II, Finfish Aquaculture. CRC Press, Boca Raton, Florida, p 43-51.

Huse, I. 1993. First feeding larval sensory perception and behavioural programming: implications for systems and procedures in culture. In: B.T. Walther & H.J. Fyhn (eds.), Physiological and Biochemical Aspects of Fish Development, Univ. Bergen, Norway, p 146-152.

Huse, I. 1994. Feeding at different illumination levels in larvae of three marine teleost species: cod, *Gadus morhua* L., plaice, *Pleuronectes platessa* L., and turbot, *Scophthalmus maximus* (L.). Aquac. Fish. Mgt. 25:687-695.

Huse, I., & P.A. Jensen. 1983. A simple and inexpensive spawning and egg collection system for fish with pelagic eggs. Aquacultural Eng. 2:165-171.

Hussain, N.A., & M. Higuchi. 1980. Larval rearing and development of the brown spotted grouper, *Epinephelus tauvina* (Forskål). Aquaculture 19:339-350.

Hussain, N.A., M. Saif & M. Ukawa. 1975. On the culture of *Epinephelus tauvina* (Forskal). Kuwait Inst. Sci. Res., Rep. No. MAB III.XI 75, 14 p.

Hussain, N., S. Akatsu & C. El-Zahr. 1981. Spawning, egg and early development, and growth of *Acanthopagrus cuvieri* (Sparidae). Aquaculture 22:125-136.

Hussain, S.M. 1992. The reproductive biology of the lantern fish *Benthosema fibulatum* from the northern Arabian Sea. Fish. Res. 13:381-393.

Hutchinson, T.H., T.D. Williams & G.J. Eales. 1994. Toxicity of cadmium, hexavalent chromium and copper to marine fish larvae (*Cypinodon variegatus*) and copepods (*Tisbe battagliai*). Mar. Env. Res. 38:275-290.

Hwang, D.-F., M.-R. Tsai, T. Yoshida & S.-S. Jeng. 1992. Toxicity and excretion of butylated hydroxytoluene (BHT) through feed in Japanese eel. Nippon Suisan Gakkaishi 58:69-74.

Hyndes, G.A., M.E. Platell, I.C. Potter & R.C.J. Lenanton. 1998. Age composition, growth, reproductive biology, and recruitment of King George whiting, *Sillaginodes punctata*, in coastal waters of southwestern Australia. Fish. Bull., U.S. 96:258-270.

Hysmith, B.T., G.C. Matlock & R.L. Colura. 1983. Effects of stocking rate and food type on growth and survival of fingerling red drum. World Maric. Soc. Spec. Pub. (3):133-141.

Ibeas, C., J.R. Cejas, R. Fores, P. Badía, T. Gómez & A.L. Hernández. 1997. Influence of eicosapentaenoic to docosahexaenoic acid ratio (EPA/DHA) of dietary lipids on growth and fatty acid composition of gilthead seabream (*Sparus aurata*) juveniles. Aquaculture 150:91-102.

Ibrahim, A., C. Shimizu & M. Kono. 1984. Pigmentation of cultured red sea bream, *Chrysophrys major*, using astaxanthin from Antarctic krill, *Euphausia superba*, and a mysid, *Neomysis* sp. Aquaculture 38:45-57.

Idler, D.R., L.W. Crim & J.M. Walsh (eds.). 1987. Proc. Third Internat. Symp. Repr. Physiol. Fish, Memorial Univ. Newfoundland, St. John's, 337 p.

Iglesias, J., G. Rodríguez-Ojea & I.E. Calcedo. 1994. Effect of the size of rotifers (*Brachionus plicatilis*) on the first feeding of turbot (*Scophthalmus maximus* L.) larvae. Eur. Aquac. Soc. Spec. Pub. (22):312-317.

Iguchi, K., & K. Maekawa. 1993. Female mate preference and male mating success of ayu fish, *Plecoglossus altivelis* (Osmeridae) under a promiscuous mating system. Ethology 95:193-201.

Iguchi, K., & M. Yamaguchi. 1994. Adaptive significance of inter- and intrapopulational egg size variation in ayu *Plecoglossus altivelis* (Osmeridae). Copeia 1994:184-190.

Iida, T., & H. Wakabayashi. 1992. Preliminary trial of vaccination against edwardsiellosis of hirame (Japanese flounder), *Paralichthys olivaceus*. Diseases in Asian Aquaculture I. Asian Fish. Soc., Manila, p 407-412.

Ikeda, S., Y. Ishibashi, O. Murata, T. Nasu & T. Harada. 1988. Optimum levels of protein and lipid in purified test diet for the Japanese parrot fish. Nippon Suisan Gakkaishi 54:151-154.

Ikeda, T. 1996. Metabolism, body composition, and energy budget of the mesopelagic fish *Maurolicus muelleri* in the Sea of Japan. Fish. Bull., U.S. 94:49-58.

Ikeda, T., S. Chuma & M. Okiyama. 1991. Identification of pelagic eggs of marine fishes by rearing method. Jpn. J. Ichthyol. 38:199-206.

Ikenoue, H., & T. Kafuku. 1992. Modern Methods of Aquaculture in Japan. Elsevier, Amsterdam, 274 p.

Ikewaki, Y., & M. Tanaka. 1993. Feeding habits of Japanese flounder (*Paralichthys olivaceus*) larvae in the western part of Wakasa Bay, the Japan Sea. Nippon Suisan Gakkaishi 59:951-956.

Imaizumi, K. 1993. Recent progress of Japan Sea-Farming Association in marine finfish hatchery activities. In: C.-S. Lee, M.-S. Su & I C. Liao (eds.), Finfish Hatchery in Asia. Tungkang Mar. Lab. Conf. Proc. (3):173-177.

Infante, J.L.Z., & C. Cahu. 1994a. Development and response to a diet change of some digestive enzymes in sea bass (*Dicentrarchus labrax*) larvae. Fish Physiol. Biochem. 12:399-408.

Infante, J.L.Z., & C.L. Cahu. 1994b. Influence of diet on pepsin and some pancreatic enzymes in sea bass (*Dicentrarchus labrax*) larvae. Comp. Biochem. Physiol. 109A:209-212.

Infante, J.L.Z., C.L. Cahu, A. Péres, P. Quazuguel & M.M. Le Gall. 1996. Sea bass (*Dicentrarchus labrax*) larvae fed different *Artemia* rations: growth, pancreas enzymatic response and development of digestive functions. Aquaculture 139:129-138.

Ingram, M. 1988. Farming rainbow trout in fresh water tanks and ponds. In: L.M. Laird & T. Needham (eds.), Salmon and Trout Farming. Ellis Horwood, New York, p 155-189.

Isaakson, A. 1991. Culture of Atlantic salmon. In: R.R. Stickney (ed.), Culture of Salmonid Fishes. CRC Press, Boca Raton, Florida, p 85-129.

Isayeva, N.M., & I.I. Kozinenko. 1992. Effects of various chemical compounds on the immune system of fishes in aquaculture. J. Ichthyol. 32:81-93.

Iseda, H., M. Ishihara, S. Sumita, M. Owaki & S. Tabata. 1982. Prevention of lordosis in the juvenile of red sea bream, *Pagrus major* reared in ponds - II. Relationship between the initial rearing conditions and the lordotic deformity. Bull. Kumamoto Pref. Inst. Fish. 1:9-17. (in Jpn.)

Ishibashi, Y., S. Ikeda, O. Murata, T. Nasu & T. Harada. 1992. Optimal supplementary ascorbic acid level in the Japanese parrot fish diet. Nippon Suisan Gakkaishi 58:267-270. (in Jpn., Eng. abs.)

Ishida, Y., & I. Hidaka. 1987. Gustatory response profiles for amino acids, glycinebetaine, and nucleotides in several marine teleosts. Nippon Suisan Gakkaishi 53:1391-1398.

Ishizaki, Y., T. Takeuchi, T. Watanabe, M. Arimoto & K. Shimizu. 1998. A preliminary experiment on the effect of *Artemia* enriched with arachidonic acid on survival and growth of yellowtail. Fish. Sci. 64:295-299.

Ismond, A.A. 1993. Designing fish farms to reduce the cost of environmental compliance. In: J.-K. Wang (ed.), Techniques for Modern Aquaculture. Am. Soc. Agric. Eng., St. Joseph, Michigan, p 187-192.

Iso, S., S. Suizu & A. Maejima. 1994. The lethal effect of hypertonic solutions and avoidance of marine organisms in relation to discharged brine from a desalination plant. Desalination 97:389-399.

Isono, R.S., Y. Itoh, H. Kinoshita & K. Kido. 1993. Acute toxicity of ozone-produced oxidants to eggs and larvae of Japanese whiting *Sillago japonica*. Nippon Suisan Gakkaishi 59:1527-1533.

Isshiki, T., K. Kawai & R. Kusuda. 1993. Detection of yellowtail ascites virus and neutralizing antibody in brood stocks of yellowtail. Fish Pathol. 28:65-69. (in Jpn., Eng. abs.)

Itzkowitz, M. 1991. Habitat selection and subsequent reproductive success in the beaugregory damselfish. Env. Biol. Fishes 30:287-293.

Iwai, T. 1967. Structure and development of lateral line cupulae in teleost larvae. In: P.H. Cahn (ed.), Lateral Line Detectors. Indiana Univ. Press, Bloomington, p 27-44.

Iwai, T. 1980. Sensory anatomy and feeding of fish larvae. In: J.E. Bardach, J.J. Magnuson, R.C. May & J.M. Reinhart (eds.), Fish Behavior and Its Use in the Capture and Culture of Fishes. ICLARM Conf. Proc. (5):124-145.

Iwai, T.Y., Jr., C.S. Tamaru, L. Yasukochi, S. Alexander, R. Yoshimura & M. Mitsuyasu. 1996. Natural spawning of captive bigeye scad *Selar crumenophthalmus* in Hawaii. J. World Aquac. Soc. 27:332-339.

Iwama, G.K. 1991. Interactions between aquaculture and the environment. Crit. Rev. Env. Control 21:177-216.

Iwama, G., & T. Nakanishi. 1996. The Fish Immune System: Organism, Pathogen, and Environment. Academic Press, San Diego, 380 p.

Iwama, G.K., J.C. McGeer & M.P. Pawluk. 1989. The effects of five fish anaesthetics on acid-base balance, hematocrit, blood gases, cortisol, and adrenaline in rainbow trout. Can. J. Zool. 67:2065-2073.

Iwama, G.K., J.D. Morgan & B.A. Barton. 1995. Simple field methods for monitoring stress and general condition of fish. Aquac. Res. 26:273-282.

Iwama, G.K., A.D. Pickering, J.P. Sumpter & C.B. Schreck (eds.) 1997. Fish Stress and Health in Aquaculture. Cambridge Univ. Press, 278 p.

Iwasaki, H. 1973. Problems in the cultivation and mass culture of marine copepods. Bull. Plankton Soc. Japan 20:72-73.

Iwata, N., K. Kikuchi, H. Honda, M. Kiyono & H. Kurokura. 1994. Fish. Sci. 60:527-531.

Izquierdo, M.S., T. Watanabe, T. Takeuchi, T. Arakawa & C. Kitajima. 1989. Requirement of larval red seabream *Pagrus major* for essential fatty acids. Bull. Jpn. Soc. Sci. Fish. 55:859-867.

Izquierdo, M.S., T. Arakawa, T. Takeuchi, R. Haroun & T. Watanabe. 1992. Effect of *n*-3 HUFA levels in *Artemia* on growth of larval Japanese flounder (*Paralichthys olivaceus*). Aquaculture 105:73-82.

Jachowski, R.L. 1970. Reproductive behavior of the emerald clingfish, *Acyrtops beryllinus* (Hildebrand and Ginsburg). Z. f. Tiersychol. 27:1100-1111.

Jackson, L., et al. 1991. The Science of Capelin: a Variable Resource. Dept. Fish. Oceans, Canada, 34 p.
Jafri, A.K., M. Al-Judaimi & K.A. George. 1981. Preliminary investigations feeding yellow-finned black porgy, *Acanthopagrus latus* (Houttuyn), with artificial feed. Aquaculture 22:117-124.
Jagoe, C.H., V.E. Matey, T.A. Haines & V.T. Komov. 1993. Effect of beryllium on fish in acid water is analogous to aluminum toxicity. Aquat. Toxicol. 24:241-256.
James, C.M. 1993. Use of rotifer chemostats in aquaculture. Ecological Studies: Analysis and Synthesis 98:253-264.
James, C.M. 1996. Technology for marine fish hatchery. Proc. Seminar on Fisheries—A Multibillion Dollar Industry. Aquac. Fdn. India, Madras, p 88-100.
James, C.M., & T. Abu Rezeq. 1989a. Intensive rotifer cultures using chemostats. Hydrobiologia 186/187:423-430.
James, C.M., & T.S. Abu-Rezeq. 1989b. Production and nutritional quality of two small-sized strains of the rotifer *Brachionus plicatilis*. J. World Aquac. Soc. 20:261-267.
James, C.M., M. Bou-Abbas, A.M. Al-Khars, S. Al-Hinty & A.E. Salman. 1983. Production of the rotifer *Brachinous plicatilis* for aquaculture in Kuwait. Hydrobiologia 104:77-84.
James, C.M., P. Dias & A.E. Salman. 1987. The use of marine yeast (*Candida* sp.) and bakers' yeast (*Saccharomyces cerevisiae*) in combination with *Chlorella* sp. for mass culture of the rotifer *Brachionus plicatilis*. Hydrobiologia 147:263-268.
James, C.M., S. Al-Hinty & A.E. Salman. 1989. Growth and ω3 fatty acid and amino acid composition of microalgae under different temperature regimes. Aquaculture 77:337-351.
Jan, R.-Q., & K.-H. Chang. 1984. Influence of feeding jumps on the egg-caring behaviour of male sergeant major *Abudefduf vaigiensis* (Pisces: Pomacentridae). Bull. Inst. Zool., Acad. Sin. 23:159-171.
Jana, B.B. 1994. Ammonification in aquatic environments: a brief review. Limnologica 24:389-413.
Jeney, G., & D.P. Anderson. 1993. Glucan injection or bath exposure given alone or in combination with a bacterin enhance the non-specific defence mechanisms in rainbow trout (*Oncorhynchus mykiss*). Aquaculture 116:315-329.
Jenkins, G.P. 1987. Comparative diets, prey selection, and predatory impact of co-occurring larvae of two flounder species. J. Exp. Mar. Biol. Ecol. 110:147-170.
Jenkins, W.E., & T.I.J. Smith. 1985. Natural and induced production of striped bass hybrids in tanks. Proc. Ann. Conf. SE Assoc. Fish Wildl. Ag. 39:255-260.
Jenkins, W.E., T.I.J. Smith, J.S. Hopkins, J.A. Heffernan & M. Baird. 1993. Intensive brackish water pond culture of striped bass hybrids with minimal water exchange. Eur. Aquac. Soc. Spec. Pub. (19):392.
Jenkins, W.E., C.B. Bridgham & T.I.J. Smith. 1997a. Closely controlled spawning of captive red drum (*Sciaenops ocellatus*) brood stock. World Aquac. '97 Book of Abs., p 227.
Jenkins, W.E., T.I.J. Smith, C.V. Sullivan & D.L. Berlinsky. 1997b. Production of southern flounder *Paralichthys lethostigma* juveniles in an outdoor nursery pond. J. World Aquac. Soc. 28:211-214.
Jennings, S., & M.G. Pawson. 1991. The development of bass, *Dicentrarchus labrax*, eggs in relation to temperature. J. Mar. Biol. Assoc. U.K. 71:107-116.
Jensen, J.O.T., W.C. Clarke, J.N.C. Whyte & W. Damon. 1992. Incubation and larval rearing of sablefish (*Anoplopoma fimbria*) and Pacific halibut (*Hippoglossus stenolepis*). Bull. Aquac. Assoc. Can. 92:49-51.
Jenson, A. 1993. Carrying capacity for nutrient salts in coastal waters. In: H. Reinertsen, L.A. Dahle, L. Jørgensen & K. Tvinnereim (eds.), Fish Farming Technology. Balkema, Rotterdam, p 151-154.
Jeon, I.G. 1995. A review of the nursery and growout culture techniques for Korean rockfish (*Sebastes schlegeli*) in Korea. In: K.L Main & C. Rosenfeld (eds.), Culture of High-value Marine Fishes in Asia and the United States. Oceanic Inst., Honolulu, p 81-90.
Jeon, I.G., K.S. Min, J.M. Lee, K.S. Kim & M.H. Son. 1993. Optimal stocking density for olive flounder, *Paralichthys olivaceus*, rearing in tanks. Bull. Nat. Fish. Res. Dev. Ag. (48):57-70. (in Kor., Eng. abs.)
Jessop, B.M. 1993. Fecundity of anadromous alewives and blueback herring in New Brunswick and Nova Scotia. Trans. Am. Fish. Soc. 122:85-98.
Jickells, T.D., & A.H. Knap. 1984. The distribution and geochemistry of some trace metals in the Bermuda coastal environment. Est. Coastal Shelf Sci. 18:245-262.
Jickells, T.D., A.H. Knap & S.R. Smith. 1986. Trace metal and nutrient fluxes through the Bermuda inshore waters. Rapp. P.-v. Réun. Cons. Int. Explor. Mer 186:251-262.
Jirsa, D.O., D.A. Davis & C.R. Arnold. 1997. Effects of dietary nutrient density on water quality and growth of red drum *Sciaenops ocellatus* in closed systems. J. World Aquac. Soc. 28:68-78.
Jobling, M. 1981. Temperature tolerance and the final preferendum--rapid methods for the assessment of optimum growth temperatures. J. Fish. Biol. 19:439-455.
Jobling, M. 1982. Some observations on the effects of feeding frequency on the food intake and growth of plaice, *Pleuronectes platessa* L. J. Fish. Biol. 20:431-444.
Jobling, M. 1988. A review of the physiological and nutritional energetics of cod, *Gadus morhua* L., with particular reference to growth under farmed conditions. Aquaculture 70:1-19.
Jobling, M. 1990. Fulfilling the impossible dream. Fish-Farmer (Jul-Aug):52-53.
Jobling, M. 1991. Arctic char, *Salvelinus alpinus*. In: R.P. Wilson (ed.), Handbook of Nutrient Requirements of Finfish. CRC Press, Boca Raton, Florida, p 1-4.
Jobling, M. 1995. Environmental Biology of Fishes. Chapman & Hall, London, 455 p.

Jobling, M., H. Tveitan & B. Hatlen. 1998. Cultivation of Arctic charr: an update. Aquac. Internat. 6:181-196.
Johannessen, T., J. Gjosæter & E. Moksness. 1993. Reproduction, spawning behaviour and captive breeding of the common wolffish *Anarhichas lupus* L. Aquaculture 115:41-51.
Johnsen, F., M. Hillestad & E. Austreng. 1993. High energy diets for Atlantic salmon. effects on pollution. Fish Nutrition in Practice, INRA, Paris, p 391-401.
Johnsen, P.B. & H.K. Dupree. 1991. Influence of feed ingredients on the flavor quality of farm-raised catfish. Aquaculture 96:139-150.
Johnson, A.G., T.D. Williams & C.R. Arnold. 1977. Chlorine-induced mortality of eggs and larvae of spotted seatrout (*Cynoscion nebulosus*). Trans. Am. Fish. Soc. 106:466-469.
Johnson, A.G., T.D. Williams, J.F. Messinger III & C.R. Arnold. 1979. Larval spotted seatrout (*Cynoscion nebulosus*) a bioassay subject for the marine subtropics. Contr. Mar. Sci. Univ. Texas 22:57-62.
Johnson, D.W., & I. Katavic. 1984. Mortality, growth and swim bladder stress syndrome of sea bass (*Dicentrarchus labrax*) larvae under varied environmental conditions. Aquaculture 38:67-78.
Johnson, D.W., & I. Katavic. 1986. Survival and growth of sea bass (*Dicentrarchus labrax*) larvae as influenced by temperature, salinity, and delayed initial feeding. Aquaculture 52:11-19.
Johnson, G.D. 1978. Development of Fishes of the Mid-Atlantic Bight: An Atlas of Egg, Larval and Juvenile Stages, Vol. 4. U.S. Fish Wildl. Serv., 314 p.
Johnson, G.D. 1984. Percoidei: development and relationships. Am. Soc. Ichthyol. Herpetol. Spec. Pub. (1):464-498.
Johnson, G., & R. Armstrong. 1997. A careful look at the Canadian fish health protection regulations. In: R. Armstrong (preparer), The Salmon Health Report. North. Aquac. Suppl., p 14-21.
Johnson, M.W., & J.B. Olson. 1948. The life history and biology of a marine harpacticoid copepod, *Tisbe furcata* (Baird). Biol. Bull. 95:320-332.
Johnson, S.C., M.L. Kent, D.J. Whitaker & L. Margolis. 1993. Toxicity and pathological effects of orally administered ivermectin in Atlantic, chinook, and coho salmon and steelhead trout. Dis. Aquat. Org. 17:107-112.
Johnson, S.K. 1990. Recognition and control of diseases common to grow-out aquaculture of red drum. Texas A&M Univ. Sea Grant TAMU-SG-90-603, p 113-130.
Johnson, W.W., & M.T. Finley. 1980. Handbook of Acute Toxicity of Chemicals to Fish and Aquatic Invertebrates. U.S. Fish. Wildl. Serv., Resour. Pub. 137, 98 p.
Johnstone, R., T.H. Simpson & A.F. Youngson. 1978. Sex reversal in salmonid culture. Aquaculture 13:115-134.
Jónasdóttir, S.H. 1994. Effects of food quality on the reproductive success of *Acartia tonsa* and *Acartia hudsonica*: laboratory observations. Mar. Biol. 121:67-81.
Jones, A. 1972. Studies on egg development and larval rearing of turbot, *Scophthalmus maximus* L., and brill, *Scophthalmus rhombus* L., in the laboratory. J. Mar. Biol. Assn. U.K. 52:965-987.
Jones, A., R. Alderson & B.R. Howell. 1974. Progress towards the development of a successful rearing technique for larvae of the turbot, *Scophthalmus maximus* L. In: J.H.S. Blaxter (ed.), The Early Life History of Fish. Springer-Verlag, New York, p 731-737.
Jones, A., R.A. Prickett & M.T. Douglas. 1981. Recent developments in techniques for rearing marine flatfish larvae, particularly turbot (*Scophthalmus maximus* L.), on a pilot commercial scale. Rapp. P.-v. Réun. Cons. Int. Explor. Mer 178:522-526.
Jones, D.A., K. Kurmaly & A. Arshad. 1987. Penaeid shrimp hatchery trials using microencapsulated diets. Aquaculture 64:133-146.
Jones, Gordon. 1998. Innovative Aquaculture Products Ltd., Skerry Bay, British Columbia, pers. comm.
Jones, P.W., F.D. Martin & J.D. Hardy, Jr. 1978. Development of Fishes of the Mid-Atlantic Bight: An Atlas of Egg, Larval and Juvenile Stages, Vol. 1. U.S. Fish Wildl. Serv., 366 p.
Jones, S., & V.B. Pantulu. 1958. On some larval and juvenile fishes from the Bengal and Orissa coasts. Indian J. Fish. 5:118-143.
Jørgensen, J.B., G.J.E. Sharp, C.J. Secombes & B. Robertsen. 1993a. Effect of a yeast-cell-wall glucan on the bactericidal activity of rainbow trout macrophages. Fish Shellfish Immunol. 3:267-277.
Jørgensen, J.B., H. Lunde & B. Robertsen. 1993b. Peritoneal and head kidney cell response to intraperitoneally injected yeast glucan in Atlantic salmon, *Salmo salar* L. J. Fish Dis. 16:313-325.
Jory, Darryl E. 1991. University of Miami, Florida, pers. comm.
Jory, D.E., E.S. Iversen & R.H. Lewis. 1985. Culture of fishes of the genus *Trachinotus* (Carangidae) in the western Atlantic: prospects and problems. J. World Maric. Soc. 16:87-94.
Joseph, M.M. (ed.). 1990. Aquaculture in Asia. Asian Fish. Soc., Indian Branch, Mangalore, India, 395 p.
Joyce, J.E., R.M. Martin & F.P. Thrower. 1993. Successful maturation and spawning of captive chinook salmon brood stock. Prog. Fish-Cult. 55:191-194.
Joyeux, J.-C., J.-L. Bouchereau & J.-A. Tomasini. 1991. Data about reproduction of *Gobius niger* (Pisces, Gobiidae) in brackish Lagoon of Mauguio—France: gonosomatic ratios, fecundities, spawning, eggs and larvae. Vie Milieu 41:97-106. (in Fr., Eng. abs.)
Ju, B. 1994. Inducing triploidy in red sea bream *Pagrus major*. Third Asian Fisheries Forum, Asian Fish. Soc., Manila, p 540-543.

Juanes, F. 1994. What determines prey size selectivity in piscivorous fishes? In: D.J. Stouder, K.L. Fresh & R.J. Feller (eds.), Theory and Application in Fish Feeding Ecology. Belle W. Baruch Library for Mar. Sci., Univ. South Carolina Press, p 79-110.

Juario, J.V. 1991. Status of finfish mariculture in the Philippines. In: Advances in Finfish and Shellfish Mariculture. Phil. Council Aquat. Mar. Res. Dev., Book Ser. (12):27-36.

Juario, J.V., M.N. Duray, V.M. Duray, J.F. Nacario & J.M.E. Almendras. 1984. Induced breeding and larval rearing experiments with milkfish *Chanos chanos* (Forskal) in the Philippines. Aquaculture 36:61-70.

Juario, J.V., M.N. Duray, V.M. Duray, J.F. Nacario & J.M.E. Almendras. 1985. Breeding and larval rearing of the rabbitfish *Siganus guttatus* (Bloch). Aquaculture 44:91-101.

Juario, J.V., M.N. Duray & J. Fuchs. 1991. Weaning of seabass, *Lates calcarifer* B., larvae to artificial diet. Eur. Aquac. Soc. Spec. Pub. (15):183.

Juell, J.-E. 1991. Hydroacoustic detection of food waste — a method to estimate maximum food intake of fish populations in sea cages. Aquacultural Eng. 10:207-217.

Jug-Dujaković, J., & B. Glamuzina. 1988. Preliminary studies of reproduction and early life history of *Diplodus vulgaris* (E. Geoffroy Saint-Hilaire 1817) in captivity. Aquaculture 69:367-377.

Juhl, T. 1992. Commercial breeding of anemonefishes. SeaScope, Aquarium Systems, Mentor, Ohio, 9(winter):1-4.

Jungalwalla, P.J. 1991. The development of an integrated saltwater salmonid farming industry in Tasmania, Australia. Can. Tech. Rep. Fish. Aquat. Sci. (1831):65-73.

Kabata, Z. 1985. Parasites and Diseases of Fish Cultured in the Tropics. Taylor & Francis, London, 318 p.

Kaeriyama, M. 1989. Comparative morphology and scale formation in four species of *Oncorhynchus* during early life. Jpn. J. Ichthyol. 35:445-452.

Kagawa, H., H. Tanaka, H. Ohta, K. Okuzawa & K. Hirose. 1995. *In vitro* effects of 17α-hydroxyprogesterone and 17α,20β-dihydroxy-4-pregnen-3-one on final maturation of oocytes at various developmental stages in artificially matured Japanese eel *Anguilla japonica*. Fish. Sci. 61:1012-1015.

Kagawa, H., H. Tanaka, H. Ohta, K. Okuzawa & N. Iinuma. 1997. Induced ovulation by injection of 17,20β-dihydroxy-4-pregnen-3-one in the artificially matured Japanese eel, with special reference to ovulation time. Fish. Sci. 63:365-367.

Kagawa, H., N. Iinuma, H. Tanaka, H. Ohta & K. Okuzawa. 1998. Effects of rearing period in seawater on induced maturation in female Japanese eel *Anguilla japonica*. Fish. Sci. 64:77-82.

Kah, O. 1986. Central regulation of reproduction in teleosts. Fish Physiol. Biochem. 2:25-34.

Kah, O., I. Anglade, E. Lepretre, P. Dubourg & D. de Monbrison. 1993. The reproductive brain in fish. Fish Physiol. Biochem. 11:85-98.

Kahan, D., G. Uhlig, D. Schwenzer & L. Horowitz. 1981/1982. A simple method for cultivating harpacticoid copepods and offering them to fish larvae. Aquaculture 26:303-310.

Kaji, T., M. Tanaka, Y. Takahashi, M. Oka & N. Ishibashi. 1996. Preliminary observations on development of Pacific bluefin tuna *Thunnus thynnus* (Scombridae) larvae reared in the laboratory, with special reference to the digestive system. Mar. Freshw. Res. 47:261-269.

Kakuno, A., & J. Koyama. 1994. Comparative studies on fish blood—II Seasonal variations of hematological parameters in marine teleostei. Bull. Nat. Res. Inst. Fish. Sci. (6):67-77.

Kalogeropoulos, N., M.N. Alexis & R.J. Henderson. 1993. Effect of dietary lipids on tissue fatty acid composition of gilthead bream (*Sparus aurata*). Fish Nutrition in Practice, INRA, Paris, p 257-267.

Kanabashira, Y., H. Sakai & F. Yasuda. 1980. Early development and reproductive behavior of the gobiid fish, *Mugilogobius abei*. Jpn. J. Ichthyol. 27:191-198.

Kanazawa, A. 1985. Essential fatty acid and lipid requirement of fish. In: C.B. Cowey, A.M. Mackie & J.G. Bell (eds.), Nutrition and Feeding in Fish. Academic Press, San Diego, California, p 281-298.

Kanazawa, A. 1991a. Ayu, *Plecoglossus altivelis*. In: R.P. Wilson (ed.), Handbook of Nutrient Requirements of Finfish. CRC Press, Boca Raton, Florida, p 23-29.

Kanazawa, A. 1991b. Puffer fish, *Fugu rubripes*. In: R.P. Wilson (ed.), Handbook of Nutrient Requirements of Finfish. CRC Press, Boca Raton, Florida, p 123-129.

Kanazawa, A. 1993a. Essential phospholipids of fish and crustaceans. Fish Nutrition in Practice, INRA, Paris, p 519-530.

Kanazawa, A. 1993b. Importance of dietary docosahexaenoic acid on growth and survival of fish larvae. In: C.-S. Lee, M.-S. Su & I C. Liao (eds.), Finfish Hatchery in Asia. Tungkang Mar. Lab. Conf. Proc. (3):87-95.

Kanazawa, A. 1995. Nutrition of larval fish. In: C. Lim & D.J. Sessa (eds.), Nutrition and Utilization Technology in Aquaculture. AOSC Press, Champaign, Ilinois, p 50-59.

Kanazawa, A. 1997. Effects of docosahexaenoic acid and phospholipids on stress tolerance of fish. Aquaculture 155:129-134.

Kanazawa, A., S.-I. Teshima & K. Ono. 1979. Relationship between essential fatty acid requirements of aquatic animals and the capacity for bioconversion of linolenic acid to highly unsaturated fatty acids. Comp. Biochem. Physiol. 63B:295-298.

Kanazawa, A., S.-I. Teshima, M. Sakamoto & A. Shinomiya. 1980. Nutritional requirements of the puffer fish: purified test diet and the optimal protein level. Bull. Jpn. Soc. Sci. Fish. 46:1357-1361.

Kanazawa, A., S. Teshima, S. Inamori, S. Sumida & T. Iwashita. 1982. Rearing of larval red sea bream and ayu with artificial diets. Mem. Fac. Fish. Kagoshima Univ. 31:185-192.

Kanazawa, A., S.-I. Teshima, S. Inamori & H. Matsubara. 1983a. Effects of dietary phospholipids on growth of the larval red sea bream and knife jaw. Mem. Fac. Fish. Kagoshima Univ. 32:109-114.

Kanazawa, A., S.-I. Teshima, T. Kobayashi, M. Iwashita & R. Uehdra. 1983b. Necessity of dietary phospholipids for growth of the larval ayu. Mem. Fac. Fish. Kagoshima Univ. 32:115-120.

Kanazawa, A., S. Koshio & S.-I. Teshima. 1989. Growth and survival of larval red sea bream *Pagrus major* and Japanese flounder *Paralichthys olivaceus* fed microbound diets. J. World Aquac. Soc. 20:31-37.

Kane, J. 1984. The feeding habits of co-occurring cod and haddock larvae from Georges Bank. Mar. Ecol. Prog. Ser. 16:9-20.

Kane, T.R. 1988. Relationship of temperature and time of initial feeding of Atlantic salmon. Prog. Fish-Cult. 50:93-97.

Karanikas, J. 1989. The spawning of the flame angel *Centropyge loriculus*. SeaScope, Aquarium Systems, Mentor, Ohio, 7(spring):1-2.

Katavic, I. 1986. Diet involvement in mass mortality of sea bass (*Dicentrarchus labrax*) larvae. Aquaculture 58:45-54.

Katayama, M. 1960. Fauna Japonica: Serranidae (Pisces). Tokyo News Service, Tokyo, 189 p + 86 plates.

Kato, T. 1994. Biotechnological researches at the prefectural fisheries experimental station in Japan. In: F. Lacanilao, R.M. Coloso & G.F. Quinitio (eds.), Aquaculture Development in Southeast Asia and Japan and Prospects for Seafarming and Searanching. SEAFDEC, Iloilo, Philippines, p 68-74.

Kattman, D.L. 1977. The way to do it from one inch to parenthood. Mar. Hob. News 5(1):1,3.

Kattman D., & D. Kattman. 1974. The courtship and spawning of *Dascyllus melanurus*. Octopus 1(8):20-23.

Kauffman, T.A., J. Lindsay & R. Leithiser. 1981. Vertical distribution and food selection of larval atherinids. Rapp. P.-v. Réun. Cons. Int. Explor. Mer 178:342-343.

Kausar, R., & J.V. Juario. 1991. Induced spawning and larval growth of the rabbit fish, *Siganus guttatus* (Osteichthyes; Siganidae). Pak. J. Sci. Ind. Res. 34:184-187.

Kaushik, S.J., & F. Médale. 1994. Energy requirements, utilization and dietary supply to salmoninds. Aquaculture 124:81-97.

Kaushik, S.J., P. Luquet, D. Blanc & A. Paba. 1989. Studies on the nutrition of Siberian sturgeon, *Acipenser baeri* I. Utilization of digestible carbohydrates by sturgeon. Aquaculture 76:97-107.

Kaushik, S.J., J. Breque & D. Blanc. 1991. Requirements for protein and essential amino acids and their utilization by Siberian sturgeon (*Acipenser baeri*). In: P. Williot (managing ed.), Acipenser. CEMAGREF, Bordeaux, p 25-39.

Kawabe, R. 1984. Spawning behavior of the bridled triggerfish, *Sufflamen fraenatus*, in the aquarium. Jpn. J. Ichthyol. 31:193-197. (in Jpn., Eng. abs.)

Kawahara, S., A.J. Shams, A.A. Al-Bosta, M.H. Mansoor & A.A. Al-Baqqal. 1997. Effects of incubation and spawning water temperature, and salinity on egg development of the orange-spotted grouper (*Epinephelus coioides*, Serranidae). Asian Fish. Sci. 9:239-250.

Kawakami, H., N. Shinohara, Y. Fukuda, H. Yamashita, H. Kihara & M. Sakai. 1997. The efficacy of lipopolysaccharide mixed chloroform-killed cell (LPS-CKC) bacterin of *Pasteurella piscicida* on yellowtail, *Seriola quinqueradiata*. Aquaculture 154:95-105.

Kawamura, G., & K. Ishida. 1985. Changes in sense organ morphology and behaviour with growth in the flounder *Paralichthys olivaceus*. Bull. Jpn. Soc. Sci. Fish. 51:155-165.

Kawazoe, I., T. Noso, S. Kuriyama, A. Akasaka & H. Kawauchi. 1988. Isolation and characterization of growth hormone from yellowtail *Seriola quinqueradiata*. Nippon Suisan Gakkaishi 54:393-399.

Kaya, C.M., A.E. Dizon, S.D. Hendrix, T.K. Kazama & M.K.K. Queenth. 1982. Rapid and spontaneous maturation, ovulation, and spawning of ova by newly captured skipjack tuna, *Katsuwonus pelamis*. Fish. Bull., U.S. 80:393-396.

Kayano, Y., & T. Oda. 1987. Spawning of red spotted grouper *Epinephelus akaara* reared in tank, 1986. Bull. Fish. Exp. Sta. Okayama Pref. 2:52-55. (in Jpn.)

Kearn, G.C., K. Ogawa & Y. Maeno. 1992a. Egg production, the oncomiracidium and larval development of *Benedenia seriolae*, a skin parasite of the yellowtail, *Seriola quinqueradiata*, in Japan. Pub. Seto Mar. Biol. Lab. 35:351-362.

Kearn, G.C., K. Ogawa & Y. Maeno. 1992b. Hatching patterns of the monogenean parasites *Benedenia seriolae* and *Heteraxine heterocerca* from the skin and gills, respectively, of the same host fish, *Seriola quinqueradiata*. Zool. Sci. 9:451-455.

Keefe, M., & K.W. Able. 1993. Patterns of metamorphosis in summer flounder, *Paralichthys dentatus*. J. Fish Biol. 42:713-728.

Keembiyehetty, C.N., & D.M. Gatlin III. 1997. Performance of sunshine bass fed soybean-meal-based diets supplemented with different methionine compounds. Prog. Fish-Cult. 59:25-30.

Keenlyne, K.D., & L.G. Jenkins. 1993. Age at sexual maturity of the pallid sturgeon. Trans. Am. Fish. Soc. 122:393-396.

Kelley, A.M.. 1946. Effect of abnormal CO_2 tension on development of herring eggs. J. Fish. Res. Bd. Can. 6:435-440.

Kelley, C., & C.-S. Lee. 1986. Artificial propagation. In: C.-S. Lee, M.S. Gordon & W.O. Watanabe (eds.), Aquaculture of Milkfish (*Chanos chanos*): State of the Art. Oceanic Inst., Honolulu, p 83-116.

Kelley, C., & C.-S. Lee. 1991. Milkfish culture in the U.S. In: J.P. McVey (ed.), CRC Handbook of Mariculture: Vol. II, Finfish Aquaculture. CRC Press Boca Raton, Florida, p 211-226.

Kelley, C.D., C.-S. Lee & C.S. Tamaru. 1987. Acceleration of maturation and double spawning in one season in the grey mullet (*Mugil cephalus*) following chronic hormone treatment. Proc. Third Internat. Symp. Repr. Physiol. Fish, Memorial Univ. Newfoundland, St. John's, p 203.

Kelley, C., P. Bass, & K. Leber. 1995. The production of Pacific threadfin, *Polydactylus sexfilis*, fry for stock enhancement and aquaculture. Aquac. '95 Book of Abs., p 151.

Kelly, K.H., & J.R. Moring. 1986. Species profiles: life histories and environmental requirements of coastal fishes and invertebrates (North Atlantic), Atlantic herring. U.S. Fish Wildl. Serv., Biol. Rep. 82(11.38), 22 p.

Kendall, A.W., Jr. & R.J. Behnke. 1984. Salmonidae: development and relationships. Am. Soc. Ichthyol. Herpetol. Spec. Pub. (1):142-149.

Kendall, W.C., & E.L. Goldsborough. 1911. Reports on the scientific results of the expedition to the tropical Pacific, in charge of Alexander Agassiz, by the U.S. Fish Commission Steamer "Albatross," from August 1899 to March 1900. 13. The shore fishes. Mem. Mus. Comp. Zool. Harvard 26(7):293-344.

Kennedy, Sarah Blain. Doctor of Veterinary Medicine, Private Practice, Vero Beach, Florida, pers. comm.

Kennedy, S.B., J.W. Tucker, Jr., C.L. Neidig, G.K. Vermeer, V.R. Cooper, J.L. Jarrell & D.G. Sennett. 1998. Bacterial management strategies for stock enhancement of warmwater marine fish: a case study with common snook, *Centropomus undecimalis*. Bull. Mar. Sci. 62: .

Kentouri, M. 1980. Elevage des larves de loup (*Dicentrarchus labrax* L.) a l'aide d'organismes du zooplancton congele: resultats preliminaires. Aquaculture 21:171-180.

Kentouri, M. 1981. Donnees preliminaires sur les facultes d'adaptation a un aliment inerte (zooplancton congele) des post-larves de 11 especes de poissons et crustace marins. Aquaculture 23:73-82.

Kentouri, M., P. Divanach, N. Papandroulakis & S. Stefanakis. 1993a. Biological performance of intensively reared sea bream (*Sparus aurata*) larvae in a pilot closed system. Eur. Aquac. Soc. Spec. Pub. (19):396.

Kentouri, M., P. Divanach & E. Maingot. 1993b. Comparaison de l'efficacite-cout de trois techniques de rationnement de la daurade *Sparus aurata*, en elevage intensif en bassins. Eur. Aquac. Soc. Spec. Pub. (18):273-283.

Kentouri, M., M. Anthouard & P. Divanach. 1993c. Response to colors in *Dicentrarchus labrax* (Moronidae), in a self-feeding and multiple choice condition. Mar. Ecol. 14:259-274.

Kerby, J.H., & R.M. Harrell. 1990. Hybridization, genetic manipulation, and gene pool conservation of striped bass. In: R.M. Harrell, J.H. Kerby & R.V. Minton (eds.), Culture and Propagation of Striped Bass and Its Hybrids. Am. Fish. Soc., Bethesda, Maryland, p 159-190.

Kerr, S.R. 1982. Estimating the energy budgets of actively predatory fishes. Can. J. Fish. Aquat. Sci. 39:371-379.

Kerstitch, A. 1977. Butterflies and angels of the Sea of Cortez. Mar. Aquarist 7(9):17-28.

Ketola, H.G. 1976. Choline metabolism and nutritional requirements of lake trout (*Salvelinus namaycush*). J. Anim. Sci. 43:474-477.

Ketola, H.G., C.E. Smith & G.A. Kindschi. 1989. Influence of diet and oxidative rancidity on fry of Atlantic and coho salmon. Aquaculture 79:417-423.

Khan, M.S., M.S. Zakaria, M.A. Ambak, M.J. Alam, M. Kashiwagi & T. Iwai. 1991. Effect of temperature on the hatching success of the eggs of Japanese whiting, *Sillago japonica* Temminck & Schlegel, during the spawning season. Aquac. Fish. Mgt. 22:317-321.

Khanaichenko, A.N., Y.E. Bitukova & N.K. Tkachenko. 1994. Experiences in rearing endemic Black Sea turbot larvae. Eur. Aquac. Soc. Spec. Pub. (22):349-358.

Kikuchi, K., & I. Sakaguchi. 1997. Blue mussel as an ingredient in the diet of juvenile Japanese flounder. Fish. Sci. 837-838.

Kikuchi, K., T. Furuta & H. Honda. 1994a. Utilization of feather meal as a protein source in the diet of juvenile Japanese flounder. Fish. Sci. 60:203-206.

Kikuchi, K., T. Furuta & H. Honda. 1994b. Utilization of soybean meal as a protein source in the diet of juvenile Japanese flounder, *Paralichthys olivaceus*. Suisanzoshoku 42:601-604.

Kikuchi, K., T. Sato, T. Furuta, I. Sakaguchi & Y. Deguchi. 1997a. Use of meat and bone meal as a protein source in the diet of juvenile Japanese flounder. Fish. Sci. 63:29-32.

Kikuchi, K., T. Sato, T. Furuta & I. Sakaguchi. 1997b. Use of meat meal as a protein source in the diet of juvenile Japanese flounder, *Paralichthys olivaceus*. World Aquac. '97 Book of Abs., p 252.

Kim, B.G., A.C. Ostrowski & C. Brownell. 1993. Review of hatchery design and techniques used at the Oceanic Institute for intensive culture of the mahimahi (*Coryphaena hippurus*) on a commercial scale. In: C.-S. Lee, M.-S. Su & I C. Liao (eds.), Finfish Hatchery in Asia. Tungkang Mar. Lab. Conf. Proc. (3):179-190.

Kim, E.-h., & T. Aoki. 1993. Drug resistance and broad geographical distribution of identical R plasmids of *Pasteurella piscicida* isolated from cultured yellowtail in Japan. Microbiol. Immunol. 37:103-109.

Kim, H.B., & J.G. Myoung. 1995. Yellowtail (*Seriola quinqueradiata*), red seabream (*Pagrus major*) and salmon (*Oncorhynchus* sp.) culture in Korea. In: K.L. Main & C. Rosenfeld (eds.) Culture of High-value Marine Fishes in Asia and the United States. Oceanic Inst., Honolulu, p 91-99.

Kim, J., K.C. Massee & R.W. Hardy. 1996a. Adult *Artemia* as food for first feeding coho salmon (*Oncorhynchus kisutch*). Aquaculture 144:217-226.

Kim, J.-Y., & Y.-J. Kang. 1992. Spawning ecology of anchovy, *Engraulis japonica*, in the southern waters of Korea. Bull. Kor. Fish. Soc. 25:331-340. (in Kor., Eng. abs.)

Kim, K.-K., I.C. Bang, Y. Kim, Y.K. Nam & D.S. Kim. 1996b. Early survival and chromosomes of intergeneric hybrids between Japanese flounder *Paralichthys olivaceus* and spotted halibut *Verasper variegatus*. Fish. Sci. 62:490-491.

Kim, S. 1992. Status of fishery and science of Bering Sea walleye pollock: (III) Biological characteristics (adult). Ocean Res. 14:171-189. (in Kor., Eng. abs.)

Kime, D.E. 1993. 'Classical' and 'non-classical' reproductive steroids in fish. Rev. Fish Biol. Fish. 3:160-180.

Kimura, R., S. Kawaguchi & S. Sugitani. 1993. Automatic feeding control system for fish. In: H. Reinertsen, L.A. Dahle, L. Jørgensen & K. Tvinnereim (eds.), Fish Farming Technology. Balkema, Rotterdam, p 217-225.

Kimura, S., & T. Kiriyama. 1992a. Natural spawning of the labrid fish, *Halichoeres poecilopterus* in a tank. Suisanzoshoku 40:81-85. (in Jpn., Eng. abs.)

Kimura, S., & T. Kiriyama. 1992b. Size, hatching rate, and incubation time of eggs of the labrid fish, *Halichoeres poecilopterus*. Suisanzoshoku 40:87-92.

Kimura, S., & T. Kiriyama. 1993. Development of eggs, larvae and juveniles of the labrid fish, *Halichoeres poecilopterus*, reared in the laboratory. Jpn. J. Ichthyol. 39:371-377.

Kimura, S., & Y. Tsukamoto. 1990. Development of larvae and juveniles of the atherinid fish, *Atherion elymus*, reared in the laboratory. Jpn. J. Ichthyol. 37:29-33.

Kimura, S., K. Tsumoto & K. Mori. 1987. Development of eggs, larvae and juveniles of the cottid fish, *Pseudoblennius cottoides*, reared in the laboratory. Jpn. J. Ichthyol. 34:346-350.

Kimura, S., K. Tsumoto & K. Mori. 1988. Development of the cottid fish, *Pseudoblennius percoides*, reared in the laboratory, with brief descriptions of juvenile *P. marmoratus* and *P. zonostigma*. Jpn. J. Ichthyol. 35:19-24.

Kimura, S., Y. Tsukamoto & K. Mori. 1989. Early developmental stages of the scorpaenid fish, *Scorpaena miostoma*, reared in the laboratory. Jpn. J. Ichthyol. 35:434-439.

Kimura, T., & M. Yoshimizu. 1991. Viral diseases of fish in Japan. Ann. Rev. Fish Dis. 1:67-82.

Kincheloe, J.W., G.A. Wedemeyer & D.L. Koch. 1979. Tolerance of developing salmonid eggs and fry to nitrate exposure. Bull. Env. Contam. Toxicol. 23:575-578.

Kindschi, G.A., & E. MacConnell. 1989. Factors influencing early mortality of walleye fry reared intensively. Prog. Fish-Cult. 51:220-226.

King, N., H. Howell, A. Tomlinson, D. Bidwell & E. Fairchild. 1997. Effect of microalgae and live diet type on the growth of first feeding winter flounders (*Pleuronectes americanus*, Walbaum). World Aquac. '97 Book of Abs., p 256.

Kingsford, E. 1975. Treatment of Exotic Marine Fish Diseases. Palmetto Publishing Co., St. Petersburg, Florida, 92 p.

Kingsford, M.J. 1985. The demersal eggs and planktonic larvae of *Chromis dispilus* (Teleostei: Pomacentridae) in north-eastern New Zealand coastal waters. N.Z. J. Mar. Freshw. Res. 19:429-438.

Kinne, O. 1976. Cultivation of marine organisms. In: O. Kinne (ed.), Marine Ecology, Vol. 3. Wiley-Interscience, London, p 19-300.

Kinne, O. (ed.). 1984. Introduction: Pisces. Vol. 4, Part 1. Diseases of Marine Animals. Biol. Anst. Helgol., Hamburg, Germany, 541 p.

Kissil, G.W. 1991. Gilthead seabream, *Sparus aurata*. In: R.P. Wilson (ed.), Handbook of Nutrient Requirements of Finfish. CRC Press, Boca Raton, Florida, p 83-88

Kissil, G.W. 1996. Aquaculture in Israel. World Aquac. 27(1):25-30.

Kissil, G.W., & W.M. Koven. 1990. Preparation of oils, enhanced in highly unsaturated fatty acid (HUFA) content, by low temperature crystallization separation, for rotifer (*Brachionus plicatilis*) enrichment. Aquaculture 88:69-74.

Kitajima, C. 1973. Experimental trials on mass cultures of copepods. Bull. Plankton Soc. Japan 20:54-60.

Kitajima, C., & Y. Tsukashima. 1982. Effects of depth of the rearing tank on the incidence of swim bladder inflation of larval red sea bream, *Pagrus major*. Bull. Nagasaki Pref. Inst. Fish. 8:137-140. (in Jpn.)

Kitajima, C., & Y. Tsukashima. 1983. Morphology, growth and low temperature- and low salinity-tolerance of sparid hybrids. Jpn. J. Ichthyol. 30:275-283.

Kitajima, C., Y. Tsukashima, S. Fujita, T. Watanabe & Y. Yone. 1981. Relationship between uninflated swim bladders and lordotic deformity in hatchery-reared red sea bream *Pagrus major*. Bull. Jpn. Soc. Sci. Fish. 47:1289-1294.

Kitajima, C., Y. Tsukashima & M. Tanaka. 1985. The voluminal changes of swim bladder of larval red sea bream *Pagrus major*. Bull. Jpn. Soc. Sci. Fish. 51:759-764.

Kitajima, C., G. Hayashida & S. Yasumoto. 1988. Early development of the laboratory-reared flounder, *Pleuronichthys cornutus*. Jpn. J. Ichthyol. 35:69-77.

Kitajima, C., M. Takaya, Y. Tsukashima & T. Arakawa. 1991. Development of eggs, larvae and juveniles of the grouper, *Epinephelus septemfasciatus*, reared in the laboratory. Jpn. J. Ichthyol. 38:47-55. (in Jpn., Eng. abs.)

Kitajima, C., T. Watanabe, Y. Tsukashima & S. Fujita. 1994. Lordotic deformation and abnormal development of swim bladders in some hatchery-bred marine physoclistous fish in Japan. J. World Aquac. Soc. 25:64-77.

Kitchell, J.F. 1983. Energetics. In: P.W. Webb & D. Weihs (eds.), Fish Biomechanics. Praeger, New York, p 312-338.

Kiyono, M., & R. Hirano. 1981. Effects of light on the feeding and growth of black porgy, *Mylio macrocephalus* (Basilewsky), postlarvae and juveniles. Rapp. P.-v. Réun. Cons. Int. Explor. Mer 178:334-336.

Kjartansson, H. 1993. Salmonid farming in Iceland. In: H. Reinertsen, L.A. Dahle, L. Jørgensen & K. Tvinnereim (eds.), Fish Farming Technology. Balkema, Rotterdam, p 333-340.

Kjesbu, O.S. 1989. The spawning activity of cod, *Gadus morhua* L. J. Fish Biol. 34:195-206.

Kjesbu, O.S., & H. Kryvi. 1993. A histological examination of oocyte final maturation in cod (*Gadus morhua* L.). In: B.T. Walther & H.J. Fyhn (eds.), Physiological and Biochemical Aspects of Fish Development. Univ. Bergen, Norway, p 86-93.

Kjesbu, O.S., J. Klungsoyr, H. Kryvi, P.R. Witthames & M.G. Walker. 1991. Fecundity, atresia, and egg size of captive Atlantic cod (*Gadus morhua*) in relation to proximate body composition. Can. J. Fish. Aquat. Sci. 48:2333-2343.

Kjørsvik, E. 1994. Egg quality in wild and broodstock cod *Gadus morhua* L. J. World Aquac. Soc. 25:22-29.

Kjørsvik, E., & I. Holmefjord. 1995. Atlantic halibut (*Hippoglossus hippoglossus*) and cod (*Gadus morhua*). In: N.R. Bromage & R.J. Roberts (eds.), Broodstock Management and Egg and Larval Quality. Blackwell, Oxford, p 169-196.

Kjørsvik, E., & S. Lonning. 1983. Effects of egg quality on normal fertilization and early development of the cod, *Gadus morhua* L. J. Fish Biol. 23:1-12.

Kjørsvik, E., A. Mangor-Jensen & I. Holmefjord. 1990. Egg quality in fishes. Adv. Mar. Biol. 26:71-113.

Klaoudatos, S., N. Tsevis & A. Conides. 1990. Energy sources during the early larval development of the European sea bass, *Dicentrarchus labrax* (L.). Aquaculture 87:361-372.

Kleiber, M. 1975. The Fire of Life: An Introduction to Animal Energetics. Krieger, Huntington, New York, 453 p.

Klein Breteler, W.C.M., & S.R. Gonzalez. 1982. Influence of cultivation and food concentration on body length of calanoid copepods. Mar. Biol. 71:157-161.

Klein-MacPhee, G. 1981. Effects of stocking density on survival of laboratory cultured summer flounder (*Paralichthys dentatus*) larvae. Rapp. P.-v. Réun. Cons. Int. Explor. Mer 178:505-506.

Klein-MacPhee, G. 1983. A Culture Manual for the Winter Flounder, *Pseudopleuronectes americanus*. Spec. Rep., Env. Res. Lab., Narragansett, Rhode Island, 65 p.

Klein-MacPhee, G., W.H. Howell & A.D. Beck. 1982. Comparison of a reference strain and four geographical strains of *Artemia* as food for winter flounder (*Pseudopleuronectes americanus*) larvae. Aquaculture 29:279-288.

Kloth, T.C. 1980. Observations on the spawning behavior of captive Florida pompano, *Trachinotus carolinus*. Copeia 1980:884-886.

Klyszejko, B., M. Mourad, G. Glebocka, W. Ciereszko & T. Sciazko. 1993. Bioacccumulation of PCB's (Aroclor 1242) and its effects on the elctrocardiogram of the eel *Anguilla anguilla* L.. Acta Ichthyologica Piscatoria 23:111-117.

Knapp, R.A., & R.R. Warner. 1991. Male parental care and female choice in the bicolor damselfish, *Stegastes partitus*: bigger is not always better. Anim. Behav. 41:747-756.

Kneib, R.T. 1993. Growth and mortality in successive cohorts of fish larvae within an estuarine nursery. Mar. Ecol. Prog. Ser. 94:115-127.

Knights, B. 1985. Energetics and fish farming. In: P. Tytler & P. Calow (eds.), Fish Energetics: New Perspectives. Johns Hopkins Univ. Press, Baltimore, p 309-340.

Knoph, M.B. 1992. Acute toxicity of ammonia to Atlantic salmon (*Salmo salar*) parr. Comp. Biochem. Physiol. 101C:275-282.

Knoph, M.B., & Y.A. Olsen. 1994. Subacute toxicity of ammonia to Atlantic salmon (*Salmo salar*) in seawater: effects on water and salt balance, plasma cortisol and plasma ammonia levels. Aquat. Toxicol. 30:295-310.

Knösche, R., & K. Schreckenbach. 1995. The influence of carbohydrates in the feed on some physiological parameters of culture fishcarp, trout and sturgeon hybrids. Aquaculture 129:253-254.

Knud-Hansen, C.F., & A.K. Pautong. 1993. On the role of urea in pond fertilization. Aquaculture 114:273-283.

Kobayashi, T., & T. Iwamoto. 1984. Early life history of the threebanded sweetlip, *Plectorhynchus cinctus*. Jpn. J. Ichthyol. 30:412-418. (in Jpn., Eng. abs.)

Kobayashi, K., K. Suzuki & Y. Shiobara. 1993. Reproduction and hermaphroditism in *Parapercis snyderi* (Teleostei, Parapercidae) in Suruga Bay, central Japan. J. Fac. Mar. Sci. Tech. Tokai Univ. 35:149-168. (in Jpn., Eng. abs.)

Koenig, Christopher C. 1992. Florida State University, Tallahassee, pers. comm.

Koenig, O. 1958. Ein Beitrag zur Fortpflanzungsbiologie von *Dascyllus trimaculatus*. Die Aquarien- und Terrarien-Zeitschrift 11:107-111.

Koh, J.-N., & Y.U. Kim. 1992. Embryonic deveopment and morphology of larvae and juveniles of parrot fish, *Oplegnathus fasciatus* (Temminck et Schlegel). Bull. Nat. Fish. Univ. Pusan 32:29-45.

Kohler, C.C., R.J. Sheehan, C. Habicht, J.A. Malison & T.B. Kayes. 1994. Habituation to captivity and controlled spawning of white bass. Trans. Am. Fish. Soc. 123:964-974.

Kohno, H., Y. Taki, Y. Ogasawara, Y. Shirojo, M. Taketomi & M. Inoue. 1983. Development of swimming and feeding functions in larval *Pagrus major*. Jpn. J. Ichthyol. 30:47-60.

Kohno, H., S. Hara & Y. Taki. 1986. Early larval development of the seabass *Lates calcarifer* with emphasis on the transition of energy sources. Bull. Jpn. Soc. Sci. Fish. 52:1719-1725.

Kohno, H., M. Duray & J. Juario. 1988. State of grouper (lapu-lapu) culture in the Philippines. Asian Aquac. 10(2):4-8.

Kohno, H., S. Diani, P. Sunyoto, B. Slamet & P.T. Imanto. 1990a. Early developmental events associated with changeover of nutrient sources in the grouper, *Epinephelus fuscoguttatus*, larvae. Bull. Pen. Perikanan, Spec. Ed. (1):51-64.

Kohno, H., P.T. Imanto, S. Diani, B. Slamet & P. Sunyoto. 1990b. Reproductive performance and early life history of the grouper, *Epinephelus fuscoguttatus*. Bull. Pen. Perikanan, Spec. Ed. (1):27-35.

Kohno, H., M. Duray, A. Ohno & Y. Taki. 1994a. Larval intervals of the seabass, *Lates calcarifer*, based on the development of swimming and feeding functions. Third Asian Fisheries Forum, Asian Fish. Soc., Manila, p 98-101.

Kohno, H., A. Ohno & T. Nakao. 1994b. Transition from endogenous nutrition to exogenous food resources in larval little tuna, *Euthynnus affinis*. Suisanzoshoku 42:165-169.

Kohno, H., A. Ohno & Y. Taki. 1994c. Why is grouper larval rearing difficult?: a comparison of the biological natures of early larvae of four tropical marine fish species. Third Asian Fisheries Forum, Asian Fish. Soc., Manila, p 450-453.

Kolkovski, S., A. Tandler & G.W. Kissil. 1993a. The effect of dietary enzymes with age on protein and lipid assimilation and deposition in *Sparus aurata* larvae. Fish Nutrition in Practice. INRA, Paris, p 569-578.

Kolkovski, S., A. Tandler, G.W. Kissil & A. Gertler. 1993b. The effect of dietary exogenous digestive enzymes on ingestion, assimilation, growth and survival of gilthead seabream (*Sparus aurata*, Sparidae, Linnaeus) larvae. Fish Physiol. Biochem. 12:203-209.

Kolkovski, S., W. Koven & A. Tandler. 1997a. The mode of action of *Artemia* in enhancing utilization of microdiet by gilthead seabream *Sparus aurata* larvae. Aquaculture 155:193-205.

Kolkovski, S., A. Arieli & A. Tandler. 1997b. Visual and chemical cues stimulate microdiet ingestion in sea bream larvae. Aquac. Internat. 5:527-536.

Kolodziejska, I., & Z.E. Sikorski. 1995. The properties and utilization of proteases of marine fish and invertebrates. Polish J. Food Nutr. Sci. 4/45:5-12.

Koltes, K.H. 1993. Aspects of the reproductive biology and social structure of the stoplight parrotfish *Sparisoma viride*, at Grand Turk, Turks and Caicos Islands, B.W.I. Bull. Mar. Sci. 52:792-805.

Kono, M., T. Matsui & C. Shimizu. 1987. Effect of chitin, chitosan, and cellulose as diet supplements on the growth of cultured fish. Nippon Suisan Gakkaishi 53:125-129.

Kopp, J.F., & G.D. McKee. 1983. Methods for Chemical Analysis of Water and Wastes. USEPA Rep. EPA-600/4-79-020.

Korn, S., & R. Earnest. 1974. Acute toxicity of twenty insecticides to striped bass, *Morone saxatilis*. Cal. Fish Game 60:128-131.

Korn, S., & S. Rice. 1981. Sensitivity to, and accumulation and depuration of, aromatic petroleum components by early life stages of coho salmon (*Oncorhynchus kisutch*). Rapp. P.-v. Réun. Cons. Int. Explor. Mer 178:87-92.

Korstad, J. 1988. A selected bibliography on the biology of *Brachionus plicatilis*, with particular reference to feeding, life history, and population dynamics. Fdn. Sci. Indust. Res., Norw. Inst. Technol., Trondheim, No. STF21 A88027, 40 p.

Korstad, J., A. Neyts, T. Danielsen, I. Overrein & Y. Olsen. 1995. Use of swimming speed and egg ratio as predictors of the status of rotifer cultures in aquaculture. Hydrobiologia 313/314:395-398.

Koskela, J., M. Jobling & J. Pirhonen. 1997. Influence of the length of the daily feeding period on feed intake and growth of whitefish, *Coregonus lavaretus*. Aquaculture 156:35-44.

Kosutarak, P., A. Kanazawa, S. Teshima, S. Koshio & S. Itoh. 1994. L-ascorbyl-2-phosphate Mg as a vitamin C source for red sea bream (*Pagrus major*) juveniles. Third Asian Fisheries Forum. Asian Fish. Soc., Manila, p 729-732.

Kosutarak, P., A. Kanazawa, S.-i. Teshima & S. Koshio. 1995a. Interactions of L-ascorbyl-2-phosphate Mg and n-3 highly unsaturated fatty acids on Japanese flounder juveniles. Fish. Sci. 61:860-866.

Kosutarak, P., A. Kanazawa, S.-i. Teshima & S. Koshio. 1995b. Interactions of L-ascorbyl-2-phosphate Mg and oxidized fish oil on red sea bream juveniles. Fish. Sci. 61:696-702.

Koumoundouros, G., G. Oran, P. Divanach, S. Stefanakis & M. Kentouri. 1997. The opercular complex deformity in intensive gilthead sea bream (*Sparus aurata* L.) larviculture. Moment of apparition and description. Aquaculture 156:165-177.

Koven, W.M., G.W. Kissil & A. Tandler, A., 1989. Lipid and *n*-3 requirement of *Sparus aurata* larvae during starvation and feeding. Aquaculture 79:185-191.

Koven, W.M., A. Tandler, G.W. Kissil, D. Sklan, O. Friezlander & M. Harel. 1990. The effect of dietary (*n*-3) polyunsaturated fatty acids on growth, survival and swim bladder development in *Sparus aurata* larvae. Aquaculture 91:131-141.

Koven, W.M., A. Tandler, G.W. Kissil & D. Sklan. 1992. The importance of *n*-3 highly unsaturated fatty acids for growth in larval *Sparus aurata* and their effect on survival, lipid composition and size distribution. Aquaculture 104:91-104.

Koya, Y., T. Matsubara & T. Nakagawa. 1994. Efficient artificial fertilization method based on the ovulation cycle in barfin flounder *Verasper moseri*. Fish. Sci. 60:537-540.

Kraak, S.B.M. 1996. A quantitative description of the reproductive biology of the Mediterranean blenny *Aidablennius sphynx* (Teleostei, Blenniidae) in its natural habitat. Env. Biol. Fishes 46:329-342.

Kraak, S.B.M., & E.P. van den Berghe. 1992. Do female fish assess paternal quality by means of test eggs? Anim. Behav. 43:865-867.

Kraljević, M., J. Jug-Dujaković & J. Dulčić. 1994. Some effects of salinity on survival of early developmental stages of gilthead sea bream (*Sparus aurata* L.). Acta Adriat. 35:37-44.

Kramer, D. 1960. Development of eggs and larvae of Pacific mackerel and distribution and abundance of larvae 1952-1956. U.S. Fish Wildl. Serv. Fish. Bull. 60:393-438.

Kraul, S. 1985. Comparative hatchery characteristics of yellowtail jack, *Seriola lalandei* and mahimahi, *Coryphaena hippurus*. Proc. Second International Conference on Warm Water Aquaculture—Finfish. Brigham Young Univ., Hawaii Campus, Laie, p 99-108.

Kraul, S. 1990a. Production of live prey for marine fish larvae. Advances in Tropical Aquaculture. AQUACOP, IFREMER, Actes Colloq. (9):595-607.

Kraul, S. 1990b. Review and current status of the aquaculture potential for the mahimahi, *Coryphaena hippurus*. Advances in Tropical Aquaculture. AQUACOP, IFREMER, Actes Colloq. (9):445-459.

Kraul, S. 1991. Hatchery methods for the mahimahi, *Coryphaena hippurus*, at Waikiki Aquarium. In: J.P. McVey (ed.), CRC Handbook of Mariculture: Volume II, Finfish Aquaculture. CRC Press, Boca Raton, Florida, p 241-250.

Kraul, S. 1993. Larviculture of the mahimahi *Coryphaena hippurus* in Hawaii, USA. J. World Aquac. Soc. 24:410-421.

Kraul, S., A. Nelson, K. Brittain & D. Wenzel. 1988. Feeding and handling requirements for hatchery culture of the mahimahi *Coryphaena hippurus*. Univ. Hawaii Sea Grant Quart. 10(3):6 p.

Kraul, S., A. Nelson, K. Brittain, H. Ako & A. Ogasawara. 1992. Evaluation of live feeds for larval and postlarval mahimahi *Coryphaena hippurus*. J. World Aquac. Soc. 23:299-306.

Kreiberg, H., G.A. Hunter, E.M. Donaldson, W.C. Clarke & I. Baker. 1987. Induced ovulation and spermiation in the Pacific herring (*Clupea harengus pallasi*) using salmon pituitary preparations and a synthetic gonadotropin-releasing hormone analogue. Aquaculture 61:155-161.

Krijgsheld, K.R., & A. van der Gen. 1986. Assessment of the impact of the emission of certain organochlorine compounds on the aquatic environment. Part I: monochlorophenols and 2,4-dichlorophenol. Chemosphere 15:825-860.

Krise, W.F. 1993. Effects of one-year exposures to gas supersaturation on lake trout. Prog. Fish-Cult. 55:169-176.

Krom, M.D., C. Porter & H. Gordin. 1985. Causes of fish mortalities in semi-intensively operated seawater ponds in Eilat, Israel. Aquaculture 49:159-177.

Kubitza, F., L.L. Lovshin & R.T. Lovell. 1997. Identification of feed enhancers for juvenile largemouth bass *Micropterus salmoides*. Aquaculture 148:191-200.

Kudo, S., & C. Teshima. 1991. Enzyme activities and antifungal action of fertilization envelope extract from fish eggs. J. Exp. Zool. 259:392-398.

Kuhlmann, D., & G. Quantz. 1980. Some effects of temperature and salinity on the embryonic development and incubation time of the turbot, *Scophthalmus maximus* L., from the Baltic Sea. Meeresforschung Rep. Mar. Res. 28:172-178.

Kuhlmann, D., G. Quantz & U. Witt. 1981. Rearing of turbot larvae (*Scophthalmus maximus* L.) on cultured food organisms and postmetamorphosis growth on natural and artificial food. Aquaculture 23:183-196.

Kuiter, R.H. 1988. Notes on hatching and raising seadragons. SeaScope, Aquarium Systems, Mentor, Ohio, 5(summer):1-3.

Kumai, H. 1995. A review of the nursery and growout culture techniques for red seabream (*Pagrus major*) and striped jack (*Caranx delicatissimus*) in Japan. In: K.L Main & C. Rosenfeld (eds.), Culture of High-value Marine Fishes in Asia and the United States. Oceanic Inst., Honolulu, p 101-111.

Kumai, H., & M. Nakamura. 1977. On the natural spawning of sand borer, *Sillago sihama* (Forskal). Mem. Fac. Agric. Kinki Univ. 10:39-43. (in Jpn., Eng. abs.)

Kungvankij, P. 1987. Induction of spawning of sea bass (*Lates calcarifer*) by hormone injection and environmental manipulation. Austral. Ctr. Internat. Agric. Res. Proc. (20):120-122.

Kungvankij, Pinij. 1992. Charoen Pokphand Feedmill Co., Bangkok, Thailand, pers. comm.

Kungvankij, P., L.B. Tiro, B.P. Pudadera & I.O. Potestas. 1986. Induced spawning and larval rearing of grouper (*Epinephelus salmoides* Maxwell). First Asian Fisheries Forum. Asian Fish. Soc., Manila, p 663-666.

Kungvankij, P., S.P. Meyers, B. Budiraharjo & M. Kadori. 1990. Growth and survival of rabbitfish *Siganus canaliculatus* on three types of formulated diets. Second Asian Fisheries Forum. Asian Fish. Soc., Manila, p 325-328.

Kuntz, A. 1913. The embryology and larval development of *Bairdiella chrysura* and *Anchovia mitchilli*. Bull. U.S. Bur. Fish. 33:3-19.

Kuntz, A. 1914. Notes on the embryology and larval development of five species of teleostean fishes. Bull. U.S. Bur. Fish. 34:409-429.

Kuntz, A., & L. Radcliffe. 1917. Notes on the embryology and larval development of twelve teleostean fishes. U.S. Bur. Fish. Bull. 35:89-134.

Kuo, C.-M. 1982. Induced breeeding of grey mullet, *Mugil cephalus* L. In: C.J.J. Richter & M.J.T. Goos (compilers). Proc. Internat. Symp. Repr. Physiol. Fish. Wageningen, Netherlands, p 181-184.

Kuo, C.-M. 1995. Manipulation of ovarian development and spawning in grey mullet, *Mugil cephalus* L. Bamidgeh 47:43-58.

Kuo, C.-M., Z.H. Shehadeh & K.K. Milisen. 1973. A preliminary report on the development, growth and survival of laboratory-reared larvae of the grey mullet, *Mugil cephalus* L. J. Fish Biol. 5:459-470.

Kuo, C.-M., Y.-Y. Ting & S.-L. Yeh. 1988. Induced sex reversal and spawning of blue-spotted grouper, *Epinephelus fario*. Aquaculture 74:113-126.

Kurokawa, T., & T. Suzuki. 1996. Formation of the diffuse pancreas and the development of digestive enzyme synthesis in larvae of the Japanese flounder *Paralichthys olivaceus*. Aquaculture 141:267-276.

Kurokawa, T., & T. Suzuki. 1997. Development of the digestive functions during larval stage of Japanese flounder *Paralichthys olivaceus*. World Aquac. '97 Book of Abs., p 266-267.

Kurokawa, T., H. Kagawa, H. Ohta, H. Tanaka, K. Okuzawa & K. Hirose. 1995. Development of digestive organs and feeding ability in larvae of Japanese eel (*Anguilla japonica*). Can. J. Fish. Aquat. Sci. 52:1030-1036.

Kurokawa, T., H. Tanaka, H. Kagawa & H. Ohta. 1996. Absorption of protein molecules by the rectal cells in eel larvae *Anguilla japonica*. Fish. Sci. 62:832-833.

Kurokura, H., H. Kumai & M. Nakamura. 1986. Hybridization between female red sea bream and male crimson sea bream by means of sperm cryopreservation. First Asian Fisheries Forum. Asian Fish. Soc. Manila, p 113-116.

Kuronuma, K., & K. Fukusho. 1984. Rearing of marine fish larvae in Japan. Internat. Dev. Res. Ctr., Ottawa, Canada, 109 p.

Kuroshima, R., & S. Kimura. 1989. Artificial seawater for testing the toxicity of pollutants to marine fish. I. Examination of the suitability by estimating the LC_{50} and acumulation of Cd for *Girella punctata* Gray. Bull. Natl. Res. Inst. Fish. Sci. (1):75-82.

Kurup, B.M., & C.T. Samuel. 1991a. Observations on the spawning biology of *Nibea albida* (Cuvier) in the Cochin Estuary. J. Mar. Biol. Assn. India 33:99-106.

Kurup, B.M., & C.T. Samuel. 1991b. Spawning biology of *Gerres filamentosus* Cuvier in the Cochin Estuary. Fish. Technol. 28:19-24.

Kusakari, M. 1991. Mariculture of kurosoi, *Sebastes schlegeli*. Env. Biol. Fishes 30:245-251.

Kusuda, R. 1992. Bacterial fish diseases in mariculture in Japan with special emphasis on streptococcicosis. Bamidgeh 44:140.

Kusuda, R., J. Yokoyama & K. Kawai. 1986. Bacteriological study on cause of mass mortalities in cultured black sea bream fry. Bull. Jpn. Soc. Sci. Fish. 52:1745-1751.

Kutty, M.N. 1981. Energy metabolism of mullets. In: O.H. Oren (ed.), Aquaculture of Grey Mullets. IBP 26, Cambridge Univ. Press, Cambridge, U.K., p 219-263.

Kuwamura, T. 1981. Diurnal periodicity of spawning activity in free-spawning labrid fishes. Jpn. J. Ichthyol. 28:343-348.

Kuwamara, T. 1983. Spawning behavior and timing of fertilization in the mouthbrooding cardinalfish *Apogon notatus*. Jpn. J. Ichthyol. 30:61-71.

Kuwamura, T., Y. Yogo & Y. Nakashima. 1993. Size-assortative monogamy and paternal egg care in a coral goby *Paragobiodon echinocephalus*. Ethology 95:65-75.

Kvenseth, P.G. 1993. Use of wrasse to control salmon lice. In: H. Reinertsen, L.A. Dahle, L. Jørgensen & K. Tvinnereim (eds.), Fish Farming Technology. Balkema, Rotterdam, p 227-232.

Kwan, S.K., T.W. Chen & L.H. Teo. 1994. The control of total ammonia and carbon dioxide levels in the packaging physiology of mollies. Third Asian Fisheries Forum, Asian Fish. Soc., Manila, p 888-891.

Kyowa, nd.a. Feeding Models for Fry Feed Kyowa. Kyowa Hakko Kogyo Co., Tokyo, Japan, 18 p.

Kyowa, nd.b. Fry Feed Kyowa A·B·C: A Blended Feed for Fish Fry. Kyowa Hakko Kogyo Co., Tokyo, 31 p.

LaBomascus, D.C., E.H. Robinson & T.L. Linton. 1987. Use of water conditioners in water-recirculation systems. Prog. Fish-Cult. 49:64-65.

Lacanilao, F., & C. Marte. 1980. Sexual maturation of milkfish in floating cages. Asian Aquac. 3:4,6.

Lahnsteiner, F., T. Weismann & R. Patzner. 1996. Cryopreservation of semen of the grayling (*Thymallus thymallus*) and the Danube salmon (*Hucho hucho*). Aquaculture 144:265-274.

Lai, K.V., & G.W. Klontz. 1980. Evaluation of Environmental and Nutritional Factors Influencing the Performance of Biofilters in Fish Rearing Systems: Final Report. Walla Walla Dist. Corps Engineers,

Walla Walla, Washington, 139 p.

Laird, L.M., & T. Needham (eds.). 1988a. Salmon and Trout Farming. Ellis Horwood, Chichester, England, 271 p.

Laird, L.M., & T. Needham. 1988b. The farmed salmonids. In: L.M. Laird & T. Needham (eds.), Salmon and Trout Farming. Ellis Horwood, Chichester, England, p 15-31.

Laird, L., & T. Needham. 1990. Sea water culture of salmonids. In: G. Barnabé (ed.), Aquaculture: Vol. 2, Ellis Horwood, New York, p 628-646.

Lall, S.P. 1989. The minerals. In: J.E. Halver (ed.), Fish Nutrition. Academic Press, San Diego, p 219-257.

Lall, S.P., & G. Olivier. 1993. Role of micronutrients in immune response and disease resistance in fish. Fish Nutrition in Practice, INRA, Paris, p 101-118.

Lam, C.W.Y. 1990. Pollution effects of marine fish culture in Hong Kong. Asian Mar. Biol. 7:1-7.

Lam, T.J. 1982. Applications of endocrinology to fish culture. Can J. Fish. Aquat. Sci. 39:111-137.

Lam, T.J. 1994. Hormones and egg/larval quality in fish. J. World Aquac. Soc. 25:2-12.

Lam, T.J., & A.D. Munro. 1987. Environmental control of reproduction in teleosts: an overview. Proc. Third Internat. Symp. Repr. Physiol. Fish, Memorial Univ. Newfoundland, St. John's, p 279-288.

Lambert, Y., J.-D. Dutil & J. Munro. 1994. Effects of intermediate and low salinity conditions on growth rate and food conversion of Atlantic cod (*Gadus morhua*). Can. J. Fish. Aquat. Sci. 51:1569-1576.

Lamm, D.R. 1984. Spawning of the coral beauty angelfish. SeaScope, Aquarium Systems, Mentor, Ohio, 1(summer):1-4.

Landesman, L. 1994. Negative impacts of coastal aquaculture development. World Aquaculture 25(2):12-17.

Landolt, M.L. 1989. The relationship between diet and the immune response of fish. Aquaculture 79:193-206.

Landsberg, J.H. 1989. Parasites and associated diseases of fish in warm water culture with special emphasis on intensification. In: M. Shilo & S. Sarig (eds.), Fish Culture in Warm Water Systems: Problems and Trends, CRC Press, Boca Raton, Florida, p 195-252.

Landsberg, J.H., K.A. Steidinger, B.A. Blakesley & R.L. Zondervan. 1994. Scanning electron microscope study of dinospores of *Amyloodinium* cf. *ocellatum*, a pathogenic dinoflagellate parasite of marine fish, and comments on its relationship to the Peridiniales. Dis. Aquat. Org. 20:23-32.

Langer, R., & Y. Zohar. 1992. Development of a novel technology for the manipulation of fish reproductive cycles; controlled release of gonadotropin releasing hormones. Mass. Inst. Tech. Sea Grant MITSG 92-1, 22 p.

Langlais, B., D.A. Reckhow & D.R. Brink. 1991. Ozone in Water Treatment: Application and Engineering. Lewis, Chelsea, Michigan, 569 p.

Langston, W.J. 1991. Toxic effects of metals and the incidence of metal pollution in marine ecosystems. In: R.W. Furness & P.S. Rainbow (eds.), Heavy Metals in the Marine Environment, CRC Press, Boca Raton, Florida, p 101-122.

Lankford, T.E., Jr. & T.E. Targett. 1994. Suitability of estuarine nursery zones for juvenile weakfish (*Cynoscion regalis*): effects of temperature and salinity on feeding, growth and survival. Mar. Biol. 119:611-620.

Laroche, W.A., W.F. Smith-Vaniz & S.L. Richardson. 1984. Carangidae: development. Am. Soc. Ichthyol. Herpetol. Spec. Pub. (1):510-522.

Larrazabal, G. 1992. Aquaculture in Spain. World Aquac. 23(4):10-16.

Larson, E.T., & A.L. Shanks. 1996. Consumption of marine snow by two species of juvenile mullet and its contribution to their growth. Mar. Ecol. Prog. Ser. 130:19-28.

Larson, G.L., F.E. Hutchins & L.P. Lamperti. 1977. Laboratory determination of acute and sublethal toxicities of inorganic chloramines to early life stages of coho salmon (*Oncorhynchus kisutch*). Trans. Am. Fish. Soc. 106:268-277.

Larsson, D.G.J., C.C. Mylonas, Y. Zohar & L.W. Crim. 1997. Gonadotropin-releasing hormone analogue (GnRH-A) induces multiple ovulations of high-quality eggs in a cold-water, batch-spawning teleost, the yellowtail flounder (*Pleuronectes ferrugineus*). Can. J. Fish. Aquat. Sci. 54:1957-1964.

Larsson, P.-O., & J. Pickova. 1993. Stock enhancement experiments with cod, *Gadus morhua*, in the Bothnian Sea - conditions and expectations. ICES Statutory Mtg., Paper C.M.1993/J:18, Session R, 10 p.

Lasker, R. 1974. Induced maturation and spawning of marine fish at the Southwest Fisheries Center La Jolla, California. Proc. World Maric. Soc. 5:313-318.

Lasker, R., H.M. Feder, G.H. Theilacker & R.C. May. 1970. Feeding, growth, and survival of *Engraulis mordax* larvae reared in the laboratory. Mar. Biol. 5:345-353.

Lau, S.R., & P.L. Shafland. 1982. Larval development of snook, *Centropomus undecimalis* (Pisces: Centropomidae). Copeia 1982:618-627.

Laurence, G.C. 1974. Growth and survival of haddock (*Melanogrammus aeglefinus*) larvae in relation to planktonic prey concentration. J. Fish. Res. Bd. Can. 31:1415-1419.

Laurence, G.C. 1975. Laboratory growth and metabolism of the winter flounder *Pseudopleuronectes americanus* from hatching through metamorphosis at three temperatures. Mar. Biol. 32:223-229.

Laurence, G.C. 1977. A bioenergetic model for the analysis of feeding and survival potential of winter flounder, *Pseudopleuronectes americanus*, larvae during the period from hatching to metamorphosis. Fish. Bull., U.S. 75:529-546.

Laurence, G.C. 1978. Comparative growth, respiration and delayed feeding abilities of larval cod (*Gadus morhua*) and haddock (*Melanogrammus aeglefinus*) as influenced by temperature during laboratory studies. Mar. Biol. 50:1-7.

Laurence, G.C., & C.A. Rogers. 1976. Effects of temperature and salinity on comparative embryo development and mortality of Atlantic cod (*Gadus morhua* L.) and haddock (*Melanogrammus aeglefinus* (L.)). J. Cons. 36:220-228.

Laurencin, F.B., J.L. Messager & G. Stephan. 1990. Two examples of nutritional pathology related to vitamin E and C deficiencies. Advances in Tropical Aquaculture. AQUACOP, IFREMER, Actes Colloq. (9):171-181.

Laurenson, L.J.B., F.J. Neira & I.C. Potter. 1993. Reproductive biology and larval morphology of the marine plotosid *Cnidoglanis macrocephalus* (Teleostei) in a seasonally closed Australian estuary. Hydrobiologia 268:179-192.

Lavens, P., & R.A.M. Remmerswaal (eds.). 1994. Turbot Culture: Problems and Prospects. Eur. Aquac. Soc. Spec. Pub. No. 22, 360 p.

Lavens, P., P. Dhert, G. Merchie, M. Stael & P. Sorgeloos. 1994. A standard procedure for the mass production of rotifers on an artificial diet. Third Asian Fisheries Forum, Asian Fish. Soc., Manila, p 745-748.

Lavens, P., P. Sorgeloos, P. Dhert & B. Devresse. 1995. Larval foods. In: N.R. Bromage & R.J. Roberts (eds.), Broodstock Management and Egg and Larval Quality. Blackwell, Oxford, 373-397.

Lazier, C.B., M. Mann & T.P. Mommsen. 1987. Estrogen receptors and regulation of vitellogenesis. Proc. Third Internat. Symp. Repr. Physiol. Fish, Memorial Univ. Newfoundland, St. John's, p 178-182.

Lazo, J.P., A. Davis & C. Arnold. 1997. The effect of dietary protein level on growth, feed and protein efficiency and survival of yellowtail snapper (*Ocyurus chrysurus*) in culture. World Aquac. '97 Book of Abs., p 273.

Lazur, A.M., & D.J. Zimet. 1995. Economic considerations of golden shiner production in Florida. IFAS, Univ. Florida, Gainesville, SS-NFREC-94-00, 13 p. (draft)

Leary, D.F., & R.T. Lovell. 1975. Value of fiber in production-type diets for channel catfish. Trans. Am. Fish. Soc. 104:328-332.

Lebedeva, L.I., & O.N. Orlenko. 1995. Feeding rate of *Brachionus plicatilis* O.F. Müller on two types of food depending on ambient temperature and salinity. Int. Revue. ges. Hydrobiol. 80:77-87.

Le Breton, A.D. 1993. Élvage de poissons marins en Méditerranée: techniques d'élevage, situation actuelle et perspectives. Rec. Méd. Vét. 170:121-128. (in Fr., Eng. sum.)

Le Breton, A. 1996. An overview of the main infectious problems in cultured seabass *Dicentrarchus labrax* and seabream *Sparus aurata*: solutions? In: B. Chatain, M. Saroglia, J. Sweetman & P. Lavens (eds.), Seabass and Seabream Culture: Problems and Prospects. Eur. Aquac. Soc., p 67-86.

LeBorgne, Y. 1990. Culture of micro-algae. In: G. Barnabé (ed.), Aquaculture, Vol. 1. Ellis Horwood, New York, p 197-206.

Leclercq, D. 1994. Turbot broodstock management: a key point to mid-term progress of the turbot industry. Eur. Aquac. Soc. Spec. Pub. (22):3-13.

Le Clus, F. 1992. Seasonal trends in sea surface temperature, dry mass per oocyte and batch fecundity of pilchard *Sardinops ocellatus* in the northern Benguela system. S. Afr. J. Mar. Sci. 12:123-134.

Lecomte-Finiger, R. 1992. Growth history and age at recruitment of European glass eels (*Anguilla anguilla*) as revealed by otolith microstructure. Mar. Biol. 114:205-210.

Le Coz, C., P. Margerit & J.P. Marion. 1990. Culturing tilapia in sea water in Martinique. In: G. Barnabé (ed.), Aquaculture: Vol. 2. Ellis Horwood, New York, p 833-840.

Lee, C.-L., & R. Hirano. 1981. Suitable food organisms and feeding schedules for the larvae of *Sillago sihama* (Forsskål). Rapp. P.-v. Réun. Cons. Int. Explor. Mer 178:533-537.

Lee, C.-L., & G.M. Weber. 1986. Effects of salinity and photoperiod on 17α-methyltestosterone-induced spermatogenesis in the grey mullet, *Mugil cephalus* L. Aquaculture 56:53-62.

Lee, C.-S. 1986. Reproduction. In: C.-S. Lee, M.S. Gordon & W.O. Watanabe (eds.), Aquaculture of Milkfish (*Chanos chanos*): State of the Art. Oceanic Inst., Honolulu, p 57-81.

Lee, C.-S., & F. Hu. 1981. Salinity tolerance and salinity effects on brood size of *Tigriopus japonicus* Mori. Aquaculture 22:377-381.

Lee, C.-S., & C.D. Kelley. 1991. Artificial propagation of mullet in the United States. In: J.P. McVey (ed.), CRC Handbook of Mariculture: Vol. II, Finfish Aquaculture. CRC Press, Boca Raton, Florida, p 193-209.

Lee, C.-S., & I-C. Liao (eds.). 1985. Reproduction and Culture of Milkfish. Oceanic Inst., Honolulu, 226 p.

Lee, C.-S., & C.S. Tamaru. 1988. Advances and future prospects of controlled maturation and spawning of grey mullet (*Mugil cephalus* L.) in captivity. Aquaculture 74:63-73.

Lee, C.-S., F. Hu & R. Hirano. 1981. Organisms suitable as food for larvae of black sea bream. Prog. Fish.-Cult. 43:121-124.

Lee, C.-S., M.S. Gordon & W.O. Watanabe (editors). 1986a. Aquaculture of Milkfish (*Chanos chanos*): State of the Art. Oceanic Inst., Honolulu, 284 p.

Lee, C.-S., C.S. Tamaru, J.E. Banno, C.D. Kelley, A. Bocek & J.A. Wyban. 1986b. Induced maturation and spawning of milkfish, *Chanos chanos* Forsskal, by hormone implantation. Aquaculture 52:199-205.

Lee, C.-S., C.S. Tamaru, C.D. Kelley & J.E. Banno. 1986c. Induced spawning of milkfish, *Chanos chanos*, by a single application of LHRH-analogue. Aquaculture 58:87-98.

Lee, C.-S., C.S. Tamaru, J.E. Banno & C.D. Kelley. 1986d. Influence of chronic administration of LHRH-analogue and/or 17α-methyltestosterone on maturation in milkfish, *Chanos chanos*. Aquaculture 59:147-159.

Lee, C.-S., C.S. Tamaru & C.D. Kelley. 1986e. Technique for making chronic-release LHRH-a and 17α-methyltestosterone pellets for intramuscular implantation in fishes. Aquaculture 59:161-168.

Lee, C.S., C.S. Tamaru, G.T. Miyamoto & C.D. Kelley. 1987. Induced spawning of grey mullet (*Mugil cephalus*) by LHRH-a. Aquaculture 62:327-336.

Lee, C.-S., et al. 1988a. Studies on the maturation and spawning of milkfish in captivity: Progress Report October 1984-December 1988. Contract No. DAN-4161-A-00-4055-00, Oceanic Inst., Honolulu, 35 p.

Lee, C.-S., C.S. Tamaru & C.D. Kelley. 1988b. The cost and effectiveness of CPH, HCG and LHRH-a on the induced spawning of grey mullet, *Mugil cephalus*. Aquaculture 73:341-347.

Lee, C.-S., C.S. Tamaru, C.D. Kelley, A. Moriwake & G.T. Miyamoto. 1992a. The effect of salinity on the induction of spawning and fertilization in the striped mullet, *Mugil cephalus*. Aquaculture 102:289-296.

Lee, C.-S., C.S. Tamaru, C.D. Kelley, G.T. Miyamoto & A.M. Moriwake. 1992b. The minimum effective dosage of 17α-methyltestosterone for induction of testicular maturation in the striped mullet, *Mugil cephalus* L. Aquaculture 104:183-191.

Lee, C.-S., P.-S. Leung & M.-S. Su. 1997. Bioeconomic evaluation of different fry production systems for milkfish (*Chanos chanos*). Aquaculture 155:367-376.

Lee, G.W.Y., & M.K. Litvak. 1996a. Weaning of metamorphosed winter flounder (*Pleuronectes americanus*) reared in the laboratory: comparison of two commercial artificial diets on growth, survival and conversion efficiency. Aquaculture 144:251-263.

Lee, G.W.Y., & M.K. Litvak. 1996b. Weaning of wild young-of-the-year winter flounder *Pleuronectes americanus* (Walbaum) on a dry diet: effects on growth, survival, and feed efficiency ratios. J. World Aquac. Soc. 27:30-39.

Lee, J.H.W., R.S.S. Wu, Y.K. Cheung & P.P.S. Wong. 1991a. Dissolved oxygen variations in marine fish culture zone. J. Env. Eng. 117:799-815.

Lee, J.H.W., R.S.S. Wu & Y.K. Cheung. 1991b. Forecasting of dissolved oxygen in marine fish culture zone. J. Env. Eng. 117:816-833.

Lee, J.Y., Y.J. Kang, S.-M. Lee & Y.J. Park. 1993a. Evaluation of protein source for Korean rockfish (*Sebastes schlegeli*) test diets. Bull. Nat. Fish. Res. Dev. Ag. 48:97-105. (in Kor., Eng. abs.)

Lee, K.-K. 1995a. Pathogenesis studies on *Vibrio alginolyticus* in the grouper, *Epinephelus malabaricus*, Bloch et Schneider. Microbial Pathogenesis 19:39-48.

Lee, M.-F., C.-F. Chang, J.-D. Huang & A.V. Schally. 1993b. Microencapsulated LH-RH analog accelerated spermiation in protandrous black porgy, *Acanthopagrus schlegeli*. Bull. Inst. Zool. Acad. Sin. 32:32-38.

Lee, P.G. 1995b. A review of automated control systems for aquaculture and design criteria for their implementation. Aquacultural Eng. 14:205-227.

Lee, S.J., S.J. Son, J.U. Lee & S.U. Park. 1995. Embryonic development and larval morphology of sweetfish, *Plecoglossus altivelis* Temminck et Schlegel. Bull. Nat. Fish. Res. Dev. Ag. 50:91-102.

Lee, S.K., D. Freitag, C. Steinberg, A. Kettrup & Y.H. Kim. 1993c. Effects of dissolved humic materials on acute toxicity of some organic chemicals to aquatic organisms. Wat. Res. 27:199-204.

Lee, S.-M., & I.-G. Jeon 1996. Evaluation of soybean meal as a partial substitute for fish meal in formulated diets for Korean rockfish, *Sebastes schlegeli*. J. Kor. Fish. Soc. 29:586-594. (in Kor., Eng. abs.)

Lee, S.M., & J.H. Yoo. 1996. Evaluation of cotton seed meal or rapeseed meal as a partial substitute for fish meal in formulated diets for Korean rockfish (*Sebastes schlegeli*). Kor. J. Anim. Nutr. Feedstuffs 20:128-135. (in Kor., Eng. abs.)

Lee, S.-M., I.-G. Jeon, J.-Y. Lee, S.-R. Park, Y.-J. Kang & K.-S. Jeong. 1996. Substitution of plant and animal proteins for fish meal in the growing Korean rockfish (*Sebastes schlegeli*) feeds. J. Kor. Fish. Soc. 29:651-662. (in Kor., Eng. abs.)

Legendre, M. 1986. Seasonal changes in sexual maturity and fecundity, and HCG-induced breeding of the catfish, *Heterobranchus longifilis* Val. (Clariidae), reared in Ebrie Lagoon (Ivory Coast). Aquaculture 55:201-213.

Leger, C. 1980. Developpements recents de la notion d'acide gras essentiel chez les poissons. Ann. Nutr. Alim. 34:207-216.

Léger, P., D.A. Bengston, K.L. Simpson & P. Sorgeloos. 1986. The use and nutritional value of *Artemia* as a food source. Oceanogr. Mar. Biol. Ann. Rev. 24:521-623.

Leibovitz, H.E., D.A. Bengtson & K.L. Simpson. 1991. Evaluation of microcapsule performance on delivering diets to silverside (*Menidia beryllina*) larvae. Eur. Aquac. Soc. Spec. Pub. (15):173-174.

Lein, I. 1991. Viability of Atlantic halibut (*Hippoglossus hippoglossus* L.) eggs exposed to seawater before fertilization. Eur. Aquac. Soc. Spec. Pub. (15):205-206.

Lein, I., & I. Holmefjord. 1992. Age at first feeding of Atlantic halibut larvae. Aquaculture 105:157-164.

Lein, I., & I. Holmefjord. 1993. Success factors in rearing of yolk sac larvae of Atlantic halibut. In: H.

Reinertsen, L.A. Dahle, L. Jørgensen & K. Tvinnereim (eds.), Fish Farming Technology. Balkema, Rotterdam, p 57-58.

Lein, I., S. Tveite, B. Gjerde & I. Holmefjord. 1997a. Effects of salinity on yolk sac larvae of Atlantic halibut (*Hippoglossus hippoglossus* L.) Aquaculture 156:291-303.

Lein, I., I. Holmefjord & M. Rye. 1997b. Effects of temperature on yolk sac larvae of Atlantic halibut (*Hippoglossus hippoglossus* L.). Aquaculture 157:123-135.

Leis, J.M. 1978. Systematics and zoogeography of the porcupinefishes (*Diodon*, Diodontidae, Tetraodontiformes), with comments on egg and larval development. Fish. Bull., U.S. 76:535-567.

Leis, J.M. 1984. Tetraodontoidei: development. Am. Soc. Ichthyol. Herpetol. Spec. Pub. (1):447-450.

Leis, J.M., & J.T. Moyer. 1985. Development of eggs, larvae and pelagic juveniles of three Indo-Pacific ostraciid fishes (Tetraodontiformes): *Ostracion meleagris*, *Lactoria fornasini* and *L. diaphana*. Jpn. J. Ichthyol. 32:189-202.

Leis, J.M., & D.S. Rennis. 1983. The Larvae of Indo-Pacific Coral Reef Fishes. New South Wales Univ. Press, Sydney, 269 p.

Leis, J.M., & W.J. Richards. 1984. Acanthuroidei: development and relationships. Am. Soc. Ichthyol. Herpetol. Spec. Pub. (1):547-551.

Lekang, O.I., & S.O. Fjæra. 1993. Fish handling in landbased fish farms. In: H. Reinertsen, L.A. Dahle, L. Jørgensen & K. Tvinnereim (eds.), Fish Farming Technology. Balkema, Rotterdam, p 389-394.

Lekang, O.I., S.O. Fjæra & J.M. Thomassen. 1996. Voluntary fish transport in land-based fish farms. Aquacultural Eng. 15:13-25.

Lemaire, P., A. Mathieu, J. Giudicelli & M. Lafaurie. 1992. Effect of diet on the responses of hepatic biotransformation enzymes to benzo(a)pyrene in the European sea bass (*Dicentrarchus labrax*). Comp. Biochem. Physiol. 102C:413-420.

Le Menn, F., J.N. Rodriguez, B. Breton & P. Copeland. 1988. Incorporation of vitellogenin in fish. In: Y. Zohar & B. Breton (eds.), Reproduction in Fish. Basic and Applied Aspects in Endocrinology and Genetics. Colloq. INRA (44):35-38.

Lemm, C.A., R.L. Herman, D.P. Lemarie & A. Arzapalo. 1993. Effects of diet and environmental salinity on the growth, mortality, and tissue structure of juvenile striped bass. J. Aquat. Anim. Health 5:294-305.

León R., J., J. Millán Q. & L. León S. 1996. Cultivo experimental del pargo dienton (*Lutjanus griseus*) en estanques. Bol. Inst. Oceanogr. Venezuela, Univ. Oriente 35:9-15.

Leong, J.C., & J.L. Fryer. 1993. Viral vaccines for aquaculture. Ann. Rev. Fish Dis. 225-240.

Leong, R. 1971. Induced spawning of the northern anchovy, *Engraulis mordax* Girard. Fish. Bull., U.S. 69:357-360.

Leong, R.J.H. 1989. Sperm concentrations and egg fertilization rates during spawning of captive anchovy, *Engraulis mordax*. Cal. Coop. Oceanic Fish. Invest. Rep. 30:136-139.

Leong, T.S. 1992. Diseases of brackishwater and marine fish cultured in some Asian countries. Diseases in Asian Aquaculture I. Asian Fish. Soc., Manila, p 223-236.

Leu, M.-Y. 1994. Natural spawning and larval rearing of silver bream, *Rhabdosargus sarba* (Forsskål), in captivity. Aquaculture 120:115-122.

Leu, M.-Y. 1997. Natural spawning and mass larviculture of black porgy *Acanthopagrus schlegeli* in captivity in Taiwan. J. World Aquac. Soc. 28:180-187.

Leu, M.-Y., & Y.-H. Chou. 1996. Induced spawning and larval rearing of captive yellowfin porgy, *Acanthopagrus latus* (Houttuyn). Aquaculture 143:155-166.

Leu, M.-Y., & C.-H. Liou. 1992. Substitution of live foods with a micro-coated diet in the feeding of larval silver bream, *Sparus sarba* (Forskål): note on swim bladder inflation. J. Fish. Soc. Taiwan 19:65-73.

Leung, L.K.-P. 1987. Cryopreservation of spermatozoa of the barramundi, *Lates calcarifer* (Teleostei: Centropomidae). Aquaculture 64:243-247.

Leung, L.K.-P., & B.G.M. Jamieson. 1991. Live preservation of fish gametes. In: B.G.M. Jamieson (ed.), Fish Evolution and Systematics: Evidence from Spermatozoa. Cambridge Univ. Press, U.K., p 245-269.

Leung, P.S., C.-S. Lee & L.W. Rowland. 1993. Production cost for a finfish hatchery: the case of striped mullet (*Mugil cephalus*). In: C.-S. Lee, M.-S. Su & I C. Liao (eds.), Finfish Hatchery in Asia. Tungkang Mar. Lab. Conf. Proc. (3):239-244.

Levine, D.Z. 1995. Ciguatera: current concepts. J. Am. Osteopathic Assn. 95:193-198.

Levings, C.D. 1994. Some ecological concerns for net-pen culture of salmon on the coasts of the northeast Pacific and Atlantic Oceans, with special reference to British Columbia. J. Appl. Aquac. 4:65-141.

Levings, C.D., A. Ervik, P. Johannessen & J. Aure. 1995. Ecological criteria used to help site fish farms in fjords. Estuaries 18:81-90.

Lewis, M.A. 1992. The effects of mixtures and other environmental modifying factors on the toxicities of surfactants to freshwater and marine life. Water Res. 26:1013-1023.

Lewis, R.J., M.J. Holmes & M. Sellin. 1994. Invertebrates implicated in the transfer of gambiertoxins to the benthic carnivore *Pomadasys maculatus*. Mem. Queensland Mus. 34:561-564.

Lewis, W.M., Jr. & D.P. Morris. 1986. Toxicity of nitrite to fish: a review. Trans. Am. Fish. Soc. 115:183-195.

Li, J.-E. 1995. A review of nursery and growout culture techniques for marine finfish in China. In: K.L Main & C. Rosenfeld (eds.), Culture of High-value Marine Fishes in Asia and the United States. Oceanic Inst., Honolulu, p 113-119.

Li, M.H., & E.H. Robinson. 1994. Catfish Feed Manufacture. Off. Agric. Commun., Miss. State Univ., State College, 14 p.
Li, M.H., & E.H. Robinson. 1996. Comparison of chelated zinc and zinc sulfate as zinc sources for growth and bone mineralization of channel catfish (*Ictalurus punctatus*) fed practical diets. Aquaculture 146:237-243.
Li, Y., S. Jin & J. Qin. 1996. Strategies for development of rotifers as larval fish food in ponds. J. World Aquac. Soc. 27:178-186.
Liang, B. 1992. Research on the culture of *Hippocampus*. J. Ocean Univ. Qingdao 22:39-44. (in Ch., Eng. abs.)
Liao, I.-C. 1981. Cultivation methods. In: O.H. Oren (ed.), Aquaculture of Grey Mullets. IBP 26, Cambridge Univ. Press, Cambridge, U.K., p 361-389.
Liao, I.-C. 1988. Status of aquaculture in Asia. Austasia Aquac. Mag. 2(8):5-7.
Liao, I.-C. 1991. Milkfish culture in Taiwan. In: J.P. McVey (ed.), Handbook of Mariculture, Vol. II, Finfish Aquaculture. CRC Press, Boca Raton, Florida, p 91-115.
Liao, I C. 1993. Finfish hatcheries in Taiwan: recent advances. In: C.-S. Lee, M.-S. Su & I C. Liao (eds.), Finfish Hatchery in Asia. Tungkang Mar. Lab. Conf. Proc. (3):1-25.
Liao, I.-C., & N.-H. Chao. 1991. Grey mullet culture in Taiwan. In: J.P. McVey (ed.), Handbook of Mariculture, Vol. II, Finfish Aquaculture. CRC Press, Boca Raton, Florida, p 117-132.
Liao, I.C., & T.-I. Chen. 1986. Milkfish culture methods in Southeast Asia. In: C.-S. Lee, M.S. Gordon & W.O. Watanabe (eds.), Aquaculture of Milkfish (*Chanos chanos*): State of the Art. Oceanic Inst., Honolulu, p 209-242.
Liao, I.C., C.P. Hung, M.C. Lin, Y.W. Hou, T.L. Huang & I.H. Tung. 1969. Artificial propagation of grey mullet, *Mugil cephalus* Linnaeus. Joint Comm. Rural Reconstr. Fish. Ser. 8:10-20. (in Ch., Eng. sum.)
Liao, I.C., Y.J. Lu, T.L. Huang & M.C. Lin. 1972. Experiments on induced breeding of the grey mullet, *Mugil cephalus* Linnaeus. In: T.V.R. Pillay (ed.), Coastal Aquaculture in the Indo-Pacific Region. Fishing News Books, Surrey, England, p 213-243.
Liao, I.-C., J.V. Juario, S. Kumagai, H. Nakajima, M. Natividad & P. Buri. 1979. On the induced spawning and larval rearing of milkfish, *Chanos chanos* (Forskal). Aquaculture 18:75-93.
Liao, I C., M.-S. Su & L.-S. Hsieh. 1994. Spontaneous spawning of red drum (*Sciaenops ocellatus*) in Taiwan. Third Asian Fisheries Forum. Asian Fish. Soc., Manila, p 211-214.
Liao, I C., M.-S. Su & S.-L. Chang. 1995. A review of the nursery and growout techniques of high-value marine finfishes in Taiwan. In: K.L Main & C. Rosenfeld (eds.), Culture of High-value Marine Fishes in Asia and the United States. Oceanic Inst., Honolulu, p 121-137.
Libey, G.S, & M.B. Timmons (eds.). 1996. Successes and Failures in Commercial Recirculating Aquaculture. Aquacultural Eng. Soc. Proc. II, 2 vols., 278 & 653 p.
Lie, Ø., & A. Mangor-Jensen. 1993. Importance of broodstock nutrition for optimal production in aquaculture. In: H. Reinertsen, L.A. Dahle, L. Jørgensen & K. Tvinnereim (eds.), Fish Farming Technology. Balkema, Rotterdam, p 35-40.
Lie, Ø., E. Lied & G. Lambertsen. 1986. Liver retention of fat and of fatty acids in cod (*Gadus morhua*) fed different oils. Aquaculture 59:187-196.
Lie, Ø., E. Lied & G. Lambertsen. 1988. Feed optimization in Atlantic cod (*Gadus morhua*): fat versus protein content in the feed. Aquaculture 69:333-341.
Lie, Ø., A. Mangor-Jensen & G.-I. Hemre. 1993. Broodstock nutrition in cod (*Gadus morhua*)—effect of dietary fatty acids. Fisk. Dir. Skr. Ser. Ernæring 6:11-19.
Liley, N.R., J.R. Cardwell & Y. Rouger. 1987. Current status of hormones and sexual behavior in fish. Proc. Third Internat. Symp. Repr. Physiol. Fish, Memorial Univ. Newfoundland, St. John's, p 142-149.
Liltved, H., H. Hektoen & H. Efraimsen. 1995. Inactivation of bacterial and viral fish pathogens by ozonation or UV irradiation in water of different salinity. Aquacultural Eng. 14:107-122.
Lim, C. 1989. Practical feeding—tilapias. In: T. Lovell, Nutrition and Feeding of Fish. Van Nostrand Reinhold, New York, p 163-183.
Lim, C. 1991a. Milkfish, *Chanos chanos*. In: R.P. Wilson (ed.), Handbook of Nutrient Requirements of Finfish. CRC Press, Boca Raton, Florida, p 97-104.
Lim, C., S. Sukhawongs & F.P. Pascual. 1979. A preliminary study on the protein requirements of *Chanos chanos* (Forskal) fry in a controlled environment. Aquaculture 17:195-201.
Lim, C., B.R. LeaMaster & J.A. Brock. 1995. Pyridoxine requirement of fingerling red hybrid tilapia grown in seawater. J. Appl. Aquac. 5:49-60.
Lim, L.C. 1991b. Larviculture of the greasy grouper (*Epinephelus tauvina* F.) and brown-marbled grouper (*E. fuscoguttatus* F.) in Singapore. Eur. Aquac. Soc. Spec. Pub. (15):321-322.
Lim, L.C. 1993. Larviculture of the greasy grouper *Epinephelus tauvina* F. and the brown-marbled grouper *E. fuscoguttatus* F. in Singapore. J. World Aquac. Soc. 24:262-274.
Lim, L.C., L. Cheong, H.B. Lee & H.H. Heng. 1985. Induced breeding studies of the John's snapper, *Lutjanus johni* (Bloch), in Singapore. Sing. J. Pri. Ind. 13:70-83.
Lim, L.C., H.H. Heng & H.B. Lee. 1986. The induced breeding of seabass, *Lates calcarifer* (Bloch) in Singapore. Sing. J. Pri. Ind. 14:81-95.
Lim, L.C., T.M. Chao & L.T. Khoo. 1990. Observations on the breeding of brown-marbled grouper,

Epinephelus fuscoguttatus (Forskal). Sing. J. Pri. Ind. 18:66-84.

Lin, D., J. Zhan, Z. Zheng, Z. Weng & Y. Shu. 1991a. Studies on the artificial propagation of the large yellow croaker, *Pseudosciaena crocea* (Richardson). J. Fujian Normal Univ. (Natural Sci.) 7:71-79. (in Ch., Eng. sum.)

Lin, H.-C., & W.A. Dunson. 1993. The effect of salinity on the acute toxicity of cadmium to the tropical, estuarine, hermaphroditic fish, *Rivulus marmoratus*: a comparison of Cd, Cu, and Zn tolerance with *Fundulus heteroclitus*. Arch. Env. Contam. Toxicol. 25:41-47.

Lin, H.-R., & R.E. Peter. 1986. Induction of gonadotropin secretion and ovulation in teleosts using LHRH analogs and catecholaminergic drugs: a review. First Asian Fisheries Forum. Asian Fish. Soc. Manila, p 667-670.

Lin, H.R., & R.E. Peter. 1992. Hormones and spawning in fishes. Abs. in program book, Third Asian Fisheries Forum. Asian Fish. Soc., Manila, p 5.

Lin, H.-R., M.-L. Zhang, S.-M. Zhang, G. Van Der Kraak & R.E. Peter. 1991b. Stimulation of pituitary gonadotropin and ovarian development by chronic administration of testosterone in female Japanese silver eel, *Anguilla japonica*. Aquaculture 96:87-95.

Lin, H.T., C.C. Chang & W.Y. Tseng. 1981. Preliminary study on the compounded feed for *Epinephelus amblycephalus*. Collection of theses, Dept. Aquac., Nat. Taiwan Coll. Mar. Sci. Tech., Keelung, Taiwan.

Lin, K.-J., R.-M. Chang, C.-Y. Liu, Y.-K. Fang, C.-L. Chen, C.-Y. Juang & J.-Y. Twu. 1988. Natural spawning of goldline sea bream *Sparus sarba* in the artificial environment and embryonic development. Bull. Taiwan Fish. Res. Inst. (45):3-16. (in Ch., Eng. abs.)

Lin, K.-J. R.-M. Chang, J.-Y. Twu & C.-Y. Liu. 1989. Experiments on the propagation of goldline sea bream *Sparus sarba* — breeder culture, natural spawning in the artificial environment of 3-year-old breeder, and hatching of fertilized eggs. Bull. Taiwan Fish. Res. Inst. (47):21-37. (in Ch., Eng. abs.)

Lin, K.-J., J.-Y. Twu & C.-H. Chen. 1990. Studies of daily spawning and quality of eggs in one goldline sea bream, *Sparus sarba*. Bull. Taiwan Fish. Res. Inst. (48):251-262. (in Ch., Eng. abs.)

Lin, L.T. 1982. Further success in induced spawning of pond-reared milkfish. China Fish. (357):17-19. (in Ch., Eng. abs.)

Lin, L.-T. 1984. Studies on the induced breeding of milkfish (*Chanos chanos* Forsskal) reared in ponds. China Fish. (378):3-29. (in Ch. Eng. abs.)

Lin, L.-T. 1985. My experience in artificial propagation of milkfish—studies on natural spawning of pond-reared broodstock. In: C.S. Lee & I.C. Liao (eds.), Reproduction and Culture of Milkfish. Oceanic Inst., Hawaii, and Tungkang Mar. Lab., Taiwan, p 185-203.

Lin, L.-T. 1991. A general introduction to the completion of a generation cycle of permit fish (*Trachinotus falcatus*). Abs. in program book. Finfish Hatchery in Asia '91. Tungkang Mar. Lab., Tungkang, Taiwan, p 44.

Lin, Lieh-Tang. 1992. Tung Hsing Aquatic Hatchery, Ping Tung Hsin, Taiwan, ROC, pers. comm.

Lin, L.T., & H.H. Lin. 1987. Studies on the induced breeding of milkfish (*Chanos chanos* Forskal) reared in ponds—Natural spawning of pond-reared broodstock in 1986. In: J.Y.L. Yu, C.M. Kuo, S.H. Shih & S.J. Chen (eds.), Fish Reproduction and Its Endocrine Control: Basic and Practical Aspects. Inst. Zool., Academia Sinica and Dept. Fish., Council Agric., Taipei, COA Fish. Ser. 7:199-210. (in Ch., Eng. abs.)

Lindberg, G.U., & Z.V. Krasyukova. 1971. Fishes of the Sea of Japan and the Adjacent Areas of the Sea of Okhotsk and the Yellow Sea. Part 3: Teleostomi. XXIX. Perciformes. In: Keys to the Fauna of the USSR, No. 99. Trans. by Israel Prog. Sci. Trans., U.S. Dept. Commerce, Nat. Tech. Info. Serv., Springfield, Virginia, 498 p.

Lindner, P., A. Eshel, S. Kolkovski, A. Tandler & S. Harpaz. 1995. Proteolysis by juvenile sea bass (*Dicantrarchus labrax*) gastrointestinal enzymes as a method for the evaluation of feed proteins. Fish Physiol. Biochem. 14:399-407.

Lippson, A.J., & R.L. Moran. 1974. Manual for Identification of Early Developmental Stages of Fishes of the Potomac River Estuary. Env. Tech. Cen., Martin Marietta Corp., Baltimore, Maryland, 282 p.

Lisovenko, L.A., & D.P. Andrianov. 1991. Determination of absolute fecundity of intermittently spawning fishes. J. Ichthyol. 31:143-155.

Little, D.C. 1994. Efficiency indicators in intensive *Oreochromis* seed production systems. Third Asian Fisheries Forum. Asian Fish. Soc., Manila, p 61-64.

Little, D.C., D.J. Macintosh & P. Edwards. 1993. Improving spawning synchrony in the Nile tilapia, *Oreochromis niloticus* (L.). Aquac. Fish. Mgt. 24:399-405.

Litvak, M.K. 1994. Winter flounder [culture]. Bull. Aquac. Assoc. Can. (94-1):4-8.

Liu, H.W., R.R. Stickney & S.D. Smith. 1991a. A note on the artificial spawning of Pacific halibut. Prog. Fish-Cult. 53:189-192.

Liu, H.W., R.R. Stickney, W.W. Dickhoff & D.A. McCaughran. 1993. Early larval growth of Pacific halibut *Hippoglossus stenolepis*. J. World Aquac. Soc. 24:482-485.

Liu, J., G. Wu, G. Xie & F. Chen. 1991b. Other species culture. In: L. Zhong (chief ed.), Pond Fisheries in China. Pergamon, Oxford, p 157-205.

Liu, J., M. Li, K. Wang, X. Wang & J. Liu. 1994. Studies on the phosphorus requirement and proper calcium/phosphorus ratio in the diet of the black sea bream (*Sparus macrocephalus*). Third Asian Fisheries Forum. Asian Fish. Soc., Manila, p 737-740.

Liu, J., K. Wang & X. Wang. 1995. Studies on the optimum protein content and the ratio of animal to plant protein in the diet of the black sea bream (*Sparus macrocephalus*). Oceanol. Limnol. Sin. 26:445-448. (in Ch., Eng. abs.)

Liu, K.K.M., & C.D. Kelley. nd.a. Milkfish (*Chanos chanos*). Oceanic Inst. Hatchery Man. Ser., Honolulu, 87 p.

Liu, K.K.M., & C.D. Kelley. nd.b. Striped Mullet (*Mugil cephalus*). Oceanic Inst. Hatchery Man. Ser., Honolulu, 87 p.

Liu, K.-M., C.-T. Chen & S.-J. Joung. 1992. Some aspects on the fishery biology of big eye *Priacanthus macracanthus* in the Tungkang waters, southwestern Taiwan. J. Fish. Soc. Taiwan 19:251-262.

Lloyd, R. 1981. Freshwater quality. In: A.D. Hawkins (ed.), Aquarium Systems. Academic Press, London, p 129-147.

Lobel, P.S. 1974. Another pygmy angel spawns. Octopus 1(7):24.

Lobel, P.S. 1975. Hawaiian angelfishes. Mar. Aquarist 6(4):30-41.

Lobel, P.S. 1978. Diel, lunar, and seasonal periodicity in the reproductive behavior of the pomacanthid fish, *Centropyge potteri*, and some other reef fishes in Hawaii. Pac. Sci. 32:193-207.

Lobel, P.S. 1991. Mating strategies of coastal marine fishes. Oceanus (fall):19-26.

Lochmann, R.T., & D.M. Gatlin III. 1993a. Essential fatty acid requirement of juvenile red drum (*Sciaenops ocellatus*). Fish Physiol. Biochem. 12:221-235.

Lochmann, R.T., & D.M. Gatlin III. 1993b. Evaluation of different types and levels of triglycerides, singly and in combination with different levels of *n*-3 highly unsaturated fatty acid ethyl esters in diets of juvenile red drum, *Sciaenops ocellatus*. Aquaculture 114:113-130.

Lochmann, R., & H. Phillips. 1996. Nutrition and feeding of baitfish. Aquac. Mag. 22(4):87-89.

Logan, S.H., W.E. Johnston & S.I. Doroshov. 1995. Economics of joint production of sturgeon (*Acipenser transmontanus* Richardson) and roe for caviar. Aquaculture 130:299-316.

Lom, J., & A.R. Lawler. 1973. An ultrastructural study of the mode of attachment in dinoflagellates invading gills of Cyprinodontidae. Protistologica 9:293-309.

Lopez, G.W. 1982. Short-term population dynamics of *Tisbe cucumariae* (Copepoda: Harpacticoida). Mar. Biol. 68:333-341.

López-Alvarado, J., & A. Kanazawa. 1994. Effect of dietary arginine levels on growth of red sea bream larvae fed diets supplemented with crystalline amino acids. Fish. Sci. 60:435-439.

López-Alvarado, J., & A. Kanazawa. 1995. Optimum levels of crystalline amino acids in diets for larval red sea bream (*Pagrus major*). ICES Mar. Sci. Symp. 201:100-105.

López-Alvarado, J., & A. Kanazawa. 1997. Effect of dietary protein sources in microdiets on feeding behavior and growth of red sea bream, *Pagrus major*, during weaning and metamorphosis. J. Appl. Aquac. 7:53-66.

López-Alvarado, J., C.J. Langdon, S.-I. Teshima & A. Kanazawa. 1994. Effects of coating and encapsulation of crystalline amino acids on leaching in larval feeds. Aquaculture 122:335-346.

Lorentzen, M., A. Maage & K. Julshamn. 1996. Manganese supplementation of a practical, fish meal based diet for Atlantic salmon parr. Aquac. Nutr. 2:121-125.

Losordo, T.M. 1991a. Engineering considerations in closed recirculating systems. Aquaculture Systems Engineering. Am. Soc. Agric. Eng., p 58-69.

Losordo, T.M. 1991b. Stratification and mixing in shallow aquaculture ponds. Adv. World Aquac. (3):396-435.

Losordo, T.M., & P.W. Westerman. 1994. An analysis of biological, economic, and engineering factors affecting the cost of fish production in recirculating aquaculture systems. J. World Aquac. Soc. 25:193-203.

Lotz, J.M., R.M. Overstreet & J.S. Franks. 1996. Gonadal maturation in the cobia, *Rachycentron canadum*, from the northcentral Gulf of Mexico. Gulf Res. Rep. 9:147-159.

Lourenço, S.O., U.M.L. Marquez, J. Mancini-Filho, E. Barbarino & E. Aidar. 1997. Aquaculture 148:153-168.

Love, M.S., A. Brooks, D. Busatto, J. Stephens & P.A. Gregory. 1996. Aspects of the life histories of the kelp bass, *Paralabrax clathratus*, and barred sand bass, *P. nebulifer*, from the southern California Bight. Fish. Bull., U.S. 94:472-481.

Lovell, T. 1989. Nutrition and Feeding of Fish. Van Nostrand Reinhold, New York, 260 p.

Loverich, G.F., & T.R. Croker. 1993. Ocean Spar® net pen system: 32 months of offshore operations. In: H. Reinertsen, L.A. Dahle, L. Jørgensen, & K. Tvinnereim (eds.), Fish Farming Technology. Balkema, Rotterdam, p 243-249.

Loverich, G.F., & K.T. Swanson. 1993. Offshore sea farms: 25 months of experience. In: J.-K. Wang (ed.), Techniques for Modern Aquaculture. Am. Soc. Agric. Eng., St. Joseph, Michigan, p 241-253.

Lubbock, R. 1975. Fishes of the family Pseudochromidae (Perciformes) in the northwest Indian Ocean and Red Sea. J. Zool., London 176:115-157.

Luberda, Z., J. Strzezek & M. Luczynski. 1993. Some proteolytic properties of hatching liquid in the sea trout, *Salmo trutta* m. *trutta* L. Fish Physiol. Biochem. 12:75-80.

Lubzens, E. 1987. Raising rotifers for use in aquaculture. Hydrobiologia 147:245-255.

Lubzens, E., A. Marko & A. Tietz. 1985. De novo synthesis of fatty acids in the rotifer, *Brachionus plicatilis*. Aquaculture 47:27-37.

Lubzens, E., A. Tandler & G. Minkoff. 1989. Rotifers as food in aquaculture. Hydrobiologia 186/187:387-400.

Lubzens, E., G. Kolodny, B. Perry, N. Galai, R. Sheshinski & Y. Wax. 1990a. Factors affecting survival of rotifers (*Brachionus plicatilis* O.F. Müller) at 4°C. Aquaculture 91:23-47.

Lubzens, E., B. Perry, S. Beddig & R. Berghahn. 1990b. Preservation of rotifers for use in aquaculture. Eur. Aquac. Soc. Spec. Pub. (11):233-249.

Lubzens, E., Y. Wax, G. Minkoff & F. Adler. 1993. A model evaluating the contribution of environmental factors to the production of resting eggs in the rotifer *Brachionus plicatilis*. Hydrobiologia 255/256:127-138.

Lubzens, E., D. Rankevich, G. Kolodny, O. Gibson, A. Cohen & M. Khayat. 1995a. Physiological adaptations in the survival of rotifers (*Brachionus plicatilis*, O.F. Müller) at low temperatures. Hydrobiologia 313/314:175-183.

Lubzens, E., O. Gibson, O. Zmora & A. Sukenik. 1995b. Potential advantages of frozen algae (*Nannochloropsis* sp.) for rotifer (*Brachionus plicatilis*) culture. Aquaculture 133:295-309.

Lubzens, E., N. Daube, I. Pekarsky, Y. Magnus, A. Cohen, F. Yusefovich & P. Feigin. 1997. Carp (*Cyprinus carpio* L.) spermatozoa cryobanks — strategies in research and application. Aquaculture 155:13-30.

Lucchetti, G.L., & G.A. Gray. 1988. Water reuse systems: a review of principal components. Prog. Fish-Cult. 50:1-6.

Ludwig, G.M. 1989. Effect of golden shiners on plankton and water quality in ponds managed for intensive production. J. World Aquac. Soc. 20:46-52.

Ludwig, G.M. 1993. Effects of trichlorfon, fenthion, and diflubenzuron on the zooplankton community and on production of reciprocal-cross hybrid striped bass fry in culture ponds. Aquaculture 110:301-319.

Ludwig, G.M. 1994. Tank culture of sunshine bass *Morone chrysops* × *M. saxatilis* fry with freshwater rotifers *Brachionus calyciflorus* and salmon starter meal as first food sources. J. World Aquac. Soc. 25:337-341.

Ludwig, G.M. 1997. Sunshine bass fry culture. Aquac. Mag. 23(5):93-97.

Lumare, F., & P. Villani. 1970. Contributo alla conoscenze delle uova e dei primi stadi larvali di *Sparus aurata* (L.). Pub. Sta. Zool. Napoli 38:364-369.

Lundberg, B., & D. Golani. 1995. Diet adaptations of Lessepsian migrant rabbitfishes, *Siganus luridus* and *S. rivulatus*, to the algal resources of the Mediterranean coast of Israel. Mar. Ecol. 16:73-89.

Lunestad, B.T., & J. Goksøyr. 1990. Reduction in the antibacterial effect of oxytetracycline in sea water by complex formation with magnesium and calcium. Dis. Aquat. Org. 9:67-72.

Luo, J., & J.A. Musick. 1991. Reproductive biology of the bay anchovy in Chesapeake Bay. Trans. Am. Fish. Soc. 120:701-710.

Luquet, P. 1991. Tilapia, *Oreochromis* spp. In: R.P. Wilson (ed.), Handbook of Nutrient Requirements of Finfish. CRC Press, Boca Raton, Florida, p 169-179.

Lutz, R.V., N.H. Marcus & J.P. Chanton. 1994. Hatching and viability of copepod eggs at two stages of embryological development: anoxic/hypoxic effect. Mar. Biol. 119:199-204.

Luzier, J.M., R.C. Summerfelt & H.G. Ketola. 1995. Partial replacement of fish meal with spray-dried blood powder to reduce phosphorus concentrations in diets for juvenile rainbow trout, *Oncorhynchus mykiss* (Walbaum). Aquac. Res. 26:577-587.

Maage, A., M. Lorentzen, M. Bjornevik & K. Julshamn. 1993. Zinc requirement of Atlantic salmon (*Salmo salar*) fry during start feeding. Fish Nutrition in Practice, INRA, Paris, p 873-882.

Macchi, G.J. 1998. Preliminary estimate of spawning frequency and batch fecundity of striped weakfish, *Cynoscion striatus*, in coastal waters off Buenos Aires province. Fish. Bull., U.S. 96:375-381.

Macer, C.T. 1974. The reproductive biology of the horse mackerel *Trachurus trachurus* (L.) in the North Sea and English Channel. J. Fish Biol. 6:415-438.

Macewicz, B.J., & J.R. Hunter. 1993. Spawning frequency and batch fecundity of jack mackerel, *Trachurus symmetricus*, off California during 1991. Cal. Coop. Oceanic Fish. Invest. Rep. 34:112-121.

Macewicz, B.J., & J.R. Hunter. 1994. Fecundity of sablefish, *Anoplopoma fimbria*, from Oregon coastal waters. Cal. Coop. Oceanic Fish. Invest. Rep. 35:160-174.

MacGregor, J.S. 1976. Ovarian development and fecundity of five species of California Current fishes. Cal. Coop. Oceanic Fish. Invest. Rep. 18:181-188.

MacGregor, R., III, M.S. Greeley, Jr., W.C. Trimble & W.M. Tatum. 1983. Seasonal variation of reproduction and fattening in gulf killifish from brackish mariculture ponds. Northeast Gulf Sci. 6:23-31.

Macintosh, D.J., & D.C. Little. 1995. Nile tilapia (*Oreochromis niloticus*). In: N.R. Bromage and R.J. Roberts (eds.), Broodstock Management and Egg and Larval Quality, Blackwell, Oxford, p 277-320.

Maciorowski, A.F., A. Henderson-Arzapalo, D.E. Roberts, Jr., R.L. Colura & C.E. McCarty. 1987. Fingerling production of common snook, *Centropomus undecimalis*, in saltwater ponds. Proc. Gulf Carib. Fish. Inst. 38:190-202.

MacKenzie, B.R., W.C. Leggett & R.H. Peters. 1990. Estimating larval fish ingestion rates: can laboratory derived values be reliably extrapolated to the wild. Mar. Ecol. Prog. Ser. 67:209-225.

MacKenzie, D.S., J. Warner & P. Thomas. 1987. Thyroid-reproductive relationships in the channel catfish, *Ictalurus punctatus*: evidence for estradiol-induced changes in plasma thyroid hormone binding. Proc.

Third Internat. Symp. Repr. Physiol. Fish, Memorial Univ. Newfoundland, St. John's, p 205.
Mackie, A.M., & A.I. Mitchell. 1985. Identification of gustatory feeding stimulants for fish--applications in aquaculture. In: C.B. Cowey, A.M. Mackie, & J.G. Bell (eds.) Nutrition and Feeding in Fish. Academic Press, San Diego, p 177-189.
MacKinnon, B.M. 1997. Sea lice: a review. World Aquac. 28(3):5-10.
MacKinnon, M.R. 1987. Rearing and growth of larval and juvenile barramundi (*Lates calcarifer*) in Queensland. Austral. Ctr. Internat. Agric. Res. Proc. (20):148-153.
MacKinnon, M.R. 1990. Status and potential of Australian *Lates calcarifer* culture. Advances in Tropical Aquaculture, AQUACOP, IFREMER, Actes Colloq. (9):713-727.
MacKinnon, M.R., S. Poole, E. Cazzola & D. Bull. 1987. Pellet-fed barramundi show potential for commercial farming. Austral. Fish. 46:42-44.
Maclean, J.L., L.B. Dizon & L.V. Hosillos (eds.). 1986. The First Asian Fisheries Forum: Proc. of the First Asian Fisheries Forum, Manila, Philippines, 26-31 May 1986. Asian Fish. Soc., Manila, 727 p.
MacMillan, J.R. 1991. Biological factors impinging upon control of external protozoan fish parasites. Ann. Rev. Fish Dis. 1:119-131.
MacMillan, R. 1995. Fish parasites, approved drugs and aquaculture. Aquac. Mag. 21(2):37-40.
MacQuarrie, D.W., J.R. Markert & W.E. Vanstone. 1978. Photoperiod induced off-season spawning of coho salmon (*Oncorhynchus kisutch*). Ann. Biol. Anim. Biochem. Biophys. 18:1051-1058.
MacQuarrie, D.W., W.E. Vanstone & J.R. Markert. 1979. Photoperiod induced off-season spawning of pink salmon (*Oncorhynchus gorbuscha*). Aquaculture 18:289-302.
Madraisau, Becky B. 1990. Micronesian Mariculture Demonstration Center, Palau, pers. comm.
Mahnken, C.V.W. 1991. Coho salmon farming in Japan. In: R.R. Stickney (ed.), Culture of Salmonid Fishes. CRC Press, Boca Raton, Florida, p 131-149.
Mahnken, C.V.W., J. Spinelli & F.W. Waknitz. 1980. Evaluation of an alkane yeast (*Candida* sp.) as a substitute for fish meal in Oregon moist pellet: feeding trials with coho salmon (*Oncorhyncus kisutch*) and rainbow trout (*Salmo gairdneri*). Aquaculture 20:41-56.
Main, K.L., & C. Rosenfeld (eds.). 1995. Culture of High-value Marine Fishes in Asia and the United States. Oceanic Inst., Honolulu, 319 p.
Main, K.L., & C. Rosenfeld (eds.). 1996. Aquaculture Health Management Strategies for Marine Fishes. Oceanic Inst., Honolulu, 280 p.
Mair, G.C., C.C. Estabillo, R.C. Sevilleja & R.D. Recometa. 1993. Small-scale fry production systems for Nile tilapia, *Oreochromis niloticus* (L.). Aquac. Fish. Mgt. 24:229-235.
Malins, D.C., & T.K. Collier. 1981. Xenobiotic interactions in aquatic organisms: effects on biological systems. Aquat. Toxicol. 1:257-268.
Malins, D.C., & H.O. Hodgins. 1981. Petroleum and marine fishes: a review of uptake, disposition, and effects. Env. Sci. Tech. 15:1272-1280.
Mallat, J. 1985. Fish gill structural changes induced by toxicants and other irritants: a statistical review. Can. J. Fish. Aquat. Sci. 42:630-648.
Malloy, K.D., & T.E. Targett. 1991. Feeding, growth and survival of juvenile summer flounder *Paralichthys dentatus*: experimental analysis of the effects of temperature and salinity. Mar. Ecol. Prog. Ser. 72:213-223.
Malone, R.F. 1992. Sludge production from recirculating systems employing expandable granular biofilters. Paper presented at 21st UJNR Aquaculture Panel Joint Meeting, Kyoto, Japan, November.
Malone, R.F., & D.G. Drennan, II. 1993. Bead filters. Alternative Aquac. Network. 11:203.
Malone, R.F., B.S. Chitta & D.G. Drennan. 1993. Optimizing nitrification in bead filters for warmwater recirculating aquaculture systems. In: J.-K. Wang (ed.), Techniques for Modern Aquaculture. Am. Soc. Agric. Eng., St. Joseph, Michigan, p 315-325.
Manabe, S., & A. Kasuga. 1989. Spawning behavior and early life history of *Epinephelus moara* in an aquarium. Rep. Yashima Aquarium, Takamatsu, Japan, 9 p. (in Jpn.)
Manci, W.E., K.G. Richardson & J.F. Bauer. 1992. Design and performance of an electronic lighting-control circuit that simulates sunrise and sunset. Prog. Fish-Cult. 54:127-129.
Maneewong, S. 1987a. Induction of spawning of sea bass (*Lates calcarifer*) in Thailand. Austral. Ctr. Internat. Agric. Res. Proc. (20):116-119.
Maneewong, S. 1987b. Research on the nursery stages of sea bass (*Lates calcarifer*) in Thailand. In: J.W. Copland & D.L. Grey (eds.), Management of Wild and Cultured Sea Bass/Barramundi (*Lates calcarifer*). Austral. Ctr. Internat. Agric. Res. Proc. (20):138-141.
Maneewong, S., & M.S. Hazza. 1992. Study on spawning of brown marble grouper, (*Epinephelus fuscoguttatus*, Forskal). Rep. UTFN/SAU/010/SAU, Fish Farming Ctr., Min. Agric. Water, Jeddah, Saudi Arabia, 12 p.
Maneewong, S., P. Akkayanont, J. Pongmaneerat & M. Iizawa. 1986. Larval rearing and development of grouper, *Epinephelus malabaricus* (Bloch and Schneider). In: Rep. of Thailand and Japan Joint Coastal Aquaculture Research Project (April 1984- January 1986). No. 2, Japan Internat. Coop. Ag., p 39-52.
Manooch, Charles S., III. 1998. National Marine Fisheries Service, Beaufort, North Carolina, pers. comm.
Manooch, C.S., III & C.L. Drennon. 1987. Age and growth of yellowtail snapper and queen triggerfish collected from the U.S. Virgin Islands and Puerto Rico. Fish. Res. 6:53-68.

Manooch, C.S., III & D.L. Mason. 1984. Age, growth, and mortality of lane snapper from southern Florida. Northeast Gulf Sci. 7:109-115.

Manooch, C.S., III & D.L. Mason. 1987. Age and growth of the Warsaw grouper and black grouper from the southeast region of the United States. Northeast Gulf Sci. 9:65-75.

Manooch, C.S., III & J.C. Potts. 1997a. Age and growth of red snapper, *Lutjanus campechanus*, Lutjanidae, collected along the southeastern United States from North Carolina through the east coast of Florida. Elisha Mitchell Sci. Soc. 113:111-122.

Manooch, C.S., III & J.C. Potts. 1997b. Age, growth and mortality of greater amberjack from the southeastern United States. Fish. Res. 30:229-240.

Manrique, R.J. 1987a. Embryonal and prolarval development of *Diplectrum formosum* (L.), (Pisces: Serranidae). Preliminary results. In: J.A.J. Verreth (ed.), Aquaculture Research in Latin America. Internat. Fdn. Sci., Lima, Peru, p 58-63. (in Sp., Eng. sum.)

Manrique, R.J. 1987b. Estado actual de la acuicultura en Venezuela. In: J.A.J. Verreth (ed.), Aquaculture Research in Latin America. Internat. Fdn. Sci., Lima, Peru, p 23-29.

Manrique, R., D.E. Jory & M. Hung-Manrique. 1988. Larval rearing, growth and survival of Atlantic spadefish (*Chaetodipterus faber*) using live and commercial diets. Memorias Reunion Anual AsoVAC, Maracay, Venezuela.

Mansueti, A.J., & J.D. Hardy, Jr. 1967. Development of Fishes of the Chesapeake Bay Region: An Atlas of Egg, Larval, and Juvenile Stages. Natural Resour. Inst., Univ. Maryland, 202 p.

Manzano, J.V., & V.B. Manzano. 1990. Pond culture of lapu-lapu. Outreach 13(1), March.

Manzano, V.B., & J.V. Manzano. 1989. You can raise fish in pens or cages. Outreach 12(1), March.

Marais, J.F.K. 1980. Aspects of food intake, food selection, and alimentary canal morphology of *Mugil cephalus* (Linnaeus, 1958), *Liza tricuspidens* (Smith, 1935), *L. richardsoni* (Smith, 1846), and *L. dumerili* (Steindachner, 1869). J. Exp. Mar. Biol. Ecol. 44:193-209.

Marais, J.F.K., & G.W. Kissil. 1979. The influence of energy level on the feed intake, growth, food conversion and body composition of *Sparus aurata*. Aquaculture 17:203-219.

Marasigan, E.T. 1994. Effects of reducing the trash fish component of the formulated diets for seabass. Third Asian Fisheries Forum, Asian Fish. Soc., Manila, p 649-652.

Marchini, S., M.D. Hoglund, S.J. Broderius & M.L. Tosato. 1993. Comparison of the susceptibility of daphnids and fish to benzene derivatives. Sci. Total Env. 1993:799-808.

Margulies, D. 1989. Size-specific vulnerability to predation and sensory system development of white seabass, *Atractoscion nobilis*, larvae. Fish. Bull., U.S. 87:537-552.

Margulies, D. 1993. Assessment of the nutritional condition of larval and early juvenile tuna and Spanish mackerel (Pisces: Scombridae) in the Panamá Bight. Mar. Biol. 115:317-330.

Marinaro, J.Y. 1991. Eggs of the *Microchirus* species (Soleidae) from the Bay of Algiers. J. Mar. Biol. Assoc. U.K. 71:117-125.

Marini, F.C. 1996. My notes and observations on raising and breeding th Banggai cardinalfish. J. MaquaCulture 4(4):2-5,27-31.

Marino, G., C. Boglione, M.G. Finoia, P. Bronzi, G. Monaco, B. Bertolini & S. Cataudella. 1991. Effects of incubation temperature on embryonic development and hatching of *Dicentrarchus labrax* (L.) eggs. Eur. Aquac. Soc. Spec. Pub. (15):230-232.

Marliave, J.B. 1977. Effects of three artificial lighting regimes on survival of laboratory-reared larvae of the sailfin sculpin. Prog. Fish-Cult. 39:117-118.

Marliave, J.B., S.J.D. Martell, D.I. Kent & H.L. Blankenship. 1997. Propagation and transplanting of lingcod (*Ophiodon elongatus*). World Aquac. '97 Book of Abs., p 337.

Marr, J.C. 1956. The "critical period" in the early life history of marine fishes. J. Cons. 21:160-170.

Mårstøl, M.J., H.J. Fyhn, O.S. Kjesbu & P. Solemdal. 1993. Free amino acid content as a potential criterion of egg quality in Atlantic cod (*Gadus morhua*). In: B.T. Walther & H.J. Fyhn (eds.), Physiological and Biochemical Aspects of Fish Development. Univ. Bergen, Norway, p 99-103.

Marte, C.L. 1990. Hormone-induced spawning of cultured tropical finfishes. Advances in Tropical Aquaculture. AQUACOP, IFREMER, Actes Colloq. (9):519-539.

Marte, C.L., & F. Lacanilao. 1986. Spontaneous maturation and spawning of milkfish in floating net cages. Aquaculture 53:115-132.

Marte, C.L., G.F. Quinitio, L.M.B. Garcia & F.J. Lacanilao. 1984. A Guide to the Establishment and Maintenance of Milkfish Broodstock. SEAFDEC-AQD, Iloilo, Philippines, Tech. Rep. No. 11, 36 p.

Marte, C.L., N.M. Sherwood, L.W. Crim & B. Harvey. 1987. Induced spawning of maturing milkfish (*Chanos chanos* Forsskal) with gonadotropin-releasing hormone (GnRH) analogues administered in various ways. Aquaculture 60:303-310.

Marte, C.L., N. Sherwood, L. Crim & J. Tan. 1988. Induced spawning of maturing milkfish (*Chanos chanos*) using human chorionic gonadotropin and mammalian and salmon gonadotropin releasing hormone analogues. Aquaculture 73:333-340.

Marte, C.L., J. Nocillado, M.N. Duray & I. Borlongan. 1994. Survival and growth of milkfish (*Chanos chanos* Forsskal) larvae reared on larval feeds as supplement to live food. Third Asian Fisheries Forum, Asian Fish. Soc., Manila, p 756-759.

Marteinsdottir, G. 1991. Early Life History Variation of the Mummichog (*Fundulus heteroclitus*): Egg Size Variation and Its Significance in Reproduction and Survival of Eggs and Larvae. Ph.D. Diss., Rutgers

State Univ., New Jersey, xii+142 p.
Marteinsdottir, G., & K.W. Able. 1992. Influence of egg size on embryos and larvae of *Fundulus heteroclitus* (L.). J. Fish Biol. 41:883-896.
Martel, Holly. 1994. New England Aquarium, Boston, Massachusetts, pers. comm.
Martin, A.M., S. Goddard & P. Bemister. 1993. Production of *Candida utilis* biomass as aquaculture feed. J. Sci. Food Agric. 61:363-370.
Martin, F.D, & G.E. Drewry. 1978. Development of Fishes of the Mid-Atlantic Bight: An Atlas of Egg, Larval and Juvenile Stages, Vol. 6. U.S. Fish Wildl. Serv., 416 p.
Martin, M. 1985. The effects of hydrogen sulphide in catfish farming. Aquac. Mag. 11:(1)39-41.
Martin, R.A. 1968. Reproduction of fishes. Salt Water Aquarium 4:34-40.
Martin, R.A., & C.L. Martin. 1971. Reproduction of the clingfish, *Gobiesox strumosus*. Quart. J. Fla. Acad. Sci. 33:275-278.
Martin-Robichaud, D.J. 1991. Culture of lumpfish (*Cyclopterus lumpus*) for roe production. Bull. Aquac. Assoc. Can. 91-3, p 83-85.
Martin-Robichaud, D.J., R.H. Peterson, T.J. Benfey & L.W. Crim. 1994. Direct feminization of lumpfish (*Cyclopterus lumpus* L.) using 17β-oestradiol-enriched *Artemia* as food. Aquaculture 123:137-151.
Martosudarmo, B. 1987. Culture trials for sea bass (*Lates calcarifer*) in floating net cages. Austral. Ctr. Internat. Agric. Res., Proc. (20):179-180.
Maruta, K., H. Miyazaki, S. Masuda, M. Takahashi, T. Marubashi, Y. Tadano & H. Takahashi. 1996. Exclusion of intestinal pathogens by continuous feeding with *Bacillus subtilis* C-3102 and its influence on the intestinal microflora in broilers. Anim. Sci. Technol. (Japan) 67:273-280.
Maruyama, I., T. Nakamura, T. Matsubayashi, Y. Ando & T. Maeda. 1986. Identification of the alga known as 'marine *Chlorella*' as a member of the Eustigmatophyceae. Jap. J. Phycol. 34:319-325.
Maruyama, Keigo. 1992. Japan Sea-Farming Association, Goto Station, Tamanoura, Nagasaki, pers. comm.
Maruyama, K., K. Nogami, Y. Yoshida & K. Fukunaga. 1994. Hatchery technology on the breeding and seed production of red spotted grouper, *Epinephelus akaara* in Japan. Third Asian Fisheries Forum, Asian Fish. Soc., Manila. p 446-449.
Mason, D.L., & C.S. Manooch III. 1985. Age and growth of mutton snapper along the east coast of Florida. Fish. Res. 3:93-104.
Masschelein, W.J. (editor). 1982. Ozonization Manual for Water and Wastewater Treatment. Wiley, New York, 324 p.
Masuda, R., K. Tsukamoto, K. Imaizumi, S. Shiozawa, S. Sekiya, T. Kogane & A. Nishi. 1995. The role of optical and chemical stimuli in the feeding behavior of striped jack *Pseudocaranx dentex*. Fish. Sci. 61:169-170.
Masuma, S., 1989. In: Ann. Rep. Japan Sea-Farming Assn., p 204-207. (in Jpn.)
Masuma, S., M. Kanematu & K. Teruya. 1990. Embryonic and morphological development of larvae and juveniles of the amberjack, *Seriola dumerili*. Jpn. J. Ichthyol. 37:164-169.
Masuma, S., N. Tezuka & K. Teruya. 1993. Embryonic and morphological development of larval and juvenile coral trout, *Plectropomus leopardus*. Jpn. J. Ichthyol. 40:333-342.
Matarese, A.C., & E.M. Sandknop. 1984. Identification of fish eggs. Am. Soc. Ichthyol. Herpetol. Spec. Pub. (1):27-31.
Matarese, A.C., W. Watson & E.G. Stevens. 1984. Blenniodea: development and relationships. Am. Soc. Ichthyol. Herpetol. Spec. Pub. (1):565-573.
Matheson, R.H., III, G.R. Huntsman & C.S. Manooch III. 1986. Age, growth, mortality, food and reproduction of the scamp, *Mycteroperca phenax*, collected off North Carolina and South Carolina. Bull. Mar. Sci. 38:300-312.
Mathews, S.B., J.B. Mock, K. Willson & H. Senn. 1976. Energy efficiency of Pacific salmon aquaculture. Prog. Fish-Cult. 38:102-106.
Mathisen, O.A. 1994. Spawning characteristics of the pink salmon (*Oncorhynchus gorbuscha*) in the eastern North Pacific Ocean. Aquac. Fish. Mgt. 25:147-156.
Matlock, G.C. 1990. The life history of red drum. Texas A&M Univ. Sea Grant TAMU-SG-90-603, p 1-21.
Matsuoka, M. 1982. Development of vertebral column and caudal skeleton of the red sea bream, *Pagrus major*. Jpn. J. Ichthyol. 29:285-294.
Matsuoka, S. 1995. A review of the nursery and growout culture techniques for flounder (*Paralichthys olivaceus*) in Japan. In: K.L. Main & C. Rosenfeld (eds.), Culture of High-value Marine Fishes in Asia and the United States. Oceanic Inst., Honolulu, p 139-145.
Matsuoka, T. 1962. On the spawning of *Pomacentrus coelestis* Jordan and Stark. The Aquacultur. 10(3):1-6.
Matsusato, T. 1975. Bacterial tuberculoidosis of cultured yellowtail. In: Proc. 3rd U.S.-Japan Meeting on Aquaculture at Tokyo, Japan, 15-16 October 1974. Spec. Pub. Fish. Ag., Japanese Gov. and Japan Sea Reg. Fish. Res. Lab., Niigata, Japan, p 115-118.
Matsuyama, M., S. Adachi, Y. Nagahama, C. Kitajima & S. Matsuura. 1988a. Annual reproductive cycle of the captive female Japanese sardine *Sardinops melanostictus*: relationship to ovarian development and serum levels of gonadal steroid hormones. Mar. Biol. 108:21-29.
Matsuyama, M., S. Adachi, Y. Nagahama & S. Matsuura. 1988b. Diurnal rhythm of oocyte development and plasma steroid hormone levels in the female red sea bream, *Pagrus major*, during the spawning season. Aquaculture 73:357-372.

Matsuyama, M., S. Adachi, Y. Nagahama, C. Kitajima & S. Matsuura. 1991a. Annual reproductive cycle of the captive female Japanese sardine *Sardinops melanostictus*: relationship to ovarian development and serum levels of gonadal steroid hormones. Mar. Biol. 108:21-29.

Matsuyama, M., T. Ishida, S. Ikeura, S. Matsui, T. Tanabe, C. Kitajima & S. Matsuura. 1991b. Induced oocyte maturation and ovulation of the Japanese sardine, *Sardinops melanostictus*, by LHRH-a. Bull. Jpn. Soc. Fish. Oceanogr. 55:301-306. (in Jpn., Eng. abs.)

Matsuyama, M., M. Hamada, T. Ashitani, M. Kashiwagi, T. Iwai, K. Okuzawa, H. Tanaka & H. Kagawa. 1993. Development of LHRH-a copolymer pellet polymerized by ultraviolet and its application for maturation in red sea bream *Pagrus major* during the non-spawning season. Bull. Jpn. Soc. Sci. Fish. 59:1361-1369.

Matsuyama, M., H. Takeuchi, M. Kashiwagi, K. Hirose & H. Kagawa. 1995. Induced gonadal development and spawning of immature red sea bream *Pagrus major* with LHRH-a administration in different ways during winter season. Fish. Sci. 61:472-477.

Matty, A.J., & K.P. Lone. 1985. Hormonal control of protein deposition. In: C.B. Cowey, A.M. Mackie & J.G. Bell (eds.) Nutrition and Feeding in Fish. Academic Press, San Diego, p 147-176.

May, R.C. 1970. Feeding larval marine fishes in the laboratory: a review. Calif. Mar. Res. Comm., Cal. Coop. Oceanic Fish. Invest. Rep. 14:76-83.

May, R.C. 1972. Effects of Temperature and Salinity on Eggs and Early Larvae of the Sciaenid Fish, *Bairdiella icistia* (Jordan and Gilbert). Ph.D. Diss., Univ. California, San Diego, 266 p.

May, R.C. 1974. Larval mortality in marine fishes and the critical period concept. In: J.H.S. Blaxter (ed.) The Early Life History of Fish. Springer-Verlag, New York, p 3-19.

May, R.C. 1975. Effects of temperature and salinity on fertilization, embryonic development, and hatching in *Bairdiella icistia* (Pisces: Sciaenidae), and the effect of parental salinity acclimation on embryonic and larval salinity tolerance. Fish. Bull., U.S. 73:1-22.

May, R.C. 1978. Marine food production: problems and prospects for Latin America. Revista de Biologia Tropical 26(suppl. 1):167-189.

May, R.C., D. Popper & J.P. McVey. 1974. Rearing and larval development of *Siganus canaliculatus* (Park)(Pisces: Siganidae). Micronesica 10:285-298.

Mayer, F.L., Jr. 1987. Acute Toxicity Handbook of Chemicals to Estuarine Organisms. USEPA Rep. 600/8-87/017, 274 p.

Mayunar. 1992. Induced spawning and larval rearing of estuary grouper. Oseana 17:69-82. (in Indones., Eng. abs.)

Mayunar, P.T. Imanto, S. Diani & T. Yokokawa. 1991. Spawning of grouper, *Epinephelus fuscoguttatus*. Bull. Penelitian Perikanan, Spec. Ed. (2):15-22. (in Indones., Eng. abs.)

Mazid, M.A., S. Sultana, M. Kamal, M.H. Hossain & S. Gheyasuddin. 1994. Preparation of feed from indigenous sources for the optimum growth of tilapia (*Oreochromis niloticus*). Third Asian Fisheries Forum., Asian Fish. Soc., Manila, p 637-640.

Mazzola, A., B. Rallo, S. Gilberto & R. Cecarelli. 1983. Report on artificial reproduction and first stage of the breeding of the sargo (*Diplodus sargus* L.). World Maric. Soc. Spec. Pub. (3):124-132.

McAlary, F.A., & W.N. McFarland. 1993. The effect of light and darkness on hatching in the pomacentrid *Abudefduf saxatilis*. Env. Biol. Fishes 37:237-244.

McCallum, I.M., & D.A. Higgs. 1989. An assessment of processing effects on the nutritive value of marine protein sources for juvenile chinook salmon (*Oncorhynchus tshawytscha*). Aquaculture 77:181-200.

McCarthy, I., E. Moksness & D.A. Pavlov. 1998. The effects of temperature on growth rate and growth efficiency of juvenile common wolffish. Aquac. Internat. 6:207-218.

McCarty, C.E. 1990. Design and operation of a photoperiod/temperature spawning system for red drum. Texas A&M Univ. Sea Grant TAMU-SG-90-603, p 44-45.

McCarty, C.E., J.G. Geiger, L.N. Sturmer, B.A. Gregg & W.P. Rutledge. 1986. Marine finfish culture in Texas: a model for the future. In: R.H. Stroud (ed.), Fish Culture in Fisheries Management. Am. Fish. Soc., Bethesda, Maryland, p 249-262.

McCoy, H.D., II. 1993. Open ocean fish farming: part one. Aquac. Mag. 19:66-74.

McCraren, J.P. 1990. Transportation and acclimation of red drum fingerlings. Texas A&M Univ. Sea Grant TAMU-SG-90-603, p 92-104.

McCraren, J.P., & J.L. Millard. 1978. Transportation of warmwater fishes. In: Manual of Fish Culture. U.S. Fish Wildl. Serv., p 43-88.

McCurdy, S.M., & B.E. March. 1992. Processing of canola meal for incorporation in trout and salmon diets. J. Am. Oil Chem. Soc. 69:213-220.

McEachron, L.W., & K. Daniels. 1995. Red drum in Texas: a success story in partnership and commitment. Fisheries 20(3):6-8.

McEllhiney, R.R. (ed.). 1985. Feed Manufacturing Technology III. American Feed Industry, Arlington, Virginia, 608 p.

McEvoy, L.-A. (Barton) 1984. Ovulatory rhythms and over-ripening of eggs in cultivated turbot, *Scophthalmus maximus* L. J. Fish Biol. 24:437-448.

McEvoy, L.-A. 1989. Reproduction of turbot (*Scophthalmus maximus* L.) in captivity. Cuadernos da Area de Ciencias Marinas, Seminario de Estudos Galegos 3:9-28.

McEvoy, L.-A., & J. McEvoy. 1991. Size fluctuation in the eggs and newly hatched larvae of captive turbot

(*Scophthalmus maximus*). J. Mar. Biol. Assn. U.K. 71:679-690.
McEvoy, L.A., & J. McEvoy. 1992. Multiple spawning in several commercial fish species and its consequences for fisheries management, cultivation and experimentation. J. Fish Biol. 41(suppl. B):125-136.
McEvoy, L., D. Holland & J. McEvoy. 1993. Effect of spawning fast on lipid and fatty acid composition of eggs of captive turbot (*Scophthalmus maximus* L.). Aquaculture 114:131-139.
McEvoy, L.A., J.C. Navarro, J.G. Bell & J.R. Sargent. 1995. Autoxidation of oil emulsions during the *Artemia* enrichment process. Aquaculture 134:101-112.
McEvoy, L.A., J.C. Navarro, F. Hontoria, F. Amat & J.R. Sargent. 1996. Two novel *Artemia* enrichment diets containing polar lipid. Aquaculture 144:339-352.
McFarland, W.N. 1960. The use of anesthetics for the handling and the transport of fishes. Cal. Fish Game 46:407-431.
McFarlane, G.A., J.Q.T. Jensen, W.T. Andrews & E.P. Groot. 1991. Egg and yolk sac larval development of Pacific halibut (*Hippoglossus stenolepis*). Internat. Pac. Halibut Comm. Tech. Rep. No. 24, 22 p.
McGeachin, R.B., R.R. Stickney & C.R. Arnold. 1977. Algae-fed brine shrimp nauplii as a food source for larval spotted seatrout. Proc. Ann. Conf. SE Assoc. Fish Wildl. Ag. 31:574-582.
McGoogan, B.B., & D.M. Gatlin III. 1997. Effects of replacing fish meal with soybean meal in diets for red drum *Sciaenops ocellatus* and potential for palatability enhancement. J. World Aquac. Soc. 28:374-385.
McGoogan, B.B., & R.C. Reigh. 1996. Apparent digestibility of selected ingredients in red drum (*Sciaenops ocellatus*) diets. Aquaculture 141:233-244.
McGowan, M.F., & F.H. Berry. 1984. Clupeiformes: development and relationships. Am. Soc. Ichthyol. Herpetol. Spec. Pub. (1):108-126.
McKenney, T.W. 1959. A contribution to the life history of the squirrel fish *Holocentrus vexillarius* Poey. Bull. Mar. Sci. 9:174-221.
McKim, J.M., G.F. Olson, G.W. Holcombe & E.P. Hunt. 1976. Long-term effects of methylmercuric chloride on three generations of brook trout (*Salvelinus fontinalis*): toxicity, accumulation, distribution, and elimination. J. Fish. Res. Bd. Can. 33:2726-2739.
McMaster, M.F. 1988. Pompano aquaculture: past success and present opportunities. Aquac. Mag. 14(3):28,30-34.
McVey, J.P. 1972. Observations on the early-stage formation of rabbitfish *Siganus fuscescens* [should be *S. canaliculatus*] at Palau Mariculture Demonstration Centre. S. Pac. Isl. Fish. Newsl. (Noumea, New Caledonia) 6:11-12.
Meade, J.W. 1985. Allowable ammonia for fish culture. Prog. Fish-Cult. 47:135-145.
Meade, J.W. 1989. Aquaculture Management. Van Nostrand Reinhold, New York, 175 p.
Meekan, M.G., M.J. Milicich & P.J. Doherty. 1993. Larval production drives temporal patterns of larval supply and recruitment of a coral reef damselfish. Mar. Ecol. Prog. Ser. 93:217-225.
Mehner, T. 1992. Diet spectra of *Pomatoschistus microps* (Krøyer) and *Pomatoschistus minutus* (Pallas)(Teleostei, Gobiidae) during first weeks after hatching. Zool. Anz. 229:13-20.
Meidi & Zhifeng. 1991. Fish disease and its control. In: L. Zhong (chief ed.), Pond Fisheries in China. Pergamon, Oxford, p 207-260.
Meier, W., M. Schmitt & T. Wahli. 1994. Viral hemorrhagic septicemia (VHS) of nonsalmonids. Ann. Rev. Fish Dis. 4:359-373.
Meilahn, C.W., D.A. Davis & C.R. Arnold. 1996. Effects of commercial fish meal analogue and menhaden fish meal on growth of red drum fed isonitrogenous diets. Prog. Fish-Cult. 58:111-116.
Meister, R.T. (ed.-in chief). 1994. Farm Chemicals Handbook '94. Meister Pub. Co., Willoughby, Ohio.
Melo, Y.C., & M.J. Armstrong. 1991. Batch spawning behaviour in lightfish *Maurolicus muelleri*. S. Afr. J. Mar Sci. 10:125-130.
Meng, L. 1993. Estimating food requirements of striped bass larvae: an energetics approach. Trans. Am. Fish. Soc. 122:244-251.
Mercer, L.P. 1984. A biological and fisheries profile of spotted seatrout, *Cynoscion nebulosus*. North Carolina Dept. Nat. Resour., Spec. Sci. Rep. No. 40, 87 p.
Mercer, L.P. 1989. Species profiles: life histories and environmental requirements of coastal fishes and invertebrates (South Atlantic): black sea bass. U.S. Fish Wildl. Serv., Biol. Rep. 82(11.99), 16 p.
Merchie, G., P. Lavens, V. Storch, U. Übel, H. Nelis, A. De Leenheer & P. Sorgeloos. 1996. Influence of dietary vitamin C dosage on turbot (*Scophthalmus maximus*) and European sea bass (*Dicentrarchus labrax*) nursery stages. Comp. Biochem. Physiol. 114A:123-133.
Merchie, G., P. Lavens & P. Sorgeloos. 1997. Optimization of dietary vitamin C in fish and crustacean larvae: a review. Aquaculture 155:165-181.
Merriner, J.V. 1976. Aspects of the reproductive biology of the weakfish, *Cynoscion regalis* (Sciaenidae), in North Carolina. Fish. Bull., U.S. 74:18-26.
Metailler, R., & C. Huelvan. 1993. Utilisation de levures dans l'alimentation du juvenile de bar (*Dicentrarchus labrax*). Fish Nutrition in Practice, INRA, Paris, p 945-948.
Métailler, R., M. Cadena-Roa & J. Person-Le Ruyet. 1983. Attractive chemical substances for the weaning of Dover sole (*Solea vulgaris*): qualitative and quantitative approach. J. World. Maric. Soc.

14:679-684.
Meyer, K.A. 1977. Reproductive behavior and patterns of sexuality in the Japanese labrid fish *Thalassoma cupido*. Jpn. J. Ichthyol. 24:101-112.
Meyers, S.P. 1985. Importance of lipid, lecithin in aquatic diets examined. Feedstuffs 57:20-21.
Meyers, S.P. 1994. Developments in world aquaculture, feed formulations, and role of carotenoids. Pure & Appl. Chem. 66:1069-1076.
Meyers, S.P. 1995. Soy lecithin and phospholipids in aquatic diets. In: C. Lim & D.J. Sessa (eds.), Nutrition and Utilization Technology in Aquaculture. AOSC Press, Champaign, Ilinois, p 105-113.
Meyers, S.P., & H.K. No. 1995. Utilization of crawfish pigment and other fishery processing by-products. In: C. Lim & D.J. Sessa (eds.), Nutrition and Utilization Technology in Aquaculture. AOSC Press, Champaign, Ilinois, p 269-277.
Micale, V., F. Perdichizzi & G. Basciano. 1996. Aspects of the reproductive biology of the sharpsnout seabream *Diplodus puntazzo* (Cetti, 1777). I. Gametogenesis and gonadal cycle in captivity during the third year of life. Aquaculture 140:281-291.
Michelson, R. 1991. Atlantic salmon restoration moves slowly; aquaculture blossoms. Aquac. Mag. 17(6):32,34-39.
Microalgal Technology Research Group. 1986. Microalgae Culture Collection 1985-1986. Solar Energy Res. Inst., Golden, Colorado, 97 p.
Middaugh, D.P. 1981. Reproductive ecology and spawning periodicity of the Atlantic silverside, *Menidia menidia* (Pisces: Atherinidae). Copeia 1981:766-776.
Middaugh, D.P., & M.J. Hemmer. 1992. Reproductive ecology of the inland silverside, *Menidia beryllina*, (Pisces: Atherinidae) from Blackwater Bay, Florida. Copeia 1992:53-61.
Middaugh, D.P., & R.L. Yoakum. 1974. The use of chorionic gonadotropin to induce laboratory spawning of the Atlantic croaker, *Micropogon undulatus*, with notes on subsequent embryonic development. Chesapeake Sci. 15:110-114.
Middaugh, D.P., J.A. Couch & A.M. Crane. 1977. Responses of early life history stages of the striped bass, *Morone saxatilis* to chlorination. Chesapeake Sci. 18:141-153.
Middaugh, D.P., M.J. Hemmer, J.M. Shenker & T. Takita. 1990. Laboratory culture of jacksmelt, *Atherinopsis californiensis*, and topsmelt, *Atherinops affinis* (Pisces: Atherinidae), with a description of larvae. Cal. Fish Game 76:4-13.
Middaugh, D.P., B.S. Anderson & M.J. Hemmer. 1992. Laboratory spawning of topsmelt, *Atherinops affinis*, with notes on culture and growth of larvae. Env. Toxicol. Chem. 11:393-399.
Middaugh, D.P., S.M. Resnick, S.E. Lantz, C.S. Heard & J.G. Mueller. 1993. Toxicological assessment of biodegraded pentachlorophenol: Microtox® and fish embryos. Arch. Env. Contam. Toxicol. 24:165-172.
Mihelakakis, A., & C. Kitajima. 1994a. Effects of salinity and temperature on incubation period, hatching rate, and morphogenesis of the silver sea bream, *Sparus sarba* (Forskal, 1775). Aquaculture 126:361-371.
Mihelakakis, A., & C. Kitajima. 1994b. The effect of feeding level on growth and feed conversion in silver sea bream, *Sparus sarba*. J. Fac. Agric. Kyushu Univ. 39:9-14.
Mihelakakis, A., & C. Kitajima. 1995. Spawning of the silver sea bream, *Sparus sarba*, in captivity. Jpn. J. Ichthyol. 42:53-59.
Miki, W., N. Toriu, Y. Kondo, M. Murakami, K. Yamaguchi, S. Konosu, M. Satake & T. Fujita. 1983. The stability of carotenoid pigments in the Antarctic krill *Euphausia superba*. Bull. Jpn. Soc. Sci. Fish. 49:1417-1420.
Miliou, H., M. Moraitou-Apostolopoulou & M. Argyridou. 1992. Biochemical composition of *Tisbe holothuriae* (Copepoda: Harpacticoida) and its differentiation in relation to developmental stages. Ophelia 36:159-166.
Millan, L.M., A. Ortega, A.G. Alcazar, R.T. Lovell & J.R. Posada. 1989. Optimum levels in substituting fish meal with soybean meal in practical diets for growing and fattening of gilthead seabream (*Sparus aurata*) Anim. Nutr. Highlights 1.
Millares, N.M., M. Borrero & E. Gonzalez. 1979. Desove inducido de *Eugerres plumieri* (patao) en condiciones de laboratorio. Revista Cubana de Investigaciones Pesqueras 4:65-87.
Miller, D.C., S. Poucher, J.A. Cardin & D. Hansen. 1990. The acute and chronic toxicity of ammonia to marine fish and a mysid. Arch. Env. Contam. Toxicol. 19:40-48.
Miller, Gary L. 1992. Coral World Bahamas, Nassau, pers. comm.
Miller, J.M., W. Watson & J.M. Leis. 1979. An Atlas of Common Nearshore Marine Fish Larvae of the Hawaiian Islands. Univ. Hawaii Sea Grant UNIHI-SEAGRANT-MR-80-02, 179 p.
Millikin, M.R. 1982a. Effects of dietary protein concentration on growth, feed efficiency, and body composition of age-0 striped bass. Trans. Am. Fish. Soc. 111:373-378.
Millikin, M.R. 1982b. Qualitative and quantitative nutrient requirements of fishes: a review. Fish. Bull., U.S. 80:655-686.
Millner, R., C. Whiting, M.G. Walker & P. Witthames. 1991. Growth increment, condition and fecundity in sole (*Solea solea* (L.)) from the North Sea and eastern English Channel. Neth. J. Sea Res. 27:433-439.
Mills, E.L., & J.L. Forney. 1981. Energetics, food consumption, and growth of young yellow perch in Oneida Lake, New York. Trans. Am. Fish. Soc. 110:479-488.

Milne, P.H. 1972. Fish and Shellfish Farming in Coastal Waters. Fishing News Books, Farnham, Surrey, England, 208 p.
Milton, D.A., & S.J.M. Blaber. 1991. Maturation, spawning seasonality, and proximate spawning stimuli of six species of tuna baitfish in the Solomon Islands. Fish. Bull., U.S. 89:221-237.
Mims, S.D., C.D. Webster, J.H. Tidwell & D.H. Yancey. 1991. Fatty acid composition of *Daphnia pulex* cultured by two different methods. J. World Aquac. Soc. 22:153-156.
Minchew, C.D., & J.D. Yarbrough. 1977. The occurrence of fin rot in mullet (*Mugil cephalus*) associated with crude oil contamination of an estuarine pond-ecosystem. J. Fish. Biol. 10:319-323.
Minkoff, G., & A.P. Broadhurst. 1994. Intensive production of turbot, *Scophthalmus maximus*, fry. Eur. Aquac. Soc. Spec. Pub. (22):14-31.
Mironova, N.V. 1974. The energy balance of *Tilapia mossambica*. J. Ichthyol. 14:431-438.
Mitchell, A.I., A. Dawson & D.F. Houlihan. 1993. Trypsin inhibitors in commercial fish food. Fish Nutrition in Practice, INRA, Paris, p 219-222.
Mitchell, H.M. 1996. Practical considerations in developing and using efficacious vaccines for aquaculture with lessons from agriculture. In: K.L. Main & C. Rosenfeld (eds.), Aquaculture Health Management Strategies for Marine Fishes. Oceanic Inst., Honolulu, p 171-182.
Mitchell, S.A. 1991. A technique for the intensive culture of *Paramecium* (group *caudatum*; Parameciidae) as a first food for ornamental fish fry. Aquaculture 95:189-192.
Mito, S. 1954. Breeding habits of a blennioid fish, *Salarias enosimae*. Jpn. J. Ichthyol. 3:144-152.
Mito, S. 1957. On the egg development and larvae of a Japanese sea bass, *Lateolabrax japonicus* (Cuvier). Sci. Bull. Fac. Agric., Kyushu Univ. 16:115-124. (in Jpn., Eng. sum.)
Mito, S. 1961. Pelagic fish eggs from Japanese waters—I. Clupeina, Chanina, Stomiatina, Myctophida, Anguillida, Belonida and Syngnathida. Sci. Bull. Fac. Agric. Kyushu Univ. 18:285-310.
Mito, S. 1963. Pelagic fish eggs from Japanese waters—III. Percina. Jpn. J. Ichthyol. 11:39-57.
Mitsch, H. 1941. Breeding of marine clown fishes {*Amphiprion percula*}. The Aquarium 10(3):48-50.
Miya, M., & T. Nemoto. 1991. Comparative life histories of the meso- and bathypelagic fishes of the genus *Cyclothone* (Pisces: Gonostomatidae) in Sagami Bay, central Japan. Deep-Sea Res. 38:67-89.
Miyaki, K., & Y. Dotsu. 1987. A scanning microscopical observation of the cryopreserved spermatozoa of the swellfishes. Bull. Fac. Fish. Nagasaki Univ. 62:17-21.
Miyashita, K., M. Hirao, E. Nara & T. Ota. 1995. Oxidative stability of triglycerides from orbital fat of tuna and soybean oil in an emulsion. Fish. Sci. 61:273-275.
Mobiha, A. 1991. The reproductive biology of eight reef fish species from the Tigak Islands of Papua New Guinea. Sci. New Guinea 17:133-142.
Mochioka, N., M. Iwamizu & T. Kanda. 1993. Leptocephalus eel larvae will feed in aquaria. Env. Biol. Fishes 36:381-384.
Moe, M.A. 1969. Biology of the red grouper *Epinephelus morio* (Valenciennes), from the eastern Gulf of Mexico. Fla. Dept. Nat. Resour. Mar. Res. Lab., Prof. Pap. Ser. No. 10, 95 p.
Moe, M.A. 1973. Breeding the clownfish, *Amphiprion ocellaris*. Salt Water Aquarium 9(2):3-14.
Moe, M.A. 1975. Propagating the Atlantic neon goby. Mar. Aquarist 6(2):4-10.
Moe, M.A., Jr. 1976. Rearing Atlantic angelfish. Mar. Aquarist 7(7)17-26.
Moe, M.A., Jr. 1977. Inside the egg of an angelfish. Mar. Aquarist 8(3):5-12.
Moe, M.A., Jr. 1989. The Marine Aquarium Reference: Systems and Invertebrates. Green Turtle Publications, Plantation, Florida, 507 p.
Moe, M.A., Jr. 1992. The Marine Aquarium Handbook; Beginner to Breeder. Green Turtle Publications, Plantation, Florida, 320 p.
Moe, M.A., Jr. 1997a. Breeding the Orchid Dottyback, *Pseudochromis fridmani*: An Aquarist's Journal. Green Turtle Publications, Plantation, Florida, 288 p.
Moe, M.A., Jr. 1997b. Spawning and rearing the large angelfish *Pomacanthus* sp. Aquarium Frontiers (May-June):14-24.
Moe, M.A., Jr. 1997c. Spawning and rearing the large angelfish *Pomacanthus* sp.: part 2. InterNet.
Moe, M.A., & F.A. Young. 1981. Spawning the jewels of the reef. Freshw. Mar. Aquarium 4:24-25,82-84.
Moe, M., R.H. Lewis & R. Ingle. 1968. Pompano mariculture: preliminary data and basic considerations. Fla. Bd. Conserv. Mar. Lab., Tech. Ser. No. 55, 65 p.
Moenich, D.R. 1990. Marine angelfish: *Holacanthus*. Aquarium Fish Mag. 2(6):59-65.
Moffatt, N.M. 1981. Survival and growth of northern anchovy larvae on low zooplankton densities as affected by the presence of a *Chlorella* bloom. Rapp. P.-v. Réun. Cons. Int. Explor. Mer 178:475-480.
Moffitt, C.A. 1991. FDA Approved Registration of Erythromycin for Treatment of Bacterial Kidney Disease (BKD) in Juvenile and Adult Chinook Salmon. Bonneville Power Administration, Portland, Oregon, Proj. No. 89-32, Ann. Rep., 44 p.
Mohapatra, B.C., & A. Noble. 1993. Changes in oxygen uptake in mullet, *Liza parsia* (Hamilton-Buchanan) exposed to dichlorvos. Fish. Technol. 30:112-114.
Mohler, J.W., K. Fynn-Aikins & R. Barrows. 1996. Feeding trials with juvenile Atlantic sturgeons propagated from wild broodstock. Prog. Fish-Cult. 58:173-177.
Moksness, E., & R. Støle. 1997. Larviculture of marine fish for sea ranching purposes: is it profitable? Aquaculture 155:341-353.

Moksness, E., J. Gjøsæter, A. Reinert & I.S. Fjallstein. 1989. Start-feeding and on-growing of wolffish (*Anarhichas lupus*) in the laboratory. Aquaculture 77:221-228.

Molinero, A., E. Gómez, J. Balasch & L. Tort. 1997. Stress by fish removal in the gilthead sea bream, *Sparus aurata*: a time course study on the remaining fish in the same tank. J. Appl. Aquac. 7:1-12.

Molsher, R.L., M.C. Geddes & D.C. Paton. 1994. Population and reproductive ecology of the small-mouthed hardyhead *Atherinosoma microstoma* (Gunther)(Pisces: Atherinidae) along a salinity gradient in the Coorong, South Australia. Trans. Royal Soc. S. Australia 118:207-216.

Montaini, E., G.C. Zittelli, M.R. Tredici, E.M. Grima, J.M.F. Sevilla & J.A.S. Pérez. 1995. Long-term preservation of *Tetraselmis suecica*: influence of storage on viability and fatty acid profile. Aquaculture 134:81-90.

Montgomery, D., & J.A. Brock. 1997. Preliminary trials on the application of hydrogen peroxide for treatment of the ectoparasites *Gyrodactylus* sp., and *Trichodina* sp., on the Chinese catfish (*Clarias fuscus*), and *Amyloodinium* sp. on the Pacific thread-fin (*Polydactylus sexfilis*). World Aquac. '97 Book of Abs. p 333-334.

Montgomery, J.L.M., & T.E. Targett. 1992. The nutritional role of seagrass in the diet of the omnivorous pinfish *Lagodon rhomboides* (L.). J. Exp. Mar. Biol. Ecol. 158:37-57.

Moon, H.Y., & D.M. Gatlin III. 1989. Amino acid nutrition of the red drum (*Sciaenops ocellatus*): determination of limiting amino acids and development of a suitable amino acid test diet. Proc. Third Internat. Symp. on Feeding and Nutr. in Fish, Toba Aug. 28-Sept 1, Japan. p 201-208.

Moon, H.Y., & D.M. Gatlin III. 1991. Total sulfur amino acid requirement of juvenile red drum, *Sciaenops ocellatus*. Aquaculture 95:97-106.

Morales-Nin, B. 1991. Parametros biologicos del salmonete de roca *Mullus surmuletus* (L. 1758), en Mallorca. Bol. Inst. Esp. Oceanogr. 7:139-147. (in Sp., Eng. abs.)

Moran, D. 1988. Species Profiles: Life Histories and Environmental Requirements of Coastal Fishes and Invertebrates (Gulf of Mexico)—Red Snapper. U.S. Fish Wild. Serv. Biol. Rep. 82(11.83). U.S. Army Corps Engineers, TR EL-82-4, 19 p.

Morgan, R.P., II, & R.D. Prince. 1977. Chlorine toxicity to eggs and larvae of five Chesapeake Bay fishes. Trans. Am. Fish. Soc. 106:380-385.

Morgan, R.P., II & R.D. Prince. 1978. Chlorine effects on larval development of striped bass (*Morone saxatilis*), white perch (*M. americana*) and blueback herring (*Alosa aestivalis*). Trans. Am. Fish. Soc. 107:636-641.

Mori, K., S. Ueyanagi & Y. Nishikawa. 1971. The development of artificially fertilized and reared larvae of the yellowfin tuna, *Thunnus albacares*. Bull. Far Seas Fish. Res. Lab. (5):219-232.

Moriarty, D.J.W. 1997. The role of microorganisms in aquaculture ponds. Aquaculture 151:333-349.

Morimoto, H. 1993. Time of maximal oocyte hydration and spawning in the Japanese sardine in Tosa Bay, southwestern Japan. Bull. Jpn. Soc. Sci. Fish. 59:7-14. (in Jpn., Eng. abs.)

Morimoto, H. 1998. Relationship between batch fecundity and egg size in Japanese sardine *Sardinops melanostictus* in Tosa Bay and off Kii Channel, southwestern Japan from 1990 to 1993. Fish. Sci. 64:220-227.

Morimura, S. 1993. Influences of the environment on fish farming. In: H. Reinertsen, L.A. Dahle, L. Jørgensen & K. Tvinnereim (eds.), Fish Farming Technology. Balkema, Rotterdam, p 155-161.

Morizane, T. 1993. The present state of finfish seed production in Ehime Prefecture. In: C.-S. Lee, M.-S. Su & I C. Liao (eds.), Finfish Hatchery in Asia. Tungkang Mar. Lab. Conf. Proc. (3):41-49.

Morrison, C.M. 1993. Histology of the Atlantic Cod, *Gadus morhua*. Part Four. Eleutheroembryo and Larva. Can. Spec. Pub. Fish. Aquat. Sci. No. 119, 496 p.

Morrison, J.A., I.R. Napier & J.C. Gamble. 1991. Mass mortality of herring eggs associated with a sedimenting diatom bloom. ICES J. Mar. Sci. 48:237-245.

Mortensen, S.H., K.Y. Borsheim, J.R. Rainuzzo & G. Knutsen. 1988. Fatty acid and elemental composition of the marine diatom *Chaetoceros gracilis* Schutt. Effects of silicate deprivation, temperature and light intensity. J. Exp. Mar. Biol. Ecol. 122:173-185.

Mortensen, S.H., Ø. Evensen, O.M. Rødseth & B.K. Hjeltnes. 1993. The relevance of infectious pancreatic necrosis virus (IPNV) in farmed Norwegian turbot (*Scophthalmus maximus*). Aquaculture 115:243-252.

Morton, S.L., D.R. Norris & J.W. Bomber. 1992. Effect of temperature, salinity and light intensity on the growth and seasonality of toxic dinoflagellates associated with ciguatera. J. Exp. Mar. Biol. Ecol. 157:79-90.

Moser, H.G., E.H. Ahlstrom & E.M. Sandknop. 1977. Guide to the Identification of Scorpionfish Larvae (family Scorpaenidae) in the Eastern Pacific with Comparative Notes on Species of *Sebastes* and *Helicolenus* from Other Oceans. NOAA Tech. Rep. NMFS Circ. 402, 71 p.

Moser, H.G., D.A. Ambrose, M.S. Busby, J.L. Butler, E.M. Sandknop, B.Y. Sumida & E.G. Stevens. 1983. Description of early stages of white seabass, *Atractoscion nobilis*, with notes on distribution. Cal. Coop. Oceanic Fish. Invest. Rep. 24:182-193.

Moser, H.G. (ed. in chief). 1984. Ontogeny and Systematics of Fishes. Am. Soc. Ichthyol. Herpetol. Spec. Pub. No. 1, 760 p.

Mosher, C. 1954. Observations on the spawning behavior and the early larval development of the sargassum fish, *Histrio histrio* (Linnaeus). Zoologica 39:141-152.

Motta, P.J., K.B. Clifton, P. Hernandez, B.T. Eggold, S.D. Giordano & R. Wilcox. 1995. Feeding relationships among nine species of seagrass fishes of Tampa Bay, Florida. Bull. Mar. Sci. 56:185-200.

Mounib, M.S. 1978. Cryogenic preservation of fish and mammalian spermatozoa. J. Repr. Fert. 53:13-18.

Mounib, M.S., P.C. Hwang & D.R. Idler. 1968. Cryogenic preservation of Atlantic cod *Gadus morhua* sperm. J. Fish. Res. Bd. Can. 25:2623-2632.

Mourente, G., & D.R. Tocher. 1992. Effects of weaning onto a pelleted diet on docosahexaenoic acid (22:6*n*-3) levels in brain of developing turbot (*Scophthalmus maximus* L.). Aquaculture 105:363-377.

Mourente, G., & D.R. Tocher. 1993. The effects of weaning on to a dry pellet diet on brain lipid and fatty acid compositions in post-larval gilthead sea bream (*Sparus aurata* L.). Comp. Biochem. Physiol. 104A:605-611.

Moyano, F.-J., G. Cardenete & M. De la Higuera. 1992. Nutritive value of diets containing a high percentage of vegetable proteins for trout, *Oncorhynchus mykiss*. Aquat. Living Resour. 5:23-29.

Moyer, J.T. 1990. Social and reproductive behavior of *Chaetodontoplus mesoleucus* (Pomacanthidae) at Bantayan Island, Philippines, with notes on pomacanthid relationships. Jpn. J. Ichthyol. 36:459-467.

Moyer, J.T., & A. Nakazono. 1978. Population structure, reproductive behavior and protogynous hermaphroditism in the angelfish *Centropyge interruptus* at Miyake-jima, Japan. Jpn. J. Ichthyol. 25:25-39.

Moyer, J.T., & M.J. Zaiser. 1981. Social organization and spawning behavior of the pteroine fish *Dendrochirus zebra* at Miyake-jima, Japan. Jpn. J. Ichthyol. 28:52-69.

Moyer, J.T., R.E. Thresher & P.L. Colin. 1983. Courtship, spawning and inferred social organization of American angelfishes (genera *Pomacanthus*, *Holacanthus* and *Centropyge*; Pomacanthidae). Env. Biol. Fishes 9:25-39.

Mudge, S.M., I.D. Goodchild & M. Wheeler. 1995. Vegetable oil spills on salt marshes. Chem. Ecol. 10:127-135.

Mueller, K.W., W.O. Watanabe & W.D. Head. 1994. Occurrence and control of *Neobenedenia melleni* (Monogenea: Capsalidae) in cultured tropical marine fish, including three new host records. Prog. Fish-Cult. 56:140-142.

Mugnier, C., A. Fostier, S. Guezou, J.-L. Gaignon & L. Quemener. 1998. Effect of some repetitive factors on turbot stress response. Aquac. Internat. 6:33-45.

Muir, P.R., & D.C. Sutton. 1994. Bacterial degradation of microencapsulated foods used in larval culture. J. World Aquac. Soc. 25:371-378.

Muiruri, R.M. 1988. Chilled and Cryogenic Preservation of Tilapia Spermatozoa. M.S. Thesis, Univ. Stirling, 57 p.

Mukhopadhyay, M.K., B.B. Ghosh & M.M. Bagchi. 1994. Toxicity of heavy metals to fish, prawn and fish food organisms of Hooghly estuarine system. Geobios 21:13-17.

Mullaney, M.D., Jr. & L.D. Gale. 1996. Ecomorphological realtionships in ontogeny: anatomy and diet in gag, *Mycteroperca microlepis* (Pisces: Serranidae). Copeia 1996:167-180.

Mullis, A.W., & J.M. Smith. 1990. Design considerations for striped bass and striped bass hybrid hatching facilities. In: R.M. Harrell, J.H. Kerby & R.V. Minton (eds.), Culture and Propagation of Striped Bass and Its Hybrids. Am. Fish. Soc., Bethesda, Maryland, p 7-16.

Munehara, H., & S. Mishima. 1986. Embryonic development, larvae and juvenile of elkhorn sculpin, *Alcichthys alcicornis*. Jpn. J. Ichthyol. 33:46-50.

Munehara, H., & K. Shimazaki. 1991. Embryonic development and newly hatched larvae of the little dragon sculpin *Blepsias cirrhosus*. Jpn. J. Ichthyol. 38:31-34.

Munilla-Moran, R., J.R. Stark & A. Barbour. 1990. The role of exogenous enzymes in digestion in cultured turbot larvae (*Scophthalmus maximus* L.). Aquaculture 88:337-350.

Munro, A.D., A.P. Scott & T.J. Lam (eds.) 1990. Reproductive Seasonality in Teleosts: Environmental Influences. CRC Press, Boca Raton, Florida, 254 p.

Munro, J., C. Audet, M. Besner & J.-D. Dutil. 1994. Physiological response of American plaice (*Hippoglossoides platessoides*) exposed to low salinity. Can. J. Fish. Aquat. Sci. 51:2448-2456.

Munro, P.D., H.A. McLean, A. Barbour & T.H. Birkbeck. 1995. Stimulation or inhibition of growth of the unicellular alga *Pavlova lutheri* by bacteria isolated from larval turbot culture systems. J. Appl. Bacteriol. 79:519-524.

Mura, G. 1995. Biometrics and fatty acid composition of the resting eggs of *Thamnocephalus platyurus* (Anostraca) in view of an eventual use as fish feed. Crustaceana 68:629-635.

Murai, M., Y. Aoki & K. Nishimura. 1984. On the egg collection of *Epinephelus fasciatus* (Forsskål). Tech. Rep. Farm. Fish 13(1):63-67. (in Jpn.)

Murashige, R., P. Bass, L. Wallace, A. Molnar, B. Eastham, V. Sato, C. Tamaru & C.-S. Lee. 1991. The effect of salinity on the survival and growth of striped mullet (*Mugil cephalus*) larvae in the hatchery. Aquaculture 96:249-254.

Murata, O., T. Harada, S. Miyashita, K.-i. Izumi, S. Maeda, K. Kato & H. Kumai. 1996. Selective breeding for growth in red sea bream. Fish. Sci. 62:845-849.

Muravjev, W., L. Buqrov & L. Bougrova. 1993. Underwater aquaculture technology in offshore areas. Eur. Aquac. Soc. Spec. Pub. (19):417.

Muroga, K. 1992. Hatchery diseases of marine fish in Japan. In: Diseases in Asian Aquaculture. Asian Fish. Soc., Manila, p 215-222.

Musgrave, G. 1976. Congratulations Gary Musgrave. Mar. Hob. News 4(7):1,6.

Mushiake, K., S. Arai, A. Matsumoto, H. Shimma & I. Hasegawa. 1993. Artificial insemination from 2 year-old cultured yellowtail fed with moist pellets. Bull. Jpn. Soc. Sci. Fish. 59:1721-1726. (in Jpn. Eng. abs.)

Mushiake, K., K. Kawano, W. Sakamoto & I. Hasegawa. 1994. Effects of extended daylength on ovarian maturation and HCG-induced spawning in yellowtail fed moist pellets. Fish. Sci. 60:647-651.

Mustafa, M.G., & H. Nakagawa. 1995. A review: dietary benefits of algae as an additive in fish feed. Bamidgeh 47:155-162.

Mustafa, M.G., T.-a. Takeda, T. Umino, S. Wakamatsu & H. Nakagawa. 1994. Effects of *Ascophyllum* and *Spirulina* meal as feed additives on growth performance and feed utilization of red sea bream, *Pagrus major*. J. Fac. Appl. Biol. Sci. 33:125-132.

Mustafa, M.G., S. Wakamatsu, T.-a. Takeda, T. Umino & H. Nakagawa. 1995. Effects of algae meal as feed additive on growth, feed efficiency, and body composition in red sea bream. Fish. Sci. 61:25-28.

Myers, J.A., & R.N. Boisvert. 1990. The economics of producing algae and bivalve seed in hatcheries. Aquaculture 86:163-179.

Myhre, P., A. Danielsberg & L. Berg. 1993. Experiments on upwelling tank systems for halibut larvae (*Hippoglossus hippoglossus* L.). In: H. Reinertsen, L.A. Dahle, L. Jørgensen & K. Tvinnereim (eds.), Fish Farming Technology. Balkema, Rotterdam, p 421-423.

Mylonas, C.C., Y. Zohar, B.M. Richardson & S.P. Minkkinen. 1995. Induced spawning of wild American shad *Alosa sapidissima* using sustained administration of gonadotropin-releasing hormone analog (GnRHa). J. World Aquac. Soc. 26:240-251.

Mylonas, C.C., Y. Magnus, A. Gissis, Y. Klebanov & Y. Zohar. 1996. Application of controlled-release, GnRHa-delivery systems in commercial production of white bass × striped bass hybrids (sunshine bass), using captive broodstocks. Aquaculture 140:265-280.

Mylonas, C.C., A. Gissis, Y. Magnus & Y. Zohar. 1997. Hormonal changes in male white bass (*Morone chrysops*) and evaluation of milt quality after treatment with a sustained-release GnRHa delivery system. Aquaculture 153:301-311.

Myouga, H., M. Yoshimizu, K. Tajima & Y. Ezura. 1995. Purification of an antiviral substance produced by *Alteromonas* sp. and its virucidal activity againts fish viruses. Fish Pathol. 30:15-22.

Myoung, J.G., P.K. Kim, S.K. Yi, J.M. Kim & H.T. Huh. 1994. Preliminary study on the seed production of small yellow croaker, *Pseudosciaena polyactis*. Third Asian Fisheries Forum. Asian Fish. Soc., Manila, p 112-115.

Myrberg, A.A. 1972. Ethology of the bicolor damselfish, *Eupomacentrus partitus* (Pisces: Pomacentridae): a comparative analysis of laboratory and field behaviour. Anim. Beh. Monogr. 5:197-283.

Myrberg, A.A., Jr., B.D. Brahy & A.R. Emery. 1967. Field observations on reproduction of the damselfish, *Chromis multilineata* (Pomacentridae), with additional notes on general behavior. Copeia 1967:819-827.

Myrseth, B. 1993. Open production systems: status and future challenges. In: H. Reinertsen, L.A. Dahle, L. Jørgensen & K. Tvinnereim (eds.), Fish Farming Technology. Balkema, Rotterdam, p 5-16.

Naas, K.E., T. van der Meeren & D.L. Aksnes. 1991. Plankton succession and responses to manipulations in a marine basin for larval fish rearing. Mar. Ecol. Prog. Ser. 74:161-173.

Naas, K.E., T. Næss & T. Harboe. 1992. Enhanced first feeding of halibut larvae (*Hippoglossus hippoglossus* L.) in green water. Aquaculture 105:143-156.

Naas, K., I. Huse & J. Iglesias. 1996. Illumination in first feeding tanks for marine fish larvae. Aquacultural Eng. 15:291-300.

Nacario, J.F. 1987. Releasing hormones as an effective agent in the induction of spawning in captivity of sea bass (*Lates calcarifer*). Austral. Ctr. Internat. Agric. Res. Proc. (20):126-128.

Nacario, J.F. 1991. Mariculture techniques for seabass, *Lates calcarifer*. In: Advances in Finfish and Shellfish Mariculture. Phil. Council Aquat. Mar. Res. Dev., Book Ser. (12):107-112.

Næss, T. 1996. Benthic resting eggs of calanoid copepods in Norwegian enclosures used in mariculture: abundance, species composition and hatching. Hydrobiologia 320:161-168.

Næss, T., K.E. Naas & O.B. Samuelsen. 1991. Toxicity of rotenone to some potential predators on marine fish larvae—an experimental study. Aquacultural Eng. 10:149-159.

Næss, T., M. Germain-Henry & K.E. Naas. 1995a. *Artemia* or wild zooplankton as first-feed for halibut larvae (*Hippoglossus hipoglossus*)—implications for abnormal pigmentation. ICES Mar. Sci. Symp. 201:201.

Næss, T., M. Germain-Henry & K.E. Naas. 1995b. First feeding of Atlantic halibut (*Hippoglossus hippoglossus*) using different combinations of *Artemia* and wild zooplankton. Aquaculture 130:235-250.

Næss, T., T. Harboe, A. Mangor-Jensen, K.E. Naas & B. Norberg. 1996. Successful first feeding of Atlantic halibut larvae from photoperiod-manipulated broodstock. Prog. Fish-Cult. 58:212-214.

Naevdal, G. 1994. Genetic aspects in connection with sea ranching of marine fish species. Aquac. Fish. Mgt. 25:93-100.

Nagahama, Y., M. Yoshikuni, M. Yamashita & M. Tanaka. 1994. Regulation of oocyte maturation in fish. In: W.S. Hoar & D.J. Randall (eds.), Fish Physiology, Vol. 13. Academic Press, New York, p 393-439.

Nagasawa, T., & T. Kobayashi. 1995. The early life history of the rockfish, *Sebastes thompsoni*

(Scorpaenidae), in the Sea of Japan. Jpn. J. Ichthyol. 41:385-396.
Nagata, W.D., & J.N.C. Whyte. 1992. Effects of yeast and algal diets on the growth and biochemical composition of the rotifer *Brachionus plicatilis* (Müller) in culture. Aquac. Fish. Mgt. 23:13-21.
Nagpal, N.K. 1994. Development of water quality and sediment criteria for PAHs to protect aquatic life. In: D.D. MacKinlay (ed.), High Performance Fish. Am. Fish. Soc., p 193-198.
Nakada, M., & T. Murai. 1991. Yellowtail aquaculture in Japan. In: J.P. McVey (ed.), Handbook of Mariculture: Vol. II, Finfish Aquaculture. CRC Press, Boca Raton, Florida, p 55-72.
Nakagawa, H. 1990. Status and problem of nutritional research on puffer fish. Fish Culture 27:113.
Nakagawa, H., H. Kumai, M. Nakamura, K. Nanba & S. Kasahara. 1984. Diagnostic studies on disease of sardine-fed yellowtail. Bull. Jpn. Soc. Sci. Fish. 50:775-782.
Nakagawa, H., S. Kasahara & T. Sugiyama. 1987. Effect of *Ulva* meal supplementation on lipid metabolism of black sea bream, *Acanthopagrus schlegeli* (Bleeker). Aquaculture 62:109-121.
Nakagawa, H., G.R. Nematipour, M. Yamamoto, T. Sugiyama & K. Kusaka. 1993. Optimum level of *Ulva* meal diet supplement to minimize weight loss during wintering in black sea bream *Acanthopagrus schlegeli* (Bleeker). Asian Fish. Sci. 6:139-148.
Nakagawa, H., J. Karakawa & M. Yamamoto. 1994. Changes in body constituents of young black sea bream, *Acanthopagrus schlegeli* (Bleeker), released into the Inland Sea of Japan. Third Asian Fisheries Forum. Asian Fish. Soc., Manila, p 880-883.
Nakagawa, H., T. Umino & Y. Tasaka. 1997. Usefulness of *Ascophyllum* meal as a feed additive for red sea bream, *Pagrus major*. Aquaculture 151:275-281.
Nakagawa, T. 1988. On the artificial propagation of *Epinephelus moara*. Saibai (48):41-44. (in Jpn.)
Nakajima, K. 1991. Effects of diet-supplemented dimethyl-B-propiothetin on growth and thrust power of goldfish, carp, and red sea bream. Nippon Suisan Gakkaishi 57:673-679.
Nakajima, H., H. Kawahara & S. Takamatsu. 1987. The breeding behavior and the behavior of larvae and juveniles of the sharksucker, *Echeneis naucrates*. Jpn. J. Ichthyol. 34:66-70.
Nakajima, K., A. Uchida & Y. Ishida. 1990. Effect of a feeding attractant, dimethyl-B-propiothetin, on growth of marine fish. Nippon Suisan Gakkaishi 56:1151-1154.
Nakajima, M., T. Takeuchi, T. Watanabe, S. Sekiya, M. Kanematsu & K. Imaizumi. 1995. Adequate and excess amounts of n-3 highly unsaturated fatty acids in the diet for juvenile striped jack, *Pseudocaranx dentex*. Suisanzoshoku 43:75-81. (in Jpn., Eng. abs.)
Nakamura, M., K. Takaii, O. Takaoka, S.-I. Furuta & H. Kumai. 1991. Rearing of Japanese sardine from hatching through juveniles. Bull. Jpn. Soc. Sci. Fish. 57:345.
Nakamura, S. 1937. Larvae and young of fishes found in the vicinity of Kominato, VII-VIII. J. Imp. Fish. Inst. (Tokyo) 32:17-25.
Nakanishi, T. 1986. Ontogenetic development of the immune response in the marine teleost *Sebasticus marmoratus*. Bull. Jpn. Soc. Sci. Fish. 52:473-477.
Nakazano, A., & H. Kawase. 1993. Spawning and biparental egg-care in a temperate filefish, *Paramonacanthus japonicus* (Monacanthidae). Env. Biol. Fishes 37:245-256.
Nakazono, A., H. Takeya & H. Tsukahara. 1979. Studies on the spawning behavior of *Chromis notata* (Temminck et Schlegel), Sci. Bull. Fac. Agr. Kyushu Univ. 34:29-37. (in Jpn., Eng. sum.)
Namdari, R., S. Abedini & F.C.P. Law. 1996. Tissue distribution and elimination of oxytetracycline in seawater chinook and coho salmon following medicated-feed treatment. Aquaculture 144:27-38.
Naqvi, S.M., & C. Vaishnavi. 1993. Bioaccumulative potential and toxicity of endosulfan insecticide to non-target animals. Comp. Biochem. Physiol. 105C:347-361.
Nardi, G. 1996. Culture of summer flounder, *Paralichthys dentatus*, at Greatbay Aquafarms. J. Shellfish Res. 15:458.
Nardi, G.C. 1997. Commercial culture of summer flounder (*Paralichthys dentatus*) at Greatbay Aquafarms. World Aquac. '97 Book of Abs., p 345.
Nash, C.E., & R.M. Koningsberger, 1981. Artificial propagation. In: O.H. Oren (ed.), Aquaculture of Grey Mullets. IBP 26. Cambridge Univ. Press, Cambridge, U.K., p 265-312.
Nash, C.E., C.-M. Kuo & S.C. McConnel. 1974. Operational procedures for rearing larvae of the grey mullet (*Mugil cephalus* L.). Aquaculture 3:15-24.
Nash, C.E., C.-M. Kuo, W.D. Madden & C.L. Paulsen. 1977. Swim bladder inflation and survival of *Mugil cephalus* to 50 days. Aquaculture 12:89-94.
Nashida, K., & S. Kanamaru. 1991. Early life stages of demersal fishes and oceanic environments in the middle of the Japan Sea. Bull. Jpn. Soc. Fish. Oceanogr. 55:218-224. (in Jpn.)
Navarro, J.C., F. Amat & J.R. Sargent. 1993. The lipids of the cysts of freshwater- and marine-type *Artemia*. Aquaculture 109:327-336.
Navas, J.M., M. Thrush, J. Ramos, M. Bruce, M. Carrillo, S. Zanuy & N. Bromage. 1995. The effect of seasonal alteration in the lipid composition of broodstock diets on egg quality in the European sea bass (*Dicentrarchus labrax*). Proc. Fifth Internat. Symp. Repr. Physiol. Fish. Univ. Texas, Austin, p 108-110.
Navas, J.M., M. Bruce, M. Thrush, B.M. Farndale, N. Bromage, S. Zanuy, M. Carrillo, J.G. Bell & J. Ramos. 1997. The impact of seasonal alteration in the lipid composition of broodstock diets on egg quality in the European sea bass. J. Fish Biol. 51:760-773.
Needham, T. 1988a. Salmon smolt production. In: L. Laird & T. Needham (eds.), Salmon and Trout

Farming. Ellis Horwood, Chichester, England, p 87-116.
Needham, T. 1988b. Sea water cage culture of salmonids. In: L.M. Laird & T. Needham (eds.), Salmon and Trout Farming. Ellis Horwood, Chichester, England, p 117-154.
Needham, T. 1990. Canadian aquaculture—let's farm the oceans. World Aquac. 21(2):76-80.
Nehr, O., J.-P. Blancheton & E. Alliot. 1996. Development of an intensive culture system for sea bass (*Dicentrarchus labrax*) larvae in sea enclosures. Aquaculture 142:43-58.
Neill, W.H. 1990. Environmental requirements of red drum. Texas A&M Univ. Sea Grant TAMU-SG-90-603, p 105-108.
Neilson, J.D., J.F. Kearney, P. Perley & H. Sampson. 1993. Reproductive biology of Atlantic halibut (*Hippoglossus hippoglossus*) in Canadian waters. Can. J. Fish. Aquat. Sci. 50:551-563.
Nel, S. 1995. Mahimahi farming information. Aquac. Eur. 19(4):39.
Nelson, G.R. 1976. Analytical method for aqueous selenite ion in the environment. Masters Thesis, Univ. South Florida, Tampa.
Nengas, I., M.N. Alexis, S.J. Davies & G. Petichakis. 1995. Investigation to determine digestibility coefficients of various raw materials in diets for gilthead sea bream, *Sparus auratus* L. Aquac. Res. 26:185-194.
Neskovic, N.K., V. Karan, I. Elezovic, V. Poleksic & M. Budimir. 1994. Toxic effects of 2,4-D herbicide on fish. J. Env. Sci. Health B29:265-279.
Neugebauer, W. 1969. So zuchten wir Korallenfische. Aquarien Magazin, December, p 483-488. (in Ger.)
Nevo, H. 1983. La Biologie et L'elevage du Turbot. Ph.D. Thesis, Univ. Paul-Sabatier de Toulouse, 96 p. (in Fr.)
New, M.B. 1986. Aquaculture diets of postlarval marine fish of the super-family Percoidae, with special reference to sea bass, sea breams, groupers and yellowtail: a review. Kuwait Bull. Mar. Sci. 7:75-151.
Nguyen, H.D., T. Mekuchi, K. Imura, T. Nakai, T. Nishizawa & K. Muroga. 1994. Occurrence of viral nervous necrosis (VNN) in hatchery-reared juvenile Japanese flounder *Paralichthys olivaceus*. Fish. Sci. 60:551-554.
Nichols, D.S., P. Hart, P.D. Nichols & T.A. McMeekin. 1996. Enrichment of the rotifer *Brachionus plicatilis* fed an Antarctic bacterium containing polyunsaturated fatty acids. Aquaculture 147:115-125.
Nichols, J.T. 1916. *Lucania parva* in the aquarium. Copeia 1916(31):37-38.
Nicholson, L.C., L.C. Woods III & J.G. Woiwode. 1990. Intensive culture techniques for the striped bass and its hybrids. In: R.M. Harrell, J.H. Kerby & R.V. Minton (eds.), Culture and Propagation of Striped Bass and Its Hybrids. Am. Fish. Soc., Bethesda, Maryland, p 141-157.
Nicolás, J.L., E. Robic & D. Ansquer. 1989. Bacterial flora associated with a trophic chain consisting of microalgae, rotifers and turbot larvae: influence of bacteria on larval survival. Aquaculture 83:237-248.
Nieland, D.L., & C.A. Wilson. 1993. Reproductive biology and annual variation of reproductive variables of black drum in the northern Gulf of Mexico. Trans. Am. Fish. Soc. 122:318-327.
Nijhof, M. 1993. Effects of body weight on diet requirements of turbot (*Scophthalmus maximus*) with special reference to feed intake and body composition. Fish Nutrition in Practice, INRA, Paris, p 501-513.
Nijhof, M. 1994a. Research on nutrition and growth in the culture of post larval turbot (*Scophthalmus maximus* L.). Eur. Aquac. Soc. Spec. Pub. (22):137-165.
Nijhof, M. 1994b. The application of commercially exploited recirculating system concepts in the culture of turbot (*Scophthalmus maximus* L.). Eur. Aquac. Soc. Spec. Pub. (22):273-282.
Nikaido, H., N. Miyabe, & S. Ueyanagi. 1991. Spawning time and frequency of bigeye tuna, *Thunnus obesus*. Bull. Nat. Res. Inst. Far Seas Fish. (28):47-74. (in Jpn., Eng. abs.)
Nilsen, F. 1995. Description of *Trichodina hippoglossi* n. sp. from farmed Atlantic halibut larvae *Hippoglossus hippoglossus*. Dis. Aquat. Org. 21:209-214.
Nilsson, E., & J.G. Cloud. 1993. Cryopreservation of rainbow trout (*Oncorhynchus mykiss*) blastomeres. Aquat. Living Resour. 6:77-80.
Nimura, Y., K. Nanba & M.I. Miah. 1994a. Apparent assimilation efficiency in *Artemia* related to body size and ingestion rate. Fish. Sci. 60:505-510.
Nimura, Y., K. Nanba & M.I. Miah. 1994b. Food utilization in *Artemia* for growth, reproduction, and maintenance. Fish. Sci. 60:493-503.
Nissling, A. 1994. Survival of eggs and yolk-sac larvae of Baltic cod (*Gadus morhua* L.) at low oxygen levels in different salinities. ICES Mar. Sci. Symp. 198:626-631.
Nissling, A., P. Solemdal, M. Svensson & L. Westin. 1994. Survival, activity and feeding ability of Baltic cod (*Gadus morhua*) yolk-sac larvae at different salinities. J. Fish Biol. 45:435-445.
Noga, E.J. 1996. Fish Disease: Diagnosis and Treatment. Mosby, St. Louis, 367 p.
Noga, E.J., S.A. Smith, J.M. Burkholder, C. Hobbs & R.A. Bullis. 1993. A new ichthyotoxic dinoflagellate: cause of acute mortality in aquarium fishes. Vet. Rec. 133:96-97.
Norberg, B., V. Valkner, J. Huse, I. Karlsen & G.L. Grung. 1991. Ovulatory rhythms and egg viability in the Atlantic halibut (*Hippoglossus hippoglossus*). Aquaculture 97:365-371.
Norcross, J.J., S.L. Richardson, W.H. Massmann & E.B. Joseph. 1974. Development of young bluefish (*Pomatomus saltatrix*) and distribution of eggs and young in Virginian coastal waters. Trans. Am. Fish. Soc. 103:477-497.
Nordlie, F.G., & D.C. Haney. 1993. Euryhaline adaptations in the fat sleeper, *Dormitator maculatus*. J. Fish

Biol. 43:433-439.

Nordmo, R., J.M. Holth & B.O. Gabrielsen. 1995. Immunostimulating effect of alginate feed in Atlantic salmon (*Salmo salar* L.) challenged with *Aeromonas salmonicida*. Molec. Mar. Biol. Biotechnol. 4:232-235.

Noriega-Curtis, P., & J. Vera Rivas. 1989. A regional survey of the aquaculture sector in Latin America. FAO/UNDP/ADCP/REP/89/39, 80 p.

Norrgren, L., T. Andersson, P.-A. Bergqvist & I. Bjorklund. 1993. Chemical, physiological and morphological studies of feral Baltic salmon (*Salmo salar*) suffering from abnormal fry mortality. Env. Toxicol. Chem. 12:2065-2075.

Norsker, N.-H., & J.G. Støttrup. 1994. The importance of dietary HUFAs for fecundity and HUFA content in the harpacticoid, *Tisbe holothuriae* Humes. Aquaculture 125:155-166.

Nortvedt, R., & S. Tuene. 1995. Multivariate evaluation of feed for Atlantic halibut. Chemometrics Intelligent Lab. Syst. 29:271-282.

Nour, A.M., E. Omar, A. Mohamed, M.H.-El Sokkary & I. El-Sokkary. 1992. Effects of supplementation of some heavy metals (cadmium and/or zinc and mercury and/or selenium) on growth performance and feed utilization of tilapia (*Oreochromis niloticus*). In: C. Phillips & D. Piggins (eds.), Farm Animals and the Environment. C.A.B. International, Wallingford, U.K., p 414.

Nowak, B.F., & S.C. Battaglene. 1996. Incidence and composition of calculi in the urinary bladder of intensively reared marine fish larvae. Mar. Freshw. Res. 47:255-260.

NRC (National Research Council). 1973. Nutrient requirements of trout, salmon and catfish. Nat. Acad. Press, Washington, D.C., 57 p.

NRC (National Research Council). 1977. Nutrient Requirements of Warmwater Fishes. Nat. Acad. Press, Washington, D.C., 78 p.

NRC (National Research Council). 1981. Nutrient Requirements of Coldwater Fishes. Nat. Acad. Press, Washington, D.C., 63 p.

NRC (National Research Council). 1983. Nutrient Requirements of Warmwater Fishes and Shellfishes. Nat. Acad. Press, Washington, D.C., 102 p.

NRC (National Research Council). 1992. Marine Aquaculture: Opportunities for Growth. Nat. Acad. Press, Washington, D.C., 290 p.

Nugegoda, D., & T.J. Lam. 1994. Treatment of fertilized eggs (embryos) with triiodothyronine (T3) enhances subsequent larval growth and development in tilapia (*Oreochromis mossambicus*). Third Asian Fisheries Forum. Asian Fish. Soc., Manila, p 852-855.

Nwadukwe, F.O. 1995. Effect of salinity on the survival of *Oreochromis niloticus* (L) in cages stocked with *Sarotherodon melanotheron* (R) in tropical mangrove brackishwater fish farm. Env. Ecol. 13:155-159.

Oakes, D., P. Cooley, L.L. Edwards, R.W. Hirsch & V.G. Miller. 1979. Ozone disinfection of fish hatchery waters: pilot plant results, prototype design and control considerations. Proc. World Maric. Soc. 10:854-870.

Ochiai, A., K. Mutsutani & S. Umeda. 1982. Development of eggs, larvae and juveniles of jack mackerel, *Trachurus japonicus*. Jpn. J. Ichthyol. 29:86-92.

O'Connell, C.P. 1981. Development of organ systems in the northern anchovy, *Engraulis mordax*, and other teleosts. Am. Zool. 21:429-446.

Oda, D. 1991. Development of eggs and larvae of California halibut *Paralichthys californicus* and fantail sole *Xystreurys liolepis* (Pisces: Paralichthyidae). Fish Bull., U.S. 89:387-402.

Oda, D.L., R.J. Lavenberg & J.M. Rounds. 1993. Reproductive biology of three California species of *Paralabrax* (Pisces: Serranidae). Cal. Coop. Oceanic Fish. Invest. Rep. 34:122-132.

Odenkirchen, E.W., & R. Eisler. 1988. Chlorpyrifos hazards to fish, wildlife, and invertebrates: a synoptic review. U.S. Fish Wildl. Serv., Biol. Rep. 85(1.13), 34 p.

Ogawa, K. 1992. *Caligus longipedis* infection of cultured striped jack, *Pseudocaranx dentex* (Teleostei: Carangidae) in Japan. Gyobyo Kenkyu 27:197-205.

Ogawa, K., H. Andoh & M. Yamaguchi. 1993. Some biological aspects of *Paradeontacylix* (Trematoda: Sanguinicolidae) infection in cultured marine fish *Seriola dumerili*. Gyobyo Kenkyu 28:177-180.

Ogburn, D.M., & N.J. Ogburn. 1994. Intensive pond culture trials of the green grouper (*Epinephelus malabaricus* Bloch et Schneider) in the Philippines. Third Asian Fisheries Forum. Asian Fish. Soc., Manila, p 74-77.

Ogino, C., N. Uki, T. Watanabe, Z. Iida & K. Ando. 1970. B vitamin requirements of carp--IV. Requirement for choline. Bull. Jpn. Soc. Sci. Fish. 36:1140-1146.

Ohta, H., & K. Takano. 1996. Testicular maturation induced by methyltestosterone in elvers of the Japanese eel *Anguilla japonica*. Fish. Sci. 62:990-991.

Ohta, H., & H. Tanaka. 1997. Relationship between serum levels of human chorionic gonadotropin (hCG) and 11-ketotestosterone after a single injection of hCG and induced maturity in the male Japanese eel, *Anguilla japonica*. Aquaculture 153:123-134.

Ohta, H., H. Kagawa, H. Tanaka, K. Okuzawa & K. Hirose. 1996a. Changes in fertilization and hatching rates with time after ovulation induced by 17, 20β-dihydroxy-4-pregnen-3-one in the Japanese eel, *Anguilla japonica*. Aquaculture 139:291-301.

Ohta, H., H. Kagawa, H. Tanaka, K. Okuzawa & K. Hirose. 1996b. Milt production in the Japanese eel *Anguilla japonica* induced by repeated injections of human chorionic gonadotropin. Fish. Sci. 62:44-49.

Ohta, H., H. Tanaka, H. Kagawa, K. Okuzawa & N. Iinuma. 1997. Artifical fertilization using testicular spermatozoa in the Japanese eel *Anguilla japonica*. Fish. Sci. 63:393-396.

Oikawa, C.K., & B.E. March. 1997. A method for assessment of the efficacy of feed attractants for fish. Prog. Fish-Cult. 59:213-217.

Oikawa, S., T. Takeda & Y. Itazawa. 1994. Scale effects of MS-222 on a marine teleost, porgy *Pagrus major*. Aquaculture 121:369-379.

Ojaveer, H., P.C. Morris, S.J. Davies & P. Russell. 1996. The response of thick-lipped grey mullet, *Chelon labrosus* (Risso), to diets of varied protein-to-energy ratio. Aquac. Res. 27:603-612.

Oka, A., N. Suzuki & T. Watanabe. 1982. Effect of fatty acids in *Moina* on the fatty acid composition of larval ayu *Plecoglossus altevelis*. Bull. Jpn. Soc. Sci. Fish. 48:1159-1162.

Okauchi, M. 1991. The status of phytoplankton production in Japan. In: W. Fulks & K.L. Main (eds.), Rotifer and Microalgae Culture Systems: proceedings of a U.S.—Asia Workshop. Oceanic Inst., Honolulu, p 247-256.

Okauchi, M., W.-J. Zhou, W.-H. Zou, K. Fukusho & A. Kanazawa. 1990. Difference in nutritive value of a microalga *Nannochloropsis oculata* at various growth phases. Nippon Suisan Gakkaishi 56:1293-1298.

Oki, D. & O. Tabeta. 1998. Age, growth and reproductive characteristics of the yellow sea bream *Dentex tumifrons* in the East China Sea. Fish. Sci. 64:191-197.

Okimasu, E., M. Matsumoto, Y. Yoshida & A. Amemura. 1992. The effect of pigments of *Rhodobacter capsulatus* on free radicals and application of the bacterium as feed to fish larvae. Nippon Suisan Gakkaishi 58:1487-1491.

Okiyama, M. (ed.). 1988. An atlas of the early stage fishes in Japan. Section on: Subfamily Sebastinae, genus *Sebastes* (including *Sebasticus marmoratus*), Trans. by L. Mottet & M. Mottet, 1989. NWAFC Processed Rep. 89-09, 34 p.

Olla, B.L., & C. Samet. 1977. Courtship and spawning behavior of the tautog, *Tautoga onitis* (Pisces: Labridae), under laboratory conditions. Fish. Bull., U.S. 75:585-599.

Olli, J.J., & Å. Krogdahl. 1994. Nutritive value of four soybean products as protein sources in diets for rainbow trout (*Oncorhynchus mykiss*, Walbaum) reared in fresh water. Acta Agric. Scand., Sec. A, Anim. Sci. 44:185-192.

Olli, J.J., Å. Krogdahl, T.S.G.A.M. van den Ingh & L.E. Brattås. 1994a. Nutritive value of four soybean products in diets for Atlantic salmon (*Salmo salar*, L.). Acta Agric. Scand., Sec. A, Anim. Sci. 44:50-60.

Olli, J.J., K. Hjelmeland & Å. Krogdahl. 1994b. Soybean trypsin inhibitors in diets for Atlantic salmon (*Salmo salar*, L.): effects on nutrient digestibilities and trypsin in pyloric caeca homogenate and intestinal content. Comp. Biochem. Physiol. 109A:923-928.

Olmedo, M., B. Alvarez-Bláquez & C. Gómez. 1995. Effect of rotifer (*Brachionus plicatilis*) density on the first-feeding of turbot larvae (*Scophthalmus maximus* L.). ICES Mar. Sci. Symp. 201:202.

Olsen, Y., J.R. Rainuzzo, O. Vadstein & A. Jensen. 1989. Kinetics of n-3 fatty acids in *Brachionus plicatilis* and changes in the food supply. Hydrobiologia 186/187:409-413.

Olsen, Y., K.I. Reitan & O. Vadstein. 1993a. Dependence of temperature on loss rates of rotifers, lipids, and ω3 fatty acids in starved *Brachionus plicatilis* cultures. Hydrobiologia 255/256:13-20.

Olsen, Y., J.R. Rainuzzo, K.I. Reitan & O. Vadstein. 1993b. Manipulation of lipids and ω3 fatty acids in *Brachionus plicatilis*. In: H. Reinertsen, L.A. Dahle, L. Jørgensen & K. Tvinnereim (eds.), Fish Farming Technology. Balkema, Rotterdam, p 101-108.

Olson, R.J., & V.P. Scholey. 1990. Captive tunas in a tropical marine research laboratory: growth of late-larval and early-juvenile black skipjack *Euthynnus lineatus*. Fish. Bull., U.S. 88:821-828.

Oorschot, R.W.A., & J.H. Boon. 1993. Mortality of marine cultured rainbow trout, *Oncorhynchus mykiss* (Walbaum), during the summer in the Netherlands. Aquac. Fish. Mgt. 24:291-298.

Oozeki, Y., A. Hagiwara, H. Eda & C.S. Lee. 1992a. Development of food selectivity of striped mullet *Mugil cephalus* during the larval stage. Nippon Suisan Gakkaishi. 58:1381.

Oozeki, Y., P.-P. Hwang & R. Hirano. 1992b. Larval development of the Japanese whiting, *Sillago japonica*. Jpn. J. Ichthyol. 39:59-66.

Oppenborn, J.B., & C.A. Goudie. 1993. Acute and sublethal effects of ammonia on striped bass and hybrid striped bass. J. World Aquac. Soc. 24:90-101.

Opstad, I., & Ø. Bergh. 1993. Culture parameters, growth and mortality of halibut (*Hippoglossus hippoglossus* L.) yolk sac larvae in upwelling incubators. Aquaculture 109:1-11.

Opstad, I., B. Strand, I. Huse & O. Garatun-Tjeldsto. 1989. Laboratory studies on the use of rotifers (*Brachionus plicatilis* O.F. Muller) as start-feed for cod larvae (*Gadus morhua* L.). Aquaculture 79:345-351.

Opstad, I., K. Boxaspen & O. Bergh. 1993. Studies on the use of rotifers (*Brachionus plicatilis*) and *Artemia*, enriched on different mixtures, as start-feed for turbot larvae (*Scophthalmus maximus*). Fish Nutrition in Practice, INRA, Paris, p 633-638.

Opuszynski, K., J.V. Shireman, F.J. Aldridge & R.W. Rottmann. 1984. Environmental manipulation to stimulate rotifers in fish rearing ponds. Aquaculture 42:343-348.

Ordonio-Aguilar, R.S., H. Kohno, A. Ohno, M. Moteki & Y. Taki. 1997. The development of grouper, *Epinephelus coioides*, larvae during changeover of energy sources. World Aquac. '97 Book of Abs., p 357.

Oren, O.H. (ed.). 1981. Aquaculture of Grey Mullets. IBP Vol. 26, 389 p.

Orhun, M.R. 1989. Early Life History of White Sea Bass *Atractoscion nobilis*. M.S. Thesis, San Diego State Univ., California, 162 p.

Ortega, A., A. Estevez, F. Linares & D.G. Carnero. 1995. Changes in the chemical composition of starved rotifers (*Brachionus plicatilis*) kept at ambient and low temperatures for extensive periods. ICES Mar. Sci. Symp. 201:112-118.

Orthoefer, F.T., S.U. Gurkin & J.D. Fisk. 1995. The use of soybean lecithin in aquaculture. In: C. Lim & D.J. Sessa (eds.), Nutrition and Utilization Technology in Aquaculture. AOSC Press, Champaign, Ilinois, p 114-129.

Orton, G.L. 1955. Early developmental stages of the California scorpionfish, *Scorpaena guttata*. Copeia 1955:210-214.

Ortubay, S., & D. Wegrzyn. 1991. Fecundacion artificial y desarrollo embrionario de *Galaxias platei* Steindachner (Salmoniformes, Galaxiidae). Medio Ambiente 11:84-89. (in Sp., Eng. abs.)

Osse, J.W.M., & J.G.M. van den Boogaart. 1995. Fish larvae, development, allometric growth, and the aquatic environment. ICES Mar. Sci. Symp. 201:21-34.

Ostrowski, A.C. 1989. Effect of rearing tank background color on early survival of dolphin larvae. Prog. Fish-Cult. 51:161-163.

Ostrowski, A.C. 1995. Nursery and growout production techniques for the mahimahi (*Coryphaena hippurus*) and Pacific threadfin (*Polydactylus sexfilis*). In: K.L. Main & C. Rosenfeld (eds.), Culture of High-value Marine Fishes in Asia and the United States. Oceanic Inst., Honolulu, p 153-166.

Ostrowski, A.C. 1997. Current status on the culture and commercial activity of the mahimahi, *Coryphaena hippurus*. World Aquac. '97 Book of Abs., p 358.

Ostrowski, A.C., & S. Divakaran. 1990. Survival and bioconversion of *n*-3 fatty acids during early development of dolphin (*Coryphaena hippurus*) larvae fed oil-enriched rotifers. Aquaculture 89:273-285.

Ostrowski, A.C., & S. Divakaran. 1991. Energy substrates for eggs and prefeeding larvae of the dolphin *Coryphaena hippurus*. Mar. Biol. 109:149-155.

Ostrowski, A.C., C. Brownell & E.O. Duerr. 1989. Growth and feeding rates of juvenile dolphins (*Coryphaena hippurus*) fed a practical diet through growout. World Aquac. 20(4):104-105.

Ostrowski, A.C., S. Divakaran, B. Kim & E.O. Duerr. 1996a. Effect of fish meal quality on biological performance of juvenile dolphin, *Coryphaena hippurus*. J. Appl. Aquac. 6:39-56.

Ostrowski, A.C., T. Iwai, S. Monahan, S. Unger, D. Dagdagan, P. Murakawa, A. Schivell & C. Pigao. 1996b. Nursery production technology for Pacific threadfin (*Polydactylus sexfilis*). Aquaculture 139:19-29.

O'Sullivan, D. 1990. Status and prospects for Australian aquaculture (1989-1990). Dosaqua, Mowbray, Tasmania, Australia, 67 p.

O'Sullivan, D. 1992. Aquaculture in Australia: part one of a two part series. Aquac. Mag. 18(4):32-47.

Otake, T. 1996. Fine structure and function of the alimentary canal in leptocephali of the Japanese eel *Anguilla japonica*. Fish. Sci. 62:28-34.

Ott, A. G. 1975. Cryopreservation of Pacific salmon and steelhead trout sperm. Ph.D. Diss., Orgeon State Univ., Corvallis, 145 p.

Otterå H., G.I. Hemre & Ø. Lie. 1994. Influence of dietary water content on feed intake, growth and survival of juvenile Atlantic cod, *Gadus morhua* L., during the weaning process. Aquac. Fish Mgt. 25:915-926.

Ottesen, O.H., & S. Bolla. 1998. Combined effects of temperature and salinity on development and survival of Atlantic halibut larvae. Aquac. Internat. 6:103-120.

Ounaïs, N. 1997. Technical and biological aspects of the ornamental fish reproduction: example of 3 species reproduced at the Aquarium of Monaco: *Amphiprion ocellaris*, *Serrasalmis nattereri*, and *Pterapogon kauderni*. Island Aquac. Trop. Aquac. Eur. Aquac. Soc., Ann. Mtg, p 220.

Ounaïs-Guschemann, N. 1989. Definition d'un modele d'elevage larvaire intensif pour la daurade *Sparus auratus*. Doctoral Thesis. Univ. d'Aix Marseille II, 184 p.

Overstreet, R.M., & R.W. Heard. 1978. Food of the red drum, *Sciaenops ocellata*, from Mississippi Sound. Gulf Res. Rep. 6:131-135.

Overton, E.B., W.D. Sharp & P. Roberts. 1994. Toxicity of petroleum. In: L.G. Cockerham & B.S. Shane (eds.), Basic Environmental Toxicology. CRC Press, Boca Raton, Florida, p 133-156.

Owen, J.M., J.W. Adron, J.R. Sargent & C.B. Cowey. 1972. Studies on the nutrition of marine flatfish. The effect of dietary fatty acids on the tissue fatty acids of the plaice *Pleuronectes platessa*. Mar. Biol. 13:160-166.

Ozawa, T., K. Kawai & I. Uotani. 1991. Stomach content analysis of chub mackerel *Scomber japonicus* larvae by quantification method. Nippon Suisan Gakkaishi 57:1241-1245.

Ozawa, T., F. Kakizoe, O. Tabeta, T. Maeda & Y. Yuwaki. 1992. Larval growth and drift of the Japanese eel *Anguilla japonica* estimated from leptocephali collection. Bull. Jpn. Soc. Sci. Fish. 58:15-23.

Ozkizilcik, S., & F.-L.E. Chu. 1994. Uptake and metabolism of liposomes by *Artemia* nauplii. Aquaculture 128:131-141.

Ozkizilcik, S., & F.-L.E. Chu. 1996. Preparation and characterization of a complex microencapsulated diet for striped bass *Morone saxatilis* larvae. J. Microencapsulation 13:331-343.

Øie, G., K.I. Reitan & Y. Olsen. 1994a. Comparison of rotifer culture quality with yeast plus oil and algal-based cultivation diets. Aquac. Internat. 2:225-238.

Øie, G., P. Makridis, K.I. Reitan & Y. Olsen. 1994b. Protein and carbon utilization of rotifers (*Brachionus plicatilis*) in first feeding of turbot larvae (*Scophthalmus maximus* L.). Aquaculture 153:103-122.

Øiestad, V., & E. Moksness. 1981. Study of growth and survival of herring larvae (*Clupea harengus* L.) using plastic bag and concrete basin enclosures. Rapp. P.-v. Réun. Cons. Int. Explor. Mer 178:144-149.

Padros, F., & S. Crespo. 1995. Swimbladder pathology during larval development of turbot (*Scophthalmus maximus* L.). ICES Mar. Sci. Symp. 201:159-162.

Painter, H.A. 1970. A review of literature on inorganic nitrogen metabolism in microorganisms. Water Res. 4:393-450.

Palawski, D., J.B. Hunn & F.J. Dwyer. 1985. Sensitivity of young striped bass to organic and inorganic contaminants in fresh and saline waters. Trans. Am. Fish. Soc. 114:748-753.

Palko, B.J., & W.J. Richards. 1969. The rearing of cowfish. Salt Water Aquarium Mag. 5(3):67-70.

Palko, B.J., G.L. Beardsley & W.J. Richards. 1982. Synopsis of the Biological Data on Dolphin-fishes, *Coryphaena hippurus* Linnaeus and *Coryphaena equiselis* Linnaeus. NOAA Tech. Rep., NMFS Circ. 443, 28 p.

Palma, A.L. 1978. Induced spawning and larval rearing of rabbitfish, *Siganus guttatus*. Philip. J. Fish. 16:95-104.

Palmer, P.J., A.W. Blackshaw & R.N. Garrett. 1993. Successful fertility experiments with cryopreserved spermatozoa of barramundi, *Lates calcarifer* (Bloch), using dimethylsulfoxide and glycerol as cryoprotectants. Reprod. Fertil. Dev. 5:285-293.

Palomera, I. 1992. Spawning of anchovy *Engraulis encrasicolus* in the northwestern Mediterranean relative to hydrographic features in the region. Mar. Ecol. Prog. Ser. 79:215-223.

Pan, B.S., C.C. Lan & T.Y. Hung. 1991. Changes in composition and proteolytic enzyme activities of *Artemia* during early development. Comp. Biochem. Physiol. 100A:725-730.

Pandian, T.J., & S.G. Sheela. 1995. Hormonal induction of sex reversal in fish. Aquaculture 138:1-22.

Pankhurst, N.W. 1994. Effects of gonadotropin releasing hormone analogue, human chorionic gonadotropin and gonadal steroids on milt volume in the New Zealand snapper, *Pagrus auratus* (Sparidae). Aquaculture 125:185-197.

Pankhurst, N.W., & J.F. Carragher. 1992. Oocyte maturation and changes in plasma steroid levels in snapper *Pagrus* (=*Chrysophrys*) *auratus* (Sparidae) following treatment with human chorionic gonadotropin. Aquaculture 101:337-347.

Paperna, I. 1977. Epitheliocystis infection in wild and cultured sea bream (*Sparus aurata*, Sparidae) and grey mullets (*Liza ramada*, Mugilidae). Aquaculture 10:169-176.

Paperna, I. 1978. Swimbladder and skeletal deformations in hatchery bred *Sparus aurata*. J. Fish Biol. 12:109-114.

Paperna, I. 1980. *Amyloodinium ocellatum* (Brown, 1931) (Dinoflagellida) infestations in cultured marine fish at Eilat, Red Sea: epizootiology and pathology. J. Fish Dis. 3:363-372.

Paperna, I. 1983. Review of diseases of cultured warm water marine fish. Rapp. P.-v. Réun. Cons. Int. Explor. Mer 182:44-48.

Paperna, I. 1984a. Chemical control of *Amyloodinium ocellatum* (Brown 1931) (Dinoflagellida) infections: in vitro tests and treatment trials with infected fishes. Aquaculture 38:1-18.

Paperna, I. 1984b. Reproduction cycle and tolerance to temperature and salinity of *Amyloodinium ocellatum* (Brown, 1931) (Dinoflagellida). Ann. Parasitol. Hum. Comp. 59:7-30.

Paperna, I. 1984c. Review of diseases affecting cultured *Sparus aurata* and *Dicentrarchus labrax*. In: G. Barnabé & R. Billard (eds.), L'Aquaculture du Bar et des Sparides. INRA, Paris, p 465-482.

Paperna, I. 1991. Diseases caused by parasites in the aquaculture of warm water fish. Ann. Rev. Fish Dis. 1:155-194.

Paperna, I., & R.M. Overstreet. 1981. Parasites and diseases of mullets (Mugilidae). In: O.H. Oren (ed.), Aquaculture of Grey Mullets. IBP 26. Cambridge Univ. Press, Cambridge, U.K. p 411-493.

Papoutsoglou, S., M.J. Costello, E. Stamou & G. Tziha. 1996. Environmental conditions at sea-cages, and ectoparasites on farmed European sea-bass, *Dicentrachus labrax* (L.), and gilt-head sea-bream, *Sparus aurata* L., at two farms in Greece. Aquac. Res. 27:25-34.

Parado-Estepa, F.D. 1991. Survival of newly-hatched larvae of *Epinephelus malabaricus* at different salinity levels. Eur. Aquac. Soc. Spec. Pub. (15):323-325.

Parauka, F.M., W.J. Troxel, F.A. Chapman & L.G. McBay. 1991. Hormone-induced ovulation and artificial spawning of Gulf of Mexico sturgeon (*Acipenser oxyrhynchus desotoi*). Prog. Fish-Cult. 53:113-117.

Parazo, M.M. 1990. Effect of dietary protein and energy level on growth, protein utilization and carcass composition of rabbitfish, *Siganus guttatus*. Aquaculture 86:41-49.

Parazo, M.M., L.M.B. Garcia, F.G. Ayson, A.C. Fermin, J.M.E. Almendras, D.M. Reyes, Jr. & E.M. Avila. 1990. Sea Bass Hatchery Operations. SEAFDEC Aquac. Ext. Man. No. 18, 38 p.

Parazo, M.M., E.M. Avila & D.M. Reyes, Jr. 1991a. Size- and weight-dependent cannibalism in hatchery-bred sea bass (*Lates calcarifer* Bloch). J. Appl. Ichthyol. 7:1-7.

Parazo, M.M., D.M. Reyes, Jr. & E.M. Avila. 1991b. Hatchery rearing of sea bass *Lates calcarifer* Bloch. Philip. Sci. 28:65-76.

Park, E.-H., H.-H. Chang, W.N. Joo, H.-S. Chung & H.-S. Kwak. 1994. Assessment of the estuarine hermaphroditic fish *Rivulus marmoratus* as a useful euryhaline species for acute toxicity tests as shown using cadmium. Can. J. Fish. Aquat. Sci. 51:280-285.

Park, H.-Y., & J.-M. Yoon. 1992. Studies on genetics and breeding in rainbow trout (*Oncorhynchus mykiss*). 7. Fertilization of fresh egg with cryo-preserved sperm and ultrastructural changes. Bull. Kor. Fish. Soc. 25:79-92. (in Kor., Eng. abs.)

Park, I.-S., H.-B. Kim, H.-J. Choi, Y.-D. Lee & J.-K. Son. 1992. Stimulation of spermiation by human chorionic gonadotropin (HCG) in mature flatfish, *Paralichthys olivaceus*. Bull. Mar. Res. Inst., Cheju Nat. Univ. 16:75-82. (in Kor., Eng. abs.)

Park, K.H., & H.D. Jeon. 1996. Enhanced resistance against *Edwardsiella tarda* infection in tilapia (*Oreochromis niloticus*) by administration of protein-bound polysaccharide. Aquaculture 143:135-143.

Park, M.-A., & S.-G. Sohn. 1996. Practical approaches to marine fish health problems in Korea. In: K.L. Main & C. Rosenfeld (eds.), Aquaculture Health Management Strategies for Marine Fishes. Oceanic Inst., Honolulu, p 69-80.

Park, M.-S. 1991. The status of mass production of live feeds in Korean hatcheries. In: W. Fulks & K.L. Main (eds.), Rotifer and Microalgae Culture Systems: Proc. of a U.S.—Asia Workshop. Oceanic Inst., Honolulu, p 287-295.

Parker, N.C. 1984. Culture requirements for striped bass. In: J.P. McCraren (ed.), The Aquaculture of Striped Bass: A Proc. Univ. Maryland Sea Grant UM-SG-MAP-84-01, p 29-43.

Parker, N. 1988. Fish culture in the United States. In: C.J. Shepard & N.R. Bromage (eds.), Intensive Fish Farming. Blackwell, Oxford, 404 p.

Parker, N.C., G.T. Klar, T.I.J. Smith & J.H. Kerby. 1990a. Special considerations in the culture of striped bass and striped bass hybrids. In: R.M. Harrell, J.H. Kerby & R.V. Minton (eds.), Culture and Propagation of Striped Bass and Its Hybrids. Am. Fish. Soc., Bethesda, Maryland, p 191-215.

Parker, S.J., A.G. Durbin & J.L. Specker. 1990b. Effects of leaf litter on survival and growth of juvenile coho salmon. Prog. Fish-Cult. 52:62-64.

Parmanne, R., & E. Kuittinen. 1991. Comparison of the fecundity of herring (*Clupea harengus* L.) in various areas of the northern Baltic Sea. Fin. Fish. Res. 12:157-164.

Parrish, K.K., & D.F. Wilson. 1978. Fecundity studies on *Acartia tonsa* (Copepoda: Calanoida) in standardized culture. Mar. Biol. 46:65-81.

Parsons, T.R., Y. Maita & C.M. Lalli. 1989. A Manual of Chemical and Biological Methods for Seawater Analysis. Pergamon, Oxford, 173 p.

Pascual, F.C., G.T. Tayo & E.R. Cruz-Lacierda. 1994. Acute toxicity of formalin to sea bass (*Lates calcarifer*) fry. Third Asian Fisheries Forum, Asian Fish. Soc., Manila, p 346-348.

Paspatis, M., & T. Boujard. 1996. A comparative study of automatic feeding and self-feeding in juvenile Atlantic salmon (*Salmo salar*) fed diets of different energy levels. Aquaculture 145:245-257.

Pastor, E., F. Riera, S. Pou, A.M. Grau & A. Grau. 1995. Summary of investigations on reproduction and larval rearing of common dentex (*Dentex dentex* (L.)). ICES Mar. Sci. Symp. 201:148-152.

Patiño, R. 1997. Manipulations of the reproductive system of fishes by means of exogenous chemicals. Prog. Fish-Cult. 59:118-128.

Pattillo, M. 1990. Selected species profiles and ELMR data sheets. Species profile: striped mullet (*Mugil cephalus*). Draft 05-25-90, 10 p.

Paul, E.A., H.A. Simonin & J. Symula. 1994. The toxicity of diquat, endothall, and fluoridone to the early life stages of fish. J. Freshw. Ecol. 9:229-239.

Pauley, G.B., K.L. Bowers & G.L. Thomas. 1988. Species Profiles: Life Histories and Environmental Requirements of Coastal Fishes and Invertebrates (Pacific Northwest)—Chum Salmon. U.S. Fish Wildl. Serv. Biol. Rep. 82(11.81). U.S. Army Corps Engineers, TR EL-82-4, 17 p.

Pauley, G.B., R. Risher & G.L. Thomas. 1989. Species Profiles: Life Histories and Environmental Requirements of Coastal Fishes and Invertebrates (Pacific Northwest)—Sockeye Salmon. U.S. Fish Wildl. Serv. Biol. Rep. 82(11.116). U.S. Army Corps Engineers, TR EL-82-4, 22 p.

Paull, R.C. 1969. Sargassum fish spawns. Salt Water Aquarium 5:77-78.

Paulsen, H. 1987. Current status of turbot culture. Proc. Third Internat. Symp. Repr. Physiol. Fish, Memorial Univ. Newfoundland, St. John's, p 85-88.

Pavlov, D.A., & E. Moksness. 1996. Repeat sexual maturation of wolffish (*Anarhichas lupus* L.) broodstock. Aquaculture 139:249-263.

Pavlov, D.A., & G.G. Novikov. 1987. On the development of biotechnology for rearing of White Sea wolffish, *Anarhichas lupus marisalbi*. I. Experience on obtaining mature sex products, incubation of eggs and rearing of the young fish. J. Ichthyol. 26:95-106.

Pavlov, D.A., K.F. Dzerzhinskiy & E.K. Radzikhovskaya. 1992. Assessing the quality of roe from White Sea wolffish (*Anarhichas lupus marisalbi*), obtained under experimental conditions. J. Ichthyol. 32:88-104.

Pedersen, B.H. 1997. The cost of growth in young fish larvae, a review of new hypotheses. Aquaculture 155:259-269.

Pedersen, B.H., K. Buchmann & M. Køie. 1993. Baltic larval cod *Gadus morhua* are infested with a protistan endoparasite in the yolk sac. Dis. Aquat. Org. 16:29-33.

Pedrazzoli, A., R. Guartatanga, A. León-Hing, L. Schwarz, M. Futagawa & H. Hosokawa. 1997. Dietary protein level and rearing density of juvenile Ecuadorian flounder *Paralichthys woolmani*. Island

Aquaculture and Tropical Aquaculture. Eur. Aquac. Soc., Ann. Mtg., p 231-232.

Peleteiro, J.B., P. Lavens, G. Rodríguez-Ojea & J. Iglesias. 1995. Relationship between egg quality and fatty acid content of various turbot broodstocks (*Scophthalmus maximus* L.). ICES Mar. Sci. Symp. 201:51-56.

Pelosi, S., P. Villani & G.C. Cozzolino. 1993. The effects of temperature on the eggs and larval development of *Dicentrarchus labrax* L. Eur. Aquac. Soc. Spec. Pub. (18):205-209.

Peng, H.-K., C.-F. Huang & F.G. Liu. 1991. Artificial rearing of broodstock and seed production of ayu (*Plecoglossus altivelis*). Abs. in program book, Finfish Hatchery in Asia '91. Tungkang Mar. Lab., Tungkang, Taiwan, p 47.

Pepin, P. 1991. Effect of temperature and size on development, mortality, and survival rates of the pelagic early life history stages of marine fish. Can. J. Fish. Aquat. Sci. 48:503-518.

Pepin, P., & R.A. Myers. 1991. Significance of egg and larval size to recruitment variability of temperate marine fish. Can. J. Fish. Aquat. Sci. 48:1820-1828.

Peppard, E.M., W.R. Wolters, J.W. Avault, Jr. & W.G. Perry. 1991. Toxicity of chelated copper to juvenile red drum *Sciaenops ocellatus*. J. World Aquac. Soc. 22:101-108.

Pereira, J.J., R. Mercaldo-Allen, C. Kuropat, D. Luedke & G. Sennefelder. 1993. Effect of cadmium accumulation on serum vitellogenin levels and hepatosomatic and gonadosomatic indices of winter flounder (*Pleuronectes americanus*). Arch. Env. Contam. Toxicol. 24:427-431.

Perera, W.M.K., C.G. Carter & D.F. Houlihan. 1995. Feed consumption, growth and growth efficiency of rainbow trout (*Oncorhynchus mykiss* (Walbaum)) fed on diets containing a bacterial single-cell protein. Brit. J. Nutr. 73:591-603.

Pérez, L., H. Gonzalez, M. Jover & J. Fernández-Carmona. 1997. Growth of European sea bass fingerlings (*Dicentrarchus labrax*) fed extruded diets containing varying levels of protein, lipid and carbohydrate. Aquaculture 156:183-193.

Pérez Mellado, Joaquín. 1992. Instituto Tecnológico del Mar en Guaymas, Mexico, pers. comm.

Perry, D.M. 1994. Artificial spawning of tautog under laboratory conditions. Prog. Fish-Cult. 56:33-36.

Perry, D.M., R. Mercaldo-Allen, C.A. Kuropat & J.B. Hughes. 1998. Laboratory culture of tautog. Prog. Fish-Cult. 60:50-54.

Perschbacker, P.W. 1985. Interactions between cultured *Fundulus grandis* (Pisces: Cyprinodontidae) and the brackishwater environment: bionomic and economic considerations. Ph.D. Diss., Texas A&M Univ., College Station, 140 p.

Perschbacher, P.W., & K. Strawn. 1983. Fertilization vs. supplementary feeding for growout of pond-raised gulf killifish. Proc. Ann. Conf. SE Assoc. Fish Wildl. Ag. 37:335-342.

Perschbacher, P.W., & K. Strawn. 1988. The sheepshead minnow *Cyprinodon variegatus*—a major pest in experimental mariculture of the gulf killifish *Fundulus grandis*. J. World Aquac. Soc. 19:113-117.

Perschbacker, P.W., D. Gonzalez & K. Strawn. 1995. Air incubation of eggs of the gulf killifish. Prog. Fish-Cult. 57:128-131.

Perschbacher, P.W., N. Stone, G.M. Ludwig & C.B. Guy, Jr. 1997. Evaluation of effects of common aerially-applied soybean herbicides and propanil on the plankton communities of aquaculture ponds. Aquaculture 157:117-122.

Person-Le Ruyet, J. 1990a. Early weaning of marine fish larvae onto microdiets: constraints and perspectives. Advances in Tropical Aquaculture. AQUACOP, IFREMER, Actes Colloq. (9):625-642.

Person-Le Ruyet, J. (coordinator). 1990b. Sole and turbot culture. In: G. Barnabé (ed.), Aquaculture: Vol. 2. Ellis Horwood, New York, p 687-734.

Person-le Ruyet, J., & P. Verillaud. 1980. Techniques d'elevage intensif de la daurade doree (*Sparus aurata* (L.)) de la naissance a l'age de deux mois. Aquaculture 20:351-370.

Person-Le Ruyet, J., F. Baudin-Laurencin, N. Devauchelle, R. Metailler, J.-L. Nicolas, J. Robin & J. Guillaume. 1991. Culture of turbot (*Scophthalmus maximus*). In: J.P. McVey (ed.), CRC Handbook of Mariculture: Vol. II, Finfish Aquaculture. CRC Press, Boca Raton, Florida, p 21-41.

Person Le Ruyet, J., J.C. Alexandre, L. Thébaud & C. Mugnier. 1993a. Marine fish larvae feeding: formulated diets or live prey? J. World Aquac. Soc. 24:211-224.

Person Le Ruyet, J., C. Fischer & L. Thebaud. 1993b. Sea bass (*Dicentrarchus labrax*) weaning and ongrowing onto Sevbar. Fish Nutrition in Practice. INRA, Paris, p 623-628.

Person-Le Ruyet, J., H. Chartois, E. Desbruyères & J.L. Thomas. 1994. Ammonia acute toxicity in turbot (*Scophthalmus maximus*) juveniles. Eur. Aquac. Soc. Spec. Pub. (22):283-291.

Person-Le Ruyet, J., H. Chartois & L. Quemener. 1995. Comparative acute ammonia toxicity in marine fish and plasma ammonia response. Aquaculture 136:181-194.

Person-Le Ruyet, J., R. Galland, A. Le Roux & H. Chartois. 1997. Chronic ammonia toxicity in juvenile turbot (*Scophthalmus maximus*). Aquaculture 154:155-171.

Persoone, G., P. Sorgeloos, O. Roels & E. Jaspers. (eds.), 1980. The Brine Shrimp *Artemia*. Vol. 3. Ecology, Culturing, Use in Aquaculture. Universa Press, Wetteren, Belgium, 428 p.

Peter, R.E., H.-R. Lin & G. Van Der Kraak. 1987. Drug/hormone induced breeding of Chinese teleosts. Proc. Third Internat. Symp. Repr. Physiol. Fish, Memorial Univ. Newfoundland, St. John's, p 120-123.

Peter, R.E., H.-R. Lin & G. van der Kraak. 1988. Induced ovulation and spawning of cultured freshwater fish in China: advances in application of GnRH analogues and dopamine antagonists. Aquaculture 74:1-10.

Peter, R.E., H.R. Lin, G. van der Kraak & M. Little. 1993. Releasing hormones, dopamine antagonists and induced spawning. Rec. Adv. Aquac. IV:25-30.

Peters, K.K., & C.M. Moffitt. 1996. Optimal dosage of erythromycin thiocyanate in a new feed additive to control bacterial kidney disease. J. Aquat. Anim. Health 8:229-240.

Petersen, C.W., & H.C. Hess. 1991. The adaptive significance of spawning synchronization in the Caribbean damselfish *Stegastes dorsopunicans* (Poey). J. Exp. Mar. Biol. Ecol. 151:155-167.

Petersen, C.W., R.P. Warner, S. Cohen, H.C. Hess & A.T. Sewell. 1992. Variable pelagic fertilization success: implications for mate choice and spatial patterns of spawning. Ecology 73:391-401.

Peterson R.H. 1991. Workshop on Biology and Culture of Striped Bass (*Morone saxatilis*). Can. Tech. Rep. Fish. Aquat. Sci. No. 1832, 66 p.

Petit, J. 1990. Water supply, treatment, and recycling in aquaculture. In: G. Barnabé (ed.), Aquaculture, Vol. 1. Ellis Horwood, New York, p 63-196.

Petrocci, C. 1995. Tails from the road: the black sea bass. Aquac. Mag. 21(2):48-49,50.

Phillips, M.J., M.C.M. Beveridge & R.M. Clarke. 1991. Impact of aquaculture on water resources. Adv. World Aquac. 3:568-591.

Phillips, P., A. Russell, J. Bender & R. Mūnoz. 1994. Management plan for utilization of a floating microbial mat with its associated detrital gelatinous layer as a complete tilapia *Oreochromis niloticus* feed system. Bioresource Technol. 47:239-245.

Phromkunthong, W., M. Boonyaratpalin & V. Storch. 1997. Different concentrations of ascorbyl-2-monophosphate-magnesium as dietary sources of vitamin C for seabass, *Lates calcarifer*. Aquaculture 151:225-243.

Pianka, E.R. 1970. On *r*- and *K*-selection. Am. Nat. 104:592-597.

Piasecki, W., & B.M. MacKinnon. 1995. Life cycle of a sea louse, *Caligus elongatus* von Nordmann, 1832 (Copepoda, Siphonostomatoida, Caligidae). Can. J. Zool. 73:74-82.

Pickering, A.D. (ed.). 1981. Stress and Fish. Academic Press, London, 367 p.

Pickering, A.D., & T.G. Pottinger. 1987a. Crowding causes prolonged leucopenia in salmonid fish, despite interrenal acclimation. J. Fish Biol. 30:701-712.

Pickering, A.D., & T.G. Pottinger. 1987b. Poor water quality suppresses the cortisol response of salmonid fish to handling and confinement. J. Fish Biol. 30:363-374.

Pickering, P.L., & G.J. Holt. 1993. Pacific oyster (*Crassostrea gigas*) trochophores as a first food for red drum (*Sciaenops ocellatus*) larvae. Proc. Texas Chap. Am. Fish. Soc. 15:27-37.

Pierce, R.H., J.M. Weeks & J.M. Prappas. 1993. Nitrate toxicity to five species of marine fish. J. World Aquac. Soc. 24:105-107.

Pierce R.J., & T.E. Wissing, 1974. Energy cost of food utilization in the bluegill (*Lepomis macrochirus*). Trans. Am. Fish. Soc. 103:38-45.

Pietsch, T.W. 1984. Lophiiformes: development and relationships. Am. Soc. Ichthyol. Herpetol. Spec. Pub. (1):320-325.

Piferrer, F., & E.M. Donaldson. 1992. The comparative effectiveness of the natural and a synthetic estrogen for the direct feminization of chinook salmon (*Oncorhynchus tshawytscha*). Aquaculture 106:183-193.

Pigott, G.M., & B.W. Tucker. 1989. Special feeds. In: J.E. Halver (ed.), Fish Nutrition. Academic Press, San Diego, p 653-679.

Piironen, J. 1987. Factors affecting fertilization rate with cryopreserved sperm of whitefish (*Coregonus muksun* Pallas). Aquaculture 66:347-357.

Pike, I.H. 1993. Freshness of fish for fish meal - effect on growth of salmon. Fish Nutrition in Practice, INRA, Paris, p 843-846.

Pillay, T.V.R. 1992. Aquaculture and the Environment. Wiley, New York, 189 p.

Pillsbury, K.S. 1985. The relative food value and biochemical composition of five phytoplankton diets for queen conch, *Strombus gigas* (Linne) larvae. J. Exp. Mar. Biol. Ecol. 90:221-231.

Pimentel, D. 1971. Ecological Effects of Pesticides on Non-target Species. U.S. Off. Sci. Technol., 220 p.

Piñon, G.C. 1994. Penetration and marketing of cultured turbot in restaurants. Eur. Aquac. Soc. Spec. Pub. (22):214-215.

Piper, R.G., I.B. McElwaine, L.E. Orme, J.P. McCraren, L.P. Fowler & J.R. Leonard. 1982. Fish Hatchery Management. U.S. Fish Wildl. Serv., Washington, D.C., 517 p.

Pironet, F.N., & F.J. Neira. 1998. Hormone-induced spawning and development of artificially reared larvae of the West Australian dhufish, *Glaucosoma hebraicum* (Glaucosomatidae). Mar. Freshw. Res. 49:133-142.

Pitcher, T.J. 1977. An energy budget for a rainbow trout farm. Env. Conserv. 4:59-65.

Pittman, K. 1996. Rearing halibut in Norway: present practices and challenges. Mar. Freshw. Res. 47:243-249.

Pittman, K., L. Berg & K. Naas. 1987. Morphological development of halibut (*Hippoglossus hippoglossus*) larvae with special reference to mouth development and metamorphosis. ICES Paper C.M.1987/F:18, 22 p.

Pittman, K., A.B. Skiftesvik & T. Harboe. 1989. Effect of temperature on growth rates and organogenesis in the larvae of halibut (*Hippoglossus hippoglossus* L.). Rapp. P.-v. Réun. Cons. Int. Explor. Mer 191:421-430.

Pittman, K., Ø. Bergh, I. Opstad, A.B. Skiftesvik, L. Skjolddal & H. Strand. 1990. Development of eggs and yolk sac larvae of halibut (*Hippoglossus hippoglossus* L.). J. Appl. Ichthyol. 6:142-160.

Planas, M. 1994. R&D on production systems. Eur. Aquac. Soc. Spec. Pub. (22):54-77.

Planes, S. 1993. Analysis of growth rhythm on the larvae and juveniles of sea-bass (*Dicentrarchus labrax*). C.R. Acad. Sci. Paris. T. 316, Ser. III, 197-203. (in Fr., Eng. abs. & abridged version)

Planko, D., & D. O'Sullivan. 1993. Simple methods for small scale use of brine shrimp (*Artemia*) nauplii - decapsulation, hatching and enrichment. Aquaculture Sourcebook pub., Univ. Tasmania, Launceston, 23 p.

Plumb, J.A. 1991. Major diseases of striped bass and redfish. Vet. Hum. Toxicol. 33:34-39.

Plumb, J.A. 1992. Disease control in aquaculture. In: Diseases in Asian Aquaculture. Asian Fish. Soc., Manila, p 3-17.

Plumb, J.A. 1996. Practical considerations on the application of diagnostic procedures to aquaculture health management. In: K.L. Main & C. Rosenfeld (eds.), Aquaculture Health Management Strategies for Marine Fishes. Oceanic Inst., Honolulu, p 183-193.

Poleo, A.B.S. 1995. Aluminium polymerization—a mechanism of acute toxicity of aqueous aluminium to fish. Aquat. Toxicol. 31:347-356.

Policansky, D., & P. Sieswerda. 1979. Early life history of the starry flounder, *Platichthys stellatus*, reared through metamorphosis in the laboratory. Trans. Am. Fish. Soc. 108:326-327.

Polo, A., M. Yufera & E. Pascual. 1991. Effects of temperature on egg and larval development of *Sparus aurata* L. Aquaculture 92:367-375.

Poole, B.M. 1978. Off-season spawning of Atlantic silversides. Prog. Fish-Cult. 40:72.

Popper, D., & N. Gundermann. 1976. A successful spawning and hatching of *Siganus vermiculatus* under field conditions. Aquaculture 7:291-292.

Popper, D., H. Gordin & G.W. Kissil. 1973. Fertilization and hatching of the rabbitfish (*Siganus rivulatus*). Aquaculture 2:37-44.

Popper, D., R. Pitt & Y. Zohar. 1979. Experiments on the propagation of Red Sea siganids and some notes on their reproduction in nature. Aquaculture 16:177-181.

Popper, D.M., S. Hassin, Y. Hanin & Y. Zohar. 1994. *Epinephelus aeneus* (Serranidae), a candidate for aquaculture. Third Asian Fisheries Forum. Asian Fish. Soc., Manila, p 71-73.

Porrello, S., F. Andaloro, P. Vivona & G. Marino. 1993. Rearing trial of *Seriola dumerili* in a floating cage. Eur. Aquac. Soc. Spec. Pub. (18):299-307.

Porter, C.W., & A.F. Maciorowski. 1984. Spotted seatrout fingerling production in saltwater ponds. J. World Maric. Soc. 15:222-232.

Porter, C.B., M.D. Krom & H. Gordin. 1986. The effect of water quality on the growth of *Sparus aurata* in marine fish ponds. Aquaculture 59:299-315.

Post, G. 1987. Textbook of Fish Health. TFH, Neptune City, New Jersey, 288 p.

Postel, R.T., M. LaDouceur, D. Holbert & M.L. Gallagher. 1996. Texture and flavor of hybrid striped bass fed soybean meal diets. J. Appl. Aquac. 5:83-91.

Poston, H.A. 1986. Vitamin requirements of finfishes: a review. Hoffman-La Roche Inc., Nutley, New Jersey, 36 p.

Poston, H.A. 1990a. Effect of body size on growth, survival, and chemical composition of Atlantic salmon fed soy lecithin and choline. Prog. Fish. Cult. 52:226-230.

Poston, H.A. 1990b. Performance of rainbow trout fry fed supplemental soy lecithin and choline. Prog. Fish. Cult. 52:218-225.

Poston, H.A. 1991. Effects of dietary aluminum on growth and composition of young Atlantic salmon. Prog. Fish.-Cult. 53:7-10.

Potthoff, T., S. Kelley, M. Moe & F. Young. 1984. Description of porkfish larvae (*Anisotremus virginicus*, Haemulidae) and their osteological development. Bull. Mar. Sci. 34:21-59.

Potthoff, T., S. Kelley, V. Saksena, M. Moe & F. Young. 1987. Description of larval and juvenile yellowtail damselfish, *Microspathodon chrysurus*, Pomacentridae, and their osteological development. Bull. Mar. Sci. 40:330-375.

Potts, G.W. 1984. Parental behaviour in temperate marine teleosts with special reference to the development of nest structures. In: G.W. Potts & R.J. Wootton (eds.), Fish Reproduction: Strategies and Tactics. Academic Press, London, p 223-244.

Potts, J.C., & C.S. Manooch III. 1995. Age and growth of red hind and rock hind collected from North Carolina through the Dry Tortugas, Florida. Bull. Mar. Sci. 56:784-794.

Potts, J.C., C.S. Manooch III & D.S. Vaughan. In press. Age and growth of vermilion snapper from the southeastern United States. Trans. Am. Fish. Soc.

Poulet, S.A., & R. Williams. 1991. Characteristics and properties of copepods affecting the recruitment of fish larvae. In: Proc. Fourth Internat. Conf. Copepoda. Bull. Plankton Soc. Jpn. Spec. Vol., p 271-290.

Pouliquen, H., H. Le Bris & L. Pinault. 1993. Experimental study on the decontamination kinetics of seawater polluted by oxytetracycline contained in effluents released from a fish farm located in a salt-marsh. Aquaculture 112:113-123.

Pouliquen, H., M. Algoet, V. Buchet & H. Le Bris. 1995. Acute toxicity of fluorescein to turbot (*Scophthalmus maximus*). Vet. Hum. Toxicol. 37:527-529.

Pounds, G.L., C.R. Engle & L.W. Dorman. 1992. Economic effects of intensification of baitfish

production. J. World Aquac. Soc. 23:64-76.
Pourriot, R. 1990. Rotifers—biology. In: G. Barnabé (ed.), Aquaculture, Vol. 1. Ellis Horwood, New York, p 213-231.
Pousão-Ferreira, P., E. Dores, O. Luís, A. Passos & L. Narciso. 1997a. Fatty acid profile of enriched *Brachionus pliactilis* on different commercial oils and emulsions for the sea bream (*Sparus aurata*) larvae. Island Aquac. Trop. Aquac. Eur. Aquac. Soc., Ann. Mtg, p 243-245.
Pousão-Ferreira, P., E. Dores & J. Amaral. 1997b. White sea bream (*Diplodus sargus*) juvenile production: perspectives for commercial aquaculture. Island Aquac. Trop. Aquac. Eur. Aquac. Soc., Ann. Mtg, p 241-242.
Powell, A.B. 1993. A comparison of early-life-history traits in Atlantic menhaden *Brevoortia tyrannus* and gulf menhaden *B. patronus*. Fish. Bull., U.S. 91:119-128.
Powell, A.B., & A.J. Chester. 1985. Morphometric indices of nutritional condition and sensitivity to starvation of spot larvae. Trans. Am. Fish. Soc. 114:338-347.
Powell, A.B., & Gordy, H.R. 1980. Egg and larval development of the spot, *Leiostomus xanthurus* (Sciaenidae). Fish. Bull., U.S. 78:701-714.
Powell, A.B., & T. Henley. 1995. Egg and larval development of laboratory-reared gulf flounder, *Paralichthys albigutta*, and southern flounder, *P. lethostigma* (Pisces: Paralichthyidae). Fish. Bull. U.S. 93:504-515.
Powell, A.B., & J.W. Tucker, Jr. 1992. Egg and larval development of laboratory-reared Nassau grouper, *Epinephelus striatus* (Pisces, Serranidae). Bull. Mar. Sci. 50:171-185.
Poxton, M.G. 1991. Incubation of salmon eggs and rearing of alevins: natural temperature fluctuations and their influence on hatchery requirements. Aquacultural Eng. 10:31-53.
Poxton, M.G., & S.B. Allouse. 1982. Water quality criteria for marine fisheries. Aquacultural Eng. 1:153-191.
Practical Aquaculture and Lake Management. 1992. March/April issue, several articles.
Prappas, J.M., L.E. Greene & R. White. 1991. Reproductive behavior of the sergeant major, *Abudefduf saxatilis*, within a closed system aquarium. Env. Biol. Fishes 31:33-40.
Prater, D.A., J.J. Bertone, T. Letonja & S.D. Vaughn. 1996. Clinical investigation of new animal drugs by veterinary practitioners. J. Am. Vet. Med. Assoc. 208:190-194.
Prentice, J.A., & P. Thomas. 1987. Successful spawning of orangemouth corvina following injection with des-Gly^{10},[D-Ala^{6}]-luteinizing hormone-releasing hormone (1-9) ethylamide and pimozide. Prog. Fish-Cult. 49:66-69.
Prentice, J.A., R.L. Colura & B.W. Bumguardner. 1989. Observations on induced maturation and spawning of orangemouth corvina. Cal. Fish Game 75:27-32.
Prijono, A., Tridjoko, I.N.A. Giri, A. Poernomo, W.E. Vanstone, C. Lim & T. Daulay. 1988. Natural spawning and larval rearing of milkfish in captivity in Indonesia. Aquaculture 74:127-130.
Pritchard, J.B. 1993. Aquatic toxicology: past, present, and prospects. Env. Health Persp. 100:249-257.
Procarione, L.S., & T.L. King. 1993. Upper and lower temperature tolerance limits for juvenile red drums from Texas and South Carolina. J. Aquat. Anim. Health 5:208-212.
Prosch, R.M. 1991. Reproductive biology and spawning of the myctophid *Lampanyctodes hectoris* and the sternoptychid *Maurolicus muelleri* in the southern Benguela ecosystem. S. Afr. J. Mar. Sci. 10:241-252.
Pryor, V.K., & C.E. Epifanio. 1993. Prey selection by larval weakfish (*Cynoscion regalis*): the effects of prey size, speed, and abundance. Mar. Biol. 116:31-37.
Pullin, R.S.V. 1972. The storage of plaice (*Pleuronectes platessa*) sperm at low temperature. Aquaculture 1:279-283.
Pullin, R.S.V., & H. Bailey. 1981. Progress in storing marine flatfish eggs at low temperatures. Rapp. P.-v. Réun. Cons. Int. Explor. Mer 178:514-517.
Pullin, R.S.V., & R.H. Lowe-McConnell. 1982. The Biology and Culture of Tilapias. ICLARM Conf. Proc. No. 7, 434 p.
Pullin, R.S.V., H. Rosenthal & J.L. Maclean (eds.). 1993. Environment and Aquaculture in Developing Countries. ICLARM, Manila, 359 p.
Pursley, M.G., & W.R. Wolters. 1994. Effect of total water hardness and chloride on survival, growth, and feed conversion of juvenile red drum *Sciaenops ocellatus*. J. World. Aquac. Soc. 25:448-453.
Quattro, J.M., & S.C. Weeks. 1991. Correlations between egg size and egg energetic content within and among biotypes of the genus *Poeciliopsis*. J. Fish Biol. 38:331-334.
Quin, B.J. 1993. The culture and harvest of zooplankton in eutrophic impoundments. Eur. Aquac. Soc. Spec. Pub. (18);31-37.
Quinitio, G.F., & J.D. Toledo. 1991. Mariculture techniques for *Epinephelus* sp. in the Philippines. Advances in Finfish and Shellfish Mariculture. Philip. Council Aquat. Mar. Res. Dev., Book Ser. No. 12/1991, p 94-106.
Quinitio, G.F., J.D. Toledo, A.T. Duller & D.M. Reyes, Jr. 1991. Survival of yolk-sac larvae of grouper (*Epinephelus suillus*) under simulated transport conditions. Eur. Aquac. Soc. Spec. Pub. (15):326-327.
Raa, J., G. Roerstad, R. Engstad & B. Robertsen. 1992. The use of immunostimulants to increase resistance of aquatic organisms to microbial infections. Diseases in Asian Aquaculture I. Asian Fish. Soc., Manila, p 39-50.
Rabalais, N.N., S.C. Rabalais & C.R. Arnold. 1980. Description of eggs and larvae of laboratory reared

red snapper (*Lutjanus campechanus*). Copeia 1980:704-708.

Rabben, H. 1991. Progress in methodology development for rearing halibut fry. Aquac. Eur. 16(2):6-12.

Rach, J.J., T.M. Schreier, G.E. Howe & S.D. Redman. 1997. Effect of species, life stage, and water temperature on the toxicity of hydrogen peroxide to fish. Prog. Fish-Cult. 59:41-46.

Rackowski, J.P., & E.K. Pikitch. 1989. Species profiles: life histories and environmental requirements of coastal fishes and invertebrates (Pacific southwest). Pacific and speckled sanddabs. U.S. Fish Wildl. Serv. Biol. Rep. 82(11.107), 18 p.

Raikova, E.V. 1994. Life cycle, cytology, and morphology of *Polypodium hydriforme*, a coelenterate parasite of the eggs of acipenseriform fishes. J. Parasitol. 80:1-22.

Rainuzzo, J.R. 1993. Fatty acid and lipid composition of fish egg and larvae. In: H. Reinertsen, L.A. Dahle, L. Jørgensen & K. Tvinnereim (eds.), Fish Farming Technology. Balkema, Rotterdam, p 43-49.

Rainuzzo, J.R., Y. Olsen & G. Rosenlund. 1989. The effect of enrichment diets on the fatty acid composition of the rotifer *Brachionus plicatilis*. Aquaculture 79:157-161.

Rainuzzo, J.R., K.I. Reitan, L. Jørgensen & Y. Olsen. 1994. Lipid composition in turbot larvae fed live feed cultured by emulsions of different lipid classes. Comp. Biochem. Physiol. 107A:699-710.

Rajaguru, A. 1992. Biology of two co-occurring tonguefishes, *Cynoglossus arel* and *C. lida* (Pleuronectiformes: Cynoglossidae), from Indian waters. Fish. Bull., U.S. 90:328-367.

Rajasilta, M., J. Eklund, J. Hanninen, M. Kurkilahti, J. Kaaria, P. Rannikko & M. Soikkeli. 1993. Spawning of herring (*Clupea harengus membras* L.) in the Archipelago Sea. ICES J. Mar. Sci. 50:233-246.

Ramírez-Sevilla, R., R. Rueda-Jasso, J.L. Ortiz-Galindo & B. González-Acosta. 1991. Methodology for the experimental culture of the rotifer *Brachionus plicatilis*. Inv. Mar. CICIMAR 6:287-290.

Ramos, J. 1986a. Induction of spawning in common sole (*Solea solea* L.) with human chorionic gonadotropin (HCG). Aquaculture 56:239-242.

Ramos, J. 1986b. Luteinizing hormone-releasing hormone analogue (LH-RHa) induces precocious ovulation in common sole (*Solea solea* L.). Aquaculture 54:185-190.

Ramsay, N., R. Stagg & D. Houlihan. 1993. Osmoregulation of developing herring, *Clupea harengus* L.: effects of chronic exposure to sublethal levels of heavy metals. In: B.T. Walther & H.J. Fyhn (eds.), Physiological and Biochemical Aspects of Fish Development. Univ. Bergen, Norway, p 263-268.

Rana, K. 1995. Preservation of gametes. In: N.R. Bromage & R.J. Roberts (eds.), Broodstock Management and Egg and Larval Quality. Blackwell, Oxford, p 53-75.

Randall, C.F., J. Duston & N.R. Bromage. 1987. Photoperiodic history and the entrainment of the annual cycle of reproduction in the female rainbow trout (*Salmo gairdneri*). Proc. Third Internat. Symp. Repr. Physiol. Fish, Memorial Univ. Newfoundland, St. John's, p 310.

Randall, D. 1991. The impact of variations in water pH on fish. Adv. World Aquac. 3:90-104.

Randall, J.E. 1987. A preliminary synopsis of the groupers (Perciformes: Serranidae: Epinephelinae) of the Indo-Pacific region. In: J.J. Polovina & S. Ralston (eds.), Tropical Snappers and Groupers: Biology and Fisheries Management. Westview Press, Boulder, Colorado, p 89-188.

Randall, J.E., & H.A. Randall. 1963. The spawning and early development of the Atlantic parrot fish, *Sparisoma rubripinne*, with notes on other scarid and labrid fishes. Zoologica 48:49-62.

Randall, J.E., G.R. Allen & R.C. Steene. 1990. Fishes of the Great Barrier Reef and Coral Sea. Univ. Hawaii Press, Honolulu, 507 p.

Rasem, B.M., C.M. James, S.A. Al-Thobaiti & M.H. Carlos. 1997. Spawning of the camouflage grouper *Epinephelus polyphekadion* (Bleeker) in the hypersaline waters of Saudi Arabia. Asian Fish. Sci. 9:251-259.

Rasmussen, K.J. 1994. Spray-dried blood in diets to Atlantic salmon (*Salmo salar* L.). Fisk. Dir. Skr. Ser. Ernæring 6:151-160.

Raymont, J.E.G. 1983. Plankton and Productivity in the Oceans. Vol. 2, Zooplankton. Pergamon, Oxford, 824 p.

Raynard, R.S., et al. 1994. Development of a vaccine for the control of sea lice (*Lepeophtheirus salmonis* and *Caligus elongatus*) in Atlantic salmon (*Salmo salar*). ICES Pap. CM 1994/F:17, 6 p.

Ré, P. 1980. The eggs and newly hatched larvae of *Abudefduf luridus* (Cuvier, 1830)(Pisces: Pomacentridae) from the Azores. Arquivos Mus. Bocage 7:109-115.

Ré, P., & J. Gomes. 1982. The eggs, newly hatched larvae and juveniles of the Azorian *Chromis* (Pisces: Pomacentridae). Bolm Soc. Port. Ciénc. Nat. 21:9-18.

Rea, M.S. (ed. in chief). 1995. Lighting Handbook: Reference & Application. Illuminating Eng. Soc. N. Am., New York, 989 p (1993 edition reprinted with corrections).

Reardon, I.S., & R.M. Harrell. 1994. Effects of varying salinities on the toxicity of potassium permanganate to larval and juvenile striped bass, *Morone saxatilis* (Walbaum). Aquac. Fish. Mgt. 25:571-578.

Reay, P.J. 1984. Reproductive tactics: a non-event in aquaculture? In: G.W. Potts & R.J. Wootton (eds.), Fish Reproduction: Strategies and Tactics. Academic Press, London, p 291-309.

Redding, T.W., A.V. Schally, T.R. Tice & W.E. Meyers. 1984. Long-acting delivery systems for peptides: inhibition of rat prostate tumors by controlled release of [D-Trp^6]luteinizing hormone-releasing hormone from injectable microcapsules. Proc. Nat. Acad. Sci., U.S. 81:5845-5848.

Reddy, P.K., & T.J. Lam. 1991. Effect of thyroid hormones on hatching in the tilapia, *Oreochromis* *mossambicus*. Gen. Comp. Endocrinol. 81:484-491.

Reddington, J.J. 1996. Potential applications of agricultural diagnostic techniques to aquaculture health management. In: K.L. Main & C. Rosenfeld (eds.), Aquaculture Health Management Strategies for Marine Fishes. Oceanic Inst., Honolulu, p 195-208.

Redmayne, P. 1992. Norway's salmon industry shatters into chaos. Seafood Leader 11:74-82.

Reed, P., & R. Francis-Floyd. 1994. Amyloodinium infections of marine fish. Fla. Coop. Ext. Serv., Inst. Food Agric. Sci., Univ. Fla., Gainesville, Fact Sheet VM-90, 4 p.

Rees, J.T., & J.M. Oldfather. 1980. Small-scale mass culture of *Daphnia magna* Straus. Proc. World Maric. Soc. 11:202-210.

Rees, R.A., & R.M. Harrell, 1990. Artificial spawning and fry production of striped bass and hybrids. In: R.M. Harrell, J.H. Kerby & R.V. Minton (eds.), Culture and propagation of striped bass and its hybrids. Am. Fish. Soc., Bethesda, Maryland, p 43-72.

Regidor, S.E., & J.R. Arthur. 1992. A survey of the parasite fauna of milkfish, *Chanos chanos* (Forskal), in the Philippines. Diseases in Asian Aquaculture. Asian Fish. Soc., Manila, p 313-322.

Reguera, B. 1984. The effect of ciliate contamination in mass cultures of the rotifer, *Brachionus plicatilis* O.F. Müller. Aquaculture 40:103-108.

Reid, B., & C.R. Arnold. 1994. Use of ozone for water treatment in recirculating-water raceway systems. Prog. Fish-Cult. 56:47-50.

Reid, G.M. 1997. Culture of ornamental fishes in zoo aquaria. Island Aquaculture and Tropical Aquaculture. Eur. Aquac. Soc., Ann. Mtg., p 257-258.

Reigh, R.C., & S.C. Ellis. 1992. Effects of dietary soybean and fish-protein ratios on growth and body composition of red drum (*Sciaenops ocellatus*) fed isonitrogenous diets. Aquaculture 104:279-292.

Reinertsen, H., L.A. Dahle, L. Jørgensen & K. Tvinnereim (eds.). 1993. Fish Farming Technology. Balkema, Rotterdam, 482 p.

Reish, D.J., P.S. Oshida, A.J. Mearns & T.C. Ginn. 1988. Effects on saltwater organisms. J. Wat. Poll. Contr. Fed. 60:1065-1077.

Reish, D.J., P.S. Oshida, A.J. Mearns & T.C. Ginn. 1993. Effects of pollution on saltwater organisms. Wat. Env. Res. 65:573-585.

Reisman, H.M. 1963. Reproductive behavior of *Apeltes quadracus*, including some comparisons with other gasterosteid fishes. Copeia 1963:191-192.

Reitan, K.I., S. Bolla & Y. Olsen. 1991. Ingestion and assimilation of microalgae in yolk sac larvae of halibut, *Hippoglossus hippoglossus* (L.). Eur. Aquac. Soc. Spec. Pub. (15):332-334.

Reitan, K.I., J.O. Evjemo, Y. Olsen & G. Øie. 1993a. Biomass growth in Atlantic halibut (*Hippoglossus hippoglossus* L.) from eggs to metamorphosis. In: H. Reinertsen, L.A. Dahle, L. Jørgensen & K. Tvinnereim (eds.), Fish Farming Technology. Balkema, Rotterdam, p 417-420.

Reitan, K.I., J.O. Evjemo, Y. Olsen, I. Salvesen, J. Skjermo, O. Vadstein & G. Øie. 1993b. Comparison of incubator concepts for yolk sac larvae of Atlantic halibut (*Hippoglossus hippoglossus* L.). In: H. Reinertsen, L.A. Dahle, L. Jørgensen & K. Tvinnereim (eds.), Fish Farming Technology. Balkema, Rotterdam, p 51-55.

Reitan, K.I., J.R. Rainuzzo, G. Øie & Y. Olsen. 1993c. Nutritional effects of algal addition in first-feeding of turbot (*Scophthalmus maximus* L.) larvae. Aquaculture 118:257-275.

Reitan, K.I., S. Bolla & Y. Olsen. 1994a. A study of the mechanism of algal uptake in yolk-sac larvae of Atlantic halibut (*Hippoglossus hippoglossus*) J. Fish Biol. 44:303-310.

Reitan, K.I., J.R. Rainuzzo & Y. Olsen. 1994b. Influence of lipid composition of live feed on growth, survival and pigmentation of turbot larvae. Aquac. Internat. 2:33-48.

Reitan, K.I., J.R. Rainuzzo, G. Øie & Y. Olsen. 1997. A review of the nutritional effects of algae in marine fish larvae. Aquaculture 155:207-221.

Render, J.H., & C.A. Wilson. 1992. Reproductive biology of sheepshead in the northern Gulf of Mexico. Trans. Am. Fish. Soc. 121:757-764.

Render, J.H., B.A. Thompson & R.L. Allen. 1995. Reproductive development of striped mullet in Louisiana estuarine waters with notes on the applicability of reproductive assessment methods for isochronal species. Trans. Am. Fish. Soc. 124:26-36.

René, F. 1974. Rearing of gilt-head *Sparus aurata*. In: J.H.S. Blaxter (ed.), The Early Life History of Fish. Springer-Verlag, New York, p 747.

Reubush, K.J., & A.G. Heath. 1997. Effects of recovery water salinity on secondary stress responses of hybrid striped bass fingerlings. Prog. Fish-Cult. 59:188-197.

Reyes-Sosa, C.F., & R. Castellanos-Molina. 1995. Nutritional evaluation of gizzard erosion positive brown fish meal in starter diets for Nile tilapia, *Oreochromis niloticus*. Aquaculture 138:323-329.

Riaza, A., & J. Hall. 1993. Large scale production of turbot. In: H. Reinertsen, L.A. Dahle, L. Jørgensen & K. Tvinnereim (eds.), Fish Farming Technology. Balkema, Rotterdam, p 147-148.

Ribeiro, R., J. Reis, C. Santos, F. Gonçalves & A.M.V.M. Soares. 1996. Spawning of anchovy *Engraulis encrasicolus* in the Mondego Estuary, Portugal. Est. Coastal Shelf Sci. 42:467-482.

Rice, J.A. 1990. Bioenergetics modeling approaches to evaluation of stress in fishes. Am. Fish. Soc. Symp. (8):80-92.

Rice, M.A., & A.Z. DeVera. 1998. Aquaculture in Dagupan City, Philippines. World Aquac. 29(1):18-24.

Richards, W.J. 1984. Gonorhynchiformes: development and relationships. Am. Soc. Ichthyol. Herpetol.

Spec. Pub. (1):138-139.
Richards, W.J., & Leis. 1984. Labroidei: development and relationships. Am. Soc. Ichthyol. Herpetol. Spec. Pub. (1):542-547.
Richards, W.J., & V.P. Saksena. 1980. Description of larvae and early juveniles of laboratory-reared gray snapper, *Lutjanus griseus* (Linnaeus)(Pisces, Lutjanidae). Bull. Mar. Sci. 30:515-522.
Richardson, K., & G. Kullenberg. 1987. Physical and biological interactions leading to plankton blooms: a review of *Gyrodinium aureolum* blooms in Scandinavian waters. Rapp. P.-v. Réun. Cons. Int. Explor. Mer 187:19-26.
Richardson, L.B., & D.T. Burton. 1981. Toxicity of ozonated estuarine water to juvenile blue crabs (*Callinectes sapidus*) and juvenile Atlantic menhaden (*Brevoortia tyrannus*). Bull. Env. Contam. Toxicol. 26:171-178.
Richmond, A. (ed.). 1986. CRC Handbook of Microalgal Mass Culture. CRC Press, Boca Raton, Florida, 528 p.
Ricker, W.E. 1979. Growth rates and models. In: W.S. Hoar, D.J. Randall, & J.R. Brett (eds.), Fish Physiology, Vol. VIII: Bioenergetics and Growth. Academic Press, Orlando, Florida, p 677-743.
Rico-Martinez, R., & S.I. Dodson. 1992. Culture of the rotifer *Brachionus calyciflorus* Pallas. Aquaculture 105:191-199.
Rieper-Kirchner, M., K. Hinz & B. Biddanda. 1991. Ingestion of microbially-synthesized organic aggregates and egestion of fecal pellets by marine harpacticoid copepods. Kieler Meeresforsch 8:257-263.
Rijnsdorp, A.D. 1991. Changes in fecundity of female North Sea plaice (*Pleuronectes platessa* L.) between three periods since 1900. ICES J. Mar. Sci. 48:253-280.
Rijnsdorp, A.D., N. Daan, F.A. van Beek & H.J.L. Heessen. 1991. Reproductive variability in North Sea plaice, sole, and cod. J. Cons. 47:352-375.
Riley, C.M., G.J. Holt & C.R. Arnold. 1995. Growth and morphology of larval and juvenile captive bred yellowtail snapper *Ocyurus chrysurus*. Fish. Bull., U.S. 93:179-185.
Rimmer, M. 1995. Barramundi farming: an introduction. Queensland Dept. Pri. Ind., Info. Ser. Q195020, 26 p.
Rimmer, M.A., & A. Reed. 1990. Effects of nutritional enhancement of live food organisms on growth and survival of barramundi/seabass *Lates calcarifer* (Bloch) larvae. Advances in Tropical Aquaculture. AQUACOP, IFREMER, Actes Colloq. (9):611-623.
Rimmer, M.A., A.W. Reed, M.S. Levitt & A.T. Lisle. 1994a. Effects of nutritional enhancement of live food organisms on growth and survival of barramundi, *Lates calcarifer* (Bloch), larvae. Aquac. Fish. Mgt. 25:143-156.
Rimmer, M.A., R.N. Garrett & M.A. Samoilys. 1994b. In vitro fertilization of eggs of coral trout, *Plectropomus leopardus* (Serranidae), collected from an aggregation site on the Great Barrier Reef, Australia. Bull. Mar. Sci. 54:356-358.
Ringø, E., & F.-J. Gatesoupe. 1998. Lactic acid bacteria in fish: a review. Aquaculture 160:177-203.
Ringo, E., E. Strøm & J.-A. Tabachek. 1995. Intestinal microflora of salmonids: a review. Aquac. Res. 26:773-789.
Robaina, L., M.S. Izquierdo, F.J. Moyano, J. Socorro, J.M. Vergara, D. Montero & H. Fernández-Palacios. 1995. Soybean and lupin seed meals as protein sources in diets for gilthead seabream (*Sparus aurata*): nutritional and histological implications. Aquaculture 130:219-233.
Robaina, L., F.J. Moyano, M.S. Izquierdo, J. Socorro, J.M. Vergara & D. Montero. 1997. Corn gluten and meat and bone meals as protein sources in diets for gilthead seabream (*Sparus aurata*): nutritional and histological implications. Aquaculture 157:347-359.
Robert, N., R. Le Gouvello, J.C. Mauviot, F. Arroyo, P. Aguirre & S.J. Kaushik. 1993. Use of extruded diets in intensive trout culture: effects of protein to energy ratios on growth, nutrient utilization and on flesh and water quality. Fish Nutrition in Practice, INRA, Paris, p 497-500.
Roberts, C.M., N. Quinn, J.W. Tucker, Jr. & P.N. Woodward. 1995. Introduction of hatchery-reared Nassau groupers to a coral reef environment. N. Am. J. Fish. Mgt. 15:159-164.
Roberts, D.E., Jr. 1987. Induced maturation and spawning of common snook, *Centropomus undecimalis*. Proc. Gulf Carib. Fish. Inst. 38:222-230.
Roberts, D.E., Jr. 1990. Photoperiod/temperature control in the commercial production of red drum (*Sciaenops ocellatus*) eggs. Texas A&M Univ. Sea Grant TAMU-SG-90-603, p 35-43.
Roberts, D.E., Jr. & R.A. Schlieder. 1983. Induced sex inversion, maturation, spawning and embryogeny of the protogynous grouper, *Mycteroperca microlepis*. J. World Maric. Soc. 14:639-649.
Roberts, D.E., Jr., B.V. Harpster, W.K. Havens & K.R. Halscott. 1976. Facilities and methodology for the culture of the southern sea bass (*Centropristis melana*). Proc. World Maric. Soc. 7:163-198.
Roberts, D.E., Jr., B.V. Harpster & G.E. Henderson. 1978. Conditioning and induced spawning of the red drum (*Sciaenops ocellata*) under varied conditions of photoperiod and temperature. Proc. World Maric. Soc. 9:311-332.
Roberts, R.J. 1982. Microbial Diseases of Fish. Academic Press, New York, 466 p.
Roberts, R.J. (ed.). 1989a. Fish Pathology. Bailliere Tindall, London, 467 p.
Roberts, R.J. 1989b. Miscellaneous non-infectious diseases. In R.J. Roberts (ed.), Fish Pathology. Bailliere Tindall, London, p 363-373.

Roberts, R.J. 1989c. Neoplasia of teleosts. In R.J. Roberts (ed.), Fish Pathology. Bailliere Tindall, London, p 153-172.

Roberts, R.J. 1989d. The mycology of teleosts. In R.J. Roberts (ed.), Fish Pathology. Bailliere Tindall, London, p 320-336.

Roberts, R.J. 1989e. The nutritional pathology of teleosts. In R.J. Roberts (ed.), Fish Pathology. Bailliere Tindall, London, p 337-362.

Roberts, R.J. 1989f. The pathophysiology and systematic pathology of teleosts. In R.J. Roberts (ed.), Fish Pathology. Bailliere Tindall, London, p 56-134.

Roberts, R.J., & A.M. Bullock. 1989. Nutritional pathology. In: J.E. Halver (ed.), Fish Nutrition. Academic Press, San Diego, p 423-473.

Robertson, D.R. 1991. The role of adult biology in the timing of spawning of tropical reef fishes. In: P.F. Sale (ed.), The Ecology of Fishes on Coral Reefs. Academic Press, San Diego, p 356-386.

Robertson, D.R. 1992. Patterns of lunar settlement and early recruitment in Caribbean reef fishes at Panamá. Mar. Biol. 114:527-537.

Robin, J.H. 1995. The importance of n-6 fatty acids in the culture of marine fish larvae. ICES Mar. Sci. Symp. 201:106-111.

Robin, J.H., M.M. Le Gall & H. Le Delliou. 1993. Comparison of three kinds of rotifer enrichments for turbot larval culture. Fish Nutrition in Practice. INRA, Paris, p 619-622.

Robinette, H.R. 1983. Species profiles: life histories and environmental requirements of coastal fishes and invertebrates (Gulf of Mexico). Bay anchovy and striped anchovy. U.S. Fish Wildl. Serv., Biol. Rep. 82(11.14), 15 p.

Robinson, E. 1991. Red drum, *Sciaenops ocellatus*. In: R.P. Wilson (ed.), Handbook of Nutrient Requirements of Finfish. CRC Press, Boca Raton, Florida, p 145-152.

Robinson, E.H., & M.H. Li. 1995. Use of cottonseed meal in aquaculture feeds. In: C. Lim & D.J. Sessa (eds.), Nutrition and Utilization Technology in Aquaculture. AOSC Press, Champaign, Ilinois, p 157-165.

Robinson, R.D., L.F. Khalil, R.N. Hall & R.D. Steele. 1992. Infection of red hybrid tilapia with a monogenean in coastal waters off southern Jamaica. Proc. Gulf Carib. Fish. Inst. 42:441-447.

Roche, H., G. Boge & G. Peres. 1993. Toxicity of colchicine towards a marine fish, *Dicentrarchus labrax*. Comp. Biochem. Physiol. 106C:371-375.

Rodehutscord, M. 1996. Response of rainbow trout (*Oncorhynchus mykiss*) growing from 50 to 200 g to supplements of dibasic sodium phosphate in a semipurified diet. J. Nutr. 126:324-331.

Rodgers, L.J., & C.G. Barlow. 1987. Better nutrition enhances survival of barramundi larvae. Austral. Fish. 46(7):30-32.

Rodríguez, C., J.A. Pérez, M.S. Izquierdo, J.R. Cejas, A. Bolaños & A. Lorenzo. 1996. Improvement of the nutritional value of rotifers by varying the type and concentration of oil and the enrichment period. Aquaculture 147:93-105.

Rodríguez, C., J.A. Pérez, M. Díaz, M.S. Izquierdo, H. Fernández-Palacios & A. Lorenzo. 1997. Influence of the EPA/DHA ratio in rotifers on gilthead seabream (*Sparus aurata*) larval development. Aquaculture 150:77-89.

Rodriguez, J., & T.R. Gregg. 1993. Considerations for the use of ultraviolet in fish culture. In: J.-K. Wang (ed.), Techniques for Modern Aquaculture. Am. Soc. Agric. Eng., St. Joseph, Michigan, p 481-485.

Rodriguez, J.N., & F. Le Menn. 1988. Vitellogenesis in a teleost fish: the sole (*Solea vulgaris* L.). In: Y. Zohar & B. Breton (eds.), Reproduction in Fish. Basic and Applied Aspects in Endocrinology and Genetics. Colloq. INRA (44):31-34.

Rodríguez-Serna, M., M.A. Olvera-Novoa & C.C. Carmona-Osalde. 1996. Nutrional value of animal by-product meal in practical diets for Nile tilapia *Oreochromis niloticus*(L.) fry. Aquac. Res. 27:67-73.

Roesijadi, G. 1992. Metallothioneins in metal regulation and toxicity in aquatic animals. Aquat. Toxicol. 22:81-113.

Rogers, W.A., & D. Xu. 1992. Protective immunity induced by a commercial *Vibrio* vaccine in hybrid striped bass. J. Aquat. Anim. Health 4:303-305.

Rombough, P.J. 1988. Respiratory gas exchange, aerobic metabolism, and effects of hypoxia during early life. In: W.S. Hoar & D.J. Randall (eds.), Fish Physiology, Vol. 11, Part A. Academic Press, San Diego, p 59-161.

Rønnestad, I., H.J. Fyhn & K. Gravningen. 1992. The importance of free amino acids to the energy metabolism of eggs and larvae of turbot (*Scophthalmus maximus*). Mar. Biol. 114:517-525.

Ronnestad, I., W.M. Koven, A. Tandler, M. Harel & H.J. Fyhn. 1994. Energy metabolism during development of eggs and larvae of gilthead sea bream (*Sparus aurata*). Mar. Biol. 120:187-196.

Ronyai, A., A. Ruttkay, L. Varadi & A. Peteri. 1991. Growth comparative trial of fingerlings of sterlet (*Acipenser ruthenus* L.) and that of its hybrid with male Siberian sturgeon (*Acipenser baeri* B.). In: P. Williot (ed.), Acipenser. CEMAGREF, Bordeaux, p 417-421.

Ronzani-Cerquiera, V. 1986. L'elevage larvaire intensif du loup, *Dicentrarchus labrax*: influence de la lumiere, de la densite en proies et de la temperature sur l'alimentation, sur le transit digestif et sur les performances zootechniques. Doctoral Thesis, Univ. d'Aix Marseille II, 191 p.

Rosas, J., T. Cabrera & J. Millan. 1997. Factors that affect the survival of mangrove snapper (*Lutjanus griseus* L. 1758) larvae. Island Aquac. Trop. Aquac. Eur. Aquac. Soc., Ann. Mtg. p 267-268.

Rosas C., J., T. Cabrera B. & J. Millán Q. In press. Aplicación de dos tipos de hormonas en la reproducción artificial del pargo de mangle *Lutjanus griseus* L. Proc. Gulf Carib. Fish. Inst. 50.

Rosenlund, G., I. Meslo, R. Rødsjø & H. Torp. 1993. Large scale production of cod. In: H. Reinertsen, L.A. Dahle, L. Jørgensen & K. Tvinnereim (eds.), Fish Farming Technology. Balkema, Rotterdam, p 141-146.

Rosenlund, G., C. Talbot, E. Santos & R. Hole. 1996. Potential gains through alimentation/nutrition. Improvements in the grow-out. In: B. Chatain, M. Saroglia, J. Sweetman & P. Lavens (eds.), Seabass and Seabream Culture: Problems and Prospects. Eur. Aquac. Soc., Oostende, Belgium, p 160-174.

Rosenlund, G., J. Stoss & C. Talbot. 1997. Co-feeding marine fish larvae with inert and live diets. Aquaculture 155:183-191.

Rosenthal, H. 1994. Aquaculture and the Environment. World Aquac. 25(2):4-11.

Rosenthal, H., & D.F. Alderdice. 1976. Sublethal effects of environmental stressors, natural and pollutional, on marine fish eggs and larvae. J. Fish. Res. Bd. Can. 33:2047-2065.

Ross, J.L., T.M. Stevens & D.S. Vaughan. 1995. Age, growth, mortality, and reproductive biology of red drums in North Carolina waters. Trans. Am. Fish. Soc. 124:37-54.

Ross, L.G., R.W. McKinney, S.K. Cardwell, J.G. Fullarton, S.E.J. Roberts & B. Ross. 1992. The effects of dietary protein content, lipid content and ration level on oxygen consumption and specific dynamic action in *Oreochromis niloticus* L. Comp. Biochem. Physiol. 103A:573-578.

Ross, R.M., T.W.H. Backman & R.M. Bennett. 1993. Evaluation of the anesthetic metomidate for the handling and transport of juvenile American shad. Prog. Fish.-Cult. 55:236-243.

Rosti, P. 1967. Breeding the royal gramma. Salt Water Aquarium 2:106-108.

Roszell, L.E., & R.S. Anderson. 1994. Inhibition of phagocytosis and superoxide production by pentachlorophenol in two leukocyte subpopulations from *Fundulus heteroclitus*. Mar. Env. Res. 38:195-206.

Roth, M., R.H. Richards, D.P. Dobson & G.H. Rae. 1996. Field trials on the efficacy of the organophosphorus compound azamethiphos for the control of sea lice (Copepoda: Caligidae) infestations of farmed Atlantic salmon (*Salmo salar*). Aquaculture 140:217-239.

Rothhaupt, K.O. 1993. Rotifers and continuous culture techniques: model systems for testing mechanistic concepts of consumer-resource interactions. Ecological Studies: Analysis and Synthesis 98:178-192.

Rotllant, J., M. Pavlidis, M. Kentouri, M.E. Abad & L. Tort. 1997. Non-specific immune responses in the red porgy *Pagrus pagrus* after crowding stress. Aquaculture 156:279-290.

Rottmann, R.W., J.V. Shireman, C.C. Starling & W.H. Revels. 1988. Eliminating adhesiveness of white bass eggs for the hatchery production of hybrid striped bass. Prog. Fish-Cult. 50:55-57.

Rottmann, R.W., J.V. Shireman & F.A. Chapman. 1991a. Determining sexual maturity of broodstock for induced spawning of fish. South. Reg. Aquac. Ctr. Pub. No. 423, 4 p.

Rottmann, R.W., J.V. Shireman & F.A. Chapman. 1991b. Techniques for taking and fertilizing the spawn of fish. South. Reg. Aquac. Ctr. Pub. No. 426, 6 p.

Rotunno, T. & R.K. Cowen. 1997. Temporal and spatial spawning patterns of the Atlantic butterfish, *Peprilus triacanthus*, in the South and Middle Atlantic Bights. Fish. Bull., U.S. 95:785-799.

Rouse, D.B., C.D. Webster & I.A. Radwin. 1992. Enhancement of the fatty acid composition of the nematode *Panagrellus redivivus* using three different media. J. World Aquac. Soc. 23:89-95.

Rowland, S.J. 1988. Hormone-induced spawning of the Australian freshwater fish Murray cod, *Macculochella peeli* (Michell)(Percichthyidae). Aquaculture 70:371-389.

Ruangpanit, N. 1987a. Biological characteristics of wildstock sea bass (*Lates calcarifer*) in Thailand. Austral. Ctr. Internat. Agric. Res. Proc. (20):55-56.

Ruangpanit, N. 1987b. Developing hatchery techniques for sea bass (*Lates calcarifer*): a review. Austral. Ctr. Internat. Agric. Res. Proc. (20):132-135.

Ruangpanit, N., & R. Yashiro. 1995. A review of grouper (*Epinephelus* spp.) and seabass (*Lates calcarifer*) culture in Thailand. In: K.L Main & C. Rosenfeld (eds.), Culture of High-value Marine Fishes in Asia and the United States. Oceanic Inst., Honolulu, p 167-183.

Rubin, J.-F. 1994. Survival and development of sea trout, *Salmo trutta* (L.), eggs in Baltic Sea saltwater. Fish. Res. 20:1-12.

Ruchimat, T., T. Masumoto, H. Hosokawa, Y. Itoh & S. Shimeno. 1997. Quantitative lysine requirement of yellowtail (*Seriola quinqueradiata*). Aquaculture 158:331-339.

Ruchon, F., T. Laugier & J.-P. Quignard. 1993. Variation saisonniere de la taille des oeufs et de la fecondite par acte de ponte de *Lipophrys pavo* (Teleostei, Blenniidae) dans une lagune Nord-Mediterraneenne (France, Mauguio). Cybium 17:197-214. (in Fr., Eng. abs.)

Rudi, H., & E. Dragsund. 1993. Localization strategies. In: H. Reinertsen, L.A. Dahle, L. Jørgensen & K. Tvinnereim (eds.), Fish Farming Technology. Balkema, Rotterdam, p 169-175.

Rueda-Jasso, R.A. 1996. Nutritional effect of three microalgae and one cyanobacteria on the culture of the rotifer *Brachionus plicatilis* Müller: 1786. Ciencias Marinas 22:313-328.

Rueter, J., & R. Johnson. 1995. The use of ozone to improve solids removal during disinfection. Aquacultural Eng. 14:123-141.

Rumsey, Gary L. 1992. Dryden, New York, pers. comm.

Rumsey, G.L., J.G. Endres, P.R. Bowser, K.A. Earnest-Koons, D.P. Anderson & A.K. Siwicki. 1995. Soy protein in diets of rainbow trout: effects on growth, protein absorption, gastrointestinal histology, and nonspecific serologic and immune response. In: C. Lim & D.J. Sessa (eds.), Nutrition and Utilization Technology in Aquaculture. AOSC Press, Champaign, Ilinois, p 166-188.

Runyan, S. 1961. Early development of the clingfish, *Gobiesox strumosus* Cope. Chesapeake Sci. 2:113-141.

Ruple, D. 1984. Gobioidei: development. Am. Soc. Ichthyol. Herpetol. Spec. Pub. (1):582-587.

Russell, F.S. 1976. The Eggs and Planktonic Stages of British Marine Fishes. Academic Press, London, 524 p.

Russo, R.C. 1985. Ammonia, nitrite, and nitrate. In: G.M. Rand & S.R. Petrocelli (eds.), Fundamentals of Aquatic Toxicology: Methods and Applications. Hemisphere Pub., Washington, DC, p 455-471.

Russo, R.C., & R.V. Thurston. 1991. Toxicity of ammonia, nitrite, and nitrate to fishes. Adv. World Aquac. 3:58-89.

Rutecki, T.L., & E.R. Varosi. 1997. Distribution, age, and growth of juvenile sablefish, *Anoplopoma fimbria*, in southeast Alaska. NOAA Tech. Rep. NMFS 130, p 45-54.

Rutledge, W.P. 1988. Fish Culture Production Strategy for Marine Fishes in Saltwater Rearing Ponds. M.S. Thesis, Southwest Texas State Univ., San Marcos, 44 p.

Rutledge, W.P., & M.A. Rimmer. 1991. Culture of larval sea bass, *Lates calcarifer* (Bloch), in saltwater rearing ponds in Queensland, Australia. Asian Fish. Sci. 4:345-355.

Ruttanagosrigit, W., Y. Musig, C.E. Boyd & L. Sukchareon. 1991. Effect of salinity on oxygen transfer by propeller-aspirator-pump and paddle wheel aerators used in shrimp farming. Aquacultural Eng. 10:121-131.

Rørstad, G., P.M. Aasjord & B. Robertsen. 1993. Adjuvant effect of a yeast glucan in vaccines against furunculosis in Atlantic salmon (*Salmo salar* L.). Fish Shellfish Immunol. 3:179-190.

Saat, T. 1993. The morphology and chronology of oocyte final maturation and fertilization in fish. In: B.T. Walther & H.J. Fyhn (eds.), Physiological and Biochemical Aspects of Fish Development. Univ. Bergen, Norway, p 71-85.

Sabates, A., & J.-M. Gili. 1991. Processes occurring during fish larvae stages as determinants of year-class strength in fish populations in the northwestern Mediterranean. Œcologia Aquatica 10:203-213.

Sabatés, A., & P. Martin. 1993. Spawning and distribution of bluefish *Pomatomus saltatrix* (L.) in the northwestern Mediterranean. J. Fish Biol. 43:109-118.

Sabaut, J.J., & P. Luquet. 1973. Nutritional requirements of the gilthead bream *Chrysophrys aurata*. Quantitative protein requirements. Mar. Biol. 18:50-54.

Sadler, K. 1982. The toxicity of ammonia to the European eel (*Anguilla anguilla* L.). Aquaculture 26:173-181.

Safford, S.E., & P. Thomas. 1987. Effects of capture and handling on circulating levels of gonadal steroids and cortisol in the spotted seatrout, *Cynoscion nebulosus*. Proc. Third Internat. Symp. Repr. Physiol. Fish, Memorial Univ. Newfoundland, St. John's, p 312.

Saifulina, E. 1991. Larval rearing of the mullet *Mugil so-iyu*. Eur. Aquac. Soc. Spec. Pub. (15):317-318.

Saitoh, S., T. Mori, M. Kusakari & K. Miura. 1991. Studies on the spawning of Japanese flounder (*Paralichthys olivaceus*) I. Induction of spawning of Japanese flounder by manipulation of water temperature. Sci. Rep. Hokkaido Fish. Exp. Sta. 36:71-79. (in Jpn., Eng. abs.)

Sakai, K., M. Nomura, F. Takashima & H. Oto. 1975. The over-ripening phenomenon of rainbow trout—II changes in the percentage of eyed eggs, hatching rate and incidence of abnormal alevins during the process of over-ripening. Bull. Jpn. Soc. Sci. Fish. 41:855-860. (in Jpn., Eng. abs.)

Sakai, M., H. Kamiya, S. Ishii, S. Atsuta & M. Kobayashi. 1992. The immunostimulating effects of chitin in rainbow trout, *Oncorhynchus mykiss*. Diseases in Asian Aquaculture. Asian Fish. Soc., Manila, p 413-417.

Sakai, T., H. Murata, M. Endo, T. Shimomura, K. Yamauchi, T. Ito, T. Yamaguchi, H. Nakjima & M. Fukudome. 1998. Severe oxidative stress is thought to be a principal cause of jaundice of yellowtail *Seriola quinqueradiata*. Aquaculture 160:205-214.

Sakamoto, T., & K. Suzuki. 1978. Spawning behavior and early life history of the porcupine puffer, *Diodon holacanthus*, in aquaria. Jpn. J. Ichthyol. 24:261-270. (in Jpn., Eng. abs.)

Sakamoto, S., & Y. Yone. 1973. Effect of dietary calcium/phosphorus ratio upon growth, feed efficiency and blood serum Ca and P level in red sea bream. Bull. Jpn. Soc. Sci. Fish. 39:343-348.

Sakamoto, S., & Y. Yone. 1979. Requirement of red sea bream for dietary Mg. Bull. Jpn. Soc. Sci. Fish. 45:57-60.

Sakaras, W. 1987. Optimum stocking density of sea bass (*Lates calcarifer*) cultured in cages. Austral. Ctr. Internat. Agric. Res. Proc. (20):172-175.

Sakata, T., & T. Kawazu. 1992. Production of hemolysin by a fish pathogenic strain of *Vibrio damsela*. Mem. Fac. Fish. Kagoshima Univ. 41:9-18.

Sako, H. 1996. Practical approaches to marine fish health problems in Japan. In: K.L. Main & C. Rosenfeld (eds.), Aquaculture Health Management Strategies for Marine Fishes. Oceanic Inst., Honolulu, p 81-90.

Saksena, V.P., & W.J. Richards. 1975. Description of eggs and larvae of laboratory-reared white grunt, *Haemulon plumieri* (Lacepede) (Pisces, Pomadasyidae). Bull. Mar. Sci. 25:523-536.

Sakurai, Y., & T. Hattori. 1996. Reproductive behavior of Pacific cod in captivity. Fish. Sci. 62:222-228.

Saleh, G., & W. Eleraky. 1993. Rearing newly hatchied larvea [sic] of *Tilapia nilotica* with natural food and artificial diets. J. Egypt. Vet. Med. Assoc. 53:97-103.

Salhi, M., M.S. Izquierdo, C.M. Hernández-Cruz, M. González & H. Fernández-Palacios. 1994. Effect of lipid and *n*-3 HUFA levels in microdiets on growth, survival and fatty acid composition of larval gilthead seabream (*Sparus aurata*). Aquaculture 124:275-282.

Salvesen, I., L. Jørgensen & O. Vadstein. 1991. Evaluation of four chemicals for surface-disinfection of marine fish eggs. Eur. Aquac. Soc. Spec. Pub. (15):406-408.

Salvesen, I., G. Øie & O. Vadstein. 1997. Surface disinfection of Atlantic halibut and turbot eggs with glutaraldehyde: evaluation of concentrations and contact times. Aquac. Internat. 5:249-258.

Samiullah, Y. 1985. Biological effects of marine oil pollution. Oil Petrochem. Poll. 2:235-264.

Samoilys, M.A., & L.C. Squire. 1994. Preliminary observations on the spawning behavior of coral trout, *Plectropomus leopardus* (Pisces: Serranidae), on the Great Barrier Reef. Bull. Mar. Sci. 54:332-342.

Samuelsen, O.B., & B.T. Lunestad. 1996. Bath treatment, an alternative method for the administration of the quinolones flumequine and oxolinic acid to halibut *Hippoglossus hippoglossus*, and *in vitro* antibacterial activity of the drugs against some *Vibrio* sp. Dis. Aquat. Org. 27:13-18.

Sanborn, H.R., & D.A. Misitano. 1991. Hormonally induced spawning of English sole (*Parophrys vetulus* Girard). J. Appl. Ichthyol. 7:15-25.

Sánchez-Velasco, L., & W. Norbis. 1997. Comparative diets and feeding habits of *Boops boops* and *Diplodus sargus* larvae, two sparid fishes co-occurring in the northwestern Mediterranean (May 1992). Bull. Mar. Sci. 61:821-835.

Sancho, E., M.D. Ferrando & E. Andreu. 1996. Physiological stress responses of *Anguilla anguilla* to fenitrothion. J. Env. Sci. Health B31:87-98.

Sanderson, G.W., & S.O. Jolly. 1994. The value of *Phaffia* yeast as a feed ingredient for salmonid fish. Aquaculture 124:193-200.

Sandifer, P.A., J.S. Hopkins, A.D. Stokes & R.D. Smiley. 1993. Experimental pond grow-out of red drum, *Sciaenops ocellatus*, in South Carolina. Aquaculture 118:217-228.

Sanjeev, S., & J. Stephen. 1992. Antibiotic sensitivity of Kanagawa-positive and Kanagawa-negative strains of *Vibrio parahaemolyticus* isolated from fishes marketed in Kochi. Fish. Technol. 29:162-165.

Santiago, C.B. 1986. Nutrition and feeds. In: C.-S. Lee, M.S. Gordon & W.O. Watanabe (eds.), Aquaculture of Milkfish (*Chanos chanos*): State of the Art. Oceanic Inst., Honolulu, p 181-207.

Santinha, P.J.M., E.F.S. Gomes & J.O. Coimbra. 1996. Effects of protein level of the diet on digestibility and growth of gilthead sea bream, *Sparus auratus* L. Aquac. Nutr. 2:81-87.

Santulli, A., & V. D'Amelio. 1986. Effects of supplementary dietary carnitine on growth and lipid metabolism of hatchery-reared sea bass (*Dicentrarchus labrax* L.). Aquaculture 59:177-186.

Santulli, A., A. Modica, A. Curatolo & V. D'Amelio. 1988. Carnitine administration to sea bass (*Dicentrarchus labrax* (L.) during feeding on a fat diet: modification of plasma lipid levels and lipoprotein pattern. Aquaculture 68:345-351.

Santulli, A., E. Puccia & V. D'Amelio. 1990. Preliminary study on the effect of short-term carnitine treatment on nucleic acids and protein metabolism in sea bass (*Dicentrarchus labrax* L.) fry. Aquaculture 87:85-89.

Sanz, A., A.E. Morales, M. de la Higuera & G. Cardenete. 1994. Sunflower meal compared with soybean meal as partial substitutes for fish meal in rainbow trout (*Oncorhynchus mykiss*) diets: protein and energy utilization. Aquaculture 128:287-300.

Sanz, F., & J. Coll. 1992. Techniques for diagnosing viral diseases of salmonid fish. Dis. Aquat. Org. 13:211-223.

Sarasquete, M.C., A. Polo & M. Yúfera. 1995. Histology and histochemistry of the development of the digestive system of larval gilthead seabream, *Sparus aurata* L. Aquaculture 130:79-92.

Sargent, J.R. 1995. Origins and functions of egg lipids: nutritional implications. In: N.R. Bromage & R.J. Roberts (eds.), Broodstock Management and Egg and Larval Quality. Blackwell, Oxford, p 353-372.

Sargent, J.R., & R.J. Henderson. 1986. Lipids. In: E.D.S. Corner & S.C.M. O'Hara (eds.), The Biological Chemistry of Marine Copepods. Clarendon Press, Oxford, p 59-108.

Sargent, J., R.J. Henderson & D.R. Tocher. 1989. The lipids. In: J.E. Halver (ed.), Fish Nutrition. Academic Press, San Diego, p 153-218.

Sargent, J.R., L.A. McEvoy & J.G. Bell. 1997. Requirements, presentation and sources of polyunsaturated fatty acids in marine fish larval feeds. Aquaculture 155:117-127.

Sarig, S. 1981. The Mugilidae in polyculture in fresh and brackish water fishponds. In: O.H. Oren (ed.), Aquaculture of Grey Mullets. IBP 26. Cambridge Univ. Press, Cambridge, U.K., p 391-409.

Saroglia, M.G., G. Scarano & E. Tibaldi. 1981. Acute toxicity of nitrite to sea bass (*Dicentrarchus labrax*) and European eel (*Anguilla anguilla*) J. World Maric. Soc. 12:121-126.

Sato, T., & K. Kikuchi. 1997. Meat meal as a protein source in the diet of juvenile Japanese flounder. Fish. Sci. 63:877-880.

Sato, M., Y. Hatano & R. Yoshinaka. 1991. L-ascorbyl 2-sulfate as a dietary vitamin C source for rainbow trout *Oncorhyncus mykiss*. Nippon Suisan Gakkaishi 57:717-721.

Satoh, H., K. Yamamori & T. Hibiya. 1992. Induced spawning of the Japanese eel. Bull. Jpn. Soc. Sci. Fish. 58:825-832.

Satoh, S., T. Takeuchi, T. Watanabe, M. Hoshi & K. Toyama. 1984. Nutritional evaluation of scrap meal as a protein source in diets for rainbow trout and carp. Bull. Jpn. Soc. Sci. Fish. 50:2077-2083.

Satsmadjis, J., & N. Friligos. 1983. Red tide in Greek waters. Vie et Milieu 33:111-117.

Saucier, M.H., & D.M. Baltz. 1993. Spawning site selection by spotted seatrout, *Cynoscion nebulosus*, and

black drum, *Pogonias cromis*, in Louisiana. Env. Biol. Fishes 36:257-272.
Saunders, R.L. 1986. The thermal biology of Atlantic salmon: influence of temperature on salmon culture with particular reference to constraints imposed by low temperature. Rep. Sotvattenslaboratoriet 63:77-90.
Scarano, G., & M.G. Saroglia. 1984. Recovery of fish from functional and haemolytic anaemia after brief exposure to a lethal concentration of nitrite. Aquaculture 43:421-426.
Scarano, G., M.G. Saroglia, R.H. Gray & E. Tibaldi. 1984. Hematological responses of sea bass *Dicentrarchus labrax* to sublethal nitrite exposures. Trans. Am. Fish. Soc. 113:360-364.
Scarpa, J. 1997. Method for the production of genetically modified bivalve trochophore larvae as a first-feed for fish larvae. World Aquac. '97 Book of Abs., p 410-411.
Schach, H. 1939. Die Künstliche Aufzucht von *Clupea harengus* L. Helgol. wiss. Meeresuntersuch 1:359-372.
Schaefer, K.M. 1987. Reproductive biology of black skipjack, *Euthynnus lineatus*, an eastern Pacific tuna. Inter-American Trop. Tuna Comm. Bull. 19:167-260.
Schaefer, K.M. 1996. Spawning time, frequency, and batch fecundity of yellowfin tuna, *Thunnus albacares*, near Clipperton Atoll in the eastern Pacific Ocean. Fish. Bull., U.S. 94:98-112.
Schauer, P.S., & K.L. Simpson. 1985. Bioaccumulation and bioconversion of dietary labeled fatty acids in *Artemia* and winter flounder (*Pseudopleuronectes americanus*). Can. J. Fish. Aquat. Sci. 42:1430-1438.
Schauer, P.S., D.M. Johns, C.E. Olney & K.L. Simpson. 1980. International study on *Artemia* IX. Lipid level, energy content and fatty acid composition of the cysts and newly hatched nauplii from five geographical strains of *Artemia*. In: G. Persoone, P. Sorgeloos, O. Roels & E. Jaspers (eds.), The Brine Shrimp *Artemia*. Vol. 3. Ecology Culturing Use in Aquaculture. Universa Press, Wetteren, Belgium, p 365-373.
Scheerer, P.D., & G.H. Thorgaard. 1989. Improved fertilization by cryopreserved rainbow trout semen treated with theophylline. Prog. Fish-Cult. 51:179-182.
Schiller, G. 1990. Spawning behavior of the decorated firefish (*Nemateleotris decora*). SeaScope, Aquarium Systems, Mentor, Ohio, 7(fall):4.
Schimmel, S.C., & D.J. Hansen. 1974. Sheepshead minnow (*Cyprinodon variegatus*): an estuarine fish suitable for chronic (entire life-cycle) bioassays. Proc. Ann. Conf. SE Assoc. Game Fish Comm. 28:392-398.
Schimmel, S.C., D.J. Hansen & J. Forester. 1974. Effects of Aroclor® 1254 on laboratory-reared embryos and fry of sheepshead minnows (*Cyprinodon variegatus*). Trans. Am. Fish. Soc. 103:582-586.
Schlüter, M., & J. Groeneweg. 1981. Mass production of freshwater rotifers on liquid wastes I. The influence of some environmental factors on population growth of *Brachionus rubens* Ehrenberg 1838. Aquaculture 25:17-24.
Schlüter, M., & J. Groeneweg. 1985. The inhibition by ammonia of population growth of the rotifer, *Brachionus rubens*, in continuous culture. Aquaculture 46:215-220.
Schmahl, G. 1991. The chemotherapy of monogeneans which parasitize fish: a review. Folia Parasitologica 38:97-106.
Schmid-Araya, J.M. 1992. The biochemical composition and calorific content of a rotifer and its algal food: comparison of a two stage chemostat and batch culture. Oecologia 92:327-338.
Schmidt, H., T. Braunbeck, M. El-Matbouli, R.W. Hoffmann, H. Mehlhorn, G. Ritter & G. Schmahl. 1997. The use of a medicated feed containing quinine for the therapy of protozoan diseases of fish. World Aquac. '97 Book of Abs., p 414.
Schmidt, T. 1995. About seahorses. SeaScope, Aquarium Systems, Mentor, Ohio, 12(winter):1-2.
Schmittou, H.R. 1993. High Density Fish Culture in Low Volume Cages. Am. Soybean Assoc., Singapore, 78 p.
Schnick, Rosalie A. 1998. National Coordinator for Aquaculture New Animal Drug Applications (administrated by Michigan State University), La Crosse, Wisconsin, pers. comm.
Schnick, R.A., & R. Armstrong. 1997. Aquaculture drug approval progress in the United States. In: R. Armstrong (preparer), The Salmon Health Report. North. Aquac. Suppl., p 22-28.
Schnick, R.A., D.J. Alderman, R. Armstrong, R. Le Gouvello, S. Ishihara, E.C. Lacierda, S. Percival & M. Roth. 1997. World wide aquaculture drug and vaccine registration progress. Bull. Eur. Assn. Fish Pathol. 17:251-260.
Schreck, C.B., M.S. Fitzpatrick & K.P. Currens. 1995. Pacific salmon (*Oncorhynchus* spp.). In: N.R. Bromage & R.J. Roberts (eds.), Broodstock Management and Egg and Larval Quality. Blackwell, Oxford, p 197-219.
Schreibman, M.P., & H. Margolis-Nunno. 1989. The brain-pituitary-gonad axis in poikilotherms. In: M.P. Schreibman & C.G. Scanes (eds.), Development, Maturation, and Senescence of Neuroendocrine Systems: A Comparative Approach. Academic Press, San Diego, p 97-133.
Schreier, T.M., J.J. Rach & G.E. Howe. 1996. Efficacy of formalin, hydrogen peroxide, and sodium chloride on fungal-infected rainbow trout eggs. Aquaculture 140:323-331.
Schreiner, W. 1972. Breeding report—clownfishes. Mar. Aquarist 3(3):31-33.
Schreitmuller, W. 1930a. Kugelfishe. Das Aquarium, Jan, p 12-16. (in Ger.)
Schreitmuller, W. 1930b. Kugelfishe. Das Aquarium, Feb, p 20-26. (in Ger.)

Schulstad, G. 1997. Design of a computerized decision support system for hatchery production management. Aquac. Eng. 16:7-25.

Scott, A.P., & S.M. Baynes. 1978. Effect of algal diet and temperature on the biochemical composition of the rotifer, *Brachionus plicatilis*. Aquaculture 14:247-260.

Scott, A.P., & S.M. Baynes. 1980. A review of the biology, handling and storage of salmonid spermatozoa. J. Fish. Biol. 17:707-739.

Scott, A.P., & A.V.M. Canario. 1987. Status of oocyte maturation-inducing steroids in teleosts. Proc. Third Internat. Symp. Repr. Physiol. Fish, Memorial Univ. Newfoundland, St. John's, p 224-234.

Scott, A.P., & C. Middleton. 1979. Unicellular algae as a food for turbot (*Scophthalmus maximus* L.) larvae—the importance of dietary long-chain polyunsaturated fatty acids. Aquaculture 18:227-240.

Scott, D.B.C. 1979. Environmental timing and the control of reproduction in teleost fish. Symp. Zool. Soc. London (44):105-132.

Scott, S.G., & N.W. Pankhurst. 1992. Interannual variation in the reproductive cycle of the New Zealand snapper *Pagrus auratus* (Bloch & Schneider)(Sparidae). J. Fish Biol. 41:685-696.

Scott, S.G., J.R. Zeldis & N.W. Pankhurst. 1993. Evidence of daily spawning in natural populations of the New Zealand snapper *Pagrus auratus* (Sparidae). Env. Biol. Fishes 36:149-156.

Scott, W.B., & E.J. Crossman. 1973. Freshwater Fishes of Canada. Fish. Res. Bd. Can. Bull. 184, 966 p.

Scudder, B.C., J.L. Carter & H.V. Leland. 1988. Effects of copper on development of the fathead minnow, *Pimephales promelas* Rafinesque. Aquat. Toxicol. 12:107-124.

Scura, E.D., & C.W. Jerde. 1977. Various species of phytoplankton as food for larval northern anchovy, *Engraulis mordax*, and relative nutritional value of the dinoflagellates *Gymnodinium splendens* and *Gonyaulax polyedra*. Fish. Bull., U.S. 75:577-583.

Segawa, S., & W.T. Yang. 1988. Population growth and density of an estuarine cladoceran *Diaphanosoma aspinosum* in laboratory culture. Bull. Plankton Soc. Japan 35:67-73.

Segner, H., V. Storch, M. Reinecke, W. Kloas & W. Hanke. 1994. The development of functional digestive and metabolic organs in turbot, *Scophthalmus maximus*. Mar. Biol. 119:471-486.

Sekiya, T., H. Murata, T. Sakai, K. Yamauchi, K. Yamashita, M. Ugawa, M. Kanai & M. Shimada. 1991. An attempt to control lipid peroxidation in the tissues of yellowtails and to enhance their biological protective ability by feeding high a-tocopherol brown meals. Nippon Suisan Gakkaishi 57:287-292.

Semple, G.P. 1991. Reproductive behaviour and early development of the honey blue-eye, *Pseudomugil mellis* Allen and Ivantsoff 1982 (Pisces: Pseudomugilidae), from the North-east Coast Division, south-eastern Queensland, Australia. Austral. J. Mar. Freshw. Res. 42:277-286.

Senoo, S., K.J. Ang & G. Kawamura. 1994a. Development of sense organs and mouth and feeding of reared marble goby *Oxyeleotris marmoratus* larvae. Fish. Sci. 60: 361-368.

Senoo, S., M. Kaneko, S.H. Cheah & K.J. Ang. 1994b. Egg development, hatching, and larval development of marble goby *Oxyeleotris marmoratus* under artificial rearing conditions. Fish. Sci. 60:1-8.

Serrano, J.A., G.R. Nematipour & D.M. Gatlin III. 1992. Dietary protein requirement of the red drum (*Sciaenops ocellatus*) and relative use of dietary carbohydrate and lipid. Aquaculture 101:283-291.

Service, R.F. 1995. Antibiotics that resist resistance. Science 270:724-727.

Shaffer, R.V., & E.L. Nakamura. 1989. Synopsis of biological data on the cobia *Rachycentron canadum* (Pisces: Rachycentridae). NOAA Tech. Rep. NMFS 82, 21 p.

Shafland, P.L., & D.H. Koehl. 1980. Laboratory rearing of the common snook. Proc. Ann. Conf. SE Assoc. Fish Wildl. Ag. 33:425-431.

Shahidi, F. 1995. Further utilization of seafood processing discards. In: C. Lim & D.J. Sessa (eds.), Nutrition and Utilization Technology in Aquaculture. AOSC Press, Champaign, Ilinois, p 278-288.

Shamsudin, L. 1994. Food value of indigenous zooplankton species collected from various coastal waters of the South China Sea for aquaculture. Third Asian Fisheries Forum, Asian Fish. Soc., Manila, p 752-755.

Shansudin, L., M. Yusof, A. Azis & Y. Shukri. 1997. The potential of certain indigenous copepod species as live food for commercial fish larval rearing. Aquaculture 151:351-356.

Shang, Y.C. 1990. Aquaculture Economic Analysis. Adv. World Aquac., Vol. 2, 211 p.

Shapiro, D.Y. 1992. Plasticity of gonadal development and protandry in fishes. J. Exp. Zool. 261:194-203.

Shapiro, D.Y., & M.B. Rasotto. 1993. Sex differentiation and gonadal development in the diandric, protogynous wrasse, *Thalassoma bifasciatum* (Pisces, Labridae). J. Zool. London 230:231-245.

Shaw, B.A., P.J. Harrison & R.J. Andersen. 1994. Evaluation of the copepod *Tigriopus californicus* as a bioassay organism for the detection of chemical feeding deterrents produced by marine phytoplankton. Mar. Biol. 121:89-95.

Shaw, E.S. 1955. The embryology of the sergeant major, *Abudefduf saxatilis*. Copeia 1955:85-89.

Shearer, K.D., A. Maage, J. Opstvedt & H. Mundheim. 1992. Effects of high-ash diets on growth, feed efficiency, and zinc status of juvenile Atlantic salmon (*Salmo salar*). Aquaculture 106:345-355.

Sheehan, E.M., P.J.A. Sheehy & R. Fitzgerald. 1994. Feeding trial, composition, oxidative stability and sensory evaluation of cultured turbot fed three commercial feeds. Proc. Nutr. Soc. 53:128A.

Shehadeh, Z.H., & J.N. Ellis. 1970. Induced spawning of the striped mullet *Mugil cephalus* L. J. Fish Biol. 2:355-360.

Sheikh-Eldin, M., S.S. De Silva, T.A. Anderson & G. Gooley. 1996. Comparison of fatty acid composition of muscle, liver, mature oocytes, and diets of wild and captive Macquarie perch, *Macquaria australasica*, broodfish. Aquaculture 144:201-216.

Shelbourne, J.E. 1971. The Artificial Propagation of Marine Fish. T.F.H., Neptune City, New Jersey, 83 p.

Shelbourne, J.E. 1974. Population effects on the survival, growth and pigment of tank-reared plaice larvae. In: F.R. Harden Jones (ed.) Wiley, New York, p 357-377.

Shelbourne, J.E. 1975. Marine fish cultivation: pioneering studies on the culture of the larvae of the plaice (*Pleuronectes platessa* L.) and the sole (*Solea solea* L.). Fish. Invest. Ser. II, Vol. 27, No. 9, 29 p.

Shelley, C. (ed.). 1993. Barramundi Farming in the Northern Territory. N. Terr. Dept. Pri. Ind. Fish., Darwin, 45 p.

Shelton, W.L. 1989. Management of finfish reproduction for aquaculture. Rev. Aquat. Sci. 1:497-535.

Shepherd, C.J., & N.R. Bromage (eds.). 1988. Intensive Fish Farming. Blackwell, Oxford, 404 p.

Shepherd, J. 1988. Fish health and disease. In: C.J. Shepherd (ed.), Intensive Fish Farming. Blackwell, Oxford, p 198-238.

Sherwood, N.M. 1987. Brain peptides in the control of fish reproduction. Proc. Third Internat. Symp. Repr. Physiol. Fish, Memorial Univ. Newfoundland, St. John's, p 5-15.

Sherwood, N.M., & B. Harvey. 1986. Topical absorption of gonadotropin-releasing hormone (GnRH) in goldfish. Gen. Comp. Endocrinol. 61:13-19.

Sherwood, N.M., L.W. Crim, J. Carolsfeld & S.M. Walters. 1988. Sustained hormone release. I. Characteristics of in vitro release of gonadotropin-releasing hormone analogue (GnRH-A) from pellets. Aquaculture 74:75-86.

Shiau, S.-Y., & C.-W. Lan. 1996. Optimum dietary protein level and protein to energy ratio for growth of grouper (*Epinephelus malabaricus*). Aquaculture 145:259-266.

Shiau, S.-Y., & H.-S. Liang. 1995. Carbohydrate utilization and digestibility by tilapia, *Oreochromis niloticus* × *O. aureus*, are affected by chromic oxide inclusion in the diet. J. Nutr. 125:976-982.

Shiau, S.-Y., & G.-S. Suen. 1992. Estimation of the niacin requirements for tilapia fed diets contining glucose or dextrin. J. Nutr. 122:2030-2036.

Shibata, N. 1983. Effect of fishing season on lipid content and composition of Antarctic krill. Bull. Jpn. Soc. Sci. Fish. 49:259-264.

Shibuno, T., I. Chiba, K. Gushima, S. Kakuda & H. Hashimoto. 1993. Reproductive behavior of the wrasse, *Halichoeres marginatus*, at Kuchierabu-jima. Jpn. J. Ichthyol. 40:351-359.

Shields, R.J., J. Davenport, C. Young & P.L. Smith. 1993. Oocyte maturation and ovulation in the Atlantic halibut, *Hippoglossus hippoglossus* (L.), examined using ultrasonography. Aquac. Fish. Mgt. 24:181-186.

Shields, R.J., N.P. Brown & N.R. Bromage. 1997. Blastomere morphology as a predictive measure of fish egg viability. Aquaculture 155:1-12.

Shim, K.F. 1988. Mass production of *Moina* in Singapore using pig waste. World Aquac. 19:59-60.

Shimeno, S. 1991. Yellowtail, *Seriola quinqueradiata*. In: R.P. Wilson (ed.), Handbook of Nutrient Requirements of Finfish. CRC Press, Boca Raton, Florida, p 181-191.

Shimeno, S., H. Hosokawa & M. Takeda. 1979. The importance of carbohydrate in the diet of a carnivorous fish. In: Proc. World Symp. Finfish Nutr. Fishfeed Technol., Hamburg 20-23 June, 1978. Vol. I, Berlin, p 127-143.

Shimeno, S., H. Hosokawa, M. Takeda & H. Kajiyama. 1980. Effects of caloric to protein ratios in formulated diets on the growth, feed conversion and body composition of young yellowtail. Bull. Jpn. Soc. Sci. Fish. 46:1083-1087.

Shimeno, S., H. Hosokawa, M. Takeda, H. Kajiyama & T. Kaisho. 1985. Effect of dietary lipid and carbohydrate on growth, feed conversion and body composition in young yellowtail. Bull. Jpn. Soc. Sci. Fish. 51:1893-1898.

Shimeno, S., T. Mima, O. Yamamoto & Y. Ando. 1993a. Effects of fermented defatted soybean meal in diet on the growth, feed conversion, and body composition of juvenile yellowtail. Nippon Suisan Gakkaishi 59:1883-1888. (in Jpn., Eng. abs.)

Shimeno, S., T. Mima, T. Imanaga & K. Tomaru. 1993b. Inclusion of combination of defatted soybean meal, meat meal, and corn gluten meal to yellowtail diets. Nippon Suisan Gakkaishi 59:1889-1895. (in Jpn., Eng. abs.)

Shimeno, S., T. Mima, H. Kinoshita & S. Kishi. 1994. Inclusion of malt protein flour to diet for fingerling yellowtail. Nippon Suisan Gakkaishi 60:521-525. (in Jpn., Eng. abs.)

Shimeno, S., H. Hosokawa, T. Masumoto, H. Ando, H. Yamanaka & M. Ukawa. 1995a. Effects of defatted soybean meal in diet on growth, feed conversion and body composition in juvenile yellowtail. Bull. Mar. Sci. Fish. Kochi Univ. (15):43-50. (in Jpn., Eng. abs.)

Shimeno, S., H. Hosokawa, T. Matsumoto, H. Ando, H. Yamanaka & M. Ukawa. 1995b. Effects of lipid contents in soybean meal diet on growth, feed conversion and body composition in juvenile yellowtail. Bull. Mar. Sci. Fish. Kochi Univ. (15):51-57. (in Jpn., Eng. abs.)

Shimeno, S., Y. Kanetaka, T. Ruchimat & M. Ukawa. 1995c. Nutritional evaulation of several soy proteins for fingerling yellowtail. Nippon Suisan Gakkaishi 61:919-926. (in Jpn., Eng. abs.)

Shimeno, S., H. Hosokawa, T. Masumoto, T. Ruchimat & S. Kishi. 1996a. Addition of combined defatted soybean meal, malt protein flour, and meat meal to yellowtail diet. Nippon Suisan Gakkaishi 62:243-247. (in Jpn., Eng. abs.)

Shimeno, S., H. Hosokawa & M. Takeda. 1996b. Metabolic response of juvenile yellowtail to dietary carbohydrate to lipid ratios. Fish. Sci. 62:945-949.

Shimeno, S., T. Ruchimat, M. Matsumoto & M. Ukawa. 1997. Inclusion of full-fat soybean meal in diet for fingerling yellowtail. Nippon Suisan Gakkaishi 63:70-76. (in Jpn., Eng. abs.)

Shimizu, A., & S. Kimura. 1991. Acute toxicity of triphenyltin chloride (TPTC) to the saltwater goby, *Chasmichthys dolichognathus* in natural seawater and artificial seawater. Bull. Nat. Res. Inst. Fish. Sci. (2):33-39.

Shimizu, C., A. Ibrahim, T. Tokoro & Y. Shirakawa. 1990. Feeding stimulation in sea bream, *Pagrus major*, fed diets supplemented with Antarctic krill meals. Aquaculture 89:43-53.

Shiobara, Y., & Y. Tanaka. 1994. Development eggs and early larvae of the gobiid fish, *Trimma grammistes*. J. Fac. Mar. Sci. Tech. Tokai Univ. (38):257-264.

Shiobara, Y., K. Suzuki & Y. Tanaka. 1990. Life history of the hairychin goby, *Sagamia geneionema*, from the coast of Suruga Bay. J. Fac. Mar. Sci. Tech., Tokai Univ. 31:25-36.

Shiogaki, M. 1981a. Life history of the gobiid fish *Pterogobius zacalles*. Jpn. J. Ichthyol. 28:70-79. (in Jpn., Eng. abs.)

Shiogaki, M. 1981b. Notes on the life history of the stichaeid fish *Opisthocentrus tenuis*. Jpn. J. Ichthyol. 28:319-328. (in Jpn., Eng. abs.)

Shiogaki, M. 1982. Life history of the stichaeid fish *Opisthocentrus ocellatus*. Jpn. J. Ichthyol. 29:77-85. (in Jpn., Eng. abs.)

Shiogaki, M. 1983. On the life history of the stichaeid fish *Chirolophis japonicus*. Jpn J. Ichthyol. 29:446-455. (in Jpn., Eng. abs.)

Shiogaki, M., & Y. Dotsu. 1971. The life history of the clingfish, *Aspasma minima*. Jpn. J. Ichthyol. 18:76-84. (in Jpn., Eng. abs.)

Shiogaki, M., & Y. Dotsu. 1972a. The life history of the gobiid fish, *Luciogobius elongatus*. Bull. Fac. Fish. Nagasaki Univ. 34:9-18.

Shiogaki, M., & Y. Dotsu. 1972b. The spawning and the larva rearing of the clingfish, *Aspasmichthys ciconiae* and the spawning of the clingfish, *Aspasma minima*. Bull. Fac. Fish. Nagasaki Univ. 34:29-33. (in Jpn., Eng. abs.)

Shiogaki, M., & Y. Dotsu. 1973a. The egg development and larva rearing of the tripterygiid blenny, *Tripterygion etheostoma*. Jpn. J. Ichthyol. 20:42-46. (in Jpn., Eng. abs.)

Shiogaki, M., & Y. Dotsu. 1973b. The spawning behavior of the tripterygiid blenny, *Tripterygion etheostoma*. Jpn. J. Ichthyol. 20:36-41.

Shiranee, P., P. Natarajan & R. Dhevendran. 1993. The role of gut and sediment bacterial flora in the nutrition of cultured pearl spot (*Etroplus suratensis*, Bloch). Bamidgeh 45:45-58.

Shirota, A. 1970. Studies on the mouth size of fish larvae. Bull. Jpn. Soc. Sci. Fish. 36:353-368.

Shokita, S., K. Kakazu, A. Tomori & T. Toma (trans. by M. Yamaguchi). 1991. Aquaculture in Tropical Areas. Midori Shobo, Tokyo, 360 p.

Shotts, E.B., Jr. & J.D. Teska. 1989. Bacterial pathogens of aquatic vertebrates. In: B. Austin & D.A. Austin (eds.), Methods for the Microbiological Examination of Fish and Shellfish. Ellis Horwood, Chichester, England, p 164-186.

Siddiqui, A.Q., A.H. Al-Harbi & A.R. Al-Najada. 1994. Durability, water absorption and water stability of two commercial fish feed pellets and their evaluation as tilapia feeds in Saudi Arabia. Arab Gulf J. Scient. Res. 12:581-594.

Sieswerda, P.L. 1977. Doing what comes naturally. Mar. Aquarist 8(1):18-28.

Sikkel, P.C. 1994. Filial cannibalism in a paternal-caring marine fish: the influence of egg developmental stage and position in the nest. Anim. Beh. 47:1149-1158.

Silva, A. 1994. Spawning of the Chilean flounder *Paralichthys microps* Gunther, 1881 in captivity. J. World Aquac. Soc. 25:342-344.

Simpson, A.C. 1971. Diel spawning behaviour in populations of plaice, dab, sprat, and pilchard. J. Cons. 34:58-64.

Sindermann, C.J. (ed.). 1977. Disease diagnosis and control in North American Marine Aquaculture. Elsevier, New York, 330 p.

Sindermann, C.J. 1990. Principal Diseases of Marine Fish and Shellfish, Volume 1: Diseases of Marine Fish. Academic Press, San Diego, 521 p.

Singh, T. 1991. Malaysia enjoys the seabass boom. INFOFISH Internat. 2/91:45-48.

Sitjà-Bobadilla, A., & P. Alvarez-Pellitero. 1992. Effect of Fumagillin treatment on sea bass *Dicentrarchus labrax* parasitized by *Sphaerospora testicularis* (Myxosporea: Bivalvulida). Dis. Aquat. Org. 14:171-178.

Sitjà-Bobadilla, A., & P. Alvarez-Pellitero. 1993. Population dynamics of *Sphaerospora dicentrarchi* Sitja-Bobadilla et Alvarez-Pellitero, 1992 and *S. testicularis* Sitja-Bobadilla et Alvarez-Pellitero, 1990 (Myxosporea: Bivalvulida) infections in wild and cultured Mediterranean sea bass (*Dicentrarchus labrax* L.). Parasitology 106:39-45.

Siwicki, A.K., D.P. Anderson & G.L. Rumsey. 1994. Dietary intake of immunostimulants by rainbow trout affects non-specific immunity and protection against furunculosis. Vet. Immunol. Immunopathol. 41:125-139.

Skaar, A., & T. Bodvin. 1993. Full-scale production of salmon in floating, enclosed systems. In: H. Reinertsen, L.A. Dahle, L. Jørgensen & K. Tvinnereim (eds.), Fish Farming Technology. Balkema, Rotterdam, p 325-328.

Skaramuca, B.s. 1994. An approach to the construction of a simple and low-cost aeration system for rotifer culture. Aquacultural Eng. 13:153-161.

Skiftesvik, A.B. 1992. Changes in behaviour at onset of exogenous feeding in marine fish larvae. Can. J. Fish. Aquat. Sci. 49:1570-1572.

Skiftesvik, A.B. 1994. Impact of physical environment on the behaviour of cod larvae. ICES Mar. Sci. Symp. 198:646-653.

Skjermo, J., & O. Vadstein. 1993. Characterization of the bacterial flora of mass cultivated *Brachionus plicatilis*. Hydrobiologia 255/256:185-191.

Skjermo, J., T. Defoort, M. Dehasque, T. Espevik, Y. Olsen, G. Skjåk-Bræk, P. Sorgeloos & O. Vadstein. 1995. Immunostimulation of juvenile turbot (*Scophthalmus maximus* L.) using an alginate with high mannuronic acid content administered via the live food organism *Artemia*. Fish Shellfish Immunol. 5:531-534.

Slater, C.H., C.B. Schreck & D.F. Amend. 1995. GnRHa injection accelerates final maturation and ovulation/spermiation of sockeye salmon (*Oncorhynchus nerka*) in both fresh and salt water. Aquaculture 130:279-285.

Smail, D.M., & A.L.S. Munro. 1989. In R.J. Roberts (ed.), Fish Pathology. Bailliere Tindall, London, p 173-241.

Smigielski, A.S. 1975a. Hormonal-induced ovulation of the winter flounder, *Pseudopleuronectes americanus*. Fish. Bull., U.S. 73:431-438.

Smigielski, A.S. 1975b. Hormone-induced spawnings of the summer flounder and rearing of the larvae in the laboratory. Prog. Fish-Cult. 37:3-8.

Smigielski, A.S., & C.R. Arnold. 1972. Separating and incubating winter flounder eggs. Prog. Fish-Cult. 34:113.

Smith, I.P., N.B. Metcalfe, F.A. Huntingford & S. Kadri. 1993. Daily and seasonal patterns in the feeding behaviour of Atlantic salmon (*Salmo salar* L.) in a sea cage. Aquaculture 117:165-178.

Smith, I.R., & R. Pestano-Smith. 1985. Social feasibility of coastal aquaculture. ICLARM Newsl. 8(3):6-8.

Smith, J.M., & D.K. Whitehurst. 1990. Tank spawning methodology for the production of striped bass. In: R.M. Harrell, J.H. Kerby, & R.V. Minton (eds.), Culture and Propagation of Striped Bass and Its Hybrids. Am. Fish. Soc., Bethesda, Maryland, p 73-77.

Smith, K.L., Jr. 1973a. Energy transformations by the sargassum fish, *Histrio histrio* (L.). J. Exp. Mar. Biol. Ecol. 12:219-227.

Smith, L.L., Jr., D.M. Oseid, I.R. Adelman & S.J. Broderius. 1976. Effect of hydrogen sulfide on fish and invertebrates. part I—acute and chronic toxicity studies. USEPA Rep. EPA-600/3-76-062a, 285 p.

Smith, P., M.P. Hiney & O.B. Samuelsen. 1994. Bacterial resistance to antimicrobial agents used in fish farming: a critical evaluation of method and meaning. Ann. Rev. Fish Dis. 4:273-313.

Smith, P.J., & M. Hataya. 1982. Larval rearing and reseeding of red sea bream (*Chrysophrys major*) in Japan. New Zealand Fish. Res. Div., Occ. Pub. No. 39, 19 p.

Smith, R.R. 1989a. Nutritional energetics. In: J.E. Halver (ed.), Fish Nutrition. Academic Press, San Diego, p 1-29.

Smith, R.R., R.A. Winfree, G.W. Rumsey, A. Allred & M. Peterson. 1995. Apparent digestion coefficients and metabolizable energy of feed ingredients for rainbow trout *Oncorhynchus mykiss*. J. World Aquac. Soc. 26:432-437.

Smith, S.J., W.O. Watanabe, J.R. Chan, D.H. Ernst & R.I. Wicklund. 1991. Hatchery production of Florida red tilapia seed in brackishwater tanks: the influence of broodstock age. Aquac. Fish. Mgt. 22:141-147.

Smith, T.I.J. 1973b. The commercial feasibility of rearing pompano, *Trachinotus carolinus* (Linnaeus) in cages. Florida Sea Grant Tech. Bull. No. 26, 62 p.

Smith, T.I.J. 1989b. The culture potential of striped bass and its hybrids. World Aquac. 20(1):32-38.

Smith, T.I.J., & W.E. Jenkins. 1987. Aquaculture research with striped bass (*Morone saxatilis*) and its hybrids in South Carolina. Proc. Ann. Conf. SE Assoc. Fish Wildl. Ag. 39:217-227.

Smith, T.I.J., W.E. Jenkins & R.V. Minton. 1990. Production of advanced fingerling and subadult striped bass and striped bass hybrids in earthen ponds. In: R.M. Harrell, J.H. Kerby & R.V. Minton (eds.), Culture and Propagation of Striped Bass and Its Hybrids. Am. Fish. Soc., Bethesda, Maryland, p 121-139.

Smith, T.I.J., W.E. Jenkins & L.D. Heyward, Sr. 1997. Hatchery production of white bass and sunshine bass using cultured brood stocks. World Aquac. '97 Book of Abs., p 431.

Smith, W.G., & M.P. Fahay. 1970. Description of eggs and larvae of the summer flounder, *Paralichthys dentatus*. U.S. Bur. Sport Fish. Wildl. Res. Rep. No. 75. 21 p.

Smith, W., P. Berrien & T. Potthoff. 1994. Spawning patterns of bluefish, *Pomatomus saltatrix*, in the northeast continental shelf ecosystem. Bull. Mar. Sci. 54:8-16.

Snell, T.W., & K. Carrillo. 1984. Body size variation among strains of the rotifer *Brachionus plicatilis*. Aquaculture 37:359-367.

Snell, T.W., & G. Persoone. 1989. Acute toxicity bioassays using rotifers. I. A test for brackish and marine environments with *Brachionus plicatilis*. Aquat. Toxicol. 14:65-80.

Snell, T.W., M.J. Childress, E.M. Boyer & F.H. Hoff. 1987. Assessing the status of rotifer mass cultures. J. World Aquac. Soc. 18:270-277.

Soderberg, R.W., & J.W. Meade. 1991. The effects of ionic strength on un-ionized ammonia concentration. Prog. Fish-Cult. 53:118-120.

Sogard, S.M., K.W. Able & M.P. Fahay. 1992. Early life history of the tautog *Tautoga onitis* in the Mid-Atlantic Bight. Fish. Bull., U.S. 90:529-539.

Soh, C.L., & T.J. Lam. 1973. Induced breeding and early development of the rabbitfish *Siganus oramin* (Schneider). Malay. Nat. J. 1973:49-56.

Solaas, F., H. Rudi, A. Berg & K. Tvinnereim. 1993. Floating fish farms with bag pens. In: H. Reinertsen, L.A. Dahle, L. Jørgensen & K. Tvinnereim (eds.), Fish Farming Technology. Balkema, Rotterdam, p 317-323.

Solar, I.I., I.J. Baker & E.M. Donaldson. 1987. Effect of salmon gonadotropin and a gonadotropin releasing hormone analogue on ovarian hydration and ovulation in captive sablefish (*Anoplopoma fimbria*). Aquaculture 62:319-325.

Solar, I.I., E. McLean, I.J. Baker, N.M. Sherwood & E.M. Donaldson. 1990. Short communication: induced ovulation of sablefish (*Anoplopoma fimbria*) following oral administration of des Gly^{10}-(D-Ala^{6})LH-RH ethylamide. Fish Physiol. Biochem. 8:497-499.

Solberg, T., & S. Tilseth. 1984. Growth, energy consumption and prey density requirements in first feeding larvae of cod (*Gadus morhua* L.). Flodevigen rapportser. 1:145-166.

Soletchnik, P., E. Thouard & M. Suquet. 1987. Synthese des donnes acquises sur l'elevage de deux poissons tropicaux en Martinique: la sarde queue jaune (*Ocyurus chrysurus*), la carangue aile ronde (*Trachinotus goodei*). IFREMER Doc. Sci. No. 9, 69 p.

Soletchnik, P., E. Thouard, D. Gallet de Saint Aurin, M. Suquet, P. Hurtaud & J.-P. Mesdouze. 1988. Overview of the studies conducted on the finfish, palometa (*Trachinotus goodei*), in Martinique, French West Indies. Proc., First Iberoamerican and Caribbean Congress on Fisheries and Aquaculture, Fundacion La Salle de Ciencias Naturales, 8-14 May 1988, Punta de Piedras, Margarita Island, Venezuela.

Soletchnik, P., M. Suquet, E. Thouard & J.P. Mesdouze. 1989. Spawning of yellowtail snapper (*Ocyurus chrysurus* Bloch 1791) in captivity. Aquaculture 77:287-289.

Soletchnik, P., E. Thouard & E. Goyard. 1990. Intensive larval rearing trials of red drum (*Sciaenops ocellata*) in Martinique (F.W.I.). Advances in Tropical Aquaculture. AQUACOP, IFREMER, Actes Colloq. (9):661-675.

Soliman, A.K., K. Jauncey & R.J. Roberts. 1987. Stability of L-ascorbic acid (vitamin C) and its forms in fish feeds during processing, storage and leaching. Aquaculture 60:73-83.

Solomon, D.J., & A.E. Brafield, 1972. The energetics of feeding, metabolism and growth of perch (*Perca fluviatilis* L.). J. Anim. Ecol. 41:699-718.

Sorensen, E.M.B. 1991. Metal Poisoning in Fish. CRC Press, Boca Raton, Florida, 374 p.

Sorgeloos, P., & P. Léger. 1992. Improved larviculture outputs of marine fish, shrimp and prawn. J. World Aquac. Soc. 23:251-264.

Sorgeloos, P., E. Bossuyt, P. Lavens, P. Léger, P. Vanhaecke & D. Versichele. 1983. The use of brine shrimp *Artemia* in crustacean hatcheries and nurseries. In: J.P. McVey (ed.), CRC Handbook of Mariculture, Vol. 1: Crustacean Aquaculture, CRC Press, Boca Raton, Florida, p 71-96.

Sorgeloos, P., P. Lavens, P.H. Léger, W. Tackaert & D. Versichele. 1986. Manual for the Culture and Use of Brine Shrimp *Artemia* in Aquaculture. Artemia Ref. Ctr., State Univ. Ghent, Belgium, 319 p.

Sorgeloos, P., P. Léger & P. Lavens. 1988. Improved larval rearing of European and Asian seabass, seabream, mahi-mahi, siganid and milkfish using enrichment diets for *Brachionus* and *Artemia*. World Aquac. 19:78-79.

Sorgeloos, P., P. Lavens, P. Leger & W. Tackaert. 1993. The use of *Artemia* in marine fish larviculture. In: C.-S. Lee, M.-S. Su & I C. Liao (eds.), Finfish Hatchery in Asia. Tungkang Mar. Lab. Conf. Proc. (3):73-86.

Southgate, P.C., & P.S. Lee. 1993. Notes on the use of microbound artificial diets for larval rearing of sea bass (*Lates calcarifer*). Asian Fish. Sci. 6:245-247.

Sövényi, J.F. 1992. Effect of fumagillin feeding on the haematopoietic organs and susceptibility of common carp, *Cyprinus carpio* L., to erythrodermatitis. J. Fish. Dis. 15:255-260.

Sower, S.A., C.B. Schreck & E.M. Donaldson. 1982. Hormone-induced ovulation of coho salmon (*Oncorhynchus kisutch*) held in seawater and fresh water. Can. J. Fish. Aquat. Sci. 39:627-632.

Sower, S.A., R.N. Iwamoto, W.W. Dickhoff & A. Gorbman. 1984. Ovulatory and steroidal responses in coho salmon and steelhead trout following administration of salmon gonadotropin and D-Ala^{6},des Gly^{10} gonadotropin-releasing hormone ethylamide (GnRHa). Aquaculture 43:35-46.

Spannhof, L., & H. Plantikow. 1983. Studies on carbohydrate digestion in rainbow trout. Aquaculture 30:95-108.

Spedicato, M.T., G. Giorgetti, G. Lembo, G. Bertoja, G. Bovo & G. Ceschia. 1993a. Introduction of fertilized eggs of *Paralichthys olivaceus* (Temminch e Schlegel) from Japan: quarantine techniques and preliminary results of rearing. Rivista Ital. Acquac. 28:25-37.

Spedicato, M.T., G. Lembo & P. Trotta. 1993b. Larval and post-larval rearing of *Paralichthys olivaceus*

(Temminck and Schlegel): the first experience in Italy. Eur. Aquac. Soc. Spec. Pub. (19):476.

Spedicato, M.T., G. Lembo & P. Trotta. 1994. Larval and post-larval rearing of *Paralichthys olivaceus* (Temminck and Schlegel): the first experience in Italy. Eur. Aquac. Soc. Spec. Pub. (22):292-301.

Spehar, R.L. 1976. Cadmium and zinc toxicity to flagfish, *Jordanella floridae*. J. Fish. Res. Bd. Can. 33:1939-1945.

Spotte, S. 1992. Captive Seawater Fishes. Wiley, New York, 942 p.

Sprengel, G., & H. Lüchtenberg. 1991. Infection by endoparasites reduces maximum swimming speed of European smelt *Osmerus eperlanus* and European eel *Anguilla anguilla*. Dis. Aquat. Org. 11:31-35.

Spyridakis, P., R. Metailler & J. Gabaudan. 1989. Studies on nutrient digestibility in European sea bass (*Dicentrarchus labrax*) 2. Effect of sodium alginate on protein and lipid digestibility. Aquaculture 77:71-73.

Squires, K. 1971. Breeding clown fish. Mar. Aquarist 2(2):30-33.

St.-Félix, C., & F. Blouin. 1997. Identification of new species in tropical marine fish farming: first data on culture of the *Elagatis bipinnulatus* (Quoy and Gaimard, 1825). Island Aquac. Trop. Aquac. Eur. Aquac. Soc., Ann. Mtg., p 272-274

Stahl, C.J., D.M. Gatlin & W.H. Neill. 1997. The effect of dissolved oxygen on the growth performance of juvenile red drum, *Sciaenops ocellatus*. World Aquac. '97 Book of Abs., p 438.

Staples, D.J., & M. Nomura. 1976. Influence of body size and food ration on the energy budget of rainbow trout *Salmo gairdneri* Richardson. J. Fish Biol. 9:29-43.

Starr, R.C., & J.A. Zeikus. 1993. UTEX—the culture collection of algae at the University of Texas at Austin 1993 list of cultures. Phycology 29(2)suppl.:1-106.

Stebbing, A.R.D., S. Soria, G.R. Burt & J.J. Cleary. 1990. Water quality bioassays in two Bermudan harbours using the ciliate *Euplotes vannus*, in relation to tributyltin distribution. J. Exp. Mar. Biol. Ecol. 138:159-166.

Steele, C.W. 1983. Acute toxicity of copper to sea catfish. Mar. Poll. Bull. 14:168-170.

Steene, R.C. 1978. Butterfly and Angelfishes of the World, Vol. 1, Australia. Wiley, New York, 144 p.

Stefanussen, D., Ø. Lie, E. Moksness & K.I. Ugland. 1993. Growth of juvenile common wolffish (*Anarhichas lupus*) fed practical fish feeds. Aquaculture 114:103-111.

Steffens, W. 1989. Principles of Fish Nutrition. Ellis Horwood, Chichester, England, 384 p.

Stein, J.R. (ed.). 1973. Handbook of Phycological Methods. Culture Methods and Growth Measurements. Cambridge Univ. Press, New York, 448 p.

Stephan, G., J.L. Messager, F. Lamour & F.B. Laurencin. 1993. Interactions between dietary alpha-tocopherol and oxidized oil on sea bass *Dicentrarchus labrax*. Fish Nutrition in Practice, INRA, Paris, p 215-218.

Stephenson, R.L. 1990. Multiuse conflict: aquaculture collides with traditional fisheries in Canada's Bay of Fundy. World Aquac. 21(2):34-45.

Steslow, F.A. 1991. Ozone kinetics in seawater. Freshw. Mar. Aquarium 14:154-158.

Stevens, E.G., A.C. Matarese & W. Watson. 1984. Ammodytoidei: development and relationships. Am. Soc. Ichthyol. Herpetol. Spec. Pub. (1):574-575.

Stevens, R.E. 1966. Hormone-induced spawning of striped bass for reservoir stocking. Prog. Fish-Cult. 28:19-28.

Stewart, J.E. 1994. Aquaculture in Atlantic Canada and the research requirements related to environmental interactions with finfish culture. Fisken Havet 13:1-18.

Stickney, R.R. 1979. Principles of Warmwater Aquaculture, Wiley, New York, 375 p.

Stickney, R.R. (ed.) 1986. Culture of Nonsalmonid Freshwater Fishes. CRC Press, Boca Raton, Florida, 201 p.

Stickney, R.R. 1988. Aquaculture on trial. World Aquac. 19(3):16-18.

Stickney, R.R. (ed.). 1991a. Culture of Salmonid Fishes. CRC Press, Boca Raton, Florida, 189 p.

Stickney, R.R. 1991b. Growout of Pacific salmon in net-pens. In: R.R. Stickney (ed.), Culture of Salmonid Fishes. CRC Press, Boca Raton, Florida, p 71-83.

Stickney, R.R. 1991c. Salmonid life histories. In: R.R. Stickney (ed.), Culture of Salmonid Fishes. CRC Press, Boca Raton, Florida, p 1-20.

Stober, Q.J., & C.H. Hanson. 1974. Toxicity of chlorine and heat to pink (*Oncorhynchus gorbuscha*) and chinook salmon (*O. tshawytscha*). Trans. Am Fish Soc. 103:569-576.

Stoeckel, J.N., & R.J. Neves. 1992. Comparison of methods for viewing the germinal vesicle in fish oocytes. Prog. Fish-Cult. 54:115-118.

Storebakken, T., & E. Austreng. 1987. Binders in fish feeds. II. Effect of different alginates on the digestibility of macronutrients in rainbow trout. Aquaculture 60:121-131.

Stoskopf, M.K. (ed.). 1993. Fish Medicine. Saunders, Philadelphia, 882 p.

Stoss, J., & E.M. Donaldson. 1982. Preservation of fish gametes. In: C.J.J. Richter & M.J.T. Goos (compilers). Proc. Internat. Symp. Repr. Physiol. Fish. Wageningen, Netherlands, p 114-122.

Stoss, J., & T. Refstie. 1983. Short-term storage and cryopreservation of milt from Atlantic salmon and sea trout. Aquaculture 30:229-236.

Stoss, J.H., & J.E. Røer. 1993. Controlled reproduction in turbot (*Scophthalmus maximus*). In: H. Reinertsen, L.A. Dahle, L. Jørgensen & K. Tvinnereim (eds.), Fish Farming Technology. Balkema, Rotterdam, p 29-34.

Støttrup, J.G. 1993. First feeding in marine fish larvae: nutritional and environmental aspects. In: B.T. Walther & H.J. Fyhn (eds.), Physiological and Biochemical Aspects of Fish Development, Univ. Bergen, Norway, 123-131.

Støttrup, J.G., & Y. Attramadal. 1992. The influence of different rotifer and *Artemia* enrichment diets on growth, survival and pigmentation in turbot (*Scophthalmus maximus* L.) larvae. J. World Aquac. Soc. 23:307-316.

Støttrup, J.G., & N.H. Norsker. 1997. Production and use of copepods in marine fish larviculture. Aquaculture 155:231-247.

Støttrup, J.G., K. Richardson, E. Kirkegaard & N.J. Pihl. 1986. The cultivation of *Acartia tonsa* Dana for use as a live food source for marine fish larvae. Aquaculture 52:87-96.

Støttrup, J.G., K. Gravningen & N.H. Norsker. 1995. The role of different algae in the growth and survival of turbot larvae (*Scophthalmus maximus* L.) in intensive rearing systems. ICES Mar. Sci. Symp. 201:173-186.

Stradmeyer, L., N.B. Metcalfe & J.E. Thorpe. 1988. Effect of food pellet shape and texture on the feeding response of juvenile Atlantic salmon. Aquaculture 73:217-228.

Strand, H.K., T.K. Hansen, A. Pedersen, I.B. Falk-Petersen & V. Øiestad. 1995. First feeding of common wolffish on formulated dry feed diets in a low water-level raceway system. Aquac. Internat. 3:1-10.

Straughan, R.P.L. 1956. Salt water fishes are spawned. The Aquarium 25:157-159.

Straughan, R.P.L. 1959a. Salt water black angelfish spawned. The Aquarium 28(7):211-212.

Straughan, R.P.L. 1959b. Salt water black angelfish spawned. Aquarium J. 30(9):338,340.

Straughan, R.P.L. 1968. First spawning of high hat. Salt Water Aquarium 4:73-74.

Straughan, R.P.L. 1973. The Marine Collector's Guide. Barnes, New York, 222 p.

Strawn, K., P.W. Perschbacher, R. Nailon & G. Chamberlain. 1986. Raising mudminnows. Texas A&M Sea Grant TAMU-SG-86-506, 8 p.

Stroud, G.J., B. Gldman & W. Gladstone. 1989. Larval development, growth and age determination in the sharpnose pufferfish *Canthigaster valentini* (Teleostei: Tetraodontidae). Jpn. J. Ichthyol. 36:327-337.

Strøm, E., & E. Ringo. 1993. Changes in the bacterial composition of early developing cod, *Gadus morhua* (L.) larvae following inoculation of *Lactobacillus plantarum* into the water. In: B.T. Walther & H.J. Fyhn (eds.), Physiological and Biochemical Aspects of Fish Development. Univ. Bergen, Norway, p 226-228.

Struffenegger, P. 1992. Sturgeon farming in California: a promising new industry. Aquac. Eur. 17(1):6-9.

Sturmbauer, C., & R. Hofer. 1985. Can amylase inhibitors from wheat reduce the digestibility of starch and the growth rate in fish. In: C.B. Cowey, A.M. Mackie & J.G. Bell (eds.), Nutrition and Feeding in Fish. Academic Press, London, p 265-267.

Sturmer, L.N. 1990. Zooplankton composition and dynamics in fingerling red drum rearing ponds. Texas A&M Sea Grant TAMU-SG-90-603, p 80-90.

Su, H.-M., M.-S. Su & I C. Liao. 1994. Fatty acid composition of the rotifer *Brachionus plicatilis* fed selected microalgae or yeast, alone or in combination with various oils. Third Asian Fisheries Forum, Asian Fish. Soc., Manila, p 207-210.

Subasinghe, R.P., & C. Sommerville. 1992. Effects of temperature on hatchability, development and growth of eggs and yolksac fry of *Oreochromis mossambicus* (Peters) under artificial incubation. Aquac. Fish. Mgt. 23:31-39.

Subasinghe, R.P., J.R. Arthur & M. Shariff (eds.). 1996. Health Management in Asian Aquaculture. FAO Fish. Tech. Pap. 360, 142 p.

Subiyanto, I. Hirata & T. Senta. 1992. The occurrence of pelagic larvae of the Japanese flounder in the Yatsushiro Sea and adjacent waters, Japan. Bull. Jpn. Soc. Sci. Fish. 58:229-234.

Sugita, H., T. Asai, K. Amanuma, T. Mitsuya, K. Hayashi, C. Maruyama & Y. Deguchi. 1994. Determination of total residual oxidants and survival of fish pathogenic bacteria in the ozonated seawater. Third Asian Fisheries Forum, Asian Fish. Soc., Manila, p 379-382.

Sugiura, S.H., F.M. Dong & R.W. Hardy. 1998. Effects of dietary supplements on the availability of minerals in fish meal: preliminary observations. Aquaculture 160:283-303.

Sukenik, A., O. Zmora & Y. Carmeli. 1993. Biochemical quality of marine unicellular algae with special emphasis on lipid composition. II. *Nannochloropsis* sp. Aquaculture 117:313-326.

Sukumasavin, N., W. Leelapatra, E. McLean & E.M. Donaldson. 1992. Orally induced spawning of Thai carp (*Puntius gonionotus*, Bleeker) following co-administration of des Gly^{10}(D-Arg^{6})sGnRH ethylamide and domperidone. J. Fish Biol. 40:477-479.

Sulaiman, A.H. 1993. Acute toxicity relationships for two species of fish using a simultaneous testing method. Sci. Total Env. Suppl. 1993:1001-1009.

Sullivan, J.A., & R.C. Reigh. 1995. Apparent digestibility of selected feedstuffs in diets for hybrid striped bass (*Morone saxatilis* ♀ × *Morone chrysops* ♂). Aquaculture 138:313-322.

Sumagaysay, N.S. 1993. Growth, daily ration, and gastric evacuation rates of milkfish (*Chanos chanos*) fed supplemental diet and natural food. J. Appl. Ichthyol. 9:65-73.

Sumagaysay, N.S., Y.N. Chiu-Chern, V.J. Estilo & M.A.S. Sastrillo. 1990. Increasing milkfish (*Chanos chanos*) yields in brackishwater ponds through increased stocking rates and supplementary feeding. Asian Fish. Sci. 3:251-256.

Sumpter, J.P. 1993. The deleterious effects of stress and their significance to aquaculture. Eur. Aquac. Soc. Spec. Pub. (18):157-165.

Sumpter, J.P., J. Carragher, T.G. Pottingher & A.D. Pickering. 1987. The interaction of stress and reproduction in trout. Proc. Third Internat. Symp. Repr. Physiol. Fish, Memorial Univ. Newfoundland, St. John's, p 299-302.

Sun, B., & J.W. Fleeger. 1995. Sustained mass culture of *Amphiascoides atopus* a marine harpacticoid copepod in a recirculating system. Aquaculture 136:313-321.

Sun, G., & Z. Liu. 1992. Research on the applicable technique of natural zooplankton on artificial breeding of red porgy *Pagrosomus major* (T and S). Mar. Sci. 4:296-300.

Sun, M., Y. Zhang & J. Cai. 1991. Preliminary study of breeding and biology of *Daphnia magna* HB, a high-temperature-resistant species. Acta Hydrobiol. Sin. 15:166-173.

Sunde, L.M., A.K. Imsland, A. Folkvord & S.O. Stefansson. 1998. Effects of size grading on growth and survival of juvenile turbot at two temperatures. Aquac. Internat. 6:19-32.

Supriatna, A., & H. Kohno. 1990. A larval rearing trial of the grouper, *Epinephelus fuscoguttatus*. Bull. Pen. Perikanan, Spec. Ed. (1):37-43.

Suquet, M., M.H. Omnes, Y. Normant & C. Fauvel. 1992. Influence of photoperiod, frequency of stripping and presence of females on sperm output in turbot, *Scophthalmus maximus* (L.). Aquac. Fish. Mgt. 23:217-225.

Suresh, A.V., & C.K. Lin. 1992. Tilapia culture in saline waters: a review. Aquaculture 106:201-226.

Sutherland, R. 1997. Review of the economics of potential systems for farmed production of Atlantic halibut. Aquac. Eur. 21(4):6-11.

Suzuki, K., & S. Hioki. 1978. Spawning behavior, eggs, and larvae of the sea bream, *Gymnocranius griseus*, in an aquarium. Jpn. J. Ichthyol. 24:271-277. (in Jpn., Eng. abs.)

Suzuki, K., & S. Hioki. 1979. Spawning behavior, eggs, and larvae of the lutjanid fish, *Lutjanus kasmira*, in an aquarium. Jpn. J. Ichthyol. 26:161-166.

Suzuki, K., K. Kobayashi, S. Hioki & T. Sakamoto. 1978. Ecological studies of the anthiine fish, *Franzia squamipinnis*, in Suruga Bay, Japan. Jpn. J. Ichthyol. 25:124-140.

Suzuki, K., S. Hioki, Y. Tanaka & K. Iwasa. 1979. Spawning behavior, eggs, larvae, and sex reversal of two pomacanthine fish, *Genicanthus lamarck* and *G. semifasciatus*, in the aquarium. J. Fac. Mar. Sci. Tech. Tokai Univ. 12:149-165. (in Jpn., Eng. abs.)

Suzuki, K., Y. Tanaka & S. Hioki. 1980. Spawning behavior, eggs, and larvae of the butterflyfish, *Chaetodon nippon*, in an aquarium. Jpn. J. Ichthyol. 26:334-341.

Suzuki, Y., M. Kobayashi, O. Nakamura, K. Aida & I. Hanyu. 1988. Induced ovulation of the goldfish by oral administration of salmon pituitary extract. Aquaculture 74:379-384.

Suzuki, Z. 1994. A review of the biology and fisheries for yellowfin tuna (*Thunnus albacares*) in the western and central Pacific Ocean. FAO Fish. Technical Pap. (336):108-137.

Svendsen, Y.S., & T. Haug. 1991. Effectiveness of formalin, benzocaine, and hypo- and hypersaline exposures against adults and eggs of *Entobdella hippoglossi* (Müller), an ectoparasite on Atlantic halibut (*Hippoglossus hippoglossus* L.). Laboratory studies. Aquaculture 94:279-289.

Swann, D.L., J.R. Riepe, J.D. Stanley, M.E. Griffin & P.B. Brown. 1994. Cage culture of hybrid striped bass in Indiana and evaluation of diets containing three levels of dietary protein. J. World Aquac. Soc. 25: 281-288.

Swanson, P., W.W. Dickhoff, D. Larsen, C.C. Mylonas & Y. Zohar. 1997. Induction of ovulation and spermiation in Pacific salmon captive broodstock using gonadotropin-releasing hormone agonist in sustained-delivery devices. World Aquac. '97 Book of Abs., p 453.

Swedmark, M., & A. Granmo. 1981. Effects of mixtures of heavy metals and a surfactant on the development of cod (*Gadus morhua* L.). Rapp. P.-v. Réun. Cons. Int. Explor. Mer 178:95-103.

Sweetman, J.W. 1992. Larviculture of Mediterranean marine fish species: current status and future trends. J. World Aquac. Soc. 23:330-337.

Sylvester, J.R., & C.E. Nash. 1975. Thermal tolerance of eggs and larvae of Hawaiian striped mullet, *Mugil cephalus* L. Trans. Am. Fish. Soc. 104:144-147.

Sylvester, J.R., C.E. Nash & C.R. Emberson. 1975. Salinity and oxygen tolerances of eggs and larvae of Hawaiian striped mullet, *Mugil cephalus* L. J. Fish Biol. 7:621-629.

Szedlmayer, S.T. 1987. Hormone-induced spawning of weakfish. Prog. Fish-Cult. 49:158-160.

Szyper, J.P. 1989. Nutritional depletion of the aquaculture feed organisms *Euterpina acutifrons*, *Artemia* sp. and *Brachionus plicatilis* during starvation. J. World Aquac. Soc. 20:162-169.

Szyper, J.P. 1991. Culture of mahimahi: review of life stages. In: J.P. McVey (ed.), Handbook of Mariculture, Vol. II, Finfish Aquaculture. CRC Press, Boca Raton, Florida, p 228-240.

Szyper, J.P., & H. Ako. 1990. Feed formulas producing improved palatability and growth with juvenile mahimahi, *Coryphaena hippurus*. World Aquac. 21(4):104-105.

Szyper, J.P., M.J. Anderson & N.H. Richman. 1991. Preliminary aquaculture evaluation of moi (*Polydactylus sexfilis*). Prog. Fish-Cult. 53:20-25.

Tabachek, J.-A.L. 1988. The effect of feed particle size on the growth and feed efficiency of Arctic charr [*Salvelinus alpinus* (L.)]. Aquaculture 71:319-330.

Tabachek, J.L., & B.G.E. de March. 1991. Research Priorities for the Culture of Arctic Charr, *Salvelinus alpinus* (L.), in Atlantic Canada. Can. Ind. Rep. Fish. Aquat. Sci. No. 211, 15 p.

Tabachek, J.L., M.J. Foster, C.E. Engel & R.N. Olson. 1993. A system for the incubation of separate groups of salmonid eggs. Prog. Fish-Cult. 55:101-105.

Tabaru, K., S. Ishibashi, A. Tanaka & K. Ookubo. 1984. *Pagrus major* larvae production. Ann. Rep. Miyazaki Pref. Aquac. Ctr., Miyazaki, Japan, p 9-25. (in Jpn.)

Tabeta, O., K. Tanaka, J. Yamada & W.-N. Tzeng. 1987. Aspects of the early life history of the Japanese eel *Anguilla japonica* determined from otolith microstructure. Nippon Suisan Gakkaishi 53:1727-1734.

Tachihara, K., & S. Kimura. 1991. Embryonic development and morphological changes with growth of the larval and juvenile ayu *Plecoglossus altivelis altivelis* of Lake Ikeda in southern Kyushu. Nippon Suisan Gakkaishi 57:789-795.

Tachihara, K., R. Ebisu & Y. Tukashima. 1993. Spawning, eggs, larvae and juveniles of the purplish amberjack *Seriola dumerili*. Bull. Jpn. Soc. Sci. Fish. 59:1479-1488. (in Jpn., Eng. abs.)

Tachihara, K., M.K. El-Zibdeh, A. Ishimatsu & M. Tagawa. 1997. Improved seed production of goldstriped amberjack *Seriola lalandi* under hatchery conditions by injection of triiodothyronine (T_3) to broodstock fish. J. World Aquac. Soc. 28:34-44.

Tacon, A.G.J. 1988. The nutrition and feeding of farmed fish and shrimp—a training manual. 3. Feeding methods. FAO, Brasilia, Brazil, GCP/RLA/075/ITA Field Doc. 7/E, 208 p.

Tacon, A.G.J. 1992. Nutritional Fish Pathology: Morphological Signs of Nutrient Deficiency and Toxicity in Farmed Fish. FAO Fish. Tech. Pap. No. 330, 75 p.

Tacon, A.G.J. 1995. Application of nutrient requirement data under practical conditions: special problems of intensive and semi-intensive fish farming systems. J. Appl. Ichthyol. 11:205-214.

Tacon, Albert G.J. 1998. Food and Agriculture Organization of the United Nations, pers. comm.

Tacon, A.G.J., & A.J. Jackson. 1985. Utilisation of conventional and unconventional protein sources in practical fish feeds. In: C.B. Cowey, A.M. Mackie & J.G. Bell (eds.) Nutrition and Feeding in Fish. Academic Press, San Diego, p 119-145.

Tacon, A.G.J., N. Rausin, M. Kadari & P. Cornelis. 1991a. The food and feeding of tropical marine fishes in floating net cages: Asian seabass, *Lates calcarifer* (Bloch), and brown-spotted grouper, *Epinephelus tauvina* (Forskal). Aquac. Fish. Mgt. 22:165-182.

Tacon, A.G.J., N. Rausin, M. Kadari & P. Cornelis. 1991b. Use of night-lights to attract food organisms into tropical marine fish cages. Aquac. Fish. Mgt. 22:183-192.

Tait, J.S. 1960. The first filling of the swim bladder in salmonoids. Can. J. Zool. 38:179-187.

Takahashi, K. 1993. Sea ranching the black rockfish, *Sebastes schlegeli*, in Japan. In: Marine Fish Culture and Enhancement, Conf. Proc., Seattle, Washington, October 4-6, 1993. Washington Sea Grant, Univ. Washington, Seattle, p 35-36.

Takaoka, O., K. Takii, M. Nakamura, H. Kumai & M. Takeda. 1995. Identification of feeding stimulants for tiger puffer. Fish. Sci. 61:833-836.

Takatsu, T., T. Nakatani, T. Mutoh & T. Takahashi. 1995. Feeding habits of Pacific cod larvae and juveniles in Mutsu Bay, Japan. Fish. Sci. 61:415-422.

Takeda, M., S. Shimeno, H. Hosokawa, H. Kajiyama & T. Kaisyo. 1975. The effect of dietary calorie-to-protein ratio on the growth, feed conversion, and body composition of young yellowtail. Bull. Jpn. Soc. Sci. Fish. 41:443-447.

Takeshita, G.Y. 1976. An angel hybrid. Mar. Aquarist 7(1):27-35.

Takeuchi, T., S. Arai, T. Watanabe & Y. Shiina. 1980. Requirement of eel *Anguilla japonica* for essential fatty acids. Bull. Jpn. Soc. Sci. Fish. 46:345-353.

Takeuchi, T., M. Toyota, S. Satoh & T. Watanabe. 1990. Requirement of juvenile red seabream *Pagrus major* for eicosapentaenoic and docosahexaenoic acids. Nippon Suisan Gakkaishi 56:1263-1269.

Takeuchi, T., Y. Shiina & T. Watanabe. 1991. Suitable protein and lipid levels in diet for fingerlings of red sea bream *Pagrus major*. Nippon Suisan Gakkaishi 57:293-299.

Takeuchi, T., M. Toyota & T. Watanabe. 1992a. Comparison of lipid and n-3 highly unsaturated fatty acid incorporations between *Artemia* enriched with various types of oil by direct method. Nippon Suisan Gakkaishi 58:277-281.

Takeuchi, T., M. Toyota & T. Watanabe. 1992b. Dietary value of *Artemia* enriched with various types of oil for larval striped knifejaw and red sea bream. Nippon Suisan Gakkaishi 58:283-289.

Takeuchi, T., Y. Shiina & T. Watanabe. 1992c. Suitable levels of n-3 highly unsaturated fatty acids in diet for fingerlings of red sea bream. Nippon Suisan Gakkaishi 58:509-514.

Takeuchi, T., Y. Shiina, T. Watanabe, S. Sekiya & K. Imaizumi. 1992d. Suitable levels of n-3 highly unsaturated fatty acids in diet for fingerlings of yellowtail. Nippon Suisan Gakkaishi 58:1341-1346.

Takeuchi, T., Y. Shiina, T. Watanabe, S. Sekiya & K. Imaizumi. 1992e. Suitable protein and lipid levels in diet for fingerlings of yellowtail. Nippon Suisan Gakkaishi 58:1333-1339.

Takeuchi, T., R. Masuda, Y. Ishizaki, T. Watanabe, M. Kanematsu, K. Imaizumi & K. Tsukamoto. 1996. Determination of the requirement of larval striped jack for eicosapentaenoic acid and docosahexaenoic acid using enriched *Artemia* nauplii. Fish. Sci. 62:760-765.

Takeyama, H., K. Iwamoto,, S. Hata, H. Takano & T. Matsunaga. 1996. DHA enrichment of rotifers: a simple two-step culture using the unicellular algae *Chlorella regularis* and *Isochrysis galbana*. J. Mar. Biotechnol. 3:244-247.

Taki, Y., H. Kohno & S. Hara. 1987. Morphological aspects of the development of swimming and feeding functions in the milkfish *Chanos chanos*. Jpn. J. Ichthyol. 34:198-208.

Takii, K., M. Ukawa, M. Nakamura & H. Kumai. 1995a. Suitable lipid level in brown fish meal diet for tiger puffer. Fish. Sci. 61:841-844.
Takii, K., M. Ukawa, M. Nakamura & H. Kumai. 1995b. Suitable sugar level in brown fish meal diet for tiger puffer. Fish. Sci. 61:837-840.
Takii, K., S. Miyashita, M. Seoka, Y. Tanaka, Y. Kubo & H. Kumai. 1997a. Changes in chemical contents and enzyme activities during embryonic development of bluefin tuna. Fish. Sci. 63:1014-1018.
Takii, K., K. Konishi, M. Ukawa, M. Nakamura & H. Kumai. 1997b. Influence of feeding rates on digestion and energy flow in tiger puffer and red sea bream. Fish. Sci. 63:355-360.
Takita, T. 1980. Embryonic development and larvae of three dragonets. Bull. Jpn. Soc. Sci. Fish. 46:1-7.
Takita, T. 1983. Embryonic and larval development of the callionymid fish, *Callionymus calliste*. Jpn. J. Ichthyol. 29:441-445.
Takita, T., & K. Nakamura. 1986. Embryonic development and prelarva of the atherinid fish, *Hypoatherina bleekeri*. Jpn. J. Ichthyol. 33:57-61.
Takita, T., & E. Okamoto. 1979. Spawning behavior of the two dragonets, *Callionymus flagris* and *C. richardsoni*, in the aquarium. Jpn. J. Ichthyol. 26:282-288. (in Jpn., Eng. abs.)
Takita, T., T. Iwamoto, S. Kai & I. Sogabe. 1983. Maturation and spawning of the dragonet, *Callionymus enneactis*, in an aquarium. Jpn. J. Ichthyol. 30:221-226.
Talbot, B., & S. Battaglene. 1991. Brine shrimp in aquaculture. New South Wales Agric. Fish., Agdex Advisory Note No. 9/90, 4 p.
Talbot, B., I. Smith & J. Piddington. 1990. Mass production of the rotifer *Brachionus plicatilis*. New South Wales Agric. Fish., Agdex Advisory Note No. 2/90, 4 p.
Talbot, C. 1993. Some biological and physical constraints to the design of feeding regimes for salmonids in intensive cultivation. In: H. Reinertsen, L.A. Dahle, L. Jørgensen & K. Tvinnereim (eds.), Fish Farming Technology. Balkema, Rotterdam, p 19-26.
Talbot, C., & R. Hole. 1994. Fish diets and the control of eutrophication resulting from aquaculture. J. Appl. Ichthyol. 10:258-270.
Tamaru, C.S., C.-S. Lee, C.D. Kelley, J.E. Banno, P.Y. Ha, K. Aida & I. Hanyu. 1988. Characterizing the stage of maturity most receptive to an acute LHRH-analogue therapy for inducing milkfish (*Chanos chanos*) to spawn. Aquaculture 74:147-163.
Tamaru, C.S., C.D. Kelley, C.-S. Lee, K. Aida, I. Hanyu & F. Goetz. 1991a. Steroid profiles during maturation and induced spawning of the striped mullet, *Mugil cephalus* L. Aquaculture 95:149-168.
Tamaru, C.S., C.-S. Lee & H. Ako. 1991b. Improving the larval rearing of striped mullet, *Mugil cephalus*, by manipulating quantity and quality of the rotifer, *Brachionus plicatilis*. In: W. Fulks & K. Main (eds.), Rotifer and Microalgae Culture Systems, Proc. of a US-Asia Workshop. Oceanic Inst., Honolulu, p 89-103.
Tamaru, C.S., H. Ako & C.-S. Lee. 1992. Fatty acid and amino acid profiles of spawned eggs of striped mullet, *Mugil cephalus* L. Aquaculture 105:83-94.
Tamaru, C.S., P. Bass, R. Murashige & C.-S. Lee. 1993a. Environmental parameters and their role in the larval rearing of striped mullet (*Mugil cephalus*). In: C.-S. Lee, M.-S. Su & I C. Liao (eds.), Finfish Hatchery in Asia. Tungkang Mar. Lab. Conf. Proc. (3):191-202.
Tamaru, C.S., H. Ako, C.-S. Lee & P. Bass. 1993b. Fatty acid composition of rotifers cultured on various combinations of phytoplankton and baker's yeast and its effect on the growth and survival of larval milkfish (*Chanos chanos*). In: C.-S. Lee, M.-S. Su & I C. Liao (eds.), Finfish Hatchery in Asia. Tungkang Mar. Lab. Conf. Proc. (3):97-108.
Tamaru, C.S., R. Murashige, C.-S. Lee, H. Ako & V. Sato. 1993c. Rotifers fed various diets of baker's yeast and/or *Nannochloropsis oculata* and their effect on the growth and survival of striped mullet (*Mugil cephalus*) and milkfish (*Chanos chanos*) larvae. Aquaculture 110:361-372.
Tamaru, C.S., R. Murashige & C.-S. Lee. 1994. The paradox of using background phytoplankton during the larval culture of striped mullet, *Mugil cephalus* L. Aquaculture 119:167-174.
Tamaru, C.S., F. Cholik, J.C.-M. Kuo & W.J. Fitzgerald, Jr. 1995. Status of the culture of milkfish (*Chanos chanos*), striped mullet (*Mugil cephalus*), and grouper (*Epinephelus* sp.). Rev. Fish. Sci. 3:249-273.
Tamaru, C.S., C. Carlstrom-Trick, W.J. FitzGerald, Jr. & H. Ako. 1996. Induced final maturation and spawning of the marbled grouper *Epinephelus microdon* captured from spawning aggregations in the Republic of Palau, Micronesia. J. World Aquac. Soc. 27:363-372.
Tamse, C.T., & R.Q. Gacutan. 1994. Acute toxicity of nifurpirinol, a fish chemotherapeutant, to milkfish (*Chanos chanos*) fingerlings. Bull. Env. Contam. Toxicol. 52:346-350.
Tan, O.K.K., & T.J. Lam. 1973. Induced breeding and early development of the marble goby (*Oxyeleotris marmorata*, Blk.) Aquaculture 2:411-423.
Tanaka, A., & S. Fujiwara. 1983. Larvae culture of *Pagrus major*. Ann. Rep. Miyazaki Pref. Aquac. Ctr., Miyazaki, Japan, p 9-32. (in Jpn.)
Tanaka, H., H. Kagawa, H. Ohta, K. Okuzawa & K. Hirose. 1995. The first report of eel larvae ingesting rotifers. Fish. Sci. 61:171-172.
Tanaka, Y., & K. Suzuki. 1991. Spawning, eggs and larvae of the hawkfish, *Cirrhitichthys aureus*, in an aquarium. Jpn. J. Ichthyol. 38:283-288.
Tanase, H., & C. Araga. 1975. Observation of breeding and taxonomy of the anemonefish, *Amphiprion*

clarkii (Bennett). J. Jpn. Assoc. Zool. Gardens Aquar. 17:16-21.

Tanasichuk, R.W., A.H. Kristofferson & D.V. Gillman. 1993. Comparison of some life history characteristics of Pacific herring (*Clupea pallasi*) from the Canadian Pacific Ocean and Beaufort Sea. Can. J. Fish. Aquat. Sci. 50:964-971.

Tandler, A. 1991. Marine aquaculture in Israel with special emphasis on larval rearing. Eur. Aquac. Soc. Spec. Pub. (15):379-380.

Tandler, A. 1993. Marine aquaculture in Israel with special emphasis on larval rearing. J. World Aquac. Soc. 24:241-245.

Tandler, A., & S. Helps. 1985. The effects of photoperiod and water exchange rate on growth and survival of gilthead sea bream (*Sparus aurata*, Linnaeus; Sparidae) from hatching to metamorphosis in mass rearing systems. Aquaculture 48:71-82.

Tandler, A., & S. Kolkovski. 1991. Rates of ingestion and digestibility as limiting factors in the successful use of microdiets in *Sparus aurata* larval rearing. Eur. Aquac. Soc. Spec. Pub. (15):169-171.

Tandler, A., & C. Mason. 1983. Light and food density effects on growth and survival of larval gilthead seabream (*Sparus aurata*, Linnaeus; Sparidae). World Maric. Soc. Spec. Pub. (3):103-116.

Tandler, A., & C. Mason. 1984. The use of ^{14}C labelled rotifers (*Brachionus plicatilis*) in the larvae of gilthead seabream (*Sparus aurata*): measurements of the effect of rotifer concentration, the lighting regime, and seabream's larval age on their rate of rotifer ingestion. Eur. Maric. Soc. Spec. Pub. (8):241-259.

Tandler, A., B.A. Berg, G.W. Kissil & A.M. Mackie. 1982. Effect of food attractants on appetite and growth rate of gilthead bream, *Sparus aurata* L. J. Fish. Biol. 20: 673-681.

Tang, H.-C., J.-Y. Twu & W.-C. Su. 1979. Preliminary report on artificial propagation of black-spotted grouper, *Epinephelus amblycephalus* Bleeker. Bull. Taiwan Fish. Res. Inst. (31):511-517. (in Ch., Eng. abs.)

Tang, Y.-A. 1964. Induced spawning of striped mullet by hormone injection. Jpn. J. Ichthyol. 12:23-28.

Taniguchi, A.K. 1981. Survival and growth of larval spotted seatrout (*Cynoscion nebulosus*) in relation to temperature, prey abundance and stocking densities. Rapp. P.-v. Réun. Cons. Int. Explor. Mer 178:507-508.

Taniguchi, M. 1983a. Effects of the food quality on the appearance of yellowtail streptococcicosis. Bull. Jpn. Soc. Sci. Fish. 49:363-366.

Taniguchi, M. 1983b. Feeding experiment of yellowtail using frozen diet. Bull. Jpn. Soc. Sci. Fish. 49:611-615.

Taranger, G.L., S.O. Stefansson & T. Hansen. 1992. Advancement and synchronization of ovulation in Atlantic salmon (*Salmo salar* L.) following injections of LHRH analogue. Aquaculture 102:169-175.

Tarazona, J.V., M.J. Munoz, G. Carbonell, M. Carballo, J.A. Ortiz & A. Castano. 1991. A toxicological assessment of water pollution and its relationship to aquaculture development in Algeciras Bay, Cadiz, Spain. Arch. Env. Contam. Toxicol. 20:480-487.

Tareen, I.U., & L.A. Hadi. 1981. Mariculture diseases and methods of control. Kuwait Inst. Sci. Res., 1981 Ann. Res. Rep. p 107-109.

Tashiro, K., & Y. Iwatsuki. 1995. Growth and feeding habits in reared Japanese centropomid fish *Lates japonicus*. Nippon Suisan Gakkaishi 61:684-688.

Tatum, W.M., W.C. Trimble & R.F. Helton, Jr. 1979. Production of gulf killifish in brackish-water ponds. Proc. Ann. Conf. SE Assoc. Fish Wildl. Ag. 32:502-508.

Tatum, W.M., J.P. Hawke, R.V. Minton & W.C. Trimble. 1982a. Alabama researchers study gulf killifish: prime culture candidate in coastal regions. Aquac. Mag. 8(4):20-25.

Tatum, W.M., J.P. Hawke, R.V. Minton & W.C. Trimble. 1982b. Production of bull minnows (*Fundulus grandis*) for the live bait market in coastal Alabama. Alabama Mar. Res. Bull. No. 13, 35 p.

Tave, D. 1990a. Effective breeding number and broodstock management: how to minimize genetic drift. Auburn Symposium on Fisheries and Aquaculture, Alabama Agric. Exp. Sta. J. No. 8-881732P, p 39-46.

Tave, D. 1990b. Effective breeding number and broodstock management: how to minimize inbreeding. Auburn Symp. Fish. Aquac., Alabama Agric. Exp. Sta. J. No. 8-881732P, p 27-38.

Tawada, S. 1989a. Development of eggs, larvae and juveniles of the grouper *Epinephelus microdon* (Bleeker) reared in the hatchery. Suisanzoshoku 37:99-103. (in Jpn.)

Tawada, S. 1989b. Spawning of brood stock of the grouper *Epinephelus microdon* (Bleeker). Suisanzoshoku 37:105-108. (in Jpn.)

Taylor, D., B.G. Maddock & G. Mance. 1985. The acute toxicity of nine 'grey list' metals (arsenic, boron, chromium, copper, lead, nickel, tin, vanadium and zinc) to two marine fish species: dab (*Limanda limanda*) and grey mullet (*Chelon labrosus*). Aquat. Toxicol. 7:135-144.

Taylor, D.S., S.A. Ritchie & E. Johnson. 1992. The killifish *Rivulus marmoratus*: a potential biocontrol agent for *Aedes taeniorhynchus* and brackish water *Culex*. J. Am. Mosquito Control Assoc. 8:80-83.

Teichman, J. (preparer). 1994. Report on the Economics of Ornamental Fish Culture in Hawaii. USDA Ctr. Trop. Subtrop. Agric., Honolulu, Contract No. 067, 99 p.

Teng, S.-K., T.-E. Chua & P.-E. Lim. 1978. Preliminary observation on the dietary protein requirement of estuary grouper, *Epinephelus salmoides* Maxwell, cultured in floating net-cages. Aquaculture 15:257-271.

Teruya, K., S. Masuma & Y. Hondo. 1992. Spawning and spawning behavior of *Plectropomus leopardus* in an aquarium. Saibaigiken 12:15-20. (in Jpn.)

Terver, D. 1971. Comportement de ponte, reproduction et developpement embryonnaire d'un poisson-clown (*Amphiprion polymnus*). La Pisciculture Francaise 26:9-23. (in Fr.)

Teshima, S.-I., S. Yamasaki, A. Kanazawa & H. Hirata. 1983. Effects of water temperature and salinity on eicosapentaenoic acid level of marine *Chlorella*. Bull. Jpn. Soc. Sci. Fish. 49:805.

Teshima, S.I., A. Kanazawa, S. Koshio & S. Itoh. 1993. L-ascorbyl-2-phosphate-Mg as vitamin C source for the Japanese flounder (*Paralichthys olivaceus*). Fish Nutrition in Practice, INRA, Paris, p 157-166.

Tesorero, L.B., & R.D. Fortes. 1994. Efficiency of the sea bass (*Lates calcarifer*) as biological control for tilapia (*Oreochromis mossambicus*) young at higher stocking density in polyculture. Third Asian Fisheries Forum. Asian Fish. Soc., Manila, p 65-68.

Thacker, S.G., & W.L. Griffin. 1994. Indoor intensive red drum aquaculture: a stochastic sensitivity analysis. J. World Aquac. Soc. 25:86-100.

Thangaraja, M. 1985. On the laboratory reared fish eggs and larvae of five species of carangids from the Vellar Estuary, Porto Novo. Mahasagar 18:477-488.

Theilacker, G.H. 1987. Feeding ecology and growth energetics of larval northern anchovy, *Engraulis mordax*. Fish. Bull., U.S. 85:213-228.

Theilacker, G., & K. Dorsey. 1980. Larval fish diversity, a summary of laboratory and field research. In: G. Sharp (rapporteur), Workshop on the Effects of Environmental Variation on the Survival of Larval Pelagic Fishes. UNESCO, Intergov. Oceanogr. Comm. Workshop Rep. (28):106-142.

Theilacker, G.H., & M.F. McMaster. 1971. Mass culture of the rotifer *Brachionus plicatilis* and its evaluation as a food for larval anchovies. Mar. Biol. 10:183-188.

Theiling, D.L., & H.A. Loyacano, Jr. 1976. Age and growth of red drum from a saltwater marsh impoundment in South Carolina. Trans. Am. Fish. Soc. 105:41-44.

Thiel, A.J. 1989a. Advanced Reef Keeping. Aardvark Press, Bridgeport, Messilla Park, New Mexico, 440 p.

Thiel, A.J. 1989b. Small Reef Aquarium Basics. Aardvark Press, Bridgeport, Connecticut, 175 p.

Thiel, A.J. 1990. Water quality management revisited. Marine Fish Monthly (Dec 5):11,15-22,31-32.

Theisen, D.D., D.D. Stansell & L.C. Woods III. 1998. Disinfection of nauplii of *Artemia franciscana* by ozonation. Prog. Fish-Cult. 60:149-151.

Thomas, P., & C.R. Arnold. 1993. Environmental and hormonal control of reproduction in sciaenid fish. Blackwell, Oxford, Rec. Adv. Aquac. IV:31-42.

Thomas, P., & N. Boyd. 1988. Induced spawning of spotted seatrout, red drum and orangemouth corvina (family Sciaenidae) with luteinizing hormone-releasing hormone analog injection. Contr. Mar. Sci. 30(suppl.):43-47.

Thomas, P., & N.W. Boyd. 1989. Dietary administration of an LHRH analogue induces spawning of spotted seatrout (*Cynoscion nebulosus*). Aquaculture 80:363-370.

Thomas, P., P.A. Copeland & J.A. Prentice. 1994. Preliminary observations on the reproductive physiology of female orangemouth corvina in captivity. J. World Aquac. Soc. 25:214-224.

Thomas, P., C.R. Arnold & G.J. Holt. 1995. Red drum and other sciaenids. In: N.R. Bromage & R.J. Roberts (eds.), Broodstock Management and Egg and Larval Quality. Blackwell, Oxford, p 118-137.

Thomas, W.H., A.N. Dodson & C.A. Linden. 1973. Optimum light and temperature requirements for *Gymnodinium splendens*, a larval fish food organism. Fish. Bull., U.S. 71:599-601.

Thomassen, J.M. 1993. A new method for control of salmon lice. In: H. Reinertsen, L.A. Dahle, L. Jørgensen & K. Tvinnereim (eds.), Fish Farming Technology. Balkema, Rotterdam, p 233-236.

Thompson, A.G. 1990. The danger of exotic species. World Aquac. 21(3):25-32.

Thompson, B.M., & J.D. Riley. 1981. Egg and larval development studies in the North Sea cod (*Gadus morhua* L.). Rapp. P.-v. Réun. Cons. Int. Explor. Mer 178:553-559.

Thompson, C.Q., & J.W. Tucker, Jr. 1989. Toxicity of the organophosphate insecticide fenthion, alone, and with thermal fog carriers, to an estuarine copepod and young fish. Bull. Env. Contam. Toxicol. 43:789-796.

Thompson, I., A. White, T.C. Fletcher, D.F. Houlihan & C.J. Secombes. 1993. The effect of stress on the immune response of Atlantic salmon (*Salmo salar* L.) fed diets containing different amounts of vitamin C. Aquaculture 114:1-18.

Thompson, P.A., P.J. Harrison & J.N.C. Whyte. 1990. Influence of irradiance on the fatty acid composition of phytoplankton. J. Phycol. 26:278-288.

Thoney, D.A., & W.J. Hargis, Jr. 1991. Monogenea (Platyhelminthes) as hazards for fish in confinement. Ann. Rev. Fish Dis. 1:133-153.

Thorburn, M.A. 1996. Practical applications of epidemiology to aquaculture. In: K.L. Main & C. Rosenfeld (eds.), Aquaculture Health Management Strategies for Marine Fishes. Oceanic Inst., Honolulu, p 209-215.

Thorgaard, G.H. 1995. Biotechnical approaches to broodstock management. In: N.R. Bromage & R.J. Roberts (eds.), Broodstock Management and Egg and Larval Quality. Blackwell, Oxford, p 76-93.

Thorgaard, G.H., M.E. Jazwin & A.R. Stier. 1981. Polyploidy induced by heat shock in rainbow trout. Trans. Am. Fish. Soc. 110:546-550.

Thorisson, K. 1994. Is metamorphosis a critical interval in the early life of marine fishes? Env. Biol. Fishes 40:23-36.

Thorogood, J., & A. Blackshaw. 1992. Factors affecting the activation, motility and cryopreservation of the spermatozoa of the yellowfin bream, *Acanthopagrus australis* (Gunther). Aquac. Fish. Mgt. 23:337-344.

Thorpe, J.E. 1980. Salmon Ranching. Academic Press, London, 441 p.

Thorpe, J.E., & C.Y. Cho. 1995. Minimising waste through bioenergetically and behaviourally based feeding strategies. Wat. Sci. Tech. 31:29-40.

Thorsen, A., & H.J. Fyhn. 1991. Osmotic effectors during preovulatory swelling of marine fish eggs. Proc., Fourth Internat. Symp. Repr. Physiol. of Fish. Norwich, U.K., p 312-314.

Thorsen, A., H.J. Fyhn & R.A. Wallace. 1993. Free amino acids as osmotic effectors for oocyte hydration in marine fishes. In: B.T. Walther & H.J. Fyhn (eds.), Physiological and Biochemical Aspects of Fish Development. Univ. Bergen, Norway, p 94-98.

Thouard, E. P. Soletchnik & J.-P. Marion. 1990. Selection of finfish species for aquaculture development in Martinique (F.W.I.). Advances in Tropical Aquaculture, AQUACOP, IFREMER, Actes Colloq. (9):499-510.

Thresher, R.E. 1984. Reproduction in Reef Fishes. T.F.H., Neptune City, New Jersey, 399 p.

Thurston, R.V., R.C. Russo, E.L. Meyn & R.K. Zajdel. 1986. Chronic toxicity of ammonia to fathead minnows. Trans. Am. Fish. Soc. 115:196-207.

Tibaldi, E., F. Tulli & D. Lanari. 1994. Arginine requirement and effect of different dietary arginine and lysine levels for fingerling sea bass (*Dicentrarchus labrax*). Aquaculture 127:207-218.

Tibaldi, E., P. Beraldo, L.A. Volpelli & M. Pinosa. 1996. Growth response of juvenile dentex (*Dentex dentex* L.) to varying protein level and protein to lipid ratio in practical diets. Aquaculture 139:91-99.

Tillman, R.E., & J.T. Surovik. 1990. Broodstock collection. Texas A&M Univ. Sea Grant TAMU-SG-90-603, p 30-32.

Tilney, R.L., & T. Hecht. 1993. Early ontogeny of *Galeichthys feliceps* from the south east coast of South Africa. J. Fish Biol. 43:183-212.

Tilseth, S., G. Blom & K. Naas. 1992. Recent progress in research and development of marine cold water species for aquaculture production in Norway. J. World Aquac. Soc. 23:277-285.

Timmons, M.B. 1994. Use of foam fractionators in aquaculture. In: M.B. Timmons & T.M. Losordo (eds.), Aquaculture Water Reuse Systems: Engineering Design and Management. Elsevier, Amsterdam, p 247-279.

Timmons, M.B. (ed.). 1995. Aquacultural Engineering and Waste Management. Proc. Aquaculture Expo VIII and Aquaculture in the Mid-Atlantic Conference, Washington, D.C., June 24-28, 1995. NE Reg. Agric. Eng. Serv., Ithaca, New York, 381 p.

Timmons, M.B., & T.M. Losordo. 1994. Aquaculture Water Reuse Systems: Engineering Design and Management. Elsevier, Amsterdam, 333 p.

Timmons, M.B., & T. Losordo (eds.). 1997. Advances in Aquacultural Engineering. Aquac. Eng. Soc. Proc. No. 3, NE Reg. Agric. Eng. Serv., Ithaca, New York, 393 p.

Tocher, D.R., G. Mourente & J.R. Sargent. 1997. The use of silages prepared from fish neural tissues as enrichers for rotifers (*Brachionus plicatilis*) and *Artemia* in the nutrition of larval marine fish. Aquaculture 148:213-231.

Toepfer, C.S., & J.W. Fleeger. 1995. Diet of juvenile fishes *Citharichthys spilopterus*, *Symphurus plagiusa*, and *Gobionellus boleosoma*. Bull. Mar. Sci. 56:238-249.

Toften, H., & M. Jobling. 1996. Development of spinal deformities in Atlantic salmon and Arctic charr fed diets supplemented with oxytetracycline. J. Fish Biol. 49:668-677.

Tojo, J., M.T. Santamarina, F.M. Ubeira, J. Estevez & M.L. Sanmartín. 1992a. Anthelmintic activity of benzimidazoles against *Gyrodactylus* sp. infecting rainbow trout *Oncorhynchus mykiss*. Dis. Aquat. Org. 12:185-189.

Tojo, J., M.T. Santamarina, D. Peris, F.M. Ubeira, S.L. Leiro & M.L. Sanmartín. 1992b. *In vitro* effect of anthelmintics on *Anisakis simplex* survival. Jpn. J. Parasitol. 41:473-480.

Toledo, Joebert D. 1992. Fisheries Department, Ministry of Industry and Primary Resources, Bandar Seri Begawan, Brunei Darussalam., pers. comm.

Toledo, J.D., & A.C. Gaitan. 1992. Egg cannibalism by milkfish (*Chanos chanos* Forsskal) spawners in circular floating net cages. Abs. in program book, Third Asian Fisheries Forum. Asian Fish. Soc., Manila, p 148.

Toledo, J.D., A. Nagai, D. Javellana & J.D. Tan-Fermin. 1990. Successive spawning of grouper, *Epinephelus malabaricus*, in tank and floating net cage. Asian Aquac. 12:8-10.

Toledo, J.D., C.L. Marte & A.R. Castillo. 1991. Spontaneous maturation and spawning of sea bass *Lates calcarifer* in floating net cages. J. Appl. Ichthyol. 7:217-222.

Toledo, J.D., A. Nagai & D. Javellana. 1992. Successive spawning of grouper *Epinephelus suillus* in a tank and a floating net cage. Abstract in program book, Third Asian Fisheries Forum. Asian Fish. Soc., Manila, p 148.

Toledo, J.D., A. Nagai & D. Javellana. 1993. Successive spawning of grouper, *Epinephelus suillus* (Valenciennes), in a tank and a floating net cage. Aquaculture 115:361-367.

Tomasso, J.R. 1994. Comparative inhibition of nitrite toxicity by chloride in temperate freshwater and euryhaline fishes. In: D.D. MacKinlay (ed.), High Performance Fish. Am. Fish. Soc., p 239-242.

Tomasso, J.R., & C.J. Kempton. 1997. Effects of water temperature on production chracteristics and blood chemistries of red drum fingerlings, *Sciaenops ocellatus*. World Aquac. '97 Book of Abs., p 249.

Tominaga, O., & K. Nashida. 1992. Spawning period and water depths of fivespot flounder *Pseudorhombus pentophthalmus* in the coastal waters of Northern Niigata Prefecture. Bull. Japan Sea Natl. Fish. Res. Inst. (42):41-47. (in Jpn., Eng. abs.)

Toner, R.C., & B. Brooks. 1975. The effects of ozone on four species of phytoplankton, crab zoea and megalops and the Atlantic silverside, *Menidia menidia*. In: Aquatic Applications of Ozone. Internat. Ozone Inst., p 154-160.

Tookwinas, S. 1990a. Larviculture of seabass (*Lates calcarifer*) and grouper (*Epinephelus malabaricus*) in Thailand. Advances in Tropical Aquaculture, AQUACOP, IFREMER, Actes Colloq. (9):645-659.

Tookwinas, S. 1990b. Review of grow-out techniques under tropical conditions: experience of Thailand on seabass (*Lates calcarifer*) and grouper (*Epinephelus malabaricus*). Advances in Tropical Aquaculture, AQUACOP, IFREMER, Actes Colloq. (9):737-750.

Tookwinas, S. 1990c. Review of knowledge on grouper aquaculture in South East Asia. Advances in Tropical Aquaculture, AQUACOP, IFREMER, Actes Colloq. (9):429-435.

Tookwinas, S., N. Srichantulk & C. Choongan. 1994. Coastal aquaculture in Thailand. In: F. Lacanilao, R.M. Colos & G.F. Quinitio (eds.), Aquaculture Development in Southeast Asia and Japan and Prospects for Seafarming and Searanching. SEAFDEC, Tigbauan, Philippines, p 84-90.

Toranzo, A.E., J.L. Barja & S. Devesa. 1994. An overview of the main infectious problems in cultured turbot: present status and future necessities. Eur. Aquac. Soc. Spec. Pub. (22):106-126.

Torrissen, O.J., & R. Christiansen. 1995. Requirements for carotenoids in fish diets. J. Appl. Ichthyol. 11:225-230.

Torrissen, O.J., J.C. Holm, G. Nævdal & T. Hansen. 1995. Aquaculture in Norway. World Aquac. 26(3):11-20.

Tortora, G.J., B.R. Funke & C.L. Case. 1992. Microbiology. Benjamin/Cummings, Redwood City, California, 810 p.

Touraki, M., S. Mourelatos, G. Karamanlidou, S. Kalaitzopoulou & C. Kastritsis. 1996. Bioencapsulation of chemotherapeutics in *Artemia* as a means of prevention and treatment of infectious diseases of marine fish fry. Aquacultural Eng. 15:133-147.

Toyama, K., M. Hoshi, Y. Takashima, S. Satoh, T. Takeuchi & T. Watanabe. 1984. Chemical evaluation of Japanese scrap fish meal as a protein source of fish foods. Bull. Jpn. Soc. Sci. Fish. 50:2065-2075.

Trant, J.M., & P. Thomas. 1987. The production of a novel maturation inducing steroid *in vitro* in the Atlantic croaker *Micropogonias undulatus*. Proc. Third Internat. Symp. Repr. Physiol. Fish, Memorial Univ. Newfoundland, St. John's, p 276.

Tribble, G.W., & H. Nishikawa. 1982. An analysis of the diets of four spatially overlapping damselfishes of the genus *Chromis*. Jpn. J. Ichthyol. 29:267-271.

Trimble, W.C. 1979. Yield trials for red drum in brackish-water ponds, 1976-1979. Proc. Ann. Conf. SE Assoc. Fish Wildl. Ag. 33:432-441.

Trimble, W.C., W.M. Tatum & S.A. Styron. 1981. Pond studies on gulf killifish (*Fundulus grandis*) mariculture. J. World Maric. Soc. 12:50-60.

Trotta, P. 1983. An indoor solution for mass production of the marine rotifer *Brachionus plicatilis* Muller fed on the marine microalga *Tetraselmis suecica* Butcher. Aquacultural Eng. 2:93-100.

Truelson, R.L., & R.P. Bio. 1994. Dissolved oxygen criteria proposed for aquatic life in British Columbia. In: D.D. MacKinlay (ed.), High Performance Fish., Am. Fish. Soc., p 205-210.

Trumble, R.J., J.D. Neilson, W.R. Bowering & D.A. McCaughran. 1993. Atlantic Halibut (*Hippoglossus hippoglossus*) and Pacific Halibut (*H. stenolepis*) and Their North American Fisheries. Can. Bull. Fish. Aquat. Sci. No. 227, 84 p.

Tsay, T.-T., & T.-C. Yu. 1991. Induced maturation and egg development of black sea bream (*Acanthopagrus latus*) reared in tank. Abs. in program book, Finfish Hatchery in Asia '91. Tungkang Mar. Lab., Tungkang, Taiwan, p 49.

Tseng, W.-y., & K.-f. Chan. 1985. On the larval rearing of the white-spotted green grouper, *Epinephelus amblycephalus* (Bleeker), with a description of larval development. J. World Maric. Soc. 16:114-126.

Tseng, W.Y., & K.L. Chan. 1982. The reproductive biology of the rabbitfish in Hong Kong. J. World Maric. Soc. 13:313-321.

Tseng, W.-Y., & S.-K. Ho. 1979. Egg development and early larval rearing of red grouper (*Epinephelus akaara*, Temminck & Schlegel). Quart. J. Taiwan Mus. 32:209-219.

Tseng, W.Y., & S.K. Ho. 1988. Grouper culture—a practical manual. Chien Cheng Publisher, Kaohsiung, Taiwan, ROC, 134 p.

Tseng, W.Y., & C.T. Poon. 1983. Hybridization of *Epinephelus* species. Aquaculture 34:177-182.

Tsevis, N., S. Klaoudatos & A. Conides. 1992. Food conversion budget in sea bass, *Dicentrarchus labrax*, fingerlings under two different feeding frequency patterns. Aquaculture 101:293-304.

Tsujigado, A., & C.-S. Lee. 1993. The design of a typical marine finfish hatchery in Japan using Mie Prefectural Fish Farming Center as an example. In: C.-S. Lee, M.-S. Su & I C. Liao (eds.), Finfish Hatchery in Asia. Tungkang Mar. Lab. Conf. Proc. (3):27-39.

Tsukamoto, K., & T. Kajihara. 1984. On the relation between yolk absorption and swimming activity in the ayu larvae *Plecoglossus altivelis*. Bull. Jpn. Soc. Sci. Fish. 50:59-61.

Tsukamoto, K., A. Umezawa, O. Tabeta, N. Mochioka & T. Kajihara. 1989. Age and birth date of

Anguilla japonica leptocephali collected in western North Pacific in September 1986. Nippon Suisan Gakkaishi 55:1023-1028.

Tsukamoto, Y., & S. Kimura. 1993. Development of laboratory-reared eggs, larvae and juveniles of the atherinind fish, *Hypoatherina tsurugae*, and comparison with related species. Jpn. J. Ichthyol. 40:261-267.

Tsutsumi, H. 1995. Impact of fish net pen culture on the benthic environment of a cove in south Japan. Estuaries 18:108-115.

Tubongbanua, E.S. 1987. Development of artificial feeds for sea bass (*Lates calcarifer*). Austral. Ctr. Internat. Agric. Res. Proc. (20):186-188.

Tucker, C.S., & J.F. Martin. 1991. Environment-related off-flavors in fish. Adv. World Aquac. 3:133-179.

Tucker, J.W., Jr. 1982. Larval development of *Citharichthys cornutus*, *C. gymnorhinus*, *C. spilopterus*, and *Etropus crossotus* (Bothidae), with notes on larval occurrence. Fish. Bull., U.S. 80:35-73.

Tucker, J.W., Jr. 1984. Hormone-induced ovulation of black sea bass females and rearing of larvae. Prog. Fish-Cult. 46:201-203.

Tucker, J.W., Jr. 1986. Sheepshead, the versatile porgy. Trop. Fish Hob. (Jan):64-65,68.

Tucker, J.W., Jr. 1987a. Sheepshead culture and preliminary evaluation for farming. Prog. Fish-Cult. 49:224-228.

Tucker, J.W., Jr. 1987b. Snook and tarpon snook culture and preliminary evaluation for commercial farming. Prog. Fish-Cult. 49:49-57.

Tucker, J.W., Jr. 1988. Growth of juvenile spotted seatrout on dry feeds. Prog. Fish-Cult. 50:39-41.

Tucker, J.W., Jr. 1989. Energy utilization in bay anchovy and black sea bass eggs and larvae. Fish. Bull., U.S. 87:279-293.

Tucker, J.W., Jr. 1992a. Feeding intensively-cultured marine fish larvae. In: G.L. Allan & W. Dall (eds.), Proc., Aquac. Nutr. Workshop. New South Wales Fish., Brackish Water Fish Cult. Res. Sta., Salamander Bay, Australia, p 129-146.

Tucker, J.W., Jr. 1992b. Marine fish nutrition. In: G.L. Allan & W. Dall (eds.), Proc., Aquac. Nutr. Workshop. New South Wales Fisheries, Brackish Water Fish Cult. Res. Sta., Salamander Bay, Australia, p 25-40.

Tucker, J.W., Jr. 1994. Spawning by captive serranid fishes: a review. J. World Aquac. Soc. 25:345-359.

Tucker, J.W., Jr. & S.R. Alshuth. 1997. Development of laboratory-reared sheepshead, *Archosargus probatocephalus* (Pisces: Sparidae). Fish. Bull., U.S. 95:394-401.

Tucker, J.W., Jr. & P.A. Barbera. 1987. Laboratory spawning of sheepshead, *Archosargus probatocephalus*. Prog. Fish-Cult. 49:229-230.

Tucker, J.W., Jr. & S.W. Campbell. 1988. Spawning season of common snook along the east central Florida coast. Florida Sci. 51:1-6.

Tucker, J.W., Jr. & B.E. Faulkner. 1987. Voluntary spawning patterns of captive spotted seatrout (*Cynoscion nebulosus*). Northeast Gulf Sci. 9:59-63.

Tucker, J.W., Jr. & W.J. FitzGerald. 1994. Induced spawning of two western tropical Pacific groupers, *Plectropomus areolatus* and *Epinephelus fuscoguttatus*, in Palau. Asian Fish. Sci. 7:57-62.

Tucker, J.W., Jr. & D.E. Jory. 1991. Marine fish culture in the Caribbean region. World Aquac. 22(1):10-14,16-17,19-27.

Tucker, J.W., Jr. & C.Q. Thompson. 1987. Dangers of using organophosphorous pesticides and diesel oil in fish ponds. Aquac. Mag. 13(3):62-63.

Tucker, J.W., Jr. & P.N. Woodward. 1995. Egg production and completion of the life cycle of belted sandfish (*Serranus subligarius*) in captivity. Bull. Mar. Sci. 56:701-707.

Tucker, J.W., Jr. & P.N. Woodward. 1996. Nassau grouper aquaculture. In: F. Arreguín-Sánchez, J.L. Munro, M.C. Balgos & D. Pauly (eds.), Biology, Fisheries and Culture of Tropical Groupers and Snappers. ICLARM Conf. Proc. (48):363-377.

Tucker, J.W., Jr., M.P. Landau & B.E. Faulkner. 1985. Culinary value and composition of wild and captive common snook, *Centropomus undecimalis*. Florida Sci. 48:196-200.

Tucker, J.W., Jr., C.Q. Thompson, T.C. Wang & R.A. Lenahan. 1987. Toxicity of organophosphorous insecticides to estuarine copepods and young fish after field applications. J. Fla. Anti-Mosquito Assn. 58:1-6.

Tucker, J.W., Jr., M.R. MacKinnon, D.J. Russell, J.J. O'Brien & E. Cazzola. 1988. Growth of juvenile barramundi (*Lates calcarifer*) on dry feeds. Prog. Fish-Cult. 50:81-85.

Tucker, J.W., Jr., J.E. Parsons, G.C. Ebanks & P.G. Bush. 1991. Induced spawning of Nassau grouper *Epinephelus striatus*. J. World Aquac. Soc. 22:187-191.

Tucker, J.W., Jr., R.M. Baptiste, F.A. Prahl & M.S. Clark. 1993a. Growth of sunshine bass on experimental dry feeds. Prog. Fish-Cult. 55:199-203.

Tucker, J.W., Jr., P.G. Bush & S.T. Slaybaugh. 1993b. Reproductive patterns of Cayman Islands Nassau grouper (*Epinephelus striatus*) populations. Bull. Mar. Sci. 52:961-969.

Tucker, J.W., Jr., P.N. Woodward & D.S. Sennett. 1996. Voluntary spawning of captive Nassau groupers *Epinephelus striatus* in a concrete raceway. J. World Aquac. Soc. 27:373-383.

Tucker, J.W.. Jr., W.A. Lellis, G.K. Vermeer, D.E. Roberts, Jr. & P.N. Woodward. 1997a. Growth of red drum, *Sciaenops ocellatus*, to maturity on experimental diets. J. Appl. Aquac. 7:93-108.

Tucker, J.W.. Jr., W.A. Lellis, G.K. Vermeer, D.E. Roberts, Jr. & P.N. Woodward. 1997b. The effects of experimental starter diets with different levels of soybean or menhaden oil on red drum (*Sciaenops ocellatus*). Aquaculture 149:323-339.

Tucker, J.W., Jr., W.E. Lellis & G.K. Vermeer. ms. The correlation between liver dry matter and liver lipid in cultured red drum (*Sciaenops ocellatus*) fed six commercial and 15 experimental diets.

Tudor, K.W., R.R. Rosati, P.D. O'Rourke, Y.V. Wu, D. Sessa & P. Brown. 1996. Technical and economical feasibility of on-farm fish feed production using fishmeal analogs. Aquacultural Eng. 15:53-65.

Tuene, S., & R. Nortvedt. 1995. Feed intake, growth and feed conversion efficiency of Atlantic halibut, *Hippoglossus hippoglossus* (L.). Aquac. Nutr. 1:27-35.

Tully, O., P. Daly, S. Lysaght, S. Deady & S.J.A. Varian. 1996. Use of cleaner-wrasse (*Centrolabrus exoletus* (L.) and *Ctenolabrus rupestris* (L.)) to control infestations of *Caligus elongatus* Nordmann on farmed Atlantic salmon. Aquaculture 142:11-24.

Tuncer, H., & R.M. Harrell. 1992. Essential fatty acid nutrition of larval striped bass (*Morone saxatilis*) and palmetto bass (*M. saxatilis* × *M. chrysops*). Aquaculture 101:105-121.

Tuncer, H., R.M. Harrell & E.D. Houde. 1990. Comparative energetics of striped bass (*Morone saxatilis*) and hybrid (*M. saxatilis* × *M. chrysops*) juveniles. Aquaculture 86:387-400.

Tuncer, H., R.M. Harrell & T.-j. Chai. 1993. Beneficial effects of *n*-3HUFA enriched *Artemia* as food for larval palmetto bass (*Morone saxatilis* × *M. chrysops*). Aquaculture 110:341-359.

Tung, I.-H. 1973. On the egg development and larval stages of the grey mullet, *Mugil cephalus* Linnaeus. Rep. Inst. Fish. Biol., Min. Econ. Affairs & Nat. Taiwan Univ. 3:187-215.

Tung, M.-C., & G.-H. Kou. 1986. Pathology. In: C.-S. Lee, M.S. Gordon & W.O. Watanabe (eds.), Aquaculture of Milkfish (*Chanos chanos*): State of the Art. Oceanic Inst., Honolulu, p 243-262.

Turk, P.E., M.E. Krejci & W.T. Yang. 1982. A laboratory method for the culture of *Acartia tonsa* (Crustacea: Copepoda) using rice bran. J. Aquaric. Aquat. Sci. 3:25-27.

Turner, C.H., & E.E. Ebert. 1962. The nesting of *Chromis punctipinnis* (Cooper) and a description of their eggs and larvae. Cal. Fish Game 48:243-248.

Turner, C.H., E. Ebert & R.R. Given. 1969. Manmade reef ecology. Cal. Dept. Fish Game, Fish. Bull. No. 146, 221 p.

Turner, R.E. 1977. Intertidal vegetation and commercial yields of penaeid shrimp. Trans. Am. Fish. Soc. 106:411-416.

Tvinnereim, K. 1993. Critical factors in fish farming in closed production systems. In: H. Reinertsen, L.A. Dahle, L. Jørgensen & K. Tvinnereim (eds.), Fish Farming Technology. Balkema, Rotterdam, p 285-288.

Twiddy, D.R. 1995. Antibiotic-resistant human pathogens in integrated fish farms. ASEAN Food J. 10:22-29.

Tytler, P., & P. Calow (eds.). 1985. Fish Energetics: New Perspectives. Johns Hopkins Univ. Press, Baltimore, 349 p.

Tzeng, W.-N. 1990. Relationship between growth rate and age at recruitment of *Anguilla japonica* elvers in a Taiwan estuary as inferred from otolith growth increments. Mar. Biol. 107:75-81.

Tzovenis, I., N. De Pauw & P. Sorgeloos. 1997. Effect of different light regimes on the docosahexaenoic acid (DHA) content of *Isochrysis* aff. *galbana* (clone T-ISO). Aquac. Internat. 5:489-507.

Uchida, K., S. Imai, S. Mito, S. Fujita, M. Ueno, Y. Shojima, T. Senta, M. Tahuka & Y. Dotsu (eds.). 1958. Studies on the eggs, larvae, and juveniles of Japanese fishes. Series 1. Second Lab. Fish. Biol., Fish. Dept., Fac. Agric. Kyushu Univ., 525 p. (in Jpn.)

Uchiyama, J.H., R.K. Burch & S.A. Kraul, Jr. 1986. Growth of dolphins, *Coryphaena hippurus* and *C. equiselis*, in Hawaiian waters as determined by daily increments on otoliths. Fish. Bull., U.S. 84:186-191.

Ueda, R., H. Sugita & Y. Deguchi. 1996. Determination of the naturally occurring bactericidal activity of sera from Japanese coastal crustaceans. Fish. Sci. 62:325-326.

Ugelstad, I., I. Lein & K. Hjelmeland. 1995. Synthesis of the digestive enzyme trypsin in halibut (*Hippoglossus hippoglossus* L.) larvae and juveniles. ICES Mar. Sci. Symp. 201:205.

Uhlig, G. 1984. Progress in mass cultivation of harpacticoid copepods for mariculture purposes. Eur. Maric. Soc. Spec. Pub. (8):261-273.

Ukawa, M., M. Higuchi & S. Mito. 1966. Spawning habits and early life history of a serranid fish, *Epinephelus akaara* (Temminck et Schlegel). Jpn. J. Ichthyol. 13:156-161 (in Jpn., Eng. sum.)

Ukawa, M., K. Takii, M. Nakamura, M. Akutsu, H. Kumai & M. Takeda. 1994. Effect of iron supplements on a soy protein concentrate diet on hematology of yellowtail. Fish. Sci. 60:165-169.

Umezawa, A., T. Otake, J. Hirokawa, K. Tsukamoto & M. Okiyama. 1991. Development of the eggs and larvae of the pike eel, *Muraenesox cinereus*. Jpn. J. Ichthyol. 38:35-40.

Umezawa, S. 1994. Overview of seafarming and searanching technology in Japan. In: F. Lacanilao, R.M. Colos & G.F. Quinitio (eds.), Aquaculture Development in Southeast Asia and Japan and Prospects for Seafarming and Searanching. SEAFDEC, Tigbauan, Philippines, p 93-104.

Underwood, E.J. 1977. Trace Elements in Human and Animal Nutrition. Academic Press, New York, 545 p.

UNDP/FAO. 1980. Fish Feed Technology. United Nations Development Programme, FAO, Rome, 395 p.

Ungson, J.R., Y. Matsuda & H. Hirata. 1995. Ranching the red sea bream in Japan. World Aquac. 26(1):6-12.

Unwin, P.D. 1986. Breeding the yellow tailed damselfish (*Glyphidodontops hemicyaneus*). SeaScope, Aquarium Systems, Mentor, Ohio, 3(winter):1.

Urho, L. 1992. Morphological and ecological differences in the development of smelt and herring larvae in the northern Baltic Sea. Sarsia 77:1-10.

Urup, B. 1994. Methods for the production of turbot fry using copepods as food. Eur. Aquac. Soc. Spec. Pub. (22):47-53.

Ushiro, M., S. Yamasaki & H. Hirata. 1980. Examination of bacteria as food for *Brachionus plicatilis*, in culture. Min. Rev. Data File Fish. Res. 1:96-106.

Usui, A. 1991. Eel Culture. Fishing News Books, Oxford, 148 p.

Vadstein, O. 1997. The use of immunostimulation in marine larviculture: possibilities and challenges. Aquaculture 155:401-417.

Vadstein, O., G. Øie, Y. Olsen, I. Salvesen, J. Skjermo & G. Skjåk-Bræk. 1993. A strategy to obtain microbial control during larval development of marine fish. In: H. Reinertsen, L.A. Dahle, L. Jørgensen & K. Tvinnereim (eds.), Fish Farming Technology, Balkema, Rotterdam, p 69-75.

Valenti, R.J. 1972. The embryology of the neon goby, *Gobiosoma oceanops*. Copeia 1972:477-482.

Valenti, R.J. 1975. Semi-closed system culture of the northern puffer, *Sphoeroides maculatus*. Proc. World. Maric. Soc. 6:479-485.

Vallin, L., & A. Nissling. 1994. Estimation of egg quality at early blastula stages in eggs from Skagerrak cod and Baltic cod (*Gadus morhua*). ICES Pap. C.M. 1994/J:23, 10 p.

van Ballaer, E., F. Amat, F. Hontoria, P. Léger & P. Sorgeloos. 1985. Preliminary results on the nutritional evaluation of ω3-HUFA-enriched *Artemia* nauplii for larvae of the sea bass, *Dicentrarchus labrax*. Aquaculture 49:223-229.

Van Coillie, R., C. Thellen & R. Schoenert. 1980. Toxicite sous-letale du cuivre chez les oeufs d'omble de fontaine: caracteristiques biochimiques de la toxicite retardee. Can. Tech. Rep. Fish. Aquat. Sci. (975):57-67.

Van Den Avyle, M.J. 1984. Species profiles: life histories and environmental requirements of coastal fishes and invertebrates (South Atlantic)—American eel. U.S. Fish Wildl. Serv. Rep. FWS/OBS-82/11.24. U.S. Army Corps Engineers, TR EL-82-4, 19 p.

Van Der Kraak, G., H.M. Dye, E.M. Donaldson & G.A. Hunter. 1985. Plasma gonadotropin, 17Beta-estradiol, and 17Alpha,20Beta-dihydroxy-4-pregnen-3-one levels during luteinizing hormone-releasing hormone analogue and gonadotropin induced ovulation in coho salmon (*Oncorhynchus kisutch*). Can. J. Zool. 63:824-833.

van der Meeren, T. 1991. Production of marine fish fry in Norway. World Aquac. 22(2):37-40.

van der Meeren, T. 1995. Vekst og overlevelse for larver og yngel av atlantiske torskebestander, komparative studier i posesystemer. Fisken og havet No. 25, 19 p. (in Nor., Eng. abs.)

van der Meeren, T., & T. Næss. 1993. How does cod (*Gadus morhua*) cope with variability in feeding conditions during early larval stages? Mar. Biol. 116:637-647.

van der Meeren, T., K.E. Jørstad, P. Solemdal & O.S. Kjesbu. 1994. Growth and survival of cod larvae (*Gadus morhua* L.): comparative enclosure studies of northeast Arctic cod and coastal cod from western Norway. ICES Mar. Sci. Symp. 198:633-645.

van der Ploeg, M., & C.S. Tucker. 1994. Seasonal trends in flavor quality of channel catfish, *Ictalurus punctatus*, from commercial ponds in Mississippi. J. Appl. Aquac. 3:121-140.

van der Ploeg, M., C.S. Tucker & C.E. Boyd. 1992. Geosmin and 2-methylisoborneol production by cyanobacteria in fish ponds in the southeastern United States. Water Sci. Technol. 25:283-290.

van der Wal, E.J., & J.A. Nell. 1986. Effect of food concentration on the survival and growth of Australian bass (*Macquaria novemaculeata*) larvae. Prog. Fish-Cult. 48:202-204.

van Heerden, E., J.H.J. van Vuren & G.J. Steyn. 1995. LC_{50} determination for malachite green and formalin on rainbow trout (*Oncorhynchus mykiss*) juveniles. Water S.A. 21:87-94.

Van Olst, J.C., & J.M. Carlberg. 1987. The Feasibility of Corvina Aquaculture for Population Enhancement and Food Production. Phase I Report. NSF/IST-87069, Nat. Tech. Info. Serv., 69 p.

van Rooij, J.M., F.J. Kroon & J.J. Videler. 1996. The social and mating system of the herbivorous reef fish *Sparisoma viride*: one-male versus multi-male groups. Env. Biol. Fishes 47:353-378.

Varner, P.W., & D.H. Lewis. 1991. Characterization of a virus associated with head and lateral line erosion syndrome in marine angelfish. J. Aquat. Anim. Health 3:198-205.

Varriale, A.M.C., A. Castelli, S. Cecchini & M. Saroglia. 1993. Study of *Diplectanum aequans* (Monogenea), parasite of sea bass (*Dicentrarchus labrax*) in intensive farming. Eur. Aquac. Soc. Spec. Pub. (18):143-153.

Vásquez-Yeomans, L., & E. Sosa-Cordero. 1992. Culture of grunion *Leuresthes tenuis* (Ayres) larvae to metamorphosis, in the laboratory. Ciencias Marinas 18:1-18.

Venizelos, A., & D.D. Benetti. 1996. Epitheliocystis disease in cultured yellowtail *Seriola mazatlana* in Ecuador. J. World Aquac. Soc. 27:223-227.

Venkataramiah, A., G.J. Lakshmi, C. Best, G. Gunter, E. Hartwig & P. Wilde. 1982. Studies on toxicity of OTEC plant components on *Eucalanus* sp. from the Gulf of Mexico. Ocean Sci. Eng. 7:353-401.

Ventling, A.R., & N.W. Pankhurst. 1995. Effects of gonadal steroids and human chorionic gonadotrophin on final oocyte maturation *in vitro* in the New Zealand snapper *Pagrus auratus* (Sparidae). Mar. Freshw. Res. 46:467-473.

Ventura, R.F., & E.M. Enderez. 1980. Preliminary studies on *Moina* sp. production in freshwater tanks. Aquaculture 21:93-96.

Verakunpiriya, V., T. Watanabe, K. Mushiake, V. Kiron, S. Satoh & T. Takeuchi. 1996. Effect of broodstock diets on the chemical components of milt and eggs produced by yellowtail. Fish. Sci. 62:610-619.

Verakunpiriya, V., K. Watanabe, K. Mushiake, K. Kawano, T. Kobayashi, I. Hasegawa, V. Kiron, S. Satoh & T. Watanabe. 1997a. Effect of krill meal supplementation in soft-dry pellets on spawning and quality of egg of yellowtail. Fish. Sci. 63:433-439.

Verakunpiriya, V., K. Mushiake, K. Kawano & T. Watanabe. 1997b. Supplemental effect of astaxanthin in broodstock diets on the quality of yellowtail eggs. Fish. Sci. 63:816-823.

Verdonck, L., M. Dehasque, J. Swings, P. Sorgeloos & P. Léger. 1991. The microbial environment of rotifer (*Brachionus plicatilis*) and *Artemia* production systems. Eur. Aquac. Soc. Spec. Pub. (15):398.

Verdonck, L., L. Grisez, E. Sweetman, G. Minkoff, P. Sorgeloos, F. Ollevier & J. Swings. 1997. Vibrios associated with routine productions of *Brachionus plicatilis*. Aquaculture 149:203-214.

Vergara, J.M., & K. Jauncey. 1993. Studies on the use of dietary energy by gilthead sea bream (*Sparus aurata* L.) juveniles. Fish Nutrition in Practice, INRA, Paris, p 453-458.

Vergara, J.M., L. Robainà, M. Izquierdo & M. De La Higuera. 1996a. Protein sparing effect of lipids in diets for fingerlings of gilthead sea bream. Fish. Sci. 62:624-628.

Vergara, J.M., H. Fernández-Palacios, L. Robainà, K. Jauncey, M. De La Higuera & M. Izquierdo. 1996b. The effects of varying dietary protein level on the growth, feed efficiency, protein utilization and body composition of gilthead sea bream fry. Fish. Sci. 62:620-623.

Verischele, D., P. Leger, P. Lavens & P. Sorgeloos. 1990. The use of *Artemia*. In: G. Barnabé (ed.), Aquaculture, Vol. 1. Ellis Horwood, New York, p 246-263.

Verlhac, V., J. Gabaudan, A. Obach, W. Schüep & R. Hole. 1996. Influence of dietary glucan and vitamin C on non-specific and spcific immune responses of rainbow trout (*Oncorhynchus mykiss*). Aquaculture 143:123-133.

Vermeer, G.K., J.H. Landsberg, A.E. Burke, N.E. Perry & S.A. Richards. Submitted. Investigation and control of *Amyloodinium* cf. *ocellatum* infestations in cultured red drum *Sciaenops ocellatus* (L.).

Verwey, J. 1930. Coral reef studies. I. The symbiosis between damselfishes and sea anemones in Batavia Bay. Treubia 12:305-366.

Veshchev, P.V. 1991. Characteristics of spawning stocks and reproduction of Volga stellate sturgeon, *Acipenser stellatus*, under new ecological conditions. J. Ichthyol. 31:121-132.

Vidy, G., & J. Franc. 1992. Saisons de presence a la cote des alevins de muges (Mugilidae) en Tunisie. Cybium 16:53-71. (in Fr., Eng. abs.)

Viette, M., P.G. Giulianini & E.A. Ferrero. 1997. Reproductive biology of scad, *Trachurus mediterraneus* (Teleostei, Carangidae), from the Gulf of Trieste. ICES J. Mar. Sci. 54:267-272.

Vilela, M.H. 1992. Mass culture and nutritional quality of the marine copepod *Tigriopus brevicornis* Müller. Bol. Inst. Nac. Invest. Pescas 17:49-60.

Villaluz, A.C. 1986. Fry and fingerling collection and handling. In: C.-S. Lee, M.S. Gordon & W.O. Watanabe (eds.), Aquaculture of Milkfish (*Chanos chanos*): State of the Art. Oceanic Inst., Honolulu, p 153-180.

Villegas, C.T. 1990. The effects on growth and survival of feeding water fleas (*Moina macrocopa* Straus) and rotifers (*Brachionus plicatilis*) to milkfish (*Chanos chanos* Forsskal) fry. Bamidgeh 42:10-17.

Villegas, C.T., O. Millamena & F. Escritor. 1990. Food value of *Brachionus plicatilis* fed three selected algal species as live food for milkfish, *Chanos chanos* Forsskal, fry production. Aquac. Fish. Mgt. 21:213-219.

Vincent, A.C.J. 1992. Prospects for sex role reversal in teleost fishes. Neth. J. Zool. 42:392-399.

Vincent, A.C.J. 1995a. Exploitation of seahorses and pipefishes. NAGA (ICLARM) (Jan):18-19.

Vincent, A. 1995b. Seahorse keeping. Breeder's Registry 3(1):1,5-6.

Vincent, A. 1995c. Seahorse keeping: feeding adults, mating, rearing young, microculturing. Breeder's Registry 3(2):1-5.

Viola, S., & Y. Arieli. 1983. Nutrition studies with tilapia hybrids. 2. The effects of oil supplements to practical diets for intensive aquaculture. Bamidgeh 35:44-52.

Viyakarn, V., T. Watanabe, H. Aoki, H. Tsuda, H. Sakamoto, N. Okamoto, N. Iso, S. Satoh & T. Takeuchi. 1992. Use of soybean meal as a substitute for fish meal in a newly developed soft-dry pellet for yellowtail. Nippon Suisan Gakkaishi 58:1991-2000.

Volkart, B. 1991. Damsel, fair damsel. Trop. Fish Hob. 39(9):52-54,56-57,60.

Volkman, J.K., & G.M. Hallegraeff. 1988. Lipids in marine diatoms of the genus *Thalassiosira*: predominance of 24-methylenecholesterol. Phytochemistry 27:1389-1394.

Volkman, J.K., S.W. Jeffrey, P.D. Nichols, G.I. Rogers & C.D. Garland. 1989. Fatty acid and lipid composition of 10 species of microalgae used in mariculture. J. Exp. Mar. Biol. Ecol. 128:219-240.

Volkman, J.K., G.A. Dunstan, S.W. Jeffrey & P.S. Kearney. 1991. Fatty acids from microalgae of the genus *Pavlova*. Phytochemistry 30:1855-1859.

von Bodungen, B., T.D. Jickells, S.R. Smith, J.A.D. Ward & G.B. Hillier. 1982. The Bermuda marine environment, Vol. III. Bermuda Biol. Sta. Spec. Pub. No. 18, 123 p.

Von Ledebur, J.T., & W. Wunder. 1938. Beitrage zur Physiologie der Schwimmblase der Fische. IV.

Beobachtungen an Stichlingen die ihre Schwimmblase nicht mit Gas fullen konnten. Z. Vergl. Physiol. 25:149-155.

Von Westernhagen, H., & H. Rosenthal. 1975. Rearing and spawning siganids (Pisces: Teleostei) in a closed seawater system. Helgol. Wiss. Meeresunters 27:1-18.

Von Westernhagen, H., & H. Rosenthal. 1976. Induced multiple spawning of reared *Siganus oramin* (Schneider)(= *S. canaliculatus* Park). Aquaculture 7:193-196.

Vuorinen, M., P.J. Vuorinen, J. Hoikka & S. Peuranen. 1993. Lethal and sublethal threshold values of aluminium and acidity to pike (*Esox lucius*), whitefish (*Coregonus lavaretus pallasi*), pike perch (*Stizostedion lucioperca*) and roach (*Rutilus rutilus*) yolk-sac fry. Sci. Total Env. Suppl. 1993:953-967.

Waas, B.P., K. Strawn, M. Johns & W. Griffin. 1983. The commercial production of mudminnows (*Fundulus grandis*) for live bait: a preliminary economic analysis. Texas J. Sci. 35:51-60.

Wagner, E.J., T. Bosakowski & S.A. Miller. 1994. Evaluation of the absorption efficiency of the low head oxygenation system. Aquacultural Eng. 14:49-57.

Wagner, R.E., W. Kotas & G.A. Yogis. 1992. Guide to Environmental Analytical Methods. Genium Publishing, Schenectady, New York.

Wagstaff, R.K. 1975. The concept and process of growing pompano in confinement. Feedstuffs (2 Jun 1975):24-27.

Wahbeh, M.I. 1997. Amino acid and fatty acid profiles of four species of macroalgae from Aqaba and their suitability for use in fish diets. Aquaculture 159:101-109.

Wajsbrot, N., A. Gasith, M.D. Krom & D.M. Popper. 1991. Acute toxicity of ammonia to juvenile gilthead seabream *Sparus aurata* under reduced oxygen conditions. Aquaculture 92:277-288.

Wajsbrot, N., A. Gasith, A. Diamant & D.M. Popper. 1993. Chronic toxicity of ammonia to juvenile gilthead seabream *Sparus aurata* and related histopathological effects. J. Fish Biol. 42:321-328.

Wakeman, J.M., & D.E. Wohlschlag. 1977. Salinity stress and swimming performance of spotted seatrout. Proc. Ann. Conf. SE Assoc. Fish Wildl. Ag. 31:357-361.

Walch, J.C. 1993. Breeding jawfish. Breeder's Registry 1(4):1-4.

Walch, J. 1994a. Breeding the royal gramma, *Gramma loreto*. Breeder's Registry 2(1):1-2.

Walch, J. 1994b. Reproduction of yellowhead jawfish in captivity. SeaScope, Aquarium Systems, Mentor, Ohio, 11(winter):1-2.

Walford, J., & T.J. Lam. 1987. Effect of feeding with microcapsules on the content of essential fatty acids in live foods for the larvae of marine fishes. Aquaculture 61:219-229.

Walford, J., & T.J. Lam. 1991. Growth and survival of seabass (*Lates calcarifer*) larvae fed microencapsulated diets alone or together with live food. Eur. Aquac. Soc. Spec. Pub. (15):184-187.

Walford, J., & T.J. Lam. 1993. Development of digestive tract and proteolytic enzyme activity in seabass (*Lates calcarifer*) larvae and juveniles. Aquaculture 109:187-205.

Walford, J., T.M. Lim & T.J. Lam. 1991. Replacing live foods with microencapsulated diets in the rearing of seabass (*Lates calcarifer*) larvae: do the larvae ingest and digest protein-membrane microcapsules? Aquaculture 92:225-235.

Walker, K.F. 1981. 13. A synopsis of ecological information on the saline lake rotifer *Brachionus plicatilis* Muller 1786. Hydrobiologia 81:159-167.

Walker, S. 1977. Walker successfully spawns five species. Mar. Hob. News 5(9):1,4.

Walker, S.D. 1991. Reproducing the Atlantic spadefish at the Tulsa Zoo. Seascope, Aquarium Systems, Mentor, Ohio, 8(winter):1-2.

Walker, S.D., & N. Herwig. 1976. Salinity and spawning. Mar. Aquarist 7(2):45-50.

Wallace, R.A., & K. Selman. 1981. Cellular and dynamic aspects of oocyte growth in teleosts. Am. Zool. 21:325-343.

Wallace, R.A., K. Selman, M.S. Greeley, Jr., P.C. Begovac, Y.-W.P. Lin, R. McPherson & T.R. Petrino. 1987. Current status of oocyte growth. Proc. Third Internat. Symp. Repr. Physiol. Fish, Memorial Univ. Newfoundland, St. John's, p 167-177.

Waller, U., K. Wieland & D. Schnack. 1993. The survival of eggs and the hatching of larvae of cod (*Gadus morhua*) at different oxygen levels. ICES Pap. CM 1993/J:21, 6 p.

Wallin, J.E., & M.J. Van Den Avyle. 1995. Interactive effects of stocking site salinity and handling stress on survival of striped bass fingerlings. Trans. Am. Fish. Soc. 124:736-745.

Walsh, D.T., C.A. Withstandley, R.A. Kraus & E.J. Petrovits. 1987. Mass culture of selected marine microalgae for the nursery production of bivalve seed. J. Shellfish Res. 6:71-77.

Walsh, M., & A.D.F. Johnstone. 1992. Spawning behaviour and diel periodicity of egg production in captive Atlantic mackerel, *Scomber scombrus* L. J. Fish Biol. 40:939-950.

Walsh, W.A., C. Swanson & C.-S. Lee. 1991. Combined effects of temperature and salinity on embryonic development and hatching of striped mullet, *Mugil cephalus*. Aquaculture 97:281-289.

Walz, N. 1993. Methods of single- and two-stage chemostat experiments. Ecological Studies: Analysis and Synthesis 98:21-29.

Wan, P.J., F. Zhang & R.J. Hron, Sr. 1995. Extraction, composition, and stability of pigments from crawfish shell waste. In: C. Lim & D.J. Sessa (eds.), Nutrition and Utilization Technology in Aquaculture. AOSC Press, Champaign, Ilinois, p 255-268

Wanakowat, J., M. Boonyaratpalin & T. Watanabe. 1993. Essential fatty acid requirement of juvenile seabass. In: Fish Nutrition in Practice, INRA, Paris, p 807-817.

Wang, J.-K. (ed.). 1993. Techniques for Modern Aquaculture. Am. Soc. Agric. Eng., St. Joseph, Michigan, 604 p.

Wang, T.C., & J.L. Bricker. 1979. 2-butanone and tetrahydrofuran contamination in the water supply. Bull. Env. Contam. Toxicol. 23:620-623.

Ware, D.M. 1975. Relation between egg size, growth, and natural mortality of larval fish. J. Fish. Res. Bd. Can. 32:2503-2512.

Ware, G.W. 1994. The Pesticide Book. Thomson, Fresno, California, 386 p.

Waring, C.P., R.M. Stagg & M.G. Poxton. 1992. The effects of handling on flounder (Platichthys flesus L.) and Atlantic salmon (*Salmo salar* L.). J. Fish Biol. 41:131-144.

Waring, C.P., M.G. Poxton & R.M. Stagg. 1997. The physiological response of the turbot to multiple net confinements. Aquac. Internat. 5:1-12.

Warner, R.R., & S.E. Swearer. 1991. Social control of sex change in the bluehead wrasse, *Thalassoma bifasciatum* (Pisces: Labridae). Biol. Bull. 181:199-204.

Warren, H.J., R.M. Harrell, J.G. Geiger & R.A. Rees. 1990. Design of rearing facilities for striped bass and hybrid striped bass. In: R.M. Harrell, J.H. Kerby & R.V. Minton (eds.), Culture and Propagation of Striped Bass and Its Hybrids. Am. Fish. Soc., Bethesda, Maryland, p 17-27.

Warrington, P.D. 1994. Development of water quality criteria for the protection of aquatic life from chlorophenols. In: D.D. MacKinlay (ed.), High Performance Fish. Fish Physiol. Assoc., Vancouver, BC, p 211-216.

Washington, B.B., H.G Moser, W.A. Laroche & W.J. Richards. 1984. Scorpaeniformes: development. Am. Soc. Ichthyol. Herpetol. Spec. Pub. (1):405-428.

Washio, M., S. Komiya & T. Takita. 1993. Maturation of the mudskipper *Boleophthalmus pectinirostris* distributed in the mud flats of the Midori River, Kumamoto Prefecture. Bull. Jpn. Soc. Sci. Fish. 59:575-580.

Wass, B.P., K. Strawn, M. Johns & W. Griffin. 1983. Commercial production of mudminnows (*Fundulus grandis*) for live bait: a preliminary economic analysis. Texas J. Sci. 35(1):51-60.

Wassink, H., & R. Brons. 1988. Eine gelungene zucht des mirakelfisches, *Calloplesiops altivelis* (Steindachner, 1903). Die Aquarien und Terrarien Zeitschrift 41(3):126-130.

Wassink, H., & R. Brons. 1990. A successful cultivation of the comet *Calloplesiops altivelis* (Steindachner, 1903). SeaScope, Aquarium Systems, Mentor, Ohio, 7(spring):1-3.

Watanabe, K., K. Sezaki, K. Yazawa & A. Hino. 1992. Nutritive fortification of the rotifer *Brachionus plicatilis* with eicosapentaenoic acid-producing bacteria. Nippon Suisan Gakkaishi 58:271-276.

Watanabe, T. 1979. Nutritional quality of living feeds used in seed production of fish. Proc. 7th Japan-Soviet Joint Symp. Aquac., Sep 1978, Tokyo. p 49-57.

Watanabe, T. 1982. Lipid nutrition in fish. Comp. Biochem. Physiol. 73B:3-15.

Watanabe, T. 1985. Importance of the study of broodstock nutrition for further development of aquaculture. In: C. B. Cowey, A. M. Mackie & J. G. Bell (eds.), Nutrition and Feeding in Fish. Academic Press, San Diego, California, p 395-414.

Watanabe, T. 1988. Intensive marine farming in Japan. In: C.J. Shepherd & N.R. Bromage (eds.), Intensive Fish Farming. Blackwell, Oxford, p 239-267.

Watanabe, T. 1990. Enrichment of live foods. In: Advances in Tropical Aquaculture. In: Advances in Tropical Aquaculture. AQUACOP, IFREMER, Actes Colloq. (9):609.

Watanabe, T., & V. Kiron. 1994. Prospects in larval fish dietetics. Aquaculture 124:223-251.

Watanabe, T., & V. Kiron. 1995. Red sea bream (*Pagrus major*). In: N.R. Bromage & R.J. Roberts (eds.), Broodstock Management and Egg and Larval Quality. Blackwell, Oxford, p 398-413.

Watanabe, T., & W. Miki. 1993. Astaxanthin: an effective dietary component for red seabream broodstock. Fish Nutrition in Practice, INRA, Paris, p 27-36.

Watanabe, T., & H. Sakamoto. 1992. Development of a new type of dry pellet for maricultured fish. Bamidgeh 44:124-125.

Watanabe, T., F. Oowa, C. Kitajima & S. Fujita. 1978. Nutritional quality of brine shrimp, *Artemia salina*, as a living feed from the viewpoint of essential fatty acids for fish. Bull. Jpn. Soc. Sci. Fish. 44:1115-1121.

Watanabe, T., C. Kitajima & S. Fujita. 1983a. Nutritional values of live organisms used in Japan for mass propagation of fish: a review. Aquaculture 34:115-143.

Watanabe, T., T. Tamiya, A. Oka, M. Hirata, C. Kitajima & S. Fujita. 1983b. Improvement of dietary value of live foods for fish larvae by feeding them on ω3 highly unsaturated fatty acids and fat-soluble vitamins. Bull. Jpn. Soc. Sci. Fish. 49:471-479.

Watanabe, T., A. Itoh, C. Kitajima & S. Fujita. 1984a. Effect of dietary protein levels on reproduction of red sea bream. Bull. Jpn. Soc. Sci. Fish. 50:1015-1022.

Watanabe, T., T. Arakawa, C. Kitajima & S. Fujita. 1984b. Effect of nutritional quality of broodstock diets on reproduction of red sea bream. Bull. Jpn. Soc. Sci. Fish. 50:495-501.

Watanabe, T., A. Itoh, A. Murakami, Y. Tsukashima, C. Kitajima & S. Fujita. 1984c. Effect of nutritional quality of diets given to broodstock on the verge of spawning on reproduction of red sea bream. Bull. Jpn. Soc. Sci. Fish. 50:1023-1028.

Watanabe, T., A. Itoh, S. Satoh, C. Kitajima & S. Fujita. 1985a. Effect of dietary protein levels and feeding period before spawning on chemical components of eggs produced by red sea bream broodstock.

Bull. Jpn. Soc. Sci. Fish. 51:1501-1509.

Watanabe, T., T. Koizumi, H. Suzuki, S. Satoh, T. Takeuchi, N. Yoshida, T. Kitada & Y. Tsukashima. 1985b. Improvement of quality of red sea bream eggs by feeding broodstock on a diet containing cutttlefish meal or on raw krill shortly before spawning. Bull. Jpn. Soc. Sci. Fish. 51:1511-1521.

Watanabe, T., S. Satoh & T. Takeuchi. 1988. Availability of minerals in fish meal to fish. Asian Fish. Sci. 1:175-195.

Watanabe, T., T. Arakawa, T. Takeuchi & S. Satoh. 1989a. Comparison between eicosapentaenoic and docosahexaenoic acids in terms of essential fatty acid efficiency in juvenile striped jack *Pseudocaranx dentex*. Bull. Jpn. Soc. Sci. Fish. 55:1989-1995.

Watanabe, T., M.S. Izquierdo, T. Takeuchi, S. Satoh & C. Kitajima. 1989b. Comparison between eicosapentaenoic and docosahexaenoic acids in terms of essential fatty acid efficacy in larval red seabream. Bull. Jpn. Soc. Sci. Fish. 55:1635-1640.

Watanabe, T., T. Takeuchi, T. Arakawa, K. Imaizumi, S. Sekiya & C. Kitajima. 1989c. Requirement of juvenile striped jack *Longirostris delicatissimus* for n-3 highly unsaturated fatty acids. Bull. Jpn. Soc. Sci. Fish. 55:1111-1117.

Watanabe, T., H. Sakamoto, M. Abiru & J. Yamashita. 1991a. Development of a new type of dry pellet for yellowtail. Nippon Suisan Gakkaishi 57:891-897.

Watanabe, T., M.-J. Lee, J. Mizutani, T. Yamada, S. Satoh, T. Takeuchi, N. Yoshida, T. Kitada & T. Arakawa. 1991b. Effective components in cuttlefish meal and raw krill for improvement of quality of red seabream *Pagrus major* eggs. Nippon Suisan Gakkaishi 57:681-694.

Watanabe, T., T. Fujimura, M.-J. Lee, K. Fukusho, S. Satoh & T. Takeuchi. 1991c. Effect of polar and nonpolar lipids from krill on quality of eggs of red seabream *Pagrus major*. Nippon Suisan Gakkaishi 57:695-698.

Watanabe, T., J. Pongmaneerat, S. Sato & T. Takeuchi. 1993a. Replacement of fish meal by alternative protein sources in rainbow trout diets. Nippon Suisan Gakkaishi 59:1573-1579.

Watanabe, T., V. Verakunpiriya, K. Mushiake, K. Kawano & I. Hasegawa. 1996a. The first spawn-taking from broodstock yellowtail cultured with extruded dry pellets. Fish. Sci. 62:388-393.

Watanabe, T., V. Kiron, & S. Satoh. 1997. Trace minerals in fish nutrition. Aquaculture 151:185-207.

Watanabe, T., R. Vassallo-Agius, K. Mushiake, K. Kawano, V. Kiron & S. Satoh. 1998a. The first spawn-taking from striped jack broodstock fed soft-dry pellets. Fish. Sci. 64:39-43.

Watanabe. W.O. 1986. Larvae and larval culture. In: C.-S. Lee, M.S. Gordon & W.O. Watanabe (eds.), Aquaculture of Milkfish (*Chanos chanos*): State of the Art. Oceanic Inst., Honolulu, p 117-152.

Watanabe, W.O. 1991. Saltwater culture of tilapia in the Caribbean. World Aquac. 22(1):49-54.

Watanabe, Wade O. 1993 & 1998. University of North Carolina at Wilmington, pers. comm.

Watanabe, W.O. 1995. Aquaculture of the Florida pompano and other jacks (family Carangidae) in the western Atlantic, Gulf of Mexico and Caribbean Basin: status and potential. In: K.L. Main & C. Rosenfeld (eds.), Culture of High-value Marine Fishes in Asia and the United States. Oceanic Inst., Honolulu, p 185-205.

Watanabe, W.O., C.-M. Kuo & M.-C. Huang. 1985b. Salinity tolerance of the tilapias *Oreochromis aureus*, *O. niloticus*, and an *O. mossambicus* × *O. niloticus* hybrid. ICLARM. Tech. Rep. No. 16, 22 p.

Watanabe, W.O., K.E. French, D.H. Ernst, B.L. Olla & R.I. Wicklund. 1989. Salinity during early development influences growth and survival of Florida red tilapia in brackish and seawater. J. World Aquac. Soc. 20:134-142.

Watanabe, W.O., D.H. Ernst, B.L. Olla & R.I. Wicklund. 1990. Aquaculture of red tilapia (*Oreochromis* sp.) in marine environments: state of the art. Advances in Tropical Aquaculture, AQUACOP, IFREMER, Actes de Colloques (9):487-498.

Watanabe, W.O., S.J. Smith, R.I. Wicklund & B.L. Olla. 1992. Hatchery production of Florida red tilapia seed in brackishwater tanks under natural-mouthbrooding and clutch-removal methods. Aquaculture 102:77-88.

Watanabe, W.O., D.H. Ernst, M.P. Chasar, R.I. Wicklund & B.L. Olla. 1993b. The effects of temperature and salinity on growth and feed utilization of juvenile, sex-reversed male Florida red tilapia cultured in a recirculating system. Aquaculture 112:309-320.

Watanabe, W.O., S.C. Ellis, E.P. Ellis, V.G. Lopez, P. Bass, J. Ginoza & A. Moriwake. 1996b. Evaluation of first-feeding regimens for larval Nassau grouper *Epinephelus striatus* and preliminary, pilot-scale culture through metamorphosis. J. World Aquac. Soc. 27:323-331.

Watanabe, W.O., E.P. Ellis, S.C. Ellis, J. Chaves, C. Manfredi, R.W. Hagood, M. Sparsis & S. Arneson. 1998b. Artificial propagation of mutton snapper *Lutjanus analis*, a new candidate marine fish species for aquaculture. J. World Aquac. Soc. 29:176-187.

Watanabe, W.O., M.W. Feeley, S.C. Ellis & E.P. Ellis. 1998c. Light intensity and salinity effects on eggs and yolk sac larvae of the summer flounder. Prog. Fish.-Cult. 60:9-19.

Water Farming Journal. 1992a. May issue, several articles.

Water Farming Journal. 1992b. June issue, several articles.

Waterman, M.P. 1995. Haddock: northern aquaculture's new rising star? Aquac. Mag. 21(2):30,32,34-36.

Watson, J.J., I.G. Priede, P.R. Witthames & A. Owori-Wadunde. 1992. Batch fecundity of Atlantic mackerel, *Scomber scombrus* L. J. Fish Biol. 40:591-598.

Watson, L.C. 1987. Spawning and hatching Atlantic tomcod. Prog. Fish-Cult. 49:69-71.

Watson, L.C. 1992. Temperature-modified prediction of ovulation for striped bass. Prog. Fish-Cult. 54:189-195.

Watson, T.A., K.L. Cooper, M.G. Rankin & E. Garay. 1995. Research and applied investigation to establish a pentachlorophenol (PCP) concentration to protect a salmon creek impacted by wood treatment operations. Can. Tech. Rep. Fish. Aquat. Sci. 2050:46.

Watten, B.J. 1994. Aeration and oxygenation. In: M.B. Timmons & T.M. Losordo (eds.), Aquaculture Water Reuse Systems: Engineering Design and Management. Elsevier, Amsterdam, p 173-208.

Weatherley, A.H., & H.S. Gill. 1987a. Growth increases produced by bovine growth hormone in grass pickerel, *Esox americanus vermiculatus* (Le Seur), and the underlying dynamics of muscle fiber growth. Aquaculture 65:55-66.

Weatherley, A.H., & Gill, H.S. 1987b. The Biology of Fish Growth. Academic Press, London, 443 p.

Weaver, Dallas E. 1995. Scientific Hatcheries, Huntington Beach, California, pers. comm.

Webb, K.L., & F.-L.E. Chu. 1981. Phytoplankton as a food source for bivalve larvae. World Maric. Soc. Spec. Pub. (2):272-291.

Webster, C.D., & R.T. Lovell. 1990a. Comparison of live brine shrimp nauplii and nonliving diets as first food for striped bass larvae. Prog. Fish.-Cult. 52:171-175.

Webster, C.D., & R.T. Lovell, 1990b. Response of striped bass larvae fed brine shrimp from different sources containing different fatty acid combinations. Aquaculture 90:49-61.

Webster, C.D., J.H. Tidwell, L.S. Goodgame-Tiu & D.H. Yancey. 1995. Evaluation of distillers grains with solubles as an alternative plant protein in aquaculture diets. In: C. Lim & D.J. Sessa (eds.), Nutrition and Utilization Technology in Aquaculture. AOSC Press, Champaign, Ilinois, p 189-198.

Wedemeyer, G.A. 1982. Effects of environmental stressors in aquacultural systems on quality, smoltification and early marine survival of anadromous fish. In: Proc. of the North Pacific Aquaculture Symposium. Alaska Sea Grant Rep. No. 82-2, p 155-169.

Wedemeyer, G.A. 1996. Physiology of Fish in Intensive Culture Systems. Chapman & Hall, New York, 232 p.

Wedemeyer, G.A., & J.W. Wood. 1974. Stress as a predisposing factor in fish diseases. U.S. Fish Wildl. Serv. Fish Dis. Leaflet 38, 8 p.

Wedemeyer, G.A., R.L. Saunders & W. C. Clarke. 1981. The hatchery environment required to optimize smoltification in the artificial propagation of anadromous salmonids. Am. Fish. Soc., Fish Cult. Sec. Pub. (1):6-20.

Weeks, N.C., M.B. Timmons & S. Chen. 1992. Feasibility of using foam fractionation for the removal of dissolved and suspended solids from fish culture water. Aquacultural Eng. 11:251-265.

Weis, J.S., & P. Weis. 1987. Pollutants as developmental toxicants in aquatic organisms. Env. Health Persp. 71:77-85.

Weis, J.S., P. Weis & J.L. Ricci. 1981. Effects of cadmium, zinc, salinity, and temperature on the teratogenicity of methylmercury to the killifish (*Fundulus heteroclitus*). Rapp. P.-v. Réun. Cons. Int. Explor. Mer 178:64-70.

Wells, R.M.G., J. Baldwin & R.S. Seymour. 1997. Low concentrations of methaemoglobin in marine fishes of the Great Barrier Reef, Australia. Mar. Freshw. Res. 48:303-309.

Welsh, W.W., & C.M. Breder, Jr. 1922. A contribution to the life history of the puffer, *Spheroides maculatus* (Schneider). Zoologica 2:260-276.

Wester, P.W., & C.J. Roghair. 1992. Teratogenic effect in the gas gland of fish induced by the fabric softener ditallow dimethyl ammonium chloride. Dis. Aquat. Org. 12:207-213.

Weston, D.P. 1991. The effects of aquaculture on indigenous biota. Adv. World Aquac. 3:534-567.

Wheatland, S.B. 1956. Oceanography of Long Island Sound, 1952-1954. VII. Pelagic fish eggs and larvae. Bull. Bingham Oceanogr. Coll. 15:234-314.

Wheaton, F.W. 1977. Aquacultural Engineering. Wiley, New York, 708 p.

Wheaton, F., J. Hochheimer & G.E. Kaiser. 1991. Fixed film nitrification filters for aquaculture. Adv. World Aquac. 3:272-303.

Wheaton, F.W., J.N. Hochheimer, G.E. Kaiser, R.F. Malone, M.J. Krones, G.S. Libey & C.C. Easter. 1994a. Nitrification filter design methods. In: M.B. Timmons & T.M. Losordo (eds.), Aquaculture Water Reuse Systems: Engineering Design and Management. Elsevier, Amsterdam, p 127-171.

Wheaton, F.W., J.N. Hochheimer, G.E. Kaiser, M.J. Krones, G.S. Libey & C.C. Easter. 1994b. Nitrification filter principles. M.B. Timmons and T.M. Losordo (eds.), Aquaculture Water Reuse Systems: Engineering Design and Management. Elsevier, Amsterdam, p 101-126.

Wheeler, P.A., & G.H. Thorgaard. 1991. Cryopreservation of rainbow trout semen in large straws. Aquaculture 93:95-100.

White, A., T.C. Fletcher, C.J. Secombes & D.F. Houlihan. 1993. The effect of different dietary levels of vitamins C and E on their tissue levels in the Atlantic salmon, *Salmo salar* L. Fish Nutrition in Practice, INRA, Paris, p 203-207.

White, B.N., R.J. Lavenberg & G.E. McGowen. 1984. Atheriniformes: development and relationships. Am. Soc. Ichthyol. Herpetol. Spec. Pub. (1):355-362.

White, G.C. 1986. The Handbook of Chlorination. Van Nostrand Reinhold, New York, 1017 p.

White, Roger L. 1993. The Living Seas, Epcot Center, Disney World, Orlando, Florida, pers. comm.

Whitehead, P.J.P., M.-L. Bauchot, J.-C. Hureau, J. Nielsen & E. Tortonese (eds.). 1984. Fishes of the North-eastern Atlantic and the Mediterranean. UNESCO, Paris, 3 vol., 1473 p.

Whitehurst, A., & H.R. Robinette. 1994. Tolerance of juvenile red drum *Sciaenops ocellatus* to rapidly decreasing water temperatures. J. World Aquac. Soc. 25:225-229.

Whitson, J., P. Turk & P. Lee. 1993. Biological denitrification in a closed recirculating marine culture system. In: J.-K. Wang (ed.), Techniques for Modern Aquaculture. Am. Soc. Agric. Eng., St. Joseph, Michigan, p 458-466.

Whyte, J.N.C. 1987. Biochemical composition and energy content of six species of phytoplankton used in mariculture of bivalves. Aquaculture 60:231-241.

Whyte, J.N.C., & W.D. Nagata. 1990. Carbohydrate and fatty acid composition of the rotifer, *Brachionus plicatilis*, fed monospecific diets of yeast or phytoplankton. Aquaculture 89:263-272.

Whyte, J.N.C., W.C. Clarke, N.G. Ginther, J.O.T. Jensen & L.D. Townsend. 1994. Influence of composition of *Brachionus plicatilis* and *Artemia* on growth of larval sablefish (*Anoplopoma fimbria* Pallas). Aquaculture 119:47-61.

Wickins, J.F., & M.M. Helm. 1981. Sea water treatment. In: A.D. Hawkins (ed.), Aquarium Systems. Academic Press, London, p 63-128.

Wicklund, R.I. 1970. Observations on the spawning of the cunner in waters of northern New Jersey. Chesapeake Sci. 11:137.

Widodo, J. 1991. Maturity and spawning of shortfin scad (*Decapturus macrosoma*)(Carangidae) of the Java Sea. Asian Fish. Sci. 4:245-252.

Wiegand, M.D. 1982. Vitellogenesis in fishes. In: C.J.J. Richter & M.J.T. Goos (compilers). Proc. Internat. Symp. Repr. Physiol. Fish. Wageningen, Netherlands, p 136-146.

Wiegand, M.D., L.W. Crim, C.E. Wilson, C.M. Hewitt & J.M. Loewen. 1987. A study of oocyte development and induction of spawning in the winter flounder following LHRH analog implantation. Proc. Third Internat. Symp. Repr. Physiol. Fish, Memorial Univ. Newfoundland, St. John's, p 104-105.

Wiles, M. 1981. Successful rearing of striped bass, *Morone saxatilis* (Walbaum), from fertilised eggs in small aquaria, with reference to environment, feeding and growth, and bacterial disease. Rapp. P.-v. Réun. Cons. Int. Explor. Mer 178:518-521.

Wilkerson, J.D. 1992a. *Amphiprion clarkii*: journey from egg to juvenile: part I. Freshw. Mar. Aquarium 15(10):17,19,22-24.

Wilkerson, J.D. 1992b. *Amphiprion clarkii*: journey from egg to juvenile: part II. Freshw. Mar. Aquarium 15(12):154,156,158,160,162.

Wilkerson, J. 1996. C-Quest Hatchery: innovations in captive ornamental marine fish culturing. Freshw. Mar. Aquarium 19(4):123-126,128,130,134,136,138.

Wilkerson, J.D. 1998. Clownfishes: A Guide to Their Captive Care, Breeding & Natural History. Micocosm Ltd, Shelburne, Vermont, 240 p.

Wilkins, M.E., & M.W. Saunders (eds.). 1997. Biology and Management of Sablefish, *Anoplopoma fimbria*. NOAA Tech. Rep. NMFS 130, 275 p.

Willey, J.D. 1977. Coprecipitation of zinc with silica in seawater and in distilled water. Mar. Chem. 5:267-290.

Williams, C.D., & E.H. Robinson. 1988. Response of red drum to various dietary levels of menhaden oil. Aquaculture 70:107-120.

Williams, K.C., & B.J. Blaney. 1992. The potential of moulds to reduce the value of feedstuffs. In: G.L. Allan & W. Dall (eds.), Proc. Aquac. Nutr. Workshop. New South Wales Fisheries, Brackish Water Fish Cult. Res. Sta., Salamander Bay, Australia, p 205-213.

Williams, S.F., & R.S. Caldwell. 1978. Growth, food conversion and survival of 0-group English sole (*Parophrys vetulus* Girard) at five temperatures and five rations. Aquaculture 15:129-139.

Williams, S., R.T. Lovell & J.P. Hawke. 1985. Value of menhaden oil in diets of Florida pompano. Prog. Fish.-Cult. 47:159-165.

Williot, P. (ed.). 1991. Acipenser. CEMAGREF, Bordeaux, 519 p.

Williot, P. 1990. Sturgeon farming. In: G. Barnabé (ed.), Aquaculture: Vol. 2. Ellis Horwood, New York, p 735-755.

Williot, P., R. Brun, T. Rouault & O. Rooryck. 1991. Management of female spawners of the Siberian sturgeon, *Acipenser baeri* Brandt: first results. In: P. Williot (ed.), Acipenser. CEMAGREF, p 365-379.

Wilson, D.P. 1958. Notes from the Plymouth Aquarium. III. J. Mar. Biol. Assn. U.K. 37:299-307.

Wilson, D.P. 1978. Territorial behaviour of male dragonets (*Callionymus lyra*). J. Mar. Biol. Assn. U.K. 58:731-734.

Winberg, G.G. 1956. Rate of metabolism and food requirements of fishes. Beloruss. State Univ., Minsk. (Fish. Res. Bd. Can., Trans. Ser. No. 194, 253 p, 1960).

Winemiller, K.O., & K.A. Rose. 1992. Patterns of life-history diversification in North American fishes: implications for population regulation. Can. J. Fish. Aquat. Sci. 49:2196-2218.

Winger, P.V., & P.J. Lasier. 1994. Effects of salinity on striped bass eggs and larvae from the Savannah River, Georgia. Trans. Am. Fish. Soc. 123:904-912.

Winton, J.R. 1991. Recent advances in detection and control of infectious hematopoietic necrosis virus in aquaculture. Ann. Rev. Fish Dis. 1:83-93.

Wise, D.J., C.R. Weirich & J.R. Tomasso. 1989. Toxicity of ammonia to red drum *Sciaenops ocellatus*

fingerlings with information on uptake and depuration. J. World Aquac. Soc. 20:188-192.
Wissing, T.E. 1974. Energy transformations by young-of-the-year white bass *Morone chrysops* (Rafinesque) in Lake Mendota, Wisconsin. Trans. Am. Fish. Soc. 103:32-37.
Withler, F.C., & L.C. Lim. 1982. Preliminary observations of chilled and deep-frozen storage of grouper (*Epinephelus tauvina*) sperm. Aquaculture 27:389-392.
Witthames, P.R., & M.G. Walker. 1995. Determinacy of fecundity and oocyte atresia in sole (*Solea solea*) from the Channel, the North Sea and the Irish Sea. Aquat. Living Resour. 8:91-109.
Wohlfarth, G. 1990. The tools of genetics in aquaculture. In: R. Flos, L. Tort & P. Torres (eds.), Mediterranean Aquaculture. Ellis Horwood, New York, p 167-180.
Wohlfarth, G.W., & G. Hulata. 1989. Selective breeding of cultivated fish. In: M. Shilo & S. Sarig (eds.), Fish Culture in Warm Water Systems: Problems and Trends. CRC Press, Boca Raton, Florida, p 21-63.
Woiwode, J.G., & I.R. Adelman. 1991. Effects of temperature, photoperiod, and ration size on growth of hybrid striped bass × white bass. Trans. Am. Fish. Soc. 120:217-229.
Wolfsheimer, G. 1957. A spawning of porcupine puffers. The Aquarium 26:288-290.
Woltering, D.M. 1984. The growth response in fish chronic and early life stage toxicity tests: a critical review. Aquat. Toxicol. 5:1-21.
Wolters, W.R., & R. DeMay. 1996. Production characteristics of striped bass × white bass and striped bass × yellow bass hybrids. J. World Aquac. Soc. 27:202-207.
Wong, H.-C., L.-L. Chen & C.-M. Yu. 1995. Occurrence of vibrios in frozen seafoods and survival of psychotrophic *Vibrio cholerae* in broth and shrimp homogenate at low temperatures. J. Food Protection 58:263-267.
Wong, P.S., & R.S.S. Wu. 1987. Red tides in Hong Kong: problems and management strategy with special reference to the mariculture industry. J. Shoreline Mgt. 3:1-21.
Woo, N.Y.S., & S.F. Chiu. 1994. Effects of nitrite exposure on growth and survival of sea bass, *Lates calcarifer*, fingerlings in various salinities. J. Appl. Aquac. 4:45-54.
Woo, N.Y.S., & S.P. Kelly. 1995. Effects of salinity and nutritional status on growth and metabolism of *Sparus sarba* in a closed seawater system. Aquaculture 135:229-238.
Woo, P.T.K. 1992. Immunological responses of fish to parasitic organisms. Ann. Rev. Fish Dis. 2:339-366.
Woods, L.C., III & C.V. Sullivan. 1993. Reproduction of striped bass, *Morone saxatilis* (Walbaum), broodstock: monitoring maturation and hormonal induction of spawning. Aquac. Fish. Mgt 24:211-222.
Woods, L.C., III, J.H. Kerby & M.T. Huish. 1985. Culture of hybrid striped bass to marketable size in circular tanks. Prog. Fish-Cult. 47:147-153.
Woods, L.C., III, J.G. Woiwode, M.A. McCarthy, D.D. Theisen & R.O. Bennett. 1990. Noninduced spawning of captive striped bass in tanks. Prog. Fish-Cult. 52:201-202.
Woods, L.C., III, R.O. Bennett & C.V. Sullivan. 1992. Reproduction of a domestic striped bass brood stock. Prog. Fish-Cult. 54:184-188.
Woods, L.C., III, C.C. Kohler, R.J. Sheehan & C.V. Sullivan. 1995. Volitional tank spawning of female striped bass with male white bass produces hybrid offspring. Trans. Am. Fish. Soc. 124:628-632.
Wootten, R. 1989. The parasitology of teleosts. In R.J. Roberts (ed.), Fish Pathology. Bailliere Tindall, London, p 242-288.
Wootton, R.J. 1982. Environmental factors in fish reproduction. In: C.J.J. Richter & M.J.T. Goos (compilers). Proc. Internat. Symp. Repr. Physiol. Fish. Wageningen, Netherlands, p 210-219.
Wourms, J.P. 1994. The challenges of piscine viviparity. Israel J. Zool. 40:551-568.
Wrobel, D. 1988. The dwarf angels of the genus *Centropyge*. SeaScope, Aquarium Systems, Mentor, Ohio, 5(spring):1,3.
Wu, C.-L., & C.-L. Kuo. 1993. Maturity and fecundity of albacore, *Thunnus alalunga* (Bonnaterre), from the Indian Ocean. J. Fish. Soc. Taiwan 20:135-152.
Wu, R.S.S. 1990. Biological and economic factors in the selection of cultured fish species and the development of a bio-economic model. Advances in Tropical Aquaculture, AQUACOP, IFREMER, Actes Colloq. (9):437-444.
Wu, R.S.S. 1994. Application of scientific research to solve practical problems in marine fish culture management. Third Asian Fisheries Forum, Asian Fish. Soc., Manila, p 164-167.
Wu, R.S.S., & N.Y.S. Woo. 1983. Tolerance of hypo-osmotic salinities in thirteen species of adult marine fish: implications for estuarine fish culture. Aquaculture 32:175-181.
Wu, R.S.S., & N.Y.S. Woo. 1984. Respiratory responses and tolerance to hypoxia in two marine teleosts, *Epinephelus akaara* (Temminck & Schlegel) and *Mylio macrocephalus* (Basilewsky). Hydrobiologia 119:209-217.
Wu, R.S.S., K.S. Lam, D.W. MacKay, T.C. Lau & V. Yam. 1994a. Impact of marine fish farming on water quality and bottom sediment: a case study in the sub-tropical environment. Mar. Env. Res. 38:115-145.
Wu, Y.V., R. Rosati, D.J. Sessa & P. Brown. 1994b. Utilization of protein-rich ethanol co-products from corn in tilapia feed. J. Am. Oil Chem. Soc. 71:1041-1043.
Wu, Y.V., R.R. Rosati, D.J. Sessa & P.B. Brown. 1995a. Evaluation of corn gluten meal as a protein source in tilapia diets. J. Agric. Food Chem. 43: 1585-1588.
Wu, Y.V., R. Rosati, D.J. Sessa & P. Brown. 1995b. Utilization of corn gluten feed by Nile tilapia. Prog. Fish.-Cult. 57: 305-309.

Wu, Y.V., R.R. Rosati & P.B. Brown. 1996a. Effect of diets containing various levels of protein and ethanol coproducts from corn on growth of tilapia fry. J. Agric. Food. Chem. 44:1491-1493.

Wu, Y.V., K. Warner, R. Rosati, D.J. Sessa & P. Brown. 1996b. Sensory evaluation and composition of tilapia (*Oreochromis niloticus*) fed diets containing protein-rich ethanol by-products from corn. J. Aquat. Food Prod. Technol. 5:7-16.

Wurts, W.A., & R.R. Stickney. 1993. Growth rates of juvenile red drum *Sciaenops ocellatus* reared on commercial salmon feed in fresh and salt water. J. World Aquac. Soc. 24:422-424.

Xu, B.-T., S. Yamasaki & H. Hirata. 1993. Supplementary *Ulva* sp. var. meal level in diet of Japanese flounder, *Paralichthys olivaceus*. Suisanzoshoku 41:461-468. (in Jpn., Eng. abs.)

Xu, D., & W.A. Rogers. 1993. Oxytetracycline residue in hybrid striped bass muscle. J. World Aquac. Soc. 24:466-472.

Yakupitiyage, A. 1991. Constraints to the use of plant fodder as fish feed in tropical small-scale tilpaia culture systems: an overview. Fish Nutrition in Practice, INRA, Paris, p 681-689.

Yamada, J., & M. Kusakari. 1991. Staging and the time course of embryonic development in kurosoi, *Sebastes schlegeli*. Env. Biol. Fishes. 30:103-110.

Yamada, K., K. Kobayashi & Y. Yone. 1980. Conversion of linolenic acid to w3-highly unsaturated fatty acids in marine fishes and rainbow trout. Bull. Jpn. Soc. Sci. Fish. 46:1231-1233.

Yamagami, K. 1988. Mechanisms of hatching in fish. In: W.S. Hoar & D.J. Randall (eds.), Fish Physiology, Vol. 11, Part A. Academic Press, San Diego, p 447-499.

Yamagami, K., T.S. Hamazaki, S. Yasumasu, K. Masuda & I. Iuchi. 1992. Molecular and cellular basis of formation, hardening, and breakdown of the egg envelope in fish. Internat. Rev. Cytol. 136:51-92.

Yamagata, Y., S. Oonaka, M. Harada & M. Niwa. 1983. Influence of concentration of dissolved oxygen on the growth of Japanese eel *Anguilla japonica*. Bull. Jpn. Soc. Sci. Fish. 49:1335-1339.

Yamaha Fishery Journal. 1989. Pioneering the culture of saltwater fishes. Fishery Journal, Yamaha Motor Co., Iwata, Shizuoka, Japan, No. 29, May, 8 p.

Yamaha Fishery Journal. 1990. The "king of fish," its history and development as a fishery. Fishery Journal, Yamaha Motor Co., Iwata, Shizuoka, Japan, No. 33, Jul, 8 p.

Yamaha Fishery Journal. 1991. The big shift from trawl fishery to fish farming and aquaculture. Fishery Journal, Yamaha Motor Co., Iwata, Shizuoka, Japan, No. 37, Oct, 8 p.

Yamaha Fishery Journal. 1992a. Internationalism comes to the Japanese eel industry. Fishery Journal, Yamaha Motor Co., Iwata, Shizuoka, Japan, No. 39, Mar, 8 p.

Yamaha Fishery Journal. 1992b. Propagation within an ecosystem. Fishery Journal, Yamaha Motor Co., Iwata, Shizuoka, Japan, No. 38, Jan, 8 p.

Yamamoto, K. 1980. Cleavage pattern and formation of the blastocoel in the egg of the eel, *Anguilla japonica*. Jpn. J. Ichthyol. 27:301-308.

Yamamoto, K. 1981. Cleavage pattern and formation of the blastocoel in the egg of the eel, *Anguilla japonica*. Jpn. J. Ichthyol. 27:301-308.

Yamamoto, K., & K. Yamauchi. 1974. Sexual maturation of Japanese eel and production of eel larvae in the aquarium. Nature 251:220-222.

Yamamoto, T.S., & W. Kobayashi. 1992. Closure of the micropyle during embryonic development of some pelagic fish eggs. J. Fish Biol. 40:225-241.

Yamamoto, T., P.A. Marcouli, T. Unuma & T. Akiyama. 1994. Utilization of malt protein flour in fingerling rainbow trout diets. Fish. Sci. 60:455-460.

Yamamoto, T., T. Unuma & T. Akiyama. 1995a. The effect of combined use of several alternative protein sources in fingerling rainbow trout diets. Fish. Sci. 61:915-920.

Yamamoto, T., T. Unuma & T. Akiyama. 1995b. Utilization of malt protein flour in fingerling Japanese flounder diets. Bull. Nat. Res. Inst. Aquac. (24):33-42.

Yamamoto, T., T. Unuma, T. Akiyama & S. Kishi. 1996. Utilization of malt protein flour in the diets for fingerling red sea bream. Fish. Sci. 62:59-63.

Yamasaki, S., & K. Tanaka. 1996. Possibility of utilizing sterile mutant *Ulva pertusa* as a bio-filter in a closed culture system of the rotifer *Brachionus plicatilis*. Fish. Sci. 62:323-324.

Yamasaki, S., D.H. Secor & H. Hirata. 1987. Population growth of two types of rotifer (L and S) *Brachionus plicatilis* at different dissolved oxygen levels. Nippon Suisan Gakkaishi 53:1303.

Yamashita, K. 1982a. Differentiation of the swimbladder structure in larvae of the red seabream *Pagrus major*. Jpn. J. Ichthyol. 29:193-202.

Yamashita, K. 1982b. Sensory cupulae found in prelarvae of red seabream *Pagrus major*. Jpn. J. Ichthyol. 29:279-284.

Yamauchi, K., & K. Yamamoto. 1982. Experiments on artificial maturation and fertilization of the Japanese eel (*Anguilla japonica*). In: C.J.J. Richter & M.J.T. Goos (compilers). Proc. Internat. Symp. Repr. Physiol. Fish. Wageningen, Netherlands, p 185-189.

Yamazaki, F. 1976. Application of hormones in fish culture. J. Fish. Res. Bd. Can. 33:948-958.

Yamazaki, F. 1983. Sex control and manipulation in fish. Aquaculture 33:329-354.

Yang, X., & T.A. Dick. 1993. Effects of dietary fatty acids on growth, feed efficiency and liver RNA and DNA content of Arctic charr, *Salvelinus alpinus* (L.). Aquaculture 116:57-70.

Yao, C., D. Hu, M. Shi & G. Hu. 1993. Studies on antibiotic 230 as a molluscicide. Jpn. J. Med. Sci. Biol. 46:103-119.

Yao, S.-J., T. Umino & H. Nakagawa. 1994. Effect of feeding frequency on lipid accumulation in ayu. Fish. Sci. 60:667-671.

Yasuda, K., & N. Taga. 1980. Culture of *Brachionus plicatilis* Müller using bacteria as food. Bull. Jpn. Soc. Sci. Fish. 46:933-939.

Ye, L.-X., D.A. Ritz, G.E. Fenton & M.E. Lewis. 1991. Tracing the influence on sediments of organic waste from a salmonid farm using stable isotope analysis. J. Exp. Mar. Biol. Ecol. 145:161-174.

Yeager, D.M., J.E. Van Tassel & C.M. Woolley. 1990. Collection, transportation, and handling of striped bass brood stock. In: R.M. Harrell, J.H. Kerby & R.V. Minton (eds.), Culture and Propagation of Striped Bass and Its Hybrids. Am. Fish. Soc., Bethesda, Maryland, p 29-42.

Yeh, S.-L., Y.-Y. Ting & C.-M. Kuo. 1987. Induction of sex reversal and spawning of grouper *Epinephelus salmonoides Epinephelus fario*. Bull. Taiwan Fish. Res. Inst. 43:143-152. (In Ch., Eng. abs.)

Yeheskel, O., & R.R. Avtalion. 1988. Artificial fertilization of tilapia eggs, a preliminary study. In: Y. Zohar & B. Breton (eds.), Reproduction in Fish. Basic and Applied Aspects in Endocrinology and Genetics. Colloq. INRA No. 44, p 169-175.

Yi, Y.-H., & Y.-J. Chang. 1994. Physiological effects of seamustard supplement diet on the growth and body composition of young rockfish, *Sebastes schlegeli*. Bull. Kor. Fish. Soc. 27:69-82. (in Kor., Eng. abs.)

Yoklavich, M.M., & E.K. Pikitch. 1989. Reproductive status of Dover sole, *Microstomus pacificus*, off northern Oregon. Fish. Bull. U.S. 87:988-995.

Yokoyama, K., Y. Kamei, M. Toda, K. Hirano & Y. Iwatsuki. 1994. Reproductive behavior, eggs and larvae of a caesionine fish, *Pterocaesio digramma*, observed in an aquarium. Jpn. J. Ichthyol. 41:261-274.

Yonamine, M. 1989. Growth and spawning of the spangled emperor, *Lethrinus nebulosus* (Forsskal) reared in the net-cage at Ishigaki Island of the Ryukyus. Suisanzoshoku 37:253-257. (in Jpn.)

Yone, Y. 1976. Nutritional studies of red sea bream. In: K.S. Price, W.N. Shaw & K.S. Danberg (eds.), Proc. First Internat. Conf. Aquac. Nutr., Univ. Delaware, Lewes/Rehoboth, p 39-64.

Yone, Y. 1978. Essential fatty acids and lipid requirements of marine fish. In: Jpn. Soc. Sci. Fish. (eds.), Dietary Lipids in Aquaculture. Koseisha-Koseikaku, Tokyo, p 43-59.

Yone, Y., M. Furuichi & S. Sakamoto. 1971. Studies on nutrition of red sea bream. III. Nutritive value and optimum content of lipids in diet. Rep. Fish. Res. Lab., Kyushu Univ., (1):49-60.

Yone, Y., M. Furuichi & K. Urano. 1986a. Effects of dietary wakame *Undaria pinnatifida* and *Ascophyllum nodosum* supplements on growth, feed efficiency, and proximate compositions of liver and muscle of red sea bream. Bull. Jpn. Soc. Sci. Fish. 52:1465-1468.

Yone, Y., M. Furuichi & K. Urano. 1986b. Effects of wakame *Undaria pinnatifida* and *Ascophyllum nodosum* on absorption of dietary nutrients, and blood sugar and plasma free amino-N levels of red sea bream. Bull. Jpn. Soc. Sci. Fish. 52:1817-1819.

Yone, Y., M.A. Hossain, M. Furuichi & F. Kato. 1986c. Effect of fermented and fermented-resteamed scrap meals on growth and feed efficiency of red sea bream. Bull. Jpn. Soc. Sci. Fish. 52:549-552.

Yoshimatsu, T., S. Matsui & C. Kitajima. 1992a. Early development of laboratory-reared keelback mullet. Nippon Suisan Gakkaishi 59:765-776.

Yoshimatsu, T., S. Matsui & C. Kitajima. 1992b. Early development of laboratory-reared redlip mullet, *Liza haematocheila*. Aquaculture 105:379-390.

Yoshimatsu, T., M. Furuichi & C. Kitajima. 1992c. Optimum level of protein in purified experimental diets for redlip mullet. Nippon Suisan Gakkaishi 58:2111-2117.

Yoshimatsu, T., M. Furuichi & C. Kitajima. 1993. Effects of dietary lipid levels on the growth, efficiency of feed utilization and body composition of young redlip mullet. J. Fac. Agr. Kyushu Univ. 37:273-281.

Yoshimatsu, T., M. Hayashi, K. Toda, M. Furuichi & C. Kitajima. 1995. Preliminary experiment on the requirement of larval redlip mullet for essential fatty acids, and the supplemental effect of *Nannochloropsis* to rearing water. Nippon Suisan Gakkaishi 61:912-918. (in Jpn., Eng. abs.)

Yoshimura, K., A. Hagiwara, T. Yoshimatsu & C. Kitajima. 1996a. Culture technology of marine rotifers and the implications for intensive culture of marine fish in Japan. Mar. Freshw. Res. 47:217-222.

Yoshimura, K., Y. Ohmori, T. Yoshimatsu, K. Tanaka & A. Ishizaki. 1996b. On the aeration method in high density culture of rotifer *Brachionus rotundiformis*. Nippon Suisan Gakkaishi 62:897-903.

Young, F.A. 1980. The spawning and larval rearing of the jackknife fish. Freshw. Mar. Aquarium 3(5):31-32,78.

Young, F.A. 1991. Rearing the Red Sea anemonefish *Amphiprion bicinctus*. Freshw. Mar. Aquarium 14(4):136,142.

Young, G., C.L. Brown, R.S. Nishioka, L.C. Folmar, M. Andrews, J.R. Cashman & H.A. Bern. 1994. Histopathology, blood chemistry, and physiological status of normal and moribund striped bass (*Morone saxatilis*) involved in summer mortality ('die-off') in the Sacramento-San Joaquin Delta of California. J. Fish Biol. 44:491-512.

Young, J.A., M.F. Capra & A.W. Blackshaw. 1992. Cryopreservation of summer whiting *Sillago ciliata* spermatozoa. Aquaculture 102:155-160.

Young, P.S., & C.E. Dueñas. 1993. Salinity tolerance of fertilized eggs and yolk-sac larvae of the rabbitfish *Siganus guttatus* (Bloch). Aquaculture 112:363-377.

Yousif, O.M., & G.A. Alhadhrami. 1993. Bioresource Technol. 45:153-155.

Yousif, O.M., G.A. Alhadhrami & M. Pessarakli. 1994. Evaluation of dehydrated alfalfa and salt bush (*Atriplex*) leaves in diets for tilapia (*Oreochromis aureus* L.). Aquaculture 126:341-347.

Yousif, O.M., M.F. Osman & G.A. Alhadhrami. 1996. Evaluation of dates and date pits as dietary ingredients in tilapia (*Oreochromis aureus*) diets differing in protein sources. Bioresource Technol. 57:81-85.

Yu, J.-P., & K. Hirayama. 1986. The effect of un-ionized ammonia on the population growth of the rotifer in mass culture. Bull. Jpn. Soc. Sci. Fish. 52:1509-1513.

Yu, J.-P., K. Hirayama & A. Hino. 1994. The role of bacteria in mass culture of the rotifer *Brachionus plicatilis*. Bull. Nat. Res. Inst. Aquac., Suppl. 1, p 67-70.

Yu, T.C., & R.O. Sinnhuber. 1979. Effect of dietary ω3 and ω6 fatty acids on growth and feed conversion efficiency of coho salmon (*Oncorhyncus kisutch*). Aquaculture 16:31-38.

Yu, T.-C., & T.-Y. Tung. 1991. Broodstock management of blackfish (*Girella punctata*). Abs. in program book, Finfish Hatchery in Asia '91. Tungkang Marine Lab., Tungkang, Taiwan, p 53.

Yu, T.-C., C.-L. Tsai, Y.-S. Tsai & J.-Y. Lai. 1991. Induced breeding of Japanese eel (*Anguilla japonica*). Abs. in program book, Finfish Hatchery in Asia '91. Tungkang Mar. Lab., Tungkang, Taiwan, p 51.

Yúfera, M. 1987. Effect of algal diet and temperature on the embryonic development time of the rotifer *Brachionus plicatilis* in culture. Hydrobiologia 147:319-322.

Yúfera, M., E. Pascual & A. Polo. 1991. The importance of prey sequencing during early growth of the gilthead seabream *Sparus aurata* L. Eur. Aquac. Soc. Spec. Pub. (15):116-118.

Yúfera, M., E. Pascual & J. Guinea. 1993. Factors influencing the biomass of the rotifer *Brachionus plicatilis* in culture. Hydrobiologia 255/256:159-164.

Yúfera, M., C. Fernández-Díaz & E. Pascual. 1995. Feeding rates of gilthead seabream (*Sparus aurata*), larvae on microcapsules. Aquaculture 134:257-268.

Yúfera, M., M.C. Sarasquete & C. Fernández-Diáz. 1996. Testing protein-walled microcapsules for the rearing of first-feeding gilthead sea bream (*Sparus aurata* L.) larvae. Mar. Freshw. Res. 47:211-216.

Zamarro, J. 1992. Determination of fecundity in American plaice (*Hippoglossoides platessoides*) and its variation from 1987 to 1989 on the tail of the Grand Bank. Neth. J. Sea Res. 29:205-209.

Zanuy, S., M. Carrillo & F. Ruiz. 1986. Delayed gametogenesis and spawning of sea bass (*Dicentrarchus labrax* L.) kept under different photoperiod and temperature regimes. Fish Physiol. Biochem. 2:53-63.

Zastrow, C.E., E.D. Houde & L.G. Morin. 1991. Spawning, fecundity, hatch-date frequency and young-of-the-year growth of bay anchovy *Anchoa mitchilli* in mid-Chesapeake Bay. Mar. Ecol. Prog. Ser. 73:161-171.

Zeldis, J.R., P.J. Grimes & J.K.V. Ingerson. 1995. Ascent rates, vertical distribution, and a thermal history model of development of orange roughy, *Hoplostethus atlanticus*, eggs in the water column. Fish. Bull., U.S. 93:373-385.

Zhang, Q., & G. Uhlig. 1991. Progress in mass cultivation of harpacticoid copepods of the genus *Tisbe*. Eur. Aquac. Soc. Spec. Pub. (15):98-100.

Zhang, Q., & G. Uhlig. 1993. Dry weight and chemical composition (CHN) in relation to population density of cultivated *Tisbe holothuriae* (Copepoda, Harpacticoida). Helgoländer Meeresunters 47:221-227.

Zhang, W. 1992. Growth and development of *Bostrichthys sinensis* larvae and juveniles. Tropic Oceanol. 11:100-104.

Zheng, Z.-A., L.-H. Cai, J.-C. Ye, Q.-R. Wang & D.-S. Xu. 1993. Studies on the artificial propagation and larval rearing of Japanese sea bass, (*Lateolabrax japonicus*). Taiwan Fish. Res. Inst. Conf. Proc. (3):177-183. (in Ch., Eng. abs.)

Zheng, F., T. Takeuchi, K. Yoseda, M. Kobayashi, J. Hirokawa & T. Watanabe. 1996. Requirement of larval cod for arachidonic acid, eicosapentaenoic acid, and docosahexaenoic acid using by their enriched *Artemia* nauplii. Nippon Suisan Gakkaishi 62:669-676.

Zhuravleva, N.G.. 1995. Mariculture and its potential along the Murmansk coast. ICES Mar. Sci. Symp. 201:187-188.

Zhuravleva, N.G., & I.I. Petrov. 1993. The method of cod larvae rearing under the climatic conditions of East Murman. Eur. Aquac. Soc. Spec. Pub. (19):492.

Zillioux, E.J. 1969. A continuous recirculating culture system for planktonic copepods. Mar. Biol. 4:215-218.

Zohar, Y. 1988. Gonadotropin releasing hormone in spawning induction in teleosts: basic and applied considerations. In: Y. Zohar & B. Breton (eds.), Reproduction in Fish. Basic and Applied Aspects in Endocrinology and Genetics. Colloq. INRA No. 44, p 47-62.

Zohar, Y. 1989. 1989. Fish reproduction: its physiology and artificial manipulation. In: M. Shilo & S. Sarig (eds.), Fish Culture in Warmwater Systems: Problems and Trends. CRC Press, Boca Raton, Florida, p 65-119.

Zohar, Y. 1991. Biotechnologies for the manipulation of reproduction in fish farming. 1991 Internat. Mar. Biotechnol. Conf., Vol. II, p 486-490.

Zohar, Y. 1996. New approaches for the manipulation of ovulation and spawning in farmed fish. Bull. Natl. Res. Inst. Aquac. Suppl. 2:43-48.

Zohar, Y., & H. Gordin. 1979. Spawning kinetics in the gilthead sea-bream, *Sparus aurata* L. after low doses of human chorionic gonadotropin. J. Fish Biol. 15:665-670.

Zohar, Y., G. Pagelson, M. Tosky & Y. Finkelman. 1987. GnRH control of gonadotropin secretion, ovulation and spawning in the gilthead seabream *Sparus aurata*. Proc. Third Internat. Symp. Repr. Physiol. Fish. Memorial Univ. Newfoundland, St. John's, p 106.

Zohar, Y., M. Harel, S. Hassin & A. Tandler. 1995. Gilt-head sea bream (*Sparus aurata*). In: N.R. Bromage & R.J. Roberts (eds.), Broodstock Management and Egg and Larval Quality. Blackwell, Oxford, p 94-117.

Zumpe, D. 1963. Uber das Ablaichen von *Thalassoma bifasciatum* (Bloch). Die Aquarien- und Terrarien-Zeitschrift 16:86-88. (in Ger.)

Zuñiga, H.N., & E.S. Acuña. 1992. Larval development of two sympatric flounders, *Paralichthys adspersus* (Steindachner, 1867) and *Paralichthys microps* (Gunther, 1881) from the Bay of Coquimbo, Chile. Fish. Bull., U.S. 90:607-620.

Zurlini, G., I. Ferrari & A. Nassogne. 1978. Reproduction and growth of *Euterpina acutifrons* (Copepoda: Harpacticoida) under experimental conditions. Mar. Biol. 46:59-64.

GLOSSARY

α-starch. Partly hydrolyzed starch.
Abcess. Swollen inflamed area of tissue containing pus.
Absorb. Attract and retain internally.
Acacia. The plant from which gum arabic is extracted.
Acetylcholinesterase. Enzyme that catalyses hydrolysis of acetylcholine to choline and acetic acid.
Acid detergent fiber. In feeds, mainly cellulose, lignin, and silica, which are very indigestible.
Acidosis. Buildup of acid or loss of alkali from the body, to a harmful level.
Acinar. Concerned with a terminal sac of an acinous type of gland (multicellular with secretory sacs).
Actin. Protein in muscle fiber that complexes with myosin to cause muscle contraction and relaxation.
Actomyosin. Complex of actin and myosin.
Adhesion. Joining of parts to each other, sometimes abnormal.
Adrenal gland. Endocrine gland near the kidney, which secretes hormones necessary for nutrient utilization.
Adsorb. Attract and retain on the surface.
Air dry. State of feed as fed, usually with about 10% moisture.
Aldehyde. Organic compound that contains an aldehyde functional group (—CHO).
Alkaline phosphatase. Enzyme that catalyzes hydrolysis and synthesis of phosphoric acid esters and transfer of phosphate groups from phosphoric acid to other substituted compounds under alkaline conditions.
Alkalosis. Buildup of alkali or loss of acid from the body, to a harmful level.
Ambicoloration. Pigmentation on the normally white, eyeless side of a flatfish.
Amino acid. Basic unit of protein (molecule contains an amino, NH_2, and a carboxyl, COOH, group).
Amylase. Enzyme that catalyzes hydrolysis of starch into simpler substances.
Anabolism. Use of nutrients in the formation and repair of body tissue (constructive metabolism).
Anemia. Subnormal number of red cells and/or volume of hemoglobin in the blood.
Anisocytosis. Excessive size variation in red blood cells.
Ankylosis. Joint immobility and consolidation.
Anorexia. Pathological loss of appetite.
Anoxia. Lack of oxygen.
Antihelmintic. Drug used against parasitic worms.
Arteriosclerosis. Thickening and loss of elasticity in arterial walls.
Artesian well. One from which water will rise higher that the level of the aquifer supplying it.
Ascites. Buildup of serous fluid in the abdominal cavity.
Ash. Minerals remaining after organic matter is incinerated.
Assimilation. Processing of food in the body (i.e., digestion, absorption, distribution, and metabolism).
Ataxia. Irregularity or lack of muscular coordination.
Atrophy. Pathological wasting of part of the body.
β-starch. Raw starch.
Bilirubin. Bile pigment from the breakdown of heme and reduction of biliverdin.
Bioaccumulation. Uptake and deposition of a contaminant in an organism.
Bioconcentration. Increasing concentration of a contaminant at a single level of a food chain or web.
Biogenic. Produced by living organisms.
Biomagnification. Increasing concentration of a contaminant at successively higher levels in a food chain or web.
Biopsy. Removal of tissue from a live body for examination.
Botulism. Food poisoning from the neurotoxin botulin, produced by *Clostridium botulinum* in spoiled food.
Bran. Outer coarse coat of grain that is separated during processing.
Cachexia. Serious general bad health and/or malnutrition.
Calcinosis. Abnormal deposition of calcium salts in tissue.
Calculus. Abnormal concretion (usually mineral salts) in the body.
Calorie. A unit of energy; by definition, the amount of heat needed to raise the temperature of 1 g of water by 1°C.
Cannula. Tube used for drawing fluid from the body.

Carboxylation. Addition of a carboxyl group (—COOH)
Carcinogen. Something that causes cancer.
Cartilage. A fibrous connective tissue.
Caseation. Necrosis resulting in a dry, cheese-like appearance.
Catabolism. Breaking down of nutrients to produce heat and work (destructive metabolism).
Cataract. Opacity of the eye lens or capsule.
Cellulose. The primary polysaccharide (which supports the cell wall) in plants; consists of chains of 2000-4000 units of ($C_6H_{10}O_5O_n$).
Cercaria. Final, free-swimming larval stage of a parasitic trematode.
Ceroid body. Small mass of a yellow to brown pigment that occurs especially in the liver and is related to the diet.
Chitin. A horny polysaccharide, a major structural component of arthropod exoskeletons.
Chitosan. Substance formed by partial deacetylation of chitin with alkali.
Cholesterol. A fatlike steroid alcohol ($C_{27}H_{45}OH$) that occurs in animal fats, bile, blood, milk, brain, nerves, liver, kidney, adrenal glands, and egg yolk; a precursor of bile acids and is used in steroid hormone synthesis.
Cholinesterase. Enzyme that catalyzes hydrolysis of acylcholine to choline and an anion.
Choroid. The vascular layer of the eye between the sclera and retina.
Ciliate. Protozoans in the phylum Ciliophora; they use cilia for propulsion.
Cladoceran. A "water flea," member of the crustacean order Cladocera; they mostly inhabit fresh and brackish water.
Coenzyme. An organic compound (usually containing phosphorus and a vitamin) that works with an enzyme for a particular purpose.
Collagen. The protein substance in the white fibers of all animal connective tissue.
Combustion. Combination of a substance with oxygen, causing liberation of heat.
Compensatory growth. Accelerated growth after a period of limited feeding.
Complete diet. One that fulfills all known nutritional requirements of the animal.
Complete protein. One with an amino acid balance necessary for normal growth.
Condensed. Made denser by removing moisture.
Congestion. Abnormal blood accumulation.
Copepod. A member of the crustacean subclass Copepoda, which contains about nine orders: Calanoida and Harpacticoida are of the most interest for feeding fish; the other orders contain parasites. Copepods probably are the most abundant multicellular animals in the world.
Cornea. Transparent anterior part of the eye.
Corticosteroid. Except for the sex hormones, steroids produced by the adrenal cortex, or synthesized ones.
Cortisol. A major corticosteroid hormone that increases with acute stress and is used as an indicator of stress.
Crossbite. Misaligment of opposing teeth.
Crude fat. Substances extracted from moisture-free material by ether (mainly fats and oils with small amounts of waxes, resins, and pigments).
Crude fiber. In feeds, cellulose, hemicellulose, and lignin. Determined by first extracting fats with ether, then boiling the remainder alternately in a weak acid and a weak base until only crude fiber and ash are left; finally ash % is subtracted.
Crude protein. True protein plus non-protein nitrogenous compounds (e.g., nucleic acids).
Crumbles. Granular feed, often made by breaking and sieving larger pieces.
Cryopreservation. Preservation of live organisms or parts of them at very low temperature.
Cyst. 1. An epithelium-lined sac or cavity, which usually contains liquid or semisolid substances and can be normal or abnormal. 2. Stage of certain animals (especially parasites) protected by a capsule-like wall.
Cytochrome P-450. A group of enzymes that act mainly in the liver to help detoxify organic contaminants. Because production of cytochrome P-450 quickly increases (days) in response to toxicants, its level can be used as an early warning sign of toxicity from organic pollutants.
Decarboxylation. Removal of a CO_2 molecule from a compound.
Demersal. Existing near the bottom of a natural body of water.
Dermatitis. Inflammation of the skin.
Dextrin. A carbohydrate resulting from the hydrolysis of starch to sugar.
Diathesis. Unusual predisposition to a particular disease.
Digestible nutrient. The portion of a nutrient that is absorbed by an animal.

Digestibility. The ratio of nutrient absorbed to nutrient ingested (%).
Disaccharide. Type of sugar that can be hydrolyzed into two monosaccharides.
Dissolved organic carbon (DOC). The carbon content of dissolved organic matter (DOM).
Dissolved solids. As opposed to particulate solids.
Dropsy. Abnormal buildup of serous fluid in tissue or a body cavity.
Dry matter. Percentage of moisture-free material.
Dry matter basis. Moisture-free basis.
Dystrophy. Any disorder caused by faulty nutrition.
EC_{50}. Concentration of a dissolved or gaseous substance that has a designated effect on 50% of a test group.
Edema. Swelling caused by excess intracellular fluid retention.
Embolism. Sudden blockage of an artery by a clot, gas bubble, or other substance carried by the blood flow.
Emphysema (in fish). Pathological accumulation of gas in tissue, especially the skin.
Emulsification. Division of larger fat or oil particles into smaller, better dispersed ones.
Endocrine gland. One that secretes directly into the blood or lymph (ductless).
Endogenous. Originating within the body (e.g., from yolk).
Endosperm. Carbohydrate portion of a seed.
Energy. The capacity to do work.
Energy feed. One high in energy and low in fiber (below 18%), usually with less than 20% protein.
Enteritis. Inflammation of the intestine.
Eosinophil. Specialized white blood cells that accumulate in response to an allergen (e.g., during parasitism).
Epicarditis. Inflammation of the outer wall of the heart.
Epilepsy. A temporary disturbance of brain function, usually recurring.
Erucic acid. A toxic monoethenoid acid ($C_{22}H_{42}O_2$) that constitutes 25-55% of rapeseed oil, and also occurs in wallflower seed and mustard seed. Canola is a rapeseed variety with low (2%) erucic acid content.
Erythropoiesis. Production of red blood cells.
Essential nutrient. One that is necessary for normal growth, maintenance, activity, health, and reproduction, but cannot be made in the body and must be obtained from exogenous sources.
Esterase. Enzyme that catalyzes hydrolysis of an ester into an alcohol and acid.
Ester. Compound produced by reaction of an acid and an alcohol with elimination of a water molecule.
Estradiol. A potent estrogen.
Estrogen. Female sex hormones.
Ether extract. Fatty substances soluble in ether.
Exogenous. Originating outside the body (e.g., from food).
Exonuclease. A type of nuclease (enzymes that split nucleic acids).
Exophthalmia. Abnormal protrusion of the eye.
Expanded. Feed produced by holding ingredients under pressure until a chosen moment when the pressure is suddenly released. Such feeds have low density and usually float in water.
Expeller process. This uses a screw press to mechanically remove oil from seeds.
Extrusion. Forcing feed ingredients through a die with pressure.
Exudation. Release of a fluid containing protein and cellular debris from blood vessels into tissues or onto their surfaces, usually during imflammation.
Fat. A naturally-occurring mixture of triglycerides; usually solid at room temperature (often used to refer to lipids in general).
Fat soluble. Soluble in fats and fat solvents but usually not in water.
Fatty acid. A monobasic aliphatic acid consisting of only carbon, hydrogen, and oxygen, which will combine with glycerol to form a fat.
Feed conversion ratio. Dry (or air-dry) weight of dry feed necessary to produce an increment (wet) of growth.
Fibrinous. Containing fibrin (insoluble protein necessary for blood clotting).
Fines. Material that will pass through a screen of specified mesh size (e.g., dust from pellets).
Flagellates. Protozoans possessing one or more flagella and classified in several phyla. Phytoflagellates are important as foods for rotifers, copepods, *Artemia*, and other filter feeders. Some zooflagellates are parasitic.
Flashing. Quick scraping against a solid object to remove an irritant (e.g., parasite), usually very obvious to an observer.

Flocculation. Formation of a colloidal suspension of discrete (usually visible) particles, as opposed to coagulation, which forms a continuous mass.
Flour. Soft, finely ground and bolted meal of cereal grains and other seeds (mainly gluten and starch of the endosperm).
Free amino acid. Amino acid not combined in a protein. Most amino acids in the body are contained in proteins; free amino acids are mainly transitional. Sometimes FAAs are included in feeds.
Free radical. Atomic, diatomic, or polyatomic molecule that has at least one unpaired electron.
Fructose. A monosaccharide ($C_6H_{12}O_6$) from honey and fruits.
Furuncle. A boil, nodule in the skin with a central core caused by bacterial infection.
Gelatinization. Completely rupturing starch granules by a combination of moisture, heat, and pressure, sometimes also with mechanical shear.
Germ. Embryo of a seed.
Glucose. A monosaccharide ($C_6H_{12}O_6$), the main internal energy source for living organisms (also called dextrose).
Glucosinolates. Compounds in rapeseed and related plants that break down into other compounds that interfere with thyroid function. Canola is a rapeseed variety with low glucosinolate content.
Glutathione peroxidase. Enzyme that catalyzes reactions that detoxify hydrogen peroxide and organic peroxides.
Glutathione. A tripeptide important in plant and animal oxidation reactions; contains glutamic acid, cysteine, and glycine, $C_{10}H_{17}N_3O_6S$. It combines with certain organic contaminants to help detoxify them, mainly in the liver.
Glutathione transferase. An enzyme that acts mainly in the liver to help detoxify certain organic contaminants. Because production of glutathione transferase increases slowly in response to toxicants, its level can be used as an index of chronic toxicity.
Gluten. Proteinaceous substance remaining after starch and germ are extracted from wheat, corn, and other grains.
Glycogen. A polysaccharide that is the main stored carbohydrate in animals; it can be broken down into glucose molecules as needed.
Glycopeptide. Type of peptide that contains carbohydrate.
Gonadotropin. Any hormone that stimulates the gonads.
Gossypol. A toxic yellow pigment in cottonseed.
Granuloma. Tumor-like nodule or mass of granulation tissue caused by chronic inflammation.
Groats. Grain from which hulls have been removed.
Gross growth efficiency. Weight (or energy) of fish gained divided by weight (or energy) of food eaten in a given time, usually on a dry:dry basis.
Heat increment. Heat loss from digestion and transformation of food.
Hemagglutinin titer. Measure of hemagglutinin, an antibody that causes agglutinization (clumping) of erythrocytes.
Hematocrit. Percent volume of red cells in whole blood.
Hematoma. Abnormal internal accumulation of blood, especially subcutaneous.
Hematopoietic. Concerning the formation of blood cells.
Hemoglobin. The oxygen-carrying pigment in red blood cells; contains amino acids and iron.
Hemolysis. Release of hemoglobin from red blood cells.
Hemorrhage. Bleeding.
Hemosiderin. An insoluble form of iron stored in tissue.
Histopathology. 1. Microscopic examination for disease signs. 2. Disease signs seen microscopically.
Homeostasis. Tendency toward normal physiological stability.
Hull. Outer covering of grain.
Hyaline. A clear, homogeneous, structureless substance in the matrix of cartilage, vitreous body, mucin, and glycogen.
Hydraulic process. This uses a hydraulic press to mechanically remove oil from seeds.
Hydrolysis. Splitting of a compound into smaller molecules by chemical addition of water.
Hydroxylation. A reaction adding at least one hydroxyl group (OH^-) to an organic molecule.
Hypercalcemia. Abnormally high blood calcium (antonym hypocalcemia).
Hypercapnia. Excess carbon dioxide in the blood (antonym hypocapnia).
Hyperchromic. Applied to red blood cells with excess hemoglobin and color (antonym hypochromic).
Hyperemia. Presence of excess blood in part of the body (antonym hyporemia).
Hyperglycemia. Abnormally high blood glucose (antonym hypoglycemia).
Hyperkalemia. Abnormally high blood potassium (antonym hypokalemia).

Hypermagnesemia. Abnormally high blood magnesium (antonym hypomagnesemia).
Hypernatremia. Abnormally high blood sodium (antonym hyponatremia).
Hyperoxia. Excess oxygen in the system (antonym hypoxia).
Hyperplasia. Increased size of an organ or tissue from an abnormal increase in the number of normal cells in normal arrangement.
Hypertrophy. Increased size of an organ or tissue from an increase in cell size.
Hyphae. Filaments that make up the mycelium of a fungus.
Hypervitaminosis. Abnormal condition resulting from vitamin overdose (antonym hypovitaminosis).
Incomplete diet. One that does not fulfill all known nutritional requirements of the animal.
Infarct. Localized necrosis from insufficient blood flow.
Inflammation. Response of tissue to injury or destruction to exclude the damaging agent and protect the tissues.
Iritis. Inflammation of the iris (pigmented membrane behind the cornea through which the pupil opens).
Ischemia. Insufficient blood flow.
Isopod. Member of the crustacean order Isopoda; usually dorsoventrally flattened; many species are carnivorous or parasitic.
Isthmus. The usually narrow, ventral structure connecting the lower jaw with the trunk of a fish.
IU (International unit). A standard unit of potency for a biological substance (e.g., vitamin, hormone, antibiotic, antitoxin) defined by the International Conference for Unification of Formulae and based on bioassays.
Joule. An international unit of work; by definition, 1 joule can maintain a 1-ampere current against 1 ohm of resistance for 1 s.
Keratin. A sulfur-containing protein, the main component of epidermis, hair, wool, hoof, horn, and the organic matrix of teeth.
Keratinization. Formation of an outer horny layer of skin.
Keratosis. A horny growth.
Ketone. An organic compound with a carbonyl group (C=O) contained in the carbon chain.
Kilogram. 1000 grams; 2.2046 pounds. 1 pound = 453.6 grams.
Kyphosis. Curvature of the spine in a dorsal direction.
Lactose. A disaccharide ($C_{12}H_{22}O_{11}$), milk sugar.
LC_{50}. Concentration of a dissolved or gaseous substance that kills 50% of the test subjects after a certain time.
LD_{50}. Dosage (weight/weight) of a substance that kills 50% of the test subjects after a certain time.
Lesion. A pathological discontinuity or dysfunction of a body part.
Leucocyte. White blood cell.
Leucopenia. Abnormally low number of white cells in the blood.
Leucosis. Abnormally high white blood cell count.
Limiting nutrient. The one most likely to result in a deficiency (e.g., limiting amino acid).
Lipids. Fats and fatlike substances, including fatty acids, neutral fats, waxes, phosphatides, sterols, and soaps.
Lipositol. Inositol.
Listless. Inactive and languorous.
Lordosis. Curvature of the spine in a ventral direction.
Lymphocyte. A non-granular white blood cell produced by lymphoid tissue that functions in immunity.
Lymphoma. Tumor of lymphoid tissue.
Macrocytic anemia. Types of anemia with abnormally large red cells (antonym microcytic)
Maltose. A disaccharide ($C_{12}H_{22}O_{11}$) resulting from hydrolysis of starch by amylase.
Meal. Material with a slightly larger particle size than a flour.
Medusa (pl. medusae). Jellyfish stage of members of the cnidarian classes Hydrozoa and Scyphozoa.
Megaloblastic anemia. Anemia with megaloblasts (immature form of an abnormal type of red blood cell) in the bone marrow.
Megalops (pl. megalopa). The last larval stage of a crab, which resembles a crab with a tail and can swim or walk.
Metabolism. All changes in nutrients after they have been absorbed.
Metabolite. Substance resulting from a metabolic process.
Metacercaria. Encysted resting or maturing stage of a parasitic trematode in an intermediate host or in the environment.
Metanauplius. Crustacean larval stage following the nauplius.

Middlings. Flour milling byproduct, combination of bran, shorts, germ, flour, and offal from the "tail of the mill."

Miliary. Resembling millet seeds.

Mill run. Ungraded (usually uninspected) milling byproduct, combination of bran, shorts, germ, flour, and offal from the "tail of the mill."

Monosaccharide. A simple sugar, which cannot be hydrolyzed.

Mucus. Viscous fluid secreted by mucous membranes and glands (mainly mucin, inorganic salts, sloughed cells, white blood cells, and water).

Mutagen. Something that causes genetic mutation.

Mycotoxin. Toxic metabolite produced by fungus.

Myocardium. Middle, muscular layer of the heart wall.

Myopathy. Any muscle disease.

Myosin. Part of muscle fiber that complexes with actin to cause muscle contraction.

Nauplius. The simplest type of crustacean larva, occurring in some members of all classes, including copepods and *Artemia*.

Necrosis. Death of cells.

Nephrocalcinosis. Precipitation of calcium phosphate in kidney tubules, causing blockage and renal insufficiency.

Net growth efficiency. Energy gained by the fish divided by the energy of food eaten less the energy of egestion, excretion, and heat increment.

Neutral detergent fiber. In feeds, acid detergent fiber plus hemicellulose, which is more or less digestible for certain animals.

Neutral fat, neutral lipid. A triglyceride (typical fat).

Neutrophil. Type of granular white blood cell with a lobed nucleus.

Nitrogen-free extract. Mainly sugars, starches, pentoses, and non-nitrogenous organic acids, determined by subtracting percent moisture, crude protein, crude fat, crude fiber, and ash from 100%.

Nodule. Circumscribed, raised, firm lesion.

Non-settleable suspended solids. Particulate matter that will sink less than 1 m/h in still water ($\approx$ particles <50 µm).

Normochromic anemia. Types of anemia with normal hemoglobin content and normally colored red blood cells.

Normocytic anemia. Anemia with decreased hemoglobin content, red cell volume, and red cell number.

Nutrient. Substance used by an organism for growth, maintenance, activity, health, and reproduction (mainly proteins, lipids, carbohydrates, vitamins, minerals, and water).

Nutrient allowance. Recommended amount to include in feed to compensate for variations in animals, feeds, and conditions (nutrient requirement plus safety margin).

Nutrient requirement. Minimum amount needed for normal growth, maintenance, activity, health, and reproduction, with no safety margin added.

Nutrient to energy ratio. Weight of a specific nutrient or class of nutrient per unit of feed energy (e.g., mg protein/kcal energy).

Oil. Any of various greasy, combustible materials (derived from animals, plants, or minerals) that are insoluble in water, soluble in certain organic solvents, and usually liquid at room temperature.

Oligosaccharide. Carbohydrate from which a small number of monosaccharides are produced when it is hydrolyzed.

Operculum (opercle). One of the main bones supporting the gill cover of fish.

Ossification. Formation of or conversion into bone.

Particulate organic carbon (POC). The carbon content of particulate organic matter (POM).

Pelagic. Occurring in the open ocean.

Pellet. Feed produced when ingredients are mixed, treated with steam, then compacted by forcing them through die openings.

Pepsin. Enzyme in gastric juice that catalyzes hydrolysis of proteins to polypeptides.

Pericardium. Sac that contains the heart and large blood vessels connected to it.

Peritonitis. Inflammation of the peritoneum (membrane lining the abdominal cavity).

Peroxidation. Formation of a peroxide (oxide containing two atoms of oxygen linked by a single bond).

Petechial. Referring to pinhead-sized hemorrhages (petechiae).

pH. Measure of acidity or alkalinity of a liquid. The strongest acid is at pH 0, neutral at pH 7, and the strongest base at pH 14.

Phospholipid. Fatlike substance that contains phosphorus, nitrogen, fatty acids, and cholesterol.

Photophobia. Abnormally low visual tolerance of light.

Photosynthesis. Primarily, the production of carbohydrates from water and carbon dioxide in plant tissue containing chlorophyll that is exposed to light.
Phytate. Salt of phytic acid (phosphorus compound in cereal grains), which binds calcium, phosphorus, zinc, and other minerals making it impossible for monogastric animals to absorb them.
Planktonic. Living organism that normally exists drifting in an aquatic environment.
Plasma (blood). Colorless, fluid part of blood without the corpuscles.
Platelet. Disc-shaped blood cell that functions in clotting.
Poikilocytosis. Presence of a type of abnormal red blood cell, oversized and without a nucleus.
Polyunsaturated fatty acid. One that has more than one double bond.
Practical feed (diet). One made from ingredients that are economical and available enough for use in production feeds.
Premix. Mixture of microingredients with a carrier, used to promote uniform mixing.
Preservative. A chemical used to prevent oxidation, fermentation, or other deterioration—especially by bacteria.
Proerythrocyte. Immature red blood cell.
Prostaglandin. One of a group of hydroxy fatty acids that can affect several physiological processes including the action of certain hormones.
Protein. Complex organic compound consisting mainly of amino acids in characteristic sequence and proportion (always contain carbon, hydrogen, oxygen, nitrogen, usually sulfur, often phosphorus).
Protein:energy ratio. Usually stated as weight/energy (i.e., mg/cal, mg/mJ).
Protein sparing. Increased utilization of dietary carbohydrate or fat for energy so that more protein can be used for growth.
Protease. A type of enzyme that hydrolyzes proteins into (smaller) polypeptides.
Prothrombin. During blood clotting, the part of plasma that changes to the enzyme thrombin, which converts the protein fibrinogen into fibrin to form the clot.
PUFA. Polyunsaturated fatty acid, one with two or more double bonds.
Pugheaded. With an abnormally short, thickened head.
Purified diet. A test diet made from standard, well-defined ingredients.
Pustule. Small, elevated, pus-containing skin lesion.
Pyloric cecum. Pouchlike extensions of the midgut that probably function mainly to increase surface area for better absorption of nutrients.
Pyruvate. Salt, ester, or anion of pyruvic acid—the end product of glycolysis.
Rancid. Describes fats decomposed by hydrolysis or oxidation.
Ration. Amount of feed eaten during a specified time interval, usually 24 h.
Rice polishings. Fine byproduct remaining after rice kernels are polished.
Red dog. Wheat flour milling byproduct, offal from the "tail of the mill" plus fine particles of bran, germ, and flour.
Respiration. The process by which living cells take in oxygen and oxidize organic compounds, causing energy, carbon dioxide, water, etc. to be released
Retina. Interior lining of the back of the eye that transforms visual stimuli into nerve impulses.
Rheotaxis. Movement preferentially with or against the flow of liquid.
Rotifers. Members of the phylum Rotifera (the smallest multicellular animals), which mainly live in fresh, brackish, and coastal waters.
Salinity. Concentration of dissolved solids in seawater, given in parts per thousand (ppt).
Saponine. One of a group of glucosides from plants that form a lasting foam when shaken with water; some are toxic.
Saturated fat. One that is completely hydrogenated; all carbon atoms associated with the maximum number of hydrogen atoms; no double bonds.
Scoliosis. Lateral curvature of the spine.
Semi-purified diet. One containing both practical ingredients (e.g., fish meal, wheat flour) and ingredients typically used only for research (e.g., casein, starch).
Septicemia. Systemic disease with pathogenic microorganisms or their toxins persisting in the blood.
Serum (blood). Colorless, fluid part of blood that remains after clotting and removal of corpuscles (i.e., plasma minus fibrinogen).
Settleable suspended solids. Particulate matter that will sink at least 1 m/h in still water ($\approx$ particles $\geq$50 µm).
Shorts. Flour milling byproduct, small particles of bran, germ, flour, and offal from the "tail of the mill."
Silage. Material that has undergone anaerobic fermentation.

Solubles. Byproduct liquid from animal or plant processing, contains dissolved and possibly fine suspended substances.
Solvent-extracted. Fat and oil have been removed by organic solvents.
Spongiosis. Edema occurring between epidermal cells.
Steatitis. Inflammation of fatty tissue.
Stress. Physical or psychological factor to which an animal does not adapt, resulting in harm.
Sucrose. A disaccharide ($C_{12}H_{22}O_{11}$), usually obtained from sugar cane or sugar beets.
Suspended solids. See total, settleable, and non-settleable suspended solids.
TAN. Total ammonia nitrogen—concentration of only the nitrogen present as ammonia, excluding the hydrogen.
TDN (total digestible nutrients). A measure of the energy value of food.
% TDN = % DCP + % DCF + % DNFE + (% DEE x 2.25)
DCP = digestible crude protein; DCF = digestible crude fiber; DNFE = digestible nitrogen-free extract; DEE = digestible ether extract.
Teratogen. Something that causes a physical defect in a developing animal.
Testosterone. The primary male hormone ($C_{19}H_{28}O_2$), released by the testes.
Tetany. Presence of localized, spasmodic muscle contractions.
Total suspended solids (TSS). All particulate matter in the water column.
Toxemia. Presence of toxins (especially bacterial) in the blood.
Transaminase. Type of enzyme that catalyzes transfer of an amino group from an amino acid to a keto acid (both an acid and a ketone), with help from the coenzyme pyridoxal phosphate, to produce another amino acid.
Transferrin. A type of beta globulin in blood serum that binds and transports iron to the bone marrow and storage areas.
Transformation. Final phase of fish larval development, during which the full complement of meristic elements (especially fin rays and scales) is completed and the body takes a form similar to that of the adult.
Transketolase. An enzyme that splits a substrate at the carbonyl group.
Triglyceride. A neutral fat or lipid; the molecule consists of three fatty acids esterified to glycerol.
Trochophore. The first larval stage of bivalve molluscs (also in other phyla), which swims, lacks a shell for about a day, and can be used as food for fish larvae.
True protein. Nitrogenous compound that can be hydrolyzed completely into amino acids.
Trypsin. A proteolytic enzyme originating in the pancreas that catalyzes hydrolysis of peptide linkages in proteins and partly hydrolyzed proteins.
Tubercle. A small, rounded mass caused by infection with *Mycobacterium* spp.
Ulcer. Defect in an organ or tissue caused by sloughing of necrotic inflammatory tissue.
Ultrasound. High-frequency sound, >20,000 hertz (cycles per second).
Unknown factors. Unidentified dietary substances that promote growth or other processes in animals. Such factors seem to exist in dried whey, animal byproducts, distiller's solubles, antibiotic fermentation residues, alfalfa, and green forages.
Unsaturated fat. One that is not completely hydrogenated and therefore has one or more double bonds.
Unsaturated fatty acid. One that contains at least one double bond.
Vacuole. A membrane-lined cavity inside a cell, for digestion, storage, secretion, or excretion.
Ventilation. Movement of water past the gills for the purpose of gas exchange.
Vitamin. A complex organic compound necessary in small amounts (usually for proper functioning of enzyme systems) and essential for normal growth, maintenance, activity, health, and reproduction.
Vitellogenin. A lipophosphoprotein, produced in the liver and incorporated into yolk proteins in the ovary.
Weaning. In fish, the transition from living to non-living food.
Xanthophyll. A type of yellow carotenoid pigment, derived from carotenes, found in certain flowers, fruits, and leaves.
Xerophthalmia. Dryness and thickening of the conjunctiva (mucus membrane covering the eyeball and eye socket) and cornea.
Zein. A combustible white to yellowish protein from corn (insoluble in water but soluble in dilute alcohol) used for microencapsulation and coating food.

Conversion Factors

To convert	to	multiply by
Acres	Hectares	0.4047
Acres	Square feet	43560
Acres	Square meters	4047
Acres	Square yards	4840
Acres-feet	Cubic feet	43560
Acres-feet	Cubic meters	1233.49
Acres-feet	U.S. gallons	325851
Atmospheres	Bars	101.32
Atmospheres	Feet of fresh water	33.932
Atmospheres	Feet of seawater	33.044
Atmospheres	Inches of mercury	29.921
Atmospheres	Millimeters of mercury	760.000
Atmospheres	Pounds/square inch	14.696
Bars	Atmospheres	0.98697
British thermal unit	Foot-pounds	777.5
British thermal unit	Kilocalories	0.252
British thermal unit	Kilogram-meters	107.5
British thermal unit	Kilowatt-hours	0.0002926
British thermal unit/minute	Horsepower	0.02356
British thermal unit/minute	Kilowatts	0.01757
Calories	Joules	4.184
Cattie (China)	Kilograms	2
Centimeters	Feet	0.03281
Centimeters	Inches	0.3937
Chain (Gunters)	Feet	66
Cubic centimeters	Cubic inches	0.06102
Cubic feet	Cubic inches	1728
Cubic feet	Cubic meters	0.02832
Cubic feet	Liters	28.32
Cubic feet	U.S. gallons	7.48052
Cubic feet of fresh water	Pounds of fresh water	62.3
Cubic feet of seawater	Pounds of seawater	64.0
Cubic feet/second	Million gallons/day	0.646317
Cubic inches	Cubic centimeters	16.387
Cubic inches	Liters	0.01639
Cubic meters	Cubic feet	35.314
Cubic meters	Cubic yards	1.308
Cubic meters	U.S. gallons	264.2
Cubic yards	Cubic meters	0.7646
Fathoms	Feet	6
Feet	Meters	0.3048
Feet of fresh water	Atmospheres	0.295
Feet of fresh water	Inches of mercury	0.8826
Feet/second	Knots	0.5921
Gallons (Imperial	Cubic inches	277.42
Gallons (Imperial)	U.S. gallons	1.20095
Gallons (Imperial)	Liters	4.546
Gallons (U.S.)	Cubic feet	0.1337
Gallons (U.S.)	Cubic inches	231.00
Gallons (U.S.)	Imperial gallons	0.83267
Gallons (U.S.)	Liters	3.785
Gallons (U.S) of fresh water	Pounds of fresh water	8.3453
Grains (troy)	Grams	0.0648

Conversion Factors (continued)

To convert	to	multiply by
Grams	Grains (troy)	15.43
Grams	Ounces	0.03527
Grams	Pounds	0.002205
Grams/liter	Parts per thousand	~1
Hectares	Acres	2.471
Hectares	Mu (China)	15
Hectares	Square feet	107600
Hectares	Square meters	10000
Horsepower	BTU/minute	42.44
Horsepower	Foot-pounds/minute	33000
Horsepower	Horsepower (metric)	1.014
Horsepower	Kilocalories/minute	10.7
Horsepower	Kilowatts	0.7457
Horsepower (boiler)	BTU/hour	33479
Horsepower (boiler)	Kilowatts	9.803
Inches	Centimeters	2.54
Inches of fresh water	Atmospheres	0.002458
Inches of mercury	Atmospheres	0.03342
Inches of mercury	Feet of fresh water	1.133
Inches of mercury	Pounds/square foot	70.73
Joules	Calories	0.2390
Kilocalories	BTU	3.968
Kilograms	Pounds	2.2046
Kilometers	Feet	3281
Kilometers	Nautical miles	0.5400
Kilometers	Statute miles	0.6214
Kilometers/hour	Knots	0.5400
Kilowatts	BTU/minute	56.92
Kilowatts	Foot-pounds/minute	44250
Kilowatts	Horsepower	1.341
Kilowatts	Kilocalories/minute	14.34
Kilowatt-hours	BTU	3415
Kilowatt-hours	Foot-pounds	2655000
Kilowatt-hours	Kilocalories	860.5
Kilowatt-hours	Kilogram-meters	367100
Liters	Cubic centimeters	1000
Liters	Cubic feet	0.03531
Liters	Cubic inches	61.024
Liters	Imperial gallons	0.2200
Liters	U.S. gallons	0.2642
Meters	Feet	3.281
Meters	Inches	39.37
Meters	Yards	1.094
Miles (nautical)	Feet	6076.1
Miles (nautical)	Kilometers	1.852
Miles (nautical)	Minutes of latitude	1
Miles (statute)	Feet	5280
Miles (statute)	Kilometers	1.609
Miles (statute)/hour	Knots	0.8690
Milligrams/liter	Parts per million	~1
Milliliters	Cubic centimeters	1
Million gallons/day	Cubic feet/second	1.547
Mils	Inches	0.001
Mu (China)	Hectares	0.0667
Ounces	Grains (troy)	437.48

Conversion Factors (continued)

To convert	to	multiply by
Ounces	Grams	28.349
Ounces	Ounces (troy)	0.9115
Ounces	Pounds	0.0625
Ounces (fluid)	Cubic inches	1.805
Ounces (fluid)	Milliliters	29.57
Pounds	Grams	453.5924
Pounds	Ounces	16
Pounds of fresh water	Cubic feet of fresh water	0.01602
Pounds of fresh water	Cubic inches of fresh water	27.68
Pounds of fresh water	U.S. gallons of fresh water	0.1198
Pounds/cubic foot	Kilograms/cubic meter	16.02
Quintal (Argentina)	Kilograms	45.94
Quintal (Brazil)	Kilograms	58.75
Quintal (Chile)	Kilograms	46.00
Quintal (metric)	Kilograms	100
Quintal (Mexico)	Kilograms	46.02
Quintal (Peru)	Kilograms	46.00
Rods	Feet	16.5
Square centimeters	Square inches	0.155
Square feet	Square meters	0.0929
Square inches	Square centimeters	6.452
Square kilometers	Acres	247.1
Square kilometers	Square statute miles	0.3861
Square meters	Square feet	10.764
Square meters	Square yards	1.196
Square statute miles	Acres	640
Square statute miles	Square kilometers	2.59
Square yards	Square meters	0.8361
Tonnes (metric tons)	Kilograms	1000
Tonnes (metric tons)	Pounds	2205
Tons (long)	Kilograms	1016
Tons (long)	Pounds	2240
Tons (short)	Kilograms	907.185
Tons (short)	Pounds	2000
Total ammonia	Total ammonia nitrogen	0.7777
Total nitrate	Nitrate nitrogen	0.2258
Total nitrite	Nitrite nitrogen	0.3043
Yards	Meters	0.9144

SPECIES INDEX, BY FAMILY

PETROMYZONTIDAE	**LAMPREYS**, 249.
Petromyzon marinus	Sea lamprey, 271.
RAJIDAE	**SKATES**
Raja eglanteria	Clearnose skate, 79.
ACIPENSERIDAE	**STURGEONS**, 5, 7, 8, 187, 197, 232, 236, 249, 254, 265, 266, 268, 270, 289, 291, 298, 353, 387, 406, 414, 504, 514, 522, 533, 574.
Acipenser baerii	Siberian sturgeon, 5, 8, 212, 240, 353, 377, 387, 420.
Acipenser brevirostrum	Shortnose sturgeon, 51, 220.
Acipenser desotoi	Gulf of Mexico sturgeon, 240, 353.
Acipenser fulvescens	Lake sturgeon, 51.
Acipenser gueldenstaedtii	Russian sturgeon, 240.
A. gueldenstaedtii × *H. huso*	Russian sturgeon × Beluga, 379, 385, 415.
Acipenser medirostris	Green sturgeon, 533.
Acipenser naccarii	Italian sturgeon, 353.
Acipenser oxyrhynchus	Atlantic sturgeon, 18, 35, 51, 218, 220, 415, 533.
Acipenser ruthenus	Sterlet, 5, 8, 240, 420.
Acipenser stellatus	Starry sturgeon, 56, 219, 220, 240, 533.
Acipenser transmontanus	White sturgeon, 35, 51, 93, 197, 198, 220, 240, 258, 267, 334, 376, 386, 394, 414, 465, 530, 533.
Acipenser spp.	(sturgeons), 198.
Huso huso	Beluga, 5, 8, 240.
H. huso × *A. ruthenus*	Bester, 205, 533.
Scaphyrhynchus albus	Pallid sturgeon, 220.
POLYDONTIDAE	**PADDLEFISHES**, 254, 504.
Polyodon spathula	American paddlefish
ELOPIDAE	**TARPONS**
Megalops atlanticus	Tarpon, 17, 218, 220, 533.
ALBULIDAE	**BONEFISHES**
Albula vulpes	Bonefish, 220.
ANGUILLIDAE	**"FRESHWATER" EELS**, 5, 7, 8, 23, 207, 211, 238, 406, 415, 494, 498, 508, 514, 533-534, 577.
Anguilla anguilla	European eel, 5, 8, 41, 51, 74, 78, 82, 129, 135, 297, 353, 376, 377, 379, 381, 384, 385, 386, 387, 415, 421, 465, 493, 530, 533, 534.
Anguilla australis	Shortfin eel, 5, 8.
Anguilla japonica	Japanese eel, 5, 8, 41, 51, 82, 204, 238, 240, 258, 259, 291, 297, 302, 353, 379, 382, 384, 388, 397, 398, 399, 400, 401, 402, 403, 408, 415, 421, 465, 492, 493, 494, 497, 501, 526, 534, 572.
Anguilla rostrata	American eel, 5, 8, 13, 31, 42, 100, 111, 127, 128, 129, 131, 218, 220, 353, 534.
MURAENIDAE	**MORAY EELS**, 577.
Muraenesox cinereus	Pike eel, 41, 271.
CONGRIDAE	**CONGER EELS**, 577.
Heteroconger spp.	(garden eels), 577.
CLUPEIDAE	**HERRINGS**, 249, 270, 291, 502.
Alosa aestivalis	Blueback herring, 43, 88, 219, 220, 477.
Alosa mediocris	Hickory shad, 220, 262.
Alosa pseudoharengus	Alewife, 219, 220.
Alosa sapidissima	American shad, 218, 220, 240, 250, 524.
Brevoortia patronus	Gulf menhaden, 13, 18, 42, 129, 220, 240, 354, 534.
Brevoortia smithi	Yellowfin menhaden, 18, 534.
Brevoortia tyrannus	Atlantic menhaden, 18, 41, 51, 88, 89, 91, 102, 220, 240, 261, 354, 493, 495, 534.
Clupea harengus	Atlantic herring, 29, 30, 37, 51, 56, 62, 63, 66, 70, 82, 97, 99, 107, 111, 219, 220, 258, 269, 304, 338, 534.

Clupea pallasii	Pacific herring, 36, 87, 219, 220, 240, 251, 258, 271, 475, 534.
Dorosoma cepedianum	Gizzard shad, 220, 260.
Dorosoma petenense	Threadfin shad, 220, 259.
Etrumeus micropus	(round herring, Indo-Pacific), 37.
Etrumeus teres	Round herring, 220, 263.
Harengula jaguana	Scaled sardine, 415, 534.
Opisthonema oglinum	Atlantic thread herring, 220.
Sardina pilchardus	European pilchard, 66, 238, 263.
Sardinella aurita	Spanish sardine, 220, 263.
Sardinops melanostictus	Japanese sardine, 41, 198, 217, 220, 240, 263, 338, 534.
Sardinops ocellatus	South African pilchard, 217.
Sardinops sagax	Pacific sardine, 21, 41.
Spratelloides delicatulus	Blue sprat, 220.
Spratelloides lewisi	220.
Sprattus sprattus	Sprat, 263.
ENGRAULIDAE	ANCHOVIES, 18, 291.
Anchoa hepsetus	Striped anchovy, 43, 51, 56, 221, 415-416.
Anchoa mitchilli	Bay anchovy, 13, 21, 47, 51, 56, 62, 63, 82, 211, 214, 217, 221, 261, 287, 424, 474, 534.
Encrasicholina devisi	221.
Encrasicholina heteroloba	221.
Encrasicholina purpurea	Nehu, 262.
Engraulis encrasicolus	European anchovy, 221, 262.
Engraulis encrasicolus ponticus	Black sea anchovy, 217, 263.
Engraulis japonicus	Japanese anchovy, 22, 42, 217, 221, 301.
Engraulis mordax	Northern anchovy, 29, 30, 44, 61, 62, 63, 88, 91, 192, 213, 217, 221, 240, 262, 263, 271, 304, 338, 354, 474, 477, 534.
CHANIDAE	MILKFISH
Chanos chanos	Milkfish, 5, 7, 8, 24, 31, 42, 51, 56, 62, 63, 78, 82, 139, 196, 198, 202, 204, 205, 211, 212, 216, 221, 238, 241, 249, 250, 251, 253, 259, 263, 269, 271, 287, 290, 291, 298, 301, 302, 303, 318, 337, 338, 354, 375, 377, 379, 384, 385, 422, 423, 424, 425, 465, 493, 496, 508, 514-515, 530, 531, 535, 572, 574.
CYPRINIDAE	CARPS, MINNOWS, 1, 240, 251, 296, 406, 470, 485, 513, 519, 530, 531.
Barbodes gonionotus	Java barb, Thai carp, 1, 4, 251.
Carassius auratus	Goldfish, 50, 100, 251, 258, 406.
Cirrhinus mrigala	Indian major carp, 395.
Cyprinus carpio	Common carp, 4, 84, 111, 135, 139, 270, 303, 376, 381, 386, 387, 388, 392, 393, 394, 396, 397, 399, 400, 401, 402, 406, 511.
Danio rerio	Zebrafish, 64, 270.
Hypophthalmichthys molitrix	Silver carp, 133.
Labeo rohita	Rohu, 1.
Notemigonus crysoleucas	Golden shiner, 575.
Phoxinus phoxinus	Eurasian minnow, 102.
Pimephales promelas	Fathead minnow, 74, 78, 88, 99, 100, 101, 102, 105, 106, 113, 114, 115, 117, 119, 123, 124, 126, 127, 128, 129, 130, 131, 132, 133, 134, 135, 136, 137, 511.
COBITIDAE	LOACHES, 406.
CHARACIDAE	CHARACINS
Prochilodus platensis	Sabalo, Curimbata, 258.
ICTALURIDAE	BULLHEAD CATFISHES, 1, 381, 404, 471, 485, 486, 487, 488, 524.
Ictalurus punctatus	Channel catfish, 77, 78, 79, 85, 87, 138, 268, 375, 378, 388, 392, 395, 396, 398, 399, 400, 401, 402, 404-405, 467, 510, 512, 524, 530.
BAGRIDAE	BAGRID CATFISHES
Chrysichthys nigrodigitatus	(bagrid catfish), 4.
CLARIIDAE	AIRBREATHING CATFISHES, 251, 519.
Heterobranchus longifilis	African catfish, Sampa, 213, 221, 241.
ARIIDAE	SEA CATFISHES, 535.

Ariopsis bonillai	New Granada sea catfish, 535.
Arius felis	Hardhead sea catfish, 99, 216, 221.
Arius herzbergii	Pemecou sea catfish, 535.
Arius proops	Crucifix sea catfish, 535.
Bagre marinus	Gafftopsail catfish, 35, 221.
Galeichthys feliceps	White sea catfish, 35.
PLOTOSIDAE	**EELTAIL CATFISHES**
Cnidoglanis macrocephalus	Cobbler, 221.
Plotosus lineatus	Striped eel-catfish, 40, 59, 577.
ESOCIDAE	**PIKES**, 487.
ARGENTINIDAE	**ARGENTINES**
Argentina silus	Greater silversmelt, 221.
OSMERIDAE	**SMELTS**, 18.
Hypomesus transpacificus	Pond smelt, 303.
Mallotus villosus	Capelin, 39, 213, 221.
Osmerus eperlanus	European smelt, 38.
Plecoglossus altivelis	Ayu, 5, 7, 8, 38, 221, 241, 258, 271, 303, 338, 354, 375, 379, 384, 404, 425, 426, 465, 495, 501, 508, 515, 535-536, 574.
GALAXIIDAE	**TROUTMINNOWS**
Galaxias platei	36.
Lovettia sealii	Tasmanian whitebait, 37, 339, 355, 536.
SALMONIDAE	**SALMONS, TROUTS**, 5, 7, 18, 70, 71, 74, 78, 80, 82, 83, 84, 87, 88, 89, 93, 99, 101, 104, 107, 121, 124, 125, 126, 127, 128, 131, 147, 148, 181, 186, 187, 191, 197, 232, 238, 240, 249, 250, 268, 269, 270, 289, 291, 296, 297, 298, 353, 355, 375, 380, 381, 384, 386, 388, 395, 399, 400, 401, 403, 404, 406, 407, 409, 426-427, 465, 467, 470, 471, 473, 485, 486, 487, 492, 493, 494, 495, 496, 498, 499, 500, 501, 502, 503, 508, 510, 515, 519, 520.
Coregonus clupeaformis	Lake whitefish, 92.
Coregonus lavaretus	Powan, 5, 8, 213, 258, 271, 465.
Coregonus muksun	Whitefish, 269.
Coregonus spp.	(whitefishes, ciscoes), 5, 8.
Hucho hucho	Danube salmon, 40.
Oncorhynchus clarki	Cutthroat trout, 74, 79, 106, 510, 511.
Oncorhynchus gorbuscha	Pink salmon, 40, 51, 88, 115, 214, 221, 271, 298.
Oncorhynchus keta	Chum salmon, 40, 51, 74, 100, 214, 221, 297, 399.
Oncorhynchus kisutch	Coho salmon, 5, 7, 8, 29, 40, 51, 67, 106, 111, 116, 119, 127, 131, 132, 133, 135, 137, 197, 202, 214, 221, 241, 258, 271, 298, 355, 400, 501, 511, 532, 536.
Oncorhynchus masou	Masu salmon, 40, 297.
Oncorhynchus mykiss	Rainbow trout, Steelhead trout, 5, 7, 8, 51, 56, 74, 78, 79, 87, 88, 92, 93, 98, 100, 101, 102, 103, 104, 105, 106, 109, 116, 119, 120, 121, 124, 125, 126, 127, 129, 130, 132, 133, 134, 135, 136, 137, 138, 139, 238, 241, 258, 267, 268, 269, 270, 289, 297, 298, 375, 377, 379, 380, 381, 386, 387, 388, 389, 392, 393, 394, 395, 396, 397, 398, 399, 400, 401, 402, 404, 406, 411, 471, 473, 477, 479, 503, 508, 510, 515, 523, 530, 532.
Oncorhynchus nerka	Sockeye salmon, 18-19, 40, 51, 137, 214, 221, 241, 477, 479.
Oncorhynchus tshawytscha	Chinook salmon, 5, 8, 22, 40, 51, 78, 80, 88, 98, 109, 129, 131, 214, 241, 271, 297, 298, 379, 381, 398, 402, 523.
Oncorhynchus spp.	(salmons, trouts), 109.
Salmo salar	Atlantic salmon, 5, 7, 8, 40, 50, 51, 56, 66, 70, 74, 78, 82, 87, 88, 89, 93, 98, 99, 101, 104, 107, 108, 110, 112, 113, 130, 131, 133, 197, 202, 241, 267, 270, 298, 376, 379, 380, 382, 385, 387, 395, 397, 399, 400, 401, 402, 405, 411, 426, 464, 492, 493, 497, 501, 505, 507, 508, 510, 515, 519, 525, 526, 536-537.
Salmo trutta	Brown trout, 5, 7, 8, 40, 51, 92, 106, 125, 127, 133, 139, 241, 258, 270, 478, 510.
Salmo spp.	(salmons, trouts), 5, 8.
Salvelinus alpinus	Arctic char, 5, 8, 40, 51, 56, 65, 214, 258, 271, 427, 465, 472, 536.
Salvelinus fontinalis	Brook trout, 5, 9, 51, 98, 101, 105, 124.

Salvelinus namaycush	Lake trout, 93, 393, 394.
Salvelinus spp.	(chars), 5, 9.
Thymallus thymallus	Grayling, 5, 9.
GONOSTOMATIDAE	**BRISTLEMOUTHS, LIGHTFISHES**
Cyclothone alba	Albino anglemouth, 221.
Cyclothone atraria	Black bristlemouth, 221.
Cyclothone pseudopallida	Hiiroonihadaka, 221.
Maurolicus muelleri	Lightfish, 217, 219, 221, 472.
MYCTOPHIDAE	**LANTERNFISHES**
Benthosema fibulatum	Hokutohadaka, 221.
Lampanyctodes hectoris	Lanternfish, 219, 221.
GADIDAE	**CODS**, 18, 298, 502.
Enchelyopus cimbrius	Fourbeard rockling, 222.
Gaidropsarus capensis	Cape rockling, 70, 74, 78, 80, 82.
Gadus macrocephalus	Pacific cod, 37, 222, 271, 302, 304.
Gadus morhua	Atlantic cod, 6, 9, 24, 29, 41, 50, 51, 56, 62, 63, 64, 65, 82, 99, 107, 132, 134, 135, 136, 137, 192, 196, 198, 202, 212, 213, 217, 218, 222, 258, 259, 269, 270, 271, 287, 292, 339, 355, 377, 379, 382, 385, 387, 398, 407, 409-410, 427, 428, 429, 465, 479, 492, 494, 501, 515, 537, 572.
Merlangius merlangus	English whiting, 213, 272.
Melanogrammus aeglefinus	Haddock, 23, 41, 51, 56, 101, 116, 213, 222, 537.
Merluccius bilinearis	Silver hake, 45, 218, 222.
Microgadus tomcod	Atlantic tomcod, 219, 222, 261, 272, 537.
Pollachias virens	Pollock, 219, 222, 272.
Theragra chalcogramma	Walleye pollock, 41, 218, 222, 352.
Urophycis chuss	Red hake, 222.
Urophycis regia	Spotted hake, 222.
Urophycis tenuis	White hake, 222.
BATRACHOIDIDAE	**TOADFISHES**
Opsanus beta	Gulf toadfish, 35.
Opsanus tau	Oyster toadfish, 35, 215, 222.
Porichthys notatus	Midshipman, 222, 263, 272.
Porichthys plectrodon	Atlantic midshipman, 219, 222.
LOPHIIDAE	**GOOSEFISHES**
Lophius americanus	Goosefish, 39, 218, 222, 259.
SYNODONTIDAE	**LIZARDFISHES**, 577.
ANTENNARIIDAE	**FROGFISHES**, 577.
Histrio histrio	Sargassumfish, 27, 40, 261, 272, 476, 478, 577.
Phrynelox scaber	Splitlure frogfish, 40, 213, 260.
GOBIESOCIDAE	**CLINGFISHES**, 577-578.
Acyrtops beryllinus	Emerald clingfish, 37, 222, 272.
Aspasma minima	37, 215, 222, 272.
Aspasmichthys ciconiae	Yellow clingfish, 272, 578.
Gobiesox rhessodon	California clingfish, 26.
Gobiesox strumosus	Skilletfish, 38, 192, 214, 222, 259, 272.
EXOCOETIDAE	**FLYINGFISHES**, 17.
Cheilopogon heterurus	Atlantic flyingfish, 36, 219, 222.
Hemirhamphus sajori	Japanese halfbeak, 36.
Hirundichthys oxycephalus	Pointed flyingfish, 37, 537.
Hyporhamphus unifasciatus	Halfbeak, 222.
BELONIDAE	**NEEDLEFISHES**
Strongylura marina	Atlantic needlefish, 222.
Tylosurus acus	Agujon, Blackfin needlefish, 35, 222.
Tylosurus crocodilus	Houndfish, 213, 222.
SCOMBERESOCIDAE	**SAURIES**
Cololabis saira	Pacific saury, 36.
Scomberesox saurus	Atlantic saury, 41, 223.
ADRIANICHTHYIDAE	**RICEFISHES**

Oryzias latipes	Japanese medaka, 64, 134.
APLOCHEILIDAE	**RIVULINS**
Rivulus marmoratus	Mangrove rivulus, 99, 107, 112, 575.
CYPRINODONTIDAE	**KILLIFISHES**, 18, 202, 289, 295, 296, 298, 355, 427-428, 465, 502, 575.
Cyprinodon nevadensis	Desert pupfish, 255.
Cyprinodon variegatus	Sheepshead minnow, 13, 36, 51, 56, 70, 75, 99, 102, 105, 109, 112, 116, 119, 123, 124, 125, 126, 127, 128, 129, 575-576.
Floridichthys carpio	Goldspotted killifish, 427.
Fundulus confluentus	Marsh killifish, 216, 223.
Fundulus diaphanus	Banded killifish, 219, 223, 260.
Fundulus grandis	Gulf killifish, 37, 51, 56, 127, 136, 196, 204, 223, 261, 269, 272.
Fundulus heteroclitus	Mummichog, 36, 51, 56, 70, 82, 99, 101, 103, 107, 109, 112, 114, 129, 130, 137, 139, 216, 575-576.
Fundulus luciae	Spotfin killifish, 216, 223, 260.
Fundulus majalis	Striped killifish, 112, 219, 223, 575-576.
Fundulus similis	Longnose killifish, 123, 124, 125, 126, 127, 128, 129, 131, 133, 134.
Fundulus spp.	(killifishes), 59.
Jordanella floridae	Flagfish, 106, 111, 118, 131.
Lucania parva	Rainwater killifish, 216, 223, 272.
ANABLEPIDAE	**FOUR-EYED FISHES**, 17.
POECILIIDAE	**LIVEBEARERS**, 17, 296, 465.
Gambusia affinis	Western mosquitofish, 13, 215, 223, 352.
Poecilia reticulata	Guppy, 13, 80, 352, 399.
Xiphophorus variatus	Variable platyfish, 13.
Poecilia spp.	(mollies), 529.
MELANOTAENIIDAE	**RAINBOWFISHES**
Melanotaenia fluviatilis	Crimsonspotted rainbowfish, 129, 130.
PSEUDOMUGILIDAE	**BLUE-EYES**
Pseudomugil mellis	Honey blue-eye, 36, 216, 223, 272, 339, 578.
ATHERINIDAE	**SILVERSIDES**, 18, 289, 537.
Atherinomorus lacunosus	Hardyhead silverside, 223.
Atherinops affinis	Topsmelt, 36, 99, 136, 192, 198, 216, 223, 262, 263, 272, 339, 537-538.
Atherinopsis californiensis	Jacksmelt, 36, 339, 537.
Atherinosoma microstoma	Smallmouth hardyhead, 223.
Atherion elymnus	Japanese silverside, 38, 273, 538.
Hypoatherina bleekeri	37, 223, 287.
Hypoatherina tsurugae	36.
Leuresthes tenuis	California grunion, 40, 58, 130, 198, 289, 339, 355, 538.
Membras martinica	Rough silverside, 223.
Menidia beryllina	Inland silverside, 70, 75, 88, 130, 133, 137, 214, 217, 223, 355, 428, 538.
Menidia menidia	Atlantic silverside, 37, 88, 89, 99, 109, 112, 116, 124, 130, 133, 211, 215, 223, 241, 259, 273, 287, 355, 538.
Menidia peninsulae	Tidewater silverside, 123, 128, 136, 137, 138.
TRACHICHTHYIDAE	**ROUGHIES**
Hoplostethus atlanticus	Orange roughy, 41.
HOLOCENTRIDAE	**SQUIRRELFISHES**
Holocentrus vexillarius	Dusky squirrelfish, 23.
GASTEROSTEIDAE	**STICKLEBACKS**, 121.
Apeltes quadracus	Fourspine stickleback, 223, 273.
Gasterosteus aculeatus	Threespine stickleback, 36, 92, 215, 223.
FISTULARIIDAE	**CORNETFISHES**
Fistularia petimba	Red cornetfish, 21, 22, 24, 41, 578.
Fistularia tabacaria	Bluespotted cornetfish, 223.
CENTRISCIDAE	**SNIPEFISHES**
Macrorhamphosus scolopax	Longspine snipefish, 43.
SYNGNATHIDAE	**PIPEFISHES, SEAHORSES**, 428, 503, 577, 578.

Hippocampus coronatus	Crowned seahorse, 273.
Hippocampus erectus	Lined seahorse, 21, 35, 215, 223.
Hippocampus fuscus	216.
Hippocampus guttulatus	Continental seahorse, 215.
Hippocampus hippocampus	European seahorse, 578.
Hippocampus histrix	Thorny seahorse, 215.
Hippocampus japonicus	Japanese seahorse, 22, 24, 52.
Hippocampus kuda/comes	Yellow or Spotted seahorse, 51, 215.
Hippocampus obtusus	Offshore seahorse, 223.
Hippocampus reidi	Longsnout seahorse, 214.
Hippocampus trimaculatus	Longnose seahorse, 52.
Hippocampus zosterae	Dwarf seahorse, 35, 216, 273.
Hippocampus spp.	(seahorses), 92, 273, 339, 356, 578.
Phycodurus eques	Leafy seadragon, 35, 215, 578.
Pyyllopteryx taeniolatus	Common seadragon, 578.
Syngnathus floridae	Dusky pipefish, 215, 223, 263.
Synganthus fuscus	Northern pipefish, 35, 216, 223.
Syngnathus louisianae	Chain pipefish, 223.
Syngnathus pelagicus	Sargassum pipefish, 223.
Syngnathus scovelli	Gulf pipefish, 428, 578.
DACTYLOPTERIDAE	**FLYING GURNARDS**
Dactylopterus volitans	Flying gurnard, 46.
SCORPAENIDAE	**SCORPIONFISHES**, 6, 7, 9, 17, 494-495, 578-579.
Dendrochirus brachypterus	Pygmy lionfish, 39, 273, 579.
Dendrochirus lunulata	39.
Dendrochirus zebra	Zebra lionfish, 261.
Inimicus japonicus	Devil stinger, 538.
Scorpaena guttata	California scorpionfish, 39.
Scorpaena miostoma	40, 263, 273.
Scorpaenichthys marmoratus	Cabezon, 107.
Sebastes diploproa	Splitnose rockfish, 52, 65.
Sebastes emphaeus	Puget Sound rockfish, 219, 223.
Sebastes flavidus	Yellowtail rockfish, 223.
Sebastes inermis	Mushroom scorpionfish, black rockfish, 224.
Sebastes marinus	Ocean perch, 219, 224.
Sebastes matsubarae	Matsubara's red rockfish, 224.
Sebastes oblongus	Oblong rockfish, 224.
Sebastes schlegeli	Kurosoi, 35, 52, 99, 101, 107, 109, 112, 134, 198, 212, 218, 224, 263, 273, 292, 339, 428, 430, 431, 515, 538, 572.
Sebastes vulpes	Fox jacopever, 223.
Sebastiscus marmoratus	Kasago, 224, 273, 382, 483, 538.
Sebastolobus altivelis	Longspine thornyhead, 29.
CONGIOPODIDAE	**HORSEFISHES**
Congiopodus spinifer	Spinenose horsefish, 82.
TRIGLIDAE	**SEAROBINS**
Prionotus carolinus	Northern searobin, 43, 224.
Prionotus spp.	(searobins), 262.
PLATYCEPHALIDAE	**FLATHEADS**
Platycephalus indicus	Bartail flathead, 44.
ANOPLOPOMATIDAE	**SABLEFISHES**
Anoplopoma fimbria	Sablefish, 47, 213, 217, 224, 241, 251, 273, 298, 339, 538.
HEXAGRAMMIDAE	**GREENLINGS**
Ophiodon elongatus	Lingcod, 39, 273, 339, 356, 538.
Pleurogrammus azonus	Atka mackerel, 7, 9.
Pleurogrammus monopterygius	Atka mackerel
COTTIDAE	**SCULPINS**, 18.
Alcichthys alcicornis	Elkhorn sculpin, 37, 538.
Blepsias cirrhosus	Silverspotted sculpin, 35.
Myoxocephalus octodecemspinosus	Longhorn sculpin, 36.
Nautichthys oculofasciatus	Sailfin sculpin, 62, 65.

Orthonopias triacis	Snubnose sculpin, 38.
Pseudoblennius cottoides	Asahi-anahaze, Sunrise sculpin, 36, 259, 273, 538.
Pseudoblennius percoides	Anahaze, 36, 538.
CYCLOPTERIDAE	**SNAILFISHES**
Cyclopterus lumpus	Atlantic lumpfish, 35, 273, 297, 340, 356, 539.
CENTROPOMIDAE	**SNOOKS**, 249, 254, 336, 464, 574, 576.
Centropomus parallelus	Fat snook, 540.
Centropomus pectinatus	Tarpon snook, 13, 541.
Centropomus undecimalis	Common snook, 5, 9, 13, 25, 47, 52, 56, 62, 84, 130, 131, 149, 175, 180, 198, 203, 204, 210, 212, 232, 235, 236, 242, 254, 255, 287, 291, 376, 377, 298, 301, 336, 340, 357, 373, 380, 385, 387, 410, 414, 418, 515, 521, 527, 529, 530, 540-542.
Lates calcarifer	Barramundi, Asian sea bass, Giant perch, 5, 7, 9, 13, 30, 31, 46, 52, 56, 62, 65, 78, 79, 82, 92, 94, 95-96, 138, 182, 186, 187, 192, 196, 198, 203, 204, 205, 206, 207, 212, 216, 224, 242, 249, 250, 253, 262, 263, 269, 273, 287, 288, 292, 298, 301, 302, 318, 334, 335, 336, 337, 341, 356-357, 373, 376, 377, 379, 380, 382, 384, 385, 392, 395, 414, 418, 430, 431-434, 465, 472, 492, 495, 514, 515, 527, 528, 531, 533, 539-540, 542, 572.
Lates japonicus	Japanese snook, 542.
Lates niloticus	Nile perch, 5, 9, 540.
PERCICHTHYIDAE	**TEMPERATE BASSES**, 68, 249, 254, 336, 464, 465, 500.
Dicentrarchus labrax	European sea bass, 6, 7, 9, 30, 41, 52, 56, 62, 65, 68, 70, 75, 77, 79, 82, 91, 92, 96, 99, 138, 148, 192, 198, 203, 205, 207, 212, 224, 238, 242, 250, 251, 253, 269, 273, 289, 292, 298, 300, 303, 305, 306, 318, 335, 336, 341, 357, 373, 375, 376, 377, 379, 381, 384, 385, 395, 397, 405, 406, 407, 434, 435, 465, 466, 485, 493, 494, 495, 508, 510, 511, 514, 515-516, 524, 530, 533, 542-543, 572.
Dicentrarchus spp.	(sea basses), 6, 7, 9, 202.
Lateolabrax japonicus	Japanese sea bass, 6, 7, 9, 25, 41, 52, 92, 242, 273, 303, 341, 359, 545.
Maccullochella peelii	Murray perch, 5, 9, 258.
Macquaria ambigua	Golden perch, 5, 9.
Macquaria novemaculeata	Australian bass, 42, 52, 56, 65, 91, 92, 94, 199, 242, 270, 300, 341, 359, 518, 545.
Morone americana	White perch, 88.
Morone chrysops	White bass, 52, 213, 224, 242, 250, 267, 268, 292, 296, 341, 358-359, 477, 543.
M. chrysops × *M. saxatilis*	Sunshine bass, 5, 7, 9, 13, 38, 70, 71, 74, 78, 80, 82, 87, 92, 93, 99, 107, 108, 149, 197, 199, 202, 203, 204, 207, 258, 292, 341, 358, 376, 377, 385, 414, 418, 435, 436, 478, 525, 533, 543-545, 572, 574.
Morone saxatilis	Striped bass, 18, 40, 52, 56, 70, 71, 74, 75, 78, 80, 82, 87, 88, 89, 92, 93, 94, 95, 99, 100, 101, 103, 107, 108, 109, 110, 111, 112, 115, 116, 117, 125, 126, 127, 128, 129, 130, 131, 132, 133, 134, 135, 136, 137, 138, 139, 197, 199, 207, 212, 218, 224, 239, 242, 250, 252, 254, 257, 258, 266, 267, 269, 270, 273, 289, 290, 292, 296, 303, 305, 307, 341, 358, 375, 379, 399, 434-435, 472, 476, 477, 503, 512, 516, 522, 524, 531, 532, 543-545.
M. saxatilis × *M. chrysops*	Palmetto bass, 5, 7, 9, 52, 56, 65, 70, 71, 74, 78, 80, 82, 87, 92, 93, 99, 107, 108, 197, 199, 202, 204, 207, 274, 292, 305, 341, 358, 387, 435-437, 499, 519, 533, 543-545, 574.
Stereolepis gigas	Giant sea bass, 31, 207, 211, 232, 239, 249.
SERRANIDAE	**GROUPERS, SEA BASSES**, 6, 7, 9, 67, 68, 91, 186, 187, 202, 255, 269, 288, 289, 290, 291, 292, 296, 298, 336, 359, 406, 437, 464, 465, 492, 493, 516, 521, 525, 527, 531, 532, 533, 545-546, 574, 579.
Centropristis striata	Black sea bass, 13, 44, 62, 63, 80, 243, 253, 274, 288, 342, 474, 548, 571, 579.
Cromileptes altivelis	Polka dot grouper, 45, 224, 274, 545.
Diplectrum formosum	Sand perch, 47, 243, 549.

Epinephelus adscensionis	Rock hind, 243, 571.
Epinephelus aeneus	White grouper, 224, 243, 549.
Epinephelus akaara	Redspotted grouper, 7, 9, 45, 56, 59, 83, 192, 199, 216, 224, 238, 243, 253, 274, 290, 342, 359, 516, 545, 546, 572.
Epinephelus amblycephalus	Blackspotted grouper, 52, 56, 199, 243, 342, 379, 548-549.
E. amblycephalus × *E. akaara*	Blackspotted grouper × red grouper, 43.
Epinephelus andersoni	Spotted rockcod, 274.
Epinephelus areolatus	Squaretail grouper, 6, 9.
Epinephelus awoara	Yellow grouper, 59, 274.
Epinephelus chlorostigma	Brownspotted grouper, 213, 224.
Epinephelus coioides	Orangespotted or Estuary grouper, 46, 62, 192, 199, 243, 274, 300, 342, 360, 504, 546, 572.
Epinephelus cruentatus	Graysby, 243.
Epinephelus diacanthus	Sixband grouper, 224, 274.
Epinephelus fario	Bluespotted grouper, 243, 253, 290.
Epinephelus fasciatus	Banded grouper, 274.
Epinephelus fulvus	Coney, 44, 224, 243.
E. fulvus × *E. guttatus*	Coney × red hind, 13.
E. fulvus × *P. furcifer*	Coney × creole-fish, 13.
Epinephelus fuscoguttatus	Brownmarbled grouper, 13, 31, 45, 94, 196, 199, 212, 243, 253, 265, 274, 301, 302, 342, 359, 546, 572.
Epinephelus guaza	Mérou, 438.
Epinephelus guttatus	Red hind, 13, 44, 224, 243, 254, 255, 571.
Epinephelus itajara	Jewfish, 548.
Epinephelus malabaricus	Malabar grouper, 45, 52, 56, 75, 83, 87, 192, 196, 199, 224, 243, 253, 263, 269, 274, 342, 359, 376, 377, 379, 385, 406, 437, 545, 546-547, 572.
Epinephelus marginatus	549.
Epinephelus moara	Kelp grouper, 192, 275, 545.
Epinephelus morio	Red grouper, 218, 224, 243, 549, 571.
E. morio × *C. striata*	Red grouper × black sea bass, 13.
Epinephelus nigritus	Warsaw grouper, 571.
Epinephelus polyphekadion	Camouflage grouper, 45, 192, 199, 224, 243, 275, 547.
Epinephelus septemfasciatus	Sevenband grouper, 45, 243, 253, 342, 549.
Epinephelus striatus	Nassau grouper, 13, 19, 22, 25, 27, 44, 67, 75, 76, 79, 92, 180, 192, 199, 203, 212, 216, 225, 232, 236, 243, 254, 255, 256, 258, 261, 264, 275, 287, 289, 291, 293, 336, 342, 359-360, 373, 376, 377, 385, 414, 418, 437, 465, 483, 511, 522, 523, 527, 528, 530, 531, 547-548, 572, 579.
Epinephelus tauvina	Greasy grouper, 6, 7, 9, 44, 52, 56, 83, 91, 92, 94, 205, 225, 243, 253, 269, 290, 336, 343, 359, 377, 379, 433, 437, 465, 547, 572.
Epinephelus tukula	Potato grouper, 275.
Epinephelus spp.	(groupers), 6, 9, 205.
Hypoplectrus gemma	Blue hamlet, 243, 275, 579.
Hypoplectrus guttavarius	Shy hamlet, 579.
Hypoplectrus indigo	Indigo hamlet, 275.
Hypoplectrus nigricans	Black hamlet, 275.
Hypoplectrus puella	Barred hamlet, 244, 275.
Hypoplectrus unicolor	Butter hamlet, 244, 261, 275, 579.
H. unicolor × *H. gemma*	Butter hamlet × blue hamlet, 579.
Hypoplectrus spp.	(hamlets), 46.
Liopropoma rubre	Swissguard basslet, 579.
Mycteroperca bonaci	Black grouper, 275, 571.
Mycteroperca microlepis	Gag, 44, 225, 244, 256, 275, 302, 549, 571.
Mycteroperca phenax	Scamp, 244, 571.
Mycteroperca venenosa	Yellowfin grouper, 275.
Paralabrax clathratus	Kelp bass, 22, 25, 217, 225, 244, 253.
Paralabrax maculatofasciatus	Spotted sand bass, 45, 244, 253, 259, 261, 275, 343, 548.
Paralabrax nebulifer	Barred sand bass, 45, 225, 244, 253, 259, 549.
Paranthias furcifer	Creole-fish, 13, 44, 244, 291.
Plectropomus areolatus	Squaretail coraltrout, 13, 46, 212, 244, 257, 266, 275, 291, 549.
Plectropomus leopardus	Leopard coraltrout, 45, 225, 244, 261, 269, 275, 287, 343, 352, 549.

Pseudanthias squamipinnis	Lyretail fairy basslet, 47.
Sacura margaritacea	Cherry bass, 46.
Serranus subligarius	Belted sandfish, 13, 47, 215, 225, 260, 261, 275, 290, 291, 438, 579.
Serranus tabacarius	Tobaccofish, 244, 579.
Serranus tigrinus	Harlequin bass, 244, 275, 579.
Siniperca chuatsi	Chinese perch [Mandarin fish], 6, 7, 9, 548.
PSEUDOCHROMIDAE	**DOTTYBACKS**, 360, 577, 579.
Ogilbyina novaehollandiae	Australian dottyback, 276, 579.
Pseudochromis aldabrensis	Neon dottyback, 276, 579.
Pseudochromis flavivertex	Sunrise dottyback, 37, 215, 276, 407, 579.
Pseudochromis fridmani	Orchid dottyback, 37, 199, 216, 225, 276, 343, 407, 579.
Pseudochromis fuscus	Brown or Yellow dottyback, 37, 276, 579.
Pseudochromis olivaceus	Olive dottyback, 276, 579.
Pseudochromis sankeyi	Striped dottyback, 276, 579.
Pseudochromis springeri	Springer's dottyback, 276, 579.
GRAMMATIDAE	**BASSLETS**, 361, 577.
Gramma loreto	Royal gramma, 37, 192, 216, 276, 343, 361, 579.
Gramma melacara	Blackcap basslet, 215, 260, 276, 343, 361, 579.
PLESIOPIDAE	**LONGFINS**
Assessor macneilli	MacNeill's mini-grouper, 35, 192, 215, 276.
Calloplesiops altivelis	Comet, 37, 215, 276, 343, 361, 577, 580.
TERAPONTIDAE	**GRUNTERS**
Therapon jarbua	Threestripe tigerfish, 46, 59, 99, 107, 244, 549.
CENTRARCHIDAE	**SUNFISHES**
Lepomis cyanellus	Green sunfish, 510.
Lepomis macrochirus	Bluegill, 131, 137, 139, 479, 511, 512, 524.
Micropterus dolomieu	Smallmouth bass, 488.
Micropterus salmoides	Largemouth bass, 4, 77, 119, 139, 406, 471, 488, 511, 524.
PERCIDAE	**PERCHES**, 487.
Perca flavescens	Yellow perch, 137, 301, 476.
Perca fluviatilis	Eurasian or European perch, 118, 478.
Stizostedion lucioperca	Zander, 110.
Stizostedion vitreum	Walleye, 68, 91, 92, 94, 125.
PRIACANTHIDAE	**BIGEYES**
Priacanthus macracanthus	Red bullseye, 212, 225.
APOGONIDAE	**CARDINALFISHES**, 580.
Apogon conspersus	35.
Apogon cyanosoma	Goldstriped cardinalfish, 260, 276.
Apogon maculatus	Flamefish, 193, 216, 276.
Apogon notatus	Blackspot cardinalfish, 225, 260.
Apogon taeniatus	Twobelt cardinalfish, 580.
Archamia zosterophora	Blackbelt cardinalfish, 225.
Foa fo (brachygramma)	Bay cardinalfish, 24.
Pterapogon kauderni	Banggai cardinalfish, 35, 276, 580.
SILLAGINIDAE	**WHITINGS, SILLAGOS**
Sillaginodes punctata	Spotted or King George whiting, 45, 212, 225.
Sillago ciliata	Sand or Summer whiting, 62, 63, 92, 269, 549.
Sillago japonica	Japanese whiting, 47, 89, 244, 343, 549.
Sillago sihama	Silver whiting, 213, 262, 276.
MALACANTHIDAE	**TILEFISHES**
Caulolatilus microps	Blueline tilefish, 571.
Malacanthus latovittatus	Blue whiting, 225, 261.
Malacanthus plumieri	Sand tilefish, 47.
POMATOMIDAE	**BLUEFISHES**
Pomatomus saltatrix	Bluefish, 9, 43, 225, 263, 472.
RACHYCENTRIDAE	**COBIA**
Rachycentron canadum	Cobia, 7, 10, 18, 42, 52, 56, 217, 225, 232, 269, 343, 549.
ECHENEIDAE	**REMORAS**
Echeneis naucrates	Sharksucker, 41, 225, 276, 549.

CARANGIDAE	**JACKS**, 187, 202, 249, 254, 290, 298, 361, 406, 464.
Alepes djedaba	Shrimp scad, 47.
Carangoides equula	Whitefin crevalle, 59.
Carangoides malabaricus	Malabar trevally, 43.
Decapterus macrosoma	Shortfin scad, 225.
Decapterus russelli	Russell's scad, 47.
Decapturus spp.	(scads), 6, 7, 10.
Elagatis bipinnulata	Rainbow runner, 551.
Gnathanodon speciosus	Banded trevally, 550.
Oligoplites saurus	Leatherjacket, 438.
Pseudocaranx dentex	Striped jack, 52, 276, 304, 305, 361, 384, 438, 440, 495, 507, 551.
Scomberoides tol	Needlescaled queenfish, 47.
Selar crumenophthalmus	Bigeye scad, 193, 225, 262, 276.
Selaroides leptolepis	Yellowstripe scad, 47.
Selene vomer	Lookdown, 276.
Seriola dumerili	Greater amberjack, 7, 8, 10, 42, 52, 205, 244, 276, 343, 440, 550, 571.
Seriola lalandi	Goldstriped amberjack, 41, 52, 92, 213, 225, 244, 276, 343, 550.
Seriola mazatlana	Mazatlan yellowtail, 52, 92, 193, 277, 293, 361, 550.
Seriola quinqueradiata	Yellowtail, 6, 7, 10, 42, 52, 56, 83, 186, 193, 199, 205, 207, 211, 225, 244, 277, 293, 301, 303, 304, 305, 306, 344, 361, 375, 376, 377, 379, 382, 385, 386, 387-388, 389, 392, 393, 394, 395, 396, 397, 400, 403, 404, 406, 407, 438-440, 465, 492, 493, 494, 495, 498, 499, 500, 505, 516, 533, 549-550, 572.
Seriola zonata	Banded rudderfish, 28.
Trachinotus blochii	Snubnose pompano, 6, 10, 44, 244, 551.
Trachinotus carolinus	Florida pompano, 25, 52, 56, 80, 83, 99, 138, 139, 212, 218, 225, 244, 253, 344, 376, 377, 385, 438, 440, 550-551, 572.
Trachinotus falcatus	Permit, 52, 57, 196, 225, 244, 277, 344, 438, 551, 572.
Trachinotus goodei	Palometa, 277, 440, 551.
Trachurus japonicus	Japanese horse mackerel, 6, 10, 45, 244, 254, 516, 551.
Trachurus mediterraneus	Mediterranean scad, 226.
Trachurus murphyi	Peruvian jack mackerel, 217.
Trachurus symmetricus	Jack mackerel, 20, 551.
Trachurus trachurus	Atlantic horse mackerel, 226.
Trachurus spp.	(jack & horse mackerels), 6, 10.
CORYPHAENIDAE	**DOLPHINS**, 68.
Coryphaena equisetis	Pompano dolphin, 41.
Coryphaena hippurus	Dolphin, 13, 19, 41, 52, 57, 66, 68, 75, 83, 93, 193, 199, 202, 213, 226, 232, 277, 288, 291, 293, 298, 300, 318, 332, 335, 344, 352, 361-362, 376, 378, 385, 440, 441, 476, 478, 516, 521, 524, 533, 552, 571, 572, 574.
LUTJANIDAE	**SNAPPERS**, 6, 7, 10, 187, 202, 210, 211, 249, 290, 293, 298, 362, 464, 552, 574.
Lutjanus analis	Mutton snapper, 46, 52, 218, 245, 344, 440, 516, 553, 571.
Lutjanus apodus	Schoolmaster, 553.
Lutjanus argentimaculatus	Asian red snapper, 6, 7, 10, 46, 199, 212, 226, 245, 263, 277, 301, 335, 344, 362, 552-553, 573.
Lutjanus argentiventris	Amarillo snapper, 46.
Lutjanus buccanella	Blackfin snapper, 226.
Lutjanus campechanus	Atlantic red snapper, 27, 45, 193, 218, 226, 277, 553, 571.
Lutjanus gibbus	Australian red snapper, 219, 226.
Lutjanus griseus	Gray snapper, 47, 53, 57, 218, 226, 245, 553, 571.
Lutjanus johnii	Golden snapper, 46, 199, 205, 212, 226, 245, 293, 344, 362, 553.
Lutjanus kasmira	Bluestriped snapper, 45, 193, 263, 277.
Lutjanus russellii	Russell's snapper, 6, 10, 59, 83.
Lutjanus stellatus	Whitespotted snapper, 45, 263, 277.
Lutjanus synagris	Lane snapper, 53, 57, 226, 245, 553, 571.
Lutjanus vittus	Blackstriped snapper, 217, 226, 260.
Lutjanus vivanus	Silk snapper, 226.
Lutjanus spp.	(snappers), 6, 10.
Ocyurus chrysurus	Yellowtail snapper, 13, 43, 53, 89, 139, 226, 245, 268, 289, 291,

345, 483, 510, 511, 553, 571.
Pterocaesio digramma — Twin-yellowstriped fusilier, 45, 261, 277.
Rhomboplites aurorubens — Vermilion snapper, 571.
GERREIDAE — **MOJARRAS**, 553, 576.
Diapterus plumieri — Striped mojarra
Eucinostomus gula — Silver jenny, 441.
Eugerres brasilianus — Striped patao, 47, 92, 199, 212, 245, 345, 362, 553.
Gerres filamentosus — Threadfin silverbelly, 226.
HAEMULIDAE — **GRUNTS**, 406, 554, 576.
Anisotremus virginicus — Porkfish, 554.
Haemulon plumieri — White grunt, 44, 441, 554, 571.
Haemulon sciurus — Bluestriped grunt, 277.
Orthopristis chrysoptera — Pigfish, 46, 226, 441, 576.
Parapristipoma trilineatum — Threeline grunt, 59, 554.
Plectorhinchus celebicus — Celebes sweetlips, 218, 226.
Plectorhinchus cinctus — Yellowspotted grunt, 46, 218, 226, 245, 253, 554.
Pomadasys commersonnii — Spotted grunter, 53.
Pomadasys hasta — Silver grunt, 59, 553.
Pomadasys sp. — (grunt)
Pseudopristipoma nigra — Brown sweetlips, 24.
SPARIDAE — **SEA BREAMS, PORGIES**, 6, 7, 10, 68, 202, 210, 249, 290, 293, 298, 336, 363, 406, 464, 465, 494, 554, 574, 576.
Acanthopagrus australis — Surf sea bream [yellowfin sea bream], 269.
Acanthopagrus berda — River bream, 6, 10, 226, 253, 263.
Acanthopagrus latus — Yellowfin sea bream, 44, 53, 245, 277, 345, 364, 446, 554, 556.
Acanthopagrus schlegeli — Black sea bream, 6, 7, 10, 45, 53, 57, 59, 62, 63, 65, 70, 71, 83, 92, 199, 213, 226, 245, 253, 269, 277, 303, 345, 352, 363, 377, 382, 407, 410-411, 443, 484, 533, 554, 555, 556, 573.
Acanthopagrus sivicolus — Okinawa sea bream, 205.
Archosargus probatocephalus — Sheepshead, 13, 45, 53, 57, 149, 180, 199, 204, 217, 226, 245, 253, 260, 287, 291, 336, 345, 364, 373, 385, 444-445, 556-557, 573.
Archosargus rhomboidalis — Caribbean sea bream, sea bream, 45, 62, 65, 199, 445, 474, 557.
Calamus nodosus — Knobbed porgy, 571.
Dentex dentex — Common dentex, 7, 10, 43, 53, 92, 193, 245, 301, 345, 377, 379, 446, 554, 558.
Dentex gibbosus — Pink dentex, 43, 193, 277.
Dentex tumifrons — Yellow sea bream, 217, 226.
Diplodus puntazzo — Sharpsnout sea bream, 45, 200, 203, 245.
Diplodus sargus — White sea bream, Sargo, 6, 10, 45, 70, 75, 79, 80, 203, 345, 558.
Diplodus vulgaris — Twoband sea bream, 43, 277, 301, 345, 557.
Diplodus spp. — (sea breams), 7, 10.
Evynnis japonica — Crimson sea bream, 6, 10, 269.
Lagodon rhomboides — Pinfish, 43, 99, 116, 126, 127, 128, 129, 130, 131, 133, 134, 135, 136, 137, 227, 245, 253, 441.
Lithognathus mormyrus — Sand steenbras, 75, 79, 80.
Pachymetopon blochii — Hottentot, 75, 83.
Pachymetopon grande — Bronze sea bream, 75, 227.
Pagrus major — New Zealand snapper, Red sea bream, 6, 7, 10, 29, 43, 44, 53, 57, 58, 59, 62, 65, 70, 75, 83, 91, 92, 94, 139, 191, 193, 200, 205, 206, 207, 213, 216, 217, 226, 227, 245, 250, 258, 261, 262, 269, 270, 277, 293, 298, 303, 304, 305, 306, 317, 332, 334, 335, 337, 345, 352, 363, 375, 377, 378, 379, 382, 384, 385, 386, 388, 392, 394, 395, 398, 399, 400, 401, 403, 404, 405, 406, 407, 441, 442-443, 472, 508, 514, 517, 524, 533, 554-555, 573.
Pagrus pagrus — Red porgy, 6, 10, 203.
Rhabdosargus haffara — Haffara sea bream, 45, 245, 346, 558.
Rhabdosargus sarba — Silver sea bream, 6, 7, 10, 43, 53, 57, 58, 59, 65, 83, 92, 200, 213, 216, 227, 262, 263, 277, 278, 346, 364, 445-446, 554, 555-556.
Sparidentex hasta — Sobaity, Bluefin porgy, 24, 53, 92, 213, 261, 278, 556, 558.
Sparus auratus — Gilthead sea bream, 6, 7, 10, 30, 44, 49, 53, 57, 61, 62, 63, 65, 66,

	75, 91, 92, 94, 95, 148, 200, 203, 205, 207, 213, 217, 238, 245, 249, 250, 251, 253, 255, 259, 260, 269, 278, 289, 300, 301, 305, 306, 318, 332, 334, 335, 346, 363-364, 373, 375, 376, 377, 379, 384, 385, 392, 395, 407, 443-444, 465, 467, 483, 485, 492, 493, 494, 502, 503, 508, 513, 514, 517, 525, 533, 554, 556, 573.
Spondyliosoma sp.	278.
Stenotomus chrysops	Scup, 43, 227.
CENTRACANTHIDAE	**CENTRACANTHIDS**
Pterosmaris axillaris	83.
LETHRINIDAE	**SWEETLIPS**
Gymnocranius griseus	Wise porgy, 46, 278.
Lethrinus choerorhynchus	Bluestreak emperor, 558.
Lethrinus hypselopterus	Masked emperor, 218, 227.
Lethrinus lentjan	Redspot emperor, 219, 227.
Lethrinus nebulosus	Spangled emperor, 59, 193, 278.
SCIAENIDAE	**DRUMS**, 7, 11, 249, 254, 290, 293, 298, 336, 364, 464, 465, 501, 525, 558, 574, 580.
Argyrosomus hololepidotus	Mulloway, 44, 213, 246, 346, 365, 514, 561.
Atractoscion nobilis	White seabass, 29, 42, 53, 62, 63, 65, 93, 193, 200, 246, 346, 365, 561.
Bairdiella chrysoura	Silver perch, 46, 227, 446.
Bairdiella icistia	Gulf croaker, 47, 53, 57, 58, 213, 246, 258.
Cynoscion nebulosus	Spotted seatrout, 13, 46, 53, 57, 58, 75, 79, 88, 96, 130, 131, 139, 149, 180, 193, 200, 203, 210, 212, 217, 227, 234, 246, 251, 253, 254, 263, 268, 278, 288, 291, 294, 335, 346, 365, 373, 377, 380, 385, 414, 418, 446, 449, 484, 509, 511, 512, 559-561, 573, 576.
Cynoscion regalis	Weakfish, 45, 53, 57, 227, 246, 262, 287, 365, 472, 561.
Cynoscion striatus	Striped weakfish, 217, 227.
Cynoscion xanthulus	Orangemouth corvina, 227, 246, 253, 262, 278, 365, 561.
Equetus acuminatus	Highhat, 193, 278, 346, 365, 580.
Equetus lanceolatus	Jackknifefish, 43, 278, 580.
Equetus punctatus	Spotted drum, 42.
Genyonemus lineatus	White croaker, 46.
Leiostomus xanthurus	Spot, 13, 46, 99, 107, 123, 124, 125, 126, 128, 129, 130, 131, 132, 133, 134, 136, 137, 138, 213, 512, 561.
Menticirrhus americanus	Southern kingfish, 13.
Micropogonias furnieri	Whitemouth croaker, 227, 465, 561.
Micropogonias undulatus	Atlantic croaker, 47, 53, 57, 246, 251, 253, 269, 450, 561.
Nibea albida	227.
Nibea albiflora	Yellow croaker, 398, 399, 400, 401.
Nibea japonica	Japanese croaker, 75, 559.
Pogonias cromis	Black drum, 13, 217, 227, 246, 262, 278, 419, 559.
Pseudosciaena crocea	Large yellow croaker, 246.
Pseudosciaena polyactis	Small yellow croaker, 42, 227, 278, 561.
Sciaenops ocellatus	Red drum, 7, 10, 13, 44, 53, 57, 66, 71, 75, 79, 83, 89, 93, 138, 139, 193, 200, 202, 203, 204, 210, 212, 216, 227, 246, 253, 261, 268, 269, 278, 289, 291, 293, 300, 335, 336, 347, 364, 373, 375, 376, 377, 379, 382, 383, 384, 385, 388, 410, 414, 419, 446-449, 501, 506, 507, 511, 513, 517, 521, 522, 524, 525, 529, 530, 532, 533, 558-559, 573.
MULLIDAE	**GOATFISHES**
Mullus surmuletus	Red mullet, 227.
Parupeneus multifasciatus	Manybanded goatfish, 262.
Upeneus bensasi	Stripefin goatfish, 46.
GLAUCOSOMATIDAE	**PEARL PERCHES**
Glaucosoma hebraicum	West Australia dhufish, 43, 246, 561.
KYPHOSIDAE	**SEA CHUBS**
Girella nigricans	Opaleye, 53, 382.
Girella punctata	Largescale blackfish, Nibbler, 43, 53, 112, 404.
Kyphosus cinerascens	Ashen drummer, 219, 227.
Kyphosus vaigiensis	Brass drummer, 43.

CHAETODONTIDAE	**BUTTERFLYFISHES**, 315, 450, 580-581, 586.
Chaetodon auriga	Threadfin butterflyfish, 450.
Chaetodon austriacus	Exquisite butterflyfish, 451.
Chaetodon capistratus	Foureye butterflyfish, 581.
Chaetodon hoefleri	Hoefler's butterflyfish, 580.
Chaetodon lineolatus	Lined butterflyfish, 450.
Chaetodon lunula	Raccoon butterflyfish, 450.
Chaetodon nippon	Blackloin butterflyfish, 47, 193, 278, 581.
Chaetodon ocellatus	Spotfin butterflyfish, 47, 450.
Chaetodon ornatissimus	Ornate butterflyfish, 451.
Chaetodon quadrimaculatus	Fourspot butterflyfish, 451.
Chaetodon trifascialis	Chevron butterflyfish, 451.
Chaetodon spp.	(butterflyfishes), 261.
Chelmon rostratus	Copperbanded butterflyfish, 450.
Forcipiger longirostris	Longnose butterflyfish, 27, 450.
Hemitaurichthys polylepis	Pyramid butterflyfish, 450.
Hemitaurichthys zoster	Black pyramid butterflyfish, 450.
Heniochus acuminatus	Common bannerfish, 450.
Heniochus diphreutes	Schooling bannerfish, 450.
POMACANTHIDAE	**ANGELFISHES**, 296, 298, 315, 451, 512, 581, 586.
Centropyge acanthops	Orangeback pygmy angelfish, 216, 279.
Centropyge argi	Cherub pygmy angelfish, 43, 215, 261, 262, 279.
Centropyge bicolor	Bicolor pygmy angelfish, 215, 262, 279.
Centropyge bispinosus	Coral beauty, 215, 216, 261, 279.
Centropyge ferrugatus	Rusty pygmy angelfish, 47, 193, 262, 279, 407-408.
Centropyge fisheri	Fisher's pygmy angelfish, 215, 279.
Centropyge flavissimus	Lemonpeel, 216, 279.
Centropyge interruptus	Japanese pygmy angelfish, 47, 193, 261, 262, 279.
Centropyge loriculus	Flame pygmy angelfish, 47, 215, 216, 262, 279, 581.
Centropyge potteri	Potter's pygmy angelfish, 261, 279.
Centropyge resplendens	Resplendent angelfish, 47.
Centropyge tibicen	Keyhole angelfish, 215, 279.
Centropyge spp.	(pygmy angelfishes), 27, 47, 232, 257, 261, 262, 295, 365-366, 581, 586.
Chaetodontoplus duboulayi	Scribbled angelfish, 44, 262, 279.
Chaetodontoplus mesoleucas	Vermiculated angelfish, 261.
Genicanthus lamarck	Freckletail lyretail angelfish, 46, 279.
Genicanthus melanospilos	Blackspot lyretail angelfish, 215, 262, 279.
Genicanthus semifasciatus	Halfbanded lyretail angelfish, 46, 279.
Holacanthus bermudensis	Blue angelfish, 47, 296.
Holacanthus ciliaris	Queen angelfish, 47, 227, 262, 296, 581.
Holacanthus passer	King angelfish, 262.
Holacanthus tricolor	Rock beauty, 47, 227, 262, 287, 366.
Pomacanthus arcuatus	Gray angelfish, 44, 227, 246, 261, 279, 366, 581.
Pomacanthus paru	French angelfish, 43, 44, 227, 246, 262, 279, 366, 581.
CICHLIDAE	**CICHLIDS**, 1, 17, 31, 67, 184, 232, 269, 289, 294, 296, 298, 366, 375, 380, 381, 392, 400, 406, 451, 452, 465, 500, 502, 513, 519, 531, 533, 540, 545, 561.
Etroplus suratensis	Pearl spot, 484.
Oreochromis aureus	Blue tilapia, 4, 53, 57, 214, 297, 298, 398, 451.
Oreochromis mossambicus	Mozambique tilapia, 4, 13, 35, 53, 57, 93, 99, 102, 107, 109, 112, 130, 279, 395, 476, 478, 479, 451, 561-562.
Oreochromis niloticus	Nile tilapia, 4, 53, 57, 78, 101, 117, 137, 193, 196, 200, 214, 227, 246, 270, 279, 347, 376, 381, 386, 397, 399, 451, 561-562.
Oreochromis spilurus	4, 193, 203, 205, 206, 214, 216, 227, 279, 562.
Oreochromis urolepis	Wami tilapia, 561-562.
O. urolepis × *O. mossambicus*	Florida red tilapia, 35, 194, 196, 200, 203, 205, 279, 303, 376, 505, 517, 561-562, 573.
Oreochromis spp.	(tilapias), 4.
Sarotherodon galilaeus	Mango tilapia, 246.
Sarotherodon spp.	(tilapias), 53.
Tilapia zilli	Redbelly tilapia, 107.

EMBIOTOCIDAE	**SURFPERCHES**, 17.
POMACENTRIDAE	**DAMSELFISHES**, 18, 61, 68, 191, 269, 290, 291, 295, 298, 366, 451-452, 521, 577, 581.
Abudefduf luridus	Canary damselfish, 38.
Abudefduf saxatilis	Sergeant major, 14, 38, 62, 64, 67, 211, 213, 227, 259, 280, 347, 366, 582.
Abudefduf sordidus	Spot pom, Blackspot sergeant, 38.
Amblyglyphidodon leucogaster	Whitebelly damselfish, 214.
Amphiprion akallopisus	Skunk clownfish, 260, 280, 582.
Amphiprion akindynos	Barrier reef clownfish, 582.
Amphiprion bicinctus	Twoband clownfish, 214, 262, 280, 347, 582.
Amphiprion chrysopterus	Orangefin clownfish, 582.
Amphiprion clarkii	Clark's clownfish, 214, 228, 260, 280, 582.
Amphiprion ephippium	Red saddleback clownfish, 260, 261, 280, 582.
Amphiprion frenatus	Tomato clownfish, 215, 259, 260, 280, 347, 582.
Amphiprion latezonatus	Lord Howe or Wideband anemonefish, 280.
Amphiprion melanopus	Cinnamon clownfish, 582.
Amphiprion nigripes	Blackfoot clownfish, 215, 259, 280.
Amphiprion ocellaris	Common clownfish, 36, 77, 80, 194, 200, 214, 260, 261, 280, 582.
Amphiprion percula	Percula clownfish, 36, 194, 200, 214, 260, 280, 582.
Amphiprion perideraion	Pink clownfish, 582.
Amphiprion polymnus	Saddleback clownfish, 280, 582.
Amphiprion rubrocinctus	Red or Australian clownfish, 582.
Amphiprion sandaracinos	Orange clownfish, 582.
Amphiprion sebae	Sebae clownfish, 260, 280, 582.
Amphiprion tricinctus	Threeband clownfish, 582.
Amphiprion spp.	(clownfishes), 228, 280, 347, 366, 452, 502, 577, 581-582, 586.
Chromis cyanea	Blue chromis, 280.
Chromis dispila	Demoiselle, 38.
Chromis insolata	Sunshinefish, Olive chromis, 280.
Chromis multilineata	Yellowedge or Brown chromis, 280.
Chromis notata	Taillight damselfish, Pearlspot chromis, 39, 259, 280.
Chromis punctipinnis	Blacksmith, 38, 228.
Chromis scotti	Purple reeffish, Purple chromis, 280.
Chromis spp.	Azorian chromis, 38.
Chrysiptera cyanea	Blue devil or Orangetail damselfish, 214, 215, 259, 263, 280, 583.
Chrysiptera hemicyanea	Yellowtail blue damselfish, 215, 259, 281.
Chrysiptera parasema	Azure damselfish, 583.
Chrysiptera traceyi	Tracey's damselfish, 260, 281.
Dascyllus albisella	Spot dascyllus, 62, 64, 200, 213, 347, 366, 582.
Dascyllus aruanas	Threestripe damselfish, 39, 62, 64, 214, 260, 281, 347, 366, 582.
Dascyllus melanurus	Blacktail humbug, 281.
Dascyllus trimaculatus	Domino damselfish, 39, 194, 213, 259, 281.
Hypsypops rubicundus	Garibaldi, 228.
Microspathodon chrysurus	Yellowtail damselfish, 25, 37, 281, 582-583.
Pomacentrus amboinensis	Pallid damselfish, 228.
Pomacentrus caeruleus	Caerulean damselfish, 215, 263, 281.
Pomacentrus coelestis	Blue or Neon damselfish, 281.
Pomacentrus nagasakiensis	Nagasaki damselfish, 36, 194, 281.
Premnas biaculeatus	Maroon or Spinecheek clownfish, 281, 582.
Stegastes dorsopunicans	Orangeback damselfish, 259.
Stegastes fuscus	Dusky damselfish, 281.
Stegastes leucorus	Whitetail damselfish, Whitetail gregory, 39, 194, 281.
Stegastes leucostictus	Beaugregory, 14, 39, 80, 228, 281.
Stegastes partitus	Bicolor damselfish, 194, 214, 259, 281.
Stegastes planifrons	Threespot damselfish, 281.
Stegastes variabilis	Cocoa damselfish, 281.
OPLEGNATHIDAE	**KNIFEJAWS**
Oplegnathus fasciatus	Striped knifejaw, 27, 45, 53, 247, 281, 304, 305, 306, 347, 352, 366, 378, 379, 385, 394, 452, 494, 495, 562.
Oplegnathus punctatus	Spotted knifejaw, 53, 281.

CIRRHITIDAE	**HAWKFISHES**
Cirrhitichthys aprinus	Spotted hawkfish, 261.
Cirrhitichthys aureus	Golden curlyfin, 46, 214, 261, 281.
Paracirrhites arcatus	Whiteline hawkfish, 262.
Paracirrhites forsteri	Freckled hawkfish, 262.
LATRIDAE	**TRUMPETERS**
Latris lineata	Striped trumpeter, 93.
MUGILIDAE	**MULLETS**, 6, 7, 11, 68, 202, 210, 211, 238, 249, 336, 366, 452, 503, 517, 533, 562, 574, 576.
Aldrichetta forsteri	Yelloweye mullet, 109.
Chelon labrosus	Thicklip mullet, 70, 99, 104, 107, 109, 115, 117, 228, 379, 404, 453.
Liza affinis	Keelback mullet, 45.
Liza aurata	Golden or Longfin mullet, 228.
Liza haematocheila	Redlip or Redeye mullet, 43, 287, 347, 377, 379, 385, 398, 452, 563.
Liza macrolepis	Largescale mullet, 563.
Liza parsia	99, 107, 128.
Liza ramada	Thinlip mullet, 11, 228, 452-453.
Liza saliens	Leaping mullet, 228.
Mugil cephalus	Striped mullet, 6, 7, 11, 14, 21, 44, 53, 57, 58, 62, 64, 75, 83, 88, 96, 99, 101, 107, 109, 112, 123, 125, 126, 127, 128, 129, 130, 131, 132, 133, 134, 135, 137, 138, 200, 204, 207, 212, 218, 228, 238, 239, 247, 250, 251, 253, 255, 263, 265, 269, 287, 289, 290, 294, 297, 298, 337, 347, 348, 366-367, 375, 382, 408, 452, 465, 484, 503, 524, 562-563, 573.
Mugil curema	White mullet, 14, 23, 45, 62, 64, 200, 228, 263, 348, 453, 563.
Mugil liza	Liza, 45, 247, 348, 453, 563.
Mugil soiuy	[Redeye mullet], 348.
Valamugil seheli	Bluespotted mullet, 218, 228.
SPHYRAENIDAE	**BARRACUDAS**
Sphyraena barracuda	Great barracuda, 218, 228.
Sphyraena borealis	Northern sennet, 22, 28, 42, 563, 583.
Sphyraena pinguis	Brown barracuda, 46.
POLYNEMIDAE	**THREADFINS**
Polydactylus plebeius	Striped threadfin, 563.
Polydactylus sexfilis	Pacific threadfin, 83, 200, 348, 367, 453, 563, 573.
LABRIDAE	**WRASSES**, 259, 315, 583, 586.
Anampses caeruleopunctatus	Bluespotted tamarin, 228, 260.
Anampses cuvier	Firefish, 262, 281.
Gomphosus varius	Birdfish, 228, 259.
Halichoeres bivitattus	Slippery dick, 14.
Halichoeres marginatus	Speckled rainbowfish, 228, 260, 261.
Halichoeres poecilopterus	Multicolorfin wrasse, 47, 213, 217, 259, 281, 348, 583.
Lachnolaimus maximus	Hogfish, 14, 23, 42, 93, 228, 260, 261, 571, 583.
Nelabrichthys ornatus	Concha wrasse, 261.
Symphodus melops	Corkwing wrasse, 134.
Symphodus roissali	Fivespot wrasse, 38.
Tautoga onitis	Tautog, 43, 228, 260, 281, 287, 367, 563.
Tautogolabrus adspersus	Cunner, 45, 228, 260.
Thalassoma bifasciatum	Bluehead wrasse, 14, 127, 129, 131, 132, 216, 228, 281, 287, 583.
Thalassoma cupido	Cupido wrasse, 47, 228, 259, 281.
Thalassoma lutescens	Green moon wrasse, 228, 259.
SCARIDAE	**PARROTFISHES**, 315, 583.
Calotomus japonicus	Whitespotted parrotfish, 47.
Cetoscarus bicolor	Bicolor parrotfish, 228, 259.
Scarus croicensis	Striped parrotfish, 41, 282.
Scarus globiceps	Violetline parrotfish, 228, 259.
Sparisoma rubripinne	Redfin parrotfish, 47, 228, 260, 287.
Sparisoma viride	Stoplight parrotfish, 43, 229, 259.
ZOARCIDAE	**EELPOUTS**

Macrozoarces americanus	Ocean pout, 35, 367.
Stichaeidae	**Pricklebacks**
Anoplarchus purpurescens	High cockscomb, 36.
Cebidichthys violaceus	Monkeyface prickleback, 453.
Chirolophis japonicus	Fringed blenny, 35, 229, 282.
Opisthocentrus ocellatus	36, 282.
Opisthocentrus tenuis	36, 229, 282, 583.
Pholidae	**Gunnels**
Pholis gunnellus	Rock gunnel, 37.
Pholis sp.	(gunnel), 25.
Anarhichadidae	**Wolffishes**, 563.
Anarhichas lupus	Common wolffish, 53, 201, 214, 282, 348, 367, 453-454, 563-564.
Anarhichas lupus marisalbi	White sea wolffish, 35, 219, 229, 258, 282, 348, 563.
Opistognathidae	**Jawfishes**, 577.
Opistognathus aurifrons	Yellowhead jawfish, 35, 215, 262, 282, 348, 367, 583.
Opistognathus gilberti	Yellowhead jawfish, 583.
Opistognathus macrognathus	Banded jawfish, 282.
Nototheniidae	**Antarctic blennies, Antarctic icefishes, Cod icefishes**, 50
Pseudotrematomus eulepidotus	229.
Trichodontidae	**Sandfishes**
Arctoscopus japonicus	Sailfin sandfish, 564.
Pinguipedidae	**Sandperches**
Parapercis cylindrica	Sharpnose weever, 48.
Parapercis sexfasciata	Sixband weever, 45.
Parapercis snyderi	Snyder's weever, 261.
Tripterygiidae	**Triplefins**
Tripterygion etheostoma	Snake blenny, 37, 38, 282, 583-584.
Clinidae	**Scaled blennies**, 17.
Acanthemblemaria aspera	Roughhead blenny, 216.
Acanthemblemaria spinosa	Spinyhead blenny, 216.
Acanthemblemaria sp.	(blenny), 282.
Chaenopsis sp.	(pikeblenny), 282.
Emblemaria sp.	(blenny), 282.
Paraclinus grandicomis	Horned blenny, 584.
Paraclinus marmoratus	Marbled blenny, 37, 282.
Blenniidae	**Combtooth blennies**, 577, 584.
Aidablennius sphynx	Hen-like blenny, Mediterranean blenny, 215, 229, 282.
Chasmodes bosquianus	Striped blenny, 37, 229.
Chasmodes saburrae	Florida blenny, 454.
Ecsenius bicolor	Bicolor blenny, 39.
Hypleurochilus geminatus	Crested blenny, 38, 229.
Hypsoblennius hentz	Feather blenny, 38, 229.
Istiblennius zebra	Rockskipper, 28.
Lipophrys pholis	Shanny, 36.
Meiacanthus grammistes	Striped blenny, 215, 259.
Meiacanthus nigrolineatus	Blackline blenny, 37, 282, 584.
Salaria fluviatilis	Freshwater blenny, 37.
Salaria pavo	229.
Salarias enosimae	38.
Ammodytidae	**Sand lances**
Ammodytes marinus	Sand lance, 26.
Ammodytes personatus	Japanese sand lance, 40.
Ammodytes tobianus	Lesser sandeel, 39.
Callionymidae	**Dragonets**, 584.
Callionymus calliste	Ornate dragonet, 47, 283.
Callionymus enneactis	Mangrove dragonet, 247.
Callionymus lyra	European dragonet, 47, 229, 283.
Callionymus maculatus	Spotted dragonet, 283.

Repomucenus beniteguri	Beniteguri, 47, 584.
Repomucenus hugenini	229.
Repomucenus richardsoni	Richardson's dragonet, 283.
Repomucenus valenciennei	260, 283.
Synchiropus splendidus	Mandarinfish, 261, 454, 584.
ELEOTRIDAE	**SLEEPERS**, 4.
Bostrichthys sinensis	Chinese ocellated gudgeon, 37.
Dormitator latifrons	Pacific fat sleeper, 4.
Dormitator maculatus	Atlantic fat sleeper, 57.
Eleotris oxycephala	39.
Mogurnda mogurnda	Purplestriped gudgeon, 36.
Nemateleotris decora	Decorated firefish, 262, 263, 283.
Oxyeleotris marmorata	Marble goby, 37, 247, 303, 349.
Parioglossus taeniatus	Taeniatus hover goby, 38.
GOBIIDAE	**GOBIES**, 18, 295, 298, 465, 483, 577, 584.
Acanthogobius flavimanus	Yellowfin goby, 36, 247.
Acentrogobius masago	39.
Amblygobius bynoensis	Bynoe's goby, 215, 263, 283.
Amblygobius rainfordi	Rainford's goby, 584.
Apocryptes bato	99.
Bathygobius soporator	Frillfin goby, 37.
Boleophthalmus pectinirostris	Bluespotted mudhopper, 229, 564.
Chaeturichthys hexanema	36.
Chasmichthys dolichognathus	Snakeskin goby, 102.
Chriolepis minutillus	Rubble goby, 283.
Coryphopterus personatus	Masked goby, 283, 584.
Cryptocentrus cinctus	Yellow prawn-goby, 283, 367.
Eutaenichthys gilli	36.
Ginsburgellus novemlineatus	Nineline goby, 584.
Glossogobius giuris	Flathead goby, 564.
Gobiodon citrinus	Lemon or Citron goby, 214, 584.
Gobiodon okinawae	Yellow or Yellow clown goby, 214, 283, 584.
Gobionellus boleosoma	Darter goby, 229, 454.
Gobiosoma bosc	Naked goby, 38, 83, 88, 229.
Gobiosoma evelynae	Sharknose goby, 283.
Gobiosoma genie	Genie's cleaner goby, 283, 584.
Gobiosoma horsti	Yellowline goby, 283.
Gobiosoma multifasciatum	Greenbanded goby, 584.
Gobiosoma oceanops	Neon goby, 37, 194, 215, 216, 229, 259, 283, 367, 584.
G. oceanops × *G. genie*	Neon goby × Genie's cleaner goby, 201.
Gobiosoma prochilus	West Indian cleaner goby, 584.
Gobiosoma xanthiprora	Yellowstriped cleaner goby, 283.
Gobius lidwilli	Midget goby, 38.
Gobius niger	Black goby, 217, 229.
Gobius paganellus	Rock goby, 70.
Gobius ruthensparri	283.
Gobiusculus flavescens	Twospot goby, 134.
Luciogobius elongatus	26.
Luciogobius guttatus	37.
Luciogobius saikaiensis	36.
Lythrypnus dalli	Bluebanded goby, 283.
Microgobius gulosus	Clown goby, 454.
Mugilogobius abei	39, 194, 283, 349, 584.
Paragobiodon echinocephalus	Pricklyhead goby, 229, 260.
Pomatoschistus minutus	Small sand goby, 302.
Pterogobius zacalles	36, 213, 229, 283.
Pterogobius zonoleucus	Flagfin goby, 37.
Quisquilius hipoliti	Rusty goby, 284.
Sagamia geneionema	Hairychin goby, 35, 284.
Sicydium japonicum	39.
Signigobius biocellatus	Twospot goby, 284.

Triaenopogon barbatus	38.
Tridentiger undicerneus	38.
Trimma grammistes	39.
Typhlogobius californiensis	Blind goby, 37.
EPHIPPIDAE	**SPADEFISHES**
Chaetodipterus faber	Atlantic spadefish, 14, 27, 42, 194, 201, 214, 229, 247, 253, 263, 284, 349, 367, 564, 584-585.
Platax pinnatus	Longfin batfish, 585.
SCATOPHAGIDAE	**SCATS**
Scatophagus argus	Spotted scat, 46, 212, 229, 247, 253, 564, 585.
SIGANIDAE	**RABBITFISHES**, 249, 290, 298, 368, 406, 454, 503, 517, 533, 564.
Siganus argenteus	Streamlined spinefoot, 47, 247.
Siganus canaliculatus	Pearly spinefoot, 7, 11, 39, 53, 57, 83, 229, 247, 284, 454, 456, 565.
Siganus chrysospilos	Yellowspotted spinefoot, 39.
Siganus fuscescens	Sandy spinefoot, Dusky spinefoot, 26, 39, 565.
Siganus guttatus	Yellowblotch spinefoot, Blotched spinefoot, 31, 39, 57, 62, 63, 65, 83, 194, 201, 212, 229, 247, 253, 284, 301, 303, 318, 337, 349, 368, 377, 378, 379, 385, 408, 454, 455, 564.
Siganus lineatus	Goldline spinefoot, 39, 218, 229, 259, 284, 349, 368, 565.
Siganus luridus	Dusky spinefoot, 39, 248.
Siganus oramin	Whitespotted spinefoot, 59, 349.
Siganus rivulatus	Black spinefoot, 39, 206, 248, 284, 287, 565.
Siganus vermiculatus	Vermiculated spinefoot, 39, 248, 284, 565.
Siganus virgatus	Blueline spinefoot, 39.
Siganus spp.	(rabbitfishes), 7, 11.
ACANTHURIDAE	**SURGEONFISHES**, 512, 585.
Acanthurus coeruleus	Blue tang, 512.
Acanthurus triostegus	Convict tang, Fiveband surgeonfish, 47.
Naso unicornis	Bluespine unicornfish, 26.
Paracanthurus hepatus	Palette tang, 512.
SCOMBRIDAE	**MACKERELS**, 18, 368, 406, 565, 574.
Acanthocybium solandri	Wahoo, 571.
Auxis rochei	Bullet mackerel, 45, 565.
Auxis thazard	Frigate mackerel, 44, 565.
Euthynnus affinis	Kawakawa, 31, 47, 218, 230, 263, 284, 301, 303.
Euthynnus alletteratus	Little tunny, 230, 565.
Euthynnus lineatus	Black skipjack, 217, 230, 261, 566.
Katsuwonus pelamis	Skipjack tuna, 43, 230, 287, 294, 301, 566, 571.
Rastrelliger kanagurta	Indian mackerel, 230.
Sarda orientalis	Striped bonito, 566.
Sarda sarda	Bonito, 230.
Scomber japonicus	Chub mackerel, 23, 42, 194, 217, 219, 230, 248, 253, 284, 301, 349, 368, 472, 565-566.
Scomber scombrus	Atlantic mackerel, 42, 194, 213, 217, 259, 284.
Scomberomorus cavalla	King mackerel, 571.
Scomberomorus maculatus	Spanish mackerel, 43, 230.
Thunnus alalunga	Albacore, 230.
Thunnus albacares	Yellowfin tuna, 43, 217, 230, 263, 287, 349, 368, 566, 571, 573.
Thunnus atlanticus	Blackfin tuna, 230, 571.
Thunnus maccoyii	Southern bluefin tuna, 6, 11, 211, 217, 230, 566.
Thunnus obesus	Bigeye tuna, 230, 263.
Thunnus thynnus	Northern bluefin tuna, 6, 11, 26, 43, 54, 83, 186, 196, 207, 230, 260, 261, 269, 284, 294, 349, 368, 493, 566, 571.
Thunnus spp.	(tunas), 202, 291, 406.
XIPHIIDAE	**SWORDFISH**
Xiphias gladius	Swordfish, 41, 230.
STROMATEIDAE	**BUTTERFISHES**
Peprilus triacanthus	Butterfish, 46, 230.
Tetragonurus atlanticus	Bigeye squaretail, 42.
ANABANTIDAE	**CLIMBING GOURAMIS**, 296.

Bothidae	**Lefteye flounders**, 298, 406.
Citharichthys arctifrons	Gulf Stream flounder, 230.
Citharichthys cornutus	Horned whiff, 23, 26, 28.
Citharichthys sordidus	Pacific sandab, 230.
Citharichthys spilopterus	Bay whiff, 454.
Citharichthys stigmaeus	Speckled sandab, 54, 103, 109, 217, 230.
Etropus crossotus	Fringed flounder, 230.
Etropus microstomus	Smallmouth flounder, 230.
Paralichthys adspersus	47, 568.
Paralichthys albigutta	Gulf flounder, 45, 568.
Paralichthys californicus	California halibut, 46, 54, 212, 248, 253, 260, 284, 287, 349, 370, 568.
Paralichthys dentatus	Summer flounder, 31, 43, 54, 116, 230, 248, 303, 349, 350, 370, 501, 508, 567, 573.
Paralichthys lethostigma	Southern flounder, 44, 57, 194, 201, 213, 230, 248, 250, 253, 284, 350, 370, 567-568, 573.
Paralichthys microps	Chilean flounder, 48, 194, 217, 284, 350, 406, 568.
Paralichthys olivaceus	Hirame, 6, 7, 11, 21, 29, 44, 54, 99, 101, 107, 109, 112, 134, 194, 201, 217, 230, 284, 294, 301, 303, 304, 306, 334, 350, 351, 368-369, 375, 394, 395, 454-456, 457, 465, 492, 494-495, 497, 517, 519, 566-567, 573.
Paralichthys woolmani	Ecuadorian flounder, 54, 194, 285, 350, 370-371, 568.
Pseudorhombus pentophthalmus	Fivespot flounder, 230.
Scophthalmus aquosus	Windowpane, 14, 42, 230, 585.
Scophthalmus maximus	Turbot, 6, 7, 11, 43, 54, 57, 58, 61, 62, 64, 65, 75, 80, 83, 93, 135, 191, 194, 197, 201, 202, 203, 207, 213, 217, 218, 231, 236, 238, 257, 258, 260, 263, 267, 268, 285, 289, 294, 300, 301, 304, 305, 306, 312, 317, 318, 335, 337, 350, 369, 375, 376, 379, 382, 385, 386, 389, 392, 395, 397, 401, 456-458, 472, 484, 485, 495, 498, 503, 505, 512, 517-518, 525, 533, 567, 573.
Scophthalmus rhombus	Brill, 41, 218.
Syacium papillosum	Dusky flounder, 219, 231.
Xystreurys liolepis	Fantail sole, 46, 568.
Pleuronectidae	**Righteye flounders**, 249, 298, 336.
Eopsetta exilis	Slender sole, 23, 28.
Eopsetta grigorjewi	Spotted halibut, 42.
Errex zachirus	Rex sole, 21.
Glyptocephalus cynoglossus	Witch flounder, 231, 287, 350, 570.
Hippoglossoides platessoides	American plaice, 41, 57.
Hippoglossus hippoglossus	Atlantic halibut, 29, 40, 54, 57, 62, 63, 64, 67, 201, 202, 213, 231, 238, 258, 263, 267, 268, 269, 270, 285, 289, 294, 300, 304, 337, 350, 353, 371, 376, 377, 379, 458-459, 484, 493, 494, 499, 505, 518, 533, 568-569, 573.
Hippoglossus stenolepis	Pacific halibut, 41, 54, 213, 248, 285, 298, 518, 569.
Kareius bicoloratus	Stone flounder, 303.
Microstomus kitt	Lemon sole, 351, 371.
Microstomus pacificus	Dover sole, 217, 219, 231.
Platichthys flesus	North Atlantic flounder, 54, 105, 297, 298, 472, 503, 525.
Platichthys stellatus	Starry flounder, 39, 248, 351, 570.
Pleuronectes americanus	Winter flounder, 38, 112, 218, 231, 238, 248, 250, 285, 287, 371, 459, 475, 476, 569.
Pleuronectes cornutus	Ridged-eye flounder, 42, 248, 285, 351, 570.
Pleuronectes ferrugineus	Yellowtail flounder, 45, 231, 248.
Pleuronectes herzensteini	Littlemouth flounder, 231.
Pleuronectes limanda	Dab, 99, 109, 115, 118, 119, 263.
Pleuronectes platessa	European plaice, 30, 41, 54, 57, 62, 63, 64, 66, 70, 83, 88, 89, 99, 104, 112, 120, 138, 219, 231, 238, 263, 268, 270, 285, 297, 298, 300, 304, 317, 351, 371, 379, 382, 459, 465, 569.
Pleuronectes vetulus	English sole, 54, 58, 231, 248, 459-460.
Pleuronectes yokohamae	Marbled flounder, 38, 57, 219, 231, 248, 258, 287, 304, 570.
Psettichthys melanostictus	Sand sole, 135.
Reinhardtius hippoglossoides	Greenland halibut, 40, 231.

Rhombosolea tapirina	Greenback flounder, 54, 301, 302, 351, 371.
Verasper moseri	Barfin flounder, 258, 285, 570.
Verasper variegatus	Spotted halibut, 41, 351.
SOLEIDAE	**SOLES, TONGUEFISHES**, 336, 406.
Achirus lineatus	Lined sole, 47, 474.
Cynoglossus abbreviatus	42, 248, 351, 570.
Cynoglossus arel	Brown tonguesole, 219, 231.
Cynoglossus capensis	Sand tonguefish, 83.
Cynoglossus lida	Shoulderspot tonguesole, 219, 231.
Cynoglossus robustus	45.
Heteromycteris capensis	Cape sole, 70, 79, 83, 87.
Microchirus ocellatus	Eyed sole, 41.
Solea bleekeri	Blackhand sole, 231.
Solea ovata	Ovate sole, 28.
Solea senegalensis	Senegal sole, 43, 351, 372, 570.
Solea solea	Common sole, 7, 11, 42, 54, 57, 62, 65, 75, 89, 103, 195, 201, 213, 217, 218, 231, 238, 248, 251, 253, 263, 285, 295, 298, 301, 304, 315, 335, 351, 371-372, 379, 385, 386, 460, 492, 505, 518, 570.
Symphurus plagiusa	Blackcheek tonguefish, 460.
Synaptura kleinii	Lace sole, 79.
Trinectes maculatus	Hogchoker, 21, 219, 231, 261, 460.
ORDER PLEURONECTIFORMES	**FLATFISHES**, 6, 11, 31, 67, 296, 298, 384, 493, 498, 508.
BALISTIDAE	**LEATHERJACKETS**, 64, 290, 298, 570, 585, 586.
Balistes vetula	Queen triggerfish, 14, 67, 211, 285, 571, 585.
Brachaluteres ulvarum	38, 259, 285.
Cantherines sp.	(filefishes), 6, 11.
Canthidermis sufflamen	Ocean triggerfish, 24, 285.
Melichthys niger	Black durgon, 285.
Monacanthus hispidus	Planehead filefish, 75, 80, 231.
Oxymonacanthus longirostris	Orangespot filefish, 215, 260, 285.
Paramonacanthus japonicus	Striped filefish, 39, 214, 231, 261, 285.
Stephanolepis cirrhifer	Thread-sailfin filefish, 7, 11, 22, 39.
Sufflamen fraenatus	Bridled triggerfish, 39, 285.
OSTRACIIDAE	**BOXFISHES**, 503, 585.
Lactoria fornasini	Shortspine cowfish, 41.
Lactophrys quadricornis	Scrawled cowfish, 27, 41, 585-586.
TETRAODONTIDAE	**PUFFERS**, 6, 7, 11, 25, 270, 289, 298, 406, 494, 586.
Canthigaster rivulata	Doubleline toby, 39, 231, 260, 285, 287.
Canthigaster rostrata	Sharpnose puffer, 586.
Canthigaster valentini	Saddled toby, 39, 215, 259.
Chilomycterus schoepfi	Striped burrfish, 40, 231, 460.
Cyclichthys orbicularis	Orbiculate burrfish, 261, 285.
Diodon holocanthus	Balloonfish, 41, 231, 263, 286, 571, 586.
Diodon hystrix	Porcupinefish, 25, 195, 586.
Lagocephalus lunaris	Rough golden puffer, 22, 39, 586.
Lagocephalus spadiceus	Half-smooth golden puffer, 571.
Sphoeroides maculatus	Northern puffer, 38, 219, 231, 286, 586.
Sphoeroides nephelus	Southern puffer, 460.
Takifugu chrysops	37, 351, 571.
Takifugu exascurus	37, 231, 248, 351, 571.
Takifugu niphobles	Grass puffer, 37, 231, 261, 270, 303.
Takifugu pardalis	Panther puffer, 37.
Takifugu porphyreus	Purple puffer, 37, 231, 248, 571.
Takifugu rubripes	Tiger puffer, 37, 54, 201, 231, 286, 287, 301, 351, 372, 377, 379, 382, 385, 398, 460, 461, 472, 518, 533, 570-571, 573.
Tetraodon cutcutia	Ocellated blowfish, 37, 215, 231, 259, 286.
MOLIDAE	**MOLAS**
Mola mola	Ocean sunfish, 17, 218, 231.